WITHDRAWN
NDSU

The Role of Phosphorus in Agriculture

The Role of Phosphorus in Agriculture.

Proceedings of a symposium held 1-3 June 1976 at the National Fertilizer Development Center, Tennessee Valley Authority, Muscle Shoals, Alabama. Cosponsored by the Tennessee Valley Authority, American Society of Agronomy, Crop Science Society of America, and Soil Science Society of America.

Editorial Committee: F. E. KHASAWNEH, chairman
E. C. SAMPLE
E. J. KAMPRATH

Symposium Planning Committee: F. E. KHASAWNEH, co-chairman
E. C. SAMPLE, co-chairman
E. O. HUFFMAN
S. R. OLSEN
ALEX POPE

Coordinating Editor: MATTHIAS STELLY

Managing Editor: RICHARD C. DINAUER

Published by: American Society of Agronomy
Crop Science Society of America
Soil Science Society of America
Madison, Wisconsin USA

1980

The cover design was prepared by Mrs. Martha Lott, Display Illustrator, Division of Agricultural Development, Tennessee Valley Authority, Muscle Shoals, Alabama.

Copyright © 1980 by the American Society of Agronomy, Inc.
Crop Science Society of America, Inc.
Soil Science Society of America, Inc.

ALL RIGHTS RESERVED UNDER THE U.S. COPYRIGHT LAW OF 1978 (P.L. 94-533)
Any and all uses beyond the "fair use" provision of the law require written permission from the publishers and/or author(s); not applicable to contributions prepared by officers or employees of the U.S. Government as part of their official duties.

American Society of Agronomy, Inc.
Crop Science Society of America, Inc.
Soil Science Society of America, Inc.
677 South Segoe Road, Madison, Wisconsin 53711 USA

Library of Congress Cataloging in Publication Data

The Role of phosphorus in agriculture.
 Papers presented at a symposium held at Muscle Shoals, Ala., June 1-3, 1976, and sponsored by the Tennessee Valley Authority, the American Society of Agronomy, and others.
 Bibliography: with each chapter
 Includes index.
 1. Phosphatic fertilizers—Congresses. 2. Plants, Effect of phosphorus on—Congresses. 3. Soils—Phosphorus content—Congresses. 4. Phosphate industry—Congresses. I. Khasawneh, F. E., 1938– II. Sample, E.C., 1929–   III. Kamprath, E.J., 1926–   IV. Tennessee Valley Authority. V. American Society of Agronomy.
S647.R64   631.8′5   80-18564
ISBN 0-89118-062-1

Printed in the United States of America

CONTENTS

	Page
FOREWORD	xii
PREFACE	xiii
AN OVERVIEW	xiv
CONTRIBUTORS	xv
CONVERSION FACTORS FOR U.S. AND METRIC UNITS	xviii

1 World Phosphate Reserves and Resources
JAMES B. CATHCART

I. Introduction	1
II. Definitions	3
III. World Phosphate Reserves and Resources	5
IV. Comparison to Other Estimates	13
LITERATURE CITED	17

2 The Phosphate Industry of the United States
JAMES B. CATHCART

I. Introduction	19
II. Economic Factors	21
III. Phosphate Deposits—Atlantic and Gulf Coastal Plains	26
IV. Phosphate Deposits—Central United States	32
V. Phosphate Deposits—Western United States	36
VI. Offshore Deposits	38
VII. Environmental Considerations	39
VIII. Conservation	40
IX. Outlook for the Future	40
LITERATURE CITED	41

3 Evaluation of Phosphatic Raw Materials
G. H. MC CLELLAN AND L. R. GREMILLION

I. Introduction	43
II. Phosphate Mineralogy	44
III. Accessory Minerals	45
IV. Texture of Phosphate Rocks	48
V. Characterization Methods	52
VI. Utilization of Characterization Data	61
VII. Characterization of Marginal Phosphate Ores	74
VIII. Future Research Needs	78
LITERATURE CITED	79

4 Phosphate Raw Materials and Fertilizers: Part I—A Look Ahead
JAMES R. LEHR

I. Introduction	81
II. Phosphate Raw Materials: Present and Future	82
III. Optional Fertilizer Uses of Phosphate Rock	90
IV. Future Outlook for Phosphate Fertilizers	109

4 Phosphate Raw Materials and Fertilizers: Part II—A Case History of Marginal Raw Materials
JOHN HOARE

I. Introduction	121
II. Exploring the Options	123
LITERATURE CITED	127

5 Sulfur Requirements of the Phosphate Fertilizer Industry
D. W. BIXBY

I. Introduction	129
II. Statistics of Past and Present Use	129
III. Projections of Future Needs	137
IV. Future Supply Picture	140
LITERATURE CITED	150

6 Phosphoric Acid Technology
NORMAN ROBINSON

I. Introduction	151
II. Principles of Manufacture by the Wet Process	152
III. Worldwide Trends in Phosphoric Acid Processes	162
IV. Suitability of World Phosphate Rocks for Phosphoric Acid Manufacture	183
LITERATURE CITED	192

7 Phosphate Fertilizers and Process Technology
RONALD D. YOUNG AND CHARLES H. DAVIS

I. Introduction	195
II. Current Products and Processes	196
III. New Process Technology	218
IV. Environmental Problems in the Phosphate Fertilizer Industry	223
LITERATURE CITED	225

8 World Phosphate Fertilizer Supply-Demand Outlook
E. A. HARRE AND K. F. ISHERWOOD

I. Introduction	227
II. Components of Fertilizer Supply-Demand	228
III. Forecasting of Fertilizer Supply-Demand	230
IV. Supply-Demand Outlook for North America	231
V. World Phosphate Market Outlook	234
VI. World Phosphate Supplies	236
VII. Problems in Assessing Future Market Outlook	238
LITERATURE CITED	238

9 Energy Requirements for the Production of Phosphate Fertilizers
WILLIAM C. WHITE AND KARL T. JOHNSON

I. Introduction	241
II. Production Capacity	242

CONTENTS

    III. Phosphate Rock 242
    IV. Sulfur... 244
    V. Wet Process Phosphoric Acid...................................... 244
    VI. Ammonium Phosphates 245
    VII. Triple Superphosphate... 246
    VIII. Normal Superphosphate 246
    IX. Environmental Protection.. 246
    X. Energy Conservation 249
    LITERATURE CITED... 250

10 Energy of Phosphate Fertilizer Applications and Food Energy Returns

WILLIAM C. BURROWS AND ORVIS P. ENGELSTAD

    I. Introduction 251
    II. Phosphate Fertilizer in the Agricultural Energy Input Spectrum 252
    III. Phosphate Fertilizer in the Agricultural Energy Output Spectrum........ 254
    LITERATURE CITED... 262

11 Reactions of Phosphate Fertilizers in Soils

E. C. SAMPLE, R. J. SOPER, AND G. J. RACZ

    I. Introduction 263
    II. Phosphate Retention by Soil Constituents 264
    III. Sequence of Events After Fertilizer Application....................... 274
    IV. Characterization of Soil-Phosphorus Reactions: Methodology 288
    V. Summary Statement and Future Research Needs 302
    LITERATURE CITED... 304

12 Agronomic Effectiveness of Phosphate Fertilizers

O. P. ENGELSTAD AND G. L. TERMAN

    I. Introduction 311
    II. Fertilizer Properties Affecting Crop Response to Applied P............. 311
    III. Soil and Management Factors Affecting Crop Response to P Fertilizers... 321
    IV. Changes in Source Needs Resulting from Rising Soil P Levels 328
    V. Source Needs in the Tropics 328
    LITERATURE CITED... 329

13 Evaluation and Utilization of Residual Phosphorus in Soils

N. J. BARROW

    I. Introduction 333
    II. Definition and Measurement of Residual Value....................... 333
    III. Factors Involved in the Decrease in Effectiveness of P Fertilizers 335
    IV. Characteristics of the Slow Reactions which Follow Adsorption 340
    V. Access of Plants to Residual P 349
    VI. Evaluation of Fertilizer Residues 352
    LITERATURE CITED... 355

14 Use and Limitations of Physical-Chemical Criteria for Assessing the Status of Phosphorus in Soils

S. R. OLSEN AND F. E. KHASAWNEH

I. Introduction 361
II. Physico-Chemical Characteristics of Soil Phosphorus with Respect to Plant Growth 362
III. Physical Chemical Criteria with Respect to Soil P Characterization 379
IV. The Integrated Overview 401
LITERATURE CITED... 404

15 Assessing Organic Phosphorus in Soils

GEORGE ANDERSON

I. Introduction 411
II. Biological Immobilization of Soil and Fertilizer P 411
III. Characterization of Soil Organic P 417
IV. Availability of Soil Organic Phosphate 424
V. Future Research Needs 427
LITERATURE CITED... 428

16 Conventional Soil and Tissue Tests for Assessing the Phosphorus Status of Soils

E. J. KAMPRATH AND M. E. WATSON

I. Introduction 433
II. Evaluation of Common Soil Test Extractants........................ 434
III. Soil Solution P as a Measure of Available P 444
IV. Use of Plant Analysis in Assessing Soil P Status...................... 447
V. Future Research Needs 464
LITERATURE CITED... 464

17 Management Considerations for Acid Soils with High Phosphorus Fixation Capacity

PEDRO A. SANCHEZ AND GORO UEHARA

I. Introduction 471
II. Geographical Distribution of Soils with High Phosphorus Fixation Capacity 473
III. Magnitude and Measurement of High Phosphorus Fixation 475
IV. High Input Strategy: Phosphorus as an Amendment................... 487
V. Low Input Management Strategy................................... 495
VI. Research Needs.. 508
LITERATURE CITED... 509

18 Use of Waste Materials as Sources of Phosphorus

L. E. SOMMERS AND A. L. SUTTON

I. Introduction 515
II. Composition of Wastes 516
III. Plant Availability of Phosphorus in Wastes 523
IV. Waste-Induced Changes in Soil Phosphorus......................... 530
V. Constraints on Application of Wastes on Agricultural Land 532
VI. Potential Phosphorus Supply in Waste Materials..................... 536
LITERATURE CITED... 538

19 Agricultural Phosphorus in the Environment

A. W. TAYLOR AND V. J. KILMER

I. Introduction	545
II. Background Levels in the Environment	546
III. Effects of Phosphate Additions	551
IV. Overview	555
LITERATURE CITED	556

20 Phosphate Nutrition of Plants—A General Treatise

P. G. OZANNE

I. Introduction	559
II. Soil Factors in Phosphate Availability	560
III. Plant Factors in Phosphate Availability	562
IV. Absorption and Translocation	568
V. Metabolic Role of Phosphorus Within the Plant	572
VI. Plant Analysis and Critical Concentrations	575
VII. Phosphorus Content and Nutritional Value	579
VIII. Phosphate Responses and Their Profitability	581
LITERATURE CITED	585

21 Soil-Plant Interactions in the Phosphorus Nutrition of Plants

STANLEY A. BARBER

I. Introduction	591
II. Mechanisms of P Supply to Plant Roots Growing in Soil	592
III. Soil Factors Influencing P Diffusion Rates in Soil	593
IV. Phosphorus Uptake Characteristics of Plant Roots	596
V. Models for P Uptake by Plant Roots from Soil	605
VI. Chemical Effect of Roots on Soil Environment and P Uptake	608
VII. Effect of P Distribution in Soil on P Uptake by Roots	610
VIII. Future Research Needs	613
LITERATURE CITED	613

22 Role of Rhizosphere Microorganisms in Phosphorus Uptake by Plants

P. B. TINKER

I. Introduction	617
II. Rhizosphere Microorganism Population	619
III. Effect of Free-Living Microorganisms on High Plants	623
IV. Taxonomy and Biology of Mycorrhizal Fungi	629
V. Factors Influencing Mycorrhiza Formation	634
VI. Effect of Mycorrhizae on Phosphorus Nutrition	636
VII. Mechanism of Phosphate Uptake by Mycorrhizal Roots	638
VIII. Other Effects of Mycorrhizal Fungi on Plants	643
IX. Practical Applications of Microorganisms in Improving Phosphate Nutrition of Host Plants	644
X. Conclusion	646
LITERATURE CITED	647

23 Interactions of Phosphorus with Other Elements in Soils and in Plants

FRED ADAMS

I. Introduction	655
II. Nitrogen	656
III. Calcium	657
IV. Magnesium	660
V. Potassium	660
VI. Aluminum	661
VII. Iron	663
VIII. Manganese	666
IX. Zinc	667
X. Copper	670
XI. Sulfate	671
XII. Molybdenum	671
XIII. Boron	672
XIV. Silicate	673
LITERATURE CITED	674

24 Phosphate Nutrition of Corn, Sorghum, Soybeans, and Small Grains

J. J. HANWAY AND R. A. OLSON

I. Introduction	681
II. Phosphorus Uptake	683
III. Phosphorus Removal and Phosphorus Recycling in Crop Residues	685
IV. Yield—Phosphorus Concentration Relationships	686
V. Factors Influencing Phosphorus Uptake and Utilization	687
VI. Relation Between Phosphorus Nutrition and Quality of Harvested Plant Parts	690
LITERATURE CITED	691

25 Phosphorus Nutrition of Cotton, Peanuts, Rice, Sugarcane, and Tobacco

L. E. NELSON

I. Introduction	693
II. Cotton	694
III. Peanut	702
IV. Rice	708
V. Sugarcane	715
VI. Tobacco	722
LITERATURE CITED	729

26 Phosphorus Nutrition of Vegetable Crops and Sugar Beets

O. A. LORENZ AND M. T. VITTUM

I. Introduction	737
II. Phosphorus Composition of Vegetables and Sugar Beets	738
III. Phosphorus Uptake Demand Patterns	744
IV. Crop Responses and Phosphorus Fertilization	746
V. Phosphorus Nutrition and Crop Quality	754
VI. Relation of Soil P Levels to Crop Residue	755
VII. Effects of Excess P	757

VIII.	Future Considerations	757
	LITERATURE CITED	759

27 Phosphorus Nutrition and Fertilization of Forest Trees

RUSSELL BALLARD

I.	Introduction	763
II.	Phosphorus Nutrition in Forest Nurseries	764
III.	Phosphorus Nutrition of Forest Trees	767
IV.	Phosphorus Fertilization of Forest Stands	780
V.	Determination of P Fertilizer Requirements	790
VI.	Future Research Needs	795
	LITERATURE CITED	796

28 Phosphorus Nutrition of Forages

D. A. MAYS, S. R. WILKINSON, AND C. V. COLE

I.	Introduction	805
II.	Relationships of Phosphorus Application Techniques to Root Morphology	808
III.	Phosphorus Nutrition of Cool-Season Grasses and Legumes	812
IV.	Phosphorus Nutrition of Warm-Season Grasses and Legumes	820
V.	Phosphorus Nutrition of Semiarid Grasslands	830
VI.	Phosphorus Cycling in Grasslands	834
	LITERATURE CITED	840

29 Relationship Between Phosphorus Nutrition of Plants and the Phosphorus Nutrition of Animals and Man

R. L. REID

I.	Introduction	847
II.	Phosphorus Requirements of Animals	849
III.	Factors Affecting the Phosphorus Content of Plants	853
IV.	Availability of Phosphorus in Diets	862
V.	Maintaining Adequate Phosphorus Nutrition in Animals	866
	LITERATURE CITED	879

Glossary of Common and Scientific Names of Plants and Other Organisms ... 887

Glossary of Mineral Compositions ... 891

Subject Index ... 893

FOREWORD

Without phosphorus in the environment no living organisms could exist. Phosphorus is present in all plant and animal tissue. It is necessary for such life processes as photosynthesis, the synthesis and breakdown of carbohydrates, and the transfer of energy within the plant. Phosphorus is taken up by the plant from the soil. Unless the soil contains adequate phosphorus or it is supplied to the soil from external sources, plant growth will be limited.

Phosphorus does not occur as abundantly in soils as does the other major nutrients, nitrogen and potassium. The content of phosphorus ranges from about 100 to 2,500 kg/ha and averages about 1,000 kg/ha in the surface 20 cm of a soil. Phosphorus occurs in both inorganic and organic forms in the soil. Only a small fraction of the total phosphorus is in a form available to plants.

Plants do not require as large quantities of phosphorus as they do nitrogen and potassium. But phosphorus is just as essential. Unlike nitrogen which can be returned to the soil by fixation from the air, phosphorus cannot be replenished except from external sources once it leaves the soil in agricultural products or by erosion.

The phosphorus contained in 9,400 kg/ha of corn grain (150 bu/acre) contains about 25 kg/ha of phosphorus or about 1/40 of that contained in the surface 20 cm of a typical American soil. Removal of phosphorus from the soil in food or fiber crops over a few decades can be a significant portion of that contained in the pedon. Thus, under many systems of farming, phosphorus must be supplied to the soil from external sources, principally as mineral phosphorus fertilizer, with smaller amounts from agricultural processing and municipal wastes.

Because of the importance of phosphorus in agriculture and because of the limited supplies in most soils, members of the American Society of Agronomy, the Crop Science Society of America, and the Soil Science Society of America believe that this monograph is a much needed treatise on the subject. Its purpose is to examine all aspects of the manufacture and of supplies of phosphorus fertilizers and the best possible management of phosphorus as a plant nutrient in agriculture. The editors and the authors are outstanding authorities in the field. The monograph will be of value to all scientists, students, and administrators who deal with phosphorus in agriculture.

June 1980

Roger L. Mitchell, *president*
American Society of Agronomy

Billy E. Caldwell, *president*
Crop Science Society of America

William E. Larson, *president*
Soil Science Society of America

PREFACE

The Role of Phosphorus in Agriculture is a compilation of papers presented at a symposium held at Muscle Shoals, Alabama, June 1-3, 1976, and cosponsored by the Tennessee Valley Authority, the American Society of Agronomy, the Soil Science Society of America, and the Crop Science Society of America. The objectives of the symposium were to assemble recognized authorities to summarize current knowledge about phosphorus as it relates to agriculture and to provide an authoritative reference work on this subject.

Much of the credit for the scope and range of topics in this book goes to the Planning Committee, which consisted of E. O. Huffman, S. R. Olsen, and Alex Pope, with F. E. Khasawneh and E. C. Sample as cochairmen. The committee spent many hours outlining a comprehensive coverage of fertilizer phosphorus and choosing a slate of authors knowledgeable about each topic. Through their efforts the book achieves the goal of covering phosphorus from the mine to the end use in fertilizers. Topics include: (i) surveys of world and U.S. phosphate ore deposits, and evaluation of these raw materials; (ii) new developments in phosphoric acid and phosphate fertilizer technology and in processing low-grade deposits; (iii) patterns in supply-demand trends of phosphate fertilizers; (iv) reactions of phosphate fertilizers in soils and methods of assessing the status of soil phosphorus, including residual and organic forms; (v) agronomic factors related to the effectiveness of various phosphate sources, both inorganic and organic; (vi) phosphate nutrition of major crops and the relationship between crop nutrition and nutrition of humans and animals; and (vii) the impact of fertilizer phosphorus on the environment. The committee felt that symposium topics should be restricted to phosphorus as a nutrient; hence, phosphorus in organophosphate pesticides is not included in the book.

The authors represent a cross section of disciplines and organizations including academic, governmental, and industrial. The Editorial Committee is grateful to these authors and to the organizations they represent for their outstanding contributions. We are especially grateful to these authors for taking additional time to ensure that coverage of their subject matter is current and up-to-date in spite of the time which has elapsed since the symposium. The committee also acknowledges the assistance of numerous anonymous reviewers.

Special recognition also goes to Richard C. Dinauer, Matthias Stelly, and other members of the headquarters office of the American Society of Agronomy for their help with the symposium and in editing and publishing this book. The assistance of Mrs. Peggy Kelley of the Tennessee Valley Authority in planning and conducting the symposium, handling most of the correspondence, and in editing the manuscripts is gratefully acknowledged.

May 1980

F. E. Khasawneh, *chairman, Editorial Committee*
Tennessee Valley Authority, Muscle Shoals, Alabama

E. C. Sample
Tennessee Valley Authority, Muscle Shoals, Alabama

E. J. Kamprath
North Carolina State University, Raleigh, N.C.

AN OVERVIEW—A LOOK AHEAD

This symposium on The Role of Phosphorus in Agriculture is most timely. New challenges are facing us of a magnitude and nature that we have never encountered before. World food and fiber needs are increasing rapidly. The quality of phosphate ores is declining so that new technology must be introduced or shortages will develop. Environmental quality concerns are appearing in mining, manufacture, and use of phosphates. Ways to conserve energy in fertilizer manufacture and use are becoming important in view of energy shortages. Farming systems are changing, and farmers are placing ever greater emphasis on inputs to increase yields and profits. Residual levels of applied phosphorus are building in many intensively farmed soils of the industrial nations on the one hand while many developing nations are having to consider farming the extremely P-deficient high P-fixing acid soils in the tropics and subtropics.

The symposium addresses itself directly to many of these challenges. Its main purpose, however, is to review in depth the present state of knowledge in all phases of phosphorus technology and manufacture, the status of raw materials, reactions and interactions of P in soils, methods for predicting P needs of crops, and nutrition of major crops and animals. Most of the symposium's reviewers also have pointed out the key problems and areas needing further research.

More research in practically all phases of phosphorus in agriculture, both short and long term, is needed and could have tremendous impact upon farmers, consumers, the fertilizer industry, and entire nations. Many of the pressing problems facing us need immediate solution. It is recognized too that research in phosphorus in agriculture involves a highly complex and difficult area.

There is no substitute for phosphorus in the production of crops and animals for food, fiber, and other essential needs. It behooves researchers, administrators, and governmental institutions alike to take necessary steps to increase research activity in phosphorus for agriculture.

July 1979

Lewis B. Nelson, *manager*
Office of Agricultural and Chemical Development
TVA, Muscle Shoals, Alabama

E. O. Huffman, *former director*
Division of Chemical Development
TVA, Muscle Shoals, Alabama

CONTRIBUTORS

Fred Adams   Professor of Soil Chemistry, Department of Agronomy and Soils, Auburn University, Auburn, Alabama

George Anderson   Head, Department of Soil Organic Chemistry, Macaulay Institute for Soil Research, Craigiebuckler, Aberdeen, Scotland

Russell Ballard   Formerly Scientist, Forest Research Institute, Rotorua, New Zealand. Now Associate Professor of Forest Soils and Director of the North Carolina State Forest Fertilization Cooperative, School of Forest Resources, North Carolina State University, Raleigh, North Carolina

Stanley A. Barber   Professor of Agronomy, Department of Agronomy, Purdue University, West Lafayette, Indiana

N. J. Barrow   Senior Principal Research Scientist, CSIRO, Division of Land Resources Management, Wembley, Western Australia

David W. Bixby   Director, Fertilizer Technology Research, The Sulphur Institute, Washington, D.C.

William C. Burrows   Senior Staff Scientist, Deere and Company Technical Center, Moline, Illinois

James B. Cathcart   Commodity Geologist, Phosphate, U.S. Geological Survey, Denver, Colorado

C. V. Cole   Soil Scientist, Science and Education Administration, Agricultural Research, U.S. Department of Agriculture, Fort Collins, Colorado

Charles H. Davis   Director of Chemical Development, Division of Chemical Development, Tennessee Valley Authority, National Fertilizer Development Center, Muscle Shoals, Alabama

Orvis P. Engelstad   Chief, Soils and Fertilizer Research Branch, National Fertilizer Development Center, Tennessee Valley Authority, Muscle Shoals, Alabama

Louis R. Gremillion   Geologist, Fundamental Research Branch, Division of Chemical Development, National Fertilizer Development Center, Tennessee Valley Authority, Muscle Shoals, Alabama. Deceased 29 December 1979

John J. Hanway   Professor, Agronomy Department, Iowa State University, Ames, Iowa

Edwin A. Harre   Agricultural Economist, Economics and Market Research Section, National Fertilizer Development Center, Tennessee Valley Authority, Muscle Shoals, Alabama

John S. Hoare   Chief Engineer, British Phosphate Commissioners, Melbourne, Victoria, Australia

Keith F. Isherwood   Head, Fertilizer Service, International Superphosphate and Compound Manufacturers Association (ISMA, Ltd), Paris, France

CONTRIBUTORS

Karl T. Johnson — Director, Member Services, The Fertilizer Institute, Washington, D.C.

Eugene J. Kamprath — Professor, Department of Soil Science, North Carolina State University, Raleigh, North Carolina

Fayez E. Khasawneh — Research Soil Scientist, Soils and Fertilizer Research Branch, National Fertilizer Development Center, Tennessee Valley Authority, Muscle Shoals, Alabama

Victor J. Kilmer — Formerly Chief, Soils and Fertilizer Research Branch, National Fertilizer Development Center, Tennessee Valley Authority, Muscle Shoals, Alabama, Now retired

James R. Lehr — Senior Scientist, Fundamental Research Branch, Division of Chemical Development, National Fertilizer Development Center, Tennessee Valley Authority, Muscle Shoals, Alabama

Oscar A. Lorenz — Professor of Vegetable Crops, Department of Vegetable Crops, University of California, Davis, California

David A. Mays — Agronomist, Soils and Fertilizer Research Branch, National Fertilizer Development Center, Tennessee Valley Authority, Muscle Shoals, Alabama

Guerry H. McClellan — Research Coordinator, Fertilizer Technology Division, International Fertilizer Development Center, Muscle Shoals, Alabama

Lyle E. Nelson — Professor and Agronomist, Department of Agronomy, Mississippi State University, Mississippi State, Mississippi

Sterling R. Olsen — Research Leader, Science and Education Administration, Agricultural Research, U.S. Department of Agriculture, Fort Collins, Colorado

Robert A. Olson — Professor, Agronomy Department, University of Nebraska, Lincoln, Nebraska

Peter G. Ozanne — Senior Principal Research Scientist, Division of Land Resources Management, Institute of Earth Resources, CSIRO, Wembley, Western Australia

Geza J. Racz — Professor, Department of Soil Science, University of Manitoba, Winnipeg, Manitoba, Canada

Robert L. Reid — Professor of Animal Science, College of Agriculture and Forestry, West Virginia University, Morgantown, West Virginia

Norman Robinson — Chief Chemical Engineer and Deputy Head of Technical Development, Fertilizer Division, Levington Research Station, Fisions, Ltd, Levington, Ipswich, Suffolk, England

Eugene C. Sample — Research Soil Chemist, Soils and Fertilizer Research Branch, National Fertilizer Development Center, Tennessee Valley Authority, Muscle Shoals, Alabama

Pedro A. Sanchez — Associate Professor of Soil Science and Coordinator for Tropical Soils Programs, Department of Soil Science, North Carolina State University, Raleigh, North Carolina

CONTRIBUTORS

Lee E. Sommers	Professor of Agronomy, Agronomy Department, Purdue University, West Lafayette, Indiana
Robert J. Soper	Professor, Department of Soil Science, University of Manitoba, Winnipeg, Manitoba, Canada
Alan L. Sutton	Associate Professor, Department of Animal Science, Purdue University, West Lafayette, Indiana
Alan W. Taylor	Research Chemist, Soil Nitrogen and Environmental Chemistry Laboratory, Science and Education Administration, Agricultural Research, U.S. Department of Agriculture, Beltsville, Maryland
Gilbert L. Terman	Agronomist, Soils and Fertilizer Research Branch, National Fertilizer Development Center, Tennessee Valley Authority, Muscle Shoals, Alabama. Deceased 31 January 1979
Philip B. Tinker	Formerly Professor of Agricultural Botany, University of Leeds. Now Head, Department of Soils and Plant Nutrition, Rothamsted Experimental Station, Harpenden, Herts, England
Goro Uehara	Professor of Soil Science, Department of Agronomy and Soil Science, University of Hawaii, Honolulu, Hawaii
Morrill T. Vittum	Professor and Head, Department of Seed and Vegetable Sciences, New York State Agricultural Experiment Station, Cornell University, Geneva, New York
Maurice E. Watson	Assistant Professor, Ohio Agricultural Research and Development Center, Wooster, Ohio
Stanley R. Wilkinson	Soil Scientist, Southern Piedmont Conservation Research Center, Science and Education Administration, Agricultural Research, U.S. Department of Agriculture, Watkinsville, Georgia
William C. White	Senior Vice President, Member Services, The Fertilizer Institute, Washington, D.C.
Ronald D. Young	Chemical Engineer, Division of Chemical Development, National Fertilizer Development Center, Tennessee Valley Authority, Muscle Shoals, Alabama

CONVERSION FACTORS FOR U. S. AND METRIC UNITS

To convert column 1 into column 2, multiply by	Column 1	Column 2	To convert column 2 into column 1, multiply by
	Length		
0.621	kilometer, km	mile, mi	1.609
1.094	meter, m	yard, yd	0.914
0.394	centimeter, cm	inch, in	2.54
	Area		
0.386	$kilometer^2$, km^2	$mile^2$, mi^2	2.590
247.1	$kilometer^2$, km^2	acre, acre	0.00405
2.471	hectare, ha	acre, acre	0.405
	Volume		
0.00973	$meter^3$, m^3	acre-inch	102.8
3.532	hectoliter, hl	cubic foot, ft^3	0.2832
2.838	hectoliter, hl	bushel, bu	0.352
0.0284	liter	bushel, bu	35.24
1.057	liter	quart (liquid), qt	0.946
	Mass		
1.102	ton (metric)	ton (U.S.)	0.9072
2.205	quintal, q	hundredweight, cwt (short)	0.454
2.205	kilogram, kg	pound, lb	0.454
0.035	gram, g	ounce (avdp), oz	28.35
	Pressure		
14.50	bar	$lb/inch^2$, psi	0.06895
0.9869	bar	atmosphere, atm	1.013
0.9678	kg(weight)/cm^2	atmosphere, atm	1.033
14.22	kg(weight)/cm^2	$lb/inch^2$, psi	0.07031
14.70	atmosphere, atm	$lb/inch^2$, psi	0.06805
	Yield or Rate		
0.446	ton (metric)/hectare	ton (U.S.)/acre	2.24
0.892	kg/ha	lb/acre	1.12
0.892	quintal/hectare	hundredweight/acre	1.12
	Temperature		
$\left(\frac{9}{5} °C\right) + 32$	Celsius −17.8C 0C 100C	Fahrenheit 0F 32F 212F	$\frac{5}{9}(°F - 32)$
	Water Measurement		
8.108	hectare-meters, ha-m	acre-feet	0.1233
97.29	hectare-meters, ha-m	acre-inches	0.01028
0.08108	hectare-centimeters, ha-cm	acre-feet	12.33
0.973	hectare-centimeters, ha-cm	acre-inches	1.028
0.00973	$meters^3$, m^3	acre-inches	102.8
0.981	hectare-centimeters/hour, ha-cm/hour	$feet^3$/sec	1.0194
440.3	hectare-centimeters/hour, ha-cm/hour	U.S. gallons/min	0.00227
0.00981	$meters^3$/hour, m^3/hour	$feet^3$/sec	101.94
4.403	$meters^3$/hour, m^3/hour	U.S. gallons/min	0.227

Plant Nutrition Conversion—P and K

P (phosphorus) × 2.29 = P_2O_5
K (potassium) × 1.20 = K_2O

Chapter 1

World Phosphate Reserves and Resources

JAMES B. CATHCART
U.S. Geological Survey
Denver, Colorado

I. INTRODUCTION

A. Phosphorus in the Lithosphere

Phosphorus (P) makes up about 0.12% of the earth's crust. It is present in all soils and rocks, in water, and in plant and animal remains; and it forms complex compounds with a wide variety of elements—about 150 minerals are known that contain at least 0.44% P (1% P_2O_5). The world's supply of P comes from mineral deposits, a nonrenewable natural resource. The phosphate of almost all minable deposits is one of the minerals of the apatite group—$Ca_{10}(PO_4,CO_3)_6(F,OH)_{2-3}$. A very small percentage, however, is mined from secondary Al phosphate deposits, in which the phosphate mineral was derived from apatite by weathering.

Most phosphate deposits contain silica in the form of quartz; other common diluting materials include calcite, dolomite, Fe-oxide minerals, and clay minerals. Some deposits contain diluting materials such as zeolites derived from the alteration of volcanic ash, glauconite, cristobalite, pyrite, and so on. Apatite must be separated from the gangue minerals, and methods of beneficiation have to be tailored to the suite of minerals in the phosphate rock. It is essential to determine the mineralogy as a first step in evaluating the economics of a deposit (see Chapt. 3 and 4).

Mining methods used to extract the rock will depend on the physical character of the rock and its geologic setting. If the rock is unconsolidated and flat-lying, open-pit mining methods can be used, whereas if the rock is consolidated and steeply dipping, some method of underground mining will have to be used.

The separated apatite mineral will, in all probability, be used to manufacture fertilizer. In this case the mineral species is most important. A carbonate fluorapatite, for example, is much easier to process than is a fluorapatite. The impurities in the apatite grain or phosphate pellet are important in determining the type and method used in manufacturing the fertilizer.

Copyright 1980 © ASA-CSSA-SSSA, 677 South Segoe Road, Madison, WI 53711, USA.
The Role of Phosphorus in Agriculture.

B. Historical Overview

Man has used natural materials—manures, vegetable material, and bones—as fertilizers since the beginnings of agriculture, but it was not until 1840 that Liebig, the German chemist, suggested that dissolving bones in sulfuric acid made the P more available to plants. At about the same time, John Bennet Lawes started to experiment with a form of acid-treated phosphate or bones. The superphosphate thus made from bones proved so popular that bones came to be in very short supply, and a search was begun to develop other sources of phosphate. Lawes was granted a patent in 1842 on the use of phosphate nodules (erroneously called *coprolites*) as the source for the phosphate. Although Lawes used phosphate nodules from Estremedura in Spain for some of his experiments, the first commercial production of phosphate rock was in 1847, when the mining of "coprolites" began in Suffolk in Great Britain, and peaked in 1876 when about 250,000 metric tons was mined.

Mining of phosphate in the United States began in 1867 in South Carolina, although the deposits were known as early as 1837. The first recorded production in that year was 6 long tons (≈ 6.2 metric tons). Deposits were discovered in Florida in 1888, in Tennessee in 1894, and in the western U.S. in 1906. Deposits in North Africa (Algeria and Tunisia) were discovered in 1873; mining began in 1889. Production from large Moroccan deposits began in 1921, although they were known as early as 1914. The guano-derived deposits in the Pacific Islands were discovered in the 1890's. Mining began in 1900 on Ocean Island, in 1906 on Nauru Island, and in 1908 in Makatea Island. The extensive deposits of igneous apatite of the Kola Peninsula were discovered in about 1930. Phosphate deposits have been discovered in the last 20 years in North Carolina; Canada; Baja California, Mexico; Brazil, Peru, and Colombia in South America; Israel, Turkey, Jordan, Iraq, and Saudi Arabia in the Near East; Angola, South Africa, and Western Sahara in Africa; India; and Australia. These discoveries were made by geologists who had a background in the geology of phosphate deposits and who used theories of origin as working hypotheses; they have added billions of metric tons to world phosphate reserves.

World production reached 1 million metric tons in 1887 and 12 million metric tons in 1938. Almost 80% of the 1938 production was used for the manufacture of ordinary superphosphate, and only about 4% for phosphoric acid. Since the end of World War II, a steady and rapid increase in production has taken place, so that in 1974 more than 110 million metric tons was produced. Much of the production went into the manufacture of high-analysis fertilizers, entailing a large production of phosphoric acid.

C. World Structure of the Phosphate Mining Industry

Of the total production of phosphate rock (PR) in 1974 (110 million metric tons), 75% was from three countries—USA (38%), USSR (20%), and Morocco (17%). Other North African countries Algeria, Tunisia, Egypt, and Western Sahara) and Middle Eastern countries (Jordan, Syria) produced about 9% of the total. The remaining 16% was spread over 30 countries throughout the world.

The USSR and Morocco have announced plans to substantially increase production capacity, and the U.S. capacity will also increase if all planned mines are completed. Thus, the percentage of the world production from these countries will increase in the future.

Morocco, the USA, and the USSR also account for about 75% of international export trade. In the decade from 1964 to 1973, a total of about 345 million metric tons was exported—112 million tons (32%) by Morocco, 99 million tons (29%) by the USA, and 51 million tons (15%) by the USSR (Notholt, 1975).

Except for the United States, the world's P reserves and resources are owned or controlled by the government of the country in which the deposit occurs. Partial private ownership may be allowed in many countries, but control still rests with the government. It seems likely that patterns of world trade in PR will not change drastically until the end of this century, when Australia may become a large producer (and exporter). Much of the Australian production probably will go to Japan, cutting sharply into United States exports of PR. This change in the export patterns probably will coincide with a projected decline in production from Florida.

The picture for the future, then, would seem to be that Morocco and the USSR will dominate the export business provided announced capacity increases are real, and that Australia will take over a part, and perhaps a large part, of the export trade of the United States. It seems likely that the United States will be using almost all production domestically and may become a net importer of PR in the 21st century.

II. DEFINITIONS

A. Phosphate Rock

In the mining industry, the term *phosphate rock* or simply, *rock,* is used in two senses. If an apatite-bearing rock is high enough in P content to be used directly to make fertilizer or as a furnace charge to make elemental P, it is called *phosphate rock*. The term is also used to designate a beneficiated apatite concentrate.

B. Phosphorite

Phosphorite is a rock term for a sediment in which a phosphate mineral is a major constituent. Modifying adjectives are used to define the rock more precisely (for example, calcareous phosphorite, sandy phosphorite, and clayey phosphorite). Where the phosphate mineral is present in less than major amounts, the rock may be designated as a phosphatic sandstone, a phosphatic limestone, etc.

C. Phosphatic Particles

The phosphate mineral in sedimentary deposits is present as rounded, oval-shaped particles that range in size from boulders to very fine sand or silt. The finer particles are referred to commercially as *concentrate* and range in size from 0.1 to 1 mm. Coarser particles, called pebbles, are generally larger than 1 mm, and the terms *oolite, nodule, granule, pellet,* and *fragment* have been applied to these particles. *Oolite* is a term for a small, rounded particle having accretionary texture. *Nodule* is used to denote a rounded, irregular phosphate grain that is generally coarse sand to fine boulder in size. *Granule* is a size term for particles that range in size from 2 to 4 mm in diameter. *Pellet* is used for a rounded, structureless, sedimentary apatite particle, ranging from silt to granule size. *Fragment* is used to designate a phosphate particle that has been broken and is, therefore, angular or subangular.

D. Ore or Matrix

Ore can be simply defined as material in the ground that can be mined and processed at a profit. *Matrix* is a term used in the Florida and North Carolina phosphate districts as nearly synonymous with ore. The derivation of this variant meaning of the term may be of some interest. In the early days of Florida mining, only the coarse phosphate fraction (pebble) was recovered. The material in which the pebble was embedded was called *matrix* by the dictionary definition. There was abundant phosphate in the matrix, but it could not be separated; and when flotation methods were finally perfected, the fraction treated by flotation was the matrix of the pebble fraction, and eventually the total rock came to be known as *matrix*.

E. Grade of Ore

Historically, the phosphate industry has expressed grade only in terms of the phosphate content of the ore, either as P_2O_5, BPL (bone phosphate of lime), TPL (triphosphate of lime), or P. BPL and TPL are both equivalent to $Ca_3(PO_4)_2$. Relations between these factors are as follows:

$$\% \ P_2O_5 \times 2.185 = \% \ \text{BPL or} \ \% \ \text{TPL}.$$

$$\% \ P_2O_5 \times 0.436 = \% \ P \text{ and, therefore,}$$

$$30.0\% \ P_2O_5 = 65.55\% \ \text{BPL or TPL} = 13.08\% \ P.$$

F. Units of Weight

Tonnages of reserves of phosphorite or PR are expressed in different ways in different parts of the world; that is, as long tons (2,240 pounds), short tons (2,000 pounds), and metric tons (1,000 kg or about 2,200 pounds), and as tons of rock of a certain grade (as in the western U.S.) or as tons of concentrate of a certain grade (as in Florida and North Carolina). All numbers in this report are in metric tons, and are expressed as metric tons of phosphate that contains more than a given content of P. Thus, figures given in the literature for the carbonatite deposit of Araxa in Brazil are given as 92 million metric tons containing from about 5 to 15% P (12 to 35% P_2O_5). If the average grade is 10% P (23% P_2O_5), then the rock is equivalent to 60 million metric tons of concentrate that contains about 15% P (35% P_2O_5), and this is the figure used for reserves at Araxa in Table 2. All other tonnage figures are reduced in the same manner.

G. Reserves and Resources

Reserves are defined as material that is minable at a profit using today's prices and technology. *Resources* are identified deposits of reasonably known extent and grade, but deposits whose contained minerals cannot be profitably recovered at today's prices. In this paper, additional resources are discussed that are not quantified because tonnage and grade data are lacking, or because they are too low in grade, too deeply buried, or contain deleterious elements. Before these deposits can be mined, new techniques of mining and processing must be devised.

III. WORLD PHOSPHATE RESERVES AND RESOURCES

A. Rationale Used in Estimating Reserves and Resources

The distinction between various classes of reserves (measured, indicated, and inferred) and classes of resources (hypothetical, speculative, etc.) has been a problem probably since the first attempt at quantification of an ore deposit was made. For this report the term *reserve* designates PR for which some physical measurements of thickness and extent have been made —by drilling, trenching, or geological mapping—and for which, therefore, reasonable knowledge of tonnage exists. Reserve also means that the material can be mined and processed today, at a profit. This implies that there is knowledge of the geologic structure so that mining plans can be made, and that there is knowledge of the chemical and mineralogic characteristics of the rock so that beneficiation plans can be made. No distinction is made herein between measured, indicated, or inferred reserves, because this distinction depends on the amount of drilling, trenching, test pits, etc., and these data, generally speaking, are not available. All of the data for reserves in the accompanying tables have been rounded to the nearest million metric tons. Production data for 1974 are from the U.S. Bureau of Mines (1975); figures have been rounded to the nearest 10,000 metric tons.

Resources are defined as material for which there is geological information as to extent and thickness, but where drilling is limited or nonexistent, or as deposits that are too low grade, contain too much deleterious elements, or cannot be mined or processed profitably with existing technology. The numbers in the tables are for those deposits for which there is some solid information. Additional, large resources exist for which information is very sketchy, or for which only geologic hypotheses indicate that a deposit may be present. Some of these areas and deposits will be discussed later.

Reserves, then, are somewhat limited in terms of years of production that remain. For example, it is likely that the better grade reserves of the central Florida phosphate district will be exhausted by the year 2000. Additional resources, however, will still be present, but these resources will cost more to produce. They probably will be minable by the year 2000.

B. Reserves and Resources by Type of Phosphate Deposit

Phosphorite deposited on the adjacent, somewhat more stable, mentary, weathering, and biologic processes. These deposits can be divided into several types: (i) marine phosphorite, (ii) apatite of igneous origin, (iii) residual deposits, (iv) phosphatized rock, and (v) guano. Reserves and resources of these types of deposits will be discussed in the above order, roughly corresponding to the order of their current economic importance.

1. MARINE PHOSPHORITE DEPOSITS

About four-fifths of the world's production of PR is from deposits of marine phosphorite. The richest and largest deposits formed in warm latitudes and in areas of upwelling in the tradewind belts on the west coasts of continents or in mediterranean seas on the equatorial side of the basin. Deposits also form on the west sides of poleward-moving warm currents on the eastern coasts of continents where there are cool coastal countercurrents. The thickest accumulations occur in areas of geosynclinal subsidence where clastic sedimentation is at a minimum—i.e., in the miogeosyncline. Phosphorites deposited in the eugeosyncline are so mixed with the thick shale, chert, and volcanic material as to be generally noneconomic. Deposits in the geosyncline are characterized by a suite of rocks that consist, in a shoreward direction, of black shale, phosphatic shale, phosphorite, dolomite, chert, and saline deposits, red or light-colored sandstone, and shale. The rocks grade laterally into one another and may be repeated vertically due to lateral shifts of environment.

The phosphate in these rocks is carbonaceous and consists of pellets, nodules, and phosphatized bone material and shell. The pellets may be directly precipitated or may form by diagenesis; some have replaced calcite

Table 1—Reserves, resources, and production of marine phosphorite (in metric tons of material containing at least 30% P_2O_5).

Location	Reserves	Resources	Production 1974†
North Africa			
Algeria	500×10^6	600×10^6	0.7×10^6
Morocco	$5,000 \times 10^6$	$35,000 \times 10^6$	19.7×10^6
Tunisia	500×10^6	800×10^6	3.8×10^6
West Africa			
Angola	20×10^6	100×10^6	
Senegal (Taiba)	100×10^6	$1,000 \times 10^6$	0.4×10^6 (Al)
(Thies)	90×10^6	$2,000 \times 10^6$	1.5×10^6 (Ca)
Western Sahara	$1,600 \times 10^6$	$15,000 \times 10^6$	2.4×10^6
Togo	100×10^6	200×10^6	2.6×10^6
Middle East			
Egypt	800×10^6	$2,000 \times 10^6$	0.5×10^6
Iran	30×10^6	100×10^6	
Iraq	60×10^6	600×10^6	
Israel	100×10^6	200×10^6	0.9×10^6
Jordan	100×10^6	200×10^6	1.6×10^6
Saudi Arabia		$1,000 \times 10^6$	
Syria	400×10^6	400×10^6	
Turkey		300×10^6	
Europe			
USSR (Kazakhstan)	250×10^6	250×10^6	
(Karatau)	700×10^6	700×10^6	11.5×10^6
(Aldan, Yakut)	500×10^6	$2,000 \times 10^6$	
Other (France, Belgium, Germany)	15×10^6	30×10^6	0.1×10^6
Asia			
Australia	500×10^6	$1,500 \times 10^6$	Trace
China	100×10^6	$1,000 \times 10^6$	3.0×10^6
India	70×10^6	200×10^6	0.1×10^6
Mongolia	250×10^6	700×10^6	‡
North Vietnam	100×10^6	400×10^6	1.2×10^6
Pakistan		150×10^6	
North America			
Mexico (Baja)		$1,000 \times 10^6$	
(Zacatecas)		140×10^6	
USA (Eastern)	$1,600 \times 10^6$	$6,000 \times 10^6$	41.4×10^6
(Western)	$6,000 \times 10^6$	$7,000 \times 10^6$	
South America			
Brazil (Bambui)	200×10^6	500×10^6	
(Olinda)		20×10^6	
Colombia		600×10^6	Trace
Peru (Sechura)		$6,100 \times 10^6$	
Venezuela	20×10^6	20×10^6	0.1×10^6
Totals	$19,705 \times 10^6$	$87,810 \times 10^6$	92.1×10^6 (83.8% of world total)

† U.S. Bureau of Mines (1975).
‡ Included in totals of USSR.

or dolomite. Individual beds are as much as several meters thick and may contain 13% P (30% P_2O_5) or more, and the beds may extend over hundreds of square kilometers.

Phosphorite deposits on the adjacent, somewhat more stable, continental shelf is associated with light-colored chert, calcareous chert, sandstone, and shale. The phosphate is pelletal, and phosphatized shell and bone material is common.

Miogeosynclinal deposits are common throughout the world; examples are the Permian Phosphoria Formation of the western U.S., Kara Tau in the USSR, Miocene deposits of the Sechura desert of Peru, the Cambrian of Australia, and others. Reserves and resources are measured in millions or billions of metric tons (Table 1).

Phosphorites formed by mixing of warm currents along the eastern coasts of continents consist of phosphatic limestone or sandstone. Black shale and chert are not found, although diatomaceous material may be present. Deposits are economic only when they have been reworked by submarine currents, reworked by winnowing by stream action on the continents, subjected to chemical weathering, or a combination of these. The deposits also have to be concentrated by some method of physical beneficiation after mining. The deposits along the east coast of the United States are the most important example of this class of deposit.

Deposits of phosphate also form in epicontinental seas on the stable continental interior, where they are associated with limestone, dolomite, shale, and glauconitic sandstone. The phosphate is in the form of nodules or grains that are very large with respect to the grain size of the matrix in which they are embedded. The phosphate beds contain only a small percentage of P; however, the nodules contain as much as 15.3% (35% P_2O_5). Such deposits are common in many places in the world. Reserves are limited because of the low tonnage of nodules per unit area, but resources may be very large (Table 1).

2. APATITE DEPOSITS OF IGNEOUS ORIGIN

Apatite deposits of igneous origin occur as intrusive masses or sheets, marginal differentiates, hydrothermal veins, pegmatites, and disseminated replacements. The largest deposits are intrusive sheets or masses that are associated with alkaline igneous rocks (nepheline syenites, pyroxenites, carbonatites, ijolites, etc.). The outstanding igneous apatite deposit in the world is that which occurs in the Khibina nepheline syenite in the Kola Peninsula of the USSR. Here, arcuate ringlike bodies of apatite-nepheline rock that contain as much as 65% apatite are found. According to Barr (1960), the largest apatite lens extends for almost 4 km along the strike and has a maximum thickness of 180 m.

The igneous complex at Phalaborwa (Palabora) in South Africa is the best example of ores of the phoscorite (apatite-serpentine-magnetite rock) and pyroxenitic apatite types. The Palabora igneous complex consists of a pyroxenite intrusive into Precambrian granite and a carbonatite intrusive into the pyroxenite. The carbonatite contains only minor concentrations of apatite, but it is surrounded by a zone of phoscorite that contains as much as 50% apatite. The pyroxenites, which cover about 15.6 km^2, contain disseminations, segregations, and veins of apatite. Large tonnages of pyroxenite that contain about 15% apatite, or more, have been blocked out, and future mining will be of this type material.

Carbonatites containing only small amounts of apatite but associated with calcite, magnetite, pyrochlore, and other minerals may be residually enriched in apatite by chemical weathering in a tropical environment. The carbonatites of Brazil offer excellent examples of this type of occurrence. The residual ore is a nondescript red-brown lateritic soil that may contain as much as 11% P (25% P_2O_5). Because the material is unconsolidated, it is simple to mine and process. These bodies may contain as much as several hundred million metric tons of weathered material averaging between 4.4 and 6.5% P (10 to 15% P_2O_5). The carbonatite deposits offer, in addition, many other mineral commodities that may be recovered as coproducts. These include Nb, the rare earths, Ti, magnetite, monazite, vermiculite, Cu minerals, and perhaps others. Of course, not all of these minerals are present at every locality, but a multiproduct mining plan may offer interesting economic possibilities.

The apatite mineral of the igneous apatites is fluorapatite, although in the weathered zone of carbonatites the mineral may be carbonate fluorapatite. Chlorapatites are known in minor amounts in some pegmatite deposits, and hydroxyfluorapatites are present in some of the pegmatite or vein deposits.

Precise data on P reserves in igneous deposits are difficult to obtain, particularly from the USSR and China. Many deposits that may have considerable reserves of apatite have not been adequately prospected or explored and reserve data, therefore, are incomplete. Table 2 shows known reserves of apatite from igneous deposits and some data on resources. For the deposits of the Khibina alkaline igneous complex in the Kola Peninsula (USSR), Deans (1968) reported that "Here there are ore bodies of apatite-

Table 2—Reserves, resources, and production of apatite of igneous origin (in metric tons of material containing at least 30% P_2O_5).

Location	Reserves	Resources	Production 1974†
Kola Peninsula, USSR	400 × 10⁶	400 × 10⁶	11.0 × 10⁶
Phalaborwa, South Africa	100 × 10⁶	1,300 × 10⁶	1.4 × 10⁶
Brazil (Carbonatites)			
Araxa	50 × 10⁶	50 × 10⁶	
Jacupiranga	12 × 10⁶		
Catalao	75 × 10⁶	75 × 10⁶	0.2 × 10⁶
Tapira	100 × 10⁶	150 × 10⁶	
Others		500 × 10⁶	
Eastern Uganda	40 × 10⁶	160 × 10⁶	0.02 × 10⁶
Finland	50 × 10⁶	100 × 10⁶	
North Korea	5 × 10⁶	30 × 10⁶	0.4 × 10⁶
Southern Rhodesia			
Dorowa	10 × 10⁶	10 × 10⁶	0.1 × 10⁶
Canada	--	40 × 10⁶	
Others	10 × 10⁶	30 × 10⁶	
Totals	852 × 10⁶	2,845 × 10⁶	13.1 × 10⁶ (11.9% of total world production)

† Production figures from U.S. Bureau of Mines (1975).

nepheline rock with reserves of the order of 2,000 million tons...". Notholt (1975) reported reserves and resources as 1,660 million metric tons containing 7.9% P (18% P_2O_5). If the apatite concentrate contains 15% P (35% P_2O_5), then these reserves contain about 800 million metric tons of apatite, and in Table 2 these are divided arbitrarily between reserves and resources.

For the deposit at Phalaborwa in South Africa, Notholt (1975) reported reserves and resources of 1,400 million metric tons containing 15% P (35% P_2O_5). Deans (1968) reported that the 600 million short tons blocked out in the area should yield about 90 million tons of concentrate. Therefore, in Table 2 I have shown reserves of 100 million metric tons, and resources of 1,300 million metric tons.

Data on deposits in Brazil (all from weathered carbonatite ores) are from various sources, and reserve numbers are thought to be reasonable. Resource data are based on geologic information, and the category marked "Others" (under Brazil) is of particular interest. At the present time, many circular bodies have been delineated by various aerial photographic techniques; and, although only a few of these bodies have been investigated, some are carbonatites, and some of these bodies are expected to contain residual concentrations of apatite. The figure given for resources, therefore, is thought to be conservative, but it is admittedly a guess.

The general heading "Others" includes several carbonatite bodies in Africa, and pegmatite apatites in Chile, Canada, Brazil, and elsewhere.

Production in 1974 was about 13 million metric tons, almost 12% of the world total.

3. RESIDUAL DEPOSITS

During chemical weathering of phosphatic limestones, calcite is preferentially removed, leaving a residually enriched deposit consisting of phosphate, quartz, and other minerals resistant to weathering. The brown-rock deposits of Tennessee are residual; calcite of the source limestones has been removed, leaving a deposit of apatite, quartz, and minor clay and Fe oxide minerals. The deposits of Thies in Senegal are residual deposits of Ca-Al phosphate (Notholt, 1975). The residual deposits are at or near the surface, and they occur on deeply irregular bedrock surfaces.

Many phosphatic limestone deposits in the world contain at least a thin residual cap at the present surface. For example, the deposits of the Phosphoria Formation in the western U.S. are leached of much of their calcite content at the surface—the surface ores are much richer in P content than is the unweathered material in the subsurface. Some of the deposits in the sandy facies in Colombia, South America, contain about 15% calcite in the deep, unweathered sections, but consist only of quartz and apatite in the weathered surface outcrops. Surface weathering here extends from 10 to 20 m below the surface.

Reserves of residual deposits are measured in thousands to millions of metric tons. They may be important deposits because of high quality and thin overburden. In 1974, production from residual deposits amounted to several million metric tons, or about 3% of the world production.

Table 3—Reserves, resources, and production of deposits derived from guano.

Location	Reserves	Resources	Production 1974†
		metric tons	
	Calcium phosphate		
Nauru Island	50×10^6		2.3×10^6
Ocean Island	20×10^6		0.5×10^6
Angaur Island	2×10^6		
Paracel Island	10×10^6		
Christmas Island	100×10^6	70×10^6	1.8×10^6
Langebaan (RSA)	7×10^6		
Minjingu	6×10^6		
Curacao	10×10^6		0.1×10^6
Others‡	10×10^6		Trace
Subtotal	215×10^6	70×10^6	
	Aluminum and iron phosphates		
Trauira ⎫	10×10^6		
Pirocaua ⎬ Brazil	10×10^6		
Itacupin ⎭	5×10^6	30×10^6	
Aruba	10×10^6		
Others§	5×10^6		
Subtotal	40×10^6	30×10^6	
Grand total	225×10^6	100×10^6	4.7×10^6 (4.3% of total world production)

† Production figures from U.S. Bureau of Mines (1975).
‡ Others include Peleliu, Fais, Rota, Saipan, Okina-daito, Makatea, Fiji, Bellona, Rennell, Walpole, Surprise, Clipperton, Bismark Archipelago, Somberero, Barbuda, and Laccadives.
§ Includes Malpelo Island, Bomi Hill (Liberia), Cayman Island, Kita daito, Redonda, Martinique, Mona Island (Puerto Rico), Navassa, and Le Gran Connetable.

PHOSPHATIZED ROCK

a. Derived from Guano—In areas of even slight rainfall, the soluble phosphate of guano is dissolved and carried downward, where it replaces or alters the underlying rock. Where the underlying rock is limestone, secondary deposits of Ca phosphate (apatite) that contain hundreds of thousands to scores of millions of metric tons are formed. Where the underlying rock is volcanic or igneous, Fe or Al phosphate deposits that contain hundreds of thousands to a few million metric tons are formed.

The most important deposits of Ca phosphate in this category are found in the coral islands of the Pacific (Nauru and Ocean Islands) and the Indian Ocean (Christmas Island). Deposits are also known in many other parts of the world (Table 3).

Some of the most interesting deposits of Fe and Al phosphate derived from guano are those of Trauira, Pirocaua, and Itacupin Islands off the northeastern coast of Brazil. Although these "islands" are now connected to the coast by low, tidal mangrove swamps, during a Pleistocene rise in sea level they were probably separated from the mainland by some tens of kilometers. The "islands" are more or less circular plugs of alkaline igneous rock that were altered by lateritic weathering to form high Fe and Al cappings (in the case of Trauira, the capping was a bauxite). The P content of the original rock was about 0.1 to 0.2%; during the lateritic weathering, the P content increased to about 1.0 to 2.0%. Phosphorus content of the Al and Fe phosphate cappings today is as much as 11% P (25% P_2O_5). The source of the P is thought to be guano, derived from a Pleistocene bird rookery.

Production of phosphate derived from guano was about 4% of the world's production of PR in 1974 (Table 3). Except for 100,000 metric tons of low-fluorine PR from Curacao, all production was from Ocean, Nauru, and Christmas Islands.

b. Derived from Phosphatic Sedimentary Rocks—The apatite mineral of sedimentary rocks (carbonate fluorapatite) is insoluble except in acid solutions. Thus, in regions of tropical or subtropical weathering where the pH of meteoric waters is below 7.0, the apatite mineral is decomposed, and the P in solution may be reprecipitated or may replace the underlying rock. Again, if the rock is limestone, the phosphatized rock will be composed of Ca phosphate; in phosphatized volcanic rock, the Al and Fe phosphate minerals variscite, barrandite, and strengite are formed; and if the underlying rock is a clay, wavellite, crandallite, millisite, and augelite may be formed (Altschuler, 1973). The deposits consist of irregular, tabular re-

Table 4—Reserves, resources, and production of guano, in material containing at least 30% P_2O_5 (ND = no data; -- = no production, 1974).

Location	Reserves and resources	Production, 1974†
	metric tons	
	Bird guano	
Argentina	‡	TR (few tons)
Chile	A few tens of thousands	19,000
Peru	Do	23,000
Ecuador	ND	--
Venezuela	‡	--
Brazil	ND	--
St. Helena	Very small	--
Ascension	Do	--
Madagascar (Malagasy)	Do	--
Seychelles Island	Small	7,000
	Bat guano§	
Philippines	50,000	1,000
Malaysia	30,000	--
Indonesia	50,000	--
Thailand	Small	--
Jamaica	Do	--
Anguilla	Do	--
Total	A few hundred thousand tons	50,000 (<0.1% of total production)

† Production figures from U.S. Bureau of Mines (1975).
‡ Small islands off the coasts of Argentina, Brazil, Ecuador, Venezuela, and Madagascar contain small tonnages of bird guano. Reserves are of little importance in the world market and production is miniscule.
§ Figures for tonnages of bat guano for Philippines, Malaysia, and Indonesia are aggregates for many bat caves. Data from Bureau of Minerals, Manila (1968).

placement of the underlying rock, or of phosphate encrustations on surfaces of joints and cracks, or of both. Deposits are small, ranging from a few thousand to millions of metric tons. Although small tonnages of this type material have been mined in the past, there has been no significant production in recent years.

5. DEPOSITS OF GUANO

Guano, the dried excrement of sea birds and bats, is a valuable fertilizer material because of its content of water-soluble P and N. Most deposits are small—containing a few hundred or a few thousand metric tons—but the deposits on the desert coasts and offshore islands of Chile and Peru originally contained hundreds of thousands of tons of "fossil" guano. Thicknesses of fossil guano on Chincha Island (Peru) were as much as 45.6 m (Warin, 1968). Deposits of bird and bat guano are known and have been mined in many places in the world (Table 4), but the only significant deposits are those on the desert coasts of Peru and Chile. Those deposits are divided into fossil red guano and fresh white guano. Almost all production today is from the white guano that contains about 15% N, 4.4 to 5.2% P (10 to 12% P_2O_5), and 1.7% K (2% K_2O). No material is exported and production is limited to the amounts produced yearly by the birds.

World production of guano in 1974 was 50 thousand metric tons— <0.1% of total phosphate production—almost all of which was from Chile and Peru.

IV. COMPARISON TO OTHER ESTIMATES

A. Past Reserve and Resource Estimates

The accuracy of any reserve estimate depends on a large number of factors, the most important of which is the amount of information the estimator has at his disposal. For example, in 1938, total world reserves were estimated to be 17.5 billion tons (Barr, 1960). Jacob (1953) estimated world reserves to be 46.7 billion metric tons, but pointed out that large quantities were not included because reasonably accurate data were not available. In 1971, the British Sulphur Corp. estimated world reserves of all grades to be 130 billion metric tons, equivalent to about 19.5 billion metric tons of P. The Institute of Ecology (1971) stated that known reserves of PR might be exhausted in 90 to 130 years, equaling only about 3.4 billion tons of P—a figure that is obviously in error.

Emigh (1972) made an estimate of world phosphate reserves to refute the low estimate made by the Institute of Ecology. In this report he stated: "Tabulating only the presently quantifiable rock deposits, I can document estimated reserves of 1,298,000 million tons about 52 times the 25,000 million tons indicated by the IOE report." The number refers to total PR in the ground, and the average P content of this material is probably about 4.4% (10% P_2O_5) or slightly less.

Wells (1975) has made a different estimate of the world P. His definitions of reserve classes are:
1) *Economic Reserves*—ores that can be mined at present costs. This is equivalent to reserves of this report.
2) *Subeconomic Deposits*—deposits that can be mined at somewhat higher costs than those for the economic deposits. This category is about equivalent to resources of this report.
3) *Low-grade Deposits*—those with even higher costs, including ores that contain from 0.44 to 2.2% P (1 to 5% P_2O_5). These low-grade deposits form the bulk of Emigh's tonnage and will be discussed later.
4) *Common Rock Deposits*—those with about 0.044% P (0.1% P_2O_5).

Category 4 of Wells will not be considered in this paper, but it is a most interesting category. He estimates (Wells, 1975) that the topmost 1.6 km of the earth's crust contains 5.9×10^{14} metric tons of P, enough if used in fertilizer to last for 0.5 million years, assuming a population that eventually reaches 20 billion. Costs to extract P from common rock probably would be very high—estimates are as much as about $11,000/metric ton of P. It seems likely, then, that there is enough P in the crust of the earth to last for an extremely long time.

In this report, the total reserves of rock containing at least 13% P [30% P_2O_5 (Tables 1-4)] are about 20,000 million metric tons. Resources are an additional 90,000 million metric tons. These figures are equivalent to about 14,300 million metric tons of P.

The total reserves and resources of PR have increased immensely over the years (and in the last 20 years in particular) due to the vigorous search for new deposits.

B. Other Resources

Emigh (1972) listed several areas in the world that contain what he called "reserves for the future." These "reserves" are about equivalent to Wells' (1975) category of low-grade deposits. These deposits have not been considered reserves in the past because of low-grade, deleterious gangue minerals, problems of mining and processing, etc. In short, they cannot be mined and processed using today's technology and with today's economics. These deposits will not be discussed here in any great detail; calculations of amounts present in a few of the areas will suffice.

Total reserves and resources for the western U.S. (Table 1) are 13,000 million metric tons, equivalent to about 2,000 million metric tons of P. Cathcart and Gulbrandsen (1973) pointed out that the Phosphoria Formation contains about 16×10^9 metric tons of P; thus, 14×10^9 metric tons of P is present as an additional "resource." Most of this material is deeply buried, is not weathered, and will require new mining and processing techniques.

In the eastern U.S., very large resources are present as phosphatic carbonate rock, such as the Hawthorn Formation of Florida. In the Ft. Meade quadrangle (Cathcart, 1966), the Hawthorn Formation was estimated, on the basis of a few drill holes, to contain 1,500 million metric tons of phosphate pellets having an average content of about 11% P (25% P_2O_5). Other quadrangles in the Florida Land-Pebble phosphate district contain similar amounts. Using this figure as a projection for the whole state, the Hawthorn Formation could consist of hundreds of billions of metric tons of phosphate pellets containing tens of billions of metric tons of P. This resource is too deeply buried to be mined by conventional methods, and processing will also require new techniques.

Large resources of PR are present offshore in the modern oceans. Deposits are known in the Atlantic Ocean off the African Coast and off the coast of the United States from Florida to North Carolina, and in the Pacific Ocean off New Zealand and off the coast of North and South America. The following computation of resources off the coast of Baja California, Mexico, illustrates the possible amounts of material. D'Anglejan (1967, Fig. 1) showed the area underlain by phosphatic sediments to be 250 km long by an average of about 20 km wide. If the phosphate-bearing sediment is 1 m thick, then about 10,000 million metric tons of phosphatic sand is present. If this sand contains 10% phosphate pellets (a conservation estimate), then 1,000 million metric tons of pellets, which may contain about 13% P (30% P_2O_5), is present. This number is purely a guess, but it indicates that the total resource of P on the sea floor must be extremely large.

Other areas in the world that contain large, but unmeasured, amounts of P include Australia; Tennessee; Alaska; Africa; the Near East; Peru, Colombia, and Brazil in South America; China; Mongolia; and the USSR.

Total resources of "low-grade rock," then, are enormous, although these materials will require new techniques of mining and processing and much higher costs than the "high-grade" rock currently being mined.

A recent issue of *Economic Geology* (March–April 1979, Vol. 74) has several papers of general interest on the geology and reserves of phosphate. These include an overview (Howard, 1979); deposits not previously described in the literature, namely, Sechura (Peru) and the Ontario carbonatite province (Cheney et al., 1979; Erdosh, 1979); a review of European and USSR deposits (Notholt, 1979; Fuller, 1979); shelf and onshore deposits of southern Africa (Fuller, 1979); Geochemistry and origin of the Georgina Basin deposits in northern Australia (Howard & Hough, 1979); and marine deposits as related to plate tectonic theory (Cook & McElhinny). In addition, recent data on world reserves can be found in a paper by Stowasser (1977), and in a National Academy of Sciences report (Comm. Acc. Elem., NAS, 1979).

C. Variables that Influence Reserve Tonnage

Reserves, as defined herein, are depleted by mining at a high annual rate. World production in 1974 was about 110 million tons, and the rate of production has been increasing and will probably continue to increase in the future. Total reserves, then, will begin to decline; but as the high-grade, low-cost reserves are depleted, prices will rise, turning more and more of resources into reserves. The search for new deposits is continuing; and, I believe, tonnages of reserves will continue to rise, at least for the immediate future.

Until the present time, however, production has come from PR high in P and low in undesirable elements—Al, Fe, Mg, etc. This premium-grade rock is rapidly being depleted throughout the world, and more and more rock being sold to manufacturers contains larger and larger amounts of the contaminant elements. This is causing problems in the production of fertilizers.

Most of the tonnage mined today (about three-fourths) is from large, low-cost, open-pit mines. In the future, more and more rock will come from much higher cost underground mines. Much of the resource tonnage (Tables 1–4) and a part of the reserve tonnage are from areas that will have to be mined by underground methods. Costs for underground mining are rising, and part of the reserve tonnage may have to be put into the resource category if costs become too high.

The geographic location of a deposit may be enough to cause the deposit to be considered as a resource rather than a reserve. For example, there are very large tonnages of PR in the Sechura Desert of Peru. All of this material is considered as a resource because of the location—in a desert on the Pacific Coast of Peru, with no port, no town, no railroad, and no fresh water. To mine the mineral, it will be necessary to build everything that involves a very high capital investment. The deposit in northern Saudi Arabia is in the same position. Both of these deposits contain large tonnages of rock that can be beneficiated using present-day techniques into a high-grade phosphate product, but they are not considered as reserves because of their geographic location. Both of these deposits, however, will certainly be mined in the future.

D. Environmental Considerations

Mines and chemical processing plants in the United States are being forced, by law, to control air and water effluents and to reclaim land after mining. Although this is not true for most other countries in the world at the present time, there is a growing concern everywhere about the environment, water and air pollution, and land reclamation.

Certain phosphate deposits that may otherwise be minable may not be mined because of environmental considerations. One example in the United States is the deposit on the Savannah River: the state of Georgia refused to grant permission to mine, because the Savannah River and adjacent tidelands would be "irreparably damaged" by the mining. Thus, at least 2,000 million metric tons of recoverable PR of about 13% P (30% P_2O_5) must be considered a resource rather than a reserve.

Demands on the mining industry in the form of very restrictive laws could make open-pit mining virtually impossible, and some legislators have proposed laws that would ban open-pit mining. It is obvious that the mining and chemical industries must make strong efforts to reclaim land as promptly as possible and to keep air and water pollution to a minimum in order to avoid serious trouble with county, state, and federal legislatures.

E. Other Variables

In November 1973, Morocco announced a rise in the price of PR from $15 to $48/metric ton. Other countries quickly followed suit, and the net result was a tripling of the price at the mines. The price continued to rise and reached a maximum of $88/ton early in 1975, when there was the beginning of buyer resistance. After 1 year the price had dropped to about $45 to $55/ton (Eng. Min. J., 1976). In Dec. 1979, the price of Florida rock was $17 to $26/ton (Eng. Min. J., 1979).

The rise and fall of the price of PR, controlled by one very large producer, influences the amount of material that can be considered as reserve or resource tonnage. At high prices, material can be mined which is considered marginal or noneconomic during times of low prices.

LITERATURE CITED

Altschuler, Z. S. 1973. The weathering of phosphate deposits—Geochemical and environmental aspects. p. 33-96. *In* E. J. Griffith, A. Beeton, J. M. Spencer, and D. T. Mitchell (ed.) Environmental phosphorus handbook. John Wiley & Sons, Inc., New York.

Barr, J. A. 1960. Phosphate rock. p. 649-668. *In* J. L. Gillson (ed.) Industrial minerals and rocks. 3rd ed. Am. Inst. Mining and Metall. Eng., New York.

British Sulphur Corporation. 1971. A world survey of phosphate deposits. 3rd ed. 180 p. British Sulphur Corp., London.

Bureau of Minerals, Manila. 1968. Mineral raw material resource for fertilizer industry in the Philippines. p. 78-80. *In* Proc. of seminar on "Sources of Mineral Raw Materials for the Fertilizer Industry in Asia and the Far East." Bangkok, Thailand, 1967; sponsored by ECAFE. U.N. Min. Res. Dev. Ser. no. 32.

Cathcart, J. B. 1966. Economic geology of the Ft. Meade quadrangle, Polk and Hardee Counties, Florida. U.S. Geol. Survey Bull. 1207. 97 p.

Cathcart, J. B., and R. A. Gulbrandsen. 1973. Phosphate deposits. p. 515-525. *In* D. A. Brobst and W. P. Pratt (ed.) United States mineral resources. U.S. Geol. Survey Prof. Paper 820.

Cheney, T. M., G. H. McClellan, and E. S. Montgomery. 1979. Sechura phosphate deposits, their stratigraphy, origin, and composition. Econ. Geol. 74:232-259.

Committee on Accessory Elements, NRC, National Academy of Sciences. 1979. Redistribution of accessory elements in mining and mineral processing, Part II: Uranium, phosphate, and alumina. NAS, Washington, D.C. p. 45-104.

Cook, P. J., and M. W. McElhinny. 1979. A reevaluation of the spatial and temporal distribution of sedimentary phosphate deposits in the light of plate tectonics. Econ. Geol. 74:315-330.

D'Anglejan, B. F. 1967. Origin of marine phosphorites off Baja California, Mexico. Marine Geol. 5:15-44.

Deans, T. 1968. Exploration for apatite deposits associated with carbonatites and pyroxenites. p. 109-119. *In* Proc. seminar on "Sources of Mineral Raw Materials for the Fertilizer Industry in Asia and Far East." Bangkok, Thailand, 1967; sponsored by ECAFE. U.N. Min. Res. Div. Series 32.

Emigh, G. D. 1972. World phosphate reserves—are there really enough? Eng. Mining J., April 1972, 173(4):90–95.

Engineering and Mining Journal. 1976. Morocco (News item) March 1976. 177(3):272

Engineering and Mining Journal. 1979. Metal and non-metal markets. (News item) December 1979. 180(12:000.

Erdosh, George. 1979. The Ontario carbonatite province and its phosphate potential. Econ. Geol. 74:331–338.

Fuller, A. O. 1979. Phosphate occurrences on the western and southern coastal areas and continental shelves of southern Africa. Econ. Geol. 74:221–231.

Howard, P. F. 1979. Phosphate. Econ. Geol. 74:192–194.

Howard, P. F., and M. J. Hough. 1979. On the geochemistry and origin of the D tree, Wonarah, and Sherrin Creek phosphorite deposits of the Georgina Basin, Northern Australia. Econ. Geol. 74:260–284.

Institute of Ecology. 1971. Man in the living environment. Report of the 1971 workshop on global ecological problems. The Inst. of Ecology, Chicago.

Jacob, K. D. 1953. Phosphate resources and processing facilities. In K. D. Jacob (ed.) Fertilizer technology and resources in the United States. Agronomy 3:117–165. Academic Press, New York.

Notholt, A. J. G. 1975. Phosphate rock, world production, trade, and resources. p. 104–119. In R. F. S. Fleming (ed.) Proc. of 1st Ind. Miner. Int. Congr. Metal Bull., Ltd., London.

Notholt, A. J. G. 1979. The economic geology and development of igneous phosphate deposits in Europe and the USSR. Econ. Geol. 74:339–350.

Stowasser, W. F. 1977. Phosphate. U.S. Department of the Interior, Bureau of Mines. Mineral Commodity Profiles MCP-2, May 1977.

U.S. Bureau of Mines. 1975. Phosphate rock. Miner. Trade Notes. 72(9):10–11.

Warin, O. N. 1968. Deposits of phosphate rock in Oceania. p. 124–131. In Proc. of the seminar on sources of mineral raw materials for the fertilizer industry in Asia and the Far East. U.N. Min. Res. Div. Ser. 32.

Wells, F. J. 1975. The long-run availability of phosphorus—Resources for the Future, Inc. The Johns Hopkins Univ. Press, Baltimore, Md. 121 p.

Chapter 2

The Phosphate Industry of the United States

JAMES B. CATHCART

U.S. Geological Survey
Denver, Colorado

I. INTRODUCTION

Phosphate rock (PR) is produced from three distinct and different forms of deposits: guano or guano-derived deposits; igneous, the so-called apatite deposits; and sedimentary deposits, usually called *phosphorites*.

Guano or guano-derived deposits, particularly those of the Pacific Islands (Ocean Island, Christmas Island, and Nauru), account for about 5% of the PR produced in the world. In the United States, a few very small bat guano deposits are known from caves in the arid southwest (some of which have been mined in the past), and small guano deposits have formed from bird rookeries in southern Florida and on some of the islands of the Hawaiian chain. The occurrences in the United States are not economically important, although some may be mined for local use in the future.

Igneous apatites are of economic importance in the Kola Peninsula (USSR), in Palabora (South Africa), and from carbonatites in Brazil and Africa. About 20% of the world's production of phosphate is from igneous apatite. As shown in Fig. 1, deposits of igneous apatite are known in the United States (for example, nelsonite deposits of Virginia; apatite-magnetite deposits of the Adirondacks of New York and New Jersey; marginal apatite differentiates at Iron Springs, Utah; and apatite-rich iron ores at Pea Ridge, Missouri). None of these deposits are being mined for their apatite content, but byproduct apatite is concentrated in processing the iron ores of Pea Ridge, Missouri. Guild (1967) estimated that, at full production of iron ore, about 100,000 metric tons of apatite per year would be concentrated as a byproduct. Basic slag (also called *Thomas meal*) is a phosphatic slag produced in smelting phosphorus-rich iron ores. It is used as a fertilizer.

Sedimentary phosphorite deposits provide most of the phosphate used in the United States and about 75% of the world's production. The best deposits formed in warm latitudes and in areas of upwelling caused by

---

Copyright 1980 © ASA-CSSA-SSSA, 677 South Segoe Road, Madison, WI 53711, USA.
The Role of Phosphorus in Agriculture.

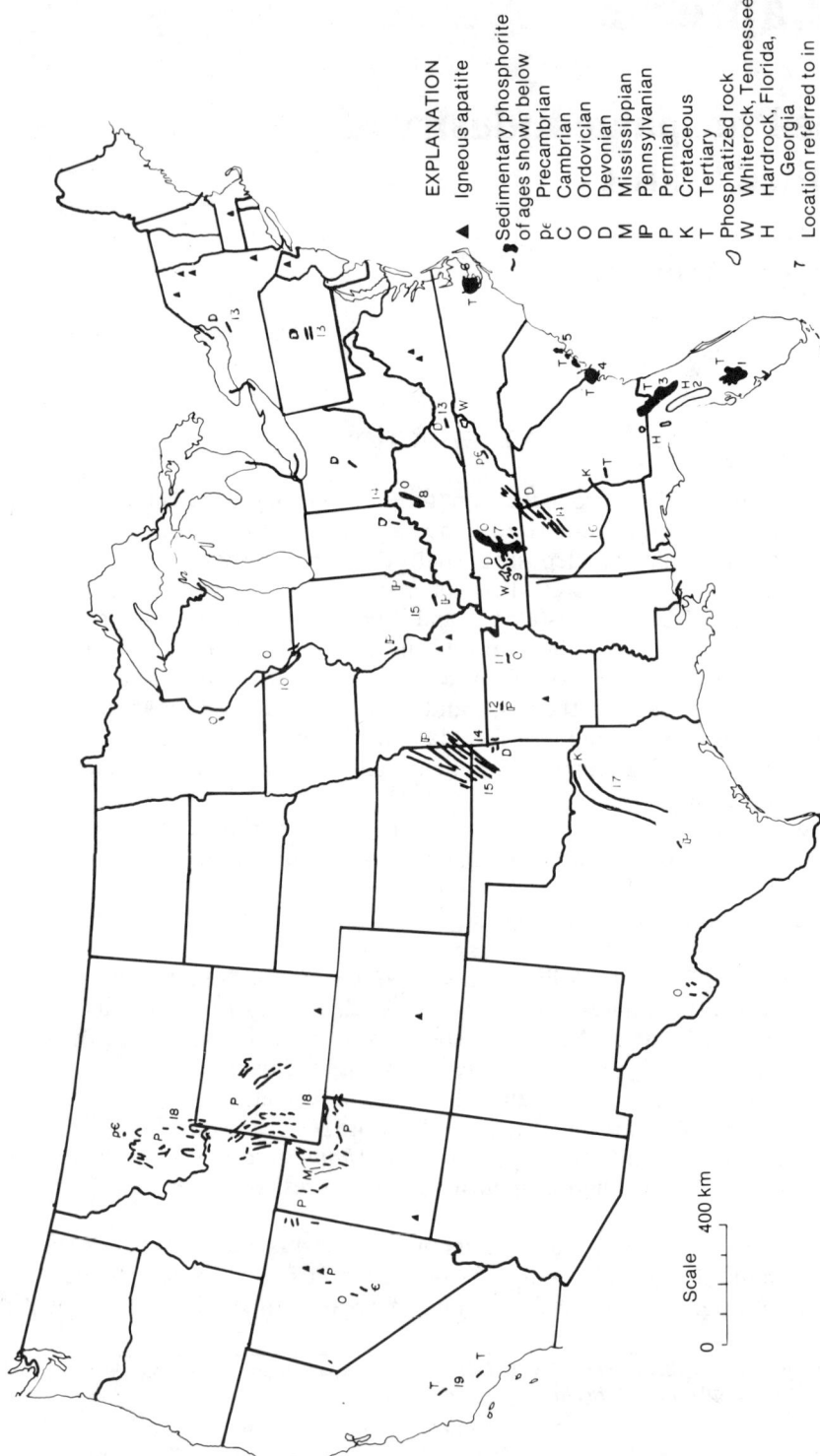

Fig. 1—Map of the United States showing distribution of phosphate deposits and occurrences.

divergence. These areas were in the tradewind belts on the west coasts of continents or in a basin along the equatorial side of mediterranean seas.

Deposits were also formed on the west sides of poleward-moving warm currents along the eastern coasts of continents where there is turbulent mixing with cool coastal countercurrents, as along the eastern coast of the United States in Miocene time.

Deposits caused by divergent upwelling are characterized by a suite of rocks that includes black shale, phosphatic shale, phosphorite, dolomite, and chert. The rocks grade laterally into one another, and the sequence may be repeated vertically because of lateral shifts caused by epeirogenic movements or eustatic sea-level changes. The thickest and best-grade accumulations of phosphorite occur where clastic sedimentation is at a minimum; that is, in the miogeosyncline.

Phosphorite that formed on the stable continental shelf adjacent to the miogeosyncline is associated with light-colored chert, sandstone, shale, and some limestone.

Phosphorites that formed in warm currents along the eastern coasts of continents consist of limestone, sandstone, diatomaceous earth, and clays or clayey sands, all containing phosphate pellets. These deposits are economic only if they have been reworked by submarine currents or enriched by weathering or both, and even then they must be upgraded by some method of beneficiation after mining to procure an economic product.

Deposits or occurrences of PR in eugeosynclinal sediments associated with chert, shale, limestone, and volcanic material have been described, but the phosphate beds are thin or too low in P content to be economic.

Marine phosphate deposits that form on the stable shelf or in the continental interior are associated with limestone, sandstone, shale, and glauconite. The phosphate is in the form of nodules that are very coarse with respect to the matrix in which they are found, and phosphate is also concentrated as lag gravels at unconformities.

II. ECONOMIC FACTORS

Factors that make a phosphate deposit economical include the obvious ones of tonnage and grade (P content), but a great many other factors are needed to determine whether a deposit can be mined and treated profitably. These factors include geologic setting, geographic location, potential use (for example, local use for direct application versus the manufacture of high-analysis chemical fertilizer for domestic use or export), availability of other raw materials (S, N, power, etc.), the physical characteristics of the deposit (hardness, size of particles, type of cementing material), and other chemical constituents of the PR (organic material, rare earths, trace metals, etc.).

The factors are interrelated, and none are absolute. To cite a most simple case, a low-grade deposit that is at or close to a potential market may be economical whereas a high-grade deposit in a remote geographic location may be uneconomical.

A. Phosphate Content

The phosphate content of the rock is of prime importance. A phosphate occurrence that contains $<4.4\%$ P (10% P_2O_5) is probably not minable under present conditions, whereas a bed containing 13% or more P (30% P_2O_5) probably is minable if other conditions are favorable. The amount of P present is interrelated with other factors. For example, a flat-lying, unconsolidated phosphorite, covered by thin overburden, might be economical if the total P content is 5.2% (12% P_2O_5), whereas a consolidated, steeply dipping bed containing 13% P or more (30% P_2O_5) might not be economical. It is not possible to cite absolute P contents that are economical, but the following minimum grades for different geologic conditions and uses for the rock are a useful guide: The minimum grade for unconsolidated material that can be concentrated by existing processes is about 5.2% P (12% P_2O_5). For consolidated material that can be upgraded by disaggregation and flotation, it is about 8% P (18% P_2O_5); for a consolidated deposit to be used for direct application to the soil, about 9% P (20% P_2O_5); for a consolidated deposit to be used as furnace feed, about 10.5% P (24% P_2O_5); to be used for acidulation, about 13% P (30% P_2O_5).

B. Tonnage

Tonnage is the most straightforward of all the economic factors. Gross tonnages of most marine phosphorites are large, but not all the material may be minable because of unfavorable structure, depth below water table, deleterious gangue minerals, thickness of the bed, thickness of the overburden, etc. Tonnages, then, must be reported in terms of potentially minable ore. Tonnage requirements for a minable deposit are a function of the total costs of the land, equipment, plant, maintenance, method of depreciation, etc.

C. Geologic Setting

The geology of the deposit will dictate the mining method and, therefore, the cost of mining. Near-surface, unconsolidated, flat-lying matrix can be mined cheaply with large, earth-moving equipment. A similar deposit that is consolidated can be mined by open-pit methods, but at a greater cost because consolidated materials must be broken before mining and crushed before processing. Steeply dipping and buried deposits must be mined by underground methods. In areas of complicated geologic structure, the phosphorite deposits may be so disrupted that mining is impossible.

D. Thickness of the Phosphorite Bed

For open-pit mining, using large equipment such as a 35-m^3 dragline, the minimum thickness of a phosphorite deposit that can be mined is about

2 m, and there is a relation of minimum thickness of the ore bed to the thickness of the overburden. This latter factor is also complex and varies with grade, degree of consolidation, etc. As a general rule, the ratio of overburden-to-ore thickness should be less than 3:1. In addition, the total thickness of material that can be mined (overburden plus ore thickness) is limited by local conditions. For example, a layer of unconsolidated material that is more than 15 m thick probably is unstable, and special mining plans have to be made for very deep deposits, regardless of favorable ratios.

E. Principal Gangue Materials

The principal gangue or diluting minerals in all marine phosphorites are quartz, carbonate minerals (calcite and dolomite), clay minerals (montmorillonite, attapulgite, illite, sepiolite, chlorite, and kaolinite), and Fe-oxide minerals (goethite, hematite, limonite). In addition, organic material is present in many deposits. These diluting materials affect processing and chemical uses of the phosphorite.

1. QUARTZ

Quartz, as silt-to-sand-size detrital grains, is present in all marine phosphorites. Quartz will have no effect on acidulation, except to reduce the amount of phosphoric acid produced per unit time as the quartz content rises and to increase the number of filters needed to remove the waste material. Costs to operate an acid plant, therefore, will rise as the amount of quartz increases. Quartz can be separated from phosphate by froth flotation, and a high-grade product can be made from a low-grade quartz-phosphate ore. Quartz may be beneficial if the rock is to be used for an electric furnace charge. The ratio of SiO_2 to CaO for a furnace charge is well established (Waggaman, 1952), and it may be possible to use a quartz PR for a furnace charge, adding only coke.

2. CARBONATE MINERALS (CALCITE AND DOLOMITE)

Calcite is an important gangue mineral in many marine phosphorites. Where the phosphorite is composed principally of calcite, apatite, and some quartz, the rock may be uneconomic because of costs of concentrating the apatite. The rock cannot be used for direct acidulation because calcite consumes acid, and the rock may not be usable as a furnace charge because of the necessity of adding large amounts of silica to flux the much more abundant Ca. Limits for CaO are not fixed for either furnace or acidulation, but the optimum ratio of CaO to P_2O_5 should be only slightly in excess of the ratio for the apatite mineral [$Ca_{10}(PO_4,CO_3)_6(F,OH)_{2-3}$; ratio CaO to P_2O_5 = 1.45–1.55]; i.e., the ratio should not exceed 1.6 to 1.

Dolomite, present in small amounts in some phosphorites, is a major gangue mineral in other marine phosphorites. Dolomite is deleterious in the same manner as calcite; however, it creates an additional problem in that only very small amounts of Mg can be tolerated in the manufacture of high-analysis phosphoric acid.

3. CLAY MINERALS

One or more of the clay minerals—montmorillonite, kaolinite, chlorite, attapulgite, or sepiolite—are locked in small amounts in the pelletal apatite of marine phosphorites. The alumina in these clay minerals is deleterious in the manufacture of wet-process acid. Alumina consumes acid and can cause problems at several points in the production, pumping, filtering, and shipping of phosphoric acid due to the formation of complex aluminum phosphates. Undissolved clays may also inhibit efficient filtration and washing of the calcium sulfate.

4. IRON MINERALS

The Fe minerals include the oxides goethite, hematite, and limonite, and the sulfide is pyrite. Iron consumes acid in wet-process acid manufacture and forms complex Fe phosphates. In the manufacture of elemental P, Fe combines with P to form the so-called ferro-phos. Ferro-phos is saved and is used to some extent, but the formation of ferro-phos causes a loss in P production. Pyrite forms poisonous H_2S gas during acidulation. Any amount of pyrite is deleterious.

5. ORGANIC MATERIAL

Organic material is present in all marine phosphorites, in the ground mass or in the phosphate pellets, or in both. The organic content of the phosphorites of Florida is low—about 0.5% (Jacob et al., 1933), whereas the organic content of the phosphorite of the western field and North Carolina is high—2 to 3% (Gulbrandsen, 1966; Rooney & Kerr, 1967). High-organic material causes foaming during acidulation, and in North Carolina the PR must be calcined at high temperature to drive off the organic material prior to acidulation.

F. Weathering

The surficial parts of all deposits are weathered to some extent. The effects of weathering on a phosphate deposit may so change the characteristics at the outcrop that beneficiation methods may have to be changed to process the fresh rock. It is essential to determine the character and tonnage of both the weathered material and the fresh material.

A sandy phosphorite, composed essentially of apatite and quartz, may be depleted in its P content by weathering by acid ground water, and the P in solution may form Fe or Al phosphate minerals in surficial samples. Thus, certain outcrop samples may be lower in P content and different mineralogically from the unweathered rock that may form most of the deposit.

In contrast, a calcareous phosphorite may be enriched in P content by removal of the more soluble carbonate. A surface sample may be high in P and completely devoid of calcite; therefore, a beneficiation scheme based on outcrop data may be erroneous.

The collection of samples that truly represents the bulk of the deposit, therefore, is of the utmost importance. Trenches must be dug to get below the zone of obvious surficial weathering, and core samples from drilling must be obtained to be sure of the character of the material below the zone of weathering.

G. Geographic Location

The simple fact of geographic location may be the most important factor in determining the economics of a phosphate deposit. Phosphate is a bulk commodity since millions of tons per year must be handled and transported. A high-grade, large-tonnage deposit that can be cheaply mined may be economical, no matter where it is located, if the cost of transport can be amortized over a long period of time and at a high tonnage rate per year.

H. Product Use

The use of the product is of great importance also. If the PR can be used only as ground rock for direct application, the material can only be used in a restricted area around the mine, because it will not stand high transport costs. If the deposit is high-grade and has large reserves, it may be possible to manufacture high-analysis phosphate chemicals at the mine and ship the product to a market area, either within the country of origin or as an export item.

I. Water

Water poses many problems in any mining operation—problems of both too little and too much. Mining below the water table means that water must be removed; and, whereas pumps are available that can handle any amount of water, cost is a problem. Furthermore, when the water pumped is a part of a domestic water supply, there is also the problem of polluting both surface and underground water.

Flotation beneficiation requires large amounts of water, and where water is scarce a deposit requiring this treatment may be uneconomical.

J. Requirements for Raw Materials Other than Phosphate

The kind and amount of raw materials used in a phosphate plant, other than the PR, depend on the process that is to be used.

Phosphate rock may be treated in a furnace with coke and silica to produce elemental P. Raw materials needed are coke, silica, and a large amount of electrical power. The amounts of silica and coke needed are given by Waggaman (1952). The charge to a furnace must be coarse grained (5 to 8 cm in diameter). The PR is mixed with enough silica to flux all of the Ca of the apatite mineral; the P is reduced by coke, and P gas is taken off the furnace and condensed to a solid P product. The calcium silicate slag is tapped at intervals and may be used as "road metal" or rock wool. Iron in the rock combines with P to form "ferro-phos," which is also tapped from the furnace at intervals and retained for possible sale.

Both ordinary superphosphate and wet-process phosphoric acid are made by treating PR with sulfuric acid. Triple superphosphate is made by treating PR with phosphoric acid; hence, sulfuric acid is the major starting acid for acidulation. Other acids (hydrochloric, nitric, or mixtures of nitric and sulfuric) can be used to acidulate PR, but are usually avoided because of environmental and production problems. Use of the other acids would become widespread only if the raw materials for manufacturing sulfuric acid became unavailable.

III. PHOSPHATE DEPOSITS—ATLANTIC AND GULF COASTAL PLAINS

A. General Geology

The Coastal Plains of the southern and eastern U.S. are underlain by sedimentary rocks of Cretaceous to Holocene age. Cretaceous rocks crop out at the landward edges of the Coastal Plains and are covered seaward by successive sequences of rocks of Tertiary and Quaternary ages. In general, the rocks are poorly consolidated and dip very gently seaward, but the seaward dips are interrupted by broad, gentle anticlinal and synclinal folds.

Phosphate pellets are widespread in sedimentary rocks of the Gulf and Atlantic Coastal Plains, but economic deposits of phosphate (central Florida, north Florida, and North Carolina) are confined to the Atlantic Coastal Plain and are known only in rocks of middle Miocene age or in younger rocks that derived much or all of their phosphate from middle Miocene rocks.

Economic phosphate deposits are in part structurally controlled. All deposits of the Atlantic Coastal Plain are in basins on the flanks of positive areas that were rising during the time of phosphate deposition. The position of the deposits suggests that the P was supplied by cool, southward-moving, nearshore ocean currents. Phosphate was probably precipitated when the cool waters, diverted by the rising positive areas, were mixed with the warmer waters corresponding to the present-day Florida Current and the Gulf Stream. The lack of economic deposits in the Gulf Coast is thought to be due to the position of the Floridan Plateau, a long-standing positive area that diverted the currents of cool water to the east, away from the Caribbean and the Gulf Coast.

All of the economic or potentially economic deposits of the Atlantic Coastal Plain have gross similarities. They are poorly consolidated or unconsolidated sedimentary rocks consisting of quartz silt and sand, phosphate pellets of silt to fine-pebble size, clay minerals, and minor amounts of carbonate minerals. The primary phosphate mineral in all of the deposits is a carbonate fluorapatite. Differences in the deposits are primarily diagenetic in origin. Thus, the deposit in North Carolina is uniform in the P content of the phosphate pellets—the deposit has not been reworked or altered by weathering, and contains no secondary phosphate or clay minerals. In contrast, the deposit in south Florida has been reworked in a marine environment, and, due to secondary enrichment, the P content of some of the phosphate particles is very high. After reworking, the Florida deposit was exposed to acid weathering, forming Al and Fe phosphate minerals and altering the original clay minerals to kaolinite.

Riggs (1979a) has speculated in detail on the origin of the phosphate deposits of Florida and has also written a detailed paper on the petrology of the deposits (Riggs, 1979b).

B. Deposits Minable Under Present Conditions

1. LAND-PEBBLE DISTRICT, POLK AND HILLSBOROUGH COUNTIES, FLORIDA (LOC. 1, FIG. 1)

The deposit is partly residual and partly reworked in a marine or continental environment and has been altered by acid weathering in the Pleistocene and Holocene. The phosphate was deposited in middle Miocene time in a shallow marine basin on the south and east flank of the Ocala Uplift and to the east of a small structure, the Hillsborough High (Cathcart, 1963). Phosphate pellets in a matrix of carbonate rock (either limestone or dolomite) of the Hawthorn Formation of middle Miocene age were concentrated and enriched by post-Miocene weathering that removed carbonate. The phosphate pellets were further concentrated and enriched when the ocean advanced over the area in Pliocene time to form the Bone Valley Formation, which is characterized by coarse phosphate nodules in a basal conglomerate and by sedimentary structures such as graded bedding, cross-bedding, and channeling.

Fossils of Pliocene age are found in the Bone Valley Formation. Further reworking in the Pleistocene (fossils of Pleistocene land mammals are common in some areas) formed linear, sinuous deposits that are probably river channels. Weathering of the phosphate-rich clayey sand in Pleistocene and Holocene times altered the Ca phosphate to Al phosphate and changed the dominant clay mineral from montmorillonite to kaolinite. The long history of reworking, enrichment, and alteration resulted in a deposit in which the phosphate pellets range in P content (from 8.7 to 17.4%—20 to 40% P_2O_5) and in amount (from a few percent to as much as 70%). The deposit, which ranges in thickness from 1 to 15 m, is minable because it is flat, is covered by thin overburden (3 to 15 m), and is unconsolidated. Thus, the phosphate can be cheaply mined and recovered by washing, screening, and froth flotation.

Minable reserves of phosphate pellets that contain >13% P (30% P_2O_5) are hundreds of millions of metric tons.

2. LAND-PEBBLE DISTRICT, HAMILTON COUNTY, NORTH FLORIDA (LOC. 3, FIG. 1)

The deposit is on the east and north flank of the Ocala Uplift and is to the east of a local high, the Barwick Arch (Sever et al., 1967; Olson, 1966a). The deposit is almost the same as the phosphorite in Polk and Hillsborough Counties. It was deposited in carbonate of the Hawthorn Formation of middle Miocene age, reworked during the Pliocene (an unnamed formation) when the phosphate was concentrated and enriched, and has been reworked to some extent in the Pleistocene. Acid weathering, although not as intense as in central Florida, has formed a thin zone of Al phosphate.

Minable reserves of phosphate pellets that contain >13% P (30% P_2O_5) are several hundred million metric tons; resources are much greater.

3. PUNGO RIVER AREA, NORTH CAROLINA (LOC. 6, FIG. 1)

The deposit is in a basin on the north flank of an unnamed high that is a part of the Cape Fear Arch (Kimrey, 1965; Gibson, 1967). Only minor clastic material was deposited in the basin from the landward side, and so the western part of the basin contains a bed of phosphate and quartz sand that ranges from 5 to about 25 m in thickness. To the east, the middle Miocene Pungo River Formation thickens to about 75 m because of the intercalation of nonphosphatic or weakly phosphatic beds of dolomite, clay, and sandy clay. Still farther east, the facies changes to clay and carbonate rock that contain no phosphate—an open-ocean facies as much as 125 m thick. Phosphate pellets deposited in the basin are round, almost spherical grains that range from 0.1 to about 1.0 mm. Coarser grains are not common; the coarse fraction (>1.0 mm) is mixed with shell material, dolomite fragments, and some quartz and is not an economic fraction. The material has not been reworked, except locally, and the overburden consists of upper Miocene and Pleistocene sediments. The overburden is 20 to 75 m thick. The phosphate pellets have not been enriched, as in the Florida deposits, and are uniform in their P content (from 11 to 13%—25 to 30% P_2O_5) and in amount of phosphate pellets (from 30 to about 50% by volume).

Reserves are hundreds of millions of metric tons; resources are billions of tons.

C. Deposits Minable Under More Favorable Conditions

1. LAND-PEBBLE DISTRICT—HARDEE, MANATEE, AND DE SOTO COUNTIES, FLORIDA (LOC. 1, FIG. 1)

This area is a southern extension of the district in Polk and Hillsborough Counties. Part of the area is minable today; part will be minable in the future. Here the overburden is thicker and much of the phosphate is a residuum of the Hawthorn Formation. The phosphate pellets contain more carbonate and somewhat less P. The amounts of recoverable pellets per unit thickness are less than in the northern area. All of these factors mean greater mining and processing costs. Mining in Hardee and Manatee Counties will probably start within a few years and will increase over the next 20 years as the deposits in Polk and Hillsborough Counties begin to phase out.

Reserves may be up to hundreds of millions of tons; resources are much larger.

2. LAND-PEBBLE DISTRICT—NORTH FLORIDA-SOUTH GEORGIA (LOC. 3, FIG. 1)

This district is an extension of the area in Hamilton County that is being mined at the present, and extends into Echols and adjacent counties in south Georgia (Sever et al., 1967; Olson, 1966b). The geology is similar to that discussed previously; the differences are in amounts of phosphate particles and in thickness of cover. The amount of recoverable phosphate (as pellets) per unit volume of rock in much of the area is low, although grades (P content) of the pellets are probably favorable. Some areas may be minable today but are in or adjacent to the Ocala National Forest, and mining leases on forest land are not being granted for environmental reasons. Total resources are not known but are probably large. Reserves are limited because of impossibility of mining at the present time.

3. SAVANNAH RIVER AREA—NORTH GEORGIA, SOUTHERN SOUTH CAROLINA (LOC. 4, FIG. 1)

Phosphate, in carbonate and clayey sand of middle Miocene age (Hawthorn Formation), is present in a basin on the north and east flanks of the Beaufort High (Heron and Johnson, 1966). Locally, this phosphate has been reworked, enriched, and concentrated in rocks of late Miocene age (Furlow, 1969). The phosphorite bed, a sandy clay or clayey sand with abundant phosphate pellets and nodules, is from 5 to 18 m in thickness and is covered by 20 to 40 m of barren material. The Savannah River cuts through the heart of the deposit, and the phosphate deposit is known to extend for at least 16 km under the sea. Reserves are not known; resources may be billions of metric tons. Because of potential damage to the Savannah River and the tidelands, mining probably will not be done here for many years.

4. CHARLESTON AREA, SOUTH CAROLINA (LOC. 5, FIG. 1)

The phosphate deposit of this area is at the base of the Ladson Formation of Pleistocene age (Malde, 1959), which rests unconformably on the Cooper Marl of Oligocene age. The phosphate member of the Ladson Formation consists of phosphate gravel, sand, and clay. The phosphate deposit is the result of reworking of the phosphatized Cooper Marl (Malde, 1959). The source of the P-rich solutions that phosphatized the Cooper Marl is probably the phosphatic Hawthorn Formation that is present as a southward-thickening wedge just to the south of the Charleston area. The first phosphate mined in the United States (in 1867) came from mines in this area. Mining continued until the mid-1920's. Large-scale mining cannot be done here; individual deposits are small and irregular, and total tonnage is small, probably some tens of millions of metric tons. Small-scale mining for local use may be possible at some time in the future.

5. NORTH CAROLINA—PUNGO RIVER AREA (LOC. 6, FIG. 1)

The only part of this deposit in the Pungo River Formation that is minable under present economic conditions is in the western one-fifth, where the overburden is thinnest and the phosphate bed is least diluted by clastic or carbonate sediments. It is probable that less than one-tenth of the total resources is in this part of the deposit. The remaining resources are covered by >30 m of overburden, and the deposit is thick (as much as 80 m). Open-pit mining methods probably cannot be used on this material, and some other method of mining will have to be devised. A method involving the injection of water into the phosphate bed and the removal of the phosphate sand as a slurry by pumping has been tried in the deeper part of the deposit and may prove to be economical.

6. HARDROCK PHOSPHATE—FLORIDA AND GEORGIA (LOC. 2, FIG. 1)

The hardrock phosphate deposits are formed when P, dissolved by acid ground water, moves downward and is precipitated on or replaces underlying limestone. The deposits are small and irregular in distribution but tend to be high in P content. Large-scale mining probably cannot be done, but small-scale mining at a low annual tonnage rate may be possible. Reserves are small, but total resources, spread over a large area from central Florida into southern Georgia, probably are hundreds of millions of metric tons.

7. RIVER-PEBBLE DEPOSITS—FLORIDA, GEORGIA, SOUTH CAROLINA

These deposits are formed when phosphate pellets, eroded from the Miocene or Pliocene deposits by modern streams, are concentrated as bars or in the flood plains of these rivers. The phosphate tends to be concentrated because of its greater specific gravity. The deposits are small, irregular in distribution, and low grade, because P is removed by the acid stream waters. They are exposed or thinly covered, so amounts of waste material are small. These deposits are found along all of the streams that drain the phosphate deposits of Florida, Georgia, and South Carolina. None are known in North Carolina because the modern streams have not penetrated the phosphate deposit.

D. Mining and Beneficiation

The phosphorite deposits of the Atlantic Coastal Plain are flat-lying bodies of unconsolidated sediments. Overburden is relatively thin and unconsolidated. These bodies are ideal for mining, using large electric draglines capable of moving thousands of tons of material per day. The largest

(55-m^3 bucket) is in North Carolina; the capacities of Florida draglines range from about 15 to 35 m^3. The machines remove overburden, then mine the ore body. The ore (called *matrix*) is dumped into a small pit at the surface, where it is broken up by hydraulic monitors, then picked up by centrifugal pumps and pumped to the beneficiation plant. At this stage in the process, the Florida matrix consists of about equal parts of recoverable phosphate particles, quartz sand, and slime (-150 mesh material). The slime fraction consists of silt and fine-sand-size quartz and phosphate particles, and one or a number of clay minerals (montmorillonite, attapulgite, and kaolinite). To make an economic product, the phosphate particles must be separated from the quartz particles and the slime fraction. The first step is sizing. The matrix is washed and fractionated, through a combination of screening, scrubbing, breaking of clay balls, and classification, to produce a pebble fraction ($+20$ mesh), flotation feed fraction ($-20+150$ mesh), and slime (-150 mesh). The screen size for pebble may vary from $+16$ to $+20$ mesh, and washer screens are changed as the deposit characteristics change. This, of course, alters the screen size of the flotation feed accordingly. Flotation feed is often further separated into coarse and fine fractions to permit selective reagentizing. Because there are virtually no quartz grains coarser than 20 mesh in the Florida deposits, the pebble fraction is a salable product. In the North Carolina deposits, the pebble fraction contains abundant carbonate grains and some quartz and must be discarded. The -150 mesh (slime) fraction is a waste product that contains considerable phosphate (average of 4.4 to 5.2% P—10 to 12% P_2O_5). The phosphate is not recovered because of particle size, but the slime fraction is being stored behind large, earth-fill dams at high cost because it cannot be dumped into the river. The flotation feed fraction is treated in flotation cells to separate the quartz and phosphate fractions. The concentrate is a high-grade phosphate product; the tailings (mostly quartz sand) are discarded as waste and are used today in land reclamation.

In Hardee, Manatee, and De Soto Counties, the southern part of the land-pebble district of central Florida, the overburden thickens, and the phosphate deposit contains fewer phosphate pellets that are somewhat lower in P content and slightly more carbonate (calcite and dolomite) than the deposit in the main part of the district (in Polk and Hillsborough Counties). These differences may result in different mining methods (one company has indicated that they may use a dredge to mine the PR under water), slightly different processing methods (for removal of carbonate minerals), and somewhat higher costs.

The deposit in North Florida (Hamilton County) is similar to the deposit in Polk and Hillsborough Counties.

The deposit in North Carolina is covered by as much as 30 m of overburden, causing higher mining costs than in Florida. When part of the overburden is removed, a dragline is put on the resulting bench, and the rest of the overburden and the ore zone is mined. Processing to separate the phosphate particles from the quartz sand and from the slime fraction is identical to the processing in Florida.

IV. PHOSPHATE DEPOSITS—CENTRAL UNITED STATES

Phosphate pellets are known in the central interior of the United States in most of the states from Texas in the south to Iowa and Wisconsin in the north. The phosphate is in rocks that range in age from Ordovician through Tertiary; however, economic deposits are known only in rocks of Ordovician age in Tennessee.

The geology of the central interior platform of the United States is basically simple. Flat or gently dipping rocks, generally carbonate or fine-grained clastics, were deposited as thin beds on the stable interior platform. The generally flat-lying beds, however, are interrupted by several major structures.

Much of the phosphate in this area of the United States is found as particles that are very coarse-grained with respect to the size of the enclosing material. Nodules of phosphate as much as several centimeters in diameter are found in very fine-grained black shales or carbonate rock. In almost all cases, the amount of P per unit volume is far too small to be economical, and only when the material is upgraded in some way, either naturally or artificially, is the phosphorite minable.

A. Deposits Minable Under Present Conditions

The brown rock deposits of Tennessee (Loc. 7, Fig. 1) are present in rocks of the following formations of Ordovician age (in ascending order): Hermitage, Bigby, Cannon, Catheys, and Leipers Limestones. Phosphate was deposited in shallow marine water on the western flank of the rising Nashville dome. The rocks are silty and sandy limestones that are phosphatic on the west flank of the dome. In each younger formation, the most phosphatic part is to the west of the next oldest formation, indicating that phosphate was being deposited at about the same water depth. Limestone on the crest and to the east of the dome is not phosphatic; thus, the source of the phosphate was the deeper sea to the west. Deep, cool currents, moving toward the rising dome, were mixed with shallow warm water, causing P to precipitate. The phosphatic limestones are exposed on the west side of the present Nashville basin and have been weathered in the modern cycle. Carbonate has been removed, enriching the rock in P. The deposits are covered by thin phosphatic soil, and are economic because they are flat-lying, poorly consolidated, and shallow.

B. Deposits Minable Under More Favorable Conditions

1. BROWN ROCK DEPOSITS—KENTUCKY (LOC. 8, FIG. 1) AND ALABAMA

All characteristics of these deposits are identical to those of the brown rock deposits of Tennessee. The Alabama deposits are in the same rocks as the Tennessee deposits; the Kentucky deposits are in the middle part of the

Lexington Limestone of Ordovician age. Some mining has been done in recent years in Limestone County, Alabama, and some mining was done many years ago in Kentucky. The Kentucky deposits are in Woodford, Fayette, Franklin, and Clark Counties in the central, bluegrass region. Most of this area is now developed into horse farms, and mining in the future is not likely because of the value of the land; but resources of a few million tons of relatively high-grade "brown rock" remain in both Alabama and Kentucky.

2. BLUE ROCK DEPOSITS—TENNESSEE (LOC. 9, FIG. 1)

The blue rock deposits of Maury, Lewis, Hickman, Perry, and Wayne Counties, central Tennessee, are the phosphatic part of the Hardin Sandstone Member of the Chattanooga Shale of Devonian age. The Hardin Sandstone Member is phosphatic only where it overlies the phosphatic parts of the Ordovician limestones. The phosphatic rocks are erratic in distribution, and the bed is generally thin—from 1 to 2 m thick where it is best developed. Locally, the overlying part of the Chattanooga Shale contains abundant coarse nodules of phosphate, and this so-called "kidney rock" was mined with the underlying blue rock. Resources are limited—about 100 million metric tons or less—and the rock would have to be mined by underground methods, which is much more expensive than open-pit mining. This is a resource for the future.

3. WHITE ROCK DEPOSITS—TENNESSEE (LOC. W, FIG. 1)

The secondary white rock deposits of Tennessee are in Perry, Decatur, and Humphrey Counties. The deposits formed where phosphate, leached by acid ground water from either the slightly phosphatic Fort Payne Chert of Mississippian age or from the "blue rock" phosphate, has replaced underlying Silurian limestones. The leaching occurred during the modern weathering cycle, and the deposits vary from almost pure precipitated apatite or completely replaced limestone to very siliceous "stony" phosphate that replaces cherty and siliceous limestone. The phosphate may also form a "breccia," in which the phosphate fills joints around irregular pieces of partly phosphatized limestone. In Johnson County, in northeast Tennessee, white rock is formed on and in the Knox Dolomite of Late Cambrian and Early Ordovician age by the same methods. The phosphate is derived from the overlying slightly phosphatic rocks.

The phosphate is deposited as joint fillings, breccia cement, and lamellar replacement. White rock may be very pure apatite and is a valuable source of coarse furnace feed, but deposits are small and erratic in distribution and total resources (Jacob, 1953) are only a few million metric tons. Small tonnages may be mined for local use but no great future can be projected for these deposits.

4. MAQUOKETA SHALE—IOWA, ILLINOIS, AND WISCONSIN (LOC. 10, FIG. 1)

Basal beds of the Upper Ordovician Maquoketa Shale in Iowa, Illinois, and Wisconsin contain abundant phosphate as pellets, nodules, and phosphatized fossils (Brown, 1966). The phosphate bed is thin (0.3 to 1.2 m) and medium in phosphate content (7.4% P—17% P_2O_5). The phosphate zone is at its best near Dubuque, Iowa, and in this area resources have been estimated to be a few million metric tons. Total resources in all of the area of outcrop of the Maquoketa Shale must be much larger. Mining here is unlikely for many years, but this may be a reserve for local use at some time in the future.

5. CASON SHALE—ARKANSAS (LOC. 11, FIG. 1)

Phosphate pellets and nodules are found associated with Mn in the Cason Shale of Ordovician age in Independence and Izard Counties in Arkansas. The phosphate-bearing beds are thin (about 1 m) and lenticular, and contain as much as 8.7% P (20% P_2O_5). The phosphate was deposited on the rising flank of the Ozark dome. Resources are limited; about 20 million metric tons was estimated by Jacob (1953). A few thousand tons was mined in the 1920's.

6. PITKIN LIMESTONE AND HALE FORMATION—ARKANSAS (LOC. 12, FIG. 1)

Lenticular calcareous phosphorite beds at the contact between the Pitkin Limestone of Mississippian age and the Hale Formation of Pennsylvanian age in northwest Arkansas were deposited in a basin on the flank of the Ozark dome. The deposit is in channel-like depressions that range in thickness from 0 to 7 m and are as much as 400 m wide and 400 to 500 m long. Resources are limited—tens of millions of metric tons. The deposit was mined in the past and may be mined for local use in the future.

Phosphate is also known in the Fayetteville Shale of Mississippian age. The beds of the black shale are thick and may contain as much as 2.6% P [6% P_2O_5 (E. E. Glick, U.S. Geolog. Surv., oral commun., 1975)]. Individual thin beds contain much greater amounts of P. Total resources may be very large, and this could be a resource for the future.

7. ORISKANY SANDSTONE—NEW YORK, PENNSYLVANIA, VIRGINIA (LOC. 13, FIG. 1)

Beds of phosphatic sandstone are known in the Oriskany Sandstone of Early Devonian age in New York, Pennsylvania, and Virginia. Deposits are thin, lenticular, and moderate in P content. Resources are limited, but the material might be mined for local use at some time in the distant future.

8. CHATTANOOGA SHALE (LOC. 14, FIG. 1)

The Chattanooga Shale of Devonian and locally Mississippian age and its equivalents, the Maury (Lower Mississippian) and New Albany (Middle Devonian to Lower Mississippian) Shales, contain sparse coarse phosphate nodules in a black-shale matrix. The rocks outcrop in Alabama, Georgia, Tennessee, Kentucky, Ohio, and Indiana. The nodules contain as much as 13% P (30% P_2O_5), but the total bed contains only 0.9 to 1.3% P (2 to 3% P_2O_5). There are no data on resources, but the amount of P must be very large because of the tremendous extent of the outcrop area. Mining under present conditions is not possible—there are too few phosphate nodules per unit volume—but there could be some mining in the distant future, when the reserves of presently minable phosphorite ae depleted.

9. PENNSYLVANIAN ROCKS—KANSAS, OKLAHOMA, MISSOURI (LOC. 15, FIG. 1)

Pennsylvanian rocks of eastern Oklahoma and Kansas and the adjacent part of western Missouri contain phosphate nodules and pellets in black shale. Phosphate is present in several shale beds that are separated by limestone. The phosphate pellets and nodules may contain as much as 13% P (30% P_2O_5); the total shale beds, only 0.9 to 1.3% P. Total resources have not been estimated but may be large; however, there are no minable reserves because of low content of P per unit area. Again, this may be a resource for the distant future.

10. CRETACEOUS ROCKS—GEORGIA, ALABAMA, MISSISSIPPI, TEXAS

Rocks of Cretaceous age contain phosphate pellets and nodules in Georgia, Alabama, Mississippi (Loc. 16, Fig. 1), and Texas (Loc. 17, Fig. 1). The phosphate pellets are associated with limestones, marls, sandstones, and shales, and in the northern parts of Alabama and into Tennessee, the phosphorite beds give way to glauconitic beds of the same age. The areal extent of the phosphate-bearing beds is large; thus the total amount of P must be large. But there is no economical material because of the very low content of P per unit volume.

11. TERTIARY ROCKS—NORTH CAROLINA, SOUTH CAROLINA, GEORGIA, TEXAS

Phosphate pellets occur in rocks of Tertiary age in North Carolina (Castle Hayne Limestone, Eocene), South Carolina (Cooper Marl, Oligocene), Georgia (Tallahatta Formation, Eocene), and Texas (Midway Group, Paleocene). The pellets are sparse but contain as much as 13% P (30% P_2O_5). No information is available as to amounts, and none of the material is economical under present conditions.

C. Mining and Beneficiation

The only deposits in this region that are being mined today are the "brown rock" deposits of the Nashville basin of Tennessee. Mining is by small draglines with 1.5- to 3-m^3 buckets (2 to 4 yd^3). The mined rock is loaded into trucks or railroad cars and hauled to the processing plants.

The PR (muck) is treated in the washing plants to remove fine material —clay- and silt-sized phosphate and quartz. The fines (-200 mesh) are a waste product and are ponded in dams to avoid pollution. Phosphate in the slime fraction is lost, but the material is available for future processing.

The coarser material ($+200$ mesh) is the product of the washing plant. Most of the production from Tennessee is used to make elemental P in electric furnaces. Fine-grained PR is treated in kilns, where it is formed into nodules by partial fusing and is sized to make an acceptable furnace charge.

V. PHOSPHATE DEPOSITS—WESTERN UNITED STATES

The major phosphate deposits in the western U.S. are in the Permian Phosphoria Formation, but phosphate occurrences are also known from rocks of Cambrian, Ordovician, and Silurian ages in Nevada (Rogers et al., 1970), from rocks of Mississippian age in Utah (Cheney, 1957), from rocks of Permian and Triassic ages in Alaska (Patton & Matzko, 1959), from rocks of Jurassic age in Montana, and in rocks of Tertiary age in California (Lowe, 1972).

A. Deposits Minable Under Present Conditions

1. PHOSPHORIA FORMATION—IDAHO, WYOMING, MONTANA, AND UTAH (LOC. 18, FIG. 1)

The deposits of the Phosphoria Formation of Permian age crop out over an area of about 350,000 km^2 in Idaho, Montana, Wyoming, Utah, and Nevada. The deposits in the eastern part of the field are of the platform type; those in the western part are geosynclinal. The rocks have been folded and faulted, with deformation more intense in the western part of the field.

The Phosphoria Formation consists of dark-colored chert, black shale (carbonaceous mudstone), and phosphorite in the area of southeast Idaho. The best phosphorite deposits are near the base of the Meade Peak Phosphatic Shale Member, but commercial deposits are also in the Retort Phosphatic Shale Member near the top of the formation. The Phosphoria in Idaho grades laterally into a more sandy sequence to the north and east in Montana and Wyoming and to a carbonate facies in Utah and Wyoming (to the south and east). The carbonate sequence grades laterally to the east into a clastic facies containing red beds (McKelvey et al., 1959).

Reserves minable under present conditions are measured in hundreds of millions of metric tons, but total resources are billions of tons. Much of the resource will not be mined until more efficient modifications of present mining methods can be devised, but the very large tonnage of PR will be a major source of supply for the future.

2. MIOCENE ROCKS OF CALIFORNIA (LOC. 19, FIG. 1)

Phosphate deposits of Miocene age in southern California are in rocks of the Santa Margarita Formation (Lowe, 1972). Potentially economic phosphate deposits are present in Ventura County, in the upper Sespe Creek and Cuyama Valley areas. Phosphate beds are also known in the Monterey Formation of middle Miocene to early Pliocene age as far north as Monterey Bay in central California. The phosphate is pelletal, and the phosphate beds contain a large component of tuffaceous or altered tuffaceous material. The P accumulated as a chemical precipitate in a coastal strait between the low-lying coastal plain and offshore islands (Lowe, 1972).

Reserves are not known, but presumably measure millions of metric tons, and total resources must be very large. There was some production in the early 1970's and production may be resumed in the near future.

B. Deposits Minable Under More Favorable Conditions

1. PHOSPHORIA FORMATION

At some time in the future, additional mining will be done in the western phosphate field. Total resources are very large and, although most of the mining will have to be done by expensive underground methods, it is likely that production from the Phosphoria Formation will eventually be the major source of PR for the United States.

2. MIOCENE—CALIFORNIA

As with the Phosphoria Formation, at some time in the future, additional mining of phosphate from rocks of Miocene and Pliocene age will be done in California. Resources are certainly large enough to support a much greater tonnage than is currently being mined.

3. PERMIAN AND TRIASSIC—ALASKA

A large potential resource of PR is present in the Brooks Range of Alaska in rocks of Permian and Triassic age. According to R. L. Detterman (written commun., 1972), beds of PR as much as 9 m thick, containing $>4.4\%$ P (10% P_2O_5), are present nearly continuously for about 190 km along strike. Beds as much as 1 m thick within this zone contain as much as 13% P (30% P_2O_5). The phosphorite beds are in a sequence of black shale, limestone, and chert; and the phosphate is pelletal, oolitic, and nodular. The area is structurally complex, and so reserves have not been measured;

but it is evident that total resources must be hundreds of millions or billions of metric tons. The geographic location, on the north slope of the Brooks Range, is such that mining is probably impossible for many years, but this area is a resource for the distant future.

C. Mining and Beneficiation

The PR from the western field is mined by both underground and open-pit methods. Underground mining methods are generally confined to the higher grade material. Open-pit mining is by tractor-drawn scrapers, power shovels, and front-end loaders—more or less standard earth-moving equipment. Blasting may be done when the rock is too hard to extract using the earth-moving machinery. The ore is moved by trucks either directly to the plant or to a railhead, where it is loaded on the railroad for shipment to the plant.

Mining of the deposits in California is by open-pit methods, using large earth-moving equipment. Total production has been small, and the rate of production in the future probably will be small.

Beneficiation depends on the type of PR produced. High-P-content ($>13.5\%$ P—31% P_2O_5) phosphorite, called *acid-grade*, is sent to plants producing wet-process phosphoric acid. The only necessary beneficiation might be the removal of excess organic material by calcining and grinding the rock to a suitable size for the acidulation. Furnace-grade rock (10.5 to 13.5% P—24 to 31% P_2O_5) is used in electric furnaces to make elemental P.

Phosphatic shales that contain from 7.8 to 10.5% P (18 to 24% P_2O_5) have to be treated to concentrate the P. Froth flotation is used to separate the phosphate particles from the gangue minerals—calcite and quartz.

At the present time, most of the production of PR from the western field is by open-pit mining, but the major part of the total resource will have to be mined by the much more expensive underground techniques.

VI. OFFSHORE DEPOSITS

A. California

Coarse phosphate nodules that may be forming in the modern ocean are present as thin surface veneers on banks offshore from southern California. They are present over large areas and, although the beds are thin, the total resource may be large. Research on mining methods has been done and some plans for mining have been made, but this is a resource for the future. The nodules may contain as much as 13% P (30% P_2O_5) and may have other minerals present in minor amounts.

B. Eastern United States

Phosphate pellets and nodules are incorporated in the sands of the modern Atlantic Ocean, from North Carolina south to the southern tip of Florida. The P is thought to be derived from onshore deposits that are Miocene in age. No precise estimates of tonnage have been made, but the areal extent is large; and, although the percent of phosphate pellets in the sediments is not large (up to about 10% by volume), the total tonnage must be enormous. Mining would be difficult, particularly where the deposit is covered with barren sand of the modern sea floor, and research on a mining method has not been done. None of the material can be classed as a reserve, although it could be a resource for the distant future.

VII. ENVIRONMENTAL CONSIDERATIONS

Environmentalists, ecologists, and other groups whose aim is to protect the environment view with concern the damages that may result when mining and processing projects, including phosphate mines, are proposed in the areas in which they live. In the past several years, open-pit mining has been strongly attacked by the environmentalists.

It is unquestionable that open-pit mining alters the environment during mining. However, mined-over land can be and is being reclaimed and returned to a variety of productive uses. State and federal laws now make it mandatory that land be reclaimed, and mining companies are required to submit environmental impact statements and to guarantee that mined land will be reclaimed.

Air- and water-pollution laws are strict and will, in all probability, become stricter, and so beneficiation and chemical processing plants must have air- and water-pollution controls.

The laws controlling land reclamation and air and water pollution increase costs of mining and processing PR. These increased costs of production will be passed on to the consumer as increases in the price of the fertilizer products.

Because of stringent state and federal laws, mined-out land in Florida and Tennessee is being reclaimed not long after mining is completed. In both states, reclaimed land is used for recreation areas (fishing, swimming, and boating in man-made lakes; golf courses and bird and wildlife sanctuaries on the land), for farming and ranching, for industrial and home-building sites—in short, for all possible land uses.

Because of the character of the land and the climate, the mined-out areas in Florida can be reclaimed quickly and, within a few years, it is nearly impossible to tell that they were mined. Such may not be the case in the Phosphoria field in the western U.S., where the character of the land and the climate are such that land reclamation will take a much longer time.

VIII. CONSERVATION

Not all parts of all phosphate deposits are minable at any given time. Some areas may be too low grade, too low in tonnage, too thin, or may contain too much clay to be minable. The nature of open-pit mining is such that a given area probably must be mined in its entirety or left and not mined at all. If a part of a deposit is low grade, for example, it could be left unmined during a time of low prices and, thus, effectively lost as a source of P. Conservation of a natural resource demands that mining be planned so that lower grade areas can be mined in the future or that the lower grade material be mined and either stockpiled for later use or blended with higher grade material to make an acceptable product. Land leased from the government must be mined so as to effect all possible conservation methods, but privately owned land has no such restrictions. Fortunately, mining companies are well aware of the situation and are, normally, practicing conservation.

IX. OUTLOOK FOR THE FUTURE

Deposits or occurrences of phosphorite are known in about 30 of the states in the United States (Fig. 1), and phosphate has been mined from at least 20 states. Mining in the United States today is from three areas—the Atlantic Coastal Plain (Florida and North Carolina), the central U.S. (Tennessee), and the western U.S. (Idaho, Montana, Wyoming, and Utah). In 1975, a few thousand metric tons of PR was mined from a deposit in northwest Arkansas, and preparations were being made to mine from deposits in southern California.

In 1974, production in the United States was 44.7 million metric tons of PR, of which 5% was from Tennessee, 13% from the western states, and 82% from the deposits of the Atlantic Coastal Plain.

The resources of PR in the United States are large enough that we need not be concerned for the immediate future. However, our high-grade and cheaply mined reserves are being depleted, and resources that are unminable at the present time will eventually have to be tapped. Several of these areas have been mentioned, previously. For example, large tonnages of PR in

However, even there, the land can be reclaimed and restored to a usable condition. Land cannot be returned to its "original" condition after mining and, if laws are written to require a return to original condition, mining probably will be impossible. It is incumbent upon the mining companies to comply with the laws for land reclamation and air and water purity, but there must be cooperation from the general public and government to insure that laws are not written that are impossible to obey. Phosphate rock is a necessary fertilizer raw material; without phosphate fertilizer our agriculture could not continue its high output for very many years.

North Carolina are not considered minable because they are too deeply buried. Mining by some method, such as sand pumping, may prove to be feasible in this case. Large reserves in the Savannah River area at the Georgia-South Carolina line and in the adjacent Atlantic Ocean cannot be mined at the present because of potential damage to the Savannah River and because, for the material offshore, no mining method is available. Large resources of PR in north Florida and south Georgia are unminable at present because of low grades and tonnages and because much of the land is National Forest, and potential damage to the environment makes it unlikely that mining will be done in the near future. The Hawthorn Formation of Florida contains a vast, low-grade, low-tonnage-per-unit-volume resource, much of which is covered by very thick overburden. A new mining method and new processing methods (to remove calcite and dolomite) will have to be worked out before any of this rock can be mined.

Resources like the Mississippian-Pennsylvanian rocks of the Midcontinent (Oklahoma, Kansas, Missouri), which contain a few tons of phosphate nodules per hectare-meter of thickness, are very large, but are impossible to mine and process using present-day methods. These may be a resource for the distant future.

In the western field, most of the resources are deep, in areas of structural complexity, and will have to be mined by underground methods. For thick beds of moderate P content that are steeply dipping, no mining method exists. Research is necessary to determine mining methods. Processing of the calcite-rich rock will pose additional problems.

Large potential resources of phosphorite are present in Alaska and form a resource for the distant future.

LITERATURE CITED

Brown, P. M. 1966. The relation of phosphorites to ground water in Beaufort County, North Carolina. Econ. Geol. 53:85-101.
Cathcart, J. B. 1963. Economic geology of the Keysville quadrangle, Florida. U.S. Geol. Survey Bull. 1128. 82 p.
Cheney, T. J. 1957. Phosphate in Utah. Utah Geol. Mineralog. Surv. Bull. 59. 54 p.
Furlow, J. W. 1969. Stratigraphy and economic geology of the eastern Chatham County phosphate deposit. Georgia Geol. Survey Bull. 82. 40 p.
Gibson, T. G. 1967. Stratigraphy and paleoenvironment of the phosphatic Miocene strata of North Carolina. Geol. Soc. Am. Bull. 78:631-649.
Guild, P. W. 1967. Phosphate. p. 219-220. *In* Mineral and water resources of Missouri. 90th Congr., 1st Sess., Senate Doc. 19.
Gulbrandsen, R. A. 1966. Chemical composition of phosphorites of the Phosphoria Formation. Geochim. Cosmochim. Acta. 30:769-778.
Heron, S. D., Jr., and H. S. Johnson, Jr. 1966. Clay mineralogy, stratigraphy, and structural setting of the Hawthorn Formation, Coosawatchie district, South Carolina. Southeast. Geol. 7:51-63.
Jacob, K. D. 1953. Phosphate resources and processing facilities. *In* K. D. Jacob (ed.) Fertilizer technology and resources in the United States. Agronomy 3;117-165. Academic Press, New York.
Jacob, K. D., W. L. Hill, H. L. Marshall, and D. S. Reynolds. 1933. The composition and distribution of phosphate rock with special reference to the United States: USDA Tech. Bull. 364. 89 p.

Kimrey, J. O. 1965. Description of the Pungo River Formation in Beaufort County, North Carolina. North Carolina Div. Min. Resour. Bull. 79. 131 p.

Lowe, D. R. 1972. The relationship between silicic volcanism and the formation of some sedimentary phosphorites. p. 217–226. *In* H. S. Puri (ed.) Proc. 7th Forum on Geol. of Ind. Minerals. Tampa, Florida, April 1971. Sponsored and published by Florida Geol. Survey, Tallahassee.

Malde, H. E. 1959. Geology of the Charleston phosphate area, South Carolina. U.S. Geol. Survey Bull. 1079. 105 p.

McKelvey, V. E., J. S. Williams, R. P. Sheldon, E. R. Cressman, T. M. Cheney, and R. W. Swanson. 1959. The Phosphoria, Park City, and Shedhorn Formations in the western phosphate field. U.S. Geol. Surv. Prof. Pap. 313-A. p. 1–47.

Olson, N. K. (ed.). 1966a. Geology of the Miocene and Pliocene Series in the north Florida-south Georgia area. Southeast. Geol. Soc., 12th Annu. Field Conf. Guidebook. 94 p.

Olson, N. K. 1966b. Phosphorite exploration in portions of Lowndes, Echols, Clinch, and Charlton Counties, Georgia. Georgia State Div. Conserv., South Georgia Minerals Program Proj. Rep. 4. 113 p.

Patton, W. W., and J. J. Matzko. 1959. Phosphate deposits in northern Alaska. U.S. Geol. Survey Prof. Pap. 302-A. 17 p.

Riggs, S. R. 1979a. Phosphorite sedimentation in Florida—a model phosphogenic system. Econ. Geol. 74:285–314.

Riggs, S. R. 1979b. Petrology of the tertiary phosphorite system of Florida. Econ. Geol. 74: 195–220.

Rogers, C. L., F. J. Kleinhampl, J. J. Ziony, and Walter Danilchik. 1970. Phosphate occurrences in Nye County and adjacent areas, Nevada. p. C49–C60. *In* Geological survey research 1970. U.S. Geol. Surv. Prof. Pap. 700-C.

Rooney, T. P., and P. F. Kerr. 1967. Mineralogic nature and origin of phosphorite, Beaufort County, North Carolina. Geol. Soc. Am. Bull. 78:731–748.

Sever, C. W., J. B. Cathcart, and S. H. Patterson. 1967. Phosphate deposits of south-central Georgia and north-central peninsular Florida: Georgia State Div. Conserv., South Georgia Minerals Program Proj. Rep. 7. 62 p.

Waggaman, W. H. 1952. Phosphoric acid, phosphates, and phosphatic fertilizers. 2nd ed. Reinhold Publ. Corp., New York.

Chapter 3

Evaluation of Phosphatic Raw Materials

G. H. MC CLELLAN AND L. R. GREMILLION

International Fertilizer Development Center, and Tennessee Valley Authority, Muscle Shoals, Alabama, respectively

I. INTRODUCTION

The phosphatic raw materials of interest to agronomists and the fertilizer industry are complex assemblages of minerals grouped under the generic heading of *phosphate rock* or *phosphorite*. *Phosphate rock* is a trade name that covers a wide variety of rock types that have widely different textures and mineral compositions, and those differences make phosphate raw material evaluation a complicated and fascinating field of mineralogical research. According to the American Geological Institute *Glossary of Geology* (Gary et al., 1972, p. 535), a *phosphate rock* (PR) is a sedimentary rock composed principally of phosphate minerals. Most commonly it is a bedded rock of marine origin composed of microcrystalline carbonate fluorapatite in the form of laminae, pellets, oolites, nodules, and skeletal and shell fragments. Aluminum and iron phosphates are usually products of weathering of calcium phosphates. Guano-derived phosphorite has been formed by replacement and characteristically has a complex phosphate mineral composition. The term *phosphorite* has been applied also to sedimentary rocks composed entirely of apatite and to igneous rocks that contain appreciable amounts of apatite. The information presented in this chapter will show that the composition and character of PR are both diversified and complex, but the discussion will be limited to those groups of minerals that are commercially valuable for their P content.

Most commercial deposits are of sedimentary marine PR, but a significant commercial amount of P is obtained also from alkaline igneous complexes and residual deposits produced by weathering of sedimentary phosphatic limestones and igneous carbonatite complexes. Some P deposits that are a combination of replacements and marine precipitates occur on elevated atolls of coral limestone (Notholt, 1975).

Nearly all these different commercial deposits have one common characteristic—they contain one or more of the minerals of the apatite group. The differences in chemical composition of the apatites (McClellan & Lehr, 1969) are reflected in the equally different properties of the P concentrates

---

Copyright 1980 © ASA-CSSA-SSSA, 677 South Segoe Road, Madison, WI 53711, USA.
The Role of Phosphorus in Agriculture.

produced from the deposits. In sedimentary marine PR deposits, the phosphate is usually a carbonate fluorapatite; in igneous deposits the phosphates have compositions nearly that of fluorapatite. In the residual deposits, Ca-Fe-Al and Fe-Al phosphates are commonly associated with the apatites. In spite of these differences, the grade of commercial phosphate rock is still expressed in terms of tricalcium phosphate, $Ca_3(PO_4)_2$, known in the trade as *bone phosphate of lime* or BPL. This term originated when tricalcium phosphate was thought to be the chief constituent of bone and PR. It is now known that both bone and PR are apatites and not tricalcium phosphate. Because of the wide use of the term BPL and the present trend toward elemental notation the conversion factors are included:

$$\% P_2O_5 = 0.4576 \times \% BPL$$

$$\% P = 0.1997 \times \% BPL.$$

Commercial PR varies in grade from about 83% BPL to about 60% BPL (17 to 12% P). About 85% of the world's annual PR production is processed to yield phosphorus and phosphoric acid which are converted into a wide variety of fertilizer materials.

II. PHOSPHATE MINERALOGY

Phosphate deposits fall into three broad classes based upon their mineral assemblages; these are Fe-Al phosphates, Ca-Fe-Al phosphates, and Ca phosphates. These three classes form a natural weathering sequence in which the stable Fe-Al phosphates represent the final stage of weathering. The order of increasing economic important for these classes is Fe-Al phosphates, Ca-Fe-Al phosphates, and Ca phosphates.

A. Fe-Al Phosphates

The most common Fe-Al phosphates are wavellite, $Al_3(PO_4)_2(OH)_3 \cdot 5H_2O$; variscite, $AlPO_4 \cdot 2H_2O$; and strengite, $FePO_4 \cdot 2H_2O$; many less common Fe-Al phosphates are described in the literature (Moore, 1973). Large deposits of Fe-Al phosphates occur in several places in the world, notably Senegal, Liberia, Brazil, and Utah. Although these phosphates are not tractable in normal wet-process phosphoric acid processes, they can be converted to fertilizers in nitric acid processes (Knudsen, 1972) or modified thermally to produce direct-application fertilizers (Anon., 1975).

B. Ca-Fe-Al Phosphates

The Ca-Fe-Al phosphates comprise an intermediate class of widely varying composition. The principal minerals are crandallite, $CaAl_3(PO_4)_2(OH)_5 \cdot H_2O$; and millisite, $(Na,K)CaAl_6(PO_4)_4(OH)_9 \cdot 3H_2O$; usually mixed either with each other or with members of the other classes of phosphate minerals. Two good examples of this class of phosphates are the Florida leached zone ore (Altschuler et al., 1956) and the Christmas Island C-zone ore (Trueman, 1965). Materials in this class are not suited for sulfuric and phosphoric acid treatments because they yield products of poor quality, but they can be used in nitric acid processes (Knudsen, 1972) or treated thermally to produce citrate-soluble direct-application fertilizers (Doak et al., 1965).

C. Calcium Phosphates

Commercial mineral phosphates in this class, known collectively as phosphate rock (PR) have the one common property that the structural arrangement of their ions shows them to belong to the broad category of apatitic minerals. Apatite is the tenth most abundant mineral in the earth's crust and it occasionally occurs in massive concentrations of economic importance. In spite of their crystal-structure similarity, however, the compositions of these apatites usually differ significantly from that of fluorapatite, $Ca_{10}(PO_4)_6F_2$, which is commonly assumed to be the phosphatic component of PR. These differences in chemical composition are reflected in the unique properties of particular commercial PR concentrates.

The different compositions of apatites reflect the conditions of their geochemical origins. Apatitic phosphates are formed under all geological conditions—igneous, metamorphic, and sedimentary environments. The apatites recovered from igneous and metamorphic rocks, including those from iron ore deposits, have commercial importance but supply only a small fraction of the world market, and that mainly for captive production (i.e., Kola, USSR; Phalaborwa, S. Africa; Araxa and Jacupiranga, Brazil).

Sedimentary apatites have been, and no doubt will continue for some time to be, the major source of commercial phosphate. Because of their widely differing modes of occurrence in geological periods ranging in age from Precambrian to Miocene, sedimentary apatites vary widely in chemical composition. In these apatites, significant amounts of Mg and Na usually have substituted for Ca, and as much as 25% of their phosphate may be replaced by carbonate plus fluoride which places these apatites in a distinct mineral class (McClellan & Lehr, 1969). There are important sedimentary phosphate deposits in North Africa (Senegal, Togo, Morocco, Algeria, Tunisia), the Near East (Jordan, Israel, Egypt), Australia (Queensland), and the United States (Florida, North Carolina, Idaho).

III. ACCESSORY MINERALS

The wide variety of geologic settings in which PR occurs has given rise to accessory mineral assemblages that range from the very simple to the very complex. In this paper, only the most important groups of accessory minerals will be mentioned. These groups have been selected because of their abundance and importance in the economic evaluation of P ores.

A. Silica

Silica is one of the most abundant gangue constituents in commercial PR. Quartz, the most common form of silica, occurs both as detrital grains and as authigenic crystals and is present in igneous, metamorphic, and sedimentary rocks. The hydrated and less well crystallized forms (chalcedony, opal, and silica gel) are more common as cements and impurities in the sedimentary phosphates. Cristobalite (high- and low-temperature forms of SiO_2) occurs in a variety of sedimentary lithologies, especially where vulcanism was common.

Silica is an important factor in the manufacture of chemical fertilizers because it is required as a flux in electric furnace operations and other thermal processes and as a defluorinating agent in the production of wet-process phosphoric acid. Its main deleterious effects in commercial processes are lowering of the ore grade, erosion of plant equipment, and interference with the grinding and beneficiation of the rock. Silica is the main impurity removed by present commercial beneficiation technology (either physical separation or flotation).

B. Silicates

A wide variety of silicate minerals occur in PR, and the phyllosilicates (layered structures) are the most important. Micas are the phyllosilicates most commonly seen in igneous and high-grade metamorphic rocks. The usual varieties include biotite, phlogopite, muscovite, vermiculite, and sericite. The low-grade metamorphic and sedimentary rocks contain practically every known clay mineral ranging from the stable kaolinites in highly weathered ores to the more reactive glauconite, chlorites, and montmorillonites in less weathered ores. Palygorskite (attapulgite), a clay mineral with a chain-type structure (inosilicate), is also a common geologic associate of sedimentary phosphates. The clay content of PR may occur as a single mineral species or, more typically, as complex mixtures of several varieties.

Other silicates are commonly found in the accessory minerals of PR but they generally are minor components. Among the more common of these minor accessory minerals are feldspar (microcline and orthoclase, most frequently; plagioclases, uncommon to rare), amphiboles, pyroxenes, feldsphathoids (nepheline), and many trace minerals.

The silicates can introduce problems in the physical beneficiation and chemical processing of PR. The texture of the silicates can affect physical beneficiation through the distribution of the apatite and nonapatite phases. The bonding between phases can control rock hardness and result in increased energy for grinding to liberate the phosphate. The presence of silicates inside the phosphate particles can present difficult separation problems. The particle size of the silicates also is an important factor. Fine particle-sized materials, usually various clay minerals, are rejected from the beneficiation process (slimes), and their disposal often presents difficult environmental problems. Chemical processing problems can result from impurities dissolved from acid-soluble silicates in acidulation processes—K, Fe, Al, and Mg are common undesirable impurities.

C. Carbonates

Carbonate-bearing ores are the most abundant type of PR deposit, presumably because apatites and carbonates accumulate in similar geochemical settings. Calcite is the most abundant carbonate mineral in PR and occurs in all rock types. Its abundance in PR ranges from the matrix in phosphatic limestones and carbonatites to a minor accessory in many ores. Dolomite is the second most common carbonate mineral in PR but is much less abundant than calcite. Dolomite also occurs in all rock types, but is much more abundant in sedimentary and metamorphic rocks. There are rocks in which dolomite is the only carbonate impurity, and the dolomite ranges in abundance from the bulk of the rock matrix to a minor accessory phase. Ankerite, magnesite, and fluorocarbonates (bastnaesite) also occur in PR and can be important accessory minerals in particular ores but generally are uncommon to rare.

Carbonates are undesirable components of PR for several reasons. They generally cannot be removed by beneficiation, and so they tend to lower the grade of the rock, consume valuable acid and cause foaming during acidulations, and contribute chemical components, particularly Mg, that are undesirable in some products. In nature carbonates are removed from PR by weathering. The only practicable process for removing carbonate minerals commercially from PR is calcination, which transforms the carbonates to oxides, followed by slaking and separation of the hydrated oxides from the phosphate. This is an expensive procedure but is in use in several commercial plants.

D. Iron and Aluminum Oxides and Hydroxides

In igneous rocks, Fe and Al oxides do not usually introduce significant problems because they are relatively inert in chemical processes and can easily be separated by beneficiation. Hematite and magnetite are common associates of phosphates in these rock types. In the sedimentary rocks, the relationship is much more complicated because the mineral species are more diverse and their identities and distribution in the ores are only poorly known. The most common Fe-Al oxides and hydroxides identified in sedimentary PR are goethite, limonite, bauxite minerals (gibbsite, boehmite, diaspore), and barbosalite. These acid-soluble minerals in the sedimentary ores are usually formed by weathering and introduce significant chemical problems in processing the phosphate ores.

E. Evaporite Minerals

Many of the important commercial phosphate deposits occur in arid and semiarid climates and have evaporites associated with the PR. Chlorides and sulfates are the most common evaporite minerals occurring in PR. Halite and gypsum are the most common mineral varieties; sylvite, carnallite, and anhydrite are less common. In one PR, $CaSO_4 \cdot 0.5H_2O$ (hemihydrate or natural plaster of Paris) was present as a pseudomorph after gypsum. The major industrial problem caused by evaporites is corrosion of metals by chlorides during acid processing. Washing with fresh water usually lowers chloride contents to satisfactory levels.

F. Other Accessory Minerals

A number of other minerals occur as minor accessories in PR. Titanium minerals, including rutile, ilmenite, anatase, and perovskite, are fairly common and can be significant in carbonatite deposits. These can be troublesome in acidulation processes and cause significant losses of P. Iron sulfides such as pyrite and marcasite are also common minor accessories in many rocks. Iron in a reduced state (ferrous iron) may present a significant corrosion problem with some types of construction materials, and it frequently creates problems in filtrations or ion-exchange operations by forming solids with undesirable physical properties. Fluorite is another minor accessory mineral in PR and is common as a minor component in rocks from the Phosphoria Formation. It is especially important because early workers frequently assigned the "excess fluorine" in PR to fluorite even though the phase could not be identified. This confusion delayed considerably the elucidation of the systematic crystal chemistry of francolites.

Organic matter in PR consists of the indigenous compounds derived from geologic plant and animal sources and sometimes the beneficiation reagents used in processing the ore. Organic compounds derived from fossilized plants and animals may cause foaming in acidulation processes and so increase reactor equipment costs. Organic matter from the PR also undesirably discolors the intermediate and final products, resulting in acids and fluid fertilizers of reduced sales appeal and market value.

IV. TEXTURE OF PHOSPHATE ROCKS

The way in which PR is processed commercially and utilized is determined to a large extent by its texture. The different origins of PR—igneous, metamorphic, and sedimentary—result in a wide range of rock textures and mineral assemblages. Igneous and metamorphic environments have produced comparatively few commercial phosphate deposits, but their importance may increase in the foreseeable future because of accelerated prospecting activity. The sedimentary rocks that form the bulk of commercial production of PR have widely different textures but usually simple mineral assemblages. This is in contrast to the igneous and metamorphic PR which have complex mineral assemblages but little textural variety. The major textural types are described briefly here to demonstrate the variety encountered in present commercial deposits and recently discovered deposits that are now being developed.

A. Igneous Rocks

Most igneous rocks that contain P in commercial concentrations are silica deficient. The famous Khibiny nepheline syenite complex on the Kola peninsula of Russia is probably the most well known and productive igneous phosphate deposit, and this area produces some 70% of Russia's phosphatic raw materials. The Phalaborwa complex in South Africa supplies phosphate from a pyroxenite associated with syenite. The rocks typically are coarsely crystalline with simple textures in which the phosphate is free from occlusions, although small amounts of magnetite, titanium minerals, and rare earth minerals occasionally occur in the apatite particles. Small amounts of byproduct phosphate are recovered from igneous Fe ore deposits in Sweden (Grangesberg) and Russia (Kovdor) and at Jefferson City, Missouri.

Carbonatites are a comparatively new igneous source of P and these rocks are characterized by the coarse textures typical of igneous deposits and may have very complicated mineral assemblages—more than 100 minerals are reported in some deposits. Carbonatite P ores have been discovered in Brazil (Jacupiranga, Araxa, and Catalao), Thailand, Canada (Nemogosenda), Uganda, Malawi, and Sri Lanka. Most of the commercial production from these ores is obtained from weathered portions of the ore bodies rather than from the unaltered rock.

Phosphate minerals are common in siliceous pegmatites and hydrothermal veins, but in these deposits the P usually is too dispersed to be of economic importance. A good example of a near-commercial deposit of this type is that in the Wiberforce-Bancroft area of Ontario, Canada; similar deposits have been reported in other areas of the world. No known pegmatite, hydrothermal vein, or other high-silica igneous P deposit is in commercial production.

About 17% of the world's production of P is from deposits of igneous apatite, and most of this is from Kola and Phalaborwa.

B. Metamorphosed Rocks

Metamorphosed rocks are transitional between the igneous and sedimentary rocks in their geologic history, texture, and mineral assemblages. They range in texture from the slightly modified shales of the low-grade regionally metamorphosed Phosphoria Formation in Idaho (USA) to the high-grade pyroxenite gneiss and marble of Templeton in Quebec, Canada. As a general rule, the metamorphism of phosphates is low grade and makes the rocks harder and the minerals more intimately mixed, but the basic sedimentary structures usually persist although massive beds of phosphate also can result. Some metamorphosed deposits are in Brazil (Patos de Minas), India (Udaipur), Pakistan (Hazara), Finland (Siilinjarvi), and northern Viet Nam.

C. Sedimentary Rocks

Sedimentary PR have a wide variety of textures that reflect complex geologic origins and histories—they may be detrital, chemical precipitates or contain significant amounts of fossil (organic) apatites. Cathcart and Gulbrandsen (1973) described three distinct genetic types of sedimentary deposits. The first type represents deposits formed by divergence upwelling with the characteristic forms of black shale, phosphatic shale, phosphorite, and in association with chert or dolomite. The phosphate in these deposits occurs mostly as pelletal particles. Some examples are the deposits in the western U.S. (Phosphoria), USSR (Kara Tau), Peru (Sechura), Colombia (Turmeque, Sardinata), Australia (Queensland), North Africa (Morocco, Algeria, Tunisia), West Africa (Senegal, Togo), and Middle East (Egypt, Israel, Jordan).

A second type comprises reworked phosphates deposited along eastern continental coasts by warm currents. The lithology is usually phosphatic limestone or sandstone. Two examples are those in the eastern U.S. (Florida, Georgia, North Carolina) and Brazil (Olinda).

A third type of deposit is formed on stable continental shelves or in continental interiors. These usually are associated with limestone, dolomite, shale, and glauconitic sandstone. One economic deposit of this type is the Mississippian sandstone mined as Tennessee "blue rock."

The textures of sedimentary PR fall into two major classes, consolidated and unconsolidated, with several subdivisions of each. The consolidated rocks can be subdivided into those cemented by silica and silicates, those cemented by carbonates, and those cemented by Fe and Al oxides. The unconsolidated rocks fall into several subdivisions that describe the size or structure or both of the apatite aggregates.

1. CONSOLIDATED ROCKS

Deposits of consolidated PR cemented by silica and silicates have been reported in increasing numbers in recent years. Silica in the form of chert and silica gel are the main cementing agents in deposits in Colombia, Pakistan, and Australia, and these ores present significant processing and beneficiation problems that are discussed later. Clay minerals are the most common silicate minerals that occur as cement, and the mineralogy of the clay-mineral assemblages is as complex and varied as that of the ores. Two examples of high-clay ores are the Peru and Saudi Arabian deposits.

EVALUATION OF RAW MATERIALS

Carbonate-cemented phosphorites are particularly important because of their abundance; there are 3 tons of carbonate-cemented phosphate for every ton of ore that can be handled by conventional milling and flotation processes. Both calcite and dolomite are common as cements; other carbonates are present only occasionally. Examples of carbonate-cemented deposits are those in Australia, Colombia, Pakistan, Turkey, India, Florida, and the western U.S.

Iron and aluminum phosphates, as well as their oxides, may occur as cementing agents. Although some of these are primary minerals, these cements usually are secondary minerals that were formed by weathering. Some examples of these types are deposits in Tennessee, Turkey, India, Senegal, and Florida.

Another class of consolidated PR comprises rocks that were formed by replacement processes in which leached soluble phosphate altered an underlying rock. These replacement ores may be quite massive or nodular in texture, depending on a number of geologic factors. Some well-known examples of this type are the Florida hard rock deposits and those of the western Pacific Islands (Nauru, Ocean, Christmas).

The detrital and chemically precipitated phosphates form another class of consolidated sedimentary rocks. In these materials, the lithologies are similar to those of siltstones and sandstones and generally are imperfectly cemented by small amounts of phosphates and accessory minerals so that they remain somewhat friable. The North African and Middle Eastern deposits are rocks of this textural type. The rocks are mostly phosphate with small amounts of evaporites, quartz, clays, and carbonates as accessories.

2. UNCONSOLIDATED ROCKS

The richest sedimentary deposits are those that have been concentrated by secondary processes such as reworking, leaching, and weathering, and these natural enrichment processes have left many PR in an unconsolidated state. Probably the best known of these unconsolidated rocks is the land-pebble phosphate typical of the central Florida deposit. Pelletal phosphates occur also in other parts of Florida and in North Carolina. Oolitic phosphates are common in North Africa and the Middle East. Organic phosphates occur in Israel, Jordan, and Angola where half or more of the P is derived from bone, teeth, and other organic fragments. Glauconitic phosphates are typical of the stable continental shelf and interior environments as typified by parts of the deposits in Tennessee, North Carolina, Baja California, and Chile. Examples of the unconsolidated replacement types of ores are those in which phosphates have replaced carbonate rocks and fossils. The deposits in Baja California are a good example of this type, and those in Angola, Israel, Florida, and North Carolina show this replacement to a smaller extent.

Many of the phosphate particles in sedimentary rocks contain occluded accessory minerals. Some of the more common ones are quartz, chert, calcite, dolomite, clays, Fe and Al oxides and hydroxides, and gypsum. The concentrations of these occluded phases vary markedly among the ores, and their distribution is a significant factor in the selection of the process for utilization of a particular ore.

Another class of sedimentary deposit that represents a large mineral resource for the distant future is the submarine phosphate on the continental shelf and in the deep ocean. These deposits are known along the southern Atlantic coast of the United States (South Carolina, Georgia, Florida), and the coasts of southern California and the Baja Peninsula, Chile, Peru, South Africa, New Zealand, and others (Notholt, 1975; Cathcart & Gulbrandsen, 1973). The nodules in these deposits contain about 13% P (30% P_2O_5), which is comparable in grade to some commercial rocks, but they usually are high in carbonates and Fe which precludes their use in fertilizer processes as long as phosphate supplies on land remain abundant and available.

D. Influence of Weathering on Texture

Weathering can change significantly the texture of any rock type, whether igneous, metamorphic, or sedimentary, and the results may be either beneficial or deleterious. For example, weathering can improve the texture by liberating the P from the matrix, as by the removal of carbonates. On the other hand, weathering can effect changes in the P mineralogy by which the apatite is converted to the less desirable Fe and Al phosphates, or it can so change the mineral assemblage that the accessories are made more soluble by the processing operation and so introduce undesirable properties in the product. In any event, weathering can cause very significant changes in the physical and chemical properties of a PR to such an extent that the rock may be rendered either more or less economic, depending on the type and extent of the alteration.

V. CHARACTERIZATION METHODS

Apatites are the most important source of P in commercial ores, and the characterization methods described in this report will concentrate on these ore types. A similar detailed study of the other classes of phosphate minerals probably would reveal systematic relationships in their compositions and properties also.

This chapter summarizes the results of examinations of apatite concentrates from samples of about 560 PR that represent almost all the commercial P deposits in the world. The examinations were made by chemical analysis, petrographic microscopy, X-ray powder diffraction, infrared spectroscopy, and electron microscopy. Although a complete characterization of an apatite might require the results of all these techniques, a rapid, preliminary characterization suitable for broad classification of samples can be made from a combination of X-ray and petrographic data.

The relationships that were established by the results of these examinations are helpful in selecting suitable methods of processing a phosphate ore from a particular deposit. The results show also that samples of ore from different parts of each major geographical deposit differ enough among themselves to require individual mineralogical characterization as well as chemical analysis for selection of the optimum processing conditions for a particular ore.

A. Chemical Analysis

Most of the PR samples examined were high-grade commercial concentrates. When only unbeneficiated ores were available, concentrates were prepared in the laboratory by screening, hand-sorting, and heavy-liquid separation. Calcined phosphates were not included in the ores examined.

A representative sample of each PR was obtained by repeated quartering and ground to -200 mesh. Free carbonate minerals were removed by extraction with Silverman's solution (Silverman et al., 1952); the extraction was repeated if residual free carbonates (usually coarse dolomite) were detected by petrographic examination. The extracted apatites were washed repeatedly with water and dried at 105°C and then examined by the several techniques mentioned above.

Of the more than 25 elements that have been reported to occur in fluorapatites (Table 1), most are present in insignificant amounts; and the apatites described in the literature are usually from igneous or metamorphic deposits. As shown previously (Lehr, 1967; McClellan & Lehr, 1969), however, the compositions of apatites in sedimentary PR can be adequately described by their contents of Ca, Na, Mg, P, CO_2, and F.

The compositions of some representative apatites in terms of these six major constituents are shown in Table 2. More extensive analytical data on the apatites in PR have been reported previously (Lehr, 1967; McClellan & Lehr, 1969; Lehr & McClellan, 1972).

Table 1—Substitutions in the apatite structure of fluorapatite; $Ca_{10}(PO_4)_6F_2$.

Constituent ion	Substituting ion
Ca^{2+}	$Na^+, Sr^{2+}, Mn^{2+}, K^+, U^{4+}, Mg^{2+}, RE^{2+,3+}$
P^{5+}	$C^{4+}, S^{6+}, Si^{4+}, As^{5+}, V^{5+}, Cr^{6+}, Al^{3+}$
F^-	OH^-, Cl^-
O^{2-}	F^-, OH^-

Table 2—Some typical francolite compositions, computed from unit-cell a dimensions.

Source	Composition, %					
	CaO	MgO	Na_2O	P_2O_5	CO_2	F
Western U.S.	55.6	0.13	0.26	40.1	1.59	4.09
Tennessee (U.S.)	55.5	0.24	0.47	38.7	2.71	4.31
Florida (U.S.)	55.5	0.36	0.72	37.1	3.95	4.56
Morocco	55.4	0.43	0.85	36.3	4.53	4.68
North Carolina (U.S.)	55.3	0.52	1.04	35.3	5.36	4.85
Tunisia	55.2	0.60	1.20	34.7	5.70	4.93

The electrostatic imbalance resulting from the substitution of planar CO_3^{2-} for tetrahedral PO_4^{3-} is only partially corrected by substitution of F^- in vacant oxygen sites so that a coupled monovalent cation substitution for Ca^{2+} is necessary to maintain electrostatic neutrality. In sedimentary apatites that were formed in marine environments, the cations most likely to replace Ca^{2+} are Na^+, Mg^{2+}, and K^+.

Gulbrandsen et al. (1966) found a statistical correlation between the Na content and degree of CO_3^{2-} substitution in apatites from Wyoming and a similar coupled substitution in an apatite from Ontario, Canada. Ames (1959) synthesized carbonate apatites under simulated marine conditions, except without F, and found a coupled substitution in which Na replaced about 10% of the Ca at the highest level of carbonate substitution. Simpson (1964) reexamined Ames' precipitation system with consideration of both K and Na and confirmed the coupled substitution of alkalies for Ca when CO_3^{2-} replaced PO_4^{3-}. He reported sodium contents of 1.2 to 1.8% Na (2 to 3% Na_2O) at the highest level of carbonate substitution (1.5% C, or 5.6% CO_2). Potassium showed little tendency to replace Ca (the ionic radius of K^+ is >130% that of Ca^{2+}, whereas the ionic radius of Na^+ is <95% that of Ca^{2+}). Potasisum also appeared to suppress the substitution of CO_3^{2-} for PO_4^{3-}. Simpson reported also that the degree of substitution of CO_3^{2-} and Na^+ increased as the pH of the precipitation medium was raised. Neither Ames nor Simpson, however, considered the effects of Mg and F on the composition of carbonate apatites. Le Geros et al. (1965, 1967) reported evidence of Mg and Na in their synthetic carbonate apatites and noted the effects of F^-, Mg^{2+}, and CO_3^{2-} on the crystallinity of the apatites.

The studies of apatites in sedimentary PR showed that the replacement of Ca^{2+} by Na^+ and Mg^{2+} is systematic, but replacement by K^+ was insignificant to nonexistent. These conclusions were obtained from results of determinations of Na, Mg, K, Ca, P, and CO_3-C in several hundred apatite concentrates. To minimize interferences from clays, feldspars, and other accessory minerals, each apatite was dissolved rapidly in warm $3N$ HCl and the filtrates were analyzed to determine the composition of the apatite.

Although K was present in nearly all the apatites, its amount was usually <0.08% K (0.1% K_2O) and its contribution to the coupled-substitution process was ignored. Sodium and Mg, however, were present in significant amounts in all the apatites, and the content of each element increased with increasing degree of CO_3^{2-} substitution as shown in Fig. 1. The degree of substitution is expressed as the mole ratio CO_3/PO_4.

The Na contents of the carbonate apatites ranged from 0.04 to 1.2% and the Mg contents from 0.03 to 0.36%. Sedimentary apatites, therefore, show the same type of substitutions that have been reported for synthetic apatites. More Mg and less Na were found in the natural apatites than in the synthetic apatites, and this difference may reflect differences in the conditions under which the apatites were formed. The synthetic apatites usually were prepared in fluoride-free systems.

As shown in Fig. 1, the Na and Mg in commercial concentrates are present mostly as constituents of the apatite and cannot be removed by

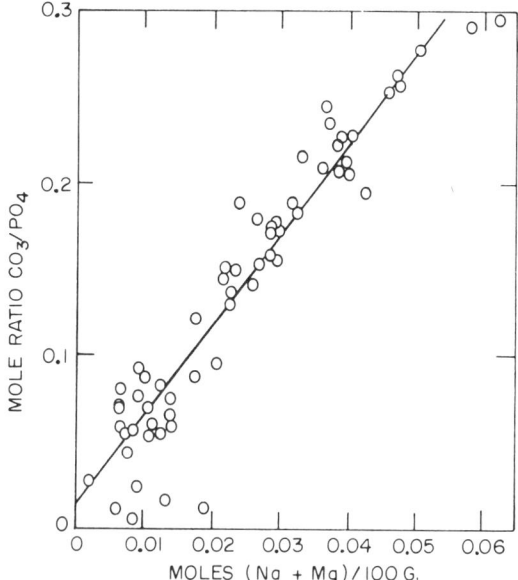

Fig. 1—Relation between degree of carbonate substitution and sum of Mg and Na contents of apatite.

beneficiation. Higher values occurred when additional Mg was contributed by bentonitic clays (western U.S.), glauconite (Baja California ore), and volcanic glass plus bentonitic clay (Peruvian ores); the additional Na in ores from Wyoming and Idaho, U.S., was the exchangeable Na in the bentonite clay associated with the phosphate.

Previous work (Lehr, 1967; McClellan & Lehr, 1969; Lehr & McClellan, 1972) has shown that the compositions of sedimentary apatites can be expressed by the generalized formula:

$$(Ca_{10-a-b}Na_aMg_b)(PO_4)_{6-x}(CO_3)_xF_{0.4x}F_2$$

in which a ranges to about 0.35, b to 0.14, and x to 1.26. Apatites with these compositions are called *carbonate apatites* or *francolites* (McConnell, 1938).

B. X-ray Characterization

The X-ray powder diffraction (XRD) patterns of the francolites (carbonate apatites) that occur in commercial phosphate rocks are typically apatitic with slight shifts in peak positions and intensities that indicate changes in the cell parameters. Changes in the unit-cell a dimentions with changes in carbonate content were reported by Maslennikov and Kavitskaya (1956), but no relationship was established between the unit-cell dimensions of the apatites and their carbonate content. Later work by Smith and Lehr

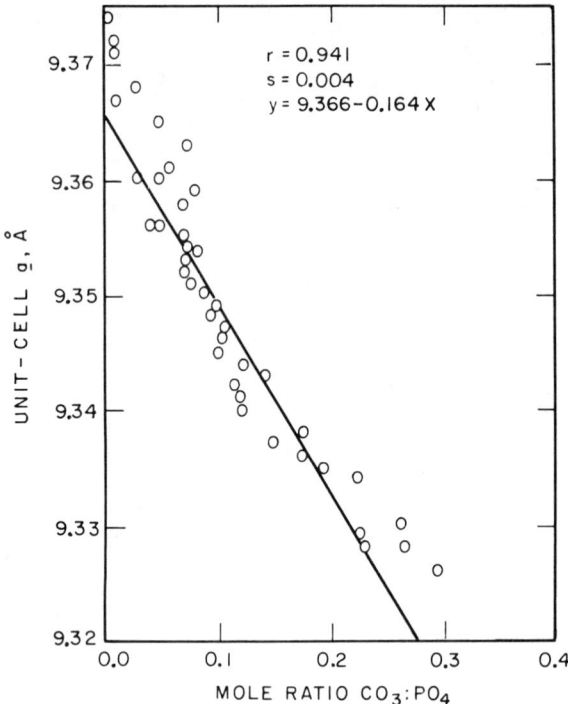

Fig. 2—Relationship between unit-cell a dimensions and mole ratio CO_3/PO_4 in apatite.

(1966) and others (Lehr, 1967; McClellan & Lehr, 1969) clearly established this relationship (Fig. 2) and defined it in terms of variations in crystal chemical compositions.

The method for determining unit-cell parameters of apatites has been described in detail elsewhere (McClellan & Lehr, 1969), and is summarized here. Basically, the procedure is to use a sample of the minus 200-mesh concentrate, free of rhombic carbonates, to record a high-resolution powder diffraction pattern over the range of 25 to 54° 2θ using CuK_α radiation. An internal standard of spinel ($MgAl_2O_4$) is used to correct for any goniometer misalignment. The cell parameters are calculated with an iterative least-squares computer program that solves a truncated Taylor series by the Newton-Raphson method for two unknowns. The apatite reflections used have the Miller indices 002, 300, 302, 310, 222, 312, 213, 321, 410, 402, and 004. The standard errors of the cell constants are usually ± 0.003 Å or less.

The results of measuring the cell parameters on nearly 500 francolite-bearing PR show that the values of a range from 9.322 to 9.376 Å and those of c from 6.877 to 6.900 Å. The larger variations in a indicate that substitutions have a greater effect along this direction, an effect confirmed by Kreidler (1967).

Apatites with a values between 9.376 and 9.421 Å are not francolites but are members of a series of fluorhydroxyapatites. Apatites from island deposits (Christmas, Nauru, Ocean) and carbonatites belong in this series, which is complex and quite different from the carbonate apatites in the more common commercial deposits. True hydroxyapatites are rare minerals. One occurrence is at Holly Springs, Georgia, and the others usually are modern organically derived materials in bones and teeth. These materials usually have a values of 9.421 Å or more, depending on their composition, and can have very complex crystal chemical structures.

C. Infrared Characterization

The infrared absorption spectrum of an apatite serves not only to identify the compositional form of the apatite but also may reveal extraneous phases present and indicate the degree of substitution of carbonate for phosphate. A comprehensive review of the absorption properties of carbonate in apatite structures was presented by Elliott[1], and several infrared studies of carbonate apatites are reported in the literature (Baddiel & Berry, 1966; Posner & Duyckaerts, 1954; Romo, 1954).

In francolites, the characteristic CO_2 absorption doublet at 1453 and 1420 cm^{-1} usually shows only small differences in relative intensity, but vary in amplitude depending upon the amount of CO_3^{2-} substitution in the particular francolite. The C–O band at 860 cm^{-1} in the apatite spectrum is significantly less intense than the corresponding band in the spectrum of either calcite (874 cm^{-1}) or dolomite (870 cm^{-1}).

With increasing carbonate substitution, the 575 cm^{-1} absorption band (P–O stretch) of fluorapatite progressively shifts to 565 cm^{-1}, and its intensity increases slightly but rarely exceeds that of the 602 cm^{-1} band. Hydroxyapatite spectra, on the other hand, usually show the 565 cm^{-1} band to be more intense than the 602 cm^{-1} band and further show weak O–H absorption bands at 3560 and 632 cm^{-1}.

For determinations of the degree of CO_3^{2-} substitution from their infrared spectra, apatite concentrates first were extracted by the Silverman procedure (Silverman et al., 1952) to remove free carbonates which might interfere by contributing background absorption. The extracted apatites were prepared for infrared examination by the KBr pellet method with 0.75 mg of sample in 300 mg of KBr compressed at 22,000 psi in an evacuated die. Spectra were recorded over the range 4000 to 250 cm^{-1} on a dual-beam spectrophotometer (Perkin-Elmer Model 521) with KBr in the reference cell.

The CO_2 index of an apatite obtained from its infrared spectrum is based on the ratio of intensities of the C–O and P–O absorptions; this ratio is directly proportional to the weight ratio CO_3/PO_4 and is independent of the concentration of apatite. As shown in Fig. 3, the C–O absorption is the average of the intensities of the 1453 and 1420 cm^{-1} bands measured from the background base at 1800 cm^{-1} to avoid interference from H_2O absorp-

---

[1] J. C. Elliott. 1964. The crystallographic structure of dental enamel and related apatites. Ph.D. Thesis, Univ. of London.

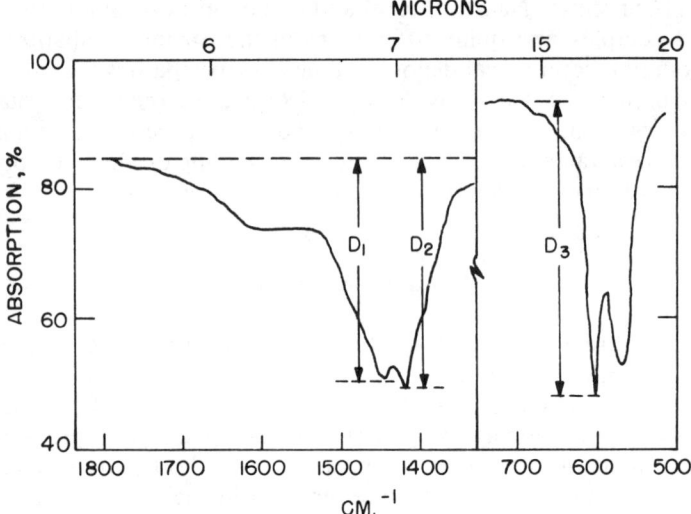

Fig. 3—Determination of CO_2 index.

tion bands at 1650 to 1640 cm^{-1}. The P-O absorption is based on the intensity of the 602 cm^{-1} band, which is consistently sharp and present in the spectra of all the apatites; the background base at 700 cm^{-1} is used to avoid interference from siliceous impurities which introduce a weak Si-O absorption at 680 to 670 cm^{-1}. The CO_2 index is defined as the ratio of intensities (Fig. 3) of the C-O and P-O absorptions by the equation:

$$CO_2 \text{ index} = 0.5\,(D_1 + D_2)/D_3.$$

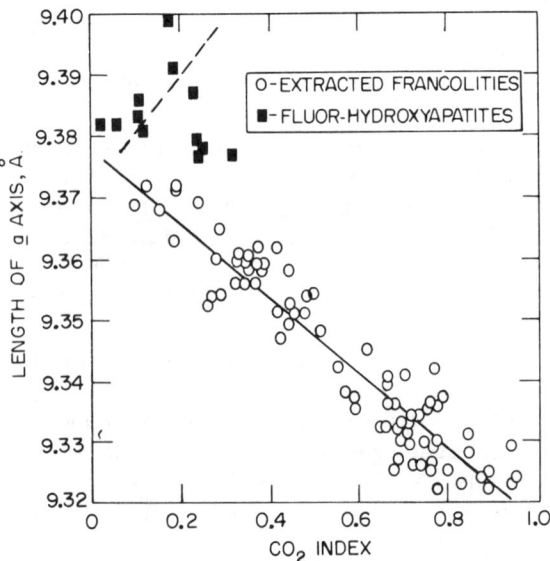

Fig. 4—Correlation of length of a axis with CO_2 index.

A correlation of the CO_2 index values with the unit-cell dimension a determined by X-ray yielded for the series of carbonate-substituted fluorapatites the linear relationship shown in Fig. 4.

Fluorine-deficient apatites that have a axes larger than that of fluorapatite, 9.372 Å, and contain OH^- in addition to CO_3^{2-} do not obey the linear solid line relationship in Fig. 4, since in these apatites CO_3^{2-} may enter hexad sites in addition to PO_4^{3-} sites. These F-deficient apatites, which are rather uncommon in sedimentary phosphorites, can be distinguished from francolites by details in their infrared spectra (Lehr, 1967).

Infrared examination of a concentrate thus yields data related to CO_3 substitution and to the unit-cell dimensions of the apatite, and hence its chemical composition may be estimated. Infrared examinations are useful for rapid screening of ore samples for homogeneity as a preliminary characterization step.

D. Petrographic Examination

The carbonate apatite in most sedimentary PR is submicrocrystalline (0.02 to 0.2 μm) and occurs in several varieties of complex aggregates. Among the common textural forms are oolitic and ovulitic pellets of marine precipitates, angular to subrounded polished grains of clastic phosphorite, replacement forms of phophatized shell, coral, and fecal pellets, and fossil vertebrate bones and teeth. The major accessory minerals (quartz, carbonates, feldspars, and heavy minerals) usually are present as discrete free mineral grains. Apatite particles frequently are stained by occlusions of finely divided Fe oxides and carbonaceous matter; in addition, colorless occlusions of clays, opaline silica, and diatom frustules occasionally are present. Although petrographic examination provides much useful information on the accessory minerals and the distribution of other impurities, the submicroscopic sizes of the carbonate apatite crystals preclude detailed optical study. Usually the only determinable property is a mean refractive index by the oil immersion method. Occluded colloidal impurities will contribute error to the measurement of refractive index of the apatite, but this interference frequently can be detected by oil immersion.

Fortunately, the refractive index of an apatite concentrate is remarkably uniform, even though the aggregates may be present in several textural forms. In a test of this uniformity, the apatite fractions of three typical phosphate ores were separated by hand sorting into recognizable textural and color groups. X-ray, infrared, and chemical examinations of the several fractions showed no significant differences in chemical composition or crystallographic properties of the apatites in the fractions, except for small variations in the amounts of occluded insoluble minerals such as Fe oxides, clays, and organic matter.

Because of this homogeneity of the apatite in most sedimentary PR, a carefully measured mean refractive index on clear fragments of the carbonate apatite that are relatively free of occluded impurity phases can be used in

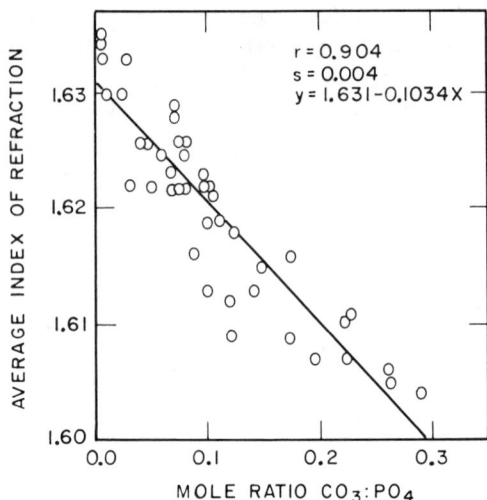

Fig. 5—Relationship between average index of refraction and mole ratio CO_3/PO_4 in apatite.

an estimate of the degree of carbonate substitution. This refractive index relationship for 36 apatite samples of variable crystal chemical composition is shown in Fig. 5. Microscopic examination thus provides a reliable measure of most apatite compositions in 5 or 10 min., but the results are less precise than those of the more complicated X-ray procedure.

Textural descriptions, which include the size, shape, and arrangement of the mineral constituents and their mode of agglomeration, are important parts of a complete evaluation of any PR and supply information that can be obtained only by petrographic analysis.

E. Interpretation of Data

The chemical compositions and unit-cell a parameters of nearly 500 francolites have been used in the development of the statistical correlations of the crystal chemical compositions with the physical and chemical properties of the francolites. These correlations have established the relationships between the composition of the francolite (as indicated in the general empirical formula) and the a-axis dimension of its unit cell, which is the most sensitive indicator of variations in crystal chemical composition.

The apatite mineral component is usually the only source of P and F in the ore and frequently contributes most, if not all, of the Mg and Na.

Sedimentary apatites have been shown to have compositions that can be represented quite adequately by a substitution series with the end-member empirical formulas (McClellan & Lehr, 1969): $Ca_{10}(PO_4)_6F_2$, fluorapatite; and $Ca_{10-a-b}Na_aMg_b(PO_4)_{6-x}(CO_3)_xF_{0.4x}F_2$, francolite. These formulas correspond to the chemical compositions shown in Table 3.

Based on data compiled for a large number of deposits, the relationship between the a unit-cell dimension and the a, b, and x parameters of the empirical formula of francolite was obtained by regression:

Table 3—Mineral composition of the end members of fluorapatite-francolite series.

Constituent, %	Fluorapatite: ($x = 0$)	Francolites: $x/(6 - x) \cong 0.30$†
CaO	55.6	55.1
P_2O_5	42.2	34.0
CO_2	0	6.3
F	3.77	5.04
Na_2O	0	1.4
MgO	0	0.7

† Maximum degree of substitution found and predicted for francolite-type apatites (McClellan & Lehr, 1969).

$$x/(6-x) = 4.90\,(9.374 - a_{obs}) \quad [1]$$

$$a = 1.327\,[x/(6-x)] \quad [2]$$

$$b = 0.515\,[x/(6-x)] \quad [3]$$

in which
 a_{obs} = length, Å, of a-axis as determined by X-ray diffraction,
 x = subscript of CO_3 in the francolite formula,
 a = subscript of Na in the francolite formula, and
 b = subscript of Mg in the francolite formula.

These regression equations permit the computation of the empirical formula of a given apatite, and hence its composition, from a measurement of the unit-cell dimension a by X-ray diffraction. Typical apatite compositions computed by this method are shown in Table 2 for six well-known geographical deposits. These compositions are congruent with those determined by detailed mineralogical analyses (Gremillion & McClellan, 1978). The compositions are arranged in the order of increasing carbonate substitution but do not span the entire range of compositions shown by the end members in Table 3.

The most significant features revealed by these compositions are the increases in the relative amounts of Mg, Na, CO_2, and F; the decrease in P_2O_5 content; and the nearly constant CaO content, as the degree of substitution in the apatite increases. It is apparent that all these compositions differ markedly from that of fluorapatite which has heretofore been assumed to be the phosphatic constituent of PR.

The application and use of these regression equations will be demonstrated in the following section on the utilization of the characterization data.

VI. UTILIZATION OF CHARACTERIZATION DATA

The characterization data on PR describe the minerals present, their relative proportions, and the manner in which they are combined. In this section, the application of the results of the detailed study of a PR will be demonstrated in several practical examples. This discussion will show how these characterization data can be used to determine the optimum method of utilization of a PR with consideration of economic, technical, and environmental factors.

A. Selection of Process Alternatives

The basic assumption will be that a new phosphate deposit has been discovered with sufficient grade and tonnage to be exploited economically. Once a phosphate deposit is located, the properties of the ore must be considered in the light of existing technology. Briefly, the principal considerations are beneficiation, direct application, thermal treatment, acidulation processes, and electric furnace technology.

B. Beneficiation

The basic beneficiation technology of PR processing consists of the two steps, liberation and separation, with success of the first step necessary for success of the second. Liberation is usually effected by size reduction or particle detachment to create free particles of the desired minerals so that they can be separated. Many ores are composed of locked particles of two or more mineral phases and liberation is measured in terms of degree, expressed as the fraction of the particles of a mineral that is free from physical combination with the gangue minerals.

In some high-grade PR, size reduction is all that is required to prepare the ore for commercial use. More often, however, the size reduction is followed by a sizing operation such as screening or wet classification to remove over- and undersize materials. With some ores this sizing is sufficient to remove enough of the free gangue minerals to produce commercial concentrates. In others, a concentration step such as washing, flotation, calcination, magnetic separation, or electrostatic separation is required. Phosphate rocks usually are concentrated by some combination of washing, flotation, and calcination.

A knowledge of the texture and mineralogy of an ore is indispensable in selecting a method for its beneficiation. This information can indicate the type and amount of grinding a particular ore may require; knowledge of the distribution of the free and occluded gangue minerals may be used to predict the properties of the finsihed concentrate and to determine when the beneficiation treatment has reached its productive end.

The characterization data may suggest uses for the particular rock. For example, a silica-cemented ore may be suitable for direct chemical extraction if lightly ground, or it might be used directly as an electric furnace feed. Nitric acid extraction of P from an ore very high in occluded Fe oxides might be suggested to take advantage of the low solubility of the undesirable Fe mineral in nitric acid. Direct acidulation may be recommended

Table 4—Mineral composition of concentrates with equivalent grade—14% P (32% P_2O_5)†.

Deposit source	Weight percent	
	Apatite‡	Gangue minerals
Western U.S.	79.8	20.2
Tennessee (U.S.)	82.7	17.3
Florida (U.S.)	86.3	13.7
Morocco	88.2	11.8
North Carolina (U.S.)	90.7	9.3
Tunisia	92.2	7.8

† Data from Lehr and McClellan (1974).
‡ Weight percent apatite = actual P_2O_5, %/theor. P_2O_5% g of apatite phase.

Table 5—Chemical composition of concentrates from representative commercial deposits.

Deposit source	BPL grade	Constituent, %†							
		CaO	P_2O_5	F	CO_2	R_2O_3	Na_2O	MgO	SiO_2
Central Florida	73	48.9	33.4	3.9	3.0	2.12	0.53	0.29	4.5
North Carolina	66	48.5	30.2	3.7	5.5	1.14	0.83	0.54	2.1
Morocco	70	51.6	32.1	4.1	5.3	0.55	0.79	0.43	1.4
Gafsa, Tunisia	63	48.3	28.8	3.4	6.3	1.22	1.30	0.59	1.8
Taiba, Senegal	82	51.2	37.4	4.0	1.7	2.06	0.20	0.06	2.9
Togo	80	52.3	36.6	4.0	1.8	1.78	0.27	0.11	1.8
Kola, Russia	83	52.0	38.2	3.1	0.2	3.14	0.50	0.06	2.0
Western Sahara	78	51.9	35.8	3.8	2.3	1.17	0.40	0.14	3.4
Angola	81	51.3	37.2	4.0	2.1	1.47	0.62	0.10	1.5
Jhamar-Kotra, India	88	54.2	40.1	3.6	0.7	0.70	0.11	0.04	1.2
Jordan	74	53.0	33.8	4.0	4.9	0.36	0.51	0.18	5.6
Oron, Israel	68	52.7	31.3	3.6	7.5	0.45	0.75	0.24	0.2
Bayovar, Peru‡	66	46.5	30.2	2.9	4.4	1.65	1.85	0.50	3.2
Algeria	63	49.3	29.0	3.6	7.4	0.70	2.00	0.81	1.0

† TVA analyses of authentic samples as received; moisture-free basis (105°C).
‡ Formerly referred to as *Sechura* deposit.

for ores containing as much as 10% free carbonates when the extra acid consumption is cheaper than a complex method of removal of the carbonates. The data may suggest the use of low-energy attrition washing to remove clay hulls from the surface of phosphate particles in another ore.

The application of different beneficiation processes to different ores with varying results is demonstrated in Table 4, in which concentrates of equal grade are shown to contain widely different amounts of apatite and gangue minerals. Knowledge of the properties of the possible final product can be extremely useful in selecting a method for processing a PR concentrate.

Because of the differences in phosphate ores, processes for their utilization cannot be arbitrarily interchanged simply because the ores have the same grade. The beneficiation of a particular PR should be custom designed, and serious problems may arise when a beneficiation process developed on one ore is applied unchanged to a different ore with no consideration for the differences in mineralogy and texture of the ores.

The compositions of 14 PR concentrates selected to represent both well-known deposits and some more recently developed deposits are compared in Table 5, in which only the major constituents are included. All these samples are authentic beneficiation concentrates and were selected for demonstration purposes only. They may or may not represent current production from the indicated locations but do illustrate the variations that occur in commercial concentrates.

C. Direct Application

One of the simplest uses of PR is for direct application as a fertilizer. Rocks used for this purpose usually are finely ground and applied as dry powders, slurries, or pellets that disaggregate readily in water. Variations in the composition of the apatites in PR, however, can affect markedly their agronomic performance.

The effectiveness of a PR as a direct-application fertilizer has been demonstrated in greenhouse and field tests and has been shown to be directly related to the degree of substitution of carbonate for phosphate in the apatite structure. The degree of carbonate substitution influences the solubility of the apatite in the rock and controls its reactivity (the rate at which P is released under favorable soil conditions). Because the reactivity of a rock is related to its agronomic effectiveness (Khasawneh & Doll, 1978), the value of a PR as a direct-application fertilizer can be determined by laboratory dissolution tests. These laboratory tests can be made rapidly and can be used as a guide in the selection of rocks to be used in greenhouse or field tests.

The use of PR as direct-application fertilizers does not enjoy the popularity that it once did. One reason for the decline in its popularity is the conflicting ways in which earlier workers expressed rock reactivity. The practice of using the ratio of solvent-soluble P_2O_5 to rock grade as a guide to agronomic effectiveness is misleading and caused many experimenters to make erroneous judgments when selecting PR to be used for direct application. The poor choice of rocks led to poor agronomic results, which, of course, discouraged further use of PR for this purpose. Phosphate rock reactivity is dependent primarily upon the type of apatite in the rock, and most of the phosphatic material from a given deposit consists of a single type of apatite whose chemical and physical characteristics are relatively uniform for that deposit. Field geology studies show, however, that the concentration of apatite in the ore varies considerably from one location in the deposit to another, and it is clear that the reactivity of a rock from a given deposit cannot be related to so variable a factor as its grade.

In recognition of this problem, Lehr and McClellan (1972) proposed that rock solubility be expressed in terms of the grade of the apatite, which is relatively constant in a given deposit, instead of the rock grade, which is not constant. Thus, when using neutral ammonium citrate as the extractant (AOAC method), it was proposed that the absolute citrate solubility (ACS) of a rock be expressed as:

$$\text{ACS} = \frac{\text{AOAC citrate-soluble P}_2\text{O}_5, \%}{\text{theoretical P}_2\text{O}_5 \text{ concentration of apatite}, \%}.$$

As the use of other solvents became more common, the absolute solubility was expressed in more general terms to include any type of solvent (McClellan, Gremillion, & Lehr, unpublished TVA data). The term *Absolute Solubility Index* (ASI) then was defined as

$$\text{ASI} = \frac{\text{solvent-soluble P}_2\text{O}_5, \%}{\text{theoretical P}_2\text{O}_5 \text{ concentration of apatite}, \%}.$$

One disadvantage of relying upon the ASI as a guide to rock selection is that the method is dependent upon the use of sophisticated X-ray analyses and computer programs that may not be available to many laboratories that are attempting to evaluate phosphate rocks as direct-application fertilizers. Recent investigations have shown, however, that unadjusted solubility data on rocks with a minimum P content of at least 8.7% (20% P_2O_5) can be used as effectively as ASI values as a guide to rock selection. Through use of this unadjusted data the need for X-ray measurements and computer calculations can be eliminated.

The three solvents most commonly used throughout the world in making solubility tests are neutral ammonium citrate, 2% citric acid, and 2% formic acid. However, because of the different properties of these solvents, the amount of P extracted from a given rock by different solvents may differ by a factor of two or more. Differences this large may cause confusion and misinterpretation in comparative evaluations of PR that are being considered for direct application. As a partial solution to this problem, equations were derived to correlate the soluble P (%) of 36 PR in neutral ammonium citrate, 2% citric acid, and 2% formic acid with the unit-cell a dimensions of the apatite in the rocks (Fig. 6). These regression equations, summarized in Table 6, can be used to convert the solubility of

Table 6—Conversion factors for solubility data based on ASI values[†].

	neutral ammonium citrate	2% Citric acid	2% Formic acid
	Without aluminum		
Neutral ammonium citrate	1	$1.259n + 7.29$	$2.416n$
2% Citric acid	$0.79c - 5.79$	1	$1.919c - 13.99$
2% Formic acid	$0.4138f$	$0.521f + 7.29$	1
	With Al-acetate		
Neutral ammonium citrate	1	$1.433n + 7.575$	$1.597n + 2.638$
2% Citric acid	$0.6977c - 5.285$	1	$1.114c - 5.802$
2% Formic acid	$0.6262f - 1.652$	$0.897f + 5.208$	1

[†] n = % P_2O_5, soluble in neutral ammonium citrate; c = % P_2O_5, soluble in 2% citric acid; and f = % P_2O_5, soluble in 2% formic acid.

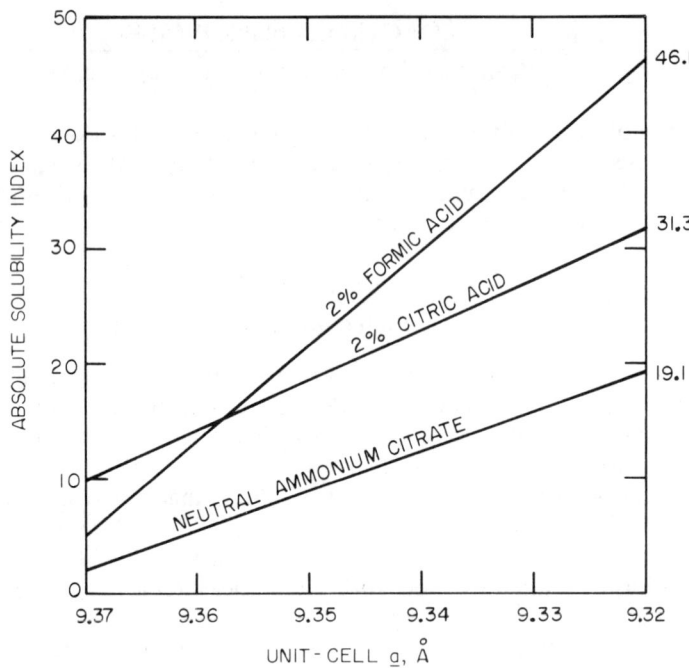

Fig. 6—Relationship between absolute solubility index and unit-cell a, Å (no soluble Al added).

an apatite in one solvent to its equivalent solubility in either of the other two solvents. In addition, the equations can be used to estimate the value of PR for direct application by comparing its solubility with the crystallographically estimated or measured solubility of a rock of known agronomic value.

In the correlation of the data it was observed that rocks from certain geographic areas had solubilities in 2% citric and 2% formic acid solutions that were consistently and significantly lower than those predicted from compositions of their apatites. It was then found that addition of small amounts of soluble Al as Al acetate increased the solubility of PR in 2% citric acid but decreased it in neutral ammonium citrate and 2% formic acid

Table 7—Regression equations relating ASI values to x, unit-cell a dimensions.

Solubilizing solution	Correlation coefficient, r	Standard error of estimate, s	Equation for y†
	Without aluminum		
ASI, neutral ammonium citrate	0.915	2.162	340.6 (9.376-x)
ASI, 2% citric acid	0.854	3.748	428.8 (9.393-x)
ASI, 2% formic acid	0.733	10.98	832.0 (9.376-x)
	With Al-acetate		
ASI, neutral ammonium citrate	0.913	2.116	330.3 (9.377-x)
ASI, 2% citric acid	0.945	2.346	473.5 (9.393-x)
ASI, 2% formic acid	0.902	3.641	527.5 (9.382-x)

† x = unit-cell a parameter. Å.

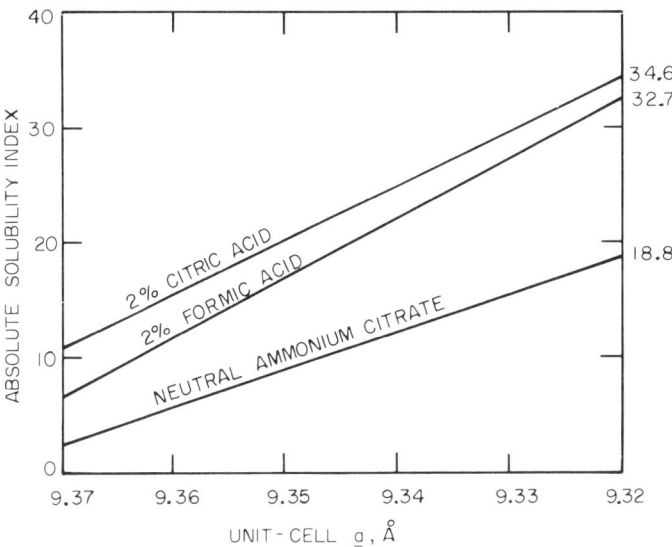

Fig. 7—Relationship between absolute solubility index and unit-cell a, Å (soluble Al acetate added to extracting solutions).

as shown by comparison of Fig. 6 and 7. The reason for these effects is not fully known, but it is thought to be related to the ability of Al to complex the fluoride that is released during the dissolution, which alters the phase equilibrium.

The addition of soluble aluminum to 2% citric acid and 2% formic acid solutions improved the correlation coefficients and the standard errors of the estimates of the solubilities to such an extent that they yielded predicted solubility values with about the same degree of reliability as a neutral ammonium citrate solution (Table 7).

The effect of soluble silica on the solubility data was studied also because it was thought that its effect on F during the dissolution would be similar to that of Al. The solubility data showed, however, that addition of silica made only a slight improvement in the correlation coefficients and standard errors of the estimates, and that the beneficial effect of soluble silica was insignificant in comparison with that of soluble Al.

D. Thermal Treatment

The next technological step beyond direct application in the utilization of P raw materials is thermal treatment. Thermal treatments fall into two general categories, calcination and thermal modification.

Calcination usually is used to upgrade ores with one or more of the following characteristics: (i) excessively high organic matter which can cause foaming and undesirable discoloration of the product in acidulation processes; (ii) highly substituted apatites that can be altered thermally to a marketable grade; and (iii) carbonate minerals that require thermal decomposition for subsequent wet chemical separation to recover the P.

Phosphate rock concentrates vary markedly in their response to thermal treatment, and their response depends largely on the composition of their apatite component. The degree of attainable upgrading increases as carbonate substitution in the apatite increases.

Thermal alteration of carbonate apatites proceeds by the reaction in which the weight loss from evolution of CO_2 and part of the F raises the P_2O_5 content of the residual calcined solids. Results calculated on the basis of such a calcination scheme for 14 phosphate concentrates are included in Table 8.

These results indicate that significant increases in grade may be effected by calcination, and this increase in grade is made even larger by losses of organic matter and bound water, which would nearly double the weight losses given in Table 8 (Lehr, 1967; J. R. Lehr and G. H. McClellan. Fluorine content and properties of commercial phosphate rocks, presented at Symp., Fluorine Sources and Technology, 164th Natl. Meet. of ACS, New York, 27 Aug.–2 Sept. 1972). Even further upgrading would result from decomposition of the carbonate minerals present, depending on their abundance, as shown in Table 9.

For estimation of the effect of calcination on carbonate ores, the amount of $CaCO_3$ must be determined. The composition of the apatite is used as the basis for the determination. Comparison of the theretical weight ratio CaO/P_2O_5 of the apatite (Table 8) with that of the apatite concentrate (Table 5), as shown in Table 9, yields an approximate distribution of the total CaO, and the CaO in excess of that required for the apatite is assigned to $CaCO_3$. The validity of this assignment is readily verified by petrographic examination of the concentrate.

The estimated amounts of free $CaCO_3$ in the concentrates in Table 9 account for as much as 70% of the gangue-mineral fractions in Table 8, and this information may be used to show where savings in acid reagents might be realized by further upgrading steps.

Most calcined PR have low reactivity because of formation of agglomerates, loss of porosity, and thermally induced growth of the crystals of the apatite mineral, all of which markedly decrease the surface area of the apatite. Calcination also lowers the citrate solubility of francolites by decreasing the substitution of carbonate for phosphate, as well as changing the physical properties of the rock.

Increases in the cost of energy have imposed economic restrictions on the use of calcination to upgrade PR, but calcination is widely used to remove carbonates and organic matter where the availability of fuel and its proximity to the PR deposit offers the most economical method of processing the rock.

Thermal modification of phosphates is discussed in detail by J. R. Lehr and J. S. Hoare in Chapter 4 in this book, and only a few of the major points are discussed here. Thermal modification of P raw materials usually involves the decomposition of PR, both apatitic and nonapatitic, with or

EVALUATION OF RAW MATERIALS

Table 8—Properties of beneficiated phosphate concentrates.

Source	Properties of apatite component			Degree of beneficiation			Effect of calcining‡	
	X-ray a obs.	Theor. P_2O_5 %	Theor. ratio CaO/P_2O_5,†	P_2O_5 grade %	Wt. % apatite	Wt. % gangue	P_2O_5 grade %	Weight loss %
Central Florida	9.340	37.2	1.48	33.4	89.8	9.2	35.0	4.6
North Carolina	9.326	35.4	1.55	30.2	85.2	14.8	32.1	5.9
Morocco	9.333	36.3	1.52	32.1	88.5	11.5	33.9	5.3
Gafsa, Tunisia	9.325	35.3	1.56	28.8	81.5	18.5	30.7	6.3
Taiba, Senegal	9.352	38.8	1.43	37.4	96.4	3.6	38.6	3.1
Togo	9.351	38.7	1.43	36.6	94.6	5.4	37.8	3.2
Kola, Russia	9.384	42.1	1.31	38.2	90.4	9.6	38.2	0
Western Sahara	9.352	38.8	1.43	35.8	92.3	7.7	36.9	3.1
Angola	9.350	38.5	1.44	37.2	96.5	3.5	38.5	3.4
Jhamar-Kotra, India	9.366	40.9	1.36	40.1	98.0	2.0	40.6	1.2
Jordan	9.333	36.6	1.52	33.8	93.1	6.9	35.7	5.3
Oron, Israel	9.336	36.7	1.51	31.3	85.4	14.6	32.7	4.3
Bayovar, Peru§	9.340	37.2	1.49	30.2	81.2	18.8	31.6	4.4
Algeria	9.325	35.3	1.56	29.0	82.1	17.9	30.9	6.1
Fluorapatite	9.374	42.2	1.32					

† Ratio CaO/P_2O_5 on weight percent basis for ease of comparisons (see discussion in text).
‡ Effect of thermal alteration of apatite only (does not include losses from organic matter, water, or decomposition of $CaCO_3$).
§ Formerly referred to as *Sechura* deposit.

Table 9—Estimations of CaCO$_3$ content of phosphate concentrates.

	Basis of calculation			CaO content (wt. %)			CaCO$_3$ content	
	Wt. ratio CaO/P$_2$O$_5$							
Product source	Theor. (A)	Actual (B)	Difference (B − A)	Total	As apatite	As CaCO$_3$	Wt. % in product	% in gangue
Central Florida	1.48	1.46	--	48.9	48.9	0	0	--
North Carolina	1.55	1.61	0.05	48.5	46.9	1.6	2.9	20
Morocco	1.52	1.61	0.09	51.6	48.8	2.8	5.0	43
Gafsa, Tunisia	1.56	1.68	0.12	48.3	44.9	3.4	6.1	33
Taiba, Senegal	1.43	1.37	--	51.2	51.2	0	0	--
Togo	1.43	1.43	--	52.3	52.3	0	0	--
Kola, Russia	1.31	1.36	0.05	52.0	50.7	1.3	2.3	24
Western Sahara	1.43	1.46	0.03	51.9	51.0	0.9	1.6	21
Angola	1.44	1.38	--	51.3	51.3	0	0	--
Jhamar-Kotra, India	1.36	1.35	--	54.2	54.2	0	0	--
Jordan	1.52	1.57	0.05	53.0	51.6	1.4	2.5	36
Oron, Israel	1.51	1.68	0.18	52.7	47.1	5.6	10.1	70
Bayovar, Peru†	1.49	1.54	0.05	46.5	44.9	1.6	2.9	15
Algeria	1.56	1.70	0.14	49.3	45.2	4.1	7.4	41

† Formerly referred to as *Sechura* deposit.

without added reagents. The purpose of the treatment is to increase the availability of the P to the soil and so make the altered rock useful as a fertilizer. Thermal treatment decreases the citrate solubility of francolites until defluorination starts at 1000 to 1200°C. In the presence of mineral impurities such as silica, sulfates, silicates, chlorides, and carbonates, thermal treatment can form new citrate-soluble compounds such as hilgenstockite $Ca_4O(PO_4)_2$, nagelschmidtite [1], silicocarnatite [2], chlorspodiosite [3], and Rhenania-type compounds.

$$[(Ca,Mg)_5(Na,K)_2(PO_4)_2(SiO_4)_2] \qquad [1]$$

$$[Ca_5(PO_4)_2SiO_4] \qquad [2]$$

$$[Ca_2Cl(PO_4)] \qquad [3]$$

The presence and distribution of accessory mineral phases thus is very important in the thermal processing of the ores.

Thermal treatment of Fe and Al phosphates at 500 to 600°C removes water from the hydrated minerals and alters the crystalline minerals to amorphous phases that have higher citrate solubilities. A modification of this technology involves reaction of the Fe and Al phosphates with limestone at higher temperatures to form citrate-soluble whitlockite, $(Ca,Mg)_3(PO_4)_2$, and this commercial process is described in Chapter 4.

The effectiveness of thermal treatments of P raw materials to increase their availability thus depends on both the calcination temperature and the type and distribution of the accessory minerals. Chemical reagents can be added in the treatment to prepare a desired product in a commercial process.

E. Acidulation Processes

Acidulation involves treating phosphatic raw materials with mineral acids to prepare water- and citrate-soluble P compounds. The most widely used processes involve treatment of apatitic PR with sulfuric acid to prepare normal superphosphate or phosphoric acid; triple or concentrated superphosphate is prepared from PR and phosphoric acid.

In the preparation of ordinary and triple superphosphates, acid-soluble Fe and Al accessory minerals (oxides, sulfides, hydroxides, silicates) may introduce technological problems. Rocks containing more than 5% R_2O_3 (Al_2O_3 + Fe_2O_3) may be difficult to process into superphosphate because of the stickiness imparted to the product by the R_2O_3 constituents (Marshall & Hill, 1952). Frazier and Lehr (1967) have shown, however, that many commercial superphosphates contain significant amounts of citrate-soluble Fe, Al, Mg, and K phosphates, which apparently had no adverse effects on their physical or agronomic properties. These studies have shown that the source and distribution of the impurity elements are critical in their effect on the physical properties of the superphosphates.

Table 10—Acidulation properties of concentrates for WPA production.

Product source	Properties of concentrate		Tons rock per ton 54% WPA	Acidulation requirements (short tons, 2,000 lb)†					
				Tons H₂SO₄ per ton			Tons sulfur per ton		
	% CaO	% P₂O₅		Rock	54% WPA	P₂O₅	Rock	54% WPA	P₂O₅
Central Florida	48.9	33.4	1.62	0.856	1.38	2.56	0.279	0.451	0.835
North Carolina	48.5	30.2	1.79	0.849	1.52	2.81	0.277	0.495	0.917
Morocco	51.6	32.1	1.68	0.903	1.52	2.81	0.294	0.495	0.917
Gafsa, Tunisia	48.3	28.8	1.88	0.845	1.58	2.93	0.276	0.516	0.956
Taiba, Senegal	51.2	37.4	1.44	0.896	1.29	2.40	0.292	0.421	0.780
Togo	52.3	36.6	1.48	0.915	1.35	2.50	0.298	0.440	0.815
Kola, Russia	52.0	38.2	1.41	0.910	1.29	2.38	0.297	0.419	0.776
Western Sahara	51.9	35.8	1.51	0.908	1.37	2.54	0.296	0.446	0.826
Angola	51.3	37.2	1.45	0.898	1.30	2.42	0.293	0.425	0.787
Jhamar-Kotra, India	54.2	40.1	1.35	0.949	1.28	2.37	0.309	0.417	0.772
Jordan	53.0	33.8	1.60	0.928	1.48	2.75	0.302	0.483	0.894
Oron, Israel	52.7	31.3	1.73	0.922	1.59	2.94	0.300	0.518	0.959
Bayovar, Peru‡	46.5	30.2	1.79	0.814	1.46	2.69	0.265	0.474	0.878
Algeria	49.3	29.0	1.86	0.863	1.61	2.98	0.281	0.524	0.970

† Assuming 100% reaction of CaO to calcium sulfate, plus a nominal excess of sulfuric acid (2.5% SO₃ loss in 54% WPA product).
‡ Formerly referred to as *Sechura* deposit.

EVALUATION OF RAW MATERIALS

In the production of wet-process phosphoric acid, PR is treated with sulfuric acid to yield a filter-grade acid. The Fe, Al, K, and Mg derived from the accessory minerals are undesirable because they slowly form sludges when the acid is concentrated (Scott et al., 1974) and also modify the crystal habit of the byproduct gypsum and decrease filtration rates (Gilbert, 1966). Carbonate minerals in the PR both consume sulfuric acid and evolve carbon dioxide which generates troublesome foam. Soluble silica minerals are useful in promoting the evolution of F from the acidulate. Iron and aluminum phosphates that have been calcined in the presence of CaO at temperatures above 1000°C have been used to prepare wet-process phosphoric acid because the high-temperature calcination renders the Fe and Al insoluble in the acid.

The acid requirements of concentrates for phosphoric acid manufacture depend upon their Ca/P ratio; weight ratios much above 2.6 (CaO/P_2O_5 of 1.6) are generally considered uneconomical for acidulation. The comparison of 14 concentrates in Table 10 (Lehr & McClellan, 1974) demonstrates how acid requirements differ from rock to rock; the requirements are based on the amount of acid for 100% conversion of the CaO to calcium sulfate and to allow for the nominal excess of free acid (1.0% S) in the merchant-grade wet-process acid (54% P_2O_5).

There is no strict linear relationship between the amounts of rock and sulfuric acid required to prepare a unit of wet-process acid because of differences in the compositions of the apatite minerals. The last column in Table 10 shows that the tons of sulfur (short ton, 2,000 lb) required per ton of P_2O_5 range from 0.772 to 0.970. The difference of 0.198 ton represents an added cost that is roughly equivalent to the value of 0.5 ton of P raw material.

Although the calculated reagent requirements in Table 10 obviously will be different for other grades of concentrate, the underlying cause for the difference in reagent requirements among PR reflects the differences in composition of their apatite components and demonstrates the significance of characterization studies in economic evaluations of proposed processes.

Citrate- and water-soluble phosphates can be prepared also from nitric acid and PR, and this process is widely used in Europe. One advantage of nitric acid processes is that they reportedly can be used to prepare commercial fertilizers from some Fe and Al phosphates (Bernard Raistrick, personal communication). Other impurities such as sulfides, titanium, uranium, and strontium, for example, have deleterious effects on different steps of this nitric acid technology.

F. Electric Furnace Methods

High energy costs have decreased the economic attractiveness of the electric furnace process, but it is still quite popular in some areas. One advantage of this process is that it can use ores of lower grades than those required in acidulation processes, generally with little or no beneficiation. In the electric furnace process, apatitic PR is reduced to elemental P at very high temperatures in the presence of coke. The Ca released from the apatite is removed as a silicate slag formed by reaction with the SiO_2 in the furnace burden. The main undesirable accessory minerals in the ore are Fe oxides and sulfides; these react with P to form ferrophosphorus (Hurst, 1961). Ferrophosphorus, Fe_2P, is a low-value material that is used in the metallurgical industry but represents a loss of 1 mole of P for every 2 moles of Fe that react and also increases the power requirement. Losses of 8% of the P as ferrophosphorus are not unusual, and rocks containing less than 3.1% Fe are preferred for furnace burden. Ferrophosphorus also reacts with the carbon electrodes in the furnace and decreases their current efficiency and service life. Magnesium, Al, and Fe also raise the viscosity of the slag and affect the performance of the furnace and cause difficulties in tapping the slag. Silicon, Mn, Ti, V, Cr, Ni, and other minor constituents from the accessory minerals are recovered in the ferrophosphorus. In addition to its formation of slag, silica (usually as quartz pebbles or chert, is beneficial in promoting liberation of F from the apatite in the furnace.

A lower temperature electric furnace product is the Mg silicophosphate formed by the fusion of low-grade PR with talc or serpentine. This product has performed well agronomically under tropical conditions where a slowly soluble P source is desired, along with soluble silica and magnesium.

The accessory minerals, particularly silica, silicates, and Fe-bearing minerals, are important in electric furnace technology. In some ores the natural mineral assemblage (for example, a silica-cemented PR) may indicate that furnace processing would be the most suitable technology. The natural accessory mineral assemblage of other ores can be supplemented to make the ores more suitable for furnace processing.

VII. CHARACTERIZATION OF MARGINAL PHOSPHATE ORES

The quality of the PR available to the phosphate industry is decreasing because of the exhaustion of high-grade reserves and the resource-conservation practice of nonselective mining. In the past, PR producers marketed the high-grade ores and processed the low grade ores in captive facilities. In those conditions the grade of the rock was the only important economic factor in comparison of raw material sources. This is no longer true. Other chemical and physical factors are becoming as important in the technical evaluation of P raw materials as P grade has been in the past. Today's more sophisticated fertilizer technology in worldwide use requires specific types of raw materials for specific processes. Different sources of commercial concentrates are not necessarily suitable for a particular process, nor is any one particular concentrate broadly suited for all processes. The economic value of a PR concentrate is increasingly dependent upon its technically feasible end uses and customer preference for some particular set of quality factors and not merely its P content (Cathcart, 1968; G. T. McBride and J. R. Camp. 1966. Resource development of the fertilizer minerals. Presented at SME-AIME Meet., 13-15 Oct., Tampa, Fla.).[2,3] These points will be discussed in Chapter 4 and are summarized here only to indicate their significance.

Because of the lack of strict specifications at present, it is difficult to define a "marginal quality" ore because the term is only relative. It can apply to ores that are either low in phosphate grade, low in chemical quality, low in physical quality, or any combination of these factors. The marginal character of a particular ore deposit also may depend upon the mining operation, beneficiation technology, or fertilizer plant process. Thus, the marginal quality depends upon a combination of process requirements and related economic factors, both of which affect the acceptability of a particular ore.

A. Physical Factors

Phosphate concentrates that consist of hard, coarsely textured particles or highly crystalline apatite such as those from Kola, Phalaborwa, and Jhamar Kotra are often unreactive in acidulation processes, especially in the manufacture of superphosphate. This type of concentrate may require expensive grinding to an acceptable particle size to be suitable for the proposed acidulation process.

Calcined concentrates also may have low reactivity because of agglomeration, loss of porosity, and thermally induced crystal growth of the apatite mineral. All these effects markedly decrease the surface area of the apatite.

In general, the preferred textures of concentrates for acidulation processes are composed of soft, porous, pelletal or sand-sized phosphate particles that are easily handled and metered mechanically. Grinding of these phosphates to the particle size required for superphosphate manufacture presents no major problem. These desired textural properties are common to many sedimentary deposits but not to igneous or metamorphic phosphate ores.

B. Chemical Factors

1. WEIGHT RATIO Ca/P

This factor ranks above all others in the evaluation of carbonate-type phosphate deposits because of its major effect on process economics. When the weight ratio Ca/P is more than 2.6 (CaO/P_2O_5 more than 1.6), the sulfuric acid requirement becomes uneconomical (Cathcart, 1968; Lehr & McClellan, 1974).[2]

---

[2] D. L. Everhart. 1971. Evaluation of phosphate rock deposits. Presented at Soc. Econ. Geol., 1-4 Mar., New York.

[3] R. St. Guilhem. 1972. Phosphate rock reserves—development of the relative importance of qualitative factors, 40th Annu. Meet. ISMA, 31 May 1972, Deauville, France.

The weight ratio Ca/P of the apatite mineral ranges up to 2.65 so that total removal of all other calcium-bearing minerals by beneficiation may be required. However, this may not be technically or economically feasible, particularly with carbonate-type P ores.

For captive use, the additional costs for H_2SO_4 reagent may be weighed against added beneficiation costs to find the most economical process modification. The additional beneficiation costs may be unavoidable, however, if the concentrate is to compete in open PR markets.

2. IRON AND ALUMINUM (R_2O_3)

These two impurities are particularly troublesome in wet-process acid and its ammoniation products. They are mainly responsible for postprecipitation sludges in phosphoric acid, scale formation in superphosphoric acid production equipment, insoluble P compounds in liquid or solid ammonium P products, and unwanted agglomeration in nongranular solid ammonium polyphosphates (Lehr, 1968). Phosphate concentrates with R_2O_3 contents above 3 or 4% are unattractive for phosphoric acid production but may be used in superphosphate manufacture.[2]

3. MAGNESIUM

The detrimental effects of Mg in the manufacture of phosphoric acid and its ammoniation products are widely recognized but costly to avoid. Magnesium occurs in concentrates mainly as residual accessory carbonates (Mg-calcite and dolomite) or as a constituent of the apatite itself (Lehr & McClellan, 1974). The apatite usually contributes most, if not all, of the Mg so that the Mg cannot be removed by physical beneficiation. The amount of Mg in apatites from different deposit sources varies widely and ranges from about 0.03 to 0.6% Mg.

In the reactor stage of phosphoric acid production, Mg precipitates F as colloidal particles that blind the gypsum filters. Magnesium also raises the viscosity of superphosphoric acid and is the primary cause of insoluble phosphate precipitates in ammonium phosphate liquid fertilizers (Lehr, 1968).

4. FLUORINE AND Si/F RATIO

Concentrates high in F usually yield phosphoric acid high in F. Reactive silica in the acidulation medium aids in removing F, but is not the controlling factor, even when the ratios Si/F appear satisfactory. The F content of phosphoric acid has a marked influence on the kinds and amounts of insoluble precipitates that form in the ammoniation products, and the concentration of residual F in liquid fertilizers is especially critical (Frazier & Lee, 1972).

5. CHLORINE

Soluble Cl is important in phosphoric acid technology because of its corrosive action on plant equipment, rather than chemical interference with manufacturing process. The usual sources of Cl in phosphate concentrates are the evaporite salts common to sedimentary ores or sea water that might have been used in beneficiation, but these sources are easily removed by fresh-water washing. In general, Cl contents higher than 0.1 or 0.2% cannot be tolerated because the corrosion rates become excessive.[2]

6. MINOR AND TRACE ELEMENTS

Minor constituents such as Mn, Fe, Zn, and Cu are beneficial in fertilizers as micronutrients, but they all contribute to post-precipitation of insoluble phosphates (Lehr, 1968). Other elements such as Cd, Pb, Cr, As, Hg, Se, and V are either toxic or potentially harmful in fertilizers. There are no rigid specifications, but concentration limits have been tabulated (Swaine, 1962). Recent interest has been focused on U and Ra, regarding both recovery and potential health hazards in manufacturing and in handling final products.

7. ORGANIC MATTER

With increasing amounts of organic matter in sedimentary ore concentrates, the foaming problem in wet-process acid manufacture adds to reactor equipment costs, and the discoloration of the products decreases their market value and sales appeal. Some high-grade rocks may require costly thermal treatment merely to alleviate the problems caused by organic matter during acidulation.

C. Combined Effects of Quality Factors

Bradley and Sweeney (1964) summarized the combined effects of physical and chemical factors on the quality and end use of PR, with specific reference to U.S. deposits as examples. Florida concentrates, because of their high grade (14 to 16% P) and chemical quality, find major use in wet-process acid manufacture. The main detrimental factors are R_2O_3 and organic matter. North Carolina concentrates have lower P contents (13 to 14% P) but have excellent physical quality and high chemical reactivity. Their major end products are phosphoric acid and ammonium phosphates. The primary causes of processing difficulties are high contents of Mg, Na, F, Fe, and organic matter. Calcination to remove organic matter is often necessary to maintain product quality.

The use of western U.S. PR is divided about equally between wet-process acid manufacture and electric furnace reduction, depending largely on chemical quality. Even in the higher-grade concentrates (~14% P) reserved for wet-process acid production, problems arise from high contents of R_2O_3, Mg, minor elements (V, Cr, Cd, As), and organic matter; the organic matter often requires calcination for its removal.

Tennessee phosphates are now largely restricted to electric furnace use because of their declining P grade and their broad range of chemical constituents undesirable in wet-process acid production. Their physical properties require an agglomeration step to make acceptable furnace burden, and minimum grade requirements (10.5% P) are met by blending ores and concentrates.

Fertilizer manufacturers are becoming more aware of the chemical factors, aside from P content of PR, with the continued development of technology and products based on wet-process and superphosphoric acids. The detrimental effects of certain chemical constituents in phosphoric acid processes, in contrast to superphosphate processes, are more widely recognized. This recognition is having an obvious effect on the relative market values of phosphate concentrates that otherwise would appear competitive on the basis of only their P grade. As St. Guilhem (1972)[3] points out, there are no immediate prospects for industry-wide specifications on the chemical factors, but it is becoming imperative to consider them in any evaluation of phosphate reserves.

In conventional processes, low-grade rock ($<12\%$ P) is presently used where the mineral diluents do not adversely affect the chemical quality. Existing plant designs may be a limiting factor, although more modern plants could handle the larger volumes of low-grade feed rock. The steady decline in grade and chemical quality of the P raw materials will make it increasingly difficult to operate existing plants and to continue to produce the current types of high-analysis materials. The shift from premium ores must be accompanied by a shift from traditional high-analysis fertilizers to lower grade materials, and this trend has already been observed in some traditional products.

In new or unconventional technology, "total matrix acidulation" offers the benefits of substantial increase in phosphate recovery (up to 90% of the mined P), elimination of expensive physical beneficiation, consolidation of accessory mineral refuse with gypsum byproduct for mine backfill, and lower energy and water requirements. This technology has been under study by the U.S. Bureau of Mines for several years (White et al., 1975) but is still largely in the exploratory stage. Use of new attack acids may allow the fertilizer process to use the differential solubility of the phosphate and accessory minerals advantageously. Another related approach would involve processes that use preconditioning (physical, chemical, or thermal) of the feed ore to allow the attack chemistry to be essentially phosphate specific. Longer range research may utilize biological activity to extract P from PR and to reclaim valuable chemical resources from fertilizer byproducts.

VIII. FUTURE RESEARCH NEEDS

The characterization of phosphatic raw materials involves the study of a variety of geologic materials described as PR. The results of these studies identify the ore type, mineral compositions, marginality factors, and textures of the rocks and serve as a guide to subsequent processing in beneficiation and fertilizer technology. The purpose of this continuing research should be to find the best possible use for a particular PR in a particular fertilizer usage situation rather than to state generalities concerning the penalties associated with the use of a valuable raw material in an unfavorable process. The custom utilization of phosphatic raw materials means that they may all have some place in fertilizer technology; the role of characterization is to find the appropriate match of raw materials and processes. This may result in the use of conventional or nonconventional processes to produce traditional or novel products that can meet the technological, agronomic, and economic requirements for the commercialization of a particular ore. The increasing importance that marginal phosphatic raw materials will have in the future indicates that characterization studies will play an increasingly important role in their technological evaluation.

LITERATURE CITED

Altschuler, F. S., E. B. Jaffee, and Frank Cuttitta. 1956. The aluminum phosphate zone of the Bone Valley Formation, Florida, and its uranium deposits. U.S. Geol. Survey Prof. Pap. 300:495-504.

Ames, L. L., Jr. 1959. The genesis of carbonate apatites. Econ. Geol. 54:829-841.

Anonymous. 1975. More phosphate from Thies. Phosphorus Potassium 75:34-36.

Baddiel, C. B., and E. E. Berry. 1966. Spectra structure correlations in hydroxy- and fluorapatite. Spectrochim. Acta 22:1407-1417.

Bradley, J., and G. Sweeney. 1964. Transportation affects phosphate output. Chem. Eng. Prog. 60(10):22-25.

Cathcart, J. B. 1968. Marine phosphorites-economic considerations. p. 295-300. *In* Proc. Seminar on Sources of Mineral Raw Materials for the Fertilizer Industry in Asia and the Far East. Miner. Resour. Dev. Ser. no. 32, U.N., New York.

Cathcart, J. B., and R. A. Gulbrandsen. 1973. Phosphate deposits in United States mineral resources. *In* D. A. Brobst and W. P. Pratt (ed.) U.S. Geol. Survey Prof. Pap. 820:515-525.

Doak, B. W., R. J. Gallaher, L. Evans, and E. B. Muller. 1965. Low temperature calcination of "C"-grade phosphate from Christmas Island. N.Z. J. Agric. Res. 8:15-29.

Frazier, A. W., and R. G. Lee. 1972. Stabilizing liquid fertilizers with fluorine compounds. Fert. Solut. 16(4):32-43.

Frazier, A. W., and J. R. Lehr. 1967. Iron and aluminum compounds in commercial superphosphates. J. Agric. Food Chem. 15:348-349.

Gary, M., R. McAfee, Jr., and C. L. Wolf (ed.). 1974. Glossary of geology. Am. Geolog. Inst., Washington, D.C.

Gilbert, R. L., Jr. 1966. Crystallization of gypsum in wet-process phosphoric acid. Ind. Eng. Chem. Process Des. Dev. 5:288-291.

Gremillion, L. R., and G. H. McClellan. 1978. The importance of chemical and mineralogical data in evaluating apatitic phosphate ores. Presented at 1978 Soc. Min. Eng. of AIME Fall Meet., Lake Buena Vista, Fla., 11-13 Sept. 1978. Preprint no. 78-B-308. Am. Inst. Min. Metall. Pet. Eng., New York.

Gulbrandsen, R. A., J. R. Kramer, L. B. Beatty, and R. E. Mays. 1966. Carbonate-bearing apatite from Faraday Township, Ontario, Canada. Am. Mineral. 51:819-824.

Hurst, T. L. 1961. Manufacture of elemental phosphorus and its major inorganic derivatives. p. 1221-1279. *In* J. R. Van Wazer (ed.) Phosphorus and its compounds. Vol. II. Interscience Publ., Inc., New York.

Khasawneh, F. E., and E. C. Doll. 1978. The use of phosphate rock for direct application to soils. Adv. Agron. 30:159-206.

Knudsen, K. C. 1972. NPK production by ion-exchange operating experience. p. 11-16. *In* Proc. ISMA Tech. Conf., Seville, Spain, 20-24 Nov. 1972.

Kreidler, E. R. 1967. Stoichiometry and crystal chemistry of apatite. Ph.D. Thesis, Penn State Univ. (Mic. no. 68-3549). Univ. Microfilm, Inc., Ann Arbor, Mich.

Le Geros, R. Z., J. P. Le Geros, and O. R. Trautz. 1965. Effect of carbonate on the lattice parameters of apatites. Nature 206:403-404.

Le Geros, R. Z., O. R. Trautz, J. P. Le Geros, and E. Klein. 1967. Apatite crystallites: Effects of carbonate on morphology. Science 155:1409-1411.

Lehr, J. R. 1967. Variations in composition of phosphate ores and related reactivity. p. 61-67. *In* Proc. 17th Ann. Meet. of Fert. Ind. Roundtable, Washington, D.C.

Lehr, J. R. 1968. Purification of wet-process acid in phosphoric acid. p. 635-686. *In* A. V. Slack (ed.) Phosphoric acid. Vol. I, Part 2. Marcel Dekker, Inc., New York.

Lehr, J. R., and G. H. McClellan. 1972. A revised laboratory reactivity scale for evaluating phosphate rocks for direct application. TVA Bull. Y-43, 36 p. TVA, Muscle Shoals, Ala.

Lehr, J. R., and G. H. McClellan. 1974. Phosphate rock: Important factors in their economic and technical evaluation. p. 194-242. *In* CENTO Symp. on Mining and Beneficiation of Fertilizer Minerals, 19-24 Nov. 1973, Istanbul, Turkey). CENTO, Ankara, Turkey.

Marshall, H. L., and W. L. Hill. 1952. Composition and properties of superphosphate. Ind. Eng. Chem. 44:1537-1540.

Maslennikov, B. M., and F. A. Kavitskaya. 1956. The phosphate substance of phosphorites. Dokl. Akad. Nauk SSSR 109:990-992.

McClellan, G. H., and J. R. Lehr. 1969. Crystal chemical investigation of nature apatites. Am. Mineral 54:1374-1391.

McConnell, D. 1938. A structural investigation of the isomorphism of the apatite group. Am. Mineral. 23:1-19.

Moore, P. B. 1973. Pegmatite phosphate: Descriptive mineralogy and crystal chemistry. The mineral. Rec. 4(May-June):103-130.

Notholt, A. J. G. 1975. Phosphate rock: World production, trade, and resources. p. 104-120. *In* Proc. of 1st Ind. Miner. Int. Congr., London, 8-9 July 1974. Metals Bull., Ltd., Surrey, England.

Posner, A. S., and G. Duyckaerts. 1954. Infrared study of the carbonate in bone, teeth, and francolite. Experientia 10:424-425.

Romo, L. A. 1954. Synthesis of carbonate-apatite. J. Am. Chem. Soc. 76:3924-3925.

Scott, W. C., G. G. Patterson, and C. A. Hodge. 1974. Status of modern wet-process phosphoric acid technology. Fert. Solut. 18(2):62, 64, 65, 68, 70, 72, 74-77.

Silverman, S., R. Fuyat, and J. Weisser. 1952. Quantitative determination of calcite associated with carbonate-bearing apatites. Am. Mineral. 37(3 & 4):211-233.

Simpson, D. R. 1964. The nature of alkali carbonate apatites. Am. Mineral. 49:363-376.

Smith, J. P., and J. R. Lehr. 1966. An X-ray investigation of carbonate apatite. J. Agric. Food Chem. 14:342-349.

Swaine, D. J. 1962. The trace element content of fertilizers, Farnham Royal, Bucks., England, Commonw. Agric. Bur. 306 p.

Trueman, N. A. 1965. The phosphate, volcanic, and carbonate rocks of Christmas Island (Indian Ocean). J. Geol. Soc. Aust. 12(2):261-283, Pls 18-20.

White, J. C., A. J. Fergus, and T. N. Goff. 1975. Phosphoric acid by direct sulfuric acid digestion of Florida land-pebble matrix. U.S. Bur. Mines, RI 8086. 12 p. (1975).

Chapter 4

Phosphate Raw Materials and Fertilizers: Part I—A Look Ahead

JAMES R. LEHR

Tennessee Valley Authority
Muscle Shoals, Alabama

I. INTRODUCTION

Perhaps no other industrial mineral commodity shows such diverse chemical and physical characteristics as the phosphorus (P) raw materials supplied to fertilizer manufacturers. Most mineral raw materials generally have more narrowly defined geochemical origins and mineral compositions. Beneficiation requirements are most stringent, yielding more highly purified concentrates, and product specifications are mutually defined by mine producers and end-product consumers.

None of these constraints apply to the phosphate rock (PR) market. With the supply of premium-grade PR product shrinking in the face of rapidly increasing demands, the changing character of P raw materials will have increasing impact on existing P fertilizer process technology, product compositions, and will very likely necessitate innovations in types of P fertilizer formulations.

Commercial mineral phosphate, known collectively as *phosphate rocks*, have but one property in common. Their P and F are contained in minerals of the apatite structure in association with Ca.

Calcium apatites constitute one of the most common and widely distributed mineral groups in the earth's crust and are the most abundant of all P-containing minerals. Occasionally, they occur in massive accumulations of economic importance in several hundred deposit locations throughout the world. These apatitic deposits occur under all geological settings in igneous, metamorphic, and sedimentary structures in an almost infinite variety of deposit characteristics.

Despite crystal structure similarities, the compositions of these apatites show significant departures from the idealized fluorapatite formula, $Ca_{10}(PO_4)_6F_2$, which has been the commonly assumed P mineral in PR.

This chapter examines the nature of commercial P raw materials as

---

Copyright 1980 © ASA-CSSA-SSSA, 677 South Segoe Road, Madison, WI 53711, USA.
The Role of Phosphorus in Agriculture.

supplied to industry, the causes of variable composition, and their collective effect on the present-day optional uses for P fertilizers. The final section deals with future trends concerning commercial P resources and the projected changes that likely may occur in traditional P fertilizers, direct-application usage, and the emergence of new P fertilizer formulations.

II. PHOSPHATE RAW MATERIALS: PRESENT AND FUTURE

A. Nature of Commercial Phosphate Raw Materials

The variable nature of commercial mineral P products can be traced to three principal factors which are summarized in Table 1.

Ore recovery methods differ widely with the type of geological structure, stratigraphic sequence of economic ore beds, overburden thickness, and other such mining technicalities. The selection of beneficiation treatments is often fixed by an allowable cost factor rather than by technical limitations, and differs for each individual mining operation. Hence, mine products do not represent a uniformly high degree of P mineral concentration, nor do they contain relatively the same kinds and amounts of accessory mineral impurities. These three interrelated factors of ore mineralogy, mining strategy, and beneficiation treatment are mainly responsible for the variable nature of commercial P products (Lehr & McClellan, 1973).

B. Influence of Mineral Assemblages on PR Quality

The principal cause of divergent mineral compositions of apatitic PR is traced to their differing geological settings and geochemical origins, as indicated in Table 1. Further variations arise from post-depositional alterations of mineral assemblages due to tectonic disturbances, chemical weathering processes, and contamination by more recent sedimentary depositions (see Chapter 3 for detailed discussion).

Table 1—Nature of commercial phosphate raw materials.

Variable types of ore resources
Igneous intrusive deposits
Metamorphic carbonatites
Sedimentary phosphates
Iron ore byproduct apatite
Variable beneficiation treatment
Simple physical methods (washing, desliming, sizing, drying)
Wet classification methods (flotation, hydrocycloning)
Thermal methods (simple calcination, calcining-slaking)
Variable types of concentrates
Wide range in P content (*10* to *16%* P, or 50 to 80 BPL†)
Wide difference in chemical composition of apatite mineral
Wide variation in types and amounts of accessory minerals
Wide variation in textural form, crystallinity, particle size

† Trade term *Bone Phosphate of Lime* expresses P content on basis of hypothetical content of tricalcium phosphate.

Listed below are the various mineralogical types of P ores that are now being exploited for commercial raw materials:

1. SEDIMENTARY DEPOSITS

    a. Siliceous—Unconsolidated or indurated mixtures of francolite-type apatite and primary silica (quartz, chalcedonic, opaline forms).

    b. Carbonate—Weathered to densely crystalline dolomitic or calcitic limestones with occluded francolite-type apatite (pelletal, fossilized, and replacement types). Pyritic minerals and organic matter are common in unweathered ores.

    c. High Fe-Al—(i) francolite-type apatite associated with clays—usually kaolinitic, montmorillonitic, illitic, or attapulgite varieties as sources of Al, Fe, Mg, and alkalies; (ii) francolite-type apatites associated with hydrous Fe-Al oxides and degraded silicates as products of intense chemical weathering; (iii) Fe-Al phosphates (crandallite, millisite, variscite-strengite, wavellite), either alone or in association with apatitic phosphate; and (iv) francolite-type apatites associated with combinations of the above types due to more complex geochemical processes.

2. IGNEOUS AND METAMORPHIC DEPOSITS

    a. Carbonatite—Coarsely crystalline varieties of fluorchlorhydroxyapatite associated with abundant calcite, dolomite, ankerite, or fluorocarbonates. Transition ore zones may also contain silicates (muscovite-biotite-vermiculite), oxides (hematite, goethite, anatase, perovskite), and accessory phosphates (Fe-Al and rare-earth varieties).

    b. Nepheline-Syenite-Pyroxinite—Coarsely crystalline fluorchlorhydroxyapatite varieties associated with pyroxinite-type silicates and felspathoids; accessory minerals may include Fe oxides and sulfides, Ti oxides, and rare-earth phosphates, among others.

    c. Apatitic Fe Ore—Coarsely crystalline fluorchlorhydroxyapatite occurring as the important accessory mineral in certain magnetite-hematite ore deposits. Accessory mineral contaminants include barite, fluorite, carbonates, and rare-earth minerals.

Virtually all of these types of P deposits are now being exploited in worldwide production with sedimentary ores providing about 87% of the supply (Notholt, 1974). While beneficiation treatments obviously vary with each ore type, both technical and economic factors restrict the degree of upgrading for each—hence no standardized product quality! Furthermore, the traditional singular objective of nearly all beneficiation methods for PR is to improve the P quality factor. Improvements in other chemical and physical quality factors are usually incidental, depending upon a rather arbitrary rejection process of specific gangue minerals in the waste mineral fraction. Since the traditional marketing parameter is still the P content of products, mine producers have little concern and incur no penalties for undesired chemical impurities retained as accessory phases in commercial P products.

Consequently, the types of phosphate concentrates currently marketed range widely with respect to both P grade and chemical quality. They no longer can be treated as mutually equivalent raw materials by producers, nor is any one product concentrate necessarily adaptable to the broad range of chemical processing options. These restraints, however, are not widely observed so that the divergent mineral quality of phosphate raw materials is an underlying cause of current problems in chemical processing and in the variable quality of fertilizer product compositions.

C. Influence of Apatite Composition on Product Quality

The compositions of the Ca apatites found in the various geological deposit types only rarely approach the stoichiometric fluorapatite composition. Collectively, the number of reported elemental substitutions in cation, anion, or halogen sites totals about 65 in various ionic states.

The most important and most common substitutions are summarized in Table 2.

Igneous and metamorphic apatites show the least departure from fluorapatite composition. Usually some substitution of Cl^-, OH^-, and CO_3^{2-} for F^- occurs. Substitutions of Mn^{2+}, Fe^{2+}, Sr^{2+}, Ba^{2+}, Mg^{2+}, and rare earths for Ca^{2+} also are common. A wide range of oxymetal anions may substitute for PO_4^{3-} but in very small amounts.

On the other hand, sedimentary apatites show the widest departures from the fluorapatite composition, but with few exceptions, these below to a distinct compositional series known as *francolite-type apatites* (McClellan & Lehr, 1969). The francolite series can be represented by the end-member compositions

$$Ca_{10}(PO_4)_6F_2 \qquad Ca_{10-a-b}Na_aMg_b(PO_4)_{6-x}(CO_3)_xF_{0.4x}(F,OH)_2$$
$$\text{(fluorapatite)} \qquad \text{(francolite)}$$

where the important substitutions are (CO_3^{2-} + F^-) replacing up to about 25% of the PO_4^{3-} and the coupled substitution of Na^+ + Mg^{2+} replacing up to about 10% of the Ca^{2+}. Other substitution processes common to apatitic structures may occur. With few exceptions, their concentrations are not

Table 2—Substitutions in apatites.

Fluorapatite, $Ca_{10}(PO_4)_6F_2$	
Ca^{2+}	Na^+, Sr^{2+}, Mn^{2+}, Fe^{2+}, K^+, U^{4+}, Mg^{2+}, R.E. (Ce, Yb, La, Pr, Y)
P^{5+}	C^{4+}, S^{6+}, Si^{4+}, As^{5+}, V^{5+}, Cr^{6+}
F^-	OH^-, Cl^-, CO_3^{2-}
O^{2-}	F^-, OH^-
Francolites, $(Ca, Na, Mg)_{10}(PO_4)_{6-x}(CO_3)_xF_y(F,OH)_2$	
Ca^{2+}	Na^+, Mg^{2+}
P^{5+}	$(CO_3)_x^{2-}$ + F_y^-
F^-	OH^-

analytically significant from a processing viewpoint. Occassionally, F-deficient francolites are found in which some (OH⁻) replacement of F⁻ occurs, as in the well-known case of the Sechura (now referred to as Bayovar) P deposit in Peru.

The compositions of the apatite mineral component, which are in no way affected by physical beneficiation treatments, have an important effect on product quality. The more highly substituted francolites have theoretical P contents approaching 15% (34% P_2O_5), which decreases product grade. More importantly, they are primary sources of Na, Mg, and F impurities that affect chemical quality of products and cause processing difficulties. Detailed discussions of apatite compositions in relation to their physical and chemical properties are available (Lehr, 1967; McClellan & Lehr, 1969; Lehr & McClellan, 1973).

D. Quality Factors of Commercial Phosphate Rocks

The cumulative sources of variability identified in the previous discussions contribute to the following set of physical and chemical quality factors, as summarized in Table 3.

This wide range of raw-material properties contributes not only to processing difficulties that have an indirect effect on product quality, but also to direct chemical modifications of product compositions. It becomes readily apparent why the principal market specification relating to P content of the PR raw material is an unreliable indicator of the P fertilizer quality obtained by some particular process or by manufacturers using the same process but raw material of equivalent P grade from different deposit sources.

Table 3—Quality factors of commercial phosphate rocks.

Physical quality factors

Texture: Hardness, porosity, cementing or coating phases
Particle size of phosphate: Coarse mesh to cryptocrystalline
Degree of crystallinity of apatite mineral
Effects of physical treatments: Raw or calcined state

Chemical quality factors

P content of PR (BPL grade)
F content of apatitic mineral
Carbonate content of apatitic mineral
CaO/P_2O_5 mole ratio (apatite + free carbonate minerals)
Fe and Al content (combined R_2O_3)
Mg content (apatite and accessory mineral sources)
Na and K (apatite and accessory mineral sources)
Organic matter (native types + beneficiation reagents)
Heavy metals (Pb, Cd, Zn, Hg) from accessory minerals
Potentially toxic elements (Se, As, Cr, V) from accessory minerals
Radionucleides (U, Th, Ra, Rn) from phosphate minerals
Miscellaneous impurities (rare-earths, Ti, Ba, Sr, S)
Chlorides (from evaporite salts and apatite substitution)
Content of inert mineral gangue (insoluble oxides, silicates)

The broader implications of these quality factors are discussed below; they are considered in more detail in later sections dealing with fertilizer product compositions and in published reports (Lehr, 1967; Lehr & McClellan, 1973; Comm. on Access. Elem., NAS, 1979).

1. EFFECTS OF QUALITY FACTORS ON FERTILIZER PROCESSES

Physical quality factors, along with certain types of chemical impurities, contribute mainly to technical problems in chemical processing. The conversion efficiency of raw phosphate to finished products is influenced by the physical quality factors and also by such chemical factors as organic matter, chlorides, free carbonates, and some solubilized metal impurities (Ba, Sr, Ti, Fe-Al, Mg, and alkalies) that interfere with chemical processing steps. This impaired efficiency of chemical processing steps indirectly affects the quality of final products (see Chapter 7).

The major group of chemical quality factors identified in Table 3 have a direct effect on the chemical and physical properties of end products because of their tendency to be passed through as soluble contaminants. Of this group of chemical impurities, those of major concern are Al, Fe, Mg, alkalies, and F.

Other metallic impurities such as the heavy metals, toxic elements, and radionucleides usually occur in concentrations that are too low to significantly affect processing steps or product grade but constitute potential sources of product adulteration that must be considered in phosphate end uses for food, fiber, and animal production.

2. *PREMIUM, NONPREMIUM,* AND *MARGINAL-GRADE* PHOSPHATE ROCKS

With increasing variability in the quality factors of commercial phosphate raw materials, a new and somewhat loose terminology is coming into usage to indicate processing quality. Raw materials are now frequently classified as *premium-grade, nonpremium-grade,* or *marginal-grade* as defined below.

Although no absolute marketing specifications presently exist, *premium-grade* PR concentrates are considered to contain at least 13.5% P (31% P_2O_5, 68 BPL), and many commercial sources were formerly in the 15 to 16% P (75 to 80 BPL) range. The term *BPL* is the marketing convention that refers to *Bone Phosphate of Lime* or the hypothetical content of tricalcium phosphate, $Ca_3(PO_4)_2$, where % P_2O_5 × 2.186 equals the *BPL* value. This corresponds to apatite concentrations of from about 82 to 95% by weight. Consequently, only a minor fraction of accessory mineral impurities remained in these concentrates so that their collective effect on chemical processing was of no major concern.

Until quite recently, nearly all the commercial phosphate raw materials were of premium quality because of intense market competition and selective mining practices to minimize production costs. Generally, their chemical compositions were such that Ca did not exceed about 29% (40%

CaO), Fe + Al was <3%, Mg <0.2% (<0.3% MgO), free carbonates <5%, and total insoluble gangue <10 to 15%, by weight. Furthermore, many of these products had been calcined to upgrade P content, evolve part of the fluorine, insolubilize some of the metal impurities, and destroy organic matter. This practice now has been largely abandoned because of uneconomical energy requirements, except in those few cases where a high content of native organic matter (2 to 4%) supplies a major fraction of the process fuel requirement.

Today, many of the P products supplied to, or used by, chemical processors are *nonpremium* grades where P content falls below 14% P (68 BPL) and may range as low as 10% P (50 BPL). Higher-grade PR products are frequently reserved for export trade, while lower grades are utilized in captive production.

Nonpremium phosphate grades, therefore, may have apatite concentrations as low as about 60% by weight with correspondingly higher concentrations of accessory mineral impurities. If these nonphosphatic mineral components are mainly insoluble SiO_2 or silicate phases, then the chemical quality of the raw material is not impaired, and the main penalty of lower P content is with regard to process economics.

However, siliceous P ores are only one of several mineralogical types that are now exploited for commercial P reagents. Because of this, the shift to nonpremium grades has been reflected in steadily rising contamination levels of such important impurities as Al, Fe, Mg, alkalies, chlorides, free carbonates, organic matter, and miscellaneous interfering metals.

The term *marginal quality* applies, therefore, to phosphate raw materials with high contents of chemical impurities, irrespective of P grade, that make them either unsuitable or of marginal value for some chemical process uses. While the principal chemical quality factor in former years was mainly that of P content, manufacturers are now more vitally concerned with this new set of chemical quality factors. Both affect process economics, but low chemical quality of raw materials imposes not only technical difficulties, but also leads to lower product quality. In fact, low chemical quality may preclude their use entirely for certain types of chemical fertilizer products.

For example, some of the marginal quality raw materials now utilized in fertilizer processing may exhibit sharp increases in acid-soluble impurities, such as 7 to 8% Fe + Al, Mg contents of 0.6 to 3%, free carbonates in excess of 10%, chloride values of 0.1 to 0.5%, and a wide range of heavy metals, toxic elements, radionucleides, and interfering metals such as Ba, Sr, Ti, and rare earths.

The use of such low chemical quality phosphates by some producers can lead to a widely varying quality and composition of traditional P fertilizer products. Thus, agronomists and agriculturalists are confronted with this new problem of evaluating these inhomogeneities in relation to actual field performance of specific fertilizer types.

This problem is likely to become acute in many of the developing agricultural regions—particularly those in the tropics—where indigenous P resources tend to be of inferior mineralogical compositions and are likely to yield, therefore, marginal-grade raw materials for chemical processing by conventional technology.

3. EFFECT OF NEW MINING STRATEGY ON PRODUCT QUALITY

The world demand for PR is increasing at an annual rate of about 10 million tons of commercial concentrate. To meet this expanding total demand, which will reach an estimated 160 million metric tons by 1977-1978, mine producers are having to abandon former "high-grading" recovery practices and adopt mining plans for maximum P recovery (Notholt, 1974). The incentive to conserve reserves by less wasteful mining practices is the sharply increased market value of phosphate products (1¢/kg of P in the early 1970's vs. about 10¢/kg of P at current market values).

These changes in mining strategy have been reflected in a general lowering of product quality due to ore-blending practices, less beneficiation treatment to reduce ore losses in waste byproducts, and marketing of non-premium-grade products of <14% P (68 BPL) that heretofore were wasted, stockpiled, or used as mine backfill.

Until recently, mining operations were largely restricted to premium quality deposits, with the USA, USSR, and North Africa (mainly Morocco) supplying the bulk of the world market requirements. The two-step price surge in the 1973-1974 period triggered an unprecedented expansion of existing mine operations and active exploitation of long-dormant P resources. New supply sources have emerged in North Africa, the Middle East, South America, and Australia; other new sources are under development (Notholt, 1974; Howard, 1979). Phosphate production from many of these newer deposit sources does not meet the premium quality standards of former commercial raw materials. A corresponding slippage is also occurring among the established producers, due to the cited changes in mining strategy.

Thus, a trend is developing wherein the special interest and technical concerns of the mine producers are being reflected in a general lowering of phosphate product quality over that formerly available to fertilizer manufacturers. The problem will be aggravated by new production from inferior-quality ore deposits that in prior years were uncompetitive in world markets.

E. Resource Exploitation Trends Vs. Product Quality

On the basis of widely reported, recent world surveys dealing with: (i) known world resources, (ii) distribution of resources by geological deposit type, and (iii) trends in worldwide production rates, there is a forewarning that continued exploitation patterns will further contribute to declining quality of P products.

The important interrelationships are diagrammed in Fig. 1, based on recent market data and authoritative estimates (Notholt, 1974, 1979; Anon., 1974a; see also Chapters 1 and 2).

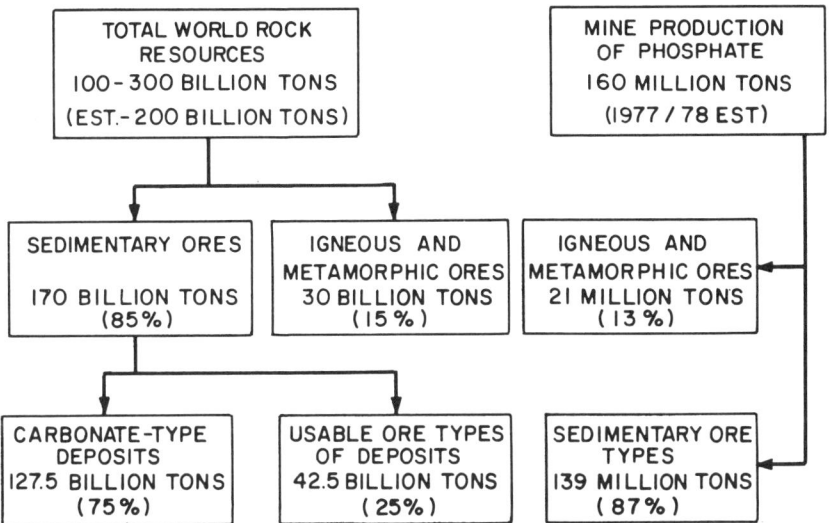

Fig. 1—Relationship of phosphate rock production quotas to phosphate rock reserves (metric tons).

In terms of total P resources, there is no foreseeable shortage on the basis of estimates in the 150- to 300-billion-metric ton range. A mean value of 200 billion tons is used in this illustration.

This total P resource estimate embraces all known geological deposits, which includes Fe-Al phosphates as well as apatitic varieties. Sedimentary phosphate deposits account for about 85% of this total, with igneous and metamorphic deposits (all types) making up the remainder. About three-fourths of the sedimentary phosphate accumulations occur in limestone matrices, which are presently excluded as economic reserves since there is no adequate beneficiation technology to upgrade such ores to commercial standards of acceptance. There are, of course, some exceptions where shallow-buried or surface-exposed deposits have undergone limited upgrading by natural weathering processes.

Nevertheless, the estimates in Fig. 1 illustrate that the potentially exploitable resources on the basis of today's economics and technology are roughly of the same order of magnitude for igneous-metamorphic ores and sedimentary ores (30 billion tons vs. 42.5 billion metric tons, respectively). Assuming about equal fractions of economic reserves for each resource type, it then becomes apparent from the current pattern of rock production that the depletion rate of sedimentary ore types is about 6.6 times that of all other sources combined.

This type of exploitation pressure will encourage even more stringent conservation measures by mine producers operating on finite reserves of sedimentary-type deposits. Thus, the trend in declining product quality has long-term implications and will become more serious as world food demands stimulate further increases in mine production of phosphate raw materials.

III. OPTIONAL FERTILIZER USES OF PHOSPHATE ROCK

Historically, the phosphatic fertilizers available to agriculture have advanced from pulverized raw natural phosphates (PR, bonemeal) to crude acidulated phosphates and thermally altered rock compositions, and, finally, to present-day, highly refined, water-soluble P compounds in a wide variety of formulations. Rapid advances in mining and beneficiation technology since 1930 provided low-cost, premium P raw materials, leading to a new generation of fertilizer processes based on phosphoric and nitric acid intermediates to prepare high-analysis products. The demand for higher-analysis, readily soluble phosphatic fertilizers by land-intensive modern agriculture and the inherent cost reductions in handling and transportation offered economic incentives for such highly refined chemical fertilizers on a worldwide production basis. Consequently, the bulk of today's worldwide P fertilizer is supplied in formulations of water-soluble salts of NH_4 or Ca [for example, concentrated superphosphate (CSP), monoammonium phosphate (MAP), diammonium phosphate (DAP), ammonium polyphosphate (APP), and nitric phosphates (NP)]. This situation is now changing due to a combination of factors as discussed in the preceding section, and a partial transition to first-generation, low-analysis phosphatic fertilizer materials as substitutes now appears unavoidable. The demands of environmentalists and conservationists, as well as changing P fertilization requirements, signal a shift away from highly soluble phosphates. Water solubility may also be deemphasized due to unusual soil or climatic factors in some of the new agricultural areas under development.

These trends prompt a reexamination of possible P fertilizer options for the future, which is discussed in the following sections in light of present-day chemical fundamentals. A historical review of prior process technology and product compositions is omitted here (Waggaman, 1969).

A. Direct-Application Fertilizers

In today's inflationary economy, especially with respect to energy costs, the direct application usage offers the lowest cost and least energy-intensive processing of raw natural phosphates. It is also the least sensitive to ore quality and chemical impurities contributed by the accessory minerals in marginal-quality PR so long as the apatite component is a reactive type.

Except for the USA, direct-application practices have been expanding rapidly in recent years (Notholt, 1974). Complete tonnage statistics are not available, but an estimated 8 to 10 million metric tons of PR is now used annually in the free world (exported rock plus utilized indigenous resources). In addition, unofficial reports of USSR rock production from marginal-quality sedimentary deposits for direct application usage indicate an additional 3 to 4 million metric tons annually.

Several factors have been responsible for deterring more widespread effective use of natural phosphates in the past. Until quite recently, all types of PR were assumed to contain the same phosphate mineral, fluorapatite, leading to the false premise that rock sources could be selected indiscriminately. Differences in reactivity among rock sources were attributed primarily to particle size and surface area factors. The detrimental effect of rock calcining on P availability was unsuspected or ignored. Furthermore, P solubility values, as expressed by the various empirical laboratory solubility procedures, were expressed as fractions of total P grade of the rock, resulting in misleading comparisons of sources (Lehr & McClellan, 1972; TVA unpublished data).

This combination of unsatisfactory selection principles made any large-scale direct application use an unpredictable and often unsatisfactory practice. The agronomic practice was largely restricted to a few widely tested phosphate rocks, mainly North African sources, of demonstrated performance.

1. MINERALOGICAL SELECTION BASIS

Recent mineral characterization studies have established that apatite compositions differ markedly among PR (Lehr, 1967; McClellan & Lehr, 1969; Lehr & McClellan, 1973; Chien, 1977; Chien & Black, 1976). Mineralogical characterizations of PR from several hundred worldwide deposits have shown that the generally assumed fluorapatite composition applies more closely to igneous and high-grade metamorphic deposit types than to sedimentary phosphates. The two well-known examples of fluorapatite varieties are Kola, USSR, and Phalaborwa, South Africa. The composition, high crystallinity, and low reactivity of such igneous apatites make them virtually useless for direct application.

Sedimentary PR contains apatitic phosphates of more complex chemical composition. They constitute a distinct series of carbonate-substituted apatites known collectively as *francolite* (see Section II-C), which can be satisfactorily represented by the end-member empirical compositions of:

$Ca_{10}(PO_4)_6F_2$ $\quad Ca_{10-a-b}Na_aMg_b(PO_4)_{6-x}(CO_3)_xF_{0.4x}(F,OH)_2$
(fluorapatite) $\quad\quad\quad\quad\quad\quad$ (francolite-type)

where the observed (and calculated) limiting value of $x/(6-x) \cong 0.3$.

These apatite compositions can be highly correlated with crystallographic unit-cell parameters, particularly the a_{obs}-axis, as determined by refined X-ray powder diffraction measurements. The following linear regression equations are satisfactory for routine analytical appraisal of apatite compositions:

$$x/(6-x) = (9.374-a_{obs})/0.204$$

$$Na_a = 1.327 \{x/(6-x)\}$$

$$Mg_b = 0.515 \{x/(6-x)\}$$

where a_{obs} is the observed unit-cell a-axis parameter determined by the refined X-ray diffraction procedure (McClellan & Lehr, 1969).

This method permits a rapid evaluation of the apatite composition in a particular PR source and provides a basis for comparing the relative reactivities of various rock sources as shown in Table 4 and Fig. 2. Results of greenhouse and field evaluations using PR characterized in this manner showed that crop response was closely related to carbonate substitution (Lehr, 1967; Lehr & McClellan, 1972).

2. LABORATORY SOLUBILITY INDEXES

Reactive phosphate rocks to be considered for direct application are those containing an apatite with a high degree of carbonate substitution. This reestablishes, in principle, the validity of laboratory solubility indexes, provided that the chosen solvent offers an adequate measure of distinction among the compositional forms of francolite-type apatites.

Neutral solvents, as employed in the AOAC neutral ammonium citrate (NAC) procedure, are relatively inefficient for distinguishing among basic Ca phosphates, and yield only a narrow solubility range in the case of francolites (0 to 3% citrate soluble P). This particular method is least able, therefore, to distinguish among the subtle compositional variations of the francolites found in sedimentary PR.

Acidic solvents such as 2% citric acid (CA) and 2% formic acid (FA) yield much wider solubility ranges, and thus can more readily differentiate apatite compositions on the basis of P solubility. Both procedures are widely employed outside the United States.

Currently, two choices for laboratory evaluation of rocks are available; each has its limitations, and no general procedure has gained international acceptance. These procedures consist of either making a direct measurement of apatite solublity in NAC, CA, or FA, or using X-ray diffraction data to estimate these solubilities. For a more detailed discussion, the reader is referred to Chapter 3. Evaluation data for a broad representation of world PR sources are compared in Fig. 3 on a relative scale to illustrate the wide divergence in rock reactivity (Lehr & McClellan, 1972).

3. TRENDS IN DIRECT USAGE OF PR

A major drawback to larger scale use of pulverized PR has been the inherent difficulties in handling, shipping, and applying such a finely divided dusty material. These problems can be alleviated by granulation using various chemical binders, but this technology is currently in the development stage. The methods of agglomeration being used, or under consideration, however, may have profound chemical effects on the apatitic component of rocks and its agronomic performance. Even though a detailed discussion of this technology is not within the scope of this chapter, a brief description is given below to point out areas of concern to the agronomist reader.

Currently, granulation of ground natural PR is promoted by a very small percentage of a salt binder such as KCl or $(NH_4)_2SO_4$, or by partial

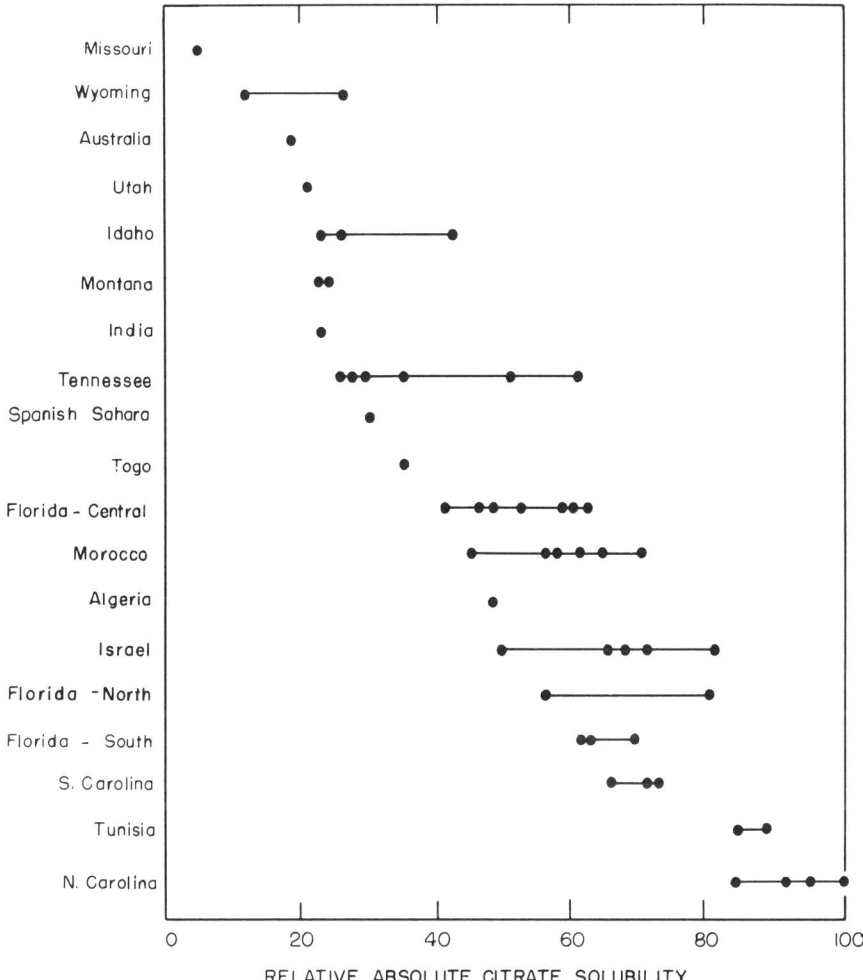

Fig. 2—Relative reactivity of phosphate rocks from various sources based on a comparison of absolute citrate solubility values; N. Carolina sample designated as 100 on solubility scale.

acidulation with H_2SO_4 or with H_3PO_4. Salt binders do not chemically alter the apatite; granules tend to readily disaggregate and redisperse in soils under rainfall action so that the end result approaches that of direct application of pulverized rock.

Partial acidulation with H_2SO_4 may lead to nonreversible agglomeration, since the cementing phases are likely to be water-insoluble Ca sulfates and phosphates.

Granulation by partial acidulation with H_3PO_4 will produce a water-soluble binder, monocalcium phosphate, which may not hinder redispersion of the agglomerated PR particles but will contribute some readily available P in addition to P supplied by the apatite phase.

Other granulation methods are being investigated that employ addi-

Table 4—Citrate solubility of phosphate rocks (Lehr & McClellan, 1972).

No.	Source	P₂O₅, % Total of Rock	P₂O₅, % Total of Apatite†	C.S.‡	Citrate solubility of P₂O₅, % Usual§	Citrate solubility of P₂O₅, % Absolute¶	Length of a axis, Å#	Mole ratio CO₃/PO₄
113	Idaho	31.3	39.0	2.0	6.4	5.1	9.354	0.088
118B	Spanish Sahara	34.8	39.0	2.7	7.8	6.9	9.354	0.060
128A	Idaho	30.3	39.7	2.3	7.6	5.8	9.358	0.078
18	Utah	28.9	39.6	1.9	6.6	4.8	9.357	0.100
14	Togo	36.6	38.7	3.1	8.5	8.0	9.351	0.075
13	N. Carolina	30.2	34.5	6.6	21.9	19.1	9.323	0.262
101	Morocco	31.6	34.8	5.1	16.1	14.6	9.325	0.236
111	Florida	33.6	37.0	3.9	11.6	10.5	9.339	0.169
98	Tennessee	27.7	37.4	3.0	10.8	8.0	9.342	0.137
24	Tennessee	35.6	39.4	2.6	7.3	6.6	9.356	0.045
15	Gafsa, Tunisia	29.2	35.0	7.0	24.0	20.0	9.326	0.285
106	Morocco	31.4	35.8	4.5	14.3	12.6	9.331	0.225
137	Florida (North)	32.7	35.6	4.5	13.8	12.6	9.330	0.178
22	S. Carolina	28.3	34.4	5.6	19.8	16.3	9.322	0.273
124	Morocco	32.1	35.0	4.8	15.0	13.7	9.326	0.214
122X	Israel	33.8	36.0	5.8	17.2	16.1	9.333	0.178
103	S. Carolina	26.4	35.5	5.2	19.7	14.7	9.329	0.292
20	Florida	30.8	36.2	4.9	15.9	13.5	9.334	0.197
16	Morocco	33.1	37.2	4.7	14.2	12.6	9.341	0.185
17	Morocco	36.9	38.2	3.9	10.6	10.2	9.347	0.098
23	Wyoming	31.1	40.2	2.4	7.7	6.0	9.361	0.061
202	Tunisia	27.4	34.8	6.6	24.1	18.9	9.325	0.289
69B	S. Carolina	33.0	34.8	5.5	16.7	16.1	9.325	0.211
102	Morocco	33.6	35.9	5.7	17.0	15.9	9.332	0.158
160	Florida	33.2	37.2	4.0	12.1	10.8	9.340	0.156
49	Florida	28.6	36.7	4.3	15.0	11.7	9.337	0.140
95	Wyoming	33.4	40.0	1.1	3.3	2.8	9.360	0.068
135	Florida (South)	31.3	35.8	5.0	16.0	14.0	9.331	0.222

(continued on next page)

Table 4—Continued.

No.	Source	P$_2$O$_5$, % Total of Rock	P$_2$O$_5$, % Total of Apatite†	C.S.‡	Citrate solubility of P$_2$O$_5$, % Usual§	Citrate solubility of P$_2$O$_5$, % Absolute¶	Length of a acis, Å#	Mole ratio CO$_3$/PO$_4$
129	Florida	33.2	36.0	4.8	14.5	13.3	9.333	0.175
97	Montana	36.9	39.4	2.1	5.7	5.3	9.356	0.056
136	Florida (South)	32.2	36.0	5.0	15.5	13.9	9.332	0.205
104	Montana	34.3	40.4	2.1	6.1	5.1	9.362	0.056
93	Tennessee	34.1	39.8	2.5	7.3	6.3	9.358	0.065
68	N. Carolina	30.5	34.8	7.2	23.6	20.7	9.324	0.264
133X	Algeria	31.2	37.2	4.0	12.8	10.8	9.340	0.212
126X	Israel	33.5	37.9	4.2	12.5	11.1	9.345	0.155
171	Tennessee	34.6	38.2	4.4	12.7	11.5	9.347	0.138
94	Tennessee	31.1	38.8	2.3	7.4	5.9	9.351	0.089
21	N. Carolina	30.1	35.2	7.6	25.2	21.6	9.327	0.259
134	Florida (South)	30.5	36.1	5.6	18.4	15.5	9.333	0.228
110	Florida	33.6	36.7	3.4	10.1	9.3	9.337	0.124
505	Missouri	34.7	42.0	0.5	1.4	1.2	9.373	0.008
469	India (J-K)	40.1	40.8	2.1	5.2	5.1	9.365	0.028
466	Florida (North)	32.4	36.3	6.6	20.4	18.2	9.334	0.184
44	Australia	39.2	41.0	1.8	4.6	4.4	9.366	0.012
465	Idaho	32.3	39.5	3.7	11.5	9.4	9.356	0.089
467	N. Carolina	29.9	34.5	7.8	26.1	22.6	9.322	0.266
464	Florida	32.7	37.9	5.3	16.2	14.0	9.345	0.120
468	Tennessee	30.7	39.6	4.5	14.9	13.7	9.358	0.061

† P$_2$O$_5$ content of apatite, calculated from length of a axis.
‡ Soluble in neutral ammonium citrate solution, AOAC method.
§ Fraction of rock P$_2$O$_5$, citrate soluble.
¶ (Citrate-scluble P$_2$O$_5$)/(P$_2$O$_5$ content of apatite).
# Determined by X-ray.

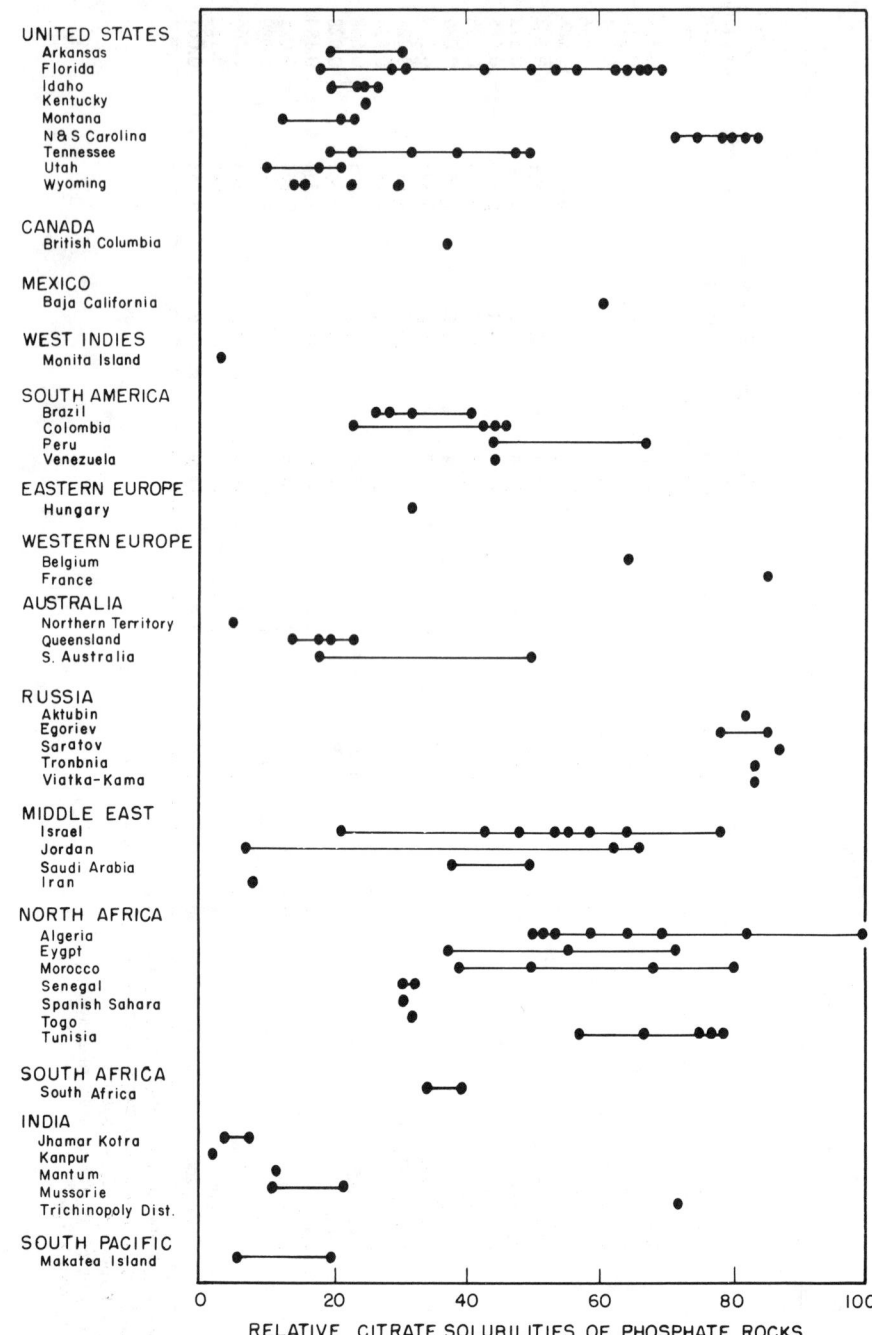

Fig. 3—Comparison of absolute citrate solubility values for phosphate rocks from various deposit sources on a relative scale based on the most reactive example.

tives of water-soluble ammonium phosphates or ammonium polyphosphate liquid grades not only to act as binders but also to increase the fraction of readily available P. These hybrid P fertilizers will have agronomic properties intermediate to those of pulverized reactive PR and the water-soluble chemical fertilizers (CSP, ammonium phosphates).

Agronomic research is needed to determine the merits of granulating PR and to guide the choice of granulation technology yielding the best product performance. Otherwise, granulation methods may evolve that provide good mechanical properties but greatly reduce the effectiveness of even the most reactive natural phosphates due to adverse chemical changes.

4. HIGH Fe-Al PR FOR DIRECT APPLICATION

In recent years, a new type of natural PR has come into commercial production and is gaining acceptance as a direct-application fertilizer. These rocks are mined from sedimentary accumulations of Fe-Al phosphate deposits which are otherwise unsuited for chemical fertilizers.

These deposits have diverse geochemical origins differing from sedimentary apatitic phosphorites but usually consist of the same group of Fe-Al minerals. This group comprises simple Fe-Al phosphates (variscite, strengite, wavellite), complex Fe-Al phosphates containing Ca, K, or Na (crandallite, wardite, millisite), and hydrous Fe-Al oxides, along with clays and usually some residual apatite.

The most noteworthy deposits are found in northwest Africa (Senegal, Liberia, Nigeria), Christmas Island in the Indian Ocean, the Florida supergene altered "Leached Zone" of the Bone Valley Formation, and the smaller well-known deposit on Curacao Island in the Caribbean Sea.

These deposits have long been ignored and considered submarginal for chemical fertilizer manufacture. While these Fe-Al phosphate materials usually rank in P grade with premium-quality apatitic ores, they require novel acidulation or caustic extraction processes to prepare conventional phosphate fertilizers, but the processes are presently uneconomical. Recently commercialized calcination technology has stimulated new interest at a time when intensified geological exploration is adding to known world resources of this under-utilized ore type. Raw mixtures of these Fe-Al phosphate minerals now can be converted to direct-application materials in pulverized or granular form by a relatively mild thermal treatment. Compared to reactive apatitic phosphates, this thermal product has much higher citrate solubility (60 to 65% of total P) and is applicable to a broader range of soil pH's.

Collectively, the structures of these Fe-Al minerals contain chemically bound H_2O. On heating to the 450 to 600°C range, they undergo dehydroxylation with resultant structural collapse to amorphous compositions. Citrate solubility approaches a maximum in this range of complete decomposition; above 600°C, recrystallization or vitrification sets in with loss of citrate solubility.

These amorphous compositions are thermodynamically metastable, which accounts for their high solubility and chemical reactivity. In soils, therefore, these phases will undergo dissolution and react to form stable crystalline compounds. Soil pH merely influences which of several possible reversion pathways will prevail.

Two commercial products of nearly identical composition now have become available for direct application usage. These are "Calciphos" prepared from C-zone Christmas Island Fe-Al PR (White, 1971b; Anon., 1974b), and "Phosphal" produced from the Thies deposit, Senegal (Anon., 1975; Atanasiu, 1971).

As with apatitic PR products, the trend is towards granulated grades using water-soluble alkali salt binders. While this achieves the desired physical characteristics, moisture can initiate an undesired chemical reversion due to interaction of the amorphous P phases with the soluble alkali binding agents. The reaction products are the same types of citrate-insoluble, crystalline, alkali-Fe and Al phosphate compounds as formed in soil-P reversion processes (Haseman et al., 1951; Lehr et al., 1967).

One cannot assume, therefore, that all granulated grades of Fe-Al phosphate materials will have consistent solid-phase compositions and solubility properties. Their variable quality will depend on granulation methods, type of salt binder, and storage history. These precautions should be carefully observed in the course of either laboratory or field evaluation studies.

B. High-Temperature Calcined Phosphates

A variety of phosphatic fertilizers can be prepared from either premium- or nonpremium-grade rock by thermally promoted reactions in the range of 1000 to 1450°C that destroy apatite structures and eliminate a substantial fraction of the fluorine, which allows recombination of $(PO_4)^{3-}$ into more reactive P compounds.

Most of these fertilizer substances have a long history of testing and use in agriculture but have become largely displaced in recent years by high-analysis, water-soluble P compounds. Internationally, there is no standard product nomenclature (Atanasiu, 1971). Typical product examples include defluorinated PR, Rhenania phosphate, basic slag, silicocarnotite-nagelschmitite, fused rock, and other so-called *calcined phosphates*. Most of the above are prepared from apatitic rocks, alone or admixed with various inorganic reagents; most of the reactions involve F evolution as an attendant step in decomposing the apatite minerals, leading to new calcium phosphate compositions.

These thermal phosphate products prepared by high-temperature decomposition reactions are not to be confused with the normal calcined grades of premium-grade apatitic rock. The latter involve a milder thermal treatment in the range of 650 to 1000°C, which does not destroy the apatite structure. It merely promotes a phase transition of francolite-type apatites to the more stable fluorapatite form, with evolution of the CO_2 and the increment of F present in coupled substitution for PO_4^{3-} in francolites (Lehr &

McClellan, 1973). Premium-grade PR products lack sufficient quantities of reactive accessory minerals to promote apatite decomposition reactions to any significant degree during calcination.

This change in apatite composition with attendant crystal growth upon calcination renders the PR too unreactive for direct application use, but this important difference between calcined and uncalcined grades of an individual PR heretofore has been frequently ignored.

Simple calcination of nonpremium grades of PR may lead inadvertently, however, to partial decomposition and reorganization of apatitic phosphate due to the much larger concentrations (up to 30 to 40%) of reactive accessory minerals. In such instances, the calcined grade may have re-enhanced reactivity over the natural apatitic PR due to the presence of newly formed calcium phosphate compositions. A future trend will be to exploit this thermal behavior of low-grade PR to increase their reactivity for direct application usage. Important exploratory research to prepare these novel direct application materials is now underway at the International Fertilizer Development Center at Muscle Shoals, Alabama.

Commercial thermal phosphate products are prepared from mixtures of commercial PR and selected reagents in the stoichiometric proportions to yield specific reaction products of either a vitreous or crystalline nature. The added reagents promote decomposition of the apatite, expelling F, and supplying constituents for recombining the Ca and P in compositions other than apatitic forms. Background discussion of apatite defluorination chemistry is available (Kreidler, 1967).

Although vast technical and patent literature exists regarding production of thermal phosphates, only a relatively few inorganic reagents have been found useful to decompose apatites. These are listed in Table 5.

The asterisk notations in Table 5 denote possible natural mineral associations in sedimentary phosphate rocks (see Chapter 3). Some types of marginal PR may thus require very little adjustment in composition to prepare some particular type of thermal phosphate, whereas these rocks might be otherwise totally unusable for either direct application or for chemical fertilizer manufacture.

Despite this wide choice of reactants, the varieties of thermal phos-

Table 5—Reactants used to decompose apatites at high temperature.

Silica and silicates	Phosphates	Halides
†Quartz, opal	H_3PO_4	†Chlorides (Na,Ca,K,Mg)
†Feldspar	Alkali phosphates	†$KMgCl_3 \cdot 6H_2O$ (carnallite)
†Olivine, serpentine (Mg silicate)	†$AlPO_4 \cdot nH_2O$	
Alkali metals	Alkaline earths	
Carbonates (Na,K)	†$CaCO_3, CaMg(CO_3)_2$	
Sulfates (Na,K,NH_4)	†$CaSO_4 \cdot 2H_2O$ (gypsum)	
†Silicates (Na,K type)	†$Mg(Na,K)_2(SO_4)_2 \cdot 4H_2O$	
Hydroxides (Na,K)		

† Denotes reactants that may occur in natural association with apatite in phosphate rocks (see Chapt. 3).

Table 6—Types of thermal phosphate products.

Idealized composition	Structural type	Reagents†
$(Ca,Mg)_3(PO_4)_2$	α,β-TCP; whitlockite; glasses	None
$Ca_4O(PO_4)_2$	Hilgenstockite; basic slags	Alkaline earths
$CaNaPO_4$		
$CaKPO_4$	Crystalline "Rhenania" compounds	Alkalies
$Ca_7P_2Si_2O_{16}$		
$(Ca,Mg)_7P_2Si_2O_{16}$	Nagelschmitite series of solid solutions	Silica, Mg silicates
$Ca_5Na_2(P,Si)_4O_{16}$		
$Ca_5P_2SiO_{12}$	Silicocarnotite series of solid solutions	Silica, silicates
$Ca_{n/2+1}(P_nO_{3n+1})$ where $n \geq 2$	Polyphosphates	Phosphates
Ca_2ClPO_4	Chlorspodiosite	Halides

† Accessory mineral components in phosphate rock may supply part or all of the necessary reactants.

phate products can be grouped for convenience on the basis of structural type and compositional properties, as shown in Table 6.

Reviews of these compositional types and their preparation chemistry have been reported by Ando (1961, 1965). Compositions in the nagelschmitite-silicocarnotite solid solution series, $Ca_3(PO_4)_2$-Ca_2SiO_4, have been described by Wolfkovich (1967). Werner (1967) has published a monograph on Rhenanian types. Fused rock and defluorinated compositions also have been described by Waggaman (1969).

Chlorspodiosite (14.9% P; 33.7% P_2O_5) is a new experimental composition based on the use of alkaline-earth chlorides as an acidic fused salt solvent for apatite (TVA unpublished research). Fluorine is precipitated as CaF_2, rather than evolved, and the reaction temperature is only about one-half that required for most thermal phosphates (600° vs. 1100 to 1400°C). The citrate solubility (NAC method) of chlorspodiosite is about 75%. Some upgrading of P_2O_5 content is possible since most metal impurities form water-soluble chlorides which can be stripped along with excess reagent from the water-insoluble Ca_2ClPO_4 by water leaching.

Since thermal phosphate products contain a primary water-insoluble phosphate phase diluted by residual inert mineral matter in the rock, they are inferior in P grade and quality to present-day high-analysis chemical fertilizers such as MAP, DAP, and CSP. On the other hand, all contain more reactive phosphate compositions of much higher citrate solubility than the precursor apatites (Atanasiu, 1971). However, their energy-intensive processing and inferior product quality in comparison to water-soluble fertilizers have restricted their importance to a relatively minor fraction of total P production. This cost factor must now be weighed against the advantages of using unbeneficiated nonpremium- or marginal-grade PR, as well as cheaper reagents than required in wet-chemical processes. Also, their slow-release characteristics and retained contents of secondary and micronutrient elements from the rock may enhance their usefulness in future agricultural practices, where such special requirements become identified.

C. Chemical Processing Options of Today

Manufactured chemical fertilizers consume well over 90% of the commercial PR now produced annually. The bulk of these chemical fertilizers are currently manufactured by wet-acidulation processes employing either H_2SO_4, H_3PO_4, or HNO_3. Production of fertilizer phosphates via the electric-furnace reduction process has diminished to insignificant tonnages due to uneconomical energy costs.

Electric furnace reduction processes yielded very high purity P intermediates from PR of inferior quality and grade. In contrast, wet-acidulation processes are highly sensitive to chemical impurities in PR materials. Hence, most acidulation processes were developed primarily to operate on premium-quality PR since those were the commercial grades that were generally available until recently.

However, the current worldwide slippage in chemical quality of commercial phosphates now is having a serious impact on all acidulation processes. A combination of problems arises with regard to processing difficulties, corrosion, precipitation losses of P, lowered production rates, and quality of final products.

As discussed earlier (Section II-D), chemical quality factors such as chlorides, organic matter, sulfides, free carbonates, and excessive contents of inert mineral diluents mainly create processing difficulties or add to manufacturing costs due to excessive reagent consumption and reduced production rates. Further discussion of these is omitted since they have only a minimal effect on final product compositions.

On the other hand, metallic impurities dissolved from the PR not only contribute to processing difficulties but, more importantly, they tend to pass through to finished products where they can markedly affect their chemical compositions and physical properties. Foremost among this troublesome group are Al, Fe, Mg, Na, and K, which are among the most common impurities in phosphate ores, especially sedimentary deposit types.

Of lesser magnitude are metal impurities such as alkaline earths (Ba, Sr), heavy metals (Pb, Cd), micronutrients (Zn, Cr, Se), toxic elements (As), radionucleides (U, Th, Ra), and interfering elements like Ti and the rare earths. Usually, the concentrations of such elements are too low to impose serious process difficulties or significantly alter product compositions. Nevertheless, the prevailing opinion is that many of these elements are undesirable constituents in final products, even in very small concentrations, because of their possible toxicity to crops or their potential risk of entering into food chains.

The compositional forms in which these low concentration metal impurities usually occur in final P products are summarized here and omitted in further discussions in individual product types. Barium and Sr may form insoluble phosphates or enter into other compositions in isomorphic substitution for Ca. Heavy metals readily precipitate as insoluble phosphates, often in combination with ammonia, resulting in some loss of soluble N and P. Radionucleides and rare earths normally enter into phosphate compositions in cation forms substituting for Ca; rare earths also form highly insoluble, anhydrous simple phosphates of the general composition, $(X)PO_4$, where X may be Ce, La, Nd, Y, or Th, among others. Elements such as Se, As, and Cr tend to substitute for PO_4 as $(X)O_4$ anions. Titanium tends to enter solution as soluble Ti sulfate or titanyl sulfate ($TiOSO_4$), which is sensitive to temperature and pH changes leading to decomposition in polyphosphate processes to form acid-insoluble TiP_2O_7 or objectionable gels of titanyl acid, $TiO(OH)_2$ or "$TiO_2 \cdot H_2O$" (Slack, 1968; Lehr, 1967; unpublished TVA data).

Sedimentary PR deposits recovered by surface mining are generally shallow-buried and substantially weathered so that concentrations of heavy metals, As, and Se, are generally insignificant due to weathering of precursor sulfide minerals. Cadmium is an important exception. Concentrations of Cd in sedimentary deposits range from 10- to about 150 ppm among geographical sources and commonly fall in the 40- to 70-ppm range; its mode of occurrence has not been established, but a major fraction can be readily evolved during normal calcination treatment. Igneous and metamorphic crystalline apatites show much wider ranges of metallic impurities, not only as accessory mineral phases but also as substituted constituents in the host apatite structure.

The following discussions of specific chemical P products will deal primarily with the compositional disturbances caused by the major metallic impurities, Al, Fe, Mg, Na, and K. Fluorine, which is contributed by the apatitic phosphate, also must be included because it participates in the formation of citrate-insoluble phosphate compounds.

1. WET-PROCESS ACID (WPA) PRODUCTION VIA H_2SO_4 ACIDULATION

During the production of merchant-grade H_3PO_4 (22 to 24% P), three precipitation processes tend to remove dissolved metallic impurities. These refer to the precipitation of byproduct sulfates and coprecipitation of metal fluorides in the attack stage (Frazier et al., 1975), and the precipitation of metal phosphate "sludges" during the concentration step (Lehr et al., 1966). Examples of fluoride precipitates that remove metal impurities are listed in Table 7. Further concentration to super acid composition causes an additional precipitation of metals in the form of insoluble pyrophosphate, tripolyphosphate, or metaphosphate salts.

The current generation of phosphoric acid processes differs widely in the distribution modes of impurities derived from similar phosphate raw

Table 7—Fluorine-derived precipitates† in WPA.

Na_2SiF_6	Na_3AlF_6
K_2SiF_6	$Ca_3(AlF_6)_2 \cdot 4H_2O$
$NaKSiF_6$	NaK_2AlF_6
$CaSiF_6 \cdot 2H_2O$	$MgNaAlF_6 \cdot 2H_2O$ (Ralstonite)
$MgSiF_6 \cdot 6H_2O$	$CaF_2\text{-}MgF_2$
	$Ca_4AlSiSO_4F_{13} \cdot 10H_2O$ (Chukhrovite)

† Precipitation processes controlled by Al, Si, F, Na, and Mg.

Table 8—Post concentration precipitates of impurities in WPA.

Compound	Impurity
$(Fe,Al)_3KH_{14}(PO_4)_8 \cdot 4H_2O$	Fe, Al, K
$FeNaH_5(PO_4)_3 \cdot H_2O$	Fe, Na
$(Fe,Al)_3(K,H_3O)H_8(PO_4)_6 \cdot 6H_2O$	Fe, Al, K
$(Na,K)_2SiF_6$	Na, K, Si, F
$Na_xMg_xAl_{2-x}(F,OH)_6 \cdot H_2O$	Na, Mg, Al, F
$Fe(H_2PO_4)_2 \cdot 2H_2O$	Fe(II)
$CaSO_4 \cdot 2H_2O\text{-}CaSO_4 \cdot 0.5H_2O$	Ca, excess SO_3
$MgSiF_6 \cdot 6H_2O$	Mg, Si, F

materials so that generalities are not warranted, even for different manufacturing plants operating on the same process. Unless the WPA products are subjected to elaborate purification treatment, significant concentrations of Al, Fe, Mg, alkalies, and F are retained to pass through to final fertilizer products.

Detailed discussions of the role of impurities in WPA production are available (Slack, 1968). Most of the important compositional types of precipitated impurities have been well characterized (Lehr et al., 1967) and are listed in Table 8. More up-to-date discussions of WPA processes appear in other chapter sections of this text.

2. AMMONIATION OF WPA

Monoammonium phosphate and DAP are the two principal product types produced by ammoniation of merchant-grade WPA. The main compositional disturbances in these products are due to Al and Fe. These cations initially precipitate during intermediate stages of ammoniation as metastable $(Al,Fe)PO_4 \cdot nH_2O$ in the form of a colloidal or cryptocrystalline gel-like phase. As such, it constitutes a diluent and contributes a water-insoluble component in products, but its highly reactive nature is reflected in high citrate solubility. Some examples of this occurrence in commercial MAP compositions are illustrated in Fig. 4, where the interstitial bonding phase of $(Fe,Al)PO_4 \cdot nH_2O$ dessicated gel in the granular MAP is in the range of 12 to 13% by weight.

This metastable form of Al and/or Fe phosphate is highly reactive in concentrated alkali electrolytes. It can react to form a variety of hydrated crystalline phosphate salts containing Fe-Al and alkalies (Haseman et al., 1951). These crystallization processes may be hastened by elevated temperatures during final ammoniation, granulation, and drying; by long retention periods; or by recycled process streams that introduce reactive crystal nuclei (seeding). This crystallization problem differs among individual processing methods and operating parameters. Most frequently, the crystallizing phase has the general composition, $(Al,Fe)(NH_4,K)(HPO_4)_2 \cdot 0.5H_2O$, although other compositional types are possible (Frazier et al., 1966; Slack, 1968). Collectively, these compounds are water-insoluble and have low citrate solubility. Important examples are listed in Table 9.

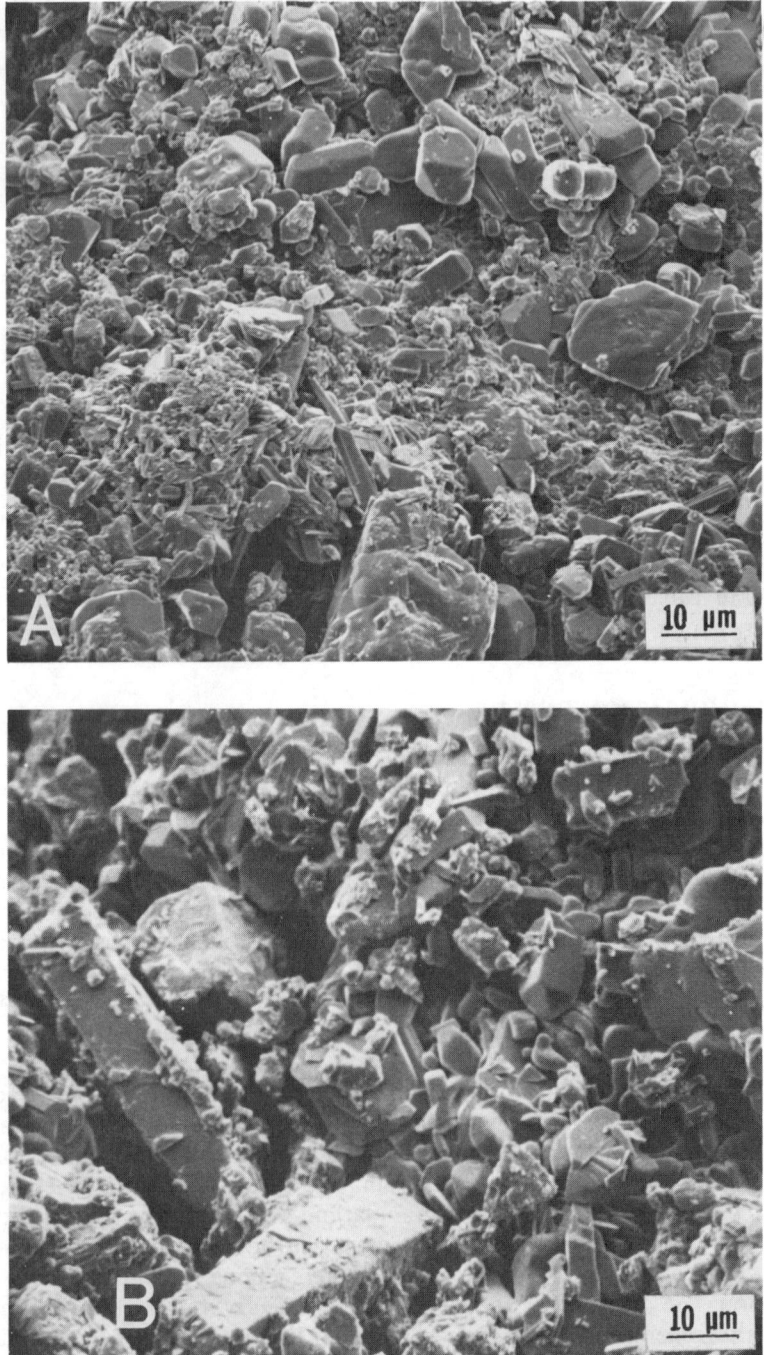

Fig. 4—Monoammonium phosphate fertilizer products as produced (*A* and *B*), and after water extraction (*C* and *D*) to show matrix of insoluble phosphate components (1000×).

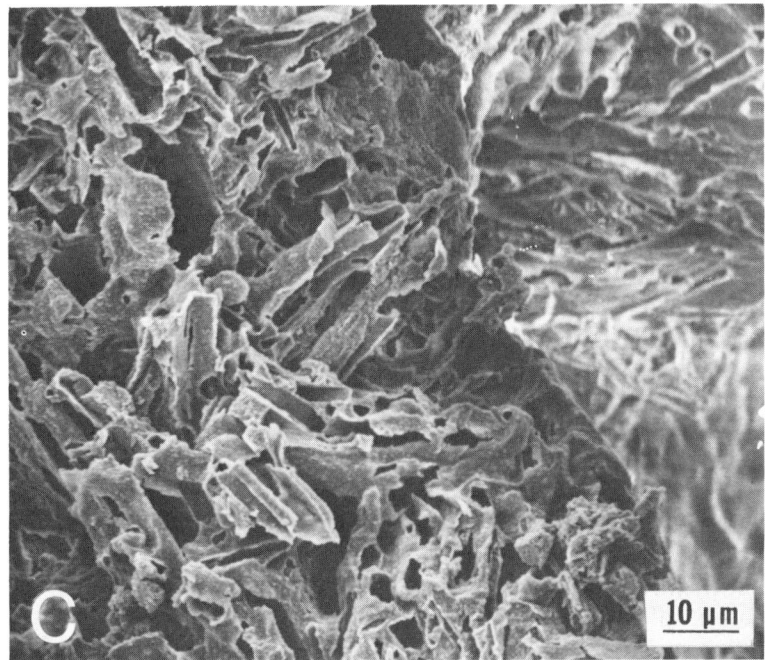

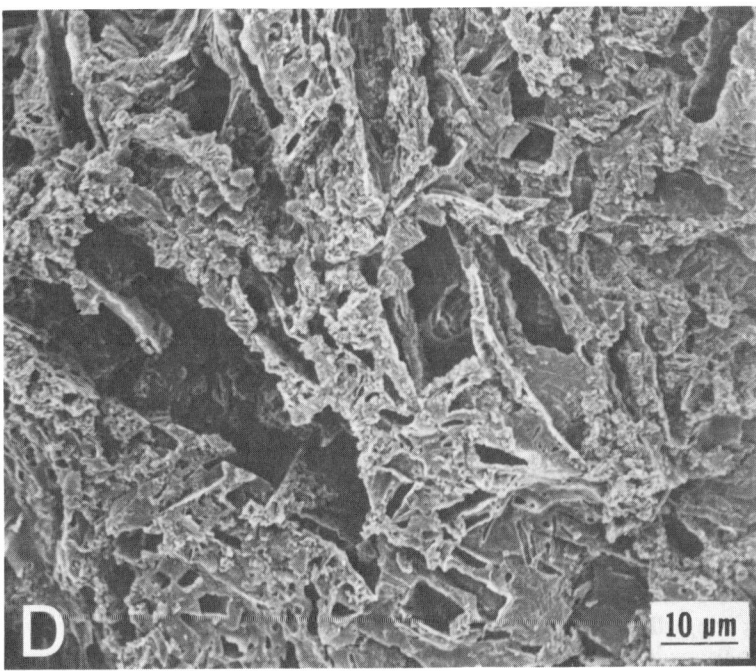

Fig. 4—continued.

Table 9—Precipitated impurities in ammoniated WPA.

Compound	Impurity
$(Fe,Al)PO_4 \cdot nH_2O$ (gel)	Fe, Al
$(Al,Fe)(NH_4,K)(HPO_4)_2 \cdot 0.5H_2O$	Al, Fe, K
$(Al,Fe)(NH_4,K)_2(HPO_4)_2F$	Al, Fe, K, F
$(Fe,Al)_3NH_4H_8(PO_4)_6 \cdot 6H_2O$	Fe, Al
$MgNH_4PO_4$ (mono- and hexahydrate)	Mg
$MgNH_4HPO_4F$	Mg, F
Fe, Al, NH_4 Polyphosphates	Fe, Al, F

High concentrations of F in the WPA intermediate favor the formation of salts such as $(Fe,Al)(NH_4,K)_2(HPO_4)_2F$, which also contributes to low water solubility and citrate-insoluble P (Ando & Akiyama, 1965). Magnesium impurity precipitates phosphate and NH_3 in several water-insoluble compositions, commonly the struvite type, $MgNH_4PO_4 \cdot 6H_2O$, or as the monohydrate salt. In the presence of F, more complex fluorophosphates of low solubility also may form (Ando et al., 1968a).

With further decline in chemical quality of phosphate raw materials, manufacturers have two options. They must either employ elaborate purification schemes or accept the change in product quality due to an increasing proportion of insoluble metal ammonium phosphate phases. In fact, the transition in MAP and DAP product compositions is already underway, and this change deserves serious consideration by agronomic researchers dealing with such formulations.

Ammonium polyphosphate formulations derived from impure WPA or the more concentrated super acid intermediates also have compositional disturbances, primarily due to Al, Fe, Mg, and F. These impurities can combine in a variety of pyro-, tripoly-, or metaphosphate salts, alone or in combination with NH_4 (Ando et al., 1968b; Slack, 1968; Frazier et al., 1972). Here, too, the change in product composition is a lowering of soluble N and citrate-soluble P. Engineering solutions to this problem may be found by new reactor technology that minimizes conversion of dissolved metal impurities into highly insoluble crystalline phases (Achorn & Salladay, 1975).

3. ROCK ACIDULATION WITH WPA OR H_2SO_4

Superphosphate compositions produced by either H_2SO_4 acidulation (ordinary superphosphate, OSP) or H_3PO_4 acidulation (CSP) are highly sensitive to dissolved metal impurities, especially Al, Fe, and alkalies, contributed by both the PR and acid reagents. The important compositions formed by these impurities are summarized in Table 10.

In OSP products, as in ammoniation of WPA, Al and Fe precipitates initially as the colloidal, metastable $(Al,Fe)PO_4 \cdot nH_2O$ phase, which creates processing difficulties (White, 1974). This gel phase is quite reactive and tends to react with alkalies or hydronium ion (H_3O^+) to form crystalline citrate-insoluble salts, such as $(Fe,Al)_3KH_8(PO_4)_6 \cdot 6H_2O$ (Ando & Lehr, 1967; Lehr et al., 1967).

Table 10—Compositional forms of impurities in superphosphates.

Compound	Impurity
$(Fe,Al)PO_4 \cdot nH_2O$ (gel)	Fe, Al
$AlPO_4 \cdot 2H_2O$	Al
$CaAlH(PO_4)_2 \cdot 6H_2O$	Al
$Ca(Al,Fe)H(PO_4)_8 \cdot 4H_2O$	Al, Fe
$(Al,Fe)_3KH_{14}(PO_4)_8 \cdot 4H_2O$	Al, Fe, K
$(Al,Fe)_3KH_8(PO_4)_6 \cdot 6H_2O$	Al, Fe, K
$CaFe_2(HPO_4)_4 \cdot 8H_2O$	Fe
$MgHPO_4 \cdot 3H_2O$	Mg
$Fe_3(PO_4)_2 \cdot 8H_2O$	Fe(II)

The Al impurity also may participate in another reaction sequence in which the gel first crystallizes and then reacts as:

$$AlPO_4 \cdot nH_2O \text{ (gel)} \rightarrow AlPO_4 \cdot 2H_2O \text{ (crystalline)} \quad [1]$$

$$AlPO_4 \cdot 2H_2O + Ca(H_2PO_4)_2 \cdot H_2O$$
$$\rightarrow CaAlH(PO_4)_2 \cdot 6H_2O + H_3PO_4. \quad [2]$$

The crystalline reaction product of step 2 is a highly insoluble phosphate of low citrate solubility (Lehr et al., 1964).

Thus, the two most important compositional disturbances in OSP products due to metal impurities are: (i) the formation of insoluble Fe and Al phosphate compositions, and (ii) residual undecomposed apatitic rock due to chemical and mechanical interferences caused by impurities during the acidulation steps.

Concentrated superphosphate compositions have undergone far more drastic changes in solid-phase composition in recent years, due partly to the use of lower quality PR but mainly to the increasing use of "sludge WPA" in substitution for clarified WPA.

Byproduct WPA "sludge acids" contain large amounts of suspended precipitates, mainly acidic phosphate salts of Fe and Al in combination with K. The principal compound is $(Al,Fe)_3KH_{14}(PO_4)_8 \cdot 4H_2O$ (Slack, 1968). During acidulation, these salts may partially dissolve and react along with Fe and Al released from the rock to form similar insoluble phosphates, such as $(Fe,Al)_3(K,H_3O)H_8(PO_4)_6 \cdot 6H_2O$ and $Ca(Al,Fe)H(PO_4)_2 \cdot 4H_2O$ (Frazier & Lehr, 1967; Ando & Lehr, 1967).

In some commercial CSP products examined in our laboratories, the content of monocalcium phosphate monohydrate was below X-ray and optical detection levels, and products consisted almost entirely of a mixture of complex, water-insoluble Fe and Al phosphates, most of which had high citrate solubility.

This is in sharp contrast to former CSP products prepared from premium quality raw materials, where the principal phosphatic component was the water-soluble phase, monocalcium phosphate monohydrate.

This transition has been occurring over the past decade, yet it has not appeared to impair agronomic performance nor customer acceptance, and has gone largely undetected by the control procedures now in force. However, the possible adverse effect may not be readily apparent on soils of high P status.

Superphosphates, therefore, will likely continue to be an acceptable outlet for nonpremium-grade PR, but a further transition from water-soluble Ca phosphate components to less soluble acidic metal phosphate components can be anticipated. Contrary to existing and somewhat arbitrary control specifications, this compositional trend may become increasingly attractive if high water solubility is deemphasized in future agricultural practices, or in special soil situations where rapid fixation is likely to occur.

4. ROCK ACIDULATION WITH HNO_3

Conventional nitric phosphate products obtained by ammoniation of nitric acid extracts of premium-grade P raw materials have traditionally comprised mixtures of water-soluble ammonium salts (phosphates, nitrates, and double salts) and water-insoluble Ca salts (phosphates and sulfates) (Ando & Lehr, 1968). The relative proportion of water-soluble P could be varied from about 40 to 70% by varying degrees of ammoniation, which converts water-soluble P to dicalcium phosphate and other more basic compounds (octacalcium phosphate, apatite).

Nonpremium-grade P raw materials, particularly those of lower chemical quality due to Fe, Al, Mg, and other soluble metal impurities, have an impact on HNO_3 acidulation analogous to other acid treatments. One mitigating factor is that HNO_3 is a less vigorous extraction of metal impurities from some accessory mineral components of PR than is H_2SO_4 or H_3PO_4.

On the other hand, nearly all of the dissolved metal impurities are retained as soluble nitrates in the acid extract up to the ammoniation stage since the only intervening precipitation stage is the removal of byproduct $Ca(NO_3)_2$ or removal of some Ca as gypsum by addition of ammonium sulfate.

Upon ammoniation, the major metal impurities (Al, Fe, Mg, alkalies) largely enter into the compositions of ammoniation products, along with any dissolved heavy metals, toxic elements, and radionucleides. Most of these metal impurities tend to combine with phosphate and ammonia to produce water-insoluble components, some of which are analogous to those described for ammonium phosphate products; important examples are summarized in Table 11.

With regard to nonpremium- or marginal-quality rock feeds, nitric processes and phosphoric acid processes are equally applicable to low quality ores where the accessory minerals are largely insoluble, such as in siliceous ore types. Nitric processes can accept rocks of higher free carbonate content since the excess HNO_3 consumed can be recovered as $Ca(NO_3)_2$ or NH_4NO_3, although this increases the N:P ratio in products. Rocks contain-

Table 11—Precipitated impurities in nitric phosphates.

Types of insoluble compounds (NH_4 and PO_4)
Fe-Al-NH_4-phosphates†
Mg-NH_4-phosphates†
Heavy metal phosphates (with NH_4)
Rare-earth phosphates
Sr, Ba-phosphates
$(NH_4,K,Na)_3AlF_6$ cryolites
Phosphates of radioactive elements

† See Table 9.

ing more than about 10% free carbonates are considered uneconomical to process, however.

If the marginal quality of the PR is due to high Fe-Al content or other such acid-soluble metal impurities, then nitric phosphate products show the same adverse compositional effects as those prepared via the WPA route. Unless such rocks are avoided, nitric phosphate fertilizer products of the future will show the same gradual transition to higher contents of water-insoluble phosphate components other than dicalcium phosphate, with a corresponding decrease in citrate solubility.

Whether such compositional disturbances, which are now beginning to appear in commercial products, have an adverse effect on the agronomic performance of nitric phosphates has yet to be determined.

IV. FUTURE OUTLOOK FOR PHOSPHATE FERTILIZERS

The first half of this decade has produced several important developments and trends that are beginning to reshape traditional patterns of P fertilizer production and usage.

Foremost among these factors is the onset of a worldwide decline in the quality of P raw materials, not only among major deposit sources, but also from newer resources of more divergent mineral composition now coming into mine production. This will have growing impact on existing chemical process technology which was developed primarily for the former premium-grade PR concentrates.

Secondly, developments in rapidly expanding subtropical and tropical agricultural markets quite possibly may identify situations requiring new phosphate formulations differing in solubility properties and composition from those chemical fertilizers now widely produced. No accurate projection of this possible trend can be made on the existing base of agronomic data, but the subject is being intensively researched by the International Fertilizer Development Center at Muscle Shoals, Alabama.

Thirdly, a trend also developed in temperate-zone agriculture during the 1972-1975 period of acute shortage followed by the steep rise in fertilizer costs. The practice of using cheap, high-analysis fertilizers at luxury application rates became no longer possible or profitable. A shift to lower cost P materials for maintenance applications took place in some countries.

This situation may be reversed if a fertilizer oversupply again develops by the late 1970's, but perhaps not before lower cost, water-insoluble P materials regain a foothold in agricultural usage.

Finally, the abrupt upsurge in costs of both energy and P raw materials presents a serious challenge to the current chemical technology base of the worldwide P industry and especially to new production facilities. Modern chemical processes are energy-intensive at all stages from raw materials preparation and chemical treatment to product drying, sizing, and granulation. These economic factors, together with a reassessment of preferred products, will influence the choice of new production capacity, particularly in developing countries.

In this early stage of transition created by the above developments, no accurate forecast of changing product compositions can be made, other than to examine the possible fertilizer options. It appears highly improbable that major short-term changes will come about in process technology, phosphate products, and agronomic usage on an international scale. Current research on new process technology is not likely to yield revolutionary technology in the near future, nor is it likely that industry will abandon existing plants and processes to adopt radically new processes.

Instead, changes in phosphate product composition and agronomic usage most likely will be gradual and localized as obsolete production facilities are retired, as new production centers emerge, and as agricultural markets develop for new P formulations. Quite probably, the developing agricultural regions may provide the proving grounds and stimulus for any new generation of P products in place of conventional chemical phosphates. Thus, the transition may be more abrupt there than in technologically developed countries which are heavily committed to production of traditional products. Aside from those considerations, an inadvertent transition in product composition is more likely to arise from the changing quality of P raw materials.

The foreseeable options for phosphatic fertilizer are reviewed in the following section, but no reliable forecast can be offered as to which will gain wide acceptance. This offers an important challenge to agronomic research to favorably influence the selection process.

A. Trends in Traditional Fertilizer Products

With the projected PR demands increasing at the rate of about 10 million metric tons/year, the shortage of premium-grade rock will become more acute, forcing even wider use of lower quality raw materials to produce P fertilizers. The unavoidable consequence of this will be a steadily declining quality of traditional product types with respect to P grade and soluble components unless dramatic breakthroughs in acid purification technology suddenly emerge. No major new resource developments or radically improved mining and beneficiation methods are likely to halt or reverse this trend in raw materials before the 1980's.

Any long-range forecast of product quality trends must consider four possible technological developments:

1) Improving mining and milling technology to restore premium quality to phosphate raw materials.
2) New or improved acidulation technology designed to process raw materials of inferior chemical quality.
3) New or improved purification processes to reject contaminants from intermediate phosphate products or process streams.
4) Modification of product specifications based on agronomic performance rather than empirical solubility criteria.

No early solution to the raw materials problem seems likely from any radically improved minerals beneficiation technology. Until very recently, mine producers needed only simple upgrading methods (washing, sizing, silica flotation) to increase P content, with little concern for other chemical quality factors. These methods now are proving inadequate to selectively reject the key impurities in lower quality ores that adversely affect product quality. Research on new physical metallurgical methods is mainly in the exploratory stage, forcing the phosphate processing industry to seek its own alternative solutions.

The chemical processing industry has made impressive modifications in acidulation technology within the limits of current plant designs to accept lower quality raw materials, as discussed in other chapters. However, the limits of engineering adjustments and cost penalties in the present generation of acidulation processes may be fast approaching. Further slippage in raw material quality increases the prospect for even higher passthrough of undesired impurities. Research on new acidulation methods is mainly in the exploratory stage so that there are no short-term prospects for new processes to mitigate the growing problem of raw material quality.

In fact, a counter trend could develop in U.S. wet-process phosphoric acid technology if the "total matrix acidulation" concept is adopted. This approach, now under investigation by TVA and the U.S. Bureau of Mines, would substitute even lower quality raw unbeneficiated phosphate ore, as mined, as process raw material in place of conventional beneficiated PR concentrates.

Industry may be forced into accepting this trade-off due to a combination of conservational, environmental, and economic factors relating to current mining and beneficiation practices and waste disposal. Except in the case of siliceous ores, this substitution of low quality raw rock for beneficiated concentrates will release a much higher concentration of metal impurities during acidulation, to be passed through to the ammonium phosphate products.

If industry is forced to adopt this processing concept, then the traditional ammonium phosphate products will show a much sharper decline in grade and quality and a marked change in solubility properties due to admixture of insoluble metal phosphate salts.

In anticipation of such processing trends, an intensified agronomic research effort is needed to determine the limits of trade-off in water-soluble

P in relation to actual performance characteristics from the agricultural viewpoint. Are water-insoluble phosphates to be avoided at all cost? The need for a more realistic set of product specifications is one of the most important problems confronting phosphate producers to seek relief from unnecessary and costly purification steps. Only agronomic research can provide the necessary guidance. Otherwise, industry's third option to cope with increasing levels of impurities is to add further costly purification steps for the intermediate products of acidulation. Most of the research and development effort has concentrated on this post-purification approach. The most widely employed methods are proprietary solvent-extraction processes and various schemes for precipitating out impurity "sludges" (Slack, 1968). Both approaches add to process costs and also create waste byproducts; neither is highly efficient in rejecting all objectionable metal impurities.

If the slippage in P raw material quality continues, then economic as well as technical limitations will determine the extent to which a pass-through of impurities to fertilizer products becomes a practical necessity in these purification schemes.

The remaining option for industry is to seek remedial adjustments in product specifications, mainly a downgrading of nutrient content and lower standards for water-soluble and citrate-soluble P content, as defined by the available phosphoric acid (APA) control procedures.

Agronomic research should play a vital part in revamping these specifications, which, as presently defined, perhaps ill-serve both industry and agriculture. For example, would a decrease to 90% APA or citrate-soluble P impair agronomic performance of products to a detectable extent, especially in view of the $< 30\%$ overall recovery of P by crops from applied sources?

In retrospect, there is little prospect for short-term developments in any of those four remedial options to provide the P industry with a wholly satisfactory solution to the deteriorating chemical quality of raw materials. The trend toward steadily decreasing quality of P products is likely, therefore, to continue as an unavoidable consequence, and its agronomic implications await evaluation. Its future impact on traditional P fertilizers type is reviewed below.

1. ORDINARY SUPERPHOSPHATES

Single superphosphate processes have a fairly wide tolerance for non-premium grades of PR, but there is a trade-off in grade due to the added burden of inert minerals, and there are problems with acidulation caused by increases in solubilized Fe and Al. The P compounds formed by metal impurities are generally citrate soluble and unlikely, therefore, to impair agronomic performance (Taylor et al., 1960). In the future, some P production capacity may shift back to OSP products to utilize low-quality raw material—a trend that may occur more rapidly in developing countries who are depending upon inferior indigenous ores and serving captive regional markets where arbitrary grade specifications do not apply.

Manufacture of OSP quite possibly may develop to be the only practical use for some marginal phosphate rocks with high contamination levels of Mg, Al, Fe, and free carbonates. It is to be emphasized that these OSP products will have different solid-phase compositions than former products comprised essentially of monocalcium phosphate and Ca sulfates (Frazier & Lehr, 1967; Ando & Lehr, 1967).

The additional incentives for a return to OSP products, aside from the raw material factor, are: (i) lower energy requirements in processing, (ii) efficient use of byproduct sulfuric acid reagents, (iii) avoidance of waste byproduct disposal, and (iv) retention of secondary and micronutrient credits in raw materials which are now largely discarded.

2. CONCENTRATED SUPERPHOSPHATES

In the past, the CSP quota of total P production has been dictated primarily by market demand. Production was based on premium-grade PR and WPA intermediates to meet high P grade and solubility requirements of products in international trade.

In more recent years, CSP production has offered an outlet for byproduct "sludge acid" from WPA production. As discussed earlier, CSP compositions have undergone a drastic change due to the use of these sludge acids, with various metal phosphates replacing the traditional monocalcium phosphate component. However, this compositional shift does not significantly alter P grade of the CSP or its content of citrate-soluble P but it markedly lowers its water-soluble P content.

From the agronomic viewpoint, this unpublicized transition in solid-phase composition of CSP over the past decade has gone virtually undetected in agricultural performance. Perhaps no other P fertilizer material has undergone such a major change in product composition as have superphosphates. This furnishes a clear example of the artificial character of existing control specifications to detect and differentiate such wide compositional variations, especially since these differences had no apparent significance on actual agricultural usage. The same may not be true, however, for other chemical P fertilizers.

The process of utilizing sludge acids has important implications. The effects on CSP composition due to Al, Fe, and alkali impurities are the same whether derived from highly impure byproduct sludge acids or from nonpremium-grade PR in the acidulation process. Either route introduces the same set of impurities and yields similar end products with similar solid-phase composition. This obviously creates an economical outlet for lower quality phosphate raw materials without undue process penalties or impaired agronomic performance of products.

As a consequence, a shift in the future to a larger production base of superphosphate-type formulations may become a virtual necessity due to the quality of available phosphate raw materials, especially in localized cases. The tolerance limit for this shift in CSP composition with respect to its content of water-insoluble Fe and Al phosphate phases in relation to agronomic performance has not been established. It appears timely, therefore, that agronomic researchers reexamine the superphosphate-soil interactions and responses in light of present-day product compositions. It also has important implications in the continuing use of CSP as a standard of reference in fertilizer evaluation experiments on the assumption that the reference compound is monocalcium phosphate.

3. AMMONIATED WPA PRODUCTS

Although faced with the problem of deteriorating raw materials quality the P industry is heavily committed to WPA processing due to major capital investments in both existing manufacturing facilities and firm expansion commitments.

As shown in Table 12, the production of WPA intermediates has accounted for about 60% of the PR consumption in the 1970's, with little change forecasted by the early 1980's (Anon., 1974a).

This worldwide major production commitment ensures a continued emphasis on high-analysis ammonium phosphate fertilizer types well into the 1980's to amortize investments, despite changes occurring in raw material quality and possible shifting requirements of phosphate products for agriculture.

The manner and rate at which individual segments of the WPA industry will adjust to the above factors will obviously differ, leading to gradual changes in the types of ammoniation products being produced. Many of the older WPA plants incapable of much further engineering adjustment to compensate for raw material quality face the choices of producing lower quality MAP or DAP grades, shifting to superphosphate-type compositions, or adopting new compositional types of N-P products, as described later.

More modern WPA production facilities are designed with greater flexibility to handle nonpremium-grade raw materials, especially by add-on purification steps for acid intermediates. There are, however, technical and economic limits to this adjustment, which differs for individual manufacturers. Without the use of purified acid, the future trend in MAP and DAP compositions will be towards increasing amounts of water-insoluble P compounds of the (Fe,Al)-, $(Fe,Al)NH_4$-, and $MgNH_4$-types. While these combinations of metal cations invariably yield salts of low water solubility, those compounds having high citrate solubility can be favored by strict process controls during the ammoniation step. Conditions of high temperatures, long retention time, and recycle loops that induce seeding must be avoided, as discussed in Section III-C. The presence of Mg impurity usually

Table 12—World production statistics of WPA.

Year	Phosphate rock		WPA production capacity, P‡	% of total production
	Total†	P‡		
1973	97.2	13.6	8.3	~60
1978	160	22.3	12.9	~58

† Million metric tons averaging 14% P (70 BPL).
‡ Million metric tons of P as merchant-grade WPA averaging 21.9 to 22.6% P.

yields citrate-soluble Mg or MgNH$_4$ phosphates if the concentration of F is low; otherwise, a series of citrate insoluble fluorophosphate compounds is possible (Ando et al., 1968a).

In the future, the grade notation of MAP and DAP products will become an imprecise measure of their actual composition due to the variability in the kinds and amounts of Al, Fe, and Mg phosphates that they contain. The agronomic implications of this compositional trend have not been evaluated but may have far reaching significance in agriculture!

Ammonium polyphosphates have been produced successfully from WPA intermediates which have been partially purified by clarification or solvent extraction, or derived from premium-grade rock. With lower quality acids, their contents of Fe, Al, Mg, and F caused the formation of highly insoluble pyrophosphate precipitates which affected citrate solubility of end products as well as created process difficulties (Slack, 1968). These problems have been partially circumvented by the new pipe reactor technology (Meline, 1974). Purification of WPA acid intermediates looms as the only option to maintain the traditional high-analysis water-soluble APP formulations. Otherwise, a corresponding tradeoff in product quality will result due to the formation of various insoluble metal pyrophosphate compositions (Frazier et al., 1966, 1972; Frazier & Lee, 1972; Ando et al., 1968b).

4. NITRIC PHOSPHATE PRODUCTS

Although nitric acid is a less vigorous solvent than H$_2$SO$_4$ for some classes of PR accessory minerals, laboratory solubility studies do not indicate a clearly demonstrated superiority of HNO$_3$ acidulation over the H$_2$SO$_4$ processes used to prepare WPA. The overriding effect appears to be corrosive attack of acidic fluoride released by apatite dissolution (unpublished TVA data). Thus, both acidulation routes release chemical impurities (notably Fe, Al, Mg, F, chlorides, alkalies, and heavy metals, among others) that create parallel problems in processing and end-product compositions.

Nitric acid processes differ from WPA processes, however, in the proportion of dissolved metal impurities that is passed through to end products. With nitric acid acidulation, metal impurities are retained as soluble nitrates until the ammoniation stage where the metal cations preferentially recombine with (PO$_4$)$^{3-}$ and ammonia to form insoluble phosphate precipitates. In WPA processes, some removal of metal impurities as sulfate and F precipitates occurs in the attack stage and by precipitation of phosphate salts during acid clarification steps (Frazier et al., 1975; Lehr et al., 1966).

Both acidulation processes can accept low-grade siliceous phosphate raw materials at a penalty of lowered production rates. Nitric acid processes can be used on PR of nonpremium grades having a high content of free carbonates, since the excess HNO$_3$ consumed by free carbonates is recovered as byproduct Ca(NO$_3$)$_2$ or as NH$_4$NO$_3$ by a reaction with added (NH$_4$)$_2$SO$_4$. Regarding the overall trend toward declining PR quality, nitric acid processes do not appear to offer marked technical advantages over WPA processes insofar as product quality is concerned. Any shift in the present balance of production via the two acidulation routes is more apt to come about because of changes in costs and availability of the respective acid reagents since HNO_3 is indirectly linked to the cost and availability of hydrocarbon feedstocks.

From the agronomic viewpoint, NP products are likely to show an increasing proportion of water-insoluble P components (Table 11). This will mainly include the metal ammonium phosphate precipitates along with the usual Ca phosphate compositions. Hence, the agronomic evaluation problem is quite similar to that discussed for ammoniated phosphoric acid products.

A second evaluation problem of potential importance concerns the fate of heavy metals, toxic metals, and radionucleides carried through to the final NP products as substituted constituents in soluble nitrates or phosphates. With decreasing chemical quality of P raw materials and especially the wider use of igneous and metamorphic apatite ores, there is a much greater risk of undesired accumulations of those metals in final fertilizer products than with WPA processes. Capricious environmental constraints may be placed on usage of such P products unless this potential problem is adequately researched by agronomists to show whether or not these substances are actually passed into agricultural products.

B. Possible Trends in New P Formulations

A technological search for alternative P fertilizer products is being stimulated by the changing raw materials situation and further incentives may arise for new products in some of the new agricultural areas now being developed.

While the worldwide impact of these factors may be one of gradual transition, the emergence of new products in a given country or geographical region may be more rapid. This forecast is based on the following considerations: (i) dependence on low-quality indigenous P resources; (ii) need for less complex, lower cost, less energy-intensive processing; (iii) need for products that retain essential nutrient elements from the precursor rock or function as carriers for same; and (iv) the necessity to revise solubility standards that now favor materials with 100% water solubility.

Some examples of possible new formulations and products are reviewed below. Most of these are in the experimental stage, but the agronomic researchers should be alert to their potential applications in agriculture.

1. DIRECT APPLICATION MATERIALS

A much wider use of apatitic PR is forecast since better selection principles are now available to choose the most reactive rock sources for agricultural use. The former practice of applying pulverized rock is being abandoned, however, in favor of granulated grades which may lead to markedly different properties, depending upon the method of granulation. The use of water-soluble salt binders, particularly ammonium phosphate, leads to hybrid granular products (see discussion in Section III, this chapter).

These hybrid types may actually bridge a gap between conventional direct-application rock products and the water-soluble manufactured chemical phosphates. Currently, developments in granulation technology outpace agronomic evaluations. Although collectively they are referred to as *granulated PR products,* it is important to recognize that the new granulation art is yielding a spectrum of new phosphate compositions which are not likely to have comparable agronomic performances. The best engineering solution may be the poorest from an agronomic viewpoint!

Phosphate rock suspensions offer an alternative to granulation as a convenient means for transporting and applying ground PR. Now under experimental development by TVA, satisfactory suspensions of high pulp density (70% rock by weight; 10% P) have been produced with the aid of polyelectrolytes, organic wetting agents, $Na_4P_2O_7$ conditioner, and attapulgite clay suspension aids (TVA unpublished data).

This approach appears to offer technical and economic advantages that may offset higher transportation costs per unit of P. Such advantages include: (i) higher efficiency and lower cost of wet grinding; (ii) ease of rock dispersion in soils compared to granule placements; and (iii) use of the aqueous suspensions as carriers of soluble primary, secondary, or micronutrients to achieve some special formulation requirement. Suspension-type products provide a handling convenience but in no way alter the inherent solubility of a particular rock source so that the selection of reactive rocks is a necessity here as in other direct-application grades.

Another new development concerns the use of nonapatitic PR for direct application. The successful commercial development of a process to convert Fe-Al mineral phosphates into direct-application goods forecasts much wider use of these neglected marginal-type resources (see discussion of Calciphos and Phospal, Section III) (Anon., 1975; Atanasiu, 1971).

2. THERMAL PHOSPHATE PRODUCTS

A much wider use of thermal P products can be reasonably forecast, based on the following considerations (Atanasiu, 1971). One or more of the process options can be adapted to a wide variety of marginal-quality PR to take advantage of reactive accessory mineral phases, which otherwise may make them unsuitable for other fertilizer production. The compositions of thermal phosphates and their physical state (glass or crystalline form) can be modified to achieve a range of desired physical and chemical properties. Also, thermal phosphate processes require a lower level of technology and simpler process facilities than wet-chemical manufacturing processes.

The main deterrent to wider use of thermal phosphates will be the continued scarcity and high cost of hydrocarbon fuels for processing. This situation may ease, however, on a regional basis.

3. CHEMICAL FERTILIZER TRENDS

a. N-P Suspension Fertilizers—Suspension fertilizers offer an outlet for ammoniation products of WPA prepared from low-quality PR. Viscous black acids having increasingly higher contamination levels of metal impurities and organic matter become unsuited for conventional liquid formulations, and possibly for some granulation processes, unless subjected to a partial purification treatment.

These suspensions, or "slurry fertilizers," consist of supersaturated solutions of ammonium phosphate in which the suspended solid MAP and DAP, along with sludge solids from the acid, are stabilized through controlled crystallization and addition of gel agents, usually attapulgite-type clays. Typical product grades are in the range of 10 to 12% N and 15 to 17% P (36 to 39% P_2O_5).

This new technology offers several obvious technical and economic advantages that may encourage its usage in the future. Such a process accepts lower quality phosphate raw materials and intermediates. The ammoniation stage is simpler, less costly, and more easily controlled. Slurry-type products retain the advantages of liquid formulations with regard to ease of storage, transportation, and field application. Their liquid base also provides a carrier for secondary and micronutrient amendments. Metal impurities from the acid which include such recognized nutrient elements as Fe, Mn, Zn, Mg, S, K, and soluble silica, are retained in the product rather than being rejected as in refined ammonium phosphate grades.

Suspension-type products will incur higher shipping costs per unit of P, however, because of their lower grade in comparison to current fertilizers. Also, their P content is distributed between water-soluble and insoluble components. Whether this factor will have an adverse effect on agronomic performance has yet to be established.

b. Urea Phosphoric Acid—The composition $CO(NH_2)_2 \cdot H_3PO_4$ has been reinvestigated in recent TVA research in connection with the development of a novel purification method for upgrading and concentrating H_3PO_4 from low-quality WPA. In this process, urea added to impure WPA dissolves to form the well-recognized molecular complex with H_3PO_4, which crystallizes out as the anhydrous solid adduct, $CO(NH_2)_2 \cdot H_3PO_4$. The complexing principle is the same as with liquid amine reagents in solvent-extraction purification schemes, but the recovery process is much simpler since the urea complex forms a stable solid phase which can be readily separated, thus avoiding liquid-liquid partitioning problems as in solvent purification methods.

The urea-phosphoric acid compound is a stable, free-flowing, water-soluble, white crystalline solid with a theoretical composition of 17.7% N and 19.6% P (44.9% P_2O_5). Technical grades of this recognized fertilizer composition recovered from impure WPA have slightly lower analyses due to a minor retention of solid impurities from the acid (unpublished TVA data).

Urea phosphoric acid recovered from impure WPA intermediates is a potential substitute for the traditional water-soluble ammoniation products of WPA such as MAP, DAP, and APP. Its main advantage is that it can be prepared from low-quality WPA without supplemental purification steps. A further advantage is that $CO(NH_2)_2 \cdot H_3PO_4$ can undergo internal decomposition, dehydration, and condensation under mild heat treatment to yield dry solid ammonium polyphosphates. This offers an alternative route to APP manufacture.

The future prospects for this process hinge on three developments: (i) increasing worldwide production and availability of urea, (ii) continued major emphasis on WPA intermediates, and (iii) the projected decline in phosphate raw material quality. A point may soon be reached where this nondestructive use of urea to recover H_3PO_4 from low-quality intermediates in the form of high-analysis N and P fertilizer products may prove to have technical and economic advantages over present WPA ammoniation processes.

4. POLYPHOSPHATE PRODUCTS DERIVED FROM IMPURE WPA

Ammonium polyphosphate formulations have been produced directly from highly impure WPA feedstock in the newly developed pipe reactor (TVA unpublished data). The pipe reactor design concept minimizes retention time of reactants which overcomes scaling problems due to formation of acid-insoluble $(Al,Fe)NH_4P_2O_7$, and also minimizes citrate-insoluble P components in product melts. Co-melts of APP with urea, ammonium nitrate, or ammonium sulfate allow N-P grade adjustments and a wider range of formulations (see discussion, Chapter 7).

Based on these experimental studies, it appears reasonably certain that APP production via relatively simple pipe-reactor technology may be extended to even lower quality feed acids than originally conceived. Thus, APP products are not likely to be phased out as future fertilizer options due to decline in the quality of phosphate raw materials.

Potassium phosphates may provide another economical outlet for impure crude WPA intermediates through the well-known, two-step thermal reaction with KCl as:

$$KCl + H_3PO_4 \stackrel{\Delta}{=} KH_2PO_4 + HCl\uparrow \qquad [3]$$

$$KH_2PO_4 \stackrel{\Delta}{=} (KPO_3)_n + H_2O\uparrow \qquad [4]$$

which yield a highly concentrated, water-soluble potassium polyphosphate to be used in formulating N-P-K compound fertilizers. Process economics depend on recovery of anhydrous HCl or chlorine as a credit against energy costs and comparatively expensive reagents (KCl and concentrated WPA). More detailed technical discussions are given in Chapter 7.

Another outlet for highly impure WPA is based on a newly developed reactor concept known as the "pipe-cross reactor" (Achorn and Salladay, 1975). The pipe-cross reactor differs from conventional pipe reactors in that impure WPA, H_2SO_4, and NH_3 are injected simultaneously at high flow rates for turbulent mixing to yield melt products, for example, 12–48–0. This reactor innovation lowers the retention time, reduces engineering costs, yields nearly anhydrous products that preclude the need for supplemental drying, and greatly minimizes the formation of citrate-insoluble P components.

These three technological improvements to produce high-analysis polyphosphate products from crude WPA intermediates demonstrate that engineering solutions may be found to soften the impact of declining phosphate raw material quality. It is unlikely, therefore, that high-analysis APP formulations will disappear in the foreseeable future, although some compositional adjustment appears inevitable. In this regard, the retention of metal impurities from WPA feeds may add to, rather than detract from, their agronomic value.

C. Conclusions Concerning Forecasted Trends

The forecasts of future P fertilizer compositions and agronomic options discussed in the preceding section are intended to illustrate possible departures from current technology. They are necessarily presumptive but not necessarily inclusive of all new technology under development.

An industry suddenly faced with declining quality of PR raw materials in the mid-1970's obviously cannot long continue to maintain the former quality standards of the traditional products which heretofore have been widely available to agriculture. The present period of engineering adjustments and add-on process steps to cope with the new problem will inevitably reach the limits in economic penalties and wasted resources. The impact will be more severe in older manufacturing facilities, perhaps hastening their obsolescence.

Agronomic research in the several defined key areas can make an important contribution to this selection process of foreseeable options for adjusting product compositions or adopting new processing routes. More intensive evaluation of new product performance in agriculture is needed, and more realistic product specifications need to be established to guide manufacturers in the selection of raw materials and processes.

Few such opportunities have emerged in recent decades for the agricultural sciences and fertilizer technologists to interact in a manner to ensure the production of high-performance P fertilizers. The wide array of options is not necessarily restricted to those of highest grade or water solubility. The final proving ground is actual agronomic performance.

Chapter 4

Phosphate Raw Materials and Fertilizers:

Part II—A Case History of Marginal Raw Materials

JOHN HOARE

The British Phosphate Commissioners, Melbourne, Australia

I. INTRODUCTION

Christmas Island is an isolated seamount situated latitude 10° 25′19″ South, longitude 105°42′57″ East. It rises from 4,500-m depths on the southern edge of the Java Trench and is an exposed portion of a northeast-trending ridge extending through the Cocos Islands to the southwest.

The basal rocks are volcanic, igneous, varying from andesite to trachybasalt interbedded with carbonate rocks from the Eocene era. Overlying these is a series of interbedded limburgites, basalts, and palogonite tuffs surmounted with a cap of Miocene limestone. This latter limestone is of biogenic origin and contains abundant foraminifera, corals, algae, molluscs, and gastropods. Carbonate rock textures are variable, ranging from dense recrystallized dolomite to reefs of shelly material with a loose, sandy composition. The Island contains large deposits of phosphate minerals varying from high-grade calcium phosphate (apatite) to lower-grade iron and aluminum phosphates (crandallite/millsite).

The surface material or upper layer phosphate is dominantly Al and Fe phosphate (C zone) and profiles into a high quality A-zone phosphate (apatite) which overlies a coral pinnacle or limestone Karrenfield base formation. The B-zone ore is intermediate and contains a variable mixture of C- and A-zone phosphates.

The A-zone ore is used for the manufacture of superphosphate and wet-process phosphoric acid, while the B- and C-zone ores contain amounts of iron and aluminum which are unacceptable in their as-mined state for normal acidulation processes.

In 1949, the mining rights were acquired by the Australian and New Zealand Governments and the Christmas Island Phosphate Commission

---

Copyright 1980 © ASA-CSSA-SSSA, 677 South Segoe Road, Madison, WI 53711, USA. *The Role of Phosphorus in Agriculture.*

Table 13—Phosphate reserves in Christmas Island.

Grade	Predominant minerals	Chemical assay, %				Available tonnage	Remarks
		P	Al	Fe	Ca		
		%				million metric tons	
A	Apatite	>15.5	1.7	0.9	34.3	20	White to cream coarse aggregate
B	Apatite, crandallite, millisite	14.5	6.6	3.6	23.6	54	Variable mixture of A- and C-grade ores
C	Crandallite, millisite, barrandite	13.1	12.6	9.8	15.7	53	Red-brown, friable clay-like
D	Crandallite, millisite, barrandite	10.2	13.2	10.5	8.6	95	Overburden "soil" or upper layer phosphate

was formed to mine and develop the P deposits. At that time, reserves of high-grade P ores were estimated to be approximately 25 million tons (25.4 million metric tons). An extensive drilling survey undertaken in 1958/1959 refined these estimates and indicated that the reserves were as follows:

	Million metric tons	P assay, %
A-grade	30	16.6
B-grade	25	15.3
C-grade	28	12.2

The revised estimates confirmed that the deposits were quite limited. A decision was made to initiate research into means to beneficiate the overlying B- and C-grade phosphates rather than to simply discard them as overburden. We shall attempt in this discussion to highlight our experience in dealing with these marginal grades of phosphate ores over the past 20 years. The nature of marginality that is increasingly encountered in the major world P deposits was discussed in the first part of this chapter. A brief look at the mineralogy of the Christmas Island deposits is given here to illustrate the point that in our efforts to utilize the B- and C-zone phosphates for fertilizer production, we have had to contend with several of the same marginality factors that will be encountered to an increasing degree on a worldwide basis in the years to come. The C-zone Fe and Al phosphates presented a research challenge of far-reaching importance because of other similar underutilized deposits elsewhere in the world.

The mineralogy of the P deposits in Christmas Island is summarized in Table 13. These data were developed from core analyses from a comprehensive drilling program that commenced in 1966. The objective was to assess the total reserves of all grades of phosphatic ore at Christmas Island. The data in Table 13 show that crandallite and millisite minerals are predominant in the C- and D-zone strata, with minor intrusions of barrandite. The crandallite and millisite minerals occur in the lateritic profiles derived by weathering of the carbonate and volcanic rocks in the presence of phosphate-bearing minerals.

The discussion which follows will focus on our research and development efforts directed towards finding means of utilizing the B- and C-zone ores as economic sources of P fertilizers. The A-zone material is a high-grade apatitic ore, quite similar to other premium-grade apatitic ores except for minor contamination by carbonate minerals and, therefore, will not be discussed further.

II. EXPLORING THE OPTIONS

A. Drying and Air Classification

Research into methods of utilizing the overburden initially concentrated on beneficiation techniques associated with upgrading B-zone ore into a product equivalent to the established marketable A-zone ore. Drying caused the disintegration of the very fine clay-like Fe and Al phosphates from the coarser Ca phosphates. The fines were separated with hot air forced into the dryers. Upgrading by this method did not effectively reduce the Fe and Al in the final product below 2.2 to 4.0%, the acceptable level for acidulation processes. With a B-zone ore containing 7 to 8% Fe + Al, nearly 60% of the rock feed would be recovered with Fe + Al content of 2.2% by removing the less than 48 mesh fines by classification. However, adhesion of the fine millisite and crandallite particles to the coarser apatite could not be overcome prior to screening. Wet-washing and screening was found far more effective than dry screening in recovering a large fraction of B-zone ore in a grade equivalent to A-zone product.

The Fe and Al in C-zone ore is chemically combined with the phosphate, and physical beneficiation methods were not expected to be applicable and results confirmed these expectations.

B. Calcination

Calcination of C-zone ore at temperatures near 430°C rendered over 70% of the P soluble in neutral ammonium citrate. The P content was increased to 12.2%, primarily due to elimination of free water on drying and loss of crystallization water above 100°C. The concentration of Fe and Al showed a corresponding increase. This process did not improve the suitability of the material for production of superphosphate or phosphoric acid. However, after fine grinding, this material was demonstrated to be quite useful for direct application (Doak et al., 1965).

The engineering aspects of this process underwent several stages of development, primarily to preclude overheating. Temperatures in excess of 700°C caused recrystallization of the dehydrated (amorphous) phosphate minerals, with considerable loss of citrate solubility.

High temperature calcination ($>1000°C$) was investigated for its effect on A-, B-, and C-zone ores and on mixtures of these with or without additional lime. The basic aim of high temperature calcination with the lime amendment is to deactivate, or render inert, the maximum amount of Fe and Al contained in the B-zone ore by changing their compositional form.

Whitlockite is formed from the phosphate in the crandallite/millisite phases combining stoichiometrically with the added Ca, while the liberated Al and Fe are converted to corundum and hematite, respectively. Some Al phosphate recrystallizes as $AlPO_4$ (cristobalite-form) which is soluble in phosphoric acid. However, the corundum and hematite produced by this thermal beneficiation treatment are relatively insoluble in H_2SO_4 in the strengths used for superphosphate manufacture.

White (1971a) concluded that acidulation of calcined (1100°C) A-zone rock resulted in a superphosphate product with physical properties comparable to that produced from Nauru rock. Calcining of B-zone ore deactivated the Fe and Al contaminants to give a product equivalent to as-mined A-zone product (3% Fe + Al). In addition, calcination fo B-zone ores with limestone significantly reduced the acid solubility of the Fe and Al contaminants from 9% (15% R_2O_3) to 1.3 or 1.6% (2 to 2.5% R_2O_3) which is even superior to the as-mined A-zone rock (White, 1971b).

Calcining C-zone ore with limestone deactivates the initial high Fe and Al content from 22% (34% R_2O_3) to 6% (9% R_2O_3), which is still unacceptably high. This residual content of active Fe and Al and the low-P content of the calcined product (12% P) combine to make it unsuitable for the manufacture of superphosphate or phosphoric acid by conventional methods. A blended mixture of 40% calcined C-zone ore and 60% rock from Nauru produced a phosphoric acid that contained 3.6% Al, 3.8% Fe, and 20% P (45% P_2O_5) as available acid soluble in neutral ammonium citrate. However, the water-soluble P content of this acid was only 15%, and triple superphosphate made from this acid and Nauru rock was sticky and hard to process and handle on a commercial scale.

Power consumption in high-temperature calcining is quite substantial and can significantly affect the economic considerations of this process in any cost/benefit analysis of the utilization of these marginal-grade phosphate ores.

C. Direct Application of Calcined C-Zone Ore

Granulated, calcined (at 550°C) C-zone ore is marketed under the trade name of "Calciphos," and the same material in the form of fines (<100 mesh) is marketed as "Citraphos." These materials have been under investigation since 1963 (some of the earlier products were calcined at lower temperatures). The early work was reported by Doak et al. (1965). Pot and field trials indicated that fine, calcined C-zone ore was nearly equivalent to superphosphate on P responsive soils. Companion laboratory investigations using x-ray diffraction, infrared absorption, and differential thermal analysis indicated that calcining destroyed the crystal structure of the minerals and resulted in an amorphous product. This also accounted for the increase in specific surface areas and in citrate solubility. Calciphos was granulated alone or with additives that might aid in the slaking down and disintegration of the granules after application. Granulation reduced effectiveness to pasture plants. Granulation with well-cured superphosphate was better than with fresh superphosphates indicating interaction between components. Additional field testing of Calciphos and Citraphos has been carried out by Mason and Cox (1969), Muller (1970), and Lipsett and Williams (1970) on pastures and wheat. Most of the results show that Calciphos has little value for short-term crops such as wheat, but is quite effective for perennial crops and pastures. Citraphos was clearly inferior to superphosphate in the first year on pastures covering a range of soils, climates, and plant species. The residual effect of Citraphos, however, improved relative to superphosphate in subsequent years.

The option of direct application for this marginal-type phosphate rock resource appears to be a feasible one, especially for pastures and where residual effects are desirable. However, the finely ground product is difficult to handle and granulation reduces its effectiveness.

D. Other Thermal and Chemical Conversion

A fundamental research program directed towards the utilization of C-zone ore was undertaken recently. Chemical, pyrometallurgical, and hydrometallurgical techniques were employed and a number of fertilizer processes have been developed and evaluated. The processes appear technically sound, though their commercial feasibility is either unfavorable or awaits further economic analysis.

1. ION EXCHANGE PROCESSES

Calcined C-zone ores are contacted with cation- and anion-exchange resins to strip off the P. The resulting acid is dilute and regeneration of the resins is costly. Generally, ion-exchange methods are substantially more costly than other alternative processes.

2. THE ALUM PROCESS

Dilute sulfuric acid leaching of 800°C calcined C-zone ore gave the most promising results. The leach liquor contained nearly all the P as phosphoric acid, excess sulfuric acid, most of the Al, and little of the Fe. Ammoniation of the liquor precipitated Al as metallurgical grade Al_2O_3 and produced an ammonium phosphate-sulfate fertilizer containing 12% P with a 74% water solubility of the P component. This process is currently assessed as being marginally uneconomic and is under further investigation and modification to improve its commercial possibilities.

3. HYDROCHLORIC ACID PROCESS

In this process, C-zone phosphate rock is leached with HCl and the clarified leach liquor is contacted with an immiscible organic solvent (such as B-butanol) to extract the phosphoric acid. Most of the alumina is removed by crystallization. A cost analysis of this process is in progress.

4. CAUSTIC LEACH PROCESS

This is patterned after the Bayer process used in extraction of alumina from bauxite (Rothbaum & Reeve, 1968). The C-zone ore is leached with caustic soda which removes most of the Al and over half of the P and leaves a red muddy residue comprised mainly of P (12%), Fe (25%), and Ca (25%). The leach liquor is filtered and crystallized in two stages, extracting pure trisodium phosphate in the first and alumnia in the second stage. The alumina is sufficiently pure for smelting to Al, and the trisodium phosphate can be converted to contaminant-free basic Ca phosphate by reaction with calcined lime.

Manufacturing costs are partially reduced by sale of alumina (23% or ore feed rate), but this is not enough to make the process economically feasible. The insoluble residue (56% of ore feed rate) contains about 12% P and efforts are underway to recover this P to help write off another portion of the cost. Hydrochloric acid attack is being evaluated as one option. Another option under investigation is the calcining of the residue with limestone to make the Fe unreactive, followed by leaching with sulfuric and recycled phosphoric acids. An additional alternate route of processing trisodium phosphate is to convert it to magnesium-ammonium phosphate. Conversion to ammonium phosphate is not technically feasible without going through the intermediates of basic calcium phosphates then phosphoric acid. In general, the caustic leach process appears uneconomic at this time unless P in the P-bearing residue can be recovered through a less costly route.

5. LIME SODA SINTER PROCESS

This process consists of sintering a slurry mixture of C-zone ore, lime, and sodium carbonate to produce Ca phosphate (whitlockite), sodium aluminate, and ferric oxide. The alumina is leached leaving a residue of Ca-phosphate and Fe-oxide. The residue is leached with sulfuric acid and recycled phosphoric acid to produce calcium sulfate and phosphoric acid. Production of phosphoric acid by this process, associated with an alumina plant of an economically viable size would be far in excess of Australian consumption capacity.

E. Concluding Remarks

The phosphate reserves on Christmas Island are limited in a more tangible way than other phosphate deposits. Relatively simple and inexpensive processes have been established on the Island to allow considerable utilization of P in the material previously considered as overburden. The current scheme consists of the following:

1) B-zone ores are effectively beneficiated by the wash-screen process. high temperature calcination holds promise of higher recoveries of a better product than with uncalcined material.

2) The C-zone ores, comprising the bulk of the Island's P deposits, are processed with simple low temperature calcination to produce a highly citrate-soluble phosphate for direct application. It is recognized that effectiveness of this material is acceptable only with certain crop and soil situations, and these considerations are recognized in the marketing of these products.

3) Chemical and thermal conversion techniques to produce alumina along with P fertilizers have been proven to be technically feasible but not economically viable in the present market.

LITERATURE CITED

Achorn, F. P., and D. G. Salladay. 1975. Production of monoammonium phosphate in a pipe-cross reactor. p. 196-211. In Proc. Annu. Meet. Trans. Fert. Ind. Round Table, 4-6 Nov. 1975, Washington, D.C.

Ando, J. 1961. Studies of calcined phosphates. Rep. no. D-20:74, Faculty of Eng., Chuo Univ., Bunkyo-ku, Tokyo, Japan.

Ando, J. 1965. Studies of chemical fertilizers. Nissin Shuppan Co., Ltd., Tokyo, Japan. 297 p.

Ando, J., and T. Akiyama. 1965. Formation of crystalline ferric ammonium phosphates during ammoniation of wet-process phosphoric acid. Kogyo Kagaku Zasshi. 68:1056-1061.

Ando, J., T. Akiyama, and M. Morita. 1968a. Magnesium ammonium phosphates, related salts, and their behavior in compound fertilizers. Bull. Chem. Soc. Japan. 41:1716-1723.

Ando, J., A. W. Frazier, and J. R. Lehr. 1968b. Insoluble compounds in ammonium polyphosphates made from wet-process phosphoric acid. J. Agric. Food Chem. 16:691-697.

Ando, J., and J. R. Lehr. 1967. Ammoniation reactions of superphosphates. J. Agric. Food Chem. 15:741-750.

Ando, J., and J. R. Lehr. 1968. Compounds in nitric phosphates. J. Agric. Food Chem. 16: 391-398.

Anonymous. 1974a. The politics of phosphate rock production. Eur. Chem. News. 18 Oct. 1974. p. 54-55.

Anonymous. 1974b. Fertilizers from Christmas Island "C-grade" phosphate. Phosphorus Potassium 74:29-31, 37.

Anonymous. 1975. Phospal—An available alternative source of P_2O_5. Phosphorus Potassium 73:34-36.

Atanasiu, N. 1971. A comparative study of the effect of water and citrate-soluble phosphatic fertilizers on yield and P uptake on tropical and subtropical soils. J. Ind. Soc. Soil Sci. 19: 119-127.

Chien, S. H. 1977. Thermodynamic considerations on the solubility of phosphate rock. Soil Sci. 123(2):117-121.

Chien, S. H., and C. A. Black. 1976. Free energy of formation of carbonate apatites in some phosphate rocks. Soil Sci. Soc. Am. J. 40:234-239.

Committee on Accessory Elements, NRC, National Academy of Sciences. 1979. Redistribution of accessory elements in mining and mineral processing, Part II: Uranium, phosphate, and alumina. NAS, Washington, D.C. p. 45-104.

Doak, B. W., P. J. Gallaher, J. Evans, and F. B. Muller. 1965. Low temperature calcination of C-grade phosphate from Christmas Island. N.Z. J. Agric. Res. 8:15-29.

Frazier, A. W., and R. G. Lee. 1972. Stabilizing liquid fertilizers with fluorine compounds. Fert. Solut. 16:32-43.

Frazier, A. W., and J. R. Lehr. 1967. Iron and aluminum compounds in commercial superphosphates. J. Agric. Food Chem. 15:348-349.

Frazier, A. W., J. R. Lehr, and E. F. Dillard. 1975. Chemical behavior of fluorine in the production of wet-process phosphoric acid. TVA Bull. Y-113. Tenn. Valley Auth., Muscle Shoals, Ala. 19 p.

Frazier, A. W., R. M. Scheib, and R. D. Thrasher. 1972. Clarification of ammonium polyphosphate fertilizer solutions. J. Agric. Food Chem. 20:138-145.

Frazier, A. W., J. P. Smith, and J. R. Lehr. 1966. Precipitated impurities in fertilizers prepared from wet-process phosphoric acid. J. Agric. Food Chem. 14:522-529.

Haseman, J. F., J. R. Lehr, and J. P. Smith. 1951. Mineralogical character of some iron and aluminum phosphates containing K or NH$_4$. Soil Sci. Soc. Am. Proc. 15:76-84.

Howard, P. F. 1979. Phosphate. Econ. Geol. 74:192-194.

Kreidler, E. R. 1967. Stoichiometry and crystal chemistry of apatite. Ph.D. Thesis. Penn. State Univ. Mic. no. 68-3549. Univ. Microfilm, Inc., Ann Arbor, Mich.

Lehr, J. R. 1967. Variations in composition of phosphate ores and related reactivity. p. 61-67. *In* Proc. 17th Annu. Meet. Fert. Ind. Round Table, Washington, D.C.

Lehr, J. R., E. H. Brown, A. W. Frazier, J. P. Smith, and R. D. Thrasher. 1967. Crystallographic properties of fertilizer compounds. Chem. Eng. Bull. no. 6, Tenn. Valley Auth., Muscle Shoals, Ala. 166 p.

Lehr, J. R., A. W. Frazier, and J. P. Smith. 1966. Precipitated impurities in wet-process phosphoric acid. J. Agric. Food Chem. 14:27-33.

Lehr, J. R., A. W. Frazier, and J. P. Smith. 1964. A new calcium aluminum phosphate, CaAlH(PO$_4$)$_2$•6H$_2$O. Soil Sci. Soc. Am. Proc. 28:38-39.

Lehr, J. R., and G. H. McClellan. 1972. A revised laboratory reactivity scale for evaluating phosphate rocks for direct applications. TVA Bull. Y-43. Tenn. Valley Auth., Muscle Shoals, Ala. 36 p.

Lehr, J. R., and G. H. McClellan. 1973. Phosphate rocks—important factors in their economic and technical evaluation. p. 194-242. *In* Proc. CENTO Symp. Mining and Beneficiation of Fertilizer Minerals, 19-24 Nov. 1973, Istanbul, Turkey. TVA Pub. no. X-235.

Lipsett, J., and C. H. Williams. 1970. Evaluation of Christmas Island C-grade phosphate as a fertilizer on some soils in southern New South Wales. Aust. J. Exp. Agric. Anim. Husband. 10:783-789.

Mason, M. G., and W. J. Cox. 1969. Calcined rock phosphate as fertilizer for pasture and cereal production in western Australia. Aust. J. Exp. Agric. Anim. Husband. 9:99-104.

McClellan, G. H., and J. R. Lehr. 1969. Crystal chemical investigation of natural apatites. Am. Mineral. 54:1374-1391.

Meline, R. S. 1974. Production of high-polyphosphate liquid fertilizer by the pipe reactor process. PTE/74/5, p. 1-23. *In* Proc. Tech. Conf. Int. Superphosphate and Compound Manufacturers Assoc., Prague, Czechoslovakia, 23-27 Sept. 1974. ISMA, Paris.

Muller, F. B. 1970. Agronomic use of calcined Christmas Island iron/aluminum phosphates. I. Field trials. N.Z. J. Agric. Res. 13:453-464.

Notholt, A. J. G. 1974. Phosphate rock: world production, trade, and resources. p. 104-120. *In* Proc. 1st Ind. Miner. Int. Congr., London, 8-9 July 1974. Metals Bull., Ltd., Surrey, England.

Notholt, A. J. G. 1979. The economic geology and development of igneous phosphate deposits in Europe and the USSR. Econ. Geol. 74:339-350.

Rothbaum, H. P., and A. J. Reeve. 1968. Recovery of alumina and phosphate from Christmas Island "C" phosphate. N.Z. J. Sci. 11:608-617.

Slack, A. V. 1968. Purification of wet-process acid. p. 637-686. Chapt. 8. *In* A. V. Slack (ed.) Phosphoric acid-part II. Marcel-Dekker, Inc., New York.

Taylor, A. W., E. L. Gurney, and W. L. Lindsay. 1960. An evaluation of some iron and aluminum phosphates as sources of phosphate for plants. Soil Sci. 90:25-31.

Waggaman, W. H. 1969. Phosphoric acid, phosphates, and phosphatic fertilizers. 2nd Ed. Hafner Publ. Co., New York. 683 p.

Werner, W. 1967. Monograph on the manufacture of Rhenania fertilizers and their properties and action. Verlay M. and H. Schaper, Hannover, W. Germany. p. 59-97.

White, M. S. 1971a. Superphosphate from Christmas Island phosphate rock. N.Z. J. Sci. 14:364-391.

White, M. S. 1971b. Calcination of Christmas Island phosphates. N.Z. J. Sci. 14:971-992.

White, M. S. 1974. The liquid phase of superphosphates. N.Z. J. Sci. 17:171-182.

Wolfkovich, S. I. 1967. Hydrothermal processing of naturally occurring phosphates. *In* Proc. no. 98 Fert. Soc. (London), 30 Nov. 1967.

Chapter 5

Sulfur Requirements of the Phosphate Fertilizer Industry

D. W. BIXBY

The Sulphur Institute
Washington, D.C.

I. INTRODUCTION

Sulfur and sulfuric acid have been closely associated with the phosphate fertilizer industry from its inception. They have grown up and matured together since that day back in 1840 when Justus von Liebig recommended that plant nutrients be added to the soil and demonstrated that the value of bones, as fertilizer, could be increased by treatment with sulfuric acid. About the same time, G. B. Lawes, working independently, applied the same treatment to bones and shortly thereafter to a Ca phosphate mineral which was discovered about that time. In his barn at Rothamsted, the first "superphosphate" factory was established in 1843.

The history of the industry is a fascinating story which has been told better and at greater length elsewhere. The association of P and S has continued, however, and we must now look at today's situation and then attempt to see what form it may take in the future.

In Fig. 1, world consumption of fertilizer P and consumption of S used to make P fertilizers are compared for the years 1950–1973 and estimated for 1980 and 1985. The close relationship between the use of the two materials is readily apparent. Similarly, the increasing ratio of S to P is illustrated, brought about by the emergence of ammonium phosphates which use wet process phosphoric acid as the sole source of P in their manufacture.

II. STATISTICS ON PAST AND PRESENT USE

A. Sulfur Use in Phosphate Fertilizers by Type

Most P fertilizers consume S at some stage of their manufacture, although there are only two products which may be said to be basic consumers—ordinary superphosphate (OSP) and wet process phosphoric acid (WPPA, Fig. 2).

Copyright 1980 © ASA-CSSA-SSSA, 677 South Segoe Road, Madison, WI 53711, USA.
The Role of Phosphorus in Agriculture.

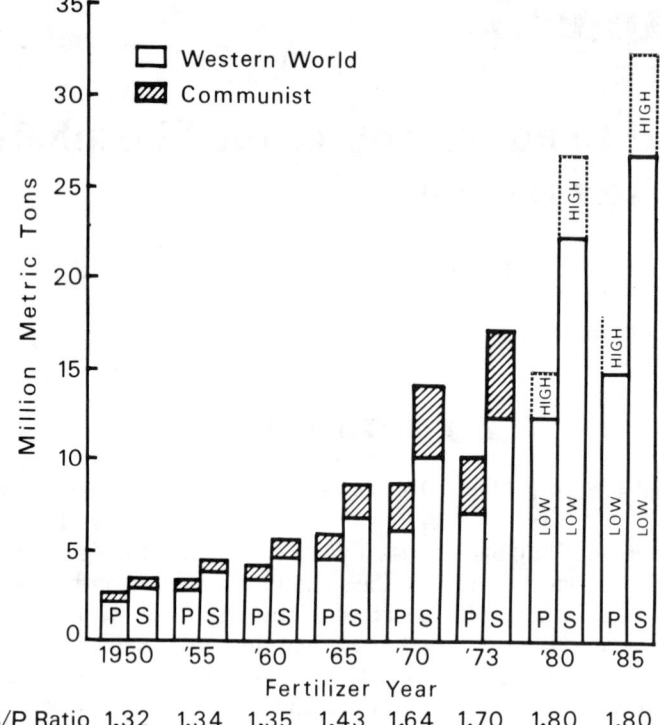

Fig. 1—World fertilizer P consumption and S consumption for phosphate fertilizer manufacturing 1950–1985.

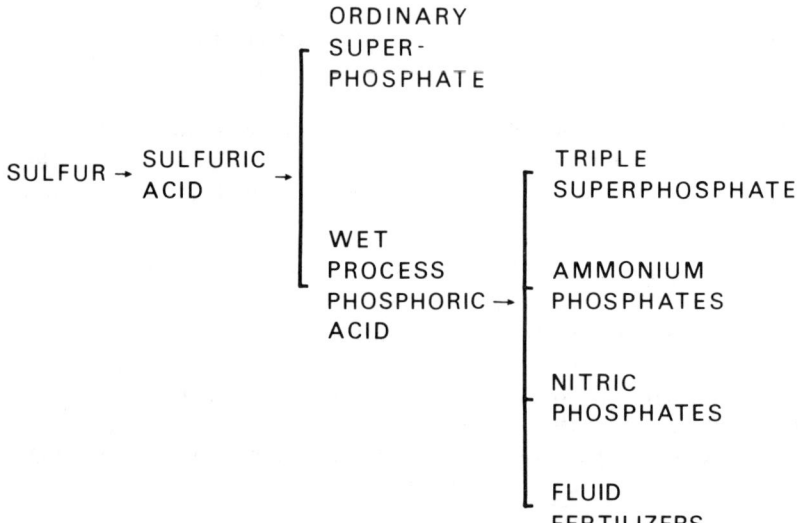

Fig. 2—Phosphate fertilizer derivatives of H_2SO_4.

1. ORDINARY SUPERPHOSPHATE

Use of P from OSP is declining as a percentage of world consumption (excluding ground rock), dropping from 62% in 1955 to 32% in 1973. However, in absolute terms, OSP is far from disappearing, having almost doubled in annual output over the same period, from 1.7 to 3.3 million metric tons of P. Production probably peaked in 1975 and is likely to begin a continuing decline. This product consumes between 1.4 and 1.5 metric tons of 3/metric ton of available P, depending on the grade of phosphate rock (PR) used, perhaps averaging 1.45 metric tons of S/metric ton of P, thus accounting for about 4.7 million metric tons/year of S use worldwide at this time.

2. WET PROCESS PHOSPHORIC ACID

Wet process phosphoric acid is now the major consumer of both PR and S in the fertilizer industry. In 1973, about 6.35 million metric tons of world P production (about 57%) was in this form, at least temporarily. Back in 1955, less than 10% of world P passed through the WPPA stage. The manufacture of WPPA consumes about 2.18 metric tons of S/metric ton of P. In 1973, this was equivalent to 13.8 million metric tons of S.

Enriched and concentrated (triple) superphosphates are secondary S consumers, deriving anywhere from 30 to 80% of the product P content from WPPA, the remainder coming from PR. Enriched superphosphate (11.8 to 13.1% P), employing a mixture of sulfuric and phosphoric acids in its manufacture, is of minor importance in the United States, and we have no data from other countries. Concentrated superphosphate (CSP) is assumed to be the predominant product, with WPPA acting as the only acidulant. If it is assumed that a theoretical 73% of the P content of this material is from WPPA, a balance is probably effected between the small amount of

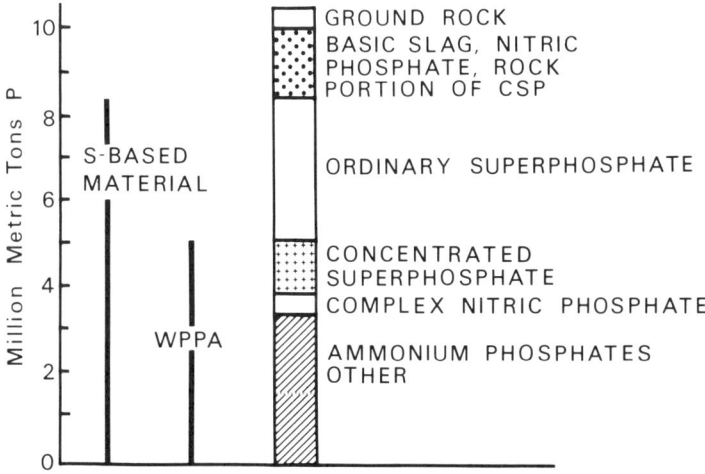

Fig. 3—World fertilizer P consumption by type 1972-1973.

enriched superphosphates and those of CSP plants which make a product containing more than 73% of its P content derived from WPPA. Thus, consumption may be assumed to average about 1.58 metric tons of S/metric ton of P in the finished material.

Concentrated superphosphate represented about 12% of world P production (excluding ground rock) in 1955 and 19.5% in 1973. Tonnage has, of course, greatly increased, from 0.35 to 1.83 million metric tons of P over the same period, equivalent to 0.55 to 2.9 million metric tons of S consumption. Little or no change is expected in world production of CSP over the next few years.

Figure 3 illustrates (i) the relationship among the principal P fertilizer materials (P basis), (ii) the portion which is S based, (iii) the share of the latter which is occupied by OSP and WPPA, and (iv) the relative share of the materials employing WPPA.

B. Forms of Sulfur Used for Phosphate Fertilizer Manufacture

It is estimated that about the equivalent of 18 million metric tons of S was consumed in the manufacture of P fertilizers in the 1973–1974 fertilizer season. This was slightly more than 39% of world S consumption. Because phosphates are the largest user of S (as H_2SO_4), it is to be expected that the forms of S consumed will reflect the mix used to make sulfuric acid generally. This, in turn, will vary according to the country and locality within the country.

In the western world, S consumption in 1974 was distributed among the various forms approximately as follows (million metric tons of S equivalent):

Brimstone	23.4
Smelter	5.5
Pyrites	5.0
Other nonbrimstone	5.5

Of the total brimstone usage, 14.2 million metric tons were recovered, 7.3 were obtained from Frasch operations, and a further 1.9 million metric tons of Frasch production were imported from Communist countries.

Within the Communist group of countries, S consumption was as follows: from brimstone approximately 5.2 million metric tons; from pyrites 5.0 million metric tons; and from other forms, including smelters, 2.3 million metric tons.

The fertilizer industry will continue to use the most economical source of an assured supply of H_2SO_4 available at any particular location, whether of their own manufacture or that of a reliable supplier. In the United States, the usual S sources are the Gulf Coast Frasch producers because of their proximity to the PR mining and processing areas. Here and elsewhere, when non-Frasch sources of S or H_2SO_4 can be obtained more economically, they will prevail in the marketplace. In Canada, for example, H_2SO_4 for the fertilizer industry is obtained from both the element as recovered from sour

SULFUR REQUIREMENTS OF THE P INDUSTRY

Table 1—Units of S used per unit of consumed P.

Calendar year	North America	Western Europe	Eastern Europe + USSR	Latin America	Africa	Asia	Communist Asia	Oceania	Western	Communist	World
1973	2.00	1.31	1.72	1.88	1.81	2.06	1.55	1.54	1.69	1.69	1.69
1974	2.01	1.34	1.76	1.89	1.83	2.07	1.55	1.54	1.74	1.73	1.74
1975	2.05	1.40	1.74	1.92	1.91	2.12	1.55	1.54	1.79	1.71	1.76
1976	2.06	1.42	1.73	2.04	2.04	2.11	1.55	1.54	1.82	1.70	1.79
1977	2.07	1.42	1.74	2.05	2.08	2.09	1.55	1.53	1.84	1.71	1.81
1978	2.07	1.42	1.75	2.05	2.09	2.09	1.55	1.53	1.84	1.72	1.81
1979	2.08	1.41	1.75	2.05	2.09	2.07	1.55	1.53	1.84	1.73	1.81
1980	2.08	1.41	1.76	2.05	2.09	2.07	1.55	1.53	1.85	1.73	1.82

gas, and byproduct acid from smelter operations. In Spain, pyrites are the major source. In the foreseeable future, pollution-abatement S is quite likely to assume a major role.

C. Effect of Geography on Sulfur Use

The preferred type of P fertilizer varies considerably with geography. Some types, such as diammonium phosphate (DAP) and monoammonium phosphate (MAP), require about 2.3 metric tons of S/metric ton of P in the product. Concentrated superphosphate requires about 1.5 metric tons of S/metric ton of P, OSP about 1.44 metric tons. Other forms of P fertilizers need no S, such as basic slag or those made exclusively with HCl or HNO_3. Thus, when estimating S needed for P fertilizer manufacture in a given region of the world, the probable product mix needs to be identified. This has been done in a TVA publication (Harre et al., 1974). Data from this publication, appropriately updated, were used to construct Table 1.

Account has been taken of such items as use of WPPA in certain nitric phosphate facilities, estimated operating rates of wet process and nitric process plants, estimated factory-to-farm losses, etc. The obtained factors now provide a basis for estimating S use for P fertilizers, either from actual P consumption data from previous years or for anticipated consumption in the near-term future. Note the relatively low factor for western Europe, due to the large basic slag use, and for Oceania, where OSP has been preferred because of its S content. In both New Zealand and Australia significant quantities of OSP fortified with elemental S are also produced.

D. Proportion of Sulfur Output Used in Phosphate Fertilizer Production

Today P fertilizers account for the major share of S consumption, over 48 and 43% of that used in the United States and the world, respectively (Table 2). In 1950, phosphates accounted for about a quarter of S usage.

Table 2—Proportion of S consumption used in P fertilizer production.

	1950	1960	1970	1974	1980
	million metric tons				
	United States				
S consumption	5.00	6.00	9.50	10.95	
P production	0.92	1.11	2.22	2.79	
S for P fertilizers	1.35	1.74	4.10	5.30	
% of total U.S. S consumption	27.0%	29.0%	43.2%	48.4%	
	World				
Estimated S consumption	13.50	23.00	42.00	46.40	53.70
P production	2.50	4.20	8.90	10.60	13.10
S for P fertilizers	3.50	6.20	13.90	18.20	23.40
% of world S consumption	25.9%	27.0%	33.1%	39.2%	43.6%

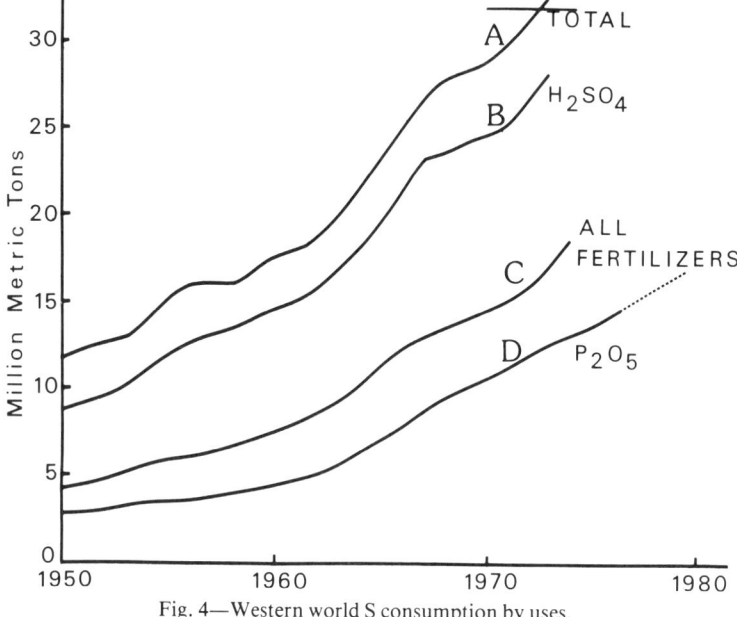

Fig. 4—Western world S consumption by uses.

In Fig. 4, the same data are shown graphically (curves A and D). Curve B shows the high proportion of S which is converted to H_2SO_4. Curve C indicates S usage for all fertilizers, with the greater part of the excess over curve D going into ammonium sulfate (AS). There are hundreds of other uses for S, the most important of which are shown in Fig. 5. The proportion of S consumed as H_2SO_4 and in nonacid forms is also illustrated in this figure.

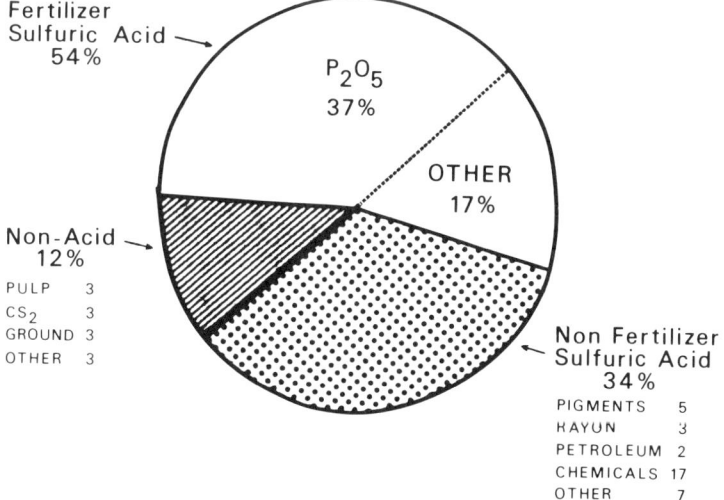

Fig. 5—Uses of S—western world 1974. Total consumption 33.9 million metric tons.

Table 3—Forecast of phosphate fertilizer consumption.

Region	Fertilizer year							% annual growth 1976/80
	1973/74	1974/75	1975/76	1976/77	1977/78	1978/79	1979/80	
	million metric tons							
North America	2.2	2.0	2.2	2.3	2.3	2.4	2.5	3.5
Western Europe	2.6	2.2	2.1	2.2	2.3	2.4	2.6	3.7
Eastern Europe	1.2	1.3	1.4	1.5	1.6	1.7	1.9	7.5
USSR†	1.2	1.2	1.3	1.4	1.5	1.6	1.7	6.0
Latin America	0.6	0.6	0.6	0.7	0.7	0.8	0.9	8.2
Africa	0.3	0.3	0.4	0.4	0.4	0.5	0.5	8.0
Oceania	0.7	0.5	0.5	0.5	0.5	0.6	0.6	5.2
Asia (non-comm.)	1.1	1.1	1.1	1.1	1.2	1.3	1.4	13.1
Comm. Asia	0.7	0.7	0.8	0.9	1.0	1.0	1.1	9.0
Total western World	7.5	6.6	6.9	7.2	7.6	8.0	8.4	5.2
Total Communist	3.1	3.3	.35	.38	4.1	4.4	.47	7.3
Total World	10.6	9.9	10.4	11.0	11.7	12.3	13.1	6.5

† Ground rock removed from P consumption figure for USSR (avg. approximately 0.4 million metric tons P), but left in for all others (approximately 0.13 million metric tons of P for rest of world).
Note: Totals may not check due to rounding off of regional data.

III. PROJECTIONS OF FUTURE NEEDS

A. Future Phosphate Consumption Forecasts

Table 3 shows a forecast of fertilizer P consumption for various regions of the world (Sulphur Institute, 1975, unpublished data). It is not as optimistic as some, nor as pessimistic as others. In these days of economic and political uncertainties, only sheer coincidence is likely to cause a forecast to turn out as predicted. The decline in P use in western Europe which began in 1973–1974 continued into the 1976 fertilizer year before showing signs of recovery. However, previous consumption patterns in this region may have been so seriously disrupted by recent high prices for P goods (although now declining significantly) that there may never be a resumption of historical growth patterns. The recent resumption of subsidies in Oceania may cause a marked reversal of the severe decline in P consumption experienced there this past year.

In any event, the forecast was made primarily to arrive at an approximation of the tonnage of S which will be used to manufacture these P fertilizer materials. Let us now take a brief look at the status of S-using processes vs. alternative processes to see if the competitive position of S is likely to continue.

B. Sulfur Processes vs. Alternative Processes

The basic purpose of the various P fertilizer processes is to render the P content of PR, generally locked up as fluorapatite, into a soluble form. This may be done by chemical or thermal means.

Chemically, PR may be treated with one of several acids to achieve P solubility. For various reasons, H_2SO_4 has assumed by far the most important position, followed by HNO_3 and HCl. Phosphoric acid itself may be used, but is considered here to be a secondary product.

1. SULFURIC ACID

Sulfuric acid has numerous advantages. It utilizes a relatively inexpensive raw material which is in plentiful supply. It need not be a byproduct of some other chemical process. When H_2SO_4 is reacted with PR, the byproduct Ca sulfate is insoluble and can be easily filtered off. This permits separation of the main product, phosphoric acid, and a still higher concentration of P with further treatment. Should the Ca sulfate be left in the product, as in OSP, it contributes plant-nutrient S. Fertilizers made with H_2SO_4 contain most of the P in a water-soluble form, an important consideration in many parts of the world. Sale of coproducts is not necessary for the viability of sulfuric acid-consuming phosphate plants. Flexibility in making several fertilizer grades is inherent to plants using H_2SO_4.

2. HYDROCHLORIC ACID

A process using HCl has been pioneered by Israeli Mining Industries, this material being a byproduct of potash processing operations in Israel. Three other small plants have operated in Japan, Brazil, and the United States. The cost of purchased HCl is such that it is impractical to consider, except under special circumstances. In addition, difficult technical and economic problems are encountered. When HCl is reacted with PR, phosphoric acid and $CaCl_2$ are formed. The latter is soluble, remaining in solution. The acid must be extracted with a water-immiscible organic solvent. The acid-solvent combination is contacted with water, with the acid entering the aqueous phase and the solvent separated and recycled. Corrosion is severe, and capital and maintenance costs are correspondingly high. About 4.6 tons of 100% HCl are required per ton of P. The relatively dilute concentration of industrial-grade HCl (31-35%) makes freight costs an important consideration in plant location.

3. NITRIC ACID

Nitric acid may also be used to dissolve PR, but separation of the resulting Ca nitrate to the degree necessary to give a commercially acceptable phosphoric acid is apparently not readily accomplished. However, nitric acid is widely used to make nitric phosphates, the acidulate being ammoniated to give a solid product containing part or all of the Ca as a citrate-soluble phosphate.

Water-soluble P can be enhanced by removal of part of the Ca nitrate by crystallization and separation in a filter or centrifuge (Odda process). Water solubility (and presumably cost) is increased to the degree that unwanted Ca nitrate is removed. Calcium nitrate can be sold as such, or converted by ammoniation into calcium-ammonium nitrate. Ammonia and CO_2 may also be reacted with Ca nitrate to produce ammonium nitrate (AN) and Ca carbonate.

Another process involves the cycling of ammonum sulfate (AS) to remove Ca. Here AS is added to the acid-rock acidulate, Ca sulfate is precipitated and filtered off, later to be reacted with NH_3 and CO_2 yielding recyclable AS and byproduct Ca carbonate. Complete water solubility of the P is achieved.

Nitric phosphates (NP) are of minor importance in the United States and Canada; however, they do occupy a prominent place in western Europe, accounting for about 0.9 million metric tons of P production. In eastern Europe and the USSR they account for perhaps 0.5 million metric tons.

The comparison of NP processes and those using H_2SO_4 is too complex to deal with adequately here. Also, such comparisons are best made in the context of a particular situation, which may favor H_2SO_4 or HNO_3, or may find these acids equally competitive. Generally, it is felt that S-based processes offer more flexibility, With N/P ratios being adjustable according to needs, whereas NP processes are constrained to N/P ratios of about 1:0.22 ($N/P_2O_5 = 2/1$). Over all, sulfuric-based processes produce more concentrated material with resulting savings in distribution costs. Water solubility is more economically attainable using sulfuric acid processes.

4. THERMAL PROCESSES

There have been several processes involving heating a mixture of PR and other reactants to obtain P in a form available to crops. They fall into three general classes: (i) those based on chemical reactions between rock and an alkali salt or Mg silicate; (ii) those based on the removal of fluorine by volatilization; and (iii) the electric furnace process based upon high-temperature acidulation with Si in the presence of carbon as a reducing agent.

The first two classes yield products of about 11% P mostly in the citrate-soluble form, and are of minor importance today. The electric furnace process, since it is the only method of making elemental P which, in turn, can be converted to highly purified phosphoric acid, remains important in the chemical industry, although representing only a minor factor in the fertilizer sector. Indeed, new methods for purification of WPPA are now enabling this source of P to make inroads into the detergent market, once solely supplied by electric furnace acid.

A few years ago, the opinion was expressed that a combination of low-cost power and increasing costs for S would soon move electric furnace phosphoric acid to a point where it could be considered competitive with WPPA for fertilizer use. A study comparing the economics of the two methods was completed in 1969 by The Sulphur Institute (Bixby et al., 1969). It was concluded that even under the most favorable circumstances for electric furnaces (use of low-grade, low-cost rock, electric power available at 3 mills, long haul of elemental P product), WPPA would be competitive at S costs in the $50 to $60/metric ton range. Today, the prospect of seeing such low power costs is dim indeed, and manpower costs for the labor-intensive electric furnace plants have escalated dramatically. Consequently, the relative economic position of furnace acid and WPPA has shifted still more in favor of the latter. It is unlikely that any further serious consideration will be given to furnace acid except for specialty fertilizer use in the foreseeable future.

C. Forecast of Sulfur Consumption for Phosphates

Using the product mix mentioned in Section II. C., and the S/P ratios developed for various regions in Table 1, we compiled Table 4. Note that this shows the amount of S expected to be consumed to produce the phosphates which will be applied to crops. In some years, such as the 1974–1975 season, considerable material may go into inventory for use in future years. This will also consume S, but there is no effective way of monitoring it. Also, there is no adequate way of accounting for the large amount of spent

Table 4—Forecast of S use in phosphate fertilizer manufacture.

Region	Fertilizer year					
	1974/75	1975/76	1976/77	1977/78	1978/79	1979/80
	million metric tons S					
North America	4.0	4.5	4.7	4.9	5.0	5.2
Western Europe	3.0	3.0	3.2	3.3	3.5	3.6
Eastern Europe	2.3	2.4	2.6	2.8	3.0	3.3
USSR	2.2	2.3	2.4	2.6	2.8	2.9
Latin America	1.3	1.3	1.4	1.5	1.7	1.8
Africa	0.6	0.7	0.8	0.9	1.0	1.0
Oceania	0.8	0.8	0.8	0.8	0.9	0.9
Asia (non-Comm.)	2.1	2.3	2.4	2.5	2.7	2.8
Comm. Asia	1.2	1.3	1.4	1.5	1.6	1.8
Total western World	11.6	12.5	13.2	13.9	14.6	15.4
Total Communist	5.6	6.0	6.4	6.9	7.4	8.0
Total World	17.3	18.4	19.6	20.8	22.1	23.4

Note: Totals may not check due to rounding off of regional data.

H_2SO_4 used by the fertilizer industry, acid which is often listed as consuming S for its original purpose, and listed again when reused in fertilizers. Granulation acid is another item not readily identifiable. Many granulation processes may consume as much as 100 kg of acid/metric ton of product to provide heat and fluidity to achieve proper granulation. The apparently diminishing role of granular complete goods in the face of bulk blends and/or fluids may be reducing the importance of this S use.

IV. FUTURE SUPPLY PICTURE

A. Sulfur Resources

The world's total S resources are vast, but only a fraction is now minable or recoverable at competitive prices. Previous estimates of these resources have differed widely for several reasons, including uncertainty about their geological nature and lack of information on ore deposits in remote parts of the world.

Estimates of identified, hypothetical, and speculative resources of S in various types of deposits in the United States, Canada, and Mexico were compiled and published in 1973 (Bodenlos, 1972). Table 5 is a condensed version of the summary of the U.S. situation. Table 6 is a similar compilation dealing with the rest of the world.

At the time, Bodenlos assumed that "recoverable" could be mined or recovered at less than about $25/metric ton; "paramarginal," between $25 and $35; and "submarginal," higher than $35/metric ton. Of course, a considerable inflation factor would now have to be introduced.

In summary, the world position on S resources is adequate, with apparent identified reserves, excluding coal and gypsum, totaling nearly 2 billion metric tons, and will sustain rates of production much higher than those at present.

Table 5—Sulfur resources of the United States.

Type of S resource	Identified resources		Hypothetical resources	Speculativa resources	Total
	Recoverable	Marginal			
	million metric tons				
1. Elemental S deposits in evaporites	150	50	100	150	450
2. H_2S in sour natural gas	10	5	185	--	200
3. Organic S in petroleum	10	245	1,000	--	1,255
4. Pyrite deposits	10	90	20	20	140
5. Elemental S in volcanic rocks	--	30	--	--	30
6. S in metallic sulfides	20	80	100	200	400
7. Gypsum	--	7,200	1,800	--	9,000
8. Organic S in tar sands	--	10	--	--	10
9. S in coal	--	21,400	19,600	--	41,000
10. S in oil shale	--	†	†	--	81,000
Total	200	29,100	22,805	370	133,485

† Amount not estimated.

Table 6—Sulfur resources outside of the United States.

Type of S resource	Identified	Probable
	million metric tons	
1. Elemental S deposits in evaporites	380	?
2. H_2S in sour natural gas	140	700
3. Organic S in petroleum	10	330
4. Pyrite deposits	550	?
5. Elemental S in volcanic rocks	100	100
6. S in metallic sulfides	260	140+
7. Gypsum	?	Vast
8. Organic S in tar sands	40†	1,800+†
9. S in coal	?	180,000
10. S in oil shale	?	200,000
11. S in seawater	Unlimited	
Total	1,480+	?

† Canada only.

A word about the cost/price relationship with respect to reserves of S might be in order. It is evident that the huge reserves of S-bearing raw material throughout the world preclude any long-term shortage of S or H_2SO_4. The situation may be likened to an inverted pyramid (Fig. 6) with the size of reserves and cost of production increasing from the bottom tip (brimstone) to the broad upper levels (anhydrite and gypsum), the as yet unexploited resources in coal, and the ultimate resources of the world's oceans. Normally, if lower-cost sources are unable to supply the entire demand, producers can move up the pyramid as higher prices make the development of other sources economically feasible. However, pollution-abatement measures, and consequent S recovery, may alter the "normal" situation by bringing S with a partially subsidized cost of production into the marketplace, in effect bypassing some of the steps in the pyramid.

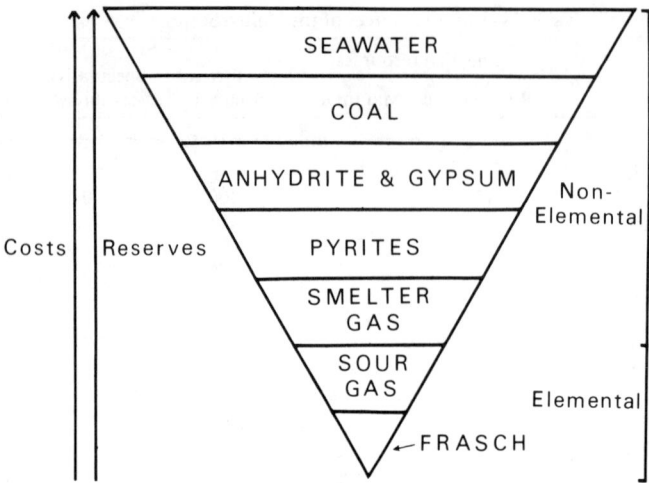

Fig. 6—The "inverted sulfur pyramid."

B. Voluntary Sulfur Supply Sources

Although natural S resources are ample, as has been indicated, evaluation of S supply requires the consideration of other factors which may be economic, logistic, or even political.

Voluntary S comes from two main sources: Frasch [mined, elemental S (brimestone)] and pyrites (Fe sulfide). In the United States, Frasch S production is expected to begin a gradual decline over the next several years, continuing until the deposits are naturally and/or economically depleted. The decline is a net reduction, with some mines being phased out, but with recent or reopened mines making up part of the difference. Barring new discoveries, however, the decade following the turn of the century should find the United States relying principally on byproduct S for its supplies of domestic origin. Pyrites are relatively unimportant in the U.S. Sulfur-supply picture and are expected to remain so.

In the rest of the world, some expansion in Frasch-process S production is expected in Poland, Russia, and Iraq. Net pyrite usage will increase

Table 7—Forecast of voluntary S supply.

	1974	1977	1980	1985
		million metric tons		
U.S.				
Frasch	8.0	7.5	7.3	5.6
Rest of world				
Frasch	10.0	12.5	14.5	18.5
Pyrites	11.0	12.0	12.0	16.0
World				
Frasch	18.0	20.0	21.8	24.1
Pyrites	11.0	12.0	12.0	16.0
Total voluntary	29.0	32.0	33.8	40.1

because of large-scale projects using indigenous material in Spain, Turkey, and the USSR. This will more than compensate for the precipitous drop in pyrite consumption by countries formerly importing S in this form, but which have now changed to brimstone. The near-term outlook for voluntary S supply is summarized in Table 7 (B. F. Newton, 1976. Forecast of world sulfur supply sources. 5th Phosphate-sulfur Symposium, John's Island, Fla., 15-16 Jan.).

C. Involuntary, Byproduct, and Pollution-abatement Sulfur

Pollution-abatement S in a form useful to the P fertilizer industry comes or will come from three principal sources: (i) industries which process fossil fuels, (ii) industries which consume fossil fuels, and (iii) metal smelting and refining operations.

The *potential* S available from these sources has been calculated for the United States and is shown in Fig. 7 (Jimeson et al., 1974). The tonnage

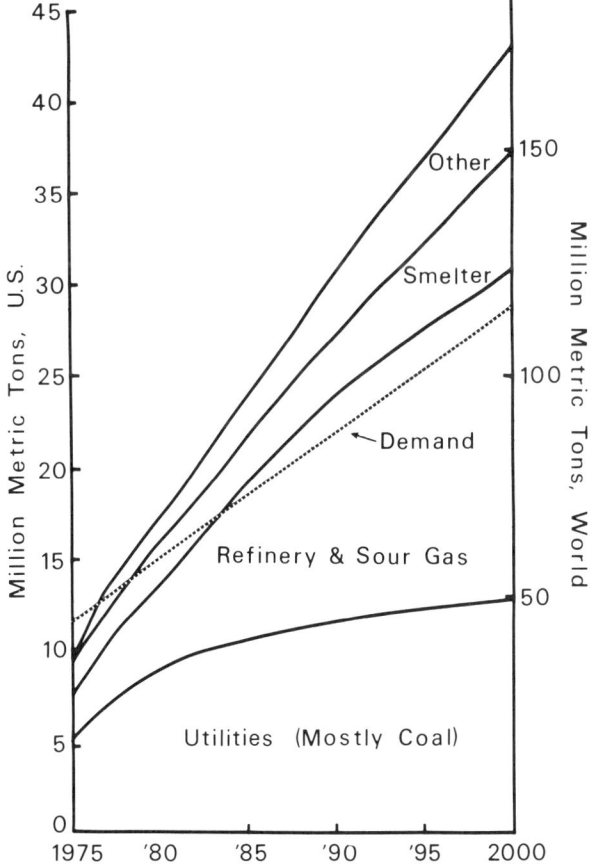

Fig. 7—Potential S available from involuntary sources.

represented is shown to almost equal U.S. demand at the time of writing, and to exceed estimated demand in about 2 years. Obviously, this has not happened, but ultimately there will be a significant impact on the S market, and in turn, upon the fertilizer industry. The paramount question is how much of the S will actually be recovered as a useful product in comparison with that ending up as an essentially useless throwaway product. An effort will be made to quantify this impact following a brief review of the individual sources.

1. INDUSTRIES WHICH PROCESS FOSSIL FUELS (AND/OR CHEMICAL RAW MATERIALS)

The following operations remove S from fuels prior to combustion.

a. Natural Gas—Gas with a significant S content (in the form of hydrogen sulfide, for example) is termed "sour." Because of their toxicity and corrosive properties, S compounds must be removed before the gas can be marketed as a fuel or used as a chemical raw material. Typically, hydrogen sulfide is removed by adsorption by various chemical agents which are regenerated. The hydrogen sulfide is then converted to elemental S.

Sulfur recovery from sour gas is not only necessary, but can be economically attractive when the price is sufficient to cover costs of manufacture, transport, and sale. However, its *production* represents a variable which is almost exclusively a function of the demand for natural gas. Thus, *capacity* can be more apparent than real.

Prior to 1950, only a small amount of S came from sour gas. At present, the main western world countries producing S from gas are Canada, France, and the United States. In 1974, they produced 6.8, 1.8, and 1.2 million metric tons, respectively, from this source, accounting for over 19% of world production of S in all forms. Of the Canadian production, about 2 million metric tons were put into inventory due to various logistical problems. These problems are being dealt with and the stockpile, which has in 1977, reached a level of about 20 million metric tons, is expected to be gradually reduced over the next several years.

Reserves of S in identified resources of U.S. sour gas are relatively small, perhaps 10 million metric tons; an additional 65 million metric tons may occur in recoverable undiscovered sources. In Canada, another 140 million metric tons may be available in identified resources, and a similar amount in those classed as hypothetical or speculative. The largest resources of such S are the extensive natural gas fields of the Middle East, which may contain about 500 million metric tons.

b. Oil—Crude oil from certain areas of the world (e.g., Venezuela and the Persian Gulf) contains significant amounts of S and is normally referred to as "sour" crude. Such oil must be desulfurized if it is to be used in combustion processes in areas where air pollution regulations are in effect. In the United States, about 1.4 million metric tons of S were recovered from refinery gases in 1974, about 12% of U.S. production. Another 2 million metric tons may have been produced elsewhere in the western world.

The potential quantities of S which may become available from refining operations are subject to considerable uncertainty. Between 1974 and 1980, announced new projects, worldwide, total nearly 6.8 million metric tons of S annually. However, historical capacity vs. production data indicates a capacity utilization ratio (CUR) of only 33%. Thus, forecast production from this source is on the order of 2.2 million additional metric tons annually by 1980. The low CUR isdue to several factors—some capacity may be standby only, some will be unused when sweet crude is available, and some may never be built.

In the United States and Canada, S in petroleum yet to be recovered amounts to perhaps 15 million metric tons, with an additional 25 million metric tons possible in marginal resources. Venezuealan oil may contribute about 30 million metric tons. However, by far the greatest part of presently known S in petroleum is contained in the vast resources of the Middle East, which perhaps amounts to 300 million metric tons.

c. **Coal**—Most coal contains S, varying from 0 to 7% by weight. The processing of coal in order to remove S before the coal is burned can go in any or all of several directions, and it is uncertain which will prevail. One pathway, the technology for which is well established, is ordinary beneficiation, consisting of crushing, separating, washing, and drying. Beneficiation removes somewhat less than 50% of the S contained in most coals, being capable of taking out the inorganic (pyritic) S only, leaving the remaining organic S untouched. Since the impure pyrites will not be a useful form of S in the foreseeable future, this source will not be considered further. However, it should be noted that for some time to come, beneficiation combined with lime flue gas desulfurization is likely to be the method of choice for reduction of stack gas emissions, particularly for steam-generating power plants.

A promising new process for producing a clean solid fuel from coal has recently been announced (Anon., 1975). Called the Battelle Hydrothermal Coal Process, it involves grinding and beneficiation followed by hydrothermal removal of S, fuel leachant separation and regeneration, and fuel drying. It is said to remove 99% of the pyritic S and 70% of the organic S, recovering it in the elemental form.

It seems certain that synthetic gas and oil produced by coal conversion (gasification and liquefaction) processes will eventually play an important role in our energy economy, supplementing or replacing natural gas and petroleum for home heating, automotive fuels, and petrochemical feed stocks. Sulfur will be an important byproduct of these operations.

About a dozen processes for coal gasification have been discussed in the literature, and a similar number for coal liquefaction, many of which produce an oil-gas combination. About 18 commercial and demonstration projects for gasification have been announced. Several liquefaction projects are in the pilot stage. In December 1977, construction has begun on only one demonstration plant, with many delays having been caused by lack of governmental policy, uncertainty about how the high sales prices of the gas or liquid will be handled, problems in obtaining capital, etc. Nevertheless, it is reasonable to assume that these problems will eventually be resolved in the coal-rich, oil-poor countries, given the pressures induced by a diminishing supply and/or increasing cost of natural oil and gas.

The various coal conversion processes differ in the method of producing heat, composition of products, configuration of reactor vessels, and mechanics of operation. However, the S in the coal appears primarily as hydrogen sulfide in the product gas. It will be recovered in the elemental form, using a process similar to the one now employed on sour natural gas.

The amount of S which will become available to industry from coal conversion processes before 1985 will be negligible. At present, the period following 1985 cannot be forecast with any degree of confidence because of the many complicating factors. All one can say is that a very large amount of S is likely to be forthcoming from coal conversion at some time following 1990. At least 200 *billion* metric tons of S are contained in coal deposits, worldwide. If conversion of a significant portion of that coal takes place in the future, then a S supply from this source is correspondingly assured.

2. INDUSTRIES WHICH CONSUME FOSSIL FUELS

The primary reference here is to operations which will be making use of flue gas desulfurization (FGD) facilities in order to comply with sulfur dioxide (SO_2) emission restrictions. The principal such industry will be electric utilities. Although there will be exceptions, other industries and nonindustrial fuel consumers, lacking the volume to justify a recovery effort, will use fuels low enough in S to ensure compliance with emission standards in their particular area.

In the United States, the near-term, commercially available options for FGD consist almost entirely of processes of the throwaway type, in which the flue gases are reacted with lime or limestone to produce a precipitate or sludge, which is then discarded. When combined with coal beneficiation, lime/limestone FGD is presently the most economical choice for utilities attempting to comply with emission standards. Of the FGD systems already installed or planned in the United States, over 70% of the plants representing 82% of capacity use lime or limestone. The S content of these sludges is of no present value to the chemical industry and will not be discussed further. It is mentioned only to correct any impression that *most* utilities will soon be entering the S marketplace.

However, recovery methods are also receiving attention. In these processes, material used to react with SO_2 in the flue gases can be regenerated for recycling. During the regeneration step, the adsorbed SO_2 is liberated at a high concentration, then treated to make H_2SO_4 or elemental S. Among these processes, the following are the most highly developed.

The *magnesium oxide process* (Chemico) uses a magnesia slurry as the gas scrubbing medium. The resulting magnesium sulfite/sulfate slurry is treated to regenerate magnesia for recycle, and SO_2 for oxidation to H_2SO_4. One test-demonstration program has been completed and a second plant has begun initial operation.

The *Wellman-Lord process* (Davy Powergas) uses sodium sulfite solution as the scrubbing medium. In a separate reactor, SO_2 is driven off and the solution recycled. A side stream rich in sodium sulfate must be discarded. Several units have operated on oil-fired boilers in Japan and three are under construction for coal-fired boilers in the United States. The SO_2 product may be used for H_2SO_4 or reduced to elemental S.

These processes and several others are in various stages of testing in U.S. coal-fired plants and may offer alternative FGD methods, but no one regenerable system dominates the field. In time, it is likely that a few basic systems will gradually prove to be superior.

Although byproduct H_2SO_4 and S from regenerable FGD systems will have commercial value which can help offset operational costs, it is unlikely that such systems will ever compete on a profit basis with the conventional industries for these products. Nevertheless, it must be expected that the problems of sludge disposal and the lure of salable byproducts will provide a sufficiently powerful incentive for the ongoing development of regenerable FGD processes, and that ultimately the supply of S available to the P fertilizer industry will be greatly augmented from this source. Although the complexities of marketing abatement FGD S and/or H_2SO_4 from widely scattered, small-scale producing points are formidable, as well as being inappropriate for discussion in this publication, it may be said that ample amounts are potentially available, should circumstances so dictate.

3. METAL SMELTING AND REFINING OPERATIONS

The practice of recovering the SO_2 content of nonferrous metal smelter waste gases is well established. It probably originated as a practical step to meet needs for H_2SO_4 and has now become a necessity to avoid atmospheric pollution in the vicinity of the smelters. The most abundant source of smelter-gas SO_2 has been Zn blende, which yields a steady supply, in contrast to the lower and fluctuating SO_2 content of gases from Pb or Cu smelters. However, convenience is no longer as important as pollution abatement, and complete S recovery in the future from the smelting and refining of all three metals is virtually certain.

At present, nonferrous smelters account for some 6% of United States S production—0.65 million metric tons. Canadian production is about the same. In the western world, S recovered as acid from smelter operations contributes about 5 million metric tons of S equivalent annually, or 13% of the estimated available supply. If the same percentage is assumed to be contributed by smelters in the Communist countries, then about 6.6 million metric tons of S are currently coming from this source.

Of course, several qualifications have to be made about smelter acid, when considered as a future source of supply for the fertilizer industry. Among them are distance from the market, increasing captive use of byproduct acid in hydrometallurgical operations, and fluctuations in production brought about by metals market conditions. However, a great deal of acid is bound to become available at a favorable price for fertilizer use. It may be added that some S in the elemental form is produced now and more may be expected as processes improve. This may be especially helpful to plants in situations too remote to ship acid economically.

What, then, is the outlook for smelter S in the future? Full utilization of present capacity and that under construction will contribute a great deal of additional S. However, in the United States, less than 25% of new smelter acid capacity announced since 1970 is being utilized. When conditions are favorable for full operation, smelter acid production could easily double.

As for the *potential* S or S equivalent available from the world's nonferrous metal sulfides, perhaps 20 million metric tons would have been recoverable before present statutory requirements for cleaner air. In addition, much of the subeconomic, identified resources, some 80 million metric tons, will eventually be recovered regardless of cost because of antipollution regulations.

Today, involuntary S (all of which may also be thought of in one sense or another as pollution-abatement and/or byproduct S) accounts for over 29% of United States S production and about 43% of world production. In 1985, the share of involuntary S is expected to increase to about 59 and 48% for the United States and the world, respectively. Our forecast at this writing for the United States and world supply of involuntary S is shown in Table 8.

Some comments about the makeup of this table may be helpful. In Canada, production of S recovered from sour natural gas has probably peaked, and will gradually decline, even with new discoveries. Logistical problems relating to the supply of Canadian S are discussed in another section of this paper. Growth in both western Europe and the United States is principally refinery byproduct. Most of the new tonnage elsewhere in the western world will be from major S recovery projects in the Middle East, notably Iran and Saudi Arabia. In the Communist countries, the principal new source is a large natural gas recovery operation in the USSR which has recently come on stream.

Table 8—Forecast of involuntary S supply.

	1974	1977	1980	1985
	million metric tons			
U.S.				
Petroleum	2.7	3.4	4.2	4.9
Oil shale	--	--	--	0.1
Utility stacks	--	--	0.1	0.5
Coal	--	--	0.05	0.5
Smelter	0.6	1.4	1.7	2.6
Rest of World				
Byproduct	14.3	15.6	17.7	20.0
Others (mostly smelter)	4.4	4.6	5.3	7.4
World				
Byproduct	17.0	19.0	22.0	26.0
Others	5.0	6.0	7.0	10.0
Total involuntary	22.0	25.0	29.0	36.0

Reasons for reduction of the present involuntary S estimate over prior estimates are: cutbacks in refinery operating levels; greater use of low-S crudes causing deferment of announced programs in Japan, western Europe, and the United States; present dominance of throwaway processes for treating utility stack gases; and delays in implementation of coal, shale, and oil-sand processing schemes primarily related to burgeoning financial problems.

D. Factors Affecting the Supply of Sulfur

Factors which may affect the supply of S or of any commodity are almost innumerable, since "everything affects everything else." Nevertheless, certain factors, while not peculiar to S, are worthy of mention. One example is the cost and/or availability of energy. In phosphate rock mining, the costs of electricity to operate a drag line and fuel to dry the rock have their effects on the cost of the finished product. Similarly, to an even greater degree, the cost and/or availability of fuel to heat water for Frasch-mining S is crucial to the process. The fuel of choice has been natural gas, forwhich the average consumption has been 215 m^3/metric ton of mined S, contributing \$7.6/metric ton to the production cost if gas is selling for \$1/28 m^3 (1,000 cu. ft.). This energy is largely returned in fertilizer processing as steam or chemical reaction energy, but still must be paid for. The detrimental effect of the unavailability of gas in the first place is obvious. Energy also may have an indirect effect on S availability as a result of government policies either encouraging or discouraging the export of natural gas or the use of fuel.

Numerous effects of logistics on S supply could be cited. The existence of smelter acid at points far removed from any noncaptive market is one. In some cases, such acid is being neutralized and thrown away as gypsum. Massive stockpiles of byproduct S are located at points with inadequate or no rail service in Canada and Iraq. Canadian S stockpiles will probably reach the level of 21 million metric tons by the end of 1978. There, lower production levels and improved rail service will probably bring about reduction of this inventory beginning around 1980.

Factors affecting S supply could be the subject of an extensive paper in itself. We have only touched upon them because the primary interest here is the relation of S to P fertilizers.

E. Forecast of Sulfur Supply for Phosphates

The sources of S for phosphates are, of course, the same as those for all industry. Sulfur will remain one of the basic building blocks of the chemical industry. Elemental S is used in the sulfite paper process; carbon disulfide is used by the textile industry and in carbon tetrachloride manufacture; and H_2SO_4, the workhorse of industry, is used for pigments, oil refining, alcohols, and in countless other ways (Fig. 5). Will the demands for these uses withdraw needed S from the manufacture of P fertilizers? The answer is no.

Some industrial applications of S phase out as processes change (steel pickling, paint pigments), but are offset by new uses (caprolactam, road construction). As a result, the overall industrial demand growth rate has been quite constant, closely following a country's GNP or a region's contribution to the GNP. At present, industrial growth rates are depressed, and although they may again increase in a year or two, they are not expected to do so at as rapid a pace as those of the fertilizer industry. The cause of the slump in the phosphate sector of the fertilizer industry—namely, high costs—has had political overtones which it is hoped will be resolved. If the price of PR does not come down, growth of P consumption in severely affected areas may also be less than historical rates for some time.

There should be ample S to take care of expected P consumption in the foreseeable future. Based on a projection of 13.1 million metric tons of P produced in 1980, about 23.4 million metric tons of S will be required in manufacturing P fertilizers (Table 2). Even if all of the projected (announced as "definite") world phosphoric acid plants should be operating at a 100% *capacity* utlization rate (12.9 million metric tons of P), the S requirement for all phosphates (WPPA and OSP) in 1980 would be only 33 to 35 million metric tons. The nonphosphate industrial demand for S in 1980 is expected to be 23 million metric tons, for a total of 56 to 58 million metric tons—well within the S supply forecast of 63 million metric tons in 1980. Thus, although regional situations may differ markedly, a sufficient supply of S for the P fertilizer industry seems assured, as is a continuing close association between S and fertilizer P.

LITERATURE CITED

Anonymous. 1975. Hydrothermal process cleans up coal. Chem. Eng. News. 53(27):24–25.
Bixby, D. W., H. L. Fike, and J. Platou. 1969. Phosphoric acid—electrothermal vs. wet process. An economic evaluation. Tech. Bull. 15. Sulphur Institute, Washington, D.C.
Bodenlos, A. J. 1972. Sulfur. U.S. Geol. Survey Prof. Paper 820.
Harre, E. A., O. W. Livingston, and J. T. Shields. 1974. World fertilizer market review and outlook. Bull. Y-70. Tennessee Valley Authority, Muscle Shoals, Ala. 68 p.
Jimeson, R. M., T. S. Needels, and L. W. Richardson. 1974. Fossil fuels and their environmental impact. Proc. Symp. energy and environmental quality, Chicago, Ill., May 10. Illinois Institute of Technology, Chicago.

Chapter 6

Phosphoric Acid Technology

NORMAN ROBINSON

Fisons, Ltd., Fertilizer Division
Levington Research Station
Levington, Ipswich, Suffolk, England

I. INTRODUCTION

This review considers the basics of "wet" process phosphoric acid technology; of necessity certain important areas of development have been omitted. No mention has been made, for example, of developments in materials of construction or in equipment design. In addition, processes not yet in commercial operation or those involving acids other than sulfuric acid have also been excluded on the grounds of limited application.

The main chemical reactions involved in the "wet" process are well understood, as are the important rock impurities which cause processing problems such as corrosion, scaling, sludge deposition, and foaming. Less clear are the factors which control crystal habit and crystal growth and thereby influence filtration rates. Also, although the zones of stability of the crystal hydrates of calcium sulfate have been well investigated, the effects of rock impurities on the kinetics of transformation need to be clarified. The criteria which influence the precipitation of complex compounds also need to be identified as these compounds can adversely affect phosphorus (P) recovery efficiencies, sludge deposition, and filtration rates.

The need to obtain this knowledge becomes more urgent when one considers the gradual depletion of world stocks of high-grade phosphates. This trend will increase the pressure on phosphoric acid plant suppliers to design plants which match the processing requirements of individual rocks or a range of rocks, depending on whether the plant is sited at or away from the rock mine. There will also be an increasing need to provide beneficiation processes, which not only consider P enrichment, but also the processing implications when the rock is used for phosphoric acid manufacture.

The dihydrate process is well established as the main "wet" process route. There will still be a place for hemihydrate processes where local circumstances provide economic advantages. The comparatively low P recovery efficiency given by this type of process, however, is a disadvantage which is likely to prevent wide-scale application.

Copyright 1980 © ASA-CSSA-SSSA, 677 South Segoe Road, Madison, WI 53711, USA.
The Role of Phosphorus in Agriculture.

The main challenge to the dihydrate process will be the two-stage hemihydrate-dihydrate processes which offer ideal performance of very high P recovery efficiencies and direct production of concentrated acid. This type of process also eases effluent problems in that a much purer gypsum is produced. As effluent regulations become tighter; this advantage could become more significant. Commercial application of the two-stage process is as yet limited and more evidence is needed on performance, reliability, and flexibility.

II. PRINCIPLES OF MANUFACTURE BY THE WET PROCESS

A. Chemistry

The so-called "wet" processes for producing phosphoric acid are all based on the replacement of Ca from fluorapatites (FA) by an acid stronger than phosphoric acid itself. Due to the relative ease of separation of calcium sulfate from phosphoric acid solutions the commercial application of the "wet" process has been almost entirely limited to the use of sulfuric acid.

The principal reaction occurring is the decomposition of fluorapatite:

$$Ca_{10}(PO_4)_6F_2 + 10H_2SO_4 + 10xH_2O \rightarrow 10CaSO_4 xH_2O + 6H_3PO_4 + 2HF$$

where $x = 0, 0.5$ to 0.7, or 2.0.

Fluorapatite is the common constituent of phosphate rock (PR) which is mined in every continent of the world. The reaction between PR and acids is essentially a surface reaction in which the rate is controlled by reaction temperature, hydrogen ion concentration, diffusion through liquid films at the surface, and the specific surface area of the rock. In the presence of sulfate ions a solid layer of calcium sulfate may also form round the rock grains, inhibiting diffusion and thereby reducing the reaction rate. The extent to which this occurs depends on the choice of sulfate concentration at the point in the reaction system where rock is added. If this concentration is high for the reactivity of the rock and reactor residence time in use, then inhibition can occur and this will be self-generating, leading to poor reaction control and high P losses. Figure 1 illustrates rock particles which have been seriously inhibited in this manner. Normally, in a well-operated phosphoric acid process, P losses arising from the coating of rock grains by calcium sulfate are low. Effective control of sulfate concentrations, however, is necessary to maintain this position, the degree of control required being dependent on the specific surface area of the rock processed and reactor residence times.

Apart from P, Ca, and F, PR contains a wide range of impurities which more often than not affect in some way the chemistry of the process and the performance of a phosphoric acid plant.

For example, calcium carbonate is often present, sometimes combined with FA as francolite, but more frequently in association with the mined

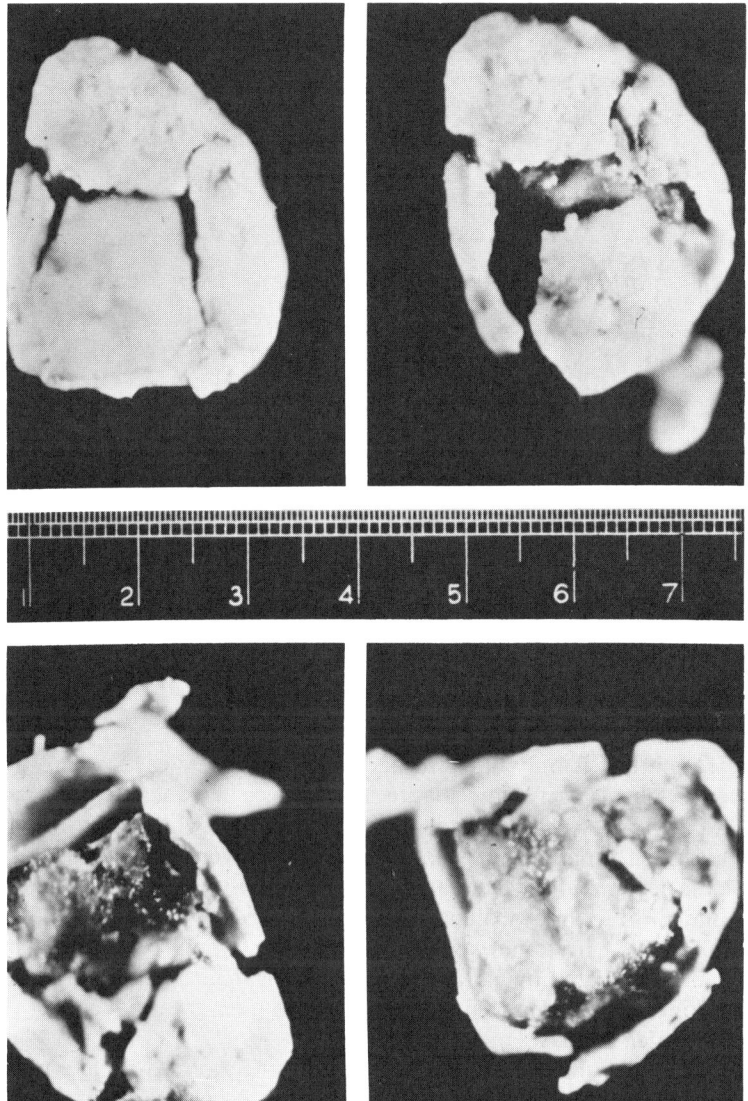

Fig. 1—"Inhibited" rock grains.

rock in the form of calcite. From either source the calcium carbonate is decomposed, precipitating calcium sulfate and liberating carbon dioxide.

$$CaCO_3 + H_2SO_4 + xH_2O \rightarrow CaSO_4 \cdot xH_2O + CO_2 + H_2O$$

where $x = 0$, 0.5 to 0.7, or 2.0.

Evolution of carbon dioxide can increase the reaction rate by causing disintegration of rock grains and providing an increase in specific surface

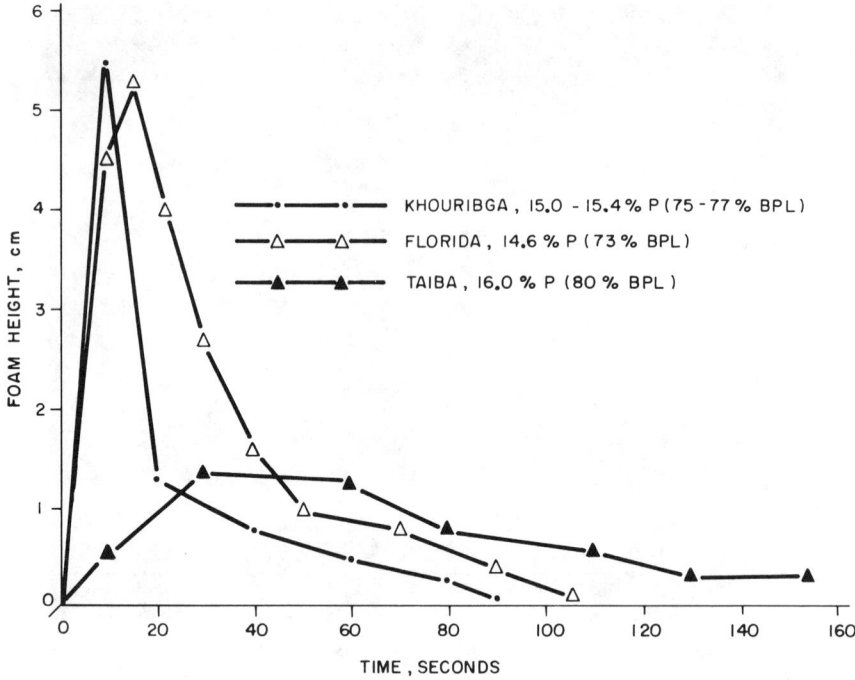

Fig. 2—Foaming properties of phosphate rocks.

area. This effect is only likely to be significant in the cases where the calcium carbonate is part of the apatite structure, as in francolite.

An adverse effect of carbon dioxide evolution, which is far more common, is the formation of foam which can prevent the wetting of rock particles, affect reaction rate and control, and also cause loss of P due to entrainment of foam by exhaust air, which is used for venting or cooling the reaction slurry. In some cases it is possible to control foaming by the turbulence generated by agitation; if this is insufficient, the only solution, apart from calcining or changing the rock source, is to make use of a defoamer, which increases operating costs.

Carbon dioxide evolution is not the only criterion which determines the extent of foam. Phosphate rocks usually contain organic matter and this can reduce the surface tension of the reaction slurry and stabilize foam caused by gas evolution. Figure 2 illustrates the point by showing the foaming characteristics of Khouribga, Florida, and Taiba PR. Normally, Khouribga PR can be processed without the need of a defoamer; currently, however, 72% BPL grade does require a defoamer. Florida PR, although having a lower carbon dioxide content, requires a defoamer because of the stabilizing influence of organic matter. Taiba PR has a low carbon dioxide content and can normally be processed without the need for a defoamer. This rock, however, contains organic matter, which imparts a high stability to the foam, and because of this a bad foaming problem can arise if Taiba is processed with any rock containing a significant amount of carbon dioxide.

Silicon is usually present in the mined rock, either as quartz or as silicates in the form of clays. Hydrofluoric acid, produced from the decomposition of the fluorapatite, reacts with the silica and the rate of this reaction depends on the type of silica present. Silicates, for example, readily react, whereas quartz reacts extremely slowly.

Under the conditions of phosphoric acid manufacture the usual reactions involving silica are:

$$SiO_2 + 4HF \rightarrow SiF_4 + 2H_2O \qquad (a)$$

$$3SiF_4 + 2H_2O \rightarrow 2H_2SiF_6 + SiO_2 \qquad (b)$$

$$H_2SiF_6 \rightarrow SiF_4 + 2HF \qquad (c)$$

In those processes producing phosphoric acid containing 13% P (30% P_2O_5) in the reactor, the main F reaction which occurs is the formation of fluosilicic acid by reaction (b). Very little F is evolved by way of reactions (a) or (c). Silica is also produced by reaction (b) and this is available for further reaction with fluoride ions or is removed with the precipitated gypsum by filtration. Most of the F which is evolved in the gaseous form occurs by reaction (c) during the concentration of the 13% P reaction acid to the 22% P (50% P_2O_5) acid in an evaporation step. The evolved F is normally recovered as fluosilicic acid by absorption in water using a scrubber with recycling facilities.

In those processes which produce phosphoric acid containing 22% P acid directly, there is no evaporation step and F is evolved from the reaction system mainly by way of reactions (a) and (c). Again, the F evolved is normally recovered as fluorsilicic acid by absorption in water.

The more common processes producing 13% P acid will contain most of the F content of the rock within the reaction system, whereas those processes producing 22% P acid in the reaction system will contain much lower F levels due to F evolution. Fluorine in the reaction acid is present as fluosilicates, fluorides, or as fluorsilicic acid. Precipitation of fluosilicates can cause processing problems, the most common being that due to scaling by Na and Na/K salts. The extent of this problem, which appears most frequently during the filtration and evaporation stages, is dependent on the concentrations of alkali metals in the rock processed.

Cationic impurities can form complex compounds with anions such as phosphate, silicate, and fluoride. The most common cationic impurities are Fe, Al, Mg, Na, and K, and these elements are mainly responsible for the formation of compounds that slowly precipitate from phosphoric acid forming sludges which complicate acid storage and shipping. Other compounds may also be produced which hinder the filtration of calcium sulfate during phosphoric acid manufacture. Magnesium fluosilicate is a well-known example of this and creates problems for rocks which have a relatively high Mg content relative to Al and alkali metals. Fortunately, this does not apply to many in commercial use at the present time; this situation could well change in the future as more low-grade rocks are exploited.

Fig. 3—Octahedral crystals (40 μm).

Another example which is more common is the formation of complexes of the type $Ca_3Al_3(RE)SO_4F_{13} \cdot 10H_2O$ where "RE" could be a mixture of rare earths and yttrium. This compound is one of an isomorphous series where Al^{3+} can be replaced by other trivalent ions or by Ca^{2+} and Si^{4+} jointly, and F^- by OH^-. Compounds of this type have been reported by several workers (Yermilova et al., 1960; Lehr et al., 1966; Coates & Woodward, 1966) and in each case the crystals could be classified in the cubic system and are octahedral in habit.

Typical crystals are shown in Fig. 3. Figure 4 illustrates a filter cloth which has been blinded by octahedral crystals causing loss of filtration capacity.

The technology of "wet" process phosphoric acid manufacture is dominated by the chemistry of calcium sulfate. This is not surprising when one considers that this is a byproduct whose crystalline form controls filtration capacity, and more often than not plant capacity, and also determines the P recovery efficiency of the process.

Calcium sulfate can precipitate in different hydrated forms during the decomposition of fluorapatite. Those hydrates having commercial im-

Fig. 4—Blinded filter cloths (400 μm).

portance are gypsum ($CaSO_4 \cdot 2H_2O$) and hemihydrate ($CaSO_4 \cdot \frac{1}{2}H_2O$). Anhydrite ($CaSO_4$), although of minor significance in the past, is not a basis for any process operating at the present time.

Various phase diagrams relating the stability of hydrated forms of calcium sulfate with reaction temperatures and acid P concentrations have been published. These range from calculated thermodynamic equilibrium curves to those determined experimentally. Such curves only provide guideli lines, as crystals produced during the manufacture of phosphoric acid are not at equilibrium thermodynamically.

Figure 5 illustrates typical curves which have been published (Dahlgren, 1960). Region I represents a thermodynamic equilibrium state and has no practical significance in "wet" process phosphoric acid manufacture. Normally, dihydrate processes operate within region II and hemihydrate processes within region III; in both these regions anhydrite is the thermodynamic equilibrium state. The time required, however, to transform dihydrate to anhydrite is very long compared to the residence time used in phosphoric acid plant reaction vessels, and in this situation dihydrate for all practical purposes is the stable phase.

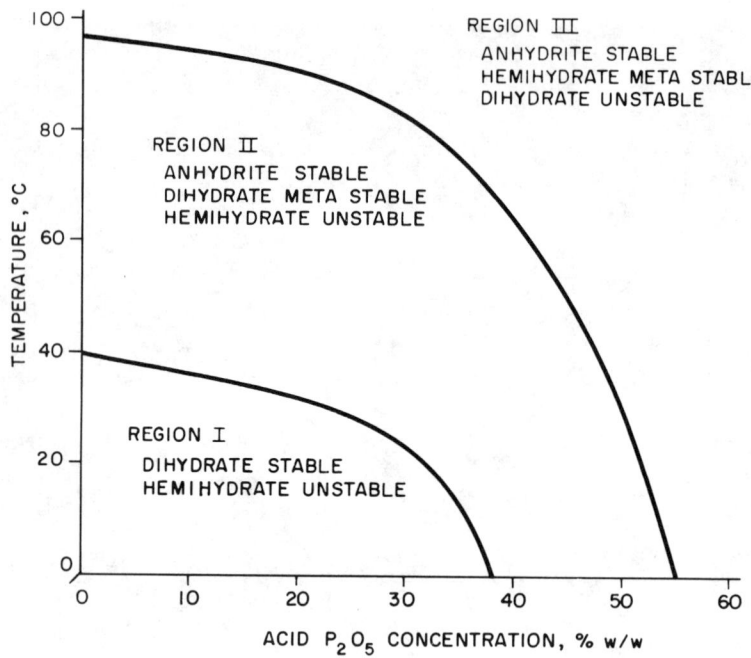

Fig. 5—Stability of calcium sulphate hydrates.

For dihydrate processes, the more significant transition, and one which must be avoided, is:

$$CaSO_4 \cdot 2H_2O \rightleftharpoons CaSO_4 \cdot \tfrac{1}{2}H_2O + 1\tfrac{1}{2}H_2O.$$

The tendency for this transition to proceed in the direction of hemihydrate is influenced by the water content of the reaction acid, reaction temperature, impurities present in the rock processed, and calcium sulfate supersaturation levels.

For the hemihydrate process the most important transition is:

$$CaSO_4 \cdot \tfrac{1}{2}H_2O \rightarrow CaSO_4 + \tfrac{1}{2}H_2O.$$

Again, the rate of transformation to anhydrite is low and for the operating conditions used for hemihydrate processes, hemihydrate is in reality the stable phase. The transition in this case, however, is irreversible and therefore must be avoided. In practice the only time when this transition may occur is during a period of long shutdown of the reaction system. This can be avoided by controlling the reaction slurry temperature during the shutdown period. The degree of control necessary is influenced by the length of the shutdown and on the nature of impurities present in the rock being processed.

Commercial dihydrate and hemihydrate processes employ reaction conditions which ensure that the desired crystal phase of calcium sulfate is precipitated. Formation of the "wrong" phase only occurs as a result of maloperation, due, for example, to instrument failure or operator error.

The most important aspects of crystallization are the size, distribution, and shape of the crystals produced, as these affect both filtration rate and P recovery efficiency. Although this is probably the most critical of all the unit operations involved in phosphoric acid manufacture, it is the least understood. This is due to the numerous factors which influence these properties, many of which interact. For example, most of the operating conditions used in the process have some influence when processing a given rock. Impurities in PR also influence crystal habit and the stability of the crystal hydrates.

B. Unit Operations

It is not possible within the scope of this review to describe in depth the unit operations involved in the manufacture of phosphoric acid. Some explanation, however, is required of the basic operations involved before specific commercial processes can be adequately described.

All modern commercial phosphoric acid processes employ common unit operations; any differences which exist arise mainly from design features and the requirements of operating within the dihydrate or hemihydrate crystal modes.

These operations involve the following considerations.

1. ROCK HANDLING

The operation of modern "wet" process phosphoric acid plants involves the continuous handling of large tonnages of PR which are shipped to the plant site in road or rail cars, ocean-going ships, or even canal barges, depending on plant location.

Off-loading the rock usually involves the use of traveling grab cranes coupled to a belt conveyor/elevator system, which either transports the rock to store, or alternatively to bottom-dumping rail cars in the case of plants which are distantly located.

Storage capacity must have sufficient reserve to ensure continuity of rock supply to the processing plant. This must take into account plant capacity, frequency and size of rock shipments, and the number of rock types to be processed.

Phosphate rock delivered from the mine has a variable particle size distribution and more often than not a grinding step is required before the rock is suitable for processing. The need for grinding depends on the reactivity of the rock, the design of the plant, and the beneficiation process required to upgrade the rock. For example, the rock may already be of sufficient fineness as a result of grinding carried out at the time to meet the requirement for flotation. On a phosphoric acid plant, various types of mills may be used. For instance, rod or ball mills are extensively used on plants located in

North and South America, while in Europe ring and roller mills are most common. In both cases air classification is usually employed.

The feed of rock to the plant must be accurately metered if satisfactory control of reaction operating conditions is to be maintained. Belt weighers incorporating load cells are the usual means of achieving this, but slurry systems which are metered by density and flux meters are also being used to avoid the drying of rock following wet grinding and/or beneficiation.

2. REACTION

The reaction system of a phosphoric acid plant normally consists of a series of agitated vessels which are connected by underflow or overflow ports. An external or submerged pump is used to recycle reaction slurry from the end to the beginning of the reaction train. Adequate mixing is necessary in the vessels where rock and sulfuric acid are added. Both flat- and pitch-bladed turbines are used, and in some cases the slurry flow generated by agitation is used to promote slurry recycle.

The main reactions occurring are exothermic, and in order to control the reaction temperature, it is necessary to transfer heat and mass by evaporation of water, either by air impingment cooling or by use of flash evaporation.

Crystallization of calcium sulfate dictates the residence time required in the reaction system, which is usually within the range of 4 to 6 hours. Close control of the many operating parameters is necessary to avoid loss of production capacity.

3. FILTRATION

The slurry from the reaction system is fed via a holding tank to a filter, where calcium sulfate is separated and washed countercurrently to complete the recovery of phosphoric acid. Many types of filters are in use, including belt, tilting-plan, and table filters which make use of a scroll for the discharge of filter cake. Some plants employ filters of different types for the separation of both hemihydrate and dihydrate. The size of filter usually determines the capacity of a given plant, as overloading will eventually be uneconomic due to incurred P losses.

4. RECRYSTALLIZATION

This operation applies only to hemihydrate processes in which recrystallization to the dihydrate form is used before disposal of calcium sulfate, or as a means of increasing P recovery efficiencies or the final purity of the calcium sulfate byproduct. Agitated vessels are the usual means of promoting recrystallization.

5. EVAPORATION

Normally acid containing 17 to 22% P (40 to 50% P_2O_5) is required for downstream processes such as those producing ammonium phosphates or superphosphates.

In dihydrate processes it is necessary to concentrate the 11 to 13% P product acid by evaporation. The evaporators used are normally forced circulation units incorporating vertical heat exchangers fitted with graphite tubes. High acid flows are maintained to minimize temperature differences and scaling of equipment. Axial flow pumps are used extensively for this purpose.

Some plants, which operate in the hemihydrate region do not require an evaporation stage, as it is possible in these cases to produce a 17 to 22% P acid directly within the reaction system.

6. FLUORINE RECOVERY

Fluorine recovery is often part of a phosphoric acid plant, either combined with the evaporation stage of a dihydrate process or with the flash cooler of a hemihydrate process. In both cases the recovered product is fluosilicic acid produced by absorption of SiF_4 and HF offgases in water.

Void spray towers are normally used which may incorporate several absorption stages in series using a recycle system. Absorption units are also used to remove F compounds from exhaust air to comply with statutory effluent requirements. In these cases it is usually uneconomic to recover F because of the quantities involved.

7. DISPOSAL OF CALCIUM SULFATE

Calcium sulfate is produced in all phosphoric acid processes and in the majority of cases this is discharged as a waste product. Disposal methods normally used involve the formation of a water slurry of calcium sulfate which is either pumped into tidal rivers or into lagoons. In the latter case calcium sulfate is used to build retaining walls in order to form a series of settling ponds which allows water to be collected for reuse within the process.

A minority of commercial phosphoric acid plants utilize calcium sulfate as a byproduct, either as a retarder in the manufacture of cement or to produce plaster block or board. In these cases, calcium sulfate filter cake is usually discharged from the phosphoric acid plant directly, rather than in the form of an aqueous slurry. The availability of natural gypsum, the capital cost involved, the marketing position regarding byproducts and local effluent restrictions are all important factors which influence the viability of utilizing calcium sulfate from "wet" process phosphoric acid plants. These are favorable in some areas, notably in Japan, France, Germany, South Africa, and Brazil.

8. SULFURIC ACID

The sulfuric acid needed for phosphoric acid manufacture is produced by the Contact process using sulfur, pyrites, or byproduct gases from smelting operations. A large quantity of high-pressure steam is also produced by this process and is used for generating power and in the evaporation step for concentrating phosphoric acid.

III. WORLDWIDE TRENDS IN PHOSPHORIC ACID PROCESSES

A. Basic Landmarks in Development

Phosphoric acid was first produced in Germany (Schucht, 1926) as long ago as the late 19th century. The reaction was carried out batchwise using 16 to 40% sulfuric acid, and the calcium sulfate was separated on filter presses. An acid containing only 3.5 to 4.4% P was produced and the process was uneconomical, giving gypsum which had poor filtration and washing properties. The Dorr Company then incorporated thickeners which were arranged in series for countercurrent washing of the gypsum, and the process was finally made continuous, giving an acid concentration of 9 to 10% P. Rotary vacuum filters were also included in addition to thickeners for removing and washing the gypsum.

Subsequently, mixing was made more efficient by using mechanical agitators rather than air sparges and the process finally became known as the *Dorr Weak Acid Process*. During the period 1917-1929 more than 25 plants of this type were in operation (West, 1930). Additional refinements included the grinding of rock with phosphoric acid in pebble mills (Heckenbleikner, 1928; Spicer, 1926) and the use of classifiers. This process made a significant advance in these early years; it is now obsolete, being superseded by more efficient processes capable of producing phosphoric acid of higher concentration.

One of the most notable landmarks in the development of phosphoric acid processes was the result of work carried out by the Swiss company, Kunstdunger Patent Verwertungs A. G. (K.P.V.) which was a combination of the Dorr Company and F. Liljenroth of Germany. In the 1920's, K.P.V. pioneered the idea of recycling reaction slurry before calcium sulfate is separated. This procedure increases crystal growth as a result of returning crystals through precipitation zones, largely by controlling supersaturation through avoiding high, local sulfate concentrations, and minimizing temperature fluctuations. This principle (Weber et al., 1936) was used by Dorr for the design of a phosphoric acid plant built in 1932 for Cominco at Trail. Continuous filters were also used and these allowed gypsum to be separated and washed countercurrently, which improved the P recovery efficiency. The use of slurry recycle and a continuous filter also allowed the acid strength to be raised from 9 to 10% P to 12 to 14% P. This concentration range still applies to modern dihydrate processes which produce the bulk of the world's supply of phosphoric acid for fertilizer use. Because of the higher acid concentration produced, the process became known as the *Dorr Strong Acid Process* and was widely adopted in the period 1930-1950.

Although there are still some plants based on the Dorr Strong Acid Process in operation, this pocess has largely been superseded by other dihydrate process designs which have evolved since 1950. One result of this has been a dramatic increase in plant capacity from 65 metric tons of P/day, which was the original capacity of the Trail plant, to capacities which have now reached 550 metric tons of P/day. Design changes have also occurred, mainly connected with the reaction vessel, filter, agitators, and cooling equipment; these are considered further in Section III-B when specific processes are covered.

The search for processes producing phosphoric acid of 17% P concentration was first carried out as early as 1927 (Lehrecke, 1935) and preceded the development of the dihydrate process.

Nordengren et al. (1955) first attempted to produce directly a 17 to 22% P acid by carrying out the reaction under pressure and at elevated temperatures. These trials were abandoned, due to corrosion and filtration problems, in favor of evaluating a process operated at atmospheric pressure and lower temperatures, i.e., 100°C. Although this process never reached commercial application, the basic studies carried out by Nordengren and his coworkers established the conditions under which the anhydrite, hemihydrate, and dihydrate forms of calcium sulfate are produced in phosphoric acid slurries.

A rather different approach to the problem was taken by the K.P.V. group under Liljenroth. Using a recrystallization technique, they produced a concentrated phosphoric acid (17% P) in addition to coarse gypsum crystals of excellent filterability.

The process (Larsson, 1933, 1939) consisted of first producing the hemihydrate, then filtering this off from 17% P acid, followed by recrystallization of gypsum, refiltering, and completion of the washing. In these early years problems were experienced due to hydration of hemihydrate, unsatisfactory materials of construction, inadequte filter designs, and unavailability of suitable filter cloths. Such complications were responsible for this process being abandoned in favor of the dihydrate route which involved slurry recirculation.

During the mid-1950's further work by Nordengren et al. led to the development of the first phosphoric acid plant using an anhydrite process, initially in the form of a prototype (7 metric tons of P/day) at Vercelli, Italy, and later as a small production plant (15 metric tons of P/day) at Landskrona, Sweden. During the period 1956–1963 (Hakansson, 1965), an 18% P acid was produced without evaporation in the Landskrona plant, but mainly while operating as a hemihydrate process. The anhydrite route was discontinued after a few years due to technical problems, one being the need to feed PR intermittently. In 1963 the plant was rebuilt to operate as a dihydrate process in order to increase product capacity.

The development of two-stage hemihydrate-dihrate processes producing phosphoric acid having concentrations of 13 to 15% P was extensively studied by several Japanese companies, notably Nissan, NKK, and Mitsubishi during the 1950's and 1960's. In these processes, hemihydrate is first produced which is recrystallized to gypsum without filtration. Phosphorus recovery efficiencies are higher than those obtained by a dihydrate process, but acid concentrations are similar.

A novel process was developed in the late 1960's by the two companies Central Glass of Japan and Societe de Prayon of Belgium. This is a two-stage process and is unique in that dihydrate is first produced, and then, after separation by centrifuge, is converted to hemihydrate, and finally filtered and washed. Acid concentrations of 14 to 17% P are claimed by this process at P recovery efficiencies of 99%. The first commercial plant by this process was a 90-metric tons of P/day unit built in 1970 for Societe Chimique Des Charbonages at Douvrin, France. Early technical problems were experienced when processing Khouribga PR and this particular plant was eventually modified to allow operation by the dihydrate route. Other plants using this process are in commercial operation processing different phosphate rocks.

The first plants to be built of commercial size which allow concentrated acids of 45 to 50% P_2O_5 to be produced without evaporation are plants based on the single-stage hemihydrate and two-stage hemihydrate-dihydrate (HDH) processes developed by Fisons. The first single-stage hemihydrate plant was installed and commissioned for Windmill Holland NV in 1970. A two-stage HDH plant which gives both a concentrated acid and high P_2O_5 recovery was installed in latter half of 1975 for RMHK Trepca in Yugoslavia, and completion of commissioning is expected early in 1976.

B. Modern Commercial Processes

Since the early 1950's a large number of phosphoric acid processes have been developed; only those which are commercially operated and have distinctive features will be considered in any detail.

Phosphoric acid processes are normally divided into groups according to the degree of hydration of the precipitated calcium sulfate. The dihydrate route was the first process to be commercially developed and these plants are still by far the most common. Plants which produce acid of similar concentration, but with the precipitation of hemihydrate and gypsum, are also used to some extent, but mainly in areas where the utilization of calcium sulfate is significant. The more recent hemihydrate or hemihydrate-dihydrate processes which produce concentrated acids without evaporation are in the early years of application, but plants of this type are likely to become more widely used, especially as results from the new plants become available. The main commercial processes may be classified as shown on Table 1.

The differences in design between the various dihydrate processes given are located mainly within the reaction system.

For all these processes the simplified flow diagram shown in Fig. 6 is usually common.

During the 1950's and 1960's the design of the reaction system changed from a series of agitated tanks arranged for cascade flow, which was the original arrangement used in the Dorr Strong Acid Process, to systems based on single tanks. The new design differed in the way the single tank was partitioned and in its method of operation.

The Saint-Gobain/Rhône-Progil Company was the first and remains the only company to offer a single reactor which has no internal partitions.

Table 1—Classification of west phosphoric acid processes.

Process type	Design features	Acid P concentration, % wt/wt	Process examples
Dihydrate	Multitank reaction system Low slurry recycle ratios	13	Dorr
	Single-tank reaction system Agitation induced slurry recycle	13	St. Gobain/Rhone Progil
	Single compartmented reaction tank High slurry recycle ratio	13	Fisons DH
	Annular reaction tank Agitation induced slurry recycle	13	Dorr
	Single reactor with draught tube Very high slurry recycle ratio	13	Swenson Isothermal
	Two-reactor loop Very high slurry recycle ratio	13	Kellog-Lopker
Hemihydrate	Multitank reaction system Very low slurry recycle ratio	20–22	Fisons HH
	Multitank reaction system Hemihydrate is not separated	13	Nissan NKK Mitsubishi
Hemihydrate-dihydrate	Multitank reaction system Hemihydrate is separated but not washed	19	Singmaster and Breyer/Heurty
	Multitank reaction system Hemihydrate is separated and washed	20–22	Fisons HDH
Dihydrate-hemihydrate	Multitank reaction system Dihydrate is separated but not washed	14–17	Central Prayon

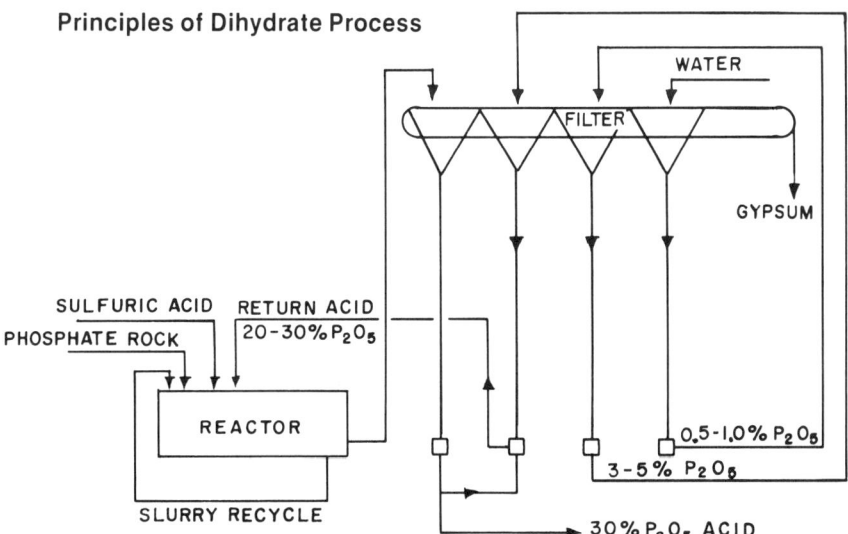

Fig. 6—Principles of dihydrate process.

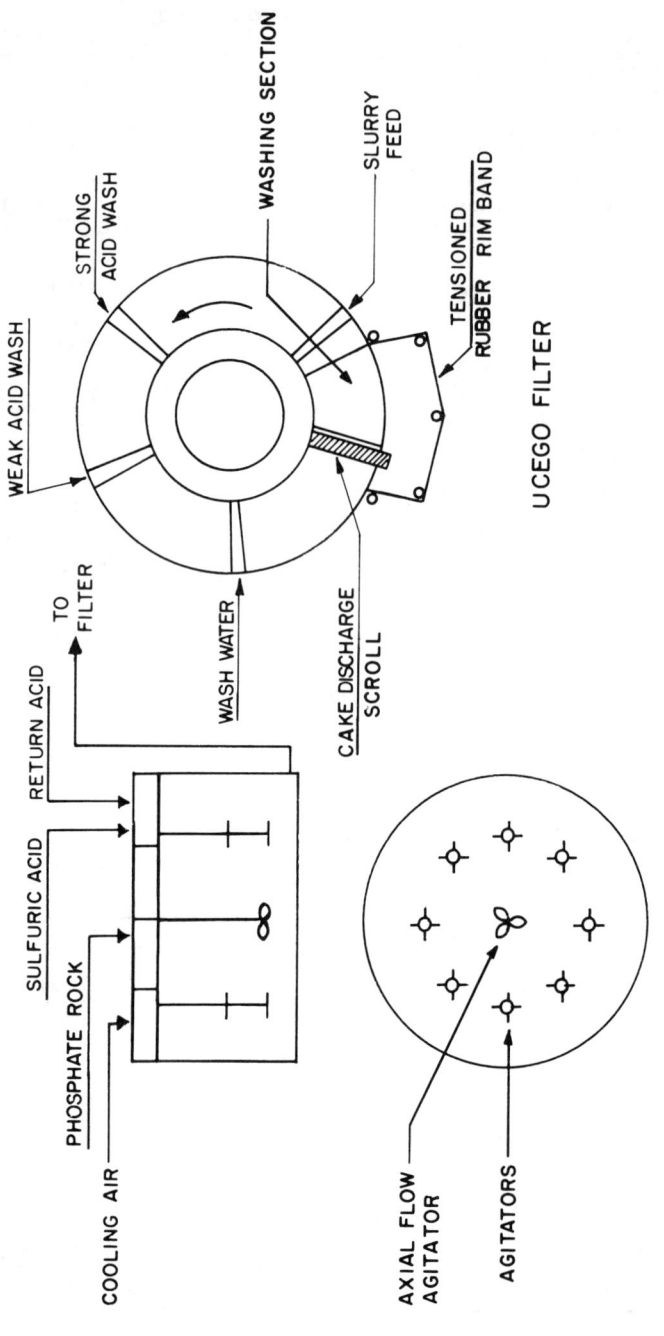

Fig. 7—Saint-Gobain Rhone-Progil dihydrate process.

The design (Fig. 7) relies on slurry movement induced by a series of agitators, mounted near the periphery of the tank wall, to provide homogeneity of the reaction slurry. Both radial and axial flow turbines are used to achieve this. Air cooling is employed and this may be supplemented by cooling the sulfuric acid feed after dilution.

A special feature of this process is the installation of the filter at a low level and the direct connection of pumps to the filtrate legs, which are less than the length required to give a true barometric seal. The usual filter installed is the UCEGO which is a rotating table design employing a flexible peripheral band to contain the filter cake (Fig. 7). This band separates from the table in order to allow the cake to be discharged by a scroll.

The use of a single reactor increases the danger of short-circuiting the reactants and subsequent decline in P recovery efficiency. The design inevitably relies on agitation to maintain suitable flow patterns within the reactor in order to minimize this danger.

The Prayon dihydrate process is operated by many companies throughout the world; one of the latest and largest examples being the 550-metric tons of P/day plant commissioned early in 1975 for Agrico at Faustina, Louisiana. This plant employs a modification introduced by Davy Power Gas involving wet grinding of PR. The design of the reaction system (Fig. 8)

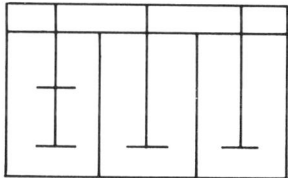

Prayon Dihydrate Process

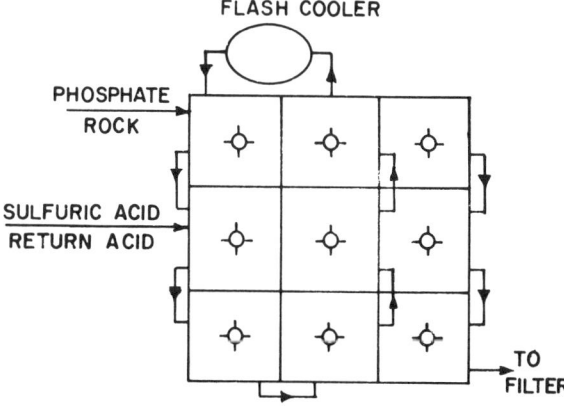

Fig. 8—Prayon dihydrate process.

follows similar lines to earlier Prayon plants in having nine (sometimes eight) reaction compartments contained within a single, lined concrete shell. The agitation system, however, has been modified in that a combination of axial and radial flow turbines are now used rather than radials alone and the slurry recycle ratio has been increased significantly.

In the Prayon process the bulk of the gypsum precipitates in the first reaction compartment. This gypsum is then allowed to stabilize in five or more reaction compartments before being recycled via a flash cooler to the first reactor. A recycle ratio of about 10:1 is normally employed. The last two vessels act as further stabilizing compartments before the gypsum slurry is filtered. Sulfuric acid is normally fed to compartment two and/or one and has for many years been diluted to 55% H_2SO_4 before addition. The use of wet grinding will no longer allow this without penalizing acid P concentration and/or P losses.

The type of operation used in the Prayon process gives gypsum having good filtration properties. Some loss of P recovery efficiency occurs due to higher coprecipitation of P compounds.

The Fisons dihydrate process (Fig. 9) features a single reaction vessel which is divided into four compartments by means of radial walls. Slurry is recycled at a ratio of about 20:1 from compartment three to compartment one, this minimizes concentration differences and provides a well-defined

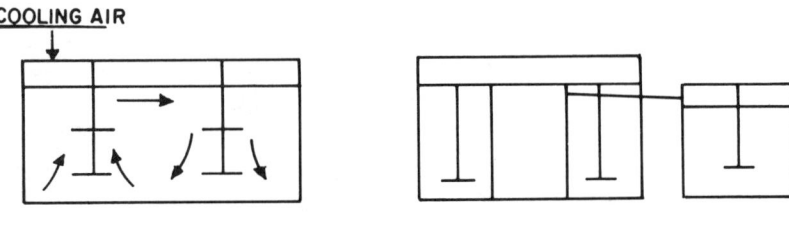

Fisons Dihydrate Process **Dorr Dihydrate Process**

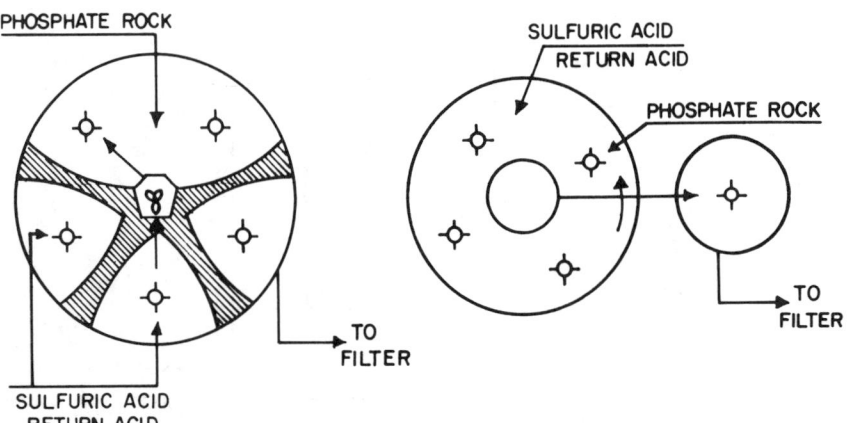

Fig. 9—Fisons dihydrate process.   Fig. 10—Dorr dihydrate process.

controllable flow pattern within the recycle system. The first reaction compartment is also twice the volume of any one of the other compartments, allowing coarsely ground and unreactive rocks to be processed, while maintaining good process control and high P recovery efficiencies. A combination of pitch-bladed agitators are used to generate a flow loop in the first reaction compartment and minimize the mixing time. Eimco tilting-pan, Landskrona belt, and UCEGO table filters have all been used with Fisons reaction system.

The method of operation used by the Fisons process increases P recovery efficiencies by reducing losses of coprecipitated insoluble P. This is achieved at the expense of a slight loss in gypsum filterability.

Commercial plants using the Fisons process are in operation in the United Kingdom, South Africa, and Brazil. Other plants are under construction in Yugoslavia.

The Dorr dihydrate process also involves a single-tank design (Fig. 10) but in this case the reaction occurs in an annulus using agitation to induce slurry recycle flow. Surplus slurry flows from the annulus to a central, concentric vessel, which in turn feeds a filter-holding tank. Either air cooling or evaporative cooling is used and may be supplemented by using precooled, diluted sulfuric acid. Tilting-pan filters, such as the Bird-Prayon or Eimco, are normally coupled to the Dorr reaction system, but Oliver and Giorgini filters have also been used on the smaller capacity plants.

The Dorr process, like the Saint-Gobain/Rhone-Progil process, relies on agitation to sustain slurry recirculation. Unlike the Saint-Gobain process, however, where there are no physical restraints, the recycle flow in the Dorr process is better defined in that it is restricted along the annulus. As with the Prayon and Fisons designs this makes the control of reaction conditions easier and more predictable.

Many plants are in operation throughout the world based on the Dorr process, the first plant was commissioned in 1967 for the Bunker Hill Company at Kellogg, Idaho.

The Kellogg-Lopker and Swenson Isothermal dihydrate processes were developed in the late 1960's and both present departures from other commercial designs involving agitated reaction vessels. Both processes involve the use of very high slurry recirculation rates to minimize temperature and concentration gradients and incorporate flash cooling within the reaction system. Each process utilizes conventional filtration equipment.

The Kellogg process (Fig. 11) employs two vessels, neither of which is agitated. One operates at atmospheric pressure and is used to dissolve the PR, and the other operates under vacuum and acts as a precipitation vessel and flash cooler. Slurry is pumped at a high rate from the dissolution tank to the evaporator before returning to the dissolver by gravity flow. Both the evaporator and dissolver have tangential inlets in order to provide a swirling motion to the slurry and aid the dissolution and mixing of the rock and sulfuric acid feeds.

Reduced utility requirements, capital, and maintenance costs are claimed (Anon., 1972a) for this process which so far has been adopted for

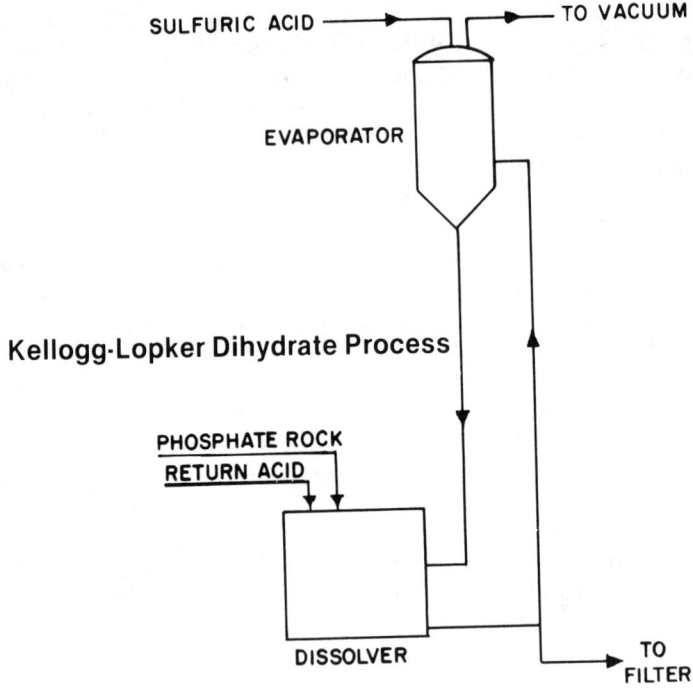

Fig. 11—Kellogg-Lopker dihydrate process.

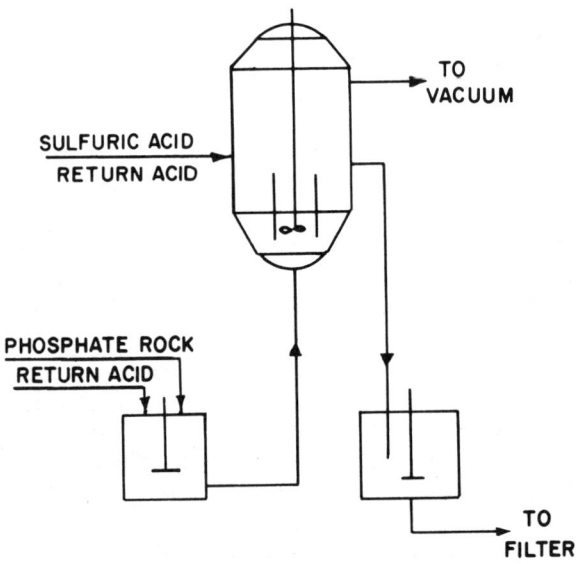

Fig. 12—Swenson isothermal dihydrate process.

one, 120-metric tons of P/day plant operated by Marchon at Whitehaven, England.

The Swenson Isothermal process (Fig. 12) incorporates a single vessel having a draught tube and submerged axial flow pump which provides a very high slurry recirculation rate. The single vessel acts as both a reactor-crystallizer and as a flash cooler. Product slurry overflows from the reactor via a barometric leg to a holding tank which supplies the filter.

Reduced utility requirements, capital, and maintenance costs are also claimed by this process which is used for a 260-metric tons of P/day plant operated by Farmland Industries, USA. Since commissioning early in 1972, various technical problems have been experienced (Salter, 1975) which include scaling of the product overflow, sulfuric acid entrainment from the use of a spray feed, failure of the reactor propellor shaft, rubber lining failures on the draught tube due to erosion, and problems controlling reactor level. Because of these problems, design changes have been made which will be incorporated into the design of a second 103-metric tons of P/day plant being built for Valley Nitrogen Producers, California.

The Fisons single-stage hemihydrate process (Fig. 13) is the only one in commercial operation that produces a concentrated acid directly without evaporation. This process involves the precipitation of hemihydrate and not gypsum.

Following extensive pilot-plant development and operation of a 17-metric tons of P/day prototype unit, the first plant (87 metric tons of P/day) using this process was commissioned for Windmill Holland NV in 1970. The reaction system comprises three vessels arranged for cascade flow and these supply slurry to a conventional three wash-stage filter, via a buffer tank. Slurry is recycled via a flash cooler from the second vessel to the first. The flow rate and sulfate concentration of the slurry are controlled so that the ratio of SO_4/CaO fed to the reactor is within the range required to achieve satisfactory hemihydrate crystal growth. Unlike a dihydrate process, there are wide differences in composition of the slurry contained in the two reaction vessels of the recycle system. Sulfate concentrations, for instance, are controlled in the second reactor within the range 1.5 to 2.3%, whereas in the first reactor an excess of Ca is maintained. Phosphoric acid containing 17 to 24% P can be produced by this process, but the normal recommended operating range is 20 to 22%. The hemihydrate filter cake is normally slurried with water in an agitated tank and pumped directly to a disposal point, either in an unchanged form or after conversion to gypsum. The method selected depends on the type of PR being processed.

The Windmill plant has now been in operation for more than 5 years and is giving production rates of 122 to 124 metric tons of P/day, which are about 40% above design. Togo PR has been the main phosphate processed, but Morocco PR has also been assessed during 1975. An average utilization of 89% has been achieved; most of the downtime results from routine maintenance on pumps, agitators, and filter grids. A small quantity of fluosilicate scale is produced in the second and third reaction vessels and represents a buildup of 2 to 3 cm over 3 years. This is insignificant compared to the

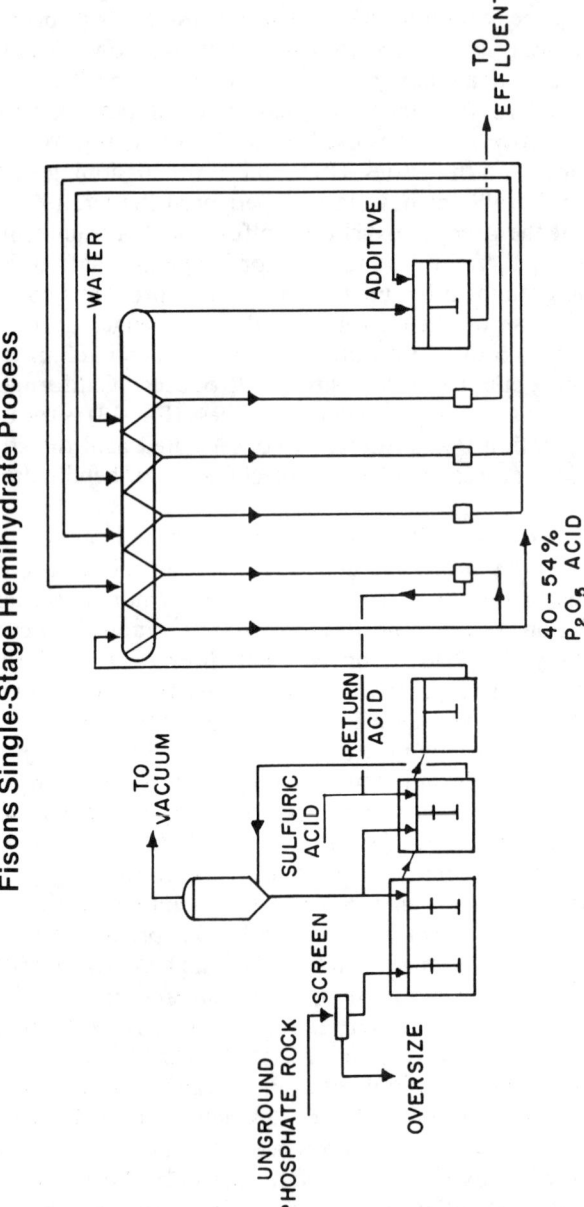

Fig. 13—Fisons single-stage hemihydrate process.

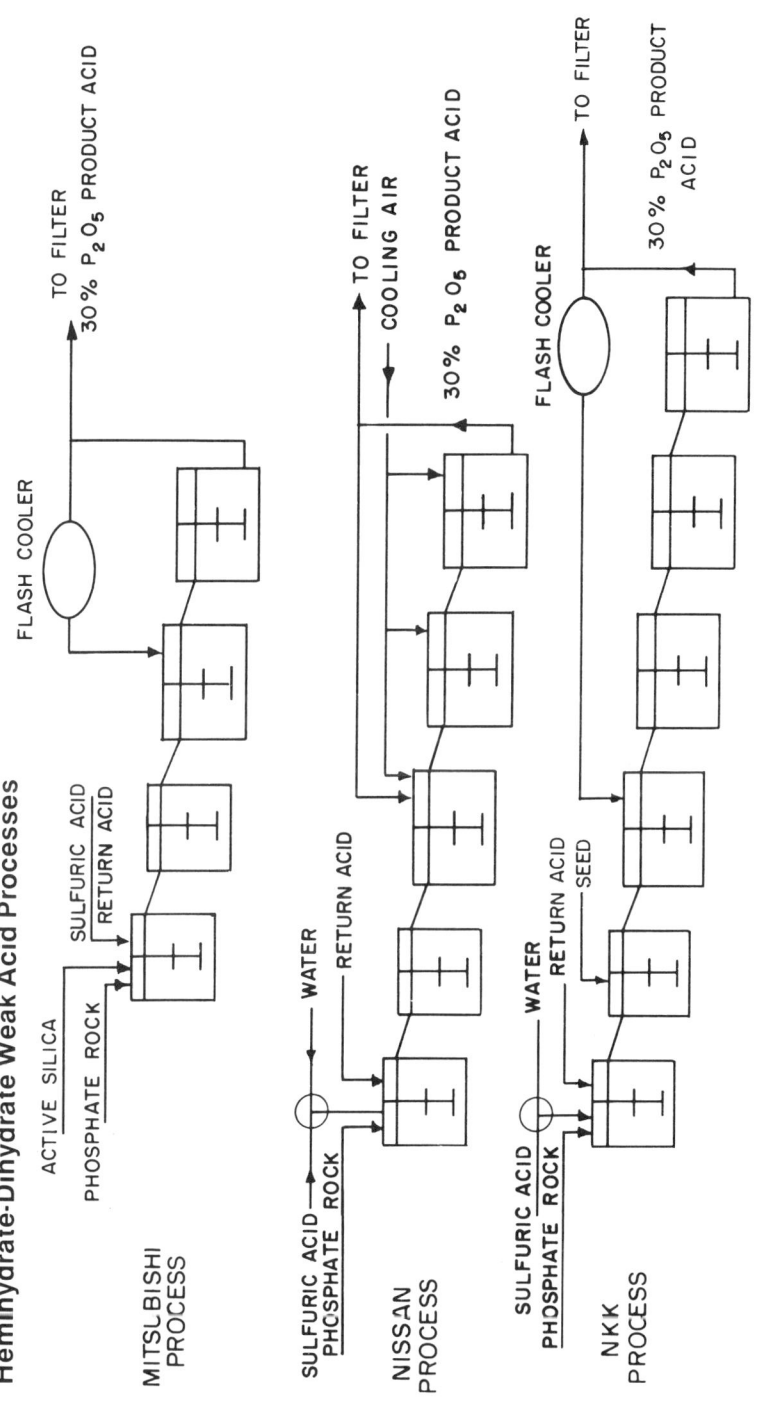

Fig. 14—Hemihydrate-dihydrate weak acid processes.

usual rate of buildup in a dihydrate process, which can be as much as 45 cm over the same period. There is no scale formation in the first reaction vessel. Scale due to calcium sulfate is completely inhibited by the use of chemical additives. The plant is normally operated to produce a 21.4% P acid in the reaction system, which gives 21% P after filtration. Recovery efficiencies of P when producing a 22% P acid are 2 to 3% lower compared to efficiencies given by the dihydrate process. For acid strengths of 20% P the efficiencies for the two processes are similar.

A two-stage hemihydrate-dihydrate process producing 13% P acids was developed in Japan during the 1950's and 1960's following detailed crystallization studies by various companies. The advantages offered by this type of process are higher P recovery efficiencies and the production of a purer gypsum suitable for the manufacture of plaster products or for use as a cement retarder. These advantages are particularly important for Japan where no indigenous supplies of gypsum or PR are available.

The Nissan, Mitsubishi, and NKK processes are well-known examples which resulted from these early studies. The three process (Fig. 14) are basically similar in that PR is first reacted using conditions which produce hemihydrate; this is then transformed to gypsum in a recrystallization stage. The processes produce a 13% P acid and there is no intermediate separation of hemihydrate. Conventional filtration equipment is used for the final separation and washing of the gypsum. The three processes differ from each other in the use of gypsum seed crystals and/or active silica as recrystallization aids.

A rather different approach has been taken by Societe de Prayon and Central Glass of Japan in that they offer a high-efficiency, two-stage pro-

Central Prayon Dihydrate-Hemihydrate Process

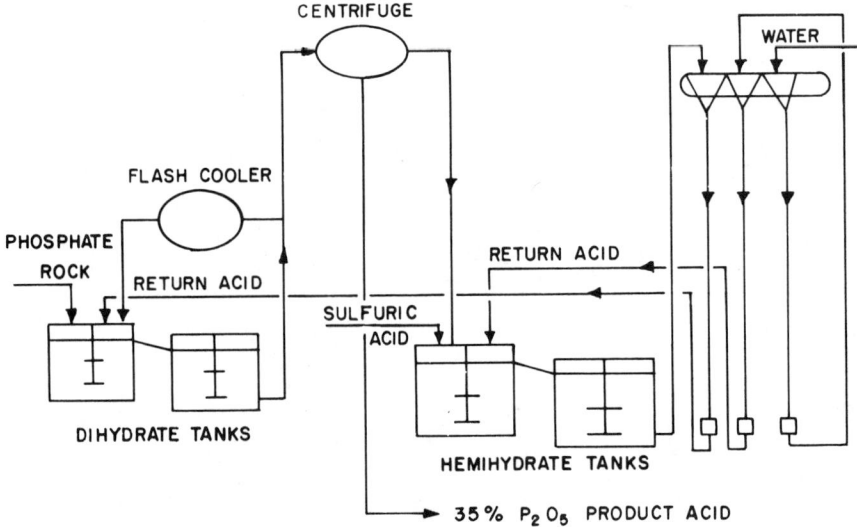

Fig. 15—Central Prayon dihydrate-hemihydrate process.

cess involving the precipitation of gypsum followed by recrystallization to hemihydrate. A centrifuge is used to separate the gypsum which is fed to the recrystallization reactors without washing. The acid concentration is 14 to 16% P and therefore requires concentration before use in the manufacture of phosphate intermediates.

A few commercial plants are using this process (Fig. 15) and in some cases technical problems have occurred with the operation of the centrifuge and with scaling during hemihydrate filtration. This type of process is particularly suited to the processing of Kola PR which is a difficult rock for the hemihydrate-dihydrate type of process.

The first two-stage hemihydrate-dihydrate process producing a more concentrated acid (19% P) was developed by Singmaster and Breyer in the 1960's. A centrifuge was used in this process to separate hemihydrate before recrystallizing to gypsum. Two or three plants were operating by this process in the late 1960's, but it is uncertain whether these are still operating, as little or no information has been published. Recently, a modification (Fig. 16) of this process has been developed by Heurty, but has not yet been commercially demonstrated.

Over the last 5 years three two-stage, hemihydrate-dihydrate processes have been developed which allow a concentrated acid to be produced with the added advantage of a high P recovery efficiency.

The Dorr HYS process (Weber et al., 1969) was used towards the end of 1969 for a 106-metric tons of P/day plant built for Rikkihappo Oy in Finland. An unusual feature of this plant was the use of a "double dump" filter on which both hemihydrate and gypsum were separated. The intermediate separation and washing ot hemihydrate used in this process allows an acid of 17 to 20% P to be produced. The Rikkihappo plant was designed to process Kola PR, which is in fact a difficult rock for a two-stage process of this type. Technical difficulties were encountered in the crystallization stage and these do not appear to have been overcome, since the plant is now operating as a dihydrate process (Anon., 1972b).

A two-stage process has recently been devleoped by Nissan which involves the intermediate separation and washing of hemihydrate before recrystallizing to gypsum. This process was first published (Miyamoto, 1975) at the end of 1974 and early in 1975 but no details of any commercial plant have yet been disclosed.

Fisons two-stage HDH process (Fig. 17) was a later development of the single-stage process already described, and involves the addition of a recrystallization stage to convert the washed hemihydrate to gypsum, which is then separated and washed on a second filter. Recovery efficiencies of 98 to 99% of the P are claimed for this process when producing 20 to 22% P acids.

The first commercial plant employing the Fisons HDH process was built for RMHK Trepca, Yugoslavia, during 1975.

The plant is designed to produce 22,000 metric tons of P/year as 22% P acid from Morocco PR. The unavailability of the correct type of PR and difficulties on procuring equipment spare parts have slowed down the ex-

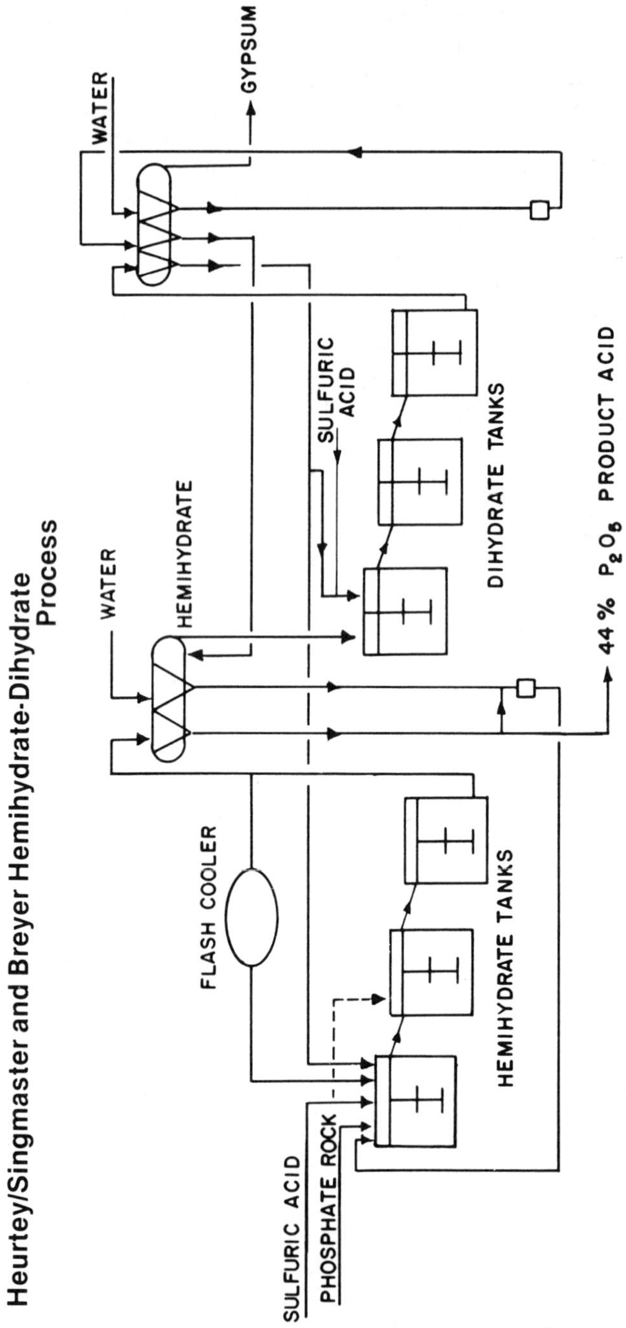

Fig. 16—Heurtey/Singmaster and Breyer hemihydrate-dihydrate process.

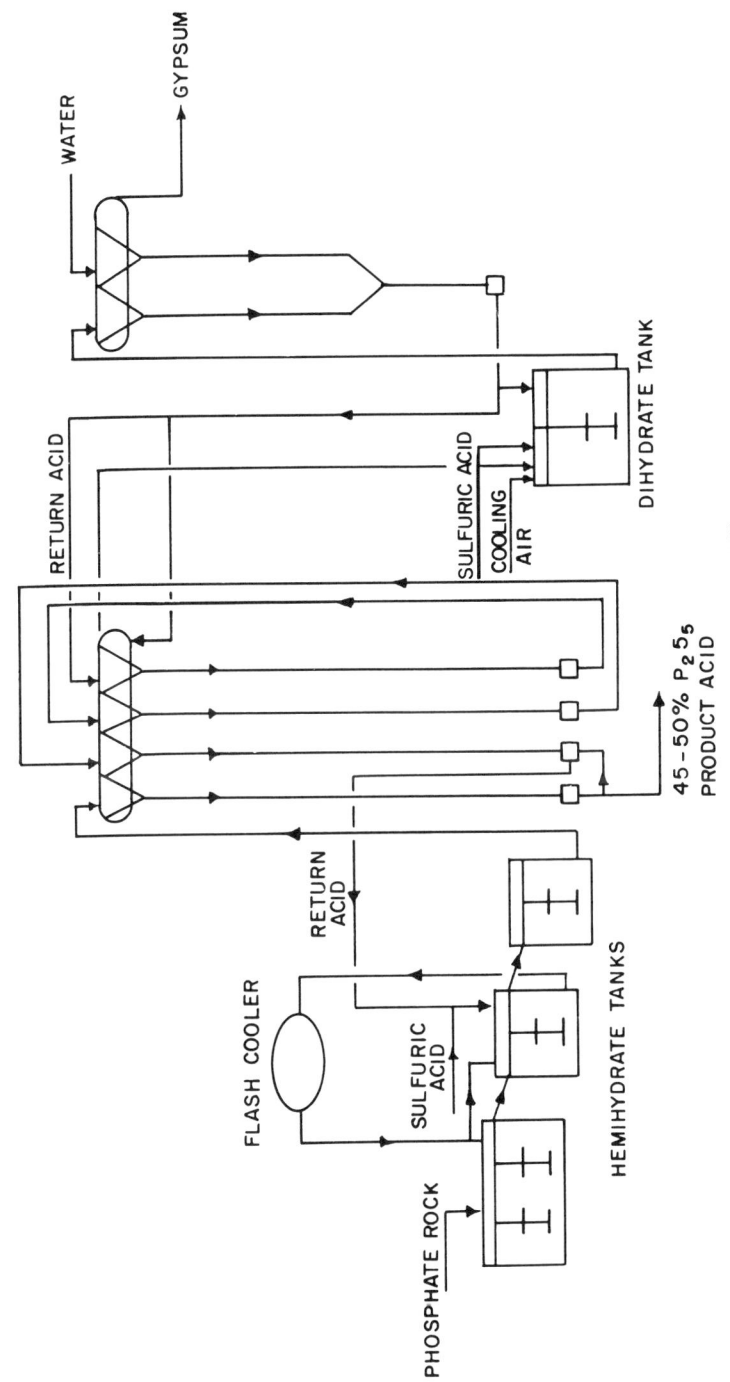

Fig. 17—Fisons hemihydrate-dihydrate (HDH) process.

pected rate of progress of commissioning, which is now expected to be completed early in 1976. Analytical data obtained so far when operating at full load are encouraging and indicate that expected performance will be achieved. Detailed results from the operation of this plant will be available during 1976.

C. Future Development Trends

1. PHOSPHORIC ACID PROCESSES

In predicting future development trends some useful guidelines can be obtained by first looking at past achievements.

Over the last 25 years notable advances have been made in the technology of the "wet" phosphoric acid process. We have seen a dramatic increase in the capacity of single-stream dihydrate plants to the point where the most economic sizes of 440 metric tons of P/day or more are becoming commonplace. There have also been changes in reaction design toward the use of a single or compartmented reaction vessel, and the incorporation in some designs of flash cooling within the reactor.

Hemihydrate and hemihydrate/dihydrate processes have also been developed and these processes, although as yet limited in application, are now commercial realities.

New materials of construction continue to be introduced which are more resistant to corrosion and erosion. We have also seen the introduction of new filter designs and other changes in equipment design.

Some important external changes are occurring which are likely to influence the future development of the "wet" process. One is the increasing tendency for phosphoric acid plants, particularly large units, to be sited at or near the location of the PR mine. Another is the depletion of world stocks of "good" PR and the ever increasing need to process lower grade rocks, which often present greater processing difficulties. There have also been dramatic increases in the cost of energy. All of these changes will inevitably affect policy decisions regarding new installations.

In general, it is expected that the attainment of a high recovery efficiency will continue to be an important consideration, because of the high significant effect on process economics of raw material prices and possible environmental or technical reasons connected with the disposal or utilization of calcium sulfate.

In the dihydrate process, future developments are likely to be toward the integration of the process with the beneficiation plant at the mine. In these cases, plants are likely to be designed for optimum utilization of a specific rock, covering a range of grades. For plants which are not located at the mine there will be a need for a more flexible process capable of handling as efficiently as possible a wide range of PR, including the lower grades. Plant capacities are likely to increase still further, but not to the same extent as previously.

The single-stage hemihydrate process is most suited to situations where steam is unavailable on established sites, or to plants located at the rock mine, where the lower P recovery efficiency given by this type of process has less economic significance.

The two-stage hemihydrate-dihydrate processes are the ones which could have the most impact on future development. Technically, they offer ideal performance, in that they provide a concentrated acid without the need for evaporation, at a much higher P recovery efficiency than is possible with a dihydrate process. The product acid and calcium sulfate are also purer, the latter being suitable for use as a cement retarder or the manufacture of plaster products.

These processes, however, need to be proved on a commercial scale and shown to be capable of handling an acceptable range of PR. Eventually, plant capacities will also need to be raised to 440 metric tons of P/day if these processes are to provide a major proportion of the world's phosphate output, at present produced by established dihydrate processes. Progress in this direction will depend on commercial results which are only just becoming available. Although preliminary, these results indicate expected performance and offer exciting prospects for this type of process over the next few years.

2. RECOVERY OF URANIUM FROM WET PROCESS ACID

It has been recognized for many years that phosphate rocks exhibit radioactivity due to the presence of trace amounts of uranium (U). Most phosphate rocks contain from 30 to 220 ppm of U, although levels ranging from 10 to 600 ppm have been reported (Swaine, 1962).

Phosphate rock	Typical uranium content, ppm
Florida	150–200
Kola	8–20
Khouribga	90–120
Gafsa	35–60
Djebel M'Dilla	60–80
Egypt	30–80
Christmas Island/Nauru	60–80
Israel	100–105

The world demand for U for medical, strategic, and energy uses has stimulated and maintained interest in finding commercially attractive ways of recovering U from phosphate rock. Encouraged by the Atomic Energy Authority several companies in the United States were involved in the early 1950's in developing methods for recovering U. However, no satisfactory process was found for selectively extracting U directly from phosphate rock or from waste gypsum. Attention was then concentrated on the extraction of U from wet process phosphoric acid which was found to contain about 90% of the original U content of the rock. During the years 1952 to 1961 several companies in the United States operated U recovery processes for varying periods. International Minerals and Chemicals was one of the first companies to build and operate a recovery process which was located at the Bonnie Chemical plant in Florida (Greek et al., 1957). Although these developments provided invaluable operating experience no extensive followup occurred, presumably for technical and economic reasons.

Early research centered around precipitation and ion exchange methods but gradually solvent extraction techniques became dominant and any large-scale application of U recovery is likely to be based on this technology. Solvent extraction is suited to continuous processing, it avoids dilution of wet process phosphoric acid, and raw material costs are claimed to be lower than those needed for ion exchange processes. The search for more efficient solvents for U still goes on and much of the credit for the ascendence of the solvent extraction method is due to the basic work carried out (McGinley, 1972) over many years by the Atomic Energy Authority's Oak Ridge National Laboratory (ORNL) in the United States. In the last few years ORNL have published (Hurst & Crouse, 1974) details of an improved process in which tetravalent U is extracted from wet process phosphoric acid (13% P) using a mixture of mono- and dioctyl phenyl phosphoric acids in an aliphatic diluent. The U in this extract is then stripped using wet process evaporated acid (24% P) containing sodium chlorate which oxidizes the U to the hexavalent state and increases the stripping efficiency. A second extraction is then carried out using di 2-ethyl hexyl phosphoric acid and trioctyl phosphine oxide from which ammonium uranyl tricarbonate is recovered after stripping with ammonium carbonate. Calcination can then be used to produce U_3O_8.

Several companies are actively studying the recovery of U from wet process phosphoric acid at the present time. Gulf Research and Development have reported (Anon., 1974a) the use of a portable pilot plant in 1974 to gather design data from several phosphoric acid plants operating in Florida. The process is based on ORNL technology but has been modified by Gulf. If results are promising a full-scale plant design was expected to follow during 1975. It was also reported in August 1974 (Anon., 1974b) that W. R. Grace and Co. and IMC were to build recovery plants at Bartow and Mulberry, respectively. The aqueous U extracts from both plants would be processed at Mulberry.

Assuming the technology is now proven, the main obstacle to large-scale development becomes one of economics and the price of U is obviously highly significant in this respect. It seems unlikely that U recovery plants will be coupled to wet phosphoric acid plants which have annual capacities of less than 65,500 metric tons of P/year if adequate returns on investment are to be met. In view of this, and the relatively high U content of Florida rock, it seems likely that the most significant expansion of U recovery processes will occur in the next few years in the state of Florida.

Conservative estimates (Facer, 1975) indicate that at least 70,000 metric tons/year of U_3O_8 could be produced from wet process acid by the year 2000 if expansion is maintained.

3. UTILIZATION OF CALCIUM SULFATE BYPRODUCT

Most phosphate rocks when used in the wet phosphoric acid process produce approximately 11 metric tons of gypsum for every tone of P produced. Generally, this gypsum is treated as a waste product and is discharged to the sea or tidal rivers where possible, or to artificial ponds or stock piles where plants are landlocked.

The potential value of byproduct gypsum has long been recognized and many companies have conducted research into possible ways of utilizing gypsum for the manufacture of plaster products (board or block), as a retarder in cement manufacture and as a source of sulfur for sulfuric acid manufacture. In most cases, however, unfavorable economics have prevented major developments and progress has been largely restricted to those areas where indigenous supplies of natural gypsum are not available or transport costs are high. This has applied particularly in Japan and to a lesser degree to some other countries such as Brazil and South Africa.

In the future some expansion in the utilization of byproduct gypsum is expected due to environmental pressures and the attraction of possible high returns on investment. Even taking an optimistic view the extent of utilization compared to the total output of byproduct gypsum is unlikely to be very significant and will not solve environmental problems posed by disposal needs. In addition, concern has been generated in recent years on the possible health hazards created by residual radioactivity present in byproduct gypsum when used for building materials. This may promote further efforts to find ways of extracting U directly from phosphate rock, rather than from wet process acid. Mitsui Toatsu Chemicals, Inc., is one of a group of Japanese companies which are reported (Anon., 1975b) to be involved in this activity. In the United Kingdom the National Radiological Protection Board in collaboration with the Building Research Establishment and Fertilizer Manufacturers has recently concluded that wet process byproduct gypsum can be safely used as a building material provided the radium content of the finished board or block does not exceed 25 pCi/g. This conclusion is unlikely to be serious in most cases; it hardly encourages, however, a development already inhibited by the need for high investment and extensive marketing.

4. POLLUTION ABATEMENT PROBLEMS

For many years the emission standards relating to effluents from fertilizer plants have been progressively tightened, and in some countries like the United States these standards have become very precise demanding high efficiency recovery techniques.

In the manufacture of wet process phosphoric acid attention has mainly concentrated on reducing fluoride levels in gaseous and liquid effluents. There have been many cases where F emissions have been blamed for damage to the environment, but lack of data and local variables such as factory location and meteorological conditions make it very difficult to establish maximum concentrations for "safe" emission standards.

Available evidence (Whalley, 1976) suggests that damage caused by atmospheric F to vegetation and animals is much more widespread than the ill effects on humans. The rate of absorption of gaseous F by plants is variable and concentrations of F as low as 0.015 to 0.04 mg/m^3 can damage the most sensitive plants such as tulips or small conifers. Normally, fluoride-containing dusts, e.g., phosphate rock, have no effect on plants, but can be toxic to grazing animals if consumed. It has been suggested (Walley, 1976) that cattle absorb about half the water-insoluble F they ingest and concentrations of available F in excess of 30 mg/kg of feed are known to be toxic.

There is a lack of uniformity in air quality standards between countries. In the United Kingdom there is no definite standard as flexibility is retained to take into account dispersion of emission, meteorological conditions, and the proximity of residential areas. A maximum 3-minute ground level concentration of 0.07 mg/m^3 is taken as a starting point in the case of hydrogen fluoride. In Germany maximum concentrations of 2 and 4 mg/m^3 are specified, based on measurements taken over 1- or 8-hour periods, respectively.

Differences also exist between emission standards set by different countries. In the United Kingdom presumptive limits are set which represent the maximum allowable emission concentrations assuming the best practicable recovery systems are used. These limits are tightened whenever there are improvements in technology, but only after discussion with industry and only after any new technology has been proved in the long term. Other countries prefer to set rigid emission limits based upon an assessment of the capabilities of current technology. In the United States the standard for fluoride emission is 0.023 kg of F/metric ton for P feed (Anon., 1975a), which is a very severe limit compared to those specified by European countries. For example, in the United Kingdom the presumptive limit is 320 mg/m^3 and an emission of about 2.0 kg of F/metric ton of P feed was indicated based upon one example of commercial practice given in the 1974 report of the Alkali Inspectorate.

The liquid effluent from a typical gypsum process phosphoric acid plant contains about 10^3 g of F/m^3 assuming fluorosilicic acid is not recovered, or about 10^2 g/m^3 if recovery is practiced. This water may be reused to scrub reaction gases before mixing with gypsum slurry prior to pumping to the sea or estuary. Alternatively, the water may be fed to a gypsum pond for cooling and reuse and some treatment with lime, for example, may be necessary to control the concentrations of soluble phosphate and fluoride.

The disposal of gypsum presents a major environmental problem when manufacturing phosphoric acid. Methods currently used include mixing with salt water and pumping the slurry to the sea, river, or tidal estuary; using barges to dump the filter cake into deep salt water; use as a land-fill material for reclamation of waste land; and conversion of the gypsum into marketable products. In some countries, notably the United States and Norway, the gypsum slurry is pumped to ponds where the gypsum settles allowing water to be reused. The American practice (Fullman & Faulkner, 1971) indicates that a gypsum pond should be at least 0.53 ha for each metric ton of P produced per day. This is much larger than an example given (Kivela, 1974) of Norwegian practice which is 0.16 ha for each metric ton/day. Gypsum deposited in ponds usually contain volatile fluoride compounds which can cause damage to vegetation in the surrounding area. Treatment of the water to reduce fluoride emission or to comply with local standards of purity in the case of disposal to rivers is sometimes practiced, but costs are high.

The ever increasing difficulty in finding environmentally satisfactory gypsum disposal methods, particularly for inland sites, has encouraged the development of processes designed to convert waste gypsum to a valuable raw material. In Europe major contributions have been made by Giullini (Foester, 1974), Knauf (Wirsching, 1971), ICI (Allen, 1975), and Rhône-Poulenc (Anon., 1975c) on processes to produce plaster products suitable for use as building materials. Commercial plants in the past have generally been small, usually less than 100,000 metric tons/year of gypsum. Recently, however, Rhône-Poulenc technology has been applied for a single-stream, 300,000-metric tons/year plant which is now in operation near Rouen, while a 400,000 metric tons/year plant using Giulini technology is being built in the USSR.

In general, the present trend of increasing environmental pressure is expected to continue. Although investment in modern pollution control technology, coupled with its diminishing return, are obvious deterrents to industry, it is the responsibility of industry to continue to develop new technology to provide maximum environmental protection. Wherever possible pollution control should be achieved by increasing process efficiency or by the sale of byproducts, rather than by the use of more energy or raw materials, as these routes are likely to inhibit progress.

IV. SUITABILITY OF WORLD PHOSPHATE ROCKS FOR PHOSPHORIC ACID MANUFACTURE

A. Basic Mineralogy

Phosphate rock deposits occur in nature either as crystalline apatite in association with igneous rocks, or more commonly as phosphorites associated with sedimentary rocks.

The world's largest deposits of mineral apatite are located in the Kola Peninsula in the USSR; other deposits may also be found in Brazil and in South Africa. Phosphorites are widely distributed, the main deposits being actively exploited are those found in Morocco and Florida.

The mining methods used are largely dependent on the depth and nature of the overburden. These may involve open cast methods using scrapers or drag lines, or it may be necessary to apply underground shaft and tunnel methods.

The raw rock generally requires beneficiating before it is suitable for "wet" process phosphoric acid manufacture. Three methods are generally employed, the choice depending on the nature of the associated impurities and minerals. The first and simplest method involves crushing and screening operations followed by density separations, using, for example, hydrocyclones. This type of operation is extensively used to beneficiate Morocco deposits. For some PR, however, an efficient separation cannot be achieved by this method, and it is then necessary to apply other techniques, such as flotation or calcination, to obtain a satisfactory material. Flotation is extensively used to beneficiate many rocks, including those found in Florida and the USSR. Calcination is less widely used, but is necessary in some cases to remove organic matter and carbonates, both of which cause processing problems. It is used mainly for PR found in Algeria, Syria, and Israel.

Phosphate rocks contain a wide range of impurities, none of which are completely removed during the beneficiation process. Silica is one of the most common impurities, originating from minerals such as quartz, clays, serpentine, pyroxenes, or micas. Carbonates are also present in the form of calcite, dolomite, or francolite. Metallic impurities include Fe in the form of magnetite or hematite, Al in the form of clays or micas, Mg in the form of dolomite, magnesite, or pyroxenes, and Ba as barite. There is also a range of minor impurities which can include rare earths, Cr, Mn, Sr, Cu, etc.

Table 2 gives typical compositions of the main PR sources used for wet process phosphoric acid manufacture.

B. The Effect of Rock Impurities on Process Performance

The impurities present in a PR generally characterize its behaviour when used to manufacture wet process phosphoric acid. Although important performance parameters, such as filterability and efficiency of P recovery, can be controlled to some extent by optimizing operating conditions, the overall performance achieved more often than not is the result of processing a given type of PR.

Many of the impurities shown in Table 2 affect processing performance in some way; the degree of influence depends not only on individual concentration levels but also on relative levels due to interactions. The effects of certain impurities are well known and understood; other effects, particularly these involving minor components, are less clear. This applies particularly to changes in crystal growth, crystal habit, and precipitation of complex compounds other than calcium sulfate.

The main effects of the more common impurities on process performance have been discussed in Section II-A. These may be summarized as follows.

The ratio of P_2O_5 to CaO is the main criterion which determines the consumption of sulfuric acid per unit weight of P_2O_5 produced. This consumption figure will also be affected if the PR contains sulfate, as this is usually combined with Ca.

The level of chloride in the PR will determine the corrosivity of the phosphoric acid produced and influence the choice of materials of construction for a plant. Fluoride also influences corrosivity, but to a lesser degree; usually this is only of significance when the rock contains high amounts of fluoride compared to silica.

Alkali metals determine in most cases the extent of scaling due to pre-

Table 2—Typical analyses of phosphate rocks.

Component, % wt/wt†	Morocco Khouribga	Morocco Yousoufia	Calcined Djebel Onk, Algeria	Gafsa, Tunisia	Phalaborwa, South Africa	Bucraa, Sahara	Africa, other Taiba	Africa, other Togo
P_2O_5	33.4	32.1	34.3	28.2	36.8	36.6	37.5	36.8
CaO	51.2	52.0	53.8	48.4	52.1	51.9	51.2	50.5
F	4.1	3.2	3.96	3.2	2.2	3.9	3.91	3.84
SiO_2	2.1	2.8	1.60	3.4	2.6	4.5	2.49	3.62
Al_2O_3	0.49	0.41	0.36	0.68	0.15	0.35	1.10	1.0
Fe_2O_3	0.21	0.31	0.32	0.32	0.27	0.14	1.10	1.14
MgO	0.28	0.40	1.03	0.6	1.06	0.06	0.02	0.06
SrO	0.08	0.08	0.15	NA§	0.4	0.03	NA	NA
CO_3-C	1.11	1.91	0.44	1.8	0.75	0.50	0.41	0.35
Organic-C	0.17	0.21	Nil	>0.5	0.03	0.05	0.46	0.12
Na_2O	0.87	0.81	0.4	1.1	0.04	0.24	0.20	0.23
K_2O	0.1	0.08	0.03	0.4	0.1	0.06	0.03	0.05
Cl	0.03	0.01	0.014	0.1	0.022	0.036	0.047	0.12
SO_4	1.8	1.4	2.4	3.9	0.2	0.53	0.1	0.2

Component	Kola, USSR	Florida, USA‡ 68% BPL	Florida, USA‡ 73% BPL	Brazil Jacupiranga	Brazil Araxa	Ruseifa, Jordan	Eastern, Syria	Oron, Israel
P_2O_5	38.7	31.6	33.4	35.7	34.9	31.3	28.9	30.9
CaO	50.7	46.5	48.3	52.8	46.8	50.9	46.4	52.9
F	3.2	3.65	3.8	1.47	2.1	3.8	3.5	3.7
SiO_2	1.13	9.5	5.3	1.5	0.25	3.6	7.4	0.83
Al_2O_3	0.69	1.26	1.21	0.35	0.46	0.38	2.6	0.15
Fe_2O_3	0.37	1.44	1.12	0.25	2.86	0.14	0.14	0.07
MgO	0.10	0.38	0.25	0.8	0.06	0.23	0.26	0.27
SrO	2.1	<0.1	<0.01	0.58	1.11	NA	NA	0.15
CO_3-C	0.02	1.0	0.82	1.33	0.22	1.43	1.55	1.81
Organic-C	0.03	0.15	0.13	0.09	0.03	0.45	0.25	0.23
Na_2O	0.13	0.64	0.6	0.26	0.23	0.7	0.56	1.15
K_2O	0.11	0.08	0.08	0.44	0.02	0.08	0.01	0.1
Cl	0.006	0.025	0.024	0.003	0.004	0.06	0.23	0.1
SO_4	Nil	0.9	0.94	1.48	3.05	0.81	1.6	1.8

† P% = 0.437 × P_2O_5%; Ca% = 0.714 × CaO%; Si% = 0.467 × SiO_2%; Al% = 0.265 × Al_2O_3%; Fe% = 0.699 × Fe_2O_3%; Mg% = 0.603 × MgO%; Sr% = 0.846 × SrO%; Na% = 0.742 × Na_2O%; K% = 0.830 × K_2O%; S% = 0.333 × SO_4%.
‡ P% = 0.20 × BPL%.
§ NA = not available.

cipitation of fluosilicates. Scaling occurs in the filter circuit and also in the reactor and can be particularly troublesome with some rocks in a dihydrate process. Normally, scaling adversely affects plant utilization rather than instantaneous production rates.

The ratio of SiO_2 to F can affect corrosion, scaling, and F recovery. In association with other impurities it can also influence crystal growth, recrystallization of hemihydrate to gypsum, and the formation of complexes which can affect filtration rates by causing the binding of filter cloths.

Carbonates influence foaming and the need for a defoamer. They can also indirectly increase the reactivity of a PR by causing disintegration of rock grains and increasing specific surface area.

Organic matter can stabilize foam generated by carbonates and also affect crystal growth and the recrystallization of hemihydrate to gypsum.

Cationic impurities, such as Fe, Al, Mg, Cr, etc., affect crystal growth and influence crystal habit. They also affect acid viscosity and the precipitation of complex compounds. All of these effects can influence filtration rates, acid purity, and the composition of P intermediates.

The reactivity of a PR can also influence the performance of a phosphoric acid plant in that it influences the grind of rock required to minimize P losses; both the rock grind and reactivity determine operating sulfate levels which in turn influence filtration rate and lattice P losses.

C. Performance of Phosphate Rocks

An ideal PR for phosphoric acid manufacture by the "wet" process should have the following features.

1) High P_2O_5 to CaO ratio—to minimize sulfuric acid consumption.
2) Low organic matter—to minimize foaming, to avoid interference with filtration, crystal growth, and recrystallization of hemihydrate to gypsum.
3) Low carbon content—to minimize foaming.
4) Low chloride content—to minimize corrosion.
5) The SiO_2 to F ratio should be sufficient to avoid a large excess of fluoride in the phosphoric acid, thereby minimizing corrosion and avoiding interference of crystal growth and the recrystallization of hemihydrate to gypsum.
6) Low alkali metal content—to minimize fluosilicate scaling.
7) Low levels of cationic impurities—to minimize loss of acid purity and loss of P water solubility of P intermediates.
8) Contain sufficient cationic impurities to produce rhombic-shaped dihydrate crystals. This is not necessary in hemihydrate processes.
9) High reactivity to minimize rock P_2O_5 losses and grinding required.
10) Be relatively soft to improve grinding rates.

None of the commercially available PR meet all the above requirements, which vary in importance depending on the type of process used and the plant design. Within the scope of this review it is only possible to give a guide to the processing characteristics of the most commonly used PR, and in the following only the "good" and "bad" characteristics are highlighted.

1. ROCKS FROM THE UNITED STATES

The major PR producers in the United States are located in Florida. There are other producers, but these operate on a much lower scale, the main areas being North Carolina, Tennessee, and the western states of Idaho, Utah, Montana, and Wyoming.

After beneficiation by flotation, Florida rocks generally give good dihydrate filtration rates, low corrosion rates, and satisfactory P recovery efficiencies. Disadvantages include foaming, fluosilicate scaling, low grinding rates, increased mill wear dur to rock hardness, and the formation of sludge in acid storage. The lower the P content of the Florida PR the more pronounced these disadvantages become. Florida PR are also suitable for hemihydrate and hemihydrate-dihydrate processes. Filtration rates in these processes are usually lower compared to rates given by other rocks of higher purity.

Rocks from the western states normally contain high levels of carbonate and organic matter and are difficult to process, unless calcined. Phosphate rocks from North Carolina are generally satisfactory, but again calcination is usually carried out to reduce levels of organic matter and carbonates.

2. ROCKS FROM SOUTH AMERICA

Deposits of PR in the form of igneous apatite and phosphorite occur in several countries on the South American continent; exploitation, however, is carried out on a small scale and all the production is internally used.

Brazil is the most significant producer, the more important deposits being carbonatites associated with a wide range of other minerals such as barite, calcite, mica, magnatite, and pyrochlore. The deposits can also be rich in metals such as columbium, titanium, thorium, etc. New flotation processes, coupled with magnetic separation, have been developed to beneficiate these ores which will eventually be used for phosphoric acid manufacture. The main deposits being exploited are those of Araxa and Tapira in the state of Minas Gerais. The Jacupiranga deposit in the state of Sao Paulo is already being used to produce phosphoric acid and phosphate intermediates.

Generally, Brazilian PR offer high P recovery efficiencies, low sulfuric acid consumptions, and satisfactory filtration rates when processed by the dihydrate route. A defoamer is usually not required. The dihydrate crystals tend toward a needle shape rather than the more common rhombic shape.

3. ROCKS FROM AUSTRALIA, NAURU, AND INDIA

Nauru and Christmas Island PR have been exploited for more than 70 years and have been a major source of P supply to Australia, New Zealand, and, to a lesser extent, Europe.

The Nauru deposit contains a high P content and only needs drying in rotary kilns before shipment. When used for phosphoric acid manufacture good efficiencies and filtration rates are achieved and the acid contains low levels of Al and Fe. The PR contains, however, organic matter, and a large excess of fluorine over silica, which can cause foaming and corrosion problems, respectively.

Although there are many known deposits of phosphate in Australia and India, exploitation is still in an early stage. The most significant deposits in Australia are those located in Queensland and the Northern Territory, which when developed could turn the country into a major producer for both the home and overseas markets. The Duchess deposit, located in South Queensland, is being used to manufacture superphosphates and trials are to be carried out by manufacturers of phosphoric acid. The exploitation of the Lady Annie deposit in North Queensland is already well advanced but further progress will depend on the future marketing position and economic factors. Lady Annie concentrate when processed by the dihydrate process gives good P recovery efficiencies, high filtration rates, and requires no defoamer. It does not have a particularly high reactivity and contains organic matter which imparts a brown coloration to the acid.

Exploitation of P deposits in India has gradually gathered momentum over the last 7 years and deposits such as the phosphorites at Udipur in the state of Rajasthan are increasingly being used for the manufacture of phosphoric acid. It has been reported (Anon., 1972c) that PR from the Jhamar Kotra deposit near Udipur gives high P recovery efficiencies, reduced problems due to foaming and scaling compared to Florida PR, and satisfactory filtration rates. The rock, however, is very hard and gives reduced grinding rates; it also has a low reactivity.

4. ROCKS FROM AFRICA

The P deposits found on the continent of Africa have been used for many decades for the manufacture of phosphoric acid. The main producer and the world's largest exporter is Morocco, which, together with the United States and the USSR, produces the bulk of the world's supply of P. A guide to the relative outputs and exports appropriate to 1974 is given in Table 3.

The P deposits in Morocco are exploited in two main areas centered near Khouribga and Youssoufia. The richest and most extensive deposits occur in the Khouribga area where the P beds alternate with marl, clay, flint, and phosphatic limestone. Both open-cast and underground mining methods are used depending on the thickness and nature of the overburden.

Khouribga PR is reactive and offers good overall performance when used to manufacture phosphoric acid by either the dihydrate or hemihy-

Table 3—Approximate levels of production of phosphate rock in 1974.

Country	Production	Exports
	million metric tons	
USA	41.2	12.9
USSR	22.9	6.5
Morocco	19.3	18.7
Tunisia	3.9	2.4
Togo	2.5	2.6
Senegal	2.0	1.8
Sahara	2.2	2.2
South Africa	1.5	--
Algeria	0.8	0.4

drate routes. Compared to Florida PR, the processing of Khouribga PR in the dihydrate process produces less fluorsilicate scaling, less sludge, and less foaming and gives a purer acid. Filtration rates, however, are lower usually by about 10 to 20%. The performance of Khouribga PR is also influenced by the grade of rock, particularly with respect to filtration rate and foaming.

The P deposits of Youssoufia contain higher contents of organic matter and are less pure compared to the best Khouribga grades. Relative to Khouribga PR, lower filtration rates are obtained when processing Youssoufia PR. Foaming problems are also increased and a defoaming agent is usually required in most processes.

Morocco PR are suitable for hemihydrate and hemihydrate-dihydrate processes. They are not ideal, however, for dihydrate-hemihydrate processes due to hydration problems which can occur during the final filtration of hemihydrate.

The Bu-Craa PR in the Sahara is the most recent deposit to be exploited and expansion in production is continuing. It is a high-grade PR giving a processing performance similar to that of Khouribga PR. Problems, however, have arisen with respect to chloride levels which have led to high corrosion rates. The increase in chloride level in the PR is the result of increased production and problems with washing in the beneficiation process; these problems are expected to be corrected in due course.

The Algerian deposits are mainly located in the Djebel-Onk area where the phosphorites are associated with limestone, marl, and coprolites. After stripping off the overburden, the ore is blasted and shipped to the beneficiation plant, where it is crushed, washed, screened, and finally calcined to remove calcite and magnesia. As mined, the PR is harder and of lower grade than those from Khouribga. The calcined PR is less reactive than Khouribga PR and contains a higher level of Mg. Good P recovery efficiencies are obtained, however, when the PR is used for phosphoric acid manufacture. Filtration rates are much lower when processing Djebel Onk PR compared to the rates given by Khouribga or Florida PR.

The Tunisian deposits are closely associated with those in Algeria, the main deposits being located in the Gafsa region. The phosphorite is associated with phosphatic limestone, marls, and clays and is beneficiated by air classification after first crushing and washing. Gafsa PR is reactive, but contains a high level of organic matter and carbonate which cause foaming problems and adversely affect filtration rates when used for phosphoric acid manufacture.

Togo's production of PR arises from a single mine located at Hahotoe. The sedimentary deposits have undergone considerable erosion and decalcification and are associated with clays and shales overburdened by sand and clay. Open-cast mining methods are used and the untreated ore is shipped to the beneficiation plant by rail. Beneficiation includes washing with sea water, classification using cyclones and hydrocyclones, washing to reduce salinity, and, finally, screening and magnetic separation. Togo PR is a high-analysis phosphate which gives high P recovery efficiencies and filtration rates in the dihydrate process. It is also suitable for hemihydrate and hemihydrate-dihydrate processes. There are no problems due to foaming and the extent of scaling due to fluosilicates is small. The main disadvantage associated with the processing of Togo PR is its high chloride content which causes corrosion problems.

The Senegal PR exploited for fertilizer use are mainly centered in the region of Taiba. The phosphate bed underlies an overburden of sand which is removed by bucket-wheel excavators before recovering the P by dragline. Beneficiation comprises washing, screening, and desliming before flotation and drying. Taiba PR is a high-analysis phosphate giving high P recovery efficiencies and filtration rates. Unlike Togo PR, Taiba PR presents no problems due to corrosion. It does contain, however, organic matter which can stabilize foam. This only presents a problem when Taiba PR is processed with a second PR containing a significant quantity of carbonate.

Phalaborwa PR is an igneous apatite found in South Africa and occurs as pyroxenite ringed by syenite and granite. The apatite is broken up by blasting and beneficiated by crushing, screening, and flotation. The beneficiated rock is dense, very hard, and abrasive. Grinding rates are low and erosion of grinding equipment can present problems unless the mills are suitably protected. Phalaborwa has a very low reactivity, but when processed in a finely ground form gives good P recovery efficiencies. There are no significant problems due to corrosion, foaming, or scaling. Filtration rates, however, are not high unless a crystal habit modifier is used.

5. ROCKS FROM THE MIDDLE EAST AND EASTERN EUROPE

Extensive reserves of PR occur in Syria in areas south and west of Palmyra. The main center is located in the Ghadir el Hamel region where the Kneifiss and Eastern PR are being exploited. Syrian rocks are hard and are associated with calcite, quartz, and organic matter. Calcination has been used to reduce the organic matter and carbonate levels in the case of the Eastern rock. In recent years both Kneifiss and Eastern rocks are being produced uncalcined. When used for phosphoric acid manufacture, Syrian rocks give P recovery efficiencies and filtration rates which are similar or lower compared to Morocco rock. Corrosion rates can be high due to the presence of chloride and defoamers are required unless the rock has been calcined.

The main P deposits in Jordan occur in the region of Ruseifa and El Hassa, where the phosphorite is found in association with limestone, chalk, chert, and marl. Both underground and open-cast mining methods are used and beneficiation includes crushing, washing, and screening operations and involves classification using hydrocyclones. Jordan rocks generally give good P recovery efficiencies when used for phosphoric acid manufacture, but filtration rates are usually lower compared to the rates given by Morocco and Florida rocks. A defoamer is also required and most grades of Jordan rock contain high chloride levels which can cause corrosion problems. Recently a new grade of rock has been produced from the El Hassa mine (14.6 to 15.0% P; 73 to 75% BPL) which has a low chloride content; this should substantially reduce corrosion rates.

The P fields in Israel are located in the Negev desert where intensive surveys of mineral resources have been in progress since the early 1950's. The main deposits which are being exploited are centered at Oron, Ha'Makhtesh, Zefa-efe, and 'En Yahau. The mining of these deposits is most established at Oron, which first began operating in 1952. The phosphorite beds occur in association with limestone or chalk and are recovered by open-cast methods. The main beneficiation process used for Oron rock involves calcination, slaking, screening, and drying to reduce the contents of carbonate and organic matter. When used for phosphoric acid manufacture, Israel rocks give good P_2O_5 recovery efficiencies but foaming and filtration rates can be variable, depending on the rock grade. Most grades give high corrosion rates due to the presence of chloride. As is the case with Jordon rocks, work is in progress in Israel to produce a low chloride rock by incorporating additional washing facilities during beneficiation.

The USSR possesses a substantial proportion of the total world's reserves of P and is the second largest producer. The main mining area is centered in the Kola Peninsula, where igneous apatite is associated with amphiboles, nepheline, and other minerals which contain Sr, Ba, and traces of rare earths. The ore is claimed by both open-cast and underground mining methods and is beneficiated by fine grinding, flotation, and cyclonic classification. Kola rock is extremely unreactive but gives good P recovery efficiencies when used to manufacture phosphoric acid by the dihydrate process. There are no problems due to corrosion, foaming, or sludge deposition, but filtration rates are lower compared to those given by Morocco rock. Fluosilicate scaling is also more evident when processing Kola rock. The performance of Kola rock is satisfactory when processed in the hemihydrate or dihydrate-hemihydrate processes. The rock is not very suitable, however, for hemihydrate-dihydrate processes, because of the slow rate of hydration of hemihydrate.

In addition to the Kola apatite deposits, there are also large resources of phosphorites. The main deposits are found in the region of the Lesser Karata Range. Open-cast mining methods are employed and flotation is used to enrich the P and reduce the level of carbonates. Kara Tau rock is used domestically by fertilizer manufacturers and for the manufacture of elemental P.

D. Research Trends

The increase in world consumption of PR will inevitably deplete the reserves of the well-established phosphates whose processing characteristics are well known and can be handled. This trend is already evident and is affecting supplies of Florida, Morocco, and Kola rocks, which together constitute the bulk of the world's exports.

Clearly, the fertilizer industry must in the longer term be prepared to process lower grade rocks and also accept more flexibility with regard to the choice of PR. Where phosphoric acid plants are located at the mine there will be a greater need to provide a "Tailor made" design to match the characteristics of a range of rock grades.

Rock producers can also contribute, by giving attention to ways of reducing the levels of harmful impurities which cause problems during phosphoric acid manufacture. Beneficiation treatments involving density separations, flotation, or calcination steps should be designed, not only to enrich the phosphate but also to minimize the levels of chloride, carbonate, and organic matter. Attempts should also be made to obtain the correct balance of reactive silica to F and ideally restrict the concentration of cationic impurities in the rock to within the range of 1 to 2%.

The effect of impurities, particularly trace elements, on crystal habit, crystal growth, filtration rate, and rates of recrystallization need to be determined and quantified wherever possible, if real progress is to be made in obtaining the best performance from a wider range of rocks.

LITERATURE CITED

Allen, M. 1975. Conversion of by-product gypsum to α-hemihydrate by ICI's process. Phosphorus Potassium 78:42–44.
Anonymous. 1972a. Kellogg details progress in phosphoric acid. Eur. Chem. News 21(526):25.
Anonymous. 1972b. Kemira Oy commissions new fertilizer units. Eur. Chem. News 22(554):6.
Anonymous. 1972c. Phosphoric acid production from Jhamar Kotra phosphate rock. Phosphorus Potassium 61:24–25.
Anonymous. 1974a. A new method for extracting uranium. Chem. Eng. News 52(16):29.
Anonymous. 1974b. Recovery of uranium from fertilizer phosphoric acid. Chem. Eng. News 52(34):16.
Anonymous. 1975a. Standard for fluorides. USEPA, Fed. Regis. 40:33154.
Anonymous. 1975b. Uranium in phosphate rock. Phosphorus Potassium 80:46.
Anonymous. 1975c. Phosphogypsum, Rhône-Poulenc transformation processes. Rhône-Poulenc Div. Chimie Minerale, Courbevoie, Paris.
Coates, R. V., and G. D. Woodward. 1966. Similarity between "Chukhrovite" and the Octahedral crystals found in gypsum in the manufacture of phosphoric acid. Nature 212—392.
Dahlgren, S. E. 1960. Calcium sulfate transitions in superphosphate. J. Agric. Food Chem. 8:411–412.
Facer, J. F. 1975. Production statistics. p. 151-158. *In* ERDA Uranium Ind. Sem., Grand Junction, Colo.
Foester, H. J. 1974. Processing in accordance with the Giullini method of synthetic gypsum from phosphoric acid production to high-grade construction materials. ISMA Tech. Conf., 23-27 Sept. 1974, Prague, ISMA, Paris.
Fullam, H. T., and B. P. Faulkner. 1971. Inorganic fertilizer and phosphate mining industries. USEPA Rep. no. 12020.
Greek, B. F., O. W. Allen, and D. E. Tynan. 1957. Uranium recovery from wet process phosphoric acid. Ind. Eng. Chem. 49:628–638.
Hakansson, R. 1965. Full scale production of semihydrate phosphoric acid with 42% P_2O_5. ISMA Tech. Conf., 14-16 Sept. 1965, Edinburgh, ISMA, London.
Heckenbleikner, I. 1928. Art of phosphoric acid manufacture. U.S. Patent 1,667,549.
Hurst, F. J., and D. J. Crouse. 1974. Recovery of uranium from wet process phosphoric acid by extracting with octylphenyl phosphoric acid. Ind. Eng. Chem. Process Des. Dev. 13:286-291.
Kivela, T. 1974. Minimising pollution from phosphate fertilizer plants including captive acid plants. Doc. ID/WG 175/12. U.N. Int. Dev. Organ. Meet., Minimising Fertiliser Plant Pollution, Helsinki, 26-31 Aug. 1974. UNIDO, Vienna.

Larsson, M. 1933. Process of leaching phosphate rock. U.S. Patent 1,916,431.
Larsson, M. 1939. Method of treating raw phosphate. U.S. Patent Re 20,994. (Original patent no. 1,902,648, dated 1933.)
Lehr, J. R., A. W. Frazier, and J. P. Smith. 1966. Precipitated impurities in wet-process phosphoric acid. J. Agric. Food Chem. 14:27-33.
Lehrecke, H. 1935. Recent experiments in the dissolution of crude phosphate by means of sulphuric acid. Tek. Tid. Uppl. c., kemi 65:81-85, 92-94.
McGinley, F. E. 1972. Latent capability from byproducts. Nucl. Ind. Sem., U.S. Atom. Energy Comm., Grand Junction, Colo. 9 p.
Miyamoto, M. 1975. Practice of new Nissan phosphoric acid process. ISMA/ANDA Tech. Conf., 22-24 Apr. 1975, Sao Paulo, Brazil.
Nordengren, S., I. Francia, and R. Nordengren. 1955. The first installation of phosphoric acid plant according to the anhydrite method at Vercelli, Italy. Fert. Soc. Proc. 33:3-22.
Salter, J. W. 1975. Development of the Swenson isothermal reactor for the production of wet-process phosphoric acid. Central Florida Sec. of Am. Inst. of Chem. Eng. 23 p.
Schucht, L. 1926. Die fabrikation des superphosphates. Vieweg und Sohn, Braunschweig, Germany.
Spicer, H. N. 1926. Lining for ball mills. U.S. Patent 1,590,655.
Swaine, D. J. 1962. The trace element content of fertilizers. Tech. Commun. no. 52. Commonw. Agric. Bur., Farnham Royal, Slough, UK.
Weber, W. C., E. J. Roberts, I. S. Mangat, and E. Uusitalo. 1969. The Dorr-Oliver HYS phosphoric acid process. Fert. Soc. Proc. no. 112:49-77.
Weber, W. C., R. W. Shafor, and E. J. Roberts. 1936. Preparation of phosphoric acid. U.S. Patent 2,049,032.
West, J. Z. 1930. A quantitative X-ray analysis of the structure of potassium dihydrogen phosphate (in English). Krist. 74:306-332.
Whalley, L. 1976. The environmental impact of gaseous emissions from the manufacture of fertilizers. Fert. Soc. Proc. no. 156:55-102.
Wirsching, F. X. 1971. Chemical gypsum as a basis for high quality building gypsum. Tonind. Ztg. Keram. Rundsch. 95:14-18.
Yermilova, L. P., V. A. Moleva, and R. F. Klevtsova. 1960. Chukhrovite, a new mineral from Central Kazakhstan. Zap Vsesoyuzn. Miner. Obshch. 89:15-25.

Chapter 7

Phosphate Fertilizers and Process Technology

RONALD D. YOUNG AND CHARLES H. DAVIS
Tennessee Valley Authority
Muscle Shoals, Alabama

I. INTRODUCTION

The first phosphate fertilizer was made of ground bones and was used widely in Europe during the early part of the 19th century. Treatment of bones with sulfuric acid was pioneered by Liebig in Germany about 1810 and soon became a common practice. Dilute sulfuric acid was used, so the product was a slurry that was distributed in wooden casks. Other nutrient materials, such as ammonium sulfate, potash, and nitrate of soda, were added later, thus producing the first fluid mixed fertilizer (Hignett, 1971).

In about 1830 the practice progressed to treatment of the more abundant pulverized phosphate rock (PR) with sulfuric acid. The solid product was called *superphosphate*. The first successful factory for production of superphosphate was established by Lawes in England in 1842. The practice grew, and by 1853 there were 14 producers in the UK. By 1862 Lawes was operating a continuous superphosphate mixer and produced 100 metric tons/day. The production or ordinary superphosphate (OSP) spread rapidly to other parts of Europe and to the USA (Hignett, 1971).

Production of concentrated or triple superphosphate (TSP) went hand in hand with the production of wet-process phosphoric acid. The first known commercial production was in Germany, and the practice gradually spread to other European countries and the USA. Triple superphosphate did not become an important phosphate fertilizer until the 1950's when several large plants were built. The TVA's production and distribution of large quantities of concentrated superphosphate (CSP) made with electric furnace acid, starting in the late 1930's, led to widespread testing and market development of this higher analysis fertilizer.

Although ammonium phosphate had long been known as an effective fertilizer, and small amounts had been produced for fertilizer use from time to time, it did not become an important fertilizer until TVA developed a practical and economical process for granular diammonium phosphate

(DAP) in 1961 (Young et al., 1962). This process proliferated, and a number of large plants for production of the 18-20-0 (18-46-0) grade were built in the USA, in Europe, and later in Asia and South America. The TVA researchers also developed a slightly modified process for production of granular monoammonium phosphate (MAP) that is predominant in Canada and Pakistan and is produced at other locations (Young & Hicks, 1967).

Development of nitric phosphate processes in which phosphate rock is acidulated with nitric acid or with nitric and sulfuric or phosphoric acid started in Europe in the early 1930's. Several process modifications were developed over the years. Popularity of nitric phosphates stemmed from the more favorable economics at many places as compared with alternative process routes using imported S. Production has been small in the USA but there are several nitric phosphate plants in Europe and also some in Asia and South and Central America (Slack, 1967a, 1967b, 1967c; Slack et al., 1967).

Ammonium polyphosphate (APP) fertilizers have become quite important commercially. Liquid and solid APP products were pioneered by TVA starting about 1955 (Siegel & Young, 1969). They form the backbone of the liquid fertilizer industry that is growing rapidly in the USA (about 2,800 plants in 1975).

Starting about 1970, several processes were developed for production of potassium phosphate fertilizers of various types. One small plant began operation in the USA in 1975.

II. CURRENT PRODUCTS AND PROCESSES

Phosphate rock mining and processing were described in a previous chapter. Fortunately this important resource is available in many places throughout the world. Recent surveys and explorations in diverse areas indicate that the world should have ample supplies of this important resource for the near future. However, it likely will be necessary to utilize P of lower grade and with high levels of objectionable impurities.

Some of the mining and beneficiation operations in Florida, North Carolina, Morocco, and USSR range in capacity from 2 million to 8 million metric tons per year. A small, but significant quantity of this PR is directly applied to agricultural fields. This practice is discussed elsewhere in this book (see Chapt. 4). The largest portion of this raw material is processed further, either chemically, thermally, or both. The objectives of all processes are (i) to increase the water solubility of P in the apatite which enhances its availability for plant uptake and (ii) to upgrade the plant nutrient content of the finished product, which cuts down on cost of shipping, handling, and storage. In addition, the fertilizer industry is always interested in alternative processes which utilize more economical schemes, consume less energy, have less adverse impact on the environment, or offer increased flexibility with respect to variability in raw materials or finished products.

A. Acidulation Processes

1. ORDINARY (SINGLE) SUPERPHOSPHATE

Most of the earlier factories were primitive, the reactants being mixed manually with hoes or paddles in kettles or vats. By 1862 Lawes was using a continuous mixer with a capacity of 100 metric tons/day. Production of OSP spread rapidly in Europe, the USA, and later in Central and South America and Asia. Most of the plants were comparatively small and provided superphosphate for formulation of mixed fertilizers to serve local markets. After about 1950 most of the NPK fertilizers formulated with OSP were granular.

Varied types of equipment have been used over the more than 130-year history of OSP production. Most of the early plants were for batch-type operation. A popular type used a shallow pan into which the acid and rock were measured, mixed with a mechanical stirrer, and dumped into a wooden or concrete "den." After the den was filled and the superphosphate was held long enough to solidify and become friable, it was excavated by a dragline or by various types of mechanical excavators. A popular type of equipment in the USA was the Steadman pan mixer and excavator (Fig. 1).

Continuous mixers and dens for OSP were developed later and resulted in a more efficient process and increased production rate. There have been various types of continuous mixers, but the lowest cost and simplest is the TVA cone mixer that has no moving parts as shown in Fig. 2. Mixing is ac-

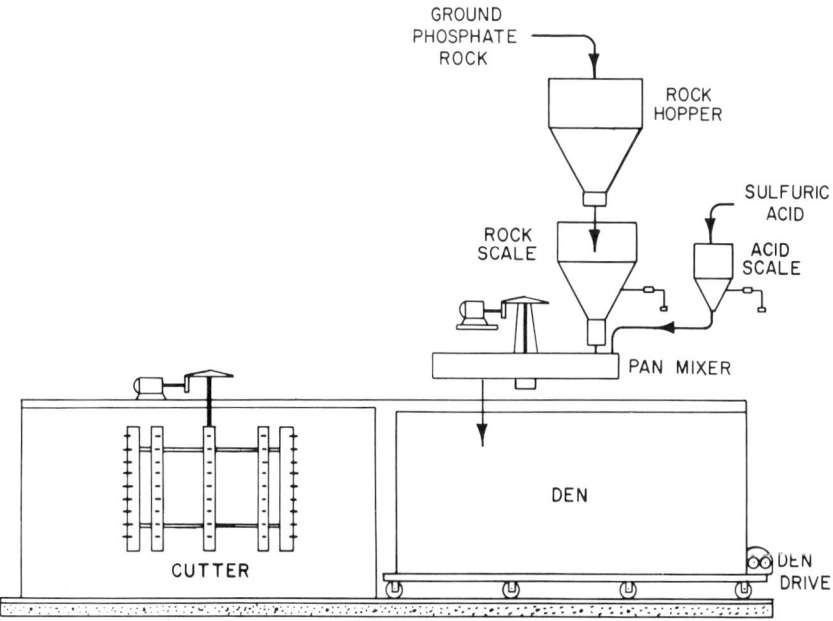

Fig. 1—Batch manufacture of ordinary superphosphate.

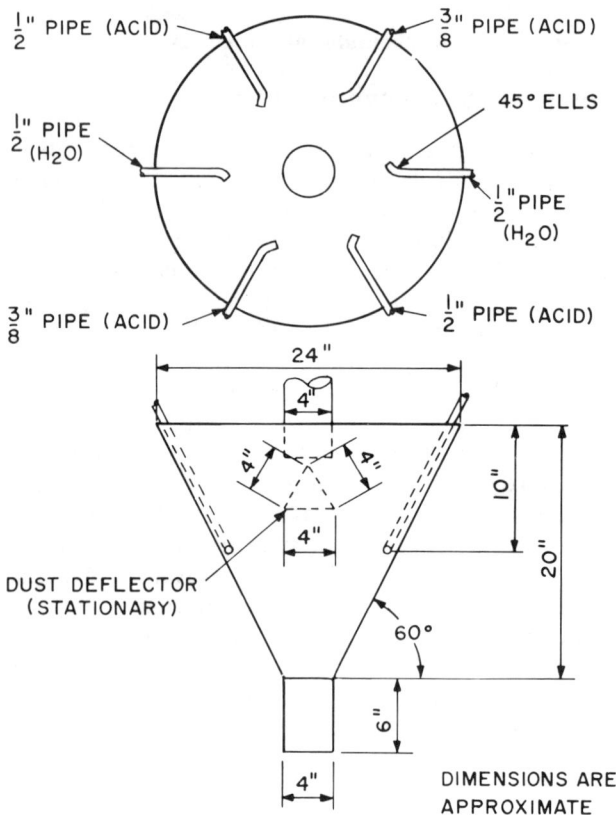

Fig. 2—Typical cone mixer for production of 25 to 30 metric tons/hour of ordinary superphosphate.

complished by the swirling action of the acid. Short, single-shaft or double-shaft pugmills also are quite widely used throughout the world for mixing of sulfuric acid and the rock prior to denning. The most widely used continuous den is the slat conveyor-type generally referred to as the *Broadfield den*. A schematic sketch of a plant using a cone mixer and Broadfield den is shown in Fig. 3. Some other types of continuous dens popular in Europe are the Moritz-Standaert and Forbis (USDA-TVA, 1964).

The reaction of PR with sulfuric acid to produce OSP can be expressed in chemical stoichiometric terms. However, in general practice the proportioning is usually based on a simpler relationship of about 0.59 to 0.61 kg of sulfuric acid (100% H_2SO_4 basis) per kg of phosphate rock containing 13 to 14% P (30 to 32% P_2O_5). The PR usually is pulverized so that 90% is <0.15 mm (minus 100 mesh) and 70% is <0.075 mm (minus 200 mesh).

Gases that are released as the superphosphate is solidifying (setting) results in a porous, friable mass in the den. Ordinary superphosphate made from typical rock will "set" in 40 to 50 min in a continuous den and can be cut and handled satisfactorily. The cycle is usually 1.5 to 2 hours in a batch den before excavating.

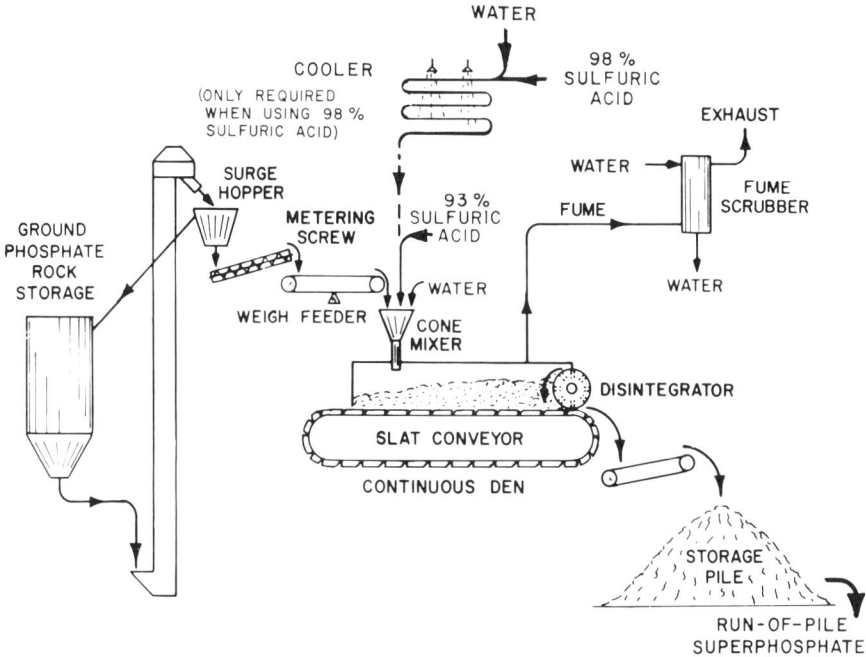

Fig. 3—Continuous process for the manufacture of ordinary superphosphate.

The superphosphate usually is held in storage piles (cured) for 4 to 6 weeks to obtain better handling properties and to allow the continuation of chemical reactions and the attainment of high conversion (95 to 96%) of P_2O_5 to citrate soluble (available) form. Usual grade of OSP made from Florida rock is 9% P (20% available P_2O_5). Use of lower grade rock results in product containing 18 or 19% available P_2O_5. An analysis of OSP made with Florida rock is shown below.

	Analysis, % by wt							
	P							
Total	Available	H_2O soluble	Ca	Free acid	SO_4-S	F	R_2O_3	Mg
8.82	8.65	7.86	20.1	3.7	9.9	1.6	1.6	0.09

Because of the low analysis of OSP (20% P_2O_5), economics favors shipping the PR (14% P, 32% P_2O_5) to local plants where the superphosphate is produced and usually processed in formulations for granular NPK fertilizers. Ordinary superphosphate of good quality absorbs ammonia readily during the granulation process in the proportion of up to about 6 kg ammonia per 20 kg of P_2O_5 in the superphosphate. Usually the maximum amount of OSP possible is incorporated in formulations for grades in which it can be used, since it is the lowest cost form of P_2O_5.

Ordinary superphosphate is granulated in large quantities in Australia and New Zealand, normally as it is discharged from the den. Pan granulators and rotary drums are used; sometimes steam is employed to promote granulation. Smaller amounts are granulated in western Europe by use of steam and water in a rotary drum.

2. TRIPLE SUPERPHOSPHATE

Triple superphosphate is made by acidulation of PR with phosphoric acid with equipment and process similar to that for OSP. Triple superphosphate (20% P; 45 to 46% P_2O_5) did not appear on the scene in any appreciable quantity until wet-process phosphoric acid was produced commercially. Production of similar concentrated superphosphate was initiated by TVA in the late 1930's by use of electric furnace phosphoric acid. Widespread agronomic testing, and market development through use of large tonnages by fertilizer manufacturers in demonstration programs, led to rapid acceptance of this much higher analysis P material. Producers of PR in the USA and other countries moved into production of wet-process acid and TSP. Logistics favored production of the higher analysis TSP near the source of rock and shipping this intermediate to local mixed-fertilizer plants near the markets.

The TVA cone mixer has been almost universally used in production of nongranular TSP. The "set time" for TSP is only 14 to 20 min, as compared with 40 to 50 min for OSP. A simple cupped conveyor belt, instead of the slat-type den for OSP, is used to hold the acidulate until the superphosphate solidifies. Production rate ranges from 40 to 50 metric tons/hour with a belt about 1.5 m wide and 30 m long. A diagram of the cone mixer and "wet-belt" system is shown in Fig. 4.

Proportioning is typically 2.4 to 2.5 kg of P from acid for each kg from rock. The rock is pulverized to about 90% minus 100 mesh and 70% minus 200 mesh. The TSP usually is cured 4 to 6 weeks prior to shipment or use at the site. Chemical analysis of a TSP made with Florida rock is shown below.

Analysis, % by wt								
P								
Total	Available	H_2O soluble	Free acid	Ca	R_2O_3	MgO	F	H_2O
20.5	20.2	18.3	3.4	13.8	3.1	0.3	2.7	4.5

Use of TSP in granular NPK fertilizer formulations, together with or in place of single superphosphate, allowed production of higher analysis grades of granular compound fertilizers. These included 12-5-10 (12-12-12), 8-7-13 (9-16-16), and 5-9-17 (5-20-20) instead of 10-5-8 (10-10-10), 6®5-10 (6-12-12), and 3-5-10 (3-12-12) using only OSP. With TSP supplying all of the P, grades as high as 3-6-11 (13-13-13) and 9-8-15 (9-18-18) can be produced.

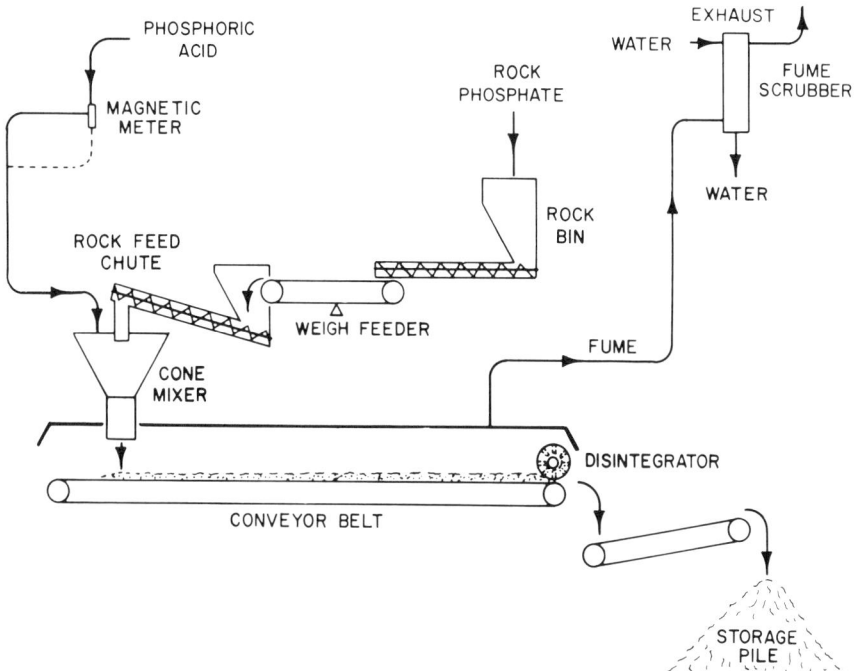

Fig. 4—Continuous process for the manufacture of triple superphosphate.

The TSP can be readily ammoniated to about 3.5 kg of ammonia per 8.7 kg of P during the granulation process.

Granular TSP is produced in large quantities for use in direct application and in bulk blends. In some processes cured TSP is granulated in a rotary drum or pan granulator using steam and water to promote granulation. In Australia, New Zealand, and the UK, the superphosphate is granulated in a pan or drum as it comes from the den.

A slurry-type process as shown in Fig. 5 is used in the USA and other countries. Pulverized PR is treated with wet-process phosphoric acid in a two-stage slurry reaction system and the slurry is sprayed into a pugmill or rotary drum for layering on recycled fines at a ratio of 10 to 12 kg of recycle/kg product. Product granules (20% P; 45% P_2O_5) are quite spherical and dense. The lower grade results from the need to decrease the acidulation ratio from the 2.45 kg of acid P per kg of rock P to about 2.25 to control free acid content and resultant prolonged stickiness. For production rate higher than about 23 metric tons/hour, a rotary drum is usually used instead of a pugmill.

Most of the granular TSP in the USA is used in bulk blending, as discussed later.

3. PHOSPHORIC ACID

Wet-process phosphoric acid (WPA, made by reaction of PR with sulfuric acid) became important in fertilizer production when TSP was estab-

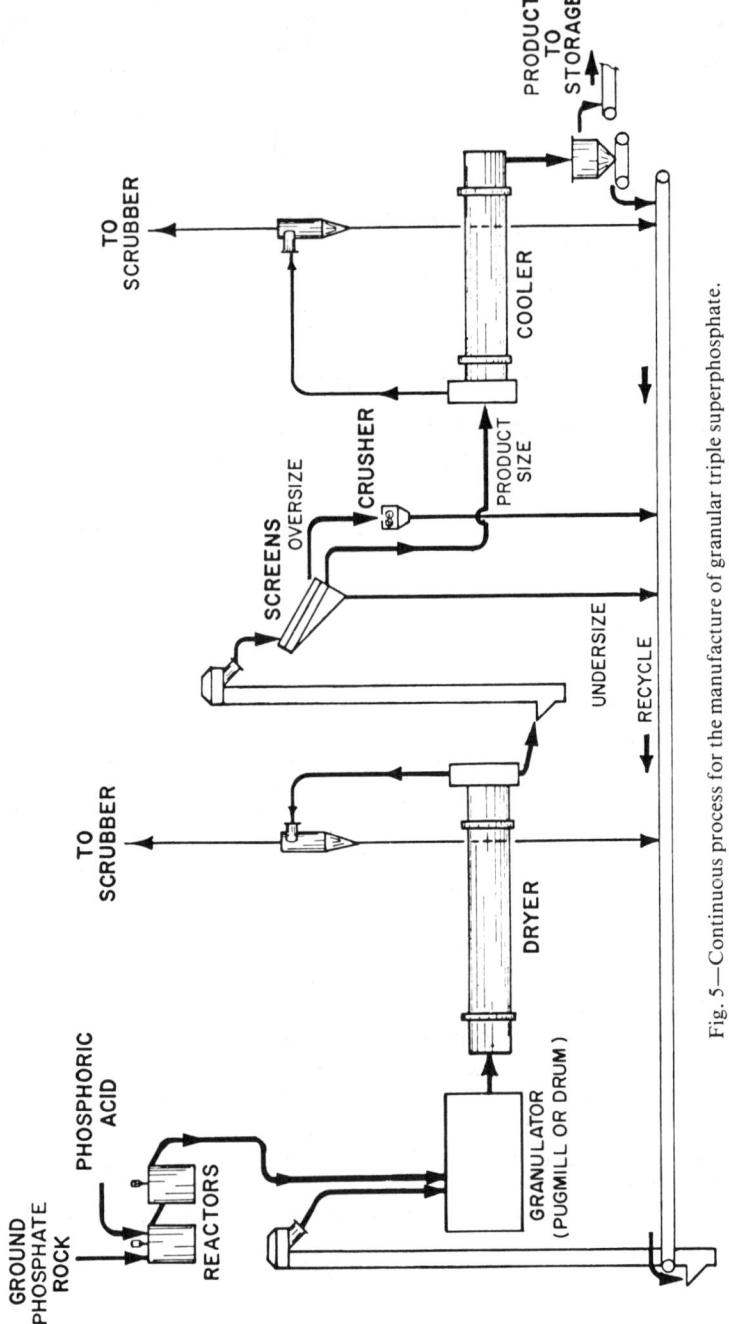

Fig. 5—Continuous process for the manufacture of granular triple superphosphate.

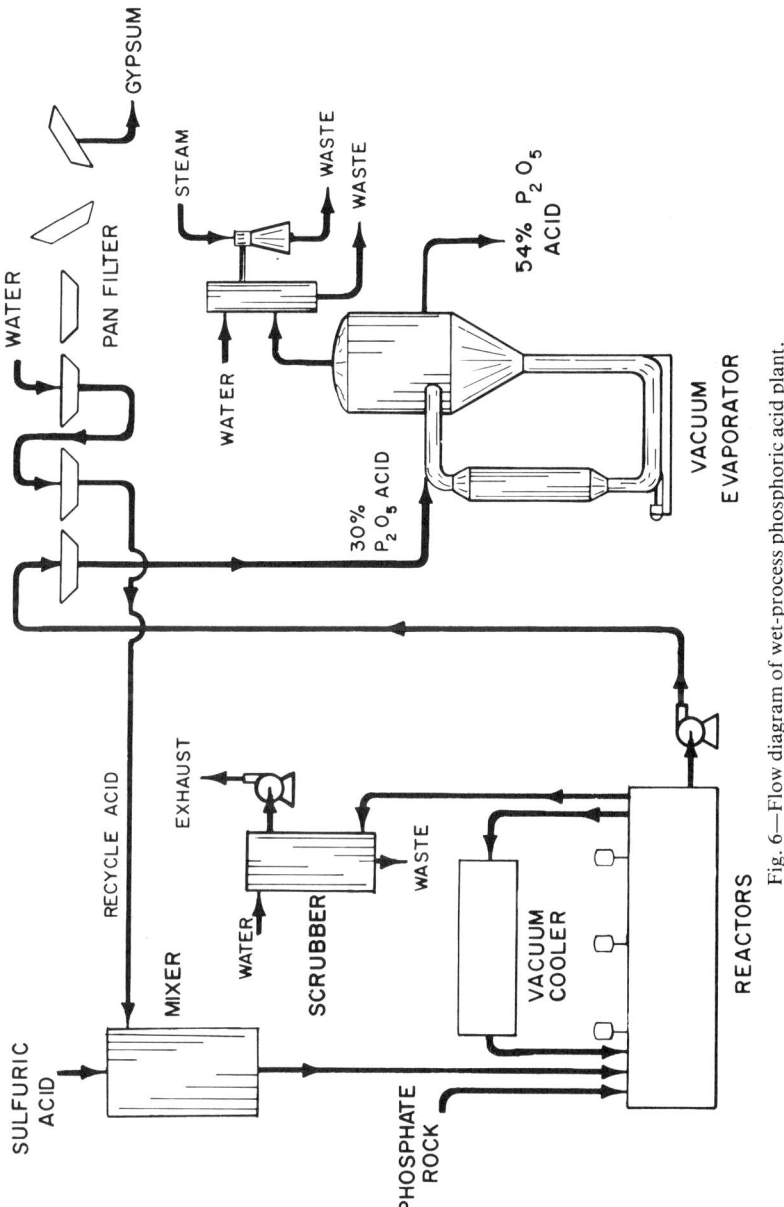

Fig. 6—Flow diagram of wet-process phosphoric acid plant.

lished as a good, economical, high-analysis fertilizer and intermediate. In the USA, most of the wet-process acid is produced in large production complexes near the site of PR deposits. A large part of this acid is used at the site in production of nongranular and granular TSP and granular ammonium phosphates. However, substantial amounts of the acid are shipped to local plants. Canada imports PR for use in production of wet-process acid and subsequent TSP and granular MAP. Most phosphate plants in Europe, Asia, and South America import PR for use in producing wet-process acid and other products. In North Africa and Mexico, wet-process phosphoric acid is produced for export.

A schematic diagram of the WPA process is shown in Fig. 6. Pulverized PR is reacted (digested) with sulfuric acid in multiple-reaction tanks or in a compartmented single tank. Raw PR ordinarily is used, but calcined rock is sometimes used when a clear, green acid for use in making liquid fertilizers is desired. Calcination is sometimes a necessary step in beneficiation. Retention time ranges from as low as 1.5 to 2 hours to as long as 10 to 12 hours. Slurry from the reaction system is filtered to remove the calcium sulfate that is precipitated in the reaction. Conditions during reaction are controlled to obtain crystals of calcium sulfate that filter at a high rate. Cooling of the reactant slurry in process is provided; a flash (vacuum) cooler often is used. Cake on the filter is washed to recover about 95% or more of the P in PR as phosphoric acid from the filter at a concentration usually ranging from 12.2 to 13.1% P (28 to 30% P_2O_5) for the dihydrate process or 18.3 to 21.8% P (42 to 50% P_2O_5) for the hemihydrate or anhydrite processes.

The most widely used process results in precipitation of calcium sulfate dihydrate (gypsum) (Kulp & Leyshon, 1968; Janikowski, 1968; Roubinet, 1968; Kaji, 1968). There are processes which are operated at higher temperatures resulting in the hemihydrate or the anhydrite precipitate (Murakami & Hori, 1968; Versteegh & Boontje, 1968). Various types of continuous filters are used, but the most popular one is the tilting-pan type shown schematically in Fig. 6. The next most popular type is the belt filter. The filter acid is concentrated to 22.7 to 23.6% P (52 to 54% P_2O_5) for merchant grade and for use in producing TSP. For ammonium phosphate production at the site, a concentration of only 17.5% (40% P_2O_5) is sufficient. Wet-process superphosphoric acid is prepared by further concentration to 29.7 to 30.6% P (68 to 70% P_2O_5) content; single-stage vacuum evaporators of the forced-circulation type are used in most superphosphoric acid plants.

4. AMMONIUM PHOSPHATES

Although ammonium phosphates did not come on the scene in significant quantities until the early 1960's, they have become the leading form of phosphate in the USA and are increasing in popularity throughout the world. Almost all new phosphate fertilizer complexes built in recent years and now being built are for production of ammonium phosphate as the major product.

Crystalline DAP was produced by TVA starting in the late 1940's, and demonstration programs showed it to be a very good high-analysis fertilizer. Smaller amounts were produced by others as a byproduct. When the comparatively simple and dependable TVA process for granular DAP of grade 18-20-0 (18-46-0) was developed in 1960-1961, it was rapidly adopted by the industry (Young et al., 1962). Many granular DAP plants have production capacity of about 50 metric tons/hour and a few as high as 70 to 100 metric tons/hour.

a. Granular DAP—A flow diagram of a typical granular DAP production unit of the TVA type that has become standard in the industry is shown in Fig. 7.

Wet-process phosphoric acid of about 17.5% P content (or a mixture of 23.6% P acid and acid from the scrubbing circuit of 12.2 to 13.1% P content) is fed to a preneutralizer. Anhydrous ammonia is sparged through open-end pipes that project through the walls of the tank. The NH_3 neutralizes the acid to an NH_3/H_3PO_4 mole ratio of about 1.4. This is in a range of maximum solubility of ammonium phosphate as shown in the solubility curve of Fig. 8. The heat of reaction evaporates considerable water, and water content of the slurry is 16 to 20%.

The slurry is pumped at a controlled rate and distributed on the bed in a rotary drum, TVA-type ammoniator-granulator. Several types of distributors for the slurry, including saw-toothed types, spray nozzles, and several open-end pipes, are used. The most commonly used metering system for preneutralized slurry is a variable-speed centrifugal pump with automatic control signal from a magnetic flowmeter. A few plants have had success recently with a magnetic flowmeter and automatic control valve of a special ball type.

Anhydrous, gaseous ammonia is sparged beneath the bed to ammoniate the slurry further to near DAP; the usual finishing NH_3/H_3PO_4 mole ratio is 1.85 to 1.90. Ammonia evolved from the preneutralizer and granulator is recovered in acid of about 13.1% P content in a scrubbing circuit. Recycled material at a rate of 5 to 7 kg per kg of product is the primary control of granulation.

Discharge from the granulator is dried with moderate heat to 82 to 88°C product temperature. Most plants screen hot and only the product fraction is cooled, since the material is not very sticky. Rotary coolers of the fluidized bed type are used. The product, with moisture content of 1.5 to 2%, does not require a conditioner. It has excellent storage and handling properties in bags or in bulk.

Materials of construction other than mild steel are required for the acid, preneutralizer, and slurry handling systems and for the scrubbing circuits. Type 316L stainless steel or rubber- and bricklined mild steel is used for the preneutralizer. Type 316L stainless is also used for the slurry pumps and piping. Fiberglass-reinforced polyester plastic and high-density polyvinyl chloride are sometimes used for wet-process acid pipes and for scrubbers that may also be of rubber-lined mild steel construction.

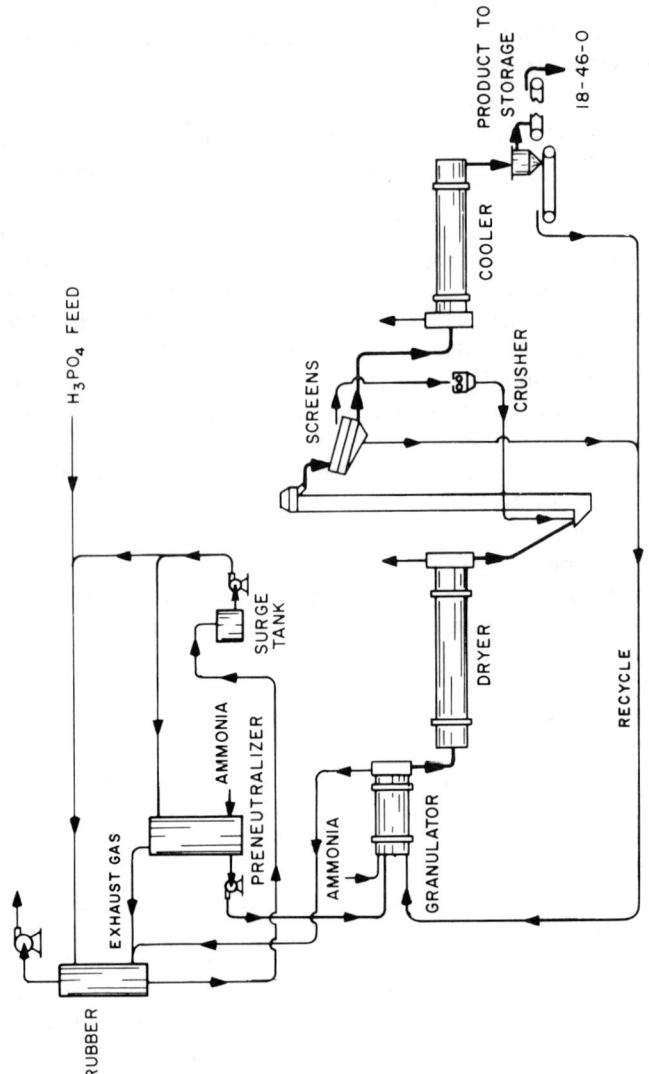

Fig. 7—Flow diagram of diammonium phosphate plant.

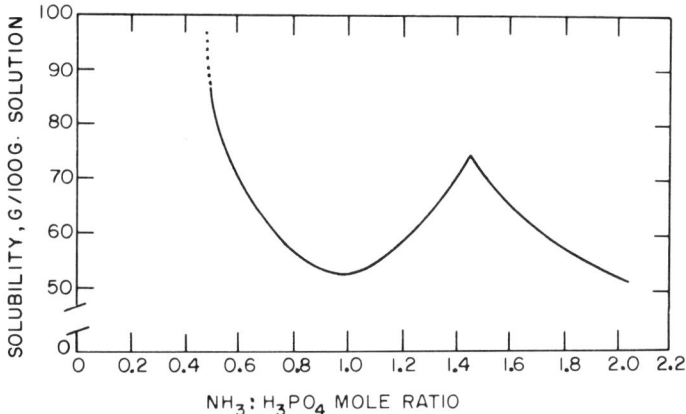

Fig. 8—Effect of NH_3/H_3PO_4 mole ratio on solubility of ammonium phosphate at 88°C.

b. Granular MAP—Granular DAP of grade 18-20-0 (18-46-0) have become household terms in the world fertilizer industry. But substantial interest in granular MAP persists, particularly where soils are mainly alkaline, as in Canada and Pakistan. Also, where the primary interest is in producing and shipping phosphate, the 11-22-0 (11-52-0) to 10-23-0 (10-54-0) grades of MAP provide higher phosphate payload.

The TVA developed two comparatively minor modifications of the granular DAP process to allow production of MAP (Young & Hicks, 1967). In one method, the preneutralizer is ammoniated to only an NH_3/H_3PO_4 mole ratio of about 0.6 and then to about 1.0 in the granulator drum. In the other procedure, the preneutralizer is ammoniated to about 1.4 as with DAP, and additional wet-process acid is distributed onto the bed in the granulator to adjust back to MAP mole ratio of about 1.0. Remainder of either process modification is the same as for DAP, but higher drying temperature can be used to increase the production rate for MAP.

c. Nongranular MAP—Starting about 1968, simple processes have been developed for production of nongranular, or powdered, MAP. The main processes were developed by Fisons in England, Scottish Agricultural Industries (SAI), Swift in the USA, and Nissan in Japan. The TVA has done some pilot-plant work in production and use of this intermediate (TVA, 1972). In the Fisons process, wet-process acid is ammoniated in a reactor at a pressure of about 2.5 kg/cm² (35 psi) and the slurry is sprayed into a fabric-enclosed tower. Product solidifies as the droplets fall downward through airflow in the tower and form very small prill-like particles. The Swift process is similar, except that a pipe reactor is used for the reaction. In the SAI process, slurry produced in a preneutralizer at about 1.4 mole ratio is fed to a pugmill-type unit and additional acid is sprayed into the material to adjust back to a mole ratio of about 1.0. The material quickly solidifies at this point of minimum solubility. The Nissan process uses a spray tower for ammoniation of the acid and solidification of the product.

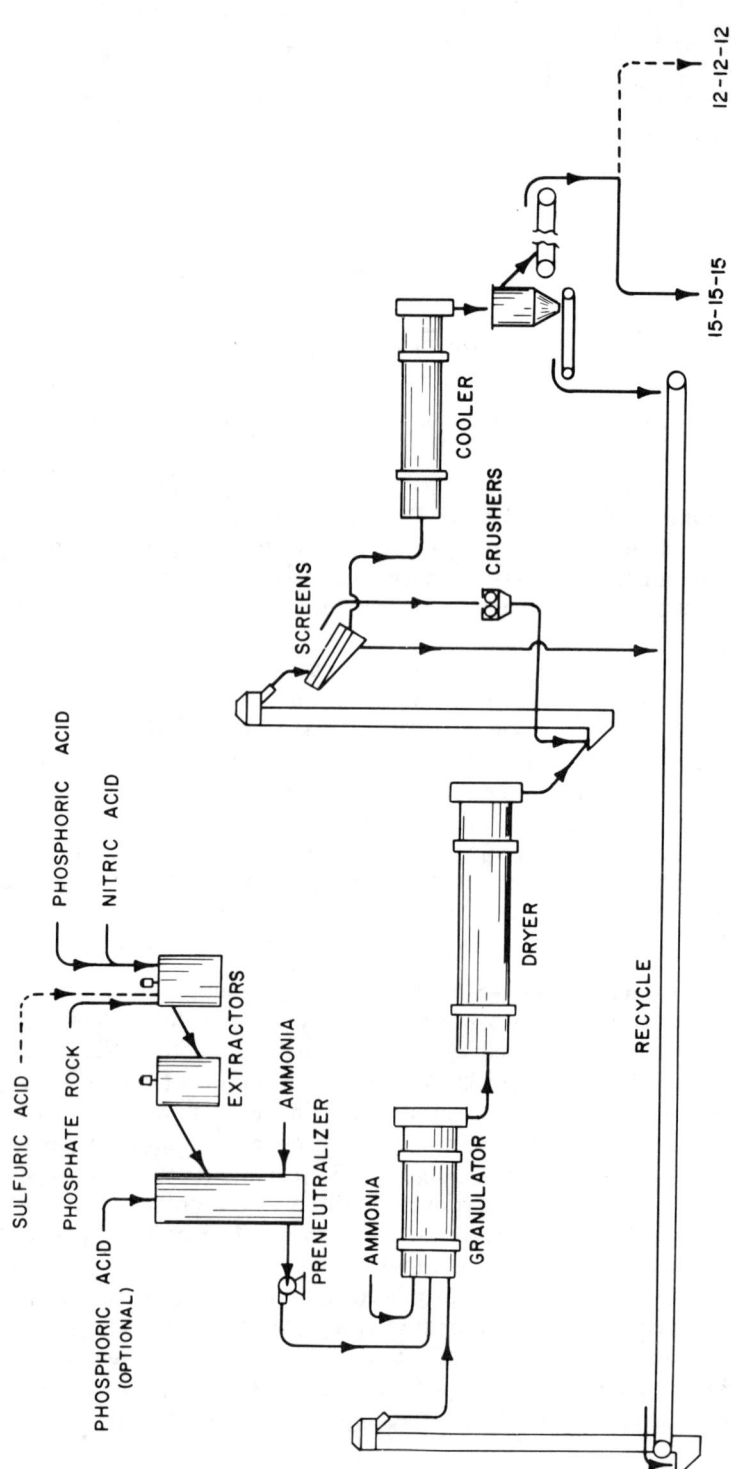

Fig. 9 — The TVA mixed acid nitrophosphate process.

Product from the various processes ranges in its moisture content from 4 to 8%, but storage and handling properties generally are good.

A number of these comparatively low-cost units have been built commercially including plants in the UK, the Netherlands, Japan, Australia, Spain, USA, Brazil, and Iran. In some locations the nongranular MAP is shipped to other plants where it is fed as a major component in production of NPK fertilizers. Major fertilizer producers in Spain make coastal and overland shipments of nongranular MAP for use in NPK granulation plants for distances up to 500 km. Nongranular MAP has been shipped and handled satisfactorily in international trade.

5. NITRIC PHOSPHATES

Fertilizers referred to as nitric phosphate or nitrophosphate are produced by acidulation of PR with nitric acid or with mixtures of nitric and sulfuric or phosphoric acids. A variety of processes and equipment has been used in Europe since the late 1930's (Davis et al., 1968; Hignett, 1966; Young, 1966). There are also a number of plants in Central and South America and Asia. In 1976 there were only three nitric phosphate plants in the USA, and these had been modified to use higher proportions of phosphoric acid and other materials.

The primary advantage of nitric phosphate processes is that little or no S is required as compared with ammonium phosphates; this is particularly important during a shortage of S, or where S must be shiped long distances over land.

In the Odda process no supplemental sulfuric or phosphoric acid is used; the nitric acid-rock extraction slurry is cooled to crystallize Ca nitrate that is removed by centrifugation. The Ca nitrate is either sold as a fertilizer or converted to coproduct ammonium nitrate. Various process modifications use a rotary drum, pugmill, or spray drum (Spherodizer) for granulation (Hignett, 1966). In recent years NPK nitric phosphate grades have been produced by a prilling process in a few European plants.

In earlier years, the main disadvantage of nitric phosphate processes was low water solubility of phosphate in the products. Use of supplemental phosphoric acid, or "deep cooling" by refrigeration to remove a higher proportion of Ca as Ca nitrate in the Odda-type processes, has increased water solubility to 60% or higher. Additional disadvantages of nitric phosphate processes now include (i) limited flexibility in N/P ratio, (ii) more N is produced than phosphate; and (iii) susceptibility of some NPK grades to "cigar burning" and decomposition if exposed to a sufficient source of heat.

A flow diagram for the TVA mixed acid nitric phosphate process is shown in Fig. 9. The TVA operated a demonstration plant of 16 to 18 metric tons/hour capacity for several years and produced fertilizers of grade 20-9-0 (20-20-0) and 26-6-0 (26-13-0).

Popular grades of nitric phosphates include 14-6-12 (14-14-14), 10-4-8 (10-10-10), 22-5-9 (22-11-11), 20-9-0 (20-20-0), and 26-6-0 (26-13-0). Growth in production of nitric phosphates has been slow since the popular processes for granular DAP and MAP started their rapid growth in the early 1960's throughout the world.

B. Compound (NPK) Fertilizers

A large part of the phosphate intermediates—ordinary and triple superphosphate, wet-process phosphoric acid, and ammonium phosphates—is used in granulation plants nearer the market as components in formulation of a variety of grades of NPK granular fertilizers (TVA, 1971b). In the most commonly used system, the solid phosphate intermediates together with potash salts are proportioned (usually in a batch weighing system) and fed in continuous flow to an ammoniator-granulator drum. Nitrogen solution and anhydrous ammonia are sparged beneath the bed of solids to provide heat of reaction and fluid phase for granulation. Material from the granulator is dried, cooled, and screened to prepare the product with grades such as 13-6-11 (13-13-13), 8-7-13 (8-16-16), 10-9-8 (10-20-10), and 6-10-19 (6-24-24) (Yates et al., 1954).

1. BULK BLENDING

Since its inception in the mid-1950's, bulk blending, the physical mixing of granular materials, has transformed the method of preparation of finished fertilizers in the USA. This system also led to widespread bulk handling, transport, and application of granular fertilizers of all types. Latest estimates indicate more than 5,000 bulk blending units in the USA in 1974. Production rate per unit ranges from 4,000 to 8,000 metric tons/year. There are a few blending units in Brazil, England, and some other countries, but this is primarily an institution and a practice of the U.S. industry. Figures from a survey in 1974 showed that about 40% of U.S. mixed fertilizer was bulk blends (TVA, 1975).

A schematic diagram of a typical bulk-blending plant is shown in Fig. 10. Granular intermediates that include granular DAP or MAP, granular TSP, prilled or granular urea or ammonium nitrate, and granular potassium chloride are used. They usually are received in rail hopper cars and unloaded by undertract screw and elevator into storage compartments at ground level. A front-end loader conveys the components to an elevator that delivers them into an overhead compartmented hopper for the separate materials. The components are discharged and weighed in sequence in a scale-mounted hopper. Then the full batch of 2 to 6 metric tons is blended in a rotary blender discharged into an applicator truck. Only a small percentage is bagged. A wide variety of grades can be produced, or products containing specified amounts of N, P, and K can be easily formulated. Other types of blenders, such as a scale-mounted ribbon mixer, are used. Although weighing and blending are batch operations, the entire operation is essentially continuous (Hignett, 1965; TVA, 1971b).

Rapid growth of bulk-blending practices and production of granular DAP went hand in hand, and each was effective in promoting the growth of the other.

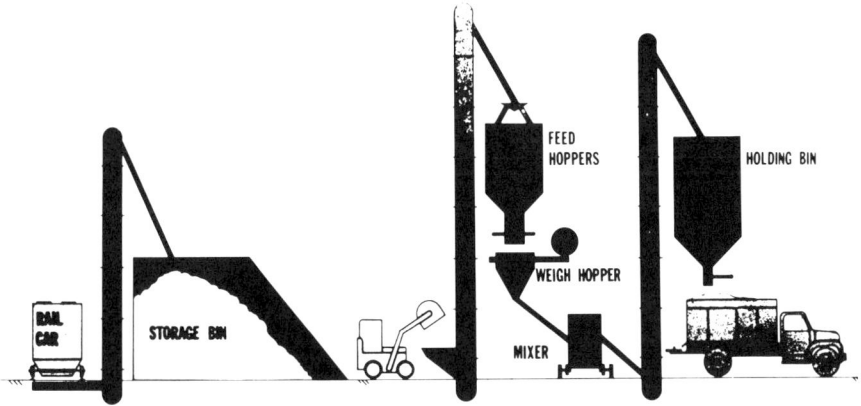

Fig. 10—A bulk blending plant.

2. LIQUID FERTILIZERS

a. General Development—The TVA pioneered the development, introduction, and marketing of liquid ammonium polyphosphates starting in about 1956. They were derived from electric furnace (thermal) phosphoric acid of high strength (32 to 34% P, 74 to 78% P_2O_5) called *superphosphoric acid*. This acid could be produced in thermal acid plants of the conventional type by adjusting water input and operating conditions. Ammoniation of superphosphoric acid with controlled addition of water produced clear base liquid fertilizers of grade 11-14-0 (11-33-0) with 50% polyphosphate content and later 11-16-0 (11-37-0) grade with polyphosphate content of 75%.

The polyphosphate content of 50 to 70% of the total P in the intermediate liquid fertilizers allowed production of more stable NP and higher analysis NPK grades by industry. Later it was found that the polyphosphates, and particularly the pyrophosphate form, would sequester Fe and Al impurities in wet-process acid. This allowed use of lower cost wet-process acid in a substantial proportion of the phosphate in liquid fertilizers. In still later work, TVA researchers developed a process for concentrating wet-process orthophosphoric acid to superphosphoric acid with a P concentration of 29 to 31% (68 to 72% P_2O_5) and with 40 to 50% of the P as polyphosphates (Fleming, 1969; Siegel & Young, 1969; Achorn & Scott, 1969; Huffman & Newman, 1970; TVA, 1971a).

The phosphate industry in the USA rapidly built acid concentrators to prepare the wet-process superphosphoric acid. The practical and economic process for liquid polyphosphates from wet-process acid became the backbone of a rapidly growing liquid fertilizer industry. The TVA continued production of thermal superphosphoric acid and 11-16-0 (11-37-0) with 75 to 80% of its P as polyphosphates through 1975. This intermediate with very high polyphosphate content was used mainly for sequestration of impurities in wet-process acid, and particularly for dealing with the problem of Mg precipitation when using acids with a Mg content of 0.24% or higher (Young et al., 1969).

b. The Pipe Reactor—In 1971 a simple, but very important, TVA development—the pipe reactor—allowed the use of wet-process acid to produce ammonium polyphosphate liquids with grades of 10-15-0 or 11-16-0 (10-34-0 or 11-37-0) with polyphosphates comprising 75 to 80% of the total P (Meline et al., 1972).

The principle of the pipe reactor is shown in Fig. 11. Wet-process superphosphoric acid with polyphosphate content of only about 20% is fed to one branch of a stainless steel pipe tee and anhydrous, gaseous ammonia to the other. The very rapid reaction in the 3 to 3.6 m of pipe extension from the tee results in temperatures of 340 to 370°C (650 to 700°F). Combined water is driven off as steam and the reaction melt contains 75 to 80% of the P as polyphosphate. The melt is quenched in cooled liquid fertilizer in a reaction tank where water and additional ammonia are added to produce 11-16-0 (11-37-0) or 10-15-0 (10-34-0) with 70 to 80% of P_2O_5 as polyphosphate. The liquid is recirculated through a cooler. This simple and low-cost development was adopted very rapidly by the liquid fertilizer industry

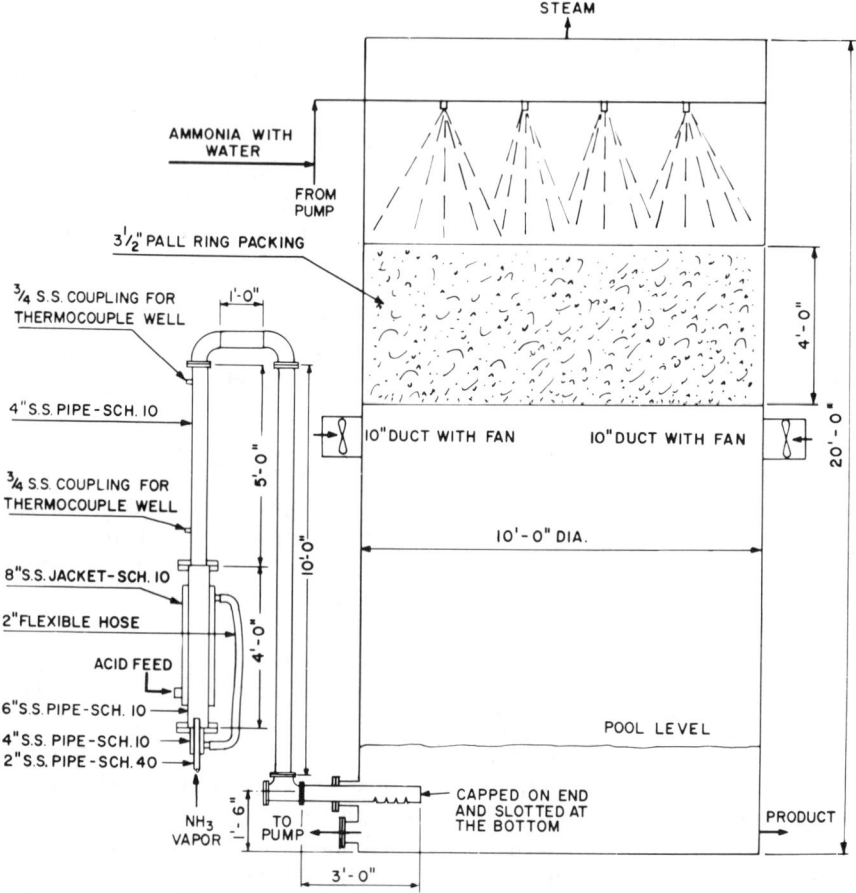

Fig. 11—The TVA pipe reactor system for liquid fertilizers of high polyphosphate content.

and opened up new vistas for high-analysis liquids of good quality made entirely from wet-process acid (Achorn & Kimbrough, 1974). There were more than 100 commercial pipe reactor units at the end of 1975. Also, the producers of wet-process superphosphoric acid were able to greatly simplify their equipment, since the second-stage concentrator, heated with high-temperature medium to attain 40 to 50% conversion to polyphosphate, could be eliminated. The product acid of low-polyphosphate content has much lower viscosity and much better properties for shipment, handling, and storage. With the advent of the pipe reactor, the liquid fertilizer industry based on wet-process acid really came of age.

c. Granular APP—The TVA developed a process for production of granular APP from electric furnace superphosphoric acid and ammonia. The process was adaptable to use of superphosphoric acid made with up to 30% of the P as wet-process orthophosphoric acid fed to the superphosphoric acid plant. The primary grade of the granular product was 15-27-0 (15-62-0). It was shipped in bulk and used primarily as the base for polyphosphate liquid fertilizers. The granules dissolve readily when mixed with water and ammoniated to a pH of about 6. A grade of 10-15-0 (10-34-0) is produced; by addition of urea-ammonium nitrate solution and potassium chloride, such grades as 7-9-6 (7-21-7) can be produced.

The process for making granular APP is simple and easily controlled. Superphosphoric acid containing 32 to 33% P (74 to 76% P_2O_5) is fed at controlled rate to a closed reactor, and anhydrous, gaseous ammonia is sparged through four equally spaced open-end pipes beneath the turbine-type agitator. The ammonia flow to the reactor is simply "by demand" to control pressure in the reactor at 2.5 to 3.5 kg/cm^2 (35 to 50 psi). The melt is withdrawn from the reactor at about 205°C (400°F) and granulated in a double-shaft pugmill with recycle at a ratio of about 3.5 to 1. The pugmill discharge is cooled and screened to obtain product of $-6+10$ mesh (Kelso et al., 1968).

The TVA researchers have also developed a "direct process" for producing liquid and granular APP (Lee et al., 1971). The heat of reaction of merchant-grade 22 to 23% P (52 to 54% P_2O_5) wet-process acid and anhydrous, gaseous ammonia in a pipe reactor (similar to that described in a previous section) drives off free and combined water to form polyphosphate melt. The melt is passed through a vapor disengager and dissolved in water with ammonia addition to produce a 10-15-0 (10-34-0) liquid with up to 40% or more of P_2O_5 as polyphosphate.

The melt from a system like that described above is granulated by TVA in a pugmill in a demonstration plant (11 to 13 metric tons/hour) with recycle to produce 11-24-0 (11-55-0) grade and with addition of 99.5% urea solution to produce 28-12-0 (28-28-0) urea-ammonium phosphate (Lee et al., 1974). The product polyphosphate level is limited to about 20 to 30% of the P because materials of higher polyphosphate content from wet-process acid are too "sticky" to granulate satisfactorily. polyphosphate content of about 20% of the P is too low to produce clear liquids of good quality. However, good suspension fertilizers can be readily produced from this intermediate.

Additional pilot-plant development work is being done by TVA to improve the quality of liquids made by the direct process.

d. Suspension Fertilizers—Suspension fertilizers of NP and NPK types are an offshoot resulting from efforts to increase the nutrient concentration of clear liquid fertilizers. Their production and use have grown steadily for the past 15 years or so, and in 1975, it was estimated that about half of the liquid fertilizer manufacturers produced suspensions. The primary advantage of suspension fertilizers is that they can be produced in grades of higher analysis than clear liquids. Suspensions can also be used as carriers of micro- and secondary nutrients and pesticides that are not sufficiently soluble to be added with clear liquids.

The ammonium polyphosphate or ammonium orthophosphate solutions are made as previously discussed for clear liquid fertilizers except that the grades are higher. A suitable gelling type clay is then added in proportion of 1.5 to 2% and thoroughly mixed, preferably with high shear-type agitation. The sequence of addition of the clay and other components is important. The properly produced suspensions will store for extended periods without serious settling difficulty if they are agitated occasionally. Experience in storage has been satisfactory with air agitation with open-end pipes extending downward to points near the bottom of the tank.

TVA pioneered the process and market development of suspensions. The first grade of polyphosphate suspension (made with electric-furnace superphosphoric acid) was 12-17-0 (12-40-0), but the product was later upgraded to 13-18-0 (13-41-0) and then changed to 13-17-0 (13-39-0) when about 40% of the phosphoric acid was fed as the low-cost merchant-grade wet-process acid (Wilbanks, 1967).

In 1975 TVA started operation of a demonstration plant for production of 18 metric tons/hour of orthophosphate suspensions with all the P supplied as wet-process acid. A two-stage reaction system was used, and operating conditions were controlled to cause the formation of a very large number of small crystals (predominantly DAP). An evaporative-type cooler was used with a practical minimum of "hold-up" in the circuit and with operation to promote rapid cooling and formation of the desired small crystals. A small amount of additional ammonia is fed to the tank when clay is added in a further effort to ensure formation of small DAP crystals.

Grade of the orthophosphate suspension has been varied slightly in efforts to establish the best N/P ratio for producing suspension with least tendency to settle. The common grade is 13-17-0 (13-38-0). The TVA demonstration plant was modified for use of a three-stage reaction system in 1978.

The N/P suspensions produced by TVA have been used in local fluid fertilizer plants to prepare final NPK suspensions such as 15-7-12 (15-15-15), 18-4-15 (18-9-18), and 10-9-17 (10-20-20) by addition of potash and urea-ammonium nitrate solution. Secondary and micronutrients and herbicides can be added also.

3. UREA-AMMONIUM PHOSPHATE

The TVA took the lead in development of processes for producing urea-ammonium phosphate (UAP). The first pilot-plant work adapted a slightly modified TVA granular DAP process. Urea was added either as concentrated solution sprayed onto the bed in the ammoniator-granulator or as prills or granules. Solid urea was incorporated much better when the prills or granules were "cracked" into smaller particles with a roll crusher. grades such as 19-8-16 (19-19-19), 28-6-12 (28-14-14), and 14-21-12 (14- The first commercial plants using this process were built in India; they use urea prills. The primary grade is 28-12-0 (28-28-0) but they also produce grades such as 19-8-16 (19-19-19), 28-6-12 (28-14-14), and 14-12-12 (14-28-14). Grades containing potash are more sensitive to melting during drying. The product has a considerably lower critical humidity than without potash, and requires greater protection from moisture.

As mentioned earlier (Young, 1975), TVA also adapted the melt preparation system of the direct process to production of urea-ammonium phosphate. A demonstration plant (16 metric tons/hour) was put in operation in 1973 and produces 28-12-0 (28-28-0) and 35-8-0 (35-17-0) grades (Lee et al., 1974). The reaction melt with 20 to 25% of P as polyphosphate is granulated with 99.5% urea melt in a pugmill. Material from the pugmill requires only cooling and sizing. The product has excellent storage properties without a conditioner. It has been stored in bulk for long periods in a storage building with humidity control. Most of the product has been shipped in bulk, and no storage problems have been encountered when the material is simply covered with a plastic sheet.

Urea-ammonium phosphate also has been produced in a pilot plant using a pipe reactor and rotary drum granulator. (See Section III-A-2 on melt-type granulation, following).

4. POTASSIUM PHOSPHATES

Potassium phosphates and processes for their production have been studied for a long time, but commercial production has been very limited and primarily experimental. In some early TVA studies, wet-process orthophosphoric acid was reacted with KCl. The reaction mixture was then heated to produce potassium metaphosphate. Grade of the product was about 0-24-31 (0-54-37). In this exploratory work, major problems included corrosion due to the moist hydrochloric acid vapor evolved and poor marketability of the byproduct HCl. In current small-scale tests at TVA, wet-process superphosphoric acid is reacted with KCl. The HCl is evolved in an essentially dry state, which greatly decreases corrosion problems and enhances prospects for its recovery and utilization. The product is a potassium ammonium polyphosphate which can be used in base liquid fertilizer solutions such as 4-10-10 (4-24-12) or 7-12-10 (7-28-12).

In 1970, Goulding Fertilizers Ltd. in Ireland reported development of a novel process for potassium phosphate (Thompson, 1971; Thompson &

Somers, 1970). During the same period, Pennzoil in the USA was developing a similar process. Raw materials for the process are PR, sulfuric acid, and KCl. The technology is similar to that for wet-process phosphoric acid; preparation of readily filterable gypsum is a key objective.

In the process flow scheme, potash is reacted with excess sulfuric acid at atmospheric pressure and elevated temperature. The Cl is evolved quantitatively as HCl, leaving a chloride-free slurry of potassium bisulfate in sulfuric acid. The HCl which is relatively pure is recovered for sale. The slurry flows to a reactor operated at atmospheric pressure and about 70°C (160°F) where it reacts with PR to form potassium phosphate and phosphoric acid. The coproduced gypsum is separated by filtration. The mother liquor is then concentrated in a vacuum evaporator. Water and small amounts of fluosilicic acid and entrained silica are evolved. Methanol is added to the product solution of monopotassium phosphate in phosphoric acid to precipitate monopotassium phosphate (KH_2PO_4). The precipitated KH_2PO_4 then is centrifuged, washed with recycled methanol, and dried. Liquor from the centrifuge is stripped of methanol in a distillation column. The methanol is recovered and recycled. The bottom stream contains a purified phosphoric acid and some residual potassium phosphate. This stream can be returned to process or removed to produce phosphates of high purity.

A small wet-process acid plant in California was modified for production of limited commercial quantities of granular potassium phosphates with this technology (E. K. Drechsel. 1973. Potassium phosphates: The new generation of SUPER Phosphates. Pennzoil Chem., Inc., Hampton, Tex.). Product grades are 9-21-13 and 5-20-25 (9-48-16 and 5-46-30).

Scottish Agricultural Industries (SAI) developed a process which involves reacting wet-process phosphoric acid and KCl in a rotary drum with heat supplied by combustion gases (Ewart & Raitt, 1969). The process has been studied on pilot-plant scale at a production rate of 45 kg/hour. The drum is designed to use an internal recycle of material that provides a hot bed for the reaction. The major part of the reaction takes place at the surface of the heated particles at a temperature of 480°C away from the heating gases. This isolates the evolved HCl and facilitates its recovery in a concentrated form. A portion of the most completely reacted material is continuously removed as product. The unique design of the drum conserves heat.

Israel Mining Industries (IMI) has a process for potassium phosphates that has been studied in small-scale tests. Potassium chloride is treated with twice the stoichiometric quantity of wet-process acid of about 22% P content in a vertical column in contact with a countercurrent flow of vaporized naphtha at 140 to 180°C. The HCl is recovered at the top of the tower. A two-phase liquid is withdrawn at the bottom of the column; it consists of condensed naphtha and the double salt $KH_2PO_4 \cdot H_3PO_4$. The naphtha separates as a floating liquid layer and is recovered for recycle. The $KH_2PO_4 \cdot H_3PO_4$ may be recovered as a solid 0-26-17 grade (0-60-20) material. By ammoniation, a product ($KH_2PO_4 \cdot NH_4H_2PO_4$) of grade 8-24-15 (8-56-18) can be produced, or pure H_3PO_4 can be removed by solvent extraction to leave a solid of 0-23-29 (0-52-35) grade.

ESPINDESA in Spain has operated a small pilot-plant unit for production of potassium phosphate (Nogueira et al., 1974). Their process is based on the direct precipitation of monopotassium phosphate (KH_2PO_4) in an organic phase that consists of an alkylamine, a modifer, and commercial kerosene. When equimolar amine and phosphoric acid are brought into contact, monoamine phosphate is formed according to the following reaction:

$$H_3PO_4 + R_3N \rightleftharpoons R_3NH \cdot H_2PO_4.$$

When this organic phase containing the amine phosphate comes in contact with solid KCl, an ionic exchange takes place by which the amine is transformed into amine chloride and the solid KCl into KH_2PO_4 which is the desired product according to the reaction:

$$R_3NH \cdot H_2PO_4 + KCl \rightleftharpoons R_3NH \cdot Cl + KH_2PO_4.$$

The amine chloride is converted into free amine for recycle by treatment with lime:

$$2R_3NH \cdot Cl + Ca(OH)_2 \rightleftharpoons 2R_3N + CaCl_2 + 2H_2O.$$

The product comes from the process as small granules. Grade of the product is about 0-21-26 (0-47-31). Suitable disposable of the soluble $CaCl_2$ is required.

Potassium phosphate of 0-11-21 (0-25-25) grade has been produced experimentally by a variation of the Rhenania phosphate process (see following section). Because of present interest in potassium phosphates, their proved agronomic potential, and the varied experimental work now going on, these fertilizers may become important in the future. They certainly seem one of the most promising new types on the horizon.

C. Thermal Processes

1. RHENANIA PHOSPHATE

Kali-Chemie in Germany has produced a calcium-sodium phosphate ($CaNaPO_4$) fertilizer for several years. Substantial tonnages have been used in Germany and exported to other European countries, to Africa, and to South America. The general formula for the reaction is:

$$Ca_{10}F_2(PO_4)_6 + 2SiO_2 + 4Na_2CO_3$$
$$= \underset{\text{Rhenania phosphate}}{(6CaNaPO_4 + 2Ca_2SiO_4)} + 2NaF + CO_2$$

The feed materials—Kola apatite, silica sand, and sodium carbonate—are proportioned, mixed, and then calcined in a rotary kiln at maximum temperature of about 1,250°C. A special high-alumina refractory lining is required. The discharged clinker is cooled and pulverized to 90% minus 0.15 mm.

Fine Rhenania phosphate is granulated in a drum by addition of water and about 0.6% of potato starch as a binder. The material discharged from the granulator contains about 12% moisture; it is dried and screened to obtain product of about $-6+12$ mesh. Magnesium and/or B are added to some products (UN, 1967).

Grade of the straight Rhenania phosphate is 12 to 13% P (28 to 30% P_2O_5). Granular mixed grades include 0-7-21 (0-15-25), 0-8-17 (0-18-20), 0-5-17 + 3Mg + 0.15B (0-12-20 + 5MgO + 0.15B) and 0-7-12 + 3Mg (0-15-15 + 5MgO).

Potassium carbonate has been substituted for sodium carbonate in the mixture before calcination in experimental tests; product grade of potassium phosphate was about 0-11-21 (0-25-25). High cost of potassium carbonate is a deterrent to commercial production and marketing of this type of fertilizer.

2. BASIC SLAG

Basic slag, a byproduct of the steel industry, has been widely used in large tonnages as a fertilizer in Europe and South America, and to a lesser extent in Asia and North America. This material, also referred to as *Thomas slag,* has a P content of about 7% (17% P_2O_5). World production of basic slag totaled about 0.6 million metric tons of P in 1962. It stayed at the level through 1971 but decreased to about 0.5 million metric tons of P in 1974.

The finely divided material is very dusty and messy to apply. There has been some attempts to granulate this material with water and a binder, and by cogranulation with soluble fertilizers.

3. FUSED Ca-Mg-PHOSPHATE

Fused Mg phosphate and Ca-Mg-phosphates are produced by an electric furnace-type operation (UN, 1967). About 750,000 metric tons of this material was produced in Japan and Taiwan in 1974. It is produced by addition of a Mg source (usually olivine or serpentine) along with PR to the furnace. The glassy product is pulverized to improve the availability of its P to plants. The product contains about 9% P and 9% Mg (20% P_2O_5 and 15% MgO).

III. NEW PROCESS TECHNOLOGY

A. Present and Near Term

1. GENERAL

A continuing steady growth in production of urea-ammonium phosphate is expected throughout the world as a natural outgrowth in the use of urea and ammonium phosphate, the world's leading N and P fertilizers. As discussed earlier, a number of processes use drum, pugmill, or pan granulation. Developmental work also has been done on prilling of this type of fertilizer.

There is a place for potassium phosphate fertilizers, perhaps as premium fertilizers in liquid and granular forms, if some recent processes described earlier prove to be fully practical and economical.

2. MELT-TYPE GRANULATION

Comparatively simple and economical melt-type granulation processes for ammonium phosphate fertilizers grew out of TVA's direct process using wet-process orthophosphoric acid. As mentioned before, a demonstration plant was built in 1973 to produce 11-24-0 (11-55-0) grade of essentially MAP with 20 to 25% of the P as polyphosphate. A pipe reactor system and pugmill were used. Since the reaction produces essentially an anhydrous melt, no drying step is required. This decreases fuel and electric power requirements, and investment for a new plant is decreased about 25%. Since the expensive dryer in a granulation plant is also the greatest source of fumes and dust, pollution problems are greatly decreased by eliminating the drying step.

A later TVA melt-process modification uses a pipe reactor and rotary drum-type granulator (Young, 1975). A schematic flow diagram is shown in Fig. 12. In the pilot-plant work, the melt system was simplified by eliminating the disengager. An extension of the pipe reactor into the drum has drilled holes for distribution of the melt onto the bed. The 2 to 2.5 kg/cm^2 (30 to 35 psi) steam pressure built up in the pipe by the reaction causes good atomization of the melt. The water vapor also disengages readily. Grades produced in the pilot plant include 11-24-0 and 19-8-16 (11-55-0 and 19-19-19). Several new or modified commercial units were built or planned for using this melt-type granulation process in early 1976.

In a still later version, a "pipe-cross" reactor allows simultaneous feeding of phosphoric and sulfuric acids (Fig. 13). In pilot-plant tests, the melt from this type of reactor has been used with other materials to produce grades such as 13-6-11 (13-13-13), 10-17-8 (10-40-10), and 6-10-20 (6-24-24). The melt is discharged onto the bed in a rotary drum through an open-end inverted ell. Several commercial plants are currently using the pipe and pipe-cross reactors (Achorn & Saladay, 1976).

3. UREA PHOSPHATE-BASED FERTILIZERS

The TVA and others have been studying processes based on the production of urea phosphate by reaction of wet-process orthophosphoric acid and urea. The crystallized urea phosphate is purified, since most of the impurities in the acid stay in the mother liquor. Urea phosphate contains 17% N and 19% P (17-44-0) and is a good fertilizer, although quite acidic. It can be reacted with ammonia to produce melts with polyphosphate content as high as 60 to 80% of the phosphate. This intermediate can be used to produce polyphosphate liquid fertilizers of very high quality approaching the purity of liquids made with electric furnace acid. The TVA considers this to

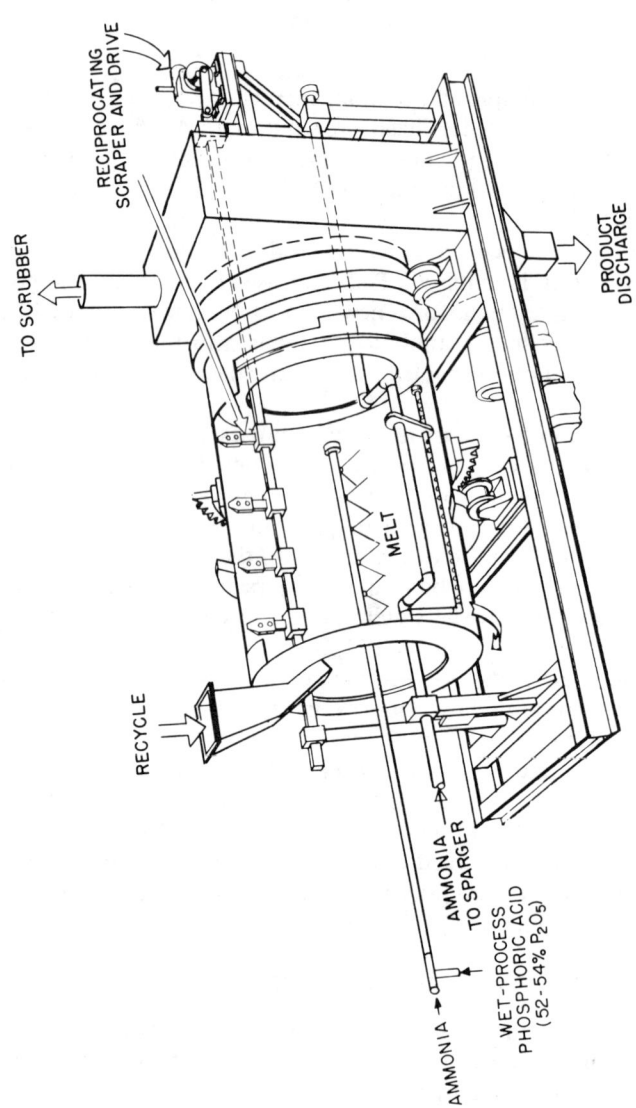

Fig. 12—Sketch of pipe reactor and rotary drum granulator.

PHOSPHATE FERTILIZERS AND PROCESS TECHNOLOGY

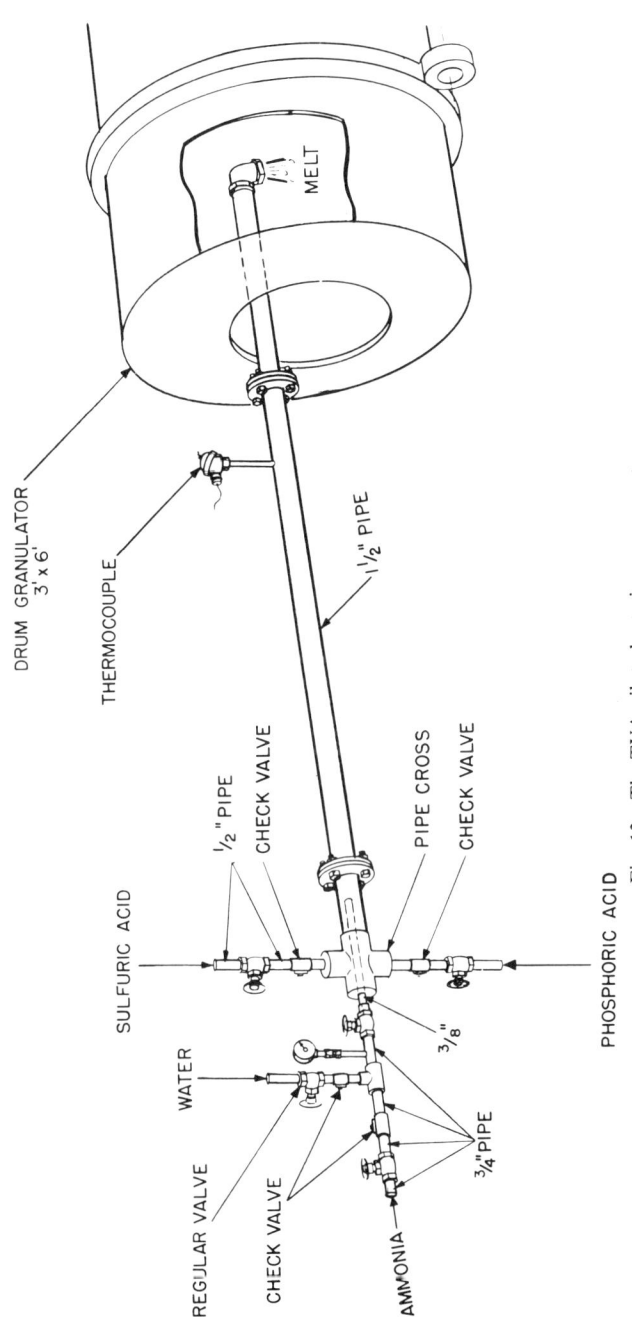

Fig. 13—The TVA pilot plant pipe-cross reactor.

be a promising route for production of liquids of high quality from merchant-grade wet-process acid (TVA, 1978). Pilot plants with production rate of 485 kg/hour of urea phosphate and for its conversion to 15-12-0 (15-28-0) polyphosphate liquid fertilizer were in operation by TVA in late 1978 (TVA, 1978).

B. Longer Term Processes and Fertilizers

1. UTILIZATION OF MARGINAL PR

Many deposits of PR throughout the world are considered of marginal quality and are not presently mined. The marginality of these deposits is due mainly to high levels of impurities such as Fe, Al, Mg, and Cl which make chemical processing by present technology uneconomical. Deposits of higher grade material, such as in Florida, will be depleted in about 20 years. The TVA started a long-range development program in 1975 aimed at new or modified processes for utilization of the marginal ores. This is considered to be by far the greatest and most urgent need and opportunity in phosphate fertilizer technology.

2. ULTRA-HIGH-ANALYSIS FERTILIZERS

The TVA research chemists have screened and studied a large number of compounds made predominantly of covalently bonded N and P. The objective is to determine possible effective fertilizers of much higher analysis than our present fertilizers. Thirty-four compounds were prepared in small quantities in the research laboratories and compared in the greenhouse as sources of N and P for corn in pot tests (Wakefield et al., 1971). The compounds evaluated included phosphonitrilic derivatives, metaphosphimates, metaphosphates, S-containing compounds, and amido- and imidophosphates. Metaphosphates, amidophosphate, thioamidophosphates, and phosphoryl triamide were excellent sources of both N and P but metaphosphimates were poor sources. Several of the S-containing compounds were toxic in early stages of growth but thereafter were good sources of nutrients. Of the phosphonitrilic compounds, only phosphoryl hexaamide was an effective source of the nutrients.

One of the most promising ultra-high-analysis compounds, phosphonitrilic hexaamide $[P_3N_3(NH_2)_6 \cdot H_2O]$, was prepared in the laboratory by ammoniation of trimeric phosphonitrilic chloride. The anhydrous compound is composed almost entirely of plant nutrients (95% as N + P or 147% as N + P_2O_5).

If suitable and economical methods for production of these compounds could be developed, the shipping, handling, and application costs would be greatly decreased. However, there likely would be difficulties in uniform application of such ultra-high-analysis fertilizers with present equipment. Thus, it is doubtful that such fertilizers will be on the commercial scene in the foreseeable future.

IV. ENVIRONMENTAL PROBLEMS IN THE PROSPHATE FERTILIZER INDUSTRY

A. Phosphate Rock Mining and Beneficiation

One concern in mining of PR is restoration of the land surface to preserve aesthetic features of the environment. This requires considerable effort, but sometimes such land is more valuable after restoration than before. In some areas, rather elaborate measures have been implemented to avoid adverse effects on ground water and other ecological aspects of the mined land.

In beneficiation of PR, the main environmental problems are containment of slimes from the operation, collection of in-plant dust from crushing and handling, and collection of particulates from stack effluents. Where calcination is used for beneficiation purposes, there are further problems in collection of particulates and fumes containing fluorine.

B. Production of Wet-Process Acid

Collection of dust from the rock drying and grinding operations is a major problem. This problem has been eliminated by wet grinding in ball mills in some of the newer plants. Fluorine removal or recovery is required to meet rigid environmental standards that have been imposed. Most plants employ closed-loop scrubbing systems using ponded water. Fluorine from the acid concentration step is now recovered as a salable product. Most of the fluorine recovered in the U.S. plants is used by the Al industry. In one instance in the United States a single processing plant has been installed to treat scrubbing liquor from three large phosphate plant complexes. This approach substantially decreases investment and operating costs per unit of product.

C. Superphosphates, Ammonium Phosphates, and Nitric Phosphates

Ordinary superphosphate production units usually are local and of medium size (18 to 23 metric tons/hour). The main pollution problem is F evolution into the plant environment and from the stacks. Perhaps the first complaints from neighbors of fertilizer plants (and realistic ones) were from those bothered by the F fumes from superphosphate plants.

Fluorine is readily removed by scrubbing with water in rather simple equipment such as towers using sprays or simple wood grids. Sprays also are used in horizontal scrubbers equipped with baffles. In the past, effluent dilute fluosilicic acid from some plants was neutralized by flowing over lump limestone or oyster shells before being discharged. Regulations on discharge have been tightening the past few years, and more thorough treatment is now required.

Some ordinary superphosphate plants use two stages of scrubbing so that good F recovery is obtained while producing fluosilicic acid of 20 to 25% concentration. Product of this concentration can be sold and shipped a limited distance for use by cities in fluoridation of water. Processes are available for production of Ca fluoride or Na fluoride from recovered F. Pooling of effluent from two or three nearby plants has been practiced to obtain output sufficient for economic viability.

Triple superphosphate is produced in large plants (45 to 55 metric tons/hour) at a complex where wet-process acid and granular ammonium phosphate also are produced. Fluorine and particulates in the plant environment and stack gases are the main problems. Pooling of the recovered F from the various production units makes economical production of salable products possible. The Swift process is widely used. This rather simple process is a particular patented version of two-step scrubbing (Parish, 1963). The scrubbing process is controlled to minimize problems with precipitated silica; concentration of the fluosilicic acid is high enough (25 to 30%) that it can be shipped economically for water fluoridation or further processing.

Scrubbing systems are required to recover ammonia and particulates in plants producing ammonium phosphates. Dilute (12 to 13% P) wet-process acid is used in primary scrubbers. Ponded water is used in closed-circuit secondary scrubbing to remove the F stripped out of the acid in the primary scrubbers. The most common scrubber used in this service is the venturi-cyclonic separator type. A low-pressure-drop (30 to 38 cm of water) venturi unit is satisfactory. Most common materials of construction are rubber-lined mild steel (with stainless lining at points of treatest wear) and fiber-glass-reinforced plastic.

Bag filters, sometimes referred to as *baghouses,* are very effective in collection of dry dust down to very small particle size. These units are very expensive and require disciplined maintenance and operation, especially if the collected dust is hygroscopic.

Phosphate fertilizer production complexes often use ponded water in closed circuit with F scrubbers. Where sufficient evaporation does not occur, the controlled effluent is neutralized before appropriate discharge. Leaks from acid pumps and minor spills must be collected and recycled. Acid storage tanks must be diked so that a massive spill caused by rupture would be contained. Comparatively inexpensive earthen dikes are acceptable.

In nitric phosphate plants there is the additional problem of controlling emission of oxides of $N(NO_x)$ within specified limits. The main emission is from the nitric acid production unit, but there is additional emission from the extraction units for nitric acid-phosphate rock attack. Several schemes (all rather costly) are being used for NO_x control in nitric acid plants. High pressure-drop scrubbing is used for removal of NO_x and particulates (aerosol type) from extraction and slurry ammoniation tanks. If the exhaust system can be essentially closed to restrict airflow, a simple and low-cost condensing-type scrubber is very effective. The simplest is an open spray tower. Noise control must be applied to the high-speed centrifugal compressor for the nitric acid plant. Enclosing the compressor with a compartment made of thick, noise-absorbing material is effective.

Control of dust in the plant enviornment (working areas) is accomplished by well-located and properly designed pick-up points.

LITERATURE CITED

Achorn, F. P., and H. L. Kimbrough. 1974. Latest developments in commercial use of the pipe reactor process. Fert. Solut. 18(4):8-9, 12, 14, 16, 20-21.

Achorn, F. P., and D. G. Salladay. 1976. Pipe-cross reactor eliminates the dryer. Farm Chem. 139(7):34, 36, 38.

Achorn, F. P., and W. C. Scott, Jr. 1969. Polyphosphates are revolutionizing fertilizers. Part III. Polyphosphates in mixtures. Farm Chem. 132(12):46-52.

Davis, C. H., R. S. Meline, and H. G. Graham, Jr. 1968. TVA mixed-acid nitric phosphate process. Chem. Eng. Proc. 64(5):75-82.

Ewart, George, and J. S. Raitt. 1969. Water-soluble polyphosphate and method of preparation. U.S. Patent 3,432,261.

Fleming, J. D. 1969. Polyphosphates are revolutionizing fertilizers. Part I. What polyphosphates are. Farm Chem. 132(8):30-36.

Hignett, T. P. 1965. Bulk blending of fertilisers: Practices and problems. Proc. Fert. Soc., no. 87, London.

Hignett, T. P. 1966. Nitrophosphate processes advantages and disadvantages. p. 92-95. In A. Spillman (ed.) Proc. 15th Annu. Meet. Fert. Ind. Round Table, 10-12 Nov. 1965, Washington, D.C. Fertilizer Ind. Round Table, Glen Arm, Md.

Hignett, T. P. 1971. History of world fertilisers and manufacturing processes. Indian Chem. Manuf. IX(4):13-17.

Huffman, E. O., and E. L. Newman. 1970. Polyphosphates are revolutionizing fertilizers. Part IV. Behavior and outlook. Farm Chem. 133(2):27-32.

Janikowski, S. M. 1968. Dihydrate processes. II. Commercial processes. B. Fisons. p. 241-252. In A. V. Slack (ed.) Phosphoric acid, Vol. 1, Part I. Marcel Dekker, Inc., New York.

Kaji, Keiji. 1968. Dihydrate processes. II. Commercial processes. F. Taki. p. 279-284. In A. V. Slack (ed.) Phosphoric acid, Vol. 1, Part I. Marcel Dekker, Inc., New York.

Kelso, T. M., J. J. Stumpe, and P. C. Williamson. 1968. Production of ammonium polyphosphate from superphosphoric acid. Commu. Fert. 116(3):10-16.

Kulp, R. L., and D. W. Leyshon. 1968. Dihydrate processes. II. Commercial processes. A. Dorr-Oliver. p. 213-240. In A. V. Slack (ed.) Phosphoric acid, Vol. 1, Part I. Marcel Dekker, Inc., New York.

Lee, R. G., M. M. Norton, and H. G. Graham. 1974. Urea-ammonium phosphate production using the TVA melt-type granulation process. Circ. Z-54, TVA, Muscle Shoals, Ala.

Lee, R. G., R. S. Meline, and R. D. Young. 1971. Pilot-plant studies of anhydrous melt granulation process for ammonium phosphate-based fertilizers. J. Ind. Eng. Chem. PDD 11(1): 90-94.

Meline, R. S., R. G. Lee, and W. C. Scott. 1972. Use of a pipe reactor in production of liquid fertilizers with very high polyphosphate content. Fert. Solut. 16(22):32-45.

Murakami, K., and S. Hori. 1968. Hemihydrate and anhydrite processes. I. Hemihydrate-dihydrate processes in Japan. A. General description. p. 287-298. In A. V. Slack (ed.) Phosphoric acid, Vol. 1, Part I. Marcel Dekker, Inc., New York.

Nogueira, E. D., J. M. Regife, and C. Gonzales. 1974. Potassium monophosphate production by solvent extraction. Proc. Int. Superphosphate and Compound Manuf. Assoc. Tech. Conf., 23-27 Sept. 1974. Prague, Czechoslavakia. Pap. no. PTE/74/4, ISMA, Ltd., Paris, France.

Parish, W. R. (to Swift and Co.). 1963. Fluorine recovery. U.S. Patent 3,091,513.

Roubinet, Jean. 1968. Dihydrate processes. II. Commercial processes. D. PSG-UCB. p. 259-272. In A. V. Slack (ed.) Phosphoric acid, Vol. 1, Part I. Marcel Dekker, Inc., New York.

Siegel, M. R., and R. D. Young. 1969. Polyphosphates are revolutionizing fertilizers. Part II. Base materials. Farm Chem. 132(9):41-47.

Slack, A. V. 1967a. It's time to consider nitric phosphates. Part 1. Background and history. Farm Chem. 130(4):28-30, 32, 34.

Slack, A. V. 1967b. It's time to consider nitric phosphates. Part 2. Process technology. Farm Chem. 130(5):24, 26, 28, 29, 62, 64.

Slack, A. V. 1967c. It's time to consider nitric phosphates. Part 4. Agronomic considerations. Farm Chem. 130(7):30, 32, 40, 42.

Slack, A. V. 1968. Dihydrate processes. II. Commercial processes. C. Prayon. P. 253–258. *In* A. V. Slack (ed.) Phosphoric acid, Vol. 1, Part I. Marcel Dekker, Inc., New York.

Slack, A. V., G. M. Blouin, and O. W. Livingston. 1967. It's time to consider nitric phosphates. Part III. Economic considerations. Farm Chem. 130(6):124–126, 128, 130, 132.

Tennessee Valley Authority. 1971a. Liquid fertilizer production and distribution. Circ. Z-27, TVA, Muscle Shoals, Ala.

Tennessee Valley Authority. 1971b. Production of compound fertilizers from intermediates in local plants. Circ. Z-30, TVA, Muscle Shoals, Ala.

Tennessee Valley Authority. 1972. New developments in fertilizer technology. 9th Demonstration, October 1972. TVA Bull. Y-50. TVA, Muscle Shoals, Ala. 72 p.

Tennessee Valley Authority. 1975. Directory of fertilizer plants in the U.S. 1974 TVA Bull. Y-87. TVA, Muscle Shoals, Ala. p. 247.

Tennessee Valley Authority. 1978. New developments in fertilizer technology. 12th Demonstr. TVA Bull. Y-136. TVA, Muscle Shoals, Ala. p. 47–51.

Thompson, W. H. 1971. New route cuts costs for potassium orthophosphates. Chem. Eng. 78(8):83–85.

Thompson, W. H., and T. N. Somers. 1970. Potassium dihydrogen phosphate manufacture. Proc. Int. Superphosphate and Compound Manuf. Assoc. Ltd. Tech. Conf., 8–11 Sept. 1970. Sandefjord, Norway. Pap. no. LTE/70/16. 121, ISMA, Ltd., Gloucester Place, London, England (also Paris, France).

United Nations. 1967. Fertilizer manual. UN Ind. Dev. Organ., New York. p. 133–135.

U.S. Department of Agriculture, Tennessee Valley Authority. 1964. Superphosphate: Its history, chemistry, and manufacture. U.S. Government Printing Office, Washington, D.C. p. 131–164.

Versteegh, P. M. R., and J. T. Boontje. 1968. Hemihydrate and anhydrite processes. II. Hemihydrate and anhydrite process in Europe. p. 331–362. *In* A. V. Slack (ed.) Phosphoric acid, Vol. 1, Part I. Marcel Dekker, Inc., New York.

Wakefield, Z. T., S. E. Allen, J. F. McCullough, R. C. Sheridan, and J. J. Kohler. 1971. Evaluation of phosphorus-nitrogen compounds as fertilizers. J. Agric. Food Chem. 19(1):99–103.

Wilbanks, J. A. 1967. Suspension and slurry fertilizers. Liquid fertilizer. p. 6–11. *In* Proc. Natl. Fert. Solut. Assoc. Round-Up, St. Louis, Mo. Natl. Fert. Solut. Assoc., Peoria, Ill.

Yates, L. D., F. T. Nielsson, and G. C. Hicks. 1954. TVA continuous ammoniator for superphosphates and fertilizer mixtures. Part I. Farm Chem. 117(7):38, 41, 43, 47–48. Part II. 117(8):34, 36–38, 40–41.

Young, R. D. 1966. Nitric phosphate processes utilizing supplemental acid. p. 67–72. *In* A. Spillman (ed.) Proc. 15th Annu. Meet. Fert. Ind. Round Table, 10–12 Nov. 1965, Washington, D.C. Fert. Ind. Round Table, Glen Arm, Md.

Young, R. D. 1975. Advantages in energy, fuel, and investment savings by melt-type granulation processes. Circ. Z-70. TVA, Muscle Shoals, Ala. p. 30.

Young, R. D., and G. C. Hicks. 1967. Production of monoammonium phosphate in a TVA-type ammonium phosphate granulation system. Commu. Fert. 114(2):26–27.

Young, R. D., G. C. Hicks, and C. H. Davis. 1962. TVA process for production of granular diammonium phosphate. J. Agric. Food Chem. 10:442–447.

Young, R. D., W. C. Scott, and R. S. Meline. 1969. Alternatives in production of ammonium polyphosphate materials for use in fluid fertilizer formulations. p. 128–135. *In* A. Spillman (ed.) Proc. 18th Annu. Meet. Fert. Ind. Round Table, 10–12 Nov. 1965, Washington, D.C. Fert. Ind. Round Table, Glen Arm, Md.

Chapter 8

World Phosphate Fertilizer Supply-Demand Outlook[1]

E. A. HARRE

Tennessee Valley Authority
Muscle Shoals, Alabama

K. F. ISHERWOOD

ISMA Ltd.
Paris, France

[1]*Editors' Note*—The authors of this chapter have insisted since the inception of the symposium that a chapter presenting forecasts of U.S. and world phosphate fertilizer supply-demand may be inappropriate in a volume of this nature. As they point out in the chapter, forecasts are only useful for a short period of time and will usually prove to be wrong as producers and consumers respond to the imbalances in the market.

However, the editors have requested that the authors revise the tables to include the latest available information and allow the reader to compare actual market developments with the original forecasts presented at the symposium. It is hoped that this will serve to illustrate (i) how market conditions may vary in relation to forecasted values, and (ii) how the relative importance of these market factors changes over time. No attempt to judge the accuracy of the forecasts of the authors is implied.

This chapter was prepared at a time when fears of world fertilizer shortages had resulted in a rapid increase in fertilizer prices and sharply lower fertilizer use in many countries. In 1977-1978 the phosphate market went through a period of oversupply, declining prices, and a lack of investment in new production facilities. But, since the chapter was revised in 1978, crop prices have increased because of weather-related crop production curtailments in some areas of the world and phosphate demand has risen sharply. Phosphates are once again in short supply, prices have moved rapidly upward, and new capacity is being added. The cyclical nature of the world phosphate market as pointed out by the authors continues, however, at a much more rapid pace than previously. Under todays market conditions forecasting becomes much more difficult and even greater judgement on the part of the forecaster is required.

I. INTRODUCTION

The fertilizer industry is cyclical in nature. It supplies a basic input to agriculture, which continually swings from undersupply to oversupply. Further, the fertilizer industry is part of the chemical industry which also is subject to

---

Copyright 1980 © ASA-CSSA-SSSA, 677 South Segoe Road, Madison, WI 53711, USA.
The Role of Phosphorus in Agriculture.

fluctuations in supply levels resulting from over-reaction to changing demand expectations or lack of adequate investment during periods of relatively slow demand growth. Thus, the P fertilizer industry is constantly changing; periods of rapidly rising demand and escalating prices typically are followed by slack demand and falling prices.

It is within this framework that this report presents a perspective on the current U.S. and world phosphate market and examines the expected state of the industry by the end of the decade. Some problems facing the market analyst are discussed in order to provide guidance to those evaluating demand and supply projections.

The report is divided into two basic parts: (i) the United States with mention of Canadian market trends and North American interrelationships; and (ii) the remainder of the world. The latter concentrates on the major phosphate markets of Europe. The contribution of the developing regions is discussed as it relates to changing world trade patterns for both phosphate rock (PR) and finished phosphatic fertilizer materials. The report begins with fertilizer pricing and its relation to agricultural prices. No attempt is made to discuss future price trends although price ratios do play an important role in the farmer's decision to fertilize his crops.

II. COMPONENTS OF FERTILIZER SUPPLY-DEMAND

A. Crop and Fertilizer Prices

The expected return for each dollar invested in fertilizer is a major factor in determining how much phosphorus (P) the farmer will purchase and apply. The effect on P fertilizer demand is illustrated by considering Table 1. The price ratio generally increased from 1965 through 1973. This meant that the farmer could afford to use higher rates of application to obtain higher yields per acre, which he did. In 1974-1975, however, fertilizer prices were high and continued to increase while prices of agricultural commodities decreased. This reversed the trend in the price ratio and signaled to the farmer that heavier use of fertilizer might not be justified. With such conditions at time of planting and without a definitive agricultural policy that indicated a better outlook at harvest, farmers cut back on P fertilizer use.

During late 1975 and early 1976, fertilizer prices declined as supplies increased. At the same time, commodity prices stabilized or increased slightly. With these improved price ratios U.S. farmers responded by increasing P use.

The current planting season holds many uncertainties for P fertilizer manufacturers. The 1978 crop year was one of record high production for almost all major crops. Wheat prices are expected to hold firm as a result of continued export demand and loan rates. But even with record export levels anticipated, carryover stocks of feedgrains are expected to increase, thus

Table 1—Relationship of concentrated superphosphate prices with prices of selected agricultural commodities in the United States.†

Year	Retail price paid concentrated superphosphate‡ (P_i)	Prices received by farmers§ (P_o)			Price ratio		
		Wheat	Corn	Cotton	Wheat	Corn	Cotton
	($/ton)	($/100 bu)	($/100 bu)	($/cwt)		P_o/P_i	
1965	80.90	135	116	29.4	1.67	1.43	0.36
1966	80.90	163	124	21.8	2.01	1.53	0.27
1967	84.10	139	103	26.7	1.65	1.22	0.32
1968	78.40	124	108	23.1	1.58	1.38	0.29
1969	74.00	125	116	22.0	1.69	1.57	0.30
1970	75.10	133	133	22.9	1.77	1.77	0.30
1971	76.60	134	108	28.2	1.75	1.41	0.37
1972	78.00	176	157	27.3	2.26	2.01	0.35
1973	87.50	395	255	44.6	4.51	2.91	0.51
1974	150.00	409	303	42.9	2.73	2.02	0.29
1975	214.00	355	254	51.3	1.66	1.19	0.24
1976	158.00	285	232	65.0	1.80	1.47	0.41
1977	146.00	229	203	60.7	1.57	1.39	0.42
1978	151.00	280	218	52.7	1.85	1.44	0.35
1979	161.00	315	227	55.1	1.96	1.41	0.34

† Source: USDA, Agricultural prices 1965–1979, monthly reports, Washington, D.C.
‡ U.S. average spring retail price paid by farmers.
§ Season average price received by farmers.

forcing prices lower. Cotton, on the other hand, should rebound sharply. Reduced plantings and poor weather made the 1978 crop one of the smallest in several years.

Overall, more P fertilizer probably will be applied in 1979 than a year earlier. But with the record crop of 1978 tending to dampen farmers' income expectations, the recovery may be modest. Fertilizer prices must react to price changes in the agricultural commodity market if price is not to be a limiting factor in the growth in demand.

B. Demand Elasticity

Before discussing the future P fertilizer supply-demand situation and outlook, several other price-related aspects of the market should be pointed out. The first of these concerns the accumulation of P in soil in areas where application rates have been in excess of the needs of the crops grown. In periods of relatively poor benefit/cost ratios, farmers may opt to "mine" their soils, and then go back to replenishing P fertility levels when price ratios are more favorable. The large-scale increase in soil testing during the past few years of shortage and rapidly increasing prices indicates that the farmer is well aware of his fertilizer needs and is responsive to short-term price movements. With the nature of the fertilizer market during the last two seasons, it appears that a reevaluation of fertilizer price elasticity studies is needed.

C. Supply Elasticity

Supply elasticity is the second factor that affects the market outlook. Depressed profits can curb investment in the necessary production and manufacturing facilities, and, if the situation warrants, lead to permanent closure of many existing plants. This forces supply toward the prevailing demand level. High net returns have the opposite effect; new plants are scheduled, older units debottlenecked or revitalized, and supply moves upward rapidly, removing any restrictions that supply may have had on consumption.

D. Availability of Raw Materials

Many studies have indicated the enormous reserve levels of P throughout the world. However, in the case of the P industry, price levels have a definite effect on PR reserves. The amount that is economically recoverable is determined by prices. Bureau of Mines estimates show that increasing PR prices from $8 to $20 per ton changes estimated recoverable reserves from 5 billion to almost 48 billion metric tons (Josephson, 1973). In 1975, PR prices for certain grades were as high as $58 to $70 per metric ton (Stowasser, 1979). As it will be shown later in the supply discussion, these price levels brought about additional investment in this phase of the P supply chain; thus the function of price in the supply-demand equation has once again helped to ensure adequate supplies.

Although these simple concepts may seem out of place in a technical discussion such as this, it is surprising how quickly they are forgotten, even by marketers and market forecasters. In periods of oversupply, there seems to be no limit to price declines; in shortage situations, price escalation is excessive. While it is doubtful that the cyclical nature of the fertilizer industry will ever be eliminated, a better awareness of the function of prices could lead to more realistic market evaluations.

III. FORECASTING OF FERTILIZER SUPPLY-DEMAND

Projecting a supply-demand balance depends on a number of interrelated factors which are not readily amenable to precise analysis. Generally, forecasting begins by extrapolating past and current trends. However, regardless of how sophisticated the statistical techniques used, the forecaster must use his own judgment in deciding whether a trend will continue unchanged, accelerate, or decline. This judgment is tempered by personal experience and knowledge of the cyclical nature of the market.

Probably the most enigmatic feature of forecasting is its purpose. *A useful forecast is one that will be proven wrong.* If a forecaster predicts a shortage, his efforts will be rewarded *only* if the market responds in time to avert the impending shortage. Thus, the purpose of forecasting is to provide a series of warning signals to bring about market corrections and hopefully avoid excessive ups and downs in the supply-demand balance. In the following sections these principles will be further illustrated when discussing the supply-demand outlook for North America and the world.

IV. SUPPLY-DEMAND OUTLOOK FOR NORTH AMERICA

Phosphate fertilizer use in the United States has increased continually over the past 30 years. The rate of increase has averaged 4.5% per year, but exceeded this level in several periods and held relatively constant for an extended period in the 1950's (Harre, 1975). The long-term upward trend was interrupted abruptly in 1974-1975 when use declined by nearly 12% (USDA, 1962-79).

From this and the previous discussion of factors bearing on the outlook for the P fertilizer market, the current situation vividly illustrates the point that forecasting market trends is extremely difficult. The market is in a period of adjustment and no clear pattern of market development is apparent. As of the end of 1975, P fertilizers in the United States were in severe oversupply. Producers' inventories were at or near record high levels. The drop in demand during 1974-1975 was the largest single decline since the depression years of the 1930's. The recovery in crop prices brought P use back to previous levels, but prices received by farmers remained well below the record levels of 1973-1974, dimming prospects for further improvement. Fall fertilizer sales in some areas were encouraging; in others, demand lacked vigor. But buying continued sluggish as many took a wait-and-see attitude toward the 1979 planting season.

The growth in the demand for plant nutrients in the United States and the relative importance of P fertilizers are shown in Table 2. Even discounting the 1974-1975 decline, the growth in the use of P fertilizers during the last 5 years had been well below that of the average growth over the longer term period indicated. Phosphate application rates have increased little in recent years for all major U.S. crops except soybeans.

Similar use data for Canada are shown in Table 3. Tracing the use pattern over time shows the relationship of fertilizer consumption to world grain price levels. In the period of depressed wheat prices on the world market of the late 1960's, fertilizer use fell sharply in the Western Provinces, where wheat is the major crop. Use recovered during the last few years as grain prices rose and farm income improved. In 1974-1975, however, consumption once again fell below the previous year's level, as it did in the United States.

Table 4 presents a forecast of P consumption for the United States and Canada. It is based on past trends in the North American P market and an evaluation of the factors that affect fertilizer use. The U.S. market has recovered from its 1974-1975 slump but is not expected to expand rapidly in the next few years. The growth rate for Canadian P materials is expected to

Table 2—Consumption of plant nutrients in the United States (USDA, 1962-1979).

Fertilizer year	N	P	K	Total
		thousand metric tons		
1961-1962	3,057	1,120	1,709	5,886
1962-1963	3,564	1,226	1,884	6,674
1963-1964	3,949	1,348	2,055	7,352
1964-1965	4,208	1,401	2,133	7,742
1965-1966	4,832	1,555	2,425	8,812
1966-1967	5,468	1,718	2,742	9,928
1967-1968	6,073	1,777	2,855	10,705
1968-1969	6,312	1,862	2,929	11,103
1969-1970	6,767	1,825	3,038	11,630
1970-1971	7,379	1,917	3,186	12,482
1971-1972	7,272	1,945	3,261	12,478
1972-1973	7,525	2,029	3,500	13,054
1973-1974	8,307	2,035	3,827	14,169
1974-1975	7,803	1,799	3,353	12,955
1975-1976	9,445	2,087	3,923	15,455
1976-1977	9,659	2,247	4,393	16,299
1977-1978	9,040	2,034	4,161	15,244
1978-1979	9,655	2,213	4,681	16,549

be around 3% per year. The range of values shown for these projections indicate that a wide degree of variation can be expected in this market.

An indication of the extent of oversupply can be seen by comparing the supply data in Table 5 with the use data in Table 3. Net domestic supply was very close to the demand levels in 1972-1973, 1973-1974, and 1975-1976 to 1977-1978. But the decline in use during 1974-1975 together with increased production pushed the available supply level well above actual use. The startup of several large phosphoric acid plants late in the 1974-1975 fertilizer season added to the potential supply level for the years ahead.

Table 3—Consumption of plant nutrients in Canada (Statistics Canada, 1962-1979).

Fertilizer year	N	P	K	Total
		thousand metric tons		
1961-1962	89	78	80	247
1962-1963	111	89	84	284
1963-1964	150	105	90	345
1964-1965	171	117	102	390
1965-1966	218	146	117	481
1966-1967	277	164	134	575
1967-1968	321	175	137	633
1968-1969	252	139	139	530
1969-1970	268	123	145	536
1970-1971	323	143	152	618
1971-1972	334	150	206	690
1972-1973	410	182	158	750
1973-1974	513	217	167	897
1974-1975	531	221	172	924
1975-1976	586	221	201	1,008
1976-1977	599	221	194	1,014
1977-1978	650	255	223	1,128

Table 4—Forecast of phosphate fertilizer consumption in North America.

Fertilizer year	United States		Canada		North America	
	Low	High	Low	High	Low	High
	thousand metric tons of P					
1975-1976	1,790	2,381	158	261	1,948	2,642
1976-1977	1,812	2,419	165	272	1,977	2,691
1977-1978	1,833	2,458	172	282	2,005	2,739
1978-1979	1,855	2,497	179	292	2,033	2,788
1979-1980	1,864	2,527	185	302	2,050	2,829

A change is taking place in the product mix of the P industry (Table 5). Normal superphosphate production started to decline 10 years ago. It is expected to continue to lose its share of the P market. Little increase is seen in the production of concentrated superphosphate (CSP). Production has been stable recently, but examination of trade patterns indicates that less is going to the U.S. market and more into the export market. Ammonium phosphates and other products based on the use of wet-process phosphoric acid have rapidly increased their share of the market at the expense of the lower analysis materials. The overall trend toward use of highly concentrated materials will continue, reflecting favorable production and transportation economics. In Canada, ammonium phosphate materials completely dominate the market (Koepke, 1971).

The trend toward greater use of phosphoric acid is shown in Table 6. Capacity has increased sharply over the past decade, with the biggest jump

Table 5—Supply of phosphate fertilizers in the United States (Mahan, 1979).

Fertilizer year	Phosphate product supply from domestic production					Total export	Total import	Net domestic supply
	Normal super-phosphate	Concentrated super-phosphate	Ammonium phosphates	Other	Total			
	thousand metric tons of P							
1961-1962	482	411	161	160	1,215	112	34	1,137
1962-1963	495	373	246	211	1,328	109	46	1,265
1963-1964	492	503	336	201	1,535	159	39	1,415
1964-1965	455	524	411	204	1,597	172	38	1,463
1965-1966	454	601	468	348	1,874	176	49	1,748
1966-1967	471	622	570	399	2,064	313	65	1,815
1967-1968	424	584	696	437	2,143	456	67	1,754
1968-1969	338	577	700	416	2,034	396	73	1,709
1969-1970	306	531	778	457	2,074	337	108	1,846
1970-1971	249	583	908	546	2,289	358	112	2,043
1971-1972	270	665	969	548	2,454	439	129	2,145
1972-1973	247	664	1,079	557	2,549	567	124	2,106
1973-1974	268	683	1,063	614	2,630	631	125	2,124
1974-1975	244	626	1,027	989	2,886	743	109	2,252
1975-1976	165	663	1,349	696	2,873	868	88	2,093
1976-1977	150	699	1,644	618	3,111	998	98	2,211
1977-1978	121	706	1,815	649	3,291	1,321	92	2,062
1978-1979	105	707	2,008	972	3,792	1,519	89	2,326

Table 6—Wet-process phosphoric acid capacity in North America and announced projects through 1980.†

Year	United States	Canada	North America
	thousand metric tons of P		
1967	2,127	355	2,482
1968	2,185	355	2,540
1969	2,350	435	2,785
1970	2,394	435	2,829
1971	2,295	435	2,730
1972	2,321	435	2,756
1973	2,570	435	3,005
1974	2,672	435	3,107
1975	3,400	435	3,835
1976	3,577	435	4,012
1977	3,715	415	4,130
1978	3,809	423	4,232
1979	3,868	423	4,291
1980	4,123	423	4,546

† Source: Capacity data from the records of Econ. and Market. Res. Sect., TVA-NFDC, Muscle Shoals, Ala., November 1979.

occurring during the 1974-1975 fertilizer season. Little capacity increase is expected in Canada since supplies are adequate to meet demand.

The capacity shown in Table 6 includes all announced plants, and should be considered as a maximum level for 1980. Supply levels are more than adequate to meet domestic demand; however, the rapid increase in exports in the past few years has maintained a fairly close supply-demand balance for the United States. With the market currently near equilibrium any further increase in the export market could lead to another round of capacity expansions in the United States.

V. WORLD PHOSPHATE MARKET OUTLOOK

From 1961-1962 to 1973-1974, world P fertilizer consumption increased at a rate which was on the average about 7% per year, slower than the rates of either N or K. Phosphorus accounted for 20% of total nutrient use in 1961-1962, but by 1974-1975, its share was down to about 15% of the market (Table 7). Further decline is expected during the remainder of the decade. Many of the reasons mentioned for the slower growth in P use in North America will also affect market growth in other major use areas.

As in the United States, the consumption of P fertilizers in Western Europe has increased steadily since 1961-1962. In some countries, however, the rate of gain in P consumption has been slowing or remaining constant, as levels of application approach the economic optimum for the crop varieties currently cultivated. The economic optimum for certain crops has been exceeded in some countries, but levels of application in Southern Europe remain substantially below recommended rates.

When prices of P fertilizers started to go up in 1973-1974, deliveries increased only slightly, despite the forward buying by farmers and merchants

Table 7—World consumption of plant nutrients (FAO, 1962-1979).

Fertilizer year	N	P	K	Total
		thousand metric tons		
1961-1962	11,554	4,654	7,207	23,415
1962-1963	13,173	4,976	7,700	25,849
1963-1964	14,942	5,577	8,365	28,884
1964-1965	16,375	6,140	9,159	31,674
1965-1966	18,828	6,577	10,194	35,599
1966-1967	21,776	7,089	10,866	39,731
1967-1968	23,918	7,451	11,719	43,088
1968-1969	26,591	7,988	12,252	46,831
1969-1970	28,677	8,272	12,922	49,871
1970-1971	31,824	8,719	13,843	54,386
1971-1972	33,324	9,280	14,612	57,216
1972-1973	35,771	9,930	15,643	61,344
1973-1974	38,900	10,700	17,300	66,900
1974-1975	38,576	9,983	16,453	65,012
1975-1976	43,140	10,629	17,915	71,684
1976-1977	45,116	11,566	19,223	75,905
1977-1978	47,768	11,988	19,350	79,106

in anticipation of further price increases. In 1974-1975, P fertilizer deliveries fell more than 16%, returning the market to 1968-1969 use levels. In some countries, the extent of the 1974-1975 decrease might have been exaggerated by forward buying in the previous year. There can be little doubt, however, that there was a real and substantial decline in the farmer use of P fertilizer materials.

In Western Europe the decline varied from less than 1% to as much as 40% below the 1973-1974 level, and was related to shifts in the agricultural product price/fertilizer price ratio. Weather had some effect but was a secondary factor. In some cases, farmers simply did not have enough cash to purchase fertilizers at their accustomed levels. Where there were increases in the prices of agricultural products, they were generally insufficient to compensate for the increased cost of fertilizers and other inputs. Total farmer expenditure for fertilizers increased sharply even though fewer tons were purchased.

There was a similar reaction in other developed countries. In Japan, consumption fell by 16%. Australian consumption showed a 47% decline because of a combination of higher fertilizer prices, removal of the P subsidy, and unfavorable commodity prices. Phosphorus use in all developed countries was 16% below the 1973-1974 level. In the centrally planned economies, use probably followed prevailing trends except in China where a reduction in PR imports took place during the year.

Consumption in some developing countries continued to expand, but India recorded a 9% drop in P deliveries and an estimated 27% decline in actual use.

The longer term repercussions of the downturn in P use are not clear yet. Advisory services in many countries are suggesting that farmers reduce rates of application on soils having high P levels. Demand for soil tests has increased in efforts to determine areas where P use can be reduced. Once these practices become established, market recovery may be slow even if economic relationships improve. Forecasts for Western Europe indicate that the 1973-1974 use level may not be regained until 1980-1981 (Isherwood & Louis, 1975).

In India, farmers were told to neglect soil maintenance applications and apply P only where a direct response could be expected. However, India has now instituted a crop and fertilizer subsidy program and P use has increased from 286,000 metric tons in 1974 to an estimated 485,000 metric tons in 1979—a 70% gain.

The effects of these trends on world food production in the future are not clear. In developing countries, with low soil fertility levels, a drop in P use is likely to cause lower yields. In India, soil fertility maps based on 1967 information indicate that 47% of the soils had a poor P status while an additional 48% tested at medium level. In Australia, the supply of N from clover depends on maintaining an adequate P level, and any extended reduction in use could thus seriously affect pasture production. In Western Europe, the situation is more complicated. Average application rates have been relatively high for several years and in many areas soil reserves are good. But there are other areas where application rates remain well below the recommended level.

VI. WORLD PHOSPHATE SUPPLIES

There should be ample supplies of P fertilizers available to meet world demand for the remainder of the 1970's. The improved investment outlook in 1974-1975 and again in 1978-1979, in addition to a trend toward self-sufficiency by some countries who were unable to obtain adequate supplies during the past few years of shortage, has brought about a substantial increase in production capacity.

Based on announced plans for the construction of additional phosphoric acid plants throughout the world, it is expected that total world capacity will increase by more than 60% by 1980. New capacity of over 2 million metric tons of P had been added in the mid-1970's, bringing the potential P supply up to an adequate level to meet world demands.

Capacity in the developing regions of the world could increase from 1.1 million to 3.3 million metric tons of P—a gain of almost 200%. Implications for future world trade patterns are obvious. Increased phosphoric acid capacity in these areas will mean greater exports of raw materials and relatively smaller tonnages of finished P materials will be exported by the traditional marketers. North African nations will be expanding capacity of processed materials for export as an alternative to exporting only PR. In total, the developing regions will increase their share of world capacity from 13 to 23% of the world phosphoric acid plants in operation if all announced plans materialize. Developed regions will decline in relative importance, but will still account for a large percentage of the total world production.

Table 8—World phosphoric acid capacity and phosphate fertilizer supply estimate.†

Region	1970	1973	1974	1975	1976	1977	1978	1979	1980
				million metric tons of P					
Developed									
North America	2.8	2.9	3.0	3.8	4.0	4.0	4.3	4.5	4.5
Western Europe	1.7	1.9	2.0	2.7	2.7	2.7	2.7	2.7	2.7
Other‡	0.6	0.7	0.7	0.7	0.7	1.0	1.0	1.0	1.1
Total	5.1	5.5	5.7	7.2	7.4	7.7	8.0	8.2	8.3
Developing									
Latin America	0.3	0.3	0.4	0.4	0.6	0.7	0.8	0.8	1.2
Africa	0.1	0.3	0.3	0.4	0.7	0.9	1.1	1.1	1.1
Asia	0.3	0.4	0.5	0.7	0.8	1.0	1.0	1.1	1.1
Total	0.7	1.0	1.2	1.5	2.1	2.6	2.9	3.0	3.4
Centrally planned economies									
East Europe-USSR	1.1	1.8	2.2	2.6	2.6	2.7	2.7	2.7	2.7
World total	6.9	8.3	9.1	11.3	12.1	13.0	13.6	13.9	14.4
World total (revised)	--	--	--	--	--	--	--	12.5	12.9
Actual fertilizer supply	8.0	9.9	10.4	10.7	10.4	11.4	12.5	--	--
Potential fertilizer supply§	--	--	--	--	13.1	13.6	14.1	14.4	14.6
Potential fertilizer supply (revised)	--	--	--	--	--	--	--	13.5	13.9
World demand¶									
Low	--	--	--	--	9.6	10.1	10.5	10.9	11.4
Midpoint	8.3	9.9	10.7	10.0	10.6	11.6	12.0	12.1	12.7
High	--	--	--	--	11.4	12.0	12.6	13.3	14.0

† Source: Harre, E. A. et al. (1975) (Demand adjusted for TVA revisions of estimates for North America and ISMA revised forecast for Western Europe.) and Harris and Harre (1979).
‡ Includes Japan, Israel, Republic of South Africa, and Oceania.
§ Includes production forecasts for normal superphosphate, basic slag, and the PR contribution to supply from CSP and nitric phosphates, plus estimated potential supply for fertilizer from phosphoric acid. Forecast 1976-1980 based on 90% operating level in developed regions, and 70% in developing regions after 2-year phasing-in period for phosphoric acid plants. Data adjusted for processing losses and nonfertilizer uses of phosphoric acid. All years adjusted for transportation and distribution losses.
¶ Actual use 1970, 1973-1978. Excludes the use of ground PR.

Total world phosphoric acid capacity, potential supply levels, and estimated world demand are shown in Table 8. Potential supply is the sum of the expected production of phosphoric acid for fertilizer and the contribution of normal superphosphate, basic slag, and the PR content of CSP and nitric phosphate materials. Allowances have been made for nonfertilizer uses of phosphoric acid and for processing and distribution losses (Harre et al., 1974).

Relating potential supply to forecasted demand shows that supply will exceed demand for the remainder of the decade, however, it will be only about 10% above demand by 1979-1980. If the expected maximum demand level is attained, world supply and demand would be in balance by the end of the decade.

Throughout the world, however, the decline in use in 1974-1975, the resulting buildup in inventories, and the rapid decline in price levels affected the longer term supply outlook. The projected imbalance in the P market could not be sustained and was less than indicated by new plant announcements and expansions. First of all, lower prices for wet-process phosphoric acid as opposed to a continued increase in the costs of producing phosphorus and thermal phosphoric acid increased efforts to substitute materials, thus siphoning off some supply from the fertilizer sector. Secondly, the less favorable investment element delayed construction programs and caused abandonment of some projects, and the closure of several production facilities.

VII. PROBLEMS IN ASSESSING FUTURE MARKET OUTLOOK

Forecasting is never easy, but it becomes extremely difficult during transition periods of the supply-demand cycle. All of the guides offered by review and extrapolation of past market performance become of little value as patterns change directions. Predicting reactions to current market conditions and the long-term effects become a matter of the analyst's judgment that no model or other analytical tool has been able to replace. There is also the tendency to overreact to the current market situation and to assume it will continue indefinitely.

The fertilizer quantities which create the difference between an oversupply and an undersupply situation often are small in relation to total demand and supply. The uncertainties in the forecasts of either demand or supply can change the assessment of the balance from one of oversupply to one of undersupply, or vice versa. If supply and demand uncertainties are combined, the results can be meaningless.

Another problem concerns the impact of forecasts. Fertilizer demand forecasts in the early to mid-1960's made by industry and international organizations for 1970 were not overly optimistic, and yet a harmful oversupply situation developed. This indicates that good forecasts are necessary but not sufficient.

Forecasted supply can, at best, refer to a period only 2 or 3 years ahead, beyond which announced new phosphoric acid capacities cannot, in general, be considered as "firm." Because of the inevitable uncertainties, it is doubtful if forecasts of the market balance will be realized. However, they should signal the directions the market is taking and indicate how these directions should be changed if an unbalanced situation is to be avoided. The forecaster must use his own view of the market outlook as he evaluates the factors that affect both supply and demand. The basic law of supply-demand and the function of price in balancing the market should be allowed to operate freely.

LITERATURE CITED

Food and Agricultural Organization. 1962-1979. Annual fertilizer review, Annu. Rep., 1962-78. United Nations, Rome, Italy.

Harre, E. A. 1975. The supply outlook for phosphate fertilizers. p. 36-44. *In* Proc. TVA Fert. Conf., Louisville, Ky. TVA Bull. Y-96, Muscle Shoals, Ala.

Harre, E. A., J. D. Bridges, and J. T. Shields. 1975. Worldwide fertilizer production facilities as related to supply and demand for the next five years. *In* A. Spillman (ed.) Proc. 25th Annual Fert. Ind. Round Table, 4-6 Nov. 1975, Washington, D.C. Fertilizer Ind. Round Table, Glen Arm, Maryland.

Harre, E. A., O. W. Livingston, and J. T. Shields. 1974. World fertilizer market review and outlook. TVA Bull. Y-70. TVA, Muscle Shoals, Ala. p. 35-37.

Harris, G. T., and E. A. Harre. 1979. World fertilizer situation and outlook—1978-1985. Int. Fert. Develop. Center, Muscle Shoals, Ala. Tech. Bull. IFDC—T-13, March 1979.

Isherwood, K. R., and P. L. Louis. 1975. The present situation and medium term global outlook for the supply/demand balances for phosphate fertilizers, phosphoric acid, phosphate rock, and sulphur. ISMA, Ltd., Paris, France (International Superphosphate and (Compound) Manufacturers' Assoc.).

Josephson, G. W. 1973. Fertilizer mineral supply and outlook. p. 34. *In* S. T. C. Orhan (ed.) CENTO Symp. on Mining and Beneficiation of Fertilizer Minerals, 19-24 Nov. 1973, Istanbul, Turkey, CENTO, Istanbul.

Koepke, W. E. 1971. Fertilizers and fertilizer minerals in Canada. Dep. of Energy, Mines, and Resour., Bull. MR 115, Ottawa, Canada.

Mahan, J. N. 1979. The fertilizer supply. USDA Agric. Stabil. & Conserv. Serv. Annu. Rep., 1963-1978. USDA, Washington, D.C.

Statistics Canada. 1962-1979. Fertilizer trade, Ann. Rep., Manufacturing and Primary Ind. Div., Ottawa, Canada.

Stowasser, W. F. 1979. Phosphate 1979. Mineral commodity profiles, Jan. 1979. U.S. Dept. of Interior, Bur. Mines, Washington, D.C.

U.S. Department of Agriculture. 1979. Agricultural prices, 1965-1979, monthly reports. Washington, D.C.

U.S. Department of Agriculture, Statistical Reporting Service. 1962-1979. Commercial fertilizers, annual reports. USDA-SRS, Washington, D.C.

Chapter 9

Energy Requirements for the Production of Phosphate Fertilizers

WILLIAM C. WHITE AND KARL T. JOHNSON
The Fertilizer Institute
Washington, D.C.

I. INTRODUCTION

Production of phosphate (P) fertilizers is essentially a process of solubilizing P ores obtained from mining operations. Each of these two major steps —mining and solubilizing P—takes its toll of energy.

Compared with capital and raw material cost inputs for mining and processing P fertilizers, expenditures for energy have been relatively small in proportion. However, with sharply rising energy prices, this component of production costs will receive increasingly closer attention by analysts in the future. Increases in cost of energy have been due in part to upgrading of the terms and conditions of energy purchase contracts, e.g., shifting from interruptible supply to firm supply contracts.

Another dimension of the energy picture for P fertilizer is the curtailment of energy supply. In 1973, phosphate rock (PR) producers in Florida reported a loss of nearly 1 million short tons of PR due to electrical power curtailment.[1] Thus, with both energy supply and cost becoming of increasing importance to the fertilizer industry, attention in the past several years has turned appropriately to compiling energy consumption data per unit of product. Survey data to date, however, have focused on total energy use by major products and little data are available on individual steps within a given process.

The objective of this paper is to present representative data on current energy use for the process involved for each major P fertilizer and for the material inputs for each product. From these data other comparisons are made, including energy per unit of P in finished products.

---

[1] W. C. White. 1975. Testimony before the Council on Environmental Quality, Washington, D.C.

Copyright 1980 © ASA-CSSA-SSSA, 677 South Segoe Road, Madison, WI 53711, USA.
The Role of Phosphorus in Agriculture.

II. PRODUCTION CAPACITY

Reference data on P production capacity in the United States for 1977 are found in Table 1. The concentration of capacity in the South Atlantic region bears directly on energy requirements for that region.

The uneven concentration of production capacity places heaviest requirements for electrical power and fuel oil in the South Atlantic and mountain regions, aside from the energy required for Frasch mining of sulfur (S).

III. PHOSPHATE ROCK

Energy use in phosphate rock mining commences with the dragline that removes the overburden. Energy use varies due to amount of overburden removed. In Florida, for example, an average of 7 m^3 of material is handled for 1 m^3 of matrix. The matrix typically consists of one-third usable rock, one-third clay, and one-third sand. Thus, there are very large volumes of solid material to handle for 1 metric ton of usable rock which will contain, if of good quality, about 14% P (32% P_2O_5 or 70% BPL).

Most of the PR mining draglines in Florida are electrically powered. Electricity is also used to power pumps that transport the matrix-water slurry to processing facilities that may be 3 to 8 km from the mine site. (In the western U.S. and Tennessee fields, trucks are used for this purpose.) Additionally, considerable power is required for beneficiation, a process of desliming and sizing the rock. For these various purposes, the electrical bill, estimated by the Florida Phosphate Council, was nearly $100 million in 1975.

Table 1—Phosphate production capacity in the United States by region for 1977 (A. D. Little, 1975).

Region	Phosphate rock		Wet process acid	Triple super-phosphate	Ammonium phosphate
			1,000 metric tons		
South Atlantic	41,400	P	2,331	789	578
Fla., N.C.		P_2O_5	5,298	1,794	1,313
E. North Central		P	142		88
Ill., Mich.		P_2O_5	323		201
W. North Central		P	76	18	141
Iowa, Kans., Mo.		P_2O_5	172	41	320
E. South Central	3,800	P	64	50	91
Miss., Tenn., Ala.		P_2O_5	145	114	207
W. South Central		P	729		637
La., Ark., Tex.		P_2O_5	1,657		1,447
Mountain	6,900	P	227	101	136
Idaho, Utah, Mont., Wyo., Ariz.		P_2O_5	515	229	310
Pacific		P	88		80
Calif., Wash.		P_2O_5	199		181
Total	52,100	P	3,656	958	1,751
		P_2O_5	8,309	2,178	3,989

Fuel for drying ranks next among energy requirements for PR. Total processing energy requirements are given in Table 2.

Data in Table 2 show consumption of fuel oil, natural gas, and electricity, in the mining and beneficiation processes of PR. Coal is used in Tennessee and in western U.S. phosphate operations, but only in small amounts.

More energy is used in drying and calcining than in mechanical handling. A. D. Little (1975) states that evaporation of 1 kg of water from PR requires 5.1×10^6 J (2,200 Btu/lb). Energy use for calcining, which is practiced only in some plants, is difficult to identify separately. One source of information (A. D. Little, 1975) contrasts energy use for PR among various regions, as shown in Table 3.

Table 2—Process consumption of energy.

Process	Energy sources and quantities per short ton of material			Total energy requirements		Comments†
	Fuel oil	Nat. gas	Electricity			
	gals	ft³	kWh	million Btu/short ton	GJ/ metric tons	
PR (14% P)	1.37	76	55.5	0.457		a
	2.52	90	50	0.623		b
				0.480		c
				0.500	0.58	d
Frasch S		6,000		6.000		a
		6,840	18.1	6.924		b
				9.000		c
				6.500	7.56	d,e
WPA‡ (23.6% P)	2.31	227	180.1	1.164		a,f
			124	0.423		b
				0.378		c
				0.700	0.81	d
DAP	2.5	1,000	38	1.490		g
(18-20-0, N-P-K)		1,500	20	1.568		b
				1.058		c
			1.500	1.75		d
Granular TSP	3.7	565	33	1.211		g
(0-20-0, N-P-K)	6		15	0.916		b
			0.460	0.460		c
				1.000	1.16	d
OSP	0.2		38	0.159		g
(0-8.7-0, N-P-K)				0.159	0.19	d

† Comments:
  a) From The Fertilizer Institute. 5 Apr. 1976. Energy Consumption Survey for July–Dec. 1975.
  b) From A. D. Little (1975). Data representative of Florida and North Carolina.
  c) From Davis (1974).
  d) Representative figure used in subsequent calculations.
  e) One metric ton S typically yields 2.9 metric tons of sulfuric acid. Process requirement for sulfuric acid, therefore, is 2.61 GJ/metric ton (2.04 million Btu/short ton sulfuric acid).
  f) This figure is weighted with energy value of steam from sulfuric acid process. Figure in 1973 survey showed 791 Btu per ton of 23.6% P acid (54% P_2O_5).
  g) From White (1973).
‡ WPA is wet process phosphoric acid.

Table 3—Energy use in PR production (A. D. Little, 1975).

Source, unit/short ton of rock	Florida North Carolina	Tennessee	Western states
Coal (lb)	--	100	4
Fuel oil (gal)	2.52	0.42	8
Natural gas (ft^3)	90	135	120
Electricity (kWh)	50	35	0
Million Btu/short ton	0.623	1.58	0.365
GJ/metric ton	0.798	2.02	0.468

Data for North Carolina rock, which is calcined, are comingled with Florida operations where essentially no calcining is practiced. In the western states calcining is more widely used but lower overall energy consumption is observed. The most plausible explanation for the relatively low energy use per unit in western states is that little beneficiation, if any, is required, thus obviating the need to evaporate water from wet-beneficiated material. In addition, energy for matrix slurry pumping is not required. For calculations throughout the paper, the value for operations in North Carolina and Florida of 0.58 GJ/metric ton (0.50 million Btu/short ton) is used for energy in producing PR.

IV. SULFUR

Most of the sulfur used in the production of fertilizer P comes from Frasch mining. In this process super-heated water (about 163°C) is pumped into the sulfur dome. The major energy requirement—supplied primarily with natural gas—is for water heating (A. D. Little, 1975).

Davis (1974) has reported that S recovered from sour natural gas and refinery gases requires little external energy, on the order of 0.35 GJ/metric ton (0.3 million Btu/short ton). He cites a figure of 35 GJ/metric ton (30 million Btu/short ton) for S obtained from desulfurization of oil. Desulfurization of oil exclusively for S recovery is expensive, energywise. For pyritic S, he estimates 0.58 GJ/metric ton (0.5 million Btu/short ton).

An energy requirement for sulfuric acid of 2.61 GJ/metric ton (2.24 million Btu per short ton of 100% H_2SO_4) was calculated on the basis of 2.9 metric tons of sulfuric acid per metric ton of S. Data on individual P products presented later in the paper clearly show sulfuric acid as the "work horse" of the phosphate industry, literally representing over half the energy requirement for each "downstream" phosphate product.

V. WET PROCESS PHOSPHORIC ACID

Electricity provides about half of the process energy for wet process phosphoric acid (Table 2). Not shown in Table 2 is the by-product steam from companion sulfuric acid plants that provides the energy for concentrating the product acid from about 14% P (32% P_2O_5) to merchant grade, 24% P (54% P_2O_5). More than enough steam as a by-product is usually

ENERGY REQUIREMENTS FOR PRODUCTION

Table 4—Product consumption of energy/metric ton of product.

Product	PR†	H_2SO_4	WPA	For process‡	For materials	Total	GJ/metric ton of product P	% of OSP
	tons of material/ton of product			GJ/metric ton of product				
WPA†	1.80	1.46	--	0.81	4.85	5.66	24.02	134
DAP†	--	--	0.85	1.75	4.81	6.56	32.66	182
TSP granular†	0.4	--	0.63	1.16	3.80	4.96	24.72	138
TSP run-of-pile†§	0.4	--	0.63	0.16	3.80	3.96	19.71	110
OSP†	0.63	0.39	--	0.19	1.38	1.57	17.96	100

† Based on the following P concentrations, in %; PR, 14; WPA, 23.6; DAP, 20.0; TSP, 20.0; and OSP, 8.7.
‡ From Table 2.
§ Calculated by reducing process energy requirement by an amount equivalent to 6 gal of fuel oil per short ton of granular TSP used in the drying process.

available from sulfuric acid plants to operate the phosphoric acid concentrator (A. D. Little, 1975). In view of the variety of plant operations and steam use, a "steam balance" is not presented.

A relatively high (67%) portion of the total energy requirement in wet process phosphoric acid manufacture is expended for S and sulfuric acid (Table 4).

Although relatively little electric furnace phosphoric acid is used in fertilizers, it is a significant end-use for PR. Davis (1974) reports that energy use in electric furnace phosphoric acid is about 72 GJ/metric ton of P (27 million Btu per short ton of P_2O_5). The energy use for the wet process acid (Table 4) is 24 GJ/metric ton of P (9 million Btu per short ton of P_2O_5), or one-third that of furnace acid. Lower per-unit costs of energy forms used in wet process phosphoric acid, in comparison to the high per-unit costs of electric energy, add to the cost advantage for sulfuric acid as an attack agent for PR.

Production of superphosphoric acid, a conversion product of wet process phosphoric acid, amounts currently to nearly 239,000 metric tons of P (600,000 short tons P_2O_5) annually. The process is essentially one of reducing the water content of phosphoric acid. Reference data on energy requirements for this process are not available, but one estimate gives an energy requirement of nearly 2.9 GJ/metric ton of P (1 million Btu per short ton P_2O_5) as superphosphoric acid made from 24% P (54% P_2O_5) wet process phosphoric acid (D. G. Mercer, Texasgulf, Inc., personal communication).

VI. AMMONIUM PHOSPHATES

Diammonium phosphate (DAP) is the leading P product tonnagewise and provides over half of the P in American fertilizers (Table 1). Yet, because phosphoric acid is the sole source of P in DAP, energy use per unit of P in diammonium phosphate is the highest of all P fertilizers, excluding P from electric furnace acid, as indicated in Table 4.

Process energy requirements for DAP (Tables 2 and 4) are less than might be expected because heat of neutralization provides much of the energy for drying and granulation. In all of these calculations, the energy requirements for NH_3 synthesis are not included, since these are more appropriately assigned to the N component of this mixed fertilizer.

Data for monoammonium phosphate (MAP) are not presented, but energy use per unit of P would be similar to that for DAP. For ammonium polyphosphate (APP), by the TVA process, there is a reduction of about 2.5 GJ/metric ton of P (1 million Btu per short ton of P_2O_5) from the DAP process, largely resulting from heat of reaction in the pipe reactor polymerizing some of the P and dehydrating the product. Despite this savings in energy, final total production costs of these two products are very similar on a nutrient basis.

VII. TRIPLE SUPERPHOSPHATE

Phosphate rock provides about 30% of the P in granular triple superphosphate (TSP), and wet process phosphoric acid the remainder. Energy requirements for TSP—process and material input—per unit of P are about 75% of that needed for DAP.

An estimate for run-of-pile TSP was obtained by deducting from the energy requirements for granular TSP an equivalent of 25 liters (6 gal/short ton) of fuel oil per metric ton of product for drying (Table 4). Thus, the granulation step costs about 5.3 GJ/metric ton of P (2 million Btu per short ton P_2O_5).

VIII. NORMAL SUPERPHOSPHATE

Solubilizing P via the normal superphosphate process has the lowest energy requirements per unit of P of all known processes. Requirements of other solid products range from 10 to 82% above normal superphosphate per unit of P (Table 4). The low energy expenditure for this product results from PR being the sole source of P, and from the relatively small quantity (0.39 metric tons) of sulfuric acid required per metric ton of normal superphosphate.

IX. ENVIRONMENTAL PROTECTION

Efforts to reduce the level of pollutants to comply with new regulations for air emissions and water discharges from P fertilizer production operations have contributed to energy consumption. Pollution abatement is not new to the U.S. fertilizer industry, but the Clean Air Act of 1970 and the 1972 amendments to the Federal Water Pollution Control Act have resulted in a concerted effort by federal and state environmental control agencies to address manufacturing point sources of air and water pollutants. In the United States, every major process from mining of PR to final product storage is currently subject to regulations prescribing the amount of pollutant which may be released. The situation in other industrialized and developing countries is variable, but the trend is towards similar regulations.

The U.S. Environmental Protection Agency (USEPA) has considered the impact of the energy required for pollution abatement in the course of developing industry-wide limitations. A summary of the energy consumed for pollution abatement for the various manufacturing processes is presented in Table 5. The USEPA was not able to obtain complete information on energy consumption for current levels of control in all cases and has, therefore, constructed model plants which are considered to be fairly representative of typical plants throughout the industry. The data presented in Table 5 are most useful in assessing the relative energy consumption for pollution abatement of the various manufacturing processes.

Air emission control for the PR processing segment consists of control of particulate emissions from rock drying, rock grinding, and rock handling. The USEPA selected as typical control systems cyclonic scrubbers for rock dryers and fabric filters (bag houses) for grinders and rock transport systems (USEPA, 1976b). The energy use for waste water effluent control was based on values for pond treatment of slimes and sand tailings to control the level of total suspended solids (USEPA, 1975).

In the wet process phosphoric acid segment, the existing control system

Table 5—Energy requirements for pollution abatement.†

Phosphate product	Air emission control	Water effluent control	Total
Phosphate rock‡			
GJ/metric ton P	0.048	0.202	0.250
GJ/metric ton P_2O_5	0.021	0.088	0.109
Million BTu/ton P_2O_5			0.094
Wet process phosphoric acid§			
GJ/metric ton P	0.030	0.009¶	0.039
GJ/metric ton P_2O_5	0.013	0.004¶	0.017
million Btu/ton P_2O_5			0.015
Superphosphoric acid			
GJ/metric ton P	0.041		0.041
GJ/metric ton P_2O_5	0.018		0.018
million Btu/ton P_2O_5			0.015
Diammonium phosphate			
GJ/metric ton P	0.183		0.183
GJ/metric ton P_2O_5	0.080		0.080
million Btu/ton P_2O_5			0.069
Run-of-pile triple superphosphate			
GJ/metric ton P	0.089		0.089
GJ/metric ton P_2O_5	0.039		0.039
million Btu/ton P_2O_5			0.034
Granular triple superphosphate			
GJ/metric ton P	0.220		0.220
GJ/metric ton P_2O_5	0.096		0.096
Million Btu/ton P_2O_5			0.083

† Individual process basis.
‡ Includes rock grinding.
§ Energy for sulfuric acid plant abatement measures not included.
¶ Gypsum pond seepage control.

for fluoride emission consists of cyclonic spray tower scrubbers which treat gases from the digester and filter vent (USEPA, 1976a). Discharge of waste water from the vast water recycling system associated with wet process phosphoric acid plants occurs normally only during periods of rainfall which exceeds the capacity of the recycling system. Therefore, treatment costs are highly variable except for the control of gypsum pond water seepage. The model system contains an estimate for power consumed for pumping water from the seepage ditch back to the recycling system (USEPA, 1974).

Since any process waste water discharge associated with the manufacture of superphosphoric acid, ammonium phosphates, and superphosphates is normally retained within the fertilizer plant complex, no energy charge for abatement is appropriate.

The energy use estimate for control of air emissions from superphosphoric acid manufacture is based on the modified submerged combustion process only. Energy use in excess of that necessary for the process itself is not required for the more commonly used vacuum evaporation process. Fluoride emission levels from the vacuum evaporation plants meet USEPA regulations without auxiliary abatement. The model system for fluoride controls at submerged combustion plants consists of three baffled spray chambers followed by an impingement scrubber (USEPA, 1976a).

In DAP production, the exhaust gases from the preneutralization reactor and the granulator are vented to a venturi scrubber for ammonia recovery. Fluorides are subsequently removed in a cyclonic spray tower using gypsum pond water as the absorbing solution. Ammonia in the flue gas from the dryer is removed in a venturi scrubber and another venturi scrubber is used for particulate removal from the air stream of the dry cyclone on the product cooling and screening equipment (USEPA, 1976a).

Gases vented from the mixer and den are pretreated in a venturi scrubber in the run-of-pile TSP plant and then combined with the storage building ventilation gases to be passed through a spray tower scrubber. Recycled gypsum pond water is the scrubber liquid in both cases (USEPA, 1976a).

The reactor and granulator off-gases from granular TSP production are passed through a venturi scrubber and cyclonic spray tower. Gypsum pond water is the scrubbing medium. Product in the dryer gases is recovered in a cyclone and the gases passed through a venturi and cyclone for fluoride removal. Gas streams from product cooling and screening operations are routed through a cyclone for product recovery and a cyclonic spray for fluoride removal. It is assumed that the storage building ventilation air is not treated (USEPA, 1976a).

Energy consumption for pollution abatement in the processes outlined above (Table 5) ranges from approximately 1% of process energy to around 6%. In most cases, the data collected by the USEPA for assessing current levels represent conditions in the 1972 to 1974 period. Increased energy consumption for pollution abatement can be expected to occur as existing facilities are retrofitted to meet abatement implementation schedules established by the pollution control agencies. If proposed air standards for existing fertilizer plants are adopted, the energy consumption may double the values shown for granular TSP process and increase by 40 to 60% the values for wet process phosphoric acid, DAP, and run-of-pile TSP. It is expected that the trend of energy consumption for pollution abatement will increase for the next few years until the effect of new plants with more efficient power utilization becomes a significant part of nation-wide production capacity.

X. ENERGY CONSERVATION

Using fertilizer consumption data for 1974, White[1] has reported that manufacture of fertilizer nutrients with representative energy values of 65.2, 26.6, and 11.2 GJ/metric ton of N, P, and K, respectively, required a total of 0.63×10^9 GJ. This energy was equivalent to 0.82% of the total energy (77×10^9 GJ) used in the United States in 1974. Of the energy expenditures for fertilizers, about two-thirds was for ammonia production. This use for ammonia was equivalent to about 2% of the 6.23×10^{11} m^3 (22 trillion ft^3) of natural gas used annually. The A. D. Little report (1975) states that electricity used for fertilizer (5 billion kWh) annually is about equivalent to 0.3% of the national total (1,900 billion kWh).

Obviously, the quantity of energy used in fertilizer production is relatively small, yet quite significant in total costs to the fertilizer industry. At $0.71 per GJ, the energy used in 1974 represented a cost of $450 million—over 1% of the consumer price of fertilizers in 1974. Thus, both factors, quantity and costs, increase the need for energy conservation.

As shown earlier, the energy requirement of Frasch S is the dominant factor in energy budgets of P fertilizers. Energy costs per unit of P relate directly to S use as long as sulfuric acid remains the primary attack agent for PR.

Substitution of by-product hydrochloric acid is one possibility discussed by Blouin (1974). It could reduce the energy input in DAP, for ex-

Table 6—Conversion factors.

1 lb coal	=	14,000 Btu
1 gal fuel oil	=	144,000 Btu
1 ft^3 natural gas	=	1,000 Btu
1 kWh	=	3,412 Btu
1 Btu	=	heat to raise temperature of 1 lb water 1°F
kcal	=	heat to raise temperature of 1 kg water 1°C
1 Btu	=	0.252 kcal
	=	1,054.8 joules (J)
Giga joule	=	10^9 joules (J)
Million Btu	=	1.055 Giga joules (GJ)

Short ton multiplied by 1.103 = metric ton
Million Btu/short ton multiplied by 1.164 = GJ/metric ton
 (1.055) (1.103) = 1.164
Million Btu/short ton P$_2$O$_5$ multiplied by 2.665 = GJ/metric ton P
 (1.055) (1.103) (2.29) = 2.665

ample, by 30%. But, this by-product acid is not plentiful and primary hydrochloric acid is more costly in energy than sulfuric acid.

Recovered S from sour natural gas and refinery gases is one of the best alternatives for energy reduction. Its process requirement is only about 0.35 GJ/metric ton (0.3 million Btu per short ton), less than 5% of that of Frasch S. According to Blouin (1974), however, recovered S currently supplies only about 20% of U.S.-produced S.

Converting from the dihydrate to hemihydrate phosphoric acid process could save much of the energy used in concentrating the acid. The dihydrate process, used throughout the United States, requires concentrating filter acid of about 13-14% to 23.6% (30-32% to 54% P_2O_5). In the hemihydrate procedure, the filter acid has 17 to 22% P (40 to 50% P_2O_5) thus little concentration is required. As mentioned previously, however, much of the heat for acid concentration is supplied by the by-product steam from the accompanying sulfuric acid. At present, acid concentration is about the best use of this steam.

Use of wet rock (8 to 10% water content) rather than dry rock (2 to 3% water) is a development in recent years that conserves energy. Five to eight percentage points of water evaporation are saved, with adjustments made to water additions in the acidulation and gypsum washing operations. Additionally, dust control problems are reduced markedly. Of course, more water incurs freight bills, and rotary dump rail hopper cars are required. Except for modifications in the grinding equipment, wet rock is processed with the same equipment at the wet process phosphoric plant site as dry rock.

LITERATURE CITED

Arthur D. Little, Inc. 1975. Economic impact of shortages on the fertilizer industry. Rep. to Fed. Energy Admin. C-77382, Cambridge, Mass.

Blouin, G. M. 1974. Effects of increased energy costs on fertilizer production costs and technology. Bull. Y-84, Natl. Fert. Dev. Center, TVA, Muscle Shoals, Ala.

Davis, C. H. 1974. Energy requirements for alternative methods for processing phosphate fertilizers. Proc. of ISMA Tech. Conf., Prague, 23-27 Sept. 1974. ISMA, Paris, France. p. PTE/74/15.

U.S. Environmental Protection Agency. 1974. Development docuemnt for effluent limitation guidelines and new source performance standards for the basic fertilizer chemicals segment of the fertilizer manufacturing point source category. EPA-440/1-74-011a. U.S. Government Printing Office, Washington, D.C.

U.S. Environmental Protection Agency. 1975. Development document for interim final effluent limitations guidelines and new source performance standards for the minerals for the chemical and fertilizer industries. Vol. II. EPA 440/1-75/095b. USEPA, Washington, D.C.

U.S. Environmental Protection Agency. 1976a. Draft guideline document: Control of fluoride emissions from existing phosphate fertilizer plants. USEPA, OAQPS, ESED, Research Triangle Park, N.C.

U.S. Environmental Protection Agency. 1976b. Standards support and environmental impact statement. An investigation of the best systems of emission reduction for the phosphate rock processing industry. Rough draft. USEPA, OAQPS, ESED, Research Triangle Park, N.C.

White, W. C. 1973. Fertilier-food-energy relationships. Am. Chem. Soc., Chicago, 28 Aug. 1973. The Fert. Inst., Washington, D.C.

Chapter 10

Energy of Phosphate Fertilizer Applications and Food Energy Returns

WILLIAM C. BURROWS AND ORVIS P. ENGELSTAD

Deere and Company, Moline, Illinois, and Tennessee Valley Authority, Muscle Shoals, Alabama, respectively

I. INTRODUCTION

"It is the business of agriculture to collect and store solar energy as food energy in plant and animal products" (Commission on Agricultural Production Efficiency, 1975). In this definition, we find the real basis for assessing the efficiency of agriculture—or of the inputs it uses. Solar energy is the only form of energy *directly* converted to food energy by plants. Other forms of energy inputs used in farming are not directly converted to biological energy. They merely change the efficiency with which the solar energy is collected and stored. Thus, many of the commonly used measures of efficiency, such as return of food energy per unit of fossil energy input, should not be used. In fact, many food crops are cultivated by the human race for their protein, mineral, or vitamin content and not for their biological energy content. We define "efficiency" in this discussion to be the ratio of biological energy content of the harvested crop parts to the incident photosynthetically active solar energy, expressed as a percentage.

We propose to show that fertilizer use, at proper rates, is a highly energy-efficient practice in food production. However, economics presently constrains the amount of fertilizer use to levels below those required for highest solar energy conversion efficiency. Greater food production directly implies more intensive agriculture—that is, greater input energy as well as greater biological energy output. The cost of inputs may be based on the energy contained in them or added in manufacture. However, the price of the agricultural produce when also based on energy content will, in most cases, rise to a degree that severe economic dislocation will be the result.

It has been estimated that U.S. agriculture uses between 2.5 and 3.0% of the total national energy budget (Council for Agricultural Science and Technology, 1973). We think it important that a distinction be made be-

---

Copyright 1980 © ASA-CSSA-SSSA, 677 South Segoe Road, Madison, WI 53711, USA.
The Role of Phosphorus in Agriculture.

tween "agriculture" and the food system of the United States. Nelson et al., (1975) have made the point that the portion of the food system energy budget allotted to production agriculture should include only that amount required to get the agricultural output to the farm gate. The additional energy required to eventually put food on the American dinner table must be assigned to a number of factors—geographic and social, among others.

Of the approximately 3.0% of the total U.S. energy budget allotted to production agriculture, slightly over one-third may be ascribed to fertilizer inputs (Nelson & Burrows, 1974), of which about 7.0% is P fertilizer. Thus, the percentage of U.S. energy in the P fertilizer used in agriculture is very small—about 0.07%.

In view of this small percentage, the concerns expressed here over efficiency may seem inordinate. However, other authors (Pimentel et al., 1973) have used returns of food energy per unit of fossil energy input to show an apparent energy advantage to the nation from decreasing fertilizer use. Either doubling or halving P fertilizer use would have an imperceptible effect on energy use in the United States. In fact, the effect on energy use in production agriculture would also be nearly imperceptible. But even though small at first, the influence on food production would be large after several years elapsed.

Feed grains (mostly maize), in relation to food crops, take the largest portion of all fertilizer energy. However, N fertilizers account for most of the input fertilizer energy for these crops. Approximately the same amount of energy, in the form of phosphate fertilizers, is expended for food crops and for feed grains. Pesticides, the other major type of agricultural chemical, represent approximately 10% of the energy budgeted to fertilizers. Again, approximately equal amounts are used for the food and feed crops.

We need to describe the conceptual size of the "envelope" which includes the energy figures that we will be using. For field operations, such as tillage and harvesting, the energy requirement is for fuel and lubricants. Energy for farm machinery includes the energy added at the factory of the farm machinery manufacturer, transport from manufacturer to farmer, and repairs and maintenance. It does not include energy used for mining and smelting of metals used in farm equipment, nor does it include energy added by manufacturers of items such as fuel pumps, tires, and generators.

After the agricultural produce passes through the farm gate, we assign the energy subsequently used to the food system, not to production agriculture (Nelson et al., 1975).

II. PHOSPHATE FERTILIZER IN THE AGRICULTURAL ENERGY INPUT SPECTRUM

A. Fertilizer from Manufacturer to Farm

The authors of the preceding chapter have reported the energy added in P fertilizer manufacture. Additionally, the energy needed to transport fertilizer to the farm site must be accounted for. Data do not exist in sufficient detail to insure extreme accuracy, and some assumptions were necessary (Hoeft & Siemens, 1975). We also had to make assumptions about fertilizer application rates. We selected maize (*Zea mays* l.) and wheat (*Triticum aestivum* L.) as typical feed and food grains, soybeans (*Glycine max* L.) for the oil crop, and cotton (*Gossypium hirsutum* L.) as the fiber crop. Fertilizer rates for soybeans and maize are those suggested for a major part of the central Corn Belt to maintain slightly better-than-average yield. An Indiana site is used for wheat response to P fertilizer. Alabama cotton completes the list. Shipment of fertilizer from manufacturer to a central storage point was assumed to be by railway. All subsequent transport of fertilizer materials was assumed to be by truck. We used Hirst's (1973) estimates of transportation energy. Our calculations show that the energy cost of diammonium phosphate (DAP) at the farm is between 2.8 and 3.0 million J/kg. This estimate includes manufacture and transport.

B. Farm Use of Fertilizer

Energy required on the farm to spread both P and K fertilizer materials from a bulk cart is about 36.5 million J/ha. If the fertilizer dealer uses a spreader truck, the amount of fuel required would probably be slightly greater. Approximately the same amount of energy is required if the fertilizer is applied with the planter (in addition to that required for the planter alone).

The amount of energy assigned to the machinery used to apply the fertilizer is slightly more difficult to estimate (Burrows & Siemens, 1974). A great deal less energy is required to manufacture, transport, and repair a bulk fertilizer spreader cart than is required for a truck used for the same purpose. However, the truck will cover more area in a season than will the cart. We have assumed that these two factors cancel each other. The machinery manufacturing energy requirement is estimated at about 1.8 million J/ha. Thus in what follows, we shall be using the energy values shown above for calculating input energy requirements for P fertilizer.

We should point out that it is not only difficult, but can be bad practice to separate a single element of the farming system such as we have done for P fertilizer. The interactions involved in the total farming system are complex. It thus is an oversimplification to separate one practice, such as fertilization, from the total system, and it is an even further simplification to attempt to separate one of the major plant nutrients from the rest because of fertilizer interactions. While recognizing the difficulty of separating P in the energy budget we shall make some estimates.

Soil management practices influence the choice of chemical and physical form, placement, and application rate of fertilizers. For example, in the Corn Belt, soil management practices vary from no-till planting to moldboard plowing (including several disking or other secondary cultivation methods). The input energy is affected by these related practices.

Other practices, particularly the use of pesticides, may also influence, or be influenced by, fertilization practice. In general, weed control (herbicides) is probably the most important. Further, insect and disease pests change the physiological status of crop plants so that the nutrient-use efficiency is changed. Thus, the efficiency of solar energy storage is changed.

Planting is another major consideration when fertilization is considered. Both pesticides and fertilizers can be applied with the planter. Also, all or only part of the fertilizer may be applied at planting.

This complexity puts a complete system analysis beyond the scope of our discussion here. Analysis of a single factor is tractable and illustrates the principles we wish to emphasize. However, before discussing the role of P fertilizer in the biological energy output which flows from agriculture, we will discuss how crops respond to P fertilization.

III. PHOSPHATE FERTILIZER IN THE AGRICULTURAL ENERGY OUTPUT SPECTRUM

A. Crop Response to Phosphate Fertilization

During the decade 1960-1969, fertilization rates increased markedly, largely as a result of bulk handling and spreading techniques. It was also during this period that many agronomists realized that soil P levels were rising quite markedly. It became increasingly difficult to find P-responsive locations necessary for experiments to compare P sources.

The management of high-P soils suddenly became more than merely of academic interest to research soil scientists. While early growth response to applied P often can be observed on soils testing high in P, final yield effects are usually insignificant. The issue now is to determine the rate and frequency of P fertilizer application necessary to maintain soil P levels over time. Indications are that some soils can supply adequate P for several years with little or no fertilizer P added (unpublished TVA data).

Because of this accumulation of P in soils, data on yield responses to multiple rates of added P are not only difficult to find, but relate to less than 50% of soils in modern agriculture. Therefore, the yield response data presented here are put into perspective by showing also the probability of receiving such response as indicated by the percentage of soil samples in the particular soil area or state that is in the low-P category.

The corn data were obtained by averaging multiyear data from the Galva-Primghar Experiment Farm in northwest Iowa and the Clarion-Webster Experiment Farm in northern Iowa.[1] The percentage of soil samples received in 1972-1973 testing low in P in these two soil areas is the same —35% of the total.

The soybean data are averages of yields over the 1972-1974 period at Orange, Virginia.[2] These data reflect the fact that soybeans generally show

---

[1] Richard W. Bohling. 1971. Effect of annual additions of phosphorus and potassium on chemical indexes and crop yields in monoculture systems. M.S. Thesis, Iowa State Univ., Ames.

[2] J. A. Lutz. (unpublished data). Virginia Polytechnic Institute and State Univ., Blacksburg.

rather small, if any, response to added P. Only 14% of soil samples tested in Virginia in 1970-1971 were low in P.

The wheat data from Indiana were obtained from a long-term rotation study at Lafayette (Barber, 1970). The percentage of soil samples testing low in P for dark-colored, fine-textured soils of northern Indiana is 26.

The cotton data were obtained by averaging yields over four seasons (1965, 1966, 1968, and 1969) on a site of Hartsells fine sandy loam in Alabama (Cope, 1970). Soil samples testing low in P for cotton in 1972-1973 in Alabama were 21% of the total submitted.

Square root or quadratic functions were fitted to the above data for convenience in subsequent calculations. In addition to the mean yields, the predicted yields are plotted as a function of P rate in Fig. 1, 2, 3, and 4 for corn, soybeans, wheat, and cotton, respectively. Also shown are the curves depicting the calculated net return per hectare at current prices of crop and fertilizer P.[3] The optimum rate is shown as well as the range in rates that would have resulted in at least 95% of the maximum net return. As indicated, net return is a rather insensitive function of fertilizer P rate. We emphasize again that the same is not true for many of the other inputs to production agriculture. For example, Fig. 1 shows the economic rate to be 33 kg/ha for corn. But any rate in a 28 kg/ha range (20 to 48 kg/ha) would have resulted in 95% or more of the maximum net return. Similar ranges in rates of 18, 20, and 18 kg of P/ha were found for soybeans, wheat, and cotton, respectively.

A high degree of precision in recommending P fertilizer rates is therefore not required; a considerable degree of variation in rate (or deviation from optimum) can be tolerated without much loss in net return. Such information may be of some comfort to farmers faced with uncertainty at the beginning of each cropping season. This assumes, of course, that the farmer could identify this range for his situation. Unfortunately, such knowledge is quite imperfect at the present time. Efficient use of fertilizer P would be enhanced by applying only where needed. This will require the proper use of soil testing services that will ensure profitable returns on the farmers' investment.

B. Biological Energy Output

We are now ready to discuss the biological energy output of crops as influenced by fertilizers. Two points should be kept in mind when discussing biological energy flow in agriculture. First, the task of agriculture is to produce food, feed, and fiber—not energy per se. This production comes directly from the process of photosynthesis, the conversion of solar energy to chemical energy. Increasing productivity means increasing the conversion

---

[3] Price of P used was: $0.96/kg of P ($0.19/lb of P_2O_5); prices of crops were as follows: corn —$0.10/kg ($2.60/bu); soybeans—$0.18/kg ($5.00/bu); wheat—$0.15/kg ($4.00/bu); and seed cotton—$0.46/kg ($0.21/lb). Prices as of the time of the symposium; these prices illustrate the basic relationships which will not change significantly even though the prices will change.

and storage of solar energy. The second point readily follows: the efficiency of agriculture should be measured by the efficiency of solar energy conversion.

Figures 1 through 4, showing the influence of P fertilizer rate on both crop and yield and net dollar return, may be recalculated to show the influence of added energy in the form of P fertilizer on solar energy conversion. Figure 5 shows the plot of energy input vs. solar energy conversion efficiency. As noted previously, efficiency is calculated by dividing biological energy in the harvested portion of the crop by the photosynthetically active solar energy falling on the field. The biological energy content of the crops may be found in standard texts such as Morrison (1951). Solar energy was estimated from average values of solar radiation during the growing season at each location, corrected to photosynthetically active wavelengths. By definition, the highest solar energy conversion efficiency (ME) is reached at the fertilizer rate that gives the maximum yield, not the economic optimum (EO) rate. For wheat and cotton, the fertilizer rate giving maximum solar energy conversion efficiency will also give at least 95% of the highest net economic return. For seed cotton, the economic optimum rate is the same as the maximum energy conversion efficiency rate. For soybeans, however, the economic optimum rate is less than half the rate for maximum energy conversion. At the prices used here,[3] use of the fertilizer rate producing the highest solar energy conversion efficiency gives an economic return for P use that is only 76% of the economic optimum.

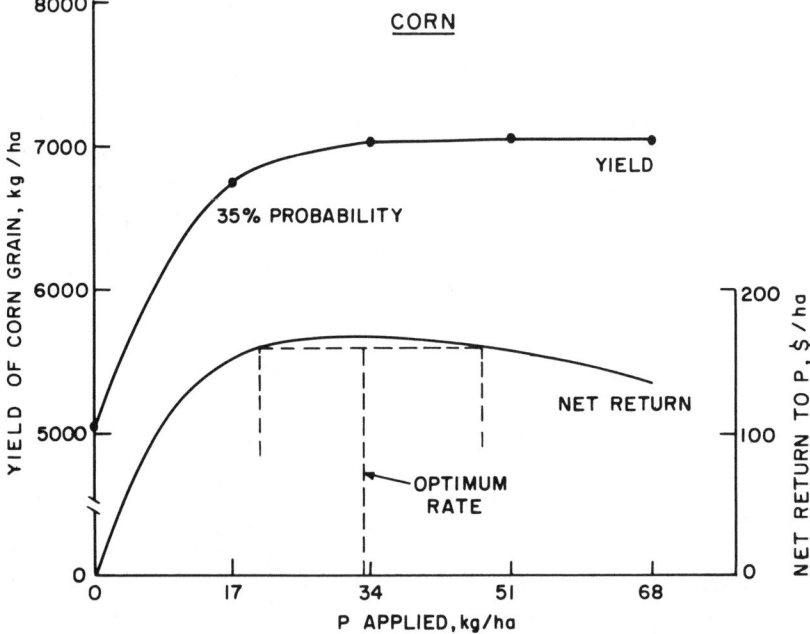

Fig. 1—Yield of corn (Iowa data)[1] and associated net return to P as affected by rate applied. Probability of yield response indicated by percentage of soil samples testing low in P in soil areas. Also shown is the optimum rate and the range in P rate resulting in at least 95% of the highest net return.

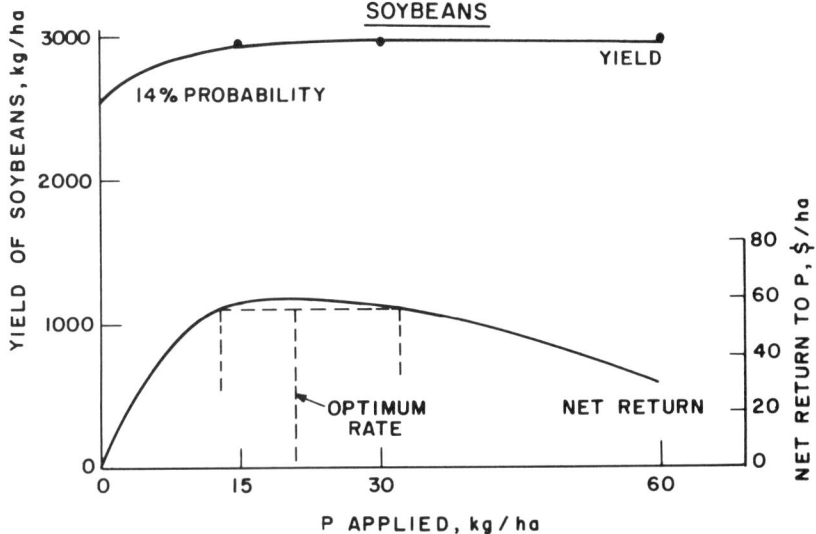

Fig. 2—Yield of soybeans (Virginia data)[2] and associated net return to P as affected by rate applied. Probability of yield response indicated by percentage of soil samples testing low in soil P in Virginia. Also shown is the optimum rate and the range in P rate resulting in at least 95% of the highest net return.

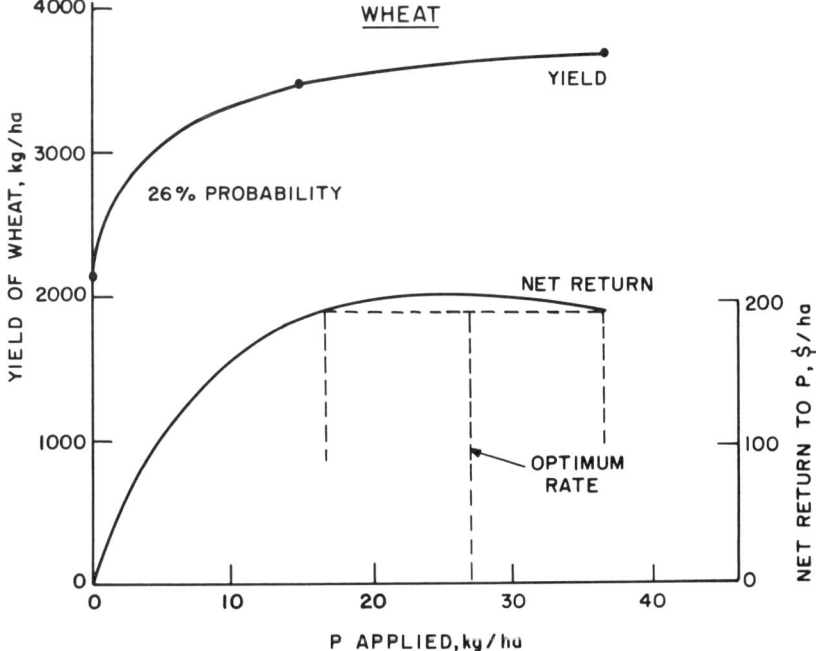

Fig. 3—Yield of wheat (Indiana data; Barber, 1970) and associated net return to P as affected by rate applied. Probability of yield response indicated by percentage of soil samples testing low in P in Indiana. Also shown is the optimum rate, and the range in P rate resulting in at least 95% of the highest net return.

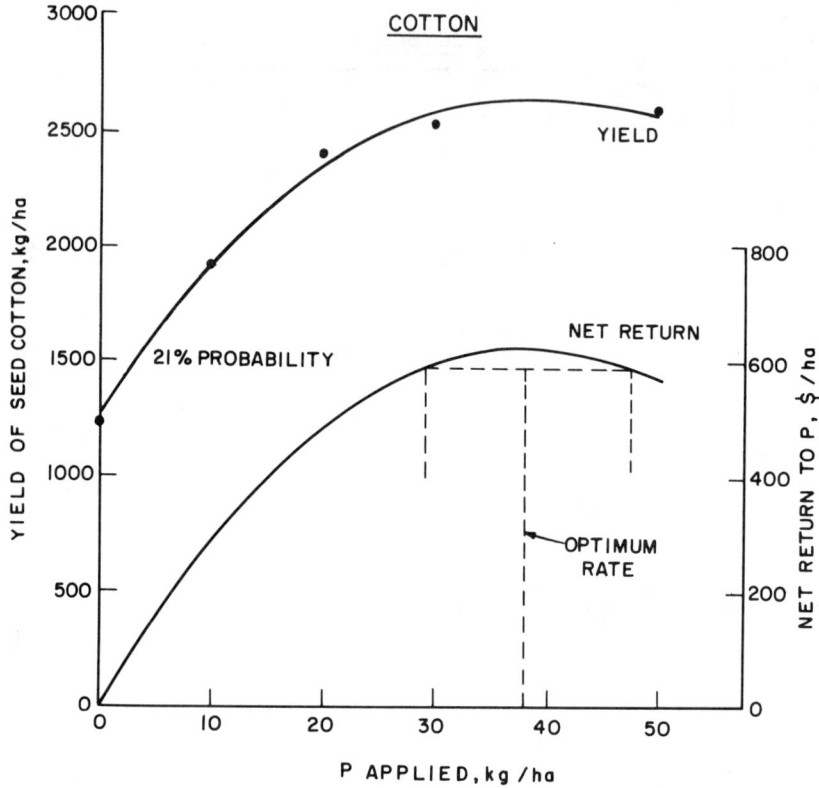

Fig. 4—Yield of cotton (Alabama data; Cope, 1970) and net return to P as affected by rate added. Probability of yield response indicated by percentage of soil samples testing low in P for cotton in Alabama in 1972-1973. Also shown is the optimum rate and the range in P rate resulting in at least 95% of the highest net return.

The maximum net energy return may be calculated in a manner similar to the maximum economic return. However, this turns out to be simply a slightly lower rate of fertilization than that giving the maximum energy conversion efficiency. For example, for soybeans the rates are 44.6 and 46.0 kg/ha, respectively.

A recap of the fertilizer rates and energy values producing the highest net economic return and highest solar energy conversion efficiency is given in Table 1.

It is obvious from the low solar energy conversion efficiencies that plants exhibit (as shown in Table 1 and Fig. 5) that any research leading to greater conversion efficiencies would have a great payoff. However, this type of research has not yet shown evidence of a breakthrough.

C. Economics of Energy Output and Input

Production agriculture responds to dollar flow—not to energy flow. Thus, as we have already noted, the economic optimum rate of fertilizer use

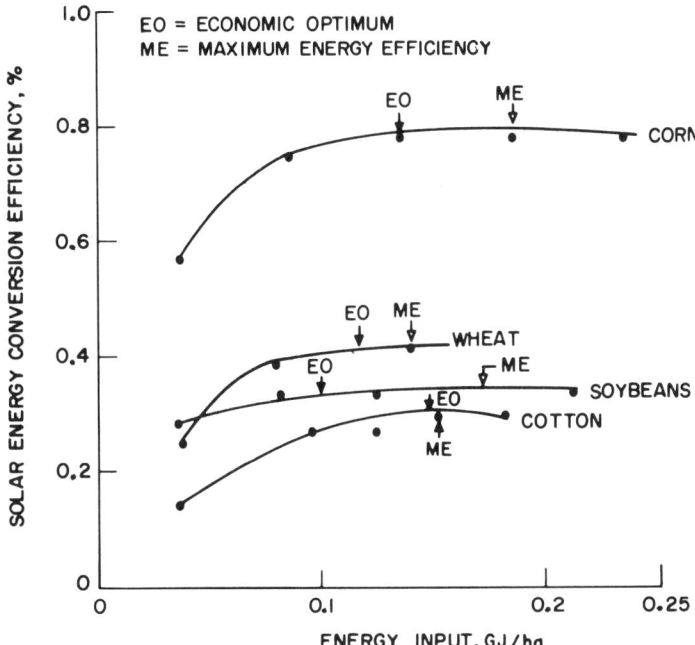

Fig. 5—Plot of energy input vs. solar energy conversion efficiency.

Table 1—Economic and energy inputs and outputs at highest net economic return and solar energy conversion efficiency.

	P fertilizer			Net return to P		Solar conversion efficiency
	Rate		Cost			
Crop	kg/ha	GJ/ha	$/ha	$/ha	GJ/ha	%
Corn						
HNER†	33.5	0.135	32.44	172.06	38.0	0.809
HSCE‡	51.0	0.186	49.34	162.74	39.4	0.818
Prob = 0.35§						
Soybeans						
HNER	21.0	0.099	20.16	61.59	8.5	0.341
HSCE	46.0	0.172	44.35	57.50	9.6	0.347
Prob = 0.14						
Wheat						
HNER	27.0	0.117	26.01	197.39	28.7	0.418
HSCE	36.0	0.143	34.56	192.81	29.2	0.4207
Prob = 0.21						
Cotton						
HNER	38.0	0.148	36.48	633.09	26.5	0.300
HSCE	39.0	0.148	37.44	632.59	26.7	0.300
Prob − 0.21						

† HNER = Highest net economic return.
‡ HSCE = Highest solar energy conversion efficiency.
§ Prob = From Fig. 1-4. Can roughly be interpreted as the proportion of crop area expected to give returns less than or equal to those shown.

Table 2—Economic comparison of returns from corn, wheat, and soybeans for different marketing bases.

Crop	Market price				Net return†	Equiv. P price
	$/bu	$/kg grain	$/kg protein	$/GJ	$/ha	$/kg
Current situation‡						
Corn	2.60	0.10	1.14	5.39	50	0.22
Wheat	4.00	0.15	1.05	7.80	147	0.32
Soybeans	5.00	0.18	0.44	9.58	19	0.39
Case 1: biological energy sold at $5.39/GJ						
Corn	2.60	0.10	1.14	5.39	50	0.22
Wheat	2.77	0.10	0.73	5.39	−20	0.22
Soybeans	2.81	0.10	0.25	5.39	−223	0.22
Case 2: protein sold at $0.44/kg						
Corn	1.00	0.04	0.44	2.07	−399	0.09
Wheat	1.67	0.06	0.44	3.25	−170	0.13
Soybeans	5.00	0.18	0.44	9.58	19	0.39
Case 3: biological energy sold at $23.30/GJ						
Corn	11.25	0.44	4.92	23.30	2480	0.96
Wheat	11.96	0.44	3.14	23.30	1229	0.96
Soybeans	12.16	0.45	1.05	23.30	808	0.96

† Costs for net return are average costs of production taken from a report on the contents of the publication, "Costs of Producing Selected Crops in the United States—1974," available from the Senate Agriculture Committee, Senate Office Building, Washington, D.C. Costs used were: Corn—$2.42/bu, Wheat—$2.92/bu, and Soybean—$4.83/bu.
‡ Current market prices are those used for Fig. 1-3 and stated in footnote.[3]

is not, in general, the rate for highest solar energy conversion efficiency. If crops are to be grown so that this efficiency is maximized, the pricing structure of crops or of inputs will have to change. We calculated three cases for the food-related crops sold on different price bases.

The first case recognizes the fact that maize is the major crop grown for the biological energy it contains. So we calculated the current price of that energy and assumed all crops should be sold for their energy value alone. This is, in fact, one logical implication of using biological energy output over fossil energy input to measure agriculture's efficiency.

Measurement of protein output per unit of fossil energy input implies the producer can be economically sound by selling the crops on a protein basis. For our second case we used the current price of the protein in soybeans, and recalculated the prices for maize and wheat.

Case 3 is based on energy price, as was case 1. But the price of biological energy in case 3 is set by the current price of fertilizer P. At $0.96/kg of fertilizer, the equivalent fossil energy in fertilizer P sells for $23.30/GJ.

The results of the three assumed cases, as well as the current situation, are shown in Table 2.

Note that, at the crop market prices used in this paper as current, wheat produces the greatest net economic return. If market price were based on biological energy output with corn as the base (case 1), corn gives the best economic return with soybeans and wheat being losing propositions. This case is interesting because, if agricultural energy efficiency is computed as the ratio of biological energy output to fossil energy input, production economics is ignored. No business—not even agriculture—will be able to exist long by producing a product at a net economic loss.

At current costs of all inputs, the market price of soybeans sold for their biological energy content would need to be about the current price, or $0.177/kg ($4.83/bu), just for the farmer to break even. We would be paying $9.25/GJ for our food energy. At this rate, corn price would rise to $4.46/bu, almost twice the current price.

Selling the crops on the basis of protein content completely reverses the previous picture (case 2). There is a drastic change in soybean price. Note that the market price of soybeans will be about five times the price of corn because of the higher protein content. Heichel (1973) has used amount of protein return per unit of fossil energy input as an efficiency rating for agricultural energy use.

Two previous examples present an economic picture related to price of production output alone. Case 3, however, considers the economics related to cost of the energy added as input energy. If the crops' biological energy were to be sold for the same unit price as the energy bought in the form of P fertilizer, the price is $23.30/GJ. For this case, corn-wheat-soybeans is the ranking of the extremely high net returns. But even more interesting are the market prices of the crops compared with "current" prices. Again, there is a rise to rather unrealistic levels.

All three cases present an artificial picture, completely out of tune with the current market structure. They illustrate that an artificially imposed pricing structure based on food energy output alone, on protein content alone, or on input energy cost alone will increase the price of food or bankrupt the farmer. Energy conservation and energy use efficiency must be an integral part of food production. But artificial standards of efficiency or marketing cannot result in a workable system.

The above discussion has been given to point out the dangers in oversimplifying a system as complex as the food production system. Both the inputs and the outputs of the system are a result of many forces—social and political—which create the demands and thus the economics of the system. In addition, the discussion also gives further basis for stating that fertilizer use is energy efficient.

Only by using those inputs—and at proper levels—that maximize the true "job" of agriculture can energy efficiency be maximized. If research can find ways to increase the efficiency of solar energy conversion and storage by crop plants at the same or at lower energy input levels, or if research can find substitutions for current energy inputs, then we can expect to see a drastic change in agriculture and its energy use patterns. This is especially true for those energy inputs which, unlike P fertilizer, have a large impact on solar energy storage. It seems a mistake to suggest that one of the lowest energy users in society today should drastically cut its use. Production would fall and food prices would rise out of all proportion to the energy conservation benefits realized.

LITERATURE CITED

Barber, S. A. 1970. Effect of rate and placement of phosphorus and potassium on a rotation, 18-year period. Purdue Univ. Res. Prog. Rep. 365, Lafayette, Ind.

Burrows, W. C., and J. C. Siemens. 1974. Determination of optimum machinery for corn-soybean farms. Trans. ASAE 17:1130-1135.

Commission on Agricultural Production Efficiency. 1975. Agricultural production efficiency. Natl. Academy of Sciences, Washington, D.C.

Cope, J. T., Jr. 1970. Response of cotton, corn, and bermudagrass to rates of N, P, and K. Auburn Agric. Expt. Stn. Circ. 181.

Council for Agricultural Science and Technology. 1973. Energy in agriculture. Council for Agric. Sci. nd Technol., Ames, Iowa.

Heichel, G. H. 1973. Comparative efficiency of energy use in crop production. Conn. Agric. Expt. Stn., Bull. 739, New Haven, Conn.

Hirst, Eric. 1973. Energy intensiveness of passenger and freight transport modes, 1950-70. ORNL-NSP-EP-44. Oak Ridge National Laboratory, Oak Ridge, Tenn.

Hoeft, R. G., and J. C. Siemens. 1975. Energy consumption and return from adding nitrogen to corn. Illinois Research, Winter. Univ. of Ill., Urbana.

Morrison, F. B. 1951. Feeds and feeding. Morrison Pub. Co., Ithaca, N.Y.

Nelson, L. F., and W. C. Burrows. 1974. Putting the U.S. agricultural energy picture into focus. ASAE Paper no. 74-1040. St. Joseph, Mich.

Nelson, L. F., W. C. Burrows, and F. C. Stickler. 1975. Recognizing productive, energy-efficient agriculture in the U.S. food system. ASAE Paper no. 75-7505. St. Joseph, Mich.

Pimental David, L. E. Hurd, A. C.Bellotti, M. J. Forester, I. N. Oka, O. D. Sholes, and R. J. Whitman. 1973. Food production and the energy crisis. Science 182:443-449.

Chapter 11

Reactions of Phosphate Fertilizers in Soils

E. C. SAMPLE

Tennessee Valley Authority
Muscle Shoals, Alabama

R. J. SOPER

University of Manitoba
Winnipeg, Manitoba

G. J. RACZ

University of Manitoba
Winnipeg, Manitoba

I. INTRODUCTION

When phosphatic (P) fertilizers are applied to soils and are dissolved by soil water or are applied as liquid fertilizers, reactions occur among the phosphate, soil constituents, and the nonphosphatic fertilizer compounds which remove P from the solution phase and render the phosphates less soluble. This phenomenon, called *P fixation* or *retention,* has been known for well over a century (Way, 1850) and has probably prompted more research than any other aspect of soil-fertilizer-plant interactions. The purpose of this review is to summarize some of the evidence for the retention of fertilizer P by soils, some factors which influence retention, some possible mechanisms through which retention may occur, and some of the methods by which retention is studied.

Although retention of P by soils plays an important role in determining the ultimate availability of fertilizer P to plants, this topic will be touched on only briefly. The agronomic effectiveness of P fertilizers is dealt with in Chapter 12 by Engelstad and Terman and the utilization of residual P is covered in Chapter 13 by Barrow in this book. The retention or immobilization of P by biological pathways is also treated in this book in Chapter 15 by G. Anderson and will not be dealt with here. Likewise, the very large body of research dealing with P relations in virgin soils or soils which have not received recent applications of fertilizer will not be reviewed extensively. Readers are referred to reviews by Larsen (1967) on various aspects of soil P

Copyright 1980 © ASA-CSSA-SSSA, 677 South Segoe Road, Madison, WI 53711, USA.
The Role of Phosphorus in Agriculture.

and by Lindsay and Vlek (1977) on the nature of native phosphate minerals in the soil environment.

An historical overview of research on P retention can be obtained from earlier reviews. Examples are those by Midgley (1940), Dean (1949), Wild (1950), Kurtz (1953), Olsen (1953), Hemwall (1957), and Mattingly and Talibudeen (1967). Research at the Tennessee Valley Authority on reactions of P in soils was summarized by Huffman (1962, 1968).

II. PHOSPHATE RETENTION BY SOIL CONSTITUENTS

In his review, Wild (1950) cited several researchers from the mid-nineteenth century who concluded that calcium carbonate and hydrous oxides of Fe and Al played key roles in P retention. They suggested that the P was either precipitated as Ca-, Fe-, or Al-phosphates or that the P was chemically bonded to these cations at the surfaces of the soil minerals. Hence, from the very early days the choices of mechanisms used to explain P retention involved precipitation or adsorption. More modern researchers have proposed several more refined mechanisms through which P may be retained by soils. These include: physical adsorption, chemisorption, anion exchange, surface precipitation, and precipitation of separate solid phases. In general, these are but special cases of precipitation or adsorption reactions.

Perhaps because the very early workers implicated specific soil constituents in P retention reactions, subsequent workers have done considerable research on the mechanisms involved in P retention by specific soil minerals. For convenience, these research efforts are discussed below according to mineral group. Though not a mineral group, soil organic matter is also included.

A. Retention by Hydrous Oxides of Iron and Aluminum

Aluminum and Fe oxides and hydrous oxides can occur as discrete compounds in soils or as coatings on other soil particles. They can also exist as amorphous Al hydroxy compounds between the layers of expandable Al silicates. Wild (1950) in his review of P retention by soil states that as early as 1866 Warington found that hydrous oxides of Fe and Al retained large amounts of P from solution and that the amount of P retained in a soil was related to hydrochloric acid-soluble Fe and Al. Wild gives considerable evidence that these compounds account for much of the P retention by acid soils. The amount of P sorbed depends upon the time of the reaction, the temperature, pH, and P concentration of the soil solution.

Despite earlier suggestions of precipitation, the mechanism for the reaction of phosphate with Fe and Al up to the time of Wild's review was thought to be one of adsorption rather than precipitation, since there did not seem to be enough Fe and Al in the soil solution to precipitate the rather large amounts of P that could be immobilized. The adsorption was postulated to be an exchange reaction between phosphate ions and hydroxyl ions associated with the metal. Wild rightly predicted that the distinction between adsorption and precipitation would be hard to resolve. A great deal of the research on P reactions with soils since that time has been devoted to solving this problem.

The mechanism for P retention in soils by Al and Fe compounds is still not resolved, but progress has been made in the past 20 years. Bache (1964) measured the sorption of P by gibbsite and hydrous ferric oxide at different pH's in buffered solutions and concluded that there were three stages of adsorption occurring at different P concentrations of the solution: (i) a high energy chemisorption of small amounts of P; (ii) precipitation of a separate phosphate phase; and (iii) a low energy sorption of P onto the precipitate. He concluded that the precipitated products would likely be $(Al,Fe)(H_2PO_4)_n(OH)_{3-n}$, and that n would be less than 1 for most soil systems. He concluded that amorphous precipitates have much higher solubilities than variscite and strengite.

The adsorption of P by kaolinite, gibbsite, and pseudoboehmite was studied by Muljadi et al. (1966a, 1966b, 1966c). They found that the complete isotherms obtained did not conform to a single isotherm type. Hsu and Rennie (1962) had similar results. The adsorption isotherms were divided into three regions similar to those found by Bache (1964). Muljadi et al. suggested that the three regions were related to the affinity of P for at least three energetically different reactive sites. They suggested that the mechanism for adsorption in regions I and II was the exchange of phosphate for OH groups associated with a positively charged Al atom $>Al\begin{smallmatrix}\nearrow OH\\ \searrow OH\end{smallmatrix}$ by surface hydrolysis of OH groups. In region I the first OH group was involved while in region II the second OH reacted. All adsorbants behaved similarly for regions I and II, the Al atoms involved being situated on the edge faces of the crystals.

Adsorption in region III was thought to arise through the penetration of P into some less crystalline region of the clay surface. Phosphate bound by kaolinite was reversible with respect to pH for all regions, but not with P concentration for region I. With the hydrous oxides, the P adsorbed in regions II and III was much less reversible with respect to concentration than for kaolinite. The authors suggested that on desorption the P bound by the hydrous oxides may undergo a phase change. Kafkafi et al. (1967) suggested that in this phase change, the P may be bonded by two Al atoms forming a stable six-membered ring.

Hingston et al. (1967, 1968) have shown that anions may be specifically adsorbed by hydrous Fe and Al oxides. The oxide surfaces contain molecules of water and hydroxide ions which are octahedrally coordinated with either Fe or Al. At the edge of the crystal some water molecules and hydroxyls will be coordinated with only one atom of Fe or Al and thus at the isoelectric point, or when the crystal has no net charge, the surface may appear like this:

$$>Al\begin{matrix}OH\\OH_2\end{matrix}$$

The Al ion shares four other O atoms within the crystal. As the pH increases, H^+ dissociates from the water molecule to form more hydroxyl groups and the charge on the surface becomes more negative. Conversely, as the pH of the solution decreases, the proportion of water molecules increases and the surface becomes more positive, as depicted by Rajan et al. (1974):

$$Al\begin{matrix}OH_2\\OH_2\end{matrix}^{1+} \underset{-H^+}{\overset{+H^+}{\rightleftarrows}} Al\begin{matrix}OH_2\\OH\end{matrix}^{0} \underset{-H^+}{\overset{+H^+}{\rightleftarrows}} Al\begin{matrix}OH\\OH\end{matrix}^{1-} \quad [1]$$

Hingston et al. (1967, 1968) suggested that phosphate is capable of exchanging with OH_2 and OH^- and becoming coordinated to the metal ion at the surface and specifically adsorbed. They indicated that, when anions are specifically adsorbed, they take up protons from either the oxide surface at pH values below the isoelectric point or from the dissociation of the acid itself when the charge on the oxide is negative. This lower average negative charge of the anion permits a coordinate link to be formed with the metallic ion. As a consequence of this, the net charge on the surface becomes more negative and the pH of the solution increases. This proposal for the mechanism of P adsorption by hydrous oxides of Fe and Al is therefore different from that of Muljadi et al. (1966b) and Hsu and Rennie (1962), who thought that the phosphate ion replaced hydroxyl ions only.

Rajan et al. (1974) made use of the fact that the pH of the solution increases with adsorption of P to determine which groups were reactive when P was adsorbed at different concentrations on a hydrous Al oxide surface. The amount of hydroxyl released or neutralized during P adsorption was measured at a constant pH of 5.1 and 6.2 using an automatic titrator. They found that at low concentrations P mainly replaced water groups, but as adsorption increased and P concentration increased, more hydroxyl groups were involved. At the adsorption maximum, P was mainly replacing hydroxyl groups. Beyond the adsorption maximum, P was still adsorbed, and it was thought that this was achieved by a breaking of a hydroxyl bonded to two Al^{3+} ions (Eq. [2]).

$$\begin{array}{c}\diagdown\!\!\!\diagup\mathrm{Al}\diagup\!\!\!\diagdown\mathrm{OH}\diagdown\!\!\!\diagup\mathrm{Al}\diagdown\end{array} \rightleftharpoons \begin{array}{c}\diagdown\mathrm{Al-H_2PO_4^{1/2-}}\diagup \\ + \\ \diagdown\mathrm{Al-OH^{1/2-}}\diagup \end{array} \underset{-\mathrm{H^+}}{\overset{+\mathrm{H^+}}{\rightleftharpoons}} \diagdown\mathrm{Al-H_2O^{1/2+}}\diagup$$

$$\diagdown\mathrm{Al-H_2O^{1/2+}}\diagup + \mathrm{H_2PO_4^{1-}} \rightleftharpoons \diagdown\mathrm{Al-H_2PO_4^{1/2-}}\diagup + \mathrm{H_2O} \quad [2]$$

In this reaction one H^+ would be consumed for every two phosphates adsorbed.

Based on infrared studies, Parfitt et al. (1975) indicated that P was specifically adsorbed by Fe oxide surfaces by replacing two adjacent surface hydroxyl ions. Two oxygen atoms of the phosphate ion are linked to two Fe ions giving a binuclear surface complex, $Fe-O-(P=O_2)-O-Fe$. The structure is similar to that suggested by Kafkafi et al. (1967), based on desorption studies, and by Hingston et al. (1967). Rajan (1975) has also suggested from calculating the number of hydroxyls released, that the adsorption of HPO_4^{2-} by hydrous Al oxide is due to a binuclear coordination of phosphate ions with two Al atoms. Thus, P may become strongly bound to hydrous oxides of Al and Fe and be relatively unavailable to plants.

One of the consequences of the specific adsorption of anions by soil colloids as described by Hingston et al. (1967, 1968) is an increase in the negative charge of the soil colloid. Several workers in the past have shown an increase in the CEC of soils with the addition of P (Davis, 1945; Coleman & Mehlich, 1948; Perkins, 1958). Recently Nakaru and Uehara (1972) studied the effect of P adsorption on the CEC of some ferruginous tropical soils which contained large amounts of hydrous Fe and Al oxides. Quite large increases in CEC of up to 100 meq/100 g were obtained, the ratio of change being approximately 0.8 meq/mmole of P added. Schalscha et al. (1972, 1974) demonstrated that the CEC of some volcanic soils as determined with KH_2PO_4 was much greater than that with KCl or NaCl, and that P adsorption increased the CEC in comparable amounts to that found by Nakaru and Uehara (1972). Sawhney (1974) reported similar results with P adsorption and concluded that the increase in CEC was due to hydrous Fe and Al oxides rather than layer silicates. All of the above authors agreed that specific adsorption of P was responsible for the increase in CEC. Rather large amounts of P are required for significant changes in CEC; however, the effects may be important around the fertilizer granule in acid soils of low CEC.

In several of the studies cited above, the authors concluded that, at low P concentrations, hydrous oxides retained P through sorption-type reactions. At higher concentrations, precipitation was suggested as the mechanism, largely because of discontinuities in conventional sorption isotherms; however, little direct evidence was offered for the existence of separate precipitated phases. A number of researchers, however, have found that hydrous oxides react with highly concentrated P solutions to form identifiable precipitates.

Haseman et al. (1950a) reacted $1.0M$ solutions of ammonium, potassium, or sodium phosphate with hydrous Fe and Al oxides and found that the oxides decomposed, forming Fe and Al phosphates as separate crystalline phases. Kittrick and Jackson (1955, 1956) also showed that concentrated P solutions decomposed hydrous oxides with the formation of separate phosphate precipitates. Lindsay et al. (1962) and Philen and Lehr (1967) reacted hydrous Fe and Al oxides with a wide variety of concentrated P solutions and identified several precipitated phosphates containing Fe and Al.

B. Retention by Alumino-Silicate Minerals

Wild (1950) reviewed numerous investigations of the reactions of phosphate with alumino-silicate clay minerals. Early researchers found appreciable variations in the amounts of P that were retained by the diverse minerals; nevertheless, Wild concluded that, when compared at nearly equal particle sizes (equal surface areas), montmorillonitic and kaolinitic clays retained similar amounts of P. This suggested to Wild that the mechanism for sorption was the same for the various clays. He cited two proposed sorption mechanisms: exchange of phosphate ions with the hydroxyl groups linked to the gibbsite layer, and as exchangeable anions which counter a possible positive charge developed by the adsorption of protons or hydroxyl groups of the gibbsite layer.

Kurtz (1953) in his review of P in acid and neutral soils stated that Ca-saturated clays retain more P than do Na-, NH_4-, or K-saturated clays. It is possible that the linkage of P to clay minerals through exchangeable Ca^{2+} or Mg^{2+} ions could contribute to P sorption. Pissarides et al. (1968) studied the adsorption of phosphorus by kaolinite, illite, and montmorillonite saturated with various cations. Phosphate at low concentrations was adsorbed in accordance with the Langmuir adsorption isotherm for all cations. Clays saturated with divalent ions adsorbed more P than clays saturated with monovalent ions. Montmorillonite had a negative adsorption when saturated with monovalent ions, but this changed to a positive adsorption when $0.1N$ NaCl was added. The authors concluded from these results that the thickness of the diffuse double layer controlled how much P was adsorbed at the surface of the clay.

There have been numerous studies of sorption by clay minerals in which one of several adsorption isotherms has been used to mathematically describe the sorption data. Because of similarities in the sorption isotherms for alumino-silicates and the hydrous oxides, several researchers have concluded that the sorption mechanisms are the same for the two mineral groups. Muljadi et al. (1966a, 1966b, 1966c) concluded that the adsorption of P was similar for kaolinite, gibbsite, and pseudoboehmite, differing only in the number of adsorption sites. When the minerals were washed, a phase change took place where the P was less reversibly adsorbed. The change was greater for the hydrous Al oxides than for kaolinite, an indication that the P would be less labile. Hingston et al. (1972) also think that the edge faces of kaolinite behave similarly to the hydrous oxides in anion adsorption.

Kuo and Lotse (1972) have proposed the adsorption of P onto kaolinite from dilute solutions of P to be the replacement of specifically adsorbed water molecules at the broken edge of the clay lattice. The P was thought to be covalently bonded to an aluminum ion. They believed that P replaced a water molecule rather than a hydroxyl group, since the pH of the solution did not significantly rise during the reaction.

Rajan and Perrott (1975) demonstrated that synthetic amorphous alumino-silicates sorbed P in much the same manner as did kaolinite.

Phosphate adsorption by some tropical soils was studied by Rajan and Fox (1975), who found that P adsorption could best be described by a binary Langmuir equation, suggesting two types of adsorption sites. In the first region P adsorption resulted in the release of sulfate and silicate and a rise in pH, which indicated that phosphate was replacing adsorbed silicate, sulfate, and hydroxyl groups. In the second region, at higher concentrations of P, there was a sharp increase in the release of silicate but with no additional sulfate, suggesting that the silicate clay minerals were being disorganized and the structural silicate displaced, allowing more P to be adsorbed.

These latter results are similar to those of Low and Black (1948, 1950), who concluded that, at high concentrations, P solutions dissolve kaolinite to release Si and Al, with subsequent precipitation of Al-P compounds. At more dilute concentrations, they believed the reaction to be one of adsorption with the replacement of surface hydroxyl groups by phosphate.

As was the case with hydrous oxides, several researchers invoke a low-concentration sorption and high-concentration precipitation mechanism to explain experimental data. Quite often there is no direct evidence for a separate precipitation phase. Researchers working with very concentrated P solutions, however, have very little difficulty in demonstrating a precipitation reaction associated with silicate clays.

Haseman et al. (1950a) reacted $1.0 M$ P solutions (as NH_4-, K-, or Na-phosphate) with illite, kaolinite, and montmorillonite and found precipitated phases to consist of Fe- and Al-phosphates containing the cation furnished by the P solution. The reactions with silicate clays were much slower than those with hydrous oxides. Kittrick and Jackson (1956) have also shown that concentrated P solutions will decompose kaolinite, resulting in precipitation of an Al-P compound. Philen and Lehr (1967) found that concentrated ammonium pyro- and tripolyphosphate solutions reacted with montmorillonite and attapulgite to form several Mg-NH_4- and Ca-NH_4-pyrophosphates as separate precipitates.

C. Retention by Soil Carbonates

The chemistry of calcareous soils is dominated by soil carbonates, thus reactions of P with pure carbonates have been studied in some detail. Boischot et al. (1950) and Cole et al. (1953) concluded that adsorption reactions dominated when dilute P solutions were added to $CaCO_3$, but that precipitation reactions dominated with more concentrated P solutions. Cole et al. (1953) found that the Langmuir adsorption isotherm described the adsorption of P by $CaCO_3$ at relatively low P concentrations. An abrupt discontinuity in the isotherm with increasing concentration suggested to the authors that precipitation was occurring. The precipitating compound was thought to be DCP or a compound with similar properties. The authors characterized the initial reaction products as having very high specific surfaces and greater solubilities than hydroxy- or fluorapatite.

The mechanism of P adsorption by $CaCO_3$ is still not well understood. Kuo and Lotse (1972) have suggested that P may replace adsorbed water molecules, bicarbonate ions, and hydroxyl ions when it is adsorbed by calcite, with adsorbing strength depending upon the solubility of the compound formed with the surface Ca ions. They showed the value of the maximum monolayer-surface saturation calculated from the Langmuir isotherm to be much less for $CaCO_3$ than for a Ca-saturated kaolinite.

Griffin and Jurinak (1973, 1974) studied the kinetics of P interactions with calcite and concluded that two reactions were occurring. The first reaction at low P concentrations consisted of the adsorption of P by the calcite surface. The second reaction was described as a nucleation process of calcium phosphate crystals where there is a surface rearrangement of amorphous phosphate into phosphate heteronuclei which supercedes crystal growth. Solubility studies indicated that at low P concentrations hydroxyapatite was formed. At higher P concentrations the solutions were saturated with respect to octocalcium phosphate. Cole et al. (1953) showed that at P concentrations beyond these, $CaHPO_4 \cdot 2H_2O$ precipitated out.

Holford et al. (1974) have demonstrated that the Langmuir equation describes the adsorption of P by soils if it is assumed that the adsorption occurs on two types of surfaces with different bonding energies. Using this two-surface Langmuir equation, Holford and Mattingly (1975) studied P adsorption by 24 calcareous soils. The high-energy adsorption surfaces were closely related to dithionite-soluble iron, which indicated that even in calcareous soils hydrous oxides are important in the adsorption of P. The low-energy adsorption was highly correlated with $CaCO_3$ surface areas and organic matter content, but not with the $CaCO_3$ content. The results suggest that P is less strongly bound to $CaCO_3$ than to the hydrous oxides and hence is more available to plants in that form.

Most researchers have had little difficulty in demonstrating separate precipitation reactions between P solutions and $CaCO_3$, even at relatively low concentrations. At high P concentrations, precipitation reactions are both copious and rapid.

Lindsay et al. (1962) found $CaHPO_4 \cdot 2H_2O$ and $CaHPO_4$ to be the initial reaction products when a saturated solution of MCP was reacted with $CaCO_3$ or $CaMg(CO_3)_2$. These same products were formed when $CaCO_3$ was added to filtrates from reactions of soils and saturated solutions of MCP. When the saturating solutions were MAP and MKP, $CaHPO_4 \cdot 2H_2O$ and $CaHPO_4$ formed, but also $Mg_3(NH_4)_2(HPO_4)_4 \cdot 8H_2O$, $MgNH_4PO_4 \cdot 6H_2O$, and $MgHPO_4 \cdot 3H_2O$. When $CaCO_3$ and $CaMg(CO_3)_2$ were added to saturated solutions of DAP and DKP, the following compounds were identified: $Ca_2(NH_4)_2(HPO_4)_3 \cdot 2H_2O$, $Ca_8H_2(PO_4)_6 \cdot 5H_2O$, $CaNH_4PO_4 \cdot H_2O$, $Ca_{10}(PO_4)_6(OH)_2$, $MgNH_4PO_4 \cdot 6H_2O$, $CaK_3H(PO_4)_2$, and $MgKPO_4 \cdot 6H_2O$. Others have found that the initial reaction product of MCP and $CaCO_3$ was $CaHPO_4 \cdot 2H_2O$ (Larsen et al., 1963). Racz and Soper (1967) studied the hydrolysis of $CaHPO_4 \cdot 2H_2O$ and $MgHPO_4 \cdot 3H_2O$ in the presence of $CaCO_3$ and $MgCO_3$. After 1 month, $CaHPO_4 \cdot 2H_2O$ hydrolyzed to form $Ca_8H_2(PO_4)_6 \cdot 5H_2O$ in the presence of $CaCO_3$, and after 3 months, $MgHPO_4 \cdot 3H_2O$ hydrolyzed to form $Mg_3(PO_4)_2 \cdot 22H_2O$ in $MgCO_3$. No other hydrolysis products were detected even after 6 months.

Philen and Lehr (1967) reacted calcite with an ammonium tripolyphosphate solution, one of the ingredients of APP fertilizers, and found that Ca-NH_4-tripolyphosphate and Ca-NH_4-pyrophosphate precipitates formed. Magnesite added to the tripolyphosphate solution resulted in precipitation of a Mg-NH_4-tripolyphosphate.

D. Soil Organic Matter and Phosphorus Retention

Humus, since it is normally negatively charged, is not thought to retain much P by itself in soils; however, in association with cations such as Fe^{3+}, Al^{3+}, and Ca^{2+} it is able to retain significant amounts (Wild, 1950). Weir and Soper (1963) prepared a complex of Fe, P, and humic acid which was water soluble and contained large amounts of ^{32}P-exchangeable P. They thought that complexes such as this are some of the forms in which Fe phosphates can be held in neutral and alkaline soils. When manure was added over a long period of time to a neutral soil, the Fe phosphate fraction according to the Chang and Jackson fractionation procedure increased in quantity as well as in the percent P exchangeable with ^{32}P (Weir & Soper, 1962). This result was in keeping with the formation of the proposed complex. Appelt et al. (1975) prepared a hydroxyl-Al-humic acid complex that was capable of adsorbing P. The amount of P adsorbed increased as the Al:OH ratio decreased, which led them to believe that the P was adsorbed by ligand exchange of phosphate for hydroxyl groups. They reasoned that humic acid could react with Al from soil minerals to form these complexes, which would give rise to new surfaces for P adsorption. The effect of an increase in organic content of the soil therefore would be to increase P adsorption rather than to decrease it by competing with P for adsorptive sites.

Many researchers have found that organic matter affects the reactions of P in soils. Moreno et al. (1960c) showed that organic matter may complex Ca ions and thus increase the P concentration in the soil solution from some of the Ca phosphates. Nagarajah et al. (1970) found that organic acids were capable of reducing the amount of P that was adsorbed by kaolinite, gibbsite, and goethite. They believed the organic acids could be adsorbed by ligand exchange on the mineral surfaces and thus compete with P for adsorption sites. Holford and Mattingly (1975) have suggested that, in calcareous soils, organic matter and P compete for the same sites for adsorption on $CaCO_3$ surfaces. They concluded that the adsorption of organic matter on these soils decreased the bonding energy of adsorbed P. Weir and Soper (1963) found that the addition of manure reduced the bonding energy of a neutral soil for P, but increased the adsorption maximum. In contrast, Appelt et al. (1975) reported that simple organic acids, fulvic acid, and humic acid had no effect on P adsorption by volcanic ash-derived soils. They concluded that for these soils P was preferentially adsorbed over the organic acids studied.

There is evidence therefore that organic matter may decrease or increase the ability of soils to adsorb P. Several researchers have reported positive relationships between the organic matter content of soils and P adsorption (Rennie & McKercher, 1959; Harter, 1969; Hinga, 1973; Lopez-Hernandez & Burnham, 1974; Holford & Mattingly, 1975). These relationships probably reflect the association of organic matter with cations such as Fe, Al, and Ca. These ions are capable of adsorbing P while still associated with organic matter, and hence a positive relationship would be expected.

As was the case with studies of P adsorption by various soil minerals, much of the research on the role of organic matter in P retention has employed dilute P solutions. While such studies are valuable in determining the effect of organic matter on P adsorption, they have generally ignored the converse question—what effects do P fertilizers have on the soil organic matter in the application microsite? Bell and Black (1970a) presented visual evidence that MAP and DAP solubilized soil organic matter. The organic matter moved with the advancing front of the fertilizer solution. Giordano et al. (1971) applied MAP or triammonium pyrophosphate (TPP) to Hartsells soil at rates to simulate the environment near a band of fertilizer. They found that up to 10% of the soil organic matter was solubilized by the fertilizers, with TPP dissolving about twice as much organic C as did MAP. A possible mechanism is that the NH_4^+ from the MAP and TPP replaced various di- and trivalent metal ions from stable metal-organic matter complexes, thus making the complexes more soluble. There are several ways these alterations may influence P retention reactions. As the fertilizer solution moves from a granule or band into the soil, the displaced di- and trivalent metal ions contribute to the concentration of soil-derived cations available for subsequent precipitation with P. Second, removal of the organic matter coatings on soil minerals may expose new surfaces which can participate in P adsorption or in mineral decomposition-P precipitation reactions. Third, as the solubilized organic matter is carried to a new location in the soil, it may reprecipitate, covering soil mineral surfaces which otherwise could have participated in P retention reactions. Little or no research has been conducted on these microsite effects of organic matter on P retention.

E. Phosphate Retention—Adsorption or Precipitation?

The research cited above constitutes a very small fraction of the studies on the mechanisms of P retention. Many important contributions have been omitted, and those cited may have been oversimplified. Evidence has been presented supporting a wide variety of sorption and precipitation mechanisms, with no real consensus as to the relative magnitudes of their contributions.

There is a rather large body of literature describing research in which soil samples or individual soil constituents were reacted with highly concentrated P solutions, usually in the molar range. The physical and chemical evidence for metathetical reactions is incontrovertible. Researchers have watched the dissolution and surface erosion of soil minerals, and they have observed precipitates forming separately and on the surfaces of the soil constituents. In these systems, contributions from adsorption reactions are completely masked. It is impossible to assess the importance of sorption when the potential sorbing surface is either being constantly eroded or occluded by surface precipitation.

There is a much larger body of literature describing research in which soil samples or individual soil constituents were reacted with relatively dilute P solutions, usually in the millimolar range. These solutions form relatively mild chemical environments in soils and induce very little breakdown of the soil minerals. Nevertheless, retention of P does occur in such systems. The retention data are often satisfactorily described by one or more of the adsorption equations over some portion of the concentration range. Despite the fact that most researchers are careful to point out that conformance to a particular adsorption model does not constitute evidence against precipitation, many readers retain a feeling that adsorption is the dominant retention mechanism. Occasionally precipitation is invoked as an explanation of a discontinuity in an adsorption isotherm, but rarely is a separate phase actually encountered or identified.

Because the experimental conditions imposed during the investigation often guarantee that the results are best explained as either adsorption or precipitation reactions, past researchers have often appeared to advocate an either/or approach to retention—either precipitation or adsorption. Many researchers today, however, view P retention as a continuum embodying precipitation, chemisorption, and adsorption, if the processes are viewed throughout the entire zone of soil influenced by a fertilizer application and through a time span encompassing an entire growing season or longer.

In practical field agriculture, highly water-soluble fertilizers are applied to soils as granules or bands, or highly concentrated solutions are applied as droplets comparable to granules. The adjacent soil is bathed in solutions of very high concentrations of P and accompanying cations. It is this environment in which P retention reactions begin, not in the very dilute solutions commonly used to quantify sorption. We will attempt to assess the various reactions entering into P retention in relation to the actual sequence of events occurring when a liquid or solid phosphatic fertilizer is applied to a soil.

III. SEQUENCE OF EVENTS AFTER FERTILIZER APPLICATION

A. Dissolution of Fertilizer Salts

Fertilizers are almost universally applied to soils having moisture contents appreciably below "field capacity," primarily because this condition is desirable before fertilization equipment is brought on the land. Even under conditions of low soil moisture, dissolution of granules of highly water-soluble P fertilizers is fairly rapid. Lawton and Vomocil (1954) found that the moisture content of granules of superphosphate in a soil at >15 atm moisture tension increased to 16.2% within 24 hours, and at tensions of 2 to 0.3 atm increased to 25 to 30% moisture. Thus, ample moisture had moved into the granules to initiate dissolution. In soils with 0.5 moisture equivalent, Lehr et al. (1959) observed rapid dissolution of granules of MCP, OSP, and CSP and theorized that moisture sufficient to initiate dissolution could be drawn into the granules from the surrounding soil by either capillarity or vapor transport. Kolaian and Ohlrogge (1959) demonstrated that vapor transport from soils at a moisture tension of about 3 atm was sufficient to initiate dissolution of a variety of fertilizer salts. They assumed that the solutions formed within and surrounding the fertilizer granules were essentially saturated with respect to the salt. That this is true in the case of MCP (the major P component of superphosphates) was demonstrated by Lindsay and Stephenson (1959a). The composition of the solution leaving a band of MCP in Hartsells soil was very close to that of the metastable triple-point solution formed during the incongruent dissolution of MCP in a pure system. Solution concentrations were affected very little by reducing soil moisture from 0.8 to 0.4 moisture equivalent.

Upon formation of the nearly saturated solution in and around the fertilizer granule or band, an osmotic potential gradient is established between the concentrated fertilizer solution and the soil water (Lehr et al., 1959). As water is drawn into this concentrated zone by vapor transport, the fertilizer solution moves into the surrounding soil. The process of inward movement of water and outward movement of solution continues to produce a nearly saturated solution so long as any of the original salt remains. Even after the salt disappears, the concentrated solution in the surrounding soil maintains an osmotic gradient attracting further inward water movement and outward solution movement. This continues until the concentration is decreased by dilution or by reaction of fertilizer and soil constituents to the level at which no osmotic gradient exists (Huffman & Taylor, 1963).

It has been postulated that all water-soluble P fertilizers produce nearly saturated solutions in soils under nonsaturated moisture conditions (Huffman & Taylor, 1963). If this is true, characterizing the solution leaving fertilizer application sites is relatively simple—the composition is very nearly

Table 1—Phosphate compounds commonly found in fertilizers and compositions of their saturated solutions.

Compound	Formula	Solution symbol	pH	P, moles/ liter	Accompanying cation, moles/liter		Reference[†]
		Composition of saturated solution					
		Highly water-soluble compounds					
Monocalcium phosphate	$Ca(H_2PO_4)_2 \cdot H_2O$	TPS	1.0	4.5	Ca	1.3	A
		MTPS	1.5	4.0	Ca	1.4	A
Monoammonium phosphate	$NH_4H_2PO_4$	MAP	3.5	2.9	NH_4	2.9	A
Monopotassium phosphate	KH_2PO_4	MKP	4.0	1.7	K	1.7	A
Triammonium pyrophosphate	$(NH_4)_3HP_2O_7 \cdot H_2O$	TPP	6.0	6.8 (3.4 P_2O_7)	NH_4	10.2	B
Diammonium phosphate	$(NH_4)_2HPO_4$	DAP	8.0	3.8	NH_4	7.6	A
Dipotassium phosphate	K_2HPO_4	DKP	10.1	6.1	K	12.2	A
		Sparingly soluble compounds					
Dicalcium phosphate	$CaHPO_4$ $CaHPO_4 \cdot 2H_2O$	DCP	6.5	~0.002	Ca	0.001	C
Hydroxyapatite	$Ca_{10}(PO_4)_6(OH)_2$	HAP	6.5	~10^{-5}	Ca	0.001	C

[†] A: Lindsay et al. (1962); B: unpublished TVA data; C: based on Farr (1950), assuming pH = 6.5 and Ca = 0.001M.

that of a saturated solution of the salt in question. Although there are hundreds of different fertilizer formulations and grades, their P is contained in a relatively few compounds. Those water-soluble P compounds listed in Table 1 account for well over 90% of the P contained in commercial fertilizers in the United States. Table 1 also shows the chemical compositions of saturated solutions of these compounds. The P concentrations supported by these compounds appear to justify the high P concentrations used in retention studies cited in Section II.

The water-soluble compounds in Table 1 are characterized by very high concentrations of P and associated cations and by a wide range of highly buffered pH values. As solutions such as these move into the first increments of soil, the chemical environment is dominated by the solution properties rather than the soil properties. Lindsay and Stephenson (1959b) reacted triple-point solution with 10 successive portions of an acid soil (pH 4.6) and a slightly calcareous soil (pH 7.6) at a 1:2 soil to solution ratio. After the 10th soil increment, the pH of the solution had risen from about 1.0 to only 2.1, and changes in solution pH were largely independent of soil pH. Lindsay and Stephenson (1959a) also found that when MCP was banded in a soil of pH 5.6, the pH of soil samples from 0 to 10 mm from the band remained below 3.0 for at least 6 weeks.

The above discussion centers largely around highly water-soluble phosphates. The principles apply equally to P compounds of low water solu-

Table 2—Composition of saturated solutions formed from selected fertilizer mixtures.

Mixture	Composition of saturated solution			Reference†
	pH	P, moles/liter	Accompanying cation, moles/liter	
Solid mixes				
$Ca(H_2PO_4)_2 \cdot H_2O/KCl/H_2O$ (wt ratio 4:1:2)	0.6	3.43	Ca 1.32 K 2.30	A
$Ca(H_2PO_4)_2 \cdot H_2O/NH_4Cl/H_2O$ (wt ratio 2.3:1:2)	0.43	2.74	Ca 1.12 NH_4 4.06	A
$Ca(H_2PO_4)_2 \cdot H_2O/NH_4NO_3/H_2O$ (wt ratio 2.3:1:2)	0.93	3.25	Ca 1.32 NH_4 3.25	A
$Ca(H_2PO_4)_2 \cdot H_2O/KNO_3/H_2O$ (wt ratio 2.3:1:2)	0.86	4.00	Ca 1.48 K 2.50	A
Liquid fertilizers				
N-P-K ($N-P_2O_5-K_2O$)				
8-10.5-0 (8-24-0)	6.5	4.26	NH_4 7.20	B
10-14.8-0 (10-34-0)	6.2	6.7	NH_4 10.0	C
8-3.5-6.6 (8-8-8)	6.2	1.4	NH_4 6.9 K 2.2	C
8-7.0-6.6 (8-16-8)	6.2	3.0	NH_4 7.5 K 2.2	C
7-9.2-5.8 (7-21-7)	6.2	4.0	NH_4 6.8 K 2.0	C

† References: A = Huffman (1962); B = unpublished TVA data; C = Scott et al. (1968).

bility, such as dicalcium phosphate and apatites. Soil moisture moves to particles of these compounds and forms a saturated solution around and in the particles. In this case, however, the saturated solutions are very dilute (Table 1), as dictated by the solubility products of such compounds, and create a very small osmotic gradient between the fertilizer solution and soil solution. Water movement to the fertilizer zone is very slow, and the solution influences only small volumes of soil around the particle. The major pathways of P movement are diffusion or mass flow in percolating rain or irrigation water. The P concentrations supported by these compounds appear to justify the low-concentration sorption studies cited previously. It should be pointed out, however, that these materials occur only rarely in U.S. fertilizers unaccompanied by highly water-soluble phosphates.

Solutions from mixed fertilizer salts may be of more importance than those from individual salts. However, solutions from mixed fertilizers are more difficult to predict. Table 2 shows the composition of solutions saturated with MCP and various nonphosphatic salts which might be encountered in mixed fertilizers. Again, the solutions are highly concentrated in P and associated cations. The pH values in these cases are very low; however, with other salt combinations they could be very high. These equilibrium solutions, however, may not represent the true composition of solutions leaving mixed fertilizers. Huffman and Taylor (1963) showed that slowly leaching a 50:50 mixture of MCP and KCl with water yielded solutions whose compositions changed constantly as successive increments were removed (Table 3). The reactions causing these changes are discussed be-

Table 3—Composition of solution and solid phases at various times during leaching of mixture of MCP + KCl at a rate of 1 ml of H_2O/hour (after Huffman & Taylor, 1963).

Time, hours	Solution composition				Solid phases present
	pH	Ca	P	K	
			moles/liter		
1	0.4	1.5	3.1	2.3	KCl major, $Ca_2KH_7(PO_4)_4 \cdot 2H_2O$ and $Ca(H_2PO_4)_2 \cdot H_2O$ minor
5	1.6	0.6	1.7	3.7	--
8	2.3	0.4	1.3	4.5	KCl major, $Ca_2KH_7(PO_4)_4 \cdot 2H_2O$ minor
10	2.5	0.3	1.2	3.9	--
12	2.5	0.4	1.5	2.6	KCl and $Ca_2KH_7(PO_4)_4 \cdot 2H_2O$ major
13	2.4	0.5	2.1	1.9	--
16	2.9	0.1	0.9	0.6	$CaHPO_4 \cdot 2H_2O$ only

low. Similar concentration changes would be expected with other combinations of salts.

Not all fertilizers are applied as solids. The use of clear liquid mixes and suspensions has increased recently, and about 15% of the mixed fertilizers applied in the USA in 1975 were fluid mixes (Hargett, 1977). The chemical environments near applications of fluid fertilizers are quite similar to those surrounding highly water-soluble solids because most fluid mixes represent nearly saturated solutions of the various component salts. The dissolution step merely takes place in a mixing tank rather than in the soil. Chemical compositions and concentrations of some representative fluid grades are shown in Table 2.

B. Precipitation Reactions Within the Application Site

As dissolution of water-soluble fertilizers proceeds in soils, conditions often favor in situ precipitation of P compounds. The soil has relatively little influence on the chemistry of the in situ reactions, serving mainly as a source of water and as a sink for the fertilizer solution (Bouldin et al., 1960). The best documented examples of such reaction products are the dicalcium phosphate (DCP) residues remaining at application sites of fertilizers containing MCP. The concentrated solution which forms within a granule of MCP (Table 1) is supersaturated with respect to DCP. During the time between formation of the solution and movement into the adjacent soil, some amount of DCP precipitates within the granule site. Lehr et al. (1959) placed MCP, OSP, or CSP granules in five soils at two levels of moisture and found that from 20 to 34% of the applied P remained as DCP after complete dissolution of the MCP component. At 1.0 moisture equivalent, the residues were predominantly $CaHPO_4 \cdot 2H_2O$ (DCPD), while at 0.5 moisture equivalent, $CaHPO_4$ (DCPA) was dominant. Based on phase-rule considerations, Brown and Lehr (1959) predicted that when DCPA was the residue about 28% of the applied P should remain, and when DCPD was the residue about 21% should remain.

Mixing of other fertilizer salts with superphosphate may alter in situ reactions and the amount of P in the granule residue (Bouldin & Sample,

1958; Brown & Lehr, 1959). Bouldin et al. (1960) placed granules containing MCP + $(NH_4)_2SO_4$ in moist soil and found that only about 2% of the added P remained as DCPA residues after 3 weeks, compared with about 21% with MCP alone. The major solid phase remaining was gypsum ($CaSO_4 \cdot 2H_2O$). At the other extreme, mixing MCP with $CaCO_3$ resulted in almost 92% of the added P remaining in the granule site in the form of DCPD and $Ca_8H_2(PO_4)_6 \cdot 5H_2O$ (octacalcium phosphate). Other salts which significantly reduced P in the residue were NH_4NO_3, NH_4Cl, KNO_3, K_2SO_4, and KCl.

To determine the sequence of reactions which might lead to reduced P in the residue, Bouldin et al. (1960) examined the residues from granules of MCP + KCl at various times after placing in moist Hartsells soil. They found the following sequence:

Time, hours	Observations
2	MCP and KCl bulk phases. $Ca_2KH_7(PO_4)_4 \cdot 2H_2O$ minor phase.
17.5	KCl and $Ca_2KH_7(PO_4)_4 \cdot 2H_2O$ major phases. MCP and DCPD minor phases.
21	$Ca_2KH_7(PO_4)_4 \cdot 2H_2O$ major phase. DCPD and KH_2PO_4 minor phases.
23.5	DCPD major phase. Abundant $Ca_2KH_7(PO_4)_4 \cdot 2H_2O$, and a trace of KH_2PO_4.
25.5	DCPD major phase. $Ca_2KH_7(PO_4)_4 \cdot 2H_2O$ minor phase.

Huffman and Taylor (1963) reported about the same sequence of events when MCP + KCl mixtures were leached slowly with water. The sequence of reactions explains why the pH and chemical compositions of the solutions leaving mixed salts vary as dissolution progresses (Table 3). Bouldin et al. (1960) found that MCP + NH_4Cl and MCP + NH_4NO_3 underwent an analogous series of intermediate reactions with NH_4^+ substituting for K^+.

Ammonium ortho- and polyphosphate fertilizers containing little or no Ca or micronutrient cations ordinarily dissolve completely in soils and leave no residues in the granule site (Bouldin & Sample, 1959; Khasawneh et al., 1974). The inclusion of other fertilizer salts with these materials, however, may lead to in situ precipitation. For example, when micronutrient sources are incorporated into these materials, a wide variety of initial reaction products may precipitate during the dissolution process. Because such inclusions normally involve relatively small additions of the micronutrient cation, such reactions immobilize only small quantities of P. Lehr (1972) gave two examples of the metathetical reactions between micronutrient compounds and ammonium phosphates which account for a large number of reaction products. For the ammonium orthophosphates and polyphosphates, respectively, general reactions were:

$$MeSO_4 + (NH_4)_2HPO_4 \rightarrow MeNH_4PO_4 + NH_4HSO_4 \qquad [3]$$

and

Table 4—Some phosphate compounds identified as in situ reaction products of solid phosphate fertilizers in soils.†

Compound	Reacting fertilizers	Reference
$CaHPO_4$	MCP, OSP, CSP	Lehr et al. (1959)
		Bouldin et al. (1960)
$CaHPO_4 \cdot 2H_2O$	MCP, OSP, CSP	Lehr et al. (1959)
$Ca_2KH_7(PO_4)_4 \cdot 2H_2O$	MCP + KCl	Bouldin et al. (1960)
$Ca_2NH_4H_7(PO_4)_4 \cdot 2H_2O$	MCP + NH_4Cl	Bouldin et al. (1960)
	MCP + NH_4NO_3	
$MnNH_4PO_4 \cdot H_2O$	$MnSO_4 \cdot H_2O$ + several ammonium	Hossner & Blanchar
$Mn(NH_4)_2P_2O_7 \cdot H_2O$	ortho- and pyrophosphates	(1968, 1970)
$Mn(NH_4)_2H_4(P_2O_7)_2 \cdot 2H_2O$		
$Mn_3(NH_4)_2(P_2O_7)_2 \cdot 2H_2O$		
$Mn(NH_4)_2P_2O_7 \cdot 2H_2O$	$MnSO_4 \cdot H_2O$ + TPP	Giordano & Mortvedt (1969)
$ZnNH_4PO_4$	$ZnSO_4 \cdot H_2O$ + several ammonium	Hossner & Blanchar
$ZnNH_4H_3(PO_4)_2 \cdot H_2O$	ortho- and pyrophosphates	(1969)
$Zn(NH_4)_2P_2O_7 \cdot H_2O$		
$Zn(NH_4)_2H_4(P_2O_7)_2 \cdot H_2O$		
$Zn(NH_4)_2(P_2O_7)_2 \cdot 2H_2O$		

† For a summary of *probable* reaction products between micronutrient cations and various phosphates, see Lehr (1972).

$$MeSO_4 + (NH_4)_4P_2O_7 + H_2O \rightarrow Me(NH_4)_2P_2O_7 \cdot H_2O + (NH_4)_2SO_4 \quad [4]$$

in which Me = Cu^{2+}, Fe^{2+}, Mn^{2+}, or Zn^{2+}. Some of the identified products of these and similar reactions are listed in Table 4, along with some products formed with Ca-phosphate fertilizers.

Depending on types of fertilizers applied, then, significant portions of the applied P can be "retained" as initial reaction products before they reach the soil. Chances of this occurring are greater with fertilizers containing MCP than with those based on the ammonium phosphates.

C. Movement of Fertilier Solution into Soil

When the concentrated fertilizer solution leaves the application site and moves into the surrounding soil, soil constituents begin to participate in the P retention reactions. The soil constituents may be altered by the solution, and the solution composition is altered by contact with the soil.

1. DISSOLUTION OF SOIL CONSTITUENTS

Some soil minerals in the first increments of soil contacted by the concentrated solution may undergo dissolution, placing relatively large quantities of reactive cations (e.g., Al, Fe, Ca) in solution. Lindsay and Stephenson (1959a) sampled the solution at 5-mm intervals from a band of MCP in acid Hartsells soil and found appreciable concentrations of Al and Fe in the fertilizer solution. When the original soil moisture was 0.4 moisture equivalent, they found Al and Fe concentrations of $0.4M$ and $0.19M$, respectively,

1 cm from the band. At higher moisture levels and at greater distances, the concentrations were less. To further study the dissolution of soil constituents, Lindsay and Stephenson (1959b) and Lindsay et al. (1959a) reacted successive increments of soil with solutions derived from MCP, both the stable and metastable triple-point solutions (TPS and MTPS). After contacting 10 increments of an acid soil, the concentrations of Fe, Al, and Mn in the TPS had reached $0.7M$, $0.2M$, and $0.009M$, respectively. With a slightly calcareous soil, concentrations in the MTPS reached about $0.2M$ Al, $0.02M$ Fe, $0.01M$ K, and $0.01M$ Mn. There was also extensive removal of Ca from soil minerals. In both soils there was evidence that considerable P was lost through precipitation reactions involving the above cations; hence, concentrations of these cations were reduced by these reactions. Lindsay et al. (1962) presented evidence of similar release of soil cations to fertilizer solutions derived from MAP, DAP, MKP, DKP, and ammoniated superphosphoric acid.

The buildup of soil-derived cations in the fertilizer solution can result from replacement of the exchangeable cations from the soil and from decomposition or dissolution of soil minerals. The quantities involved cannot be accounted for by exchangeable cations, so dissolution of soil minerals is thought to be the major contributor of cations. Early evidence that this can occur was supplied by Low and Black (1948), who found that treating kaolinite with ammonium phosphate solutions (up to 1.5 molal) released significant amounts of Si. They reasoned that the Si release was accompanied by proportionate amounts of Al. Kittrick and Jackson (1955, 1956), using an electron microscope, confirmed that P solutions induced decomposition (phosphatolysis) of hydrous oxides and kaolinite. Some degree of phosphatolysis has been demonstrated for a variety of minerals found in soils, including kaolinite, montmorillonite, attapulgite, illite, goethite, gibbsite, calcite, and magnesite (Haseman et al., 1950a; Philen & Lehr, 1967; Hashimoto et al., 1969).

2. INITIAL SOIL-FERTILIZER PHOSPHATE REACTION PRODUCTS

As successive increments of soil are contacted by the moving front of the fertilizer solution, dissolving increasing amounts of Fe, Al, Mn, Ca, Mg, and other soil-derived cations, the solution becomes supersaturated with respect to a variety of P compounds. These compounds slowly precipitate in the soil matrix. The nature of the precipitating compound depends on the kinds and amounts of cations and anions supplied by both the fertilizer and the soil, pH, and soil moisture. There have been very few studies in which researchers have successfully isolated reaction products from the soil body and identified them. Most of the compounds identified as probable soil-fertilizer reaction products have been obtained from simulations of the chemical environment near a fertilizer application, as discussed in a later section.

Lindsay et al. (1962) added soil and soil constituents to saturated solutions of the common fertilizers and identified 30 crystalline P compounds in addition to colloidal precipitates of variable composition. The addition of

MCP in a saturated solution to an acid soil and a calcareous soil dissolved considerable quantities of Fe and Al. Upon standing, the filtrates from these systems formed precipitates of colloidal $(Fe,Al,X)PO_4 \cdot nH_2O$. Addition of NH_4 or K salts to the filtrates resulted in precipitation of DCPD and either $NH_4(Al,Fe)_3H_8(PO_4)_6 \cdot 6H_2O$ or $K(Al,Fe)_3H_8(PO_4)_6 \cdot 6H_2O$. When sufficient $Al(OH)_3$ was added to raise the pH to above 2, the precipitated solids were DCPD and either $Al_5(NH_4)_3H_6(PO_4)_8 \cdot 18H_2O$ or $Al_5K_3H_6(PO_4)_8 \cdot 18H_2O$ (taranakites). Reaction of saturated MAP solutions with the soils yielded mainly DCPD and the NH_4-taranakite, while MKP solutions reacted to form the K-taranakite. When DAP solutions were reacted with the soils precipitates formed were $Al_2NH_4(PO_4)_2OH \cdot 8H_2O$, $Ca_2(NH_4)_2(HPO_4)_3 \cdot 2H_2O$, $MgNH_4PO_4 \cdot 6H_2O$, and colloidal apatite. The addition of various soil constituents, such as $CaCO_3$, $Al(OH)_3$, $Fe_2O_3 \cdot H_2O$, and MgO, to saturated solutions of MCP, MAP, MKP, and DAP resulted in precipitation of a wide variety of P compounds. These are included in Table 5.

Racz and Soper (1967) found DCPD and DMPT to be initial reaction products of pelleted MKP added to neutral and calcareous soils. DCPD was formed in soils having a water-soluble Ca to Mg ratio of approximately 1.5 or greater. DCPD and/or DMPT formed in soils having a water-soluble Ca to Mg ratio of less than 1.5. Hinman et al. (1962) found DCPD and DCPA to be dominant reaction products when MCP was added to a Saskatchewan calcareous soil.

Bell and Black (1970c) identified reaction products formed when MCP, MAP, and DAP were allowed to move into columns of slightly acid and calcareous soils from layers of the solid P salts placed at the bottom. DCPD was the dominant initial reaction product with MCP in all soils. There were minor amounts of DCPA. With MAP the dominant product was DCPD, the amount increasing with increasing amounts of Ca in the soil. $MgNH_4PO_4$ was also found when the amount of exchangeable Mg was

Table 5—Solubility products of some soil-phosphorus fertilizer reaction products.

Phosphate	pKsp	Reference
$AlPO_4 \cdot 2H_2O$	21.5–22.5†	Taylor & Gurney (1964a)
$Al_2NH_4(PO_4)_2OH \cdot 2H_2O$	57.0†	Taylor & Gurney (1964b)
$Al_2K(PO_4)_2OH \cdot 2H_2O$	55.0†	Taylor & Gurney (1964b)
$Al_5(NH_4)_3H_6(PO_4)_8 \cdot 18H_2O$	175.5†	Taylor & Gurney (1961)
$Al_5K_3H_6(PO_4)_8 \cdot 18H_2O$	178.7†	Taylor & Gurney (1961)
$CaHPO_4$	6.66	Farr (1950)
$CaHPO_4 \cdot 2H_2O$	6.56	Moreno et al. (1960a)
$Ca_8H_2(PO_4)_6 \cdot 5H_2O$	93.81	Moreno et al. (1960b)
$Ca_{10}(PO_4)_6(OH)_2$	111.82	Farr (1950)
$Ca_{10}(PO_4)_6F_2$	120.86	Huffman (1962)
$CaAlH(PO_4)_2 \cdot 6H_2O$	39.0†	Taylor & Gurney (1964c)
$FePO_4 \cdot 2H_2O$	35.35†	Huffman & Taylor (1963)
$MgHPO_4 \cdot 3H_2O$	5.82	Taylor et al. (1963a)
$Mg_3(PO_4)_2 \cdot 8H_2O$	25.20	Taylor et al. (1963a)
$Mg_3(PO_4)_2 \cdot 22H_2O$	23.10	Taylor et al. (1963a)
$MgNH_4PO_4 \cdot 6H_2O$	13.15	Taylor et al. (1963b)
$MgKPO_4 \cdot 6H_2O$	10.62	Taylor et al. (1963b)

† Not corrected for complex ions or value uncertain.

high. With the alkaline carrier, DAP, $Ca(NH_4)_2(HPO_4)_2 \cdot H_2O$, $Ca_8H_2(PO_4)_6 \cdot 5H_2O$, $CaHPO_4 \cdot 2H_2O$, and $MgNH_4PO_4 \cdot 6H_2O$ were identified but no product was dominant or appeared in all soils. DCPD was the most abundant product in high-Ca soils except for one soil where only $Ca_8H_2(PO_4)_6 \cdot 5H_2O$ was found. There was no evidence that Al or Fe phosphates were formed in any of the soils. In general, it can be concluded that DCPD is the most likely and often the dominant initial reaction product when water-soluble orthophosphate fertilizers are added to alkaline or calcareous soils (Olsen & Flowerdale, 1971). In soil containing large amounts of Mg, $MgHPO_4 \cdot 3H_2O$ is also frequently found.

Much less research has been done to identify the initial reaction products of the ammonium polyphosphate fertilizers in soils. Lindsay et al. (1962) reacted a concentrated APP solution, about half orthophosphate and half pyrophosphate, with a calcareous soil and found $Ca(NH_4)_2P_2O_7 \cdot H_2O$ to be a precipitating phase. No compounds were identified from reactions with an acid soil. However, when MgO or $CaMg(CO_3)_2$ was added to the APP solution, $Mg(NH_4)_2P_2O_7 \cdot 4H_2O$ precipitated, suggesting that this is a probable reaction product in soils with high Mg contents.

Philen and Lehr (1967) reacted concentrated solutions of a variety of ammonium pyrophosphates and tripolyphosphates with soil minerals and identified several probable reaction products. With kaolinite, no reaction products formed. Montmorillonite reacted with the polyphosphates to form three Mg-NH_4-pyrophosphates and three Ca-NH_4-pyrophosphates. Attapulgite reacted with the pyrophosphate to yield the same three Mg-NH_4-pyrophosphates, but yielded no precipitates with the tripolyphosphate solutions. Goethite and gibbsite reacted with the solutions to yield Fe-NH_4-pyrophosphates and Al-NH_4-pyrophosphates, respectively. It is assumed that polyphosphates reacting in soils will yield compounds similar to these, depending on which of the soil minerals are present.

Predicting the types of reaction products formed from polyphosphate fertilizers is complicated by competing hydrolysis reactions. Polyphosphates hydrolyze to orthophosphates according to the following reactions, using the polyphosphoric acids as example.

$$\text{Pyrophosphate} \quad (HO)(OH)P(=O)-O-P(=O)(OH)(OH) + H_2O \rightarrow 2\ (HO)(OH)P(=O)(OH) \quad [5]$$

$$\text{Tripolyphosphate} \quad (HO)(OH)P(=O)-O-P(=O)(OH)-O-P(=O)(OH)(OH) + H_2O \rightarrow$$

$$(HO)(OH)P(=O)(OH) + (HO)(OH)P(=O)-O-P(=O)(OH)(OH) \quad [6]$$

Longer chained polyphosphates undergo similar hydrolysis, with stepwise breakdown to produce orthophosphates and various shortened polyphosphate fragments. The shortened fragments, themselves, are then subject to the same hydrolysis reactions. In pure, sterile aqueous solutions these reactions are extremely slow at normal room temperature and at near neutral pH values. However, in soils, there are numerous agents which catalyze these reactions, making them much more rapid.

Sutton and Larsen (1964) and Sutton et al. (1966) found that half of the pyrophosphate added to samples of over 200 soils hydrolyzed to orthophosphate in a few days to a few weeks. They found that the general level of biological activity in the soil was the factor exerting the most influence on rate of hydrolysis. Karl-Kroupa et al. (1957) had already demonstrated that cellular extracts from a variety of microorganisms and higher plants could accelerate the hydrolysis reaction by factors of 10^5 to 10^6. They also reported that colloidal gels could catalyze the reaction, increasing rates by a factor of 10^4 to 10^5. Gilliam and Sample (1968) found appreciable hydrolysis of pyrophosphate in soils that had been steam-sterilized. Hydrolysis was more rapid in a fine-textured soil than in a coarse-textured soil. Hydrolysis in the same soils without sterilization was severalfold faster, presumably because of microbiological activity.

When polyphosphate fertilizers are applied to soils there is the opportunity for precipitation or adsorption of the polyphosphate species per se and also of the orthophosphate ions produced through hydrolysis. Very little direct work has been done to identify the reaction products of the secondary orthophosphates. It is presumed they will undergo reactions quite similar to those already discussed.

A summary of some of the types of compounds which might precipitate in soils surrounding applications of the common phosphatic fertilizers is given in Table 6. Although the list is quite long, it represents investigations with a very limited number of different soils. It is probable that the list could be added to by including more soil types in such studies. The summary should be useful in predicting the types of initial reaction products to expect with any given soil-fertilizer combination, if the nature of the fertilizer material and the general chemical and mineralogical makeup of the soil are known.

Most researchers involved in identifying initial precipitation products have tended to ignore any contribution from adsorption reactions. This is understandable in view of the everchanging chemical environment surrounding a fertilizer application site. Nevertheless, as the fertilizer solution moves through the soil, becoming more and more dilute because of precipitation and contact with soil moisture, a point should be reached beyond which adsorption dominates and separate-phase precipitation is negligible. With current knowledge, it is impossible to estimate the location of the transition zone. Very few researchers have been able to establish the presence of precipitated reaction products at varying distances away from a fertilizer application using direct methods. Bell and Black (1970a) identified traces of $CaHPO_4 \cdot 2H_2O$ and $MgNH_4PO_4 \cdot 6H_2O$ in soils at distances of 3 to 5 cm from applications of MAP. They made no judgments, however, as to

Table 6—Summary of compounds formed from reaction of phosphate fertilizers with soils or soil constituents.

Compound	Mineral name	Reference†
$AlPO_4 \cdot 2H_2O$	Variscite	9, 12, 24
$AlPO_4 \cdot 2H_2O$	Metavariscite	12
$Al(NH_4)_2H(PO_4)_2 \cdot 4H_2O$	--	30
$Al_2(NH_4)_2H_4(PO_4)_4 \cdot H_2O$	--	12
$Al_5(NH_4)_3H_6(PO_4)_8 \cdot 18H_2O$	NH_4-taranakite	8, 9, 10, 19, 27
$AlNH_4PO_4OH \cdot 2H_2O$	--	19, 27
$AlNH_4PO_4OH \cdot 3H_2O$	--	16
$Al_2NH_4(PO_4)_2OH \cdot 2H_2O$	--	16
$Al_2NH_4(PO_4)_2OH \cdot 8H_2O$	--	19
$AlKH_2(PO_4)_2 \cdot H_2O$	--	16
$Al_5K_3H_6(PO_4)_8 \cdot 18H_2O$	K-taranakite	8, 9, 10, 19, 28, 29
$Al_2K(PO_4)_2OH \cdot 2H_2O$	Leucophosphite	12
$AlKPO_4OH \cdot 0.5H_2O$	--	16
$AlKPO_4OH \cdot 1.5H_2O$	--	16
$Al_2K(PO_4)_2(F,OH) \cdot 3H_2O$	Minyulite	16
$CaHPO_4$	Monetite	2, 6, 8, 9, 10, 14, 18, 19, 24
$CaHPO_4 \cdot 2H_2O$	Brushite	1, 2, 6, 8, 9, 10, 14, 18, 19, 22, 24, 25
$Ca_8H_2(PO_4)_6 \cdot 5H_2O$	Octocalcium phosphate	2, 17, 19, 22, 25
$Ca_{10}(PO_4)_6(OH)_2$	Hydroxyapatite	1, 17, 19
$Ca_{10}(PO_4)_6F_2$	Fluorapatite	16
$CaAlH(PO_4)_2 \cdot 6H_2O$	--	19, 28
$CaAl_6H_4(PO_4)_3 \cdot 20H_2O$	--	19
$CaNH_4PO_4 \cdot H_2O$	--	19
$Ca(NH_4)_2(HPO_4)_2 \cdot H_2O$	--	2, 11, 19
$Ca_2NH_4H_7(PO_4)_4 \cdot 2H_2O$	NH_4-Flatt's salt	3
$Ca_2(NH_4)_2(HPO_4)_3 \cdot 2H_2O$	--	19
$CaKPO_4 \cdot H_2O$	--	16
$CaK_3H(PO_4)_2$	--	19
$Ca_2KH_7(PO_4)_4 \cdot 2H_2O$	K-Flatt's salt	3
$CaFe_2H_4(PO_4)_4 \cdot 5H_2O$	--	28
$CaFe_2H_4(PO_4)_4 \cdot 8H_2O$	--	19
$Ca_3Mg_3(PO_4)_4$	--	25
$FePO_4 \cdot 2H_2O$	Strengite	12, 19, 24
$FePO_4 \cdot 2H_2O$	Metastrengite	12, 19
$Fe_3(PO_4)_2 \cdot 8H_2O$	Vivianite	18
$FeNH_4(HPO_4)_2$	--	12, 19
$Fe_3NH_4H_8(PO_4)_6 \cdot 6H_2O$	--	12, 19, 20
$Fe_3KH_8(PO_4)_6 \cdot 6H_2O$	--	12, 19, 20, 28
$Fe_2K(PO_4)_2OH \cdot 2H_2O$	K-leucophosphite	7, 12
$MgHPO_4 \cdot 3H_2O$	Newberryite	9, 10, 19, 22, 25, 26
$Mg_3(PO_4)_2 \cdot 4H_2O$	--	22
$Mg_3(PO_4)_2 \cdot 22H_2O$	--	2, 22, 25
$MgNH_4PO_4 \cdot 6H_2O$	Struvite	2, 9, 10, 19
$Mg(NH_4)_2(HPO_4)_2 \cdot 4H_2O$	Schertelite	16
$Mg_3(NH_4)_2(HPO_4)_4 \cdot 8H_2O$	Hannayite	19
$MgKPO_4 \cdot 6H_2O$	--	19
$Mg_2KH(PO_4)_2 \cdot 15H_2O$	--	26
$Al(NH_4)_2P_2O_7OH \cdot 2H_2O$	--	13, 21
$Ca_2P_2O_7 \cdot 2H_2O$	--	19
$Ca_2P_2O_7 \cdot 4H_2O$	--	15, 19, 23
$Ca_3H_2(P_2O_7)_2 \cdot 4H_2O$	--	4, 5
$Ca(NH_4)_2P_2O_7 \cdot H_2O$	--	19, 21, 23, 26
$Ca_3(NH_4)_2(P_2O_7)_2 \cdot 6H_2O$	--	21, 23
$Ca_5(NH_4)_2(P_2O_7)_3 \cdot 6H_2O$	--	23
$CaNH_4HP_2O_7$	--	15
$Ca_2NH_4H_3(P_2O_7)_2 \cdot 3H_2O$	--	21

(continued on next page)

Table 6—Continued.

Compound	Mineral name	Reference†
CaK$_2$P$_2$O$_7$	--	23
Ca$_3$K$_2$(P$_2$O$_7$)$_2$•2H$_2$O	--	23
Ca$_2$K$_2$(P$_2$O$_7$)$_3$•6H$_2$O	--	15, 23
Ca$_2$KH$_3$(P$_2$O$_7$)$_2$•3H$_2$O	--	23
CaNa$_2$P$_2$O$_7$•4H$_2$O	--	24
Fe(NH$_4$)$_2$P$_2$O$_7$•2H$_2$O	--	21
Mg(NH$_4$)$_2$P$_2$O$_7$•4H$_2$O	--	19, 21
Mg(NH$_4$)$_6$(P$_2$O$_7$)$_2$•6H$_2$O	--	21
Mg(NH$_4$)$_2$H$_4$(P$_2$O$_7$)$_2$•2H$_2$O	--	21
Ca(NH$_4$)$_3$P$_3$O$_{10}$•2H$_2$O	--	21

† Numbers indicate the following references: 1. Beaton et al. (1963); 2. Bell & Black (1970c); 3. Bouldin et al. (1960); 4. Brown & Lehr (1964); 5. Brown et al. (1957); 6. Brown & Lehr (1959); 7. Cole & Jackson (1950); 8. Das & Datta (1968); 9. Das & Datta (1969a); 10. Das & Datta (1969b); 11. Frazier et al. (1964); 12. Haseman et al. (1950b); 13. Hashimoto et al. (1969); 14. Hinman et al. (1962); 15. Lehr et al. (1964b); 16. Lehr et al. (1967); 17. Lehr & Brown (1958); 18. Lehr et al. (1959); 19. Lindsay et al. (1962); 20. Lindsay & Stephenson (1959b); 21. Philen & Lehr (1967); 22. Racz & Soper (1967); 23. Savant & Racz (1973); 24. Sonoda & Fujiwara (1965); 25. Strong & Racz (1970); 26. Subbarao & Ellis (1975); 27. Tamimi et al. (1964); 28. Taylor & Gurney (1965a); 29. Taylor & Gurney (1965b); 30. Taylor et al. (1965).

how much of the total P retained at these distances was accounted for by these reaction products.

Knowing both the total retained P at a given distance and the sorption capacity of a soil should allow a judgment as to the maximum possible contribution of adsorption reactions. Using data from Bouldin and Sample (1959), one can calculate the amount of water-insoluble P retained at varying radial distances from a granule of MAP imbedded in Hartsells and Webster soils. After 3 weeks' reaction in Hartsells soil, the water-insoluble P contents ranged from 1.5 mg P/g soil immediately adjacent to the granule to 0.25 mg P/g soil at a distance of 20 mm. No fertilizer P was detected beyond 23 mm. With Webster soil, the P contents were from 21.5 mg P/g soil at zero distance to 1.1 mg P/g at 10 mm. No P movement was detected beyond 10.4 mm. Sample (1972) estimated the adsorption maxima of similar samples of Hartsells and Webster soils to be 0.33 and 0.18 mg P/g soil, respectively, using a two-site Langmuir adsorption equation. Under a rigid interpretation of the adsorption maximum, these data would indicate that adsorption could account for 20%, at most, of P retention by Hartsells soil adjacent to the granule and 60% at a distance of 15 mm. In Webster soil, adsorption could account for <1% of the P retained adjacent to the granule and only 16% at 10 mm. The authors of this review do not feel that the adsorption maximum obtained from Langmuir equations is sufficiently unambiguous for this type of exercise.

A severe deterrent to efforts to assess adsorption in the presence of precipitation can be illustrated with data from Lindsay and Stephenson (1959a). They sampled the moving fertilizer solution at various distances from a band of MCP in Hartsells soil using filter paper inclusions. After 1 day, the solution leaving the band was 3.7M with respect to P. No P was detected at a distance of 5 mm. After 2 days, the P concentration at 5 mm was

1.9M, apparently having been depleted somewhat by precipitation. The first increments of solution reaching the 10-mm filter paper had been depleted to 0.01M P. At this concentration, adsorption reactions might well be expected. By the 4th day, however, the solution reaching the 10-mm point had increased to 1.6M P. If P had been adsorbed at lower concentrations, the sorbed P was then bathed in a much more concentrated solution. This same pattern was followed for all soil increments 5 mm and beyond.

3. CHANGES IN REACTION PRODUCTS WITH TIME

The reaction products discussed above are formed in the initial stages of the reaction of soluble phosphate fertilizers with soils. These products are metastable and with time will change into more stable and less soluble P compounds. Some of them, however, may persist for sufficient time to act as good sources of P for plants.

Lehr and Brown (1958) observed in a greenhouse experiment that in calcareous soils DCPD changed to OCP and colloidal apatite in a period of 4 months. In acid soils the DCPD merely eroded with time. According to Mattingly and Talibudeen (1967), DCPD can break down in two ways:

$$8CaHPO_4 \cdot 2H_2O \rightarrow Ca_8H_2(PO_4)_6 + 2H_3PO_4 + 16H_2O \qquad [7]$$

or

$$CaHPO_4 \cdot 2H_2O \rightarrow Ca^{2+} + HPO_4^{2-} + 2H_2O. \qquad [8]$$

They suggest that in acid and neutral soils the phosphate is removed from solution by the soil constituents faster than OCP can be formed and hence the second equation would be operative. Lehr and Brown (1958) believed that the sequence of reaction changes of MCP in neutral and calcareous soils was MCP → DCPD → OCP → HAp. Talibudeen (1958) found that the labile P from MCP added to a calcareous soil decreased by one-third over a period of 3 months. Bell and Black (1970c) found in slightly acid and calcareous soils that $MgNH_4PO_4$ dissolved to form $Mg_3(PO_4)_2 \cdot 22H_2O$ and $Ca(NH_4)_2(HPO_4)_2 \cdot H_2O$ changed to a residue of DCPD. In a highly calcareous soil the initial reaction product of the addition of MCP was OCP. In other soils OCP was formed with time from DCPD. $MgNH_4PO_4 \cdot 6H_2O$ occurred as a reaction product in one soil from the addition of MAP, but disappeared after 48 weeks.

In another study Bell and Black (1970b) found that DCPD changed to OCP in 4 weeks in soils having a pH value above 7.9 and in 44 weeks in soils having pH values above 6.9. The rate of transformation increased with increasing pH. There was no evidence that OCP changed to more basic phosphates. Strong and Racz (1970) treated four soils, which were saturated with different amounts of Ca and Mg, with MKP and DKP and incubated them for periods up to 15 months. DCPD was formed in all soils, but gradually changed to OCP in all except one, where it persisted for 15 months. $MgHPO_4 \cdot 3H_2O$, $Mg_3(PO_4)_2 \cdot 22H_2O$, and $Ca_3Mg_3(PO_4)_4$ were formed in soils having a Ca/Mg ratio of less than 0.64. DMPT generally persisted for

15 months, but there was evidence that some DMPT and/or DCPD changed to the less soluble Mg and Ca phosphates with time. These changes were more rapid for DKP than for MKP.

In alkaline and calcareous soils DCPD has frequently been shown to change to OCP with time (Withee & Ellis, 1965; Racz & Soper, 1967; Strong & Racz, 1970; Bell & Black, 1970b; Sadler, 1973). The rate of change has been shown to be affected by several factors. Increasing the pH or temperature or the presence of $CaCO_3$ has resulted in a more rapid change of DCPD to OCP (Bell & Black, 1970b). Moreno et al. (1960c) showed that organic matter could inhibit expected precipitation of OCP from solutions at pH values above 6.4. They also found that Fe and Al oxides were active in removing P from solutions even at pH values close to neutral, and this would also help to prevent the formation of OCP. Bell and Black (1970b) reported that the conversion of DCPD to OCP was decreased in the presence of much Mg.

In contrast to the work of Bell and Black (1970b) and Strong and Racz (1970), Larsen et al. (1964) could show no new crystalline phase when DCPD was incubated for periods up to 26 months in acid and alkaline soils. Also, Racz and Soper (1967) found no new phases when DCPD was incubated in four calcareous soils for periods up to 6 months. Moreno et al. (1960c) found that soils saturated with DCPD remained saturated for periods as long as 1 month.

In acid soils the initial reaction products such as the taranakites and amorphous Fe and Al phosphates, as well as the Ca phosphates, are thought to change with time to variscite-like and strengite-like crystalline compounds. Taylor et al. (1963c) incubated possible acidic phosphate-soil reaction products with an acid soil for 10 months and found that $Fe_3KH_8(PO_4)_6 \cdot 6H_2O$ and $CaFe_2H_4(PO_4)_4 \cdot 5H_2O$ changed to strengite residues in this period of time. Amorphous Fe and Al phosphates and $Al_5K_3H_6(PO_4)_8 \cdot 18H_2O$ were more stable and were present at the end of the incubation period. Juo and Ellis (1968), working with pure systems, demonstrated that the crystallization of colloidal Fe phosphate to form strengite was much more rapid than the crystallization of colloidal Al phosphate to form variscite. Hsu (1974) suggested that variscite was probably not the stable compound in soils, based on studies with $AlCl_3$ and NaH_2PO_4 solutions in nonsoil systems (Pa Ho Hsu. 1974. Stability of aluminum phosphates. Agron. Abstr. p. 121). Based on a later similar study, Hsu (1979) suggested that variscite may not develop in soils, at all (Pa Ho Hsu. 1979. Crystallization of variscite at room temperature. Agron. Abstr. p. 148). However, Das and Datta (1969a) found that taranakites and traces of variscite formed when $NH_4H_2PO_4$ and KH_2PO_4 were reacted with acid soils. Sonoda and Fujiwara (1965) also identified variscite as a reaction product when MCP was reacted with bentonite and clay separates from alluvial and volcanic ash soils. Huffman (1962) suggested that some of the amorphous Fe and Al phosphate reaction products may be considered to be strengite and variscite, based on the fact that electron diffraction patterns of some amorphous phosphates were quite similar to the X-ray diffraction patterns of crystalline strengite and variscite.

Little is known of the alterations in soil of many of the initial reaction products in Table 6. Some reasonable speculation as to the ultimate reaction products can be made based on knowledge of soil properties. The soil property most widely used to predict the ultimate products is soil pH, recognizing that pH is itself a reflection of degree of weathering, minerals present, exchangeable cations, and a multitude of other factors. In nature, the ultimate products are thought to be hydroxy- and fluorapatites in alkaline and calcareous soils (Bassett, 1917; Arnold, 1950; Clark, 1955; Hsu & Jackson, 1960), and variscite and strengite in acid to neutral soils (Hsu & Jackson, 1960; Lindsay & Moreno, 1960; Chakravarti & Talibudeen, 1962). Jackson and his coworkers (Chang & Jackson, 1958; Hsu & Jackson, 1960) have shown that as a soil weathers with an accompanying decrease in pH, Ca phosphates change to amorphous and crystalline Al phosphates, which in turn change to Fe phosphates. It is natural to assume that, with time, some portion of the initial reaction products will change to the ultimate products mentioned above, depending on soil pH and mineralogy.

During the alteration process P is not quantitatively transferred from the initial product to the secondary product. As the product dissolves there are a number of pathways that P can take. Plant roots may absorb a portion; percolating rainwater or irrigation water may move it to a new site; and cultivation may place the reaction zone in a new setting, where P may diffuse into previously unreacted soil. That portion of the P not taken up by plants and not reprecipitated is subject to adsorption, which may well be the dominant retention mechanism at this stage. Again, assessing the magnitude of sorption reactions is made very difficult by the probable presence of several different P compounds and the persistence, for relatively long time periods, of concentration gradients from the center of a reaction zone to the outer extremities of movement of fertilizer P.

The overall effects of the slow reactions between soil and P, whether by adsorption or precipitation, have been discussed thoroughly by Barrow and his coworkers. They derived a mathematical expression describing the decrease with time of the relative effectiveness of P for plant growth (Barrow, 1974), decrease in P concentration in soil solution (Barrow & Shaw, 1975a), and decrease in the proportion of P exchangeable with P^{32} (Barrow & Shaw, 1975b). These are discussed in more detail by Olsen and Khasawneh (Chapt. 14 in this book). The homogeneous mixture of soil and P used by these researchers does not, however, adequately simulate the concentration gradients existing within the fertilizer reaction zones under field application conditions.

IV. CHARACTERIZATION OF SOIL-PHOSPHORUS REACTIONS: METHODOLOGY

In attempts to characterize the reactions of P fertilizers in soils, researchers have reacted soil with P solutions ranging in concentration from saturated to extremely dilute. The solutions and soils have generally been mixed homogeneously. In rare cases, the solutions have been applied to the surfaces of soil columns to simulate the concentration gradients encountered in practical situations. In studies employing concentrated solutions, solid phase reaction products have often been isolated and identified. A wide variety of techniques has been used in the identification process. In studies involving very dilute solutions, where no new solid phases were evident, several mathematical models have been employed to describe the adsorption processes. In the following discussion, some of the most widely used methods of studying P reactions in soils are reviewed.

A. Phosphate Application

Several methods which simulate the environment adjacent to a fertilizer granule or fertilizer band in a phosphated soil have been developed. These techniques include applying P to soils as saturated solutions, application of granular P salts in a band, and application of single granules. To simulate the environment farther removed from the granule or band, more dilute solutions have been used.

1. SATURATED PHOSPHATE SOLUTIONS

Lindsay et al. (1962) outlined a procedure for reacting soil with saturated P solutions. Saturated solutions were prepared by reacting P salts with water for 2 days at 25°C. Three hundred grams of soil was treated with 150 ml of P solution and the soil and P solution allowed to react for the desired period of time. The suspensions were filtered and the filtrates analyzed for ionic composition. Aliquots of the filtrates were stored to await precipitation of solid phases. After varying periods of storage, the solid phases were identified. It was assumed that the compounds formed were indicative of the types of reactions which would have occurred had the soil and solution remained in intimate contact. This technique has also been used for reacting various soil constituents with P (Lindsay et al., 1962; Taylor et al., 1965).

A similar technique was used by Lindsay and Stephenson (1959c) and Lindsay et al. (1959a) to simulate the actual events occurring during dissolution of a granule of MCP in soil. As the granule dissolves, metastable triple-point solution (MTPS) continues to emerge until all of the MCP disappears. As a result, the first increment of soil adjacent to the granule is repeatedly contacted by increments of new MTPS which have not passed through soil. Beyond this first increment, fresh soil is contacted by MTPS solution which has already passed through a layer of soil. To simulate the first event, Lindsay and Stephenson (1959c) reacted a given weight of soil with several successive volumes of fresh MTPS. To simulate the secondary events, filtrates of the MTPS already reacted with a given weight of soil were successively reacted with fresh soil (Lindsay & Stephenson, 1959c; Lindsay et al., 1959a). The ratio of soil to solution was maintained constant. Samples of filtrates from each successive reaction were removed for chemical analysis and storage for formation of solid phases.

The techniques described above adequately simulate reactions near a fertilizer granule, particularly when successive reactions are conducted. The technique has the advantage of providing filtrates that can be analyzed for ionic concentration. The solid phases formed in the filtrates are relatively free of soil contamination, aiding in identification and characterization of the solid phases. The technique, however, does result in some differences in reaction products as compared with reaction products obtained in soils treated with bands or granules of P placed into soil columns. Racz and Soper (1967) found octacalcium phosphate and/or trimagnesium phosphate predominated as reaction products in the filtrates of base-saturated and carbonated soils treated with saturated solutions of DAP and DKP. Dicalcium phosphate dihydrate and/or dimagnesium phosphate trihydrate formed in the filtrates of soils treated with saturated solutions of MAP and MKP. Placement of granular P in small openings in columns of the same soils resulted in the formation of dicalcium phosphate dihydrate and/or dimagnesium phosphate trihydrate in the soil near the placement site for all the above phosphates. The pH of the filtrates of the soil-saturated P solutions were markedly different; pH varied from 5.6 to 6.5 for the monobasic phosphates and from 7.7 to 9.3 for the dibasic phosphates. The pH of the soil near the dibasic phosphate granule after 3 weeks of incubation varied from 8.0 to 8.7, and was only about 0.3 to 0.5 pH units greater than for the monobasic phosphates. Thus, it appears the differences in the kinds of reaction products formed as a result of method of P application were partly due to differences in pH encountered with the two approaches.

2. GRANULE OR BAND APPLICATION

Several workers (Bouldin et al., 1960; Lehr et al., 1959; Hinman et al., 1962) simulated the reaction zone of fertilizer granules by placing fertilizer granules into cavities made in soil columns. Usually about 100 to 200 g of soil was placed into containers and wetted to desired moisture levels. Cavities of about 5 mm in diameter and 3 mm depth were formed in the center of soil columns and preweighed samples of granular P salts placed into the opening and the soil firmed in place adjacent to the granule. After incubation for the desired period of time, the reaction products per se or soil containing the reaction products were removed from the granule site or adjacent to the granule for identification. Bouldin et al. (1960) found a fine-tipped glass tube equipped with a glass-wool pad and hooked to a vacuum line under gentle suction aided in recovery of the residue at the pellet site with little or no contamination of the sample with soil.

More recently, Bell and Black (1970c) described a technique in which fertilizer salts were placed at the bottom of soil columns and the soil allowed to react with the band of P salt. Bell and Black (1970a) also used this technique, but with inclusions of a strip of glass fiber filter paper, a strip of Plexiglass with reagent grade $CaCO_3$ embedded into the surface, or inclusions of calcite chips to study reactions with individual soil constituents. The inclusions were recovered after incubation and examined for P species. In general, the P precipitates formed in the paper inclusions appeared to mirror the P precipitates formed in the surrounding soil. The use of the paper inclusions facilitated the identification of the reaction products, particularly when X-ray diffraction was used for identification. Distance from the placement site at which P precipitates could be detected was usually greater with paper than with soil, and in instances sensitivity with X-ray diffraction with paper strips was as good as for optical methods.

Bell and Black (1970b) also cut cylinders longitudinally in halves and placed P salts between the cylinder halves containing soil. The P salts were placed onto the soil surface or enclosed in glass fiber filter paper envelopes prior to placement. After incubation, the soil near the placement site and residue remaining in the filter paper envelope were examined for phosphate species. The use of filter paper envelopes to study transformations of P salts in soil is similar to a technique used by Racz and Soper (1967) in which dicalcium phosphate dihydrate or dimagnesium phosphate trihydrate were placed between filter papers separating layers of soil in plastic cylinders.

The above techniques simulate the reaction zone of P granules placed in a fertilizer band extremely well, and soil at various distances from the placement site or band can be examined for P species. Difficulties, however, do arise with contamination of the P solid phases with soil. Also, in certain instances, reaction products may be present in small quantities and not recovered in sufficient quantities for detection and identification.

3. APPLICATION OF DILUTE SOLUTIONS

In studying sorption-type reactions, it is customary to place samples of soil (or soil constituents) in contact with solutions of known P concentrations. The soil-solution suspensions are agitated intermittently or continuously for predetermined periods, after which the solution and soil phases are separated by centrifugation, filtration, or both. The amount of P retained by the soil is estimated by determining the decrease in P concentration of the solution phase during the reaction period.

An upper limit of concentrations employed in these studies is imposed by the sensitivity of the analytical method used and the soil-to-solution ratio. Most researchers choose upper limits much more dilute than that imposed by the analytical method, and this decision is related to the concentration range over which a chosen mathematical model satisfactorily describes the data. The various models are discussed later.

B. Phosphate Species Identification

1. PHYSICAL METHODS

One of the important reactions of phosphate with soil is the formation of crystalline solid phases; thus, physical methods such as X-ray diffraction, optical microscopy, and infrared spectroscopy offer excellent means of detection and identification of the solid phases formed. Determination of optical properties and measurement of diffraction and absorption characteristics of the compounds can lead to positive identification. The various physical methods used in identification each have their own advantages and disadvantages. In most instances, a combination of several methods is most desirable in detection and identification of the P species formed.

Methods of sample collection and handling can alter P reaction products, and thus care need be exercised during collection and handling of samples. Washing of solid residues can give rise to change in chemical properties of reaction products since reaction products dissolve incongruently. Also, washing can result in loss of soluble or relatively soluble constituents. Excessive grinding of samples can destroy crystallinity, and drying at elevated temperatures can result in loss of water of hydration. Due to the above, it is best that steps in handling of reaction products be monitored for changes in the form of reaction products. Microscopic methods are very suitable for this purpose.

Reaction products in filtrates of saturated P solutions reacted with soils are easily handled. The products can be filtered, washed briefly with water, rinsed with an organic solvent such as acetone, and air-dried. These products can then be identified without further preparation except for grinding for X-ray diffraction analysis. Obtaining P reaction products formed in the soil after treatment with phosphate granules is more difficult. At times, some concentration of the reaction products is necessary for detection.

a. Microscopic Methods—Optical microscopy provides a means of identifying solid phases through determination of shape, size, color, fluorescence, crystal habit, crystal angles, refractive index, and optical behavior under plane-polarized light. Microscopy has been widely used in identifying phosphate-soil reaction products and is a very useful adjunct to X-ray diffraction analysis. Microscopy can be used to monitor sample preparation and to note or guard against changes in composition during handling or preparation of samples. The microscope can be used prior to X-ray analysis to observe the crystallinity of compounds since amorphous materials yield no X-ray pattern.

The optical microscope is very useful in identifying P species when the optical constants and crystallographic properties of the solid phases are determined. A thorough knowledge of the fundamental principles of optical crystallography is required, however. The principles and their application and techniques used in obtaining optical constants and crystallographic properties have been described (Bloss, 1961; Zussman, 1967). Optical descriptions of most soil-P reaction products have been provided by Lehr et al. (1967).

Specific solid components in a mixture and minor phases can be identified using optical microscopy. However, in some instances, measuring optical properties may not be sufficient to identify certain reaction products. For example, the optical properties of $CaAlH(PO_4)_2 \cdot 6H_2O$ are very similar to those of $CaHPO_4 \cdot 2H_2O$, but their X-ray powder diffraction patterns are markedly different (Lehr et al., 1964a). Also, when crystals are too small for identification, X-ray or infrared appear to be more promising methods of identification.

The range of application of the microscope to analytical problems was extended with the use of the electron microscope. A modern electron microscope offers magnification of $100,000\times$, whereas an excellent optical microscope can attain a magnification of about $2,000\times$. Photographic enlargement of the images can further magnify them up to 10-fold. Juo and Ellis (1968) observed the particle size and shape of amorphous and crystalline phosphates using the parlodion film and metal vapor shadowing techniques. Kittrick and Jackson (1955), using very small Fe oxide particles and an $Al(OH)_3$ film on electron microscope screens, observed the formation of Fe and Al phosphate crystals when the screens were treated with $1M$ KH_2PO_4 or NaH_2PO_4. They observed that the Fe and Al phosphate crystals grew with a concurrent decrease in Fe oxide or Al hydroxide film. In other studies, Kittrick and Jackson (1954), using the electron microscope, were able to observe the formation of Al phosphate crystals when kaolinite was reacted with phosphate.

b. Infrared Methods—Infrared spectroscopy has proved a valuable ancillary to X-ray and optical methods in characterization of compounds. This method involves dispersion of a polychromatic infrared beam of light and passing the polychromatic light through the sample, which absorbs portions of the light. The sample absorbs specific wavelengths of the radiation depending on the vibrational and rotational energies of the bonds in the molecule. Since various functional groups are composed of definite atomic configurations, they absorb characteristic wavelengths and thus provide a means of analysis via these functional groups. The wavelength of the infrared radiation absorbed permits qualitative detection of the functional group or molecule; the intensity of the radiation provides a quantitative measure of the functional group present.

Use of infrared for identification of phosphate species has the advantage of not being critically dependent upon crystallite size as are X-ray and optical methods, and very small amounts of material are needed for identification. Beaton et al. (1963) found X-ray diffraction methods failed as a means of identifying some reaction products because the reaction products were too finely divided. However, infrared absorption proved to be a satisfactory method.

Infrared spectra provide a characteristic "fingerprint" of each compound which can be compared with spectra of known compounds for identification. Even if a positive match of spectra is not obtained with that of a known compound, the spectrum of the unknown can give some clues to its general character or identity. Infrared spectra can also provide clues about chemical composition, structure, and bonding. Arlidge et al. (1963) used infrared analysis to study the structure of some Al and Fe phosphates and provided evidence for the absence of either mono- or dihydrogen phosphate ions in these compounds. They found no absorption band at 3.8 μm associated with mono- or dihydrogen phosphates. The presence of water molecules in the Al phosphates was established by absorption bands near 6 μm. The infrared spectra were used to identify the type of phosphate ion present, to distinguish between structural hydroxyl groups and water of crystallization, and to reveal the presence of ammonium ions in these materials. Parfitt et al. (1975) used infrared analysis to obtain a structural model for surface reactions between Fe oxides and phosphate ions. Spectra of the phosphated Fe oxides were interpreted as follows: One phosphate ion replaced two surface hydroxyl ions or H_2O molecules. Two of the oxygen atoms of the phosphate ion were coordinated, each to a different Fe^{3+} ion, resulting in a binuclear surface complex of the type $Fe-O-(P=O_2)-O-Fe$.

Infrared spectra are not as dependent upon crystallinity as are X-ray or optical methods, and thus infrared analysis is more useful than X-ray or optical methods for characterizing amorphous materials. However, irregularities in crystalline structure cause a broadening of infrared absorption bands. Juo and Ellis (1968) found the P-O stretching bond in crystalline Al and Fe phosphates to occur at a lower frequency than for amorphous forms.

Infrared is best utilized when concentrates or isolated crystals of an unknown are analyzed. Spectra of samples containing several unknowns are complex and difficult to interpret. Concentrates or isolated crystals of an unknown can be obtained by handsorting, using a binocular microscope. A high resolution, double-beam spectrophotometer is usually required to produce meaningful spectra of most inorganic compounds. Newly introduced instruments have built-in computer capabilities which should facilitate the analysis of mixtures. Spectra of a variety of pure compounds can be stored in the computer memory. The computer can then perform stepwise subtractions of individual spectra from the spectrum of the mixture. This technique has not yet been used in the identification of reaction products.

Two sample mounting techniques have been described by Lehr (1972) and Lehr et al. (1967). The two mounting techniques, briefly described, were as follows: pellet or pressed disk method—a mixture of spectrographic grade KBr or CsI (KBr is most widely used as a matrix phase) and the unknown (0.25 to 0.50% concentration) are intimately mixed and ground under an inert liquid (ethyl or methyl alcohol or acetone). The organic liquid is evaporated and the dry mixture compressed at 1,400 to 1,500 kg/cm^2 in a vacuum die into a transparent wafer. A wafer of pure KBr or other matrix phase is prepared for mounting in the reference beam. The second method described involves Nujol mulls or split mulls of Fluorolube or Nujol. The unknown is finely ground in the mulling agent which is spread on or pressed between KBr wafers. The second method is utilized when the unknown reacts with KBr or atmospheric moisture during grinding.

After analysis the spectra of the unknowns are compared with spectra of known compounds. Reference spectra for a wide variety of P species have been described by Lehr et al. (1967).

c. X-ray Diffraction Analysis—Powder diffraction techniques using ground samples provide a "fingerprint" type of identification, and the identity once found is quite absolute if multiple lines are present. The identification of the unknown is made by comparing the parameters obtained ("d" spacings) with those in literature (Lehr et al., 1967). The American Society for Testing Materials (ASTM) publishes an extensive file of X-ray diffraction data of known compounds. If a comparison cannot be found in the literature, one can attempt to synthesize a compound having the same diffraction pattern, and attempt to identify the synthesized compound by other means. One drawback of this approach is that if the identity is not found, the pattern gives no clues of what the unknown might be. Also, the sample must be crystalline.

X-ray analysis has been widely used to identify soil-P reaction products in calcareous or base-saturated soils. These soils usually form crystalline Ca and/or Mg phosphates which are relatively easy to identify using X-ray analysis (Racz & Soper, 1967; Bell & Black, 1970a). However, Bell and Black (1970a) found optical methods provided greater sensitivity than X-ray analysis for detecting newly formed phosphates in soil columns. The optical method was found to be more sensitive in most instances; however, in a few instances identification was accomplished by X-ray diffraction but not by petrographic methods. Frazier and Taylor (1965) found a combination of X-ray and infrared absorption examination appeared the most promising method of identifying complex Al phosphates, particularly when the crystals were too small for detailed optical examination. X-ray patterns of two taranakites, (K or $NH_4)_3Al_5H_6(PO_4)_8 \cdot 18H_2O$, were similar and could not be distinguished. Savant and Racz (1973) also noted a disadvantage of X-ray analysis when they were unable to distinguish between $Ca_2NH_4H_3(P_2O_7)_2 \cdot 3H_2O$ and $Ca_2KH_3(P_2O_7)_2 \cdot 3H_2O$ in soils treated with pyrophosphate salts. Juo and Ellis (1968) found newly prepared synthetic colloidal Fe and Al phosphates amorphous to X-ray analysis. After digestion for 40 days at 105°C for the Al phosphate and 12 days' digestion for the Fe phosphate, crystalline $AlPO_4 \cdot 2H_2O$ and $FePO_4 \cdot 2H_2O$ formed, which could be characterized by X-ray diffraction. X-ray diffraction analysis has also been used to study the reaction of P with silicate clay. Kodama and Webber (1975) used X-ray diffraction analysis to study interlayer material of montmorillonite treated with P. X-ray diffraction showed an increase in basal spacing of montmorillonite upon P addition. The interlayer spacing was calculated to be 7.0 Å, coinciding with the edge dimension of an orthorhombic anhydrous $AlPO_4$.

As shown by the above examples, X-ray diffraction offers an excellent means of studying P reactions with soil or soil constituents.

d. Thermal Analysis—Thermogravimetric and differential thermal analysis can be used to detect the presence of water in P reaction products and the amount of water present. Thermogravimetric analysis (TGA) reveals the change in weight of the substance with change in temperature. This technique has limited applications as a change in weight is needed with each thermal event and many structural changes take place thermally without a change in weight. Despite these limitations, however, TGA curves can be used to determine the amount or molecules of water contained in a substance. Differential thermal analysis (DTA) reveals either absorption of heat (endothermal) or emission of heat (exothermal) when a sample is heated. Water of crystallization is indicated by exothermic peaks. Frazier and Taylor (1965) obtained DTA curves for some NH_4-Al phosphates. They found that effects of the atmosphere and of dilution on the thermal behavior of these phosphates made DTA unsuitable for their identification. Arlidge et al. (1963) utilized DTA and TGA to study the loss of water of crystallization and loss of hydroxyls in some Al and Fe phosphates. Thus, although TGA and DTA curves are usually unsuitable for identification of compounds, the information obtained using this technique provides valuable information on bound water and structural hydroxyl groups.

2. CHEMICAL METHODS

a. Chemical Analysis—Chemical characterization, even when all constituents are determined, usually does not lead to positive identification of the compound present because P reaction products are seldom pure, occur as mixtures, and frequently are contaminated with soil constituents. Chemical analysis can, however, provide evidence for the presence of constituents, such as micronutrient elements, in P reaction products when obtained relatively free of soil contamination. Chemical analysis has been widely used to analyze compounds prepared in pure form to obtain empirical formulas and to ascertain if the compound was prepared in pure form (Lehr et al., 1964a; Racz & Soper, 1967). These analyses are usually conducted in conjunction with physical analyses such as X-ray diffraction or infrared to fully characterize or identify the compound. Chemical analysis has also been widely used to characterize the filtrates and solid phases formed when saturated P solutions are reacted with soil (Lindsay & Stephenson, 1959a, 1959b, 1959c; Lindsay et al., 1962; Taylor et al., 1965). Analysis of the filtrates and solid phases gives information on ionic composition of the solution and solid phases.

b. Solubility Products—The solubility of soil-P fertilizer reaction products has been summarized by Huffman (1962). The solubility products of some of the major soil-P fertilizer reaction products are shown in Table 5.

The use of solubility products to identify soil-P fertilizer reaction products has many limitations. The solubility product of a pure sample must first be determined or obtained from the literature with which the solubility of the unknown can be compared. A serious limitation is the possibility of nonequilibrium conditions. Also, temperature variations can cause changes in solubility. The apparent or approximate values for some solubility products, calculated from ionic concentrations rather than ionic activities, are useful only in very dilute solutions. Soil-P fertilizer reaction products are seldom pure and can contain trace quantities of metals which can markedly affect their solubility. Huffman (1962) indicated that the surface area of many phosphates can alter their solubilities. Also, reaction of soil with P fertilizers results in many instances in more than one P compound being formed. The solubility of the most soluble compound will then determine the ionic concentrations.

Identification of soil-P fertilizer reaction products using solubility products involves the determination of ionic acitivites. Generally, the activity of H^+, activity of a phosphate species, and activity of metal ions need to be determined. A major limiting factor in determining ionic activities of the metals and phosphate species is the formation of ion-pairs or complex-ions in solution (Adams, 1971; Lindsay et al., 1959b; Webber & Racz, 1970; Chughtai et al., 1968). Not accounting for the formation of ion-pairs or complex-ions in calculation of ionic activities can result in rather large variations in pKsp values for a particular compound equilibrated in various electrolytes (Webber & Racz, 1970). The ionic activity of metals in soil solution needs to be corrected for ion-pair or complex-ion formation such as $CaHPO_4^0$, $MgHPO_4^0$, $CaSO_4^0$, $CaCO_3^0$, $Fe(OH)2^+$, $Fe(OH)_2^+$, $Al(OH)_2^+$, etc. In addition to the formation of the above complexes or ion-pairs, metals in soil solution may be complexed by soluble organic matter (Moreno et al., 1960c). The various phosphate species in solution are complexed or form ion-pairs as well. For example, HPO_4^{2-} forms ion-pairs with Ca and Mg. The activity of various phosphate species in soil solution can also be affected by soil-P adsorption reactions and reactions with soluble organic matter containing metals.

The Debye-Huckel equation is usually used in calculating activity coefficients. This dictates that the concentration of all major ions in solution be known in order that a good approximation of ionic strengths of solutions be obtained. Methods used for correction of ion-pair or complex-ion formation have been discussed by Lindsay et al. (1959b), Webber and Racz (1970), Chughtai et al. (1968), and Taylor et al. (1963a). The lack of information on the ion-pairs or complexes formed in various systems makes correcting for ion-pairs or complex-ion formation difficult. Also, in many instances the dissociation constants for ion-pairs or complexes are not available, particularly for reactions between soluble organic matter and metals. Dissociation constants for ion-pairs or complexes can be obtained from Sillen and Martell (1964).

The determination of the solubility product of a soil-fertilizer reaction product or soil containing the reaction product offers a means of identification. However, only when the compound is present in a relatively pure form can much emphasis be placed on solubility products as a means of positive identification. In most instances, solubility products as a means of identifying unknown reaction products only provide evidence to supplement physical measurements. A knowledge of the solubility or solubility product is, however, useful in predicting the solubility behavior of reaction products. Highly useful information on the availability of soil-P reaction products can be obtained by determining their solubility under various conditions.

c. Phosphate Potentials—The application of phosphate potentials to identify mineral forms of P in soils has been widely used (Lindsay & Moreno, 1960; Taylor & Gurney, 1962; Weir & Soper, 1963). Diagrams relating the phosphate potential, $pH_2PO_4 + 1/2pCa$, to lime potential, $pH - 1/2pCa$, for the Ca phosphates occurring in neutral and alkaline soils have been constructed based on the phosphates being stoichiometrically formed from MCP and $Ca(OH)_2$. The relationships for saturated solutions are:

for CaHPO$_4\cdot$2H$_2$O

$$(pH - 1/2pCa) - (pH_2PO_4 + 1/2pCa) = 0.65 \text{ (Moreno et al., 1960a)} \quad [9]$$

for Ca$_8$H$_2$(PO$_4$)$_6\cdot$5H$_2$O

$$5(pH - 1/2pCa) - 3(pH_2PO_4 + 1/2pCa) = 11.70 \quad [10]$$

(Moreno et al., 1960b)

and for Ca$_{10}$(PO$_4$)$_6$(OH)$_2$

$$7(pH - 1/2pCa) - 3(pH_2PO_4 + 1/2pCa) = 14.7 \text{ (Russell, 1973)}. \quad [11]$$

Similar expressions for the Mg phosphates were derived using the chemical potentials of Mg(OH)$_2$ and monomagnesium phosphate. The relationships for saturated solutions are:

for MgHPO$_4\cdot$3H$_2$O

$$(pH - 1/2pMg) - (pH_2PO_4 + 1/2pMg) = 1.38 \text{ (Racz \& Soper, 1968)} \quad [12]$$

for Mg$_3$(PO$_4$)$_2\cdot$8H$_2$O

$$2(pH - 1/2pMg) - (pH_2PO_4 + 1/2pMg) = 6.93 \quad [13]$$

(Taylor et al., 1963a)

and for Mg$_3$(PO$_4$)$_2\cdot$22H$_2$O

$$2(pH - 1/2pMg) - (pH_2PO_4 + 1/2pMg) = 7.98 \quad [14]$$

(Taylor et al., 1963a)

For acid soils, similar equations have been derived in terms of the chemical potential of the metal hydroxide and phosphoric acid. The relationship between the Al(OH)$_3$ and phosphoric acid potentials for Al(PO$_4$)$\cdot$2H$_2$O is:

$$3(pH - 1/3pAl) - (pH + pH_2PO_4) = -2.48 \text{ (Russell, 1973)}. \quad [15]$$

The relationship between the Fe(OH)$_3$ and phosphoric acid potential for Fe(PO$_4$)$\cdot$2H$_2$O is:

$$3(pH - 1/3pFe) - (pH + pH_2PO_4) = -6.3 \text{ (Russell, 1973)}. \quad [16]$$

The values calculated for the constants in the above equations vary depending upon pKsp value selected, values selected for ionization constants of phosphoric acid, and other factors affecting the determination and calculation of solubility products.

A graphical representation of the phosphate potential vs. metal hydroxide potential can be obtained using the above relationships. In order to determine the mineral form of soil phosphate, the hydroxide and phosphate potential of the soil solution are measured and related graphically to the equations above. The measurement of ionic concentrations and ionic activities required to calculate the chemical potentials is subject to the errors discussed earlier; i.e., corrections in ion activities for ion-pair and complex-ion formation should be conducted.

The use of chemical potentials to identify forms of P in soils has limitations. Obtaining equilibrium conditions between soil and soil extract prior to determining the chemical potentials is difficult. Time required for equilibration can in instances be very long. The ionic activities of phosphate and the metal in soil solution are affected by adsorption or desorption reactions. The lack of solubility data on all soil-fertilizer P reaction products also limits its application.

It has been a practice, by some workers, to use $0.01M$ $CaCl_2$ as the equilibrating solution in P solubility studies conducted on soils. This procedure has the advantages of maintaining a flocculated soil-water suspension and maintaining a relatively constant lime potential ($pH - 1/2pCa$) in soils. However, Racz and Soper (1968) found the solubility of dimagnesium phosphate to be markedly decreased in $0.01M$ $CaCl_2$ due to precipitation of P as Ca phosphates. The use of $0.01M$ $CaCl_2$ as an equilibrating solution would thus preclude the detection of P compounds such as dimagnesium phosphate.

Although the use of chemical potentials to identify forms of P in soils has limitations, it does provide a means of identifying native forms of P in soils which cannot be concentrated sufficiently for identification by physical methods. Measurement of chemical potentials has also been used to study changes in applied P over long periods of time. Information on the solubility and perhaps the availability of various P compounds as affected by variations in soil pH, addition of salts, and other factors can be obtained from graphical representations of chemical potentials. Chemical potentials can also be used in conjunction with other studies, such as adsorption isotherm studies, in which the calculation of chemical potentials can provide information on the presence of solid phases.

C. Characterization of Adsorption Reactions

The removal of P from dilute solutions by soils or soil constituents has been described by several different adsorption equations. The Freundlich, Langmuir, and Temkin equations, or modifications of these, have been used most frequently. Olsen and Khasawneh (Chapt. 14) and Barrow (1978) discuss the limitations of these and other adsorption models. A brief outline of the various equations is given below; for a more complete treatment the reader is referred to the above sources. As stated by Olsen and Khasawneh, all are based on the fundamental equation

$$q = f(c) \qquad [17]$$

where q is the quantity of P adsorbed at P concentration c.

One of the earliest used in soil studies is the Freundlich equation,

$$q = ac^b \qquad [18]$$

in which q is the amount of P adsorbed per unit weight of soil, c is the P concentration in solution, and a and b are constants which vary from soil to soil. It was probably first used in soil studies by Russell and Prescott (1916). Originally the Freundlich was empirically derived, but Barrow (1978) and Olsen and Khasawneh (Chapt. 14 in this book) cite several researchers who showed that the Freundlich equation corresponded to an adsorption model in which energy of adsorption decreased as amount of adsorption increased, a condition often ascribed to P adsorption by soils.

The Langmuir equation was first applied to adsorption of P by soils by Olsen and Watanabe (1957), although Cole et al. (1953) had previously used it to describe adsorption by $CaCO_3$. Although there are several linear forms, they are all derived from the basic expression

$$q = kbc/(1 + kc) \qquad [19]$$

in which: q and c are as in the Freundlich equation, b is the "P adsorption maximum," and k is a constant related to bonding energy. The equation implies that all increments of sorbed P are held with the same bonding energy (constant k) and that there is a maximum sorption capacity that will not be exceeded regardless of increasing concentration. Neither of these conditions is met in soil-P systems.

The Temkin equation, as proposed for use in soil-P systems by Bache and Williams (1971), also implies that the energy of adsorption decreases as the amount of P sorbed increases. In the middle range of P sorption the equation may be expressed as

$$q/b = (RT/B)\ln Ac \qquad [20]$$

in which A and B are constants and b, c, and q are as in the Langmuir equation.

All three equations require that equilibrium conditions exist, a state that is rarely achieved in soil-P adsorption studies (Kurtz et al., 1946; Rajan & Fox, 1972; Sample, 1972; Barrow & Shaw, 1975a). Another assumption common to the three equations is that the adsorption is reversible; however some portion of the P adsorbed by soils is almost always irreversibly adsorbed. Despite these and other disadvantages, the three equations have been useful in describing the relationship between c and q over limited ranges of concentrations. The range of conformity rarely reaches 100 ppm P for any of the equations.

One anomaly of the use of these sorption equations is that, over the concentration ranges in which they adequately describe the relationship between c and q, they offer no clues as to the mechanisms of adsorption of

P by soils. They do not show whether hydrous oxides of Fe or Al, silicate clays, $CaCO_3$, or $MgCO_3$ dominate the adsorption reactions. They do not indicate whether adsorption involves hydroxyl replacement, silica replacement, or bicarbonate replacement. The instances of nonconformity of sorption data to the various equations have led to more speculation on mechanisms than those of conformity. Discontinuities in sorption isotherms led Muljadi et al. (1966a) to propose three different adsorption sites on kaolinite and aluminum oxides, led Bache (1964) to suggest a three-stage process involving chemisorption, precipitation, and adsorption of P onto the precipitate, and led Rajan and Fox (1975) to propose two types of adsorption sites, one involving anion exchange and one at higher concentrations involving structural disruption of silicate clays.

Discontinuities in the simple adsorption expressions at relatively low P concentrations have led a number of researchers to modify the equations, thus extending the range of conformity. Gunary (1970) added a square-root term to the Langmuir expression and extended the concentration range over which adsorption was described. Fitter and Sutton (1975) improved the range of suitable fit of the Freundlich adsorption equation by adding a term related to the pool of native soil P. Several researchers have found that a multisurface Langmuir equation (Langmuir, 1918) is applicable to P sorption by soils over wider concentration ranges than the single-surface Langmuir equation. Equations similar to

$$q = \frac{k'b'c}{1 + k'c} + \frac{k''b''c}{1 + k''c} \quad [21]$$

have been used by Sample (1972), Syers et al. (1973), Holford et al. (1974), and Holford and Mattingly (1975). The superscripts refer to two adsorption surfaces. Although these modifications of the various adsorption equations have extended their range of usefulness, they offer few more clues than the originals about the actual mechanisms of adsorption.

Most of the adsorption models do not account for varying adsorption maxima at different pH values and solution concentrations, nor for varying bonding energies at different levels of sorbed P. Bowden et al. (1973) and Bowden et al. (1977) have derived a mechanistic equation which predicts that the increased negative charge with increasing P adsorption will produce decreasing bonding energies. Their model also postulates a common adsorption maximum at all pH values and ionic strengths (Barrow, 1978). Their approach has been applied to adsorption of P on Fe oxide surfaces, but has not yet been applied to whole soil systems.

The adsorption studies which appear to offer the greatest promise in predicting mechanisms are those which go beyond the mathematical description of the relationship between sorbed P and P concentration. Those studies cited under Sect. II in which concomitant changes in other solution properties (e.g., release of various anions and cations) are monitored are examples. The coupling of various physical measurements with traditional sorption studies also offers promise. One example is the work of Parfitt et al. (1975) who used infrared spectra to study the bonding of P to Fe oxide surfaces.

V. SUMMARY STATEMENT AND FUTURE RESEARCH NEEDS

Research over the past several decades has established that both adsorption and precipitation reactions are important in P retention by soils. Several major groups of soil constituents have been identified as important agents in P retention and plausible mechanisms have been proposed to explain their role in P adsorption. A large number of P compounds have been implicated as possible initial and secondary precipitation reaction products. Despite this progress, it is almost impossible to quantitatively assess the relative contributions of precipitation and adsorption processes under conditions simulating practical fertilizer applications to soils. We feel that both interpretation of past research on P retention and planning of future research can benefit from increased attention to the actual sequence of events occurring after fertilizer application. Accordingly, a brief summary of these events is given below, followed by some suggested areas for future work.

A. Summary of Events After Fertilization

In practical agriculture, highly water-soluble P compounds or highly concentrated fluid fertilizers are applied to soils. The application practices, at least in the more developed agricultural regions, involve placing granules or bands of the compounds in soils; hence, the system is a heterogeneous mixture of concentrated fertilizer zones surrounded by masses of relatively unaffected soil. This is true even with broadcast application of granules or fluid fertilizers.

As the water-soluble P compounds dissolve in soil moisture, highly concentrated solutions form and move into the soil adjacent to the application site. Depending on P source, the P concentration may range from about 1.5 to more than 6 moles/liter and concentrations of the accompanying cation may be as high as 10 to 12 moles/liter. Solution pH may range from about 1 to 10.

During the dissolution process, some initial reaction products may form directly in the granule or band. With superphosphates, from 20 to 30% of the applied P may remain at the granule site as dicalcium phosphate. With ammonium phosphates, in situ precipitation is not serious; however, inclusion of certain micronutrients may result in formation of precipitates within the application site.

The solution leaving the application site creates a chemical environment which induces appreciable dissolution of soil constituents, placing large amounts of soil-derived cations in the solution. These include exchangeable cations and cations released through degradation of hydrous oxides, silicate clays, soil carbonates, and soil organic matter. As the solution moves through the soil, concentrations of fertilizer- and soil-derived ions reach levels which exceed the solubility products of a variety of P compounds; consequently, some of these compounds precipitate in the soil matrix. These reactions begin within a few millimeters of the application site and probably continue on a diminishing scale through much of the zone of soil influenced by fertilizer P.

As new increments of soil are contacted, the fertilizer solution becomes increasingly dilute because of reactions with soil and mixing with soil moisture. At some point, degradation-reprecipitation reactions should decrease in importance and adsorption-like reactions should dominate. These reactions involve chemical bonding of the P to various cations at the surfaces of soil minerals. Similar reactions may well play an important role in P retention even in the concentrated zones; however, the complexity of the system has been a deterrent to sorption studies in these zones.

The initial reaction products and the initially adsorbed P may be metastable and change forms with time. As water from rain or irrigation invades the P reaction zone, some of the initial products undergo dissolution. A portion of the dissolved P reprecipitates as more stable reaction products; the remainder may be taken up by plant roots or adsorbed by soil constituents. This slow dissolution of initial reaction products probably furnishes most of the fertilizer P taken up by plants during the cropping season. The initially adsorbed P may be replaced and moved to new adsorption sites, or it may become more firmly attached to its original site. The general trend with time is that both the precipitated and adsorbed forms of fertilizer P slowly become more stable and support lower concentrations of P in the soil solution surrounding them.

B. Future Research Needs

Attention in the future should be given to simulating P applications under practical conditions. Phosphate sources, rates, and placements should adequately reflect those actually used in the field. It should be recognized that mixed, multinutrient fertilizers are quite common. A comparatively small amount of research has been done to characterize the solutions formed with mixtures of the common fertilizer materials.

The solutions leaving granules or bands of the primary P compounds under static moisture conditions have been characterized fairly well. Little work has been done to determine the nature of the solution under conditions simulating rainfall or irrigation at varying times after application.

Although numerous studies have been made of the distribution of total P in soil surrounding granules or bands of fertilizers in soil, little is known about the actual concentrations of P and soil- and fertilizer-derived cations and anions in the solution phase at varying distances from the application. Knowledge of these concentrations and their changes with time would be valuable in estimating contributions of precipitation (simultaneous diminution of concentrations of several ion species) and adsorption (reduction in P concentration in exchange for soil anions).

The identities of individual soil constituents decomposed by the concentrated fertilizer solution have been deduced mainly from indirect evidence. Likewise, the nature of the precipitated P compounds has been deduced mainly from indirect evidence. Few attempts have been made to isolate individual soil minerals from the soil-fertilizer reaction zones and to determine the degree of their alteration or decomposition. Few attempts have been made to isolate, identify, and quantify the various precipitated P compounds in different regions of the reaction zone. Few attempts have been made to isolate from the reaction zone hydrous oxides, silicate clays, or carbonates to determine the amounts of P associated with their surfaces or the types of bonds responsible for P retention at their surfaces. One deterrent to such studies has been the lack of adequate separation techniques. Because of their relatively high solubility and surface area, the precipitated compounds may disappear in the course of isolating clay-size fractions by conventional aqueous sedimentation techniques. Adaptation of nonaqueous separation techniques would facilitate isolation of precipitated and mineral-adsorbed P.

The slow reversion of initial reaction products to more stable forms, whether precipitated or adsorbed, is well documented in terms of decreases in overall P extractability or P availability to plants. The transformations of the individual reaction products over time have been studied in very few experiments, and these have usually involved static moisture conditions. The influence of percolating rainfall or irrigation water on the transformations has been largely ignored. Almost nothing is known about the effect of dispersing the reaction zone by cultivation on subsequent transformations of sorbed and precipitated P.

A major need in future research is the recognition that, in practical agriculture, P application results in a reaction zone grading from extremely concentrated to extremely dilute regions. Research dealing exclusively with homogeneous mixtures of soil and P solutions, whether extremely concentrated or extremely dilute, cannot adequately characterize the soil-P reaction zone.

LITERATURE CITED

Adams, F. 1971. Ionic concentrations and activities in soil solutions. Soil Sci. Soc. Am. Proc. 35:420-426.

Appelt, H., N. T. Coleman, and P. F. Pratt. 1975. Interactions between organic compounds minerals, and ions in volcanic-ash-derived soils: II. Effects of organic compounds on the adsorption of phosphate. Soil Sci. Soc. Am. Proc. 39:628-630.

Arlidge, E. Z., V. C. Farmer, B. D. Mitchell, and W. A. Mitchell. 1963. Infra-red, X-ray and thermal analysis of some aluminum and ferric phosphates. J. Appl. Chem. 13:17-27.

Arnold, P. W. 1950. The nature of precipitated calcium phosphate. Trans. Faraday Soc. 46: 1061-1072.

Bache, B. W. 1964. Aluminum and iron phosphate studies relating to soils. II. Reactions between phosphate and hydrous oxides. J. Soil Sci. 15:110-116.

Bache, B. W., and E. G. Williams. 1971. A phosphate sorption index for soils. J. Soil Sci. 22: 289-301.

Barrow, N. J. 1974. The slow reactions between soil and anions: I. Effects of time, temperature, and water content of a soil on the decrease in effectiveness of phosphate for plant growth. Soil Sci. 118:380-386.

Barrow, N. J. 1978. The description of phosphate adsorption curves. J. Soil Sci. 29:447-462.

Barrow, N. J., and T. C. Shaw. 1975a. The slow reactions between soil and anions: 2. Effect of time and temperature on the decrease in phosphate concentration in soil solution. Soil Sci. 119:167-177.

Barrow, N. J., and T. C. Shaw. 1975b. The slow reactions between soil and anions: 3. The effect of time and temperature on the decrease in isotopically exchangeable phosphate. Soil Sci. 119:190-197.

Bassett, H. 1917. The phosphates of calcium. IV. The basic phosphates. J. Chem. Soc. III: 620-642.

Beaton, J. D., T. L. Charlton, and R. Speer. 1963. Identification of soil-fertilizer reaction products in a calcareous Saskatchewan soil by infrared absorption analysis. Nature 197: 1329-1330.

Bell, L. C., and C. A. Black. 1970a. Comparison of methods for identifying crystalline products produced by interaction of orthophosphate fertilizers with soils. Soil Sci. Soc. Am. Proc. 34:579-582.

Bell, L. C., and C. A. Black. 1970b. Transformation of dibasic calcium phosphate dihydrate and octacalcium phosphate in slightly acid and alkaline soils. Soil Sci. Soc. Am. Proc. 34:583-587.

Bell, L. C., and C. A. Black. 1970c. Crystalline phosphates produced by interaction of orthophosphate fertilizers with slightly acid and alkaline soils. Soil Sci. Soc. Am. Proc. 34: 735-740.

Bloss, F. D. 1961. An introduction to the methods of optical crystallography. Holt, Reinhart, and Winston, Inc., New York.

Boischot, P., M. Coppenet, and J. Hebert. 1950. The fixation of phosphoric acid on calcium carbonate in soils. Plant Soil 2:311-322.

Bouldin, D. R., J. R. Lehr, and E. C. Sample. 1960. The effect of associated salts on transformations of monocalcium phosphate monohydrate at the site of application. Soil Sci. Soc. Am. Proc. 24:464-468.

Bouldin, D. R., and E. C. Sample. 1958. The effect of associated salts on the availability of concentrated superphosphate. Soil Sci. Soc. Am. Proc. 22:124-129.

Bouldin, D. R., and E. C. Sample. 1959. Laboratory and greenhouse studies with monocalcium, monoammonium, and diammonium phosphates. Soil Sci. Soc. Am. Proc. 23:338-342.

Bowden, J. W., M. D. A. Bolland, A. M. Posner, and J. P. Quirk. 1973. Generalized model for anion and cation adsorption at oxide surfaces. Nature 245:81-83.

Bowden, J. W., A. M. Posner, and J. P. Quirk. 1977. Ionic adsorption on variable charge mineral surfaces. Theoretical-charge development and titration curves. Aust. J. Soil Res. 15:121-136.

Brown, E. H., and J. R. Lehr. 1964. Vitreous calcium metaphosphate—some properties of its aqueous systems. J. Agric. Food Chem. 12:201-204.

Brown, E. H., J. R. Lehr, J. P. Smith, W. E. Brown, and A. W. Frazier. 1957. Crystalline intermediates in the hydrolytic degradation of calcium polymetaphosphate. J. Phys. Chem. 61:1669-1670.

Brown, W. E., and J. R. Lehr. 1959. Application of phase rule to the chemical behavior of monocalcium phosphate monohydrate in soils. Soil Sci. Soc. Am. Proc. 23:7-12.

Chakravarti, S. N., and O. Talibudeen. 1962. Phosphate equilibria in acid soils. J. Soil Sci. 13:231-240.

Chang, S. C., and M. L. Jackson. 1958. Soil phosphorus fractions in some representative soils. J. Soil Sci. 9:109-119.

Chughtai, A., R. Marshall, and G. H. Nancollas. 1968. Complexes in calcium phosphate solutions. J. Phys. Chem. 72:208-211.

Clark, J. S. 1955. Solubility criteria for the existence of hydroxyapatite. Can. J. Chem. 33: 1696-1700.

Cole, C. V., and M. L. Jackson. 1950. Solubility equilibrium constant of dihydroxy aluminum dihydrogen phosphate relating to a mechanism of phosphate fixation in soils. Soil Sci. Soc. Am. Proc. 15:84-89.

Cole, C. V., S. R. Olsen, and C. O. Scott. 1953. The nature of phosphate sorption by calcium carbonate. Soil Sci. Soc. Am. Proc. 17:352-356.

Coleman, N. T., and A. Mehlich. 1948. Some chemical properties of soils as related to their cation exchange-anion exchange ratios. Soil Sci. Soc. Am. Proc. 13:175-178.

Das, D. K., and N. P. Datta. 1968. Reaction products from phosphate fertilizers in red and laterite soils of India. Indian J. Agric. Sci. 38:382-390.

Das, D. K., and N. P. Datta. 1969a. Products of interaction of fertilizer phosphorus in acid soil of Tripura and alluvial calcareous soil of Bihar. J. Indian Soc. Soil Sci. 17:119-124.

Das, D. K., and N. P. Datta. 1969b. Reaction products from phosphate fertilizers in black soils of India. Indian J. Agric. Sci. 39:676-683.

Davis, L. E. 1945. Retention of phosphates by soils. Soil Sci. 59:175-190.

Dean, L. A. 1949. Fixation of soil phosphorus. Adv. Agron. 1:391-411.

Farr, T. D. 1950. Phosphorus: Properties of the element and some of its compounds. Chem. Eng. Rep. no. 8, TVA, Wilson Dam, Ala.

Fitter, A. H., and C. D. Sutton. 1975. The use of the Freundlich isotherm for soil phosphate sorption data. J. Soil Sci. 26:241-246.

Frazier, A. W., J. R. Lehr, and J. P. Smith. 1964. Calcium ammonium orthophosphate. J. Agric. Food Chem. 12:198-201.

Frazier, A. W., and A. W. Taylor. 1965. Characterization of taranakites and ammonium aluminum phosphates. Soil Sci. Soc. Am. Proc. 29:545-547.

Gilliam, J. W., and E. C. Sample. 1968. Hydrolysis of pyrophosphate in soils: pH and biological effects. Soil Sci. 106:352-357.

Giordano, P. M., and J. J. Mortvedt. 1969. Phosphorus availability to corn as affected by granulating manganese with ortho- and pyrophosphate fertilizers. Soil Sci. Soc. Am. Proc. 33:460-463.

Giordano, P. M., E. C. Sample, and J. J. Mortvedt. 1971. Effect of ammonium ortho- and pyrophosphate on Zn and P in soil solution. Soil Sci. 111:101-106.

Griffin, R. A., and J. J. Jurinak. 1973. The interaction of phosphate with calcite. Soil Sci. Soc. Am. Proc. 37:847-850.

Griffin, R. A., and J. J. Jurinak. 1974. Kinetics of the phosphate interaction with calcite. Soil Sci. Soc. Am. Proc. 38:75-79.

Gunary, D. 1970. A new adsorption isotherm for phosphate in soil. J. Soil Sci. 21:72-77.

Hargett, N. L. 1977. 1976 Fertilizer summary data. Bull. Y-112, Natl. Fert. Dev. Center, TVA, Muscle Shoals, Ala.

Harter, R. 1969. Phosphorus adsorption sites in soils. Soil Sci. Soc. Am. Proc. 33:630-632.

Haseman, J. F., E. H. Brown, and C. D. Whitt. 1950a. Some reactions of phosphate with clays and hydrous oxides of iron and aluminum. Soil Sci. 70:257-271.

Haseman, J. F., J. R. Lehr, and J. P. Smith. 1950b. Mineralogical character of some iron and aluminum phosphates containing potassium and ammonium. Soil Sci. Soc. Am. Proc. 15:76-84.

Hashimoto, Isao, J. D. Hughes, and O. D. Philen, Jr. 1969. Reactions of triammonium pyrophosphate with soils and soil minerals. Soil Sci. Soc. Am. Proc. 33:401-405.

Hemwell, J. B. 1957. The fixation of phosphorus by soils. Adv. Agron. 9:95-112.

Hinga, G. 1973. Phosphate sorption capacity in relation to properties of several types of Kenya soil. E. Afr. Agric. Forest. J. 38:400-404.

Hingston, F. J., R. J. Atkinson, A. M. Posner, and J. P. Quirk. 1967. Specific adsorption of anions. Nature 215:1459-1461.

Hingston, F. J., R. J. Atkinson, and A. M. Posner. 1968. Specific adsorption of anions on goethite. Int. Congr. Soil Sci., Trans. 9th (Adelaide, Aust.) I:669-678.

Hingston, F. J., A. M. Posner, and J. P. Quirk. 1972. Anion adsorption by goethite and gibbsite. I. The role of the proton in determining adsorption envelopes. J. Soil Sci. 23:177-192.

Hinman, W. C., J. D. Beaton, and D. W. L. Read. 1962. Some effects of moisture and temperature on transformation of monocalcium phosphate in soil. Can. J. Soil Sci. 42:229-239.

Holford, I. C. R., R. W. M. Wedderburn, and G. E. G. Mattingly. 1974. A Langmuir two-surface equation as a model for phosphate adsorption by soils. J. Soil Sci. 25:242-254.

Holford, I. C. R., and G. E. G. Mattingly. 1975. The high and low-energy phosphate absorbing surfaces in calcareous soils. J. Soil Sci. 26:407-417.

Hossner, L. R., and R. W. Blanchar. 1968. An insoluble manganese ammonium pyrophosphate found in polyphosphate fertilizer residues. Soil Sci. Soc. Am. Proc. 32:731-733.

Hossner, L. R., and R. W. Blanchar. 1969. The utilization of applied zinc as affected by pH and pyrophosphate content of ammonium phosphates. Soil Sci. Soc. Am. Proc. 33:618-621.

Hossner, L. R., and R. W. Blanchar. 1970. Manganese reactions and availability as influenced by pH and pyrophosphate content of ammonium phosphate fertilizers. Soil Sci. Soc. Am. Proc. 34:509-512.

Hsu, P. H., and M. L. Jackson. 1960. Inorganic phosphate transformation by chemical weathering in soils as influenced by pH. Soil Sci. 90:16-24.

Hsu, P. H., and D. A. Rennie. 1962. Reactions of phosphate in aluminum systems. I. Adsorption of phosphate by X-ray amorphous aluminum hydroxide. Can. J. Soil Sci. 42:197-209.

Huffman, E. O. 1962. Reactions of phosphate in soils: Recent research by TVA. Proc. Fert. Soc. (London) vol. 71. 48 p.

Huffman, E. O. 1968. Behaviour of fertilizer phosphates. Int. Congr. Soil Sci., Trans. 9th (Adelaide, Aust.) II:745-754.

Huffman, E. O., and A. W. Taylor. 1963. The behavior of water-soluble phosphate in soil. J. Agric. Food Chem. 11:182-187.

Juo, A. S. R., and B. G. Ellis. 1968. Chemical and physical properties of iron and aluminum phosphates and their relation to phosphorus availability. Soil Sci. Soc. Am. Proc. 32:216-221.

Kafkafi, U., A. M. Posner, and J. P. Quirk. 1967. Desorption of phosphate from kaolinite. Soil Sci. Soc. Am. Proc. 31:348-353.

Karl-Kroupa, E., C. F. Callis, and E. Seifter. 1957. Stability of condensed phosphates in very dilute solutions. Ind. Eng. Chem. 49:2061-2062.

Khasawneh, F. E., E. C. Sample, and Isao Hashimoto. 1974. Reactions of ammonium ortho- and polyphosphate fertilizers in soil: I. Mobility of phosphorus. Soil Sci. Soc. Am. Proc. 38:446-451.

Kittrick, J. A., and M. L. Jackson. 1954. Electron microscope observations of the formation of aluminum phosphate crystals with kaolinite as the source of aluminum. Science 120:508-509.

Kittrick, J. A., and M. L. Jackson. 1955. Rate of phosphate reaction with soil minerals and electron microscope observations on the reaction mechanism. Soil Sci. Soc. Am. Proc. 19:292-295.

Kittrick, J. A., and M. L. Jackson. 1956. Electron microscope observations of the reactions of phosphate with minerals, leading to a unified theory of phosphate fixation in soils. J. Soil Sci. 7:81-88.

Kodama, H., and M. D. Webber. 1975. Clay-inorganic studies. II. Hydroxy aluminum phosphate-montmorillonite complex. Can. J. Soil Sci. 55:225-233.

Kolaian, J. H., and A. J. Ohlrogge. 1959. Principles of nutrient uptake from fertilizer bands: IV. Accumulation of water around the bands. Agron. J. 51:106-108.

Kuo, S., and E. G. Lotse. 1972. Kinetics of phosphate adsorption by calcium carbonate and Ca-kaolinite. Soil Sci. Soc. Am. Proc. 36:725-729.

Kurtz, L. T. 1953. Inorganic phosphorus in acid and neutral soils. In W. H. Pierre and A. G. Norman (ed.) Soil and fertilizer phosphorus in crop nutrition. Agronomy 4:59-88.

Kurtz, Touby, E. E. DeTurk, and R. H. Bray. 1946. Phosphate adsorption by Illinois soils. Soil Sci. 61:111-124.

Langmuir, I. 1918. The adsorption of gases on plane surfaces of glass, mica, and platinum. J. Am. Chem. Soc. 40:1361-1402.

Larsen, Sigurd. 1967. Soil phosphorus. Adv. Agron. 19:151-210.

Larsen, S., D. Gunary, and J. R. Devine. 1964. Stability of granular dicalcium phosphate dihydrate in soil. Nature 204:1114.

Larsen, S., D. J. Parton, and Inga-Lisa Svensson. 1963. Reaction between monocalcium phosphate and calcium carbonate. Nature 197:317.

Lawton, Kirk, and J. A. Vomocil. 1954. The dissolution and migration of phosphorus from granular superphosphate in some Michigan soils. Soil Sci. Soc. Am. Proc. 18:26-32.

Lehr, J. R. 1972. Chemical reactions of micronutrients in fertilizers. p. 459-503. In J. J. Mortvedt, P. M. Giordano, and W. L. Lindsay (ed.) Micronutrients in agriculture. Soil Sci. Soc. Am., Madison, Wis.

Lehr, J. R., E. H. Brown, A. W. Frazier, J. P. Smith, and R. D. Thrasher. 1967. Crystallographic properties of fertilizer compounds. Chem. Eng. Bull. no. 6. Tenn. Valley Auth., Muscle Shoals, Ala.

Lehr, J. R., and W. E. Brown. 1958. Calcium phosphate fertilizers. II. A petrographic study of their alteration in soils. Soil Sci. Soc. Am. Proc. 22:29-32.

Lehr, J. R., W. E. Brown, and E. H. Brown. 1959. Chemical behavior of monocalcium phosphate monohydrate in soils. Soil Sci. Soc. Am. Proc. 23:3-7.

Lehr, J. R., A. W. Frazier, and J. P. Smith. 1964a. A new calcium aluminum phosphate, $CaAlH(PO_4)_2 \cdot 6H_2O$. Soil Sci. Soc. Am. Proc. 28:38-39.

Lehr, J. R., O. P. Engelstad, and E. H. Brown. 1964b. Evaluation of calcium ammonium and calcium potassium pyrophosphates as fertilizers. Soil Sci. Soc. Am. Proc. 28:396-400.

Lindsay, W. L., A. W. Frazier, and H. F. Stephenson. 1962. Identification of reaction products from phosphate fertilizers in soils. Soil Sci. Soc. Am. Proc. 26:446-452.

Lindsay, W. L., J. R. Lehr, and H. F. Stephenson. 1959a. Nature of the reactions of monocalcium phosphate monohydrate in soils: III. Studies with metastable triple-point solution. Soil Sci. Soc. Am. Proc. 23:342-345.

Lindsay, W. L., and E. C. Moreno. 1960. Phosphate phase equilibria in soils. Soil Sci. Soc. Am. Proc. 24:177-182.

Lindsay, W. L., M. Peech, and J. S. Clark. 1959b. Determination of aluminum ion activity in soil extracts. Soil Sci. Soc. Am. Proc. 23:266-269.

Lindsay, W. L., and H. F. Stephenson. 1959a. Nature of the reactions of monocalcium phosphate monohydrate in soils: I. The solution that reacts with the soil. Soil Sci. Soc. Am. Proc. 23:12-18.

Lindsay, W. L., and H. F. Stephenson. 1959b. Nature of the reactions of monocalcium phosphate monohydrate in soils: II. Dissolution and precipitation reactions involving iron, aluminum, manganese, and calcium. Soil Sci. Soc. Am. Proc. 23:18-22.

Lindsay, W. L., and H. F. Stephenson. 1959c. Nature of the reactions of monocalcium phosphate monohydrate in soils: IV. Repeated reactions with metastable triple-point solution. Soil Sci. Soc. Am. Proc. 23:440-445.

Lindsay, W. L., and P. L. G. Vlek. 1977. Phosphate minerals. p. 639-672. *In* J. B. Dixon and S. B. Weed (ed.) Minerals in soil environments. Soil Sci. Soc. Am., Inc., Madison, Wis.

Lopez-Hernandez, Danilo, and C. P. Burnham. 1974. The covariance of phosphate sorption with other soil properties in some British and tropical soils. J. Soil Sci. 25:196-206.

Low, P. F., and C. A. Black. 1948. Phosphate induced decomposition of kaolinite. Soil Sci. Soc. Am. Proc. 12:180-184.

Low, P. F., and C. A. Black. 1950. Reactions of phosphate with kaolinite. Soil Sci. 70:273-290.

Mattingly, G. E. G., and O. Talibudeen. 1967. Progress in the chemistry of fertilizer and soil phosphorus. p. 157-290. *In* Martin Grayson and E. J. Griffith (ed.) Topics in phosphorus chemistry. Vol. 4. Interscience Publ., New York.

Midgley, A. R. 1940. Phosphate fixation in soils—a critical review. Soil Sci. Soc. Am. Proc. 5:24-30.

Moreno, E. C., W. E. Brown, and G. Osborn. 1960a. Solubility of dicalcium phosphate dihydrate in aqueous systems. Soil Sci. Soc. Am. Proc. 24:94-98.

Moreno, E. C., W. E. Brown, and G. Osborn. 1960b. Stability of dicalcium phosphate dihydrate in aqueous solutions and solubility of octocalcium phosphate. Soil Sci. Soc. Am. Proc. 24:99-102.

Moreno, E. C., W. L. Lindsay, and G. Osborn. 1960c. Reactions of dicalcium phosphate dihydrate in soils. Soil Sci. 90:58-68.

Muljadi, D., A. M. Posner, and J. P. Quirk. 1966a. The mechanism of phosphate adsorption by kaolinite, gibbsite, and pseudoboehmite. Part I. The isotherm and the affect of pH on adsorption. J. Soil Sci. 17:212-229.

Muljadi, D., A. M. Posner, and J. P. Quirk. 1966b. The mechanism of phosphate adsorption by kaolinite, gibbsite, and pseudoboehmite. Part II. The location of the adsorption sites. J. Soil Sci. 17:230-237.

Muljadi, D., A. M. Posner, and J. P. Quirk. 1966c. The mechanism of phosphate adsorption by kaolinite, gibbsite, and pseudoboehmite. Part III. The effect of temperature on the adsorption. J. Soil Sci. 17:238-247.

Nagarajah, S., A. M. Posner, and J. P. Quirk. 1970. Competitive adsorption of phosphate with polygalacturonate and other organic anions on kaolinite and oxide surfaces. Nature 228:83-84.

Nakaru, T., and G. Uehara. 1972. Anion adsorption in ferriginous tropical soils. Soil Sci. Soc. Am. Proc. 36:296-300.

Olsen, S. R. 1953. Inorganic phosphorus in alkaline and calcareous soils. p. 89-122. *In* W. H. Pierre and A. G. Norman (ed.) Soil and fertilizer phosphorus in crop nutrition. Academic Press, Inc., New York.

Olsen, S. R., and A. D. Flowerday. 1971. Fertilizer phosphorus interactions in alkaline soils. p. 153-185. *In* R. A. Olson, T. J. Army, J. J. Hanway, and V. J. Kilmer (ed.) Fertilizer technology and use. Soil Sci. Soc. Am., Inc., Madison, Wis.

Olsen, S. R., and F. S. Watanabe. 1957. A method to determine a phosphorus adsorption maximum of soil as measured by the Langmuir isotherm. Soil Sci. Soc. Am. Proc. 21:144-149.

Parfitt, R. L., R. J. Atkinson, and R. St. C. Smart. 1975. The mechanism of phosphate fixation by iron oxides. Soil Sci. Soc. Am. Proc. 39:837-841.

Perkins, A. T. 1958. Effect of phosphate on the cation exchange capacity of minerals and soils. Soil Sci. Soc. Am. Proc. 22:509-511.

Philen, O. D., Jr., and J. R. Lehr. 1967. Reactions of ammonium polyphosphates with soil minerals. Soil Sci. Soc. Am. Proc. 31:196-199.

Pissarides, A., J. W. B. Stewart, and D. A. Rennie. 1968. Influence of cation saturation on phosphorus adsorption by selected clay minerals. Can. J. Soil Sci. 48:151-157.

Racz, G. J., and R. J. Soper. 1967. Reaction products of orthophosphates in soils containing varying amounts of calcium and magnesium. Can. J. Soil Sci. 47:223-230.

Racz, G. J., and R. J. Soper. 1968. Solubility of dimagnesium phosphate trihydrate and trimagnesium phosphate. Can. J. Soil Sci. 48:265-269.

Rajan, S. S. S. 1975. Adsorption of divalent phosphate on hydrous aluminum oxide. Nature 253:434-436.

Rajan, S. S. S., and R. L. Fox. 1972. Phosphate adsorption by soils. I. Influence of time and ionic environment on phosphate adsorption. Comm. Soil Sci. Plant Anal. 3:439-504.

Rajan, S. S. S., and R. L. Fox. 1975. Phosphate adsorption by soils. II. Reactions in tropical acid soils. Soil Sci. Soc. Am. Proc. 39:846-851.

Rajan, S. S. S., and K. W. Perrott. 1975. Phosphate adsorption by synthetic amorphous aluminosilicates. J. Soil Sci. 26:257-266.

Rajan, S. S. S., K. W. Perrott, and W. M. H. Saunders. 1974. Identification of phosphate-reactive sites of hydrous alumina from proton consumption during phosphate adsorption at constant pH values. J. Soil Sci. 25:438-447.

Rennie, D. A., and R. B. McKercher. 1959. Adsorption of phosphorus by four Saskatchewan soils. Can. J. Soil Sci. 39:64-75.

Russell, E. J., and J. A. Prescott. 1916. The reaction between dilute acids and the phosphorus compounds of the soil. J. Agric. Sci., Camb. 8:65-110.

Russell, E. W. 1973. Soil conditions and plant growth. 10th ed. Longman Group Ltd., London. p. 561-562.

Sadler, J. M. 1973. Influence of applied phosphorus on the nature and availability of inorganic phosphorus in a catenary sequence of Saskatchewan soils. Ph.D. Thesis. Univ. of Saskatchewan, Saskatoon, Sask., Univ. Microfilms, Ann Arbor, Mich. Acc. no. 74-9524.

Sample, E. C. 1972. Factors affecting phosphate retention parameters derived using the Langmuir adsorption equation. Ph.D. Thesis. North Carolina State Univ. at Raleigh. Univ. Microfilms. Ann Arbor, Mich. Diss. Abstr. 34:2406B, 1973.

Savant, N. K., and G. J. Racz. 1973. Reaction products of applied polyphosphates in some Manitoba soils. Can. J. Soil Sci. 53:111-117.

Sawhney, B. L. 1974. Charge characteristics of soils as affected by phosphate sorption. Soil Sci. Soc. Am. Proc. 28:159-160.

Schalscha, E. B., P. F. Pratt, T. Kinjo, and J. Amara. 1972. Effect of phosphate salts as saturating solutions in cation-exchange capacity determinations. Soil Sci. Soc. Am. Proc. 36:912-914.

Schalscha, E. B., P. F. Pratt, and D. Soto. 1974. Effect of phosphate adsorption on the cation exchange capacity of volcanic ash soils. Soil Sci. Soc. Am. Proc. 38:539-540.

Scott, W. C., J. A. Wilbanks, and M. R. Burns. 1968. Production and use of fluid fertilizers made with wet-process superphosphoric acid. Fert. Solut. 12:6-14.

Sillen, L. G., and A. E. Martell. 1964. Stability constants. Chem. Soc. Spec. Publ. no. 17. London.

Sonoda, Yoji, and Akio Fujiwara. 1965. Studies on the availability of mixed fertilizers. IV. The movement of the chemical reactions of fertilizer elements in soils. Tohoku (Japan) J. Agric. Res. 16:67-78.

Strong, J., and G. J. Racz. 1970. Reaction products of applied orthophosphate in some Manitoba soils as affected by soil calcium and magnesium content and time of incubation. Soil Sci. 110:258-262.

Subbarao, Y. V., and Roscoe Ellis, Jr. 1975. Reaction products of polyphosphates and orthophosphates with soils and influence on uptake of phosphorus by plants. Soil Sci. Soc. Am. Proc. 39:1085-1088.

Sutton, C. D., and S. Larsen. 1964. Pyrophosphate as a source of phosphorus for plants. Soil Sci. 97:196-201.

Sutton, C. D., D. Gunary, and S. Larsen. 1966. Pyrophosphate as a source of phosphorus for plants: II. Hydrolysis and initial uptake by a barley crop. Soil Sci. 101:199-204.

Syers, J. K., M. G. Browman, G. W. Smillie, and R. B. Corey. 1973. Phosphate sorption by soils evaluated by the Langmuir adsorption equation. Soil Sci. Soc. Am. Proc. 37:358-363.

Tamimi, Y. N., Y. Kanehiro, and G. D. Sherman. 1964. Reactions of ammonium phosphate with gibbsite and with montmorillonitic and kaolinitic soils. Soil Sci. 98:249-255.

Talibudeen. P. 1958. Iostopically exchangeable phosphorus in soils. III. The fractionation of soil phosphorus. J. Soil Sci. 9:120-129.

Taylor, A. W., A. W. Frazier, and E. L. Gurney. 1963a. Solubility products of di- and trimagnesium phosphates and the dissociation of magnesium phosphate solutions. Trans. Faraday Soc. 59:1585-1589.

Taylor, A. W., A. W. Frazier, and E. L. Gurney. 1963b. Solubility products of magnesium ammonium and magnesium potassium phosphates. Trans. Faraday Soc. 59:1580-1584.

Taylor, A. W., and E. L. Gurney. 1961. Solubilities of potassium and ammonium taranakites. J. Phys. Chem. 65:1613-1616.

Taylor, A. W., and E. L. Gurney. 1962. Phosphate equilibria in an acid soil. J. Soil Sci. 13:188-197.

Taylor, A. W., and E. L. Gurney. 1964a. Solubility of variscite. Soil Sci. 98:9-13.

Taylor, A. W., and E. L. Gurney. 1964b. The dissolution of basic potassium and ammonium aluminum phosphates. Soil Sci. Soc. Am. Proc. 28:289-290.

Taylor, A. W., and E. L. Gurney. 1964c. The dissolution of calcium aluminum phosphate $CaAlH(PO_4)_2 \cdot 6H_2O$. Soil Sci. Soc. Am. Proc. 28:63-64.

Taylor, A. W., E. L. Gurney. 1965a. Precipitation of phosphate by iron oxide and aluminum hydroxide from solutions containing calcium and potassium. Soil Sci. Soc. Am. Proc. 29:18-22.

Taylor, A. W., and E. L. Gurney. 1965b. Precipitation of phosphate from concentrated fertilizer solution by soil clays. Soil Sci. Soc. Am. Proc. 29:94-95.

Taylor, A. W., E. L. Gurney, and A. W. Frazier. 1965. Precipitation of phosphate from ammonium phosphate solutions by iron oxide and aluminum hydroxide. Soil Sci. Soc. Am. Proc. 29:317-320.

Taylor, A. W., E. L. Gurney, and J. R. Lehr. 1963c. Decay of phosphate fertilizer reaction products in an acid soil. Soil Sci. Soc. Am. Proc. 27:145-148.

Way, J. T. 1850. On the power of soils to absorb manure. J. Roy. Agric. Soc. Engl. 11:313-379.

Webber, M. D., and G. J. Racz. 1970. Soluble complexes in the systems dicalcium phosphate dihydrate or dimagnesium phosphate trihydrate equilibrated with aqueous salt solutions. Can. J. Soil Sci. 50:243-253.

Weir, C. C., and R. J. Soper. 1962. Adsorption and exchange studies of phosphorus in some Manitoba soils. Can. J. Soil Sci. 42:31-42.

Weir, C. C., and R. J. Soper. 1963. Solubility studies of phosphorus in some calcareous Manitoba soils. J. Soil Sci. 14:256-261.

Wild, A. 1950. The retention of phosphate by soil. A review. J. Soil Sci. 1:221-238.

Withee, L. V., and Roscoe Ellis, Jr. 1965. Change of phosphate potentials of calcareous soils on adding phosphorus. Soil Sci. Soc. Am. Proc. 29:511-514.

Zussman, J. 1967. Physical methods in determinative mineralogy. Academic Press, New York.

Chapter 12

Agronomic Effectiveness of Phosphate Fertilizers

O. P. ENGELSTAD AND G. L. TERMAN
Tennessee Valley Authority
Muscle Shoals, Alabama

I. INTRODUCTION

A surprising amount of research has been conducted over the years on the agronomic evaluation of P fertilizers. Early experiments were rather crude, but yield responses were often large enough to compensate for lack of design sophistication. In more recent years, such research involving modern experimental designs has often been frustrated by lack of significant yield response to added P on most commercially farmed soils. This diminished frequency of yield response has been a result of accumulation of P fertilizer residues.

The various factors discussed below that affect crop response to P fertilizers are primarily important where soils are low in available P. Whereas low P soils may no longer be dominant, they are still of significance in the agriculture and forestry of the United States. They are of even greater importance to agriculture in the developing world.

II. FERTILIZER PROPERTIES AFFECTING CROP RESPONSE TO APPLIED P

Several summary papers have been published in the last 15 years relating crop response to the chemical and physical properties of applied P. These include van Burg (1963), Mattingly (1963), Seatz and Stanberry (1963), Terman et al. (1964), Terman (1971), and others. A handbook of agronomic principles and practices of fertilizer evaluation is also available (Terman & Engelstad, 1976). Therefore, only brief summaries of results concerning effects of various properties will be given.

Copyright 1980 © ASA-CSSA-SSSA, 677 South Segoe Road, Madison, WI 53711, USA.
The Role of Phosphorus in Agriculture.

A. Chemical Compounds Present and Their Solubility

Chemical compounds and nutrient percentages present in commonly used P sources are given in Table 1. The various compounds undergo many types of chemical reactions with soil constituents. The nature of the products formed in these reactions and their availability to crops depend on the P source, soil properties, and time. Crystallographic properties of various fertilizers have been described by Lehr et al. (1967a).

Most of the P fertilizers made in processes involving acidulation of phosphate rock (PR) are largely water soluble. These include NH_4, K, and Na orthophosphates; ordinary superphosphate (OSP), enriched OSP, and concentrated superphosphate (CSP); ammonium polyphosphate (APP) and urea-ammonium polyphosphate (UAPP); and H_3PO_4, $H_4P_2O_7$, and various solutions. Most of the remaining P in predominately water-soluble phosphates is soluble in neutral ammonium citrate solution, the extractant used in the USA and several other countries to determine chemically "available" P (AOAC, 1975).

Some phosphates only sparingly soluble in water are citrate soluble. These are used as fertilizers in much smaller tonnages, and include dicalcium phosphate (DCP), tricalcium phosphate (TCP), basic slag, Rhenania phosphate, and serpentine phosphate. Nitric phosphates (NP) and ammoniated OSP and CSP are citrate soluble, but vary from low to high in water solubility. High temperature processes formerly used to produce limited amounts of Ca and K polyphosphates and fused tricalcium, serpentine, and Rhenania phosphates have been largely abandoned because of high energy costs.

Regardless of the source, unacidulated PR was formerly considered to contain the mineral, fluorapatite, which is largely citrate insoluble. Differences in availability were thought to be related largely to particle size and hardness. Some precipitated apatites, complex alkali Al and Fe phosphates, and metal pyrophosphates are also very low in citrate solubility. It is now recognized that PR sources vary widely in chemical substitution in the apatite structure, chemical reactivity, and availability of the P to crops (Armiger & Fried, 1957; Caro & Hill, 1956; Lehr, 1967; Lehr et al., 1967b; Lehr & McClellan, 1972).

Lehr and McClellan proposed an absolute citrate solubility index (ACS) as follows:

$$ACS = \frac{\text{AOAC citrate solubility in \%}}{\text{theoretical } P_2O_5 \text{ (\%) of apatite}}.$$

This index is in closer agreement with chemical and physical characteristics of various PR sources than AOAC citrate solubility alone and ranges from near 0 for unreactive apatites to a maximum of 23% for the most reactive PR sources having the greatest degrees of carbonate substitution in the apatite structure. Both AOAC citrate solubility and ACS are related to agronomic availability of P in various rocks.

Table 1—Compounds present in N and NP commercial and experimental fertilizers.

Fertilizer material	Representative N-P-K grades†	Water solubility of P, %	Major compounds present
P sources			
Phosphoric acid:			
Wet-process	24	100	H_3PO_4
Super acid	34.5	100	H_3PO_4, polyphosphoric acids
Superphosphates:			
Ordinary (OSP)	9	85	$Ca(H_2PO_4)_2 \cdot H_2O$, $CaSO_4 \cdot 2H_2O$
Triple (TSP)‡	20	87	$Ca(H_2PO_4)_2 \cdot H_2O$
Concentrated (CSP)	21	90	$Ca(H_2PO_4)_2 \cdot H_2O$
Dicalcium phosphate (DCP)	21	3	$CaHPO_4$, $CaHPO_4 \cdot 2H_2O$
Rhenania phosphate	14	<2	Ca silico-phosphates
Serpentine phosphate	10	<2	Ca Mg silico-phosphates
Basic slag	4	<2	Ca silico-carnotite
Phosphate rock (PR)	14	<1	Carbonate apatite
NP sources			
Ammoniated ordinary superphosphate (AOSP)	6	35	$NH_4H_2PO_4$, $CaHPO_4$, reprecipitated apatite, $CaSO_4 \cdot 2H_2O$
Ammoniated concentrated superphosphate (ACSP)	21 21	50 50	$NH_4H_2PO_4$, $CaHPO_4$, $(NH_4)_2HPO_4$, reprecipitated apatite
Ammonium phosphate nitrate (APN)	4 11 6	100 100 100	$NH_4H_2PO_4$, NH_4NO_3, $(NH_3)_2HPO_4$ $(NH_4)_2HPO_4$, NH_4NO_3 $NH_4H_2PO_4$, NH_4NO_3
Ammonium phosphate sulfate (APS)	21 17 9	>90 >90 >90	$NH_4H_2PO_4$, $(NH_4)_2SO_4$
	21	90	$NH_4H_2PO_4$, $(NH_4)_2HPO_4$, $(NH_4)_2SO_4$
Diammonium phosphate (DAP)	23 20	100 >95	$(NH_4)_2HPO_4$ $(NH_4)_2HPO_4$
Ammonium polyphosphate (APP)	26	100	$NH_4H_2PO_4$, $(NH_4)_3HP_2O_7$, longer chain polyphosphates
Urea ammonium phosphate (UAP)	7 13 15	100 100 100	$CO(NH_2)_2$, $(NH_4)_2HPO_4$, $NH_4H_2PO_4$
Urea ammonium polyphosphate (UAPP)	8 13 19	100 100 100	$CO(NH_2)_2$, $(NH_4)_3HP_2O_7$, $NH_4H_2PO_4$
Nitric phosphate (NP)	9	40	$CaHPO_4$, $NH_4H_2PO_4$, $Ca(NO_3)_2$, reprecipitated apatite

† Convert % P to % P_2O_5 by multiplying by 2.29.
‡ Triple (TSP) and concentrated superphosphates (CSP) are used interchangeably.

B. Chemical Evaluation of Phosphates

It was noted more than 125 years ago that the water-soluble P in OSP, made by H_2SO_4 acidulation of PR, tended to revert to less soluble basic phosphates. Chemists found that the reverted P was readily soluble in ammonium citrate solution, whereas the unacidulated PR was only slightly soluble. Agronomists also found that the reverted P was more available to plants than was unacidulated PR.

More recently, the P in various fertilizers solubilized by neutral normal ammonium citrate has been officially termed "available P" in the USA and in several other countries. Development and publication of standardized methods of analysis, which are necessary for the successful operation of fertilizer control laws, are done by the Association of Official Analytical Chemists (AOAC, 1975). These are also the official methods in Canada, Mexico, and several other countries. In the USA, regulations have been adopted by each of the 50 states, not by the federal government.

Dissolution of P in water or alkaline ammonium citrate is the basis for available P in certain other countries. The use of 2% citric acid was developed specifically for evaluation of basic slag. As mentioned above, the P in various PR sources has been found to vary widely in citrate solubility.

Few new developments in chemical extraction methods have been made since the reviews of Terman et al. (1962, 1964), Mattingly (1965), Cooke (1966), and Terman (1971). Kind of crop, level of soluble soil P, amount of fertilizer P applied, soil pH, soil texture, climate, and other factors affect the relationship between crop yields and "available P" in fertilizers.

Certain condensed polyphosphates have been found to be less soluble in ammonium citrate than in water. This caused some problems until an alternative AOAC procedure adopted in 1964 combined, without filtration, a water extract of the fertilizer sample and a subsequent citrate extract for the determination of available P content. Some highly condensed ammonium phosphates show low available P in water and citrate because of slow dissolution. Plant growth with such highly condensed phosphates, however, was found to be equal to that for readily water-soluble P (unpublished TVA data). This indicates that rates of dissolution and hydrolysis in the soil are adequate to release P for crop growth. Extraction with 2% $NaNO_3$ or $NaCl$ (Harris, 1963) has been proposed for chemical evaluation of K polyphosphates.

C. Granule Size—P Solubility Relationships

Numerous investigators have found that early crop response increases with increase in granule size of water-soluble P fertilizers applied to acid soils low in available P (analogous to banding). Whether such early growth responses persist to mature harvest depends on nonfertilizer factors (discussed later).

For water-soluble P fertilizer granules up to about 6 mm in diameter, effectiveness is related to the amount of water-soluble P per granule which in turn determines the volume of soil affected by P (Taylor & Terman, 1964). For example, Sample and Taylor (1964) found that the P in 6-mm granules of fertilizers having 14 and 70% P water solubility diffused into 4.2 and 20.6 cm^3 volumes of soil, respectively. Crop response to larger granules depends on the probability of roots finding the very few diffusion zones (or a fertilizer band) at a given rate of P application (Moreno, 1959; van Burg, 1963). A normal field application of granular water-soluble P fertilizer affects 2% or less of soil in the root zone.

For alkaline soils, granulation or banding of Ca phosphates has little or no agronomic advantage. This may not be true for ammonium phosphates, with which local acidification develops in the fertilizer band as the NH_4-N is nitrified. This may temporarily increase availability of P and some micronutrients.

Despite some commercial claims to the contrary, solutions, suspensions, and solid P fertilizers having similar AOAC solubility and similarly placed in the soil at a given P application rate, show similar availability to crops. Agronomic effects of spray application of fluid fertilizer in fine droplets vs. "dribble" application (large drops) are analogous to those with fine and granular solid P fertilizers.

In contrast to results with water-soluble P fertilizers, AOAC water-insoluble P fertilizers should usually be fine (<20 mesh or <1 mm) and be mixed well with either acid or alkaline soils. Agronomic availability of water-insoluble P compounds is a function of granule surface area (Bouldin et al., 1960). Very little agronomic evidence has been obtained to show that hardness of water-insoluble P granules has appreciable or long lasting effects. More porous granules, however, have a greater granule surface area per unit of applied P. The same is true for granulation of other salts or even inert materials with the P source. However, granulating NH_4-N with water-insoluble P may increase P solubility and the volume of the P diffusion zone. Presence of NH_4-N has also been found by many investigators to increase P uptake by crops.

Granule size—P solubility effects in general tend to be obscured at higher available P levels in soils. At high soil-P levels, yield response is unlikely with any applied P fertilizer. Early growth response may still occur, however, especially with short-season vegetable crops.

D. Specific P Source Effects

1. AMMONIUM VS. CALCIUM PHOSPHATES

Granulated ammonium phosphates or NH_4-N granulated with other P sources may have the following results as compared with Ca phosphates: (i) supply N which may be deficient for growth, (ii) in alkaline soils, temporarily acidify the soil adjacent to the P granule or band and increase P availability and uptake, (iii) increase the P granule diffusion volume for water-insoluble P, and (iv) increase the granule surface area for water-insoluble P per unit of applied P. The stimulatory effect of NH_4-N on P availability has been attributed by various investigators to low soil pH in the rhizosphere, stimulation of root growth, increased metabolic activity of the plant, and to other causes. The net effect on P-deficient soils is for ammonium phosphates to result in appreciably higher early growth response and occasionally in higher final crop yields than Ca phosphates of similar P water solubility (Terman, 1971). Such benefits tend to be obscured on soils more adequate in available P, especially in field experiments. The residue of

DCP (20% or more of the applied P) which usually remains from superphosphate granules (Brown & Lehr, 1959) reduces immediate P availability.

Compounds such as urea and diammonium phosphate (DAP) temporarily increase pH of the adjacent soil. In such an alkaline environment, there is an inhibition by NH_4-N of complete oxidation to NO_3-N. As a result, temporary accumulation of NO_2-N may occur, which is toxic to plant growth. In addition, NH_3-N tends to be formed, which is also toxic and may result in loss of N to the atmosphere. The net result is that DAP or urea-based fertilizers such as urea ammonium phosphate (UAP) or UAPP may give poorer response in some alkaline soil conditions than do monoammonium phosphate (MAP) and superphosphates. In acid soils, ammonium phosphates (especially DAP) and urea may also result in seedling toxicity if placed with or near germinating seeds. Part of the problem may be an imbalance of Ca by NH_4 (Bennett & Adams, 1970). Atmospheric loss of N from urea is also a problem with surface application on both acid and alkaline soils and from DAP on alkaline soils.

2. AMMONIATED SUPERPHOSPHATES

In the 1930 to 1960 period ammonia was used extensively in the production of ammoniated superphosphates. Since then production of these fertilizers has declined, along with that of OSP. The production of CSP, on the other hand, has become fairly constant since the late 1960's. Use of ammoniation-granulation processes has also decreased because of pollution problems. Increased production of ammonium phosphate solid fertilizers, liquids, slurries, and bulk blends has more than offset the decline in superphosphates. However, ammoniation of superphosphates is still important locally, particularly for the production of lower analysis NPK grades.

Granule size-solubility relationships (see Section II, C) for various ammoniated superphosphates were summarized by Terman et al. (1964) and Wright et al. (1963). With OSP, water solubility of the P decreases progressively with increase in degree of ammoniation as a result of formation of DCP and more basic Ca phosphates. With CSP, water solubility declines to about 50% with use of 0.35 to 0.45 kg of ammonia per kg of P, but then increases with higher use of ammonia as a result of formation of a higher proportion of ammonium phosphates than of DCP.

3. POLYPHOSPHATES

Interest in polyphosphates, used here to include metaphosphates, pyrophosphates, and longer-chain condensed phosphates, grew rapidly following the development at TVA of superphosphoric acid in the 1950's. The major use of this acid has been to react with ammonia to produce high-analysis fluid ammonium phosphate fertilizers (10-34-0 to 13-42-0 $N-P_2O_5-K_2O$ grades to replace 8-24-0 made by ammoniating orthophosphoric acid). These high-analysis fluid fertilizers reduce transportation and application costs and maintain much higher concentrations of Fe, Mn, and Zn in solution. More recently, various polyphosphate solutions and suspensions have been produced locally in small plants by means of the heat produced by reacting ammonia and ortho acid in simple pipe reactors (Meline et al., 1972). Solid APP can also be produced by various methods.

Many claims have been made that solid APP or fluid polyphosphates are agronomically superior to ammonium orthophosphates. However, most results in the USA and from other countries show equal availability of the P to crops grown on most soils (Terman & Engelstad, 1966; Terman, 1975). Some favorable results with APP on neutral to alkaline soils may have been caused by appreciable amounts of Fe or Zn in the polyphosphates. Some commercial fluid fertilizers have been found to contain up to 30 ppm of Cu, 8,900 of Fe, 300 of Mn, and 300 of Zn as impurities (unpublished TVA data). Most of the APP produced by TVA was made with electric furnace superphosphoric acid and usually contained <5 ppm of Cu, $<2,000$ of Fe, <50 of Mn, and <25 of Zn.

Polyphosphate fertilizers have also been found to be superior to orthophosphates as carriers of Zn (Mortvedt & Giordano, 1967), and of Fe (Mortvedt & Giordano, 1970, 1971), but were poorer as carriers of Mn (Giordano & Mortvedt, 1969). Polyphosphates, which must be hydrolyzed to be used effectively by plants, have been shown to be poorer than orthophosphates at low temperatures (Engelstad & Allen, 1971).

Calcium polyphosphate was produced experimentally by TVA for several years. Production was discontinued because of lack of agronomic advantages over superphosphate and because of high energy costs. Potassium polyphosphate has also been produced experimentally by TVA and others. It has not attained commercial production because of high energy costs, and the problem of byproduct HCl disposal.

4. ULTRA-HIGH-ANALYSIS N-P COMPOUNDS

The highest analysis N-P grade fertilizers which have been produced commercially include DAP (18-20-0 and 21-23-0) and APP (15-27-0). The TVA has produced experimentally several compounds of much higher analyses (Allen, 1970; Wakefield et al., 1971; Terman, 1971). These include products made by gas phase reaction of NH_3, P, and S up to 42% N, 51% P, and 3% S, which were essentially inert (Terman and Allen, 1967). Products hydrolyzed by boiling water were about equal as sources of N and P for corn as ammonium nitrate (AN) and CSP. A second group of compounds included linear polyamides and thiopolyamides (Wakefield et al., 1971). The P in two monomers, $PO(NH_2)_3$ and $PS(NH_2)_3$, was similar in availability to P in CSP. Long-chain compounds decreased in effectiveness with increase in chain length.

A third group of compounds included two metaphosphimates (saturated ring structure 16-26-0 and 27-30-0) in which both the N and P were unavailable, and phosphonitrilic hexaamide (unsaturated ring structure, 50-37-0) in which N and P were readily available.

As indicated above, all of these compounds are experimental, and no known processes are now available to produce them at costs competitive with present commercial fertilizers.

5. COATING GRANULAR WATER-SOLUBLE PHOSPHATES

The chief objective of coating granules of water-soluble phosphates is to reduce fixation of applied P by the soil, thereby increasing its availability for crop growth. Negative results have usually been obtained, however. Terman et al. (1970) reported no response of a first crop of flooded rice to P in S-coated CSP, but the P became available to a second crop after degradation of the coating. Allen and Mays (1971) found that insufficient P was released from S-coated DAP for early growth of forage sorghum and resulted in lower total yields than did uncoated DAP.

The reason for poor response to coated soluble P fertilizers is probably that a diffusion zone of high P concentration does not develop, as is also the case with granular water-insoluble P fertilizers. As a result, the crop is unable to obtain enough P for rapid early growth, and lower yields usually result. The soil itself rapidly reverts soluble P to less soluble "slow-release" P compounds. Thus, coating of granules of soluble P fertilizers is largely unnecessary and usually reduces P availability.

6. UNACIDULATED PHOSPHATE ROCK

Ground PR has been used in direct applications to supply P for crops for more than 150 years. In the USA, the use of PR was especially promoted in Illinois by C. G. Hopkins, in Kentucky by George Roberts, and by others. Direct application of PR was recommended largely in crop rotations involving legumes. Tennessee and Florida PR were the chief sources. Unfortunately, it was not recognized until recently that these are not the most satisfactory sources for direct application. However, no PR source is effective on soils having pH levels above 5.5 to 6.0.

Discovery of new PR ore deposits in several countries recently has led to renewed interest in PR both for direct application and for production of acidulated fertilizers. The TVA has had a project to characterize all of the known world deposits (Lehr, 1967; Lehr et al., 1967b). Various PR deposits range from rather pure igneous apatites to those highly substituted by CO_3^-, Cl^-, and other anions (Lehr & McClellan, 1972). Citrate solubility and ACS are rather highly correlated with degree of substitution in the apatite structure. Highly substituted PR sources are also softer, so that grinding them to pass a given screen size results in a finer product, e.g., North African and North Carolina PR are both finer and higher in citrate solubility than are Idaho, Florida, Tennessee, and other sources.

Several pot experiments on P-deficient soils were conducted to evaluate a number of PR sources characterized at TVA. These sources varied in AOAC citrate solubility from 1% of the total P content for a Missouri apatite to 26% for North Carolina PR. These sources were 0 to 84% as effective as CSP for rice with increase in citrate solubility (Engelstad et al., 1974). In this experiment, the pots were flooded at 3 weeks after fertilizer application and seeding.

A second group of PR sources varying in citrate solubility from 5% of the total P for a Quebec apatite to 24% for Gafsa PR and 25% for North

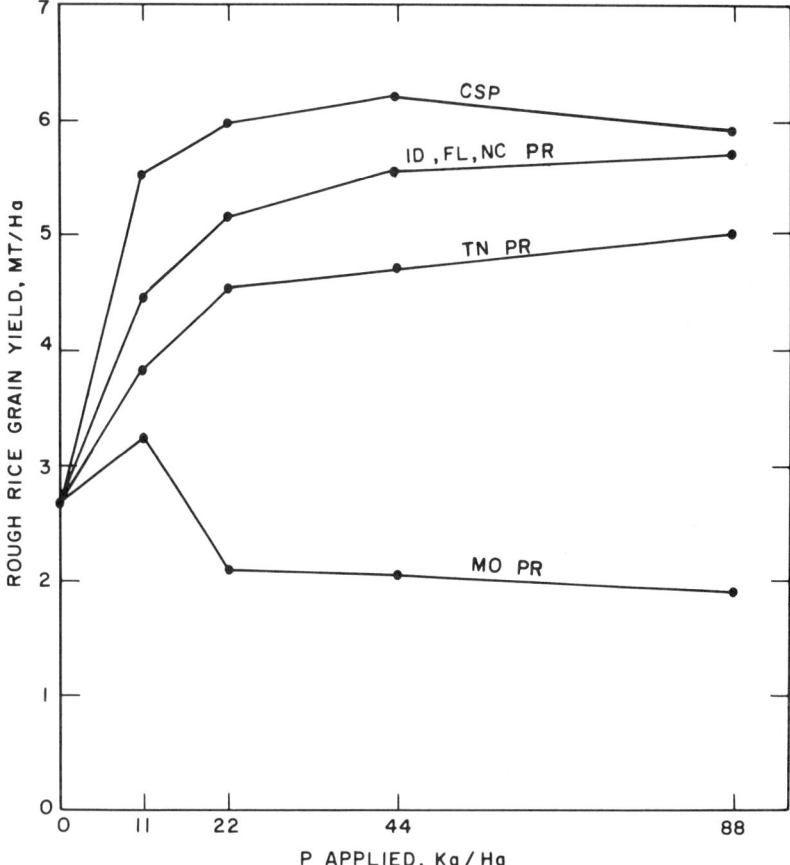

Fig. 1—Yield of rough rice at Klong Luang Rice Experiment Station, Thailand, as affected by rate and source of applied P (crop 1). The phosphate rocks (PR) are designated by the state in which they originated in the U.S.

Carolina PR was evaluated for slash pine. Both dry matter yield and P uptake increased with citrate solubility (Bengtson et al., 1974). Response of corn to a third set of PR sources also increased with higher citrate solubility (Terman et al., 1969).

Results from a cooperative field experiment with rice in Thailand (Engelstad et al., 1974) are shown in Fig. 1. Average yield responses increased curvilinearly with citrate solubility for all PR sources except Missouri apatite.

7. PARTIALLY ACIDULATED PR

Acidulation of PR with less than the stoichiometric amount of H_2SO_4 or H_3PO_4 required to produce fully acidulated OSP or CSP has been proposed at various times to conserve S and reduce product cost. Nongranular, *fine,* partially acidulated products were found by McLean and Wheeler

Table 2—Relative effectiveness of North Carolina PR products for rice.†

P source	Total P, %	Mesh size	Experiment 1 Crop 1	Experiment 1 Crop 2	Experiment 2 Crop 1	Experiment 2 Crop 2	Experiment 3 Crop 1
CSP (88%) water soluble	20.7	−6+9	100	27	100	62	100
Phosphate rock	13.0	−6+9	4	17	0	43	16
		−100	25	45	56	50	70
Urea-PR (18-8-0)	7.4	−6+9	9	15	0	36	28
Partially acidulated PR (20% water soluble)	11.1	−6+9	--	--	39	42	36
		−100	--	--	57	46	--

† Unpublished TVA data.
‡ Based on average increase in uptake of P from 50, 100, and 200 mg of P applied/pot over no applied P (CSP for crop 1 = 100).

(1964), McLean et al. (1965), and McLean and Logan (1970) to be more effective P sources on very acid soils than was fine superphosphate. However, *granular,* partially acidulated products and granulated mixtures of fine Florida PR and CSP were found by Terman and Allen (1967) to be inferior to CSP alone. Effectiveness of these granular products increased with increase in water-soluble P in the various products. Results in more recent pot experiments (Table 2) show much poorer response of rice to granular, partially acidulated North Carolina PR than to the fine, nonacidulated product (unpublished TVA data).

These contrasting results illustrate the granule size–P water solubility relationships previously described. Granulation tends to increase the effectiveness of the water-soluble P component, but that of the unacidulated PR is greatly reduced. Presumably, nongranular, partially acidulated phosphates are useful under the same acid soil conditions under which fine, unacidulated PR is effective.

8. GRANULATION OF PR ALONE AND WITH OTHER MATERIALS

Acidulation of PR in the soil after granulation with elemental S and various acids has also been studied at TVA and elsewhere. Granular PR-S products were more effective than PR granulated alone, but all were much poorer than water-soluble P sources (Terman et al., 1969). Urea nitrate and oxalic acid granulated with PR were more effective than was citric acid. Granulation of fine PR greatly reduced its effectiveness for supplying P to corn. Both granulated PR and PR-urea were also unsatisfactory P sources for rice (Table 2).

Granulation of fine PR into porous granules which are much easier to apply, but which disintegrate on contact with moisture, has been suggested for forest fertilization in P-deficient areas. No results are yet available, but this method should give results comparable to surface application of fine PR. Application of suspensions of fine PR is also being studied to avoid the dustiness of fine, dry PR.

E. Limiting Yield Differences Among P Sources

Crop yields usually increase curvilinearly with increasing amounts of water-soluble P applied to a P-deficient soil. A limiting yield is reached at some high rate of application, beyond which yields level off or even decrease at higher rates. The particular limiting yield level attained depends on the levels of other growth limiting factors.

Yield response to granules of less available P sources is usually linear with amount applied. In most comparisons, one would have to extrapolate to extremely high application rates of some water-insoluble P sources to reach the same limiting yield level obtained with water-soluble P. This means that in the range of practical application rates, limiting yields with granular water-insoluble acidulated phosphate or with less soluble PR sources are lower (Engelstad et al., 1974; Terman, 1971). This markedly affects the interpretation of relative effectiveness values based on regression calculations. Terman and Engelstad (1976) suggest an economic evaluation of yields with P sources that give different limiting yields.

No fully adequate explanation of higher limiting yields with soluble P sources has been advanced. However, the lack of sufficiently high concentrations of soluble P from less soluble sources to stimulate early growth response is the probable explanation (Terman, 1971). Such differences in limiting yields, as with other P effects, tend to be obscured as P residues accumulate in soils and increase the available P levels.

III. SOIL AND MANAGEMENT FACTORS AFFECTING CROP RESPONSE TO P FERTILIZERS

While the chemical and physical properties of phosphate fertilizers have important effects on crop response, various soil and management factors are equally important. Since soluble P reacts quickly with soil components and is therefore quite immobile, placement to minimize loss of availability (solubility) and maximize access by plant root systems is the desired goal. Needless to say, the relative importance of these factors varies with soil and crop.

A. Soil pH

Generally, P availability is greatest in the soil pH range of 5.5 to 7.0. Soil pH affects the availability of applied P by influencing the nature of ensuing P reactions, and thereby the reaction products formed.

Soil pH can also affect the growth of plant root systems and their ability to take up P. While this factor is usually not considered as part of the P availability question per se, it nevertheless has practical implications for the utilization of both soil and applied P.

B. Application Method

1. BROADCASTING

Fertilizer is considered broadcast when applied over the entire soil surface. While most fertilizers are applied in this way (with subsequent incorporation), broadcasting may also include topdressing on growing crops. However, broadcast P is generally applied prior to planting since the growing plant needs P early in its development.

Phosphate broadcast and plowed under is positioned in thin layers between the turned furrows that become more vertical with depth. Disc harrowing is effective for mixing the fertilizer only with the surface 5 to 6 cm of soil. The pattern resulting after plowing and disc harrowing is shown schematically in Fig. 2. Deeper application via plowing has been shown to be advantageous during periods of dry weather.

Even though mixing of fertilizer with the entire plow layer is incomplete, the common system of broadcasting and incorporating fertilizer P prior to planting results in deeper and more intimate mixing with the soil than is true for other methods. With water-soluble P, formation of reaction products of lower solubility proceeds at a faster rate than for placement in more concentrated zones. The crop must, therefore, rely increasingly on P supplied by these reaction products with time. For phosphates of low solubility, however, broadcast application is the most effective method since it encourages a higher dissolution rate.

Generally, acidulated P materials to be broadcast on acid to neutral soils do not need to be high in water-soluble P. Field data in Table 3 (averaged within arbitrarily selected solubility groupings) show that water solubility of broadcast P fertilizer on acid to neutral soils is not important (Webb & Pesek, 1959). However, water solubility of P fertilizer broadcast on calcareous soils was found to be quite important (Webb et al., 1961).

At higher rates of P, application is needed less frequently. In a rotation experiment, Barber (1969) broadcast P fertilizer once every 4 years at rates of 98 and 196 kg of P/ha over a 16-year period. Both rates were effective in maintaining yields through the fourth year following each application. He concluded that more flexibility in P application is possible without seriously affecting yield, provided that the P is plowed under or mixed deeply into the soil.

With the advent of chisel plow tillage, studies were conducted by

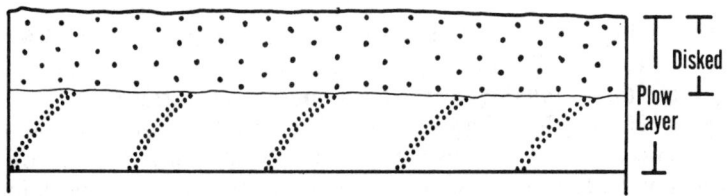

Fig. 2—Distribution of broadcast P after plowing and discing operations (schematic).

Table 3—Corn yield response to broadcast and incorporated P fertilizers as affected by soil reaction and content of water-soluble P. Data averaged over 15 and 30 kg of applied P/ha and two or more sites per year.

Water-soluble P % of available P	Acid to neutral sites			Alkaline (calcareous) sites	
	1953	1954	1956	1958	1959
	kg/ha			kg/ha	
0– 25	920	527	552	1,364	755
26– 50	967	559	439	--	678
51– 75	973	--	--	--	--
76–100	1,017	590	684	1,773	979
Difference (5% prob.)	NS	NS	NS	*	*
Number of sites	3	4	2	2	2
Calculated from:	Webb and Pesek, 1959			Webb et al., 1961	

Cihacek et al. (1974) to compare alternative P application techniques for corn. These were chisel-broadcast, chisel place (at 18- to 20-cm depth), chisel-row band, and moldboard-broadcast. Results of 3 years of experiments in Nebraska and Illinois showed the moldboard incorporation of broadcast P to be the most effective combination. It was concluded, however, that tillage effects were more important than placement of P.

2. ROW AND BAND PLACEMENTS

Early rates of application for row crops were quite low, and fertilizers were applied in or near the seed row. Later, with heavier rates of application, fertilizers were banded beside the seed row (such as 5 cm to the side and 5 cm below) to reduce germination injury. The aim was to apply all the fertilizer needs in this way without causing seedling damage. However, in commercial farming areas, more and more fertilizer has recently been applied before planting via bulk handling and spreading techniques. This relieves time and labor problems during the critical planting operation. With the major portion of the plant nutrients applied broadcast before planting, some farmers felt that a return to a low rate of fertilizer in the seed row was needed for early growth stimulation. This row or starter placement in combination with plow-down fertilizer was known as "pop-up" fertilization (Burson, 1968). Pop-up fertilizers are generally higher in P than in N and K. They may be either liquid or solid, and are usually applied with greater precision than was true for earlier seed row fertilization.

For both row and band applications, P should be largely in water-soluble form for providing the desired growth stimulation. Webb and Pesek (1958) studied the effect of a range of water-soluble P content from 2 to 100% of the available P placed 2.5 cm to the side and 4 cm below corn seed planted in 100- by 100-cm hills. Table 4 shows the data grouped into solubility ranges or classes as in Table 3. The increase in yield with higher water-soluble P content is quite marked.

With soil P levels rising on many soils, interest in placement methods is found primarily in new fertilizer use areas and where soil temperature is low. With the current trend toward earlier corn planting, low soil tempera-

Table 4—Corn yield response to hill-placed P fertilizers varying widely in water solubility. Data averaged over 7.5 and 15 kg of applied P/ha and three or more sites per year (Webb & Pesek, 1958).

Water-soluble P % of available P	1952	1953	1954
		kg/ha	
0– 25	320	358	622
26– 50	471	524	716
51– 75	--	--	--
76–100	559	647	910
Number of sites	6	4	3

tures at planting time are a more frequent occurrence. Even in such circumstances, results from banding and pop-up have not been particularly consistent. Bates (1971) reported that in only 2 out of 22 field experiments in Ontario, Canada, did corn yield increases result from pop-up fertilizer. Where moderate soil temperatures prevail at planting, there has been less interest in banding and pop-up applications. Yield increases are apparently too transitory to justify the added time and labor required during planting time. While early growth responses often occur even on high P soils, such effects very often disappear as the root system develops. Also, Barber and Olson (1968) stated that corn can absorb adequate P from band or row applications only for the first 4 weeks after planting.

Small grains in cooler regions are usually fertilized by means of drill attachments which position the fertilizer near the seed. This causes few problems since row widths are fairly narrow (17 to 35 cm) and the fertilizer concentration in the row is quite low.

The usual practice of banding no more than one-half the broadcast rate implies that banding results in higher effectiveness per unit of applied P. Rudd and Barrow (1973) found that superphosphate row-placed for wheat in Australia was about twice as effective as that applied broadcast at seeding. Prummel (1956) found the same for such crops as pulses, corn, and small grains on P-deficient soil in the Netherlands. Studies by Welch et al. (1966) showed that banded P for row crops can be higher in effectiveness than broadcast P at lower rates of application. However, highest yields were obtained with a combination of banded and broadcast P. This agrees with the findings of Ham et al. (1973) for soybeans and of Barber (1958) for corn that banding alone on low P soils is inadequate and that supplementary broadcast P is needed to reach top yields. Banded P at lower rates tends to maximize return on the investment in fertilizer P, while broadcast P usually gives the highest return per hectare.

Banding an immobile nutrient such as P generally becomes less important as soil P reserves rise over time. The exceptions might include soils quite low in temperature at planting and also short-season vegetable crops. Barber (1958) found that yield response to banded P decreased with rising soil P levels. Soil buildup and maintenance have become more dominant for P fertilization in heavy use areas.

3. STRIP APPLICATION

The concept of applying fertilizer P in strips was suggested by Barber (1974) as a compromise between the extremes of broadcast and row placement. He found that surface placement of fertilizer in narrow strips before plowing was more effective than either banding or broadcast-plowing treatments alone. The strip treatment resulted in 8 to 10% of the plow layer being affected by fertilizer P after plowing. Presumably, the width of the strips could be adjusted according to the needs of the crop and/or the capacity of the soil to fix P in less soluble forms. The seed row could be centered in the middle of treated strips, if desired.

4. EFFECTIVENESS OF SURFACE APPLICATIONS (AS RELATED TO NO-TILL SYSTEMS)

Cropping systems in which soil tillage is not used (no-till) necessarily require surface application of fertilizer P without incorporation. Considering that P is quite immobile and that soil surfaces are often rather dry under conventional systems during the growing season, it might be expected that such application would not be very effective. However, some special surface soil conditions prevail with no-till systems. With crop residue or a killed sod on the surface, moisture levels are markedly higher there because of reduced evaporation (Blevins et al., 1971; Jones et al., 1969).

Uptake of applied P has been shown by Singh et al. (1966) to be unaffected by tillage method. Belcher and Ragland (1972) concluded that P surface-applied in a no-till system was equal in effectiveness to P incorporated into the soil. Several workers have shown no-till corn yields to be equal to or higher than those obtained by conventional tillage (Shear & Moschler, 1969; Moschler & Martens, 1975; Triplett & Van Doren, 1969). These findings are consistent with those resulting from P topdressed on legumes and grass crops. Stanford et al. (1950) found that topdressing of P was quite satisfactory for alfalfa, Ladino clover, and orchardgrass.

While increased soil moisture has been shown to enhance P uptake by corn seedlings (Watanabe et al., 1960), many of the reported increases in corn yield due to the no-tillage system are likely a result of increased moisture availability to the crop. The P source used for surface application should be largely water soluble.

5. FLUID VS. SOLID FORMS OF P

The term *fluid fertilizers* is used here to include both solutions (liquids) and suspensions. For a valid comparison of fluid and solid fertilizer P, the P should be supplied in the same chemical compounds in both cases and be similarly placed, that is, have comparable contact with soil and proximity to developing root systems.

While claims for superiority of solution fertilizers over solid forms are often made, the trials producing such results usually do not comply with the above requirements. For example, solid sources may contain compounds that have different citrate and water solubilities. The initial growth effects might be quite different. Then, too, as mentioned previously for polyphosphate-orthophosphate comparisons, dissimilar micronutrient contents may be important enough to influence the results on some soils.

Lathwell et al. (1960) summarized the results of a number of field experiments conducted to compare P sources in solution and solid form. These field experiments were conducted in several southeastern states and in Iowa and New York, involving corn, wheat, oats, and cotton. The authors concluded that P in solution form is as satisfactory as in comparable solid sources, but likely to be superior to those solid materials which contain a large proportion of water-insoluble P. For the most part, they suggest that the price per unit of P applied in the field should be the primary criterion in choosing between solid and solution forms of P.

Suspension fertilizers do not require that the P compounds be water soluble. Therefore, one would need to know the compounds contained before judging the effectiveness of the suspension fertilizer. The P applied can in fact be quite insoluble. Finely ground PR for direct application can be applied in suspension form rather conveniently and avoid dust problems as well. In this way the material can be applied in finely divided state as it should be for greatest effectiveness; granulation of solid PR for ease of handling results in a marked decrease in agronomic effectiveness.

Many fluid fertilizers have a fairly high content of polyphosphate. These nonortho forms of P have particular advantages for solution fertilizers, since polyphosphates possess the ability to sequester common impurities that originate from wet-process acid (Meline et al., 1972).

It has been suggested by proponents of fluid fertilizers that polyphosphates applied as fluids also offer agronomic advantages. However, as indicated above, there is little evidence to support this contention.

6. FOLIAR FEEDING

Research on foliar application of P has been justified on one or more of the following bases:

1) High fixation of P applied to soil (low recovery of applied P).
2) High cost of P fertilizer during periods of shortage.
3) Seeking crop yields higher than those attainable with soil-applied P alone.

While it has been shown that foliar-applied P is absorbed by leaves of various plants (Silberstein & Wittwer, 1951; Fisher & Walker, 1955), the amount absorbed at any one spraying is limited. Spray solutions must be quite dilute to avoid injury to the leaves. Therefore, repeated applications are necessary for sufficient nutrient absorption by the leaves to correct deficiencies.

The results from experiments have been inconsistent. Several workers have found increases in effectiveness of P through foliar application in pot experiments (Datta & Vyas, 1967; De Datta & Moomaw, 1965). The latter workers noted a direct relationship between response of sugarcane to foliar applied P and the capacity of the soils to immobilize P. Bouma (1969), working with subterranean clover grown in nutrient solution-sand culture, concluded that foliar applications of P offer little hope as a practical means of correcting P deficiencies.

In three field experiments with cotton, Lancaster and Savatli (1965) found that frequent foliar applications of P increased the yield where P deficiencies existed, but not beyond what could be obtained with soil application alone. They concluded that foliar-applied P was not necessary for top yields and was not a practical alternative to soil application.

It is expected that some research will continue with foliar application of P, particularly where fixation of soil-applied P is high. Also, foliar application of N-P-K-S fertilizers after initiation of the reproductive phase of growth has produced some marked yield increases in crops such as soybeans.

7. PHYTOTOXIC EFFECTS

The primary possibilities for plant injury from soil-applied P is with row or seed placement. Even in such circumstances the frequency of injury is quite low. Various investigators have found P sources containing primarily monocalcium phosphate to be fairly harmless at usual rates of application (Cummins & Parks, 1961; Olson & Dreier, 1956; Dubetz et al., 1959). Multinutrient fertilizers may be quite a different matter, however. For example, NH_4-N in DAP can be injurious to germinating seedlings (Bennett & Adams, 1970). Likewise, ammonia given off from UAP also can be toxic if the fertilizer is placed too close to the seed row.

C. Crop Species

Many rapidly growing crops respond to soluble P fertilizer during early stages of growth, even at fairly high soil P levels. This is because the need for P exceeds the capacity of the young root system to absorb soil P. Persistence of growth responses to added P during later growth stages depends in part upon soil P levels, and also on crop species and other factors. Species differences involve such parameters as rate of growth, length of growth period, and degree of root proliferation. Generally, longer season crops, such as corn, outgrow early growth responses to applied P, with little effect on final yield. Root development of such crops usually provides uptake of soil P adequate for later growth. Olson et al. (1962) concluded that corn and sorghum are quite efficient in utilization of soil P as compared with small grains.

For short-season crops, such as certain vegetables, growth responses to added P tend to persist until harvest. Root development is often inadequate for P uptake during the short growing period.

The implications of these relationships for P fertilizers are that water solubility and placement in or near the seed row are more important for short- than for long-season crops. Most any P source and application method would be satisfactory for long-term crops, particularly if soil P level is rather high.

IV. CHANGES IN SOURCE NEEDS RESULTING FROM RISING SOIL P LEVELS

It is now obvious that soil P levels have been rising generally in commercial farming areas in the United States and in other countries as a result of accumulated fertilizer residues. Yield responses to applied P have become quite infrequent, indicating that these accumulated reserves are supplying P to the crop (Engelstad & Terman, 1966; Terman 1957). Soil test summaries also show this trend. Table 5 shows percentages of total soil samples received by university testing laboratories in several states testing high in extractable P.

Accumulations of P in soils influence choice of P sources. On high P soils where crop response to applied P is no longer expected, current applications are justified only for maintenance purposes. Rate of dissolution of the applied P is not important, and therefore, sources of lower solubility could be considered for this use. The main consideration would be that the source used should eventually react with the soil to supply P. Exceptions include potatoes and short-season vegetable crops, which may respond to water-soluble P, even on high P soils.

V. SOURCE NEEDS IN THE TROPICS

Proposals for special P sources for the tropics are based on the assumption that soils found there are uniquely different from those found in temperate regions. While some soils in the tropics are more heavily weathered than any found in temperate regions, soil variability in the tropics is so extreme that it is unwise to generalize. In fact, certain soil orders are found in both temperate and tropical regions.

Fixation of applied P is high in certain soils of the tropics, such as those derived from volcanic ash; however, such soils are of rather minor occurrence in the tropics. Generally, P fixation seems to have been over em-

Table 5—Percentages of total soil samples testing high in soil P in several states (university laboratories only).

	High in soil P, of total samples	Period covered
Alabama	48	1972–73
Colorado	27	1974–75
Georgia	40	1972
Iowa	28	1972–73
Mississippi	50	1971–72
Nebraska	32	1966–72
North Carolina	64	1974
Ohio	16	1971–72
Oklahoma	46	1972–73
Tennessee	40	1967–70
Virginia	58	1970–71

phasized for the soils of that region (Greenland, 1973). In fact, P fixation in tropic and temperate regions seems to be similar, as stated by Russell (1974, p. 222):

> There has been a considerable body of work on the phosphate manuring of tropical crops. At one time the statement was made that tropical soils fixed phosphate strongly, in a way different from that of temperate soils. In general, experimental work has not borne out this earlier belief, and in fact most tropical soils hold phosphate in much the same way as temperate soils. Where they often differ is that their initial phosphate status is much lower than was found on well-farmed temperate soils. This has the consequence that the residual effect of a small dressing of placed phosphate is often small compared with its effect in the season of application; and it may take a number of small applications for the phosphate level of the soil in bulk to be raised.

However, research is needed in the tropics on effective use of conventional P sources. This includes studies of optimum rates, solubility, granulation, and placement for the various combinations of climate, soil, and cropping conditions that occur there.

LITERATURE CITED

Allen, S. E. 1970. New forms of fertilizer phosphorus—agronomic evaluation. Phosphorus Agric. 55:25-35.

Allen, S. E., and D. A. Mays. 1971. Sulfur-coated fertilizers for controlled release: agronomic evaluation. J. Agric. Food Chem. 19:809-812.

Armiger, W. H., and Maurice Fried. 1957. The plant availability of various sources of phosphate rock. Soil Sci. Soc. Am. Proc. 21:183-188.

Association of Official Analytical Chemists (AOAC). 1975. Official methods of analysis. 12th ed. AOAC, Washington, D.C.

Barber, S. A. 1958. Relation of fertilizer placement to nutrient uptake and crop yield. I. Interaction of row phosphorus and the soil level of phosphorus. Agron. J. 50:535-539.

Barber, S. A. 1969. Flexibility in applying phosphorus and potassium. Crops Soils 21(9):16-17.

Barber, S. A. 1974. A program for increasing the efficiency of fertilizers. Fert. Solutions 18(2):24-25.

Barber, S. A., and R. A. Olson. 1968. Fertilizer use on corn. p. 163-188. In L. B. Nelson (ed.) Changing patterns in fertilizer use. Soil Sci. Soc. of Am., Madison.

Bates, T. E. 1971. Response of corn to small amounts of fertilizer placed with the seed: II. Summary of 22 field trials. Agron. J. 63:369-371.

Belcher, C. R., and J. L. Ragland. 1972. Phosphorus absorption by sod-planted corn (*Zea mays* L.) from surface-applied phosphorus. Agron. J. 64:754-756.

Bengtson, G. W., E. C. Sample, and S. E. Allen. 1974. Response of slash pine seedlings to P sources of varying citrate solubility. Plant Soil 40:83-96.

Bennett, A. C., and Fred Adams. 1970. Calcium deficiency and ammonia toxicity versus separate causal factors of $(NH_4)_2HPO_4$—injury to seedlings. Soil Sci. Soc. Am. Proc. 34:255-259.

Blevins, R. L., Doyle Cook, S. H. Phillips, and R. E. Phillips. 1971. Influence of no-tillage on soil moisture. Agron. J. 63:593-596.

Bouldin, D. R., J. D. DeMent, and E. C. Sample. 1960. Interaction between dicalcium and monoammonium phosphates granulated together. J. Agric. Food Chem. 8:470-474.

Bouma, D. 1969. The response of subterranean clover (*Trifolium subterranean* L.) to foliar applications of phosphorus. Aust. J. Agric. Res. 20:435-445.

Brown, W. E., and J. R. Lehr. 1959. Application of phase rule to the chemical behavior of monocalcium phosphate monohydrate in soils. Soil Sci. Soc. Am. Proc. 23:7-12.

Burson, Paul M. 1968. Fertilizer placement for corn. Minn. Sci. 24:10-12.
Caro, J. H., and W. L. Hill. 1956. Characteristics and fertilizer value of phosphate rock from different fields. J. Agric. Food Chem. 4:684-687.
Cihacek, L. J., D. L. Mulvaney, R. A. Olson, L. F. Welch, and R. A. Wiese. 1974. Phosphate placement for corn in chisel and moldboard plowing systems. Agron. J. 66:665-668.
Cooke, G. W. 1966. Phosphorus and potassium fertilisers: their forms and their places in agriculture. Proc. Fert. Soc. (London) 92:1-45.
Cummins, D. G., and W. L. Parks. 1961. The germination of corn and wheat as affected by various fertilizer salts at different soil temperatures. Soil Sci. Soc. Am. Proc. 25:47-49.
Datta, N. P., and K. K. Vyas. 1967. Isotopes in plant nutrition and physiology. p. 371-376. *In* Proc. Symp. Isotopes in Plant Nutrition and Physiology, IAEA/FAO, Vienna.
De Datta, S. K., and J. C. Moomaw. 1965. Availability of phosphorus to sugar cane in Hawaii as influenced by various phosphorus fertilizers and methods of application. Exp. Agric. 1: 261-270.
Dubetz, S., R. L. Smith, and G. C. Russell. 1959. The effect of fertilizers and osmotic pressure on germination. Can. J. Soil Sci. 39:157-164.
Engelstad, O. P., and S. E. Allen. 1971. Ammonium pyrophosphate and ammonium orthophosphate as phosphorus sources: effects of soil temperature, placement, and incubation. Soil Sci. Soc. Am. Proc. 35:1002-1004.
Engelstad, O. P., A. Jugsujinda, and S. K. De Datta. 1974. Response by flooded rice to phosphate rocks varying in citrate solubility. Agron. J. 38:524-529.
Engelstad, O. P., and G. L. Terman. 1966. Fertilizer nitrogen: Its role in determining crop yield levels. Agron. J. 58:536-539.
Fisher, E. G., and D. R. Walker. 1955. The apparent absorption of phosphorus and magnesium from sprays applied to the lower surface of McIntosh apple leaves. Proc. Am. Soc. Hortic. Sci. 65:17-24.
Giordano, P. M., and J. J. Mortvedt. 1969. Response of several corn hybrids to level of water soluble zinc in fertilizers. Soil Sci. Soc. Am. Proc. 33:145-148.
Greenland, D. J. 1973. Soil factors determining responses to phosphorus and nitrogen fertilizers used in tropical Africa. Afr. Soils XVII:99-108.
Ham, G. E., W. W. Nelson, S. D. Evans, and R. D. Frazier. 1973. Influence of fertilizer placement on yield response of soybeans. Agron. J. 65:81-84.
Harris, F. J. 1963. Potassium metaphosphate: a novel method of manufacture and a summary of its behavior as a fertilizer. Proc. Fert. Soc. (London) 76:1-48.
Jones, J. Nick, Jr., J. E. Moody, and J. H. Lillard. 1969. Effects of tillage, no tillage, and mulch on soil water and plant growth. Agron. J. 61:719-721.
Lancaster, J. D., and Z. A. Savatli. 1965. Foliar application of phosphorus for cotton. Miss. Agric. Exp. Stn. Bull. 708, Mississippi State, Miss.
Lathwell, D. J., J. T. Cope, Jr., and J. R. Webb. 1960. Liquid fertilizers as sources of phosphorus for field crops. Agron. J. 52:251-254.
Lehr, J. R. 1967. Variations in composition of phosphate ores. p. 61-67. Proc. 17th Ann. Meeting, Fert. Ind. Round Table. Washington, D.C.
Lehr, J. R., and G. H. McClellan. 1972. A revised laboratory reactivity scale for evaluating phosphate rocks for direct application. TVA Bull. Y-43. Muscle Shoals, Ala. 36 p.
Lehr, J. R., E. H. Brown, A. W. Frazier, J. P. Smith, and R. D. Thrasher. 1967a. Crystallographic properties of fertilizer compounds. TVA Chem. Eng. Bull. no. 6. Muscle Shoals, Ala. 166 p.
Lehr, J. R., G. H. McClellan, J. P. Smith, and A. W. Frazier. 1967b. Characterization of apatites in commercial phosphate rocks. p. 29-44. *In* Colloq. Int. Phosphates Mineraux Solides, Toulouse, France.
Mattingly, G. E. G. 1963. The agricultural value of some water and citrate soluble fertilisers: an account of recent work at Rothamsted and elsewhere. Proc. Fert. Soc. (London) 75: 57-97.
Mattingly, G. E. G. 1965. Evaluation of phosphate fertilizer by solubility tests. NAAS Advisory Papers no. 3., Ministry of Agric. Fisheries & Food, London.
McLean, E. O., and T. J. Logan. 1970. Sources of phosphorus for plants grown in soils with differing phosphorus fixation tendencies. Soil Sci. Soc. Am. Proc. 34:907-911.
McLean, E. O., and R. W. Wheeler. 1964. Partially acidulated rock phosphate as a source of phosphorus to plants: I. Growth chamber studies. Soil Sci. Soc. Am. Proc. 28:545-550.
McLean, E. O., R. W. Wheeler, and J. D. Watson. 1965. Partially acidulated rock phosphate as a source of phosphorus to plants: II. Growth chamber and field corn studies. Soil Sci. Soc. Am. Proc. 29:625-628.

Meline, R. S., R. G. Lee, and W. C. Scott. 1972. Use of a pipe reactor in production of liquid fertilizers with very high polyphosphate content. Fert. Solutions 16(3):32-45.

Moreno, E. C. 1959. Probability theory applied to fertilizer granule-size effects. Soil Sci. Soc. Am. Proc. 23:326-327.

Moschler, W. W., and D. C. Martens. 1975. Nitrogen, phosphorus, and potassium requirements in no-tillage and conventionally tilled corn. Agron. J. 39:886-891.

Mortvedt, J. J., and P. M. Giordano. 1967. Crop response to zinc oxide applied in liquid and granular fertilizers. J. Agric. Food Chem. 15:118-122.

Mortvedt, J. J., and P. M. Giordano. 1970. Crop response to iron sulfate applied with fluid polyphosphate fertilizers. Fert. Solutions 14(4):22-27.

Mortvedt, J. J., and P. M. Giordano. 1971. Response of grain sorghum to iron sources applied alone or with fertilizers. Agron. J. 63:758-761.

Olson, R. A., and A. F. Dreier. 1956. Fertilizer placement for small grains in relation to crop stand and nutrient efficiency in Nebraska. Soil Sci. Soc. Am. Proc. 20:19-24.

Olson, R. A., A. F. Dreier, C. A. Hoover, and H. F. Rhoades. 1962. Factors responsible for poor response of corn and grain sorghum to phosphorus fertilization: I. Soil phosphorus level and climatic factors. Soil Sci. Soc. Am. Proc. 26:571-574.

Prummel, J. 1956. Placement of fertilizers. Int. Congr. Soil Sci., Trans. 6th (Paris) IV:167-171.

Rudd, C. L., and N. J. Barrow. 1973. The effectiveness of several methods of applying superphosphate on yield response by wheat. Aust. J. Exp. Agric. Anim. Husb. 13(63):430-433.

Russell, E. W. 1974. The role of fertilizers in African agriculture. p. 213-238. *In* V. Hernandez Fernandez (ed.) Fertilizers, crop quality and economy. Elsevier Scientific Publishing Company, Amsterdam, Oxford, New York.

Sample, E. C., and A. W. Taylor. 1964. Rapid, nondestructive method for estimating rate and extent of movement of phosphorus from fertilizer granules in soil. Soil Sci. Soc. Am. Proc. 28:296-297.

Seatz, L. F., and C. O. Stanberry. 1963. Advances in phosphate fertilization. p. 155-187. *In* M. H. McVickar, G. L. Bridges, and L. B. Nelson (ed.) Fertilizer technology and usage. Soil Sci. Soc. Am., Madison, Wis.

Shear, G. M., and W. W. Moschler. 1969. Continuous corn by the no-tillage and conventional tillage methods: a six-year comparison. Agron. J. 61:524-527.

Silberstein, O., and S. H. Wittwer. 1951. Foliar application of phosphatic nutrients to vegetable crops. Proc. Am. Soc. Hortic. Sci. 58:179-190.

Singh, T. A., G. W. Thomas, W. W. Moschler, and D. C. Martens. 1966. Phosphorus uptake by corn (*Zea mays* L.). Agron. J. 58:147-150.

Stanford, G., C. McAuliffe, and Richard Bradfield. 1950. The effectiveness of superphosphate topdressed on established meadows. Agron. J. 42:423-426.

Taylor, A. W., and G. L. Terman. 1964. The nature and distribution of fertilizer phosphate in soils. Int. Congr. Soil Sci., Trans. 8th (Bucharest, Romania) IV:451-457.

Terman, G. L. 1957. Variability in phosphorus rate and source experiments in relation to crop and yield levels. Agron. J. 49:271-276.

Terman, G. L. 1971. Phosphate fertiliser sources: agronomic effectiveness in relation to chemical and physical properties. Proc. Fert. Soc. (London) 123:1-39.

Terman, G. L. 1975. Agronomic results with polyphosphate fertilizers. Phosphorus Agric. 65: 21-26.

Terman, G. L., and S. E. Allen. 1967. Response of corn to phosphorus in under-acidulated phosphate rock and rock-superphosphate fertilizers. J. Agric. Food Chem. 15:354-358.

Terman, G. L., and O. P. Engelstad. 1966. Crop response to nitrogen and phosphate in ammonium polyphosphate. Commer. Fert. 112(6):30, 32-33, 36, 40.

Terman, G. L., and O. P. Engelstad. 1976. Agronomic evaluation of fertilizers: principles and practices. TVA Bull. Y-21, rev. Muscle Shoals, Ala. 45 p.

Terman, G. L., S. E. Allen, and O. P. Engelstad. 1970. Response by paddy rice to rates and sources of applied phosphorus. Agron. J. 62:390-394.

Terman, G. L., D. R. Bouldin, and J. R. Webb. 1962. Evaluation of fertilizers by biological methods. Adv. Agron. 14:265-319.

Terman, G. L., W. M. Huffman, and B. C. Wright. 1964. Crop response to fertilizers in relation to content of "available" phosphorus. Adv. Agron. 16:59-100.

Terman, G. L., V. J. Kilmer, and S. E. Allen. 1969. Reactivity of phosphate rocks with acids, salts, and soils in relation to effectiveness for crops. Fert. News 14(8):41-45.

Triplett, G. B., Jr., and D. M. Van Doren. Jr. 1969. Nitrogen, phosphorus, and potassium fertilization for non-tilled maize. Agron. J. 61:637-639.

van Burg, P. F. J. 1963. The agricultural evaluation of nitrophosphates with particular reference to direct and cumulative phosphate effects, and to interaction between water solubility and granule size. Proc. Fert. Soc. (London) 75:5-54.

Wakefield, Z. T., S. E. Allen, J. F. McCullough, R. C. Sheridan, and J. J. Kohler. 1971. Evaluation of phosphorus-nitrogen compounds as fertilizers. J. Agric. Food Chem. 19: 99-103.

Watanabe, F. S., S. R. Olsen, and R. E. Danielson. 1960. Phosphorus availability as related to soil moisture. Int. Congr. Soil Sci. Trans. 7th (Madison, Wis.) III:450-454.

Webb, J. R., and J. T. Pesek. 1958. An evaluation of phosphorus fertilizers varying in water solubility: I. Hill applications for corn. Soil Sci. Soc. Am. Proc. 22:533-538.

Webb, J. R., and J. T. Pesek. 1959. An evaluation of phosphorus fertilizers varying in water solubility: II. Broadcast applications for corn. Soil Sci. Soc. Am. Proc. 23:381-384.

Webb, J. R., Kalju Eik, and J. T. Pesek. 1961. An evaluation of phosphorus fertilizers applied broadcast on calcareous soils for corn. Soil Sci. Soc. Am. Proc. 25:232-236.

Welch, L. F., D. L. Mulvaney, L. V. Boone, G. E. McKibben, and J. W. Pendleton. 1966. Relative efficiency of broadcast versus banded phosphorus for corn. Agron. J. 58:283-287.

Wright, B., J. D. Lancaster, and J. L. Anthony. 1963. Availability of phosphorus in ammoniated ordinary superphosphate. Miss. Agric. Exp. Stn. Tech. Bull. 52:1-35.

Chapter 13

Evaluation and Utilization of Residual Phosphorus in Soils

N. J. BARROW

CSIRO
Wembley, Western Australia

I. INTRODUCTION

Many soils are deficient in P, but this is not the only reason for continued use of P fertilizers. If it were, all that would be needed is one adequate application of fertilizer followed by small applications to replace P removed in agricultural products. This is not the case. Phosphate applications do not remain fully effective and further applications are therefore needed. Even though previously applied P may be less effective than freshly applied P, it nevertheless has value. This value may be partly associated with effects other than those on the P status per se. There may be such beneficial physical effects as an increase in the water-holding capacity of the soil and a decrease in the modulus of rupture (Lutz & Haque, 1975). There may also be increases in the organic matter content of the soil due to stimulation of plant growth. However, this chapter is mainly concerned with the direct effects of previous P applications on P status. The approach is first to consider reasons for the decline in effectiveness, then to consider mechanisms by which plants might gain access to the residues, and finally to consider field evaluation of fertilizer residues.

II. DEFINITION AND MEASUREMENT OF RESIDUAL VALUE

For effective discussion of the factors affecting residual value, it is important that an agreed index be used. One index is the period over which an application remains adequate. Thus, Kamprath (1967) showed that an application of 687 kg of P/ha was adequate for corn 7 to 9 years later. Even though he concluded that the residual value was marked, his results do not preclude the possibility that there was a several-fold loss of availability over this period; conceivably the level of application was so high that there was adequate P present despite a loss of availability. It would seem to be more informative to measure the current availability of a previous application.

Copyright 1980 © ASA-CSSA-SSSA, 677 South Segoe Road, Madison, WI 53711, USA.
The Role of Phosphorus in Agriculture.

However, there is no way to measure an absolute value of availability (Black & Scott, 1956). All that can be done is to measure the availability relative to that of another fertilizer. Thus, the current effect of a fertilizer may be compared with the current effect of fresh superphosphate and expressed as fresh superphosphate equivalents (Mattingly & Widdowson, 1963; Mattingly, 1968; Devine et al., 1968). Alternatively, though less satisfactorily, the current effect may be compared with the original effect after adjustment for seasonal differences (Arndt & McIntyre, 1963). In such cases, the property compared is the effectiveness of a fertilizer dressing rather than the availability—removal of P in agricultural produce will reduce the future effectiveness of the fertilizer dose, though it will not, of itself, reduce the availability coefficient of the P remaining. Hence, a general term to describe such measurements is relative effectiveness.

If relative effectiveness is to be measured, the fertilizer must have a measurable effect—there must be a response. In general, the larger the response, the better the response curve can be defined, and hence, the more precise the measurement of relative effectiveness. However, the precision of the measurement has seldom been indicated and this is partly because of the difficulty of assigning confidence regions when nonlinear response curves are used (Barrow & Campbell, 1972).

The response measured may be in yield or in uptake of P. Since the purpose of applying fertilizer is usually to increase yield, it could be argued that this is the better measure. However, uptake may continue at levels of fertilizer beyond that required for maximum yield, and by measuring uptake it may be possible to obtain a measure of response over a wider range of P levels. Under pot conditions, the two measures give very similar indications of residual value (Barrow & Campbell, 1972), but they may give different values in the field. Mattingly and Widdowson (1963) found that residual values measured from uptake were greater than those measured from yield. They suggested that the crop continued to take up P from residues buried by plowing.

It is implied in the measurement of relative effectiveness that the effects of the previously applied fertilizer can be duplicated by freshly applied fertilizer. This may not always be so. In experiments in England it has been found that larger yields of potatoes and sugar beets can be obtained from residues of previous fertilizer than from freshly applied superphosphate (Mattingly, 1971). This may be partly due to movement of phosphate through a greater volume of soil and perhaps partly due to effects of previous applications on other soil properties.

The ultimate measure of the residual value of a fertilizer is its ability to help grow a desired product. But such measures are tedious, time-consuming, and often imprecise. Consequently, some workers have used more convenient measures. For example, Larsen et al. (1965) sought an indication of the change in status of previously applied fertilizer on 24 sites by measuring the L-value (this is isotopic dilution technique which requires growth of plants in a glasshouse); Fox and Kamprath (1970) measured the effects of previous applications of phosphate on the P sorption isotherms and used these to assess the current status; and Power et al. (1964) and Fitter (1974) measured changes through time in the P extracted by $NaHCO_3$. The merit of such approaches depends on the closeness of the relation between the measure used and the ability of the soil to grow the desired product. The relation is usually far from perfect. Laboratory extractions cannot be expected to always give the correct balance to the numerous factors which affect uptake of P (Williams & Knight, 1963), and may dissolve soil P which is not taken up by plants (Mattingly & Pinkerton, 1961). Furthermore, such measures may indicate trends which are not apparent when plant growth is used as the criterion. Thus I found that there was a slight trend for the relative effectiveness of previously applied phosphate to increase as the buffering capacity of the soil increased (Barrow, 1973), whereas Fitter (1974) and Barrow and Shaw (1976b) found that P extracted by $NaHCO_3$ decreased more quickly in soils of high buffering capacity. A soil test, such as the bicarbonate test, is an attempt to integrate the several factors that control P status into a single value. Factors involved include buffering capacity of the soil for P, amount of P present, and period of contact between soil and P. The test does not necessarily mirror the separate effects of any of these; hence, the change with time may not indicate the change in status. The bicarbonate test involves adding another specifically adsorbed anion, and usually, increasing the pH. This is more effective in removing phosphate from weakly buffered soils, and hence the decline in P extracted with time on such soils may be less than the decline in effectiveness of the P for plant growth.

III. FACTORS INVOLVED IN THE DECREASE IN EFFECTIVENESS OF P FERTILIZERS

A. Removal of P in Agricultural Products

The amount of P removed from a field is very dependent on the kind of product exported. At one extreme, if wool is the only product, the amount of P removed is negligible. However, even in this case, there may be removal of P from part of the area and accumulation in other parts because of the tendency for sheep to camp in specific areas (Hilder, 1964). The other extreme may well be repeatedly cut, tropical pasture yielding about 20 metric tons of dry matter/ha and removing 75 kg of P/year (Younge & Plucknett, 1966). Annual removal on irrigated alfalfa is also high and may reach 60 kg of P/ha (Leamer, 1963). Between these extremes, the amount removed can usually be estimated after reference to tables for average P concentration in produce (Spector, 1956). In countries in which P fertilizers are widely used, the amount of P removed in produce is, on the average, only a small proportion of the amount added. For example, for the whole of Australia, the amount removed from farms is now less than 20% of that applied in fertilizer (Gifford et al., 1975). Since the amount removed includes produce from unfertilized areas, the amount removed for the fertilized areas may be less than this. Larsen (1974) estimates that for Denmark the amount removed is only about 10% of that applied. Thus, on the average, removal of P in products is not the major cause of the decrease in effectiveness, though it may be very important in some cases.

B. Losses of P by Leaching and Erosion

In most soils, P is retained close to the site of application and the main mechanisms for loss are physical erosion of the topsoil and loss of solution P in surface runoff. These losses are normally small. In experimental catchments in Iowa, contour-planted to corn, annual losses of P in surface water averaged less than 2% of the annual application rate (Schuman et al., 1973). Most of this was as sediment. Losses from level terraces were even lower. However, in some soils, appreciable P may be moved from the topsoil by leaching. This usually occurs on soils which are sandy and which, therefore, have both a low capacity to retain water and a low buffering capacity for phosphate. Thus, appreciable leaching loss of P has been reported in sandy soils in Florida by Neller et al. (1951), in Dorset (UK) by Mattingly (1970), in Tasmania by Paton and Loneragan (1960), and in Western Australia by Ozanne and Shaw (1961). The removal may be as high as 80% of the applied P (Neller et al., 1951; Ozanne & Shaw, 1961). However, the P may not be completely removed from the profile; Mattingly (1970) reported that much of the P removed from the topsoil was present in the B horizon. Liming of the soil often decreases the loss of P (Neller et al., 1951; Chaiwanakupt & Robertson, 1976) and may result in improved residual value of P (Paton & Loneragan, 1960).

C. Immobilization in Organic Matter

When legume-based pastures are sown on previously P-deficient soil, an appreciable accumulation of organic matter may result. Until this buildup approaches equilibrium, or until it is depleted by, say, cultivation, it is a sink for plant nutrients including P. This process has received much emphasis in Australia because many soils were very P deficient in the virgin state, biological productivity was low, and the organic matter level of the soil was low. The capacity for change in the level of organic matter is therefore high. Some values for the annual accumulation of organic P are: 2.8 kg/ha in the Southern Tablelands of New South Wales (Donald & Williams, 1954); 1.5 kg/ha at Kybybolite, South Australia (Russell, 1960a, 1960b); and 1.7 kg/ha on the coastal plain of Western Australia (Barrow, 1969). Since the annual rate of application of P in these areas may range from 5 to 20 kg/ha, the accumulation of organic P can represent a significant loss of P.

Over long periods, and for soils on which the P deficiency is not as extreme, accumulation of organic P under pasture may be less important. On Park Grass at Rothamsted, on plots maintained between pH 6.2 and 6.5

and fertilized with P for over 100 years, only 1% of the P which accumulated in the surface soil was organic (Oniani et al., 1973). For plots maintained at pH 4.5 the equivalent value was 6%. Accumulation of organic P is also unimportant in farming systems in which there is little long-term change in the organic matter content of the soil. Mattingly et al. (1974) found that between 1888 and 1959 there had been little change in the organic P content of plots in continuous wheat or barley even in those plots fertilized with farmyard manure.

D. Mixing and Diffusion of Phosphate

When a P fertilizer is applied to soil, its distribution is initially heterogeneous. If roots are distributed at random through the soil, heterogeneity can only stimulate uptake if the relation between concentration of added phosphate and uptake is curved upwards. In this case the decrease in the proportion of the roots in contact with the fertilizer is more than compensated for by the increased uptake at the zones of concentration. An upward curving relation could arise from two components. One is a threshold concentration for uptake; the other arises from the increase in the effective diffusion coefficient because of the curved relation between the total diffusible phosphate and the solution concentration. However, opposing these effects is the relation between solution concentration of phosphate and the rate of uptake by the root. As the concentration increases, the rate of uptake does not increase proportionally, but approaches a maximum (Asher & Loneragan, 1967; Keay et al., 1970; Khasawneh & Copeland, 1973; Barrow, 1975). This latter effect would be expected to become dominant at high concentrations, such as those which might be expected in a fertilizer band. This would suggest that banding would reduce uptake. But roots are not distributed at random and often proliferate near zones of high concentration. The change in the proportion of the roots which are near the fertilizer is an important component in making heterogeneity favorable. However, the ability to develop roots in zones of high concentration may differ between plants (Strong & Soper, 1973). Banding of fertilizer may have additional advantages, such as placing the fertilizer in a region which is rapidly reached by a germinating seed or in a zone of the soil which does not readily dry out. It may also have disadvantages, such as restricting the exploration of the rest of the soil and thus reducing the plant's access to other nutrients and to water.

In time, the initial heterogeneity of applied fertilizer tends to decrease. This is partly due to the slow diffusion of P away from its original site, but it can also be due to destruction of a fertilizer band by subsequent cultivation. Thus, Williams and Simpson (1965) reported that cultivation of clover pastures appeared to induce P deficiency, and they showed that mixing of the top 10 cm of soil decreased P uptake. The extent to which these processes can account for the decrease in effectiveness with time depends on the extent to which the initial heterogeneity increased effectiveness, and no general figure can be given.

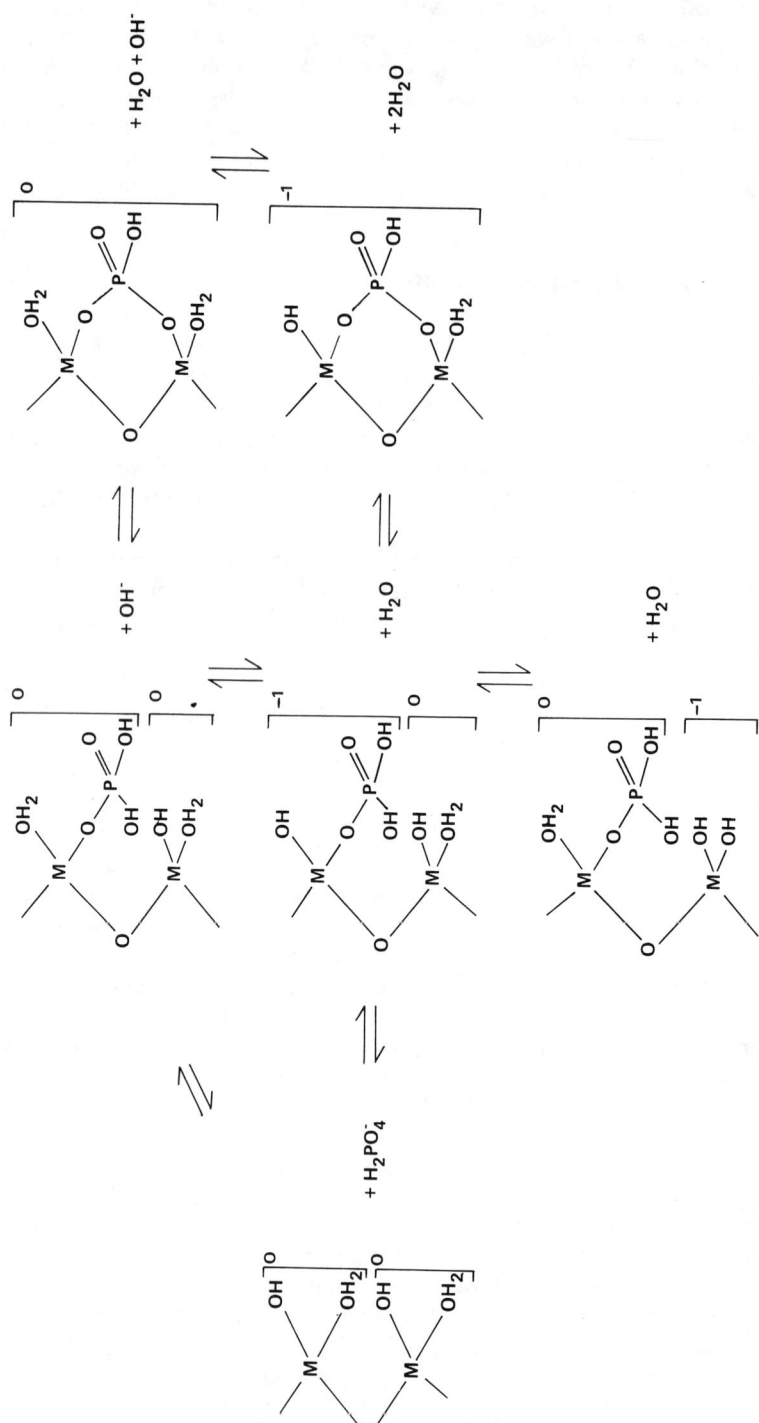

Fig. 1—Changes in charge and possible changes in configuration following adsorption of phosphate on the surface of a metal oxide.

E. Slow Reactions Between Soil and Phosphate

When a P fertilizer is added to soil, a complex sequence of reactions follows. These reactions are described in detail by Soper and Sample in Chapt. 11 of this book, and only a broad outline will be given here. In general, three zones may be recognized near a fertilizer granule. The central zone contains the residue of the P fertilizer. For example, for monocalcium phosphate fertilizer it would be largely dicalcium phosphate. Around this central zone there is a zone into which the concentrated solution resulting from the solution of the fertilizer has moved. This solution may dissolve Ca, Fe, and Al ions from the soil. Phosphate may react with these ions in solution and form precipitates. This zone grades into an outer zone in which concentrations are lower and the P reactions are mainly with atoms on the surface of soil particles, rather than atoms in solution, and can therefore be regarded as adsorption. The chemical mechanisms involved in this adsorption have been reviewed by Mott (1970) and by Mattingly (1975). For the present it is sufficient to note that the phosphate is closely and chemically bonded to the surface of Fe and Al oxides by specific chemical bonds. The reaction may be regarded as partly a displacement of water molecules and partly a displacement of hydroxyls, so that the negative charge conveyed to the surface is usually lower than the charge on the anion (Fig. 1). In a calcareous soil, there will also be reactions with the $CaCO_3$. These have been reviewed by Mattingly (1975). Some of the P will be adsorbed on the $CaCO_3$ surface and some will be precipitated from solution, probably as octocalcium phosphate, and deposited on the adsorbed layer. This P and that on Fe and Al surfaces must approach equilibrium with each other. Hence, if there are decreases in the concentration of solution P in equilibrium with the Fe and Al surfaces, a dissolution of some of the P on the $CaCO_3$ surface would be expected. In this respect the $CaCO_3$ might have an effect similar to that of adding a sparingly soluble fertilizer.

There are differences of opinion about the relative importance of the inner zone of precipitation and the outer zone of adsorption. Some 20 years ago Hemwall (1957), in a review of phosphate reactions, emphasized the importance of precipitation reactions. However, the high concentrations of the inner zone may not affect a very large volume of soil. Further, the concentration in the inner zone falls as the phosphate continues to move outward by diffusion (Williams, 1971), and some of the precipitates in the inner zone may therefore dissolve. The resolution of this argument is not the subject of this paper. It is concerned with the long-term changes which reduce the effectiveness of P fertilizers. These changes occur when granular fertilizers are used (Allen et al., 1954; Terman et al., 1960; Massey et al., 1970; Barrow, 1974b). They could therefore be partly due to changes occurring in the zone of precipitation and slow movement of P away from this zone. But the effectiveness of phosphate fertilizers is also reduced when powdered fertilizer is used (Devine et al., 1968)—indeed, Terman et al. (1960) found that the decrease was greater with fine than with granular superphosphate. Furthermore, similar decreases occur when the phosphate is supplied as resin P (Larsen & Gunnary, 1964) or as solutions of potassium phosphate (Barrow et al., 1977). In these cases incongruent dissolution would not have occurred, and when solutions are mixed through the soil, heterogeneity would be reduced. This suggests that changes which follow the initial adsorption reactions are a very important component in the decrease in effectiveness of phosphate fertilizers. These will be considered in more detail in the next section.

IV. CHARACTERISTICS OF THE SLOW REACTIONS WHICH FOLLOW ADSORPTION

A. The Existence of the Slow Reactions

When considering the continuing reaction between soil and P, many workers have thought in terms of a model in which the P is present in at least three different categories. One of these is phosphate in the soil solution and another is phosphate adsorbed on the soil particles and in quasi-equilibrium with the solution (Wiklander, 1950; Williams, 1952; Larsen, 1967, 1974; Barrow & Shaw, 1975a). The third category comprises P which is variously described as fixed on the surface layers of the soil particles (Wiklander, 1950; Williams, 1952), nonlabile (Larsen, 1967, 1974, or in a form which is not in direct equilibrium with the solution (Barrow & Shaw, 1975a). The variety of phrases used to describe this third category is probably a reflection of uncertainty about its nature, but in all cases the essence is the implication that the P is more firmly held and is less readily available to plants. When any previous equilibrium between these categories is disturbed by adding P fertilizer, it is envisaged that reaction between soil and P in the solution takes place in two steps—a rapid step in which some of the phosphate is adsorbed, and a slower step in which some of the phosphate is converted into a more firmly held form.

The two-step model was initially proposed to explain the continuing reaction between soil and phosphate as observed in the laboratory. There are two lines of evidence which suggest that this model is realistic. One is based on the effects of temperature. Provided conditions have been chosen so that the second step has become so slow that it can be ignored, high temperatures favor high concentrations of P in solution (Barrow & Shaw, 1975a). Thus the adsorption step is exothermic as is required on thermodynamic grounds (Hayward & Trapnell, 1964). If, on the other hand, conditions are chosen so that the effect of temperature on the first step are eliminated, high temperatures favor low concentrations in solution (Barrow & Shaw, 1975a). This may be interpreted as an effect of temperature in increasing the rate of a reaction which had not yet reached equilibrium. This is consistent with the effects of high temperatures of incubation of soil plus P in decreasing its subsequent effectiveness for plant growth (Robinson, 1942; Beaton & Read, 1963; Engelstad & Allen, 1971; Barrow, 1974b).

The second line of evidence is based on the effects of concentration. The distribution of P between the solution and the adsorbed phase is usually described by a curve such that each successive increase in concentration results in a smaller increase in the amount adsorbed; that is, the adsorption isotherm is curved. In contrast there appears to be a linear relation between the concentration of P in the adsorbed phase and the amount changed into the firmly held form. This assumption was used to describe results for changes in plant response (Barrow & Campbell, 1972), in solution concentration (Barrow & Shaw, 1975a), in isotopically exchangeable P (Barrow & Shaw, 1975b), and in the P desorbed by $CaCl_2$ (Barrow & Shaw, 1975c) or by $NaHCO_3$ (Barrow & Shaw, 1976a).

B. Rates of the Slow Reactions

The slow reaction between soil and phosphate seems to be very slow indeed. In a pot trial, Devine et al. (1968) found that, after 1 year's contact, the effectiveness of powdered single superphosphate on four soils averaged 58% of that of fresh superphosphate (Fig. 2). After 2 years, it was only 38% as effective and after 3 years, 20%. These figures are of the same order as those found in several field trials even though some of the other factors listed in Section III would have also operated (Trumble & Donald, 1938; Arndt & McIntyre, 1963; Mattingly, 1971).

An important aspect of the rate of the reaction is the shape of the relation between time and the proportion of the phosphate remaining effective. If this can be specified, the relation can be described by quoting one or more numbers. Further, some indication of the long-term fate of the phosphate can be gained by extrapolation. There is some evidence that the rate can be described by a geometric progression with the amount remaining effective decreasing by a constant fraction each year. This may also be expressed by saying that the half-life or period for half-decay is constant. When this relation holds, a straight line is obtained when the logarithm of the proportion remaining effective is plotted against time. Such a graph describes the results from the first few crops of Arndt and McIntyre (1963; Fig. 2) and it also describes the results, averaged over four soils, of Devine et al. (1968). Such a relation was assumed by Larsen et al. (1965) to describe the change in L-value over a period of 5 years in a range of soils. However, as they state, their results were variable, and it seems difficult to justify any particular equation from them.

While a geometric progression may describe fairly well the rate of change of effectiveness over the first couple of years, there is reason to question whether it is appropriate over long periods. If it were, repeated application of P could not build up a large reserve. If a is an annual rate of application and r the proportion remaining effective at the end of 1 year, the maximum reserve which could be accumulated is $a/(1 - r)$. As r appears to be about 0.5–0.6 (see above), large reserves could not accumulate. Further, if applications then ceased, there would be a decline in effectiveness which

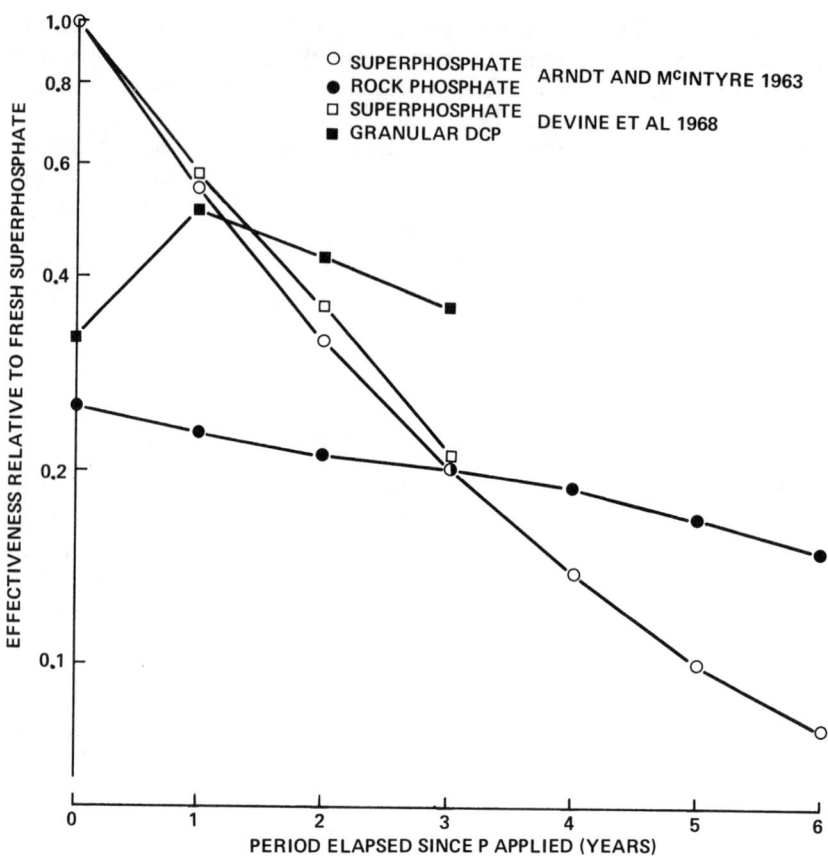

Fig. 2—Change in effectiveness of an application of 125 kg/ha of superphosphate or of rock phosphate for successive annual crops of sorghum at Katherine, Northern Territory, Australia, and changes in the effectiveness of superphosphate and of granular dicalcium phosphate (DCP) with time of contact averaged over four soils. In both cases, fresh superphosphate is scaled as unity.

would be just as rapid as for freshly applied P. These predictions do not seem to coincide with observations. Thus, Piper and de Vries (1964) found that when P had been applied at a fairly low rate to a P-deficient soil in South Australia over 23 years, wheat could be grown for 5 years before a response developed. Even after a period of 50 years had elapsed without fertilizer on the exhaustion land at Rothamsted, plots which had been fertilized over a prior period of 45 years yielded twice as much as previously unfertilized plots (Warren, 1956). Such long-term studies suggest that the decline in effectiveness ultimately becomes slower than would occur if a geometric progression persisted. This trend is visible in the results of Arndt and McIntyre (1963; Fig. 2), even though they removed all of the sorghum plants (including the roots in the top soil) from their plots so that there would have been appreciable losses of available P in addition to the continuing slow reaction.

The results of such long-term studies seem to be consistent with a study

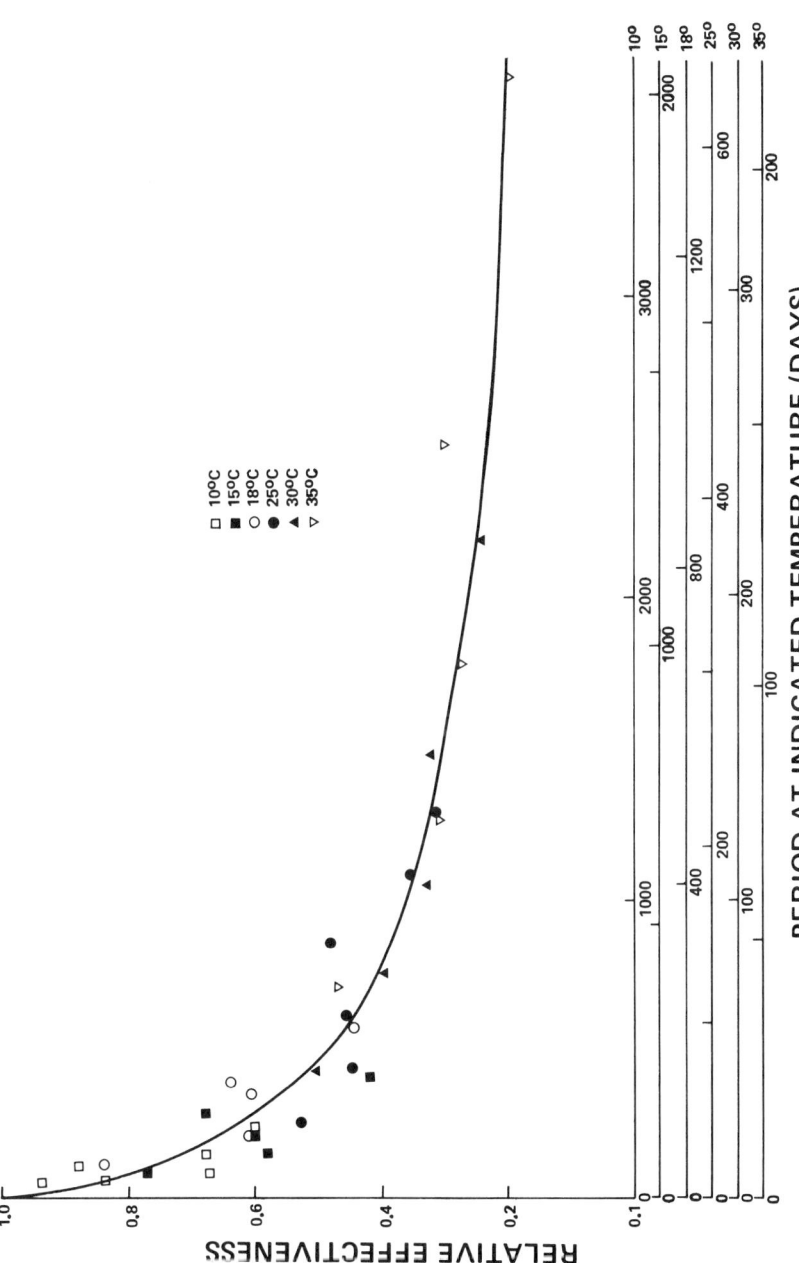

Fig. 3—Effect of period and temperature of incubation of soil plus phosphate on its subsequent effectiveness for plant growth relative to freshly applied phosphate (Barrow, 1974b). Results for the differing temperatures have been shown on different time scales using parameters of fitted equations to calculate equivalent periods.

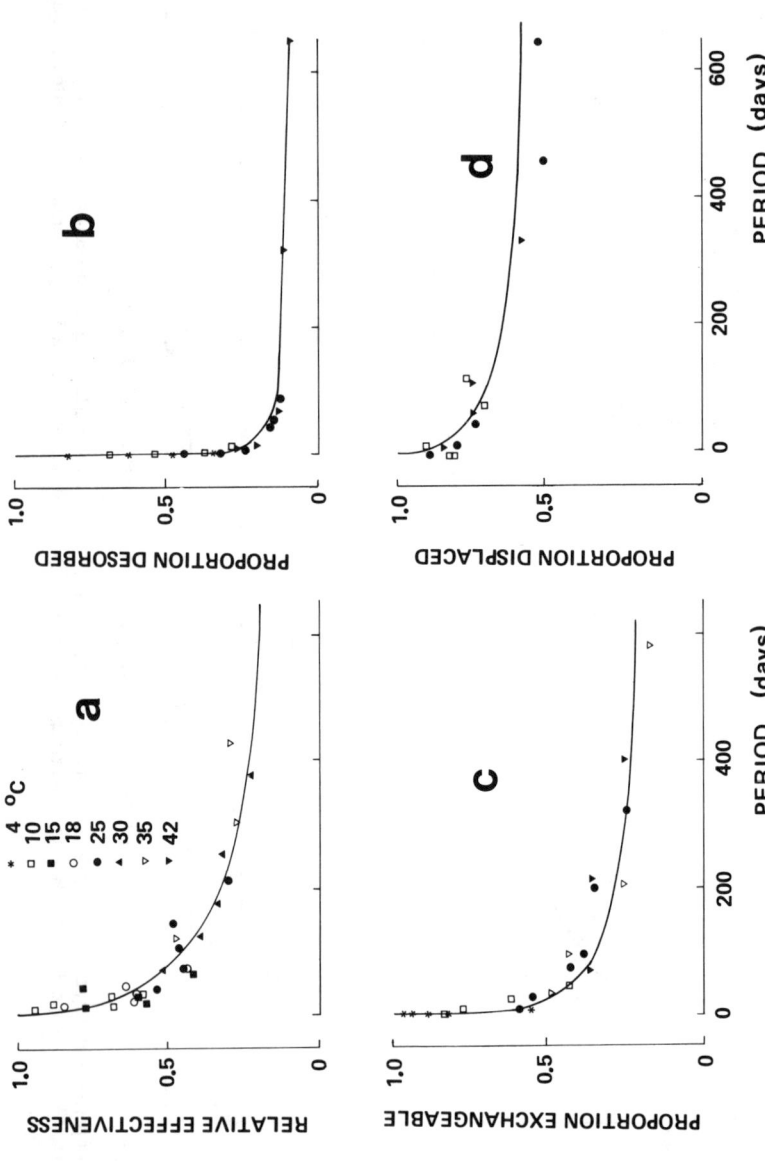

Fig. 4—Comparison of the effects of time and temperature of incubation of soil plus phosphate on: (a) relative effectiveness; (b) the proportion of added P which could be desorbed in 24 hours by 0.1M CaCl₂ solution at very wide solution/soil ratio (Barrow & Shaw, 1975a, 1975c); (c) the proportion which remained isotopically exchangeable in 24 hours (Barrow & Shaw, 1975b); and (d) the proportion which could be displaced in 16 hours by NaHCO₃ solution, after adjustment for secondary adsorption (Barrow & Shaw, 1976a). In each case the time scale has been adjusted to an equivalent period at 25°C.

in which advantage was taken of the increased rate of reaction between soil and phosphate at high temperatures to give a large range of effective periods (Barrow, 1974b; Barrow & Shaw, 1974). As Fig. 3 shows, the period required for the relative effectiveness to drop from 1 to 0.5 was shorter than the period required for the drop from 0.5 to 0.25. Thus a geometric progression did not apply. In addition, once effectiveness had declined to about a fifth of that of fresh phosphate, further decline appeared to be slow. These results were described by the equation

$$y = (1 + k\,t)^{-b} \qquad [1]$$

where y is the proportion of the phosphate remaining effective, t is the period elapsed, and k and b are coefficients. Equations of similar form were found to describe closely the changes in several other properties which would also reflect the slow reaction between soil and P (Fig. 4). Other equations have been proposed to describe the continuing reaction between phosphate and both hematite and gibbsite by Kuo and Lotse (1973) and between phosphate and lake sediments by Kuo and Lotse (1974). Equation [1] appears to be appropriate to describe the results presented by Enfield et al. (1976) for reaction between P and soils, but was not considered among the model equations used by them.

C. Differences Between Soils

One of the difficulties of generalizing from studies of residual value is uncertainty about the importance of differences among soils. Can a particular result be expected on other soils, or is it a consequence of some special property of the soil used? There have been surprisingly few studies of this important question. Larsen et al. (1965) studied the change in L-value on 24 sites spread through Britain and thus subject to variation in factors other than soil. The only soil property they found to be correlated with the rate of change was soil acidity. Decline was faster on soils of high pH; however, in a subsequent pot study on four soils which ranged in pH from 4.7 to 7.1, there was no consistent effect of pH (Devine et al., 1968). Changing the pH by liming appears to decrease the initial effectiveness of P fertilizers, but to also decrease the subsequent change (Terman et al., 1960). Differences in the relative effectiveness of previously applied P were observed in a group of 22 soils in a pot trial by Barrow (1973). The soils were selected to test the hypothesis that the relative effectiveness of P would decline more quickly in soils of high buffering capacity; however, the only effect observed was a slight and barely significant trend in the opposite direction. It seemed that, in this group of soils, differences in buffering capacity were caused by differences in the amount of adsorbing surface present rather than by differences in its nature, and subsequent changes followed a similar path in all soils.

Thus, there is at present no clear evidence of which soil properties, if any, are important. This may be partly because such studies tend to use local soils, and these are inevitably of limited diversity. Perhaps what is needed are studies in which soils from widely different regions are compared. A hypothesis which may be worth testing is that differences may be associated with the nature of the adsorbing surface present, such as, for example, differences in the proportion of Fe and Al.

D. Reversibility of the Slow Reactions

When soluble P is added to a soil, any pre-existing equilibrium will be disturbed, and there will be a movement of P to the adsorbed form and thence to the firmly held form. When P is removed from the soil solution, for example by plants, there will tend to be a movement in the opposite direction. To what extent will there be a movement from the firmly held to the adsorbed form; that is, to what extent are the slow reactions reversible? Since the forward reaction is slow, and any equilibrium is seemingly towards the firmly held form, the back reaction may also be slow. Hence, testing for full reversibility may involve periods of years. Leamer (1963) found that after 4 years of alfalfa and 1 of sorghum about two-thirds of the P applied at a rate of 235 kg/ha had been recovered in the crops. Subsequent crops responded to phosphate, but there was a slow increase in the proportion of the original fertilizer recovered up to almost 80% after 9 years. At lower levels of application, the recovery was greater and was virtually complete after 9 years. Such results suggest that, given time, the reaction is reversible though the last stages may be very slow. This conclusion presumes that the behavior observed in this soil could occur on other soils and, because of uncertainty about the differences between soils, must be treated with some caution.

In the laboratory, the reversibility of the slow reaction may be studied by reducing the concentration of P in the solution phase. This may be by increasing the volume of solution and thus obtaining a desorption isotherm or by introducing an anion exchange resin. Given that a time scale of years is needed in the field, it is not surprising that desorption measured in the laboratory over periods of hours or days is incomplete. Consequently desorption isotherms differ from adsorption isotherms as was observed, for example, by Kafkafi et al. (1967). The most extensive data for rate of desorption have been obtained in studies in which anion exchange resins were used. Since the resin induces a low concentration in the solution, it seems likely that release from the soil is the limiting factor rather than uptake by the resin. Generally the rate of desorption decreases with time, and for most published results, P adsorbed by resin is approximately proportional to $t^{0.3}$ (Amer et al., 1955; Moser et al., 1959; Elrashidi et al., 1975). Release is therefore described by a curve of gradually decreasing slope. The shape is such that after about 48 hours further release is slow, but seems to be still continuing. A similar result was obtained by Olsen (1975) for the desorption of phosphate in the presence of EDTA. He found that release was approximately proportional to the cube root of time.

The continuing release to resins and to EDTA differs from some results obtained when soil or phosphated oxides were equilibrated with finite volumes of solution (Vaidyanathan & Nye, 1970; Hingston et al., 1974). In these cases desorption was reported to have ceased within 24 hours. Such a result would indicate that an irreversible change in the adsorbed phosphate had occurred. However, Hingston et al. (1974) measured desorption after only a brief period of prior contact, and their results suggest an initial release of P followed by a continuing adsorption. This effect becomes less marked as the period of prior contact is increased (Munns & Fox, 1976). After a long period of prior contact, the results may be more consistent with resin studies (Barrow & Shaw, 1975c). In this work, it was found that at low volumes (and thus relatively high P concentrations in solution), desorption ceased within a few hours. As the volume was increased and P concentration thereby reduced, desorption continued for longer periods, until at "infinite dilution", it was proportional to $t^{0.3}$.

It seems likely that the rate of the back reaction would also increase with temperature, and thus could be studied more conveniently in the laboratory. Evidence for this has been obtained by Cooke and Hislop (1968) and by Evans and Jurinak (1976). In both cases, anion exchange resin was used, and therefore part of the effect might be on the resin rather than soil.

Both for experiments involving plant uptake of P and for laboratory studies of desorption, the property measured is the P released to the solution; this may be the resultant of two processes. It is not easy to calculate from such measurements whether the P released was from the adsorbed form or whether some was firmly held. There is no obvious point at which there is a change of rate of release; rather, there is a continuing fall in rate with continuing removal. Instead of attempting to measure the effects of a treatment on the amount of firmly held P, it might be more useful to measure the effects on the rate of desorption. For example, it has been shown that rate of desorption decreases with increasing period of prior contact (Barrow & Shaw, 1975c). The concomitant decrease in the effectiveness of P for plants might then be viewed as due to a decrease in the rate of desorption. Effects of rate of desorption have not been considered in most models of P uptake by roots, as it was not considered to be important in limiting uptake.

The rate of desorption of P may be increased if the conditions are modified. Soil tests which include another specifically adsorbed anion or which modify the pH, increase the rate of desorption and therefore extract more P in a given time than, say, water. Such observations have led to suggestions that plants might excrete anions, such as citrate or malate, and thus desorb phosphate near their roots. However, in a recent review, Tinker and Sanders (1975) calculated that the amounts required are likely to be higher than a plant could provide. Another argument is that such compounds would be good substrates for many soil organisms, and so would be unlikely to survive for long enough to be effective. There has been less emphasis on the effects of cations on desorption, but there is some evidence that these could be important. Vaidyanathan and Talibudeen (1970) found that rate of desorption was increased if a Na cation exchange resin was shaken with soils both with and without anion exchange resin. The effect increased as soil pH increased from 4 to 7—suggesting that it might be connected with replacement of soil Ca by Na from the resin, This would be consistent with the results of Lehr and Van Wesemael (1956), who found that uptake of P by wheat seedlings from soil was greater when N was supplied as $NaNO_3$ than as $Ca(NO_3)_2$.

E. Mechanism of the Slow Reaction

Many of the studies on the decline in effectiveness of P fertilizers with time have been concerned with the characteristics of the process, that is, with the outputs from the "black box." Perhaps, if we understood how the contents of the black box worked, we would be in a better position to extrapolate from our empirical studies. It seems likely that the slow reactions which follow adsorption are an important component of the black box. What can be said about their mechanism?

The activation energy appears to be about 84 kJ/mole (Barrow, 1974b; Barrow & Shaw, 1975a, 1975b, 1976a). Such a value would normally indicate a fairly fast reaction at soil temperatures. The slowness of the reaction might arise because it takes place on a surface, and it is difficult for a sufficiently energetic molecule to successfully rearrange its activation energy into the new degrees of freedom. It seems unlikely that the reaction involves vacating the original adsorption site and diffusion to a new site. One line of evidence for this is that the buffering capacity for further additions of phosphate is reduced (Barrow, 1974a), suggesting that the site is still occupied. Another line is that the proportion changed seems to be independent of level of addition (Section IV. A.), whereas the effective diffusion coefficient would increase with level of addition. The lack of effect of level of addition on the proportion changed also suggests that interaction of adsorbed phosphate molecules is not involved. In this case the reaction might be expected to follow first-order kinetics. It seems possible that this might be reconciled with the observed kinetics (Eq. [1]) by postulating that there is a suite of first-order reations, but with different reaction rates. Probert and Larsen (1972) have discussed implications of Eq. [1] and its relation to a series of first-order reaction terms, and Barrow and Shaw (1975b) have presented a general equation for such a series.

A mechanism which is often invoked to explain the fast and slow reactions is that the initial adsorption reaction is through one of the oxygen atoms of the phosphate and the second reaction involves a link through a second oxygen atom (Fig. 1), thus giving a ring structure. This mechanism was proposed by Kafkafi et al. (1967) to explain their inability to desorb P from soil by washing. It was also used by Atkinson et al. (1972) to explain the slow isotopic exchange of adsorbed P. There is now evidence from infrared absorption that this structure is formed when phosphate reacts with Fe oxides (Parfitt et al., 1975), though seemingly it was formed fairly quickly. Further evidence for this mechanism comes from the work of Hingston et al. (1974). They showed that phosphate could not be readily desorbed from goethite, but fluoride could, and suggested that this was because fluoride was unable to form a second link. More recent work, however, raises doubts that formation of a second link is indeed the mechanism of the slow reaction. Barrow and Shaw (1978) found that fluoride reacted with a range of soils in much the same way as phosphate—with increasing time and temperature, its concentration in the solution decreased; it was desorbed more slowly; and the capacity of the soil to adsorb either further fluoride or phosphate was reduced. They suggested a mechanism represented by the lower reaction in Fig. 1 to account for the similarity between the behavior of fluoride and phosphate. The negative charge induced by the adsorption of the anion moves from the adsorption site to an adjacent site by migration of protons. Further movement to nearby sites could give rise to the postulated suite of reaction rates. Similar movement to nearby sites could also occur on sites in which the phosphate formed a second link, though this is not shown in Fig. 1. Since the surface retains the negative charge, further adsorption is reduced. Desorption of P from the neutral sites so formed may be by reaction with hydroxyls, as in the top reaction of Fig. 1, or by reaction with other specifically adsorbed anions, as in some soil tests. It may also be by reverse migration of charge thus giving rise to the observed slow, but continuing desorption. The migration of charge might be affected by the closeness of approach of the cation which balances the negative charge. This varies between cations (Shainberg & Kemper, 1966), and this might explain the observed effects of cations (Section IV. D.). An alternative explanation for the role of cations has recently been published by Helyar et al. (1976a, 1976b). They suggest that Ca, among the common soil cations, has a specific effect because it forms a complex with the adsorbed phosphate molecules.

V. ACCESS OF PLANTS TO RESIDUAL P

A. Possible Mechanisms

The question of whether some plants have better access to residual P than others is obviously important on land on which much P has been applied in previous years. Before considering this question, it is important to specify it more carefully. It is not, do plant roots differ in the efficiency with which they take up P from soil–there is much evidence that they do (see Chapt. 20 and 21 of this book). Rather the question is, are there plants with any special mecahnisms for reversing the processes which lead to a decline in effectiveness with time and which therefore have a disproportionately good access to residual P? Possible mechanisms are considered next.

In areas in which leaching of P is an important cause of loss of effectiveness, perennial plants would be expected to have an advantage over annual plants, as they would have a developed root system early in the season when the P was applied. Further, deep-rooted plants would have an advantage over shallow-rooted plants. Where accumulation of organic P is a cause of loss of effectiveness, plants which could hydrolyze some of the organic P may have an advantage. Plants differ in their production of exocellular phosphatases (McLachlan, 1976), but the importance of these differences is uncertain. In McLachlan's experiments, phosphatase production was inversely related to the ability of the plants to obtain phosphate from soil. If plants differ in the extent to which they are benefited by an initial heterogeneous distribution (Section III. D.), it follows that there will be differences in the extent to which they are disadvantaged by the loss of this heterogeneity. These possibilities, however, seem unimportant when compared to the benefits which would accrue if a plant could gain access to some of the firmly held P. One mechanism which might help it do this is to have uptake zones which remained active over long periods and thus gain access to the slowly released phosphate. Desorption varies, however, with time according to the function $K\ t^{0.3}$ where t is the period of desorption (Section IV. D.). Increasing the period of contact between soil and P reduces the value of K. Increasing the period of desorption would not give any special advantage on soils on which K was low compared with soils on which K was high; the ratio of amounts desorbed would remain the same. And in both cases, desorption appears to become very slow after about 2 days, so this does not appear to be a very effective strategy. An alternative strategy would be to modify the root environment so that the reduction in K with period of contact was not as marked. This might be by excreting organic acids (Vancura, 1964). But some workers have not detected organic acids (McLachlan, 1976), others have found them unimportant (Johnston & Olsen, 1972), and still others have asked whether sufficient acid could be excreted to have much effect (Tinker & Sanders, 1975). Plants may also modify the pH near their roots as a result of differences in the balance of cations and anions taken up. McLachlan (1976) showed that buckwheat had a high uptake of Ca and tended to decrease the pH near its roots. The changes in pH might be of importance of themselves, but a high uptake of Ca has also been considered important in obtaining P from sparingly soluble Ca phosphates (Drake & Steckel, 1955; Johnston & Olsen, 1972). It has been suggested (Section IV. D.) that depletion of Ca near a soil surface might increase desorption, so a high uptake of Ca might of itself help a plant gain access to firmly held P. In many soils, however, the uptake of water by plants carries Ca to the surface of the root faster than it is absorbed (Barber et al., 1963), and Ca accumulates near the root. One of the roles of mycorrhyzae may be to extend the zone of P uptake to outside this zone of Ca accumulation. However, current evidence is that vesicular arbuscular mycorrhizae do not give the host plant any special access to firmly held phosphate (Mosse, 1973).

Many plants have long, thin roots which, together with mycorrhizas, seem well adapted to gather P from the soil by diffusion. This anatomy brings a large proportion of the soil P within diffusive range of the root.

However, it would also permit any modifying agents produced by the root to diffuse away from the root without building up to a high concentration. In contrast, some plants have special structures, such as the multiple-branched, "proteoid" roots (Lamont, 1972; Trinick, 1977) or the similar "dauciform" roots of sedges (Lamont, 1978). These structures do not seem well-adapted to gather P by diffusion. It has been suggested (W. K. Gardner, personal comm.) that they are, instead, well-adapted to decrease diffusion of modifying agents away from the root. Further investigation of these structures from this viewpoint could be fruitful.

B. Experimental Approach

Several approaches to this problem have been used. One is to mix ^{32}P through soil and test whether plants differ in the specific activity of their P. The argument is that plants which have access to additional pools of P would have a lower specific activity. There are experimental difficulties due to differences in seed P, but after allowing this, Nye and Foster (1958) found no difference in the specific activity of a range of crop plants. Of course, such experiments are inevitably confined in the range of plants tested and it remains possible that had different plants been tested, differences would have been found.

Another approach which may be relevant is a comparison of the ability of certain species to use sparingly soluble compounds. There is evidence, for example, that plants differ in their ability to obtain P from rock phosphate (Fried, 1953; Drake & Steckel, 1955). A factor in such an ability may be a plant's capability to provide a good sink for Ca and thus increase the rate of solution. Such observations of good uptake from sources of low solubility, however, do not necessarily show that the plants concerned will have a disproportionate advantage in using residual phosphate.

A third approach is a direct comparison of the availability of fresh and residual P using response curves. Barrow and Campbell (1972) used this technique to compare the residual value to clover with that to ryegrass and found no difference. Although ryegrass is more efficient at obtaining phosphate than clover, the decline in relative effectiveness with period of prior contact was similar for both species. A practical difficulty with this approach is the need to plan experiments well in advance in order to give the residual P treatment time to react with the soil. In their experiments on residual P, Devine et al. (1968) used periods of prior contact of up to 3 years. This problem can be circumvented by using temperatures of 70°C to increase the rate of the reaction. Appreciable decrease in effectiveness then occurs within a few days (Barrow et al., 1977). A more important difficulty is that this approach is limited to comparisons among plants with similar growth rates. This is because, during the period in which the plants are grown, the availability of both the freshly applied and the residual P continues to decline, but the change in the fresh P is faster (Section IV. B.). Hence, the longer the period of growth, the smaller the difference will become. In some cases the result may cause what appears to be improved access to residual P with increasing period of growth (Barrow & Campbell, 1972), whereas it may largely be a decreased access to fresh P. Because of these effects it may not be possible to use this technique to compare the residual value of P to plants of widely differing growth rates.

C. Current Status

The foregoing indicates that there is little information on whether there are any plants with a disproportionate access to firmly held P. If such plants could be discovered, and their P uptake mechanism understood or utilized, the world's reserves of phosphate could be increased greatly. However, most evidence suggests that plants have a proportionate access. This means that plants which are capable of growing at low levels of freshly applied phosphate (Chapt. 20 of this book) would also be capable of growing at low levels of firmly held P. Such plants could provide a means of using both freshly applied and residual P more effectively.

VI. EVALUATION OF FERTILIZER RESIDUES

A. Sources of Phosphate

The initial effectiveness of phosphate fertilizers is strongly affected by their solubility and granule size (Terman et al., 1960). Especially for sparingly soluble fertilizers, such as dicalcium phosphate, effectiveness can be reduced by granulation (Terman et al., 1960; Devine et al., 1968; Mattingly et al., 1971). This may be interpreted as due to a decrease in the amount going into solution. Similarily, liming may induce a decrease in the initial effectiveness (Terman et al., 1960). This may be due to adsorption and precipitation on $CaCO_3$ (Section III. E.) and thus to a decrease in the amount remaining in solution. Published results appear to be consistent with the idea that the effectiveness of all soluble P fertilizers decreases with time in a similar manner so that high initial effectiveness is followed by a markedly decreased effectiveness. This also seems to apply to condensed phosphates: polyphosphates (Miner & Kamprath, 1971), pyrophosphates (Englestad & Allen, 1971), and powdered metaphosphates (Mattingly et al., 1971) decrease in effectiveness in much the same way as orthophosphates. With less soluble fertilizer, however, the initial effectiveness is lower, but the decline is less marked. Devine et al. (1968) found that the effectiveness of granular dicalcium phosphate was greater 1 year after application than when freshly applied, and even after 3 years it was still slightly better (Fig. 2). Similarly, rock phosphate, though initially less effective than superphosphate, declined in effectiveness more slowly (Arndt & McIntyre, 1963; Fig. 2). And the decrease in initial effectiveness as a result of liming (Terman et al., 1960) was followed by a less marked decrease in effectiveness after 6 months contact.

B. Synthesis into a Model of P Status

As summarized so far in this article, the literature on residual value seems to have two contrasting strands. In one, there is emphasis on the decline in effectiveness over the first few months, or years, after application and, at least by implication, on the need for repeated applications. In the other strand there is emphasis on the continuing uptake of P for several years after application and on the long-term recovery of added P. Russell (1973) has attempted to resolve this dichotomy in terms of phosphate pools. Provided enough P has been added to raise the level of the pool appreciably, there will be a long-continued residual effect. Further, the reason for the initially greater effectiveness of soluble P fertilizers is that they give rise to several pools in which the level is high, but this level drops with time as the P equilibrates with the general soil pool. The present article differs from this interpretation in two ways. One is that, provided residual value is measured in terms of relative effectiveness rather than the period for which an effect lasts (Section II), residual value seems to be little affected by the level of application (Section IV; Arndt & McIntyre, 1963). The other is that the decline in effectiveness with time of contact does not appear to depend on having an initially heterogeneous distribution (Section III. E.), and hence the idea of several, small, full pools does not seem appropriate. The thesis developed in this article is that the dichotomy can be resolved better as outlined in Section IV. B.—that the initial rate of reaction is not maintained. In order to illustrate this, the data quoted have been built into a simple model to calculate the changes in phosphate status under several fertilizer regimes (Fig. 5).

The model assumes that the changes in phosphate status are as described by Eq. [1]. The value of the k parameter of the equation was chosen to simulate the first 2 years' results of Devine et al. (1968), and the value of the b parameter was chosen as close to that obtained by Barrow (1974b). The level of applied P is specified in arbitrary units and the level of P required to overcome deficiency is not specified. These would be affected by the buffering capacity of the soil for phosphate—the higher the buffering capacity, the more P required (Ozanne & Shaw, 1968). Phosphorus applied and P required would also be affected by other factors, such as the crop grown. It is assumed that 0.1 units of P are removed each year in agricultural produce when the total available P is greater than one unit and 10% of the available P when the total is less than one unit. The proportion removed would, in practice, vary inversely with the buffering capacity of the soil. The output of the model shows that regular annual applications of P result in a steady increase in P status. There is, however, always a decrease in status between one year and the next. In absolute amounts, this increases with time. Equilibrium would be reached when this decrease was equal to the annual rate of addition. This decrease may not be reflected in yield if the status exceeds requirement. If annual applications cease, the fall-off is fairly quick if only a few applications have been made, but after, say, 10 years,

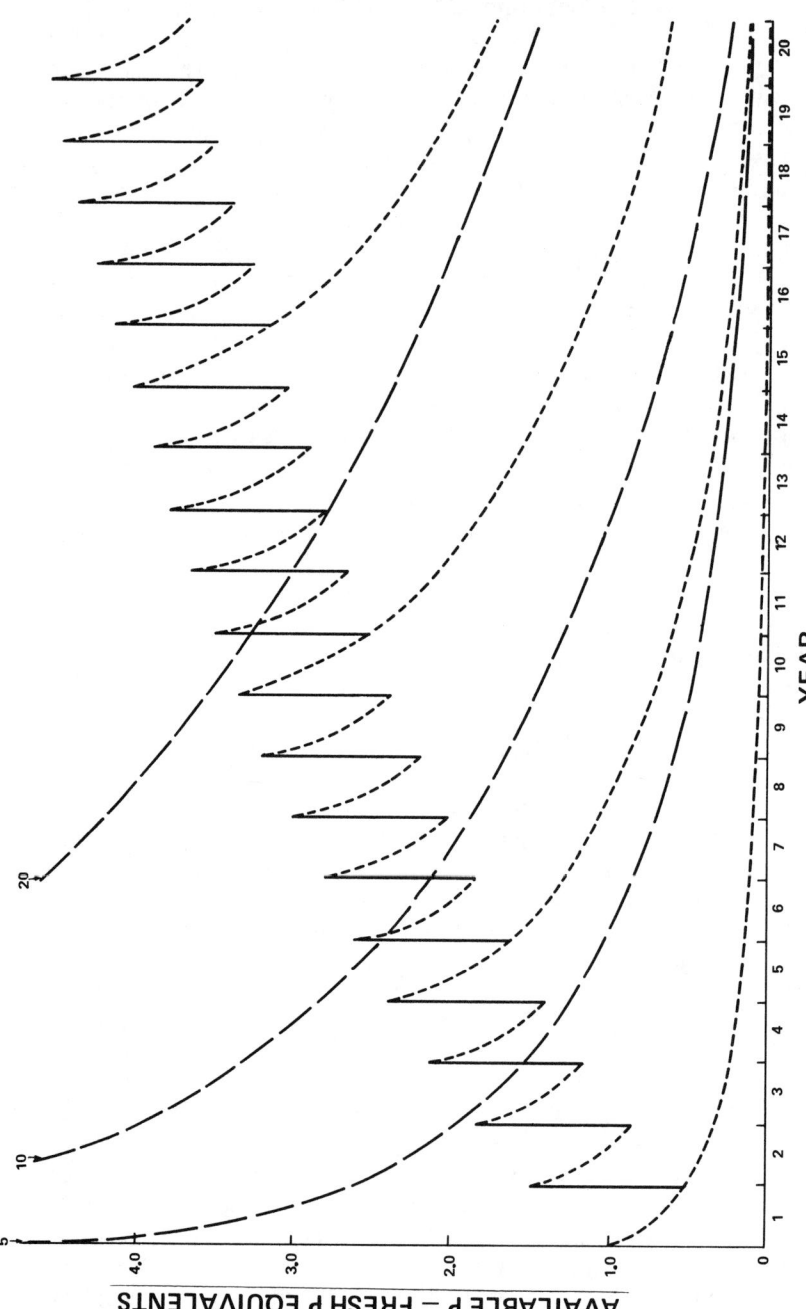

Fig. 5—Output of a model which calculates changes in available P after given fertilizer treatments. Inputs to the model are either annual applications of one arbitrary unit of P, or initial applications of the indicated arbitrary units. The broken lines indicate subsequent decreases in available P. The assumptions of the model are specified in greater detail in the text.

a high status is maintained for several years. A high status can also be achieved and maintained for several years by one heavy application as was observed by Younge and Plucknett (1966) and by Kamprath (1967). The model is thus consistent with the observed behaviour of phosphate fertilizers even though it has deliberately been kept fairly simple. More sophisticated models have been described by Russell (1977) and by Barrow and Carter (1978). These models include removal of P in produce and accumulation of P in organic matter and were fitted to field data.

ACKNOWLEDGEMENTS

I would like to thank Dr. G. E. G. Mattingly of Rothamsted Research Station for helping to plan this article and for making many helpful suggestions for revision of an early draft.

LITERATURE CITED

Allen, S. E., R. J. Spear, and M. Maloney. 1954. Utilization of phosphate as influenced by plant species and by placement and by time of application. Soil Sci. 77:67–73.

Amer, F., D. R. Bouldin, C. A. Black, and F. R. Duke. 1955. Characterization of soil phosphorus by anion exchangeable resin adsorption and ^{32}P equilibration. Plant Soil 6:391–408.

Arndt, W., and G. A. McIntyre. 1963. The initial and residual effects of superphosphate and rock phosphate for sorghum on a lateritic red earth. Aust. J. Agric. Res. 14:785–795.

Asher, C. J., and J. F. Loneragan. 1967. Response of plants to phosphate concentration in solution culture. II. Rate of phosphate absorption and growth. Soil Sci. 103:311–318.

Atkinson, R. J., A. M. Posner, and J. P. Quirk. 1972. Kinetics of isotopic exchange of phosphate at the α-FeOOH—aqueous solution interface. J. Inorg. Nucl. Chem. 34:2202–2211.

Barber, S. A., J. M. Walker, and E. H. Vasey. 1963. Mechanisms for the movement of plant nutrients from the soil and fertilizer to the plant root. J. Agric. Food Chem. 11:204–207.

Barrow, N. J. 1969. The accumulation of soil organic matter under pasture and its effects on soil properties. Aust. J. Exp. Agric. Anim. Husb. 9:437–444.

Barrow, N. J. 1973. Relationship between a soil's ability to adsorb phosphate and the residual effectiveness of superphosphate. Aust. J. Soil Res. 11:57–63.

Barrow, N. J. 1974a. Effect of previous additions of phosphate on phosphate adsorption by soils. Soil Sci. 118:82–89.

Barrow, N. J. 1974b. The slow reactions between soil and anions. 1. Effect of time, temperature and water content of a soil on the decrease in effectiveness of phosphate for plant growth. Soil Sci. 118:380–386.

Barrow, N. J. 1975. The response to phosphate of two annual pasture species. II. The specific rate of uptake of phosphate its distribution and use for growth. Aust. J. Agric. Res. 26:145–156.

Barrow, N. J., and N. A. Campbell. 1972. Methods of measuring residual value of fertilizers. Aust. J. Exp. Agric. Anim. Husb. 12:502–510.

Barrow, N. J., and E. D. Carter. 1978. A modified model for evaluating residual phosphate in soil. Aust. J. Agric. Res. 29:1011–1021.

Barrow, N. J., and T. C. Shaw. 1974. Factors affecting the long-term effectiveness of phosphate and molybdate fertilizers. Commun. Soil Sci. Plant Anal. 5:355–364.

Barrow, N. J., and T. C. Shaw. 1975a. The slow reactions between soil and anions. 2. Effect of time and temperature on the decrease in phosphate concentration in the soil solution. Soil Sci. 119:167–177.

Barrow, N. J., and T. C. Shaw. 1976b. The slow reactions between soil and anions. 2. The effects of time and temperature on the decrease in isotopically exchangeable phosphate. Soil Sci. 119:190-197.

Barrow, N. J., and T. C. Shaw. 1975c. The slow reactions between soil and anions. 5. Effects of period of prior contact on the desorption of phosphate from soils. Soil Sci. 119:311-320.

Barrow, N. J., and T. C. Shaw. 1976a. Sodium bicarbonate as an extractant for soil phosphate 1. Separation of the factors affecting the amount of phosphate displaced from soil from those affecting secondary adsorption. Geoderma 16:91-107.

Barrow, N. J., and T. C. Shaw. 1976b. Sodium bicarbonate as an extractant for soil phosphate. 3. Effects of the buffering capacity of a soil for phosphate. Geoderma 16:273-283.

Barrow, N. J., and T. C. Shaw. 1978. The slow reaction between soil and anions. 6. Effect of time and temperature of contact on fluoride. Soil Sci. 124:265-278.

Barrow, N. J., N. Malajczuk, and T. C. Shaw. 1977. A direct test of the ability of vesicular-arbuscular mycorrhiza to help plants take up fixed soil phosphate. New Phytol. 78:269-276.

Beaton, J. D., and D. W. L. Read. 1963. Effects of temperature and moisture on phosphorus uptake from a calcareous Saskatchewan soil treated with several pelleted sources of phosphorus. Soil Sci. Soc. Am. Proc. 27:61-65.

Black, C. A., and C. O. Scott. 1956. Fertilizer evaluation. I. Fundamental principles. Soil Sci. Soc. Am. Proc. 20:176-179.

Chaiwanakupt. P., and W. K. Robertson. 1976. Leaching of phosphate and selected cations from sandy soils as affected by lime. Agron. J. 68:507-511.

Cooke, I. J., and J. Hislop. 1963. Use of anion-exchange resin for the assessment of available soil phosphate. Soil Sci. 96:308-312.

Devine, J. R., D. Gunary, and S. Larsen. 1968. Availability of phosphate as affected by duration of fertilizer contact with soil. J. Agric. Sci. 71:359-364.

Donald, C. M., and C. H. Williams. 1954. Fertility and productivity of a podzolic soil as influenced by subterranean clover (*Trifolium subterraneum* L.) and superphosphate. Aust. J. Agric. Res. 5:664-687.

Drake, M., and J. E. Steckel. 1955. Solubilization of soil and rock phosphate as related to root cation exchange capacity. Soil Sci. Soc. Am. Proc. 19:449-450.

Elrashidi, M. A., A. Van Diest, and A. M. El Damaty. 1975. Phosphorus determination in highly calcareous soils by the use of an anion exchange resin. Plant Soil 42:273-286.

Enfield, C. G., C. C. Harlin, and B. E. Bledsoe. 1976. Comparison of five kinetic models for orthophosphate reactions in mineral soils. Soil Sci. Soc. Am. J. 40:243-248.

Engelstad, O. P., and S. E. Allen. 1971. Ammonium pyrophosphate and ammonium orthophosphate as phosphorus sources: effects of soil temperature, placement and incubation. Soil Sci. Soc. Am. Proc. 35:1002-1004.

Evans, R. L., and J. J. Jurinak. 1976. Kinetics of phosphate released from a desert soil. Soil Sci. 121:205-211.

Fitter, A. M. 1974. A relationship between phosphorus requirement, the immobilization of added phosphate, and the phosphate buffering capacity of colliery shales. J. Soil Sci. 25:41-50.

Fox, R. L., and E. J. Kamprath. 1970. Phoshate sorption isotherms for evaluating the phosphate requirements of soil. Soil Sci. Soc. Am. Proc. 34:902-907.

Fried, M. 1953. The feeding power of plants for phosphates. Soil Sci. Soc. Am. Proc. 17:357-359.

Gifford, R. M., J. D. Kalma, A. R. Aston, and R. J. Millington. 1975. Biophysical constraints in Australian food production. Implications for population policy. Search 6:212-223.

Hayward, D. O., and B. M. W. Trapnell. 1964. Chemisorption. 2nd ed. Butterworths, London.

Helyar, K. R., D. N. Munns, and R. G. Burau. 1976a. Adsorption of phosphate by gibbsite. I. Effects of neutral chloride salts of calcium, magnesium, sodium and potassium. J. Soil Sci. 27:307-314.

Helyar, K. R., D. N. Munns, and R. G. Burau. 1976b. Adsorption of phosphate by gibbsite. II. Formation of a surface complex involving divalent cations. J. Soil Sci. 27:315-323.

Hemwall, J. B. 1957. The fixation of phosphorus by soils. Adv. Agron. 9:95-112.

Hilder, E. J. 1964. The distribution of plant nutrients by sheep at pasture. Proc. Aust. Soc. Anim. Prod. 5:241-248.

Hingston, F. J., A. M. Posner, and J. P. Quirk. 1974. Anion adsorption by goethite and gibbsite. II. Desorption of anions from hydrous oxide surfaces. J. Soil Sci. 25:16-26.

Johnston, W. B., and R. A. Olsen. 1972. Dissolution of fluorapatite by plant roots. Soil Sci. 114:29-36.

Kafkafi, U., A. M. Posner, and J. P. Quirk. 1967. The desorption of phosphate from kaolinite. Soil Sci. Soc. Am. Proc. 31:348-353.

Kamprath, E. J. 1967. Residual effect of large applications of phosphorus on high phosphorus fixing soils. Agron. J. 59:25-27.

Keay, J., E. F. Biddiscombe, and P. G. Ozanne. 1970. The comparative rates of phosphate absorption by eight annual pasture species. Aust. J. Agric. Res. 21:33-44.

Khasawneh, F. E., and J. P. Copeland. 1973. Cotton root growth and uptake of nutrients: relation of phosphorus uptake to quantity, intensity, and buffering capacity. Soil Sci. Soc. Am. Proc. 37:250-254.

Kuo, S., and E. G. Lotse. 1973. Kinetics of phosphate adsorption and desorption by hematite and gibbsite. Soil Sci. 116:400-406.

Kuo, S., and E. G. Lotse. 1974. Kinetics of phosphate adsorption and desorption by lake sediments. Soil Sci. Soc. Am. Proc. 38:50-54.

Lamont, B. 1972. The morphology and anatomy of proteoid roots in the genus Hakea. Aust. J. Bot. 20:155-174.

Lamont, B. 1978. The root systems of sedges. Aust. Plants 9:259-261.

Larsen, S. 1967. Soil phosphorus. Adv. Agron. 19:151-210.

Larsen, S. 1974. Food. Neth. J. Agric. Sci. 22:270-274.

Larsen, S., and D. Gunary. 1964. The determination of labile soil phosphate as influenced by the time of application of labelled phosphate. Plant Soil 20:135-142.

Larsen, S., D. Gunnary, and C. D. Sutton. 1965. The rate of immobilization of applied phosphate in relation to soil properties. J. Soil Sci. 16:141-148.

Leamer, R. W. 1963. Residual effects of phosphorus fertilizer in an irrigated rotation in the south west. Soil Sci. Soc. Am. Proc. 27:65-68.

Lehr, J. J., and J. C. Van Wesemael. 1956. Variations in the uptake by plants of soil phosphate as influenced by sodium nitrate and calcium nitrate. J. Soil Sci. 7:148-155.

Lutz, J. F., and I. Haque. 1975. Effects of phosphorus on some physical and chemical properties of clays. Soil Sci. Soc. Am. Proc. 39:33-36.

Massey, D. L., R. W. Sheard, and M. H. Miller. 1970. Availability of reaction products of fertilizer phosphorus to alfalfa and bromegrass seedlings. Can. J. Soil Sci. 50:141-149.

Mattingly, G. E. G. 1968. Evaluation of phosphate fertilizers. II. Residual value of nitrophosphates, Gafsa rock phosphate, basic slag and potassium metaphosphate for potatoes, barley and swedes grown in rotation, with special reference to changes in soil phosphorus. J. Agric. Sci. 70:139-153.

Mattingly, G. E. G. 1970. Residual value of basic slag, Gafsa rock phosphate and superphosphate in a sandy podzol. J. Agric. Sci. 75:413-418.

Mattingly, G. E. G. 1971. Residual value of phosphate fertilizer on neutral and calcareous ground. p. 1-9. In Residual value of applied nutrients. Tech. Bull. 20, Ministry of Agriculture, Fisheries and Food, HMSO, London.

Mattingly, G. E. G. 1975. Labile phosphate in soils. Soil Sci. 119:369-375.

Mattingly, G. E. G., and A. Pinkerton. 1961. Some relationships between isotopically exchangeable phosphate, soil analysis and crop growth in the greenhouse. J. Sci. Food Agric. 12:772-777.

Mattingly, G. E. G., and F. W. Widdowson. 1963. Residual value of superphosphate and rock phosphate on an acid soil. 1. Yields and phosphorus uptakes in the field. J. Agric. Sci. 60:399-407.

Mattingly, G. E. G., Margaret Chater, and A. E. Johnston. 1974. Experiments made on Stackyard field, Woburn, 1876-1974. III. Effect of NPK fertilizers and farmyard manure on soil carbon, nitrogen and organic phosphorus. Rothamsted Exp. Stn. Rep. for 1974, Part 2:61-77.

Mattingly, G. E. G., A. Penny, and Marie Blakemore. 1971. Evaluation of phosphate fertilizers. III. Immediate and residual values of potassium metaphosphate and magnesium ammonium phosphate for potatoes, radishes, barley and ryegrass. J. Agric. Sci. 76:131-141.

McLachlan, K. D. 1976. Comparative phosphorus responses in plants to a range of available phosphorus situations. Aust. J. Agric. Res. 27:323-341.

Miner, G., and E. Kamprath. 1971. Reactions and availability of banded polyphosphate in field studies. Soil Sci. Soc. Am. Proc. 35:927-930.

Moser, U. S., W. H. Sutherland, and C. A. Black. 1959. Evaluation of laboratory indexes of absorption of soil phosphorus by plants. Plant Soil 10:356-374.

Mosse, B. 1973. Advances in the study of vesicular-arbuscular mycorrhiza. Annu. Rev. Phytopathol. 11:171-196.
Mott, C. J. B. 1970. Sorption of anions by soils. Society Chemistry and Industry, London. Monograph no. 37:40-52.
Munns, D. N., and R. L. Fox. 1976. The slow reaction which continues after phosphate adsorption: kinetics and equilibrium in some tropical soils. Soil Sci. Soc. Amer. J. 40:46-51.
Neller, J. R., D. W. Jones, N. Gammon, and R. B. Forbes. 1951. Leaching of fertilizer phosphorus in acid sandy soil as affected by lime. Circ. Fla. Univ. Agric. Exp. Stn. No. S-32.
Nye, P. H., and W. N. M. Foster. 1958. A study of the mechanism of soil-phosphate uptake in relation to plant species. Plant Soil 11:338-352.
Olsen, R. A. 1975. Rate of dissolution of phosphate from minerals and soils. Soil Sci. Soc. Am. Proc. 39:634-639.
Oniani, O. G., Margaret Chater, and G. E. G. Mattingly. 1973. Some effects of fertilizers and farmyard manure on the organic phosphorus in soils. J. Soil Sci. 24:1-9.
Ozanne, P. G., and T. C. Shaw. 1961. The loss of phosphorus from sandy soils. Aust. J. Agric. Res. 12:409-423.
Ozanne, P. G., and T. C. Shaw. 1968. Advantages of the recently developed phosphate sorption test over the older extractant methods for soil phosphate. Int. Congr. Soil Sci., Adelaide, Trans. 9th 2:273-280.
Parfitt, R. L., R. J. Atkinson, and R. St. C. Smart. 1975. The mechanism of phosphate fixation by iron oxides. Soil Sci. Soc. Am. Proc. 39:837-841.
Paton, D. F., and J. F. Loneragan. 1960. An effect of lime on residual phosphorus in soil. Aust. J. Agric. Res. 11:524-529.
Piper, C. S., and M. P. C. deVries. 1964. The residual value of superphosphate on a red-brown earth in South Australia. Aust. J. Agric. Res. 15:234-272.
Power, J. F., D. L. Grunes, G. A. Reichman, and W. O. Willis. 1964. Soil temperature effects on phosphorus availability. Agron. J. 56:545-548.
Probert, M. E., and S. Larsen. 1972. The kinetics of heterogenous isotopic exchange. J. Soil Sci. 23:76-81.
Robinson, R. R. 1942. Phosphorus fixation as affected by soil temperature. J. Am. Soc. Agron. 34:301-306.
Russell, E. W. 1973. Soil conditions and plant growth. 10th ed. Longman, London. p. 595.
Russell, J. S. 1960a. Soil fertilizer changes in the long-term experimental plots at Kybybolite, South Australia. I. Changes in pH, total nitrogen, organic carbon and bulk density. Aust. J. Agric. Res. 11:902-926.
Russell, J. S. 1960b. Soil fertility changes in the long-term experimental plots at Kybybolite, South Australia. II. Changes in phosphorus. Aust. J. Agric. Res. 11:927-947.
Russell, J. S. 1977. Evaluation of residual nutrients in soils. Aust. J. Agric. Res. 28:461-475.
Schuman, G. E., R. G. Spomer, and R. F. Piest. 1973. Phosphorus losses from four agricultural watersheds in Missouri valley loess. Soil Sci. Soc. Am. Proc. 37:424-427.
Shainberg, I., and W. D. Kemper. 1966. Hydration status of adsorbed cations. Soil Sci. Soc. Am. Proc. 30:707-713.
Spector, W. S. 1956. Handbook of biological data. W. B. Saunders Co., Philadelphia and London.
Strong, W. M., and R. J. Soper. 1973. Utilization of pelleted phosphorus by flax, wheat, rape and buckwheat from a calcareous soil. Agron. J. 65:18-21.
Terman, G. L., J. D. Dement, L. B. Clements, and J. A. Lutz. 1960. Crop response to ammoniated superphosphates and dicalcium phosphate as affected by granule size, water solubility, and time of reaction with the soil. J. Agric. Food Chem. 8:13-18.
Tinker, P. B. H., and F. E. Sanders. 1975. Rhizosphere microorganisms and plants nutrition. Soil Sci. 119:363-368.
Trinick, M. J. 1977. Vesicular-arbuscular infection and soil phosphorus utilisation in Lupinus spp. New Phytol. 28:297-304.
Trumble, H. C., and C. M. Donald. 1938. The relation of phosphate to the development of seeded pasture on a podzolised sand. CSIR, Aust. Bull. 116, Melbourne.
Vaidyanathan, L. V., and P. H. Nye. 1970. The measurement and mechanism of ion diffusion in soils. IV. The effect of concentration and moisture content in the counter-diffusion of soil phosphate against chloride ion. J. Soil Sci. 21:15-26.
Vaidyanathan, L. V., and O. Talibudeen. 1970. Rate processes in the desorption of phosphate from soils by ion-exchange resins. J. Soil Sci. 21:173-183.

Vancura, V. 1964. Root exudates of plants. I. Analysis of root exudates of barley and wheat in their initial phases of growth. Plant Soil 21:231-248.

Warren, R. G. 1956. N. P. K. residues from fertilizers and farmyard manure, in long-term experiments at Rothamsted. Proc. Fert. Soc. (London) 37:3-24.

Wiklander, L. 1950. Kinetics of phosphate exchange in soils. Ann. R. Agric. Coll. (Sweden) 17:407-423.

Williams, C. H. 1971. Reaction of surface-applied superphosphate with soil. II. Movement of the phosphorus and sulfur into the soil. Aust. J. Soil Res. 9:95-106.

Williams, C. H., and J. R. Simpson. 1965. Some effects of cultivation and waterlogging on the availability of phosphorus in pasture soils. Aust. J. Agric. Res. 16:413-427.

Williams, E. G. 1952. Evaluating the phosphorus status of soils. Int. Soc. Soil Sci., Trans. Comm. II & IV (Dublin, Ireland, 21-31 July 1952) 1:31-47.

Williams, E. G., and A. H. Knight. 1963. Evaluations of soil phosphate status by pot experiments, conventional extraction methods and labile phosphate values estimated with the aid of phosphorus-32. J. Sci. Food Agric. 14:555-563.

Younge, O. R., and D. L. Plucknett. 1966. Quenching the high phosphorus fixation of Hawaiian latosols. Soil Sci. Soc. Am. Proc. 30:635-655.

Chapter 14

Use and Limitations of Physical-Chemical Criteria for Assessing the Status of Phosphorus in Soils

S. R. OLSEN

*Science & Education Administration, AR, USDA, and
Colorado State University
Fort Collins, Colorado*

F. E. KHASAWNEH

*Tennessee Valley Authority
Muscle Shoals, Alabama*

I. INTRODUCTION

The assessment of soil P is a goal that is pursued in a variety of ways and for a variety of reasons. On the one hand, several soil-test methods have evolved over the years for the purpose of making fertilizer recommendations. This involves a certain degree of assessing soil P, either directly or indirectly, but emphasis is placed on correlating soil-test results to crop response as measured in field experiments. However, assessment is often attempted for the primary purpose of understanding the fundamental principles that govern soil-phosphate-plant interactions. Here interest is focused on the chemical nature of soil P compounds, their physical and chemical interactions and transformations, and the factors that affect the rate and direction of such transformations.

In the traditional soil-test methods, crop response is correlated with the fraction of soil P that is extracted by a given procedure. Understanding the chemical nature of soil P extracted by soil test methods is desirable but not essential. Such an understanding undoubtedly helps in improving soil test methods, and may even explain why certain extractants give results that are better correlated with crop response than others. It is towards this end that the two approaches of assessing soil P merge. While this paper will not concern itself directly with soil test methods and philosophy, as that is adequately covered in Chapter 16 by E. J. Kamprath and M. E. Watson, a considerable amount of overlap in our literature necessarily exists.

Copyright 1980 © ASA-CSSA-SSSA, 677 South Segoe Road, Madison, WI 53711, USA.
The Role of Phosphorus in Agriculture.

Several review articles and textbook chapters have been written recently on certain aspects of the subject matter (e.g., Larsen, 1967; Black, 1968; Mattingly & Talibudeen, 1967). It is not our intention here to duplicate these efforts, but rather to concentrate on more recent developments, both conceptual and methodological, particularly since the middle 1960's.

It is customary to describe soil P in terms of the following relationship (Larsen, 1967):

$$\text{Soil solution P} \rightleftharpoons \text{labile soil P} \rightleftharpoons \text{nonlabile P} \qquad [1]$$

where equilibrium is rapidly established between labile and soil solution P, whereas true equilibrium is seldom, if ever, established between the labile and nonlabile pools of soil P. This compartmentalization of soil P is largely arbitrary, since exact boundaries between compartments cannot be precisely delineated. Definition of the labile and nonlabile components is deliberately postponed until later in this chapter, largely because of some confusion that has arisen in literature regarding these terms. Suffice it here to say that equilibrium between labile P and solution P occurs very fast as compared with that between the labile and nonlabile pools. Therefore, quasi-equilibrium conditions develop rapidly between the first two components whereas equilibrium develops only slowly, if at all, between the two latter ones, which in certain cases reach steady state conditions.

The soil solution component of soil P has several characteristics, which are not unique to P but apply also to several other nutrients. First, it is the medium from which plants absorb several nutrients, including P. The soil solution can be partly, though not totally, separated from soil, i.e., it can be sampled; it is the only phase of soil P that is amenable to detailed and precise physico-chemical characterization. Apparent constancy of the composition of the soil solution is not sufficient proof of equilibrium conditions, nor is equilibrium the sole requirement for constant chemical make-up. "Apparent constancy" occurs also when rate processes that affect the chemical composition of the soil solution reach steady state. Equilibrium never exists in an agricultural soil on the microscale level, and only seldom on the macroscale level on disturbed soil that is brought into the laboratory. Nevertheless, lack of true equilibrium does not diminish the value of solution P characterization, because measuring changes in the soil solution composition gives data that are pertinent to the kinetics of change in the labile and the nonlabile fractions of soil P.

II. PHYSICO-CHEMICAL CHARACTERISTICS OF SOIL PHOSPHORUS WITH RESPECT TO PLANT GROWTH

A. The Intensity Factor

Because the soil solution is the medium from which P is absorbed by plant roots, and because the composition of the soil solution is largely determined by the solid phase, its characterization provides data that can be used for agronomic and physico-chemical inferences. Inferences related to plant growth are inherently predicated on a relationship that defines the process of P uptake by root surfaces in terms of the chemical composition of the soil solution. Such a relationship is usually defined in terms of an *intensity factor* and a relative factor (Khasawneh, 1971). The intensity factor is a measure of the gradient in the electrochemical potential of the phosphate ions across the absorbing surfaces of plant roots. The relative factor describes the effect of other ions in solution on P uptake. Except for OH^-, no anions compete with P uptake, and thus, the relative factor is equal to 1.

1. MEASUREMENT OF INTENSITY

The intensity factor has been measured and expressed in many ways by many investigators, and this has caused occasional confusion in the literature.

a. Molar Concentration of P—Here the concentration refers to *total analyzable* P in solution, and includes all species of orthophosphate ions as well as ion pairs in association with other cations. Quite often the concentration is expressed in ppm rather than in molar units.

b. Molar Activities of Specified Phosphate Ions, Such as of $H_2PO_4^-$ or HPO_4^{2-}—In the absence of ion pair formation, phosphates in solutions with a pH range of 3 to 8 exist mainly as $H_2PO_4^-/HPO_4^{2-}$. This ratio is determined by pH and the thermodynamic dissociation constants of the polybasic acid.

c. The Chemical Potential of $H_2PO_4^-$.

d. The Monocalcium Phosphate Potential.

Although the least adequate, the first method is the most widely used expression of P intensity. Persistence in its usage is attributed to convenience and simplicity; it is what the analyst reports, not complicated by further computations. Whether or not such over-simplification is sufficiently justified will be discussed later.

The other methods are based upon *calculated* activities of phosphate ion species. There have been a few attempts to design a single ion electrode for phosphates, but the results are not yet conclusive. The subject matter has been reviewed by Buck (1976). Activities of phosphate ions cannot be satisfactorily *measured* yet; they have to be *calculated*.

2. ACTIVITY CALCULATIONS

There are several methods for calculating activities in a solution of mixed electrolytes. Two of the most commonly used are:

a. An Iterative Method (Adams, 1971)—The starting point is a near-complete analysis of solution, especially ions in quantities large enough to affect ionic strength, ions that associate significantly, and pH. The iteration cycle consists of using an initial value for ionic concentrations and ionic strength to calculate ion activities. Ion-pair concentrations are computed next, then these are subtracted from initial ionic concentrations to provide a revised set of concentrations and the cycle is repeated. These computations require a computer facility.

b. The Electrical Conductivity Method (Ponnamperuma et al., 1966; Griffin & Jurinak, 1973c; Marion & Babcock, 1976)—The ionic strength is calculated from a measurement of the electrical conductivity of the solution rather than by iteration. Ions of special interest and ions that participate in ion-pair formation should be measured, along with solution pH. Using the "measured" ionic strength, concentrations of ion pairs are subtracted from total P concentration, and the remainder is partitioned between $H_2PO_4^-$ and HPO_4^{2-} according to pH.

In either of the above methods, the starting point is total P concentration. In addition to the two dominant protonated forms $H_2PO_4^-$ and HPO_4^{2-}, soluble soil P exists in a variety of ion pairs, mainly with Ca^{2+} and Mg^{2+} and to a lesser extent with Mn^{2+}, Fe^{2+}, Fe^{3+}, Al^{3+}, and Zn^{2+}. A partial listing of stability constants for these ion pairs is shown in Table 1. Complexes of orthophosphates with organic ligands originating from dissolved organic matter have been postulated, but their role in complex-ion formation is neither well documented nor well understood.

Activity coefficients are usually calculated with the Debye-Huckel equation of the form:

$$\log f_i = -Az^2 I^{1/2}/(1 + a_i B I^{1/2})$$

where f is the single ion activity coefficient of ion i; z, its valence; a, the ion size parameter; and I, the ionic strength; A and B are constants dependent on temperature and the dielectric constants of the solvent medium. For water at 25°C, the value of A is 0.509 and of B is 0.329, if a is measured in angstrom units (1 Å = 10^{-8} cm). Values of a for different ions can be found in many standard texts; a partial list was reproduced by Adams (1971).

The above equation is adequate in solutions with ionic strengths up to $0.05M$, which probably covers most soil solution situations. At higher concentrations, the interaction between ionic charges and the dipolar moments

Table 1—Dissociation constants of selected ion pairs.

Reaction	pK†	Reference
$CaH_2PO_4^+ \rightleftharpoons Ca^{2+} + H_2PO_4^-$	0.71	Gregory et al., 1970
$CaHPO_4^0 \rightleftharpoons Ca^{2+} + HPO_4^{2-}$	2.41	Gregory et al., 1970
$MgHPO_4^0 \rightleftharpoons Mg^{2+} + HPO_4^{2-}$	2.91	Taylor et al., 1963
$FeH_2PO_4^+ \rightleftharpoons Fe^{2+} + H_2PO_4^-$	2.7	Nriagu, 1972b
$FeHPO_4^0 \rightleftharpoons Fe^{2+} + HPO_4^{2-}$	3.6	Nriagu, 1972b
$FeH_2PO_4^{2+} \rightleftharpoons Fe^{3+} + H_2PO_4^-$	5.43	Nriagu, 1972a
$MnHPO_4^0 \rightleftharpoons MN^{2+} + HPO_4^{2-}$	2.58	Sillen & Martell, 1964
$ZnHPO_4^0 \rightleftharpoons Zn^{2+} + HPO_4^{2-}$	3.3	Nriagu, 1973
$PbHPO_4^0 \rightleftharpoons Pb^{2+} + HPO_4^{2-}$	3.1	Nriagu, 1972c
$H_3PO_4^0 \rightleftharpoons H^+ + H_2PO_4^-$	2.15	Wagman et al., 1968
$H_2PO_4^- \rightleftharpoons H^+ + HPO_4^{2-}$	7.21	Wagman et al., 1968
$HPO_4^{2-} \rightleftharpoons H^+ + PO_4^{3-}$	12.35	Wagman et al., 1968

† Negative logarithm of the dissociation constants.

of water molecules increases, resulting in a salting-out effect, and the resulting change in the dielectric constant is proportional to the total ionic concentration. The final form of the Debye-Huckel equation, therefore, is (Glasstone, 1946):

$$\log f_i = [-Az_i^2/(1 + a_iB I^{1/2})] + \alpha' I$$

where α' is an empirical constant.

In solutions of mixed electrolytes, the a_i parameter is no longer single-valued, and an average value is more appropriate. However, this average cannot be evaluated by other independent methods. The alternative is to fit experimental data to the above equation and find the best values of a_i and α' by least squares. However, this equation is not sensitive to variations in a_i; any loss of fit of experimental data to the equation caused by variation in a_i is compensated for by changes in α' (Prue, 1966). Thus a_i can be arbitrarily chosen at approximately the correct magnitude, and then the value of α' that gives the best fit is determined. If a_i is chosen so that $a_iB = 1$, then we can write

$$\log f_i = [-Az_i^2 I^{1/2}/(1 + I^{1/2})] + \alpha I^{1/2}.$$

This is the Davies equation, where α was determined to be 0.20 for many electrolytic solutions.

In solutions of mixed electrolytes, and in solutions more concentrated than 0.05 molar units of ionic strength, the Davies equation is probably more appropriate than the extended form of the Debye-Huckel equation.

The equation to calculate ionic strength from electrical conductivity is

$$I = \lambda \cdot EC$$

where λ is 0.016 according to Ponnamperuma et al. (1966), 0.013 according to Griffin and Jurinak (1973c), and 0.0144 according to Marion and Babcock (1976). Also, I is in molar units, and EC is in mmhos/cm at 25°C. These are empirical equations. The first was developed for soil "extracts" obtained from rice paddies (moisture content substantially above saturation, with standing water to cause anerobic conditions); the other two were developed for soil extracts at saturation and at 1/5 soil water ratio, respectively. This factor, plus the discrepancy in the three populations of soils, provides an explanation for the disparity in the values of λ. Furthermore, the data of Griffin and Jurinak and of Marion and Babcock were corrected for ion-pair formation, whereas the Ponnamperuma data were not. Even though these equations are empirical, they do offer an alternative to a complicated iteration procedure, and the calculations can be done on a desk-top calculator, whereas iterations require a computer facility.

The iteration procedure of Adams (1971) could be streamlined at a savings in computer time and a reduction in number of iterations needed if the ionic strength for the first cycle is calculated from electrical conductivity data rather than from measured concentrations.

Table 2—Comparison of estimates of the activity of $H_2PO_4^-$ as calculated by two methods.

Soil		pH	Ca	P	Adams[†]	Larsen[‡]	Relative error[§]
			—mM—		—M—		%
Unidentified[†]		6.26	18.65	0.150	6.00×10^{-5}	1.00×10^{-4}	67
		6.26	73.50	0.050	1.59×10^{-5}	2.38×10^{-5}	50
Lucedale sl[¶]	1	5.16	7.16	2.76	2.19×10^{-3}	2.70×10^{-3}	23
		5.18	9.93	5.35	4.13×10^{-3}	5.19×10^{-3}	26
		5.21	9.67	5.48	4.24×10^{-3}	5.31×10^{-3}	25
	2	5.67	12.32	1.82	1.25×10^{-3}	1.65×10^{-3}	32
		5.65	12.36	6.97	1.42×10^{-3}	1.86×10^{-3}	31
		5.69	12.42	2.14	1.46×10^{-3}	1.93×10^{-3}	32
	3	6.00	4.65	1.85	1.32×10^{-3}	1.63×10^{-3}	23
		6.00	4.84	1.90	1.34×10^{-3}	1.67×10^{-3}	25
		6.03	4.91	2.10	1.45×10^{-3}	1.82×10^{-3}	26
	4	5.92	6.80	1.43	0.99×10^{-3}	1.26×10^{-3}	27
		5.98	7.11	1.74	1.16×10^{-3}	1.50×10^{-3}	29
		5.94	7.69	1.84	1.23×10^{-3}	1.60×10^{-3}	30
Malbis sl[¶]	1	5.26	6.89	0.315	2.47×10^{-4}	3.06×10^{-4}	24
		5.45	15.4	3.46	2.44×10^{-3}	3.23×10^{-3}	32
	2	6.02	2.87	0.229	1.75×10^{-4}	2.04×10^{-4}	17
		5.78	7.59	2.93	2.09×10^{-3}	2.65×10^{-3}	27
	3	5.95	5.49	0.164	1.15×10^{-4}	1.45×10^{-4}	26
		5.69	14.5	1.99	1.30×10^{-3}	1.78×10^{-3}	37
	4	6.05	25.3	0.061	2.72×10^{-5}	4.47×10^{-5}	64
		6.01	20.9	1.01	5.13×10^{-4}	7.77×10^{-4}	51

[†] Adams (1971).
[‡] Larsen (1965).
[§] Values calculated by the Adams method are presumed correct.
[¶] Data from Bennett and Adams (1976).

Other procedures for calculating $H_2PO_4^-$ and HPO_4^{2-} activities have been used by various investigators. These range from simply applying the $H_2PO_4^-/HPO_4^{2-}$ equilibrium equation, disregarding all ion pairs, to equations that took only the $CaHPO_4^0$ ion-pair into account, but disregarded other ion pairs. Such an equation was proposed by Larsen (1965) and subsequently used by others (e.g., Jensen, 1970, 1971):

$$(H_2PO_4^-) = [P]/[1 + 10^{pH-pK_2} + (Ca) \times 10^{pH-pK_2 + pK_{Ca}}]$$

where K_2 is the second dissociation constant of H_3PO_4 and K_{Ca} the dissociation constant of the ion pair $CaHPO_4^0$. In addition to ignoring ion-pairs other than $CaHPO_4^0$, this equation disregards activity coefficients of $H_2PO_4^-$ and HPO_4^{2-}. In an attempt to compare this equation with the rigorous iteration scheme of Adams (1971), the activity of $H_2PO_4^-$ was computed with Larsen's (1965) equation from data published by Adams (1971) and Bennett and Adams (1976). The results are shown in Table 2.

3. ACTIVITIES AND POTENTIALS

From the standpoint of plant uptake, the activity of $H_2PO_4^-$ plays a greater role than does that of HPO_4^{2-} in determining P absorption by roots, for two reasons: (i) the $H_2PO_4^-$ ion is by far the more prevalent of the two in solutions of pH lower than 6, and (ii) plant roots absorb $H_2PO_4^-$ nearly 10 times faster than HPO_4^{2-} (Hagen & Hopkins, 1955).

Several investigators have, therefore, correctly chosen to relate indexes of plant growth to the activity of $H_2PO_4^-$. However, when soil pH is larger than 7.2 ($pK_2 = 7.2$), (HPO_4^{2-}) becomes larger than ($H_2PO_4^-$), with the divalent form becoming the dominant ion accounting for absorption at pH values higher than 8 (Hendrix, 1967). In the discussion to follow, phosphate activity in solution refers to the $H_2PO_4^-$ unless otherwise indicated.

Ionic potentials are, by definition, functions of ionic activities. The term *potential* has a rather specific connotation in physical chemistry, but in soil literature, the term *phosphate potential* has been used rather loosely.

White and Beckett (1964), trying to resolve this ambiguity, defined three terms: (i) the chemical potential of $H_2PO_4^-$; (ii) the electrochemical potential of $H_2PO_4^-$; and (iii) the "Schofield potential" of soil phosphate. The first two terms were defined in the same terminology used in standard physical chemistry textbooks.

The chemical potential (μ) of an ion is related to its activity, a, by the equation:

$$\mu = \mu^\circ + RT \ln a$$

where μ° refers to an arbitrary standard state, R is the gas constant (8.314 joules deg^{-1} $mole^{-1}$), and T is the temperature in degrees Kelvin. The electrochemical potential (η) is given by

$$\eta = \mu + zF\psi$$

where F is the Faraday and ψ the electric field potential on an ion of z valence. It is seen from the above equations that the measurement of η is always linked to a measurement of ψ, and neither is a constant quantity in a system of charged soil or clay particles.

The electrochemical potential, however, is a constant quantity for a system at equilibrium, a situation which can occur *only* if μ changes with ψ. Thus μ of any ion in a given soil clay-water system is not single-valued but is variable in space with respect to distance from charged surfaces. The inability to obtain a meaningful measurement of chemical potentials of single ions in soil suspensions has been overcome by measuring relative relationships of ion pairs as follows: (i) for ions of like charges, differences between chemical potentials are measured, and (ii) for ions of opposite charges, sums of chemical potentials are measured.

This is essentially the approach adopted by Schofield and his students. For a more detailed exposé on the first method, the reader is referred to Beckett (1972). The second method was described by White and Beckett (1964) and will be discussed further.

Let us consider a soil-water suspension, and concentrate on the properties of $H_2PO_4^-$ ions. Since Ca^{2+} is the predominant ion in many soils, let us choose Ca^{2+} as the counter ion for $H_2PO_4^-$. Next, we choose two points in space with respect to distance from charged surfaces of soil clays, and let the points be referred to as (') and ("), then at constant temperature and pressure, we can write:

$$1/2\eta_{Ca} = 1/2\mu'_{Ca} + F\psi' = 1/2\mu''_{Ca} + F\psi''$$

and

$$\eta_{H_2PO_4} = \mu'_{H_2PO_4} - F\psi' = \mu''_{H_2PO_4} - F\psi''.$$

Notice that η_{Ca} and $\eta_{H_2PO_4}$ are single-valued throughout the system. When like sides of these identities are summed, we obtain:

$$1/2\,\eta_{Ca} + \eta_{H_2PO_4} = 1/2\,\mu'_{Ca} + \mu'_{H_2PO_4} = 1/2\,\mu''_{Ca} + \mu''_{H_2PO_4}.$$

The conclusion is obvious: Even though the chemical potential of either Ca or H_2PO_4 vary in space within the soil-water system, their sum ($1/2\,\mu_{Ca} + \mu_{H_2PO_4}$) *does not vary* and, furthermore, that sum equals the sum of their electrochemical potentials expressed as $1/2\,\eta_{Ca} + \eta_{H_2PO_4}$. One of these points in space can be chosen at which the electrostatic fields of charged clay particles have decayed to zero, as is the case for the separated clear solution phase of a soil suspension. The other point can be chosen to refer to the solid phase of soil, $\bar{\mu}$, where the bar denotes the solid phase. In either case the sum is invariant and is a property of the entire soil suspension system. This is the Schofield phosphate potential. It is not required that a solid phase of the composition $Ca(H_2PO_4)_2$ be present, nor that Ca^{2+} and $H_2PO_4^-$ be chemisorbed on soil surfaces in the same ionic ratio.

The concept of adding chemical potentials of ions of opposite charge is sound under the most rigorous and exacting thermodynamic standards. Difficulties arise, however, when we prepare a soil suspension that exerts some disturbances on the surface chemical potentials of Ca^{2+} and $H_2PO_4^-$, and we ignore these disturbances and equate soil suspension data with soil solution data. The term *soil solution* is used here to refer to the "quasi-equilibrium solution of electrolytes that occurs in the soil under unsaturated moisture conditions" (Pearson, 1971). In the words of White and Beckett (1964) "The measurement of phosphate potential is not as straightforward as is sometimes supposed. There are several reasons why simply shaking a soil in dilute calcium chloride solution may give rise to values which differ considerably from the original potential of the soil." Basically, the reasons enumerated by White and Beckett are all related to procedures which alter $\bar{\mu}_{Ca}$, $\bar{\mu}_{H_2PO_4}$, or both.

Whenever moist soil is suspended in enough water such that the aqueous phase ceases to be a soil solution and becomes a "soil extract," this dilution will be accompanied by an adjustment in exchangeable cations that is often, but not always, in line with Schofield's Ratio Law (Khasawneh & Adams, 1967; Moss, 1963). The magnitude of this adjustment depends on the extent of dilution, and changes in $\bar{\mu}_{Ca}$ are therefore dependent on the extent of dilution. On the other hand, if the soil is suspended in a solution with the same composition as that of the original soil solution, the surface chemical potentials of ions should remain unchanged. The analyst, however, does not know a priori the composition of the soil solution. An approximation, therefore, is adopted. For many agricultural soils, a $0.01M$ $CaCl_2$ solution and a soil/solution ratio of 1:10 seems to be a satisfactory approximation; *For many others, it is not.* The conditions under which the above approximation is valid can be easily deduced—Ca should be the predominant exchangeable cation, and the exchange capacity should be mostly pH-independent and reasonably high, thus minimizing salt adsorption (Khasawneh & Adams, 1967).

For soils which can be suspended in $0.01M$ $CaCl_2$ without significantly disturbing $\bar{\mu}_{Ca}$, the procedure is reduced to determining a P concentration in this $0.01M$ $CaCl_2$ solution such that $\bar{\mu}_{H_2PO_4}$ is not disturbed either, as evidenced by absence of phosphate adsorption in the "null method," as discussed by White and Beckett (1964).

The criterion of constancy of the lime potential (pH $- \frac{1}{2}$ pCa) is not sufficient proof that $\bar{\mu}_{Ca}$ is not significantly disturbed. Constancy of the lime potential attests only to an unchanged relationship between surface H and surface Ca (ratio of exchangeable H to square root of exchangeable Ca). This relationship can remain constant if changes in surface Ca are accompanied by changes in surface H. Other criteria and other difficulties often encountered in the measurement of phosphate potential were discussed by White and Beckett (1964), White (1964), and Larsen (1967), such as attainment of equilibrium in the soil suspension, temperature, anaerobic conditions, and other soil handling and processing. All of these difficulties are manageable and can be overcome rather easily. The measurement of Schofield phosphate potential of such soils in suspensions with $0.01M$ $CaCl_2$ is therefore feasible, and often the most convenient way.

There are, however, soils for which this method would not be valid. When such soils are suspended in $0.01M$ $CaCl_2$, drastic changes are induced in the surface chemical potential of Ca. Acid soils with Al as the predominant exchangeable cation, intensely weathered soils that have mostly pH-dependent charges, or other saline and alkaline soils fall in this category. Such soils are quite widespread worldwide, and the indiscriminant practice of measuring phosphate potentials in suspensions with $0.01M$ $CaCl_2$ is an excercise in futility. For these soils, there is only one recourse—making the measurement in a displaced soil solution.

This alternative is not without its own difficulties. Larger samples of soil are needed, and the process of displacing the soil solution is slow and tedious. Saturation extracts, or even 1:1 soil-water extracts, provide reasonable compromises. The dilution effect in these extracts represents minor perturbations of $\bar{\mu}_{Ca}$ and $\bar{\mu}_{H_2PO_4}$ when compared with the effect of $0.01M$ $CaCl_2$ in such soil systems. The problems associated with turbid extracts have been solved (in our laboratory) by high-speed centrifugation, preliminary filtration through regular filter paper, and subsequent filtration through 0.2 μm Millipore® filter membranes. Analytical problems associated with submicrogram quantities of P in aliquots as large as 100 ml have been resolved also.

4. RELATION OF PLANT GROWTH TO P INTENSITY

In the preceding section we discussed methods of calculating activities of $H_2PO_4^-$ from measurements of [P] in soil solutions or soil extracts. We further discussed methods of calculating intensity expressions other than activities such as pH_2PO_4 or $\frac{1}{2} pCa + pH_2PO_4$. In this section the relation of P intensity, expressed in one or more of the above formulations, to plant growth will be discussed briefly. A more detailed discussion was given by Khasawneh (1971).

The subject of ion absorption by plant roots has been reviewed by several authors, and for broader coverage the reader is referred to Hope (1971), Gauch (1972), MacRobbie (1971), Epstein (1972), Higinbotham (1973), and Hodges (1973). It is generally accepted that anions are actively transported across the plasma membrane into the cytoplasm (i.e., against a gradient in their electrochemical potential). On the other hand, cations, except perhaps K^+, are absorbed into the cytoplasm down an electrochemical potential gradient, but also seem to be actively transported back across the membrane and into the external solution (efflux).

The kinetics of ion absorption by plant roots were first interpreted in a manner analogous to enzyme kinetics, and the Michaelis-Menten equation was adapted for that purpose by Epstein and Hagen (1952). Over wide ranges in solute concentrations, however, the process of absorption does not seem to saturate and the simple Michaelis-Menten kinetics fail. Absorption in the low concentration range (0.1 mM for many ions) does saturate, and can be adequately described by Michaelis-Menten kinetics (Epstein, 1972). Absorption in the lower range is referred to as *mechanism I,* and that in the higher range as *mechanism II* (Epstein, 1972). Recent developments in enzyme kinetics have again been invoked to account for failure of the Michaelis-Menten analogy (Hodges, 1973). The concept of negative cooperativity for single enzymes is invoked. This concept visualizes enzymes as made up of a number of subunits (protomers) possessing identical binding sites for a particular substrate or ligand. Substrate binding to one protomer, however, alters the Michaelis-Menten constants (k_m) of other subunits. With negative cooperativity, k_m increases; with positive cooperativity, it decreases. For ion uptake, negative cooperativity prevails. As substrate concentration increases, k_m for ion absorption increases and so does V_{max}, the maximum rate of absorption. Saturation does not develop. For a detailed mathematical description of the concept of cooperativity, see Rubinow (1975).

In the case of P uptake, there are two distinct possibilities with respect to P concentration, or more accurately P activity in soil solutions. For many soils, P concentrations in soil solutions are less than 0.5 ppm or less than $2 \times 10^{-5} M$. This is within the range of mechanism I, where Michaelis-Menten kinetics are adequate, especially where activity of $H_2PO_4^-$ would be even less than $2 \times 10^{-5} M$. The other situation develops around fertilizer granules or bands. Khasawneh et al. (1974) measured concentrations substantially higher than 1 mM (31 ppm P) at distances of up to 50 mm away from a diammonium phosphate band. Thus when seed is placed 50 mm below and

50 mm to the side of a band, the young seedling roots are certain to grow into soil zones where P concentration is such that kinetics of mechanism II become pertinent.

Regardless of which kinetics apply, the electrochemical potential of $H_2PO_4^-$ determines its rate of absorption, even though its entry is termed active, i.e., against a gradient in its electrochemical potential.

In this discussion we will not give details of the energetics of P uptake. For such, the reader is referred to review articles by Hodges (1973) and Higinbotham (1973), and for some dissenting but highly controversial views, the reader is also referred to Nissen (1974). In this discussion, we wish to assert that the electrochemical potential of P in the soil solution is *the intensity factor that determines the instantaneous rate of P absorption by a plant root.* Having made this assertion, the following statements must be added (i) Since the chemical potential and electrochemical potential of ionic P in soil solutions are identical, intensity of P becomes directly related to the molar activities of ionic P in the soil solution, i.e., of $H_2PO_4^-$, HPO_4^{2-}, or linear combination of both. (ii) While the electrochemical potential of ionic P, i.e., $H_2PO_4^-$ and HPO_4^{2-}, determines the instantaneous rate of absorption of these ions, it is not the only soil factor which governs the pattern of P supply to plant roots or its accumulation in a growing plant. The quantity of soil P, the buffering capacity, and the diffusion coefficient of P are additional *soil* factors which govern the supply pattern of P to plant roots (Khasawneh, 1971; Barber, 1976). (iii) Concentrations of P, [P], in the soil solution may be more relevant to the *supply* of soil P than the molar activities of the ions $H_2PO_4^-$ and/or HPO_4^{2-}. This statement and that in (i) appear to be contradictory, but they are not. The proportions of the various ionic and ion-pair species that make up the total [P] are always changing in the ambient soil solution, and if one species, such as $H_2PO_4^-$, is absorbed faster than the other species, then ion pairs such as $CaHPO_4$ and $CaH_2PO_4^+$ will dissociate, and ions like HPO_4^{2-} will hydrolyze to provide more $H_2PO_4^-$ ions. The net effect is that other forms of soluble P act as a reserve for the ionic species which is absorbed faster than the others. These reactions, dissociation and hydrolysis, are nearly instantaneous as compared with the rate process of diffusion in soil and ionic absorption by roots. Thus, while the rate of uptake is primarily a function of the molar activity of $H_2PO_4^-$, the supply pattern for growth is probably related to [P]. There are no data to indicate whether or not ion pairs such as $CaHPO_4$ and $CaH_2PO_4^+$ are absorbed by roots. However, considering the specificity of ion uptake with respect to energy relations and sites of absorption, it is doubtful that ion pairs are absorbed in the undissociated form. Published data either relate plant growth to [P] or $(H_2PO_4^-)$, as discussed below, but seldom compare the two (Wild, 1964).

Studies pertaining to the relation of P uptake to P intensity can be classified into the following categories:

a. Solution Culture Studies—Usually short-term experiments, where interaction of uptake and growth is minimum. Two types of experiments have been used—experiments with excised roots, and experiments with whole plants.

Both types are usually designed to relate the *rate of uptake* to intensity factors. Until recently, experiments with whole plants were usually terminated after a few hours of uptake, due to excessive depletion of the nutrient solution. This difficulty has been overcome in two ways. Asher et al. (1965) developed a method of flowing solution culture, and grew plants at maintained specified concentrations of P for as long as 28 days. Claassen and Barber (1974) developed another approach where the depletion curves of an initial concentration were utilized for calculating uptake parameters. The latter method is short-term but has advantages. Plants of different ages and different root systems can be used, and the plants do not have to be starved for P, or otherwise unduly preconditioned by exposing them to abnormal levels of nutrients, before they are used for such experiments. The plants can be healthy and normal and the only constraint on growth is that imposed by experimental design.

Studies with excised roots have been reviewed and compiled by Fried and Broeshart (1967). The process of P absorption has been described in terms of a dual mechanism, which was first thought to be due to the two major ionic species of the phosphate ions $H_2PO_4^-$ and HPO_4^{2-}. This hypothesis was not substantiated by later experiments. Research by Andrew (1966), Carter and Lathwell (1967), and Edwards (1968; 1970) showed conclusively that absorption of $H_2PO_4^-$ alone was dual, with one mechanism operating at a low concentration range ($k_m < 1.0 \times 10^{-5}M$) and the other operating at a higher concentration range ($k_m > 1 \times 10^{-4}M$). All such data, unfortunately, were reported as [P] rather than as molar activities of $H_2PO_4^-$.

Experiments with whole plants by Russell and Martin (1953, for 24 hours), Loneragan and Asher (1967), Asher and Loneragan (1967), and Edwards (1970) indicated a continuous hyperbolic type of relationship between rate of uptake and P concentration. The functions were continuous so that even if dual mechanisms did operate, they must have overlapped to such an extent as to appear as single-valued functions. Another salient feature of these data is that maximum growth was obtained at a much lower concentration than thought possible for solution culture, and that this value varied among plant species and ranged from 1 to 24 μM (0.03 to 0.7 ppm).

b. Soil Solution Studies—Unlike the previous studies, these types of experiments relate total plant uptake of P, rather than P absorption rates, to P concentrations in soil solution or soil extracts. Concentrations are measured at the start of the experiment, at harvest time, or averaged for the entire growth period (Soltanpour et al., 1974; Hossner et al., 1973; Fox & Kamprath, 1970; Olsen & Watanabe, 1970; Ozanne & Shaw, 1968; Woodruff & Kamprath, 1965; Beckwith, 1965). Maximum yield or P uptake was obtained at variable P concentrations, ranging from 2 to 24 μM (Table 3). This variability is partly due to differences among crop species, and partly an indication that while intensity of P is the main factor that determines uptake, additional factors come into play in soil systems. These other factors were listed as quantity and buffering capacity by Khasawneh (1971) and as capacity and P diffusion by Olsen and Flowerday (1971).

Table 3—Minimum concentration of P in soil solution or soil extract needed to give maximum growth (critical P concentration).

Crop	Critical [P], μM	Reference
Cotton	2	Khasawneh & Copeland, 1973
Sorghum sudangrass	22	Soltanpour et al., 1974
Barley	24	Olsen & Watanabe, 1970
Rice	3	Hossner et al., 1973
Millet	20	Woodruff & Kamprath, 1965
	$CaCl_2$ extracts	
Sorghum sudan	12	Soltanpour et al., 1974
Millet	6	Fox & Kamprath, 1970
Oats	10	Ozanne & Shaw, 1968
"Most plants"	6	Beckwith, 1965

It should be noted here that data from such experiments give hyperbolic relationships between yield (or total P uptake) and P concentrations that are quite similar to, and often mistakenly equated with, the hyperbolae which describe the relationship of P absorption rates to P concentration in solution culture studies. Actually, very little data exist, and for good reasons, on the relationship of P absorption rates by roots *growing in soil* and P concentration in soil solution. The main difficulty is that P concentration in the soil solution near the root-soil interface varies in time and space as theorized by Olsen and Kemper (1968). However, these authors described situations that led to the development of a steady-state concentration at the root-soil solution interface, where P concentration is substantially below, yet related to, the initial concentration in the bulk soil solution. If development of such a steady-state is typical of what happens at root surfaces, then it would be valid to relate absorption rates to bulk soil solution concentrations, as was done by Khasawneh and Copeland (1973) for cotton roots.

Additional soil studies have been conducted from a slightly different angle, although basically related to the intensity factor, except in these experiments intensity was expressed in terms of phosphate potential. Since the original papers by Schofield (1955) and Aslyng (1954), several authors have related plant growth (or plant uptake) to Schofield's phosphate potential (White & Beckett, 1964; Jensen, 1970; Barrow, 1967). It will be recalled that this potential is the sum of the electrochemical potentials of soil P and soil Ca. So what is the rationale for using Schofield's phosphate potential to measure the availability of P to plants? Although this question was adequately answered by White and Beckett (1964), we would like to give a slightly different perspective on the matter.

It was shown earlier that measurement of the Schofield phosphate potential by the null method in suspensions of $0.01M$ $CaCl_2$ was valid *only for certain soils*. Specifically, these are soils whose $\bar{\mu}_{Ca}$ is not significantly changed by suspending it in $0.01M$ $CaCl_2$. In other words, these soils vary only slightly in their Ca status. The suspending medium, at a 1:10 soil/solution ratio, provides a constant Ca concentration and a constant ionic strength, hence constant a_{Ca}. In theory and in practice, therefore, we have deliberately set ½ pCa of (½ pCa + pH_2PO_4) as constant. The Schofield's phosphate potential, therefore, is a measure of $a_{H_2PO_4}$ in a uniform Ca environment which validates the comparison. Again, this argument does not presume that a solid phase of monocalcium phosphate exists in soil, nor does it presume that phosphate and Ca are taken up by the plant in the stochiometric ratio of monocalcium phosphate, as interpreted by Fried and Broeshart (1967). This rationale, like that of White and Beckett (1964), not only justifies the validity of its use, but also sets some stringent constraints on it. When properly measured and used, it is a valid expression of P *intensity*.

Now that we have equated this potential to a constant plus pH_2PO_4, we find that it has faired no better than other formulations of P intensity as a measure of P availability. The potential buffering capacity, therefore, was invoked by many authors to help explain why there was no uniform phosphate potential that would always give maximum growth (e.g., Mattingly et al., 1963; Mattingly, 1965).

B. The Quantity Factor

When P intensity is diminished by withdrawal of P from solution, solid phase P goes into solution to replenish the loss. The quantity of solid phase P that acts as a reserve has been called by many authors the *quantity factor*. It will become clear later on in this section that this definition is oversimplified. This definition has been confounded further in that some authors have called soil P reserves the *capacity factor* (Fried & Broeshart, 1967; Mattingly, 1965; Olsen & Flowerday, 1971). The recent trend has been toward adopting the word *quantity* to describe these reserves, and to assign *capacity* to gradients which relate quantity to intensity.

It is recalled that the generalized depiction of soil P was given by Eq. [1] after Larsen (1967), where solution P was related to two forms of P, both of which were in the solid phase. Equilibrium between labile P and solution P was described as being substantially faster than between labile and nonlabile P. Labile P, by definition of the word *labile,* readily exchanges with solution P, and when P intensity is decreased, solution P is quickly replenished by P from this *labile pool.* Depletion of labile P usually causes nonlabile P to become labile again, but at a very slow rate. In other words, *reserves of soil P* refers to both the labile and nonlabile forms of soil P, the first directly and with a high dissociation rate, and the latter indirectly, but with a low dissociation rate; thus the oversimplified nature of that definition.

The division of P in the soil's solid phase into the labile and nonlabile forms comes about from a kinetic consideration. From a mechanistic point of view, P in the soil's solid phase can be classified by yet another way into adsorbed P and crystalline P. The first refers to P adsorbed on active surfaces in the soil, and the second to distinct P compounds either formed as reaction products, or inherently present in the soil matrix. The two types of categorization (i.e., labile vs. nonlabile and adsorbed vs. crystalline) are not synonymous, although a great deal of overlap exists between the two. These and other categorizations of soil P were discussed by Beckett and White (1964), Mattingly and Talibudeen (1967), and more recently by Mattingly (1975).

1. QUANTITY FACTOR BY ISOTOPIC EXCHANGE

Basically, labile P is the fraction of soil P which is isotopically exchangeable with ^{32}P within a specified time. Isotopic exchange is usually a rapid process in the first few hours, and decreases to a slow rate of exchange for days thereafter (Amer et al., 1955). The measurement of labile P, therefore, should specify the time period allowed for equilibrium to proceed.

The use of ^{32}P in soil P research was reviewed by Mattingly (1957), and its use to measure the labile fraction of soil P was reviewed by Fried (1964). There have been several terms developed and adapted to the various approaches taken by different workers; e.g., the *E value, L value, surface P, isotopic dilution factor,* etc. These terms were discussed by Fried (1964), Larsen (1967), and Fried and Broeshart (1967). Isotopically exchangeable P, sampled by a growing plant over the span of a growing season is called the *L value* (Larsen 1950, 1952; Fried, 1964). This definition was further refined by correcting for seed P, and the addition of a stipulation requiring the attainment of isotopic equilibrium (Larsen, 1967). The *E value* is the laboratory equivalent of the *L value,* measured over a shorter but specified time. It is usually designated as E_t, where t refers to the time of equilibration (Larsen & Widdowson, 1971). The effects of time, addition of carrier, presence of extracting chemicals, organic anions, exchangeable base saturation, temperature, time of preequilibration, and soil drying before equilibration were reviewed by Fried and Broeshart (1967). Surface P is measured by the initial rapid stage of isotopic dilution which is presumed to involve only the P on surfaces of soil particles, as indicated by a close relationship between this isotopically exchangeable P and total soil surface area (Olsen, 1953). Larsen (1967) does not consider this sufficient proof of the surface P concept, since the number of minute P-containing crystals in soil are also related to total surface area.

The E_t and L values are both based on the law of isotopic dilution. Another tracer-derived parameter that is sometimes confused with the L value is the A value. These two values, however, were clearly described and contrasted by Fried and Broeshart (1967), and also by Larsen (1967, p. 188) who summarized the differences between the two values in the form of a table.

It is evident, therefore, that labile P does not represent a precisely defined and a clearly distinct phase of solid phase P, but one that has arbitrary boundaries of time and other procedural factors. Any loss of precision in defining labile P is paralleled by an equal uncertainty in defining the remainder of solid phase P. Beckett and White (1964) distinguished two major categories of labile P—one held on soil's "net-exchange sites" and the other on surface isotope-exchange sites. Phosphates in the first category can be exchanged by OH^- or other $H_2PO_4^-$ ions, and the degree of coverage (or of dissociation) of these sites is proportional to the intensity of $H_2PO_4^-$ in solution. The second category is comprised of phosphate ions which are part of crystal surfaces, which readily exchange isotopically, and which dissociate from such surfaces upon the removal of an equivalent amount of complementary cation. Further, these ions are sorbed on surfaces in a manner not too different from salt sorption, where both the cation and the anion are involved, except in this case, the surface phenomenon seems to be similar to an extension of the underlying crystal lattice. Each of these categories were further subdivided to include subclasses of P which are not easily accessible to permit quick equilibrium, the physical impediment being occlusion in the first category and deep embedment within crystal lattices in the other.

Such compartmentalization has been indicated from numerous other experiments on rates of isotopic exchange. This will be discussed further in the section on kinetic methods of characterizing soil P.

If we interpret labile P as the fraction which readily participates in isoionic isotope exchange, the question arises as to which of the labile P fractions controls P intensity in solution. Is it P on the net exchane sites, in which case the relation between solution P and that part of labile P can be adequately described by an adsorption equation? Or is it P on the surface exchange sites, in which case the principle of solubility products will govern the relation between labile P and solution P? Or do both sites operate simultaneously?

Murrman and Peech (1969a, 1969b) indicated that P concentration in solution and the rate at which it was replenished was related to labile P and not to crystalline P. They also showed a considerable variation in labile P and solution P with pH, and the solution composition did not conform to the solubility product of any known crystalline phosphate. On the other hand, Vaidyanathan and Talibudeen (1968) did not find a close correlation between labile P and soluble P. The apparent discrepancy is most likely related to the measurement of labile P and to the history of P fertilization of the soils used. If the terminology of Beckett and White (1964) is used, P on net-exchange sites and on surface-exchange sites represents forms of labile P with decreasing rate of exchange, which when extended further towards lower rates of exchange will eventually merit a change of name to nonlabile P.

2. QUANTITY FACTOR AS MEASURED BY ADSORPTION PARAMETERS

Adsorption experiments are all predicated on the validity of one fundamental equation:

$$q = f(c)$$

where q is the quantity of adsorbed P and c is the P concentration in solution. This equation assumes that equilibrium is attainable and gives a mathematical formalism to the relation described earlier between labile P and solution P. It has been expressed in several different forms such as, for example, the Langmuir, the Freundlich, and the Temkin equations. These equations describe the nature of $f(c)$. Let us consider the Langmuir equation as an example:

$$q = kbc/(1 + kc)$$

where b is the adsorption maximum and k is a constant related to energy of adsorption. If we have determined k and b for a given soil, then a measurement of c will enable us to compute the quantity factor q. The underlying assumption is, of course, reversibility of the adsorption reaction; this assumption is common to all adsorption equations. However, nearly all experimental evidence is to the contrary. Whatever the relationship between q and c, it does not seem to be a wholly reversible one, and the adsorption isotherm is not the same as the desorption isotherm. The lack of complete reversibility of adsorption reactions throws some doubt on the utility of adsorption parameters for the purpose of measuring q. These doubts are somewhat alleviated if desorption rather than adsorption parameters are used to calculate q.

Traditionally, adsorption experiments measure the amount of P that disappears from an equilibrating solution, and then relate what has disappeared to what is left in solution. What is removed is usually designated as q, without any effort to determine if this q is totally or only partly in reversible equilibrium with c. There have been few measurements that indicated only a fraction of q was isotopically exchangeable over a relatively short period (Kafkafi et al., 1967; Ryden & Syers, 1977; White & Taylor, 1977). This is similar to the partitioning of labile P into net-exchangeable and surface-exchangeable fractions. This parallelism goes even further; adsorption isotherms usually shift toward more adsorption with time, higher temperatures, and with increasing P concentration in the equilibrium solution. Furthermore, adsorption isotherms have recently been mathematically resolved into two or more isotherms,[1] thus applying the concept of compartmentalization to adsorbed P (Sample, 1972; Holford et al., 1974; Ryden et al., 1977). Are the compartments of P deduced from adsorption studies the same ones arrived at by isotopic exchange? The evidence indicates they are. Beckett and White (1964) in defining the net-exchange sites, stipulated that coverage of these sites was proportional to the P concentration in solution, or in other words, coverage of these sites conforms to a Langmuir type relationship. They further reported that for the particular soil they studied, the net-exchange sites represented only one-fifth of the labile P. Holford et al. (1974), on the other hand, obtained dual isotherms from their adsorption data, and the high-energy sites represented about one-third of the adsorption maxima for the soils they studied. Similar data were reported by Sample (1972).

3. QUANTITY FACTOR BY RESIN EXCHANGE

Extraction of soil P by anion resins was first suggested by Amer et al. (1955), and later used by several authors (Cooke & Hıslop, 1963; Hislop & Cooke, 1968; Zunino et al., 1972). Anion resins, usually saturated with Cl⁻,

---
[1] Adsorption isotherms have been graphically resolved earlier (Shapiro & Fried, 1959; Syers et al., 1973).

provide a sink for soil P, thus simulating a withdrawal of P from solution and from the labile P without exerting a harsh and destructive influence on soil. Amer et al. (1955) theorized that this P removal from solution is more analogous to P withdrawal by plant roots than is the process of isotopic exchange where there is no net removal of P from the system. The analogy is, of course, correct; but, plant roots deplete other ions from the soil solution besides $H_2PO_4^-$, which is certain to affect the process of P removal. Vaidyanathan and Talibudeen (1970) showed that the anion resin in the Cl$^-$ form removed only a part of the isotopically exchangeable P, but the two fractions were highly correlated. When the resin system included a cation resin (saturated with a monovalent cation), the amount of resin-extractable P increased but did not exceed two-thirds of the pool of labile P. In addition to the finding that resin-extractable P was related to labile P, it was also found to be related to Olsen's bicarbonate extraction (Hislop & Cooke, 1968), and to crop uptake (Moser et al., 1959; Cooke & Hislop, 1963). It is apparent, however, that resin-extractable P, even when using mixed anion and cation exchange resins, represents only a fraction of labile P, and that resin systems seem to mobilize P from the labile pool only. Resin P and labile P are highly correlated, but not equivalent. Resin P, therefore, is a valid measure of the quantity factor, with the provision that the "units" of measurement are on a different scale from that of isotope-exchangeable P.

4. QUANTITY FACTOR BY CHEMICAL EXTRACTANTS

Several chemical extractants have been developed and used over the years. They range from acids, organic and inorganic complexing agents, to alkali and even water. Procedures for using these reagents are reviewed in more detail in another chapter (see Chapt. 16 in this book). These extractants were developed primarily for soil test purposes, and extractable P was usually correlated with crop growth and response to fertilizer additions. The extractants generally are not selective; they often extract all or part of the labile P plus undefined proportions of other forms of soil P. The chemical nature of P extracted by any one extractant varies among soils and depends on the mechanism of extraction, i.e., whether extraction is accomplished by H$^+$ of dilute acids; by complexing certain cations with reagents containing citrate, lactate, tartrate, or F$^-$; or by anion exchange with OH$^-$ or HCO_3^-. The NaHCO$_3$ (pH 8.5) extractant of Olsen et al. (1954) seems to extract surface P in calcareous soils. The Bray-Kurtz dilute acid-fluoride extractant for slightly acid to neutral soils (P_1 and P_2) seems to dissolve some Ca-bound as well as some Al-bound P, and these two fractions seem to be highly correlated to labile P (Tandon & Kurtz, 1968).

For a group of closely related soils, there would exist a certain degree of correlation between labile P and P extractable by several of these extractants (Moser et al., 1959; Thompson et al., 1960). It is this relationship that gives a valid reason for measuring the quantity factor via chemical extractants. Another reason is related to the presence of rate processes that regulate the transformation of many metastable forms of solid phase P towards the more stable ones. When soil has not been recently fertilized, these transformations have either reached equilibrium or a steady-state of change, and in both cases labile P will be highly correlated to nearly any other measure of solid phase P. On the other hand, when soil samples are brought in from recently fertilized fields, the soils will be in a state of disequilibrium, and any relationship that holds between the various measures of the quantity factor will be a transient one which rapidly changes with time.

III. PHYSICAL CHEMICAL CRITERIA WITH RESPECT TO SOIL P CHARACTERIZATION

A. Equilibrium Methods

The solubility product principle has been used frequently to explain the chemical behavior of soil P. Solubility-product constants must be known in this approach and much valuable information about them has become available since 1950. These constants may be expressed as solubility isotherms, for example, as a function of pH, and the isotherms indicate the characteristic behavior and relative stability of phosphate compounds.

1. SOLUBILITY ISOTHERMS

Solubility isotherms for known phosphate compounds have been graphically expressed in various ways. Aslyng (1954) plotted the relationship between the potential of $Ca(H_2PO_4)_2$, expressed as $\frac{1}{2}$ pCa + pH_2PO_4, and the potential of $Ca(OH)_2$, expressed as pH − $\frac{1}{2}$ pCa. One isotherm represents a compound for variable concentrations of Ca^{2+}, $H_2PO_4^-$, HPO_4^{2-}, PO_4^{3-}, pH, and ionic strength. Lindsay and Moreno (1960) plotted the relationship between pH_2PO_4 and pH for various phosphate compounds with conditions specified for other compounds (gibbsite, goethite, and fluorite to control the activity of Al, Fe, and F, respectively) at known levels and Ca at an arbitrary level. The isotherms for Ca phosphates would change for different levels of Ca. Examples of these two approaches are shown in Fig. 1 and 2.

a. Uses—Solubility isotherms are particularly useful for determining relative stabilities of phosphate compounds in soils at various pH values. If a calculated solubility point falls above a given isotherm, the soil solution is supersaturated with respect to that compound and undersaturated if the point falls below the isotherm. Precipitation of the compound would be indicated for the supersaturated condition and dissolution of the compound for the undersaturated solution.

Liming an acid soil would tend to precipitate gibbsite and dissolve variscite until a new equilibrium level of pH_2PO_4 is reached. If the pH increased to 7, the soil solution would be supersaturated with respect to Ca phosphates and precipitation in these forms would be expected.

Long-term applications of manure, P-containing organic wastes, and inorganic P fertilizers add P amounts in excess of crop needs. As a result,

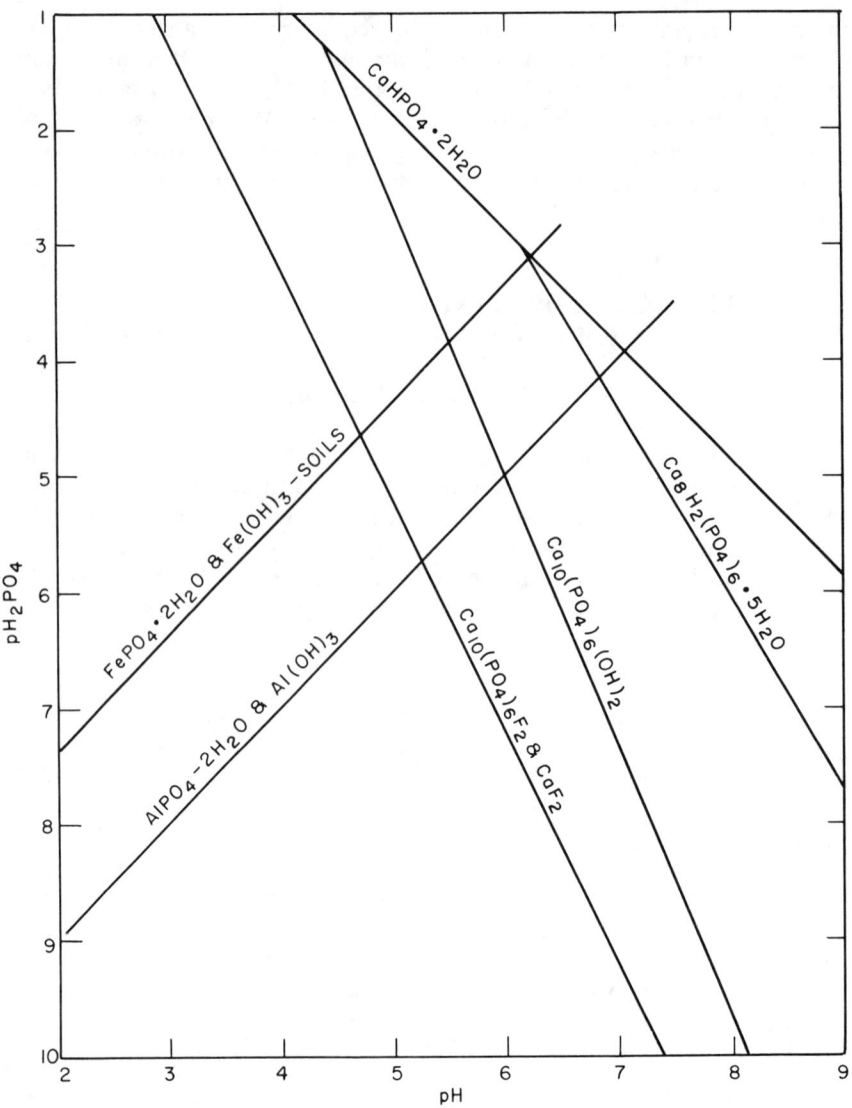

Fig. 1—Solubility isotherms for indicated crystalline phases. Activity of Ca was arbitrarily set at pCa = 2.50. Thermodynamic solubility products for the indicated phases are given by the following pK:dicalcium phosphate dihydrate = 6.60, octocalcium phosphate = 93.96, hydroxyapatite = 116.40, fluorapatite = 120.86, fluorite = 9.84, variscite = 22.52, strengite = 26.43, gibbsite = 33.96, and ferric hydroxide (soil) = 39.5. The second and third dissociation constants of H_3PO_4 are given by pK_2 = 7.21 and pK_3 = 12.35.

residual P accumulates and available P increases. Solubility isotherms have been quite useful in indicating the solid phase Ca phosphate that controls soluble P levels. Such information may be useful in adjusting fertilizer management practices and in evaluating the availability of the residual P. For example, if a fertilizer practice has raised soluble P levels to be in equilibrium with octocalcium phosphate (OCP), other data suggest that the soil

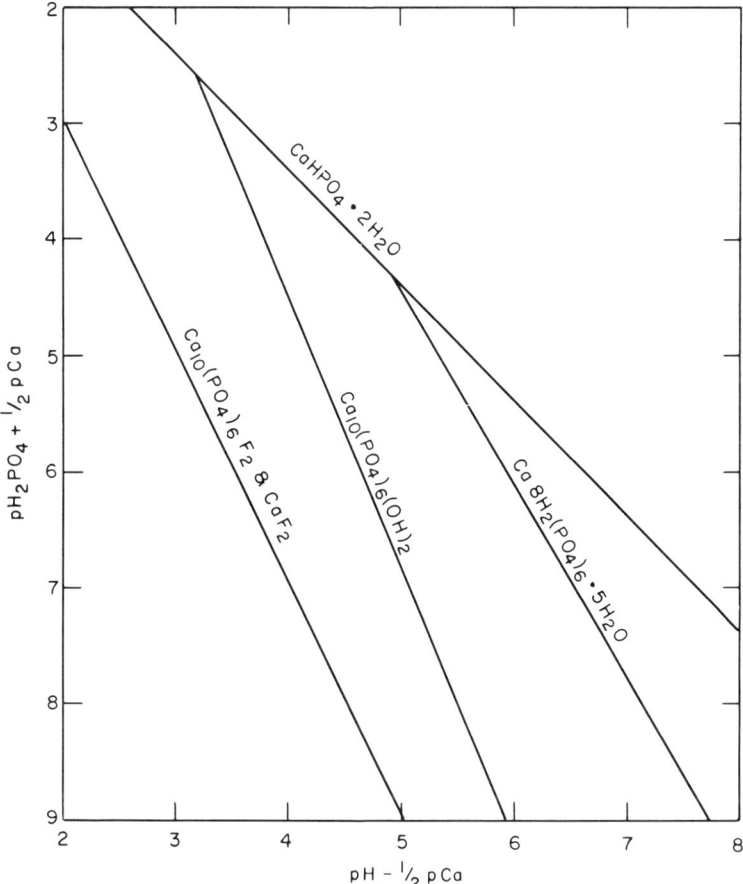

Fig. 2—Solubility isotherms of the indicated calcium phosphates expressed in terms of monocalcium phosphate and lime potentials. The activity of Ca does not need to be arbitrarily set at any value.

P status is high and exceeds the levels known to be adequate for crop growth.

b. Limitations—Chemical transformations in which one solid phase disappears and others form in soil occur very slowly (Lindsay & Moreno, 1960). Often, ionic products are assumed to represent solubility products when they actually represent super- or undersaturated conditions. Even more seriously, soil scientists have tended to think of crystalline phases in soil as pure stoichiometric crystals, when in fact such purity is difficult to obtain in an environment as heterogeneous as soil. Isomorphous substitution occurs widely in soil minerals, and crystalline forms of soil P are likely to be as imperfect as the other soil minerals, whether of geologic or pedogenic origin. Solubility product isotherms, therefore, need to be defined in terms of ranges, where the line that describes the stoichiometric mineral represents an end-member in an isomorphic substitution series of minerals.

The investigator usually does not know, a priori, whether data points that do not fall on lines of known compounds are supersaturated or undersaturated with respect to these known compounds, or that they actually fall on an isotherm of a yet undetermined, imperfect, isomorphically substituted compound.

For example, Wier et al. (1971) observed a range in values for the solubility product constant of fluoroapatite which depended on the proportion of the solids dissolved in reaching a saturated solution. They attributed the range in solubilities to a range in degree of crystallinity of the solids. In carbonate apatites, where CO_3 substitutes partially for PO_4, a constant solubility product is not exhibited (Chien & Black, 1976; Khasawneh & Doll, 1978).

For the purpose of assessing soil P, solubility product isotherms can be excellent tools in delineating the *kind* of P compound that influences P in the soil solution and possibly the pattern of supplying P to a growing plant. However, solubility product isotherms cannot measure the *quantity* of such compounds. This is a very serious limitation.

2. ADSORPTION-DESORPTION ISOTHERMS

a. The Freundlich Equation—In general, adsorption reactions occur rapidly in comparison with slower dissolution and precipitation reactions involving crystalline solid phases. The Freundlich equation has often been used to describe adsorption reactions in the form,

$$q = ac^b \qquad [2]$$

where q is the amount of P adsorbed per unit weight of soil, c is the concentration of P in solution, and a and b are constants that vary among soils. This isotherm, as first proposed, is empirical and the constants have no physical meaning. However, the isotherm has been derived for gases if one assumes that a precursor is formed first, which rapidly establishes equilibrium with the bulk adsorbing gas, and if it is further assumed that the rate-determining step is the conversion of the precursor to a more strongly adsorbed state (Crickmore & Wojciechowski, 1977). Halsey and Taylor (1947) showed that a broad exponential distribution of adsorption energies also lead to Freundlich behavior. A generalized Freundlich isotherm was proposed by Sips (1950) for adsorption of gases on metal surfaces, but this isotherm has not been utilized in soil research. This isotherm is written as

$$\theta = [c/(a + c)]^n$$

where θ is degree of surface coverage, c the concentration of the adsorbate, and a and n are constants, with $n \leq 1$. At low c, this reduces to the familiar Freundlich equation, but in this form avoids the difficulty of q increasing indefinitely as c increases. The energy of adsorption for this isotherm decreases exponentially with increasing saturation of the surface.

Fitter and Sutton (1975) found that a slight modification of one parameter of the Freundlich isotherm improved its applicability to soil P sorption data. The modification consisted of including the P which must be removed to reduce the concentration to zero. This term was closely related to resin-extractable P and they added this value to the experimentally sorbed P. By relating concentration to the total sorbed P (native pool plus experimentally adsorbed), it was possible to reduce deviations from linearity over the lower end of the concentration range. In a group of 29 soils the fit was linear for most soils over the approximate range 0.1 to 100 μM of P which covers the concentrations most relevant to studies of plant uptake of P. The exponent representing the slope of the quantity/intensity relation, log (Q/I), is a function of the sorption process. This exponent from the modified isotherms was closely related to exchangeable Al in 14 acid soils (pH <5) and to exchangeable Ca in 15 near neutral to calcareous soils (pH >5).

Although altering the labile pool size in any one soil does not affect the isotherm slope at a given concentration (Bache & Williams, 1971) the concentration maintained by a soil will be a function of the total amount of P (native-labile plus added) with which the soil solution is in equilibrium. Conventional plots of the Q/I relationship which set concentration against only the P sorbed from experimental solution, therefore, ignore the contribution from the native pool making comparisons between soils less meaningful (Fitter & Sutton, 1975). It is pertinent to indicate here that the Q/I relation, used by Olsen et al. (1965) and by Olsen and Watanabe (1970), included the native-labile plus P sorbed from fertilizer.

Bache and Williams (1971) discussed other complications related to soil P sorption isotherms. It is not possible to define rigorously or to measure unequivocally the amount of solid-phase P in equilibrium with the solution. They emphasized that there was no single-valued amount of isotopically exchangeable P because of a continuing slow rate of isotopic exchange. It would seem necessary, therefore, that native labile P be measured under the same conditions of time, temperature, and shaking procedure.

b. The Langmuir Equation—The Langmuir isotherm has been used also to describe sorption data in soils (Olsen & Watanabe, 1957). It is usually expressed as

$$q = kbc/(1 + kc).$$

This isotherm has constants which, at least when applied to the adsorption of gases on solids, have a physical meaning. One constant, b, is the maximum amount of gas that can be adsorbed as a monolayer, and the other constant, k, is related to the energy of bonding. The same equation often applies to the adsorption of liquids and ions from solutions by solids but the same rigorous, theoretical basis is not as fully developed. Adamson (1960) lists three principal postulates of the Langmuir isotherm: (i) the energy of adsorption is constant, which implies uniform sites and no interaction between adsorbed molecules; (ii) adsorption is on localized sites, which implies no translational motion of adsorbed molecules in the plane of the surface; (iii) the maximum adsorption possible corresponds to a complete monomolecular layer.

Larsen (1967) discussed the unlikely condition that all of these postulates will hold for P in soil. The energy of P adsorption is likely to be constant only within a narrow concentration range. The restriction to a monolayer seems unlikely, especially at higher concentrations where some sort of lattice structure begins to form.

Olsen and Watanabe (1957) observed that P adsorption in acid and alkaline soils followed the Langmuir isotherm. The adsorption maximum was correlated with the surface area and clay content of the soils. They stressed that agreement of the P adsorption with the Langmuir isotherm did not necessarily imply any specific adsorption or reaction mechanism. Larsen (1967) referred to Langmuir isotherms for P adsorption in 120 soils, and curvilinear relationships were found in most soils. Deviations from linearity were also observed at higher concentrations of P by Olsen and Watanabe (1957).

c. The Temkin Equation—Bache and Williams (1971) proposed the Temkin equation which is obtainable from the Langmuir equation, but which also implies that the energy of adsorption decreases linearly with increasing surface coverage. For the middle range of surface coverage, this equation gives

$$q/b = (RT/B) \ln Ac \qquad [3]$$

where A and B are constants. A plot of q against log c gives a straight line. The results for four Scottish soils gave gentle curves, but they approached linearity over certain limited concentration ranges.

It is interesting to note here that plots of sorbed P vs. log c had been used prior to Bache and Williams' paper, for example, by Beckwith (1965), Fox et al. (1968), and Fox and Kamprath (1970), although association of such plots with the Temkin equation was not explicitly recognized by them. Such plots have been used quite extensively recently (Gardner & Jones, 1973; Rajan & Fox, 1972; Nishimoto et al., 1977).

d. General Discussion—Adsorption of P on kaolinite, gibbsite, and pseudoboehmite was studied in detail by Muljadi et al. (1966a, 1966b, 1966c). Adsorption occurred mainly at low to medium concentrations of P, whereas at higher concentrations dissolution and precipitation reactions occurred with the formation of new phases. A range in pH from 3 to 10 was studied in their experiments. The adsorption isotherms were divided into three regions corresponding to at least three energetically different reactive sites. Region I with the highest energy of adsorption appeared to have sites for P associated with the edge face of the crystals or with the exchange site of kaolinite. The basic mechanism involved the grouping $-Al(OH)_2$, which is common to the oxides and kaolinite. The adsorption sites for region II were considered to be located also on the edge face of the crystals or the exchange site of kaolinite, but the adsorption energy was lower. Adsorption in region III appeared to arise through penetration of KH_2PO_4 into a less crystalline region of the clay surface. The adsorption isotherms of P on kaolinite were reversible with respect to concentration in regions II and III. The isotherms for the hydrated oxides were largely irreversible with respect to concentration which indicated a phase change during desorption. Muljadi et al. (1966b) suggested that one of the OH groups on $-Al(OH)_2$ was the reactive site for region I and the second OH groups was the site for region II. One-half of the Al on exchange sites of kaolinite seemed to be reactive sites for P. Calculations of solubility products gave no evidence for taranakite or variscite formation.

Desorption of P from kaolinite was also investigated by Kafkafi et al. (1967). Initially all of the P adsorbed by kaolinite was isotopically exchangeable with ^{32}P. Presumably, this P is held to the surface by one bond. On washing, some of the P became fixed in a nonexchangeable form, but when the amount of P that remained labile with ^{32}P was plotted against solution concentration the original isotherm was preserved. Thus, washing transformed some of the adsorbed P into a form having a greater affinity for the surface. The mechanism they suggested was the formation of a partly covalent bond between the phosphate anion and the surface, or the linkage of phosphate to two Al atoms to form a stable six-membered ring. Kyle et al. (1975) obtained additional evidence based on isotopic exchange that P formed a stable six-membered ring structure on gibbsite surfaces.

Hingston et al. (1974) measured the desorption isotherms of phosphate, selenite, and fluoride for gibbsite and goethite samples. The desorption of the anions varied between complete reversibility and almost complete irreversibility. When the isotherm was irreversible upon washing, OH^- was desorbed (or H^+ adsorbed) in preference to the desorption of the specifically adsorbed anion, whereas when the isotherm was reversible the specifically adsorbed anion was desorbed.

Irreversibility appears to involve the nature of the adsorption complex at the surface of the hydrous oxide. Where only monodentate ligands form, as with fluoride adsorption, the isotherm is reversible, whereas bridging or multidentate ligands and the formation of ring structures at the surface cause irreversibility. Hingston et al. (1974) postulated that the formation of six-membered ring structures, as illustrated below, would be expected to give considerably greater stability to the Fe-phosphate complex at the surface than for complexes in which the structure was monodentate.

Parfitt et al. (1975) used infrared spectroscopic techniques to obtain a structural model of the surface reaction between Fe oxides and phosphate ions. Two surface OH^- ions (or water molecules) are replaced by one phosphate ion. Two of the oxygen atoms of the phosphate ion are coordinated, each to a different Fe^{3+} ion, resulting in a binuclear surface complex of the type Fe-O-(P = O_2)-O-Fe. This coordination structure was found for the surfaces of goethite, hematite, lepidocrocite, B-ferric hydroxide, and amorphous ferric hydroxide gel. Similar reactions were suggested for gibbsite, allophane, and hydroxyaluminum species, but unfortunately the spectra of phosphate adsorbed on these minerals are obscured by absorption bands of the minerals.

$$\begin{bmatrix} \text{OH}_2 \\ \backslash / \\ \text{Fe} \text{---} \text{O} \\ / \quad \backslash \quad \text{O} \\ \text{O} \quad \quad \text{P} \overset{\|}{} \\ \backslash \quad / \quad \backslash \\ \text{Fe} \text{---} \text{OH} \quad \text{O} \quad \text{O} \\ / \backslash \\ \text{OH}_2 \end{bmatrix}^{2-} \rightleftarrows \begin{bmatrix} \text{OH}_2 \\ \backslash / \\ \text{Fe} \text{---} \text{O} \\ / \quad \backslash \quad \text{O} \\ \text{O} \quad \quad \text{P} \overset{\|}{} \\ \backslash \quad / \quad \backslash \\ \text{Fe} \text{---} \text{O} \quad \text{O} \\ / \backslash \\ \text{OH}_2 \end{bmatrix}^{-} + \text{OH}^{-}$$

Reversibly adsorbed phosphate (*labile*)   Irreversibly adsorbed phosphate (*inert*)

Hingston et al. (1974) suggested that the more reversible behavior of anions adsorbed on gibbsite than those adsorbed on goethite indicated that the surface complexes between Al^{3+} and anions are more labile than those involving Fe^{3+} and anions.

Wilson (1968) found that P was adsorbed on mica crystals only on the broken edges of freshly cleaved untreated mica. Phosphate is adsorbed on the 001 face of mica crystals treated with Al or Fe chlorides but mica treated with $CaCl_2$ did not adsorb more P than untreated mica. Desorption of P was much slower than sulfate desorption from mica surfaces. This difference was the result of the slower rate of P diffusion into and out of a thin homogeneous phase of hydrous oxide polymer deposited on the negatively charged mica surface.

3. TWO-SURFACE EQUATIONS

The traditional method of presenting Langmuir adsorption data for a single surface is to plot c/q vs. c according to

$$c/q = (c/b) + (1/kb) \qquad [4]$$

where q = amount absorbed per unit weight, b = maximum monolayer adsorption capacity, and k = adsorption/desorption equilibrium constant related to the bonding energy. If the line is not linear, more than one surface may be implicated. Another linear transformation of Langmuir's equation is actually better suited than Eq. [4] to detect deviation from single-surface Langmuir adsorption (Syers et al., 1973), especially at low concentrations. This transformation is as follows:

$$q = b - q/kc. \qquad [5]$$

High correlation coefficients associated with the linear transformations of the Langmuir equation are not the proper measure of goodness-of-fit. Regressing c/q on c (for Eq. [4]) or q on q/c (for Eq. [5]) does not give a least-squares fit of the data to the Langmuir hyperbola (q on c). Nonlinear regression methods would give a true least-squares fit of the data, and would generate the proper statistics.

Sample (1972) and Syers et al. (1973) evaluated P sorption by soils using the Langmuir equation. When the sorption data were plotted according to the conventional transformation of the Langmuir equation, as in Eq. [4], nonlinear relationships were obtained, and the authors inferred that at least two populations of sites with widely different affinities for P were present. They used two surface equations to describe their data. Shapiro and Fried (1959) had derived a kinetic equation for P-release and P-retention reactions of soil-P which indicated at least two forms of soil P, i.e., a rapidly released form which diminished quickly with time, and a more slowly released form which was relatively constant with time. Muljadi et al. (1966a) used a three-surface equation to describe P adsorption on clay minerals. Adsorption on two surfaces was described by the Langmuir equation, and on the third by a term linear in concentration. Gunary (1970) included a square-root term in the Langmuir equation and improved linearity for P adsorption by 24 soils. This term implied that the soil will "adsorb a little phosphate firmly, a slightly greater amount less firmly," and so on, until a limiting value is reached when all the components of the P adsorption system are saturated.

Holford et al. (1974) studied 41 soils (pH >5.0) and found the double Langmuir equation was an excellent model for describing P adsorption from solutions having a P concentration $>10^{-3}M$. Their equation, and that used by Sample (1972), was the same as that which had been proposed by Langmuir (1918) for adsorption of a gas on more than one surface:

$$q = [k'b'c/(1+k'c)] + [k''b''c/(1+k''c)]. \qquad [6]$$

The superscripts refer to the two adsorption surfaces.

Sample (1972) and Holford et al. (1974) used the method of least squares to calculate the four parameters of Eq. [6]. Syers et al. (1973) resolved the curved line into two straight-line components by a graphical method after Hofstee (1952) and Shapiro and Fried (1959). Recently, an iterative method was adopted by this group (Ryden et al., 1977) for a triple-surface equation. Muljadi et al. (1966a) estimated the linear surface by graphical inspection, and then resolved the other two after assuming that one reached saturation at a low concentration.

Nonlinear regression methods are required for fitting data to Eq. [6]. Such methods require computer processing, and dual or multiple isotherms are fitted with little more time and effort than is required for a single isotherm. However, investigators should recognize that the number of observations (q vs. c) should be increased for each additional surface to permit reliable estimates of the k and b terms.

Rajan and Perrott (1975) applied a two-surface Langmuir equation to characterize P adsorption by synthetic amorphous aluminosilicates. A deviation at high P adsorption values suggested the presence of more than two types of adsorption sites. At concentrations below about 10 mmol/liter, P exchanged mainly with aquo and hydroxo ligands and with adsorbed silicate. At higher concentrations P was adsorbed on sites arising from the disruption of hydroxy-aluminum polymers and by displacement of structural silicate.

Griffin and Jurinak (1973a) plotted adsorption data for P on calcite according to Eq. [4] and observed two distinct linear sections. They interpreted this data as signifying two different mechanisms or regions for adsorption. Previous work had not shown a two-slope Langmuir plot for calcite (Cole et al., 1953; Kuo & Lotse, 1972), but Griffin and Jurinak (1973a) obtained considerable data in the equilibrium P concentration range of <20 μM (0.6 ppm). Their data for P adsorption on calcite gave linear plots according to the BET equation which implied a multilayer adsorption phenomenon. They interpreted their data as indicating the onset of heterogeneous nucleation of calcium phosphate crystallites on the calcite surface. A plot of selected points on a solubility diagram showed that solutions in the high-energy region of the isotherm were supersaturated with respect to hydroxyapatite and that solutions from the low energy were either saturated or supersaturated with respect to octocalcium phosphate, a more soluble mineral species than hydroxyapatite.

4. USES AND LIMITATIONS

Adsorption isotherms have been used to diagnose P requirement of soils and the loading capacity for P in waste disposal sites (Fox & Kamprath, 1970; Ryden et al., 1973). Desorption isotherms are necessary to calculate diffusion coefficients and diffusion rates of P to plant roots (Nye, 1968; Olsen & Watanabe, 1970; Brewster et al., 1975a, 1975b).

Bache and Williams (1971) considered the P sorption isotherm as a fundamental soil property. They studied two samples of the same soil from a field experiment, but one had received more P fertilizer than the other. When P sorbed (q) in a laboratory experiment was added to the isotopically exchangeable P (E) initially present in the samples to give total sorbed P, i.e., ($E + q$), the plots of this sum vs. log c gave identical curves. This experiment illustrated that the sorption isotherm itself is a characteristic common to both samples, while the position on it depends on the initial P status and is altered by fertilizer application. Barrow (1974a) tested this concept in six soils but the curves did not coincide.

Procedures for obtaining isotherms have the disadvantage that a complete sorption curve is a cumbersome way of presenting results and requires much time and effort. Bache and Williams (1971) suggested simpler and quicker methods which would eliminate the need of ^{32}P and require only one or two sorption measurements. As a basis for comparing the simpler methods, they considered the slope of the adsorption isotherm as its most distinctive single-value feature. Since the isotherms are usually curved, the slope varies with position on the isotherm, so the slopes should be compared for different soils at the same equilibrium concentration. They selected a P concentration of $10^{-4}M$, because it was within a convenient experimental range and on a part of the curve where more accurate determinations of q and c were possible. They considered the isotherm slope, $dq/d\log c$ as a reference index against which others could be compared.

Several single-point methods were tested by Bache and Williams (1971) against this reference index. The highest correlation was obtained when the equilibrium concentration was taken into account in the quotient $q/\log c$ from a single addition of 150 mg of P as $KH_2PO_4/100$ g of soil suspended in approximately $0.02M$ KCl solution in the ratio of 1:20 and shaken for 18 hours. This quotient was suggested as a simple yet adequate way of indicating a fundamental soil property—its P sorption index.

Fox and Kamprath (1970) proposed using the adsorption isotherm to predict fertilizer P needs of soils. The method consisted of determining the amount of P needed to increase the P concentration in solution to 0.2 ppm (6 μM). They recognized a limitation of this approach in that a concentration of 0.2 ppm was reached on soils of low buffer capacity by the addition of insufficient P for good plant growth. Another limitation is that this approach considers only one factor, namely concentration or intensity of P, and disregards the other factors that affect P uptake rates by roots. The intensity level which is needed to obtain an adequate rate of uptake would vary among soils, depending on other soil properties.

Holford and Mattingly (1976a) used a Langmuir two-surface equation to derive the maximum buffer capacity of a soil P system, defined as the change in quantity of adsorbed P per unit change in concentration of solution P. The buffer capacity at any concentration is shown in Eq. [7].

$$dq/dc = [k'b'/(1+k'c)^2] + [k''b''/(1+k''c)^2]. \quad [7]$$

To overcome the problem of a constantly varying buffer capacity, they obtained the maximum slope of the adsorption isotherm as c tends to zero:

$$\lim_{c \to 0} (dq/dc) = k'b' + k''b'' \quad [8]$$

Both k and b determine the maximum buffer capacity, and it is independent of P saturation. The maximum buffer capacity accounted well for the changes that occurred in the intensity and quantity of labile P in a group of eight related soils following P fertilization.

Fitter (1974) observed a close correlation between the P requirement of colliery shales and a combination of $NaHCO_3$-soluble P and the P buffering capacity measured by the method of Ozanne and Shaw (1967).

Nye (1968) noted that values of P buffer power, dC/dC_e, derived from the relation between ^{32}P exchangeable P (C) and the P concentration in the soil pore solution (C_e) were much greater than those derived by solution desorption. In a related study, solution desorption values were obtained by suspending soils in 0.006 to $0.125M$ $CaCl_2$ at a wide range of soil to solution ratios (Drew & Nye, 1970). They agitated these suspensions for 18 hours and then determined P in solution and the quantity of P desorbed from the soil. Nye (1968) and Drew and Nye (1970) indicated reasons favoring the use of the slope of the desorption isotherm as a measure of P buffer power.

Brewster et al. (1975b) measured desorption isotherms determined at a high solution/soil ratio using an anion exchange resin to remove P from the solution. They observed an apparent continuity of resin desorption and adsorption isotherms on soils equilibrated with added P for 6 months, but not by dilution with excess solution. Solution desorption methods that dilute soil minerals with large volumes of neutral salt solutions may in some way "fix" adsorbed P (Kafkafi et al., 1967; Hingston et al., 1974). Brewster et al. (1975b) indicated that the use of resins was a more desirable technique than dilution with large volumes of electrolyte solution; however, they suggested that buffer powers for well-equilibrated soils could be more easily estimated by extrapolation of the relation between ΔP and $\ln C_e$ from adsorption isotherms conducted at low solution/soil ratio. They caution that applicability of these isotherms to the transport of P around roots may need modification, especially for undisturbed, structured soils.

B. Kinetic Methods

The discussion in Section II-B was concluded with the assertion that nearly all measures of the quantity factor were highly correlated for equilibrium soils. We would like to emphasize here that certain agricultural soils are nearly always in a state of disequilibrium with regard to P transformation. Soils that have been intensively cropped and fertilized with optimum P rates for many years belong in this group, because equilibrium is precluded by periodic addition of P fertilizers. Virgin soils, forested soils, range soils, and soils under less intensive farming practices fall in a different category, where either quasiequlibrium or steady-state conditions prevail with respect to soil P and its transformation. Thus, two types of soil situations emerge. Soils with continuous disequilibrium and soils with a steady state of change. Each is comprised of large acreages and each is economically important in its own way.

The extent of disequilibrium of the first soil category depends on the frequency and amplitude of the periodic disturbances, which are related to the intensity of soil-fertilizer reactions. Granular, water-soluble P sources, especially when band placed, generate zones of intense reactions with soil due to the high concentration of the fertilizer solution emanating from the granule or the band. These reactions are discussed elsewhere (see Chapt. 11).

With slightly water-soluble sources of P, the intensity of the reactions with soil is very mild, because the fertilizer solution emanating from these sources is several orders of magnitude more dilute than is that from water-soluble sources. Such mild reactions do not cause excessive disequilibrium and heterogeneity in soil P. Reaction products formed near the granule site of water-soluble sources are not stable end-products of P in soil, but usually continue to change metathetically. The initial reaction products, which exhibit varying degrees of crystallinity, dissolve only slowly and usually incongruently. The process is governed by P diffusion away from, and cations toward these reaction sites. At first, the reacted soil is only a small fraction of the total, the actual value depending on granule size and placement patterns. When fine granules (e.g., <0.25 mm diam, −60 mesh) are broadcasted and plowed under, the volume of soil affected per unit of P is larger than with coarse granules (e.g., 2.36 to 3.35 mm, $-6 + 8$ mesh), and the intensity of the reaction is lessened. Zones of disequilibrium in P transformation are small and are smoothed out rapidly, especially if the soil is well mixed.

Many of the initial reaction products and the subsequent transformation products have been identified and well characterized for disequilibrium soils, but there is little information on the kinetics involved. Nearly all the information on kinetics of soil P transformation has been obtained from soils that were deliberately selected to be equilibrium soils. If the experiment called for adding P, it was added with dilute solutions, usually less than 10 ppm, but sometimes as much as 100 ppm. An example of avoiding disequilibrium soils is the recommendation by Beckett and White (1964) that field soils be stored in a moist condition for a few months before being used for P potential determinations. There is now a considerable accumulation of recent research on P kinetics in laboratory soils, which are equilibrium soils, with relatively dilute solutions of P. These data will be reviewed next; but it remains a challenge to extend the kinetic approach beyond the steady state or equilibrium soils to the disequilibrium soils of field agriculture.

1. ADSORPTION-DESORPTION RATE

Adsorption processes were discussed earlier within the constraint of reaching equilibrium, a constraint which cannot be adhered to rigorously. Given more time, adsorption isotherms usually would shift toward more sorption. The rapid attainment of an apparent equilibrium has come to be known as the *fast reaction* (<10 days), while the continuing adsorption thereafter is called the *slow reaction*. Although these designations are arbitrary, they are often quite useful.

In general, kinetic investigations of soil P can be rationalized on the grounds that they add a different perspective to the overall problem of assessing the status of soil P. Interest in this phase of P research is stimulated by the realization that P adsorption on and desorption from soil colloids are only two of a series of rate processes that govern P uptake by plant roots and its subsequent utilization in plant growth.

Kinetic studies of P sorption on laboratory soils have proceeded along two fundamentally different premises. The first is an extension of the adsorption isotherm experiment, where *approach* to equilibrium or apparent equilibrium is investigated. The adsorption and desorption rates are considered with the intent of resolving them out of the time curve describing approach to equilibrium. In the second line of investigation, the experimenter is interested only in the rate of disappearance of P from solution in adsorption experiments, or alternately, in the desorption rate of P to a sink, but not in the two simultaneously. Approach to equilibrium will be considered first.

A second-order kinetic equation was developed by Kuo and Lotse (1972) for P sorption by $CaCO_3$ and Ca-kaolinite. The differential equation was developed along the same lines used for deriving the Langmuir adsorption equation. They assumed that adsorption would be proportional to solution concentration and to concentration of vacant adsorption sites, and that desorption would be proportional to the concentration of occupied sites. At equilibrium, these assumptions yield the familiar Langmuir equation. Single-surface Langmuir adsorption, therefore, is an implied prerequisite for the applicability of their equation. In two subsequent papers, they found that P sorption on hematite and gibbsite (Kuo & Lotse, 1974a) and on lake sediments (Kuo & Lotse, 1974b) was better described by the Freundlich equation. They modified the Freundlich equation to empirically incorporate a time-dependent factor and derived an equation which described the rate of disappearance of P from solution. This treatment, therefore, falls under the second approach discussed above, since it does not attempt to resolve the rates of adsorption and desorption from the concentration-time curve. More will be said about these two papers in a later section.

The kinetics of P interaction with calcite were investigated by Griffin and Jurinak (1974) who used a second-order rate equation of the form

$$- \mathrm{d}c/\mathrm{d}t = k_1 c^2$$

for the initial 10 min of the reaction (their Eq. [5]) and a pseudo-first-order rate equation for longer periods (up to 4 hours, their Eq. [1]). In the initial stages of exposing calcite to a P solution, desorption could be ignored with adequate justification. The authors, however, did not adequately justify their contention that the reaction was second-order in c. In contrast, Kuo and Lotse (1972) assumed that $- \mathrm{d}c/\mathrm{d}t$ would be proportional to solution concentration *and* to concentration of vacant adsorption sites. It is not readily apparent why the concentration of vacant adsorption sites can be expressed in terms of P concentration in solution, as proposed by Griffin and Jurinak (1974).

Langmuir adsorption was also assumed by Lindstrom et al. (1970) in their derivation of a differential equation for simultaneous adsorption and desorption. They used the resulting isotherm (in differential form) to describe sorption of certain herbicides on illite and silica gel. A computer program was necessary to numerically integrate the differential equation and to yield an adsorption isotherm and a least-squares estimate of the two-rate constants. Griffin and Jurinak (1973b) attempted to fit their P-calcite data to this model and concluded that the model was not valid for endothermic processes.

Novak and Adriano (1975) tested four mechanistic models of P reaction kinetics in an Ottawa loamy fine sand, and concluded that a second-order reaction rate equation based on Langmuir kinetics gave a better fit of the experimental data than the other three models, especially in the first 3 hours of reaction. Their Langmuir rate equation was similar to that of Kuo and Lotse (1972). For reaction periods longer than 3 hours, however, none of the models was adequate.

Kuo and Lotse (1972) had observed that the rate constants for P adsorption in Ca-kaolinite increased with increasing P concentration, but offered no explanation. According to Bronsted's theory of ionic interactions, the authors pointed out that the rate constants should have decreased with increasing P concentration. Novak and Adriano (1975) pointed out that they did indeed, if surface area of adsorbent had been factored out of the k of Kuo and Lotse (1972).

More recently, McLaughlin et al. (1977) employed an altogether different approach to describe the kinetics of P sorption by hydrated ferric oxide gels. They measured adsorption isotherms after varying shaking times, and then partitioned each into three simultaneous Langmuir isotherms. They observed that adsorption in region I was the only one that changed with time. When adsorption in this region was subtracted, the residual isotherms were invariant with time up to 28.7 days. Other work by this group indicated that P is chemisorbed in region I and that the reaction neutralizes positively charged surfaces ($-OH_2^+$) and displaces H_2O by ligand exchange (Ryden et al., 1977). It remains to be seen if the kinetic model developed by McLaughlin et al. (1977) can be extended to soil colloids, or to variation in time of undisturbed incubation as contrasted to variation in shaking time.

We can conclude from the foregoing discussion that there are basic, inherent limitations to this "approach to equilibrium" method which preclude meaningful or consistent results. Langmuir adsorption is seldom exhibited by soils, except over narrow ranges of concentration. Likewise, true equilibrium is never attained. The obvious corollary is that it would not be possible to extract adsorption and desorption rate constants from the concentration-time curves that would not be limited to the same narrow ranges of concentration, or be subject to the same approximation engendered by neglecting the slow reaction. The data discussed above confirm this anticipation quite vividly.

2. DESORPTION BY RESINS

Anion exchange resins (AER) have been used extensively to measure the rate of P release from soils and as an index of the quantity of available P in soils (Amer et al., 1955; Moser et al., 1959; Cooke & Hislop, 1963; Cooke, 1966; Gunary & Sutton, 1967; Vaidyanathan & Talibudeen, 1970). Amer et al. (1955) showed that if an AER is added to a soil suspension under appropriate conditions, P uptake by the resin occurs at a rate which is independent of the properties of the resin and dependent only on the rate of dissolution of P from the solid phase of the soil. Temperature control is important because the P release reaction has a relatively large temperature coefficient (Cooke & Hislop, 1963).

The extensive work cited above indicated that the rate of P release from the solid phase to resins could be measured and was highly correlated to crop growth. Cooke (1966) measured the release rate of phosphate ions from the solid phase with an AER and assigned a numerical parameter to the release rate. This parameter was highly correlated with P uptake by ryegrass in greenhouse trials from a wide range of soils. A lower correlation was found between P uptake and the equilibrium P concentration in solution prior to resin addition. Gunary and Sutton (1967) suggested that the resin method provided a good measure of the buffer capacity and nearly as good a measure of an intensity/kinetic complex, which included an interrelated complex of intensity, rate of release, and diffusion factors.

Vaidyanathan and Talibudeen (1970) studied the mechanisms controlling the transfer of phosphate ions to AER. The labile P pool was the source of all the desorbed P. Release of P was measured by an AER with and without a cation exchange resin (CER) present. The observed differences in the amounts of P sorbed by the AER with and without the CER was attributed to the degree of dissociation and not to the rate of dissociation of the labile soil P. They suggested that P uptake by the AER was rate-limiting with bulk or 'particle' diffusion in the resin beads as the major rate-limiting factor, and that the amounts taken up were determined by the P concentration maintained in the soil solution. If the results of Vaidyanathan and Talibudeen (1970) apply generally to soil-resin systems, previous studies should be reexamined where uptake of P by an AER was assumed to be measuring the release rate of solid-phase P to the solution.

In general, resin desorption data appear to be adequately represented by a sum of simultaneous first-order reactions, with correspondingly different reaction rate constants. It was postulated that each first-order reaction term represented a quantity of adsorbed P with a given energy of desorption (Amer et al., 1955). These pools of soil P were interpreted as groupings of sorption sites and not as distinct chemical compounds or physical states of one compound. However, critical analysis of prolonged desorption data revealed that the number of first-order terms required to describe the data adequately increased with increasing desorption times. Development of the concept of compartmentalizing soil P from such data was also paralleled by isotopic exchange data, as described in the next section. It must be remembered, however, that such compartmantalization is more of a mathematical artifact and a convenience than a quantitative description of the physical realities of desorption.

3. ISOTOPIC EXCHANGE METHODS

A strong interest in this topic continues because many authors have suggested that the fraction of total soil P which is readily accessible to isotopic exchange may represent the P available to plants (White, 1976). Methods based on reaction rates have been used to distinguish rapidly from slowly exchangeable forms (Vaidyanathan & Talibudeen, 1968, 1970). Rates of isotopic exchange, however, may be limited by diffusion of P to an exchange site (Wilson, 1968). Isotopic exchange of soil P has been resolved into a fast reaction, essentially complete within 24 hours, and a slow one which may continue for many days (McAuliffe et al., 1948; Jose & Krishnamoorthy, 1972; White, 1976).

Reaction kinetics of ^{32}P exchange in soil have been investigated in several ways. The earliest approach was to use a linear combination of exponential terms analogous to a series of simultaneous first-order reactions (Newman & Newman, 1958; Arambarri & Talibudeen, 1959) as given below

$$x = 1 - a_1\exp(-k_1t) - a_2\exp(-k_2t) - a_3\exp(-k_3t)\ldots - a_n\exp(-k_nt) \quad [9]$$

where x is the fraction of the tracer in the adsorbed state at time t, k_i are adsorption rate constants, and a_i are constants such that $\Sigma\, a_i = 1$. The process of fitting experimental data to the above equation is quite cumbersome. Probert and Larsen (1972) adopted a two-constant formulation, which was originally proposed by Edgington (1965) to approximate the sum of exponential terms. Edgington's formulation is

$$1 - x = [(t+\gamma)/\gamma]^{-b} \qquad [10]$$

where γ and b are constants. Probert and Larsen (1972) reported that Eq. [10] adequately described ^{32}P exchange data in several soils. The reader should note that neither of the two equations above makes a reference to adsorption or desorption rates per se. The equations simply describe the rate of disappearance of the tracer from solution and its exchange with solid phase P. Presumably this involves both adsorption and desorption reactions, and the algebraic sum of both gives the net adsorption rate of the isotope. It should also be noted that for both of these equations equilibrium is asymptotically approached with increasing time.

Kinetics of isotopic exchange of P have also been investigated for soil-related materials such as goethite, gibbsite, hydroxyapatite, and lake-bed sediments. Li et al. (1972) resolved a curve describing ^{32}P exchange in lake-bed sediments into three first-order simultaneous reactions. On the other hand, Kukura et al. (1973) applied the McKay equation to isotopic exchange on well-crystallized, aged hydroxyapatite. This equation had been developed for homogeneous systems, but their data indicated the exchange reaction was heterogenous. Atkinson et al. (1972) and Kyle et al. (1975) used a modified form of the Elovich equation to describe isotopic exchange on goethite and gibbsite surfaces, respectively. This modified equation expresses x as a function of t as follows:

$$\phi = (1/\beta) \ln [(t+\tau)/\tau] + \phi_0$$

where

$$\phi = [x/(1-x)] \cdot [(1-x_\infty)/x_\infty],$$

x and x_∞ being the fractions of the tracer in the adsorbed state at times t and at equilibrium, respectively, ϕ_0 the value of ϕ at zero time, and β and τ are constants (Kyle et al., 1975). This equation presumes a heterogenous distribution of adsorption energies, where the activation energy *increases linearly with coverage.* All the P adsorbed on goethite was isotopically exchangeable and the data could be represented by the Elovich equation (Atkinson et al., 1972). For gibbsite, however, the equation fitted the data only after subtracting a very slowly exchanging component (Kyle et al., 1975). The proportion of this component varied, and was largest at low solution pH. Probert and Larsen (1972), on the other hand, reported that the Elovich equation was not suitable for ^{32}P isotope exchange data in soils.

4. LONG-TERM EFFECTS AND TIME-TEMPERATURE INTERACTIONS

Although continued interest has been shown in the amount of P measured by isotopic exchange over short periods of time, i.e., 1 or 2 days, the long-term effects are quite relevant, especially to prevailing cultural and fertilization practices. Jose and Krishnamoorthy (1972) summarized various studies which indicated the importance of the rapid exchange processes and stressed that equilibration time should be selected so that the contribution of the slow rate processes is negligible. While the slower reactions may be irrelevant to measuring labile P, they are quite relevant to the problem of evaluating the residual effectiveness of added P. Vaidyanathan and Nye (1970) found that less than one-fourth of the long-term ^{32}P-exchangeable P (equilibrated for 12 weeks) contributed to diffusion of P to a resin paper. Barrow and Shaw (1975b, 1975c) have shown that slow reactions result in a tighter bond between P and soil surfaces; with increasing period and temperature of contact, P became less ready to exchange with isotopically labeled P. Their evidence did not suggest that part of the P was converted to a nonexchangeable fraction—rather that increasing time of contact decreased the rate at which the previously added P reacted with the isotope.

Barrow (1974b) was able to describe the effects of time and temperature by what he termed an empirical equation

$$d\alpha/dt = k(1-\alpha)^n \quad [11]$$

where α was defined as the proportion of added P converted into a form which was ineffective for plant uptake, or a form that did not participate in the equilibrium with solution P (Barrow & Shaw, 1975a), or a form that was not isotopically exchangeable (Barrow & Shaw, 1975b). The rate constant k was a function of temperature, and n was a constant. Barrow (1974b) pointed out that this equation was similar to Edgington's equation for crystal growth. The two are similar indeed (e.g., see Probert and Larsen, 1972, where x of their Eq. [2] is the same as α in Barrow's equations). The similarity may be readily apparent by inspecting the differential equations of each.

Edgington's is normally written as:

$$dc/dt = -k'c^m \quad [12]$$

where change in solution concentration is traced in time. Equation [11] traces α in time, where α refers to the fraction which is not in solution. It can be shown that the two differential equations[2] are identical by relating c to α, especially that the fraction which disappears from solution plus that which remains in solution adds up to 1.

---

[2] The Elovich equation is dissimilar to either of the above equations [11 or 12]. In differential form, the Elovich equation (in terms of α or c) is:

$$d\alpha/dt = k_1 \exp(-b\alpha) \text{ or } dc/dt = -k_2 \exp(hc)$$

where k_1, b, k_2, and h are constants (Winter, 1965).

Barrow and Shaw (1975a) observed that the rate constant, k, was dependent on temperature according to the Arrhenius equation, and that the activation energy for the slow reaction was around 20 kcal/mol. A 10°C increase in temperature, therefore, would cause approximately a threefold increase in the rate of the slow reaction. These authors incorporated the effects of level of added P, time, and temperature into one expression that they derived from Eq. [11]; this expression is

$$P_S = \beta_0 \cdot (P_a^{\beta_1}/t^{\beta_2}) \cdot \exp(\beta_3/T) \qquad [13]$$

where P_S is P concentration in soil extract after time, t, which is the time of prior contact between soil and P, and T is the temperature of incubation in degrees K, and P_a is the level of P addition in concentration units (same as P_S). The terms β_0, β_1, β_2, and β_3 are constants. This relationship indicates that log P_S will be a linear function of log P_a, log t, and $1/T$. Barrow's data indicated that variation in temperature of incubation can be transformed into equivalent variation in incubation time at a chosen temperature. With this transformation, Eq. [13] was applicable over a range in equivalent time on the temperature-time surface that spanned nearly four orders of magnitude ($\sim 10^{-2}$ to 10^2 day equivlents).

Another equation was similarly derived for isotopically exchangeable P (P_e):

$$P_e = t_e^{b_2} \cdot P_a [1 + kt/b_1]^{-n_1} \qquad [14]$$

where t_e is the shaking time as distinguished from t, the incubation time (Barrow & Shaw, 1975b), and b_1 and b_2 are constants.

It is appropriate at this point to recall that Kuo and Lotse (1974a) derived an equation to describe the rate of disappearance of P from solution in suspensions of hematite and gibbsite. The starting point was the Freundlich equation and the final expression was

$$q = k c_0 t^{1/m} \qquad [15]$$

where q is μg of P adsorbed per g of solid, c_0 is the initial P concentration in solution, and t is reaction time (shaking time, not incubation time), and k and m are constants (not to be confused with those of Eq. [12] above). Note the similarity between Eq. [14] and [15]. In both equations, the "reversibly" adsorbed phase is proportional to a fractional power of shaking time. Barrow and Shaw (1975b) commented about this similarity and observed that the activation energies reported by Kuo and Lotse (1974a) were too low (1.9 kcal/mol vs. 16 to 25 kcal/mol as reported by Barrow and Shaw), and were erroneously calculated—it is log k_2 of Eq. [2], not log k of Eq. [1] (numbers refer to Kuo and Lotse 1974a paper) that should be plotted against $1/T$. In addition, Kuo and Lotse erroneously defined k_1 of their Eq. [2]—the correct definition being $k^m = k_1[1 - \exp(-k_2 t)]$. A plot of ln k vs. $1/T$ gives $(1/m) \cdot E_a$ rather than E_a. Multiplying the values reported by Kuo and Lotse by m to correct for this error (1/0.08 for gibbsite and 1/0.12

for hematite) would alter the activation energies to 24 and 16 kcal/mol, respectively. Although these activation energies are of the same magnitude as the values reported by Barrow and Shaw (1975b), the latter are derived from the effect of incubation temperatures whereas those of Kuo and Lotse relate to shaking or equilibration temperatures.

The equation of Kuo and Lotse (1974a) was applicable to desorption of P from phosphated hematite and gibbsite to solutions of $EDTA^{2-}$ or F^-. It was also applicable to adsorption of P by lake sediments and to its subsequent desorption to $EDTA^{2-}$ and F^- solutions (Kuo & Lotse, 1974b). Again, the activation energy for P sorption by lake sediments is in error as described above.

The equation of Kuo and Lotse has not been tested in soil yet. On the other hand, Edington's equation, as adopted by Barrow and Shaw (1975a) has been tested in a limited range of soils.

It is not certain at this point how these ideas will hold for different soils. Barrow and Shaw (1975a) pointed out that the relationship expressed by Eq. [13] will vary among soils largely because of variation in B_0. In addition to that, however, it can be deduced from their data that soils with varying levels of native P will also behave differently. For example, their data on isotopically exchangeable ^{32}P (Barrow & Shaw, 1975b) showed that the proportion of the added P which remained isotopically exchangeable was independent of the quantity of added P. The soil used for this study was extremely deficient to the point that isotopically exchangeable P of the untreated soil was nearly zero. Therefore, when P_e was plotted against levels of P addition (P_a), they obtained a series of straight lines passing through the origin. For a soil whose P_e is substantially different from zero, the relationship between P_e and P_a might still be linear, but it would not pass through the origin. For such a soil, $P_e = e_0 + f \cdot P_a$, and P_e/P_a will no longer be independent of P_a, but, in fact, will be inversely proportional to P_a, and more so with higher e_0, the isotopically exchangeable P of the untreated soil. That P_e/P_a was independent of P_a was basic to the development of Eq. [13] and [14]. It is therefore necessary that these theories be tested with other soils.

On the other hand, Eq. [14] quite adequately explains the data of Larsen and Widdowson (1971). They used a calcareous soil with e_0 of 62 ppm, and P was added only at one level. They measured isotopically exchangeable P in monthly intervals for 5 months using 24- and 500-hour shaking times. The ratio P_e (24-hour)/P_e (500-hour) remained constant throughout the 5-month period of aging. According to Eq. [14], this ratio should remain constant, and should be equal to $(24/500)^{b_2}$, regardless of the aging time.

C. Phosphate Fractionation Methods

Fractionation of soil P through the use of selective chemical solvents has been attempted by several investigators since the turn of the century, culminating in the well-known Chang and Jackson fractionation scheme

(Chang & Jackson, 1957). The underlying assumptions here are that inorganic soil P consists of varying proportions of three discrete classes of compounds, namely, phosphate of Fe, Al, and Ca, some of which could be occluded or enclosed within coatings of Fe oxides and hydrated oxides. The fractionation schemes cannot differentiate between individual compounds within a class. Thus, octocalcium phosphate, hydroxyapatite, and fluoroapatite, if all are present, will be lumped in the Ca-phosphate class.

The Chang-Jackson scheme has been criticized and modified repeatedly by several researchers. The main criticism has centered on the reliability of separation between Al and Fe phosphates. A brief chronology of these modifications is given below. The reader should refer to the original papers for a more extensive discussion.

As outlined by Chang and Jackson (1957) and Jackson (1958), the original scheme consisted of shaking 1 g of soil in 50 ml of $1N$ NH_4Cl for 30 min, centrifuging, and discarding the supernatant solution. Next, the residue would be shaken with 50 ml of neutral $0.5N$ NH_4F for 1 hour. This treatment was claimed to extract Al phosphates. After centrifuging and washing with a saturated NaCl solution, the soil residue would be extracted with 50 ml of $0.1N$ NaOH for 17 hours. This treatment was supposed to extract Fe phosphates. After washing with NaCl solution again, the residue was extracted with 50 ml of $0.5N$ H_2SO_4 for 1 hour to obtain Ca phosphates. Finally, the residue was extracted with 40 ml of $0.3M$ Na citrate (pH 7.3) to which 1 g of $Na_2S_2O_4$ was added for 15 min in a water bath at 80 to 90°C. The last treatment yielded the so-called "reductant soluble" phosphates.

Chang and Jackson realized that neutral NH_4F was capable of extracting some Fe phosphates in addition to the Al phosphates. To correct that, they suggested that 10% of the NaOH-extractable Fe phosphate should be subtracted from the P in NH_4F and allocated to the Fe-phosphate fraction. The net error was claimed to be within a "few ppm."

The first modification was proposed by Glenn et al. (R. E. Glenn, P. H. Hsu, M. L. Jackson, and R. B. Corey. 1959. Flow sheet for soil phosphorus fractionation. Agron. Abstr. p. 9). Alkaline NH_4F (pH 8.2) was substituted for the neutral solution to minimize extraction of Fe phosphates; the reductant soluble P was extracted prior to the H_2SO_4 treatment to preclude extraction of occluded Fe and Al phosphates by the acid reagent. These modifications were later formalized by Petersen and Corey (1966) and the analytical scheme facilitated to permit simultaneous fractionation of a large number of soils.

Fife (1959) also advocated an alkaline NH_4F reagent (pH 8.5) and called attention to the problem of resorption of the P released by NH_4F (from Al phosphates) on Fe oxides. Initially, the choice of pH 8.5 for the NH_4F reagent was predicated on eliminating this resorption, but later work by Fife (1962) demonstrated that resorption persisted at pH 8.5. Fife continued to favor this pH for different reasons. Less Fe-bound P was extracted by NH_4F at pH 8.5, and the solution was better buffered at this pH than at pH 7. This buffering action was more effective at wide solution/soil ratios; at narrow ratios, pH increased especially in soils that contained substantial amounts of allophane, and the increased pH reduced the extractability of Al-bound P (Fife, 1962). In such soils the initial pH of NH_4F should be reduced accordingly.

The problem of resorption was addressed by using a correction factor. Fife (1962) assumed that resorption of NH_4F-released P wold be equivalent to fractional resorption of an additional increment of P (1 ppm in the NH_4F solution) under the same conditions of extraction. The correction factor is based on percentage of recovery of this added P and is used to adjust NH_4F-extractable P to resorption-free basis. Smith (1965), however, reported that fractional resorption of added P decreased as the concentration of added P increased. A correction factor was obtained by plotting recovery against P addition and extrapolating back to zero addition. Williams et al. (1967) adopted Fife's correction method, with the realization that if fractional resorption was not a linear function of added P, then neither correction would be satisfactory. Bromfield (1970) arrived at the same conclusion by measuring sorption of known amounts of P by synthetic Fe oxides in a $0.5N$ NH_4F (pH 8.5). Soils, however, may not behave like synthetic Fe oxides, especially since such oxides in soil are partially phosphated prior to fluoride extraction (Smith, 1972).

Further modifications were introduced by Williams et al. (1967). They retained the pH 8.2 level of NH_4F as given by Petersen and Corey (1966), but extended the shaking period from 1 to 24 hours. The $0.1N$ NaOH extractant was made in $1N$ NaCl to facilitate centrifugation, and the reducing extractant was buffered with $NaHCO_3$ (Mehra & Jackson, 1960) to minimize dissolution of apatite and other less basic forms of Ca-P. A second extraction with NaOH was added prior to the acid extraction of Ca-P. The remaining P was further fractionated into residual organic P (by ignition) and residual inorganic P (by fusion).

In addition to these modifications, Williams et al. (1967) called attention to the incompatibility of the NH_4F reagent with calcareous soils. Later work indicated that the problem stemmed from reaction of NH_4F with calcite to form CaF_2 which strongly sorbed P (Williams et al., 1971). Syers et al. (1972) reported that P sorbed on CaF_2 was subsequently recovered in the reductant soluble and the Ca-P fractions, leading to overestimation of these two fractions and compensating underestimation of Fe-P and Al-P fractions. These complications were sufficiently serious to warrant a complete revamping of the fractionation procedure specifically for calcareous soils (Williams et al., 1971; Syers et al., 1972). The NH_4F reagent was omitted altogether, as no attempt was made to distinguish between Fe- and Al-bound P. The fractionation procedure consisted of sequential extraction with (i) $0.1N$ NaOH to recover both Fe- and Al-bound P, (ii) $1N$ NaCl twice, (iii) citrate-bicarbonate (CB) to remove P sorbed by carbonate surfaces during the NaOH extraction, (iv) citrate-dithionite-bicarbonate (CDB) to remove reductant soluble P, and (v) $1N$ HCl to remove Ca-bound P. The authors recommended that extraction be carried out with and without P addition to the NaOH extractant (155 µg of P/g of soil) to aid in tracing and correcting for resorption. It is worth noting here that problems encountered with calcareous soils seem to be related to free calcite. The NH_4F reagent did not react with dolomite to produce CaF_2 or MgF_2 (Syers et al., 1972).

Given the uncertainty that persists in delineating Al-bound from Fe-bound P through the use of NH_4F, we are not certain that fractionation schemes are selective enough to give reliable data, whether the soils are calcareous or noncalcareous. There is yet another uncertainty that clouds the basic concepts of all fractionation methods, especially with regard to heavily fertilized soils where intermediate reaction products persist, and where such products cannot be simplistically classified as Ca-, Fe-, and Al-P. Lindsay et al. (W. L. Lindsay, P. F. Pratt, F. L. Blair, and A. W. Frazier. 1968. Effectiveness of the Chang and Jackson procedure for extracting well-characerized phosphorus compounds from soils. Agron. Abstr., p. 84) reported that the Petersen and Corey version (1966) of the fractionation scheme was satisfactory for stable reactions products such as variscite, strengite, and hydroxyapatite, but could not distinguish between metastable products such as dicalcium phosphate, octocalcium phosphates, taranakites, and amorphous Fe and Al phosphate.

Analysis of discrete phosphate grains in several cultivated and uncultivated Connecticut soils with an electron microprobe indicated that composition of precipitated soil P was quite variable and consisted of Fe, Al, Si, and, in some cases, Ca, intimately mixed with P. None of the grains examined showed a composition similar to strengite or variscite (Sawhney, 1973).

In view of the limitations of fractionation methods as discussed above, investigators should exercise discretion in the use and interpretation of such data.

IV. THE INTEGRATED OVERVIEW

A. Status of P Supply

Ability to predict the status of P supply in soils would improve the precision of fertilizer recommendations and utilization of fertilizer P. It would allow the farmer to utilize available reserves, if present, and to reduce the pollution potential connected with excessive P supplies. Prediction of P-supplying power of a soil depends on measurement of various parameters affecting P uptake by plant roots, a knowledge of the interrelationships among these factors, and their changes with time. Computer simulation models have been described to account for most of the known variables (Stewart et al., 1973; Helyar & Munns, 1975). Only the plant can measure the amount of available P in soils. Chemical tests and simulation models must be evaluated by correlation with appropriate plant observations.

The plant's P requirement for optimum growth, the nature of its root system, and the P status of the soil are all interrelated factors. Thus, a knowledge of the mechanism of P uptake by roots will assist the interpretation of soil chemical tests and aid in developing useful tests that function for a wider range of soil characteristics. Khasawneh and Copeland (1973) reported that P uptake by cotton was a singular function of root length regardless of P treatment or harvest date. Newman and Andrews (1973) also found a good correlation between P uptake by winter wheat and root growth irrespective of root density or plant age. Phosphorus uptake during a period was more closely correlated with root growth during that period than with the total amount of roots on the plant.

Brewster et al. (1976) observed a difference between rape and onions in their responsiveness to the level of soil P. The greater ability of rape to extract P from poor soil appeared to be related to its long and abundant root hairs, but could not be wholly explained by the roots and their hairs acting as simple sinks for P.

The inadequacy of using a single parameter to describe the P status of a range of soils has been evident for some time. At least three factors interact with each other to determine the functional relationship between P uptake and the status of P in the soil. An intensity factor describes the concentration of P in the soil solution (I), or it may be expressed more appropriately as the activity of the ion in solution. A quantity term (q) is a measure of the amount of P in reserve associated with the solid phase. A buffering capacity term, $\Delta q/\Delta I$, measures the resistance of the soil system to changes in I. Khasawneh (1971) designated a replenishment factor as the result of interaction of these three mutually dependent parameters—intensity, quantity, and buffering capacity. The replenishment factor influences the supply pattern in several ways. Khasawneh (1971) postulated that P uptake would be proportional to quantity and intensity; but, in soils of equal P intensity, P uptake would be proportional to buffering capacity, whereas in soils of equal quantities of P, uptake will be inversely proportional to the buffering capacity. Phosphorus uptake data by ryegrass from greenhouse experiments on 24 soils confirmed these concepts (Holford & Mattingly, 1967b). Results on rate of P uptake by cotton showed that a term combining the factors of intensity, quantity, and buffering capacity was a better measure of soil P status than either quantity or intensity alone (Khasawneh & Copeland, 1973). Phosphorus uptake by ryegrass from a range of soils was better correlated when intensity and quantity factors were considered together (Gunary & Sutton, 1967).

The kinetic model of soil P developed by Helyar and Munns (1975) included P diffusion to roots and several soil and plant factors. The quantity factor was treated as adsorbed P occurring mainly on the surfaces of soil minerals. Since a continuum of reaction rates probably exists between the various adsorbed and crystalline phosphates, their distinction on a kinetic basis is difficult. Thus, it was convenient to define adsorbed P as a phase which exchanges with solution P rapidly—within 24 hours. This approach eliminated a need to define a rate equation for the exchange between solution and adsorbed phases, since the exchange is adequately characterized by the sorption isotherm measured at 24 hours. These authors included many effects of soil properties on solution P in their model by defining the effects of pH, Fe and Al oxide contents, and solution Ca concentration on the isotherms. Their model takes into account the rate of immobilization of the rapidly reacting P forms. In acid soils this immobilization may be an important cause of depletion of available soil P. Such information would be useful in assessing the value of pH amendments, and in deciding the relative merits of frequent small applications of P fertilizers vs. infrequent large applications.

Under conditions where the rate of P removal is high or continued for a long period, the release of nonlabile P may be significant and should be taken into account (Larsen & Sutton, 1963). Grunes et al. (1955) found that soil cropped for 38 years, with addition of fertilizer P, had a similar level of labile P to that in a comparable uncropped virgin soil. Mobilization of nonlabile P in the cropped soil must have kept pace with its removal by crops. Organic P declined from the virgin to the cropped soil, and it probably contributed some to P removal. However, available P in the soil was not correlated with organic P. Vaidyanathan and Talibudeen (1965) measured the extent to which nonlabile P in a soil can restore a given loss of labile P via crop removal. Significant recoveries occurred during 8 weeks of incubation of moist soil, but a full recovery was not observed where large removals of P occurred by treatment of the soil with anion plus cation exchange resins.

B. Concluding Remarks

The physical-chemical methods of assessing soil P as discussed in this chapter complement soil tests, aid in interpreting soil test data, and often play a role in choosing one test in preference to another. The utility of physical-chemical data, however, far exceeds this limited application and extends to the timing and placement methods of P fertilizers, their chemical composition, water solubility, granular nature, and several other aspects of soil-fertilizer-plant interactions. To be complete and useful, a physical-chemical assessment of soil P should be such that it can be combined with soils and crops data to generate answers on sources and rates of P fertilizers to optimize crop production under a variety of soil and crop combinations, with the attendant variation in climate and farming practices. For such a goal, however, physical-chemical data will have to become increasingly predictive rather than just descriptive. Research is needed to find ways of incorporating static measurements of such parameters as quantity and intensity with kinetic parameters to give a clearer and more definitive measure of P supply. Such research should be extended to disequilibrium soils characterized by localized P reaction zones rather than limited to homogenized equilibrium laboratory soils. Research is also needed to find ways and means of utilizing residual P in high-P soils without sacrificing yield potentials. Physical-chemical measurements, for example, have indicated that high P levels in calcareous soils are characterized by the presence of OCP, and that when soils are cropped down to the level at which they begin to respond to P additions, the soil solution data fall off the OCP solubility isotherm. What is needed then is a way to measure how much OCP was there initially, not just whether or not it was there. Furthermore, it is desirable to know (before the fact) how many crops can be grown on such soils before P additions become profitable. Or, alternatively, would the strategy of cropping down residual P be less profitable than that of applying low maintenance rates? The answers to these and other questions are likely to be more forthcoming from an interdisciplinary approach than from a purely physical-chemical approach. Agronomic data on crop growth, rooting patterns, nutrient demands, etc., are just as essential as the seemingly more sophisticated physical-chemical measurements of soil P. It is hoped that the merging of efforts in these and other related botanical and plant physiological disciplines would be attempted in earnest in the very near future.

LITERATURE CITED

Adams, F. 1971. Ionic concentrations and activities in soil solutions. Soil Sci. Soc. Am. Proc. 35:420–426.
Adamson, A. W. 1960. Physical chemistry of surfaces. p. 470–471. Wiley and Sons (Interscience Publishers), New York.
Amer, F., D. R. Bouldin, C. A. Black, and F. R. Duke. 1955. Characterization of soil phosphorus by anion exchange resin adsorption and ^{32}P-equilibration. Plant Soil 6:391–408.
Andrew, C. S. 1966. A kinetic study of phosphate absorption by excised roots of *Stylosanthes humilis, Phaseolus lathyroides, Desmodium uncinatum, Medicago sativa,* and *Hordeum vulgare.* Aust. J. Agric. Res. 17:611–624.
Arambarri, P., and O. Talibudeen. 1959. Factors influencing the isotopically exchangeable phosphate in soils. Plant Soil 11:343–354.
Asher, C. J., and J. F. Loneragan. 1967. Response of plants to phosphate concentration in solution culture: I. Growth and phosphorus content. Soil Sci. 103:225–233.
Asher, C. J., P. G. Ozanne, and J. F. Loneragan. 1965. A method for controlling the ionic environment of plant roots. Soil Sci. 100:149–156.
Aslyng, H. C. 1954. The lime and phosphate potentials of soils; the solubility and availability of phosphates. p. 1–50. *In* Roy. Vet. Agric. Coll. Yearbook (reprint). Copenhagen, Denmark.
Atkinson, R. J., A. M. Posner, and J. P. Quirk. 1972. Kinetics of isotopic exchange of phosphate at the α-FeOOH-aqueous solution interface. J. Inorg. Nucl. Chem. 34:2201–2211.
Bache, B. W., and E. G. Williams. 1971. A phosphate sorption index for soils. J. Soil Sci. 22:289–301.
Barber, S. A. 1976. Efficient fertilizer use. p. 13–29. *In* F. L. Patterson (ed.) Agronomic research for food. Am. Soc. Agron. Spec. Pub. no. 26, Madison, Wis.
Barrow, N. J. 1967. Relationship between uptake of phosphorus by plants and other phosphorus potential and buffering capacity of the soil—an attempt to test Schofield's hypothesis. Soil Sci. 104:99–106.
Barrow, N. J. 1974a. Effect of previous additions of phosphate on phosphate adsorption by soils. Soil Sci. 118:82–89.
Barrow, N. J. 1974b. The slow reactions between soil and anions: 1. Effects of time, temperature, and water content of a soil on the decrease in effectiveness of phosphate for plant growth. Soil Sci. 118:380–386.
Barrow, N. J., and T. C. Shaw. 1975a. The slow reactions between soil and anions: 2. Effect of time and temperature on the decrease in phosphate concentration in the soil solution. Soil Sci. 119:167–177.
Barrow, N. J., and T. C. Shaw. 1975b. The slow reactions between soil and anions: 3. The effect of time and temperature on the decrease in isotopically exchangeable phosphate. Soil Sci. 119:190–197.
Barrow, N. J., and T. C. Shaw. 1975c. The slow reactions between soil and anions: 5. Effects of period of prior contact on the desorption of phosphate from soils. Soil Sci. 119:311–320.
Beckett, P. H. T. 1972. Critical cation activity ratios. Adv. Agron. 24:379–412.

Beckett, P. H. T., and R. E. White. 1964. Studies on the phosphate potential of soils. III. The pool of labile inorganic phosphate. Plant Soil 21:253–282.

Beckwith, R. S. 1965. Sorbed phosphate at standard supernatant concentration as an estimate of the phosphate needs of soils. Aust. J. Exp. Agric. Anim. Husb. 5:52–58.

Bennett, A. C., and F. Adams. 1976. Solubility and solubility product of dicalcium phosphate dihydrate in aqueous solutions and soil solutions. Soil Sci. Soc. Am. J. 40:39–42.

Black, C. A. 1968. Soil-plant relationships, 2nd Ed. John Wiley and Sons, Inc., New York.

Brewster, J. L., K. K. S. Bhat, and P. H. Nye. 1975a. The possibility of predicting solute uptake and plant growth response from independently measured soil and plant characteristics. III. The growth and uptake of onions in a soil fertilized to different initial levels of phosphate and a comparison of the results with model predictions. Plant Soil 42:197–226.

Brewster, J. L., K. K. S. Bhat, and P. H. Nye. 1976. The possibility of predicting solute uptake and plant growth response from independently measured soil and plant characteristics. V. The growth and phosphorus uptake of rape in soil at a range of phosphorus concentrations and a comparison of results with the predictions of a simulation model. Plant Soil 44:295–328.

Brewster, J. L., A. N. Gancheva, and P. H. Nye. 1975b. The determination of desorption isotherms for soil phosphate using low volumes of solution and an anion exchange resin. J. Soil Sci. 26:364–377.

Bromfield, S. M. 1970. The inadequacy of corrections for resorption of phosphate during the extraction of aluminum-bound soil phosphate. Soil Sci. 109:388–390.

Buck, R. P. 1976. Ion selective electrodes. Anal. Chem. 48:23R–39R.

Carter, O. G., and D. J. Lathwell. 1967. Effects of temperature on orthophosphate absorption by excised corn roots. Plant Physiol. 42:1407–1412.

Chang, S. C., and M. L. Jackson. 1957. Fractionation of soil phosphorus. Soil Sci. 84:133–144.

Chien, S. H., and C. A. Black. 1976. Free energy of formation of carbonate apatites in some phosphate rocks. Soil Sci. Soc. Am. J. 40:234–239.

Claassen, N., and S. A. Barber. 1974. A method for evaluating the influence of concentration on the uptake rate of nutrients. Plant Physiol. 54:564–568.

Cole, C. V., S. R. Olsen, and C. O. Scott. 1953. The nature of phosphate sorption by calcium carbonate. Soil Sci. Soc. Am. Proc. 17:352–356.

Cooke, I. J. 1966. A kinetic approach to the description of soil phosphate status. J. Soil Sci. 17:56–64.

Cooke, I. J., and J. Hislop. 1963. Use of an anion-exchange resin for the assessment of available soil phosphorus. Soil Sci. 96:308–312.

Crickmore, P. J., and B. W. Wojciechowski. 1977. Kinetics of adsorption on energetically heterogeneous surfaces. J. Chem. Soc. Farad. Trans. I. 73:1216–1223.

Drew, M. G., and P. H. Nye. 1970. The supply of nutrient ions by diffusion to plant roots in soil. III. Uptake of phosphate by roots of onion, leek, and rye-grass. Plant Soil 33:545–563.

Edgington, D. N. 1965. A proposed mechanism for the uptake of radioactive tracers by an *in vitro* hydroxy apatite system. Radiat. Res. 25:257–268.

Edwards, D. G. 1968. The mechanism of phosphate absorption by plant roots. Int. Congr. Soil Sci. Trans. 9th (Adelaide, Aust.) II:183–190.

Edwards, D. G. 1970. Phosphate absorption and long-distance transport in wheat seedlings. Aust. J. Biol. Sci. 23:255–264.

Epstein, E. 1972. Mineral nutrition of plants: Principles and perspectives. Wiley, New York.

Epstein, E., and C. E. Hagen. 1952. A kinetic study of the absorption of alkali cations by barley roots. Plant Physiol. 27:457–474.

Fife, C. V. 1959. An evaluation of ammonium fluoride as a selective extractant for aluminum-bound soil phosphate: II. Preliminary studies on soils. Soil Sci. 87:83–88.

Fife, C. V. 1962. An evaluation of ammonium fluoride as a selective extractant for aluminum-bound soil phosphate: III. Detailed studies on selected soils. Soil Sci. 93:113–123.

Fitter, A. H. 1974. A relationship between phosphorus requirements, the immobilization of added phosphate, and the phosphate buffering capacity of colliery shales. J. Soil Sci. 25:41–50.

Fitter, A. H., and C. D. Sutton. 1975. The use of the Freundlich isotherm for soil phosphate sorption data. J. Soil Sci. 26:241–246.

Fox, R. L., and E. J. Kamprath. 1970. Phosphate sorption isotherms for evaluating the phosphate requirements of soils. Soil Sci. Soc. Am. Proc. 34:902–907.

Fox, R. L., D. L. Plucknett, and A. S. Whitney. 1968. Phosphate requirements of Hawaiian latosols and residual effects of fertilizer phosphorus. Int. Congr. Soil Sci. Trans. 9th (Adelaide, Aust.) II:301-310.

Fried, M. 1964, "E," "L," and "A" values. Int. Congr. Soil Sci. Trans. 8th (Bucharest, Romania) IV:29-39.

Fried, M., and H. Broeshart. 1967. The soil-plant system in relation to inorganic nutrition. Academic Press, New York.

Gardner, B. R., and J. P. Jones. 1973. Effects of temperature on phosphate sorption isotherms and phosphate desorption. Comm. Soil Sci. Plant Anal. 4:83-93.

Gauch, H. G. 1972. Inorganic plant nutrition. Dowden, Hutchinson, and Ross, Inc., Stroudsburg, Pa.

Glasstone, S. 1946. Textbook of physical chemistry, 2nd Ed. D. Van Nostrand Co., Inc., Princeton, N.J.

Gregory, T. M., E. C. Moreno, and W. E. Brown. 1970. Solubility of $CaHPO_4 \cdot 2H_2O$ in the system $Ca(OH)_2-H_3PO_4-H_2O$ at 5, 15, 25, and 37.5°C. J. Res. Natl. Bur. Stand. 74A: 461-475.

Griffin, R. A., and J. J. Jurinak. 1973a. The interaction of phosphate with calcite. Soil Sci. Soc. Am. Proc. 37:847-850.

Griffin, R. A., and J. J. Jurinak. 1973b. Test of a new model for the kinetics of adsorption-desorption processes. Soil Sci. Soc. Am. Proc. 37:869-872.

Griffin, R. A., and J. J. Jurinak. 1973c. Estimation of activity coefficients from the electrical conductivity of natural aquatic systems and soil extracts. Soil Sci. 116:26-30.

Griffin, R. A., and J. J. Jurinak. 1974. Kinetics of the phosphate interaction with calcite. Soil Sci. Soc. Am. Proc. 38:75-79.

Grunes, D. L., H. J. Haas, and S. H. Shik. 1955. Effect of longtime dryland cropping on available phosphorus of Cheyenne fine sandy loam. Soil Sci. 80:127-138.

Gunary, D. 1970. A new adsorption isotherm for phosphate in soil. J. Soil Sci. 21:72-77.

Gunary, D., and C. D. Sutton. 1967. Soil factors affecting plant uptake of phosphate. J. Soil Sci. 18:167-173.

Hagen, C. E., and H. T. Hopkins. 1955. Ionic species in orthophosphate absorption by barley roots. Plant Physiol. 30:193-199.

Halsey, G., and H. S. Taylor. 1947. The adsorption of hydrogen on tungsten powders. J. Chem. Phys. 15:624-630.

Helyar, K. R., and D. N. Munns. 1975. Phosphate fluxes in the soil-plant system: A computer simulation. Hilgardia 43(4):103-130.

Hendrix, J. E. 1967. The effect of pH on the uptake and accumulation of phosphate and sulfate ions by bean plants. Am. J. Bot. 54:560-564.

Higinbotham, N. 1973. Electropotentials of plant cells. Annu. Rev. Plant Physiol. 24:25-46.

Hingston, F. J., A. M. Posner, and J. P. Quirk. 1974. Anion adsorption by goethite and gibbsite. II. Desorption of anions from hydrous oxide surfaces. J. Soil Sci. 25:16-26.

Hislop, J., and I. J. Cooke. 1968. Anion exchange resin as a means of assessing soil phosphate status:. A laboratory technique. Soil Sci. 105:8-11.

Hodges, T. K. 1973. Ion absorption by plant roots. Adv. agron. 25:163-207.

Hofstee, B. H. J. 1952. On the evaluation of the constants V_m and K_m in enzyme reactions. Science 116:329-331.

Holford, I. C. R., and G. E. G. Mattingly. 1976a. A model for the behavior of labile phosphate in soil. Plant Soil 44:219-229.

Holford, I. C. R., and G. E. G. Mattingly. 1976b. Phosphate adsorption and plant availability of phosphate. Plant Soil 44:377-389.

Holford, I. C. R., R. W. M. Wedderburn, and G. E. G. Mattingly. 1974. A Langmuir two-surface equation as a model for phosphate adsorption by soils. J. Soil Sci. 25:242-255.

Hope, A. B. 1971. Ion transport and membranes, a biophysical outline. Butterworth, London.

Hossner, L. R., J. A. Freeouf, and B. L. Folsom. 1973. Solution phosphorus concentration and growth of rice (*Oryza sativa* L.) in flooded soils. Soil Sci. Soc. Am. Proc. 27:405-408.

Jackson, M. L. 1958. Soil chemical analysis. Prentice-Hall, Inc., Englewood Cliffs, N.J.

Jensen, H. E. 1970. The phosphate potential and phosphate capacity of soils. Plant Soil 33:17-29.

Jensen, H. E. 1971. Phosphate solubility in Danish soils equilibrated with solutions of differing phosphate concentrations. J. Soil Sci. 22:261-266.

Jose, A. I., and K. K. Krishnamoorthy. 1972. Isotopic exchange of phosphates in soil: E value. Soils Fert. 35:620-627.

Kafkafi, U., A. M. Posner, and J. P. Quirk. 1967. Desorption of phosphate from kaolinite. Soil Sci. Soc. Am. Proc. 31:348-353.

Khasawneh, F. E. 1971. Solution ion activity and plant growth. Soil Sci. Soc. Am. Proc. 35: 426-436.

Khasawneh, F. E., and F. Adams. 1967. Effect of dilution on calcium and potassium contents of soil solutions. Soil Sci. Soc. Am. Proc. 31:172-176.

Khasawneh, F. E., and J. P. Copeland. 1973. Cotton root growth and uptake of nutrients: Relation of phosphorus uptake to quantity, intensity, and buffering capacity. Soil Sci. Soc. Am. Proc. 37:250-254.

Khasawneh, F. E., and E. C. Doll. 1978. The use of phosphate rock for direct application to soils. Adv. Agron. 30:159-206.

Khasawneh, F. E., E. C. Sample, and I. Hashimoto. 1974. Reactions of ammonium ortho- and polyphosphate fertilizers in soil: I. Mobility of phosphorus. Soil Sci. Soc. Am. Proc. 38: 446-451.

Kukura, M., L. C. Bell, A. M. Posner, and J. P. Quirk. 1973. Kinetics of isotope exchange on hydroxyapatite. Soil Sci. Soc. Am. Proc. 37:364-466.

Kuo, S., and E. G. Lotse. 1972. Kinetics of phosphate adsorption by calcium carbonate and Ca-kaolinite. Soil Sci. Soc. Am. Proc. 36:725-729.

Kuo, S., and E. G. Lotse. 1974a. Kinetics of phosphate adsorption and desorption by hematite and gibbsite. Soil Sci. 116:400-406.

Kuo, S., and E. G. Lotse. 1974b. Kinetics of phosphate adsorption and desorption by lake sediments. Soil Sci. Soc. Am. Proc. 38:50-54.

Kyle, J. H., A. M. Posner, and J. P. Qurirk. 1975. Kinetics of isotopic exchange of phosphate adsorbed on gibbsite. J. Soil Sci. 26:32-43.

Langmuir, I. 1918. The adsorption of gases on plane surfaces of glass, mica, and platinum. J. Am. Chem. Soc. 40:1361-1402.

Larsen, S. 1950. Studies on the uptake of phosphorus in plants with radiophosphorus as an indicator. K. Veterinaerog Landhohøgskole, Copenhagen, Denmark.

Larsen, S. 1952. The use of P-32 in studies on the uptake of phosphorus by plants. Plant Soil 4:1-10.

Larsen, S. 1965. The influence of calcium chloride concentration on the determination of lime and phosphate potentials of soil. J. Soil Sci. 16:285-278.

Larsen, S. 1967. Soil phosphorus. Adv. Agron. 19:151-210.

Larsen, S., and C. D. Sutton. 1963. The influence of soil volume on the absorption of soil phosphorus by plants and on the determination of labile soil phosphorus. Plant Soil 18: 77-84.

Larsen, S., and A. E. Widdowson. 1971. Ageing of phosphate added to soil. J. Soil Sci. 22:5-7.

Li, W. C., D. E. Armstrong, J. D. H. Williams. R. F. Harris, and J. K. Syers. 1972. Rate and extent of inorganic phosphate exchange in lake sediments. Soil Sci. Soc. Am. Proc. 36: 279-285.

Lindsay, W. L., and E. C. Moreno. 1960. Phosphate phase equilibria in soils. Soil Sci. Soc. Am. Proc. 24:177-182.

Lindstrom, F. T., R. Haque, and W. R. Coshow. 1970. Adsorption from solution. III. A new model for the kinetics of adsorption-desorption processes. J. Phys. Chem. 74:495-502.

Loneragan, J. F., and C. J. Asher. 1967. Response of plants to phosphate concentration in solution culture. II. Rate of phosphate absorption and its relation to growth. Soil Sci. 103: 311-318.

MacRobbie, E. A. C. 1971. Fluxes and compartmentation in plant cells. Annu. Rev. Plant Physiol. 22:75-96.

Marion, G. M., and K. L. Babcock. 1976. Predicting specific conductance and salt concentration in dilute aqueous solutions. Soil Sci. 122:181-187.

Mattingly, G. E. G. 1957. The use of the isotope ^{32}P in recent work on soil and fertilizer phosphorus. Soils Fert. 20:59-68.

Mattingly, G. E. G. 1965. The influence of intensity and capacity factors on the availability of soil phosphate. p. 1-9. In Soil phosphorus. Minist. of Agric., Fisher., and Food Tech. bull. no. 13, H. M. Stat. Office, London.

Mattingly, G. E. G. 1975. Labile phosphate in soils. Soil Sci. 119:369-375.

Mattingly, G. E. G., R. D. Russell, and B. M. Jephcott. 1963. Experiments on cumulative dressings of fertilizers in calcareous soils in South-West England: II. Phosphorus uptake by ryegrass in the greenhouse. J. Sci. Food Agric. 14:629-637.

Mattingly, G. E. G., and O. Talibudeen. 1967. Progress in the chemistry of fertilizer and soil phosphorus. p. 157-290. *In* M. Grayson and E. J. Griffith (ed.) Topics in phosphorus chemistry, Vol. 4. Interscience Publ., New York.

McAuliffe, C. D., N. S. Hall, L. A. Dean, and S. B. Hendricks. 1948. Exchange reactions between phosphates and soils: Hydroxylic surface of soil minerals. Soil Sci. Soc. Am. Proc. 12:119-123.

McLaughlin, J. R., J. C. Ryden, and J. K. Syers. 1977. Development and evaluation of a kinetic model to describe phosphate sorption by hydrous ferric oxide gel. Geoderma 18: 295-307.

Mehra, O. P., and M. L. Jackson. 1960. Iron oxide removal from soils and clays by a dithionite-citrate system buffered with sodium bicarbonate. Proc. 7th Natl. Conf. Clays, 1959. 5:317-327.

Moser, U. S., W. H. Sutherland, and C. A. Black. 1959. Evaluation of laboratory indexes of absorption of soil phosphorus by plants: I. Plant Soil 10:356-374.

Moss, P. 1963. Some aspects of the cation status of soil moisture. I. The ratio law and soil moisture content. Plant Soil 18:99-113.

Muljadi, D., A. M. Posner, and J. P. Quirk. 1966a. The mechanism of phosphate adsorption by kaolinite, gibbsite, and pseudoboehmite. Part I. The isotherm and the affect of pH on adsorption. J. Soil Sci. 17:212-229.

Muljadi, D., A. M. Posner, and J. P. Quirk. 1966b. The mechanism of phosphate adsorption by kaolinite, gibbsite, and pseudoboehmite. Part II. The location of the adsorption sites. J. Soil Sci. 17:230-237.

Muljadi, D., A. M. Posner, and J. P. Quirk. 1966c. The mechanism of phosphate adsorption by kaolinite, gibbsite, and pseudoboehmite. Part III. The effect of temperature on the adsorption. J. Soil Sci. 17:238-247.

Murrman, R. P., and M. Peech. 1969a. Effect of pH on labile and soluble phosphate in soils. Soil Sci. Soc. Am. Proc. 33:207-210.

Murrman, R. P., and M. Peech. 1969b. Relative significance of labile and crystalline phosphates in soil. Soil Sci. 107:249-255.

Neuman, W. F., and M. W. Neuman. 1958. The chemical dynamics of bone mineral. Univ. Chicago Press, Chicago, Ill.

Newman, E. I., and R. E. Andrews. 1973. Uptake of phosphorus and potassium in relation to root growth and root density. Plant Soil 38:49-69.

Nishimoto, R. K., R. L. Fox, and P. E. Parvin. 1977. Response of vegetable crops to phosphorus concentrations in soil solution. J. Am. Soc. Hortic. Sci. 102:705-709.

Nissen, Per. 1974. Uptake mechanisms: Inorganic and organic. Annu. Rev. Plant Physiol. 25:53-79.

Novak, L. T., and D. C. Adriano. 1975. Phosphorus movement in soils: soil-orthophosphate reaction kinetics. J. Environ. Qual. 4:261-266.

Nriagu, J. O. 1972a. Solubility equilibrium constant of strengite. Am. J. Sci. 272:476-484.

Nriagu, J. O. 1972b. Stability of vivianite and ion-pair formation in the system $Fe_3(PO_4)_2$-H_3PO_4-H_2O. Geochim. et Cosmochim. Acta 36:459-470.

Nriagu, J. O. 1972c. Lead orthophosphate. I. Solubility and hydrolysis of secondary lead orthophosphate. Inorg. Chem. 11:2499-2503.

Nriagu, J. O. 1973. Solubility equilibrium constant of α-hopeite. Geochim. Cosmochim. Acta 37:2357-2361.

Nye, P. H. 1968. The use of exchange isotherms to determine diffusion coefficients in soil. Int. Congr. Soil Sci. Trans. 9th (Adelaide, Aust.) I:117-126.

Olsen, S. R. 1953. Inorganic phosphorus in alkaline and calcareous soils. p. 89-122. *In* W. H. Pierre and A. G. Norman (ed.) Soil and fertilizer phosphorus in crop nutrition. Academic Press, Inc., New York.

Olsen, S. R., C. V. Cole, F. S. Watanabe, and L. A. Dean. 1954. Estimation of available phosphorus in soils by extraction with sodium bicarbonate. USDA Circ. 939.

Olsen, S. R., and A. D. Flowerday. 1971. Fertilizer phosphorus interactions in alkaline soils. p. 153-185. *In* R. A. Olson, T. J. Army, J. J. Hanway, and V. J. Kilmer (ed.) Fertilizer technology and use, 2nd Ed. Soil Sci. Soc. Am., Madison, Wis.

Olsen, S. R., and W. D. Kemper. 1968. Movement of nutrients to plant roots. Adv. Agron. 20: 91-151.

Olsen, S. R., W. D. Kemper, and J. C. van Schaik. 1965. Self-diffusion coefficients of phosphorus in soil measured by transient and steady-state methods. Soil Sci. Soc. Am. Proc. 29:154-158.

Olsen, S. R., and F. S. Watanabe. 1957. A method to determine a phosphorus adsorption maximum of soils as measured by the Langmuir isotherm. Soil Sci. Soc. Am. Proc. 21: 144-149.

Olsen, S. R., and F. S. Watanabe. 1970. Diffusive supply of phosphorus in relation to soil textural variations. Soil Sci. 110:318-327.

Ozanne, P. G., and T. C. Shaw. 1967. Phosphate sorption by soils as a measure of the phosphate requirement for pasture growth. Aust. J. Agric. Res. 18:601-612.

Ozanne, P. G., and T. C. Shaw. 1968. Advantages of the recently developed phosphate sorption test over the older extractant methods for soil phosphate. Int. Congr. Soil Sci. Trans. 9th (Adelaide, Australia) II:273-280.

Parfitt, R. L., R. J. Atkinson, and R. St. C. Smart. 1975. The mechanism of phosphate fixation by iron oxides. Soil Sci. Soc. Am. Proc. 39:837-841.

Pearson, R. W. 1971. Introduction to symposium—the soil solution. Soil Sci. Soc. Am. Proc. 35:417-420.

Petersen, G. W., and R. B. Corey. 1966. A modified Chang and Jackson procedure for routine fractionation of inorganic soil phosphates. Soil Sci. Soc. Am. Proc. 30:563-565.

Ponnamperuma, F. N., E. M. Tianco, and T. A. Loy. 1966. Ionic strengths of the solutions of flooded soils and other natural aqueous solutions from specific conductance. Soil Sci. 102:408-413.

Probert, M. E., and S. Larsen. 1972. The kinetics of heterogenous isotopic exchange. J. Soil Sci. 23:76-81.

Prue, J. E. 1966. Ionic equilibria. Pergamon Press, Oxford.

Rajan, S. S. S., and R. L. Fox. 1972. Phosphate adsorption by soils. I. Influence of time and ionic environment on phosphate adsorption. Comm. Soil Sci. Plant Anal. 3:493-504.

Rajan, S. S. S., and K. W. Perrott. 1975. Phosphate adsorption by synthetic amorphous alumino silicates. J. Soil Sci. 26:257-266.

Rubinow, S. I. 1975. Introduction to mathematical biology. John Wiley and Sons, New York.

Russell, R. S., and R. P. Martin. 1953. A study of the absorption and utilization of phosphate by young barley plants. I. The effect of external concentration on the distribution of absorbed phosphate between roots and shoots. J. Exp. Bot. 4:108-127.

Ryden, J. C., J. R. McLaughlin, and J. K. Syers. 1977. Mechanisms of phosphate sorption by soils and hydrous ferric oxide gel. J. Soil Sci. 28:72-92.

Ryden, J. C., and J. K. Syers. 1977. Desorption and isotopic exchange relationships of phosphate sorbed by soils and hydrous ferric oxide gel. J. Soil Sci. 28:596-609.

Ryden, J. C., J. K. Syers, and R. F. Harris. 1973. Phosphorus in runoff and streams. Adv. Agron. 25:1-45.

Sample, E. C. 1972. Factors affecting phosphate retention parameters derived using the Langmuir adsorption equation. Ph.D. Thesis. North Carolina State Univ., Raleigh, N.C. Univ. Microfilms, Ann Arbor, Mich. (Diss. Abstr. 34:2406B, 1973).

Sawhney, B. L. 1973. Electron microprobe analysis of phosphate in soils and sediments. Soil Sci. Soc. Am. Proc. 37:658-660.

Schofield, R. K. 1955. Can a precise meaning be given to "available" soil phosphorus? Soils Fert. 18:373-375.

Shapiro, R. E., and M. Fried. 1959. Relative release and retentiveness of soil phosphates. Soil Sci. Soc. Am. Proc. 23:195-198.

Sillen, L. G., and A. E. Martell. 1964. Stability constants of metal-ion complexes. Spec. Publ no. 17. The Chemical Society, London.

Sips, Robert. 1950. On the structure of a catalyst surface. II. J. Chem. Phys. 18:1024-1026.

Smith, A. N. 1965. Distinction between iron and aluminum phosphate in Chang and Jackson's procedure for fractionating inorganic soil phosphorus. Agrochimica 9:162-168.

Smith, A. N. 1972. Corrections for resorption of phosphate during the extraction of aluminum-bound soil phosphate. Soil Sci. 113:55-56.

Soltanpour, P. N., F. Adams, and A. C. Bennett. 1974. Soil phosphorus availability as measured by displaced soil solutions, calcium-chloride extracts, dilute-acid extracts, and labile phosphorus. Soil Sci. Soc. Am. Proc. 38:225-228.

Stewart, J. W. B., B. J. Halm, and C. V. Cole. 1973. Nutrient cycling: 1. Phosphorus Tech. Rep. no. 40, Can. Comm. Int. Biol. Prog. (Matador Project).

Syers, J. K., M. G. Browman, G. W. Smillie, and R. B. Corey. 1973. Phosphate sorption by soils evaluated by the Langmuir adsorption equation. Soil Sci. Soc. Am. Proc. 37:358-363.

Syers, J. K., G. W. Smillie, and J. D. H. Williams. 1972. Calcium fluoride formation during extraction of calcareous soils with fluoride: I. Implications to inorganic P fractionation schemes. Soil Sci. Soc. Am. Proc. 36:20-25.

Tandon, H. L. S., and L. T. Kurtz. 1968. Isotopic exchange characteristics of aluminum- and iron-bound fractions of soil phosphorus. Soil Sci. Soc. Am. Proc. 32:799-802.

Taylor, A. W., A. W. Frazier, E. L. Gurney, and J. P. Smith. 1963. Solubility products of di- and trimagnesium phosphates and the dissociation of magnesium phosphate solutions. Trans. Faraday Soc. 59:1585-1589.

Thompson, E. J., A. L. F. Oliveira, U. S. Moser, and C. A. Black. 1960. Evaluation of laboratory indexes of absorption of soil phosphorus by plants. II. Plant Soil 13:28-38.

Vaidyanathan, L. V., and O. Talibudeen. 1965. A laboratory method for the evaluation of nutrient residues in soils. Plant Soil 23:371-376.

Vaidyanathan, L. V., and O. Talibudeen. 1968. Rate-controlling processes in the release of soil phosphate. J. Soil Sci. 19:342-353.

Vaidyanathan, L. V., and O. Talibudeen. 1970. Rate processes in the desorption of phosphate from soils by ion-exchange resins. J. Soil Sci. 21:173-183.

Vaidyanathan, L. V., and P. H. Nye. 1970. The measurement and mechanism of ion diffusion in soils. VI. The effect of concentration and moisture content on the counter-diffusion of soil phosphate against chloride ion. J. Soil Sci. 21:15-27.

Wagman, D. D., W. H. Evans, V. B. Parker, I. Halow, S. M. Bailey, and R. H. Schumm. 1968. Selected values of chemical thermodynamic properties. Natl. Bur. of Stand. Tech. Note 270-3, U.S. Dep. of Commerce.

White, R. E. 1964. Studies on the phosphate potentials of soils. II. Microbial effects. Plant Soil 20:184-193.

White, R. E. 1976. Concepts and methods in the measurement of isotopically exchangeable phosphate in soil. Phosphorus in Agric. no. 67. ISMA Ltd., 1, Ave. Franklin D. Roosevelt, 75008, Paris. p. 9-16.

White, R. E., and P. H. T. Beckett. 1964. Studies on the phosphate potentials of soils. I. The measurement of phosphate potential. Plant Soil 20:1-16.

White, R. E., and A. W. Taylor. 1977. Reactions of soluble phosphate with acid soils: The interpretation of adsorption-desorption isotherms. J. Soil Sci. 28:314-328.

Wier, D. R., S. H. Chien, and C. A. Black. 1971. Solubility of hydroxyapatite. Soil Sci. 111: 107-112.

Wild, A. 1964. Soluble phosphate in soil and uptake by plants. Nature 203:326-327.

Williams, J. D. H., J. K. Syers, R. F. Harris, and D. E. Armstrong. 1971. Fractionation of inorganic phosphate in calcareous lake sediments. Soil Sci. Soc. Am. Proc. 35:250-255.

Williams, J. D. H., J. K. Syers, and T. W. Walker. 1967. Fractionation of soil inorganic phosphate by a modification of Chang and Jackson's procedure. Soil Sci. Soc. Am. Proc. 31: 736-739.

Wilson, A. T. 1968. The chemistry underlying the phosphate problem in agriculture. Aust. J. Sci. 31:55-61.

Winter, E. R. S. 1965. The kinetics of adsorption on a nonuniform surface. J. Catalysis 4:134-139.

Woodruff, J. R., and E. J. Kamprath. 1965. Phosphorus adsorption maximum as measured by the Langmuir isotherm and its relationship to phosphorus availability. Soil Sci. Soc. Am. Proc. 29:148-150.

Zunino, H., M. Aquilera, and P. Peirano. 1972. A modified resin exchange method for measurement of available phosphate in soils derived from volcanic ash. Soil Sci. 114:404-405.

Chapter 15

Assessing Organic Phosphorus in Soils

GEORGE ANDERSON
The Macaulay Institute for Soil Research
Aberdeen, Scotland

I. INTRODUCTION

The presence of P in organic form in soils was first recorded over a century ago. Since then numerous investigators have examined this fraction, studying its quantity, its nature, and its properties in relation to soil fertility. There is an extensive literature on the subject, and this review seeks to summarize our present state of knowledge regarding the factors influencing the conversion of P to organic forms in soil, the nature of the phosphate esters which accumulate, and the conditions under which the esters contribute to the P nutrition of crops.

II. BIOLOGICAL IMMOBILIZATION OF SOIL AND FERTILIZER P

A. Biological Immobilization of P in Virgin Soils

When a soil first begins to develop, the only P present is in inorganic form, derived from the parent rocks. Gradually, plants, microorganisms, and animals become established, all of which require P as an essential major nutrient, converting some of it to organic forms within their cells. When the organisms die and decompose, both inorganic and organic P are returned to the soil, initiating a biological P cycle (Fig. 1).

Most naturally occurring organic forms of P are esters of orthophosphoric acid, and numerous mono- and di-esters have been characterized. Derivatives of a phosphonic acid, containing a C–P bond, have been detected in certain organisms, for example rumen protozoa (a composite of members of *Diplodinium, Isotricha, Ophryoscolex, Dasytricha,* and *Entodinium*) (Horiguchi & Kandatsu, 1959), sea-anemones (*Anthopleura elegantissima, Metridium dianthus*), and the pond snail (*Heterogen longispira*),

---

Copyright 1980 © ASA-CSSA-SSSA, 677 South Segoe Road, Madison, WI 53711, USA.
The Role of Phosphorus in Agriculture.

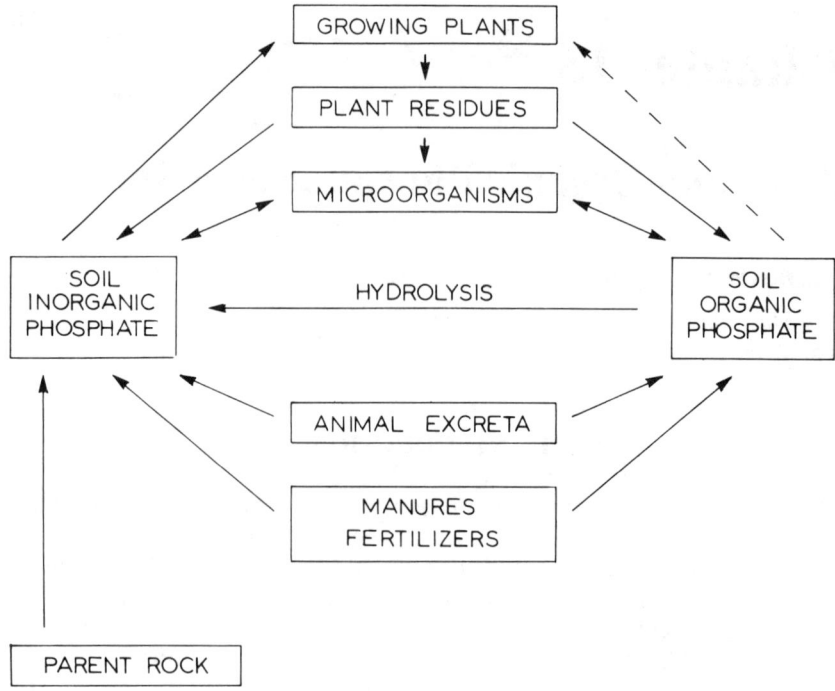

Fig. 1—Phosphorus cycle in soils.

but have not yet been reported in soils.[1] Some phosphate esters, such as nucleic acids, nucleotides, and sugar phosphates, are essential to life and occur in all living cells, though their proportion of the total P may vary widely. Others, for example teichoic acids and inositol phosphates, have been detected in some organisms only, and only in selected tissues of these organisms.

A wide range of esters, differing in nature, quantity, and stability, will constantly be released in the soil from dead organisms. Some esters will be quickly broken down and have only a transient existence, while others will be stabilized and thus accumulate to a relatively high level. The relative proportions of the esters in soil therefore bear little relationship to those in the living organisms from which they were derived.

The extent to which the esters accumulate will depend both on their rate of production and their rate of decomposition. The growth of the organisms synthesizing the esters, and the other materials from which soil organic matter is produced, will be dependent on many factors, including a favorable temperature and a ready supply of water and essential nutrients. Some nutrients, such as C and N, can be obtained from the atmosphere, but others can be derived only from the soil parent material. One of the most important of them is P, and when it occurs in small amounts, production of

---

[1] Since this chapter was prepared, phosphonate has been detected by nuclear magnetic resonance spectrometry in two New Zealand virgin soils (R. H. Newman and K. R. Tate, Symposium on NMR held at Australian National University, Canberra, 1978).

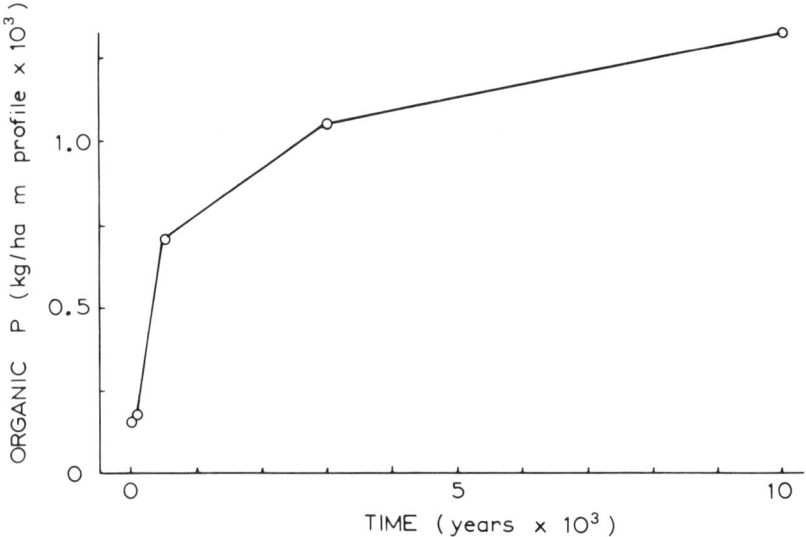

Fig. 2—Accumulation of organic phosphorus with time in a New Zealand soil (Syers & Walker, 1969).

the whole soil organic matter may be limited (Kaila, 1956a; Walker & Adams, 1958).

One example of the accumulation of organic P with time can be seen in a chronosequence of weakly weathered soils in New Zealand (Syers & Walker, 1969). The soils have been developed on wind-blown beach sand and five different stages of development can be distinguished, spanning a period of 10,000 years. The present beach sand is regarded as representing time zero and it is assumed that the parent materials of the soils representing the other stages had the same initial P composition. The organic P content increased steeply at first (Fig. 2), but the rate of increase dropped with time, a trend also followed by organic C and N. The net gain in organic P during the whole period was 1,050 kg of P/ha/m-depth in the profile. In the same time, total P dropped from about 5,000 to about 3,000 kg/ha. Initially, the distribution of total P down the profile was very uniform, but the organic P accumulation occurred mainly in the surface horizons and, as P was redistributed or lost by leaching, the total P also decreased with depth.

Eventually, as P continues to be lost from the profile, the organic P level will decline also. The time taken for this to happen will depend on the rate of weathering and leaching (Walker & Adams, 1958).

B. Changes in Organic P Content of Agricultural Soils

When a virgin soil is brought into cultivation the environment is immediately altered. Plowing, for example, increases aeration and exposes the organic matter to a more vigorous microbial attack, resulting in a net loss. The organic P is not always mineralized at the same rate as the C or N, how-

Table 1—Organic P in the surface horizon (15-cm depth) of uncultivated and cultivated soils in New Jersey (calculated from data of Van Diest, 1968).

Soil type	Cropping system	Soil organic P (kg/ha)	
		Uncultivated	Cultivated
Sandy loam	pasture	98	281
Loamy sand	vegetables and winter-cover	168	367
Sandy loam	vegetables	119	209
Sandy loam	peach orchard	444	750
Sand	peach orchard	290	87
Silt loam	potatoes/wheat	117	333

ever, and the C/organic P and N/organic P ratios in the cultivated soils often differ from those in the corresponding virgin soils (Barrow, 1961).

Application of inorganic fertilizers, by encouraging vigorous plant growth, can subsequently lead to an increase in soil organic matter and its phosphatic components. Because fertilizer nutrients are so readily available, the rate of accumulation of soil organic matter is often very much greater than in soils developing under virgin conditions. For example, Jackman (1964) examined the organic P under New Zealand pastures that had received superphosphate at an annual rate of about 300 kg/ha for periods ranging from 0 to 30 years. The soils accumulated organic P at a mean rate of up to 11 kg of P annually in the top 15 cm, but the rate of increase dropped with time, and a steady state was almost attained in most of them within 30 years. Comparison of uncultivated with cultivated coastal plain soils in New Jersey (Van Diest, 1968) has indicated similar trends (Table 1), where increases in organic P were thought to have occurred over a span of about 30 years during which considerable P fertilizer applications had been made. On this assumption the mean annual rate of increase in organic P ranged from about 3 to 10 kg/ha in the surface 15 cm. Several other observations of this kind have been recorded, for example, in Finland (Kaila & Missilä, 1956) and Germany (Owssia et al., 1966). In areas which have been receiving heavy fertilizer applications for much longer periods of time, such as parts of the United Kingdom, a steady state is now likely to have been attained.

In the case of organic manures, changes in soil organic P will to some extent depend on whether the material is left on the surface of the soil or is plowed in. The importance of the organic P content of the decomposing material in determining whether net synthesis or net mineralization initially occurs was demonstrated by Kaila (1949) in incubation tests with composts and feces. She concluded that no net mineralization of organic P occurs until its content is above a certain critical level which varies considerably in different materials, and on average is about 0.2% of the dry matter. Eventually, as decomposition proceeds, net mineralization will occur in all materials, but some of the organic P will find its way into the soil and part of the increase in soil organic P under pasture is likely to be derived from such sources (Bromfield, 1961).

When organic manures deficient in P are incorporated in the soil, the proliferating microorganisms assimilate the extra P they require from the

Table 2—Effects of farmyard manure on the organic P in continuously cropped soils in southern England (Dean, 1938; Oniani et al., 1973).

Soil type	Crop	Annual FYM addition, metric ton/ha	Soil organic P, $\mu g/g$	
Acidic sandy loam	Barley	from 1843	sampled 1888	sampled 1927
		none	185	200
		25	220	310
Calcareous clay loam	Wheat	from 1843	sampled 1893	sampled 1936
		none	80	90
		35	100	250
Calcareous clay loam	Roots	from 1856	sampled 1958	--
		none	112	--
		35	130	--

soil, increasing the amount of organic P and reducing the amount of inorganic P available to plants. Crop deficiencies can consequently occur initially unless inorganic P is added to the soil along with material of this kind (Chouchak, 1929). As the freshly synthesized energy sources are themselves attacked, most of the phosphate esters in the material will eventually undergo mineralization, but some will be selectively adsorbed by soil colloids in such a way that they are removed from the cycle in a stable form.

Examples of the accumulation of soil organic P in English soils treated with farmyard manure are shown in Table 2. In one of these soils, all of which were continuously cultivated, there was an increase of 150 ppm in organic P in a period of 43 years, though none accumulated when only inorganic P fertilizers were applied. In contrast, organic manuring of a soil in Germany over a period of 30 years did not affect the organic P level, whereas mineral P fertilizers gave an increase (Owssia et al., 1966).

C. Factors Affecting the Equilibrium Level of Organic P in Soils

When soils are receiving a plentiful supply of nutrients and organic residues, they will gradually accumulate organic matter up to an equilibrium level where input and decomposition balance. Although the organic P is well correlated with C and N (Schollenberger, 1920), substantial variations occur in the C/organic P and N/organic P ratios; the factors affecting the equilibrium levels of organic P do not necessarily affect the C or N levels to the same extent. For example, differences in pH, drainage, and cultivation can all influence the C/organic P ratios (Barrow, 1961). Nevertheless, most of the factors affecting the organic P levels do have some corresponding effects on the other organic matter components. The initial effects of cultivation on the organic matter content of virgin soils have already been noted, and the frequency of cultivation and the cropping system can cause similar changes in organic P levels. Thus, in soils that are frequently plowed, the rate of organic matter breakdown is accelerated and the equilibrum level in such soils is lower than in, for example, soils under permanent grass (DeTurk, 1938).

In very acid soils the activity of microorganisms is suppressed and such soils usually contain more organic matter in their surface horizons than comparable soils of higher pH. Liming acid soils decreases organic P (Damsgaard-Sörensen, 1946; Kaila, 1948) by creating a more favorable environment for microbial attack and probably also by increasing the solubility and thus the accessibility of some of the esters. In soils that are only slightly acid, liming has no significant effect, and in some calcareous soils, the organic P rises with calcium carbonate content (Sen Gupta & Cornfield, 1962).

The stability of phosphate esters in soils has been attributed to sorption by soil colloids, and consequently, they are found predominantly in the clay and silt fractions (Williams & Saunders, 1956). Also pertinent to this, clay-rich soils tend to have higher organic P levels than soils of lighter texture. The nature of the clay minerals may also be important; for example, the equilibrium level of organic P in some New Zealand pasture soils is higher when allophane is present (Jackman, 1964). In a range of cultivated Scottish soils, amorphous hydroxides of Al and Fe are probably the most important agents stabilizing the organic components. There is a high correlation between organic C and oxalate-soluble Al; the latter is thought to be largely present as humate complexes containing a fairly constant proportion of the soil organic matter (Williams et al., 1958).

In the same soil types very striking differences in organic P level are associated with drainage (Williams and Saunders, 1956). Comparison of pairs of freely drained and imperfectly drained soils, derived from the same parent materials and cultivated and cropped in the same way, has shown the organic P levels to be much higher in the freely drained samples. Williams and Saunders attributed the differences to both input and output. Good drainage gave better crop growth and higher returns of residues containing organic P. In addition, the freely drained soils contained higher amounts of oxalate-soluble Al which tended to retard decomposition.

In temperate climates net transformations of organic P are usually slow, although seasonal variations have been observed in some instances (Dormaar, 1972). In warmer climates more rapid transformations are likely.

D. Amounts of Organic P in Soil

The levels of organic P which have been reported in soils throughout the world vary enormously, ranging from virtually zero to over 0.2%. The distribution of organic and total P in profiles of a number of soils of various types and geographical origins is shown in Fig. 3. Because large differences can occur within short distances, these profiles do not necessarily typify extensive areas in the regions represented.

In most mineral soils there is a sharp decrease down the profile, but with peats there are often increases over considerable depths (Kaila, 1956b), and the proportion of organic P to inorganic P also increases.

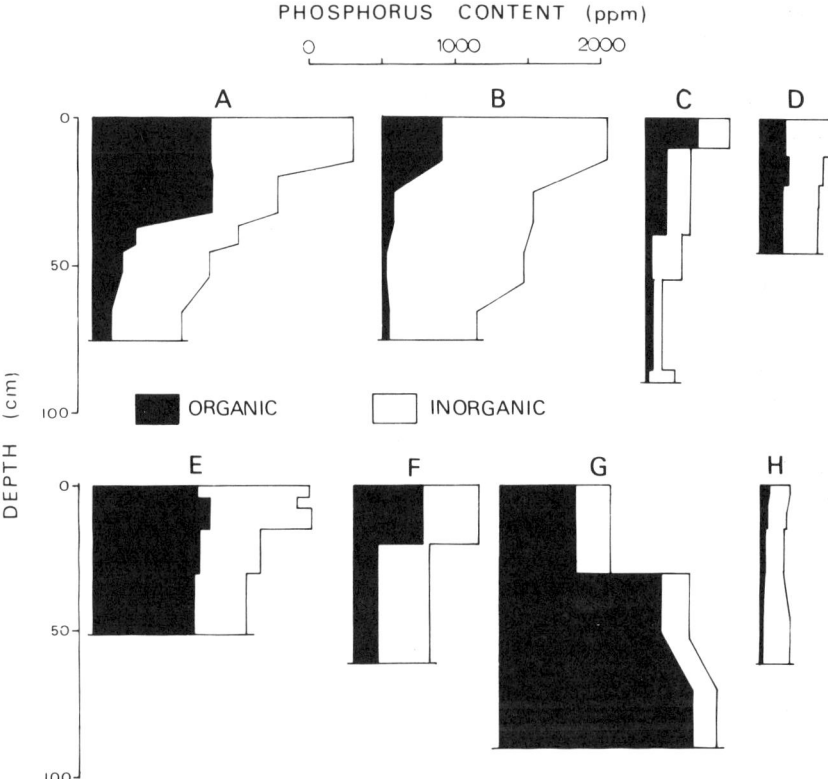

Fig. 3—Distribution of phosphorus in various soil profiles.
  A & B. Freely and poorly drained cultivated clay loams of the Insch Association, Scotland (Williams & Saunders, 1956).
  C. Uncultivated Koputaroa soil developed on windblown sand, New Zealand (Syers & Walker, 1969).
  D. Uncultivated Dawes silt loam, Nebraska (Allaway & Rhoades, 1951).
  E. Uncultivated Pima calcareous clay loam, Arizona (Fuller & McGeorge, 1951).
  F. Cultivated Orthic Deep Black, Melfort, Saskatchewan (R. B. McKercher, 1966. The distribution and significance of various categories of organic and inorganic phosphorus in soils. Ph.D. Thesis. Univ. of Aberdeen).
  G. Uncultivated *Carex globularis* pine bog, northern Finland (Kaila, 1956b).
  H. Leached forest soil, Ibadan, Nigeria (Nye & Bertheux, 1957).

III. CHARACTERIZATION OF SOIL ORGANIC P

A. Outline of Analytical Procedures

A number of procedures can be used to extract organic P from biological tissues and to fractionate it into its various components. Some are based on the method of Schmidt and Thannhauser (1945) in which sequential extraction enables measurements to be made of phospholipids, nucleic acids, phosphoprotein, and acid soluble esters which can be further subdivided by chromatography or electrophoresis (Bieleski & Young, 1963). General fractionations of this kind have not, however, proved successful with soils due to interference by other organic materials and by inorganic components.

The techniques which have given most information about the nature of soil organic P can conveniently be divided into two categories. In the first the soil is treated with reagents which are well suited to the extraction of a particular class of ester known to occur in microorganisms, plants, or animals. Many investigations have been carried out to establish the optimum conditions for extracting and isolating such esters as the phospholipids, the inositol phosphates, and the nucleic acids, so that the total amounts of each can be assessed. To be successful the procedure must break the links, if any, between the esters and other soil components, yet cause minimum breakdown or alteration of the esters themselves.

The second approach, used as yet in very few investigations, aims to establish the association that exists between the phosphate esters and other soil components. Here the whole organic matter is fractionated, for example, on a molecular weight basis, and attempts are made to characterize each of the fractions as fully as possible. A given ester may, in this case, appear in more than one fraction.

Total organic P in soil can be measured by extracting the soil with acid and alkali and determining the inorganic and total P in the extracts, the organic P being obtained by difference. Alternatively, ignited and unignited soil samples are extracted with acid and the difference in the amounts of inorganic P in the extracts is taken as a measure of organic P. The methods in most common use have been reviewed by Anderson (1975).

B. Inositol Phosphates

Inositol phosphates are released in soil at a much slower rate than many other esters, but they are quickly stabilized and have accumulated to such an extent in some soils that they constitute more than half of the organic P and about one quarter of the total P.

Inositol is a homocyclic sugar-like compound, $C_6H_{12}O_6$, which can form a series of phosphate esters ranging from monophosphates up to a hexaphosphate. It can exist in several stereo-isomeric forms (Angyal & Anderson, 1959), and phosphate esters of *myo-*, *scyllo-*, *neo-* and *chiro-*inositol have all been characterized in soil (Cosgrove, 1962; Cosgrove & Tate, 1963). Esters of *myo-*inositol are the most common in nature and, as far as the writer is aware, only *myo-*inositol phosphates have as yet been isolated from plants, animals, or pure microbial cultures. However, other isomers have been detected in the apparently undecomposed surface litter horizons of soils in Arizona, under ponderosa pine and velvet mesquite (L'Annunziata & Fuller, 1971). L'Annunziata (1975) has suggested a possible mechanism for the conversion of *myo-*inositol phosphates to other isomeric forms involving several enzymatic reactions.

The best-known ester of this group, and the one occurring in soil in greatest amount, is *myo-*inositol hexaphosphoric acid (phytic acid), which occurs widely in nature. It is a very stable substance, especially in alkaline solution, but it is gradually hydrolyzed in acid media, with an optimum rate near pH 4, to give a range of intermediate inositol phosphates and ultimately the parent inositol. It is also hydrolyzed by a group of enzymes called the phytases.

The ester forms a number of very insoluble salts and also forms strong complexes with proteins and with some metal ions. For example, Fe and Al salts are precipitated over a wide range of acid concentrations and insoluble calcium salts are formed in alkaline solution (Jackman & Black, 1951; Anderson, 1963). Some of these are much more resistant to enzyme attack than are the more soluble ester salts (Wrenshall & Dyer, 1941; Casida, 1959). Inositol hexaphosphate is strongly sorbed by clay minerals such as montmorillonite and by finely divided sesquioxides (Goring & Bartholomew, 1950; Anderson & Arlidge, 1962). Other *myo*-inositol phosphates are also sorbed, the degree of sorption decreasing as the number of phosphate groups decreases (Anderson & Arlidge, 1962).

Inositol phosphates are readily extracted from plant materials with dilute acid. Alkali is more effective for soils, but even after extraction, some of the soil esters are present in complex form. They can be liberated by hypobromite oxidation or alkaline hydrolysis, precipitated as Fe or Ba salts at appropriate pH values, converted to soluble forms, and finally fractionated by anion-exchange chromatography (Smith & Clark, 1951; Cosgrove, 1963; McKercher & Anderson, 1968a). The penta- and hexaphosphates of all the isomeric inositols can be measured in one fraction and usually constitute the bulk of the inositol phosphates in soil (McKercher & Anderson, 1968a). Separation of soil inositol phosphates from other phosphates has also been achieved, without breaking complexes or precipitating the esters, by direct adsorption on anion-exchange resins (Martin, 1964). Tinsley and Özsavasci (1974) found that they could extract the inositol phosphate from soil with HCl if cupferron was added to complex cations which form insoluble salts of the ester. They then devised a two-stage procedure, first extracting inorganic P with HCl and HF, containing $TiCl_4$ to suppress extraction of organic P. Treatment with HCl and cupferron then extracts inositol phosphates which can be precipitated and isolated by ion-exchange chromatography.

By extracting organic matter from soil with a chelating resin, and fractionating by gel chromatography, Moyer and Thomas (1970) obtained three fractions of differing molecular weight, two of which contained inositol phosphates. One of these was estimated to have a molecular weight between 1,000 and 50,000. Steward and Tate (1971) fractionated an alkaline extract by gel chromatography, under different conditions to those of Moyer and Thomas, and obtained inositol phosphates in only one fraction. No treatment to break complexes was required, so this technique may provide another method for measuring total soil inositol phosphate.

The stability of inositol phosphates in soil has been attributed to the factors already considered in relation to total organic P and is likely to depend primarily on the degree of sorption or precipitation. McKercher and

Anderson (1968a) found that the amounts of inositol penta- and hexaphosphate in a range of Canadian soils are significantly correlated with the orthophosphate sorption capacity, suggesting that the same active sites sorb both phosphate forms. This has been confirmed for a selection of acidic Scottish soils by directly comparing the sorption of inositol hexaphosphate and orthophosphate (Anderson et al., 1974). Maximum sorption of the added ester is highly correlated with the amounts of Al and Fe extracted with an acid oxalate solution, but it is noteworthy that the amounts of naturally occurring inositol penta- and hexaphosphate in the soil are more closely related to the Fe alone.

The inositol phosphates in soil are probably derived from many different sources. Fresh animal manures analyzed by Peperzak et al. (1959), for example, contained from 0.04 to 1.03% of P in this form. The esters also occur in plant materials with concentrations in the seeds, but little information is available about the amounts occurring in roots, leaves, and other plant debris deposited in soil. A considerable amount of the soil *myo*-inositol hexaphosphate is probably synthesized in situ by microorganisms and, although phosphate esters of the other inositol isomers have not yet been isolated from microbial cultures, they too are likely to be microbially produced. Attempts to verify this have had mixed success. Caldwell and Black (1958a) added inorganic and organic nutrients to sand, clay-sand mixtures, and soil parent materials, and inoculated with soil microorganisms. After incubating these systems for several months they obtained evidence that *myo*-inositol hexaphosphate and an isomer (the *scyllo*-form) had been synthesized. In a similar experiment Cosgrove (1964) detected *myo*-inositol hexaphosphate, but none of its isomers. The proportions of the isomers reported in soils as far apart as Canada, the United Kingdom, and Australia are similar (Cosgrove, 1967; McKercher & Anderson, 1968b).

Inositol phosphates were shown to accumulate in a variety of Australian soils when organic matter was built up under leguminous pasture receiving superphosphate (Williams & Anderson, 1968), but the proportion in the newly formed organic P was in most cases slightly lower than that in the organic P of the virgin soils. Two explanations have been suggested for this. The decrease could reflect changes in the nature of the organic residues from which the organic matter was derived, or alternatively, if the inositol phosphates are less easily broken down than the other phosphate esters, it may reflect the greater mean age of the organic matter in the virgin soils. Inositol penta- and hexaphosphate ranged from 2.2 to 20.5 ppm P in the virgin soils and 6.8 to 46.5 ppm in the improved pastures, with a mean increase of 87%. The pastures had developed over periods ranging from 6 to 32 years. At Rothamsted, grass plots receiving inorganic fertilizers for over 100 years contained 83 ppm P in the form of these esters compared with 40 ppm in unfertilized plots. In continuously cultivated plots, with and without heavy dressings of farmyard manure or superphosphate, the levels in each case were about 40 ppm of P (Oniani et al., 1973).

Values reported in other soils vary widely. A tea soil in Ceylon contains 2 ppm P as esters compared with 54 ppm in a tea soil in Georgia, U.S.S.R. (Oniani et al., 1973). Agricultural soils in Scotland commonly contain be-

Table 3—Amounts of inositol phosphates in the surface 20 cm of a cultivated loam in Scotland (calculated from data of Anderson & Malcolm, 1974).

Inositol	Amount of P, kg/ha
Hexaphosphates	562
Pentaphosphates	84
Tetraphosphates	24
Triphosphates	7
Diphosphates	trace
Monophosphates	not detected

tween 100 and 400 ppm, constituting about 50% of the organic P in the soil (Anderson et al., 1974). The contents of inositol phosphates in a typical soil in Aberdeenshire are shown in Table 3. The amounts of the esters decrease rapidly as the number of phosphate groups in the molecule decreases, in keeping with the fact that the degree of sorption is also dependent on the number of phosphate groups (Anderson & Arlidge, 1962). In a wide range of soils from the United States, Canada, and Australia, inositol hexaphosphates, or the penta- and hexaphosphates together, account on the average for less than 20% of the soil organic P (Caldwell & Black, 1958b; McKercher & Anderson, 1968a; Williams & Anderson, 1968).

C. Nucleic Acids

The rate of addition of nucleic acids to soils is likely to be much greater than that of inositol phosphates, but in most soils they are broken down more quickly.

Nucleic acids occur in all living things and exist in two distinct chemical forms—ribonucleic acid (RNA) and deoxyribonucleic acid (DNA) (Chargaff & Davidson, 1955). Each consists essentially of a chain of sugar units, either ribose or deoxyribose, joined by phosphate ester "bridges." To each sugar molecule is linked a nitrogenous base derived from either purine or pyrimidine. The purine bases, adenine and guanine, and the pyrimidine base, cytosine, are present in both types of nucleic acid, but the pyrimidines, uracil and thymine, occur only in RNA and DNA, respectively. The units containing only one molecule of sugar linked to one molecule of the nitrogenous base are called nucleosides, the phosphate derivatives of which are nucleotides.

For a time it was believed that at least half of the organic P in soils was present in nucleic acids, but when specific methods of identification and measurement were applied, much lower values were found. The methods used have been reviewed by Anderson (1967). Such analyses have been largely confined to soils with high levels of organic P, however, and it is possible in soils where organic P is very low, but microbial numbers are high, that the nucleic acid P may indeed constitute a high proportion of it. As yet it has not proved possible to isolate pure nucleic acids from soils, and measurements have usually been based on the amounts of nucleotides or purine and pyrimidine derivatives that can be released by hydrolysis of soil organic matter fractions.

Adams et al. (1954) used chromatography to examine the fulvic acid fraction of two Iowa soils for the presence of ribonucleotides and concluded that not more than 1.2 and 6 ppm P were present as such, equivalent to 0.2 and 1.8% of the total organic P, respectively. Guanine, adenine, cytosine, and thymine have all been released from humic acid by hydrolysis, and after isolation by ion-exchange and paper chromatography, they can be measured by ultraviolet spectrometry (Anderson, 1961). From their amounts Anderson calculated that DNA or related polynucleotides in a selection of Scottish agricultural soils accounted for 5 to 19 ppm P, representing up to 2.4% of the soil organic P. The relative proportions of the bases indicated that the nucleic acid had been derived predominantly from microbial rather than plant or animal sources. Further evidence of the presence of DNA was provided by the isolation from soil hydrolysates of nucleotides containing deoxyribose (Anderson, 1970). Nucleotides derived from RNA have not been detected in these Scottish soils, and although uracil has been isolated from hydrolysates, it is thought to have been produced by de-amination of cytosine derived from DNA.

If the equilibrium level of DNA in soils is in fact much higher than that of RNA, the reason is not clear. Although the ratio of RNA to DNA in microorganisms varies considerably with both the nature of the organism and its stage of growth, the ratio is usually greater than one and the input of RNA to the soil is likely to be at least as rapid as that of DNA. Both RNA (Goring & Bartholomew, 1952) and DNA (Greaves & Wilson, 1969) are sorbed by the clay mineral montmorillonite, particularly in acid solution where the solubility of nucleic acids is low. Analysis by X-ray diffraction has shown that under acid conditions the nucleic acids penetrate the interlayer spaces of the clay (Greaves & Wilson, 1969). In incubation tests both forms of nucleic acid are degraded to similar extents, the rate being lower at pH values below about 5.4 (Greaves & Wilson, 1970). It has been suggested that the purine and pyrimidine bases obtained from soil humic acid may be present, not in DNA, but in related "limit" polynucleotides remaining as very stable residues after enzyme attack on DNA (Anderson, 1961).

D. Phospholipids

Like nucleic acids, phospholipids occur in all living matter and their rate of deposition in the soil must be high.

Lipids have been defined as naturally occurring compounds which are insoluble in water, but soluble in "fat" solvents such as ether or chloroform, are actual or potential esters of fatty acids, and are utilized by living organisms (Bloor, 1925). Phospholipids, in addition, contain P and some of the most common are derivatives of glycerol.

The phospholipid content of biological material is usually measured by extracting with ether after an initial treatment with ethanol to release protein-bound lipid (Lovern, 1955) and determining P in the extract. The presence of large amounts of clay interferes with the extraction (Goring & Bartholomew, 1949), but pretreatment with a mixture of HCl and HF has been used to reduce this effect (Hance & Anderson, 1963a). Extraction is also more effective if successive treatments with organic solvents are used (Hance & Anderson, 1963a; Kowalenko & McKercher, 1970).

Most soil phospholipid values obtained in this way fall within the range 0.2 to 14 ppm P (Kowalenko & McKercher, 1971a), representing less than 5% of the organic P, but as much as 14% of the organic P has been accounted for in the B horizon of a chernozem in Alberta (Dormaar, 1970).

Chromatographic examination of the extracted materials and their hydrolysis products has indicated that phosphatidyl choline (lecithin) and phosphatidyl ethanolamine are the predominant phospholipids in soils (Hance & Anderson, 1963b; Dormaar, 1970; Kowalenko & McKercher, 1971b).

An important group of phospholipids, containing *myo*-inositol, is known as phosphoinositides (Lovern, 1955). Examples of this group occur in many animal and plant tissues and in a variety of microorganisms. Anderson and Malcolm (1974) isolated two inositol-containing phosphate esters from alkali extracts of soils and these, on hydrolysis, released glycerol, *myo*-inositol, and *chiro*-inositol. It is possible that the esters were derived from phosphoinositides, although *chiro*-inositol has not yet been reported in these lipids.

E. Other Esters

Microorganisms are probably the major source of much of the remainder of the organic P in soils. Bacterial cell walls, for example, contain a number of very stable esters whose rate of breakdown in soil may be relatively slow.

Teichoic acids, which are of common occurrence in the walls of Gram-positive bacteria, consist essentially of ribitol phosphate or glycerol phosphate polymers and also contain amino-acid or sugar side chains (Armstrong et al., 1960). Attempts by Anderson and Hance (1963) to find teichoic acids or related esters in soil fulvic acid were unsuccessful, but more recently, Halstead and Anderson (1970) obtained chromatographic evidence that ribitol was present in the hydrolysate of an ester fraction.

After fractionating alkaline extracts of several Australian soils by gel chromatography, Steward and Tate (1971) isolated material of high molecular weight that contained a major part of the soil organic P. Partial hydrolysis released five phosphate esters which could be distinguished by electrophoresis. One of the two esters present in the greatest amount contained a reducing sugar and also responded to a colorimetric test for uronic acids.

Phosphorylated carboxylic acids other than uronic acids have been isolated from alkaline extracts of a number of Scottish agricultural soils (Anderson & Malcolm, 1974). Four esters with a C/P ratio of approximately 7 or 8 to 1 were separated by chromatography and, although they were not identified, two had properties suggesting that they might be related to 2-keto-3-deoxy sugar acids. One source of such acids is the lipo-polysaccharide fraction in the cell walls of Gram-negative bacteria (Osborn, 1969).

IV. AVAILABILITY OF SOIL ORGANIC PHOSPHATE

A. Experiments with Pure Esters

Plants growing in water culture or in sand readily assimilate P added in the form of inositol hexaphosphate, nucleic acid, lecithin, and other esters known or thought to occur in soils (Whiting & Heck, 1926; Weissflog & Mengdehl, 1933), but contrasting observations have been made with regard to the mechanisms involved. Rogers et al. (1940) examined P uptake by corn and tomato plants from aqueous solutions over a period of 18 hours and concluded that inositol hexaphosphate and lecithin were absorbed directly, whereas RNA and its nucleotides were first dephosphorylated by enzymes on the surface of the plant roots. They were unable to detect any attack on inositol hexaphosphate or lecithin by plant enzymes or microorganisms within the test period. Tests with ^{32}P-labeled inositol hexaphosphate led Flaig et al. (1960) to conclude that the rate of dephosphorylation was the limiting factor governing P uptake from this ester, but found that the effect on barley yields was not the same in sterile and nonsterile cultures. Martin (1973) grew wheat plants for 14 days in nutrient solutions containing ^{32}P-labeled inositol hexaphosphate and found that the translocation of radioactivity to the plant tops occurred to the same extent in the presence or absence of microorganisms in the root environment. He concluded that, although many soil microorganisms show phytase activity (Szember, 1960; Greaves & Webley, 1965), they are unlikely to increase the dephosphorylation of inositol hexaphosphate above the rate attributable to plant enzymes.

In soil, however, the availability of the esters is decreased, sometimes to a very low level (Bertramson & Stephenson, 1942), and is likely to depend on the nature of the ester, the level of addition, the sorption capacity and pH of the soil, and possibly, the nature of the crop. The P in inositol hexaphosphate can be utilized in some instances (Fardeau et al., 1968) and is apparently unavailable in others. Thus, Martin and Cartwright (1971) could not detect any uptake of ^{32}P by ryegrass grown for 4 weeks in a calcareous sandy soil or a lateritic podzol receiving 200 ppm P in this form. Laboratory tests with a number of acidic soils from Scotland have shown that sorption of inositol hexaphosphate is complete up to a certain level of addition which varies with the soil (Fig. 4; Anderson et al., 1974). Above this level the absolute level of sorption decreases, probably due to the formation of soluble complexes of the ester with Fe and Al. These maximum levels of sorption vary with pH and are about thrice as high at pH 3 than at pH 6, but even at pH 6 are at least 10 times higher than the amounts of the inositol phosphates naturally present in these soils. The sorption is apparently dependent on the contents of amorphous Fe and Al, although the long-term stability of the esters is probably influenced to a greater extent by the Fe alone. Although orthophosphate is sorbed at the same sites, the ester is preferentially sorbed and depresses the sorption of orthophosphate. This effect is particu-

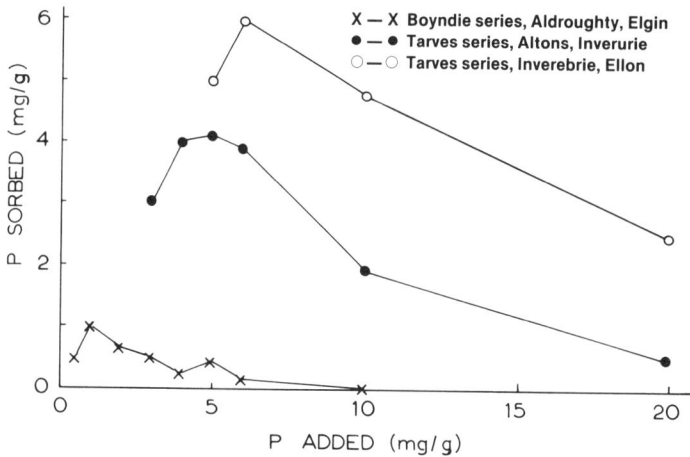

Fig. 4—Sorption of inositol hexaphosphate at pH 6 by three sandy loams from Scotland (Anderson et al., 1974).

larly marked at high levels of addition of the ester and is not diminished when soluble complexes of the ester are formed. Thus, even when the ester is itself virtually unavailable, it may have an indirect beneficial effect by increasing inorganic P availability.

The P in RNA is released in inorganic form more readily than that in inositol hexaphosphate and incubation tests have indicated that the rate is probably more dependent on soil pH than on sorption capacity (Wrenshall & Dyer, 1941; Bower, 1949; McConaghy, 1960; Greaves & Wilson, 1970). In pot experiments McConaghy (1960) showed that ryegrass grown in a basaltic loam of pH 4.8 took up less P from RNA than from superphosphate, both added at approximately 30 or 60 ppm P, but he could find no chemical evidence that any RNA remained in the soil 6 months after application.

B. Experiments with Indigenous Soil Materials

Half or more of the organic P in many soils remains uncharacterized and the properties of this fraction cannot be examined with reference compounds. Nevertheless, the availability of the native soil organic P has been assessed in a number of ways.

The availability of some fractions has been measured directly in water culture experiments. Pierre and Parker (1927) found that soil solutions from 20 sites in southern and midwestern USA contained much more organic P than inorganic P, but when rapidly growing corn, soybean or buckwheat plants were transferred to these solutions they depleted only the inorganic P over a 24-hour test period. In a similar experiment some of the organic P extracted with aqueous ammonia from an Iowa soil was utilized by corn (Rogers et al., 1940). Wild and Oke (1966) examined the organic P extracted with $CaCl_2$ from a calcareous soil in southern England, and separated the material into three fractions by ion-exchange chromatography. All the fractions provided P to clover seedlings grown over a period of 35 days, with the availability ranging from 31 to 93% of equivalent inorganic P treatments. The dry matter yield obtained with one of the fractions was lower than would be expected from the P uptake and this was attributed to the presence of a toxic substance interfering with the plant metabolism.

The significance of soil organic P in crop nutrition has also been tested by relating organic P measurements to crop values obtained in pot or field experiments. The greatest effects have been noted in the tropics where the soil organic P can apparently supply a high proportion of a crop's P requirements. In several regions of Africa very poor relationships are found between crop responses to P fertilizers and various inorganic P soil test measurements which are satisfactory in other areas. Friend and Birch (1960) observed this with a range of East African wheat soils and found that a better correlation was obtained with total organic P measurements. The water-soluble organic P did not give a good relationship with crop values, however, nor did it break down in laboratory incubation tests. It was assumed that a fairly constant proportion of the total organic P was being mineralized and that this was the main source of P for the plant. Once mineralized, however, the P released would still be subject to sorption and this too could affect availability. Friend and Birch therefore related the crop responses to the product of soil organic P and inorganic P retention capacity, and calculated that responses to P on these soils would be unlikely when the total soil organic P exceeded 640 ppm. Parallel experiments on grassland soils gave similar results, the critical level of organic P in this case being 675 ppm. In both Ghana and southern Nigeria close relationships have been noted between the growth of cocoa and soil organic P (Smith & Acquaye, 1963; Omotoso, 1971). In all but one of 12 acidic soils examined in Nigeria, cocoa responded to fertilizer P, the response decreasing as the organic P increased up to a level of 300 ppm. A soil containing 445 ppm organic P showed no response.

In temperate regions the contribution of soil organic P is very much smaller, and the availability of inorganic P is the main factor influencing crop responses. Nevertheless, there is good evidence that the organic P has some effect. Field experiments carried out over several years in southeastern Norway showed that yield responses of grass-clover leys were related to total soil organic P in soils with pH values greater than 5.5, but not in more acid soils (Semb & Uhlen, 1954). In Iowa, relationships have been noted not with total soil organic P, but with organic P mineralized either by chemical oxidation or during incubation (Van Diest & Black, 1959a, 1959b). In alkaline soils there were significant regressions of P uptake by sorghum grown in the greenhouse, on organic P mineralized by both oxidation and incubation. With acid soils significant regressions were obtained only when the incubation values were multiplied by the fractional recovery of added inorganic P, thus evaluating the labile quantity of mineralized organic P. Sekhon and Black (1968) later considered the possibility that the relationship occurred not because of any direct effect of organic P, but indirectly because CO_2 released during mineralization of organic matter increased the availability of inorganic P in the soil. On testing this hypothesis, however, they could find no evidence that the organic P effect was not direct.

The third type of evidence indicating that organic P makes a contribution to plant nutrition has been obtained by monitoring levels of soil organic P in the presence and absence of plants and by examining seasonal changes. In Japan, Hayashi and Takijima (1955) found in pot experiments that the organic P levels in a virgin soil and a paddy soil dropped more rapidly when maize or rice were grown than when the soils were uncropped. They attributed this to accelerated mineralization during cropping. A similar observation was made in Iowa by Sekhon and Black (1969) who cropped six soils in the greenhouse with oats followed by sorghum over a period of about a year, reducing the mean extractable organic P by about 7% compared to the level in uncropped soils. An attempt by Thompson and Black (1970) to show that the increased mineralization was caused by phosphatases from the plant roots was not successful. In New Zealand pastures, the availability of soil P is usually greater in spring than in autumn and winter. Because this is not reflected in laboratory measurements of available inorganic P, Saunders and Metson (1971) have suggested that organic P mineralization is at that time sufficiently rapid to sustain high pasture growth rates. A pronounced accumulation of organic P in the Ah horizons of Dark Brown Chernozems in southern Alberta has been observed to occur during the winter months (Dormaar, 1972). In unfertilized field plots growing alfalfa, the organic P content decreased in spring with crop growth, but began to increase again in the autumn.

V. FUTURE RESEARCH NEEDS

Organic P in soil should not be considered as P lost or P wasted. Evidence already presented suggests that it should be looked upon more as a reservoir whose level rises under some conditions and falls under others. Its importance may depend not only on its ability to supply P to crops but also on the part it plays as an integral component of soil organic matter. In this role it may have various indirect beneficial effects on plant nutrition. There is still much to be learned about the magnitude of its contribution to fertility.

It is not yet clear to what extent plants differ in their ability to use phosphate esters. There is evidence that the phosphatase or phytase activities of roots vary considerably, but does this influence the amounts used by different crops in a given soil under normal agricultural conditions in the field? It also appears that in soils of temperate regions there is a critical pH below which the release of P from organic forms is too slow to make a detectable contribution to plant requirements. Does this pH vary much in different regions? Is there a corresponding pH effect in tropical soils where availability is more rapid? Is there a temperature effect even within temperate regions so that availability can vary from year to year?

More information is also required about the nature of the esters. In some soils only a small percentage of the organic P has been characterized, and at best, nearly 40% remains to be identified. Most of it occurs in material of very high molecular weight, but otherwise little is known about it. The reasons for the pronounced qualitative differences in the organic P of soils from different regions also require investigation.

Further research is also required on the possible effects of phosphate esters on the physical properties of different soils. The reactions of inositol hexaphosphate, for instance, suggest that it could have a beneficial effect on soil structure. This could also be expected of phosphorylated derivatives of polyuronic acids or related acidic polysaccharides.

The most pressing need is for work to be done in as many regions as possible because it has already been shown that environmental factors can have a profound influence on both the nature and properties of the organic P in soils.

LITERATURE CITED

Adams, A. P., W. V. Bartholomew, and F. E. Clark. 1954. Measurement of nucleic acid components in soil. Soil Sci. Soc. Am. Proc. 18:40-46.

Allaway, W. H., and H. F. Rhoades. 1951. Forms and distribution of phosphorus in the horizons of some Nebraska soils in relation to profile development. Soil Sci. 72:119-128.

Anderson, G. 1961. Estimation of purines and pyrimidines in soil humic acid. Soil Sci. 91:156-161.

Anderson, G. 1963. Effect of iron/phosphorus ratio and acid concentration on the precipitation of ferric inositol hexaphosphate. J. Sci. Food Agric. 14:352-359.

Anderson, G. 1967. Nucleic acids, derivatives, and organic phosphates. p. 67-90. In A. D. McLaren and G. H. Peterson (ed.) Soil biochemistry. Marcel Dekker, New York.

Anderson, G. 1970. The isolation of nucleoside diphosphates from alkaline extracts of soil. J.Soil Sci. 21:96-104.

Anderson, G. 1975. Other organic phosphorus compounds. p. 305-331. In J. E. Gieseking (ed.) Soil components, Vol. I, Organic components. Springer Verlag, New York.

Anderson, G., and E. Z. Arlidge. 1962. The adsorption of inositol phosphates and glycerophosphate by soil clays, clay minerals and hydrated sesquioxides in acid media. J. Soil Sci. 13:216-224.

Anderson, G., and R. J. Hance. 1963. Investigation of an organic phosphorus component of fulvic acid. Plant Soil 19:296-303.

Anderson, G., and R. E. Malcolm. 1974. The nature of alkali-soluble soil organic phosphates. J. Soil Sci. 25:282-297.

Anderson, G., E. G. Williams, and J. O. Moir. 1974. A comparison of the sorption of inorganic orthophosphate and inositol hexaphosphate by six acid soils. J. Soil Sci. 25:51-62.

Angyal, S. J., and L. Anderson. 1959. The cyclitols. Adv. Carbohydr. Chem. 14:135-212.

Armstrong, J. J., J. Baddiley, and J. G. Buchanan. 1960. Structure of the ribitol teichoic acid from the walls of *Bacillus subtilis*. Biochem. J. 76:610-621.

Barrow, N. J. 1961. Phosphorus in soil organic matter. Soils Fert. 24:169-173.

Bertramson, B. R., and R. E. Stephenson. 1942. Comparative efficiency of organic phosphorus and of superphosphate in the nutrition of plants. Soil Sci. 53:215-227.

Bieleski, R. L., and R. E. Young. 1963. Extraction and separation of phosphate esters from plant tissues. Anal. Biochem. 6:54-68.

Bloor, W. R. 1925. Biochemistry of the fats. Chem. Rev. 2:243-300.

Bower, C. A. 1949. Studies on the forms and availability of soil organic phosphorus. Iowa Agric. Exp. Stn. Res. Bull. 362.

Bromfield, S. M. 1961. Sheep faeces in relation to the phosphorus cycle under pastures. Aust. J. Agric. Res. 12:111-123.

Caldwell, A. G., and C. A. Black. 1958a. Inositol hexaphosphate: 2. Synthesis by soil microorganisms. Soil Sci. Soc. Am. Proc. 22:293-296.

Caldwell, A. G., and C. A. Black. 1958b. Inositol hexaphosphate: 3. Content in soils. Soil Sci. Soc. Am. Proc. 22:296-298.

Casida, L. E. 1959. Phosphatase activity of some common soil fungi. Soil Sci. 87:305-310.

Chargaff, E., and J. N. Davidson. 1955. The nucleic acids. Academic Press, New York.

Chouchak, D. 1929. The contest between cultivated plants and soil microorganisms for their mineral nutrition; effect of dried blood on fertilizer phosphate. (In French). C. R. Acad. Sci. 189:262-264.

Cosgrove, D. J. 1962. Forms of inositol hexaphosphate in soils. Nature (London) 194:1265-1266.

Cosgrove, D. J. 1963. The chemical nature of soil organic phosphorus. 1. Inositol phosphates. Aust. J. Soil Res. 1:203-214.

Cosgrove, D. J. 1964. An examination of some possible sources of soil inositol phosphates. Plant Soil 21:137-141.

Cosgrove, D. J. 1967. Metabolism of organic phosphates in soil. p. 216-228. In A. D. McLaren and G. H. Peterson (ed.) Soil biochemistry. Marcel Dekker, New York.

Cosgrove, D. J., and M. E. Tate. 1963. Occurrence of *neo*-inositol hexaphosphate in soil. Nature 200:568-569.

Damsgaard-Sörensen, P. 1946. Studies on the soils phosphoric-acid content. 4. The organically combined phosphorus. (In Danish). Tidsskr. Planteavl. 50:653-675. From Soils Fert. 10:131 (1947).

Dean, L. A. 1938. An attempted fractionation of the soil phosphorus. J. Agric. Sci. 28:234-236.

DeTurk, E. E. 1938. Changes in the soil of the Morrow plots which have accompanied long-continued cropping. Soil Sci. Soc. Am. Proc. 3:83-85.

Dormaar, J. F. 1970. Phospholipids in chernozemic soils of southern Alberta. Soil Sci. 110:136-139.

Dormaar, J. F. 1972. Seasonal pattern of soil organic phosphorus. Can. J. Soil Sci. 52:107-112.

Fardeau, J. C., D. Delille, and C. Abramovici. 1968. Utilization of phytin by plants. p. 555-564. (In French). Proc. of symp. use of isotopes and radiation in soil organic matter studies. International Atomic Energy Agency/FAO Vienna.

Flaig, W., G. Schmid, E. Wagner, and H. Keppel. 1960. The uptake of phosphorus from inositol hexaphosphate. (In German). Landwirtsch. Forsch. Sonderh. 14:43-48. From Soils Fert. 23:345 (1960).

Friend, M. T., and H. F. Birch. 1960. Phosphate responses in relation to soil tests and organic phosphorus. J. Agric. Sci. 54:341-347.

Fuller, W. H., and W. T. McGeorge. 1951. Phosphates in calcareous Arizona soils: 2. Organic phosphorus content. Soil Sci. 71:45-49.

Goring, C. A. I., and W. V. Bartholomew. 1949. Microbial products and soil organic matter. 2. The effect of clay on the decomposition and separation of the phosphorus compounds in microorganisms. Soil Sci. Soc. Am. Proc. 14:152-156.

Goring, C. A. I., and W. V. Bartholomew. 1950. Microbial products and soil organic matter. 3. Adsorption of carbohydrate phosphates by clays. Soil Sci. Soc. Am. Proc. 15:189-194.

Goring, C. A. I., and W. V. Bartholomew. 1952. Adsorption of mononucleotides, nucleic acids, and nucleoproteins by clays. Soil Sci. 74:149-164.

Greaves, M. P., and D. M. Webley. 1965. A study of the breakdown of organic phosphates by micro-organisms from the root region of certain pasture grasses. J. Appl. Bacteriol. 28:454-465.

Greaves, M. P., and M. J. Wilson. 1969. The adsorption of nucleic acids by montmorillonite. Soil Biol. Biochem. 1:317-323.

Greaves, M. P., and M. J. Wilson. 1970. The degradation of nucleic acids and montmorillonite-nucleic acid complexes by soil microorganisms. Soil Biol. Biochem. 2:257-268.

Halstead, R. L., and G. Anderson. 1970. Chromatographic fractionation of organic phosphates from alkali, acid, and aqueous acetylacetone extracts of soils. Can. J. Soil Sci. 50:111-119.

Hance, R. J., and G. Anderson. 1963a. Extraction and estimation of soil phospholipids. Soil Sci. 96:94-98.

Hance, R. J., and G. Anderson. 1963b. Identification of hydrolysis products of soil phospholipids. Soil Sci. 96:157-161.

Hayashi, T., and Y. Takijima. 1955. Studies on utilization of soil organic phosphorus by crop plants. 3. Reduction of organic phosphorus content in soils caused by cropping. (In Japanese). J. Sci. Soil Manure, Japan. 26:215-218. From Soils Fert. 19:134 (1956).

Horiguchi, M., and M. Kandatsu. 1959. Isolation of 2-aminoethane phosphonic acid from rumen protozoa. Nature (London) 184:901-902.

Jackman, R. H. 1964. Accumulation of organic matter in some New Zealand soils under permanent pasture. 2. Rates of mineralisation of organic matter and the supply of available nutrients. N. Z. J. Agric. Res. 7:472-479.

Jackman, R. H., and C. A. Black. 1951. Solubility of iron, aluminum, calcium and magnesium inositol phosphates at different pH values. Soil Sci. 72:179-186.

Kaila, A. 1948. On the organic phosphorus in cultivated soils. (In Finnish). Valt. Maataloustieteellinen Julk. no. 129.

Kaila, A. 1949. Biological absorption of phosphorus. Soil Sci. 68:279-289.

Kaila, A. 1956a. Phosphorus in virgin peat soils. Maataloustieteellinen Aikakauskirja. 28:142-167.

Kaila, A. 1956b. Phosphorus in various depths of some virgin peat lands. Maataloustieteellinen Aikakauskirja. 28:90-104.

Kaila, A., and H. Missilä. 1956. Accumulation of fertilizer phosphorus in peat soils. Maataloustieteellinen Aikakauskirja. 28:168-178.

Kowalenko, C. G., and R. B. McKercher. 1970. An examination of methods for extraction of soil phospholipids. Soil Biol. Biochem. 2:269-273.

Kowalenko, G. C., and R. B. McKercher. 1971a. Phospholipid P content of Saskatchewan soils. Soil Biol. Biochem. 3:243-247.

Kowalenko, C. G., and R. B. McKercher. 1971b. Phospholipid components extracted from Saskatchewan soils. Can. J. Soil Sci. 51:19-22.

L'Annunziata, M. F. 1975. The origin and transformations of the soil inositol phosphate isomers. Soil Sci. Soc. Am. Proc. 39:377-379.

L'Annunziata, M. F., and W. H. Fuller. 1971. Soil and plant relationships of inositol phosphate stereoisomers; the identification of D-chiro- and muco-inositol phosphates in a desert soil and plant system. Soil Sci. Soc. Am. Proc. 35:587-595.

Lovern, J. A. 1955. The chemistry of lipids of biochemical significance. Methuen, London.

Martin, J. K. 1964. Soil organic phosphorus. 1. Methods for the extraction and partial fractionation of soil organic phosphorus. N. Z. J. Agric. Res. 7:723-735.

Martin, J. K. 1973. The influence of rhizosphere microflora on the availability of ^{32}P-myo-inositol hexaphosphate phosphorus to wheat. Soil Biol. Biochem. 5:473-483.

Martin, J. K., and B. Cartwright. 1971. The comparative plant availability of ^{32}P myo-inositol hexaphosphate and KH$_2$^{32}PO$_4$ added to soils. Commun. Soil Sci. Plant Anal. 2:375-381.

McConaghy, S. 1960. Soil phosphates, with special reference to organic forms and their availability to plants. Agric. Prog. 35:82-93.

McKercher, R. B., and G. Anderson. 1968a. Content of inositol penta- and hexaphosphates in some Canadian soils. J. Soil Sci. 19:47-55.

McKercher, R. B., and G. Anderson. 1968b. Characterization of the inositol penta- and hexaphosphate fractions of a number of Canadian and Scottish soils. J. Soil Sci. 19:302-310.

Moyer, J. R., and R. L. Thomas. 1970. Organic phosphorus and inositol phosphates in molecular size fractions of a soil organic matter extract. Soil Sci. Soc. Am. Proc. 34:80-83.

Nye, P. H., and M. H. Bertheux. 1957. The distribution of phosphorus in forest and savannah soils of the Gold Coast and its agricultural significance. J. Agric. Sci. 49:141-159.

Omotoso, T. I. 1971. Organic phosphorus contents of some cocoa growing soils of southern Nigeria. Soil Sci. 112:195-199.

Oniani, O. G., M. Chater, and G. E. G. Mattingly. 1973. Some effects of fertilizers and farmyard manure on the organic phosphorus in soils. J. Soil Sci. 24:1-9.

Osborn, M. J. 1969. Structure and biosynthesis of the bacterial cell wall. Annu. Rev. Biochem. 38:501-538.

Owssia, I., E. Wilberg, and G. Michael. 1966. Effect of 30-year mineral fertilizing and organic manuring on the status of P forms in a "Filder-loam" soil. (In German). Z. Pflanzenernaeh. Dueng. Bodenkd. 113:159-169.

Peperzak, P., A. G. Caldwell, R. R. Hunziker, and C. A. Black. 1959. Phosphorus fractions in manures. Soil Sci. 87:293-302.

Pierre, W. H., and F. W. Parker. 1927. Soil phosphorus studies. 2. The concentration of organic and inorganic phosphorus in the soil solution and soil extracts and the availability of the organic phosphorus to plants. Soil Sci. 24:119-128.

Rogers, H. T., R. W. Pearson, and W. H. Pierre. 1940. Absorption of organic phosphorus by corn and tomato plants and the mineralizing action of exo-enzyme systems of growing roots. Soil Sci. Soc. Am. Proc. 5:285-291.

Saunders, W. M. H., and A. J. Metson. 1971. Seasonal variation of phosphorus in soil and pasture. N. Z. J. Agric. Res. 14:307-328.

Schmidt, G., and S. J. Thannhauser. 1945. A method for the determination of deoxyribonucleic acid, ribonucleic acid, and phosphoproteins in animal tissues. J. Biol. Chem. 161:83-89.

Schollenberger, C. J. 1920. Organic phosphorus content of Ohio soils. Soil Sci. 10:127-141.

Sekhon, G. S., and C. A. Black. 1968. Uptake of phosphorus by plants in relation to carbon dioxide production and organic phosphorus mineralization in soils. Plant Soil 29:299-304.

Sekhon, G. S., and C. A. Black. 1969. Changes in extractable organic phosphorus in soil in the presence and absence of plants. Plant Soil 31:321-327.

Semb, G., and G. Uhlen. 1954. A comparison of different analytical methods for the determination of potassium and phosphorus in soil based on field experiments. Acta Agric. Scand. 5:44-68.

Sen Gupta, M. B., and A. H. Cornfield. 1962. Phosphorus in calcareous soils. 2. Determination of the organic phosphorus content of calcareous soils and its relation to soil calcium carbonate content. J. Sci. Food Agric. 13:655-658.

Smith, D. H., and F. E. Clark. 1951. Anion-exchange chromatography of inositol phosphates from soil. Soil Sci. 72:353-360.

Smith, R. W., and D. K. Acquaye. 1963. Fertilizer responses on peasant cocoa farms in Ghana: a factorial experiment. Emp. J. Exp. Agric. 31:115-123.

Steward, J. H., and M. E. Tate. 1971. Gel chromatography of soil organic phosphorus. J. Chromatogr. 60:75-82.

Syers, J. K., and T. W. Walker. 1969. Phosphorus transformations in a chronosequence of soils developed on wind-blown sand in New Zealand. 1. Total and organic phosphorus. J. Soil Sci. 20:57-64.

Szember, A. 1960. Influence on plant growth of the breakdown of organic phosphorus compounds by micro-organisms. Plant Soil 13:147-158.

Thompson, E. J., and C. A. Black. 1970. Changes in extractable organic phosphorus in soil in the presence and absence of plants. 3. Phosphatase effects. Plant Soil 32:335-348.

Tinsley, J., and C. Özsavasci. 1974. Studies of soil organic phosphorus using titanic chloride. Int. Congr. Soil Sci., Trans. 10th (Moscow) 2:332-340.

Van Diest, A. 1968. Biological immobilization of fertilizer phosphorus. 1. Accumulation of soil organic phosphorus in coastal plain soils of New Jersey. Plant Soil 29:241-247.

Van Diest, A., and C. A. Black. 1959a. Soil organic phosphorus and plant growth. 1. Organic phosphorus hydrolyzed by alkali and hypobromite treatments. Soil Sci. 87:100-104.

Van Diest, A., and C. A. Black. 1959b. Soil organic phosphorus and plant growth. 2. Organic phosphorus mineralized during incubation. Soil Sci. 87:145-154.

Walker, T. W., and A. F. R. Adams. 1958. Studies on soil organic matter. 1. Influence of phosphorus content of parent materials on accumulations of carbon, nitrogen, sulphur, and organic phosphorus in grassland soils. Soil Sci. 85:307-318.

Weissflog, J., and H. Mengdehl. 1933. Studies on the phosphorus metabolism of higher plants. 3. Uptake and utilization of organic phosphoric acid compounds by the plant. (In German). Planta 19:182-241.

Whiting, A. L., and A. F. Heck. 1926. The assimilation of phosphorus from phytin by oats. Soil Sci. 22:477-493.

Wild, A., and O. L. Oke. 1966. Organic phosphate compounds in calcium chloride extracts of soils: Identification and availability to plants. J. Soil Sci. 17:356-371.

Williams, C. H., and G. Anderson. 1968. Inositol phosphates in some Australian soils. Aust. J. Soil Res. 6:121-130.

Williams, E. G., and W. M. H. Saunders. 1956. Distribution of phosphorus in profiles and particle-size fractions of some Scottish soils. J. Soil Sci. 7:90-108.

Williams, E. G., N. M. Scott, and M. J. McDonald. 1958. Soil properties and phosphate sorption. J. Sci. Food Agr. 9:551-559.

Wrenshall, C. L., and W. J. Dyer. 1941. Organic phosphorus in soils. 2. The nature of the organic phosphorus compounds. A. Nucleic acid derivatives. B. Phytin. Soil Sci. 51:235-248.

Chapter 16

Conventional Soil and Tissue Tests for Assessing the Phosphorus Status of Soils

E. J. KAMPRATH AND M. E. WATSON

*North Carolina State University
Raleigh, North Carolina, and
Ohio Agriculture Research & Development Center
Wooster, Ohio, respectively*

I. INTRODUCTION

The analysis of soils and plants to determine their content of nutrients essential for plant growth had its beginnings in the nineteenth century. Early attempts to characterize the supply of nutrients available for plant growth were based on a total analysis of the soil. This approach was not of much use for accurately determining on which soils plants would respond to fertilization. Daubeny in 1845 introduced a significant concept when he suggested that nutrients existed in "active" and "dormant" forms. He proposed that carbonic acid be used to extract the active forms. This provided the conceptual framework for the development of soil test extractants to measure that portion of the soil P which was available to plants.

The idea of analyzing plant tissue for the purpose of soil nutrient status evaluation probably dates back to the middle of the 19th century (Weinhold, *in* Smith, 1962). It was then used essentially as a biological testing approach for soil fertility evaluation. This was based on the idea that the plant itself would be the best indicator of the nutrient status of the medium in which it was growing. With the finding that application of superphosphate increased crop yields in certain soils, interest was stimulated in the use of soil tests and plant analysis to identify those soils deficient in phosphate (P).

During the past 50 years soil scientists and agronomists have conducted hundreds of experiments to evaluate the P status of soils. It is not our intent to present a comprehensive review of all the research dealing with soil tests and plant analysis for P. Instead we will try to indicate some of the important principles which need to be taken into account when evaluating and interpreting results of soil tests and plant analysis for the purpose of assessing the P status of soils.

Copyright 1980 © ASA-CSSA-SSSA, 677 South Segoe Road, Madison, WI 53711, USA.
The Role of Phosphorus in Agriculture.

The reader is referred to a number of excellent reviews for additional information on soil tests and plant analysis for assessing the P status of soils. The history of the development of soil tests for P is well documented by Dyer (1894), Russell and Prescott (1916-1917), and Anderson and Noble (1937). Excellent reviews on various aspects of soil tests for P are those of Nelson et al. (1953), Bingham (1962), Olsen and Dean (1965), and Thomas and Peaslee (1973). The extensive topic of plant analysis has been recently reviewed by Smith (1962) and Munson and Nelson (1973). Specific aspects of plant analysis have been discussed by Aldrich (1973), Jones and Steyn (1973), Ulrich and Hills (1973), Humbert (1973), Sabbe and Mackenzie (1973), Small and Ohlrogge (1973), Ward et al. (1973), Jones and Eck (1973), Geraldson et al. (1973), Kenworthy (1973), Martin and Matocha (1973), and Leaf (1973).

II. EVALUATION OF COMMON SOIL TEST EXTRACTANTS

A landmark paper on measuring the available P in soils and relating the results to a need for phosphate fertilizers was that of Dyer (1894). He theorized that an ideal extracting solution would be one that had a pH similar to that of root sap. His studies with a large number of species indicated that the acidity of root sap was equivalent to a 1% citric acid solution. Dyer then tested the suitability of the 1% citric acid solution for estimating available P using soil samples from the Rothamsted plots which had a known record since 1852 of fertilizer applications and crop yields. He concluded that if a soil contained less than 0.01% phosphoric acid (44 ppm P) soluble in 1% citric acid solution, it stood in immediate need of phosphatic manure.

Another significant step in the development of soil tests for P was the work of Russell and Prescott (1916-1917). They found that dilute solutions of strong acids such as HNO_3 and HCl extracted less P than weak acids, such as citric and oxalic acid at the same concentration. The citrate and oxalate ions reduced the adsorption of P from solutions by soils as compared with HCl and HNO_3. They also found that less P was extracted in 24 hours with dilute (0.05 to 0.1N) HCl or HNO_3 than in 10 min because of readsorption of the P.

Since 1945 there have been a number of significant developments in the area of soil chemistry and soil fertility which have had a marked impact on our understanding and interpretation of soil tests for P. The availability of ^{32}P for experimental work provided for an independent measure of labile P with which the amounts of soil P extracted by various reagents could be correlated. In the late 1950's procedures were developed for fractionating soil P into various chemical forms. At the same time extensive studies were conducted by TVA and others to determine the nature of the reaction products formed in different soils when P fertilizers were added. In the 1960's principles were developed which provided a better understanding of the relationship between soil P and plant availability. Attention was focused on the effect of buffering capacity, intensity (soil solution concentration), and diffusion on availability of soil P for plant growth. All of these new developments have provided the framework for a better understanding and interpretation of P soil tests.

A. Assessing P Availability

1. OBJECTIVES OF P SOIL TESTS

The objectives of soil tests for P can be listed as (i) grouping of soils into classes for the purpose of making fertilizer P recommendations, (ii) prediction of the probability of getting a profitable response to application of fertilizer P, and (iii) providing an index of the amount of P a soil can supply.

a. Grouping of Soils into Classes for the Purpose of Making Fertilizer P recommendations—The simplest grouping of soils for the purpose of making fertilizer P recommendations consists of two categories: deficient and adequate soil P levels. The soil P level separating the deficient soils from the adequate soils has been called the critical soil test P level (Cate & Nelson, 1965). Generally, soils are grouped into at least three categories for the purpose of making fertilizer recommendations: low, medium, and high levels of soil test P (Fitts & Nelson, 1956). Rates of fertilizer P would be highest at the low soil test level and would decrease as the soil test level increased. With the use of computers for interpreting soil tests, equations have been developed which give P fertilizer rates as a function of the specific soil P level.

b. Prediction of the Probability of a Profitable Response to Application of Fertilizer P—The probability of response concept, as the basis of interpreting soil tests, has been discussed by Fitts and Nelson (1956). Soil test results are classified according to the percentage of soils in a given range of P content which give a profitable response to application of P fertilizers. Levels of soil P at which a high percentage of the soils give a profitable response to fertilizer P would be classified as low and those at which a low percentage give a profitable response would be classified as high in P.

c. Index of Available P Supply in Soils—In recent years there has been increasing interest in bringing the available P level to that point where only maintenance applications are required. This requires knowledge about the optimum level of available soil P which is required for different soils and crops. Information is also needed as to how much fertilizer P has to be added to soils to raise the available soil P to the optimum level. In order for this approach to be successful, one has to have knowledge about the soil chemical properties and the reactions between fertilizer P and the soil.

2. CRITERIA FOR A GOOD SOIL TEST

A good soil test should meet the following three criteria (i) the extractant used should extract all or a proportionate part of the available form or forms of a nutrient from soils with variable properties, (ii) the amount of nutrient extracted should be measured with reasonable accuracy and speed, and (iii) the amount extracted should be correlated with the growth and response of each crop to that nutrient under various conditions (Bray, 1948).

3. SOIL FACTORS AFFECTING P UPTAKE

Factors affecting the supply of P to plants are the amount of soil P (*quantity*), the concentration of soil solution P (*intensity*), and movement of P to the roots (*diffusion*) (Gunary & Sutton, 1967). Over 90% of the variation in the optimum rates of fertilizer P for tobacco and corn were explained by taking into account the intensity and buffering capacity values for each soil (Salmon, 1973). In any assessment of the available P status of soils by chemical tests one needs to consider the relationship among quantity, intensity, and diffusion, and factors influencing these components of P supply to plants.

The buffering capacity of acid and neutral soils is a function of the amounts and crystallinity of hydrated oxides of Fe and Al. In calcareous soils the amounts of exchangeable Ca and $CaCO_3$ determine the P buffering capacity.

Phosphorus in the soil solution is in equilibrium with a portion of the soil P generally designated as *labile P* (Mattingly, 1965). The intensity or concentration of P in the soil solution is determined by the percent saturation of the P adsorption capacity.

One of the factors influencing P diffusion in a soil is the concentration gradient across the root surfaces. Soil texture is another factor affecting diffusion (Olsen & Watanabe, 1963). As the clay content increases, the diffusion coefficients increase due to a decrease in tortuosity and an increase in buffering capacity.

Ideally, a P soil test would take into account both intensity and quantity. In practice, however, soil tests characterize either the quantity or the intensity factor. Weak extractants such as water and $CaCl_2$ evaluate the intensity factor while strong extractants such as acids, complexing ions, and alkaline-buffered solutions measure the quantity factor. The most satisfactory strong extractant would be the one whidh measures the portion of the soil P controlling the soil solution P.

B. Common Soil Test Extractants and Mechanism of P Removal

1. CLASSIFICATION OF SOIL TEST EXTRACTANTS

Soil test extractants can be placed into various categories related to the chemical nature of the extracting solutions. In certain cases the anions in the solution may have a specific effect either because of their anion replacing ability or their reaction with the cation associated with P. A list of the commonly used soil test extractants is given in Table 1.

a. Dilute Concentrations of Strong Acids—The common acids used for extracting P are HCl, HNO_3, and H_2SO_4. The concentrations of acid

Table 1—Common soil test extractants used for available P.

Common name	Extractant	Soil/solution ratio	Reference
Bray I	0.025N HCl + 0.03N NH$_4$F	1:10	Bray & Kurtz (1945)
Bray II	0.1N + 0.03N NH$_4$F	1:17	
North Carolina	0.05N HCl + 0.025N H$_2$SO$_4$	1:4	Sabbe & Breland (1974)
Troug	0.002N H$_2$SO$_4$ buffered at pH 3 with (NH$_4$)$_2$SO$_4$	1:100	Troug (1930)
Citric acid	1% citric acid	1:10	Dyer (1894)
Egner	0.02N Ca lactate + 0.02N HCl	1:20	Egner et al. (1960)
Morgan	0.54N HOAc + 0.7N NaOAc pH 4.8	1:10	Morgan (1941)
Olsen	0.5M NaHCO$_3$ pH 8.5	1:20	Olsen et al. (1954a)

vary all the way from 0.002N to 0.075N. The pH of the solutions generally range from 2 to 3. The nitrate and chloride ions have very little effect on the extraction, but the sulfate tends to reduce readsorption of the P dissolved by the acid solutions (Nelson et al., 1953).

b. Dilute Concentrations of Strong Acids plus a Complexing Ion— The complexing ion used most often is fluoride, which forms strong complexes with the Al ion. The fluoride compound generally used is NH$_4$F. The most common extracting solutions of this kind are mixtures of HCl and NH$_4$F. Concentrations of acid vary from 0.025 to 0.1N while the concentration of NH$_4$F is generally 0.03N (Fitts, 1956).

c. Dilute Concentrations of Weak Acids—The most common weak acids or salts of weak acids used for extracting P are citric, lactic, and acetic. One of the first extractants used was a 1% citric acid solution (Dyer, 1894). A solution consisting of calcium lactate and HCl is commonly used in western Europe (Egner et al., 1960). Acetic acid buffered with NaOAc at pH 4.8 was developed for the soils of the northeastern U.S. (Morgan, 1941). A number of modifications of this solution have been developed in which NH$_4$OAc is used to buffer the solution at pH 4.8.

The organic anions of these solutions tend to influence the extraction in two ways. The anions such as citrate and lactate form complexes with the polyvalent cations and thus release P. The organic anions also can replace adsorbed P and prevent its readsorption.

d. Buffered Alkaline Solutions—The most commonly used buffered alkaline solution for extracting P is NaHCO$_3$, which originally was developed for calcareous soils (Olsen et al., 1954a). The HCO$_3^-$ is quite effective in replacing adsorbed P and Na reduces the activity of Ca in solution. A modification of this method was developed by Colwell (1963) in which a soil-to-solution ratio of 1:100 is used.

2. MECHANISM OF P REMOVAL BY EXTRACTANTS

There are basically four reactions by which P is removed from the solid phase. These are solvent action of acids, anion replacement, complexing of cations binding P, and hydrolysis of cations binding P.

a. Solvent Action of Acids—The acid solutions used to extract P generally have a pH of 2 to 3. This provides sufficient H ion activity to dissolve calcium phosphates. Acid solutions will also solubilize some of the aluminum phosphates and iron phosphates. The order of greatest solubility in acid solutions is Ca-P > Al-P > Fe-P (Thomas & Peaslee, 1973).

b. Anion replacement—Phosphorus adsorbed on surfaces of $CaCO_3$ and hydrated oxides of Fe and Al can be replaced by other anions such as acetate, citrate, lactate, sulfate, and bicarbonate (Dean & Rubins, 1947; Olsen et al., 1954a). When the organic anions and sulfate are present in acid solutions they reduce readsorption of P.

c. Complexing of Cations Binding P—Fluoride ions are very effective in complexing Al ions and in this manner releasing P from Al-P (Chang & Jackson, 1957). Calcium is precipitated by F ions and therefore the P present in soils as $CaHPO_4$ will be extracted by solutions containing F ions (Thomas & Peaslee, 1973). The organic anions, citrate and lactate, also complex Al ions.

d. Hydrolysis of cations Binding P—Extracting solutions containing OH ions extract P from Al-P and Fe-P due to the hydrolysis of the Al and Fe. Thus $NaHCO_3$, which is buffered at pH 8.5, is very effective in extracting Al-P and to some extent Fe-P (Tyner & Davide, 1962).

C. Relationship Between Forms of Soil P and Plant Growth and P Extracted by Soil Tests

1. RELATIONSHIP BETWEEN FORMS OF SOIL P AND PLANT GROWTH

Since the development of fractionation procedures for soil P (Chang & Jackson, 1957) it has been possible to study the relationship between forms of soil P and plant growth. With information about which forms of soil P are related to growth and P uptake on a given soil it is possible to select soil tests which would reflect the amounts of those forms of P in the soil. Aluminum P was highly correlated with plant growth in Canadian (Halstead, 1967), North Dakota (Zubriski, 1971), and Virginia (Martens et al., 1969) soils. Both Al-P and Ca-P were found to contribute to uptake of P in Michigan soils (Susuki et al., 1963) and in an Alfisol in Virginia (Martens et al., 1969). The response to applications of superphosphate by pasture plants in New South Wales, Australia was negatively correlated with the amount of Al-P in the soil (McLachlan, 1965). About 40% of the Al-P extracted by NH_4F from heavily fertilized, highly weathered soils had equilibrated with ^{32}P, as contrasted to only 20% of the Fe-P extracted with NaOH (Dunbar & Baker, 1965). One can conclude that for acid to neutral soils Al-P is the primary source of plant P along with any $CaHPO_4$ that may be present. Therefore, soil tests which extract Al-P and $CaHPO_4$ should give a good measure of the plant available P in acid and neutral soils.

The source of P in calcareous soils, which is correlated with plant growth, is surface P which is assumed to be adsorbed on the surfaces of $CaCO_3$ (Olsen et al., 1954a). The $NaHCO_3$-extractable P in calcareous soils is highly correlated with surface P. Extraction with $NaHCO_3$ removes about half of the surface P.

2. RELATIONSHIP BETWEEN FORMS OF SOIL P AND P EXTRACTED BY SOIL TESTS

Numerous studies have been made correlating the amount of P extracted by a soil test with the various forms of soil P. The highest correlations obtained between the various soil tests and the form of soil P for a number of widely different regions are given in Table 2.

The P extracted from a wide range of soils with the Olsen ($0.5M$ $NaHCO_3$) and Bray I ($0.025M$ HCl + $0.03N$ NH_4F) solutions was primarily correlated with the Al-P form (Table 2). Based on the mechanisms of P extraction by $NaHCO_3$ and NH_4F, which were discussed previously, it is logical that these two extractants would primarily remove proportionate amounts of the Al-P form from soils.

The dilute acid extractants of North Carolina ($0.05N$ HCl + $0.025N$ H_2SO_4) and Truog ($0.002N$ H_2SO_4) will dissolve Ca-P and in soils containing appreciable amounts of Ca-P this is the main form extracted (Table 2). However, in soils containing only small amounts of Ca-P, the Al-P is the principal form of soil P removed.

The Egner extractant ($0.02N$ Ca lactate + $0.02N$ HCl) will remove both Ca-P and Al-P (Balerdi et al., 1968; Grigg, 1965). The HCl dissolves Ca-P while the lactate complexes Al and thereby releases P from Al-P.

Table 2—Correlation of soil tests with the form of soil P.

Region	Morgan NaOAc + HOAc	Truog $0.002N$ H_2SO_4	North Carolina $0.05N$ HCl + $0.025N$ H_2SO_4	Bray I $0.025N$ HCl + $0.03N$ NH_4F	Olsen $0.5M$ $NaHCO_3$	Reference
			form of P and correlation coefficient, r			
Bangladesh	Ca-P 0.79	Ca-P 0.90	Ca-P 0.88	Al-P 0.73	Fe-P 0.78	Ahmed & Islam (1975)
British Columbia	Ca P 0.14	--	Ca-P 0.65	Al-P 0.66	Al-P 0.64	John (1972)
Brazil	--	--	Al-P 0.82	Al-P 0.82	Al-P 0.74	Cajuste & Kussow (1974)
Michigan	--	Al-P 0.55	--	Al-P 0.99	Al-P 0.99	Susuki et al. (1963)
New South Wales	--	--	--	Al-P 0.88	Al-P 0.93	McLachlan, 1965
North Carolina			Al-P 0.96			Shelton & Coleman, 1968
North Dakota	--	--	--	Al-P 0.98	Al-P 0.94	Zubriski, 1971

Table 3—Correlation of amount of P extracted by soil tests with other parameters of P availability.

Region	Availability parameter	Bray I 0.025N HCl + 0.03N NH$_4$F	Olsen 0.5M NaHCO$_3$	Truog 0.002N H$_2$SO$_4$	N. Carolina 0.05N HCl + 0.025N H$_2$SO$_4$	Egner 0.02N Ca lactate + 0.02N HCl	Morgan	Reference
			correlation coefficient, r					
U.S.	A value	0.73	0.94	--	0.55	--	0.17	Fitts (1956) and Olsen et al. (1954a)
Colorado	A value	0.94	0.96	--	--	--	--	Olsen et al. (1954b)
Iowa								
Calcareous	A value	0.95	--	--	--	--	--	Smith & Pesek (1962)
Noncalcareous	A value	0.94	--	--	--	--	--	
Alabama								
Clay loam	% yield	0.91	0.91	--	0.91	--	--	Welch et al. (1957)
Sandy loam		0.72	0.73	--	0.80	--	--	
Minnesota								
Calcareous	P uptake	0.23	0.73	--	--	--	0.86	Blanchar & Caldwell (1964)
Noncalcareous		0.86	0.85	--	--	--	0.98	
New Jersey	P uptake	0.83	0.84	--	0.94	0.82	0.46	Van Diest (1963)
North Dakota	Dry matter	0.94	0.93	--	--	--	--	Zubriski (1971)
Ohio	P uptake	0.91	0.87	0.34	--	--	--	Thompson & Pratt (1954)
Pennsylvania	% P	0.74	--	--	--	--	--	Baker & Hall (1967)
Brazil	P uptake	0.89	0.87	--	0.78	--	--	Cajuste & Kussow (1974)
Canada	P uptake	0.63	0.66	--	--	--	0.31	John et al. (1967)
Central America	P uptake	0.86	0.87	--	0.85	0.95	--	Balerdi et al. (1968)
India								
Calcareous	P uptake	-0.10	0.73	--	--	--	--	Bhan & Shanker (1973)
Jamaica	P uptake	0.72	0.67	0.75	--	--	0.43	Weir (1962)
Scotland	Yield	0.46	0.69	0.63	--	0.73	--	Williams & Knight (1963)

The data of Chang and Juo (1963) indicate that the proportion of the various forms of soil P present will influence the correlation between the P extracted by a given method and the form of soil P. Generally it can be said that the alkaline solutions and NH_4F will preferentially extract Al-P and the acid solutions will preferentially extract Ca-P.

D. Suitability of Various Soil Tests For Assessing P Availability

1. CORRELATIONS OF SOIL TESTS WITH GROWTH AND P AVAILABILITY

The suitability of a soil test for predicting the P status of a soil can be evaluated by correlating the P extracted with plant growth parameters such as yield, % yield, P uptake, and P concentration (Fitts & Nelson, 1956) or with estimates of labile P such as the A or L value (Larsen, 1967). The ideal soil test would be one which is not influenced by differences in the chemical and mineralogical properties of soils. The results of a number of correlation studies with various extracting solutions are given in Table 3.

The Olsen ($NaHCO_3$) and the Bray I ($0.025N$ HCl + $0.03N$ NH_4F) methods have been studied more widely than other extracting solutions. The Olsen method was quite satisfactory across a wide range of soil conditions for estimating P availability. The Bray I method was also quite good in estimating P availability except in certain instances with calcareous soils. Where it is necessary to test soils with a wide range of chemical properties these two methods appear to be the most suitable.

The North Carolina method ($0.05N$ HCl + $0.025N$ H_2SO_4) appeared to be quite satisfactory when used for soils which had a pH of 7 or less (Fitts, 1956). The Egner method ($0.02N$ Ca lactate + $0.02N$ HCl), was as effective as the Olsen and Bray I methods. The least satisfactory method for estimating P availability across a wide range of soils was the Morgan method (NaOAc + HOAc).

2. SOIL FACTORS INFLUENCING SUITABILITY OF SOIL TESTS FOR ASSESSING P AVAILABILITY

a. Soil pH—The comprehensive study on P soil tests by the National Soil Test Work Group provided excellent information on how various soil properties influenced the correlations of soil test P with A value estimates of plant-available P (Fitts, 1956). Soil pH had a definite effect on the suitability of dilute acid extractions such as the North Carolina method in extracting labile P. Very poor correlations were obtained between dilute acid-extractable P and labile P with soils whose pH under natural conditions is 7 or higher. This could be due to two things; namely (i) exhaustion of the dilute acid by the high cation exchange capacity or by the reaction of $CaCO_3$ with the acid, and (ii) extraction of plant unavailable Ca-P compounds.

The suitability of the Bray I method on calcareous soils has been considerably improved by changing the soil-to-solution ratio from 1:7 or 1:10

to 1:50 (Smith et al., 1957; Blanchar & Caldwell, 1964). At the 1:7 soil-to-solution ratio the acid is consumed in part or entirely by the $CaCO_3$ and only small amounts of P are extracted.

b. Soil Mineralogical Properties—The North Carolina method gave much better correlations with estimates of labile P and percent yield on soils which had predominantly kaolinitic clay minerals as compared with soils which had 2:1 type clay minerals (Fitts, 1956). The soils with kaolinitic clays are more highly weathered and have lower CEC than soils with 2:1 type clays. Much better correlations were obtained on clayey Oxisols from Brazil with the North Carolina extractant when the soil-to-solution ratio was changed from 1:8 to 1:25 (Cajuste & Kussow, 1974). The soils with high clay and Fe oxide content would tend to neutralize the acid extracting solution and reduce the amount of P extracted.

Phosphorus extracted with the Bray I solution substantially decreased as the clay content of soils increased above 20% (Pratt & Garber, 1964). This was due to increased exhaustion of the Bray I reagent because of reaction with the clay.

When soils with widely varying clay contents are being tested for P by the North Carolina method or the Bray I method it may be necessary to use two different calibration scales or to use wider soil-to-solution ratios. This will give a better measure of the P buffering capacity of the clayey soils when compared with the sandy soils.

E. Critical Soil Test Levels and Recommendations

1. CRITICAL SOIL TEST LEVELS

The soil test levels above which little or no response to fertilizer P is obtained are given in Table 4. A level of 10 ppm $NaHCO_3$ extractable P was found to be adequate for wheat, alfalfa, and cotton (Olsen et al., 1954a). This critical level was found to apply to many neutral to calcareous soils. Higher critical levels have been found in Australia for $NaHCO_3$-extractable P when a 1:100 soil-to-solution ratio is used instead of a 1:20 ratio. A greater amount of P is extracted by increasing the solution-to-soil ratio and accounts for the two to three times higher critical levels.

The critical levels for the $0.025N$ HCl + $0.03N$ NH_4F extractant are generally around 25 ppm (Table 4). This agrees with the level of 30 ppm P used as the break between medium and high for midwestern U.S. soils (Thomas & Peaslee, 1973).

The critical level for the $0.05N$ HCl + $0.025N$ H_2SO_4 extractant (North Carolina) is influenced by soil texture (Table 5). The critical level on clayey soils is around 10 ppm P while on sandy soils the critical level is 20 to 25 ppm P. Although the critical soil test vlaues are less on the clayey soils than the sandy soils, the clayey soils have higher P buffering capacity than the sandy soils (Table 5). The lower amount of P extracted from the clayey soil than from the sandy soil is probably due to exhaustion of the acid extraction by the higher exchange capacity soil, or resorption of extracted P. As dis-

Table 4—Critical levels of soil test P for various crops.

Soil test	Crop	Region	Critical P level	Reference
			ppm	
$NaHCO_3$	Wheat	Bolivia	7	Waugh & Manzano (1971)
		India	11	Gattani & Seth (1973)
		India		
		Black soils	10	
		Alluvial	5	Goswami et al. (1971)
		U.S.		
		Calcareous	10	Olsen et al. (1954a)
		New South Wales	30(1:100)	Colwell (1963)
	Clover	New South Wales	25(1:100)	Spencer et al. (1969)
	Guinea grass	Queensland	20(1:100)	Bruce & Bruce (1972)
$0.025N$ HCl + $0.03N$ NH_4F	Wheat	India	11	Gattani & Seth (1973)
		Mexico	25	Ortega (1971)
	Corn	Mexico	25	Ortega (1971)
	Corn & soybeans	U.S.	30	Thomas & Peaslee (1973)
$0.05N$ HCl + $0.025N$ H_2SO_4	Corn	Alabama		
		Sandy loam, clay loam	15	Rouse (1968)
		North Carolina		
		Silty clay loam	8	Kamprath (1967)
	Coastal bermuda	Alabama		
		Loamy sand, sandy loam	25	Jordan et al. (1966)
	Millet	North Carolina		
		Silty clay loam	10	
		Fine sandy loam	17	Woodruff & Kamprath (1965)
	Pine	Southeastern U.S.	4 to 6	Pritchett & Gooding (1975)

Table 5—Effect of texture on the amount of fertilizer P required to bring soil to the critical soil test level (data from Woodruff & Kamprath, 1965).

Soil texture	Fertilizer P	Soil test P
	ppm	
Cecil fine sandy loam	52	17
Georgeville silty clay loam	86	10

cussed previously, by increasing the solution-to-soil ratio more P can be extracted from the clayey soils.

2. FERTILIZER P RECOMMENDATIONS

When soil test P is above the critical level, recommendations for fertilizer P are for maintenance of adequate soil P or to provide a starter effect.

Table 6—Increase in soil test P (0.05N HCl + 0.025N H$_2$SO$_4$) of a Portsmouth soil with varying rates of fertilizer P over a 7-year period (Kamprath, 1964).

Fertilizer P		Soil test P		
Annual	Total	Initial	Final	Increase
kg/ha		ppm		
19	133	18	25	7
38	266	18	38	20
57	399	18	58	40

If the soil test is below the critical level, the rate of fertilizer P recommended is sufficient to supply adequate P for the plant and to result in some buildup of available soil P when applied at that rate for several years.

Relatively little information is available as to changes in soil test P in the field when various rates of fertilizer P are applied over a period of time. An application of 4 kg P/ha was required to raise the 0.025N HCl + 0.03N NH$_4$F-extractable P of medium-textured Mollisols and Alfisols 1 kg/ha (Peck et al., 1971). With sandy loam to clay loam textured Ultisols an application of 5 to 6 kg P/ha was required to raise the 0.05N HCl + 0.025N H$_2$SO$_4$-extractable P 1 kg/ha, but on clayey textured soils 12 kg P/ha were required to change the soil test values 1 kg P/ha (Rouse, 1968).

The effect of increasing rates of fertilizer P on changes in soil test P were studied for a corn-soybean rotation on a Portsmouth fine sandy loam (Table 6). Annual rates of 19, 38, and 57 kg P/ha increased the soil test levels 7, 20, and 40 ppm P over a 7-year period. When the rates of fertilizer P were doubled and tripled the relative increases of soil test P were approximately 3 and 6 times greater, respectively, than the lowest rate. This points out that the P reaction products formed are more soluble as the rate of fertilizer P is increased. Similar results have been obtained with adsorption isotherms which indicate that the increases in soil solution P are a curvilinear function of P additions rather than a linear function (Fox & Kamprath, 1970).

Thus, in trying to predict how much fertilizer P is needed to change the soil test P, information is needed about the P adsorption characteristic of the soil and the initial level of soil test P.

III. SOIL SOLUTION P AS A MEASURE OF AVAILABLE P

Plants take up P from the soil solution and the concentration of P influences uptake (Wild, 1964). In recent years there has been considerable interest in determining the critical level of soil solution P required for optimum growth. The amount of P in the soil solution is a function of the quantity factor and the percentage of the adsorption capacity which is saturated by P. The theoretical development of the quantity-intensity relationships of P have been discussed in detail by S. R. Olsen and F. E. Khasawneh in Chapter 14 of this book. The presentations in the following sections will deal with the use of quantity-intensity relationships for evaluating the P availability of soils and predicting the amount of fertilizer P needed for optimum growth.

A. Phosphate Adsorption Capacity as Influenced by Soil Texture and its Effect on Soil Solution P

The P adsorption capacity of acid soils is influenced by the amounts of hydrated oxides of Fe and Al, and of calcareous soils by the amount of exchangeable Ca and $CaCO_3$. The soil constituents which determine the P adsorption capacity are often associated with the clay fraction. Soil texture, therefore, can often be used as a relative index of the P adsorption capacity, particularly with soils which have similar chemical properties. The P adsorption maxima for a group of acid soils were 18, 104, and 342 ppm of P for sand, fine sandy loam, and silty clay loam textures, respectively (Woodruff & Kamprath, 1965), and for a group of calcareous soils the P adsorption maxima were 105, 127, and 246 ppm P for fine sandy loam, silt loam, and clay textures, respectively (Olsen & Watanabe, 1957).

The effect of soil texture on the concentration of soil solution P resulting from application of fertilizer P to acid soils and calcareous soils is shown in Table 7. A given amount of fertilizer P resulted in a much greater increase in soil solution P on coarse-textured soils as compared with fine-textured soils. Much higher amounts of fertilizer P were required on acid soils than on calcareous soils with similar clay contents to give comparable increases in soil solution P. Acid soils were found to adsorb 2.17 times more P per unit surface area of soil than calcareous soils (Olsen & Watanabe, 1957).

B. Critical Levels of Soil Solution P for Plant Growth

The level of soil solution P required for optimum plant growth on a wide range of soils is given in Table 8. A higher concentration of soil solution P, approximately three times more, is required on coarse-textured soils than on fine-textured soils. The higher requirement on sandy soils is proba-

Table 7—Effect of soil texture of acid and calcareous soils on amount of fertilizer P required to increase soil solution P.

Soil property	Soil texture	Fertilizer P	Soil solution P	
			No fertilizer	Fertilizer
			ppm	
Calcareous†	Clay	35	0.017	0.053
	Silty clay loam	35	0.034	0.161
	Fine sandy loam	20	0.045	0.305
Acid‡	Silty clay loam	200	0.01	0.07
	Very fine sandy loam	75	0.01	0.20

† Data from Olsen and Watanabe, 1970.
‡ Data from Fox and Kamprath, 1970.

Table 8—Critical levels of soil solution P for plant growth.

Region	Crop	Extractant	Critical P level, ppm	Reference
Western U.S.				
Clay	Barley	Water	0.10	Olsen & Watanabe, 1963
Silty clay loam			0.16	
Fine sandy loam			0.35	
California	Small grain	Water	0.40	Martin & Mikkelsen, 1960
Idaho				
Silt loam	Sweet corn	Water	0.13	Jones & Benson, 1975
Texas	Rice	Water	0.10	Hossner et al., 1973
Southeastern U.S.				
Silty clay loam	Millet	$0.01M$ $CaCl_2$	0.07	Fox & Kamprath, 1970
Very fine sandy loam			0.20	
Sandy loam	Sudangrass	Water	0.68	Soltanpour et al., 1974
		$0.01M$ $CaCl_2$	0.37	
Western Australia	Pasture	$0.01M$ $CaCl_2$	0.2 to 0.3	Ozanne & Shaw, 1967
Hawaii	Corn	$0.01M$ $CaCl_2$	0.06	Fox et al., 1974
	Sweet potatoes		0.10	
	Lettuce		0.40	

bly due to two factors which influence P availability, diffusion rate, and buffering capacity.

Diffusion rate of P is less in sandy soils than in clayey soil (Olsen & Watanabe, 1963). Therefore, a higher concentration of solution P is required in a sandy soil as compared with a clayey soil in order for the same amount of P to reach the root.

The P buffering capacity of sandy soils is less than that of clayey soils. Consequently, soil solution P decreases more in a sandy soil than in a clayey soil for a given amount of P taken up by plants. Therefore, soil solution P in sandy soil has to be higher than in clayey soils in order to supply the same amount of P to the plant. Olsen and Watanabe (1963) found that P uptake was a function of the parameter $(b\ Dp)^{1/2}$ in which b is a measure of the buffering capacity and Dp is the P diffusion coefficient for the soil.

The plant species also has an effect on the level of soil solution P required for maximum growth. Corn, sweet potatoes, and lettuce required different levels of soil solution P for 95% of maximum growth (Table 8) (Fox et al., 1974). These differences could be due to the size of the root system and the rate of growth.

C. Measurement of Soil Solution P

The soil solution P has been estimated by extraction with water and $0.01M$ $CaCl_2$. Water-soluble P has been measured using soil-to-solution ratios of 1:1.25 (Olsen & Watanabe, 1970), 1:10 (Olsen & Dean, 1965), and 1:60 (van der Paauw, 1971). The P concentration in solution usually increases as the amount of soil increases per unit volume of water. A saturation extract would be more closely related to the P concentration found in soil solution, but this method of extraction is not suited to routine purposes.

A $0.01M$ $CaCl_2$ solution has also been used to measure the soil solution P concentration (Aslyng, 1964). Concentration of P in $CaCl_2$ extracts is generally one-third to one-half that of water extracts (Olsen & Watanabe, 1970; Soltanpour et al., 1974). Concentrations of Ca were less in the water extracts as compared with the $CaCl_2$ extracts and resulted in higher P concentrations in the water extracts.

D. Use of Adsorption Isotherms

Adsorption isotherms have been used to obtain information about both the quantity and intensity aspects of P availability (Beckwith, 1964; Ozanne & Shaw, 1967; Fox & Kamprath, 1970). The amount of P adsorbed at a standard supernatant concentration has been used as an estimate of the fertilizer P required for optimum growth. White and Haydock (1968) found that 67% of the variation in yield of pasture in 16 Queensland sites could be accounted for by the amount of P sorbed from an equilibrium solution. Determination of adsorption isotherms is not suited for routine testing purposes. However, this approach can serve as a means for characterizing soils as to their P buffering capacity and grouping together those which are similar for the purpose of making fertilizer recommendations.

IV. USE OF PLANT ANALYSIS IN ASSESSING SOIL P STATUS

The primary intent here is to evaluate the use of plant tissue analysis for the purpose of differentiating the soil P concentrations important to plant growth and final yield. To make this evaluation it is necessary to discuss certain principles basic to the use of plant analysis. Relationships between the concentration of P in the plant tissue and in the soil as well as between the concentration of P in plant tissue and plant growth or yield are discussed. The importance of the concept of "critical" concentration and minimum concentration of P in plant tissue with respect to the assessment of soil P is discussed. Factors affecting P concentrations and critical concentrations of P and the possible processes affected by the alteration of P concentration in plant tissue are discussed.

Plant analysis is conventionally defined as the determination of the quantity of inorganic nutrient element present in plant tissue. Inherent in this definition is the interpretative aspect of the results of these analysis for the ultimate purpose of attaining or maintaining maximum yield. The reasons for using plant tissue analysis for diagnostic purposes are that: (i) the plant is the ultimate measure of the truly available nutrient, and (ii) it will indicate nutrient availability regardless of soil type. In many cases, the tendency has been to make greater use of plant analysis than soil analysis, especially through the use of crop-logging methods (Richards & Bevege, 1972). The use of plant tissue analysis in crop-logging aims at the prevention of nutrient disorders through constant monitoring of the nutrient status of the plant. The plant itself may be the best indicator of its own nutritional well being and, indirectly, a good indicator of soil fertility. However, nutrient concentration values in a plant may often be misleading and may not indicate useful information about the nutrient status of the plant and consequently of the soil.

Measurement of the correlations between plant growth and nutrient concentration in the plant and between plant and soil nutrient concentration must be made and evaluated. Considerations of the effects of plant growth rate, nutrient uptake rate, and nutrient translocation rate on the nutrient concentration in the plant, independent of soil nutrient concentrations, must be made. To know just the nutrient concentration in the plant is not necessarily to know the status of the nutrient in the soil.

A. Relationship Between P Concentration of Plant Tissues and Growth

1. RELATIONSHIP DEFINED

The relationship between quantity of nutrient and cellular growth has been expressed by Boss (1964). He developed the equation $W = kX^n$, where W is the fresh weight per cell, X is the nutrient quantity, and n and k are constants. Under a defined environment, cell weight and nutrient quantity are related to each other by a constant quantitative factor. The concept that the three variables, growth, nutrient quantity in the whole plant, and nutrient concentration of certain tissues, are related directly has been used in studies of plant-soil relationships for many years. However, the relationship between plant growth and nutrient concentration in plant tissues is of greater importance with respect to the use of plant analysis as a practical diagnostic tool. Only a portion of the plant is needed for this use, whereas the whole plant is needed to measure the total quantity of nutrient absorbed per plant.

Nutrient concentration is usually considered to be the weight of nutrient per unit weight of dry tissue. In some cases, the fresh plant tissue weight is used. According to Jones (1970), element absorption and plant growth should parallel each other during most of the vegetative growth period if growing conditions are normal. Exceptions to this would be shortly after germination, during the reproductive period, and preceding senescence.

2. CRITICAL AND MINIMUM P CONCENTRATIONS

The relationship between plant growth and nutrient concentration in plant tissues was recognized, described, and separated into several components by Macy (1936) and more recently by Ulrich and Hills (1967). An important part of this relationship is the region where growth or yield approaches a maximum. The nutrient concentration in the plnat at 90% of maximum growth has often been termed the *critical* concentration. It is expected that nutrient concentration values less than the critical value will in-

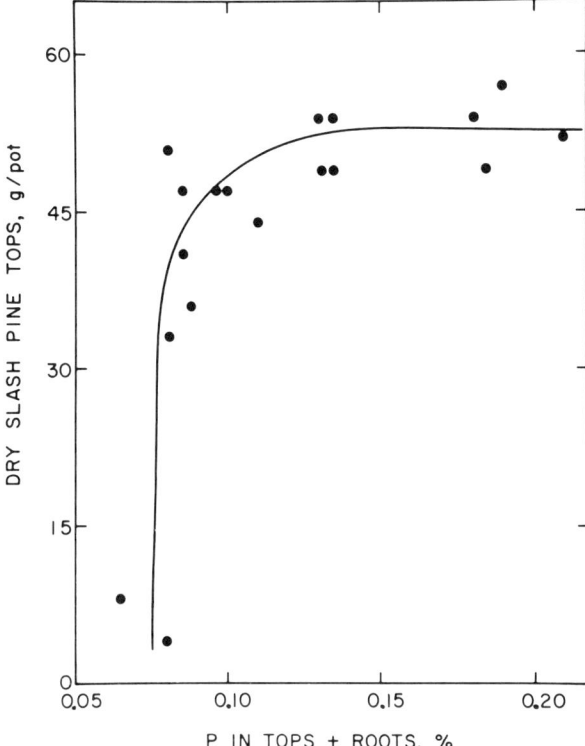

Fig. 1—Relationship between dry weight of slash pine seedlings and P concentration in tops plus roots (Terman & Bengtson, 1973).

dicate a deficient condition with respect to the growth of the plant. Values greater than the critical concentration values indicate nutrient sufficiency. The concept of a critical concentration of P in certain plant tissues is of major importance for the use of plant analysis as a diagnostic tool, whether to evaluate directly the P status of the plant or indirectly the P status of the soil. It should be remembered that the critical concentration of the nutrient is directly related to plant growth by definition; whereas nutrient concentration per se is not.

A practical example of the relationship between growth and P concentration in the whole plant is found in work by Terman and Bengtson (1973, Fig. 1). From this relationship, the critical concentration of P in tops plus roots of slash pine seedlings was estimated to be in the range from 0.09 to 0.15% of dry weight. Also a minimum concentration of P existed at about 0.07% and this concentration was sufficient to produce about 45 g of dry matter per pot. Terman et al. (1972b) developed the idea of "minimum" concentration. However, there appears to be some confusion about whether or not the plant growth rate is reduced when the P concentration is at the "minimum" as defined by Terman and associates. The minimum concentration appears similar to what Loneragan (1968a) defined as the *functional nutrient requirement*. His term refers to the nutrient concentration within the organism which can sustain its metabolic functions at rates which do *not* limit growth.

The inflection of the growth vs. tissue P concentration function probably occurs because some other factor begins to limit growth. As the maximum yield is approached, the rate of dry matter production decreases and eventually becomes zero, even though accumulation of P into the plant apparently continues, resulting in increased P concentration. It would appear that the minimum concentration of P is sufficient for further dry matter production if some other growth factors or factor were corrected. Possibly the great value of the concept of critical concentration is that it indicates that when growth is being limited some factor other than P is causing the limitation, assuming the genetic limit has not been reached. A few of the numerous examples of the relationship between growth and P concentration of whole plants and in plant tissues are found in the recent work of Bould and Parfitt (1973), De Mooy and Pesek (1969), Walker and Peck (1972), Baker and Tucker (1973), Terman et al. (1972a, 1972b), Jones et al. (1972), and Voss et al. (1970).

Critical and minimum concentration of P established in pot-type experiments may be of little use when applied to plants growing under field conditions. Because of the limited volume of soil in greenhouse pots, root growth is restricted and development of mature forage and grain yields seldom occurs. Consequently the critical and minimum concentration values obtained from pot experiments cannot be related to these mature yields. Some of the pitfalls in interpreting results from pot experiments are discussed in a paper by Terman (1974).

B. Relationship Between P Concentration of Plant Tissue and Soil P Concentration

The functional relationship between plant growth or yield and P concentration in the plant tissue does not, by itself, directly reveal the P status of the soil in which the plant is growing. To evaluate the status of the P in the soil via plant tissue analysis a high positive correlation between percent P in the tissue and some measure of soil P must exist. Two parameters of soil P which can be measured are concentration of soil P removed with soil test extractants or the concentration of P in the equilibrium soil solution surrounding the plant root. If the correlation between the concentration of P in the plant tissue and the concentration of P in the soil is nearly 1.0 then plant analysis can be used to estimate the status of soil P.

A main premise involved in soil-plant relationships is that nutrient deficiencies in plants can be corrected by altering the nutrient regime of the growth medium. It has been verified many times that more nutrient will be taken into the plant when more is added to the growth medium. However, it has not always been so clear whether or not the P concentration in the plant will correlate highly with the concentration of P in the growth medium.

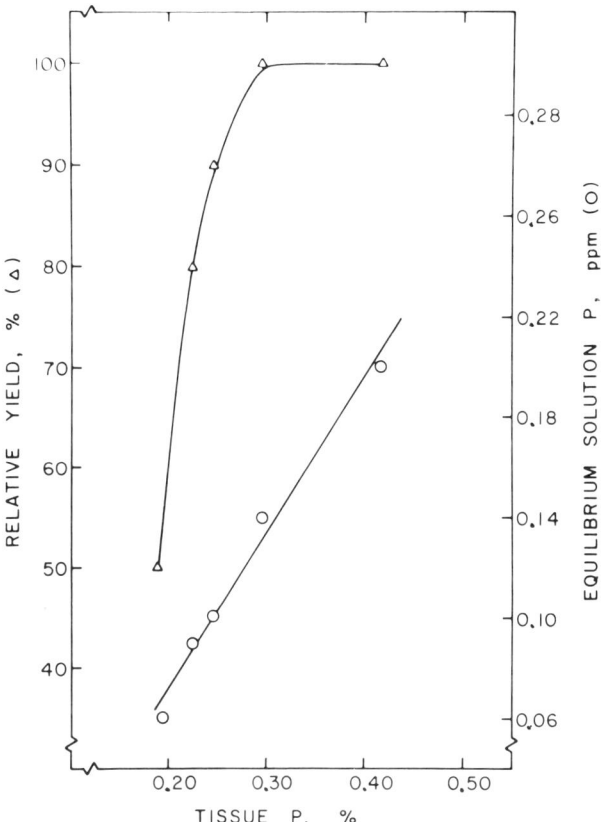

Fig. 2—Relationship among P in sweet corn leaves, relative yield, and equilibrium soil solution P (Jones & Benson, 1975).

1. EQUILIBRIUM SOLUTION P

In studying the phosphate sorption isotherms of high P fixing soils, Jones and Benson (1975) showed that the relationship between percent P in leaves of sweet corn at early tasseling and P concentration of equilibrium soil extracts was linear and highly correlated (Fig. 2). The critical concentration of P in the leaf tissue was from 0.27 to 0.29% and the corresponding equilibrium P concentration was about 0.12 to 0.13 ppm. Hossner et al. (1973) found a similar high positive linear correlation between percent P in the tissue of rice and the soil solution P concentration. At 90% of maximum yield, the rice tissue contained 0.25% P while the corresponding soil solution contained 0.10 ppm P.

A study by Fox and Kamprath (1970) showed that the relationship between P concentration in the plant and in the equilibrium soil extract was curvilinear over the range of P concentration from 0.01 to 1.8 ppm P (Fig. 3). This range was much greater in the two previously cited studies. However, relatively good linearity existed from 0.01 to 0.2 ppm P in equilibrium soil extracts. Plant growth had reached maximum at near 0.55% P but ap-

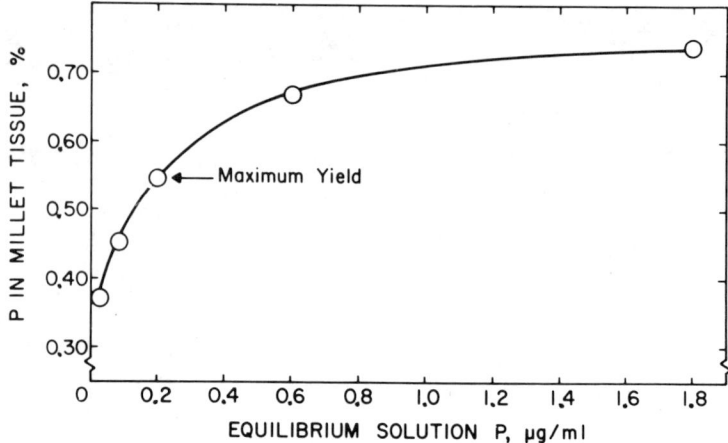

Fig. 3—Relationship between concentration of P in millet tissue and concentration of P in equilibrium soil solution (Fox & Kamprath, 1970).

parently the uptake rate of P, though greatly reduced, was not completely stopped, and the percent P continued to increase. It would appear that a percent P level in the tissue was approached which may correspond to a near toxic level (0.8% P) for these plants, as indicated by slight yield reduction at the high solution P concentration.

Asher and Loneragan (1967) also have shown that when the P concentration of the solution was increased from 0.04 to 0.2 μM a two- to fourfold increase occurred in the P concentration in tops of several plant species. However, the concentration of P in the tissue was not the same for each species.

2. PHOSPHORUS REMOVED BY CHEMICAL EXTRACTANTS

A recent study by Ballard and Pritchett (1975) showed that the relationship between the concentration of P extracted from the soil by different extractants correlated highly with the concentration of P in the tissue of pine seedlings (Table 9). The largest simple linear correlation coefficient was +0.931 for $0.5M$ $NaHCO_3$ (pH 8.5).

Table 9—Association of soil test extractants for P and concentration of P in slash pine seedlings after growth of 1 year (Ballard & Pritchett, 1975).

P extractant	Mean extractable P	P in tops
	ppm	r^2
H_2O	2.3	0.537*
NH_4OAc (pH 4.8)	4.4	0.847**
$0.002N$ H_2SO_4	8.9	0.821**
$0.5M$ $NaHCO_3$ (pH 8.5)	12.3	0.931**
$0.05N$ HCl + $0.025N$ H_2SO_4	15.2	0.925**
$0.03N$ NH_4F + $0.025N$ HCl	41.3	0.895**

* Significant at 5% level.
** Significant at 1% level.

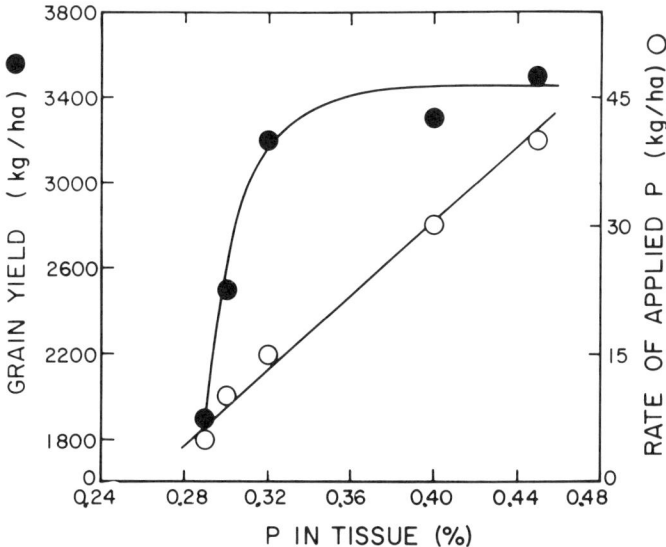

Fig. 4—Relationship of P in wheat tissue, yield and rate of applied P (Baker & Tucker, 1973).

3. PHOSPHORUS ADDED TO SOIL

Baker and Tucker (1973) reported that the concentration of P in wheat tissue was linearly related to the rate of P added to soil (Fig. 4), even though the yield-percent P relationship was not. Research by Powell and Webb (1974) showed percent P of corn leaf tissue was linearly correlated with rates of fertilizer P and that the major variability in percent P in the leaf was accounted for by the applied fertilizer variables. However, the linear correlation coefficient for the regression of grain yield on the 3-year average of P concentration in the leaf was 0.33 at one experimental location and 0.80 at another location.

4. OTHER FACTORS

Not only is applied P important to the concentration of P in plant tissue but in some cases other soil factors may prove important. De Mooy and Pesek (1969) found that percent P in soybean leaf tissue increased with P application and that 95% of the variation in percent P could be explained by variations in soil P and soil Ca, their squares, and the interactions of P, K, and Ca. The differential effects of fertilization on leaf P concentration developed progressively with time during the growth period. De Mooy and Pesek (1969) also cited earlier work which indicated that the soybean plant could concentrate P under high application rates of fertilizer P to the extent of creating P toxicity. They concluded that 1% P in soybean leaves was the threshold value for P toxicity. This toxicity would occur if growth was not proceeding fast enough to dilute the high concentrations of P in the tissues.

C. Factors Modifying P Concentration of Plants

1. EFFECTS OF PLANT GROWTH

Evidence exists that establishes the relationship between plant growth and P concentration in plant tissue and between concentration of P in tissue and concentration of soil solution P, at least for certain conditions. However, the use of plant analysis to assess the P status of soil largely depends on how well these relationships hold over different conditions which affect growth. Can the rate of plant growth alter the P concentration in the plant so that this concentration does not correlate highly with the concentration of P in soil solution?

It has been noted by many and especially more recently by Terman et al. (1972b) that indeed P concentration in plant tops decreased with increase in plant growth when another growth-limiting factor, other than P, was corrected. This is illustrated in Fig. 5 where high rates of N resulted in increased growth, causing dilution of the P concentration at a particular P rate (dotted lines). It is evident that the relative growth rate was greater than the relative P accumulation rate in the plant tops. Also, if the P concentration at 90% of the maximum yield is considered the critical concentration, then at the 480 N level, the critical concentration of P would be nearly the same as the minimum concentration of P (0.10% P). Furthermore, even though growth was greater when N was increased, the critical concentration of P did not appear to decrease greatly. It would appear that the critical concentration value of P would be a good indicator of P deficiency and an especially useful indicator to indicate the existence of a growth-limiting factor as long as the concentration of P is greater than the minimum concentration value. These authors suggested that estimations of critical concentrations of P could be made from the minimum concentration values. Also according to their results, the minimum concentration of P in corn forage may be a better indicator of the nutrient status of the plant because of its slight variability with age of the plant. They found large decreases in the critical concentration of P with age of the plant.

In another study, Terman et al. (1972a), found that yields of dry corn forage increased when the P concentration in the forage was at or near a minimum of 0.10% P. The production of corn forage was not limited by P at this minimum concentration but became limited because of insufficient Zn. The P concentration in the corn forage continued to increase because the yield increase was slowed and then stopped by inadequate Zn.

These studies show that P concentration in the plant tissue can be diluted due to increased plant growth, but no information was found relating P concentration values at different growth levels to the P concentration in soil solution. It is likely that the concentration of soil solution P would not change; consequently, the correlation between percent P in plant tissue and P concentration in soil solution would be low. However, the change in the percent P in the plant tissue may not be due to dilution of the plant P through increased growth, but might instead be the result of a decrease in

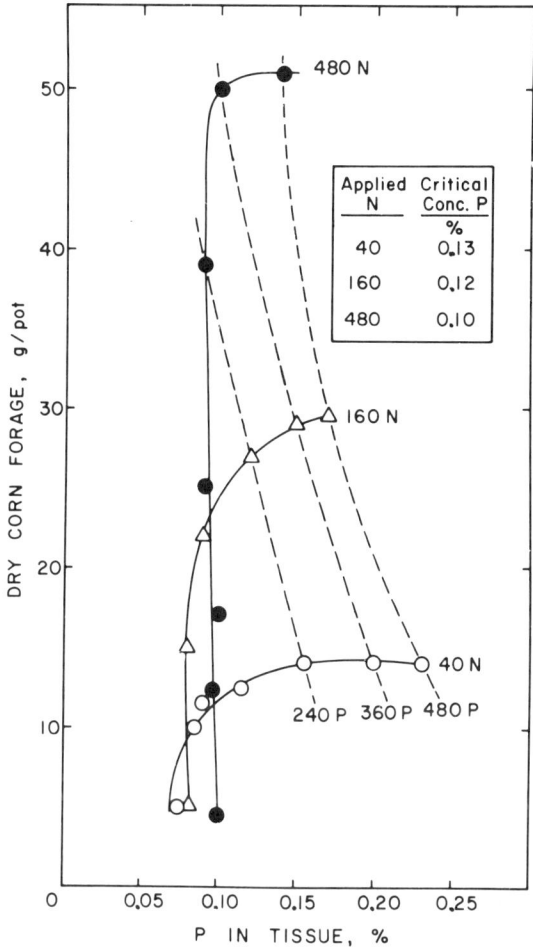

Fig. 5—Yield of dry corn forage as a function of P concentration at different applied N levels (Terman et al., 1972).

the concentration of the P in the soil solution surrounding the root. This could be true especially if the P concentration in the soil solution was initially great enough to maintain the percent P in the plant above a minimum concentration of P. If this were the case, then a decreae in percent P in the plant would correspond with the decrease in the P concentration in the soil solution and the correlation would hold up.

2. CONSTANCY OF THE CRITICAL CONCENTRATION OF P

It has been long known that nutrient concentrations in the plant are dynamic and are affected by many factors. Measuring only plant P concentration without evaluating plant growth can lead to meaningless interpretation when the purpose of the measurement is to establish guidelines to separate plants deficient in P from those sufficient in P. The use of the critical concentration does relate growth and P concentration and provides a basis for the evaluation of the P status in both plants and soils. Therefore, the important question for the establishment of this basis is: Does the critical concentration of P remain constant when other factors are varied? The evidence suggests both a yes and no answer.

Bates (1971) reviewed the factors that may cause the critical concentration of a nutrient to change. He concluded that nutrient concentrations and environmental factors can markedly alter the nutrient content of plants. Other factors, according to Bates, are age of tissue, fraction of nutrient measured, and cultivar effects. Bates also cited several references that discuss how P concentrations in plant tissue change when environmental factors change.

a. Effects of Other Nutrients—Some results of studies by Bould and Parfitt (1972) indicate that an increase in N not only caused an increase in the grams of black currant tissue but also caused the critical concentration of P to increase (Fig. 6). A third nutrient or other growth factor may have limited yields. This is likely because the rate of yield increase was not great with addition of N and because P concentration was increasing faster than yield. For unrestricted growth, it would be expected that the function would show a greater magnitude of increase in the vertical direction, before the point of inflection. However, this experiment was conducted in a nutrient

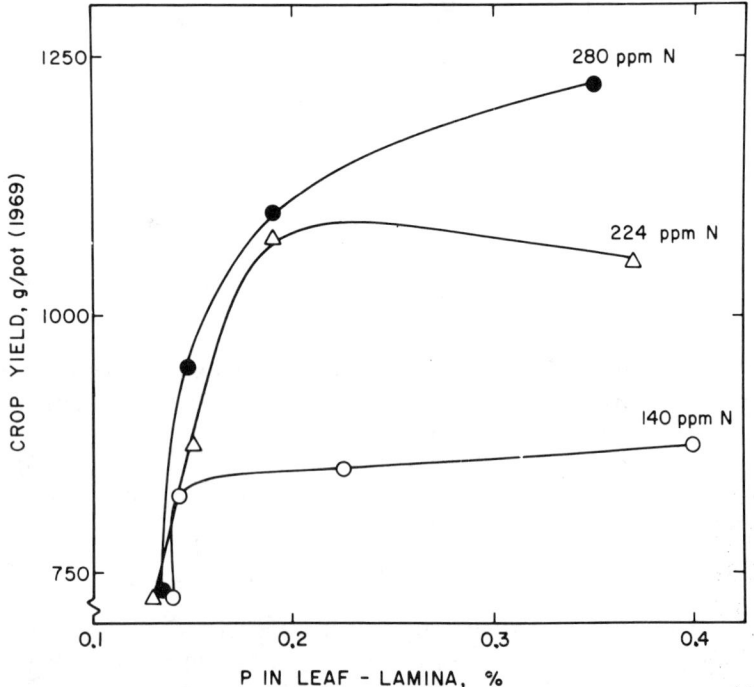

Fig. 6—Relationship between black currant yield and concentration of P of the leaf-lamina at different N concentrations in solution (Bould & Parfitt, 1972).

solution culture and the availability of P may be different than in soil. In the solution culture, the concentration of P at the root surface was probably greater than it would be in soil, resulting in greater continuous uptake of P.

Other studies have shown that there was little if any change in the critical concentration of P when N concentration in the plant was increased (Walker & Pesek, 1967). In this study, the range of P concentration was from 0.15 to 0.35% of the dry matter yield of bluegrass while the N concentration varied from 2.5 to 4.5% of dry matter. The critical concentration of P appeared to be near 0.25% and did not change with a change in the N concentration. However, the regression equation used contained a PK interaction term. This interaction term caused the critical concentration of P to be dependent on the concentration of K and 0.25% P was only considered critical at the critical concentration of K.

The idea that for maximum growth, the nutrients should be in proper balance with each other has been around for a long time. Shear et al. (1946) suggested that at any level of nutritional intensity there exists a nutritional balance at which "optimum" growth for that intensity level will result. The maximum growth and yield will result only when the proper balance of nutrient elements occurs in combination with the optimum intensity.

Bould (1964) also thought that the optimum concentration of one element varies with concentration of other nutrient elements, especially if the other elements are not at optimum concentrations. He indicated that the nutrient balance becomes more critical as the maximum yield is approached.

Results of studies by Dumenil (1961) led him to conclude that the N-P nutrient balance was most important only at or near maximum yield. He concluded that the critical concentration of P is not a narrow range of values but includes a wide range of values, dependent upon N.

Ulrich and Hills (1973) considered the critical level of nutrients to be constant, regardless of stage of growth, provided that the tissue of the same physiological age was sampled for analysis.

If all nutrients were present in the growth medium at concentrations not limiting growth, then maximum yield would only occur as a result of some non-nutritional growth factor limiting yield. It would then seem that there would exist only a narrow range in the critical concentration of P in tissue of the same physiological age. In this case, the critical concentration of a nutrient can be determined only by some non-nutritional factor causing a limitation of growth.

b. Importance of Plant Tissue Age—The type of tissue sampled may be an important factor with respect to constancy of critical concentration of P. Mature whole plants that contain relatively more mature parts than young whole plants may show greater change in the critical concentration of P than do younger plants. In the mature whole plant there may be less P per unit of mature tissue than per unit of young tissue. Consequently, for plants growing under conditions where other nutrients are not limiting growth, the critical concentration of P for the whole mature plant may be less than for the whole younger plant. The younger plant would reflect a more uniform P concentration when growth is limited. This is demonstrated by data obtained by Jones et al. (1972). Their work showed that the critical concentration of P decreased with the age of subclover tissue and decreased proportionately more in stems (46%) than in leaves (21%). Thus, when plant growth was limited at a younger growth stage, the critical concentration of P was greater than when growth was limited at an older growth stage. If leaves of the same physiological age rather than the whole plant had been sampled, the change in critical concentration of P may not have occurred.

c. Dry Matter Concentration—Friis-Nielsen (1973) suggested that much of the ambiguity characteristic of relationships between nutrient concentrations and yield could be reduced if dry matter concentrations were also considered. Results from his research have shown that nutrient concentrations, based on dry matter compared with those based on fresh matter, were relatively lower at high dry matter concentrations (upper leaves), and relatively higher at low dry matter concentrations (lower leaves). Also, he concluded that at the same level of nutrient, nutrient concentrations are identical in leaves of the same physiological age, irrespective of the chronological age of the whole plant.

d. Cultivar Affect—As discussed by Bates (1971), not only may nutrient interactions cause critical concentrations of nutrients in the plant to vary, but genetic factors tend to complicate the use of plant tissue analysis. This appears true for P in that different cultivars of a plant species may have different critical concentrations of P. Studies by Terman et al. (1975) suggested that not only the P concentration was different at the same yield level of different corn hybrids, but also it appeared that the critical concentration of P might be different. This cannot be conclusively stated since true maximum growth was not attained but the dilution effect was different with hybrids. Also, whatever factor started to limit growth did not affect each hybrid the same, i.e., growth of hybrid A leveled off before hybrid B. Work by Rivard and Bandel (1974) showed that P concentrations in field corn cultivars were significantly different, but there was no indication that the critical concentration of P was different.

3. INFLUENCE OF RATE PROCESSES ON TISSUE P CONCENTRATIONS

The P concentration is a quotient, i.e., a ratio of weight of P/unit weight of dry matter. Therefore, anything that can cause either the numerator or denominator to change in a nonparallel manner will either decrease or increase the resulting quotient. Percent P in the plant tissue can be perceived as a function of the following processes:

%$P = f$(P supply rate, P absorption rate, P translocation rate, P retranslocation rate, rate of P interaction with other nutrients, plant growth rate.)

These mechanisms probably are interrelated with each other.

a. Uptake by Roots—According to Brewster and Tinker (1972) the important question is not only whether there is sufficient available P in the soil but also whether the root system of the crop, growing in that soil, is able to absorb the P at a rate that will sustain the maximum growth rate of the crop. If rate of P uptake does not keep pace with the growth rate then P concentration in the tissue will decrease. The uptake rate of P may be sufficiently greater than the diffusion rate of P in the soil. If this happens, then the concentration of P around the root will be depleted and, according to these authors, the diffusion process then sets an upper limit to the uptake rate.

Recently, C. D. Raper, Jr. (personal communication, 1976) observed that if the relative plant growth rate (g dry matter $\cdot$ g^{-1} dry matter $\cdot$ day^{-1}) and the relative accumulation rate of P (mg $\cdot$ mg^{-1} $\cdot$ day^{-1}) from nutrient solutions were parallel then the P concentration did not change. Raper concluded that a high initial external concentration of P was more important for obtaining maximum growth of plants than was the external concentration once the plant has adjusted to its environment. When the concentration of P in the external solution was changed there was a lag period in the relative growth rate and relative accumulation rate of P before new constant rates were established. As long as these two rates were parallel the percent P in the tissue will not change; however, if new rates were established because of change in external P concentration, then the percent P changed if the parallelism was not maintained. If another nutrient limited the plant growth by reducing the relative growth rate then the percent P increased, probably because the relative accumulation rate of P was not affected in the same instant, or possibly because a longer lag period for readjustment was required.

Work by Loneragan (1968b) showed that the P flux into plants was greatly increased by very small increases in P cocnentration at very low P concentrations of a flow nutrient culture. The shape of the relationship between the influx of P into the plant and P concentration in the solution was not linear and the rate of increase of P influx began to decrease rather sharply near 4 μM of P in solution.

The amount of P absorbed may be regarded as a function of both size of the root system and rate of P absorption per unit quantity of roots. Work by Loneragan and Asher (1967) indicated that the effect of external P concentration on the rate of P absorption per unit weight of roots was similar to its effect on the total amount of P absorbed. They showed that the rate of P absorption per unit weight of roots increased more than proportionately with the first increment between 0.04 to 0.2 μM P in the external solution. This proportionality decreased as the concentration of P increased from 0.2 to 1 μM. The P absorption was proportionately much less from 1 to 5 μM or from 5 to 24 μM P concentrations. They also showed that the ability of individual species to absorb P from relatively high concentrations was not closely correlated with their ability to absorb P at lower concentrations. It was the conclusion of these investigators that the rate of P absorption per unit weight of roots necessary for maximum growth depends on (i) the initial supply of P in the seedling, (ii) the efficiency with which both seedling P and absorbed P are utilized in the processes of growth, (iii) the size of root system relative to the whole plant, and (iv) the rate of growth the plant can achieve when P is nonlimiting. The latter factor depends on genetic constitution and environmental factors such as supply of nutrients, light, temperature, and carbon dioxide concentration. Factors other than P in the environment of the root may affect root growth and consequently uptake of P. Blair et al. (1972) were able to show that not only was P uptake affected by the P concentration in solution surrounding the root but also was affected by the N form in this same solution. In addition to the effects from N forms in the solution, the status of the N in the plant was related to P uptake from the solution.

According to Loneragan (1968b) a plant with a high nutrient requirement may require a low concentration of nutrient in the culture solution as long as the plant has a high rate of nutrient absorption. Thus, the P concentration in the tissue sampled may be predominantly correlated with uptake rate rather than just P concentration of the solution. Loneragan also indicated that the relationship of growth to a nutrient concentration depends not only on conditions of nutrient supply or rate of nutrient absorption but also on the translocation rates from root to shoot and on the mobility of the nutrient in the phloem.

b. Plant Growth Rate—A main factor affecting rate of P absorption by a plant is not only the P concentration in the surrounding solution but also the relative growth rate of the whole plant. Loneragan (1968b) showed that the relative growth rate is nonlinear across the full range of the P flux into the plant (Fig. 7). Beyond about 3 μg P/g fresh roots per day, the P flux into the plant was greater even though the rate of the relative growth rate was decreasing. Greatest amount of growth was realized from the first increment of P flux.

According to White (1971), the demand for P was associated with the rate of plant growth. Growth rate appeared to have a marked influence on the rate of P uptake at deficient to optimum concentration of P in solution. At low concentrations of external P (0 to 10 μM) it appeared that the demand for P created by the plant's growth was the predominate influence on the rate of P uptake. However, at external P concentrations of 40 μM the rate of uptake was primarily dependent on only the external P concentration. The author suggested that the influx of P was regulated by the turnover rate of the pool of inorganic P in the cytoplasm and by the rate of P transport to the shoot.

c. Effects of P Concentration in External Solution—Research by Green et al. (1973) showed that P could accumulate to toxic levels in barley leaves if the barley plants were first allowed to become P deficient. Continuous exposure of the plants to the high external P solution, i.e., not allowing plant to become P deficient, did not result in toxic concentrations of P in the leaves. These authors suggested that the nondeficient plant has regulatory mechanisms that limit excessive P uptake or accumulation. This regulatory mechanism was not able to work in plants that were deficient in P. A source-sink control mechanism and not just transpirational uptake and ac-

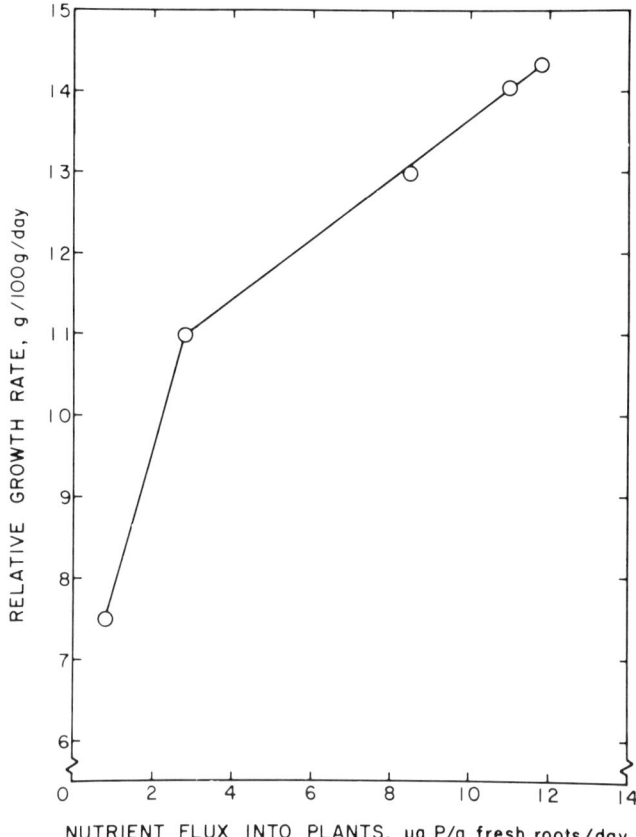

Fig. 7—Relationship between relative growth rate of plants and nutrient flux into plants (Loneragan, 1968b).

cumulation were thought to be involved. Possibly, though, P dilution due to continued growth was also important. Growth did not occur where P was deficient. In other work by Green and Warder (1973) similar results were found working with wheat. These workers thought that the P uptake by P-deficient wheat responded more to conditions within the plant, particularly the P concentration in the tissue, than to the concentration of P in the nutrient solution.

d. Interrelationships of the Plant Processes—Undoubtedly the uptake of P, translocation of P, plant growth rate and the concentration of P in the external solution are interrelated with respect to the effect on P concentration in the plant tissue. The concentration of the P in the plant tissue may be a good indicator of soil P status if these processes are operating in a parallel manner. However, in cases where the plant growth rate becomes limited, or rapidly increases, then these processes are probably operating in a non-parallel manner and the concentration of P in the plant will not correlate highly with soil P concentration which has not changed. Since P concentrations in a plant or plant tissue can change because of nonparallel changes in aspects of growth and of P accumulation, great care should be taken when interpreting the results of plant analysis. Definition of stages of growth, sampling tissue of the same physiological age, and proper sampling and handling techniques are all very important in obtaining the maximum use of plant analysis as a practical diagnostic tool. Whenever possible, plant analysis and soil analysis should be used to supply information for resolving particular plant or crop P disorders.

e. Using Models for Prediction Purposes—Nye et al. (1975) believed that correlation methods of evaluating plant growth have certain limitations. They listed these limitations as follows: (i) applicable only to the particular range of conditions of soil, climate, and crop for which the experiments were made, results cannot be extrapolated; (ii) cannot test theories about individual mechanisms involved; and (iii) the relationship between growth and other relevant factors become extremely complex when growth depends nonlinearly on many factors. Consequently, Nye et al. (1975) have attempted to develop models that would predict P uptake and plant growth response from independently measured soil and plant characteristics. These investigators were primarily interested in how the different soil-plant mechanisms work in providing maximum plant growth.

Similarly, Scaife and Smith (1973) proposed a model in which the basic sequence was: soil nutrient status-plant nutrient status-growth rate-yield, instead of just soil nutrient status-yield. The scheme of the model proposed by these authors was basically as given in Fig. 8. The aspects of plant growth as well as the interrelationships of these aspects are shown. Scaife and Smith arugued that since the decrease in P concentration at the root sur-

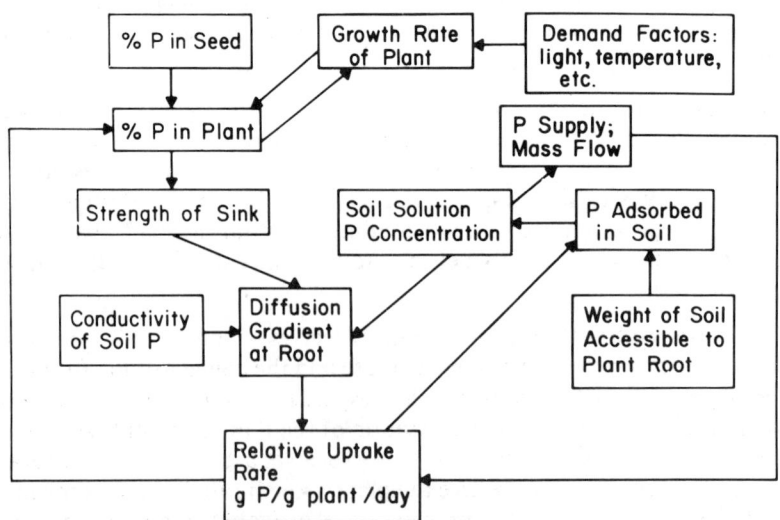

Fig. 8—Diagram of a model indicating relationships of the different processes involved in plant growth and P utilization; note processes affecting % P of the plant (Scaife & Smith, 1973).

face is caused by the plant, and could result from a suboptimal mean soil solution concentration or from impeded transport through the soil, the actual concentration of P at the root surface is only indirectly due to the supply. The P supply would be more reliably predicted from the plant nutrient status. This would be an indication of the degree to which supply is keeping pace with demand. However, from Fig. 8 it can be seen that many factors can affect percent P in the plant without directly affecting the status of soil P.

When all the necessary factors are placed into the model and the model precisely predicts the true plant growth, then it may be possible to use the model in a different way, i.e., to predict the concentration of P in solution surrounding the root. The soil solution P might then be considered a function of the different aspects of plant growth. To evaluate the status of soil P, necessary parameters such as relative plant growth rates, percent P in tissue, uptake rate of P, percent P in the seed, and probably others could be measured and used in the model to find out whether the soil P status is sufficient to provide maximum growth. Thus, possibly a better estimation of soil P status could be obtained than is now possible by measuring only the P concentration in plant tissue.

D. Fertilizer Recommendations

The ultimate objective for determining the concentration of P in plant tissue is to correct or prevent P deficiency. A high correlation between the relationship of plant growth and P concentration in plant tissue, as previously mentioned, is required to accomplish this. Unfortunately, in many cases, the variability in the P concentration does not allow a high degree of accuracy in the prediction of the increments of plant growth or yield. Consequently, the critical concentration range of P in the plant is usually the criterion that is used in diffrentiating P deficiency from P sufficiency and the decision to simply apply P or not to apply P is usually made. This probably is especially true for annual crops growing under the field environment where many uncontrolled variables are affecting the crop. Calibration of the functional relationship may be more precise and meaningful for tree crops or for crops grown under highly controlled environments. Fluctuations in the critical concentration of phosphorus, as caused by factors independent of the soil, can restrict the practical utilization of plant analysis. This restriction may be more important for annual crops that are highly sensitive to their environment. Plant tissue analysis and soil analysis should be used in a complementary fashion when the P status of plant and soil are to be evaluated. Knowledge of the growth environment and growth characteristics of the crop as well as the results from analytical determinations must all be used in making a fertilizer recommendation. Experience in the application of this knowledge in developing a fertilizer recommendation is an invaluable asset.

V. FUTURE RESEARCH NEEDS

A. Soil Testing

There is a continuing need for data on response of crops to P fertilization as related to soil test levels particularly as new varieties are developed and yield levels increase. Long-term studies with various cropping systems should be conducted to determine the amounts of fertilizer P required to reach and maintain the critical soil test level on various soils. Since it is impossible to conduct field studies on all soil series, laboratory and greenhouse studies should be done to determine which soils can be grouped together because of similar chemical properties as regards P reactions and availability.

In certain areas of the United States soils have a very high level of available P because of many years of P fertilization. Research is needed to determine how much P can be supplied for plants at these high soil test levels and how long an adequate supply could be maintained.

B. Plant Analysis

There are a number of areas in which research would be very helpful in improving the use of plant tissue analysis as a tool for diagnosing the P status of soils. More information is needed on how various growth factors affect the correlation between percent P in the tissue and plant growth. Studies should be conducted to determine the minimum concentration of P required in plant tissue for optimum growth when other growth factors have been adequately supplied.

There is very little information on the concentration of P in plant tissue at the critical soil test level and how the percent may change with age of the plant. Research on this point would provide greater insight into the relationship between percent P in the plant and available soil P.

Basic studies on rates of P absorption, P translocation, and P accumulation by plants as related to the external P supply and the effects of other growth factors on these rate processes would provide the information needed to better interpret plant analysis.

LITERATURE CITED

Ahmed, B., and A. Islam. 1975. Extractable phosphate in relation to the forms of phosphate fractions in some humid tropical soils. Trop. Agric. (Trinidad) 52:113–118.

Aldrich, S. R. 1973. Plant analysis: Problems and opportunities. p. 213–248. *In* L. M. Walsh and J. D. Beaton (ed.) Soil testing and plant analysis. Rev. ed. Soil Sci. Soc. Am., Madison, Wis.

Anderson, M. S., and W. M. Noble. 1937. Comparison of various chemical quick tests on different soils. USDA Misc. Pub. 259.

Asher, C. J., and J. F. Loneragan. 1967. Response of plants to phosphate concentration in solution culture: I. Growth and phosphorus content. Soil Sci. 103:225-233.

Aslyng, H. C. 1964. Phosphate potential and phosphate status of soils. Acta. Agric. Scand. 14:261-285.

Baker, D. E., and J. K. Hall. 1967. Measurement of phosphorus availability in acid soils of Pennsylvania. Soil Sci. Soc. Am. Proc. 31:662-667.

Baker, J. M., and B. B. Tucker. 1973. Critical N, P, and K levels in winter wheat. Comm. Soil Sci. Plant Anal. 4:347-358.

Balerdi, F., L. Muller, and H. W. Fassbender. 1968. A study of phosphorus in soils of Central America. III. Comparison of five chemical analysis methods for available phosphorus. Turrialba 18:348-360.

Ballard, R., and W. L. Pritchett. 1975. Evaluation of soil testing methods for predicting growth and response of *Pinus elliottii* to phosphorus fertilization. Soil Sci. Soc. Am. Proc. 39:132-136.

Bates, T. E. 1971. Factors affecting critical nutrient concentrations in plants and their evaluation: A review. Soil Sci. 112:116-130.

Beckwith, R. S. 1964. Sorbed phosphate at standard supernatant concentration as an estimate of the phosphate needs of soils. Aust. J. Exp. Agric. Anim. Husb. 5:52-58.

Bhan, C., and H. Shanker. 1973. Correlation of available phosphorus values obtained by different methods to phosphorus uptake by paddy. J. Indian Soc. Soil Sci. 21:177-180.

Bingham, F. T. 1962. Chemical soil tests for available phosphorus. Soil Sci. 94:87-95.

Blair, G. J., C. P. Mameril, and M. H. Miller. 1972. Effect of nitrogen status on short-term phosphorus uptake. Comm. Soil Sci. Plant Anal. 3:23-27.

Blanchar, R. W., and A. C. Caldwell. 1964. Phosphorus uptake by plants and readily extractable phosphorus in soils. Agron. J. 56:218-221.

Boss, M. L. 1964. Equational relations between nutrient quantities and cellular growth. New Phytol. 63:47-54.

Bould, C. 1964. Leaf analysis in relation to raspberry nutrition. p. 54-67. *In* C. Bould et al. (ed.) Plant analysis and fertilizer problems. IV. Am. Soc. Hortic. Sci., Michigan State Univ., East Lansing, Mich. 423 p.

Bould, C., and R. I. Parfitt. 1972. Leaf analysis as a guide to the nutrition of fruit crops: IX. Effects of initial and supplementary levels of N and P on black currants (*Rives nigrum* L.) grown in sand culture. J. Sci. Food Agric. 23:959-968.

Bould, C., and R. I. Parfitt. 1973. Leaf analysis as a guide to the nutrition of fruit crops: X. Magnesium and phosphorus sand culture experiments with apple. J. Sci. Food Agric. 24:175-185.

Bray, R. H. 1948. Correlation of soil tests with crop response to added fertilizers and with fertilizer requirement. p. 53-86. *In* H. B. Kitchen (ed.) Diagnostic techniques for soils and crops. Am. Potash Inst., Washington, D.C.

Bray, R. H., and L. T. Kurtz. 1945. Determination of total, organic and available forms of phosphorus in soils. Soil Sci. 59:39-45.

Brewster, J. L., and P. B. H. Tinker. 1972. Nutrient flow rates into roots. Soils Fert. 35:355-359.

Bruce, R. C., and I. J. Bruce. 1972. The correlation of soil phosphorus analysis with response of tropical pastures to superphosphates on some north Queensland soils. Aust. J. Exp. Agric. Anim. Husb. 12(55):188-194.

Cajuste, L. J., and W. R. Kussow. 1974. Use and limitations of the North Carolina method to predict available phosphorus in some Oxisols. Trop. Agric. (Trinidad). 51:246-252.

Cate, R. B., and L. A. Nelson. 1965. A rapid method for correlation of soil test analyses with plant response data. Int. Soil Test. Ser. Tech. Bull. 1. North Carolina State Univ. Agric. Exp. Stn., Raleigh, N.C.

Chang, S. C., and M. L. Jackson. 1957. Fractionation of soil phosphorus. Soil Sci. 84:133-144.

Chang, S. C., and A. S. R. Juo. 1963. Available phosphorus in relation to forms of phosphorus in the soils. Soil Sci. 95:91-96.

Colwell, J. D. 1963. The estimation of phosphorus fertilizer requirements of wheat in southern New Wales by soil analysis. Aust. J. Exp. Agric. Anim. Husb. 3:190-197.

Daubeny, C. G. B. 1845. Memoirs on the rotation of crops and on the quantity of inorganic matters abstracted from the soil by various plants under different circumstances. Roy. Soc. (London) Phil. Trans. 135:179-253.

Dean, L. A., and E. J. Rubins. 1947. Anion exchange in soils: I. Exchangeable phosphorus and the anion-exchange capacity. Soil Sci. 63:377-387.

De Mooy, C. J., and John Pesek. 1969. Growth and yield of soybean lines in relation to phosphorus toxicity and phosphorus, potassium and calcium requirements. Crop Sci. 9:130-134.

Dumenil, L. 1961. Nitrogen and phosphorus composition of corn leaves and corn yield in relation to critical levels and nutrient balance. Soil Sci. Soc. Am. Proc. 25:295-298.

Dunbar, A. D., and D. E. Baker. 1965. Use of isotopic dilution in a study of inorganic phosphorus fractions from different soils. Soil Sci. Soc. Am. Proc. 29:259-262.

Dyer, B. 1894. On the analytical determination of probable available mineral plant food in soils. Trans. Chem. Soc. 65:115-167.

Egner, H., H. Riehm, and W. R. Domingo. 1960. Untersuchungen uber die chemishe bodenanalyse als grundlage fur die beurteilung des nahrstoffzustandes der boden. II. Chemische extraktions-methoden zur phosphor—und kaliumbestimmung kungl. Lantbrukshoegsk. Ann. 26:204-209.

Fitts, J. W. 1956. Soil tests compared with field, greenhouse and laboratory results. N.C. Agric. Exp. Stn. Tech. Bull. 121.

Fitts, J. W., and W. L. Nelson. 1956. The determination of lime and fertilizer requirements of soils through chemical tests. Advan. Agron. 8:241-282.

Fox, R. L., and E. J. Kamprath. 1970. Phosphate sorption isotherms for evaluating the phosphate requirements of soils. Soil Sci. Soc. Am. Proc. 34:902-907.

Fox, R. L., R. Nishimoto, J. R. Thompson, R. S. De la Pena. 1974. Comparative external phosphorus requirements of plants growing in tropical soils. Int. Congr. Soil Sci., Trans. 10th (Moscow) 4:232-239.

Friis-Nielsen, B. 1973. Growth, water and nutrient status of plants in relation to patterns of varitions in concentrations of dry matter and nutrient elements in base-to-top leaves: II. Relations between distribution of concentrations of dry matter and nutrient element in tomato plants. Plant Soil 39:675-686.

Gattani, P. D., and S. P. Seth. 1973. Phosphorus soil test correlation studies in Rajasithan. J. Inc. Soc. Soil Sci. 21:373-375.

Geraldson, C. M., G. R. Klacan, and O. A. Lorenz. 1973. Plant analysis as an aid in fertilizing vegetable crops. p. 365-379. *In* L. M. Walsh and J. D. Beaton (ed.) Soil testing and plant analysis. Rev. ed. Sol Sci. Soc. Am., Madison, Wis.

Green, D. G., W. S. Ferguson, and F. G. Warder. 1973. Accumulation of toxic levels of phosphorus in the leaves of phosphorus-deficient barley. Can. J. Plant Sci. 53:241-246.

Green, D. G., and F. G. Warder. 1973. Accumulation of damaging concentrations of phosphorus by leaves of Selkirk wheat. Plant Soil 38:567-572.

Goswami, N. N., S. R. Bapat, and V. N. Pathak. 1971. Studies on the relationship between soil tests and crop responses to phosphorus under field conditions. Proc. Int. Symp. Soil Fert. Eval. (New Delhi) 1:351-359.

Grigg, J. L. 1965. Inorganic phosphorus fractions in South Island soils and their solubility in commonly used extracting solutions. N.Z. J. Agric. Res. 8:313-326.

Gunary, D., and C. D. Sutton. 1967. Soil factors affecting plant uptake of phosphate. J. Soil Sci. 18:167-173.

Halstead, R. L. 1967. Chemical availability of native and applied phosphorus in soils and their textural fractions. Soil Sci. Soc. Am. Proc. 31:414-419.

Hossner, L. R., J. A. Freeouf, and B. L. Folsom. 1973. Solution phosphorus concentration and growth of rice (*Oryza sativa* L.) in flooded soils. Soil Sci. Soc. Am. Proc. 37:405-408.

Humbert, R. 1973. Plant analysis as an aid in fertilizing sugar crops: Part II. Sugarcane. p. 289-298. *In* L. M. Walsh and J. D. Beaton (ed.) Soil testing and plant analysis. Rev. ed. Soil Sci. Soc. Am., Madison, Wis.

John, . K. 1972. Extractable phosphorus related to forms of P and other soil properties. J. Sci. Food Agric. 23:1425-1433.

John, M. K., A. L. van Ryswyk, and J. L. Mason. 1967. Effect of soil order, pH, texture, and organic matter on the correlation between phosphorus in alfalfa and soil test values. Can. J. Soil Sci. 47:157-161.

Jones, J. B., Jr. 1970. Physiological bases for plant analysis. p. 11-23. *In* Frances Greer (ed.) Proc. from a Sym. on Plant Analysis. Int. Minerals and Chem. Corp., Skokie, Ill.

Jones, J. B., Jr., and H. V. Eck. 1973. Plant analysis as an aid in fertilizing corn and grain sorghum. p. 349-364. *In* L. M. Walsh and J. D. Beaton (ed.) Soil testing and plant analysis. Rev. ed. Soil Sci. Soc. Am., Madison, Wis.

Jones, J. B., Jr., and W. J. A. Steyn. 1973. Sampling, handling, and analyzing plant tissue samples. p. 249-270. *In* L. M. Walsh and J. D. Beaton (ed.) Soil testing and plant analysis. Rev. ed. Soil Sci. Soc. Am., Madison, Wis.

Jones, J. P., and J. A. Benson. 1975. Phosphate sorption isotherms for fertilizer P needs of sweet corn (*Zea mays*) grown on a high phosphorus fixing soil. Comm. Soil Sci. Plant Anal. 6:465-477.

Jones, M. B., J. E. Ruckman, and P. W. Lawler. 1972. Critical levels of P in subclover (*Trifolium subterraneum* L.). Agron. J. 64:695-698.

Jordan, C. W., C. E. Evans, and R. D. Rouse. 1966. Coastal Bermudagrass response to applications of P and K as related to P and K levels in the soil. Soil Sci. Soc. Am. Proc. 30: 477-480.

Kamprath, E. J. 1964. Optimum soil fertility levels for corn and soybeans. Plant Food Rev. 10(3):4-6.

Kamprath, E. J. 1967. Residual effect of large applications of phosphorus on high phosphorus fixing soils. Agron. J. 59:25-27.

Kenworthy, A. L. 1973. Leaf analysis as an aid in fertilizing orchards. p. 381-392. *In* L. M. Walsh and J. D. Beaton (ed.) Soil testing and plant analysis. Rev. ed. Soil Sci. Soc. Am. Madison, Wis.

Larsen, Sigurd. 1967. Soil phosphorus. Adv. Agron. 19:151-210.

Leaf, A. L. 1973. Plant analysis as an aid in fertilizing forests. p. 427-454. *In* L. M. Walsh and J. D. Beaton (ed.) Soil testing and plant analysis. Rev. ed. Soil Sci. Soc. Am., Madison, Wis.

Loneragan, J. F. 1968a. Nutrient requirements of plants. Nature 220:1307-1308.

Loneragan, J. F. 1968b. Nutrient concentration, nutrient flux, and plant growth. Int. Congr. Soil Sci. Trans. 9th (Adelaide, Aust.) II:16-21.

Loneragan, J. C., and C. J. Asher. 1967. Response of plants to phosphate concentration in solution culture: II. Rate of phosphate absorption and its relation to growth. Soil Sci. 103:311-318.

Macy, P. 1936. The quantitative mineral nutrient requirements of plants. Plant Physiol. 11: 749-764.

Martens, D. C., J. A. Lutz, and G. D. Jones. 1969. Forms and availability of phosphorus in selected Virginia soils as related to available phosphorus tests. Agron. J. 61:616-621.

Martin, W. E., and J. E. Matocha. 1973. Plant analysis as an aid in the fertilization of forage crops. p. 393-426. *In* L. M. Walsh and J. D. Beaton (ed.) Soil Testing and plant analysis. Rev. ed. Soil Sci. Soc. Am., Madison, Wis.

Martin, W. E., and D. S. Mikkelsen. 1960. Grain fertilization in California. Calif. Agric. Exp. Stn. Bull. 775.

Mattingly, G. E. G. 1965. The influence of intensity and capacity factors on the availability of soil phosphate. Tech. Bull. Minist. Agric. Fish. 13:1-9.

McLachlan, K. D. 1965. The nature of available phosphorus in some acid pasture soils and a comparison of estimating procedures. Aust. J. Exp. Agric. Anim. Husb. 5:125.

Morgan, M. F. 1941. Chemical soil diagnosis by the universal testing system. Conn. Agric. Exp. Stn. Bull. 450.

Munson, R. D., and W. L. Nelson. 1973. Principles and practices in plant analysis. p. 223-248. *In* L. M. Walsh and J. D. Beaton (ed.) Soil testing and plant analysis. Rev. ed. Soil Sci. Soc. Am., Madison, Wis.

Nelson, W. L., A. Mehlich, and E. Winters. 1953. The development, evaluation and use of soil tests for phosphorus availability. *In* W. H. Pierre and A. G. Norman (ed.) Soil and fertilizer phosphorus. Agronomy 4:153-158. Am. Soc. of Agron., Madison, Wis.

Nye, P. J., J. L. Brewster, and K. K. S. Bhat. 1975. The possibility of predicting solute uptake and plant growth response from independently measured soil and plant characteristics. I. The theoretical basis of the experiment. Plant Soil 42:161-170.

Olsen, S. R., C. V. Cole, F. S. Watanabe, and L. A. Dean. 1954a. Estimation of available phosphorus in soils by extraction with sodium bicarbonate. USDA Circ. 939.

Olsen, S. R., and L. A. Dean. 1965. Phosphorus. *In* C. A. Black (ed.) Methods of soil analysis. Part 2. Chemical and microbiological properteis. Agronomy 9:1035-1049. Am. Soc. of Agron., Madison, Wis.

Olsen, S. R., and F. S. Watanabe. 1957. A method to determine a phosphorus absorption maximum of soils as measured by the Langmuir Isotherm. Soil Sci. Soc. Am. Proc. 21: 144-149.

Olsen, S. R., and F. S. Watanabe. 1963. Diffusion of phosphorus as related to soil texture and plant uptake. Soil Sci. Soc. Am. Proc. 27:648-653.

Olsen, S. R., and F. S. Watanabe. 1970. Diffusive supply of phosphorus in relation to soil textural variations. Soil Sci. 110:318-327.

Olsen, S. R., F. S. Watanabe, H. R. Casper, W. E. Larson, and L. B. Nelson. 1954b. residual phosphorus availability in long-time rotations on calcareous soils. Soil Sci. 78: 141-151.

Ortega, E. 1971. Correlation and calibration studies of chemical analysis in soils and plant tissues for nitrogen and available phosphorus. J. Indian Soc. Soil Sci. 19:147-153.

Ozanne, P. G., and T. C. Shaw. 1967. Phosphate sorption by soils as a measure of the phosphate requirement for pasture growth. Aust. J. Agric. Res. 18:601-612.

Peck, T. R., L. T. Kurtz, and H. L. S. Tandon. 1971. Changes in Bray P-1 soil phosphorus test values resulting from applications of phoshorus fertilizer. Soil Sci. Soc. Am. Proc. 35: 595-597.

Powell, R. D., and J. R. Webb. 1974. Effects of high rates of fertilizer N, P, and K on corn (*Zea mays* L.), leaf nutrient concentrations. Comm. Soil Sci. Plant Anal. 5:93-104.

Pratt, P. F., and M. J. Garber. 1964. Correlations of phosphorus availability by chemical tests with inorganic phosphorus fractions. Soil Sci. Soc. Am. Proc. 28:23-26.

Pritchett, W. L., and J. W. Gooding. 1975. Fertilizer recommendations for pines in the southeastern Coastal Plain of the United States. Fla. Agric. Exp. Stn. Bull. 774.

Richards, B. N., and D. I. Bevege. 1972. Principles and practices of foliar analysis as a basis for crop-logging in pine plantations. I. Basic considerations. Plant Soil 36:109-119.

Rivard, C. C., and V. A. Bandel. 1974. Effect of variety on nutrient composition of field corn. Comm. Soil Sci. Plant Anal. 5:229-242.

Rouse, R. D. 1968. Soil test theory and calibration for cotton, corn, soybeans and Coastal bermudagrass. Auburn Univ. Agric. Exp. Stn. Bull. 375.

Russell, E. J., and J. A. Prescott. 1916-1917. The reaction between dilute acids and the phosphorus compound of the soil. J. Agric. Sci. 8:65-110.

Sabbe, W. E., and H. L. Breland. 1974. Procedures used by state soil testing laboratories in the southern region of the United States. South. Coop. Ser. Bull. 190.

Sabbe, W. E., and A. J. Mackenzie. 1973. Plant analysis as an aid to cotton fertilization. p. 299-313. *In* L. M. Walsh and J. D. Beaton (ed.) Soil testing and plant analysis. Rev. ed. Soil Sci. Soc. Am., Madison, Wis.

Salmon, R. C. 1973. Effects of initial phosphate intensity and sorption or buffering capacity of soil on fertilizer requirements of different crops grown in pots or in the field. J. Agric. Sci. 81:39-46.

Scaife, M. A., and R. Smith. 1973. The phosphorus requirement of lettuce: II. A dynamic model of phosphorus uptake and growth. J. Agric. Sci. (Cambridge) 80:353-361.

Shear, C. B., H. L. Crane, and A. T. Myers. 1946. Nutrient element balance: A fundamental concept in plant nutrition. Proc. Am. Soc. Hort. Sci. 47:239-248.

Shelton, J. E., and N. T. Coleman. 1968. Inorganic phosphorus fractions and their relationship to residual value of large applications of phosphorus on high phosphorus fixing soils. Soil Sci. Soc. Am. Proc. 32:91-94.

Small, H. G., Jr., and A. J. Ohlrogge. 1973. Plant analysis as an aid in fertilizing soybeans and peanuts. p. 315-327. *In* L. M. Walsh and J. D. Beaton (ed.) Soil testing and plant analysis. Rev. ed. Soil Sci. Soc. Am., Madison, Wis.

Smith, C. M., and J. T. Pesek. 1962. Comparing measurements of the effect of residual fertilizer phosphorus in some Iowa soils. Soil Sci. Soc. Am. Proc. 26:563-566.

Smith, F. W., B. G. Ellis, and J. Grova. 1957. Use of acid-fluoride solutions for the extraction of available phosphorus in calcareous soils and in soils to which rock phosphates has been added. Soil Sci. Soc. Am. Proc. 21:400-404.

Smith, P. F. 1962. Mineral analysis of plant tissues. Annu. Rev. Plant Physiol. 13:81-108.

Soltanpour, P. N., Fred Adams, and A. C. Bennett. 1974. Soil phosphorus availability as measured by displaced soil solutions, calcium chloride extracts, dilute-acid extracts and labile phosphorus. Soil Sci. Soc. Am. Proc. 38:225-228.

Spencer, K., D. Bouma, and D. V. Moye. 1969. Assessment of the phosphorus and sulphur status of subterranian clover pastures. 2. Soil tests. Aust. J. Exp. agric. Anim. Husb. 9: 320-328.

Susuki, A., K. Lawton, and E. C. Doll. 1963. Phosphorus uptake and soil tests as related to forms of phosphorus in some Michigan soils. Soil Sci. Soc. Am. Proc. 27:401-403.

Terman, G. L. 1974. Amounts of nutrients supplied for crops grown in pot experiments. Comm. Soil Sci. Plant Anal. 5:115-121.

Terman, G. L., and G. W. Bengtson. 1973. yield-nutrient concentration relationships in Slash and Loblolly pine seedlings. Soil Sci. Soc. Am. Proc. 37:445-450.

Terman, G. L., P. M. Giordano, and S. E. Allen. 1972a. Relationships between dry matter yields and concentrations of Zn and P in young corn plants. Agron. J. 64:684-687.

Terman, G. L., P. M. Giordano, and N. W. Christensen. 1975. Corn hybrid yield effects on phosphorus, manganese, and zinc absorption. Agron. J. 67:182-184.

Terman, G. L., J. C. Noggle, and O. P. Engelstad. 1972b. Concentrations of N and P in young corn plants as affected by various growth-limiting factors. Agron. J. 64:384-388.

Thomas, G. W., and D. E. Peaslee. 1973. Testing soils for phosphorus. *In* L. M. Walsh and J. D. Beaton (ed.) Soil testing and plant analysis. Revised ed. Soil Sci. Soc. Am., Madison, Wis.

Thompson, L. F., and P. F. Pratt. 1954. Solubility of phosphorus in chemical extractions as indexes to available phosphorus in Ohio soils. Soil Sci. Soc. Am. Proc. 18:467-470.

Truog, Emil. 1930. Determination of the readily available phosphorus of soils. J. Am. Soc. Agron. 22:874-882.

Tyner, E. H., and J. G. Davide. 1962. Some criteria for evaluating phosphorus tests for lowland rice soils. p. 625-634. *In* G. J. Neale (ed.) Trans. Comm. IV and V Int. Soc. of Soil Sci., Nov. 1962, Palmerston North, New Zealand. Soil Bureau, P.B., Lower Hutt, New Zealand.

Ulrich, A., and F. J. Hills. 1967. Principles and practices of plant analysis. p. 11-24. *In* Soil testing and plant analysis. Part II. SSSA Spec. Publ. Ser. no. 2, Soil Sci. Soc. of Am., Madison, Wis.

Ulrich, A., and F. J. Hills. 1973. Plant analysis as an aid in fertilizing sugar crops: Part I. Sugar beets. p. 271-288. *In* L. M. Walsh and J. D. Beaton (ed.) Soil testing and plant analysis. Rev. ed. Soil Sci. Soc. Am., Madison, Wis.

Van Diest, A. 1963. Soil test correlation studies on New Jersey soils: I. Comparison of seven methods for measuring labile inorganic soil phosphorus. Soil Sci. 96:261-266.

Van der Paauw, F. 1971. Effective water extraction method for the detemination of plant-available soil phosphorus. Plant Soil. 34:467-481.

Voss, R. E., J. J. Hanway, and L. C. Dumenil. 1970. Relationship between grain yield, N, P, and K concentrations for corn (*Zea mays* L.) and the factors that influence this relationship. Agron. J. 62:726-727.

Walker, W. M., and T. R. Peck. 1972. A comparison of the relationship between corn yield and nutrient concentration in whole plants and different plant parts at early tassel at two locations. Comm. Soil Sci. Plant Anal. 3:513-523.

Walker, M. W., and John Pesek. 1967. Yield of Kentucky bluegrass (*Poa pratensis*) as a function of its percentage of nitrogen, phosphorus, and potassium. Agron. J. 59:44-47.

Ward, R. C., D. A. Whitney, and D. G. Westfall. 1973. Plant analysis as an aid in fertilizing small grains. p. 329-348. *In* L. M. Walsh and J. D. Beaton (ed.) Soil testing and plant analysis. Rev. ed. Soil Sci. Soc. Am., Madison, Wis.

Waugh, D. L., and A. Manzano. 1971. The correlation of phosphorus response with soil analysis in tall and dwarf wheat varieties in Bolivia. Proc. Int. Symp. Soil Fert. Eval. (New Delhi) 1:377-382.

Welch, L. F., L. E. Ensminger, and C. M. Wilson. 1957. The correlation of soil phosphorus with the yield of Ladino clover. Soil Sci. Soc. Am. Proc. 21:618-620.

Weir, C. C. 1962. Evaluation of chemical soil tests for measuring available phosphorus on some Jamaican soils. Trop. Agric. (Trinidad) 39:67-72.

White, R. E. 1971. Studies on mineral ion absorption by plants: II. The interaction between metabolic activity and the rate of phosphorus uptake. Plant Soil 38:509-523.

White, R. E., and K. P. Haydock. 1968. Phosphate availability and phosphate needs of soils under Siratro pastures as assessed by soil chemical tests. Aust. J. Exp. Agric. Anim. Husb. 8:561-568.

Wild, A. 1964. Soluble phosphate in soil and uptake in plants. Nature 203:326-327.

Williams, E. G., and A. H. Knight. 1963. Evaluation of soil phosphate status by pot experiments conventional extraction methods and labile phosphate values estimated with the aid of phosphorus 32. J. Sci. Food Agric. 14:555-563.

Woodruff, J. R., and E. J. Kamprath. 1965. Phosphorus absorption maximum as measured by the Langmuir isotherm and its relationship to phosphorus availability. Soil Sci. Soc. Am. Proc. 29:148-150.

Zubriski, J. C. 1971. Relationships between forms of soil phosphorus, some indexes of phosphorus availability and growth of sudan grass in greenhouse trials. Agron. J. 63:421-425.

Chapter 17

Management Considerations for Acid Soils with High Phosphorus Fixation Capacity

PEDRO A. SANCHEZ

North Carolina State University, Raleigh, North Carolina

GORO UEHARA

University of Hawaii, Honolulu, Hawaii

I. INTRODUCTION

Phosphorus fixation is generally understood as the transformation of soluble forms of P into less soluble ones after they react with the soil. Although this is a common occurrence in soils, it becomes a major management consideration in certain soils because of the large amounts of fertilizer P needed to meet crop requirements. Phosphorus fixation is most frequently a problem in acid soils high in finely divided sesquioxides, less frequently a problem in calcareous soils, and rarely a problem in soils that are neither high in free sesquioxides nor calcareous. A P-deficient soil which can be corrected by an application of 20 to 50 kg of P/ha is not a problem soil, but a soil which requires 300 to over 1,000 kg of P/ha needs a different management for profitable crop production. A considerable proportion of the world's potentially arable but presently unused land has this limitation.

The purpose of this chapter is to examine the occurrence and magnitude of high P fixation and the management practices designed to cope with this situation. No attempt is made to give a full literature review on P fixation, rather we attempted to illustrate management concepts with specific examples. Our discussion is confined to acid soils where the problem is most severe. For example, Olsen and Watanabe (1957) reported that representative acid soils of the United States fixed more than twice the amount of added P per unit of surface area than a similar group of neutral or calcareous soils. They also found that the P fixed was held with five times more bonding energy in acid soils as compared to calcareous soils.

Copyright 1980 © ASA-CSSA-SSSA, 677 South Segoe Road, Madison, WI 53711, USA.
The Role of Phosphorus in Agriculture.

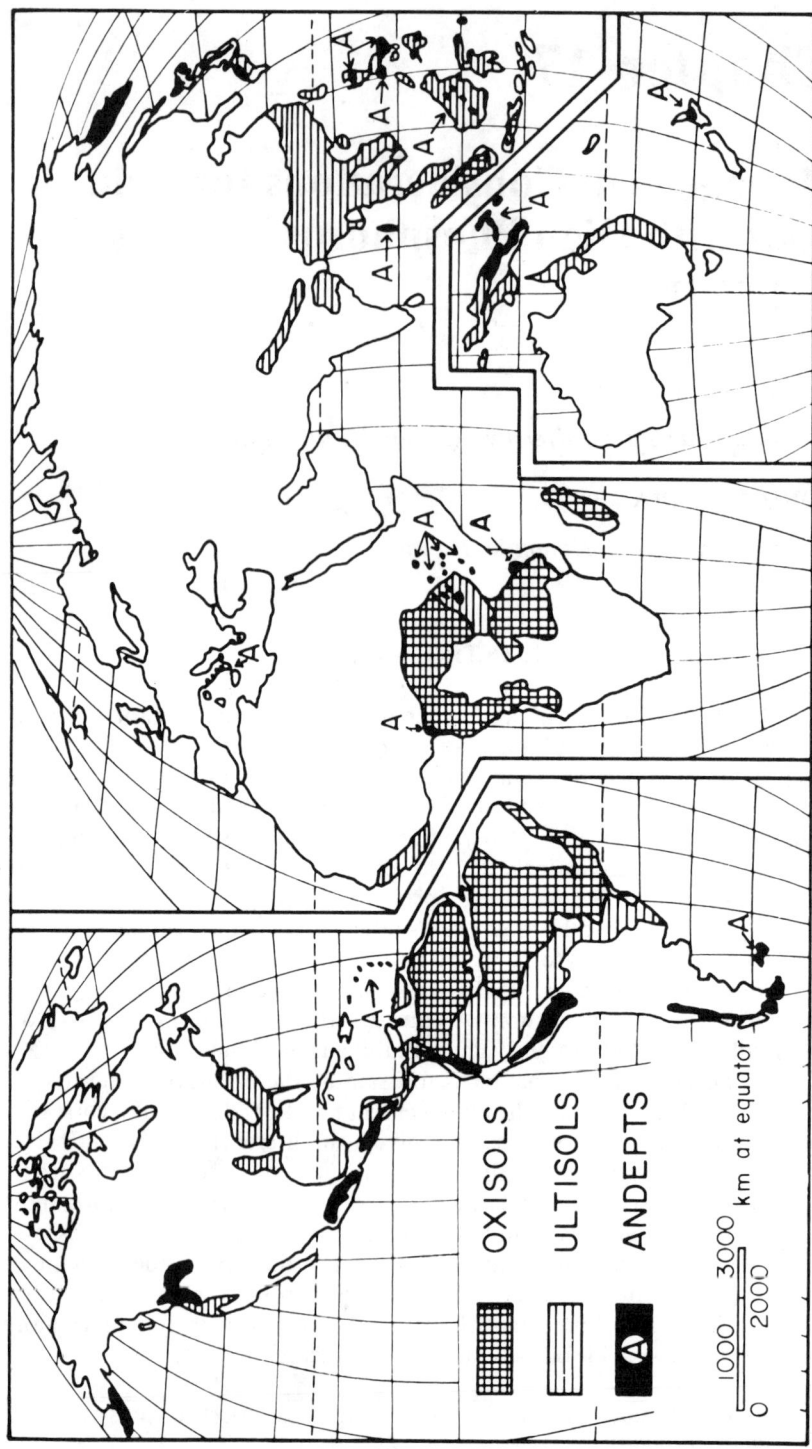

Fig. 1—Geographical distribution of Oxisols, Ultisols and Andepts in the world.

II. GEOGRAPHICAL DISTRIBUTION OF SOILS WITH HIGH PHOSPHORUS FIXATION CAPACITY

Acid soils which fix large quantities of P are invariably medium- to fine-textured soils high in oxides and hydroxides of Fe and Al. Although P fixation is not a classification criterion at any level of Soil Taxonomy or other soil classification systems, there are several broad soil groups which are notoriously high in P fixation capacity: the suborder Andepts, the orders Oxisols and Ultisols, and certain rhodic or oxic Alfisols and Inceptisols. Soils classified into these categories but with sandy surface texture are excluded. Selected properties and taxonomy of some of these soils have been described in recent review articles by Calhoun (1974), Fox (1974b), Keng and Uehara (1973), and Yuan (1973).

The worldwide distribution of Andepts, Oxisols, and Ultisols is shown in Fig. 1, based on information compiled from maps by the Soil Conservation Service (1972), the published portions of the *World Soil Map* (FAO-UNESCO, 1971, 1975), and other sources. Due to the small scale (1:50 million) only major areas where such soils are dominant are identified. Soils with low P fixation capacity are included within these mapping units; they are either coarse-textured Andepts, and Ultisols or other soils which exist in association with them. Because of these limitations it is not possible to estimate the area affected by high P fixation. Such estimations would be possible in regions where detailed information on surface texture and mineralogy are available. Nevertheless, Fig. 1 provides an overview of the location of soils with high P fixation and shows that they cover a large part of the world.

A. Andepts

Andepts, or Andosols, are relatively young soils derived from volcanic ash which have developed sufficiently to have a cambic horizon with the clay fraction dominated by x-ray amorphous colloids. These soils are found almost exclusively in areas of recent volcanic activity. Andepts are extensive in the rim around the Pacific Ocean from New Zealand northwards through Indonesia, Papua New Guinea, the Philippines, Japan, and the Kamchatka peninsula in Asia, and southwards from Alaska, the U.S. Pacific Northwest, Mexico, Central America, Colombia, Ecuador, Peru, Bolivia, and Chile to Tierra del Fuego in the Americas. Andepts do not cover all these regions but only discrete areas with recent volcanic activity. Thus, Andepts are not found in large areas of Indonesia, the Philippines, Japan, and the mountain backbone of the Western Hemisphere. Their occurrence in Europe and Africa is limited to small but locally important areas in Italy, Spain, Cameroon, Madagascar, and along the Rift Valley through Ethiopia, Kenya, Uganda, Tanzania, Rwanda, Burundi, and eastern Zäire. Andepts are abundant in many volcanic islands in the Pacific, Atlantic, and

Indian Oceans, and the Lesser Antilles in the Caribbean. Many Andept areas are under intensive production, in spite of the large quantities of P fixed as Al phosphate by allophane.

B. Oxisols

Oxisols also fix very large amounts of P because of the high contents of sesquioxides and generally medium to fine texture. Oxisols are also known as Ferralsols, Ferralitic soils, and in the Brazilian soil classification system as Latosols. Oxisols occupy large areas in the world's oldest geomorphic land surfaces as the vast Guyanan and Brazilian shields in South America and parts of Central Africa surrounding the Congo basin. They are also found in smaller extensions in Madagascar, Southeast Asia, and certain Pacific Islands. Their occurrence is limited to the tropics, except for small areas in South Africa and southern Brazil. These are the classic tropical soils with large quantities of crystalline or amorphous Fe and/or Al oxides. Phosphorus may be fixed as Fe or Al phosphates depending on the relative contents of these two elements and range of soil pH.

The vast majority of the Oxisols of the interior of South America and Africa are presently under savanna or forest vegetation. High P fixation is one of the principal limiting factors preventing their full utilization. This problem, however, has been recognized and partially solved in intensively cropped areas of Hawaii and southern Brazil, where Oxisols have become productive.

C. Ultisols

High P fixation is also encountered in clayey, sesquioxide-rich Ultisols. These soils, also known as Red Yellow Podzolic soils, Acrisols, and Dystric Nitosols, are characterized by the presence of an argillic horizon with low base saturation. There are many Ultisols with sandy topsoil texture which do not fix high amounts of P. Many of them, however, have been eroded, exposing the clayey, sesquioxide-rich argillic horizon, which is capable of fixing large quantities of P. In uneroded Ultisols having high clay contents in the topsoil, considerable quantities of P can be fixed in the form of Fe and Al phosphates.

Ultisols occupy large areas in southeastern United States, Mexico, Central America, the upper Amazon basin, Bolivia, Paraguay, and Brazil. In Africa their occurrence in large areas is limited to parts of Senegal, Guinea, Sierra Leone, Liberia, and the Ivory Coast in West Africa, and a region surrounding Lake Victoria in Uganda, Kenya, Tanzania, and eastern Zäire. The largest continuous block of Ultisols is found in Asia, extending from southeastern China through most of the southeast Asian mainland and islands, except for large valleys where rice is grown. The York Peninsula of Australia, parts of eastern and northern India, southern Japan, and southern Korea, are covered by Ultisols.

As in the case of Oxisols, there are large areas of Ultisols in the tropics covered by forest and savanna vegetation, where P fixation is a major limiting factor for optimum crop production on medium- to fine-textured topsoils. Ultisols are intensively managed and very productive soils in southeastern United States and southeastern China.

D. Other Soils

A fourth major group of soils with high P fixation capacity consists of the clayey, sesquioxide-rich Alfisols which differ from the Ultisols only in a higher subsoil base status. These Alfisols belong to rhodic great groups or subgroups, or to oxic subgroups. They are known as Eutric Nitosols and Rhodic Luvisols in the FAO legend, as Ferruginous soils in the French system, as Eutrophic Red Yellow Podzolics and Terra Roxa Estruturada in the Brazilian system, and as Reddish Brown Lateritic soils in the 1938 USDA system. They occur in large areas of Central America and the Greater Antilles, South India, Sri Lanka, East Africa, and in smaller areas of southeastern United States. Again, only those with acid, loamy, or clayey topsoils will fix large quantities of P.

Oxic subgroups of Inceptisols with properties close to Oxisols or Alfisols are also high P fixers. These soils are not identified in Fig. 1.

E. Summary

Soils with high P fixation capacity, therefore, occur in both the tropical and temperate latitudes. Many of them are already under intensive management, but vast areas in the tropics have hardly been touched. The only common property of soils with high P fixation capacity is their fine topsoil texture and high sesquioxide and/or x-ray amorphous colloid contents.

III. MAGNITUDE AND MEASUREMENT OF HIGH PHOSPHORUS FIXATION

A. Mechanisms Involved

Phosphorus is fixed into slightly soluble forms by precipitation and sorption reactions with Fe and Al compounds and crystalline and x-ray amorphous colloids of low silica-sesquioxide ratios present in acid soils. The relative importance of these mechanisms varies with soil properties. A short summary follows; the reader is also referred to the chapters by Soper, Racz, and Sample (Chapt. 11) and by Olsen and Khasawneh (Chapt. 14) in this book as well as review articles by Wild (1949), Dean (1949), Larsen (1967), Dabin (1974), and Kamprath (1974) for more detailed information.

1. PRECIPITATION REACTIONS

When a superphosphate granule is added to an acid mineral soil, water moves into the granule and dissolves some of the monocalcium phosphate (MCP), forming dicalcium phosphate (DCP) and a solution saturated with respect to both MCP and DCP. The solution emerging from the granule has a pH of 1.0 to 1.5, a P concentration of 4 to 4.5M and a Ca concentration of about 1.4M (Lindsay & Stephenson, 1959). It dissolves adjacent clay minerals, releasing Al and Fe ions which are then precipitated by the phosphate ions. These precipitates vary in chemical composition, but many are sparingly soluble although with a higher solubility product than variscite or strengite (Lindsay et al., 1962; Bache, 1964). They gradually release $H_2PO_4^-$ ions into the soil solution for several years.

In acid soils which contain significant quantities of exchangeable Al, P fertilizers can also be precipitated by this cation. Exchangeable Al, however, must first be displaced into the soil solution by the basic cations contained in P fertilizers and hydrolyzed before being able to fix P. The hydroxy Al then reacts with phosphate anions forming Al phosphate precipitates (Coleman et al., 1960; Hsu and Rennie, 1962). The following is an example of one of the possible reactions:

a. Cation Exchange:

$$\text{clay surface}\begin{bmatrix} Al^{3+} \\ Al^{3+} \end{bmatrix} + 3\,Ca^{2+} \rightleftharpoons \text{clay surface}\begin{bmatrix} Ca^{2+} \\ Ca^{2+} \\ Ca^{2+} \end{bmatrix} + 2\,Al^{3+}$$

b. Hydrolysis:

$$Al^{3+} + 2\,H_2O \rightleftharpoons Al(OH)_2^+ + 2H^+$$

c. Precipitation:

$$Al(OH)_2^+ + H_2PO_4^- \rightleftharpoons \underline{Al(OH)_2H_2PO_4} \qquad (Ksp = 10^{-29})$$

Consequently 1 mol of hydrolyzed exchangeable Al precipitates 1 mol of the orthophosphate ion. Since not all exchangeable Al is likely to be hydrolyzed under field conditions, the probable contribution of this mechanism is lower than this figure suggests.

2. SORPTION REACTIONS

Phosphate ions in the soil solution enter into ligand exchange reactions with hydroxyls in the surface of sesquioxide particles or films (Breeuwsma, 1973). Phosphate ions in the soil solution originate from the release of inorganic P or from P fertilizers. Due to their amphoteric behavior, oxide surfaces may exhibit net negative, net positive, or net zero charge (Mattson, 1931). Three reactions are given below using an Fe oxide surface as an example at pH values above, at an below the zero-point of charge (ZPC):

a. Above the ZPC-oxide Surface Negatively Charged:

[Structural diagram showing Fe oxide surface with $-O^-$ group reacting with $H_2PO_4^-$ to form Fe–O–P(=O)(OH)–O^- + OH^-]

b. At the ZPC-oxide Surface with Net Zero Charge:

[Structural diagram showing Fe oxide surface with $-OH$ group reacting with $H_2PO_4^-$ to form Fe–O–P(=O)(OH)–OH + OH^-]

c. Below the ZPC-oxide Surface with Net Positive Charge:

$$\begin{array}{c}\text{Fe structure with OH, O, H}^+\end{array} + H_2PO_4^- \rightleftharpoons \begin{array}{c}\text{Fe structure}\end{array} - O - \overset{O}{\underset{OH}{\overset{\|}{P}}} - OH + H_2O$$

In soils with variable charge colloids, the sign and magnitude of the surface charge depend on pH. Below the ZPC, the surface hydroxyls are protonated and phosphate ions then displace water ($-OH_2$). At and above ZPC, they displace hydroxyls, and an increase in pH may often be detected.

The fate of the fixed P depends on concentration and time factors. At relatively low concentrations of < 3 ppm P in solution, the amounts fixed depend directly on the quantity of P added. At higher rates, P is sorbed on sites arising from the disruption of hydroxy Al gels and by displacement of structural silica caused by the large P applications (Rajan & Perrott, 1975).

These mechanisms also vary with time. Although it is common to fix over 90% of the P within the first hour of contact, reactions may last for several days or months. Barrow (1974b) suggests that these slower reactions are a consequence of diffusion of P into inner surfaces of hydroxides, particularly amorphous hydroxides, and perhaps of rearrangement of the crystal structure as a result of the original fixation.

The interactions between the various mechanisms of P fixation are not well understood because of their complexity and the different conditions under which they have been studied. In spite of the voluminous research on the chemistry of P fixation, the relative influence of the different mechanisms and their interactions is still a matter of speculation (Perrott et al., 1974). Supporting evidence for other possible mechanisms have been proposed (Parfitt et al., 1975). Understandably, there is also confusion in terminology. The authors do not wish to add to it and will use the term fixation collectively to describe both sorption and precipitation mechanisms.

B. Measurements with P-sorption Isotherms

Phosphorus fixation has been estimated by a wide variety of methods involving widely diverging amounts of P added and extracted. One of the most widely used methods is an adaptation of the Langmuir isotherm originally developed to describe the adsorption of a monomolecular layer of gas on a solid surface. Olsen and Watanabe (1957) suggested its application for measuring P fixation and observed that P adsorption maxima could be calculated from the isotherms using the Langmuir equation. A summary of adsorption maxima of several Oxisols, Ultisols, and Andepts has been compiled by Leal and Velloso (1973a). The percent saturation of the adsorption maximum appeared to be a useful parameter similar to percent Al saturation used in liming recommendations. Unfortunately neither adsorption maxima nor their percent saturation is easily related to plant growth.

An improvement of this concept was introduced by Beckwith (1965), who suggested that the isotherm data be plotted in terms of straightforward quantity-intensity relationships, with the amount of P fixed as the quantity factor plotted on the Y axis, and the P remaining in solution as the intensity factor, plotted on the X axis on a logarithmic scale. Beckwith established the relationship with plant growth by assuming that a level of 0.2 ppm P in solution is adequate for plant growth. A subsequent study by Ozanne and Shaw (1968) supported the selection of 0.2 ppm P as the critical level.

Fox and Kamprath (1970) proposed an analytical procedure for developing the isotherms. A known amount of soil is equilibrated with different rates of MCP dissolved in $0.01M$ $CaCl_2$ and shaken for 30 min twice a day for 6 days, after which the P remaining in solution is measured. The difference between the amount added and that remaining in solution is the P which is plotted on the Y axis as a function of the ppm of P remaining in solution on the logarithmic X axis. The amount of P fixed to give a solution concentration of 0.2 ppm P is then considered the agronomically relevant estimate of P fixation.

Although a $0.01M$ $CaCl_2$ solution is probably different from actual soil solutions in acid soils, this method is considered a reasonable approximation from the analytical standpoint (Rajan and Fox, 1972). Values of equilibrium solution concentrations of more than 1 ppm P are not agronomically relevant and are excluded from further discussion. The term "P fixed" will be used henceforth to mean the amount of P adsorbed to provide 0.2 ppm P in solution according to the Fox and Kamprath (1970) procedure or minor deviations thereof. More recent results (Fox et al., 1974) suggest that plants differ in their external P requirement so that the 0.2 ppm P cannot be considered to be optimum for all crops.

Figure 2 shows several representative examples of fixation isotherms encompassing the known range in topsoils. Some of the isotherms follow essentially straight lines, but others show increasing slopes at the higher concentration which are thought to indicate rearrangements in the oxide structure which result in additional fixation at high concentrations of added P (Rajan & Fox, 1972). The slope of the isotherm provides an estimate of the P buffering capacity of the soil.

The amounts fixed at the standard concentration vary from 0 to 2,800 ppm P, which implies that the amounts of P needed to provide satisfactory crop growth varies from 0 to 5,600 kg P/ha. Although these curves represent a continuum, the authors consider soils that require additions of

150 ppm P or more to provide 0.2 ppm P in solution to be high P-fixing soils.

C. Factors Affecting the Amounts of Phosphorus Fixed

The principal factors affecting P fixation are the clay mineralogy, clay content, x-ray amorphous colloid content, exchangeable Al, and soil organic matter. Results of studies to estimate the individual contribution of these factors by successively stripping the soil of exchangeable Al, Al oxides, Fe oxides, and other components (Fassbender, 1969; Syers et al., 1971; Moura Filho et al., 1972; Biddappa and Rao, 1973; Ospina, 1974) suggest that they are not additive and that they interact with each other. Correlation or regression equations can be established, but they do not always reflect cause-effect relationships. Their combined effect can be best appreciated by comparing certain soils shown in Fig. 2.

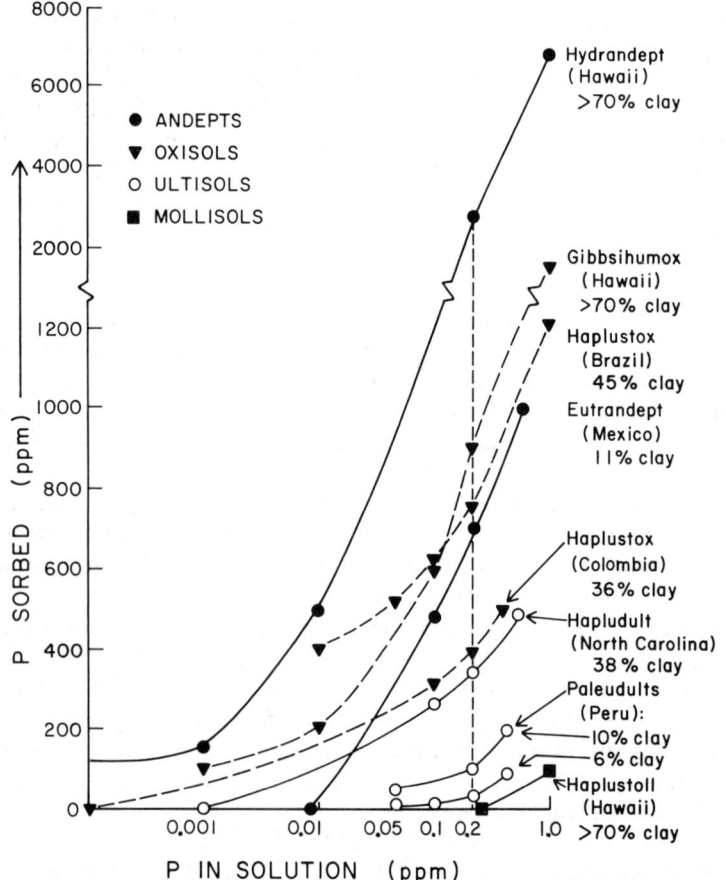

Fig. 2—Examples of P sorption isotherms determined by the method of Fox and Kamprath (1970). Sources: Rivera[1], North Carolina State University (1973), Fox (1974b).

1. CLAY MINERALOGY

The influence of clay mineralogy is dramatically illustrated by the three soils from Hawaii, all of which have more than 70% clay. The Mollisol fixed no P within the agronomic range because of the small quantity of free sesquioxides and the preponderance of montmorillonite and kaolinite. This is a typical example of soils dominated by layer silicate minerals which pose no serious P-fixation problems. The Oxisol consists primarily of well-crystallized sesquioxides, gibbsite, geothite, and hematite. It fixed about 900 ppm P. The Hydrandept is composed primarily of x-ray amorphous colloids and finely divided gibbsite and geothite. It fixed about 2,800 ppm P at the standard concentration.

Among soils with high contents of sesquioxides, it is well established that, the less crystalline they are, the higher their P fixation capacity is because of greater surface area (Colwell, 1959; Pratt et al., 1969; Fox et al., 1971; Galindo et al., 1971; Kamprath, 1973; Fox and Benavides, 1974).

2. CLAY CONTENT

Among soils of similar clay mineralogy, P fixation increases with increasing clay content. Figure 2 shows that the isotherm slopes of the Ultisols and Oxisols become steeper as texture becomes finer. Among the Ultisols, the sandy Paleudult from Peru with 6.4% clay fixed 24 ppm P, the one with 10% clay fixed 100 ppm P, but the Hapludult from North Carolina with 38% clay fixed 340 ppm P. These figures are representative of the sandy Ultisols of the Coastal Plain and the loamy or clayey Ultisols of the Piedmont regions of southeastern United States (Woodruff & Kamprath, 1965; Fox & Kamprath, 1970).

Similar relationships are found among Oxisols, which by definition have at least 12% clay in the topsoil. The loamy Haplustox from the Llanos Orientales of Colombia with 38% clay fixed 390 ppm P. Another Haplustox from the Cerrado of Brazil with 45% clay fixed almost twice that amount, 750 ppm P. A Gibbsihumox from Hawaii with more than 70% clay fixed 900 ppm P. These relationships do not follow a straight line because of variation in the amount and degree of crystallinity of the oxide minerals. Positive correlations have been reported between clay contents and P fixed in a group of Oxisols and Ultisols in southern Brazil (Syers et al., 1971) and with nine Oxisols of central Brazil (Leal & Velloso, 1973a). Topsoil texture, therefore, is an important parameter of P fixation for areas with soils of relatively similar mineralogy.

3. X-RAY AMORPHOUS COLLOID CONTENTS IN ANDEPTS

Because of their high contents of x-ray amorphous colloids, texture is often meaningless in Andepts. Phosphorus fixation, however, is positively correlated with x-ray amorphous colloid content and with surface area (Schalscha et al., 1973, 1974; Benavides, 1974). Andepts are young soils, but can develop considerable P fixation capacity rather rapidly. For example, Fox (1974a, b) studied a sequence of volcanic deposits of known age along Irazú Volcano in Costa Rica. He measured 70 ppm P fixed in a 4-year-old deposit near the crater and very high levels in older deposits down the slope. Phosphorus fixation increased as the proportion of silica in the x-ray amorphous fraction decreased with weathering (Fox et al., 1971).

Andepts as a group are the highest P fixers as shown in Fig. 2. Because of their generally low bulk densities (0.4 to 0.7 g/cm^3), however, the amounts indicated on the Y axis of the isotherms cannot be multiplied by a factor of two for conversion to kg P/ha as in most other soils. When an adjustment is made for bulk density the P requirements become much lower. For example, the actual P requirement of the Hydrandept shown in Fig. 2 is about half of what the Y axis indicates (Fox et al., 1974).

4. EXCHANGEABLE ALUMINUM

It was mentioned before that 1 meq of exchangeable Al per 100 g of soil when hydrolyzed may fix up to 102 ppm P. The neutralization of exchangeable Al via liming therefore can decrease P fixation. Correlations between P fixation and exchangeable Al have been reported (Syers et al., 1971; Udo & Uzo, 1972), but are not always entirely satisfactory because of incomplete hydrolysis. This subject will be discussed in further detail in a later section of this chapter.

5. SOIL ORGANIC MATTER

In soils with high contents of sesquioxides, organic radicals can block exposed hydroxyls on the surfaces of Fe or Al oxides and thus decrease P fixation capacity. This is one reason why topsoils of Oxisols and Andepts with texture and clay mineralogy similar to their subsoils fix considerably less P than the subsoil layers. Examples of this relationship are provided by Fox and Kamprath (1970), Moshi et al. (1974), Fox and Benavides (1974), and Rivera.[1] In these cases the main difference between the two horizons was the generally higher organic matter of the topsoils. Direct correlations between soil organic matter content and P fixation are often reported in the literature (Fassbender, 1969; Hinga, 1973; Leal and Velloso, 1973a), but these correlations do not necessarily reflect a cause-effect relationship. They probably reflect increases in finely divided sesquioxides with increasing levels of organic matter. Humic acids and Al hydroxides form complexes in Andepts, but the P fixation properties of these complexes depend on the properties of the Al hydroxides per se (Appelt et al., 1975).

6. FLOODING

Flooding decreases P fixation slightly within the agronomically relevant range because of the transformation of ferric to more soluble ferrous hydroxides under reduced conditions (Patrick et al., 1974). In acid

---

[1] C. Rivera. 1971. Phosphate fixation by tropical soils. M.S. Thesis, Soil Science Department, North Carolina State Univ., Raleigh, North Carolina.

soils flooding also increases pH towards neutrality which results in the precipitation of exchangeable Al. Flooding acid soils also decreases Al-bonded P and increases the proportion of Fe-bonded P because of the abundance of Fe^{2+} in the soil solution (Chang & Lin, 1970). Although the availability of P often increases in flooded rice soils, the changes in P fixation do not appear to be major.

D. Factors Affecting the Optimum Concentration in Solution

One advantage of the P-sorption isotherms is that the relationship between quantity and intensity factors is shown. The assumption that 0.2 ppm P in the soil solution is a reasonable approximation of the intensity factor has been subjected to recent scrutiny which revealed that the actual level varies with crop species, stage of crop growth, and with soil properties related to the diffusion of P to plant roots. Fox et al. (1974) defined the "external P requirement" as the concentration in the soil solution which produces 95% of the maximum crop yields. External P requirements have been determined in experiments with different rates of P fertilizers known to give a certain level of solution concentration previously determined via P-sorption isotherms.

1. STAGE OF CROP GROWTH

Table 1 summarizes the limited information available. At early growth stages, the external P requirements are generally higher than when the crops are carried to maturity. For example, corn required 0.2 ppm P at early stages but only 0.06 ppm P at maturity. The pasture legume *Desmodium aparines* also required 0.2 ppm P for establishment but only 0.01 ppm P after the second cut (Fox et al., 1974). These results reflect the well-known fact that plants require more P at early stages of growth. Pot experiments harvested at early growth stages, therefore, are likely to overestimate external P requirements in the field.

2. CROP SPECIES

The differences between species are also striking. Upland rice requires much less P than corn, sorghum, or millets. This confirms field observations on the generally lower needs for P in rice vs. corn in Latin America (Kamprath, 1973; Salinas & Sanchez, 1976). Two cool season vegetable crops, Chinese cabbage and lettuce, differ in their external P requirements by an order of magnitude. Sweet potatoes also have low requirements. Fox ct al. (1974) found that a concentration of only 0.003 ppm P was needed to achieve 75% of the maximum yield of sweet potatoes in Hawaii. The reason for the differences between species is not well understood. Varietal differences within species exist, but have not been quantified in terms of external P requirements.

Table 1—Summary of reported external P requirements (P in solution required to produce 95% maximum yield) of different crops.

Crop	Growth stage	Field or pot	Soils	External P requirement, ppm	Reference
Corn	Early	Field	Ustox, Hydrandept	0.20	Fox et al. (1974), Fox (1974b)
	Maturity	Field	Ustox, Hydrandept	0.06	Fox et al. (1974), Fox (1974b)
Sorghum	Maturity	Field	Hydrandept	0.05	Fox et al. (1974), Fox (1974b)
Desmodium aparines	Estab.	Field	Hydrandept	0.20	Fox et al. (1974), Fox (1974b)
	Second cut	Field	Hydrandept	0.01	Fox et al. (1974), Fox (1974b)
Sweet potato	Maturity	Field	Ustox	0.10	Fox et al. (1974), Fox (1974b)
Cabbage	Maturity	Field	Andept	0.04	Fox et al. (1974), Fox (1974b)
Lettuce	Maturity	Field	Andept	0.40	Fox et al. (1974), Fox (1974b)
Millet	Early	Pots	Ultisols	0.20	Fox and Kamprath (1970)
Millet	Early	Pots	Hydrandepts	0.02–0.06	Rajan (1973)
Millet	Early	Pots	Oxisols, Dystrandept	0.06	Rajan (1973)
Rice (flooded)	Early	Pots	Vertisols	0.10	Hossner et al. (1973)
Rice (upland)	Maturity	Pots	Oxisol	<0.1	Smyth and Sanchez (1980a, b)

3. SOIL TEXTURE AND WATER CONTENT

The external P requirement is higher in sandy soils than in finer-textured soils because of the restricted movement of P through thin water films around sand particles (Woodruff & Kamprath, 1965; Baldovinos & Thomas, 1967).

4. PHOSPHORUS FIXATION CAPACITY

The external P requirements are also higher in soils with low P fixation capacity than those with high, because of the limited buffering capacity factor of sandy soils, Histosols, and other low-fixing soils (Woodruff & Kamprath, 1965; Fox & Kamprath, 1970). For example, Fox (1969) reported maximum growth of kikuyu grass at 0.4 ppm P in a Hawaiian soil which fixed 200 ppm P, but the same crop required >1 ppm P in solution when grown in a soil which fixed only 70 ppm P.

5. ESTABLISHING EXTERNAL PHOSPHORUS REQUIREMENTS

The external P requirement of a particular crop, therefore can be calculated for soils of relatively similar fixation capacity. This is shown in Fig.

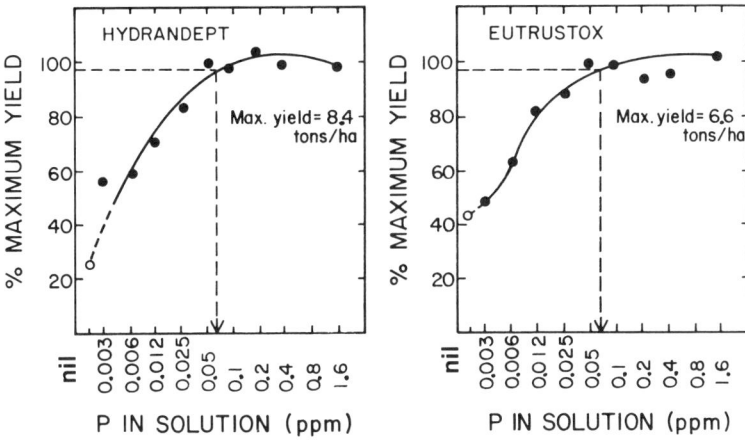

Fig. 3—Establishment of external P requirements of corn in two field experiments conducted in Hawaii. Source: Fox and Benavides (1974).

3 for corn grown in two high-fixing soils of Hawaii. Different external P requirements probably exist for the same crop on low-fixing soils and perhaps for cultivars of the same crop.

Although the concept seems well established, the amount of actual information is very limited. External P requirements remain to be established for most crops, particularly those believed to have low requirements such as cassava, and on soils of low and high P-fixing capacities. When these requirements are established for specific soil-crop situations, the amount of P needed to support 95% of the maximum yield can be estimated with the isotherms.

E. Need for Routine Methods of Estimating Phosphorus Fixation

The isotherms, although theoretically sound, are very much limited to research laboratories. The time requirement of 6 days and the analytical precision required to determine fractions of a ppm prevent their routine use in soil testing laboratories. One alternative is to use a soil test extraction procedure after the 6-day equilibration. Mendez and Kamprath (1978) tested this possibility and found that the amount of fixed P required to reach the established soil-test critical level (18 ppm P with the dilute double acid extraction) resulted in near maximum early growth of millet in small pots of Oxisols and Andepts of Panama. Figure 4 shows his results with arrows pointed at the amounts of P required to provide the critical soil test level. The points at which maximum yields were obtained are marked with an "X" in Fig. 4, showing close agreement with the soil test.

Straightforward soil tests, however, are not capable of estimating P fixation capacity. The established critical levels only separate those soils deficient in P from those that are not. An alternative amenable to routine soil

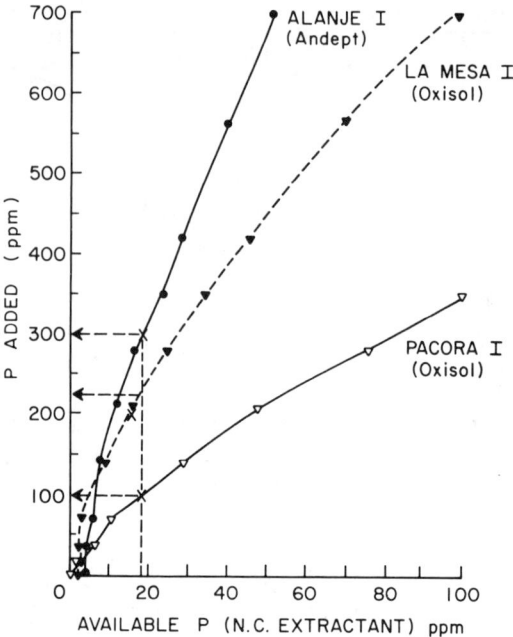

Fig. 4—Use of a soil test extraction as the intensity factor in Oxisols and Andepts of Panama. Drawn from data of Mendez and Kamprath (1978).

testing laboratories was explored by A. Cordero and G. W. Miner in Costa Rica (A. Cordero and G. S. Miner. 1974. Phosphorus fixation curves. Unpublished paper, Laboratorio de Suelos, Ministerio de Agricultura y Ganadería, San José, Costa Rica. 5 p.). Soils were tested in the conventional way and those reading below the critical level were incubated with an established amount of added P (280 ppm) from 2 to 7 days, after which they were extracted again with the conventional soil tests. Those soils reading below 50 ppm P were designated as high P fixers and those above as low fixers. In effect, they reversed the direction of the arrows in Fig. 4 and used only one level of added P instead of several. Two days' incubation gave equal results to 7 days, making this technique more applicable to routine laboratory procedures. These two approaches seem promising, but they remain to be correlated with field tests. Research in this direction is of considerable significance.

It would be even more desirable to identify high P fixation directly in the field. Buol et al. (1975) suggested the use of the NaF field test to identify allophane developed by Fieldes and Perrott (1966) as an indicator of P fixation in Andepts, and soil color hues redder than 5YR and granular structure as indicators of high P fixation in Oxisols. The NaF test seems successful, but a quick estimation in Oxisols, Ultisols, and similar soils with the proposed parameters has not been satisfactory. Pope and Buol (1976) showed that when clay contents are considered, the reliability of these predictions improve.

F. The Solution: Two Management Alternatives

The previous section has indicated that very large amounts of P need to be applied to certain acid soils in order to meet crop needs. Over the years, two main management strategies have emerged. The first or high input option consists of a heavy initial investment followed by substantial residual effects for several years. In addition to solving the problem immediately, heavy P applications also serve to act as a soil amendment, producing beneficial changes in soil physical and chemical properties in highly weathered Oxisols. The second strategy involves a low input option. It has been traditionally based on applying P in bands to satisfy the fixation capacity in a small soil volume. Recently other components have been added, such as selecting cultivars tolerant to low available P levels, decreasing the soils' capacity to fix P, and using sources of P cheaper than superphosphates. These alternatives are discussed separately in the following two sections, although they are not mutually exclusive.

IV. HIGH INPUT STRATEGY: PHOSPHORUS AS AN AMENDMENT

In certain Andept areas of Japan and in Oxisols of Latin America, agronomists were convinced that these soils did not respond to P fertilization, even though crops showed severe P deficiency symptoms, until they applied rates in the order of 500 to 1,000 kg P/ha and obtained dramatic yield responses. This approach has been described by Younge and Plucknett (1964, 1966) as "quenching" the high P fixation capacity of soils and has been tested in Hawaiian Oxisols where rates in the order of 1,000 kg P/ha are sometimes broadcast and incorporated deeply into the soil. These massive applications were supplemented with annual maintenance rates of about 60 kg P/ha and have supported high yields of pastures and crops. The heavy initial investment is considered a capital investment which can be amortized over a period of several years while the residual effect lasts. The heavy P rates also had a favorable effect on certain soil properties. The salient features of this approach are discussed in the following sections.

A. Residual Yield Effects

The main advantage of this approach is that the fixed P is gradually released over a period of several years at rates sufficient to support adequate crop growth. Knowledge of the exact duration of the residual effect is of major practical importance. Very few field experiments however, have been conducted over sufficiently long time to fully evaluate the residual effect of large initial broadcast applications on soils with high P fixation capacity. Noteworthy among those conducted are a 12-year experiment on an Oxisol from Hawaii (Younge & Plucknett, 1966; Fox et al., 1968; 1971), a 10-year experiment on an Ultisol from North Carolina (Shelton et al., 1961; Kamprath, 1967; Shelton & Coleman, 1968; Fox & Kamprath, 1970), and a group of 15 experiments ranging from 5 to 9 years' duration on Oxisols and poorly drained soils of Madagascar (Truong et al., 1973; Velly & Roche, 1973). The following discussion is based primarily on these experiences.

In all cases, a measurable residual effect was observed with initial rates of application of at least 175 kg P/ha for as long as the experiments were conducted. The higher the initial rates, the more pronounced the residual effects were in terms of crop response.

Figure 5 shows the response of a continuously grown grass-legume pasture on the Hawaiian Oxisol. It shows that the P rates needed to reach maximum yields increased with time. During the first 3 years 300 kg P/ha was sufficient. From the fourth year on, maximum yields were obtained at the highest rate applied (1,320 kg P/ha) with indications that even higher initial rates may have increased yields further in the 10th and 11th year.

A similar relationship was observed in the North Carolina and Madagascar experiments, but the duration of the residual effect of different rates of applied P varied with soil properties. Figure 6 shows two representative results of the Madagascar data plotted as relative yields as a function of time. The maximum yields attained are uniformly high as a result of the intensive management practices. In the Matsiara Oxisol, which was very low in available P and presumably a high fixer, only the two highest rates were able to maintain at least 80% of the maximum yields for several years. In the Iboaka Oxisol, however, near maximum yields were achieved with the lowest rates of 44 and 88 kg P/ha during the first 2 years, after which their effectiveness decreased markedly. The 176 kg P/ha rate gradually decreased in effectiveness with time. The 440 kg P/ha rate became superior after the fourth year. This soil had a higher level of available P and showed a lower response to P than the previous one. Truong et al. (1973) suggested

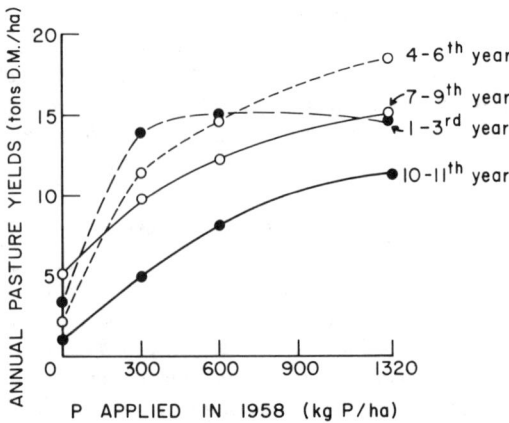

Fig. 5—Residual effect of massive P applications in a Gibbsihumox of Hawaii. Adapted from Fox et al. (1971).

that the Matsiara soil had a lower buffering capacity factor than the Iboaka, thus much higher P rates were needed in the Matsiara soil to approach maximum yields during the first 2 years.

These examples also illustrate the need to conduct such studies for several years. The performance of initial application cannot be predicted from one or two consecutive crops. Unfortunately, most of the available information is limited to 1 or 2 years of data. It should be emphasized that even the long-term studies have not fully determined the residual effects of high rates of application in high P-fixing soils. Fortunately some of these studies are being continued.

Another point that deserves emphasis is that relatively high rates of P applications are needed to produce a strong residual effect. Russell (1973)

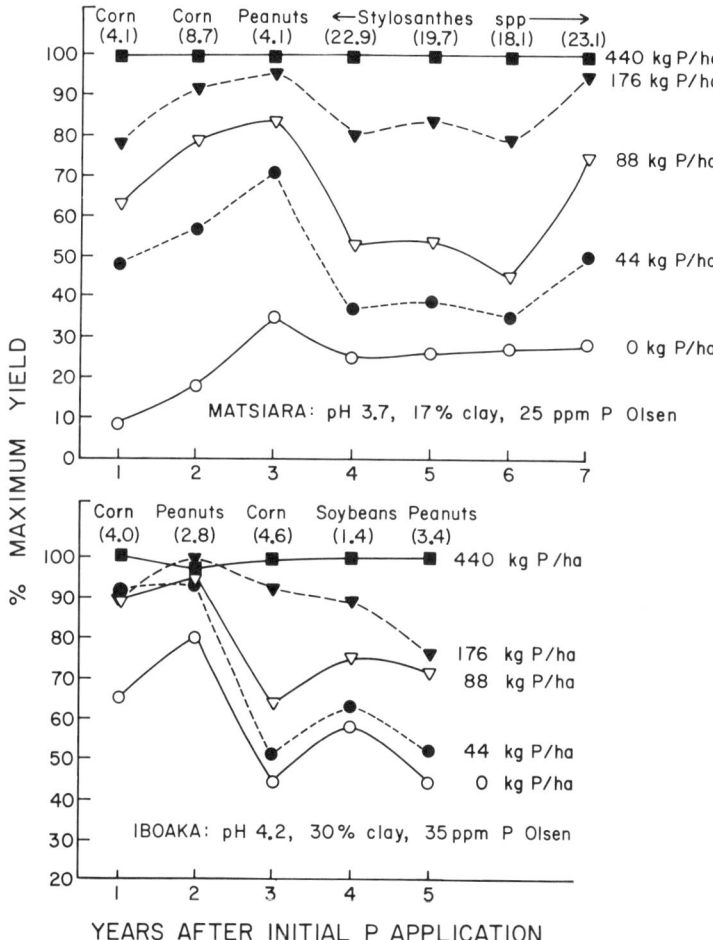

Fig. 6—Residual effects of P applications in two Oxisols from Madagascar. Calculated from data of Truong et al. (1973) and Velly and Roche (1974). Soil pH were measured in $1N$ KCl. Numbers in parentheses are maximum yields in tons/ha.

commented that there was little residual effect of P in tropical soils. His statements were based on rather low rates applied in many parts of Africa. Marginal residual effects from low P applications have also been reported on Andepts of Colombia (Monsalve & Lotero, 1972) and in Australia (Bruce, 1968; McClelland, 1968).

B. Decreases in Phosphorus Fixation Capacity

One of the benefits of heavy initial rates of application is the decrease in the soil's P-fixation capacity. Table 2 shows that the P required to give 0.2 ppm P in solution 9 or 10 years after the initial applications decreased considerably in the Hawaii and North Carolina long-term experiments. Similar results were obtained by Salmon (1973) in Rhodesia and by Barrow (1974a) in Australia within a 3-year period. The decrease in P fixation is proportional to the rate applied. Barrow (1974a), working in a range of 300 to 1,200 ppm of added P, observed that P sorbed was inversely proportional to the previous applied rate.

Table 2 also shows an estimate of the residual efficiency defined as the decrease in P fixed to attain 0.2 ppm P in solution divided by the estimated P remaining in the soil. The latter parameter is calculated by subtracting the amount of P taken up by the plants from the amount applied. The residual efficiency ranged from 20 to 80% in these experiments, decreasing with increasing rate of applied P. The residual efficiency was higher in the Hawaiian Oxisol, probably because of the year-round pasture growth, as compared with growing one crop of wheat or corn per year in temperate North

Table 2—Estimates of residual effects of heavy initial P applications in an Oxisol from Hawaii and in an Ultisol from North Carolina. Adapted from Fox et al. (1968) and Fox and Kamprath (1970).

P applied originally	Estimated residual P after 9 or 10 years (P app. − P uptake)	P requirement to give 0.2 ppm in solution	Residual efficiency†
———————————— kg/ha ————————————			%
Oxisol, Hawaii (9 years)			
0	0	1,131	--
336	224	952	80
672	459	829	66
1,344	818	605	64
Ultisol, North Carolina (10 years)			
0	0	630	--
175	105	580	48
350	220	520	50
700	540	490	20

† Residual efficiency = $\dfrac{P_0 \text{ req.} - P_x \text{ req.}}{\text{Estimated residual P}} \times 100$

in which
P_0 req. = P requirement of year 0
P_x req. = P requirement after × years.

Carolina where annual tillage may have diluted the available P due to greater mixing. Fox (1974b) observed that when 2,800 kg P/ha was applied during 3 years, the P sorbed by a Hydrandept to give 0.1 ppm in solution decreased from 3,000 to 500 ppm P.

C. Increases in Cation Exchange Capacity

A further advantage of massive P applications is an increase in cation exchange capacity (CEC). This can be readily explained by an equation which relates surface charge density to soil solution pH. The expression is (Parks, 1967):

$$\sigma = \frac{K\epsilon}{4\pi} 0.059 \,(\text{pH}_o - \text{pH})$$

where σ is the surface charge density, K is the inverse of the double layer thickness and varies with counter ion valence and salt concentration, ϵ is the dielectric constant of the medium, and pH_o is the pH corresponding to the zero-point of charge (ZPC).

Adsorbed P increases cation retention by displacing the ZPC to a lower value, thus making ($\text{pH}_o - \text{pH}$) and the surface charge more negative. Since the CEC is the product of surface charge density and specific surface

Table 3—Summary of some increases in cation exchange capacity due to P applications calculated from the literature.

Reference	Location	Soil material	Range of P added or fixed, ppm	ΔCEC, meq/100 g per 100 ppm P added	CEC† method
Murphy (1939)	California	Pure kaolinite	4,746–21,004	0.12	2
Davis (1945)	Louisiana	Hammond:			
		Unlimed (pH 5.3)	167	0.19	3
		Limed (pH 7)	167	0.35	3
Aomine and Yoshinaga (1955)	Japan	Allophane clay	300–400	0.06–0.19	2
Mekaru and Uehara (1972)	Hawaii	Subsoils of Oxisols and Andepts	423–2,326	0.26	3
Schalscha et al. (1974)	Chile	Subsoils of Andepts	1,200–87,110	0.19	1
Stoop (1974)	Hawaii	Oxisol topsoils	100–800	0.19	4
Sawhey (1974)	Conn.	Inceptisol: pH 5	140	0.12	5
		pH 6.5	140	0.20	5
		pH 7.5	140	0.25	5
Juo and Maduakor (1974)	Nigeria	Alfisols and Ultisol top and subsoils	310	0.20–0.90	2
Smyth and Sanchez (1980a, b)	Brazil	Oxisol topsoil (unlimed)	380–540	0.07	4
		Oxisol topsoil (limed)	380–540	0.28	4

† CEC methods: Buffered extractants at pH 6 (1), NH_4OAc at pH 7 (2), pH 8.2 (3). Effective CEC as sum of exch. Ca + Mg + K + Al (4). Unbuffered extraction with $1N$ $CaCl_2$ and $1N$ NaCl (5).

S, then CEC = $S\sigma$. Phosphorus sorption results in the largest increase in CEC in soils with high specific surface, which fix the largest quantities of P.

Although the existence of such relationships has been known for over 40 years (Mattson, 1931; Prince & Toth, 1936; Toth, 1937; Murphy, 1939), little attention has been given to its management implications until recently (Mekaru & Uehara, 1972; Keng & Uehara, 1974; Uehara & Keng, 1975).

Table 3 shows the increases in CEC calculated from the literature under a variety of soil materials and for a range of P applications. The increases in CEC per 100 ppm P added (ΔCEC) generally cluster between 0.1 to 0.2 meq/100 g. The magnitude of these increases is greater in oxidic than kaolinitic soils, and montmorillonitic soils exhibit still smaller increases (Coleman & Mehlich, 1948). The increases in CEC are also greater if lime is applied to soils with high percentage Al saturation (Davis, 1945; Smyth & Sanchez, 1980a, b) than in soils with relatively low percentage Al saturation (Stoop, 1974). The magnitude of the increase will also be dependent on the extent of sorption vs. precipitation mechanisms. Stoop (1974) indicated that if all the P added is precipitated, the increase in CEC is likely to be negligible.

Although increases in CEC with P applications can be measured in most soils, they are agronomically most important in soils with low effective CEC. Figure 7 compares the increases in CEC in two clayey Oxisols, a Gibbsihumox from Hawaii with a relatively high effective CEC of 3.9 meq/100 g, and a Haplustox from Brazil with an effective CEC of 1.3 meq/100 g. The highest P and lime rates increased the CEC from 3.9 to 7.2 meq/100 g in the Gibbsihumox, but almost tripled it in the Haplustox (from 1.3 to 3.1 meq/100 g). Given the excellent granular structure of these soils and the great susceptibility to leaching, doubling CEC is of great practical value. Due to the relatively high CEC of most high P-fixing Andepts, CEC increases in such soils are of less practical importance.

Heavy P applications are also likely to increase the organic matter con-

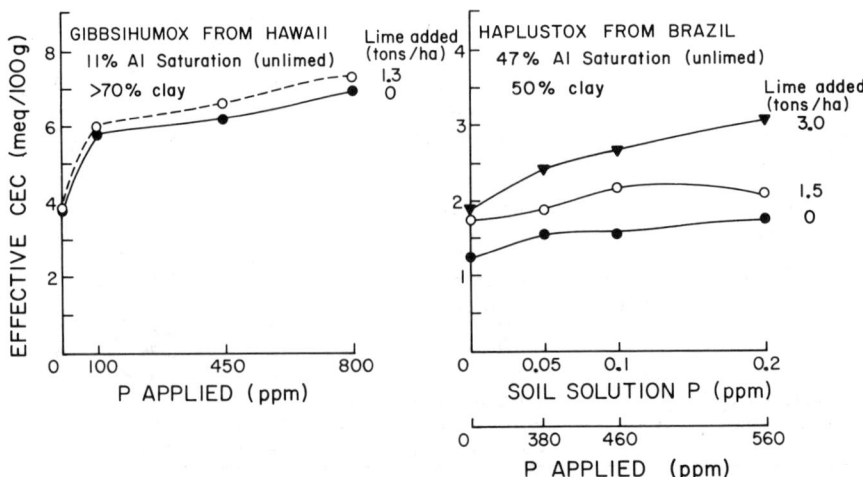

Fig. 7—Effects of P applications on increases in CEC in two Oxisols. Calculated from data of Stoop (1974), and Smyth and Sanchez (1980).

tent of the soil because of the greater amount of root decomposition which results from increased crop production. Vicente-Chandler et al. (1964) have reported this effect on intensively fertilized Oxisols and Ultisols of Puerto Rico. Organic matter increases CEC, most noticeably in low CEC Oxisols.

D. Increases in Soil pH

Heavy P applications sometimes increase the pH of soils high in sesquioxides for the same reasons indicated in the previous section. This has been referred to as "liming with P." These increases become larger as the amorphous sesquioxide content of the soils increase, but they are also dependent on the accompanying cation. These relationships are shown in Table 4 based on the work of Stoop (1974). The increases in pH are more marked in the high-fixing Andepts and Oxisols, but are nonexistent in the low-fixing montmorillonitic Mollisols. The pH increases were less when Ca phosphates rather than NH_4 phosphates were applied because Ca^{2+} is specifically adsorbed on the oxides' surfaces and NH_4^+ is not. Coleman and Mehlich (1948) and Stoop (1974) emphasized that the cation vs. anion exchange relationships determine the changes in pH when P is added. Decreases in anion exchange capacity due to P fixation increase pH, while increases in CEC decrease pH because of specific adsorption of Ca in oxide systems.

Liming with P is not a management objective *per se* because of the expense involved, but is a consequence of high rates of P application which should be taken into consideration.

E. Improvement of Soil Physical Properties

Massive initial P applications have improved the physical properties of high-fixing Ultisols in North Carolina. Lutz et al. (1960) observed that plots of Georgeville silt loam which did not receive P were very hard and difficult

Table 4—Effects of source and rate of P application on the pH of four Hawaiian soils. Cation levels at zero P are balanced with respect to Ca and NH_4 for each soil. Adapted from Stoop (1974).

Soil	P added	Source of P	
		Ca-P	NH_4-P
	ppm	pH in H_2O	
Hydrandept	0	4.2	4.4
	3,000	4.8	5.4
Gibbsihumox	0	4.9	5.1
	1,500	5.4	5.8
Eutrustox	0	4.9	5.0
	750	5.2	5.5
Haplustoll	0	6.5	6.4
	375	6.6	6.5

Table 5—Effects of initial P applications on wheat yields and improvement of physical properties 1 year after application on an Ultisol (Georgeville silty clay loam) from North Carolina. Recalculated from Lutz et al. (1966).

P applied	Wheat grain yields	Bulk density at 15 cm	Resistance to penetration	Average water content during summer	Al-P: Fe-P ratio
kg P/ha	metric tons/ha	g/cm^3	kg/cm^3	%	
0	1.50	1.26	13.4	29	0.3
175	2.06	1.17	6.8	34	1.0
350	2.02	1.09	9.7	34	0.6
700	1.98	1.18	7.8	35	1.0

to plow, while those which received from 175 to 700 kg P/ha were moist, friable, and easy to plow. Field and laboratory analyses showed that P applications decreased bulk density and resistance to penetration, and increased the average topsoil moisture content during the summer (Table 5). All three parameters were correlated with wheat yields.

The reasons for this improvement were examined in three investigations. Lutz et al. (1966) related these changes to improvements in soil aggregation caused by increased CEC and in the increase in Al-bonded phosphates. Similarly, Samra and Biswas (1974) observed that increases in aggregation in a lateritic red soil of India were related to the cementing action of Al, Fe, and reductant-soluble phosphates. Thien (1976), working with a low P-fixing Mollisol, observed that the increases in aggregate stability upon addition of phosphoric acid were related to increases in Al-P compounds. No similar results have been reported in Oxisols and Andepts which generally have good granular structure.

F. Advantages and Limitations

Massive initial P applications represent an attractive management strategy because the fixation problem is substantially reduced for some period. Farmers can expect a residual effect for many years, an increase in cation retention in low CEC soils, and an improvement in physical properties. When the high initial investments are amortized over several years at reasonable interest rates, the economic implications are also very attractive. Heavy applications are also a hedge against inflation. Heavy rates applied prior to 1972, when P fertilizer prices almost tripled, will continue to supply P and serve as an amendment for several years. The combined application of P and lime transforms naturally infertile, but physically excellent, acid Oxisols into highly productive soils. This has been the basis for successful agriculture in Hawaii where Oxisols are as productive as Vertisols and Mollisols in spite of the higher pH and lower P-fixing capacities of the latter soils.

It is difficult to justify development of extensive areas of high P-fixing soils if P fixation is treated strictly as a plant nutrition problem. If, instead, the fixed P is treated as an amendment for rejuvenating highly weathered soils, and therefore as a capital investment, it may yet be possible to persuade land use planners and agronomists to change their outlook on the role of P in agricultural development.

The main limitation of this strategy is the high initial investment required and the assurance of long-term use of the land by farmers. Both constraints are very serious in many developing countries. The finite supply of phosphate rock resources creates an additional global concern. At present consumption rates, the supply will last for several centuries, but if high P rates are applied to the 1 billion ha of Oxisols and Ultisols of the tropics, these estimates will have to be revised.

V. LOW INPUT MANAGEMENT STRATEGY

The traditional way to cope with high P fixation is to apply the fertilizer in bands in order to minimize the volume of soil with which it will react. The high cost of superphosphate and other energy-dependent inputs has led to exploring additional ways of managing high P-fixing soils with limited capital resources, particularly in small farming systems in the tropics. Salinas and Sanchez (1976) summarized a three-point strategy to overcome high P fixation under these conditions: (i) increase the efficiency of P fertilization by selecting more effective placement methods, sources, and rates of application; (ii) decrease P fixation capacity with relatively cheaper amendments such as lime or silicate; and (iii) decrease crop requirements by selecting varieties or species tolerant to low levels of available P in the soil. These points are discussed and expanded in the following sections.

A. Increased Efficiency of Phosphorus Fertilization

1. IMPROVED PLACEMENT METHODS

Banded P applications are more efficient than massive broadcast applications in certain high-fixing soils. Table 6 shows a comparison between two alternatives on the North Carolina Ultisol shown in Fig. 2 as requiring 340 ppm P to provide 0.2 ppm in solution. Kamprath (1967) found that

Table 6—Yields of corn on a North Carolina Ultisol with high P fixation capacity 7 years after either an initial heavy P application or small annual maintenance applications. Source: Kamprath (1967).

Initial application	Annual maintenance applications (banded)	
	None	22 kg P/ha
kg P/ha	metric tons/ha	
0	1.7	5.7
175	3.9	6.2
350	5.5	6.8
700	7.0	7.1

$LSD_{0.05} = 1.3$ metric tons/ha.

similar corn yields were obtained by annual banded applications of 22 kg P/ha for 7 years (providing a total of 154 kg P/ha) as were obtained by an initial application of 350 kg P/ha. Banding, therefore, saved more than half of the P requirement.

In soils with extremely high fixation capacity and *very low levels of available P,* the results are completely different. Studies by Yost et al. (1979) on a Brazilian Oxisol which requires 750 ppm P at the standard solution concentration (Fig. 2) indicate that banded applications are inferior to broadcast applications for the first corn crop. The available P in this soil is so low that root development was limited to the regions where P was applied. In the banded treatments, the very limited root development around the bands caused the plants to be less resistant to periods of moisture stress. This detrimental effect disappeared in subsequent crops because tillage operations gradually mixed the banded P with larger volumes of soil. With time, the cumulative amounts of banded applications approach the effect of high initial broadcast applications. In cropping systems which do not involve repetitive tillage, such as pastures, this cannot be expected.

In many cases the best alternative is probably a combination of an initial broadcast application followed by small annual banded maintenance applications. A basal application of 140 kg P/ha with a band application of 35 kg P/ha before each crop (including the first one) produced about 85% of the maximum yields in four consecutive corn crops in the Oxisol from Brazil previously cited. The concept of an initial amendment application followed by annual maintenance applications has been applied successfully in Africa, using low broadcast rates in the order of 33 to 44 kg P/ha in low-fixing soils and from 130 to 440 kg P/ha in high-fixing soils, such as those of Madagascar (Fig. 6; Pichot & Roche, 1972).

A compromise between banding and broadcasting is to broadcast in narrow strips and plow in. Barber (1974) found such strip applications effective in low-fixing soils of Indiana, with the fertilizer coming in contact with 10 to 30% of the topsoil volume. Strip application with different band widths deserves study in high P fixing soils.

Broadcast applications on the soil surface without incorporation may have merit in some situations. Howeler (1974) reports that surface placement is more effective than incorporation for rice grown in flooded Oxisols of the Llanos Orientales of Colombia because of their high capacity to fix P and perhaps because of the superficial root mat that rice produces in flooded environments. Surface broadcast applications were found to be more effective than banded applications for pastures grown on Andepts in Colombia (Michelin et al., 1974).

2. USE CHEAPER SOURCES OF PHOSPHORUS

Economic efficiency of P fertilization may also be increased by using water-insoluble sources which are substantially cheaper than superphosphates or ammonium phosphates. The two main sources are phosphate rocks (PR) and thermally altered sources, such as basic slags and the Rhenania phosphates.

a. **Direct Application of Phosphate Rock**—In acid soils that fix large quantities of P, direct application of PR is often more effective and economical than superphosphate. Phosphate rocks are reactive in acid soils and usually cost one-third to one-fifth per unit of P as superphosphate. In addition to savings in cost, the savings in energy consumed during the acidulation process to convert PR into superphosphate is of relevance. The effectiveness of the direct use of PR depends on its solubility, fineness, time of reaction, and soil pH.

When PR's are classified according to their absolute citrate solubility (ACS) (Lehr & McClellan, 1972), their agronomic effectiveness can be predicted. When the ACS of the best materials (North Carolina, Tunisia, and Algerian rocks) is given an index of 100%, sources with relative citrate solubility above 60% can be recommended for direct application. These are largely concentrated in North Africa, the Soviet Union, and southeastern United States. In tropical countries only certain deposits in Peru, Mexico, and India have citrate solubilities above 60%, according to Lehr and McClellan's data.

The tropical literature offers abundant examples indicating the desirability of using high-quality PR instead of superphosphates in acid soils (Motsara & Datta, 1971; Awan, 1971; Sanchez & Mata, 1972; Engelstad et al., 1974; Velly & Roche, 1972; North Carolina State University, 1974) and the poor performance of low solubility PR in acid soils (Alvarez et al., 1965; Neme & Lovadini, 1967; Viegas et al., 1970; Miranda et al., 1970; Bryan & Andrew, 1971; Michelin et al., 1974). Many of these comparisons, however, have been conducted without regard to long-term residual effects. The effectiveness of these low citrate-soluble sources relative to that of the high citrate-soluble sources increases with time. Table 7, adapted from the work of Velly and Roche (1973), compares the effect of Hyper Reno, a high citrate-soluble PR from North Africa, with a low citrate-soluble PR from Taiba, Senegal. The superior behavior of the source with higher solubility is very marked, but the difference narrows with time. Similar effects are being observed with the slowly soluble Araxá PR in Oxisols of Brazil where its effectiveness increased from the first to the second year after application (North Carolina State University, 1974). Consequently, direct application

Table 7—Yields of three successive crops in an acid hydromorphic soil from Mahitsy, Madagascar, obtained with applications of a phosphate rock of high solubility (Hyper Reno) and one of low solubility (Taiba). Adapted from Velly and Roche (1973).

Source	P added	Grain yields			Percent of maximum yield		
		Rice	Wheat	Rice	Rice	Wheat	Rice
	kg P/ha	— metric tons/ha —			— % —		
Check	0	5.0	0.4	4.1	72	18	68
Taiba	66	5.4	1.0	4.9	77	45	86
	132	5.5	1.3	5.7	78	59	93
Hyper Reno	66	6.4	1.3	5.8	91	59	96
	132	7.1	2.3	6.1	100	100	100

of low solubility PR sources could be combined with high solubility materials to satisfy the requirements of the first few crops.

A second major consideration is fineness of the materials. The effectiveness of all PR increases with increasing fineness, in contrast to an opposite effect in water-soluble sources (Terman & Engelstad, 1971). While granulation is effective in decreasing P fixation in superphosphate (Fassbender & Molina, 1969), PR generally required particle size smaller than 100 mesh to react effectively. For example, Barnes and Kamprath (1975) found that commercial North Carolina PR with only 32% of the particles smaller than 65 mesh was 65 to 83% as efficient as when ground to 100 to 115 mesh.

A third major consideration is the time of reaction. High citrate-soluble sources, such as North Carolina PR, require from 60 to 90 days to react in Ultisols of this state (Barnes & Kamprath, 1975). Very little information is available on the time required by different sources in other high fixing soils. Incorporating PR in advance may enhance its availability to the first crop.

A fourth management consideration is the relationship with liming. Phosphate rock requires an acid soil environment in order to release P into the soil solution. In Ultisols of North Carolina, for example, the effectiveness of high quality PR decreased dramatically if the soil pH increased above 5.2 (Barnes & Kamprath, 1975). Many crops cannot grow well at such low pH values because of Al or Mn toxicity. An answer to this problem is to apply the PR several months ahead of liming in order for it to react at low pH. This might work well if the first crop to be planted is relatively tolerant to Al. Upland rice is such an example. Lime can then be added prior to planting a crop such as soybean, which is more sensitive to Al. The time required for lime to react in acid soils is less than that needed for the high solubility PR.

In countries with mostly low solubility PR deposits, such as Brazil, an appropriate alternative might be to broadcast the PR and band superphosphates or other soluble sources in order to provide P while the PR slowly dissolves. This strategy has proven successful in Africa with the use of locally available low citrate-soluble sources (Pichot & Roche, 1972).

An alternative that seems of little value is partial acidulation of PR. Although this process has increased yield responses (Panda & Misra, 1970; McLean & Logan, 1970), the resulting material has very poor physical properties (Engelstad and Russel, 1975).

Many acid, high P-fixing soils are also deficient in S. A change from ordinary superphosphate (OSP) to PR will eliminate the effects of the S content of OSP. Under these conditions a combination of elemental S or other S sources with PR would be required. Bromfield (1975) observed that a mixture of S and PR was as effective as OSP and superior to PR alone in Alfisols of northern Nigeria.

Local cost of production data will determine whether direct applications of PR will be economically advantageous over the use of superphosphates. In many cases a higher rate of P is needed with PR than with superphosphates for maximum yield response, while in others the same rates of both sources are equally effective.

b. Thermally-altered and Rhenania-type Phosphates—Another group of potentially cheaper sources of P for acid, high-fixing soils includes basic slag and fused Mg phosphates, both water-insoluble products of thermal alteration. These types of fertilizers have been used primarily in Europe, but their potential in tropical areas with P-fixation problems is receiving increased attention, particularly as steel industries develop and where there are cheaper sources of energy. A recent review by Atanasiu (1971) showed that these sources were more effective than water-soluble sources in many acid soils of the tropics.

Basic slag (called "Escorias Thomas" in Latin America) is a byproduct of steel manufacture from Fe ore high in P. It has an average content of 4 to 8% P and 32% Ca, mostly as Ca silico-phosphates and Ca silicates. It has been found to be nearly as effective as superphosphates at the same rates of P application in Oxisols of Brazil (Pereira et al., 1974) and Colombia (Ortega & Guerrero, 1972; Howeler, 1974; Spain, 1974), in Andepts of Colombia (Michelin et al., 1974) and Costa Rica (Fassbender & Molina, 1969), and in acid sulfate soils of Thailand (Engelstad et al., 1974). The effectiveness of basic slag is illustrated in Table 8 where the cumulative pasture production during 3 years after the basal application of triple superphosphates (TSP), PR, and basic slag for an Andept of Colombia is compared. Basic slag produced nearly as much total dry matter as TSP and was definitely superior to the high citrate-soluble Florida PR and the low citrate-soluble local FR-22 PR. Basic slag also has a liming effect because the Ca silicates hydrolyze readily to Ca hydroxide. This effect is seen in Table 8 where basic slag increased the pH to 5.5, decreased the exchangeable Al content, and doubled the exchangeable Ca content of the soils 3 years after it was applied. These results plus the lower cost of basic slag compared to superphosphate in Colombia (Spain, 1974) suggest its use.

The Rhenania phosphates are produced by sintering PR of low citrate solubility with silica and soda ash. When serpentine or Mg silicates are

Table 8—Effect of superphosphates, phosphate rock of high solubility (Florida) and low solubility (FR-22) and basic slag initial applications on the cumulative yield of elephant grass (*Pennisetum purpureum*) and topsoil properties at the end of 3 years in an Andept from Popayan, Colombia. Adapted from Michelin et al. (1974).

Source of P	Rate	Cumulative dry matter production	Yield increase over check	Efficiency	pH	Exchangeable Al	Ca	Avail. P (Bray II)
	kg P/ha	metric tons/ha	kg	kg DM/kg P		— meq/100 g —		ppm
Check	0	53	--	--	5.2	1.7	3.6	10
TSP	44	71	18,000	409	5.2	1.4	2.8	17
TSP	88	72	19,000	216	5.1	1.7	2.8	9
Florida PR	88	61	8,000	91	5.2	1.3	3.6	7
FR-22 PR	88	54	1,000	11	5.2	1.6	2.8	12
Basic slag	88	68	15,000	170	5.5	0.6	6.0	8

fused with PR to give Ca or Mg silicophosphates, the product is called *fused magnesium phosphate* (FMP). These products vary in composition from 10 to 12% P, 20 to 30% Ca, and 0 to 8% Mg. They were developed in Germany after World War I in response to the shortage of S needed to produce superphosphates. Rhenania phosphates are being manufactured in certain countries, such as Brazil and India, in order to transform PR of low citrate solubility into highly reactive materials for their acid, high P-fixing soils. These sources have been found to be as effective as or more effective than superphosphates in high P-fixing Oxisols of Brazil (Miranda et al., 1970; Muzilli et al., 1971; Gargantini & dos Santos, 1971; Pereira et al., 1974; North Carolina State University, 1974), in Ultisols of Nigeria with high P fixation (Obigbesan & Kuhn, 1974), in Andepts and related volcanic soils of Costa Rica and Ethiopia (Fassbender and Molina, 1969; Werner, 1969; Atanasiu, 1971) and New Zealand (Sinclair, 1975), and in "lateritic" soils of South India, Liberia, and Congo (Werner, 1969; Atanasiu, 1971; Shetty et al., 1973). They are more effective in unlimed acid soils than in limed ones (Gargantini & dos Santos, 1971; Obigbesan & Kuhn, 1974) and, like basic slag, also act as a liming agent. Ongoing experiments on Oxisols in Brasil indicate that an application of 152 kg P/ha as "Termofosfato" decreased Al saturation from 70 to 38%, while no such change was observed with an equal rate of TSP which produced similar pasture yields. Another

The main disadvantage of the Rhenania phosphates is their high cost of production. In Brazil, for example, the price per kilogram of P is almost equal to that of TSP. Although the liming effect and Si content may make their use more profitable, the high cost of Rhenania phosphates is a major limiting factor. In areas with ample supply of cheap energy for thermal alteration, the situation may be different.

Production of thermally altered phosphates is sometimes suited to small fertilizer plants employing intermediate technology. While fertilizer plants with production capacity as low as 50,000 metric tons/year may not be feasible in the industrialized nations, developing countries may find it profitable and appropriate to use intermediate technology which depends on utilization of local resources and skills. Unlike superphosphates, production of thermally altered P does not require S or sulfuric acid plants. In addition high silica and/or low citric acid soluble PR may be used.

3. IMPROVED SOIL TEST INTERPRETATIONS

Efficiency of P fertilization may also be increased by use of more realistic methods for determining the optimum rates required. Use of the linear response and plateau model proposed by Waugh et al. (1973) instead of the traditional curvilinear regressions has resulted in an actual reduction of recommended rates of P without major differences in yield in a set of 30 corn experiments conducted on Oxisols in the state of Minas Gerais, Brazil (North Carolina State University, 1973). Use of the linear response and plateau model is considered more realistic because quadratic models can exaggerate the optimum rates of fertilizer applications (Anderson & Nelson, 1975).

B. Decreased Phosphorus Fixation Through Lime and Silicate Applications

A second component of this strategy is to decrease the soil's P-fixation capacity by applying less expensive amendments such as lime and silicates.

1. LIME

Considerable controversy exists in the literature regarding whether or not liming decreases P fixation. Part of the problem is the difficulty in separating direct and indirect effects of liming using only plant data. Liming acid soils to pH 5.5 or 6.0 eliminates Al and Mn toxicity and supplies Ca. These effects generally result in increases in plant growth and P uptake. When the changes in P-fixation capacity are measured, the effect of liming is found to depend on soil properties.

Liming acid soils of high Al saturation to a pH of 5.5 to neutralize the exchangeable Al decreases P fixation as measured by sorption isotherms or Langmuir adsorption maxima. This has been observed in Oxisols from Panama (Mendez & Kamprath, 1978) and Brazil (Leal & Velloso, 1973a, b; Vasconcelos et al., 1975), in Ultisols of North Carolina (Woodruff & Kamprath, 1965), and in Andepts of Reunion Island (Truong et al., 1974). In these examples original pH values ranged from 4.4 to 5.2 and Al saturation levels varied from 60 to 83%.

Liming has little or no influence in decreasing P fixation in soils with higher pH values (ranging from 5.0 to 6.0), but with low Al saturation (0 to 45%). These results have been observed in Oxisols from South Africa (Reeve & Sumner, 1970), Brazil (Leal & Velloso, 1973b), Hawaii (Fox & Benavides, 1974), in Ultisols from North Carolina (Woodruff & Kamprath, 1965), and in Andepts from Reunion (Truong et al., 1974).

Liming to pH values near or above neutrality may increase rather than decrease P fixation because of the formation of relatively insoluble Ca phosphates. This has been found in Oxisols from Hawaii (Fox et al., 1974), Ultisols from Florida (Robertson et al., 1954), as well as in many low P-fixing soils with predominantly layer-silicate mineralogy (Stoop, 1974; Mokwunye, 1975).

The beneficial effects obtained in liming acid soils to a pH of about 5.5 is attributed to the precipitation of exchangeable and hydroxy Al as Al hydroxides which fix less P and to the lower bonding energy between sorbed P and oxide surfaces as their negative charge increases with increasing pH (Kamprath & Foy, 1971). Leal and Velloso (1973b) showed that a significant proportion of the P sorbed at pH 4 was released when the pH was increased to 7 in the Oxisols of Brazil.

An example of decreased P fixation through liming is illustrated in Fig. 8 with an Oxisol from Panama. The top graph shows the decrease in P fixation when this Oxisol was limed to pH 5.5. The bottom graph shows that less than half the P application rate was needed in the limed soil to approach maximum yields of millet obtained in the unlimed soil.

Liming acid soils to a pH of 5.5 to 6.0, therefore, decreases but does not eliminate P fixation. The magnitude of this decrease appears to be greater the more acid the soil was originally. Woodruff and Kamprath (1965) observed that the adsorption maximum was reduced by liming to about 50% in Ultisols with initial pH values of 4.4 to 4.8. Leal and Velloso (1973b) observed a 25% decrease in the adosrption maxima of Oxisols with 66 to 74% Al saturation, but only a 14% decrease in Oxisols with Al saturation of 29 to 46%. When adsorption isotherms are used, the reduction in P adsorbed decreases as P in the equilibrium soil solution increases. Table 9 shows results for an Oxisol from Brazil where liming to neutralize exchangeable Al decreased P fixation by 41% at 0.03 ppm P but only by 11% at 0.2 ppm P.

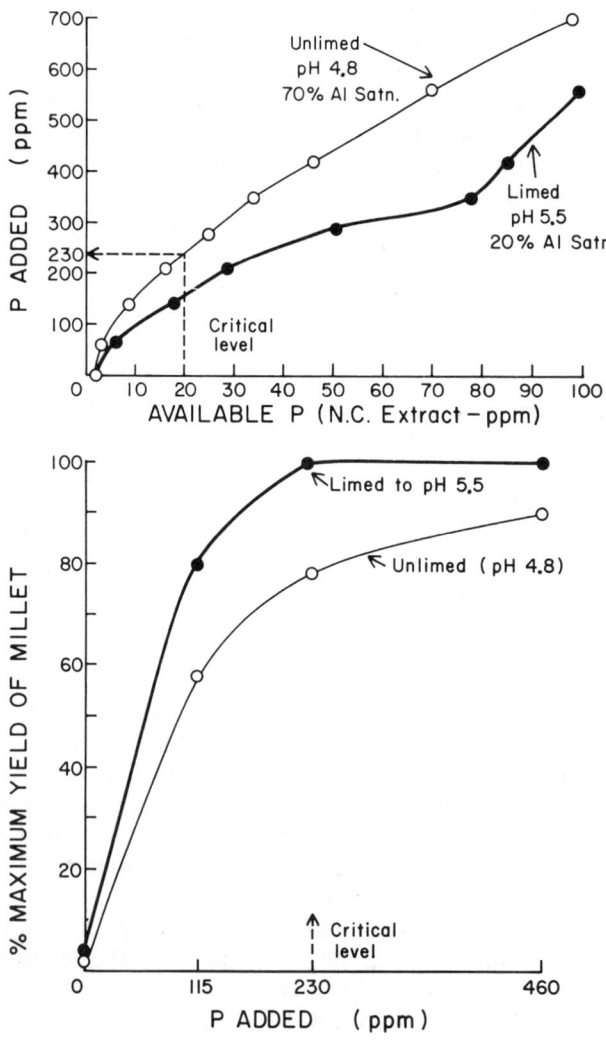

Fig. 8—Effects of liming on P fixation and response in an Oxisol from Panama. Adapted from Mendez and Kamprath (1978).

Table 9—Decrease in P fixation by lime and silicate applications sufficient to neutralize exchangeable Al in a clayey Oxisol from Brazil with an original pH of 4.6, 1.45 meq exchangeable Al/100 g and 80% saturation. Source: Smyth and Sanchez (1980 a, b)

	P fixed to give			Decrease in P fixed		
	0.03	0.10	0.20	0.03	0.10	0.20
	ppm P in solution			ppm P in solution		
	——— ppm P ———			——— % ———		
None	230	325	415	--	--	--
Lime (1.5 metric tons/ha)	135	275	370	41	15	11
Calcium silicate (1.8 metric tons/ha	125	265	355	46	18	14

The nature of the accompanying cation also affects the decrease in P fixation. Stoop (1974), working with Andepts and Oxisols of Hawaii, observed smaller decreases with Ca phosphate than with Na, K, or NH_4 phosphates. Calcium is specifically adsorbed by the oxide surfaces while Na^+, K^+, and NH_4^+ are not.

2. SILICATE APPLICATIONS

Silicate applications, usually as Ca or Na silicates, also decrease P fixation and increase P uptake by crops in certain acid Oxisols and Andepts. In some cases the increased P uptake may result from a direct growth response to Si in soils very deficient in this element, but in many others, there is a direct effect of decreasing P fixation. Silva (1971) reviewed most of the Hawaiian work and concluded that silicate anions may replace phosphates from the sorption sites and increase the availability of P. Table 10, based on the work of Roy et al. (1971), shows that the magnitude of this decrease also varies with soil properties, being more marked in soils with the lowest P-fixing capacity—the Tropohumult and the Eutrorthox. Table 9 also shows similar reductions in P fixation with $CaSiO_4$ and $CaCO_3$ when both were applied at rates sufficient to neutralize the exchangeable Al in an Oxisol from Brazil which was not deficient in Si.

The Si content of basic slag and the Rhenania phosphates may also offer the advantage of decreasing P fixation. This effect has not been measured. Even irrigating with water high in Si has been known to decrease P fixation in Hawaii (Fox et al., 1974). One important implication of this ef-

Table 10—Effect of calcium silicate applications on the decrease in P fixation of Hawaiian soils measured as the amount of fixed P required for 0.2 ppm of P in solution. Source: Roy et al. (1971).

Soil	ppm Si added		% decrease
	0	500	
	——— ppm P fixed ———		
Tropohumult	187	100	47
Eutrorthox	425	250	41
Gibbsihumox	725	550	24
Hydrandept	1,150	1,050	9

fect is that P sorption isotherms based on soils prior to receiving these amendments are likely to overestimate the amount of P required. In order to eliminate this bias, isotherms should be constructed after lime or silicate applications.

C. Select Species and Varieties Tolerant to Low Phosphorus Availability

The third component of the low-input management strategy consists of adapting plants to the soil's limitations rather than the traditional approach of adapting the soil to meet crop requirements. Many, and probably most, of the traditional food crops grown by subsistence farmers are those which can withstand high soil acidity and low P. Although their yield and nutritional value may be low, traditional varieties selected over the centuries represent optimum genotypes for conditions of low-input, subsistence agriculture.

A number of root crops, including cassava, yam, and tannier, appear to possess high tolerance to soil acidity and P stress. In great contrast to the common subsistence crops, most of the high-yielding grains are moderately to highly sensitive to exchangeable Al and low P concentrations in the soil solution. However, a wide range of tolerance to these stresses is exhibited among some cultivars. On the other extreme, the grain crops cultivated by the subsistence farmer produce low but dependable yields under adverse conditions.

A concerted effort is now being made to combine high yield and quality with high tolerance to soil-related stresses. Three commonly encountered and interrelated soil stresses are high soluble Al, low Ca, and low P. These stresses frequently occur together in high P-fixing soils in which soil fertility and not water stress is the major limiting agronomic constraint. Most of the efforts have been geared towards selecting species and cultivars tolerant to high levels of Al. Specific genes controlling Al tolerance have been identified, allowing plant breeders to select tolerant cultivars and decrease investments in lime (Foy, 1974). Active research is being conducted on Oxisols of high P-fixation capacity in the Llanos of Colombia (Spain et al., 1975) and the Cerrado of Brazil (North Carolina State University, 1974), where differences have been found in tolerance to low P availability. The following discussion summarizes the evidence of differential tolerance to P stress, the possible mechanisms involved inside the plants and in the rhizosphere, and the practical implications.

1. DIFFERENCES IN INTERNAL AND EXTERNAL REQUIREMENTS

Australian scientists working with Ultisols of Queensland found important differences in the percentage of P in plant tissue of several tropical pasture species required to achieve maximum yields (Andrew & Robins, 1969; 1971). The percentage of P in plant tops above which no further positive growth response occurred was defined by them as the "internal phosphorus requirement." Some of the results are shown in Table 11. The legumes stylo and centro require about 0.17% P for maximum growth while

Table 11—Plant P concentrations associated with maximum growth of several pasture species in Queensland, Australia. Adapted from Andrew and Robins (1969, 1971).

Species	Internal P requirement, % P
Legumes	
Townsville stylo	0.17
Centro	0.16
Greenleaf	0.22
Perennial soybean	0.23
Alfalfa	0.25
Grasses	
Pangolagrass	0.16
Molassesgrass	0.18
Guineagrass	0.19
Kikuyugrass	0.22
Rhodesgrass	0.23
Dallisgrass	0.25

perennial soybean and alfalfa require about 0.24% P. The first two species are native of regions with soils low in available P, while the last two are native of regions with soils high in available P and other nutrients. The same situation occurs with forage grasses. Molassesgrass and guineagrass are adapted to acid soils with low P availability, while rhodesgrass and dallisgrass are native to more fertile soils.

More information about the internal P requirements of crop species or cultivars within species is needed. Data for cassava, rice, corn, beans, potatoes, sweet potatoes, and other crops is being developed.

Differential tolerance to P stress can also be measured in terms of external P requirements as defined by Fox et al. (1974) and discussed previously in the section on sorption isotherms. Some of the available data presented in Table 1 show a 10-fold difference between two similar vegetable crops. This table also confirms the general observations that the recommended rates of P for upland rice are much lower than those for corn in Latin America (Kamprath, 1973). Corn has an external P requirement of 0.06 ppm P while in upland rice it is lower than 0.03 ppm in Oxisols.

The relationship between the internal and external P requirements needs to be established through joint measurement of both soil and plant P. The available information is quite limited, but ongoing work in several regions is expected to narrow this gap within the next few years.

2. DIFFERENCES IN PHOSPHORUS ABSORPTION AND TRANSLOCATION RATES

Salinas and Sanchez (1976) reviewed the literature on factors affecting varietal and species differences in tolerance to P stress and concluded that tolerance to low P and Al toxicity are intimately related. They suggested that the ability of a plant to absorb and translocate P in the presence of high levels of Al in the roots determines its tolerance to low available P. Subsequent laboratory and field work in progress has proven this assumption to be true for several crops grown in Oxisols of Brazil.

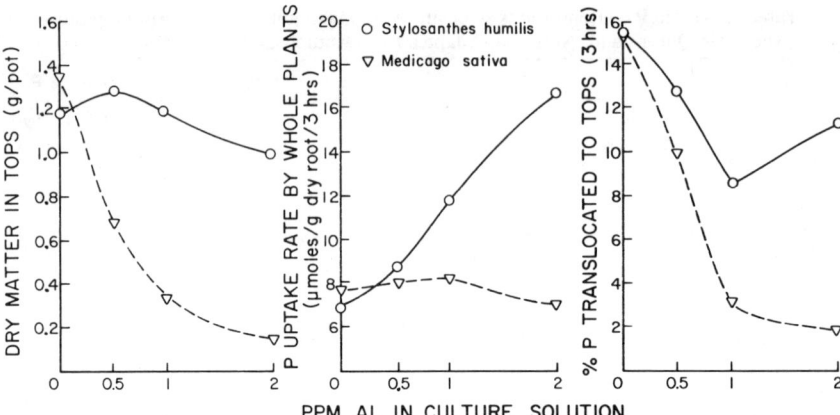

Fig. 9—Effects of Al on growth, P adsorption rate, and P translocation rates in two pasture legumes. Stylo is tolerant to P stress and alfalfa is not. Adapted from Andrew and Vanden Berg (1973).

Figure 9 compares a pasture legume considered tolerant to Al and low P, stylo, with alfalfa, which is considered susceptible to both factors (Andrew & Vanden Berg, 1973). In the absence of Al, both species produced about the same dry matter and had similar P uptake and translocation rates. In the presence of Al, alfalfa suffered from severe dry matter production losses while stylo was unaffected. Stylo absorbed more P as Al increased and translocated more than twice as much P to the tops than did alfalfa in the presence of high levels of Al. Phosphorus is known to be precipitated by Al around certain root cells (Rasmussen, 1968; McCormick & Borden, 1972). Consequently, tolerance to low P appears to be related to the rate at which this element can escape precipitation by being translocated to the tops.

3. EFFECTS IN THE RHIZOSPHERE

Another possible mechanism occurs in the rhizosphere. In a review paper, Hanotiaux and Heck (1973) suggested that plants differ in their ability to alter their rhizosphere so that different forms of solid phosphates are dissolved and absorbed by roots. Subsequent work showed that sugar beets removed more Ca-bonded phosphates than did potatoes from hydroponic sand culture, while potatoes removed more Fe-bonded phosphates than did sugar beets from the same system. Deist et al. (1971) also observed that dicotyledonous crops utilized PR better than monocotyledonous crops and attributed this to the larger amounts of Ca absorbed by the dicots. Similar comparisons with other crops and on other soils are needed to fully understand this possibility.

It is fairly well established that endomycorrhizal infection of certain plant roots increases the plant's ability to absorb P, particularly in soils deficient in this element (Gray & Gerdemann, 1969). Mycorrhizae and other rhizosphere organisms can increase P availability although the mechanisms are not completely understood (Rovira & Davey, 1974; Gerdemann, 1974).

Although most of these data are derived from tree crops, there is evidence with food crops which indicate a response to *Endogone* mycorrhiza inoculation. Studies by Murdoch et al. (1967) with corn and Ross and Gilliam (1973) with soybeans indicate such a response in soils well supplied with P, but it is not likely that a response to inoculation will occur if the soil is extremely deficient in P, unless fertilizer P is applied. The possibility of mycorrhiza inoculations in high P-fixing soils deserves more investigation, because it may be an additional means of decreasing P fertilizer requirements.

4. PRACTICAL IMPLICATIONS

The use of varieties more tolerant to low levels of available P will result in a more efficient use of fertilizer P needed, rather than in the complete elimination of fertilization. Koyama and Chammek (1971) obtained about the same yield of a tolerant rice cultivar with an application rate of 22 kg P/ha as with 44 kg P/ha when a sensitive variety was used in Thailand. Preliminary data by Salinas and Sanchez (1976) indicate that maximum P absorption rate by 'Bluebonnet 50' rice occurred at 0.2 ppm P in culture solution while the 'IR-5' showed the same absorption rate at 0.05 ppm P.

Plants tolerant to low P are likely to have lower P concentrations in their tissues. Their nutritive value may be lower than other cultivars or species. Australian pasture agronomists have recognized these differences and have used direct P supplementation to cattle in the form of salts to offset this deficiency. There is little information of a similar nature on the nutritional quality of food crops.

If the plant breeder succeeds in isolating and developing plant genotypes which combine high yield and quality with tolerance to soils low in P and Ca and high in soluble Al, then rapid and low-cost development of sizable tracts of marginal lands will be possible.

A great deal of time and effort can be saved by making selections on soils precisely adjusted for P, Ca, and Al levels. In fact, in a breeding program, it would be profitable to test new cultivars over a range of soil P, Ca, and Al levels. A high-yielding wheat or maize cultivar which requires high Ca and low Al and has an external P requirement of 0.05 ppm P has no place in the acid, high P-fixing soils of the world. But cultivars with the same or even slightly lower yield potential which can attain 95% of maximum yield when the soil solution P concentration is 0.01 ppm P and the Al saturation is >60% are urgently needed to bring about a sustained green revolution in the regions of the world where increased food production is most needed.

It is grossly unfair to expect the subsistence farmer to give up cassava and yam for more nutritious crops if the new introductions have not been selected to produce adequate yields under low soil fertility. To achieve this goal, the soil scientist must provide the plant breeder with a range of soil conditions in which to make plant selections. The soil conditions must be precisely defined so that new cultivars selected for a particular soil condition may be distributed to any region where this condition prevails or can be attained economically.

In the past, soil scientists have been placed in the unenviable position of constantly manipulating soils to meet the needs of a crop. The time has come, particularly in the humid tropics where the cost of manipulating soils is high, for soil scientists to establish the degree to which soils may be reasonably and economically improved and then work with plant breeders to develop cultivars which can perform adequately under these conditions.

D. Advantages and Limitations

The low-input strategy appears to be of greater relevance than the high-input stretegy to tropical regions with limited capital. Combining more efficient P fertilization practices with the use of cheaper sources of P and with varieties selected for tolerance to P stress is probably the most logical strategy for both small and large farm operations.

Disadvantages of the low-input strategy, compared with the high, are the lack of an amendment effect which occurs with massive applications and the likelihood of a lower yield potential. In the last analysis, the choices are a matter of economics and public policy.

VI. RESEARCH NEEDS

Although P is by far the most thoroughly studied plant nutrient in agriculture, much remains to be learned about the management of soils with a high fixation capacity. The problem is that much of the laboratory work has not been adequately tested in the field. While the mechanisms of P fixation are certainly not adequately understood, the more urgent research needs are in the applied rather than in the basic aspects.

The internal and external P requirements of most crops and of cultivars within each crop need to be defined and properly quantitied. The basic concepts and mechanisms for quantification are available; much of the data remain to be gathered.

Although the adsorption isotherm is a good tool, a simpler way of characterizing P fixation by routine and rapid methods needs to be developed.

More long-term studies on the residual effects of P applications are needed. They should be conducted in well-characterized sites considered representative of areas where high P fixation is a problem. Both water-soluble and water-insoluble sources and combinations of the two need to be compared. Such trials will permit the quantification of much of the information needed on the behavior of PR, basic slags, and Rhenania phosphates, including the time of reaction, the length of the residual effects, and their influence on soil properties.

Incorporating tolerance to P stress as a breeding objective is also needed. Plant breeders and soil scientists working together can make a major contribution to this effort.

It was regrettable that no proper comparison between the two alternate strategies could be presented. The available information was so fragmented that the combined effects of the different components of each strategy could not be quantified. Field studies designed to compare the alternative strategies and combinations or modifications thereof need to be conducted to provide suitable agronomic and economic bases for recommendations to farmers.

LITERATURE CITED

Alvarez, R., J. C. Ometto, J. Ometto, and A. Wutke. 1965. Adubação de cana-de-açúcar. Bragantia 24:97–107.

Anderson, R. L., and L. A. Nelson. 1975. A family of models involving intersecting straight lines and concomitant experimental designs useful in evaluating response to fertilizer nutrients. Biometrics 31:303–318.

Andrew, C. S., and M. F. Robins. 1969. The effect of phosphorus on the growth, chemical composition, and critical phosphorus percentages of some tropical pasture legumes. I. Growth and critical percentages of phosphorus. Aust. J. Agric. Res. 20:655–674.

Andrew, C. S., and M. F. Robins. 1971. The effect of phosphorus on the growth, chemical composition and critical phoshorus percentages of some tropical pasture grasses. Aust. J. Agric. Res. 22:693–703.

Andrew, C. S., and P. J. Vanden Berg. 1973. Influence of aluminum on phosphate sorption by whole plants and excised roots of some pasture legumes. Aust. J. Agric. Res. 24:341–451.

Aomine, S., and N. Yoshinaga. 1955. Clay minerals of some well-drained volcanic ash soils in Japan. Soil Sci. 79:349–358.

Appelt, H., N. T. Coleman, and P. F. Pratt. 1975. Interactions between organic compounds, minerals and ions in volcanic-ash derived soils. II. Effects of organic compounds on the adsorption of phosphate. Soil Sci. Soc. Am. Proc. 39:628–630.

Atanasiu, N. 1971. A comparative study on the effect of water and citrate soluble phosphatic fertilizers on yield and P-uptake on tropical and subtropical soils. J. Indian Soc. Soil Sci. 19:119–127.

Awan, A. B. 1971. Estudio comparativo de roca fosfatada y superfosfato triple como fuentes de fósforo para los cultivos. Rev. Agric. Cuba 4:55–61.

Bache, B. W. 1964. Aluminum and iron phosphate studies related to soils. II. Reaction between phosphates and hydrous oxides. J. Soil Sci. 15:110–116.

Baldovinos, S. F., and G. W. Thomas. 1967. The effect of soil clay on phosphorus uptake. Soil Sci. Soc. Am. Proc. 31:680–682.

Barber, S. A. 1974. A program for increasing the efficiency of fertilizers. Fert. Solutions 18: 24–25.

Barnes, J. S., and E. J. Kamprath. 1975. Availability of North Carolina rock phosphate applied to soils. North Carolina Agr. Exp. Stn. Tech. Bull. 229.

Barrow, N. J. 1974a. The effect of previous additions of phosphate on phosphate adsorption by soils. Soil Sci. 118:82–89.

Barrow, N. J. 1974b. The slow reactions between soil and anions. 1. Effects of time, temperature, and water content of a soil on the decrease in effectiveness of phosphate for plant growth. Soil Sci. 118:380–386.

Beckwith, R. S. 1965. Sorbed phosphate at standard supernatant concentration as an estimate of phosphorus needs of soils. Aust. J. Exp. Agric. Anim. Husb. 5:52–58.

Benavides, G. E. 1974. Adsorción de fósforo en suelos volcánicos de Nariño. Agric. Trop. (Colombia) 27(3):5–19.

Biddappa, C. C., and B. V. V. Rao. 1973. Studies on the relationship between sesquioxides, phosphorus contents, and phosphorus fixing capacity of coffee soils of South India. J. Indian Soc. Soil Sci. 21:155–159.

Breeuwsma, A. 1973. Adsorption of ions on hematite (α-Fe_2O_3). Meded. Landbowhogenschool Wageningen 73-1.

Bromfield, A. R. 1975. Effects of rock phosphate-sulfur mixture on yield and nutrient uptake of groundnuts (*Arachis hypogaea*) in northern Nigeria. Exp. Agric. 11:265–272.

Bruce, R. C. 1968. Leaching and residual value of superphosphate and rock phosphate under high rainfall conditions. Qld. J. Agric. Anim. Sci. 25:77-79.

Bryan, W. W., and C. S. Andrew. 1971. Value of Nauru rock phosphate as a source of phosphorus for some tropical pasture legumes. Aust. J. Exp. Agr. Anim. Husb. 11:532-535.

Buol, S. W., P. A. Sanchez, R. B. Cate, Jr., and M. A. Granger. 1975. Soil fertility capability classification. p. 126-141. *In* E. Bornemisza and A. Alvarado (ed.) Soil management in tropical America. North Carolina State University, Raleigh.

Calhoun, F. G. 1974. Taxonomy of Oxisols, Ultisols and Andepts. Soil Crop Sci. Soc. Florida Proc. 33:108-111.

Chang, T. C., and H. C. Lin. 1970. The influence of soil moisture conditions on the fixation of phosphates. Soils Fert. Taiwan. 1969:75.

Coleman, N. T., and A. Mehlich. 1948. Some chemical properties of soils as related to their cation exchange-anion exchange ratios. Soil Sci. Soc. Am. Proc. 13:175-178.

Coleman, N. T., J. T. Thorup, and W. A. Jackson. 1960. Phosphate sorption reactions that involve exchangeable aluminum. Soil Sci. 90:1-7.

Colwell, J. D. 1959. Phosphate sorption by iron and aluminum oxides. Aust. J. Appl. Sci. 10: 95-103.

Dabin, B. 1974. Evolution des phosphates en sols acides des régions tropicaux. Bull. Association Francaise pour l'Etude du Sol No. 2:87-194.

Davis, L. E. 1945. Retention of phosphates by soils. II. Effect of drying and of anions and cations on the cation exchange properties of soils. Soil Sci. 59:175-190.

Dean, L. A. 1949. Fixation of soil phosphorus. Adv. Agron. 1:391-411.

Deist, J., P. G. Marais, R. B. A. Harry, and C. F. G. Heyns. 1971. Relative availability of rock phosphate to different plant species. Agrochemophysica 3:35-40.

Engelstad, O. P., A. Jugsujinda, and S. K. DeDatta. 1974. Response by flooded rice to phosphate rocks varying in citrate solubility. Soil Sci. Soc. Am. Proc. 28:524-529.

Engelstad, O. P., and D. A. Russel. 1975. Fertilizers for use under tropical conditions. Adv. Agron. 27:175-208.

FAO-UNESCO. 1971. Soil map of the World. Vol. IV: South America. Unesco, Paris.

FAO-UNESCO. 1975. Soil map of the World. Vol. II: North America, Vol. III: Central, Vol. VI: Africa. Unesco, Paris.

Fassbender, H. W. 1969. Estudio del fósforo en suelos de America Central. IV. Capacidad de fijación de fósforo y su relación con caracteristicas edáficas. Turrialba 19:497-505.

Fassbender, H. W., and R. Molina. 1969. Influence of lime and silicate applications on the effect of phosphate fertilizers applied to soils derived from volcanic ash in Costa Rica. p. C.2.1-C.2.12. *In* Panel on soils derived from volcanic ash in Latin America. Interamerican Institute of Agricultural Sciences, Turrialba, Costa Rica.

Fieldes, M., and Q. Perrot. 1966. The nature of allophane in soils. III. Rapid field and laboratory test for allophane. N.Z. J. Sci. 93:623-629.

Fox, R. L. 1969. Fertilization of volcanic ash soils in Hawaii. p. C.6.1-C.6.13. *In* Panel on soils derived from volcanic ash in Latin America. Interamerican Institute of Agricultural Sciences, Turrialba, Costa Rica.

Fox, R. L. 1974a. Examples of anion and cation adsorption by soils of tropical America. Trop. Agric. (Trinidad) 51:200-210.

Fox, R. L. 1974b. Chemistry and management of soils dominated by amorphous colloids. Soil Crop Sci. Florida, Proc. 33:112-119.

Fox, R. L., and E. J. Kamprath. 1970. Phosphate sorption isotherms for evaluating the phosphate requirements of soils. Soil Sci. Soc. Am. Proc. 34:902-907.

Fox, R. L., and S. T. Benavides. 1974. El fósforo en Oxisoles. Suelos Ecuat. 6:137-175.

Fox, R. L., D. L. Plucknett, and A. S. Whitney. 1968. Phosphate requirements of Hawaiian Latosols and residual effects of fertilizer phosphorus. Int. Congr. Soil Sci., Trans. 9th (Adelaide) 2:301-310.

Fox, R. L., S. M. Hasan, and R. C. Jones. 1971. Phosphate and sulfate sorption by Latosols. Proc. Int. Symp. Soil Fert. Eval. 1:857-864.

Fox, R. L., R. K. Nishimoto, R. S. Thompson, and R. S. de la Pena. 1974. Comparative external phosphorus requirements of plant growing in tropical soils. Int. Congr. Soil Sci., Trans. 10th (Moscow) 4:232-239.

Foy, C. E. 1974. Effects of aluminum on plant growth. p. 601-642. *In* E. W. Carson (ed.) The plant root and its environment. Univ. Press of Virginia, Charlottesville.

Galindo, G. G., C. Olbuin, and E. B. Schalscha. 1971. Phosphate-sorption capacity of clay fractions of soils derived from volcanic ash. Geoderma 7:225-232.

Gargantini, H., and D. dos Santos. 1971. Competicão de fertilizantes fosfatados em soja. Bragantia 30:117-124.

Gerdemann, J. W. 1974. Mycorrhizae. p. 205-217. *In* E. W. Carson (ed.) The plant root and its environment. Univ. Press of Virginia, Charlottesville.

Gray, L. E., and J. W. Gerdemann. 1969. Uptake of phosphorus-32 by vesicular-arbuscular mycorrhiza. Plant Soil 30:415-422.

Hanotiaux, G., and J. P. Heck. 1973. L'aggresivite des especes vegetales vis-a-vis des formes du phosphore du sol. I. Definition du probleme. Pedologie 23:27-38.

Hinga, G. 1973. Phosphate sorption capacity in relation to properties of several types of Kenya soil. East Afr. Agric. For. J. 38:400-404.

Hossner, L. R., J. A. Freehouf, and B. L. Folsom. 1973. Solution phosphorus concentration and the growth of rice in flooded soils. Soil Sci. Soc. Am. Proc. 37:405-408.

Howeler, R. H. 1974. La fertilización fosfórica del arroz de riego y de secano. Suelos Ecuat. 6:245-263.

Hsu, P. H., and D. A. Rennie. 1962. Reactions of phosphate in aluminum systems. II. Precipitation of phosphate by exchangeable aluminum on a cation resin. Can. J. Soil Sci. 42: 210-221.

Juo, A. S. R., and H. O. Maduakor. 1974. Phosphate sorption of some Nigerian soils and its effect on cation exchange capacity. Commun. Soil Sci. Plant Anal. 5:479-497.

Kamprath, E. J. 1967. Residual effects of large applications of phosphorus on high fixing soils. Agron. J. 59:25-27.

Kamprath, E. J. 1973. Phosphorus. p. 138-161. *In* P. A. Sanchez (ed.) A review of soil research in tropical Latin America. North Carolina Agric. Exp. Stn. Tech. Bull. 219.

Kamprath, E. J. 1974. Aspectos químicos y formas minerales del fósforo del suelo en regiones tropicales. Suelos Ecuat. 6:1-18.

Kamprath, E. J., and C. D. Foy. 1971. Lime-fertilizer-plant interactions in acid soils. p. 105-151. *In* R. A. Olsen et al. (ed.) Fertilizer technology and use. 2nd ed. Soil Sci. Soc. Am., Madison, Wis.

Keng, J., and G. Uehara. 1974. Chemistry, mineralogy and taxonomy of Oxisols and Ultisols. Soil Crop Sci. Soc. Florida, Proc. 33:119-126.

Koyama, T., and C. Chammek. 1971. Soil-plant nutrition studies on tropical rice. I. Studies on varietal differences in absorbing phosphorus from soil low in available phosphorus. Soil Sci. Plant Nutr. 17:115-116.

Larsen, S. 1967. Soil phosphorus. Adv. Agron. 19:151-210.

Leal, J. R., and A. C. X. Velloso. 1973a. Adsorcão de fosfato em Latossolos sob vegetação de Cerrado. Pesq. Agropec. Bras. (Ser. Agron.) 8:81-88.

Leal, J. R., and A. C. X. Velloso. 1973b. Dessorção de fosfato adsorbido em Latossolos sob vegetacão de Cerrado. II. Reversibilidade da isoterma de adsorção de fosfato em relação ao pH da solução em equilibrio. Pesq. Agropec. Bras (Ser. Agron.) 8:89-92.

Lehr, J. R., and G. H. McClellan. 1972. A revised laboratory scale for evaluating phosphate rocks for direct application. TVA Bull. Y-43.

Lindsay, W. L., and H. F. Stephenson. 1959. Nature of the reactions of mono-calcium phosphate in soils. I. The solution that reacts with the soil. II. Dissolution and precipitation reactions involving iron, aluminum, manganese, and calcium. Soil Sci. Soc. Am. Proc. 23: 12-22.

Lindsay, W. L., A. W. Frazier, and H. F. Stephenson. 1962. Identification of reaction products from phosphate fertilizers in soils. Soil Sci. Soc. Am. Proc. 26:446-452.

Lutz, J. F., R. Garcia-Lagos, and H. G. Hilton. 1960. The effect of phosphate fertilizers on some physical properties of the soil. Int. Congr. Soil Sci., Trans. 7th (Madison) 1:241-248.

Lutz, J. F., R. A. Pinto, R. Garcia-Lagos, and H. G. Hilton. 1966. Effect of phosphorus on some physical properties of the soil: II. Water retention. Soil Sci. Soc. Am. Proc. 30: 433-437.

Mattson, S. 1931. The laws of colloidal behavior. V. Ion adsorption and exchange. Soil Sci. 31:311-331.

McClelland, V. F. 1968. Superphosphate on wheat: the cumulative effect of repeated applications on yield response. Aust. J. Agric. Res. 19:1-8.

McCormick, L. H., and F. Y. Borden. 1972. Phosphate fixation by aluminum in plant roots. Soil Sci. Soc. Am. Proc. 36:799-802.

McLean, E. O., and T. J. Logan. 1970. Sources of phosphorus for plants grown on soils with different phosphorus fixation tendencies. Soil Sci. Soc. Am. Proc. 34:907-911.

Mekaru, T., and G. Uehara. 1972. Anion adsorption in ferruginous tropical soils. Soil Sci. Soc. Am. Proc. 36:296-300.

Mendez, J., and E. J. Kamprath. 1978. Liming of Latosols and the effect on P response. Soil Sci. Soc. Am. J. 41:86-88.

Michelín, A., L. A. León, and A. Ramírez. 1974. Uso eficiente de fertilizantes fosfatados en suelos ácidos. Suelos Ecuat. 6:265-287.

Miranda, L. T. de, G. P. Viegas, E. S. Freire, and T. Igue. 1970. Adubação do milho. XXVII. Ensaios com diversos fosfatos. Bragantia 29:301-308.

Mokwunye, U. 1975. The influence of pH on the adsorption of phosphate by soils from the Guinea and Sudan savannah zones of Nigeria. Soil Sci. Soc. Am. Proc. 39:1100-1102.

Monsalve, S. A., and J. Lotero. 1972. Efecto residual del fósforo en un suelo negro en Antioquia. Rev. Inst. Colomb. Agropecu. 7:159-171.

Moshi, A. O., A. Wild, and D. J. Greenland. 1974. Effect of organic matter on the charge and phosphate adsorption characteristics of Kikuyu red clay from Kenya. Geoderma 11:275-285.

Motsara, M. R., and N. P. Datta. 1971. Rock phosphate as a fertilizer for direct application in acid soils. J. Indian Soc. Soil Sci. 19:107-113.

Moura Filho, W., S. W. Buol, and E. J. Kamprath. 1972. Studies on a Latosol Roxo (Eutrustox) of Brazil: Phosphate reactions. Experientiae 13:235-247.

Murdoch, C. L., J. A. Jackobs, and J. W. Gerdemann. 1967. Utilization of phosphorus sources of different availability by mycorrhizae and nonmycorrhizal maize. Plant Soil 27:329-334.

Murphy, H. F. 1939. The role of kaolinite in phosphate fixation. Hilgardia 12:343-382.

Muzilli, O., J. C. S. Rispoli, and N. A. Costa. 1971. Efeitos da adubação fosfatada em solos ácidos do sul do Parana. Bol. Univ. Fed. Paraná 9:1-7.

Neme, N. A., and L. A. C. Lovadini. 1967. Efeitos de adubos fosfatados e calcário na produção de forragem de soja perenne (*Glycine javanica*) em terra de Cerrado. Bragantia 26:365-371.

North Carolina State University. 1973. Agronomic-economic research on tropical soils. Annual report for 1973. Soil Sci. Dep., N.C. State Univ., Raleigh. 190 p.

North Carolina State Unversity. 1974. Agronomic-economic research on tropical soils. Annual report for 1974. Soil Sci. Dep., N.C. State Univ., Raleigh. 230 p.

Obigbesan, G. O., and H. Kuhn. 1974. Vergleichende Untersuchungen uber die Wirksmakeit verschiedener Phosphatdungemittel auf humiden tropischen Boden Westafrikas. Tropenlandwirt. 75:49-57.

Olsen, S. R., and F. S. Watanabe. 1957. A method to determine a phosphorus adsorption maximum of soils as measured by the Langmuir isotherm. Soil Sci. Soc. Am. Proc. 21:144-149.

Ortega, E. J., and R. R. Guerrero. 1972. Efectos de tres fuentes de fertilizantes fosfatados en el comportamiento de las fracciones de fósforo en un Latosol de Nariño, Colombia en relación a la acumulación de fósforo en avena. Turrialba 22:420-430.

Ospina, O. 1974. El fósforo en los Andosoles. Suelos Ecuat. 6:97-136.

Ozanne, P. G., and T. C. Shaw. 1968. Advantages of the recently developed phosphate sorption tests over the older extractant methods of soil phosphates. Int. Soil Sci. Congr., Trans. 9th (Adelaide) 2:273-380.

Panda, N., and U. K. Misra. 1970. Use of partially accidulated rock phosphate as a possible means of minimizing phosphate fixation in acid soils. Plant Soil 33:225-234.

Parfitt, R. L., R. J. Atkinson, and R. St. C. Smart. 1975. The mechanism of phosphate fixation by iron oxides. Soil Sci. Soc. Am. Proc. 39:837-841.

Parks, G. A. 1967. Aqueous surface chemistry of oxides and complex oxide minerals. Advances in Chemisry Series. 67:121-160.

Patrick, W. H., Jr., R. D. Delaune, and D. A. Antie. 1974. Transformation of added phosphate in flooded soil. Int. Congr. Soil Sci., Trans. 10th (Moscow) 4:296-304.

Pereira, J., J. M. Braga, and R. F. Novais. 1974. Efeitos de fontes e doses de fósforo na adubação da soja em un solo sob campo cerrado. Ceres 21:227-246.

Perrot, K. W., A. G. Langdon, and A. T. Wilson. 1974. Sorption of phosphate by aluminum and iron (III) hydroxy species on mica surfaces. Geoderma 12:223-231.

Pichot, J., and P. Roche. 1972. Phosphore dans les sols tropicaux. Agron. Trop. (France) 27:939-965.

Pope, R. A., and S. W. Buol. 1976. Improving the "i" modifier. p. 245-250. *In* Agronomic research on tropical soils. Annual report for 1975. North Carolina State University, Raleigh.

Pratt, P. F., F. F. Peterson, and C. S. Holzley. 1969. Quantitative mineralogy and chemical properties of a few soils from Sao Paulo, Brazil. Turrialba 19:491-496.

Prince, A. L., and S. J. Toth. 1936. The effect of phosphates on the cation exchange capacities of certain soils. Soil Sci. 42:281-290.

Rajan, S. S. S. 1973. Phosphorus adsorption characteristics of Hawaiian soils and their relationships to equilibrium phosphorus concentration required for maximum growth of millet. Plant Soil 39:519-532.

Rajan, S. S. S., and R. L. Fox. 1972. Phospahte adsorption by soils. I. Influence of time and ionic environment on phosphate adsorption. Commun. Soil Sci. Plant Anal. 3:493-504.

Rajan, S. S. S., and K. W. Perrott. 1975. Phosphate adsorption by synthetic amorphous aluminosilicates. J. Soil Sci. 26:257-266.

Rasmussen, H. P. 1968. Entry and distribution of aluminum in *Zea mays*: Electron microprobe X-ray analysis. Planta 81:28-37.

Reeve, N. G., and M. E. Sumner. 1970. Effects of alluminum toxicity and phosphorus fixation on crop growth on Oxisols from Natal. Soil Sci. Soc. Am. Proc. 34:263-267.

Robertson, W. K., J. K. Neller, and F. D. Bartlett. 1954. Effect of lime on the availability of phosphorus in soils of high and low sesquioxide contents. Soil Sci. Soc. Am. Proc. 18: 184-186.

Ross, J. P., and J. W. Gilliam. 1973. Effect of *Endogone* mycorrhiza on phosphorus uptake by soybeans from inorganic phosphates. Soil Sci. Soc. Am. Proc. 37:237-239.

Rovira, A. D., and C. B. Davey. 1974. Biology of the rhizosphere. p. 153-204. *In* E. W. Carson (ed.) The plant root and its environment. Univ. Press of Virginia, Charlottesville.

Roy, A. C., M. Y. Ali, R. L. Fox, and J. A. Silva. 1971. Influence of calcium silicate on phosphate solubility and availability in Hawaiian Latosols. Proc. Int. Symp. Soil Fert. Eval. (New Delhi) 1:757-765.

Russell, E. W. 1973. Some agricultural problems of semi-arid areas. p. 121-135. *In* R. P. Moss (ed.) The soil resources of tropical Africa. Cambridge Univ. Press, Oxford.

Salinas, J. G., and P. A. Sanchez. 1976. Soil-plant relationships affecting varieties and species differences in tolerance to low available soil phosphorus. Ciência e Cultura (Brazil) 28(2): 156-168.

Salmon, R. C. 1973. Phosphate intensity and capacity in some Rhodesian soils. Changes in a sandy tobacco soil due to previous fertilizer additions and subsequent cropping. Rodesian J. Agric. Res. 11:119-121.

Samra, J. S., and T. D. Biswas. 1974. Role of different forms of phosphate in the mechanism of soil agregation. J. Indian Soc. Soil Sci. 22:6-12.

Sánchez, C., and A. Mata. 1972. Efecto del método de colocación, fuente y dósis de fósforo sobre el rendimiento de maní y frijol en un suelo franco arenoso de sabana. Fitotecnia Lat. 8(3):78-84.

Sawhney, B. L. 1974. Charge characteristics of soils affected by phosphate sorption. Soil Sci. Soc. Am. Proc. 38:159-160.

Schalscha, E., C. Nieto, and F. T. Bingham. 1973. Fijación de fosfatos en suelos alofánicos de Chile. Agric. Tec. (Chile) 33:81-86.

Schalscha, E., P. F. Pratt, and D. Soto. 1974. Effect of phosphate adsorption on the cation exchange capacity of volcanic ash soils. Soil Sci. Soc. Am. Proc. 38:539-540.

Shelton, J. E., and N. T. Coleman. 1968. Inorganic phosphorus fractions and their relationships to residual value of large applications of phosphorus on high phosphorus fixing soils. Soil Sci. Soc. Am. Proc. 32:91-94.

Shelton, J. E., N. T. Coleman, and W. H. Rankin. 1961. Rate of phosphorus required to build up phosphorus in a Georgeville soil. Proc. Soil Sci. Soc. North Carolina 4:11-18.

Shetty, K. S., G. Ramanathan, K. K. Krishnamoorthy, J. Helkiah, and S. Vadivelu. 1973. Thermophos, a new phosphatic fertilizer and its comparative efficiency. Madras Agric. J. 60:953-956.

Silva, J. A. 1971. Possible mechanisms for crop response to silicate applications. Proc. Int. Symp. Soil fert. Eval. (New Delhi) 1:805-814.

Sinclair, A. G. 1975. Reaction of fused calcium-magnesium phosphate and superphosphate on a highly phosphate-fixing soil. II. Placement effects. N.Z. J. Exp. Agric. 3:111-116.

Smyth, T. J., and P. A. Sanchez. 1980a. Niveis criticos de fósforo para arroz de sequeiro em um Oxisolo dos Cerrados. Rev. Bras. Ciência Solo 4(1) (in press).

Smyth, T. J., and P. A. Sanchez. 1980b. Effects of lime, silicate, and phosphorus applications to an Oxisol in phosphorus sorption and ion retention. Soil Sci. Soc. Am. J. 44(3) (in press).

Soil Conservation Service. 1972. Soils of the world (map). Soil Geography Unit. Soil Conservation Service, USDA, Washington.

Spain, J. M. 1974. La fertilizacion fosfórica de praderas en suelos álicos. Suelos Ecuat. 6: 235-244.

Spain, J. M., C. A. Francis, R. H. Howeler, and F. Calvo. 1975. Differential species and varietal tolerance to soil acidity in tropical crops and pastures. p. 308-329. *In* E. Bornemisza and A. Alvarado (ed.) Soil management in tropical America. North Carolina State University, Raleigh.

Stoop, W. A. 1974. Interactions between phosphate adsorption and cation adsorption by soils and implications for plant nutrition. Ph.D. Thesis. Univ. of Hawaii, Honolulu. 203 p. (Diss. Abstr. 75-5042) University Microfilms, Ann Arbor, Michigan.

Syers, J. K., T. D. Evans, J. D. H. Williams, and J. T. Murdock. 1971. Phosphate sorption parameters of representative soils from Rio Grande do Sul, Brazil. Soil Sci. 112:267-275.

Terman, G. L., and O. P. Engelstad. 1971. Agronomic evaluation of fertilizers. Principles and practices. TVA Bull. Y-21.

Thein, S. J. 1976. Stabilizing soil aggregates with phosphoric acid. Soil Sci. Soc. Am. J. 40: 105-108.

Toth, S. J. 1937. Anion adsorption by soil colloids in relation to charge in free iron oxides. Soil Sci. 44:299-314.

Truong, B., J. Pichot, and S. Burdin. 1973. Etude des effets residuels du phosphore dans deux sols ferralitiques par diverses methodes analytiques (chimiques et isotopique). Agron. Trop. (France) 28:147-155.

Truong, B., R. Bertrand, S. Burdin, and J. Pichot. 1974. Contribution a l'etude du phosphore dans les sols derives de roches volcaniques de l'Ile de la Reunion (Mascaregnes). Action due carbonate et du silicate de calcium. Agron. Trop. (France) 29:663-674.

Udo, E. J., and F. O. Uzu. 1972. Characteristics of phosphorus adsorption by some Nigerian soils. Soil Sci. Soc. Am. Proc. 36:879-883.

Uehara, G., and J. Keng. 1975. Management implications of soil mineralogy in Latin America. p. 351-363. *In* E. Bornemisza and A. Alvarado (ed.) Soil management in tropical America. North Carolina State Univ., Raleigh.

Vasconcelos, C. A., J. M. Braga, R. F. Novais, and O. Pinto. 1975. Fósforo em dois Latossolos do Estado de Mato Grosso. II. Dessorção de fosfatos. III. Relaçoes entre planta, solo e fosforo. Ceres 22:22-49, 62-73.

Velly, J., and P. Roche. 1973. Comparaison de quelques engrais phosphates sur riz irrigue sur divers types de sols de Madagascar. Agron. Trop. (France) 29:593-606.

Vicente-Chandler, J., R. Caro-Costas, R. W. Pearson, F. Abruña, J. Figarella, and S. Silva. 1964. The intensive management of tropical forages in Puerto Rico. Univ. P. R. Agr. Exp. Sta. Bull. 187.

Viegas, G. P., L. T. de Miranda, and E. S. Freire. 1970. Adubação do milho. XXVI. Ensaios com diversos fosfatos. Bragantia 29:191-198.

Waugh, D. L., R. B. Cate, Jr., and L. A. Nelson. 1973. Discontinuous models for rapid correlation interpretation and utilization of soil analysis and fertilizer response data. Int. Soil Fert. Eval. Impr. Program. Tech. Bull. 7. North Carolina State Univ., Raleigh.

Werner, W. 1969. Die Bedeutung kalk-und silikathaltiger Phosphatdunger fur die Dungung vol Latosolen. Tropenlandwirt. 70:57-61.

Wild, A. 1949. The retention of phosphate by the soil. A review. J. Soil Sci. 1:221-228.

Woodruff, J. R., and E. J. Kamprath. 1965. Phosphorus adsorption maximum as measured by the Langmuir isotherm and its relationship to phosphorus availability. Soil Sci. Soc. Am. Proc. 29:148-150.

Yost, R. S., E. J. Kamprath, E. Lobato, and G. C. Naderman, Jr. 1979. Phosphorus response of corn on an Oxisol as influenced by rates and placement. Soil Sci. Soc. Am. J. 43:338-343.

Younge, O. R., and D. L. Plucknett. 1964. Lay on the fertilizer. Hawaii Farm Sci. Nov. 1964: 15-16.

Younge, O. R., and D. L. Plucknett. 1966. Quenching the high phosphorus fixation of Hawaiian Latosols. Soil Sci. Am. Proc. 30:653-655.

Yuan, T. L. 1973. Chemistry and mineralogy of Andepts. Proc. Soil Crop Sci. Soc. Fla. 33: 101-108.

Chapter 18

Use of Waste Materials as Sources of Phosphorus[1]

L. E. SOMMERS AND A. L. SUTTON
Purdue University
West Lafayette, Indiana

I. INTRODUCTION

In recent years, there has been increased emphasis on the use of agricultural land for disposal of industrial, municipal, and animal wastes. Although land application of many waste materials has been a well-established practice for many years, additional information is needed concerning the fate of nutrients and other components in wastes when added to soils to maintain environmental quality and minimize soil deterioration. Two basic approaches exist relative to the application of waste products on land: (i) a disposal approach—the material is applied at the maximum rate at which the soil is capable of assimilating the waste; and (ii) a recycling approach—the amount of waste added is commensurate with the capacity for plants growing on the soil to remove the nutrients added. Several excellent review articles have been published concerning the application of wastes on agricultural land (Agricultural Research Service, 1974; Bouwer & Chaney, 1974; Elliott & Stevenson, 1977; Sopper & Kardos, 1973b; Loehr, 1973; Nat'l. Assoc. State Univ. and Land Grant Colleges, 1973). The chemical composition of waste materials has been reviewed by McCalla et al. (1977) and Peterson et al. (1971, 1973).

In this review, we will attempt to cover the following areas: (i) chemical composition of waste products; (ii) plant availability of P added to soils in wastes; (iii) reactions of P added to soils in wastes; (iv) constraints on use of wastes as a P source; and (v) potential use of P in wastes as a resource in crop production.

---

[1] Contribution from the Purdue University Agriculture Experiment Station, Journal Paper no. 6273, West Lafayette, Indiana.

Copyright 1980 © ASA-CSSA-SSSA, 677 South Segoe Road, Madison, WI 53711, USA.
The Role of Phosphorus in Agriculture.

II. COMPOSITION OF WASTES

A. Waste Water

Waste waters are byproducts of municipal sewage treatment plants, agricultural processing plants, and industries. The composition of secondary effluents from municipal sewage treatment plants is influenced by the type of treatment employed (Table 1). In comparison with secondary effluents, waste waters from vegetable and fruit processing plants are more variable in chemical composition (Table 2). For the majority of waste waters, total P concentrations are <40 mg/liter with inorganic P constituting the bulk of total P present. Thus, waste waters contain significant amounts of P when effluent discharge into natural waters is considered, but they are a dilute source of fertilizer P. Land application has been utilized by various industries as a waste disposal method to eliminate installation of expensive biological treatment facilities, which are used on a seasonal basis only. In contrast, land application of secondary effluent from municipal sewage treatment plants has been viewed as an alternative to tertiary treat-

Table 1—Composition of effluents from municipal waste water treatment plants (from Pound and Crites, 1973).

Component	Primary	Type of secondary treatment		
		Trickling filter	Activated sludge	Oxidation pond
		mg/liter		
Total N	37	16	23	23
Total P	11	13	13	7
Na	329	267	192	257
K	22	14	20	14
Ca	96	80	52	92
Mg	34	50	37	48
SAR†	7.5	5.6	5.0	5.2

† SAR, Sodium Absorption Ratio, $Na^+/(Ca^{2+} + Mg^{2+})^{1/2}$.

Table 2—Comparison of solids, N and P contents for waste waters from selected vegetable and fruit processing plants (from Soderquist, 1975).

Material processed	Total solids	Total N	P		
			Total	Soluble inorganic	Soluble total
			mg/liter		
Greenbean-A	4,830	112	22	15	18
Greenbean-B	9,450	217	36	24	30
Cherries-A	13,600	82	19	8	11
Cherries-B	18,600	74	35	29	31
Pears	28,450	131	24	18	19
Corn	22,850	318	91	56	70
Blackberries	2,480	19	4	2.6	3.2

ment systems, allowing existing treatment facilities to meet the newly established water quality standards. With the land application approach, the waste water is renovated before entering ground water through utilization of nutrients by a growing crop and retention of particulates (e.g., microbes, suspended solids) by soil colloids as the water percolates through the soil profile. Common systems for applying waste waters to land include spray irrigation, overland flow, and flood irrigation. Due to the dilute concentration of plant nutrients in waste waters, utilization of these wastes in agriculture is commonly limited to land in close proximity to the treatment or processing plant. For example, 55 cm/ha of effluent would be required to supply 50 kg of P/ha with an effluent containing 10 mg of P/liter.

B. Municipal Wastes

The predominant types of municipal solid wastes include refuse and sewage sludge. In some cases, refuse and sewage sludge, alone or in combination, may be composted to yield a more desirable product from the standpoint of odor, stability, and ease of handling. The N, P, and K concentrations in composts and refuse are shown in Table 3. Total P levels range from <0.1 to 0.4% P, suggesting that from 12 to 50 metric tons/ha would be needed to supply 50 kg of P/ha. However, the amount of N required by the crop must be considered in determining acceptable application rates to prevent NO_3^- leaching and contamination of ground water. Furthermore, acceptable rates will vary depending on the metal concentrations in the material.

Sludges are generated in nearly all sewage treatment processes. The composition of sewage sludge is dependent upon the type of treatment process, the composition of the waste water entering the treatment plant, the method of sludge handling, and the efficiency and mode of digester operation. Trickling filter sludge, activated sludge, and other aerobic sludges are generated during secondary treatment of waste water. Primary sludge, alone or in combination with secondary sludge, is commonly subjected to anaerobic digestion. Thus, distinctly different materials are often

Table 3—Composition of municipal composts and refuses.

Type of material	N	P	K	Reference
		% dry weight		
Compost	1.3	0.26	0.97	Terman et al. (1973)
Compost	1.4	0.35	0.37	Mortvedt & Giordano (1975)
Compost	2.0	0.26	0.38	Hortenstine & Rothwell (1969)
Compost	0.4	0.08	0.58	Egawa (1974)
Compost	0.5	0.13	0.17	Egawa (1974)
Compost	1.1	0.23	0.95	Egawa (1974)
Refuse	0.6	0.08	0.81	King et al. (1974)
Refuse	0.76	0.16	0.24	Volk & Ullery, 1972 (unpublished report)

Table 4—Composition of sewage sludges produced by municipal waste water treatment plants (adapted from Sommers, 1977).

Type of digestion	Component	N†	Minimum	Maximum	Median	Mean
Anaerobic	N, %	85	0.5	17.6	4.2	5.0
	P, %	86	0.5	14.3	3.0	3.3
	K, %	86	0.02	2.6	0.3	0.5
	Zn, ppm	108	108	27,800	1,860	3,370
	Cu, ppm	108	85	10,100	1,000	1,420
	Ni, ppm	85	2	3,520	85	400
	Cd, ppm	98	3	3,400	16	106
Aerobic	N, %	16	3.5	7.6	5.4	5.5
	P, %	16	1.8	5.5	3.8	3.6
	K, %	15	0.3	1.1	0.7	0.7
	Zn, ppm	19	730	14,900	1,130	2,100
	Cu, ppm	19	280	2,600	960	970
	Ni, ppm	10	2	390	8	47
	Cd, ppm	19	6	2,170	11	125
Activated	N, %	21	0.5	7.5	4.5	4.6
	P, %	21	1.1	3.8	2.4	2.4
	K, %	21	0.08	0.5	0.3	0.3
	Zn, ppm	33	108	5,500	1,800	2,110
	Cu, ppm	33	85	2,900	975	920
	Ni, ppm	30	10	1,700	46	172
	Cd, ppm	32	5	435	130	160
All‡	N, %	191	0.1	17.6	3.3	3.9
	P, %	189	0.1	14.3	2.3	2.5
	K, %	192	0.02	2.6	0.3	0.4
	Zn, ppm	208	100	27,800	1,740	2,790
	Cu, ppm	205	84	10,400	850	1,210
	Ni, ppm	165	2	3,520	80	320
	Cd, ppm	189	3	3,410	16	110

† N, number of samples.
‡ Includes above types of sludges plus other miscellaneous sludges (e.g., primary). All data are on an oven-dry weight basis.

described by the general term *sewage sludge*. As shown in Table 4, N, P, K, and metal concentrations in sewage sludges are quite variable. This variability is apparent not only among the different types of sludges but also within a given type of sludge. For a given type of sludge, the variability in chemical composition arises from different proportions of industrial and domestic sewage entering the treatment plant. Metal concentrations are generally elevated in sludges from industrialized cities. In addition to the variability in composition of sewage sludge from one source to another, it has also been established that the composition of sewage sludge at a given treatment plant may vary with time (Sommers et al., 1976).

In general, sewage sludges contain from 2 to 4% P; however, the forms of P present remain to be characterized. Preliminary data suggest that in anaerobically digested sludges water-soluble orthophosphate ranges from 1 to 10 mg/liter, with the majority of total P being associated with the solids (Sommers et al., 1972). Even though sludges are commonly referred to as organic wastes, it is interesting to note that from 70 to 90% of the total P is present in inorganic forms (Sommers et al., 1976). Most inorganic P in anaerobically digested sewage sludges can be extracted with relatively dilute acidic reagents (Scott & Horlings, 1975; Sommers et al., 1972, 1976). Because of the relatively high concentrations of Ca, Fe, Al, and Mn in sludge (Sommers et al., 1972, 1976; Page, 1974), inorganic P may be either sorbed onto amorphous hydrous oxides or precipitated as metal phosphate solid phases. Inositol phosphates have been detected in activated sludges (Cosgrove, 1973); however, most organic P compounds present in sewage sludge have not been characterized. It is likely that a part of sludge organic P originates from microbial cells and their degradation products.

C. Animal Wastes

Animal wastes commonly include feces, urine, spilled feed, waste water, bedding, and feedlot runoff water. Management of animal wastes, including collection, storage, treatment, and utilization and/or disposal, has been an important area of research because of larger and more intensified livestock operations and greater public concern about animal wastes as a source of environmental pollutants. Therefore, research has been conducted to develop efficient and economical methods to either preserve and utilize nutrients for crop production or treat and dispose of wastes while maintaining environmental quality.

The chemical composition of animal wastes varies greatly. Factors contributing to this variation include (i) species of animal; (ii) ration fed to the animal; (iii) housing and waste management system; (iv) amount of bedding, water, and feed spillage; and (v) climate. Phosphorus is excreted mainly in the feces of herbivores, such as ruminants and the horse, while omnivores, such as swine, excrete significant amounts of P in both urine and feces (Table 5). For fowl, of course, all P excreted is contained in fecal droppings. Table 6 shows a summary of selected chemical components in freshly excreted manure from various farm animals. Differences in nutrient composition among species can be attributed to differences between the physiology and/or ration of the animals. Differences in physiological mechanisms for water retention and excretion result in horse, poultry, and sheep manures containing less moisture than manures from swine and beef or dairy cattle. Manures from ruminant animals are generally lower in P

Table 5—Distribution of P in feces and urine of cattle, horses, sheep, and swine (from Azevedo and Stout, 1974).

Species	Total phosphorus	
	Feces	Urine
	% excreted	
Cattle	97.3	2.7
Horses	100.0	Trace
Sheep	93.8	6.2
Swine	83.1	16.9

Table 6—Composition of fresh animal wastes.[†]

Source		Solids	N	P	K	Na	Ca	Mg	S	Cu	Zn
					%					mg/kg	
Cattle:											
Beef	Range	12-27	2-8	0.5-1.6	2-4	0.1-2.8	0.6-1.4	0.4-0.7	0.4-0.6	23-33	68-100
	Mean	16	4.2	0.9	2.6	0.8	0.8	0.5	0.5	28	84
Dairy	Range	10-16	3-4	0.4-0.7	2-3	0.1-1.3	1.3-1.7	0.3-0.7	0.2-0.5	28-33	83-133
	Mean	14	3.5	0.6	2.4	0.5	1.5	0.5	0.3	30	108
Swine	Range	6-28	2-10	0.6-2.5	1-6	0.6-2.9	0.3-3.2	0.3-0.5	0.5-0.8	18-163	215-805
	Mean	13	5.2	1.5	3.2	1.5	2.0	0.4	0.6	90	546
Sheep	Range	24-29	4-6	0.4-0.9	2-4	--	0.8-2.5	0.3-0.8	0.3-0.4	--	--
	Mean	26	4.4	0.6	3.0	0.8	1.7	0.5	0.3	16	81
Poultry											
Layers	Range	24-29	3-6	0.9-2.3	1-2	0.5-0.9	3.4-6.4	0.5-1.1	--	18-71	120-330
	Mean	26	4.8	1.8	1.8	0.7	5.5	0.7	0.5	40	225
Broilers	Range	--	3-6	1.1-2.1	1-2	0.5-0.9	1.4-2.1	0.4-1.1	--	18-105	125-253
	Mean	25	4.4	1.7	1.9	0.7	1.9	0.7	0.4	46	168
Turkeys	Range	22-27	5-6	0.6-2.0	1-3	--	--	--	--	--	--
	Mean	25	5.4	1.2	1.9	0.5	2.8	0.6	--	70	520
Horses	Range	21-40	2-3	0.3-0.5	1-2	--	0.7-2.9	0.3-0.4	0.2-0.3	--	--
	Mean	30	2.4	0.4	1.5	--	1.9	0.3	0.2	19	56

[†] All values expressed as % or mg/kg dry weight. Compilation of data obtained from the following sources: Powers et al., 1975; Azevedo & Stout, 1974; Graber, 1974; Menzies & Chaney, 1974; Proctor. 1965; Walsh et al., 1975; Miner & Smith, 1975; Sutton, unpublished data; Muehling, 1969; Sutton et al., 1974, 1975; Kornegay et al., 1974; Holland et al., 1975; Hashimoto, 1974; Berry et al., 1974; Horton et al., 1975; Nelson et al., 1955.

than manures from poultry or swine because ruminants have the ability to extract organically bound P from feeds (Azevedo & Stout, 1974).

Feed additives, such as Cu, As, Zn, Fe, antibiotics, sulfa drugs, etc., are included in animal rations for disease prevention and growth promotion. The level of Zn and Cu in manure is a reflection of the amount present in the ration (Kornegay et al., 1974; Menzies and Chaney, 1974). In addition to influencing manure composition, addition of As to swine rations has reduced the dry matter content of waste in anaerobic deep pits (Brumm et al., 1977). For ruminants, the amount and composition of waste produced can be influenced by the physical form and relative digestibility differences of high roughage rations compared with high concentrate rations (Grub et al., 1969; Azevedo & Stout, 1974; Fisher, 1974; Frecks & Gilbertson, 1973). The nutrient content of wastes will be diluted by additions of bedding and water, whereas feed spillage adds a concentrated source of nutrients (El-Sabban et al., 1969). The type of housing, feedlot, and waste treatment and storage system will also influence waste composition (Powers et al., 1975; Vanderholm, 1975; Adriano, 1975; Sutton et al., 1975).

Nitrogen content, more than any other waste component, is sensitive to management between the time of excretion and land application. Loss of N from animal wastes is generally greatest under partial-treatment systems, both aerobic and anaerobic, such as oxidation ditches, digesters, and lagoon systems (Miner & Smith, 1975; Jones et al., 1971; Vanderholm, 1975). A summary of available literature suggests that N losses during manure storage and application can range from <1 to 90%. In contrast, trace metals and P are not generally affected by different waste-management systems. Both P and trace metals tend to form sparingly soluble compounds through precipitation or sorption reactions and thus, dewatering of wastes will not influence the P and trace metal content of the solids. In contrast, the high water solubility of Na^+, K^+, SO_4^{2-}, and H_3BO_3 results in their enrichment in the liquid portion of wastes rather than the solid phase.

Powers et al. (1975) indicated that most P in animal wastes is bound to, or is a part of, relatively insoluble compounds. Other studies have shown that the bulk of P in liquid waste is associated with suspended or sludge solids (Mutlak et al., 1975; Loehr, 1969; Howell et al., 1974). A relatively constant soluble P concentration was found in established swine lagoon systems operating from 6 to 11 years (Booram et al., 1975). Other nutrients and chemical oxygen demand (i.e., organic C) tended to increase during this time, and 89% of the total P was present in the lagoon sludge mass. In these lagoons, an apparent equilibrium was established between soluble orthophosphate and P compounds contained in the sludge, possibly $MgNH_4PO_4 \cdot 6H_2O$. This P solid phase was often found precipitated in waste recycling plumbing systems. Ferguson et al. (1973) indicated that several Ca^{2+}- and Mg^{2+}-P compounds could be formed under conditions existing in lagoons.

The forms of organic P in animal wastes are not well defined. In a comparison of fresh manures from several species of animals, Peperzak et al. (1959) found that organic P constituted approximately 30% of the total P

Table 7—Phosphorus fractions in fresh animal manures (adapted from Peperzak et al., 1959).

Species[†]	Total P	Inorganic P	Organic P				
			Total	IHP[‡]	Other[§]	Alcohol soluble	Residual
			—————— mg P/g ——————				
Poultry (17)	28.52	22.63	5.89	3.51	0.71	0.12	1.55
Sheep (19)	11.92	7.53	4.39	0.20	2.21	0.05	1.93
Swine (34)	10.95	9.08	1.87	0.07	1.41	0.05	0.34
Horse (40)	7.25	6.89	0.36	0.08	0.08	0.06	0.14
Beef steer (28)	11.52	7.37	4.15	1.16	1.38	0.09	1.52
Dairy cow (44)	6.08	5.10	0.98	0.09	0.50	0.03	0.36
Mean by classes of livestock	9.62	6.98	2.64	0.52	1.03	0.07	1.02

[†] Numbers in parenthesis are sample numbers employed by Peperzak et al. (1959); mean based on additional data presented by authors.
[‡] IHP, inositol hexaphosphate.
[§] Other signifies the difference between IHP-P and acid-soluble P.

(Table 7). Phospholipid- and inositol hexaphosphate-P were detected in all manures, but most of the organic P was of unknown chemical structure. Since manure begins to decompose at the site of deposition, changes in the forms of P can occur in a relatively short time. McCalla et al. (1970) discussed the general end-products of manure decomposition under aerobic and anaerobic conditions. In both aerobic and anaerobic decomposition, organic P compounds are hydrolyzed, releasing orthophosphate. Salter and Schollenberger (1939) found that composting of horse and cow manure mixtures resulted in a 32% increase in extractable P. Concurrently, total and available N, organic matter, and the C/N ratio decreased in the manure. Wells et al. (1969) noted an increase of total P as a percentage of dry matter during a 10-day aerobic composting of cattle feedlot waste. In a long-term study, Peperzak et al. (1959) found that both total and inorganic P concentrations of manure piles increase with time for the first 10 years of stockpiling, whereas the P concentrations tended to decrease after 15 to 20 years.

Decomposition-induced changes of P forms in liquid waste systems have not been completely documented. However, Loehr (1971) noted an increase in inorganic P during anaerobic and aerobic decomposition of liquid animal wastes. Most researchers reported P concentrations in waste generally list total P or orthophosphate, with very few indicating any separation or fractionation of P components. The P concentrations in waste from liquid-holding tanks prior to land disposal have been shown to vary considerably within and among different species of animals (Townshend et al., 1969). Soluble P, expressed as a percentage of total P, averaged 43, 18, 54, and 46% for swine, poultry, beef cattle, and dairy cattle, respectively. Their data also indicated that the length of retention time and management of the waste system affected the distribution of P in animal wastes. Norstadt et al. (1971) found a 33% loss of orthophosphate in anaerobic lagoons. However, Vanderholm (1975) concluded that any precipitation of P in lagoon systems is not a true loss as the nutrients can be recovered if the lagoon bottom sludges are removed. In liquid-holding systems where the total waste mass was recovered, virtually all P is recovered for utilization (Howell et al., 1974; Brumm et al., 1977). Vanderholm (1975) commented that even though some losses are likely in open-lot and lagoon systems, there were insufficient data to estimate average losses. True losses of P can occur if manures are subjected to leaching in a storage area. Potential leaching losses may approach 58% of the total P for uncovered outdoor storage of solid manure (Salter & Schollenberger, 1939). With continued leaching of dairy cattle feces, a 50% reduction in total P has been obtained (Azevedo & Stout, 1974).

III. PLANT AVAILABILITY OF PHOSPHORUS IN WASTES

A. Waste Water

Effluents from municipal treatment plants and from selected industrial processing plants (e.g., vegetables, animal products, etc.) have been applied to agricultural land resulting in a twofold benefit: (i) renovation of the waste water, and (ii) increasing yields of agronomic crops grown on the disposal site. A recent review by Bouwer and Chaney (1974) summarizes the various systems, benefits, and problems encountered with waste water applications on agricultural land.

In a waste water irrigation project initiated in 1963 at the Pennsylvania State University to evaluate the efficacy of utilizing secondary effluent for production of agronomic crops and forest species, Sopper and Kardos (1973a) showed that increased yields of many crops can be obtained using spray irrigation of secondary effluent at weekly rates of 2.5 to 5 cm/ha. The amount of P applied, yield, and P uptake by crops growing on soils irrigated with secondary effluent are shown in Table 8. Soils were cropped to

Table 8—Phosphorus uptake by crops irrigated with secondary effluent (from Hook et al., 1973).

Year	Agronomy area†				Reed canarygrass area			
	Effluent applied	P in effluent	P applied	P uptake	Effluent applied	P in effluent	P applied	P uptake
	cm	mg/liter	—— kg/ha ——		cm	mg/liter	—— kg/ha ——	
1963	122	9.70	120	30	--	--	--	--
1964	168	8.55	146	21	91	8.55	79	--
1965	147	6.95	104	31	203	6.95	143	46
1966	163	5.35	89	36	198	7.70	155	38
1967	132	6.75	87	28	239	7.70	187	64
1968	102	7.10	73	--	249	8.45	214	54
1969	81	6.55	54	49	254	4.20	108	54
1970	81	4.15	34	40	218	4.05	101	64
Total	996	--	707	235	1,452	--	987	320

† Crops: 1963-1965, corn-small grains-legumes; 1966-1967, alfalfa-corn; 1968-1971, corn.

either agronomic species or Reed canarygrass. In comparison with fertilizer applications, large amounts of P were added to soils from secondary effluent, and a large proportion of the added P was utilized by the crop growing on the disposal site. Data from the early stages of the project when a variety of agronomic crops were grown indicated that silage corn and Reed canarygrass were the most efficient species for removing N and P from secondary effluent. As shown by the 1970 P-uptake data for the "Agronomy Area," P applied in previous years was available for plant uptake if the amount of P required by the crop exceeded P applied in secondary effluent. Similar yield responses to secondary effluent have been obtained for other crops (Day, 1973).

B. Municipal Wastes

A significant proportion of the P in municipal composts applied to soil is available to plants. In a greenhouse study, grain sorghum yields increased as the rate of compost application increased from 8 to 64 metric tons/ha (Hortenstine and Rothwell, 1973). Increased yields were attributed to increasing levels of plant-available N from compost additions rather than to P or K, even though 36 to 288 kg of P/ha was added by the compost. Since N was the factor limiting yield, added P utilized by the sorghum decreased from 25 to 9% as the compost application rate increased from 8 to 64 metric tons/ha. In a similar study using a low N compost (2% N), yields of oats, turnips, radishes, and pearl millet forage were decreased at rates <32 metric tons/ha. Yield decreases were attributed to N deficiency resulting from immobilization of N during decomposition of organic C added by the compost (Hortenstine & Rothwell, 1969). However, forage yields and P uptake were increased when compost was applied at rates of 128 and 512 metric tons/ha. Terman et al. (1973) conducted a series of greenhouse experiments to evaluate the availability of compost N, P, and K to corn and tall fescue. In a soil containing low levels of available N, compost addition induced N deficiency in corn forage. Addition of soluble N increased yields, but a portion of the added N was immobilized during decomposition of compost organic C. The N, P, and K in compost were 16, 71, and 64% as available as nutrients added in NH_4NO_3, concentrated superphosphate (CSP), and K_2SO_4, respectively. In addition to providing N, P, and K, compost applied at 36 metric tons/ha served as a liming material, increasing soil pH by 1 to 2 units. Experiments utilizing a P-deficient soil indicated that P uptake and crop yield increased with compost additions; however, the availability of P in compost was lower than in an equivalent amount of P added as superphosphate. This is expected if hydrolysis of the organic P in compost is necessary before plant uptake of P.

Field studies have indicated positive yield responses to application of municipal composts. With supplemental N and K fertilization, yields of sweet corn and bush beans were either unchanged or increased by compost addition at rates ranging from 56 to 448 metric tons/ha (Giordano et al., 1975). The effect of compost and supplemental N on yield and P uptake by

Table 9—Uptake of P by forage sorghum from soils treated with compost and N (from Mays et al., 1973).

Compost added	NH$_4$NO$_3$-N added	P added in compost	Total yield (1969-1971)	P uptake†
metric tons/ha	kg/ha			metric tons/ha	kg/ha
0	0	0	27.8	55.6
23	0	79	31.9	63.8
36	0	90	34.3	68.6
41	0	124	35.5	71.0
82	0	249	37.3	74.6
164	0	498	40.8	81.6
327	0	992	45.9	91.8
0	90	0	36.3	72.6
46	90	159	42.5	85.0
0	180	0	40.5	81.0
46	180	159	45.0	90.0
164	180	498	49.3	98.6

† Assuming P in plant tissue averaged 0.2%.

forage sorghum is shown in Table 9. Although P uptake increases with increasing amounts of compost added, the data suggest that N availability limited yields and, consequently, P utilization. Additional studies have indicated that forage sorghum, common bermudagrass, rye, and corn yields are increased by applying composts at rates ranging from 9 to 112 metric tons/ha (Mays et al., 1973; King et al., 1974). With all nonleguminous crops, greater yields are obtained if N fertilizer is applied with the compost. However, yield depression may be possible at compost rates >400 metric tons/ha (Mays et al., 1973). In addition to studies of agronomic crops, compost applications have been shown to improve the physical and chemical properties of strip-mine spoil, facilitating revegetation.

Application of sewage sludge on agricultural land is not a new practice but it is receiving increased emphasis because of economic and environmental advantages, as compared with alternative disposal methods. In a recent review of sewage sludge application practices, Carroll et al. (1975) summarized data indicating that 21% of the treatment plants in the United States use land application of sludges as a disposal method. Land application appears to predominate in the North Central States, with approximately 65% of the treatment plants using this approach. The potential benefits and problems arising from application of sewage sludge on agricultural land have been discussed in several reviews (Dean & Smith, 1973; Melsted, 1973; Chaney, 1973; Hinesly & Sosewitz, 1969; Walker, 1975; Page & Chang, 1975; Sopper & Kardos, 1973a; Kirkham, 1974).

Early studies indicated that increased crop yields resulted from sewage sludge application, but additional N, P, or K was needed in order to obtain maximum yield (Muller, 1929; Anderson, 1955). More recently, field studies have indicated increased yields when sewage sludge has been applied to the following crops: corn (Singh et al., 1975; Hinesly et al., 1972; Jones et al., 1975), Coastal bermudagrass and rye (King & Morris, 1972a, 1972b;

Table 10—Effect of annual sewage sludge additions on corn grain yields (Hinesly et al., 1972).

Year	Maximum sludge applied		Sludge rate, % of maximum			
			0†	25	50	100
	metric tons/ha			kg/ha		
1968	53		4,420	6,410	7,610	7,460
1969	48		9,520	9,930	10,000	10,040
1970	71		5,880	7,950	8,100	9,170
1971	99		6,440	6,910	7,360	8,370
Total	271	Average	6,570	7,810	8,270	8,750

† Plots received 188 kg P and 166 kg K/ha per year in 1969-1971.

Coker, 1966), sweet corn (Giordano et al., 1975; Singh et al., 1975), winter wheat (Sabey & Hart, 1975), fescue (Boswell, 1975), and potatoes (Dowdy & Larson, 1975a). Application rates of sludge ranging from 49 to 99 metric tons/ha have consistently supported corn grain yields significantly higher than fertilized plots (Table 10). In 1968, the control was not fertilized which resulted in a greater difference between corn yields on control and sludge plots than in 1969-1971 where the control received NPK. Surface application of liquid-digested sewage sludge to Coastal bermudagrass sod which was subsequently seeded to rye resulted in increased yield and P uptake (Table 11). compared with the check and fertilized plots. However, when application rates reached 266 metric tons/ha, slight yield depressions occurred. Sludge applications of >87 metric tons/ha resulted in yields exceeding those obtained by conventional NPK fertilization. Data in Table 11 are typical for the amounts of P that are added to soils in sewage sludge. Obviously, the percentage of added P entering the plant will decrease when P is applied at these high rates. In most situations, P is not the growth-limiting nutrient, and other nutrients, especially N, must be added in order to obtain maximum yields. Even though sufficient N is added to most soils in sewage sludge to support maximum crop yield, the efficiency of N utilization by the crop is considerably lower for N in sewage sludge than in conventional fertilizer materials (King & Morris, 1972a; Coker, 1966). When sludge is surface-applied, ammonia volatilization appears to be a significant loss

Table 11—Effect of sewage sludge on yield and P uptake of forage rye and Coastal bermudagrass (from King and Morris, 1973).

Sludge applied	P applied	Yield†	P uptake	
			kg/ha	% of applied
metric tons/ha	kg/ha			
0	0	2,320	4	--
0+NPK	174	23,760	56	32
44	382	19,400	53	14
87	765	28,820	80	10
133	1,172	30,380	82	7
266	2,343	26,850	88	4

† Total yield for Coastal bermudagrass (King & Morris, 1972a) and rye (King & Morris, 1972b) grown in 1970 and 1971. Liquid sludge was applied in 1969 and 1970.

mechanism. However, there are insufficient data to evaluate adequately the magnitude of ammonia losses under field conditions, although laboratory studies have indicated that NH_3 volatilization may range from 11 to 67% of NH_4^+-N in surface-applied sludges (Ryan & Keeney, 1975). In addition, soil water and oxygen relations following liquid sludge applications may result in significant losses of N through denitrification and/or nitrate leaching.

Most of the P present in sewage sludge appears to be available for plant uptake. However, there is a lack of data using either P-deficient soils and/or sewage sludge plus N treatments to evaluate P availability. Furthermore, nearly all sewage sludges are low in K and, thus, K may be the growth-limiting nutrient in some cases. An additional complicating factor in evaluating the effect of sewage sludge on crop yield is the presence of metals which may be essential at low concentrations but toxic at elevated levels. In several of the studies cited above, yield reductions have been encountered if high rates of sewage sludge were applied to soils. The extreme variability in metal composition of different sludges (Table 4) precludes any general statement concerning the application rates that will result in decreased crop yields. This aspect of sludge application on land will be discussed in a later section.

The accumulation of P with continued application of sewage sludge may result in yield depressions caused by excessive P or by P-metal interactions. The well-known P-induced Zn deficiency may occur in sludge-treated soils; however, most sludges contain sufficient Zn concentrations to overcome P-Zn interactions. Examples of P-Zn problems in waste-treated soils are discussed by Olsen and Barber (1977). Evidence from soils receiving annual applications of sewage sludge suggests that reduced yields of soybeans may be caused by P toxicity (Hinesly et al., 1976). Corn yields have not been influenced by the same sewage sludge treatments (T. D. Hinesly, personal communication). Furthermore, soybean yield depressions have occurred sporadically, indicating that a specific set of soil-plant-climatic conditions must be present before P toxicity symptoms occur. In view of these data, it may be advisable to set a limit for the maximum amount of P that can be applied to soil. Sufficient data are not available to assess accurately the impact of excessive P applied to soils in waste materials.

C. Animal Wastes

Numerous studies have been conducted evaluating the effects of animal wastes on plant yield and composition. Obviously, the impact of wastes on crop yields is directly related to the rate of application in conjunction with the NPK content of the waste. Beneficial effects of waste application can be evaluated from crop yield and uptake data collected during the year of application. In addition, the organic nature of N and P in wastes dictates considering the release of plant nutrients in subsequent years as a potential benefit of using animal wastes in crop production.

Greater total amounts of nutrients, especially N, must be added in animal wastes in order to obtain yields similar to those obtained with commercial fertilizers. In the year of application, manure N is 20 to 50% as effective as N in commercial fertilizers in promoting crop yields (Azevedo & Stout, 1974). In comparison with solid manures, liquid slurries of animal wastes contain elevated levels of inorganic N resulting in N utilization efficiencies of 44 to 84% relative to commercial N sources (Herriott & Wells, 1962; Herriott et al., 1963, 1965). In contrast to N, the plant availability of P in animal wastes and in commercial materials is similar when equivalent amounts of P from either source are applied to soils (Azevedo & Stout, 1974). However, the variables included in most studies only allow a comparison of varying waste application rates to a single rate of fertilizer P.

Annual applications of animal wastes at rates supplying sufficient N, P, and K result in crop yields comparable to those obtained with commercial fertilizer materials (Cooke, 1976). Phosphorus uptake by and yields of pearl millet and rye were evaluated in soil fertilized with NPK or annual rates of dairy cattle manure ranging from 22 to 267 metric tons/ha (Mugiwra, 1976). Total yields and P uptake tended to remain constant at rates >44 metric tons/ha (Table 12); however, N and K uptake continued to increase with increasing amounts of manure applied. Annual application of fertilizer P was 50 kg/ha whereas the manure-treated plots received on the average, 218 and 2,643 kg of P/ha in 22 and 267 metric tons of manure/ha, respectively. With beef cattle manure, corn silage yields increased with increasing amounts of ordinary superphosphate (OSP) or manure-P, whereas the same treatments tended to decrease corn grain yields (Table 12; Vitosh et al., 1973).

In the first year of waste application, a linear relationship has been observed between amount of beef waste applied and P uptake by corn,

Table 12—Effects of manure on crop yield and P uptake.

Crop	Annual application rate	Yield	P applied	P uptake	Comments
	— metric tons/ha —		—kg/ha—		
Pearl millet	0+OSP	63.5	150	132	Mugiwra, 1976
+ forage rye	22	65.6	224	203	Data from 3 consecutive
	44	74.5	448	247	years of dairy cattle
	89	79.2	908	254	manure application and
	178	79.3	1,816	253	cropping
	267	80.5	2,723	248	
Corn grain	0+OSP	46.3	177	125	Vitosh et al., 1973
	0+OSP	42.7	842	115	Yields and estimated P
	22	47.8	302	129	uptake after 9 annual
	45	47.0	605	127	applications of beef
	67	46.1	907	124	cattle manure
Corn silage	0+OSP	91.9	177	147	As above
	0+OSP	101.4	842	203	
	22	97.3	302	165	
	45	100.1	605	190	
	67	105.6	907	222	

whereas a quadratic relationship existed the second year after application (Wallingford et al., 1974). Reddell (1974) applied high rates (>500 metric tons/ha) of cattle manure to soil cropped to sorghum and corn and noted a quadratic effect on yield, but no specific trend for the P content of the forage. In a study with high application rates of poultry manure, elevated concentrations of P, K, Na, Ca, and Mg in the soil were related to decreased corn yields, indicating that salt accumulation was the primary cause of yield reductions (Shortall & Liebhardt, 1975). In addition, P and Mn concentrations in leaf tissue were correlated with reduced yields. Soil salinity from high applications of animal wastes has been shown by several researchers to reduce yields, especially in arid or semiarid regions (Murphy et al., 1972; O'Callaghan et al., 1971; Powers et al., 1974; Tiarks et al., 1974).

Hensler et al. (1971) summarized research showing the effects of waste handling system, bedding materials, and other management factors on the recovery of N, P, and K by plants. Dry matter yields and recovery of N, P, and K were similar for fresh and fermented liquid wastes, but materials treated in aerobic systems depressed yields and nutrient uptake. Carbonaceous materials (bedding sources) in various manure systems had little effect on corn yields and recovery of P. Increasing the amounts of manure increased total dry matter yields of corn but decreased the nutrient recovery.

The amount of P applied to soils in animal wastes often exceeds the amount of P needed for crop growth. Thus, the residual value of manure P is of considerable importance. Research conducted at the Rothamsted Experiment Station has included evaluating crop yields on soils receiving farmyard manure for more than 100 years. A 4-year experiment with wheat, barley, beans, and sugar beets was conducted on soils treated with farmyard manure for 65 years and then subjected to annual treatments of 14 metric tons/ha of farmyard manure (Williams & Cooke, 1970). Soils previously treated with manure supported higher crop yields than comparable soils receiving twice as much P in the form of OSP. Beans, barley, and sugar beets yielded more with manure plus N than with commercial fertilizer. Wheat yields were similar for manure and fertilizer treatments. In other long-term studies, the residual P from manure has maintained or increased crop yields (Warren & Johnston, 1962, 1966; Mattingly et al., 1969). As expected, the residual effect of P in manure has been demonstrated to increase with increasing rate of manure application.

More recent studies suggest that continued use of animal wastes not only provides P for plant growth but also enhances availability of soil or fertilizer P. In field plots receiving combinations of N, P, and manure, increased yields of cotton and barley were the result of manure supplying P and maintaining the added P in forms more available to plants than an equivalent amount of P added in commercial fertilizers (Abbott & Tucker, 1973). The authors suggested that manure enhanced microbial activity in soils, resulting in greater solubility and thus mobility of native soil P. Based on leaf P concentrations, applying manure at 2- or 3-year intervals at 22 metric tons/ha (67 to 112 kg of P/ha) appeared to insure adequate P availability. Similar results relative to yields and P uptake have been obtained for corn grain and silage (Vitosh et al., 1973), crop rotations (MacLeod et al., 1960; Bishop et al., 1962), and oats and sorghum (Overman et al., 1971). Representative data for the effects of manure on crop yield and P uptake are shown in Table 12.

IV. WASTE-INDUCED CHANGES IN SOIL PHOSPHORUS

During application of wastes to soils, the amounts of P added are often in excess of those used in a conventional fertility program to supply adequate P for crop growth. Thus, considerable interest has been directed toward evaluating the forms of P in soils treated with wastes and determining the applicability of P sorption isotherms and plant-available P extractants to waste-amended soils. Furthermore, the ability to predict P retention characteristics for soils under consideration as a waste disposal site is needed to minimize the potential for P contamination of ground waters.

Data from the Pennsylvania State University waste water irrigation project indicate that minimal amounts of P are leached from soils irrigated with waste water for 9 to 11 years. As summarized by Kardos and Hook (1976), <1% of added P was leached below the 120-cm depth in a clay loam soil cropped to corn. In Reed canarygrass plots, <0.1% of the added P was leached. Suction lysimeter samples obtained from soil profiles indicated that most of the effluent P was retained in the upper 30 cm of a clay loam soil profile, and soluble P levels at a 120-cm depth were comparable to control plots not irrigated with effluent. However, in a sandy loam soil, significant amounts of P moved to a depth of 120-cm. In addition to soil differences, i.e., less P sorption sites associated with Fe and Al oxides, forest species were grown on the sandy loam soil and, thus, P was recycled in the ecosystem while P was removed from the agronomic crop site during harvest. Bray P_1-extractable P increased in the upper 30 cm of the clay loam profile, whereas extractable P was increased at the 90- and 120-cm depths in the sandy loam profile. Fractionation of inorganic P indicated that most P added to the soils in waste water was extractable with NH_4F, suggesting that Al components in the soil were involved in retention of P added in waste water (Hook et al., 1973).

Since P retention by soils is a major advantage of utilizing spray irrigation as an alternative for disposal of secondary effluent, considerable emphasis has been placed on utilization of sorption isotherms in order to predict soil types that would be amenable to receiving waste water. In general, P sorption isotherms (e.g., Langmuir) tend to underestimate the capacity for soils to sorb P under waste water irrigation conditions (Ellis, 1973; Sawhney & Hill, 1975). It has been suggested that wetting and drying cycles, occurring after P has been sorbed, release Fe, Al, Ca, and other cations to expose or generate new surfaces for equilibration with the soil solution and thus create new sites for P sorption. The regeneration of P sorption sites has been demonstrated in soils irrigated with milk and vegetable processing waste water for 10 to 20 years (Adriano et al., 1975). The data indicated that

Table 13—Extractable P levels in soils amended with waste materials.

Waste added	Waste application Rate	Waste application P added	P extracted†	Reference
	metric tons/ha	kg/ha	µg/g	
None	0	0	16	Hinesly et al., 1972
Sewage sludge	43	1,885	51	Extractant: 0.1N HCl
	86	3,770	148	
	172	7,540	376	
None	0	0	3.5	King et al., 1974
Sewage sludge	11	327	7.0	Extractant: 0.5M
Refuse	188	151	3.5	NaHCO$_3$
Refuse and sludge	199	478	7.0	
Refuse and sludge	398	956	10.5	
Refuse and manure	192	274	5.2	
None	0	0	6	Mattingly et al., 1969
Manure	368	797	12	Extractant: 0.5M
Manure + CSP	368	1,983	41	NaHCO$_3$
Manure + CSP	299	2,531	38	
None (+OSP)	0	177	81	Vitosh et al., 1973
None (+OSP)	0	842	152	Extractant: Bray P$_1$
Manure	22	302	110	
	45	605	133	
	67	907	146	
None	0	0	4.8	
Manure	22	67–112	6.9	Abbott & Tucker, 1973
	44	134–224	8.4	Extractant: 0.5M
	58	174–290	10.4	NaHCO$_3$
	93	281–470	19.3	

† Soil samples from 0–15 cm.

concentrations of Bray P_1-P in the soils were twice the P level predicted from sorption isotherms.

In contrast to continual addition of small amounts of P in waste water, a single application of compost, refuse, sewage sludge, or animal waste may result in substantial increases in soil P. As summarized by Page and Pratt (1975), most research data obtained to date support the premise that P is relatively immobile in the majority of soils with P remaining in the zone of waste incorporation. Even though increases in 0.5M NaHCO$_3$-extractable P have been detected to depths of >3 to 4 m in sandy soils underlying sewage sludge ponds (Lund et al., 1976), P generally accumulates in the plow layer of soils treated with waste products. Representative data for P additions and extractable P levels in soils treated with waste products are presented in Table 13. It is apparent that significant increases in soil P occur after typical rates of waste application to soils. For surface-applied sewage sludge, most of the P applied remains in the sludge crust that accumulates on the soil surface. For example, King and Morris (1973) observed that from 49 to 55% of the P added in sewage sludge applied on Coastal bermudagrass sod was present in the sludge crust after 2 years of crop growth. In fact, P in the sludge crust was approximately three times the level found in the 0- to 15-cm soil increment. The results of these experiments demonstrated that P was

Table 14—Soluble P and total P in Barnfield plots, Rothamsted, treated with manure from 1843–1959.†

Plot	Annual rate		Total P	Soluble P	
	Manure	P		$0.01M$ $CaCl_2$	$0.5M$ $NaHCO_3$
	metric tons/ha	kg/ha	μg/g	μg/mliter	μg/g
8/0	0	0	780	0.019	23
1/0	31.4	0	1,240	0.422	83
2/0	31.4	33	1,950	0.476	140
2/N	31.4	33	1,840	0.640	132
6/0	0	33	1,220	0.061	66

† Data adapted from Olsen and Barber (1977) and Warren and Johnston (1962). All soils were pH 7.5 when sampled.

relatively immobile in soils, even though the pH of these soils was relatively acid (pH 5 to 6) after sludge application. In most animal waste application studies, increased P levels in soils are not encountered below the 30-cm depth (Mathers & Stewart, 1974; Wallingford et al., 1974; Chesnin & Anderson, 1975; Sutton et al., 1974).

The effect of short- and long-term applications of manures on soil P has been reviewed by Olsen and Barber (1977). In general, annual applications of manure and superphosphate result in increased levels of $0.01M$ $CaCl_2$- and $0.5M$ $NaHCO_3$-extractable P. In most studies, manure-treated soils tend to support a higher level of soluble P than soils treated with an equivalent amount of superphosphate. Data for $0.01M$ $CaCl_2$- and $0.5M$ $NaHCO_3$-extractable P in soils treated with manure, OSP, and manure + OSP are shown in Table 14. Obviously, manure treatment altered P equilibria even though comparable total P additions were made in manure or OSP. Equilibrium calculations indicated that P in soils treated with manures was in equilibrium with octacalcium phosphate rather than dicalcium phosphate or hydroxyapatite. In general, a linear relationship existed between increases in $NaHCO_3$-extractable P and total P due to manure applications, realizing that the absolute amounts of $NaHCO_3$-extractable P will be a function of soil properties. The sorption and mobility characteristics of soluble organic P compounds in soils treated with manure have been implicated as altering the inorganic P forms in soils, resulting in increased levels of $NaHCO_3$- and $CaCl_2$-extractable P. Applications of manure, with and without OSP, have resulted in soil organic P increases of 18 to 44 μg of P/g in the upper 0- to 23-cm layer (Oniani et al., 1973). Neither manure nor OSP influenced the inositol penta- and hexaphosphate content of soil.

V. CONSTRAINTS ON APPLICATION OF WASTES ON AGRICULTURAL LAND

Utilization of waste materials for the production of agricultural crops requires judicious management and a knowledge of the potential problems that may be encountered after repeated applications of waste materials. In view of these potential problems, U.S. federal and state regulatory agencies have attempted recently to develop guidelines for recommending the rates which waste materials may be applied to agricultural land.

The problems encountered during waste application can be divided into the following categories (i) nitrate leaching, (ii) metal accumulation, and (iii) salt accumulation. Additional considerations include alterations in soil physical properties associated with soluble salts and the presence of pathogens, especially viruses. The significance of pathogens added to soil in wastes has not been completely resolved (Burge, 1974; Miller, 1973). Furthermore, accumulation of P can be a problem with waste application; however, in most instances one of the following constraints will limit utilization of wastes in crop production.

A. Nitrate Leaching

Contamination of ground water with NO_3^- is a potential problem in soils treated with rates of waste materials supplying N in excess of the crop requirement. An example of NO_3^- leaching was illustrated by spray irrigation of food processing waste where a crop was not harvested and removed (Adriano et al., 1975). Under these conditions, NO_3^--N in subsurface waters ranged from 7 to 16 and 2 to 44 mg of N/liter, constituting from 65 to 76% of the total N added to the soils. The potential for NO_3^- leaching in soils receiving sewage sludge and other types of wastes has been demonstrated in several field projects (Sabey & Hart, 1975; King & Morris, 1972c; Hinesly et al., 1972; Manges et al., 1972; Olsen et al., 1970). Even though denitrification may account for part of the N loss from soils receiving wastes, there does not appear to be significant denitrification occurring at lower depths in the soil profile, indicating that microbial energy sources (i.e., organic C) are not moving simultaneously with NO_3^-. Thus, NO_3^- leached below the rooting zone will likely contaminate subsurface waters and subsequently pose a threat to human and animal health. In general, NO_3^- leaching is a short-term problem and can be overcome by applying an amount of waste that will supply a level of available N equivalent to the N demand of the crop being grown. However, a part of the organic N added in wastes will be mineralized for several years after waste application. If wastes are applied annually, the contribution of residual N mineralized to inorganic N should be taken into account. Current estimates indicate that from 5 to 25% of the residual organic N may be mineralized each year for 1 to 3 years after waste application.

B. Metal Accumulation

Accumulation of metals in soils is a long-term problem and may result in decreased crop yields for significant periods of time after waste application has ceased. Copper and Zn are essential micronutrients for crops, and thus application of wastes at appropriate rates may actually result in increased yields due to the addition of micronutrients; however, excessive levels of these metals are toxic to plants, resulting in yield depressions. In addition to Cu and Zn, waste materials may contain high levels of Ni and Cd, two metals that are not essential for plants and are toxic at relatively low concentrations not only to the plant but also to humans and animals consuming the final plant product. Metal problems are most often encountered with municipal (i.e., sewage sludge and composts) and industrial wastes, but they should also be considered for wastes excreted from animals fed rations containing metal additives. Most field studies using sewage sludge indicate that Cu, Zn, and Ni are the most common metals causing yield decreases for a wide variety of crops (Patterson, 1971; Lunt, 1955; Giordano et al., 1975; Terman et al., 1973; King & Morris, 1972b; Webber, 1972). Continued application of sewage to agricultural land at the Berlin and Paris sewage farms has resulted in excessive levels of soil Cu and Zn (Rohde, 1962).

In addition to the obvious economic implications caused by yield reductions, a more significant problem may be the low level enrichment of Cd in the grain and other edible parts of crops grown on soils treated with waste products. Field experiments have indicated that Cd uptake is enhanced by sewage sludge applications for vegetables (Dowdy & Larson, 1975b), corn tissue and grain (Hinesly et al., 1972, 1976; Jones et al., 1975), and fescue (Boswell, 1975). More detailed studies conducted under greenhouse conditions have confirmed the fact that Cd uptake can occur for a wide range of crops including cereals, forages, and vegetables (Kirkham, 1975; Jones et al., 1973; Dowdy and Larson, 1975a; Dudas & Pawluk, 1975; Cunningham et al., 1975a, 1975b, 1975c; Bingham et al., 1975, 1976; Bradford et al., 1975). Because metals are more available to plants in acid soils, metal uptake by crops is enhanced when acidic soils are treated with sludge (Dowdy & Larson, 1975b; Chaney, 1973).

Experiments conducted by amending a low Cd sewage sludge with increasing levels of Cd and then growing a variety of crops under greenhouse conditions indicate that leafy vegetables accumulate more Cd than do cereals and grasses. This type of data gives information concerning the relative susceptibility of plants to Cd (Table 15). Cunningham et al. (1975a, 1975b, 1975c) conducted a series of experiments to evaluate interactions among types of sludge, metals, and yields of corn and rye. Their data indicated that Zn, Cu, and Ni were more consistently related to yield decreases than were Cd or Cr. In fact, increasing Cr levels reduced the toxicity of Zn, Cu, and Ni. It appears that a metal, added to soil as a metal salt or as a metal salt-sludge mixture, is more available at a given metal level than the metal indigenously present in sludge (Cunningham et al., 1975c). Large differences between Zn availability from $ZnSO_4$, compost, and sludge have been demonstrated by Giordano et al. (1975). Based on these results, the following levels of soil Zn would result in a 100-ppm level of Zn in corn forage tissue: 27 kg of $ZnSO_4$-Zn/ha = 45 kg of sludge-Zn/ha = 90 kg of compost-Zn/ha. Thus, it may not be feasible to simulate high metal sludges by amending sludges with metals or to predict plant availability of metals from results obtained by adding metal salts to soils. These and other studies

Table 15—Cd uptake by plants from soil treated with CdSO₄-amended sewage sludge.†

Crop species	Yield component	Soil Cd at 25% yield decrement		Tissue Cd level at 25% yield decrement	
		Cd added	DTPA-Cd‡	Diagnostic leaf	Edible
		μg/g			
Spinach	Shoot	4	2.4	75	75
Soybean	Dry bean	5	3.0	7	7
Curlycress	Shoot	8	4.8	70	80
Lettuce	Head	13	7.8	48	70
Corn	Kernel	18	10.8	35	2
Carrot	Tuber	20	12.0	32	19
Turnip	Tuber	28	16.8	121	15
Field bean	Dry bean	40	24.0	15	2
Wheat	Grain	50	30.0	33	12
Radish	Tuber	96	57.6	75	21
Tomato	Ripe fruit	160	96.0	125	7
Zucchini squash	Fruit	160	96.0	68	10
Cabbage	Head	170	102.0	160	11
	Forages				
Sudangrass	--	15	11	9	--
Alfalfa	--	30	22	24	--
White clover	--	40	29	17	--
Tall fescue	--	95	71	37	--
Bermudagrass	--	145	107	43	--

† From Bingham et al., 1975; forage data from Bingham et al., 1976.
‡ DTPA-Cd, Cd extracted with DTPA procedure developed by Lindsay (1972).

have indicated that the DTPA-extraction procedure originally developed by Lindsay (1972) to evaluate Fe, Zn, and Mn availability in calcareous soils may also be applicable to assess the availability of waste-borne metals added to soils.

C. Salt Accumulation

Reduced crop yields can result when waste materials add excessive amounts of soluble salts to soils. This problem generally arises from high application rates of animal wastes or secondary effluent, especially in arid or semiarid regions. Reduced yields due to salinity effects have been observed in several studies with animal wastes (Wallingford et al., 1975; Mathers & Stewart, 1974). Excessive salt accumulation also can decrease seed germination and deteriorate soil structure due to excess Na. The presence of soluble salts and/or organic materials resulted in decreased germination and emergence for several crops planted in soils treated with sewage sludge, animal waste, or compost (Hortenstine & Rothwell, 1969; Sabey & Hart, 1975; Cunningham et al., 1975a; Molina et al., 1971). For secondary effluents and waste waters from processing plants the SAR (sodium absorption ratio) and electrical conductivity are useful parameters to evaluate potential problems arising from salt accumulation in soils.

VI. POTENTIAL PHOSPHORUS SUPPLY IN WASTE MATERIALS

Estimates of amounts of various waste materials produced can be used to evaluate the impact of wastes as a P source for crop production (Table 16). In 1972, it was estimated that municipal waste water treatment facilities in the USA generated approximately 28.4 billion liters of waste water per day. Assuming that a typical waste water contains 10 mg of P/liter, the total output of P in secondary effluent amounts to 1×10^5 metric tons/year, or $\sim 5\%$ of the total P demand in 1973. Assuming that 1×10^6 liters of treated waste water generates 0.24 metric tons of sludge, then 5×10^4 metric tons of P would be available from sewage sludge, which amounts to $\sim 2\%$ of the 1973 P demand. Municipal refuse, most of which enters landfills, could provide approximately 27% of the fertilizer P. Based on the average annual excretion of P by farm animals, and the latest census of farm animals in the United States, 2.22 million metric tons of P is excreted in wastes yearly by farm animals, or 110% of the P used. If 80 to 85% of excreted P were collectable and available for plant utilization, P from animal wastes would be equal to the amount of P used annually. It is obvious that P in wastes could supply part of the annual P demand of agriculture. However, utilization of wastes in agriculture is mainly limited to a relatively small land area contiguous to a municipality, industry, or livestock enterprise due to costs involved in transportation and application.

Significant amounts of P are contained in wastes from agriculture, industry, and municipalities and can be utilized in the production of food and fiber. In comparison to conventional fertilizer materials, all wastes are a dilute source of fertilizer P, resulting in the necessity to apply relatively large volumes of wastes to satisfy the P requirement of crops. To maintain the integrity of soil and ground water resources, it is essential to utilize

Table 16—Theoretical supply of P in wastes for use in agriculture.

Description	Municipal wastewater†	Sewage sludge‡	Refuse§	Animal waste¶
Amount (million metric tons/year)	10,400	2.5	272	1,480
P content (%)	0.001	2.0	0.2	0.15
P supply (million metric tons/year)	0.10	0.05	0.54	2.22
Percent of fertilizer P used#	5.1	2.5	27	110

† 1972 estimate of municipal effluent discharged in U.S.
‡ Assumes 0.24 metric tons of sludge produced per 10^6 liter of wastewater treated.
§ Estimated amount of municipal, commercial and institutional solid waste.
¶ Estimated from amount of P excreted per animal and number of animals in U.S. Amount expressed on wet weight basis.
# Based on P fertilizer usage (2.03×10^6 metric tons) in 1973 (Harre, 1975).

Table 17—Suggested maximum sludge metal applications for cropland.†

Metal	Soil cation exchange capacity (meq/100 g)‡		
	0-5	5-15	>15
	maximum metal addition, kg/ha		
Pb	500	1,000	2,000
Zn	250	500	1,000
Cu	125	250	500
Ni	50	100	200
Cd	5	10	20

† Metal limits adapted by NC-118 and W-124 Regional Research Committee in cooperation with Agricultural Research Service and Forest Service, USDA.
‡ Cation exchange capacity determined on soil prior to sludge application by pH 7 ammonium acetate procedure.

proper management when applying wastes to soils. For wastes applied to soils in nonarid regions, the amount applied is limited on an annual basis by either (i) N required by the crop or (ii) Cd applied (2 kg/ha). At the present time, there is considerable disagreement among both researchers and regulatory agencies concerning an annual 2 kg of Cd/ha application.[2] Nevertheless, the basic approach presented for limiting annual additions by either N or Cd appears valid. The total amount of waste applied is limited by the cumulative amounts of Pb, Zn, Cu, Ni or Cd (Table 17).[3] Due to a greater N than P requirement of crops and the approximately equal N and P levels in most wastes, it is apparent that waste application rates based on N will provide P in excess of that required for crop growth. The following diagram depicts a rationale that can be used in applying wastes on agricultural land.

---

[2] The U.S. Environmental Protection Agency recently published the *Criteria for Classification of Solid Waste Disposal Facilities and Practices* (Federal Register 34:53438-53468, 13 Sept. 1979) wherein Cd limitations on agricultural land are included along with additional criteria for utilization of sewage sludges and other solid wastes on cropland used for growing food-chain crops. In essence, these regulations involve (i) maintainence of the solid waste-soil mixture at pH 6.5 or above at the time of each waste application; (ii) an annual application of Cd from solid wastes of ≤0.5 kg/ha on cropland used for production of tobacco, leafy vegetables, or root crops grown for human consumption; and (iii) annual Cd applications on other food chain crops of 2 kg/ha until 30 June 1984, 1.25 kg/ha from 1 July 1984 to 31 Dec. 1986, and 0.5 kg/ha beginning 1 Jan. 1987. Cumulative application of Cd from solid waste for soils of pH 6.5 or above (either natural or maintained at this level through lime addition) of 5, 10, and 20 kg Cd/ha for soils with cation exchange capacities of <5, 5 to 15, and >15 meq/100 g, respectively. Additional regulations were also included in the *Criteria* for applying Cd on soils where nonhuman food-chain crops are being grown.

[3] The U.S. EPA Criteria cited above only considers Cd additions to agricultural land. Additional guidance information from EPA (*Municipal Sludge Management: Environmental Factors*, MCD-28, U.S. EPA, Washington, D.C. EPA 4301/9-77-004) and a North Central-Western Regional Publication (*Application of Sludges and Wastewater on Agricultural Land: A Planning and Educational Guide*, B. D. Knezek and R. H. Miller (ed.) North Central Regional Research Public. No. 235, Ohio Agric. Res. & Development Center, Wooster, Ohio) have utilized the same values quoted herein; however, they have not been incorporated into federal regulatory documents.

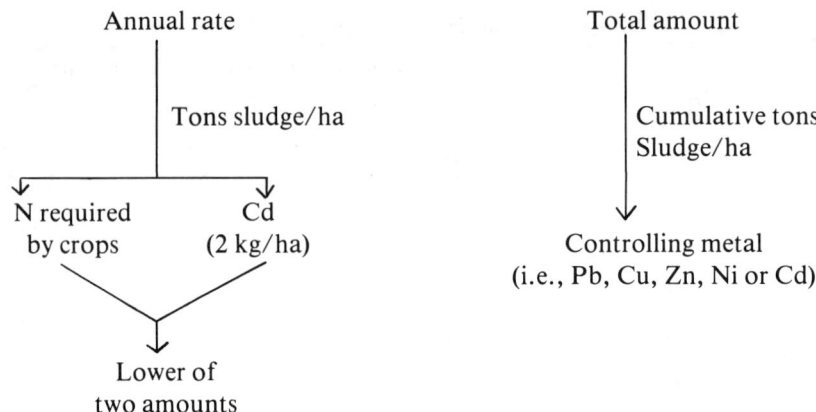

For wastes with low N contents (e.g., compost, refuse), a metal limit may be exceeded at a rate providing insufficient N for crop growth, resulting in a need for supplemental N fertilization. The above approach attempts to preclude NO_3^- contamination of ground water and metal accumulation in soils. Although this approach was developed for sewage sludge, it should be applicable to all wastes when N or metals rather than salts are limiting application rates.

LITERATURE CITED

Abbott, J. L., and T. C. Tucker. 1973. Persistence of manure phosphorus availability in calcareous soil. Soil Sci. Soc. Am. Proc. 37:60–63.

Adriano, D. C. 1975. Chemical characteristics of beef feedlot manures as influenced by housing type. p. 347–350. *In* Managing livestock wastes. Proc. Int. Symp. on Livestock Wastes. Am. Soc. Agric. Eng., St. Joseph, Mich.

Adriano, D. C., L. T. Novak, A. E. Erickson, A. R. Wolcott, and B. G. Ellis. 1975. Effect of long term land disposal by spray irrigation of food processing waste on some chemical properties of the soil and subsurface water. J. Environ. Qual. 4:242–248.

Anderson, M. S. 1955. Sewage sludge for soil improvement. USDA Circ. No. 972.

Agricultural Research Service, USDA. 1974. Factors involved in land application of agricultural and municipal wastes. Agricultural Research Service, USDA, Washington, D.C.

Azevedo, J., and P. R. Stout. 1974. Farm animal manures: an overview of their role in the agricultural environment. Calif. Agric. Exp. Stn. & Ext. Ser. Manual 44.

Berry, J. G., A. L. Sutton, and J. R. Carson. 1974. The production rate and composition of manure from growing turkeys. p. 153–158. *In* Processing and management of agricultural waste. Proc. 1974 Cornell Agric. Waste Management Conf., Rochester, N.Y.

Bingham, F. T., A. L. Page, R. J. Mahler, and T. J. Ganje. 1975. Growth and cadmium accumulation of plants grown on soils treated with cadmium-enriched sewage sludge. J. Environ. Qual. 4:207–211.

Bingham, F. T., A. L. Page, R. J. Mahler, and T. J. Ganje. 1976. Yield and cadmium accumulation of forage species in relation to cadmium content of sludge-amended soil. J. Environ. Qual. 5:57–60.

Bishop, R. F., L. B. MacLeod, L. P. Jackson, C. R. MacEachern, and E. T. Goring. 1962. A long-term field experiment with commercial fertilizers and manure. II. Fertility levels and crop yields in a rotation of potatoes, oats, and hay. Can. J. Soil Sci. 42:49–60.

Booram, C. V., T. E. Hazen, and R. J. Smith. 1975. Trends and variations in an anaerobic lagoon with recycling. p. 537–540. *In* Managing livestock waste. Proc. Int. Symp. on Livestock Wates. Am. Soc. Agric. Eng., St. Joseph, Mich.

Boswell, F. C. 1975. Municipal sewage sludge and selected element application to soil: Effect on soil and fescue. J. Environ. Qual. 4:267–273.

Bouwer, H., and R. L. Chaney. 1974. Land treatment of wastewater. Adv. Agron. 26:133-176.

Bradford, G. R., A. L. Page, L. J. Lund, and W. Olmstead. 1975. Trace element concentrations of sewage treatment plant effluents and sludges: Their interactions with soils and uptake by plants. J. Environ. Qual. 4:123-127.

Brumm, M. C., A. L. Sutton, V. B. Mayrose, J. C. Nye, and H. W. Jones. 1977. Effect of arsanilic acid in swine diets on fresh waste production, composition, and anaerobic decomposition. J. Anim. Sci. 44:521-531.

Burge, W. D. 1974. Pathogen considerations. In Factors involved in land application of agricultural and municipal waste. Agricultural Research Service, USDA, Washington, D.C.

Carroll, T. E., D. L. Maase, J. M. Genco, and C. N. Ifeadi. 1975. Review of landspreading of liquid municipal sewage sludge. EPA-670/2-75-049. Environ. Protection Agency, Washington, D.C.

Chaney, R. L. 1973. Crop and food chain effects of toxic elements in sludges and effluents. p. 129-141. In Recycling municipal sludges and effluents on land. Nat'l. Assoc. of State Univ. and Land-Grant Colleges, Washington, D.C.

Chesnin, L., and F. N. Anderson. 1975. Manure: long-term study shows its value to western soils. p. 24-26. Farm Ranch and Home Quarterly, Fall. Univ. of Nebraska, Lincoln.

Chumbley, C. G. 1971. Permissable level of toxic metals in sewage used on agricultural land. A.D.A.S. Advisory paper no. 10. 12 p.

Coker, E. G. 1966. The value of liquid digested sewage sludge. I. The effect of liquid sewage sludge on growth and composition of grass-clover swards in South-east England. J. Agric. Sci., Camb. 67:91-97.

Cooke, G. W. 1976. Long-term fertilizer experiments in England: The significant of their results for agricultural science and for practical farming. Ann. Agron. 27:503-536.

Cosgrove, D. J. 1973. Inositol polyphosphates in activated sludge. J. Environ. Qual. 2:483-485.

Cunningham, J. D., D. R. Keeney, and J. A. Ryan. 1975a. Yield and metal composition of corn and rye grown on sewage sludge-amended soil. J. Environ. Qual. 4:448-454.

Cunningham, J. D., D. R. Keeney, and J. A. Ryan. 1975b. Phototoxicity and uptake of metals added to soils as inorganic salts or in sewage sludge. J. Environ. Qual. 4:460-462.

Cunningham, J. D., J. A. Ryan, and D. R. Keeney. 1975c. Phototoxicity in and metal uptake from soils treated with metal-amended sewage sludge. J. Environ. Qual. 4:455-460.

Day, A. D. 1973. Recycling urban effluents on land using annual crops. p. 155-160. In Recycling municipal sludges and effluents on land. Nat'l. Assoc. State Univ. and Land-grant Colleges, Washington, D.C.

Dean, R. B., and J. E. Smith, Jr. 1973. The properties of sludge. p. 39-47. In Recycling municipal sludges and effluents on land. Nat'l. Assoc. of State Univ. and Land-Grant Colleges, Washington, D.C.

Dowdy, R. H., and W. E. Larson. 1975a. The availability of sludge-borne metals to various vegetable crops. J. Environ. Qual. 4:278-282.

Dowdy, R. H., and W. E. Larson. 1975b. Metal uptake by barley seedlings grown on soils amended with sewage sludge. J. Environ. Qual. 4:229-232.

Dudas, M. J., and S. Pawluk. 1975. Trace elements in sewage sludges and metal uptake by plants grown on sludge-amended soils. Can. J. Soil Sci. 55:239-243.

Egawa, T. 1974. The use of organic fertilizers in Japan. Food and Agricultural Organization of the United Nationa, A.G.L./T.M.O.F./74/19.

Elliott, L. F., and F. J. Stevenson (editors). 1977. Soils for management of organic wastes and wastewaters. Am. Soc. of Agron., Madison, Wis. 650 p.

Ellis, B. G. 1973. The soil as a chemical filter. p. 46-70. In W. E. Sopper and L. T. Kardos (ed.) Recycling treated municipal wastewater and sludge through forest and cropland. The Pennsylvania State Univ. Press, Univ. Park, Pa.

El-Sabban, F. F., T. A. Long, R. F. Gentry, and D. E. H. Frear. 1969. The influence of various factors on poultry litter composition. p. 340-346. In American waste management. Proc. Cornell Agric. Waste Management Conf., Syracuse, N.Y.

Ferguson, J. F., D. Jenkins, and J. Eastman. 1973. Calcium phosphate precipitation at slightly alkaline pH values. J. Water Pollution Control Fed. 45:620-631.

Fisher, L. J. 1974. Influence on feeding system, digestibility of ration and proportion of concentrate consumed on the quantity and quality of excreta voided by lactating cows. p. 283-290. In Processing and management of agricultural waste. Proc. 1974 Cornell Agric. Waste Management Conf., Rochester, N.Y.

Frecks, G. A., and C. B. Gilbertson. 1973. The effect of ration on engineering properties of beef cattle manure. Paper No. 73-442 presented at the summer meeting, Am. Soc. Agric. Eng., Lexington, Ky., June 17-20.

Giordano, P. M., J. J. Mortvedt, and D. A. Mays. 1975. Effect of municipal waste on crop yields and uptake of heavy metals. J. Environ. Qual. 3:394-399.

Graber, R. 1974. Agricultural animals and the environment. Monolith for feedlot waste management regional ext. project. Oklahoma State Univ., Stillwater.

Grub, W., R. C. Albin, D. M. Wells, and R. Z. Wheaton. 1969. The effect of feed, design and management on the control of pollution from beef cattle feedlots. p. 217-224. *In* Animal waste management. Proc. Cornell Agric. Waste Management Conf., Syracuse, N.Y.

Harre, E. A. 1975. The supply outlook for phosphate fertilizers. p. 36-44. T.V.A. Fertilizer Conf. Proc., Louisville, Ky., TVA, Muscle Shoals, Ala.

Hashimoto, A. G. 1974. Characterization of white leghorn manure. p. 141-152. *In* Processing and management of agricultural waste. Proc. 1974 Cornell Agric. Waste Management Conf., Rochester, N.Y.

Hensler, R. F., W. H. Erhardt, and L. M. Walsh. 1971. Effect of manure handling systems on plant nutrient cycling. p. 254-257. *In* Livestock waste management and pollution abatement. Proc. Int. Symp. on Livestock Wastes. Am. Soc. Agric. Eng., St. Joseph, Mich.

Herriott, J. B. D., and D. A. Wells. 1962. Gülle as a grassland fertilizer. J. Brit. Grassland Soc. 17:167-170.

Herriott, J. B. D., D. A. Wells, and P. Crooks. 1963. Gülle as a grassland fertilizer. Part II. J. Brit. Grassland Soc. 18:339-344.

Herriott, J. B. D., D. A. Wells, and P. Crooks. 1965. Gülle as a grassland fertilizer. Part III. J. Brit. Grassland Soc. 20:129-138.

Hinesly, T. D., and B. Sosewitz. 1969. Digested sludge disposal on cropland. J. Water Pollution Control Fed. 41:822-830.

Hinesly, T. D., R. L. Jones, J. J. Tyler, and E. L. Zigler. 1976. Soybean yield responses and assimilation of Zn and Cd from sewage sludge-amended soil. J. Water Pollution Control Fed. 48:2137-2152.

Hinesly, T. D., E. L. Zigler, and R. L. Jones. 1972. Effects on corn by application of heated anaerobically digested sludge. Compost Sci. 12(4):26 30.

Holland, M. R., E. T. Kornegay, and J. D. Hedges. 1975. Nutritive value of swine feces for swine. p. 214-217. *In* Managing livestock wastes. Proc. Int. Symp. on Livestock Wastes. Am. Soc. Agric. Eng., St. Joseph, Mich.

Hook, J. E., L. T. Kardos, and W. E. Sopper. 1973. Effects of land disposal of wastewaters on soil phosphorus relations. p. 200-219. *In* W. E. Sopper and L. T. Kardos (ed.) Recycling treated municipal wastewater and sludge through forest and cropland. Pennsylvania State Univ. Press, Univ. Park, Pa.

Hortenstine, C. C., and D. F. Rothwell. 1969. Evaluation of composted municipal refuse as a plant nutrient source and soil amendment on Leon fine sand. Proc. Soil Crop Sci. Soc. Florida. 29:312-319.

Hortenstine, C. C., and D. F. Rothwell. 1973. Pelletized municipal refuse compost as a soil amendment and nutrient source for sorghum. J. Environ. Qual. 2:343-345.

Horton, M. L., J. L. Halbeisen, J. L. Wiersma, A. C. Dittman, and R. M. Luther. 1975. Land disposal of beef wastes: climate, rates, salinity and soil. p. 258-260. *In* Managing livestock wastes. Proc. Int. Symp. on Livestock Wastes. Am. Soc. of Agric. Eng., St. Joseph, Mich.

Howell, E. S., M. R. Overcash, and F. J. Humenik. 1974. Unaerated lagoon response to loading intensity and frequency. Paper No. 74-4515 presented at Winter meeting, Am. Soc. Agric. Eng., Chicago, Ill., Dec. 10-13.

Jones, D. D., D. L. Day, and A. C. Dale. 1971. Aerobic treatment of livestock wastes. Illinois Exp. Stn. Bull. 737.

Jones, R. L., T. D. Hinesly, and E. L. Zigler. 1973. Cadmium content of soybeans grown in sewage sludge amended soil. J. Environ. Qual. 2:351-353.

Jones, R. L., T. D. Hinesly, E. L. Zigler, and J. J. Tiler. 1975. Cadmium and zinc content of corn leaf and grain production by sludge-amended soil. J. Environ. Qual. 4:509-514.

Kardos, L. T., and J. E. Hook. 1976. Phosphorus balance in sewage effluent treated soils. J. Environ. Qual. 5:87-90.

King, L. D., and H. D. Morris. 1972a. Land disposal of liquid sewage sludge: I. The effect on yield, *in vivo* digestibility and chemical composition of coastal bermudagrass (*Cynodon dactylon* L. Pers). J. Environ. Qual. 1:325-329.

King, L. D., and H. D. Morris. 1972b. Land disposal of liquid sewage sludge: II. The effect on soil pH, manganese, zinc, and growth and chemical composition of rye (*Secale cereale* L.). J. Environ. Qual. 1:425-429.

King, L. D., and H. D. Morris. 1972c. Land disposal of liquid sewage sludge: III. The effect on soil nitrate. J. Environ. Qual. 1:442-446.

King, L. D., and H. D. Morris. 1973. Land disposal of liquid sewage sludge: IV. Effect of soil phosphorus, potassium, calcium, magnesium, and sodium. J. Environ. Qual. 2:411-414.

King, L. D., L. A. Rudgers, and L. R. Webber. 1974. Application of municipal refuse and liquid sewage sludge to agricultural land: 1. Field study. J. Environ. Qual. 4:361-366.

Kirkham, M. B. 1974. Disposal of sludge on land: Effect on soils, plants and ground water. Compost Sci. 15(2):6-10.

Kirkham, M. B. 1975. Uptake of cadmium and zinc from sludge by barley grown under four different sludge irrigation regimes. J. Environ. Qual. 4:423-426.

Kornegay, E. T., J. D. Hedges, D. C. Martens, and C. Y. Kramer. 1974. Effect on soil and plant mineral levels following application of manure from swine fed high dietary copper. p. 129-134. VPI & State Univ. Livestock Res. Rep. 158.

Lindsay, W. L. 1972. Inorganic phase equilibria of micronutrients in soils. p. 41-57. *In* J. J. Mortvedt, P. M. Giordano, and W. L. Lindsay (ed.) Micronutrients in agriculture. Soil Sci. Soc. Am., Madison, Wis.

Loehr, R. C. 1969. Animal waste—A national problem. J. San. Eng. Div. Am. Soc. Civil Eng. 95:189-221.

Loehr, R. C. 1971. Liquid waste treatment. p. 54-78. *In* Agricultural wastes: principles and guidelines for practical solutions. Proc. 1971 Cornell Agric. Waste Management Conf., Syracuse, N.Y.

Loehr, R. C. 1973. Pollution implications of animal wastes—a forward oriented review. Water Pollution Control Research Series. EPA 13040—07/78. Environmental Protect. Agency, Washington, D.C.

Lund, Z. F., B. D. Boss, and F. E. Lowry. 1975. Dairy cattle manure—Its effect on rye and millet forage yield and quality. J. Environ. Qual. 4:195-198.

Lund, L. J., A. L. Page, and C. O. Nelson. 1976. Nitrogen and phosphorus levels in soils beneath sewage disposal ponds. J. Environ. Qual. 5:26-30.

Lunt, H. A. 1955. Digested sewage sludge for soil improvement. Conn. Agric. Exp. Sta. Bull. 622.

MacLeod, L. B., R. F. Bishop, L. P. Jackson, C. R. MacEachern, and E. T. Goring. 1960. A long-term field experiment with commercial fertilizers and manure. I. Fertility levels and crop yields in a rotation of swedes, oats, and hay. Can. J. Soil Sci. 40:136-145.

Manges, H. L., L. S. Murphy, and E. H. Goering. 1972. Disposal of beef feedlot wastes onto cropland. Paper no. 72-961 presented at the Winter Meeting, Am. Soc. Agric. Eng., Chicago, Ill., 11-15 Dec.

Mathers, A. C., and B. A. Stewart. 1974. Corn silage yield and soil chemical properties as affected by cattle feedlot manure. J. Environ. Qual. 3:143-147.

Mattingly, G. E. G., A. E. Johnston, and M. Chater. 1969. The residual value of farmyard manure and superphosphate in Saxmundham rotation II experiment, 1899-1968. p. 91-112. Rep. Rothamsted Exp. Sta. for 1969, p. 2.

Mays, D. A., G. L. Terman, and J. C. Duggan. 1973. Municipal compost: Effects on crop yields and soil properties. J. Environ. Qual. 2:89-92.

McCalla, T. M., L. R. Frederick, and G. L. Palmer. 1970. Manure decomposition and fate of breakdown productions in soil. p. 241-255. *In* T. L. Willrich and G. E. Smith (ed.) Agricultural practices and water quality. Iowa State Univ. Press, Ames.

McCalla, T. M., J. R. Peterson, and C. Lue-Hing. 1977. Properties of agricultural and municipal wastes. p. 11-43. *In* L. F. Elliott and F. J. Stevenson (ed.) Soils for management of organic wastes and wastewaters. Am. Soc. Agron., Madison, Wis.

Melsted, S. W. 1973. Soil-plant relationships (some practical consideration in waste management). p. 121-128. *In* Recycling municipal sludges and effluents on land. Nat'l. Assoc. of State Univ. and Land-Grant Colleges, Washington, D.C.

Menzies, J. D., and R. L. Chaney. 1974. Waste characteristics. p. 18-36. *In* Factors involved in land application of agricultural and municipal wastes. Agricultural Research Service, USDA, Washington, D.C.

Miller, R. H. 1973. Soil microbiological aspects of recycling sewage sludges and waste effluents on land. p. 79-90. *In* Recycling municipal sludges and effluents on land. Nat'l. Assoc. of State Univ. and Land-Grant Colleges, Washington, D.C.

Miner, J. R., and R. J. Smith. 1975. Livestock waste management with pollution control. Midwestern Planning Service-19, North Central Regional Res. Publ. 222.

Molina, J. A. E., O. C. Braids, T. D. Hinesly, and J. B. Cropper. 1971. Aeration-induced changes in liquid sewage sludge. Soil Sci. Soc. Am. Proc. 35:60-63.

Mortvedt, J. J., and P. M. Giordano. 1975. Response of corn to zinc and chromium in municipal wastes applied to soil. J. Environ. Qual. 4:170-174.

Muehling, A. J. 1969. Swine housing and waste management—a research review. Illinois Agric. Exp. Stn. Bull. Ag. Eng. 873.

Mugiwra, L. M. 1976. Effect of dairy cattle manure on millet and rye forage and soil properties. J. Environ. Qual. 5:60-65.

Müller, J. F. 1929. The value of raw sewage sludge as fertilizer. Soil Sci. 28:423-432.

Murphy, L. S., G. W. Wallingford, W. L. Powers, and H. L. Manges. 1972. Effects of solid beef feedlot wastes on soil conditions and plant growth. p. 449-464. *In* Waste management research. Proc. 1972 Cornell Agric. Waste Management Conf., Syracuse, N.Y.

Mutlak, S. M., A. D. McKelvie, and K. Robinson. 1975. The yield response of grass to aerobically stabilized swine waste. p. 274-276. *In* Managing livestock waste. Proc. Int. Symp. on Livestock Wastes. Am. Soc. Agric. Eng., St. Joseph, Mich.

Nat'l. Assoc. of State Univ. and Land-Grant Colleges. 1973. Recycling municipal sludges and effluents on land, Washington, D.C.

Nelson, A. B., R. W. MacVicar, W. Archer, Jr., and J. C. Meiske. 1955. Effect of a high salt intake on the digestibility of ration constituents and on nitrogen, sodium, and chloride retention by steers and wethers. J. Anim. Sci. 14:825-830.

Norstadt, R. A., L. B. Baldwin, and C. C. Hortenstine. 1971. Multi-stage lagoon systems for treatment of dairy farm waste. p. 77-80. *In* Livestock waste management and pollution abatement. Proc. Int. Symp. on Livestock Waste. Am. Soc. Agric. Eng., St. Joseph, Mich.

O'Callaghan, J. R., K. A. Pollock, and V. A. Dodd. 1971. Land spreading of manure from animal production units. J. Agric. Eng. Res. 16:280-300.

Olsen, R. J., R. F. Hensler, and O. J. Attoe. 1970. Effect of manure application, aeration, and soil pH on soil nitrogen transformations and on certain soil test values. Soil Sci. Soc. Am. Proc. 34:222-225.

Olsen, S. R., and S. A. Barber. 1977. The effect of waste application on soil phosphorus and potassium. p. 197-215. *In* L. F. Elliott and F. J. Stevenson (ed.) Soils for management of organic waste and wastewaters. Am. Soc. of Agron., Madison, Wis.

Oniani, O. G., M. Chater, and G. E. G. Mattingly. 1973. Some effects of fertilizers and farmyard manure on the organic phosphorus in soils. J. Soil Sci. 24:1-9.

Overman, A. R., C. C. Hortenstine, and J. M. Wing. 1971. Growth response of plants under sprinkler irrigation with dairy waste. p. 334-337. *In* Livestock waste management and pollution abatement. Proc. Int. Symp. on Livestock Wastes. Am. Soc. Agric. Eng., St. Joseph, Mich.

Page, A. L. 1974. Fate and effects of trace elements in sewage sludge when applied to agricultural land. EPA-670/2-74-005. Environ. Prot. Agency, Cincinnati, Ohio. 98 p.

Page, A. L., and A. C. Chang. 1975. Trace elements and plant nutrient constraints of recycling sewage sludges on agricultural land. Second Nat'l. Conf. on Water Reuse: Water's Interface with Energy, Air, and Solids. Chicago, Ill.

Page, A. L., and P. F. Pratt. 1975. Effects of sewage sludge or effluent application to soil on the movement of nitrogen, phosphorus, soluble salts, and trace elements to groundwaters. Second Nat'l. Conf. on Municipal Sludge Management and Disposal. Anaheim, Calif.

Patterson, J. B. E. 1971. Metal toxicity arising from industry. Technol. Bull. Ministry of Agric. Food, Fish and Agric. Develop. Advisory Services, Cambridge, England. 21:193-207.

Peperzak, P., A. G. Caldwell, R. R. Hunziker, and C. A. Black. 1959. Phosphorus fractions in manures. Soil Sci. 87:293-302.

Peterson, J. R., T. M. McCalla, and G. E. Smith. 1971. Human and animal wastes as fertilizers. p. 557-596. *In* R. A. Olson, T. J. Army, J. J. Hanway, and V. J. Kilmer (ed.) Fertilizer technology and use. Soil Sci. Soc. Am., Madison, Wis.

Peterson, J. R., Cecil Lue-Hing, and D. R. Zenz. 1973. Chemical and biological quality of municipal sludge. p. 26-34. *In* W. E. Sopper and L. T. Kardos (ed.) Recycling treated municipal wastewater and sludges through forest and cropland. The Pennsylvania State University Press, University Park, Pa.

Pound, C. E., and R. W. Crites. 1973. Characteristics of municipal effluents. p. 49-61. *In* Recycling municipal sludges and effluents on land. Nat'l. Assoc. of State Univ. and Land-Grant Colleges, Washington, D.C.

Powers, W. L., G. W. Wallingford, L. S. Murphy, D. A. Whitney, H. L. Manges, and H. E. Jones. 1974. Guidelines for applying beef feedlot manure to fields. Kansas Agric. Ext. Bull. C-502.

Powers, W. L., G. W. Wallingford, and L. S. Murphy. 1975. Research status on effects of land application and animal wastes. Environ. Prot. Agency. Tech. Ser. EPA-660/2-75-010.

Proctor, D. E. 1965. Amounts, composition, characteristics and pollutional properties of animal manures. Paper presented at Western Am. Soc. Animal Sci., Ft. Collins, Colo., July 14.

Reddell, D. L. 1974. Forage and grain production from land use for beef manure disposal. p. 464-483. *In* Processing and management of agricultural waste. Proc. 1974 Cornell Agric. Waste Management Conf., Rochester, N.Y.

Rohde, G. 1962. The effects of trace elements on the exhaustion of sewage-irrigated land. Inst. of Sewage Purification J. 1962; 581-585.

Ryan, J. A., and D. R. Keeney. 1975. Ammonia volatilization from surface applied sewage sludge. J. Water Poll. Cont. Fed. 47:386-393.

Sabey, B. R., and W. E. Hart. 1975. Land application of sewage sludge: I. Effect on growth and chemical composition of plants. J. Environ. Qual. 4:252-256.

Salter, R. M., and C. J. Schollenberger. 1939. Farm manure. Ohio Agric. Exp. Sta. Bull. 605.

Sawhney, B. L., and D. E. Hill. 1975. Phosphate sorption characteristics of soils treated with municipal wastewater. J. Environ. Qual. 4:342-346.

Scott, D. S., and H. Horlings. 1975. Removal of phosphates and metals from sewage sludges. Environ. Sci. Technol. 9:849-855.

Shortall, J. G., and W. C. Liebhardt. 1975. Yield and growth of corn as affected by poultry manure. J. Environ. Qual. 4:186-191.

Singh, R. N., R. F. Keefer, and D. J. Horvath. 1975. Can soils be used for sewage sludge disposal? Compost Sci. 15(2):22-25.

Soderquist, M. R. 1975. Characterization of fruit and vegetable processing wastewaters. Water Resources Research Instit., Oregon State Univ., W.R.R.I.-28.

Sommers, L. E. 1977. Chemical composition of sewage sludges and analysis of their potential use as fertilizers. J. Environ. Qual. 6:225-231.

Sommers, L. E., D. W. Nelson, J. E. Yahner, and J. V. Mannering. 1972. Chemical composition of sewage sludge from selected Indiana cities. Ind. Acad. of Sci. 82:424-432.

Sommers, L. E., D. W. Nelson, and K. J. Yost. 1976. Variable nature of chemical composition of sewage sludge. J. Environ. Qual. 5:303-306.

Sopper, W. E., and L. T. Kardos. 1973a. Vegetation responses to irrigation with treated municipal wastewater. p. 271-294. *In* W. E. Sopper and L. T. Kardos (ed.) Recycling treated municipal wastewater and sludge through forest and cropland. The Pennsylvania State Univ. Press, Univ. Park, Pa.

Sopper, W. E., and L. T. Kardos. 1973b. Recycling treated municipal wastewater and sludge through forest and cropland. The Pennsylvania State Univ. Press, Univ. Park, Pa.

Sutton, A. L., D. W. Nelson, V. B. Mayrose, and J. C. Nye. 1974. Effect of liquid swine application on soil chemical composition. p. 503-514. *In* Processing and management of agricultural waste. Proc. 1974 Cornell Agric. Waste Management Conf., Rochester, N.Y.

Sutton, A. L., D. H. Bache, J. C. Nye, A. C. Dale, D. D. Jones, V. B. Mayrose, M. P. Plumlee, and L. B. Underwood. 1975. A waste management system for a 2500-head swine operation—A case study. p. 177-180. *In* Managing livestock wastes. Proc. Int. Symp. on Livestock Wastes. Am. Soc. Agric. Eng., St. Joseph, Mich.

Terman, G. L., J. M. Soileau, and S. E. Allen. 1973. Municipal compost: Effects on crop yields and nutrient content in greenhouse pot experiments. J. Environ. Qual. 2:84-88.

Tiarks, A. E., A. P. Mazurak, and L. Chesnin. 1974. Physical and chemical properties of soil associated with heavy applications of manure from cattle feedlots. Soil Sci. Soc. Am. Proc. 38:826-830.

Townshend, A. R., K. A. Reichert, and J. H. Nodwell. 1969. Status report on water pollution control facilities for farm animal wastes in the Province of Ontario. p. 131-149. *In* Animal wastes management. Proc. 1969 Cornell Agric. Waste Management Conf., Syracuse, N.Y.

Vanderholm, D. H. 1975. Nutrient losses from livestock waste during storage, treatment and handling. p. 282–285. *In* Managing livestock wastes. Proc. Int. Symp. on Livestock Waste. Am. Soc. Agric. Eng., St. Joseph, Mich.

Vitosh, M. L., J. F. Davis, and B. D. Knezek. 1973. Long-term effects of manure, fertilizer, and plow depth on chemical properties of soils and nutrient movement in a mono-culture corn system. J. Environ. Qual. 2:296–299.

Walker, J. M. 1975. Sewage sludge—Management aspects for land application. Compost Sci. 16(2):12–21.

Wallingford, G. W., L. S. Murphy, W. L. Powers, and H. L. Manges. 1974. Effect of beef-feedlot lagoon water on soil chemical properties and growth and composition of corn forage. J. Environ. Qual. 3:74–78.

Wallingford, G. W., L. S. Murphy, W. L. Powers, and H. L. Manges. 1975. Disposal of beef-feedlot manure: effects of residual and yearly applications on corn and soil chemical properties. J. Environ. Qual. 4:526–531.

Walsh, L. M., R. F. Hensler, and E. E. Schulte. 1975. Manage manure for its value. Wisconsin Agric. Ext. Bull. A1672.

Warren, R. G., and A. E. Johnston. 1966. Hoosfield continuous barley. p. 320–338. *In* Rothamsted Rep. for 1966, Harpenden, Herts, England.

Webber, J. 1972. Effects of toxic metals and sewage on crops. J. Water Pollution Control Fed. 71:404–413.

Wells, D. M., R. C. Albin, W. Grug, and R. Z. Wheaton. 1969. Aerobic decomposition of solid wastes from cattle feedlots. p. 217–224. *In* Animal waste management. Proc. Cornell Agric. Waste Management Conf., Syracuse, N.Y.

Williams, R. J. B., and G. W. Cooke. 1970. Results of the rotation I experiment at Saxmundham, 1964–69. p. 68–97. *In* Rothamsted Rep. for 1970, Part 2, Harpenden, Herts, England.

Chapter 19

Agricultural Phosphorus in the Environment

A. W. TAYLOR

Science & Education Administration, AR, USDA
Beltsville Agricultural Research Center
Beltsville, Maryland

V. J. KILMER

Tennessee Valley Authority (retired)
Muscle Shoals, Alabama

I. INTRODUCTION

Phosphorus (P) invariably occurs naturally in the form of phosphate salts and minerals. Since P is in no way toxic, except when present as a moeity in molecules of certain nonpersistent pesticides, the environmental problems associated with P are concerned exclusively with the control of unwanted fertility levels in natural waters. Even here the direct impact is not due to the element itself but to the excess unwanted plant growth.

Since P is essential to all forms of terrestial life, the element is widely distributed over the surface of the earth in biologically available forms cycling within plants, animals, soil, and water. This biological P exists at a higher free-energy level than that of the forms present in most geological formations or unfertilized soils. Maintenance of this cycle depends upon energy input derived from photosynthesis, which is employed in the adsorption and accumulation of phosphate by plants either through roots or from water in which they float. In the absence of such input, the amount of high-energy available P will decline as it is removed by precipitation (or adsorption) on mineral substrates, acting as sinks responsible for the leakage from the biocycle.

When there is insufficient available P, biological productivity is limited and must be restored by addition of available P in the form of fertilizer or animal waste. Since the rate of leakage from the system also increases as the amount is raised, such additions are not permanent and any long-term increase in fertility can only be maintained by successive fertilization. In addi-

---

Copyright 1980 © ASA-CSSA-SSSA, 677 South Segoe Road, Madison, WI 53711, USA.
The Role of Phosphorus in Agriculture.

tion to losses in availability due to chemical "fixation," or reversal to less available forms, fertile fields may lose P by direct removal in water or by bulk removal of soil in the form of sediment. It is these losses which give rise to problems of environmental management. It should be noted, however, that even where the most acute difficulties arise there is no fundamental threat to the general environment. The difficulties are solely due to the change in productivity of a part of the biosphere which men desire to keep at a low level for aesthetic, recreational, or economic reasons.

II. BACKGROUND LEVELS IN THE ENVIRONMENT

Before considering the nature and effect of agricultural P inputs it is desirable to examine the natural levels that would exist in an essentially undisturbed system. Concentrations of P in lakes and streams of undisturbed ecosystems were never zero. Phosphorus is essential for life of any form, and must be present in any ecosystem that contains living tissue. The amount of available P may, however, be quite small where the productivity is low, particularly where the system is almost completely closed and a very small fraction of the circulating nutrient is exported.

Even in low productivity systems, such as native forest, the amount of P that leaks or is exported into streams and lakes may be inconsequential when viewed in terms of its effect on the fertility of the land, but of major importance in terms of its effect on water bodies. To illustrate, a storm event that results in water flow (surface and subsurface) equal to 10 ha-cm needs to dissolve only 30 g of P/ha to give a P concentration of 0.03 ppm. This loss would have no measurable effect on the P fertility status of the soil, but would raise the P concentration of the water draining this field to eutrophication threshold levels. It is necessary that we keep this double perspective in mind as we review the subject matter in the following sections.

Quantitative data on P concentrations in undisturbed systems is surprisingly sparse, particularly before the last decade. A series of studies published since 1970 have considerably clarified the subject.

A. Subsurface Flow

The contributions of geological sources were shown by Thomas and Crutchfield (1974) who found soluble P concentrations between 0.01 and 0.05 ppm in samples taken from streams draining five agricultural watersheds on sandstone and shale rock in central and west Kentucky. Streams draining the most intensively cultivated watersheds carried concentrations below 0.03 ppm. Levels ranging from 0.30 to 0.35 ppm were found in streams draining areas of high phosphate limestone rock. Comparison of the results, obtained in 1971 and 1972, with earlier data (McHargue & Peter, 1921) showed that general levels have remained unchanged for over 50 years, despite an 9- to 10-fold increase in P fertilization.

Other studies of subsoil water composition have produced somewhat varied results. In Boone County, Iowa, Baker et al. (1975) found soluble orthophosphate concentrations below 0.038 ppm without seasonal trends: total P concentrations, which included organic and possibly some sorbed P were generally higher, reaching 0.182 ppm. In studies in New York, Zwerman et al. (1972) found concentrations consistently below 0.01 ppm. Other data have been summarized by Ryden et al. (1973), who noted the difficulties of interpreting many experimental results obtained from artificial drains, particularly those below fertilized fields. Artificial drains increase rates of infiltration, and percolation reduces the residence time of the water during which phosphate adsorption takes place, increases the downward movement of organic matter, and produces sediment. These complications make the interpretation of variable data such as that presented by Logan and Schwab (1976) rather uncertain. Recent studies by Duxbury and Peverly (1978) and Miller (1979) have illustrated the importance of organic matter in accellerating the downward movement of P, particularly in highly organic soils. Prediction of the amounts, however, requires a detailed knowledge of the profile characteristics since interactions of the P with inorganic components of the soil can be important (Duxbury & Peverly; 1978).

Although adequate data on concentrations in natural subsoil drainage remain meager the general indications are that, except where water moves through a rock of naturally high P content or contains leachate from decaying organic matter, concentrations may be expected to be below about 0.05 ppm.

B. Surface Flow

The amount of P carried by surface runoff under natural conditions may be assessed in terms of that in runoff from natural grassland such as prairie, or woodland.

White and Williamson (1973) frequently found soluble P concentrations above 0.10 ppm in water collecting in natural surface drainage areas in South Dakota. Concentrations varied between sites, reflecting variations in topography and vegetation: at one site water derived from snow melt contained over 1 ppm. Similar data were obtained by Timmons and Holt (1977) in a 5-year study of nutrient losses in surface runoff from a native prairie: annual weighted ortho-P concentrations were 0.18 ppm, and organic-P concentrations were almost twice this figure. The overall annual losses ranged up to 0.25 kg P/ha depending on precipitation. Between 70 and 90% of the overall loss was organic P. The importance of snowmelt runoff was very evident. The importance of P release from dried or frozen vegetation, particularly when runoff moves over frozen ground, has been examined by Timmons and Holt (1970), who showed that a single leaching of frozen alfalfa could release up to 0.65 kg/ha of total soluble P, of which 80% was inorganic. Under the same treatment bluegrass released about 0.3 kg/ha, but losses from oat stubble were small. Field observations reported by Burwell et al. (1975) confirmed these data, showing that alfalfa hay plots released up to 0.3 kg/ha of soluble P in snowmelt runoff.

Measurements in runoff from an undisturbed aspen-birch forest by Timmons et al. (1977) indicated total P concentrations between 0.10 and 0.45 ppm: about half of this was in organic form. Overall losses averaged 0.22 kg/ha over the 3 years studied. Almost all of this was in snowmelt runoff.

In a 1-year study of losses from a deciduous Minnesota forest, Singer and Rust (1975) found concentrations up to 0.61 ppm with an overall loss of 0.09 kg/ha. The observed October peak was attributed to rainfall leaching of freshly fallen leaves and a March peak to snowmelt leaching. Owing to the higher volume of runoff, 90% of the annual loss was in March. No losses were observed from December through March because runoff stopped during frozen conditions. In comparable studies of pine forested watersheds in Mississippi Schreiber et al. (1976) and Duffy et al. (1978) found relatively little seasonal variation in P concentrations, although a maximum observed in November could be associated with leaf-fall: the amount of P lost was, therefore, mainly controlled by the volume of runoff water. Total soluble P losses averaged 0.088 kg/ha with 0.21 kg/ha carried by sediment: mean total solution P concentration was 0.027 ppm. Taylor et al. (1971) observed a similar pattern of loss from unfertilized woodland near Coshocton, Ohio: dissolved P concentrations ranged from 0.020 to 0.050 ppm, but variations in water-flow were the controlling variable. Highest concentrations were found in late summer when runoff was negligible: almost the entire overall loss was in February through May.

In a survey of other studies of P runoff from woodlands, Ryden et al. (1973) noted that seasonal variations in dissolved P concentrations appeared to be minor, but values above the "critical" limit of 0.03 ppm are not uncommon. It is, therefore, evident that background levels of P in the natural undisturbed environment are often close to the levels regarded as the limit for eutrophication. These background levels do, however, show considerable variation which is often imposed by climate and rainfall patterns as much as P levels in soils and plant species. In colder climates the importance of P release and transport from vegetation during snowmelt leaching can be a factor of major importance: this is also significant in agricultural as well as natural ecosystems.

C. Analytical Problems

Assessment of the significance of these data are difficult, since it is not known what fraction of P is biologically available. The use of different analytical procedures may not always give strictly comparable results, particularly when some soluble P is present in organic forms that may not be detected by a simple molybdate analytical procedure. This question has been critically discussed by Ogelesby and Schaffner (1975) who presented data showing that summer biological activity in lakes can be correlated with the total soluble P loading of the epilimnion. *Total P* was defined as that measured by molybdenum blue complex formation following a persulfate oxidation: this does not include phosphorus carried by sediment particles.

Application of this criterion to data obtained from runoff studies suggests that the amounts of biologically active soluble P transported may be somewhat higher than the analyses indicate if the method did not include an oxidation step.

It is thus evident that the quantities of soluble P removed in runoff water from unfertilized land or vegetation are not negligible and may at times be high enough to contribute significantly to accelerated eutrophication. Since, as with many other natural phenomena, the process is sporadic, individual observations must be evaluated in the light of prevailing conditions. Hydrologic observations, in particular, are of at least equal importance to the chemical data.

D. Transport on Sediments

The rapid and extensive adsorption of water-soluble P by almost all soils is the basis of a large amount of literature. Although many correlations have been obtained between the amount available to crops and that extracted by various chemical methods, no single universal index of "available soil phosphate" has been found, nor, in view of the variety of soil types and the diversity of crops, is one to be expected. Since much of the P present in soils is in highly insoluble and stable forms, total P is valueless as an indicator of biological worth. The fraction that is present in sufficiently labile adsorbed forms may be considered to be of that order of up to 5 to 10% of the total.

When erosion occurs, soil moving to lake and stream water carries adsorbed P with it. Determination of the biological importance of this P raises questions similar to, but more complex than those concerning biological availability in soils. The additional complexities are due to the more dynamic character of sediment transport, including mixing, particle sorting, exchanges of P between sediment and water, and the physical removal of P from the lake biotic zone by sediment deposition.

Before the late 1960's most research on erosion losses was concerned with the decreases in soil fertility due to topsoil erosion. This work, reviewed by Barrows and Kilmer (1963) and Taylor (1967), paid no attention to the biological significance of P in sediments. Simple calculations suggest that the total P carried by sediments is much larger than that in solution. As an example, runoff water containing 0.01% by weight of sediment which itself contains 1,000 ppm of P will contain a "total concentration" of 0.1 ppm, while the amount in true solution might be about 0.05 ppm. Data on sediment and runoff from several watershed studies (Hanway & Laflen, 1974; White & Williamson, 1973; Burwell et al., 1975) show that ratios of total P/solution P can exceed 200 to 400 when large amounts of sediment are present, although values up to 50 to 100 are more common. In a detailed examination of nutrient transport in Fall Creek, New York, Johnson et al. (1976) found that about 78% of the total P lost between September 1972 and April 1974 was carried by suspended particulate material. About 4% of this, or 3% of the whole, was in labile forms readily exchangeable with the dissolved P.

The quantity of P transported by the solid phase depends upon P content of the sediment, the concentration of solid material, and the water flow. None of these are independent of the hydrologic conditions. Owing to preferential removal of finer particles and their longer retention in suspension, the total P content of the solids will normally be higher than that of the parent soil—the "enrichment ratio". In the Fall Creek studies Johnson et al. (1976) found that the percent P of the solids tended to be more variable at low sediment loads, tending towards a uniform value of about 0.1% as the sediment load increased above 1000 mg/liter. The P load on the solid phase thus tended to be more uniform in more intense storms.

Since the concentration of total suspended solids varies widely within individual hydrologic events as well as from event to event, estimation of the particulate transport presents major difficulties unless a large number of samples are taken and analyzed. The uncertainties involved are well exemplified in the unique statistical analyses of the Fall Creek data presented by Johnson et al. (1976). In a further analysis of the same data, Bouldin et al. (1975) pointed out that, while long-term estimates (but not necessarily predictions) of sediment delivery can be made with some confidence if water-flow data are available, estimates of delivery over shorter times (monthly or yearly) are very uncertain.

The problem of estimating the impact of adsorbed P is further compounded by uncertainty as to the amount that remains accessible in the aqueous environment. Since most particulate material settles rather rapidly from the upper layers of lakes deep enough to show thermal stratification, the P will also be removed and have little effect on eutrophication unless the lake muds are stirred. The principal effect of available P carried by moving sediments will then be to act as a reserve to increase the concentration of dissolved P by desorption while the sediment remains in suspension. If it is assumed, as with soils, that about 10% of the total adsorbed P is potentially available, this will correspond to amounts between approximately 2 and 5 times that present in solution, depending on the sediment load and dissolved P concentration.

Much evidence also exists that sediment plays a considerable role in reducing the concentrations of soluble P in flowing streams and controlling the levels of the natural background. Much stream sediment is not derived from topsoil or eroding land surfaces, but from subsoil material eroded by the "gouging" of stream banks. Such material is usually P deficient and has a high adsorption capacity. In measurements of the adsorption isotherms of soils and sediments from a watershed in central Pennsylvania, Taylor and Kunishi (1971) showed that stream-bank sediments could readily reduce soluble P concentrations in large volumes of water to 0.01 ppm or less. Kunishi et al. (1972) demonstrated the practical consequences of this in measurements of P distribution between solution and sediment during the course of hydrologic events: concentrations of 0.10 to 0.13 ppm of soluble P in runoff water from fertilized fields were reduced to 0.009 ppm by adsorption on suspended sediment during the 12 hours required for downstream movement between sampling stations. Similar decreases in P concentrations with distance down flowing streams have been reported by Bouldin et al. (1975), Schuman et al. (1973), Keup (1968), and Gessner (1960). The stabilizing effect of sediments on soluble P concentrations in streamwater in relation to water flow has also been discussed in detail by Gburek and Heald (1974).

It is thus clear that P transport by sediment particles is affected by a number of factors which make any precise estimate of its significance impossible. On a unit area basis, the total amounts removed may greatly exceed the amount in solution in the associated runoff, although only a small fraction is likely to be present in available form. Much of the chemically available P may be carried through the flowing system and then removed from the biologically active zone of lake waters by sediment deposition. The introduction of P-deficient sediment by streambank erosion provides a large capacity sink for P adsorption which will frequently reduce soluble P concentrations below the 0.03- to 0.05-ppm level, resulting in environmental cleansing and a reduction in eutrophication. It is likely that this is a principal factor reducing the high concentrations which may be found in water draining woodland and natural vegetation to levels which control the biological activity in many oligotrophic lakes.

III. EFFECTS OF PHOSPHATE ADDITIONS

A. Fertilizer Applications and Soil Management

The purpose of fertilizer applications is to increase the amount of biologically available P in the soil. Although the efficiency of this practice is often limited by the chemical reactions that cause reversion of an appreciable fraction of the added P to inactive forms, both the amount of available adsorbed P and the solution concentration must be increased. The amounts that can be lost in sediment or dissolved in runoff water are thus also increased. The extent to which this may happen is illustrated by the data of Romkens and Nelson (1974) summarized in Table 1. These results show an increase in soluble and available P in direct proportion to the fertilizer added. No such relationship was found for the organic or total P which, being much larger, were dominated by the native amounts in the soil: the differences between these probably reflected the separation of different fractions of colloid material under differing erosion conditions imposed in the experiment. The organic and total P fractions were much more sensitive to the rate of application of artificial rainfall. Companion laboratory

Table 1—Increases in P concentration in runoff water and sediments by fertilization of Russell silt loam (Romkens & Nelson, 1974).†

Fertilizer applied	Dissolved P	P content of sediment		
		Available	Organic	Total
kg P/ha	ppm	ppm		
0	0.07	14.6	152	558
56	0.24	35.4	99	446
113	0.44	57.6	106	461

† Available P measured by Bray no. 1 extraction; total P by perchloric digestion; organic P by increase on ignition.

studies demonstrated the feasibility of predicting the effects of fertilization on potential losses of soluble and available P from individual fields by simple laboratory tests if runoff and erosion could be predicted.

Translation of such potential losses into direct predictions of possible losses, even from individual fields, requires a detailed understanding of the impact of soil management practices on both runoff and erosion. Data obtained by Schuman et al. (1973) on loess soils in the Missouri Valley, summarized in Table 2, show that while both concentrations in runoff and available P losses were increased by fertilization of contour plowed corn on two comparable watersheds (treatments 1 and 2, Table 2), other management practices had much larger effects. Although sediment losses from level terraced corn and a bromegrass pasture, both fertilized, were greatly reduced, the total amount of available P was not correspondingly less. On the bromegrass watershed, surface application of fertilizer and release of P from vegetation during leaching by snowmelt gave large increases in concentration. The effects of management were also investigated by Romkens et al. (1973): comparisons of five tillage practices clearly demonstrated that these outweighed fertilizer application rates in controlling P losses. Tillage systems that left either fertilizer residues or highly fertilized topsoil exposed to the action of surface runoff caused the largest losses and concentration of soluble P. In other studies, Alberts et al. (1978) showed that total P losses, including P adsorbed on sediments were highest between April and June when bare soil was exposed to erosion: concentrations in water were highest in the winter, reflecting surface leaching of decaying surface vegetation. Other data emphasizing the importance of farming systems

Table 2—Available P loss in solution and sediment from four watersheds on Missouri Valley loess under different management between 1969 and 1971 (Schuman et al., 1973).

Water-shed	Management	Fertilizer application	Available P loss		
			Solution	Sediment	Total
		kg P/ha	ppm	kg/ha per year	
1	Contour plowed corn	97	0.021	0.17	1.05	1.22
2	Contour plowed corn	39	0.016	0.11	0.58	0.69
3	Bromegrass pasture	39	0.071	0.21	0.07	0.29
4	Level terraced corn	97	0.052	0.05	0.08	0.13

and soil management have been presented by Stoltenberg and White (1953) and Burwell et al. (1975).

Overall consideration of this information indicates that while the effect of fertilizer application on P losses from individual fields or uniformly treated watersheds can be identified, detailed interpretation of such losses is difficult since they are greatly influenced by the infiltration and runoff characteristics, soil erodibility, and management practices which are not easily quantified. In an analysis of 4 years of losses of nutrients from six watersheds in Oklahoma, Menzel et al. (1978) showed that annual variations may be as great as those caused by different land uses or fertilizer additions. These authors emphasize the need for long-period records in comparing nutrient (and sediment) discharges from different management practices. Extending such observations to the prediction of the impact of fertilizer application over large land areas presents even greater problems. Amounts of soluble and available P reaching streams are greatly influenced by mixing of sediments and adsorption of soluble P as discussed in the previous section. Prediction of P behavior thus requires accurate estimation of the pathway of the phosphate-bearing sediment from an eroding field, and also the source and chemical behavior of the sediment with which it is mixed and the rate at which it is transported.

For similar reasons, interpretations of observed P loadings of lake and streamwaters in relation to different sources that exist within a watershed or drainage area are equally difficult. Even where straight comparisons between adjacent areas can be made, only the broadest conclusions are possible. In a comparison of adjacent woodland and farmland in central Ohio, Taylor et al. (1971) concluded that concentrations in water draining mixed farmland were about 50% higher than from unfertilized woodland. Since one-half the farmland was in permanent pasture, and the bulk of P release was found in the later winter and early spring, the contribution from vegetation could not be separated from that due to fertilizer application. A more comprehensive analysis of P transport in Fall Creek, New York, was presented by Bouldin et al. (1975) and Johnson et al. (1976). This stream drains a 330 km^2 area of diverse use, including woodland, a mixed agriculture with dairy farming and cropland and a rural and semi-urban population of about 12,000 people. Sufficient data was available to estimate P inputs and outputs. Fertilizer input to the watershed was estimated to be about 70% of the P cycling within the watershed due to the application of animal waste. Fertilizer usage was about half the total input, the balance being due to human sources and the import in animal feedstuffs. Although subject to some uncertainties, which were fully discussed, the analysis suggested that about 45% of the stream load could be attributed to nonhuman activities, 35% to domestic sewage, and 25% to farming. One of the principal difficulties in the analysis was the estimation of "transmission losses" due to adsorption by stream sediments. Observed loadings indicated that these losses were erratic due to flushing of stream sediments during larger hydrologic events.

B. Phosphorus in Animal Wastes

The contribution from animal wastes can under some circumstances represent a significant fraction of the P circulating in agricultural systems. The annual excretion of P ranges from about 45 g/kg of body weight for swine to about 17 to 20 g/kg for cows, horses and sheep (Millar & Turk, 1955). The impact of animal wastes upon P flows depends very greatly upon management. Data on the composition of animal manures (Taylor, 1967) show that the total P content is about the same or slightly higher than that of fertile topsoil, although, because the substrate is mainly undigested plant material, a greater fraction may be in the water-soluble form. Chichester et al. (1979) showed that concentrations of P in runoff were not increased by summer grazing of pasture in Ohio, but where animals were pastured throughout the year, winter damage to the soil surface caused both increased runoff and nutrient discharge emphasizing the adverse effect of this difference in management practice.

Where faecal matter is deposited into farm ponds or streams the direct effect may be noticeable. Spreading of animal manure upon cultivated fields may also increase P concentrations in runoff where the waste is not plowed down and is left exposed to rain or meltwater moving over frozen or saturated soil. The effect of weather on P transported in this way was demonstrated by Klausner et al. (1976) who observed soluble P losses of over 9 kg/ha from 100-metric ton surface applications of dairy waste under adverse conditions: these were exceptional, however, and considerable differences were found in different years. Variations due to weather and alternate management practices were also observed by Young and Mutchler (1976) in Minnesota. Fall applications of manure to alfalfa gave the worst losses: fall applications to fall-plowed cornfields gave much smaller losses, again emphasizing the effects of better management.

More severe problems may arise where there are local, high density animal populations in feedlots or barnyards close to streams. Actual losses will depend upon management practices, feedlot design, location, and the time the animals are confined. Where direct discharge may take place from a paved lot the problem may be regarded as a point source and alleviated by collection of both solid and liquid waste in retention systems for subsequent disposal by land spreading.

Nutrient losses in runoff from a small unpaved beef-cattle barnlot typical of many in the eastern U.S. were measured by Edwards et al. (1972). Deposition of 64 to 100 metric tons (wet) of waste on the barnlot resulted in P concentration up to 20 ppm in the discharge, with highest levels during October through January. Annual total losses ranged between 3.4 and 1.1 kg P over a 3-year period. The results show that such farming operations can act as significant sources of P for streamwater, but the final overall impact on water quality can only be estimated by an analysis of the adsorption capacity of the sediments in the stream below the point of discharge (Taylor & Kunishi, 1971).

C. Phosphorus in Urban Runoff

Data on the amounts of soluble and adsorbed P carried by runoff from paved urban areas was reviewed by Ryden et al. (1973) who noted that significant amounts, reaching 3.3 kg/ha per year, have been reported. Leaching of leaf trash and debris appears to be the main source, as in many agricultural systems, but the problem is aggravated by the impermability of the surface which leads to complete removal of the P without the amelioration due to downward movement of water moving into soil. Where urban streets and paved areas are permitted to discharge runoff directly into lakes or stream waters, the lack of infiltration capacity represents an acceleration of the natural cycle by which P tends to move from land and vegetation, and must lead to an increase in the rate of eutrophication of the recipient water.

IV. OVERVIEW

Water transport is the dominant pathway by which P moves through the environment, either in water-soluble form or in adsorbed forms carried by moving sediment. The chemistry of the adsorption process is complex, and the fraction of the P carried in this way that is biologically available is difficult to predict. On balance the effect of sediments is beneficial, particularly where they are derived from P-deficient subsoils, since they reduce the biological activity and delay the rate of movement. Much P carried into lakes by sediment may be also inactivated by depositional removal from the biotic zone.

The largest potential sources of P are the erosion of heavily fertilized fields, animal feedlots and urban areas. Since removal from all these sources is dependent upon the energy supplied by surface water movement, which is an erratic process controlled by rainfall, prediction of the amounts released at particular locations will be wholly dependent upon hydrologic forecasts of surface water movement. In general the temporal changes in water volume greatly exceed the changes in concentration of P, so that adequate hydrologic data interpretation is an absolute prerequisite for the interpretation of P transport.

Studies of P distribution in lakes and streams have demonstrated that it reflects a highly dynamic system. Interpretation of the impact of a particular P source on water quality at a downstream location must, therefore, take into account effects of dilution, adsorption, and contributions from both other sources and the natural background as well as the actual amount of P that may be released by the source at a particular time. Large amounts of chemical and hydrologic data, that can be handled by computer techniques are essential. It is doubtful, however, such complex analyses will lead to any improvement over presently known methods for reduction of the impact of particular sources. Since water transport is the primary mechanism, farm and feedlot operations consonant with good conservation practices that reduce runoff and erosion are likely to remain the only economically viable methods of control. Other methods, which involve chemical processes for removal of P from large bodies of water may prove prohibitively expensive in relation to the potential benefits. General studies have shown that natural background levels of P are frequently close to those accepted as limits for eutrophication, and future research efforts should concentrate on the evaluation of the efficiency of the simpler and most economic practices which can be used to prevent these limits from being exceeded.

LITERATURE CITED

Alberts, E. E., G. E. Schuman, and R. E. Burwell. 1978. Seasonal runoff losses of nitrogen and phosphorus from Missouri Valley loess watershed. J. Environ. Qual. 7:203-208.

Baker, J. L., K. L. Campbell, H. P. Johnson, and J. J. Hanway. 1975. Nitrate, phosphorus and sulfate in subsurface drainage water. J. Environ. Qual. 4:406-412.

Barrows, H. L., and V. J. Kilmer. 1963. Plant nutrient losses from soils by water erosion. Adv. Agron. 15:303-316.

Bouldin, D. R., A. H. Johnson, and D. A. Lauer. 1975. Influence of human activity on export of P and N from Fall Creek. p. 61-120. *In* K. S. Porter and R. J. Young (ed.) Nitrogen and phosphorus: Food production, waste, and the environment. Ann Arbor Science Publ., Ann Arbor, Mich.

Burwell, R. E., D. R. Timmons, and R. F. Holt. 1975. Nutrient transport in surface runoff as influenced by seasonal cover and seasonal periods. Soil Sci. Soc. Am. Proc. 39:523-528.

Chichester, F. W., R. W. Van Keuren, J. L. McGuinness. 1979. Hydrology and chemical quality of flow from small pastured watersheds: II. Chemical quality. J. Environ. Qual. 8:167-171.

Duffy, P. D., J. D. Schreiber, D. C. McClurkin, and L. L. McDowell. 1978. Aqueous and sediment-phase phosphorus yields from five southern pine watersheds. J. Environ. Qual. 7:45-50.

Duxbury, J. M., and J. H. Peverly. 1978. Nitrogen and phosphorus losses from organic soils. J. Environ. Qual. 7:566-570.

Edwards, W. M., E. C. Simpson, and M. H. Frere. 1972. Nutrient content of barnlot runoff water. J. Environ. Qual. 1:401-405.

Gburek, W. J., and W. R. Heald. 1974. Soluble phosphate output of an agricultural watershed in Pennsylvania. Water Resour. Res. 10:113-118.

Gessner, F. 1960. Investigations of the phosphate economy of the Amazon. Int. Rev. Hydrobiol. 45:339-345.

Hanway, J. J., and J. M. Laflen. 1974. Plant nutrient losses from tile-outlet terraces. J. Environ. Qual. 3:351-356.

Johnson, A. H., D. R. Bouldin, E. A. Goyette, and A. H. Hedges. 1976. Phosphorus loss by stream transport from a rural watershed: Quantities, processes, and sources. J. Environ. Qual. 5:148-157.

Keup, L. E. 1968. Phosphorus in flowing waters. Water Res. 2:373-386.

Klausner, S. D., P. J. Zwerman, and D. F. Ellis. 1976. Nitrogen and phosphorus losses from winter disposal of dairy manure. J. Environ. Qual. 5:47-49.

Kunishi, H. M., A. W. Taylor, W. R. Heald, W. J. Gburek, and R. N. Weaver. 1972. Phosphate movement from an agricultural watershed during two rainfall periods. J. Agric. Food Chem. 20:900-905.

Logan, T. J., and G. O. Schwab. 1976. Nutrient and sediment characteristics of tile effluent in Ohio. J. Soil Water Conserv. 31:24-27.

McHargue, J. J., and A. M. Peter. 1921. The removal of mineral plant-foot by natural drainage waters. Kentucky Agr. Exp. Sta. Bull. 237.

Menzel, R. G., E. D. Rhoades, A. E. Olness, and S. T. Smith. 1978. Variability of annual nutrient and sediment discharges in runoff from Oklahoma cropland and rangeland. J. Environ. Qual. 7:401-406.

Millar, C. E., and L. M. Turk. 1955. Soil fertility. John Wiley & Sons, New York.

Miller, M. H. 1979. Contribution of nitrogen and phosphorus to subsurface drainage water from intensively cropped mineral and organic soils in Ontario. J. Environ. Qual. 8:42-48.

Ogelsby, R. T., and W. R. Schaffner. 1975. The response of lakes to phosphorus. p. 25-27. *In* K. S. Porter and R. J. Young (ed.) Nitrogen and phosphorus: Food production, waste, and the environment. Ann Arbor Science Publ., Ann Arbor, Mich.

Romkens, M. H. M., and D. W. Nelson. 1974. Phosphorus relationships in runoff from fertilized soils. J. Environ. Qual. 3:10-13.

Romkens, M. H. M., D. W. Nelson, and J. V. Mannering. 1973. Nitrogen and phosphorus composition of surface runoff as affected by tillage method. J. Environ. Qual. 2:292-295.

Ryden, J. C., J. K. Syers, and R. F. Harris. 1973. Phosphorus in runoff and streams. Adv. Agron. 25:1-45.

Schreiber, J. D., P. D. Duffy, and D. C. McClurkin. 1976. Dissolved nutrient losses in storm runoff from five southern pine watersheds. J. Environ. Qual. 5:201-205.

Schuman, G. E., R. G. Spomer, and R. F. Piest. 1973. Phosphorus losses from agricultural watersheds on Missouri Valley loess. Soil Sci. Soc. Am. Proc. 37:424-427.

Singer, M. J., and R. H. Rust. 1975. Phosphorus in surface runoff from a deciduous forest. J. Environ. Qual. 4:307-310.

Stoltenberg, N. I., and J. L. White. 1953. Selective loss of plant nutrients by erosion. Soil Sci. Am. Soc. Proc. 17:406-410.

Taylor, A. W. 1967. Phosphorus and water pollution. J. Soil Water Conser. 22:228-231.

Taylor, A. W., W. M. Edwards, and E. C. Simpson. 1971. Nutrients in streams draining woodland and farmland near Coshocton, Ohio. Water Resour. Res. 7:81-89.

Taylor, A. W., and H. M. Kunishi. 1971. Phosphate equilibria on stream sediment and soil in a watershed draining an agricultural region. J. Agric. Food Chem. 19:827-831.

Thomas, G. W., and J. D. Crutchfield. 1974. Nitrate-nitrogen and phosphorus content of streams draining small agricultural watersheds in Kentucky. J. Environ. Qual. 3:46-49.

Timmons, D. R., and R. F. Holt. 1970. Leaching of crop residues as a source of nutrients in surface runoff water. Water Resour. Res. 6:1367-1375.

Timmons, D. R., and R. F. Holt. 1977. Nutrient losses in surface runoff from a native prairie. J. Environ. Qual. 6:369-373.

Timmons, D. R., E. S. Verry, R. E. Burwell, and R. F. Holt. 1977. Nutrient transport in surface runoff and interflow from an aspen-birch forest. J. Environ. Qual. 6:188-192.

Young, R. A., and C. K. Mutchler. 1976. Pollution potential of manure spread on frozen ground. J. Environ. Qual. 5:174-179.

White, E. M., and E. J. Williamson. 1973. Plant nutrient concentrations in runoff from fertilized cultivated erosion plots and prairie in eastern South Dakota. J. Environ. Qual. 2: 453-455.

Zwerman, P. J., T. Grewling, S. D. Klausner, and D. J. Lathwell. 1972. Nitrogen and phosphorus content of water from tile drains at two levels of management and fertilization. Soil Sci. Soc. Am. Proc. 36:134-137.

Chapter 20

Phosphate Nutrition of Plants— A General Treatise

P. G. OZANNE

CSIRO, Wembley, Western Australia

I. INTRODUCTION

In this chapter it is intended that the reader will find a broad general treatment of the particular and peculiar role of P in plant nutrition. The main emphasis is on those aspects of P nutrition which are not only important but which can be manipulated by man to his advantage. Recent reviews which treat various aspects of plant nutrition in much more detail than can be encompassed in this chapter include those by Barley (1970), Harley (1971), Epstein (1972), Bieleski (1973), Peel (1974), and the other sections of this book.

Phosphorus is indispensable for all forms of life because of its genetic role in ribonucleic acid and function in energy transfers via adenosine triphosphate. In natural ecosystems P is usually the life-limiting element due to its low availability. The prevalence of Al, Fe, and Ca which link P in highly insoluble compounds ensures this. Of the P in the soil-plant-animal system, commonly over 90% is in the soil. Of this less than 10% enters the plant-animal life cycle. Within the soil the P can be considered to be in three fractions—the organic matter which may hold up to 50% of the total soil P, the insoluble inorganic fraction, and a small, very variable part, that is soluble and can be absorbed by plants.

The central role of applied P in the nutrition of plants becomes evident when the 17 essential elements found in plants are considered. Carbon and oxygen are universally present and available from the atmosphere. Of the remaining 15 elements that are taken up from the soil, H is always abundantly available if there is enough water for plant growth. Calcium and Mg deficiencies are not frequent, except in very acid or sandy soils. Sulfur is continually supplied to the soil in rainfall, especially in industrialized regions. Sulfur is also readily released from soil organic matter by mineralization, and certain P fertilizers may contain 5 to 11% S as a residue from sulfuric acid used in the manufacturing process.

Although frequently deficient, K is readily recycled from organic residues, and fertilizer K is generally quite available to plants. Nitrogen is com-

Copyright 1980 © ASA-CSSA-SSSA, 677 South Segoe Road, Madison, WI 53711, USA. *The Role of Phosphorus in Agriculture.*

monly deficient, but legume crops can fix their needs largely from the air, while legume-containing pastures can fix not only their own requirements, but sometimes also those of cereal crops grown in alternate years.

The cost of supplying the micronutrients required by crops, and the additional Se, I, and Na required in forages by animals, is usually small. Also, Cu dressings may have a residual effect for 10 years or so, while Zn is often supplied in variable quantities as an inadvertent component of phosphate fertilizers (Ozanne et al., 1965b).

The situation with P is different. The available supply in soils is often deficient for the growth of commercial plants. This deficiency can generally be overcome only by applying P fertilizers. Phosphorus is not recycled in rainfall nor readily released from organic residues. Soluble P fertilizers applied to the soil are very rapidly changed to less-soluble compounds which, with time, become less and less available to plants.

II. SOIL FACTORS IN PHOSPHATE AVAILABILITY

A. Soil Solution P and Phosphate Buffering Capacity

The ability of 90 surface soils to sorb P from dilute solutions (1 to 24 ppm P) of monocalcium phosphate (MCP) has been examined (Ozanne, 1962). I found a close linear correlation ($r - 0.96$) between the amount of P sorbed and the acetic-acid-soluble Al and oxalate-soluble Fe. Of course, MCP is usually applied as a solid; however, Saunders (1964) found a close correlation between the P retained from solution and that retained from solid $Ca(H_2PO_4)_2 \cdot 2H_2O$.

Commonly the concentration of the soil solution is around 0.05 ppm P (Barber et al., 1963) and is seldom higher than 0.3 ppm (Fried & Shapiro, 1961) in soil not influenced by fertilizer P. The P responses of over 40 areas of pasture were measured by Ozanne and Shaw (1967). We found that no further yield response could be obtained to P applied to soils in which the solution concentrations were 0.2 to 0.3 ppm P. Earlier, Asher et al. (1965) developed a technique for growing plants in rapidly flowing culture solutions maintained at very low but constant nutrient concentration. This technique was used by Asher and Loneragan (1967) to grow eight species at five levels of P concentration. Six of the eight made maximum growth in solutions of 0.15 ppm P or less while the remaining two made 90% of maximum growth at around 0.30 ppm. As a result a soil test was proposed (Ozanne & Shaw, 1967) in which the amount of P required on a given area was based on the amount sorbed by soil samples in reaching equilibrium with a solution containing 0.3 ppm P.

The ability of a soil to resist changes in the concentration of the equilibrium solution as P is added or removed, is known as the *phosphate buffering capacity*. A useful technique for measuring it is described by Ozanne and Shaw (1968). There is a strong link between it and the equilibrium P concentration needed for maximum plant growth. Rajan (1973) found that soil solution concentrations required by millet varied from 0.02 to 0.6 ppm

P. Soils of high buffering capacity produced maximum growth of millet at low solution concentrations.

The phosphate buffering capacity is an intrinsic characteristic of soils that is of great importance in plant nutrition. Also, it is a reasonably stable one. I have found little difference in the buffering capacities of soil on two fertilizer treatments applied to a sandy loam under pasture. One treatment has received no P, and the other 40 kg P/ha as double superphosphate annually for 9 years (unpublished data). The available P status, of course, now differs very widely between the two treatments.

A word of warning is perhaps needed here. A wide range of techniques for measuring the buffering capacity of soils have been described. But the relationship between P concentration in solution and the P adsorbed is not a smooth curve in many soils. Hence phosphate buffering capacities measured using data well outside the equilibrium concentration range of 0.01 to 0.30 ppm P in solution (i.e., the range commonly encountered by plant roots) may be inappropriate for work in plant nutrition. Buffering capacities measured within this range may be successfully used not only to quantitatively predict the amount of P required, but also as a correction factor in conjunction with soil tests based on extractable P (Ozanne & Shaw, 1968).

B. Phosphate Distribution Down the Profile

Not only is the concentration of P in the soil solution usually low, but P also moves very slowly, mostly by diffusion. The diffusion coefficient for P in soil solution was found by Olsen et al. (1962) to be $< 10^{-2}$ that in water. Barber et al. (1963) reported values for the rate of diffusion of ^{32}P in soil as low as 4×10^{-11} cm^2 sec^{-1}, while Lewis and Quirk (1965) gave the mean velocity of phosphate ions in Seddon soils as only 0.04 mm/day at an applied P level of 100 ppm.

As a result of this very slow movement in soil, P tends to accumulate close to the surface. The organic P in plant residues and animal feces is normally returned to the soil surface, and incorporated into the topsoil only to a limited extent by insects, worms, fungi, etc., unless the land is cultivated. Much of the P returned to the soil by the plant-animal system is in the organic form, i.e., about 40% of the P in plant residues (Jones & Bromfield, 1969) and 60 to 70% in animal feces (Bromfield, 1961; Barrow, 1975). Barrow has shown that the inorganic P in feces is available to plants but needs to be incorporated into the soil.

In topdressed areas, the application of inorganic P fertilizer dominates the amount of P entering the soil. It, too, is largely applied to the soil surface on forage and pastures under conditions such that the bulk of the P is retained in the surface 2 to 3 cm of soil (Ozanne & Shaw, 1967; Brownlee et al., 1975). When P is plowed in or drilled into the soil it seldom reaches a depth of more than 10 to 15 cm.

This concentration of virtually all the plant available P near the surface in many soils forms a hazard where surface erosion may remove valuable amounts of P. As we shall see in the next section it also provides a major problem in plant nutrition.

III. PLANT FACTORS IN PHOSPHATE AVAILABILITY

A. Root Characteristics

1. ROOT DISTRIBUTION DOWN THE PROFILE

Whether soluble or insoluble P fertilizers are applied to soil, they are retained close to the surface. Largely as a result, plant roots proliferate in this P-rich surface zone and a relatively small proportion penetrates to the subsoil, where the bulk of the soil water, potentially available to plants, is stored. Twelve annual pasture species were grown in the field and the distribution of their roots down the soil profile studied by Ozanne et al. (1965a). After 4 months' growth, half or more of the roots of all species were present in the top 10 cm of the soil, even though all had roots to a depth of at least 1.5 m. Similar results have been obtained for perennial species by Jacques (1943), Troughton (1961), and others.

This proliferation of roots near the surface was shown to be largely due to P (Sewell & Ozanne, 1970). We collected a fertile topsoil and filled deep cylinders uniformly with it. During the course of filling, P was applied in a band at 15 cm deep in some cylinders and to the surface of others. Wheat and subterranean clover were then grown separately in the cylinders. In both species, when the P was banded at 15 cm instead of applied to the surface, root development at the 15-cm depth was trebled.

As a result of this proliferation of roots near the surface and sparse development lower down in the soil profile, the surface soil rapidly dries out in times of moisture stress. The soil P consequently becomes unavailable. Plants may cease growth or die from a combination of P deficiency and an inadequate rate of water uptake. At high levels of applied P the rate of movement could be increased 1000-fold. This raises the possibility that some of the yield responses obtained with high rates of applied P may well be due to increases in the amount of available water. We have shown (P. G. Ozanne & P. L. Sewell, 1975, unpublished data) that under conditions of intermittent drought, trebling the rate of surface-applied superphosphate or placement at the 10-cm depth gave equal responses in yields of wheat.

There appear to be four ways in which more adequate quantities of P for root and top growth may be supplied to the moist subsoil. The first, and perhaps most time-honored way, is just to apply very high levels of P to the surface soil. The second is to place the P directly into the subsoil. A third method, not yet well developed, is the application of P in a chemical form that does not react strongly with the soil but can move down the soil profile with rain or irrigation water. Lastly there are occasional reports of the successful application of P as a foliar spray. Phosphate absorbed through the leaves may be translocated down to the roots in sufficient quantities to remove any growth limitations that exist. Work to ascertain whether this occurs has not yet been done.

Curiously, although root proliferation is stimulated in a P-rich zone, the addition of P leads to a reduction in the actual amount of roots per unit weight of tops (Ozanne et al., 1969; Christie & Moorby, 1975). Indeed the total weight of roots may even be reduced when the weight of plant tops shows a response to applied P (Asher & Loneragan, 1967). It seems that photosynthates translocated downward from the leaves to the roots are trapped and utilized for root growth in the zone where P is least limiting. Similar marked increases in root growth and preferential accumulation of dry matter have been shown in localized zones of high nitrate supply by Drew et al. (1973) and Drew and Saker (1975) working with barley. Further, the relatively greater reduction in top growth than root growth under conditions of P deficiency may be partly due to the more favorable P status of the roots which have first opportunity to metabolize absorbed P. Also there is probably an evolutionary advantage in survival and growth in having relatively large and extensive root systems under conditions of low available P supply.

2. ROOT LENGTH AND MYCORRHIZA

The importance of root length in the uptake of P adsorbed by soil has been hypothesized by Barley (1970). Also, Andrews and Newman (1970), showed that reducing root length by pruning reduced P uptake, but did not significantly decrease uptake of N. Root length is, of course, a major determinant of the absorbing surface area, and data showing this were presented by Christie (1975). This may well account for the exceptional ability of plants like *Banksia* to absorb P. These plants have proteoid roots forming a dense mat in the surface soil (Jeffrey, 1967).

Recently, studies on the symbiotic associations of roots have been intensified. The long-known beneficial effects of the ectotrophic mycorrhizae, commonly found on trees, are largely due to the enhanced P supply to the host plant (Bowen, 1973). More recent work on the endotrophic mycorrhizae, in which the fungus does not form an external sheath, but lives within the root, has shown similar benefits in supplying P to a range of herbs and grasses (Khan, 1974). The most common of the endotrophic mycorrhizae are those formed by fungi belonging to the genus *Endogone* (Harley, 1971; Nicolson, 1975). However, vesicular-arbuscular (VA) mycorrhizae of a similar kind are formed by a number of other genera. Under conditions of adequate P supply, development of the mycorrhizae is reduced and the host plant shows little benefit from inoculation. Under P-deficient conditions, however, Khan (1975) found grain yields of wheat to be increased threefold by inoculation of the plants with spores from a VA mycorrhiza. Similar responses have been obtained by a number of workers in several countries and on a wide range of species.

There appears to be a similarity between the infection of leguminous plants by the *Rhizobia* bacteria and the infection of a much wider range of plants by the VA mycorrhizae. In the case of legumes and *Rhizobia* a symbiotic association is set up which allows the host plant to grow in soils very low in available N, and large yield responses to inoculation may result; however, no benefit from infection occurs if high levels of fertilizer N are applied. Much the same result in regard to P is produced by infection with VA mycorrhizae.

There ae many strains of *Rhizobia* capable of infecting a given legume, but they vary widely in ability to fix atmospheric N. In the same way it is likely that large variations exist in the ability of VA mycorrhizae to supply the host plant with P. The host plant may well be a successfully nodulated legume capable of fixing large amounts of N. Well-developed nodules containing *Rhizobia* and tissue containing the hyphae of mycorrhizal infection may occur on the same root (Daft & El-Giahmi, 1975). The growth and amount of N fixed by a legume are strongly influenced by its P supply. So the effectiveness of the strain of VA mycorrhiaza may determine not only the P supply to the host but also the amount of N fixed by it, and, indirectly, the yield of the N-dependent biomass.

So far, in contrast to *Rhizobia*, it has proved difficult to culture VA mycorrhizae on synthetic media in the absence of the host plant. But the technical problems in producing effective spores for inoculation of a field crop have now been largely overcome. Also, once an area has been successfully infected it should remain so for a period of years.

In the P-deficient soils in which the mycorrhizae are most effective in increasing P uptake, diffusion is more likely to limit adsorption. Bieleski (1973) attempted to calculate and quantify the increased absorbing surface of a root infected with fungal hyphae as compared with one uninfected. His figures show a likely increase of 5 to 20 times in inorganic P uptake. About half the P in the surface soil where the mycorrhizae form is organic in nature. The hyphae are likely to be able to decompose some of this, and so may also be able to tap organic P sources not available to the uninfected plant. Bieleski (1974) has shown that, as a response to P deficiency, the simple floating water plant *Spirodela* can increase its externally located phosphatase activity by a factor of up to 250 times. Similarly, some uninfected plants may be able to absorb a significant amount of P from the organic compounds in the soil.

B. Specific Differences

1. ROOT DEVELOPMENT

Differences in the levels of P supply needed by various species have been shown a number of times. Several factors can cause the variations. Perhaps the most simple is the case of trees which may not respond to moderate levels of applied P but do respond to heavy dressings. The reason for this is that their absorbing roots are below the depth to which the fertilizer penetrates. Sometimes the differences in requirement can be largely explained by just the mass and distribution of roots present. For example, when I compared the levels of P needed by three species growing in the field, all the applied P was retained in the top 10 cm of soil. The root weights in this layer for grass, clover, and lupins were 2.8, 2.2, and 0.8 metric tons/ha, respectively, while the applied P needs were in the reverse order of 40, 49, and 65 kg/ha, respectively (Ozanne & Petch, 1978).

Measuring the ratio of the tops present to the roots that supply them with P is more helpful in understanding the different requirements of some plants. When eight species studied in a pot experiment by Ozanne et al. (1969) were ranked in order of applied P requirement, six of the species fell in the same order as their top/root ratio; i.e., the larger the tops relative to their roots, the higher the rate of applied P needed for near maximum yield.

Measurement of just the root weight present in the fertilized zone is not fully adequate. Root length is better. However, root length should be related to the amount of P it is required to absorb. In a field experiment carried out on virgin land (P. G. Ozanne & K. M. W. Howes, 1972, unpublished data), we grew four species in pure stands, two in a mixed stand, and one with the yields varied by two levels of N. From the six rates of P applied to all species treatments, the fertilizer levels needed for 90% of maximum yield were measured. They ranged from 55 to 200 kg of P/ha. Neither root weights, root/top ratios, root lengths in the fertilized zone, nor total P taken up by the tops could be related to the differences found. However, when the length of roots present in the fertilized zone was divided by the amount of P needed in the tops for near maximum growth, the level of applied P fertilizer demanded by each different species treatment was found to rank in exactly the reverse order to the length of roots available to absorb each gram of P utilized. With increasing length of roots available to absorb each unit of P needed in the plant tops, the level of P fertilizer required decreased.

Lewis and Quirk (1967), working with wheat, suggested that it was not only the root length, but also the length of the root hairs that may determine P absorption from soil. They found that P was depleted from around the root in a cylinder corresponding in dimensions to the root length and the diameter from root hair tip to tip. In a field experiment (P. G. Ozanne & K. M. W. Howes, 1971, unpublished data) the total volume of this root hair cylinder was measured to a depth of 10 cm under a pure stand of annual ryegrass. It was found to be somewhat greater than the total pore space available to water in this soil layer. The zones of influence of the roots overlapped, and soil P may have been completely exploited under these conditions.

2. ROOT COMPETITION

By contrast, similar measurements of the total volume of the root hair cylinders under a pure stand of subterranean clover showed that the roots exploited only about 25% of the pore space in the soil in the fertilizer zone. These measurements may show why the applied fertilizer requirement for near maximum yield of the ryegrass was only about half that of the subterranean clover. The above measurements also suggest that when the ryegrass and clover were grown together at moderate levels of soil P, root competition for uptake of P occurred. This suggestion is supported by data showing that the applied P required by the ryegrass-clover mixture was increased about 25% for an increase of 50% in stand density. By contrast an increase of 80% in the stand density of clover grown alone had no effect on the level of applied P needed.

An experiment was conducted (P Lapins, 1975, personal communication) in which annual grass and clover were compared when grown with the tops intermingled but the roots in the same or separate compartments of soil. When the roots were in the same compartment and competing for adsorption, P uptake by the clover was significantly reduced.

The large volume that the root hair cylinder may occupy in the fertilizer zone of the soil is further demonstrated by data from Evans (1970). Working with perennial ryegrass grown in pots he measured root lengths of up to 1,816 m in 2.7 liters of sand culture. If we assume a root hair cylinder of 1.5 mm in diameter (Lewis & Quirk, 1965, 1967) and 1,800 m long, then the volume of the total root hair cylinder would have been up to 3 liters in a sand volume of only 2.7 liters, showing that considerable overlap and competition between roots may occur.

Work in the field with two perennial pasture species, browntop grass and white clover, was carried out by Jackman and Mouat (1972a). They showed that the level of applied P required by white clover for maximum yield was increased about threefold when it was grown in a mixed stand with browntop. Also the presence of browntop reduced the P concentration slightly in the white clover.

Although they apparently made no measurements of root length or root hairs, Jackman and Mouat (1972a) give data showing the distance between the tips of roots of perennial ryegrass and white clover. When grown in soil in pots the root tips of the clover were 6.3 mm apart, while those of the ryegrass were only 2.0 mm apart at the same depth of soil. The number of root tips, of course, represented the number of growing points exploring new sources of P. Not only do the number of root tips vary, but so do their rates of elongation. Asher and Ozanne (1966) measured the rates of 12 annual pasture species and found the main roots grew at from 0.8 to 3.2 cm/day.

3. SUSCEPTIBILITY TO INFECTION BY VESICULAR-ARBUSCULAR MYCORRHIZAL FUNGI

The formation of endotrophic mycorrhizae and their beneficial effects on P uptake have already been discussed. However, their different effects on different species are worth a mention. Khan (1974) examined 89 species from 43 families and found *Endogone* spores in the rhizosphere soil and VA mycorrhizae in halophytes and xerophytes, but not in hydrophytes. Among the 17 families in which one or more species had mycorrhizae present were the Papilionaceae, Solanaceae, Compositae, and Gramineae. No mycorrhizae were found in 26 families including the Oleaceae, Zygophyllaceae, Euphorbiaceae, Chenapodiaceae, Caryophyllaceae, and Palmae.

A short, succinct review of our knowledge of mycorrhizae is presented by Guttay (1975). In it he points out that most annuals would be in midgrowth before they had much mycorrhizal development. So the useful function of mycorrhizae in P nutrition is subject to this limitation, in contrast to the more rapid development of N fixation in legumes infected with *Rhizobia*. However, in P-deficient soils, the presence of endotrophic mycorrhizae was found by Crush (1974) to strongly stimulate the nodulation and growth of tropical legumes. These were much more depenent on the mycorrhizae for growth than were the temperate species. Even within the temperate species, when white clover and perennial ryegrass were grown together, mycorrhizae were found to preferentially stimulate the growth of the legume. This could have been due to the greater inherent ability of ryegrass to take up soil P, as compared with white clover, when neither species is inoculated with VA mycorrhizae.

4. APPLIED P REQUIREMENT

So far we have considered reasons why plants may differ in their P needs. The level of applied P required to grow a particular species is of major economic importance. Hence we may benefit from looking at the magnitude of some of the differences that have been found between species well adapted to the same climate, the same soils, and in most cases, the same management. For simplicity, only data on plants growing without interspecies competition will be discussed.

When Asher and Lonergan (1967) compared eight annual pasture species in flowing culture solutions, they showed a range of 10-fold in the P concentrations needed for maximum growth between the most demanding and least demanding species. The highest concentration needed (0.76 ppm P) was by flatweed, a composite. Next highest was required by barrel medic (a legume), while silver grass made maximum growth at the lowest concentration needed (0.03 ppm P) by any of the eight species. A somewhat similar range of concentrations was reported by Fox et al. (1974) who measured the P levels in the soil solutions of a field experiment. They found that for 95% of maximum yield, head lettuce needed 0.4 ppm, sweet potato 0.1 ppm, and corn 0.06 ppm P.

Working with plants growing in soil in pots, Ozanne et al. (1969) compared eight annual pasture species and found greater than a 10-fold range in applied P requirement. The plant with the lowest need was again the silvergrass, while cupped clover had the highest requirement of applied P for near maximum yield. The widely sown annual legume subterranean clover needed four times more applied P than the commonly sown annual wimmera ryegrass.

In further work in a field experiment (Ozanne et al., 1976), comparing two crop and two annual pasture species, a range of nearly fourfold was found in the level of applied P required. Under field conditions, subterranean clover only needed about twice the level of wimmera ryegrass, not four times as in pots. Also the requirements of wheat and subterranean clover seemed to be similar.

It is curious, however, that under conditions of acute P deficiency in the field, subterranean clover can successfully invade wimmera ryegrass. The composites flatweed and capeweed can also, although all three showed a higher level of requirement for maximum yield than the grasses when grown in the two greenhouse experiments mentioned above.

With perennial species, Jackman and Mouat (1972b) showed that browntop grass needs less than half the applied P required by white clover when both were grown in soil in pots. Andrew and Robins (1969a and 1971) examined the P requirements of 9 tropical grasses and 10 tropical legumes in soil in pots. Their data show that, unlike the temperate species discussed, tropical legumes need about the same low level of applied P as do the grasses.

When it comes to differences between similar grasses or between similar legumes, Asher and Loneragan (1967) showed a threefold range in P concentrations required by temperate annuals in flowing culture solutions. However, Christie and Moorby (1975), also with culture solutions, showed a nearer 10-fold range in the concentrations of P needed for maximum yield of three semiarid perennial grasses. In soil in pots Ozanne et al. (1969) found little difference between two grass genera, but found up to a threefold difference in requirements of applied P within the three pasture legumes, all of which were *Trifolium* species.

In a field comparison of narrow leafed lupin and sand plain lupin, both grown as a crop, a difference of 15% in P requirement was found by Rahman and Gladstones (1974). Further examination of these two species (P. G. Ozanne & Ann Petch, 1977, unpublished) has confirmed this difference and shown it to be somewhat greater.

With 19 tropical pasture species studied in pots, Andrew and Robins (1969a, 1971) showed only a two- to threefold range in applied P requirement, which was similar for the grasses and legumes. Within the genera *Phaseolus* and *Desmodium* the legume species differed by < 30% in P requirement.

These experiments show that wide differences exist in the P requirements of plants well suited to the same environment and often the same management. Big differences occurred between plants in the same genus in some cases. This work shows the need to take P supply into account in plant introduction, plant selection, and plant breeding studies. Also, it indicates the large differences in applied P that may be needed for maximum efficiency in production from different crops and pastures.

IV. ABSORPTION AND TRANSLOCATION

A. Zone of the Root

Phosphorus is absorbed by the plant root mainly in the form of the monovalent dihydrogen phosphate ion $H_2PO_4^-$. Unlike N and S in nitrate and sulfate, the P atom in phosphate is not reduced in the plant to a lower oxidation state.

The zone of entry of the phosphate ion into the plant root has been much discussed in recent times. Use of ^{32}P in studying short-term accumulation, often with excised roots, has shown that the root tip accumulates a relatively high concentration of P (Bowen & Rovira, 1971). This may be followed by a zone of lesser accumulation where the cells are elongating, then a second region of higher concentration where the root hairs are developed. Such measurements, however, may not give a true picture of the region of the root through which P enters on its way to the tops of growing, transpiring plants.

Working with the roots of corn seedlings, Burley et al. (1970) concluded that all regions of the root are effective in absorbing P, but that translocation to the shoot was correlated with the development of functional xylem. In the region within 5 cm of the root tip little or no translocation of ^{32}P was found, due to the absence of xylem. In further studies Ferguson and Clarkson (1975) found that progressive suberization and endodermal thickening along the roots of corn had little effect on the radial movement of P into the vascular tissue. But in all parts of the root, inhibitors of respiration or low temperature (4°C) reduced uptake and xylem translocation of P by $>90\%$. Also Clarkson et al. (1968) used a stream of air to dry out a section of root so that the cortex collapsed and the endodermis was discolored. The dried part of the root resembled roots present in dry surface soil. Although growth of the root continued, absorption from the desiccated segment was reduced by $>90\%$.

B. Movement into the Root

To absorb P from soil solutions of 0.05 to 0.15 ppm P (see Section II) plants must expend considerable metabolic energy. Asher and Loneragan (1967) showed that six out of the eight species grown made maximum yields in a culture solution containing 0.15 ppm P. When their data are used to calculate the concentration of P in the root sap, a value of 650 ppm P is obtained; i.e., the roots have concentrated the culture solution some 4,200 times. Hence, the plants must have a very effective pump for accumulating P into the symplasm and into the cell vacuoles. Most of the phosphate ions entering the root are taken up by the root hairs or outermost cell layers. The inner cortical cells contribute only when either the transpirational flow or the phosphate ion concentration is very high.

The various possible pathways, complex carriers, and enzymes involved in phosphate ion accumulation are reviewed in detail by Anderson (1973), Bieleski (1973), Clarkson (1974), and others. It appears that accumulation in the vacuoles of the root cells tends to compete with the processes leading to accumulation in the xylem and subsequent transport to the top of the plant. Phosphate passing radially across the root to the endodermis may move in the transpiration stream through the "outer free space" located in the walls of the epidermal and cortical cells (Epstein, 1972) and the intercellular films of moisture. Phosphate taken up into the symplasm moves by diffusion and protoplasmic streaming. Movement of P across the cortex is restricted at the endodermis by the casparian strip, the hydrophobic layer encircling the stele. Further movement into the stele and to the xylem can only take place in the symplasm through the plasmodesmata which remain free and intact even after layers of suberin have been deposited on the walls of the endodermis.

On reaching the stele the P is unloaded from the symplasm into the xylem. Perhaps another pump of some kind is involved, but this time it need not act against a concentration gradient, as the concentration of P in the cytoplasm of the cortical cells is higher than in the xylem stream (Greenway & Klepper, 1968). These workers also showed that there is a marked effect of water flux across the root and into the xylem, not on the rate of P absorption, but on the rate at which previously absorbed P is transferred to the xylem, and thus upward to the shoot. The transportation stream, by inducing more rapid movement of P across the symplast, may dilute the concentration within it and so favor transfer to the shoots rather than accumulation in the vacuoles of the root cells.

C. Translocation

Once P is absorbed it is rapidly translocated to all parts of the plant. Clarkson (1974) showed that P absorbed by a section of a seminal root of barley was distributed throughout the whole root system and the tops in 20 hours. The labeled P moved upward in the xylem and downward in the phloem. Similar results were obtained if the ^{32}P was applied as an injection to one of the mature leaves. A vigorous circulation of ^{32}P in the xylem and phloem was observed.

These results are similar to those found by Biddulph et al. (1958) working with bean plants. They found ^{32}P supplied to the roots to be absorbed and translocated mainly into the young leaves. Subsequently as these matured, most of their ^{32}P was retranslocated to leaves which had developed after the supply of labeled P to the roots had been discontinued. Thus the ^{32}P gradually moved up the plant as it grew.

Accumulation of P by plant roots is the net result of processes which lead to both absorption and loss. Appreciable leakage has been reported by Bieleski (1973). Net ion absorption is closely regulated by the growth of the plant in circumstances where the supply of ions is not limiting (Clarkson, 1974). When Asher and Loneragan (1967) grew eight species in flowing culture solutions, five of the species showed no growth response over the concentration range of 0.15 to 0.77 ppm P. Of these five, four also showed no increase in the mean rates of P absorption per unit weight of roots despite the fivefold increase in P concentration of the solution bathing the roots (Loneragan & Asher, 1967).

In an experiment by Christie and Moorby (1975) three grass species were grown in culture solutions containing five levels of P ranging from 0.003 to 30.00 ppm. Large yield responses to P were obtained in all three species and total P taken up was related to yield rather than to the concentration in the tissues. The major factor determining P uptake appeared to be the demand of the tops or their ability to dilute out absorbed P through photosynthesis and growth. The level of P in the culture solution required for 90% of maximum yield of a particular species was related to the ability of the roots to absorb and translocate P rather than the ability of the tops to function at a lower internal concentration.

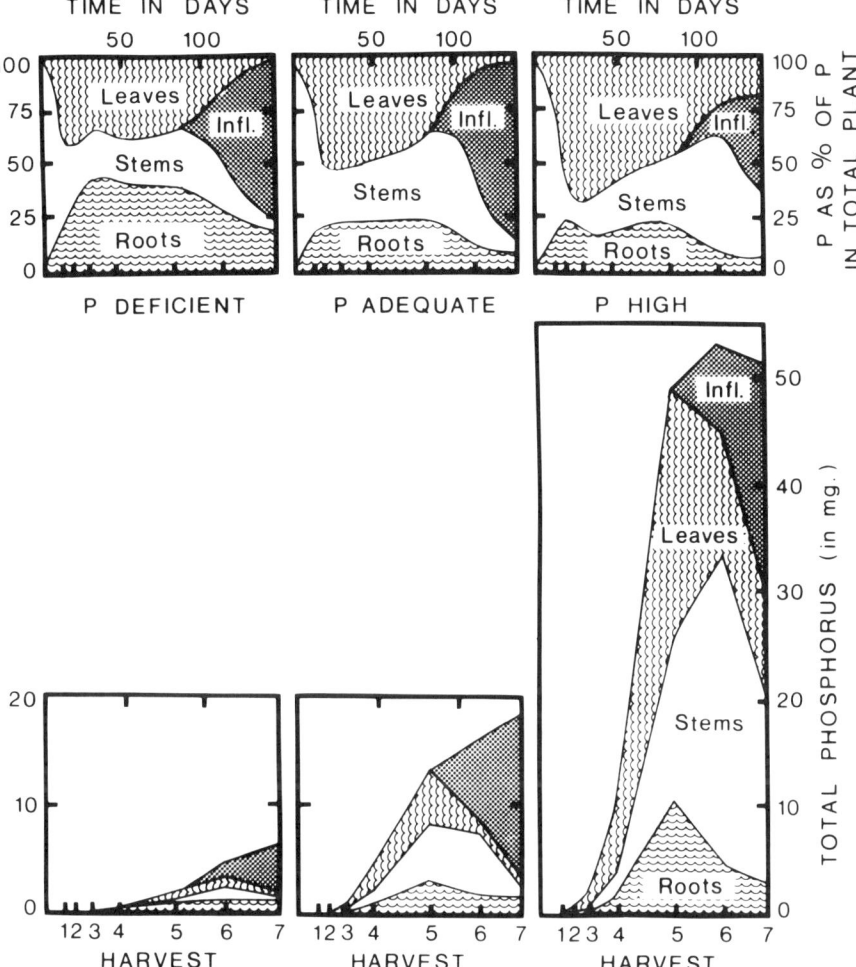

Fig. 1—The intake of P by oats and its distribution within the plant. Above: The P present in different plant parts expressed as a percentage of the total P in the plant. Below: The absolute P content of the whole plant and its parts.

D. Distribution

The quantitative distribution of absorbed P during the life cycle of the oat plant is shown in Fig. 1 (Williams, 1948). Data are from plants grown in pots containing 14 kg of sand. Phosphate uptake and distribution were measured at seven harvests—the first when the first leaf had appeared, the third at the commencement of tillering, and the last at full maturity of the plants when all the leaves were dead. The levels of P supply were such as to give three treatments—one producing severely P-deficient plants, one giving a P supply adequate for maximum yield, and one giving a high level of luxury intake.

With P-deficient plants, the roots retained a large part of the absorbed P, the P needed for inflorescence development coming mainly from the stems and leaves by retranslocation. With an adequate P supply the roots retained a smaller proportion of the total, but once again most of the P in the stems and leaves was retranslocated to the developing heads of grain. When the plants had a high P supply, the leaves and stems accumulated a large proportion of the absorbed P and retained much of it until senescence.

At the third harvest, the low, adequate, and high P treatments gave leaves containing 0.14, 0.74, and 2.27% P, respectively. At maturity, however, the range in the heads of grain was much less, being 0.17, 0.19, and 0.29% P in the low, adequate, and high P treatments. Note that these concentrations are for the total inflorescences, not just the grain.

Since this work by Williams there have been a number of similar studies although rarely so complete. From them a similar picture has emerged, for example, the data presented by Berger (1962) for maize. In early vegetative growth the plant leaves accumulate the major part of the absorbed P reaching relatively high concentration levels. As the inflorescences develop and the plant goes from the vegetative to the reproductive phase, leaf accumulation decreases and translocation of P from the roots may exceed the rate of absorption. These trends are shown in the study by Keay et al. (1970) for eight annual pasture species harvested at three stages of growth.

Measurements on wheat by Koehler (1976) showed that during the 20 days from the milk stage to mature grain the plants showed an overall loss of 7 kg of P/ha, while at the same time 36 kg of P/ha were translocated to the grain. Hanway (1975) reviewed the processes that take place in soybean during seed filling. Root growth stops, P uptake ceases, and P is translocated from the leaves and other vegetative parts to the seeds. This nutrient depletion of the leaves results in a decreasing rate of photosynthesis and in leaf senescence.

Attempts have been made to maintain the P supply to the leaves and developing seed of annual crops by spraying the foliage with P compounds, e.g., Barel (1975) working on corn and soybean, and Schultz (1975) working on wheat. Usually little or no yield response has been obtained. However, Garcia and Hanway (1976) sprayed the foliage of soybeans during the seed-filling period using balanced nutrient solutions containing not only P but also N, K, and S. Yield increases of up to 35% were obtained.

V. METABOLIC ROLE OF PHOSPHORUS WITHIN THE PLANT

A. Phosphate Compounds

More than 75% of the P moving in the xylem is inorganic. Phosphate being retranslocated in the phloem is much more similar in form to that of other tissues, being present as enzymes, proteins, and ribonucleic acid, although the concentration of adenosine triphosphate (ATP) is abnormally high. Data from Bieleski (1973) indicate that a "model" plant tissue might contain 0.004% P as desoxyribonucleic acid (DNA), 0.04% P as ribonucleic acid (RNA), 0.03% as lipid P, 0.02% as ester P, and 0.13% as inorganic P, all on a dry-weight basis. The absolute levels depend, of course, on the level of P supply to the plant, but the ratio of the organic P components of vegetative tissues is fairly constant.

Varying the P supply to agricultural plants may cause the total P concentration in the tissues to vary from about 0.1 to 1.0%. Working with *Spirodela,* Bieleski (1968) found that transferring the plants to a P-deficient medium decreases the organic P content by only 4-fold but the inorganic P concentration in the tissue by 40-fold.

Excess inorganic P may be stored in the vacuoles. Alternately the cells may synthesize a storage compound. Polyphosphate, common in lower plants, has been identified in six or more species of higher plants where it may account for 1 to 4% of the total P present (Nassery, 1969). However, phytic acid, the hexaphosphate ester of *myo*-inositol is the most common P reserve in reproductive organs. It accounts for more than 70% of the total P in rice (Mukherji et al., 1971), and wheat (Nahapetian & Bassiri, 1976), and occurs in relatively large amounts in potatoes. Unlike other P esters, there is no turnover of phytic acid in developing seeds. In the maturing and desiccating seed, phytic acid forms in the absence of phytase. Then, in the imbibing, germinating seed, phytase synthesis and rapid phytic acid hydrolysis takes place.

B. Phosphate and Inheritance

Phosphate performs a vital function in the life cycle of the plant in the nucleic acids of genes and chromosomes carrying the genetic material from cell to cell and seed to seed. This function is not easily influenced by P nutrition, but Durrant (1974 and earlier papers) describes how high levels of P supply to an inbred line of flax may produce a dwarf characteristic that is inherited for 10 or more generations. Also, Hill and Perkins (1969) found that an inbred line of tobacco grown with high P was shorter than the same cultivar grown with a low level of P. This dwarf characteristic was then retained through five generations irrespective of P supply.

Varying the level of P supply to the plant has relatively little effect on the total P concentration in the seeds. However, P deficiency may reduce seed numbers, viability, and seed size. Small seeds need to germinate nearer the soil surface than larger ones, and their roots elongate less rapidly so that they are more prone to die from drought (Ozanne & Asher, 1965). Also small seeds produce small seedlings that have a competitive disadvantage with other plants, at least for the first 2 to 3 weeks of growth.

Although P supply does not commonly affect heritable characteristics, genetic inheritance can influence P uptake. Bernard and Howell (1964) showed that in soybean the resistance to P toxicity at high levels of supply can be governed by a single pair of genes, while Barber et al. (1967) have shown genetically controlled differences in P accumulation in corn.

C. Phosphate in Photosynthesis

Through the cooperation of two photoreactions, the light energy absorbed by chlorophyll is used to reduce nicotinamide adenine dinucleotide phosphate (NADP) and to synthesize adenosine triphosphate (ATP). Evolution of O_2 accompanies these events (Knaff & Arnon, 1969). The NADPH and ATP so formed then serve as energy donors to drive many biosynthetic reactions throughout the plant (Arnon, 1967).

Throughout the plant, P compounds play a major part in the formation of many substances. For example, the sugar nucleotides, such as ATP, have an intimate role in the synthesis of starch, while the enzyme phosphorylase functions in its degradation. A detailed discussion of these processes is given by Manners (1973). Wiskich (1975) has shown that the oxidation of malate and citrate by isolated mitochondria from cauliflower and red beet are stimulated by the addition of inorganic P.

The influence of the level of inorganic P supply on photosynthesis and related processes has been studied recently by a number of workers. Bouma (1975) showed the overall effect. He found that leaves detached from P-deficient subterranean clover plants had a low rate of photosynthesis. When the petioles were placed in solutions containing from 0.70 to 6.0 mM of inorganic P, a marked increase in photosynthesis to levels as high as or higher than those obtained with leaves detached from plants grown with adequate P supply was observed. No stimulation of photosynthesis was found by placing non-P-deficient leaves in inorganic solutions.

The reduction in photosynthesis caused by P deficiency may be partly due to accumulation of polysaccharides. Photosynthetic starch formation by leaf discs of spinach and related species was found by Sheu-Hwa et al. (1975) to increase 10-fold when the cytoplasmic orthophosphate was sequestered by adding mannose to form mannose phosphate. This accumulation of polysaccharides in P-deficient plants is often accompanied by an accumulation of anthocyanins. The red or purple color they give leaves or stems is one of the common visual symptoms of P deficiency.

Inadequate P supply was also found to lower the rate of photosynthesis and the activity of various enzymes (Avdeeva & Andreeva, 1974). They found that the rate of photosynthesis was reduced more severely in corn, a C_4-species which has a four carbon atom photosynthetic pathway, than in bean, a C_3-species. Related results were obtained when Christie and Moorby (1975) compared two perennial semiarid grasses. Buffel grass, a C_4-species, showed a much larger response to higher external P levels in relative growth rate and net assimilation rate than did mulga grass, a C_3-species. To make maximum yield the C_4-species also needed a 100 times higher concentration of P in the culture solution than did the C_3-species.

D. Phosphate and N Fixation

A further function of applied P is to increase N fixation by nodulated legumes. If big yield increases to applied P are obtained, then correspondingly large or even larger increases in the amount of atmospheric N fixed may result. Working with nine tropical and one temperate pasture legumes, Andrew and Robins (1969b) found good positive correlations between the concentrations of P and N in nine of the species. Increasing the levels of applied P beyond the rates giving maximum yields gave increasing N concentrations in six species. Maximum N fixation required a higher P supply than maximum dry matter production.

This work was repeated (P. G. Ozanne, K. M. W. Howes, & A. Petch, 1976, unpublished data) by applying six levels of P to subterranean clover growing in the field. Again the concentration of N in the plants responded to higher levels of applied P than did dry matter production. Also, when grass was grown with the clover in the absence of applied N, the grass responded in yield to much higher levels of P than when grown alone and supplied with N fertilizer. Apparently grass growing with a legume tends to respond to the level of applied P giving maximum N fixation, rather than to the level of P required by the grass for its own requirements.

VI. PLANT ANALYSIS AND CRITICAL CONCENTRATIONS

A. Reasons for Sampling

Usually plant samples are taken for chemical analysis so that the current P status of the crop can be assessed. This information may be needed to estimate the profitability of a fertilizer application now or in the near future and to estimate what the rate should be. Alternately the analytical information may be of value in determining the P status of the soil and the need for fertilizer applications on succeeding crops.

Other uses of chemical analysis may be to indicate the quality of the produce, for example its probable keeping characteristics during storage or perhaps its nutritive value as a feed for animals.

B. Likelihood of Useful Values

Samples taken from a short-term crop are likely to be less variable and more useful than those from a long-term perennial stand. In the short-term crop, planting depth, spacing, age, competition, and other physical and chemical variables are usually better controlled. Also, the crop is generally of high genetic uniformity. However, tissue analysis has often been found of greater utility in "perennial" crops, such as fruit trees, while short-term crops are serviced best by soil tests.

If the stand to be sampled is a mixture of species, then the presence of a good indicator plant is helpful. For example, if the source of N for a pasture depends largely on a legume present, then it is probably the best plant to sample. It is not only least likely to be restricted by N deficiency, but also often has the highest requirement for available P.

To be useful, the analyses obtained must be related to those of other crops, other experiments, different areas, etc. So a major consideration must be standardization of sampling times, plant parts, and crop history.

C. When to Sample

Sampling time depends to some extent on the purpose for which the samples are taken. If the analyses are required to determine the need for a fertilizer application for the current crop, then the sampling must be done in time to allow the necessary fertilizer equipment to pass over the crop without damage and in time for the plants to respond before harvest or maturity. The P concentration in the sampled material is also related to the chronological or phenological age of the plants, such as days since emergence, leaf number, commencement of tillering, etc. Analyses of the tops of subterranean clover were made at 5 monthly intervals and at 9 levels of applied P by Ozanne and Howes (1971a). At all levels of applied P there was a consistent decrease in the concentration of P in the plant tops with time.

Management practices may also have a considerable effect on P content of plants, and should be borne in mind when sampling. In the work above on subterranean clover, we found that, although light defoliation had little effect, close grazing increased the P concentration present in the pasture by 50% or more. Other management practices more relevant to crops would be the time since irrigation, or the time since the last dressing of N or P fertilizer. Burns et al. (1974), working on alfalfa, found no effect of the time of day when sampling was carried out, but found a difference of from 0.33 to 0.37% P associated with a rise in maximum daily temperature from 24 to 29°C.

D. What to Sample

In some situations, as in pastures, the easiest plant part to sample may be the total tops above, say, 5 cm. Under many grazing situations this material is all younger than 4 weeks, consists mainly of young mature leaves, and so is of value as a fairly consistent, repeatable sample. But in crops, leaves, petioles, or stems are normally selected. Working with alfalfa, Rominger et al. (1975) found considerable difference between stem and leaf samples. Also, they found a consistent increase in P concentration of both stems and leaves in seven consecutive samples taken from stubble height to 70 cm. For these reasons it is usually better to select a particular leaf for analysis as is done for corn or tomatoes. Sobulo et al. (1975) found the fifth leaf of tomato plants sampled at the early flowering stage (7 to 8 weeks after sowing) to be the most useful sample. This leaf had a critical P concentration of 0.4% while the value for the third leaf was usually higher but less responsive.

Phosphorus is a mobile element in the plant, tending to accumulate in the older leaves under conditions of high P supply, but to be retranslocated from the old to the young leaves under conditions of P deficiency. For this reason it is sometimes useful to sample both old and young leaves. If the old leaves are higher in P than the young, then luxury consumption is likely. On the other hand, if the old leaves are much lower in P than the young, then it is likely that P deficiency is associated with the marked retranslocation.

E. What to Analyze

In foliage analyses for P, as in all the previous references given, the percentage of total P is usually determined. Water-soluble or acid-soluble P fractions are also useful indicators and vary much more widely than total P between conditions of deficiency and luxury consumption. However, the usefulness of their wide range is marred somewhat by fluctuations that give a relatively high error. Other possibilities are to gauge the P status of a plant by the activity of one of the many P-containing enzymes, or to measure the buildup of polysaccharides in the leaves when photosynthesis is proceeding with P supply at an inadequate level.

These measurements may be informative for special purposes; but, total P determinations are useful in that they indicate the total amount of P actually present in the plant tops per unit area. This can then be related to the amount required for a given yield of grain and also in some situations to the fertilizer requirements per unit area. If analyses of N, K, and S are also carried out, then the ratios of these to total P present can indicate whether P is likely to be the limiting element. In many healthy tissues the ratio of P/S/N/K is roughly 1:1:10:10 and a wide variation from this usually indicates a nutrient imbalance.

F. Use of Analyses

When analyses have been carried out, how can the results be used to determine the degree of nutrient stress and resulting yield depression? It is desirable to have a response in which yields are related to tissue concentrations of P. Such a response curve may have the general form shown in Fig. 2, although a complete range from acute deficiency to toxicity is seldom obtained in one experiment. Usually, only the part of the curve showing diminishing to no response is found in field trials. Figure 2 is a compositive curve drawn from data obtained in a number of field experiments with different annual species from which the youngest mature leaf was sampled just prior to flowering. The values of P concentration are given as a general guide only, and as described earlier, will vary widely depending on species, age, plant part analyzed, etc.

The "critical concentration" of P in a plant part is normally regarded as the minimum concentration associated with near maximum yield. Although it cannot usually be precisely determined, it can have considerable practical value. It is of more use to know the shape of the yield response curve in relation to the concentration of P in the selected tissue. Such curves

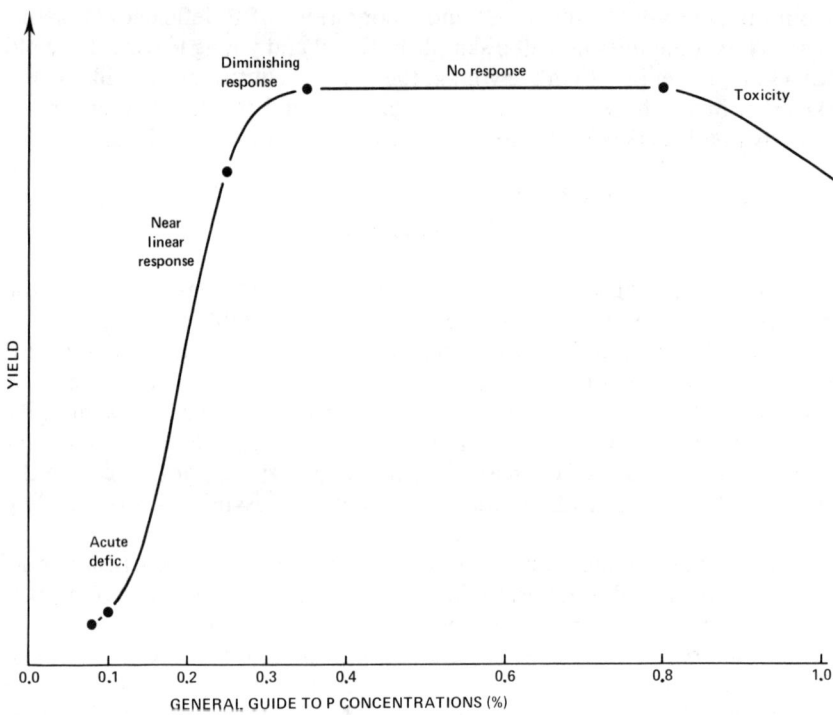

Fig. 2—The general relationship between plant yield and the concentration of P in the tissue.

are usually obtained from pot or field experiments. Because of the influence of the variables discussed earlier, the growing conditions used to establish response curves should approximate the conditions under which they are to be used as closely as possible.

The usefulness and stability of the critical percentage of total P is well shown in the work of Andrew and Robins (1969a & 1971). At the preflowering stage they found the critical concentrations (i.e., the minimum tissue concentration for near maximum yield) for nine tropical grasses, nine tropical pasture legumes, and alfalfa all to fall in the range of 0.16 to 0.25% P. Values for individual species were much less variable still. One varied by only 0.02% over six soil types, while five others varied by 0.03% or less over three or more soil types. The mean critical concentration found for alfalfa was 0.24% P which compares well with other published values—0.23% P (Larsen et al., 1952; Stivers & Ohlrogge, 1952), and 0.27% P (Bear & Wallace, 1950).

Having obtained P analyses, and with a knowledge of the critical concentrations relevant to the samples obtained, what can be ascertained? Usually plant analysis will show if P is deficient or adequate, but there is a fairly wide grey area of possible incipient deficiency between the two levels. Plants in this category may or may not respond to additional applied P. However, if the plants analyzed show symptoms which might be due to P deficiency or P toxicity, then the tissue analyses will almost always confirm or discount the diagnosis.

Even if the P concentrations found clearly indicate P deficiency, it does not follow that remedying the deficiency by a fertilizer application will be profitable. But the reverse situation does hold true. If a P deficiency is suspected but foliar analysis shows the plant to be well supplied with P, then it will certainly be profitable not to apply P alone. This situation is fairly commonly encountered due to the very considerable residual effects of P applied in previous years. Alternate interpretations are possible. If the tissue concentrations are high in P, but yields per hectare are low, it may be that the restricted growth and high P analyses may both be due to N deficiency. In this case the application of an NP fertilizer may be worthwhile. Finding the N/P ratio in the tissues as discussed earlier will aid this diagnosis. Deficiencies of K, S, or other nutrient elements may similarly give high P concentrations, but low yields which respond to applications of compound fertilizers containing P.

VII. PHOSPHORUS CONTENT AND NUTRITIONAL VALUE

A. Phosphorus Concentration in the Feed

The P nutrition of plants has long been known to be closely related to the nutritional value of the vegetation grown. Over 50 years ago Theiler et al. (1924) described the occurrence of P deficiency in ruminants and its associated syndrome. A more recent review was done by Underwood (1966), who concluded that P deficiency is the most widespread and economically important of all mineral disabilities affecting grazing livestock. This topic is discussed in more detail in Chapter 29.

Underwood and other writers have indicated that concentrations of 0.17 to 0.25% P in the available feed are required by the grazing animal. As these levels may be above the critical concentrations required by the pasture, especially toward maturity, it is easy to grow lush feed that is too low in P for animal health (Ozanne & Howes, 1971a). This problem is compounded in annual pastures by the drop in P concentration that takes place once the pasture dries off. We found that subterranean clover pastures adequately supplied with P for near maximum dry matter production contained 0.19% P at maturity. The residues providing the dry summer feed, however, had fallen to 0.09% P 5 months later. Even lower values have been published for tropical legumes (Playne & Haydock, 1972).

B. Phosphorus Imbalances

Compounding the problems of the P nutrition of pastures are the imbalances that may arise. Wise et al. (1963) found that for healthy growth of animals the ratio of total P to total Ca should fall in the range of 1:1 to 1:7.

This range is frequently exceeded. In grains, the low Ca levels frequently give a P/Ca ratio >1:1. Grasses also tend to be low in Ca compared to legumes. In a grazing experiment on irrigated pastures receiving high rates of fertilizer and giving high production of dry matter per hectare, alfalfa was found to have a P/Ca ratio of 1:1.84. Pangola grass had a P/Ca ratio of 1.22:1 and lovegrass 1.23:1 (Ozanne & Arnold, 1970).

Wide imbalances due to high Ca may occur in legumes. In a mature green subterranean clover pasture grown with a low P supply, the P/Ca ratio was 1:10 with a content of 0.10% P and 1.09% Ca (Ozanne & Howes, 1971b). Even when the pasture had received a rate of P much higher than needed for maximum dry matter production, the dry pasture residues at the end of summer contained only 0.10% P and 1.16% Ca. This wide ratio builds up in the dry feed because about half of the P is present in a water-soluble form subject to leaching, but little or none of the Ca is water soluble.

C. Nutritional Value

Reports in the literature note that application of a P fertilizer to a pasture have made it more attractive to cattle (Staten, 1949) and to sheep (Playne, 1972). When we applied nine levels of P to 45 areas of a subterranean clover based pasture, we found that grazing sheep could readily distinguish between them and showed their preference by heavily grazing the high P plots (Ozanne & Howes, 1971b). A second experiment using 100 plots gave similar results. Both in the green stages and in the dry summer residues the sheep showed a clear preference for pasture grown with high levels of applied P. The preference extended in intensity to levels of P well above those needed for maximum dry matter production. But we have so far been unable to identify the substance causing the selection.

Preference by itself does not prove nutritive value. However, when we fed sheep in pens with the dry summer residues of subterranean clover pasture grown with deficient, adequate, or high levels of applied P, feed intake and body weight gains were closely related to the level of applied P (Ozanne et al., 1976). Addition of an inorganic P supplement to the low-P feed gave a small increase in intake and body weight gain.

A similar experiment using the tropical pangola grass was carried out (M. C. Rees & D. J. Minson, 1975, unpublished data). They found that neither a high P content nor an inorganic P supplement increased feed intake, but that both increased digestibility. There is then no doubt that the P nutrition of plants is important in animal nutrition or that the levels needed in feed by the animal may be considerably higher than those required for healthy growth of the crop or pasture. Part of the beneficial effect of higher P levels in feed is that they maintain the animal in positive P balance. In addition, high levels of P in the plant are apparently associated with some other characteristics which makes feed more palatable and more digestible. Just what these changes are is not yet clear.

VIII. PHOSPHATE RESPONSES AND THEIR PROFITABILITY

A. Factors Influencing the Response

Understanding the P nutrition of plants has as its major goal the prediction of responses. In the preceding sections of this chapter, I have discussed a number of factors which influence the yield obtainable from a given application of P fertilizer. Among the soil components were the phosphate buffering capacity, the distribution of P down the profile, the reactivity of the form of fertilizer applied, and the residual value of previous applications.

Among the plant factors are those related to root characteristics including the presence or absence of mycorrhizae, other species differences, moisture stress, limitations of other nutrients, and the effects of defoliation. The influence of P on animal production was briefly mentioned. These and other factors are relevant to the response obtainable to a given P application and are discussed in more detail by Bennett and Ozanne (1973).

B. Relating Crop Response to Applied P

1. YIELD DETERMINANTS

The response of a crop or pasture to applied P has been described by a number of mathematical models. Currently relationships in use are empirical, although some have parameters which can be interpreted biologically. The more commonly used models have been reviewed by Heady (1960), FAO (1966), and Cooke (1975). One of these is the quadratic curve in which the yield (y) and available P (x) are related as follows:

$$y = z + bx + cx^2$$

where a, b, and c are constants. Another similar equation is the square root formula in which

$$y = a + bx^{1/2} + cx.$$

Among the older relationships used in a variety of forms is the exponential response curve put forward by Mitscherlich (1930) and in several of his earlier papers. The Mitscherlich model was modified by Bray (1958) to allow the inclusion of data from soil tests, and further modified by Campbell and Keay (1970) to make the equation more flexible in fitting a wider range of response curves.

A simple form of the Mitscherlich equation may be written as:

$$Y = A\,[1 - B \cdot \exp(-CX)].$$

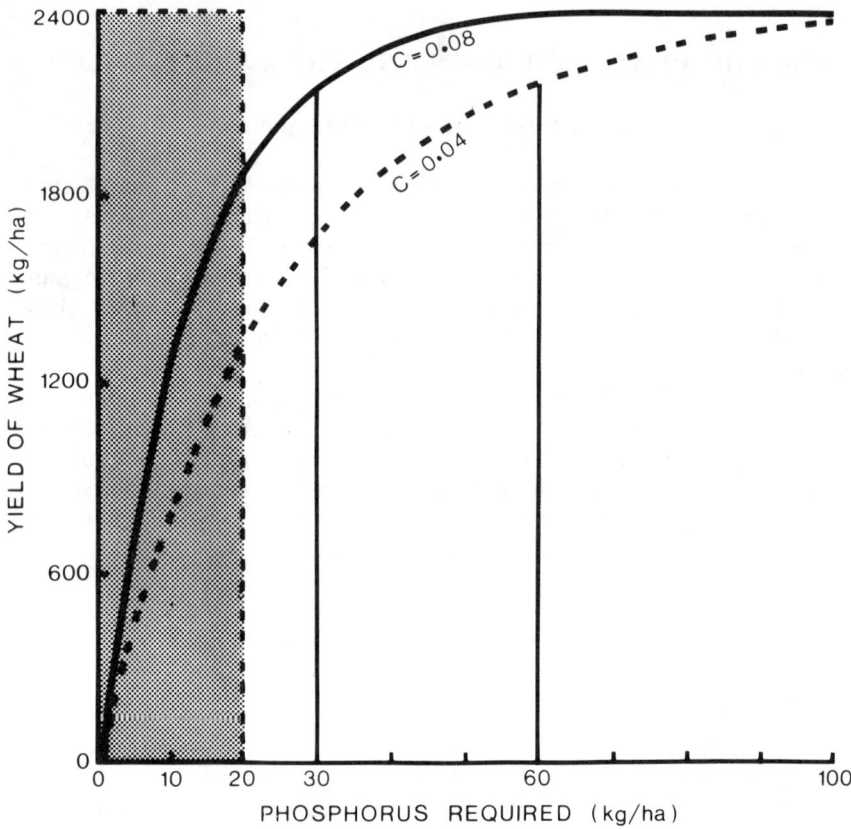

Fig. 3—Response curves of crop yield to level of available P. The shaded area represents P already present in the soil, while the vertical lines indicate P required for 90% of maximum yield. The curves represent the relationship $Y = A[1 - B \cdot \exp(-CX)]$.

The yield (Y) at any given level of available P (X) is determined by the maximum yield (A) reached when P is not limiting, the maximum obtainable response (B) expressed as a fraction of A, and the curvature (C) of the response curve. Two curves showing the relationship of Y to X are shown in Fig. 3.

2. MAXIMUM YIELD (A)

In the equation given above, A is the maximum yield determined by climate and soil fertility factors other than P. It incorporates such factors as management, weeds, insect damage, and plant disease. Different crops will, of course, have different A values, and length of growing season, temperature, and local rainfall will modify these. Usually the optimum rate of P to apply is relatively insensitive to errors in estimating A, which in Fig. 3 has arbitrarily been given the value of 2,400 kg/ha.

3. THE RESPONSIVENESS (B)

The responsiveness of the crop or forage to applications of P is represented by B. It is calculated from the difference between the maximum yield possible (A) and the yield actually obtained without extra P expressed as a fraction of A, i.e. $B = (A - Y_0)/A$. Thus B values can range from zero to one. The actual value depends on the P status of the unfertilized soil, and if this is extremely deficient, the response curve will have a B value of 1. However, a naturally fertile soil may have a value as low as 0.3, meaning that 70% of maximum yield is obtained from the virgin soil without P fertilizer.

Except in soils rich in native P, the amount available at any given time depends on how much has been added and on the time it has actually spent in contact with the soil. When recent large applications of P have been applied, B will be low.

Soils also vary in their phosphate buffering capacity, and for a given quantity of applied P, a soil with a high sorption capacity will have a higher B value. In Fig. 3 the response curves can represent those for an extremely P-deficient pair of soils with B values of 1.0. But after fertilizer applications have raised the available P level to 20 kg/ha, the B value of the soil of lower buffering capacity has fallen to 0.2, while the one of higher capacity still has a B value of 0.45.

Some plant species are better able to extract residual soil P or can grow better at lower P concentrations than others. For a given set of soil conditions such plants have a lower B value. For example in Fig. 3 the response curve showing a B value of 0.2 could be that of a grass and the other curve of $B = 0.45$ that of a clover with a higher available P requirement. Values of B appear to be largely independent of the value of A.

4. THE CURVATURE (C) OF RESPONSE

The value of C has a stronger influence on the optimum fertilizer rate than A or B. Consequently it should receive more attention. Also, C can vary independently of A or B as shown in Fig. 3.

Two broad sets of factors determine the curvature. One is the amount of applied P that remains available to plants, and this is largely determined by the phosphate buffering capacity of the soil. The other is the plant's ability to take up P and make use of it.

The range of C values found is from about 0.01 to 0.20. In Fig. 3 the effect on the shape of the response curve is shown for only the two values $C = 0.08$ and $C = 0.04$. The P required for 90% of the maximum yield (A) is shown as being 30 and 60 kg/ha for $C = 0.08$ and 0.04, respectively. This relationship is such that for any given value of the yield Y, expressed as a percentage of the maximum A, the P required is inversely proportional to the value of C (Anon., 1974).

From the equation given it follows that when $Y = 90\%$ of A, and $B = 1.0$, then $C = 2.3/X$. This means of course that C is directly related to X when X is the amount of P required for 90% of maximum yield. If this value of X can be found from the data of any experiment, then so can C, and the whole response curve can be constructed. Alternately, by comparing different C values, the efficiency of use of P can be compared in different situations, with other plant cultivars, with different management, etc.

C. Determining Profitability

The response curve method of making fertilizer decisions has been described in detail by Bowden and Bennett (1974), and at a more applied level by Robertson et al. (1975). To determine the optimum economic rate of fertilization, one needs to predict the parameters of the response curve to P in the specified situation. Then the rate of fertilizer scale must be converted to dollars, and likewise the crop yield scale to dollars worth of saleable product. Determination of the point at which marginal cost equals marginal return requires estimates of future prices for inputs and outputs. This is a relatively simple matter for the fertilizer and a cash crop, but more difficult for stock feed.

Some economic aspects of the optimum rate of P usage have been discussed by FAO (1966), Kennedy et al. (1973), and Helyar and Godden (1976). In particular these papers bring out the importance of taking into account the economic value of residual P from previous fertilizer dressings, and the value in future years of a current application of P. There have also been attempts to take into account the increased responses to P when other nutrients are applied at the same time. Colwell and Stackhouse (1970) and earlier workers have developed response surfaces to simultaneous field applications of varying rates of N, P, K, and S; but use of such data is necessarily complex. A recent comprehensive review of response surface modelling is given by Mead and Pike (1975).

Several factors influencing the optimum economic rate of P application have not yet been adequately examined in the field. Three examples are as follows:
1) The effects of cultivation versus minimum tillage on the residual value of applied P.
2) The sometimes close relationship between N fixation by legumes and applied P up to levels higher than those giving maximum yield of dry matter.
3) The influence of high rates of P on quality aspects of plant production, including the responses sometimes obtained in animal growth.

LITERATURE CITED

Anderson, W. P. 1973. Ion transport in plants. Academic Press, London, New York, and San Francisco.

Andrew, C. S., and M. F. Robins. 1969a. The effect of P on the growth and chemical composition of some tropical legumes. I. Growth and critical percentage of P. Aust. J. Agric. Res. 20:665-674.

Andrew, C. S., and M. F. Robins. 1969b. The effect of P on the growth and chemical composition of some tropical legumes. II. N, Ca, Mg, K, and Na contents. Aust. J. Agric. Res. 20:675-685.

Andrew, C. S., and M. F. Robins. 1971. The effect of P on the growth, chemical composition, and critical P percentages of some tropical pasture grasses. Aust. J. Agric. Res. 22:693-706.

Andrews, Rosalie E., and I. Newman. 1970. Root density and competition for nutrients. Oecol. Plant. 5:319-334.

Anonymous. 1974. Deciding how much super to use. Rural research, CSIRO 86:4-13.

Arnon, D. I. 1967. Photosynthetic activity of isolated chloroplasts. Physiol. Rev. 47:317-358.

Asher, C. J., and J. F. Loneragan. 1967. Response of plants to phosphate concentration in solution culture. I. Growth and phosphorus content. Soil Sci. 103:225-233.

Asher, C. J., and P. G. Ozanne. 1966. Root growth in seedlings of annual pasture species. Plant Soil 24:423-436.

Asher, C. J., P. G. Ozanne, and J. F. Loneragan. 1965. A method for controlling the ionic environment of plant roots. Soil Sci. 100:149-156.

Avdeeva, T. A., and T. F. Andreeva. 1974. Effect of conditions of phosphorus nutrition on photosynthesis and activity of carboxylating enzymes in bean and corn plants. Soviet Plant Physiol. 21:753-757.

Barber, S. A., J. M. Walker, and E. H. Vasey. 1963. Mechanisms for the movement of plant nutrients from the soil and fertilizer to the plant root. J. Agric. Food Chem. 11:204-207.

Barber, W. D., W. I. Thomas, and D. E. Baker. 1967. Inheritance of relative phosphorus accumulation in corn (Zea mays L.). Crop Sci. 7:104-107.

Barel, D. 1975. Foliar application of phosphorus compounds. Ph.D. Thesis. Iowa State Univ., Univ. Microfilms, Ann Arbor, Mich. (Diss. Abstr. 76-1820).

Barley, K. P. 1970. The configuration of the root system in relation to nutrient uptake. Adv. Agron. 22:159-201.

Barrow, N. J. 1975. Chemical form of inorganic phosphate in sheep feces. Aust. J. Soil Res. 13:63-67.

Bear, F. E., and A. Wallace. 1950. Alfalfa: its mineral requirements and chemical composition. New Jersey Agric. Exp. Stn. Bull. No. 748.

Bennett, D., and P. G. Ozanne. 1973. Deciding how much superphosphate to use. Ann. Rep. 1972 Div. Plant Indust., CSIRO, Australia. p. 45-47.

Berger, J. 1962. Maize production and the manuring of maize. Centre d'Etude de l'Azote 5 Geneva.

Bernard, R. L., and R. W. Howell. 1964. Inheritance of phosphorus sensitivity in soybeans. Crop Sci. 4:298-299.

Biddulph, O., S. Biddulph, R. Cory, and H. Koontz. 1958. Circulation patterns for phosphorus, sulphur, and calcium in the bean plant. Plant Physiol. 33:293-300.

Bieleski, R. L. 1968. Effect of phosphorus deficiency on levels of phosphorus compounds in *Spirodela*. Plant Physiol. 43:1309-1316.

Bieleski, R. L. 1973. Phosphate pools, phosphate transport, and phosphate availability. Ann. Rev. Plant Physiol. 24:225-252.

Bieleski, R. L. 1974. Development of an externally-located alkaline phosphatase as a response to phosphorus deficiency. p. 165-170. *In* R. L. Bieleski, A. R. Ferguson, and M. M. Cresswell (ed.) Mechanisms of regulation of plant growth. Bull. 12, Royal Soc. N.Z., Wellington.

Bouma, D. 1975. Effects of some metabolic phosphorus compounds on rates of photosynthesis of detached phosphorus-deficient subterranean clover leaves. J. Exp. Bot. 26:52-59.

Bowden, J. W., and D. Bennett. 1974. The DECIDE model for predicting superphosphate requirements. p. 6.1-6.36. *In* Proc. Symp. Phosphorus in Agric. November 1974. Australian Inst. of Agric. Sci. (Victorian Branch), Melbourne.

Bowen, G. D. 1973. Mineral nutrition of ectomycorrhizae. p. 151-205. *In* G. C. Marks, and T. T. Kozlowski (ed.) Ectomycorrhizae—their ecology and physiology. Academic Press, New York.

Bowen, G. D., and A. D. Rovira. 1971. Relationship between root morphology and nutrient uptake. p. 293-303. *In* R. M. Samish (ed.) Recent advances in plant nutrition. Vol. 1. Gordon & Breach Science Publishers, New York.

Bray, R. H. 1958. The correlation of a phosphorus soil test with the response of wheat through a modified Mitscherlich equation. Soil Sci. Soc. Am. Proc. 22:314-317.

Bromfield, S. M. 1961. Sheep feces in relation to the phosphorus cycle under pastures. Aust. J. Agric. Res. 12:111-123.

Brownlee, H., B. J. Scott, R. D. Kearins, and J. Bradley. 1975. Effects of topdressed superphosphate on the sheep and pasture production of dryland lucerne in central western New South Wales. Aust. J. Exp. Agric. Anim. Husb. 15:475-483.

Burley, J. W. A., F. I. O. Nwoke, G. L. Leister, and R. A. Popham. 1970. The relationship of xylem maturation to the absorption and translocation of ^{32}P. Am. J. Bot. 57:504-511.

Burns, J. C., C. L. Rhykerd, C. H. Noller, and K. R. Cummings. 1974. Influence of nitrogen, phosphorus, and potassium fertilization on the mineral concentrations of *Medicago sativa* L. II. Diurnal and day to day fluctuations. Comm. Soil Sci. Plant Anal. 5:515-529.

Campbell, N. A., and J. Keay. 1970. Flexible techniques in describing mathematically a range of response curves of pasture species. p. 332-334. *In* M. J. T. Norman (ed.) Proc. 11th Int. Grassland Congr. (Surfers Paradise, Queensland), April 1970. University of Queensland Press, St. Lucia.

Christie, E. K. 1975. Physiological responses of semiarid grasses. II. The pattern of root growth in relation to external phosphorus concentration. Aust. J. Agric. Res. 26:437-446.

Christie, E. K., and J. Moorby. 1975. Physiological responses of semiarid grasses. I. The influence of phosphorus supply on growth and phosphorus absorption. Aust. J. Agric. Res. 26:423-436.

Clarkson, D. T. 1974. Ion transport and cell structure in plants. McGraw-Hill, London.

Clarkson, D. T., J. Sanderson, and R. S. Russell. 1968. Ion uptake and root age. Nature 220:805-806.

Colwell, J. D., and K. M. Stackhouse. 1970. Some problems in the estimation of simultaneous fertilizer requirements of crops from response surfaces. Aust. J. Exp. Agric. Anim. Husb. 10:183-195.

Cooke, G. W. 1975. Fertilizing for maximum yield. Crosby Lockwood Staples, London.

Crush, J. R. 1974. Plant growth responses to vesicular-arbuscular mycorrhiza. VII. Growth and nodualtion of some herbage legumes. New Phytol. 73:743-749.

Daft, M. J., and A. A. El-Giahmi. 1975. Effects of *Glomus* infection on three legumes. p. 581-592. *In* F. E. Sanders, Barbara Mosse, and P. B. Tinker (ed.) Endomycorrhizas. Academic Press, London.

Drew, M. C., and L. R. Saker. 1975. Nutrient supply and the growth of the seminal root system in barley. II. Localized, compensatory increases in lateral root growth and rates of nitrate uptake when nitrate supply is restricted to only part of the root system. J. Exp. Bot. 26:79-90.

Drew, M. C., L. R. Saker, and T. W. Ashley. 1973. Nutrient supply and the growth of the seminal root system in barley. I. The effect of nitrate concentration on the growth of axes and laterals. J. Exp. Bot. 24:1189-1202.

Durrant, A. 1974. The association of induced changes in flax. Heredity 32:133-143.

Epstein, E. 1972. Mineral nutrition of plants: Principles and perspectives. John Wiley & Sons, Inc., New York.

Evans, P. S. 1970. Root growth of *Lolium perenne* L. I. Effect of plant age, seed weight, and nutrient concentration on root weight, length, and number of apices. N.Z. J. Bot. 8:344-356.

Ferguson, I. B., and D. T. Clarkson. 1975. Ion transport and endodermal suberization in the roots of *Zea mays*. New Phytol. 75:69-79.

Food and Agriculture Organization. 1966. Statistics of crop responses to fertilizers. Food Agric. Org. United Nations, Rome.

Fox, R. L., R. K. Nishimoto, J. R. Thompson, and R. S. de la Pena. 1974. Comparative external phosphorus requirements of plants growing in tropical soils. Int. Congr. Soil Sci., Trans. 10th (Moscow, Russia) IV:232-239.

Fried, M., and R. E. Shapiro. 1961. Soil plant relationships in ion uptake. Ann. Rev. Plant Physiol. 12:91-112.

Garcia, R. L., and J. J. Hanway. 1976. Foliar fertilization of soybeans during the seed-filling period. Agron. J. 68:653-657.

Greenway, H., and B. Klepper. 1968. Phosphorus transport to the xylem and its regulation by water flow. Planta (Berlin) 83:119-136.

Guttay, A. J. R. 1975. Fungus helps plants grow. Crops Soils Mag. 27(9):14-17.

Hanway, J. J. 1975. Interrelated development and biochemical processes in the growth of soybean plants. p. 5-15. In L. D. Hill (ed.) Proc. World Soybean Res. Conf., Univ. Illinois, Urbana, Ill., August 1975. Interstate Printers & Publishers, Inc., Danville, Ill.

Harley, J. L. 1971. Mycorrhiza. Oxford University Press, London.

Heady, E. O. 1960. Status and methods of research in economic and agronomic aspects of fertilizer response and use. Natl. Acad. Sci. Natl. Res. Coun., Washington, D.C.

Helyar, K. R., and D. P. Godden. 1976. Soil phosphate as a capital asset. In G. J. Blair (ed.) Reviews in rural science 3: Improving the efficiency of phosphorus utilization. Univ. of New England, Armidale, N.S.W.

Hill, J., and Jean M. Perkins. 1969. The environmental induction of heritable changes in *Nicotiana rustica*. Effects of genotype-environment interactions. Genetics 61:661-675.

Jackman, R. H., and M. C. H. Mouat. 1972a. Competition between grass and clover for phosphate. I. Effect of browntop (*Agrostis tenuis* Sibth) on white clover (*Trifolium repens* L.) growth and nitrogen fixation. N.Z. J. Agric. Res. 15:653-666.

Jackman, R. H., and M. C. H. Mouat. 1972b. Competition between grass and clover for phosphate. II. Effect of root activity, efficiency of response to phosphate, and soil moisture. N.Z. J. Agric. Res. 15:667-675.

Jacques, W. A. 1943. Root development in some New Zealand pasture plants. II. Perennial ryegrass (*Lolium perenne*), cocksfoot (*Dactylis glomerata*) and white clover (*Trifolium repens*). Effect of fertilizer placement on the yield of roots and herbage. N.Z. J. Sci. Tech. A Sect.: 25:91-117.

Jeffrey, D. W. 1967. Phosphate nutrition of Australian heath plants. II. The importance of proteid roots in *Banksia* (Proteaceae). Aust. J. Bot. 15:403-411.

Jones, O. L., and S. M. Bromfield. 1969. Phosphorus changes during the leaching and decomposition of hayed-off pasture plants. Aust. J. Agric. Res. 20:653-663.

Keay, J., E. F. Biddiscombe, and P. G. Ozanne. 1970. The comparative rates of phosphate adsorption by eight annual pasture species. Aust. J. Agric. Res. 21:33-44.

Kennedy, J. O. S., I. F. Whan, R. Jackson, and J. L. Dillon. 1973. Optimal fertilizer carryover and crop recycling policies for a tropical grain crop. Aust. J. Agric. Econ. 17:104-113.

Khan, A. G. 1974. The occurrence of mycorrhizas in Halophytes, Hydrophytes and Xerophytes, and of *Endogone* spores in adjacent soils. J. Gen. Microbiol. 81:7-14.

Khan, A. G. 1975. The effect of vesicular arbuscular mycorrhizal associations on growth of cereals. II. Effects on wheat growth. Ann. Appl. Biol. 80:27-36.

Knaff, D. B., and D. I. Arnon. 1969. A concept of three light reactions in photosynthesis by green plants. Proc. Natl. Acad. Sci. U.S.A. 64:715-722.

Koehler, F. E. 1976. Plant food taken up by 108 Bu/A wheat while it grows. Better Crops Plant Food 60(1):16-18 (Spring-Summer).

Larsen, W. E., L. B. Nelson, and A. S. Hunter. 1952. The effect of phosphate fertilization upon the yield and composition of oats and alfalfa grown on phosphate deficient Iowa soils. Agron. J. 44:357-361.

Lewis, D. B., and J. P. Quirk. 1965. Diffusion of phosphate to plant roots. Nature 205:765-766.

Lewis, D. G., and J. P. Quirk. 1967. Phosphate diffusion in soil and uptake by plants. III. ^{31}P-movement and uptake by plants as indicated by ^{32}P-autoradiography. Plant Soil 26:445-453.

Loneragan, J. F., and C. J. Asher. 1967. Response of plants to phosphate concentration in solution culture: II. Rate of phosphate absorption and its relation to growth. Soil Sci. 103: 311-318.

Manners, D. J. 1973. Starch and inulin. p. 176-197. In L. P. Miller (ed.) The process and products of photosynthesis, Phytochemistry Vol. 1. Van Nostrand Reinhold Co., New York.

Mead, R., and D. J. Pike. 1975. A review of response surface methodology from a biometric viewpoint. Biometrics 31:803-851.

Mitscherlich, E. A. 1930. Die Bestimmung des Dungerbedurfnisses des Bodens. Paul Parey, Berlin.

Mukherji, S., B. Dey, A. K. Paul, and S. M. Sircar. 1971. Changes in phosphorus fractions and phytase activity of rice seeds during germination. Physiol. Plant 25:94-97.

Nahapetian, A., and A. Bassiri. 1976. Variations in concentrations and interrelationships of phytate, phosphorus, magnesium, calcium, zinc, and iron in wheat varieties during two years. J. Agric. Food Chem. 24:947-950.

Nassery, H. 1969. Polyphosphate formation in the roots of *Deschampsia flexuosa* and *Urtica dioica*. New Phytol. 68:21-23.

Nicolson, T. H. 1975. Evolution of vesicular-arbuscular mycorrhizas. p. 25-34. *In* F. E. Sanders, Barbara Mosse, and P. B. Tinker (ed.) Endomycorrhizas. Academic Press, London.

Olsen, S. R., W. D. Kemper, and J. D. Jackson. 1962. Phosphate diffusion to plant roots. Soil Sci. Soc. Am. Proc. 26:222-227.

Ozanne, P. G. 1962. Some nutritional problems characteristic of sandy soils. p. 139-143. *In* G. J. Neale (ed.) Trans. Joint Meeting Comm. IV, V, Int. Soc. Soil Sci. (Palmerston North, 13-22 Nov. 1962) Soil Bureau, P.B., Lower Hutt, New Zealand.

Ozanne, P. G., and G. W. Arnold. 1970. Differences in inorganic nutrient levels for pasture and animal response. 1(d) 32-35. *In* T. C. Miller (ed.) Proc. Aust. Plant Nutr. Conf., Mt. Gambier, Sept. 1970. CSIRO, Australia.

Ozanne, P. G., and C. J. Asher. 1965. The effect of seed potassium on emergence and root development of seedlings in potassium-deficient sand. Aust. J. Agric. Res. 16:773-784.

Ozanne, P. G., C. J. Asher, and D. J. Kirton. 1965a. Root distribution in a deep sand and its relationship to the uptake of added potassium by pasture plants. Aust. J. Agric. Res. 16: 785-800.

Ozanne, P. G., and K. M. W. Howes. 1971a. The effects of grazing on the phosphorus requirement of an annual pasture. Aust. J. Agric. Res. 22:81-92.

Ozanne, P. G., and K. M. W. Howes. 1971b. Preference of grazing sheep for pasture of high phosphate content. Aust. J. Agric. Res. 22:941-950.

Ozanne, P. G., and K. M. W. Howes. 1974. Estimating the phosphate requirement of pastures. 2(a):15-18. *In* J. D. Colwell (ed.) Proc. Aust. Soil Sci. Conf., Melbourne. Feb. 1974. CSIRO, Australia.

Ozanne, P. G., K. M. W. Howes, and Ann Petch. 1976. The comparative phosphate requirements of four annual pastures and two crops. Aust. J. Agric. Res. 27:479-488.

Ozanne, P. G., J. Keay, and E. F. Biddiscombe. 1969. The comparative applied phosphate requirements of eight annual pasture species. Aust. J. Agric. Res. 20:809-818

Ozanne, P. G., and Ann Petch. 1978. Effect of species and cultivation on the responses to phosphate of annual pastures and crops. Aust. J. Agric. Res. 29:225-233.

Ozanne, P. G., D. B. Purser, K. M. W. Howes, and I. N. Southey. 1976. The effect of phosphorus concentration in dry feed on voluntary intake and liveweight gain in sheep. Aust. J. Exp. Agric. Anim. Husb. 16:353-360.

Ozanne, P. G., and T. C. Shaw. 1967. Phosphate sorption by soils as a measure of the phosphate requirement for pasture growth. Aust. J. Agric. Res. 18:601-612.

Ozanne, P. G., and T. C. Shaw. 1968. Advantages of the recently developed phosphate sorption test over the older extractant methods for soil phosphate. Int. Congr. Soil Sci., Trans. 9th (Adelaide, South Australia) II:273-280.

Ozanne, P. G., T. C. Shaw, and D. J. Kirton. 1965b. Pasture responses to traces of zinc in phosphate fertilizers. Aust. J. Exp. Agric. Anim. Husb. 5:29-33.

Peel, A. J. 1974. Transport of nutrients in plants. Butterworths, London.

Playne, M. J. 1972. Nutritional value of Townsville stylo (*Stylosanthes humilis*) and of spear grass (*Heteropogon contortus*)—dominant pastures fed to sheep. 2. The effect of superphosphate fertilizer. Aust. J. Exp. Agric. Anim. Husb. 12:373-377.

Playne, M. J., and K. P. Haydock. 1972. Nutritional value of Townsville stylo (*Stylosanthes humilis*) and of spear grass (*Heteropogon contortus*)—dominant pastures fed to sheep. 1. The effect of plant maturity. Aust. J. Exp. Agric. Anim. Husb. 12:365-372.

Rahman, M. S., and J. S. Gladstones. 1974. Differences among *Lupinus* species in field response to superphosphate. Aust. J. Agric. Anim. Husb. 14:1-26.

Rajan, S. S. S. 1973. Phosphorus adsorption characteristics of Hawaiian soils and their relationships to equilibrium phoshorus concentrations required for maximum growth of millet. Plant Soil 39:519-532.

Robertson, G. A., J. W. Bowden, and N. J. Halse. 1975. DECIDE—How much superphosphate. An outline of the DECIDE method for recommending superphosphate application rates for crops and pastures. West. Aust. J. Agric. 16:34-38.

Rominger, R. S., D. Smith, and L. A. Peterson. 1975. Yields and elemental composition of alfalfa plant parts at late bud under two fertility levels. Can. J. Plant Sci. 55:69–75.

Saunders, W. M. H. 1964. Phosphate retention by New Zealand soils and its relationship to free sesquioxides, organic matter, and other soil properties. N.Z. J. Agric. Res. 8:30–57.

Schultz. J. E. 1975. Effect on wheat yield and grain phosphorus content of deep placement and foliar application of phosphate. Agric. Record 2:51–53.

Sewell, P. L., and P. G. Ozanne. 1970. The effect of modifying root profiles and fertilizer solubility on nutrient uptake. Sect. 1(d), p. 6–9. *In* T. C. Miller (ed.) Proc. Aust. Plant Nutr. Conf., Mt. Gambier, Sept. 1970. CSIRO, Australia.

Sheu-Hwa, C., D. H. Lewis, and D. A. Walker. 1975. Stimulation of photosynthetic starch formation by sequestration of cytoplasmic orthophosphate. New Phytol. 74:383–392.

Sobulo, R. A., A. A. Fayemi, and A. Agboola. 1975. Nutrient requirements of tomatoes (*Lycopersicon esculentum*) in S.W. Nigeria. II. Foliar analysis for assessing N, P, and K requirements. Exp. Agric. 11:137–143.

Staten, H. W. 1949. Palatability trials of winter pasture crops, and effect of phosphate fertilizers on palatability. Okla. Agric. Exp. Stn. Tech. Bull. No. T-35.

Stivers, R. K., and A. J. Ohlrogge. 1952. Influence of phosphate and potassium fertilization of two soil types on alfalfa yield, stand, and content of these elements. Agron. J. 44:618–621.

Theiler, A., H. H. Green, and P. J. Dutoit. 1924. Phosphorus in the livestock industry. J. Dep. of Agric. Union of S. Afr. 8:460.

Troughton, A. 1961. Studies on the roots of leys and the organic matter and structure of the soil. Emp. J. Exp. Agric. 29:165–174.

Underwood, E. J. 1966. The mineral nutrition of livestock. Publ. Commonw. Agric. Bur., Food Agric. Org. of the U.N., Rome.

Williams, R. F. 1948. The effects of phosphorus supply on the rates of intake of phosphorus and nitrogen and upon certain aspects of phosphorus metabolism in gramineous plants. Aust. J. Sci. Res. (Series B) I:333–341.

Wise, M. B., A. L. Ordoveza, and E. R. Barrick. 1963. Influence of variation in dietary calcium:phosphorus ratio on performance and blood constituents of calves. J. Nutr. 79:79–84.

Wiskich, J. T. 1975. Phosphate-dependent substrate transport into mitochondria. Plant Physiol. 56:121–125.

Chapter 21

Soil-Plant Interactions in the Phosphorus Nutrition of Plants[1]

STANLEY A. BARBER
Purdue University
West Lafayette, Indiana

I. INTRODUCTION

Phosphorus moves through the soil-plant root interface during the process of absorption, hence the chemical and biological reactions occurring in this zone play an important role in determining P flux into the root. Since P flux into the root is a dynamic process, thermodynamic equilibrium of P with the soil within the soil zone next to the root rarely occurs. Hence, evaluation of P available to the plant with models based on kinetic parameters of soil P should be more realistic than those based on thermodynamic equilibrium for determining the availability of soil and fertilizer P for plant uptake.

In this chapter, the approach used is to discuss the mechanisms that supply P to plant roots growing in soil and to relate them to the P absorption characteristics found in plant root systems. Because diffusion is probably the dominant mechanism for the supply of P to the roots of plants growing in many soils, those soil and plant properties that influence diffusive flux of P to the root are discussed. The role of all factors, both plant and soil, that affects the rate of P uptake by the plant can be interpreted by the use of mathematical models. The several models for predicting P uptake from the soil that have been developed are discussed. The plant root can change the physical, chemical, and biological environment of the soil adjacent to the root. The degree of these changes and their effect on P flux into the root are considered. The ultimate use of new knowledge on P flux within the soil-plant root interface is the development of more efficient methods of P fertilization. Examples of such developments will be discussed.

---

[1]Journal paper no. 6259, Purdue University, Agricultural Experiment Station, West Lafayette, IN 47907. Contribution from the Department of Agronomy.

Copyright 1980 © ASA-CSSA-SSSA, 677 South Segoe Road, Madison, WI 53711, USA. *The Role of Phosphorus in Agriculture.*

II. MECHANISMS OF P SUPPLY TO PLANT ROOTS GROWING IN SOIL

When plants are grown in stirred nutrient solution, the motion of the solution replenishes the P absorbed from solution by the root at or near its surface. When roots grow in soil, only small amounts of P will be found at the root surface initially, and after this amount is absorbed, the P is replenished by P moving to the root by mass-flow and diffusion. Plant roots absorb water and this causes a convective flow of water (the soil solution) towards the root. The soil solution, therefore, carries P to the root by mass-flow. If mass-flow plus that intercepted initially are not sufficient to supply the requirement of the plant, the P concentration in solution at the root surface will be reduced as a result of P absorption by the root. This will create a P concentration gradient radiating perpendicularly to the root axis. Phosphorus will diffuse toward the root along this gradient. The proportion of P that is supplied by each mechanism will depend on the size of the root system, the P absorption characteristics of the root, the rate of water absorption by the root, and the levels of adsorbed and solution P within the soil.

Plant roots of annual crops have a volume that is usually less than 1% of the soil volume (Barber et al., 1963). Hence, the roots will contact less than 1% of the available P in the soil and this amount is usually a small percentage of the plant requirement. The amount that reaches the root without diffusing or moving by mass-flow will likely be less than the amount of available P in a volume of soil equal to the root volume. Phosphorus moving to the root by mass-flow depends on the P concentration of the soil solution. Water absorption can be calculated either in terms of the average flux to the root, about 1×10^{-7} ml cm^{-2} sec^{-1}, or it can be measured by the transpiration ratio which is measured as grams of water transpired per gram of plant produced. The latter relation, which ranges from 200 to 600 g of water per gram of plant, is useful for estimating the relative amount of supply that may be transported to the root by mass-flow. Assume a plant has a transpiration ratio of 400 and a P concentration at harvest of 0.2% P. If we divide the percent P in the plant by 400, we obtain the average P concentration needed in the soil solution to supply all the P to the plant by mass-flow. In this case, it would be 0.0005% or 5 μg/g.

Barber et al. (1963) measured P in saturation extracts of 135 soils and found values ranging from 0.01 to 1.2 μg/g, with the majority of the samples in the range 0.02 to 0.08. If we assume a soil with a P concentration of 0.05 μg/ml in soil solution, mass-flow would supply on the average only 1% of the P used by the plant. In many soils diffusion is the mechanism for the supply of 90 to 98% of the P absorbed by the root.

Evidence indicating that diffusion was an important mechanism for P supply to plant roots growing in soil has been obtained by autoradiographs (Lewis & Quirk, 1967; Bhat & Nye, 1973), which show the depletion of P about the root and the formation of a P concentration gradient extending perpendicular to the root. The concentration gradient predicted from knowledge of the rate of P diffusion in the soil was similar to that obtained experimentally.

Diffusion of P through the soil to the root is the dominant mechanism governing the supply of P to roots growing in all except soils extremely high in P. Because diffusion is important, all of the factors which govern the rate of P diffusion to the root and the extent of root growth are important in determining P availability to plants growing in soil.

III. SOIL FACTORS INFLUENCING P DIFFUSION RATES IN SOIL

Phosphorus diffusion occurs mainly through soil water, since diffusion through the solid phase is extremely slow and the phosphate ion, being negatively charged, would not likely diffuse along the negatively charged surfaces of soil particles. Phosphorus diffusion through soil is much slower than in pure water for three reasons: (i) soil water occupies only part of the soil so that the cross-sectional area for diffusion is less; (ii) the diffusion path is tortuous because the water is present as films around soil particles; and (iii) most of the diffusible P is adsorbed on soil surfaces which equilibrates with and buffers the small amount of P in soil solution that is diffusing.

The diffusion coefficient for $H_2PO_4^-$ in water at 25°C is 0.89×10^{-5} cm^2 sec^{-1} (Parsons, 1959). At pH's between 4.0 and 6.5 most of the inorganic P is in this form. Assuming that P diffuses mainly through solution, Nye (1968) proposed Eq. [1] for calculating the effective diffusion coefficient, D_e, for diffusion of an ion in soil

$$D_e = D\theta f \frac{dC_l}{dC}, \qquad [1]$$

where D is the diffusion coefficient in water, θ is the volumetric moisture content of the soil, f is a tortuosity factor and dC_l/dC is the inverse of the differential buffer capacity of the soil. Thus, θ, f, and dC_l/dC are the three principal factors which influence the size of D_e. They will each be discussed separately.

A. Volumetric Moisture

The percentage by volume of the soil that is occupied by soil water determines the average fraction of cross-sectional area through which P can diffuse. Hence, net diffusive flux should be directly related to the fractional volumetric moisture content. At low moisture contents, an additional factor, a more rigid structure of water, may reduce the diffusive flux near particle surfaces. Since $H_2PO_4^-$ would be less concentrated near soil surfaces because of repulsion of like charges, the net effect for P is rather small. The magnitude of volumetric moisture content in soils supporting vigorous plant growth varies from 0.05 for very sandy soils to 0.50 for heavy clay soils. For many loams and silt loams, values range from 0.15 to 0.30 depending on their moisture status.

B. Tortuosity

Tortuosity of the diffusion path depends on the thickness of water films and on the fineness of soil particles present. Tortuosity increases greatly as the moisture content of a soil decreases. Increasing tortuosity decreases f. Tortuosity is usually estimated by measuring the diffusion rate of a nonadsorbed ion such as Cl^- or of tritiated water. Values of 0.20 to 0.35 have been obtained on loam and silt loams soils that are near field capacity in moisture content.

Bulk density affects tortuosity of the diffusion path. Warncke and Barber (1972) measured Cl^- ion diffusion in five silt loam soils, each at four bulk densities and three moisture levels. There was little difference among soils. The relation of average values of f to soil bulk density and θ is shown in Fig. 1. There was less tortuosity or greater diffusive flux at a bulk density of 1.3 than at higher or lower bulk densities. When a soil of low bulk density was compressed, it gave greater continuity of the liquid phase. However, beyond a bulk density of 1.3, tortuosity of the diffusion path was increased because of an increased volume of solids around which the ions would need to move.

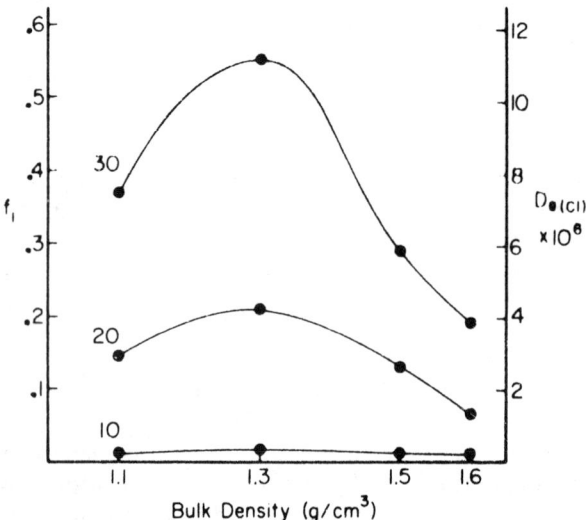

Fig. 1—Average influence of soil bulk density on the chloride diffusion coefficient and on tortuosity values calculated from $D_{e(Cl)}$ for five silt loam soils at three moisture levels. Values on curves are percent moisture (w/w).

C. Buffering Capacity

For P that is strongly adsorbed on the surface of soil particles, buffering capacity has a greater influence on the size of the effective diffusion coefficient than does volumetric moisture content or tortuosity. Equilibrium between P in solution and P adsorbed on soil surfaces can be determined by adsorption or desorption isotherms. The buffering capacity of a soil is usually highest when adsorbed P is low, and it decreases as more P is added to soil because a greater proportion of the added P remains in solution. Desorption curves conducted under conditions similar to those which occur near roots growing in soil would be the preferable way to measure buffering capacity. Bhat and Nye (1974a) found that the amount of P desorbed was related to the logarithm of the level of P in the equilibrium solution. Adepetu (1976) found values for the buffering capacities as high as 4,000 where small amounts of P (20 μg/g) were added to a high P adsorbing soil, and as low as 4 where high rates (500 μg/g) of P had been added to a soil that did not adsorb as much P. Values of 100 to 500 are common. This means that D_e is 0.01 to 0.002 of D in water due to the adsorption of P by the soil. Increasing the buffering capacity of P on the solid phase for P in solution reduces D_e, and consequently also reduces the average distance for diffusion.

D. Effect of Temperature

The effect of temperature on diffusion in solution can be calculated from the Stokes-Einstein equation $D = kT/6\pi r\eta$, where T, k, r and η are the absolute temperature, Boltzman constant, ionic radius, and viscosity, respectively. Temperature influences η. At 20°C, η is 1.002 while at 30°C it is 0.7975 centipoise (Weast, 1969). Thus, the effect of T on D is primarily due to the effect on η. Increasing T from 20 to 30°C would increase D by 30%.

Changing the rate of diffusion in solution may not be the only effect of temperature, however. Changes in temperature may change the amount of P found in solution as well as the amount of P on the solid phase that will equilibrate with solution P. Little experimental data is available on this subject. The effect of temperature on diffusion was evaluated indirectly from data on the effect of temperature on P in solution and on isotopically exchangeable P (Sutton, 1969). Sutton reviewed the subject and reported that increasing temperature increased isotopically exchangeable P. Also increasing temperature increased P in solution 1 to 2% for each degree rise in temperature. Since both the adsorbed and solution phase P were increased, any change in buffering capacity would be due to differential rates of increase of each, and this may vary with soil. However, the available data indicated that some increase in the diffusion coefficient probably did occur, but the magnitude of the increase was not known. An increase in levels of solution and adsorbed P would cause an increase in the amount of P diffusing, because an increase in concentration gradient may occur even if the diffusion coefficient was not affected; hence, increasing temperature will increase the supply of P to the root by diffusion.

IV. PHOSPHORUS UPTAKE CHARACTERISTICS OF PLANT ROOTS

Phosphorus absorbed by plant roots is either in the $H_2PO_4^-$ or HPO_4^{2-} form. The amount of each form present in solution depends on soil solution pH. At pH 7.22 there are equal amounts of $H_2PO_4^-$ and HPO_4^{2-}. Below this pH, $H_2PO_4^-$ is the main form; hence, in many soils it is the dominant form and because of this it has been the subject of most investigations.

Absorption rates for $H_2PO_4^-$ by plant roots have been found to follow Michaelis-Menten kinetics. The equation describing P flux into the root as influenced by P concentration in solution is:

$$I = I_{max} \frac{C}{K_m + C}, \qquad [2]$$

where I_{max} is the maximum rate of P uptake, C is the concentration of P in solution, and K_m is the Michaelis-Menten constant which is C where $I = 0.5$ I_{max}. Values for I_{max} and K_m are obtained by measuring I for various values of C. Commonly this has been done using excised roots and uptake times of 20 min or less. Much of the influx measurement has been with solutions of pH 6.0 or less, so that $H_2PO_4^-$ has been the P form measured. However, studies have indicated that separate carriers operate for the two P forms and the value of I_{max} for the $H_2PO_4^-$ form is approximately 10 times that for the HPO_4^{2-}. Hence, when pH increases, P influx may decrease because of the change in P form present. At P concentrations usually found in soils, HPO_4^{2-} was absorbed much more slowly than $H_2PO_4^-$ (Hendrix, 1967; Hagen & Hopkins, 1955).

Claassen and Barber (1974) developed a convenient procedure for determining the nutrient uptake characteristics of plant roots of intact plants. They measured the rate of nutrient depletion from solution to obtain uptake rate versus ion concentration and determined the uptake parameters by fitting a curve to the data using least squares. Plant roots do not completely deplete the nutrients from a solution, but only reduce it to a minimum concentration, C_{min}. Hence when fitting the curve, a value for efflux had to be subtracted so that the curve would go to C_{min}. The data could then be used to calculate net influx, I_n, vs. concentration of the ion in solution. An example of net P influx by 18-day-old corn vs. P concentration in solution is shown in Fig. 2.

The Michaelis-Menten relation usually fits the data for influx vs. C when C is less than 50 μM. Where higher concentrations are used there are increases in I_{max} and K_m as C increases. Nissen (1974) explains these changes in terms of multiphasic uptake relationships where there are specif-

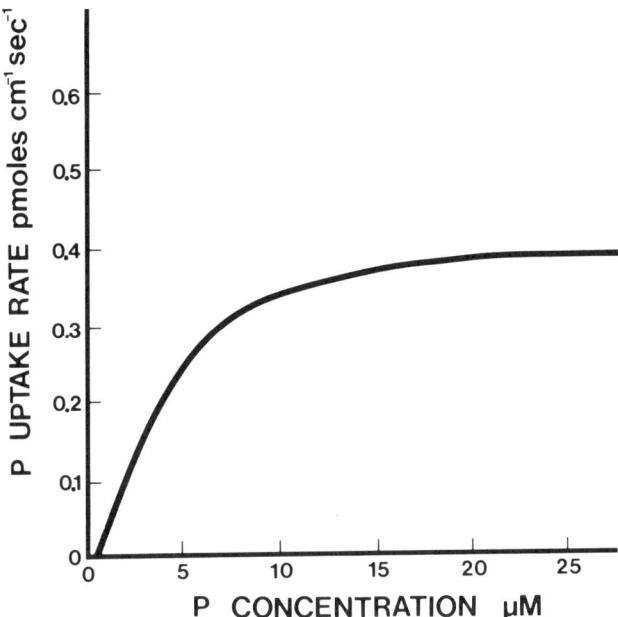

Fig. 2—The relation between P concentration in solution and net P influx for 18-day-old corn plants.

ic C values beyond which I_{max} and K_m change. Hodges (1973) explains these differences as changes in the nature of the binding site as C increases. In soil systems, the P level in soil solution rarely exceeds 20 μM except near recently applied bands of fertilizer. Since P diffuses to the root, the P level at the root will usually not exceed 2 μM except for brief periods when P absorption first begins. Therefore, there are few different phases for P absorption from soil because of the limited range of P concentration that will be present about the root.

A. Change in P Absorption with Plant Age

The effect of corn plant age on the P absorption kinetics was investigated by Jungk and Barber (1975) and that of soybean plant age by Edwards and Barber (1976). Jungk and Barber (1975) found that I_{max} reached a maximum of 0.55 pmol cm^{-1} sec^{-1} for plants about 28 days old then decreased rapidly, so that at 50 days I_{max} was only one-fourth the value for 28 days. Edwards and Barber (1976) found that I_{max} for P uptake by soybean roots varied from 0.09 pmol cm^{-1} sec^{-1} on 18-day-old plants to 0.044 pmol cm^{-1} sec^{-1} on 69- to 74-day-old plants or about a 50% reduction which was less than that found for corn. The I_{max} for soybeans was 0.2 to 0.3 that for corn.

Mengel and Barber (1974b) calculated P influx into corn roots growing in the field using data from sequential harvests. They found that the average

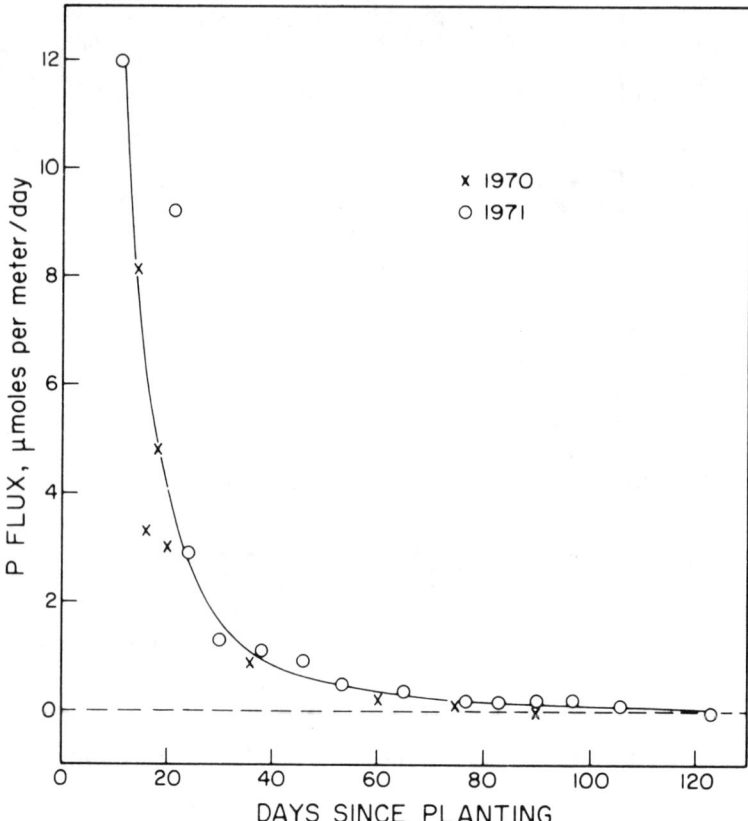

Fig. 3—The relation between average P flux into the root per unit of root length and plant age for corn grown in the field in 1970 and 1971 at the Purdue Agronomy farm, Lafayette, Indiana.

P influx rate varied from 11 to 0.1 μmol m^{-1} day^{-1} for a decrease of about 100-fold between 20-day-old and 80-day-old corn plants (Fig. 3). Warncke and Barber (1974) conducted a similar type experiment with corn in the greenhouse and observed that net influx of P varied from 5 to 0.2 μmol m^{-1} day^{-1} or about a 25-fold decrease. Barber (1974) obtained a fourfold decrease in P uptake rate between 8- and 19-day-old soybean seedlings, but with 7- to 18-day-old corn plants the uptake rate was rather constant.

The reduction in P uptake rate as the plant ages might be attributed to a reduced P influx by older or suberized roots. However, when a plant is in the vegetative stage of growth, root length frequently increases logarithmically with time, so that at any time a large fraction of the roots are not more than a few days old. The large decreases in P influx with increased plant age cannot be attributed solely to reduced uptake by older roots. Calculations indicate there also must be a reduced P uptake rate by younger roots.

While there was a large reduction in I_{max} as the plant aged, K_m did not vary with corn plant age in Jungk and Barber's (1975) experiments. An average value of 3 μM was found for K_m. Phosphorus efflux from corn roots decreased along with decreased I_{max} values. For corn, C_{min} decreased with age from 0.3 μM for 14-day-old to 0.1 μM for 52-day-old plants. With soybeans, Edwards and Barber (1976) found K_m increased from 2 to 17 μM as plant age increased from 14 to 75 days and C_{min} increased from 0.04 μM to 0.17 μM. So, while there are differences in the P absorption characteristics by plant roots according to plant age, these differences varied with species; therefore, each species needs to be investigated individually.

B. Change with Root Age

The absorption rate of P by the root cells may vary as the individual cells grow older. If this is the case net P influx would be different immediately back of the root tip from that at some distance further along the root where the root cells are older. Ferguson and Clarkson (1975) investigated these differences for P uptake from solution by roots of intact plants. They found that for corn roots P was actively adsorbed by portions of roots up to 20 cm from the tip even though the endodermis was completely suberized. In the basal region of the root, P influx was restricted. Their research indicates the P was adsorbed actively over a long period of time. A similar study of P influx using different techniques was conducted by Bowen and Rovira (1977). They also were able to show that, while the P absorption rate of wheat roots was somewhat higher in the apical 3 cm, P uptake continued over a distance of 25 cm from the tip (Rovira and Bowen, 1968).

Bhat and Nye (1974a) used autoradiographs to study the P uptake characteristics of rape roots growing in soil. They found that P influx over the entire root was relatively constant for the first 8 days for roots growing in soil except that very little P was absorbed by the apical 1.5 to 4 cm of the root. After 8 days, P was absorbed at a reduced rate.

Temperature influences the relative rate of P influx by various portions of plant roots. Bowen (1970) evaluated P influx into *Pinus radiata* roots at 14 to 25°C. At 14°C the greatest P influx occurred several centimeters behind the apex, while at 25°C P influx was greatest for the apical centimeter and decreased sharply with distance from the apex.

The data obtained from measurements of P absorption along plant roots indicate that P influx remains relatively constant for the first 8 days of root growth and that absorption by the zone immediately behind the root tip is often similar to uptake from sections up to 15 to 20 cm from the root tip. Phosphorus was absorbed in older roots even though the endodermis was suberized.

C. Change with Soil Depth

The effect of soil on the P uptake characteristics of plant roots will depend both on the structure, tilth, fertility, and water relations of the soil which affects root penetration and on the rooting characteristics of the plant. Usually the surface 15–20 cm of soil is naturally higher in P than the subsoil because P is transported to the surface by plant growth over years of soil development. In addition, P fertilization has frequently increased surface soil P fertility. It is difficult to add P to subsoils since P does not readily leach through most soils, and it is difficult to incorporate P fertilizer mechanically into subsoil horizons.

The distribution of corn roots in a Chalmers silty clay loam was measured by intervals during the growth of a corn crop (Mengel and Barber, 1974a). Thirty-four days after planting, 60% of the corn root system was in the 0- to 15-cm layer. However, the proportion present in this layer steadily decreased until at 125 days after planting only 30% was in the 0- to 15-cm layer. This change in root distribution in the soil profile shows that during later stages in the growth of the plant a large proportion of the P being absorbed may be coming from the subsoil. The amount of P absorbed from each horizon will depend greatly on the relative available P levels of each layer and on the physical and chemical nature of the subsoil which influences root penetration and development.

D. Evaluating the Significance of Root Hairs for P Uptake by Roots

The significance of root hairs will depend on whether they significantly increase the amount of P per plant that will diffuse to their surface over that diffusing to the main root surface. A model root that is characteristic of corn (Bouldin, 1961) and P diffusion characteristic of levels found in soil are used here to estimate the role of root hairs in P flux to the root. I assume that root hairs do not increase P influx into roots when P levels are maintained at the root surface as occurs in stirred solutions. I further assume the total contribution of root hairs is in providing a greater surface area for P to diffuse through the soil to the root, but does not increase uptake where high P levels are maintained at the root. The root is assumed to have the following characteristics: radius 0.03 cm; root hair radius 0.0075 cm; root hair length 0.03 cm; root hair density, 500/cm of root length (Bouldin, 1961).

The root hairs will be most important when all of the P in the root hair cylinder (the cylinder of soil penetrated by the root hairs) cannot readily reach the root or root hair surface by diffusion. In this model the average distance between root hairs is 0.0059 cm at the root surface and 0.0145 cm at the root hair tips. So if ions in this soil diffuse less than 0.003 to 0.007 cm, soil zones supplying each root hair will not overlap and root hairs would act independently in serving as a sink for P influx. Absorption of ions which diffuse greater distances than the root hair length of 0.03 cm would be affected much less by the presence of root hairs.

The average linear distance an ion will diffuse in time t with a diffusion coefficient of D is $\sqrt{2Dt}$. Using this relation the data in Table 1 were calculated for average diffusion in 4 days. Since the distance increases with $\sqrt{t}$, the distance for 1 day would be one-half these distances and the distance in 16 days would be double these values.

Table 1—Influence of size of the effective diffusion coefficient for ion diffusion in soil on the average distance the ion diffuses linearly in 4 days.

Diffusion coefficient, $cm^2 sec^{-1}$	Average diffusion distance in 4 days, cm
1×10^{-6}	0.832
1×10^{-7}	0.262
1×10^{-8}	0.083
1×10^{-9}	0.026
1×10^{-10}	0.0083
5×10^{-11}	0.0059

Phosphate diffusion in soils is usually slow and values of D_e are usually in the range 1×10^{-8} to 5×10^{-11}. The lowest values of D_e shown occur on soils with the lowest levels of solution P and highest buffering capacities. On these soils with D_e of 5×10^{-11} $cm^2 sec^{-1}$, the average distance for diffusion in 4 days is of the order of 0.006 cm which is approximately one-half the average distance between root hairs in the model used here. Hence, up until about 4 days, root hairs should contribute materially to P uptake by the root. On the other hand for soils high in plant available P and with a D_e of 1×10^{-8} $cm^2 sec^{-1}$ the average diffusion distance in 4 days would be 0.08 cm. This is almost three times the root hair length and more than 10 times the distance between root hairs. Hence, almost all of the available P in the soil in the root hair cylinder could reach the root within 1 hour. Since the average diffusion distance is much greater than the length of the root hairs, the main effect of the presence of root hairs would be absorption of P in the root hair zone and an increase in the effective radius of the roots.

Increasing the root radius from 0.03 cm to 0.06 cm will increase the effective surface area for diffusion to the root from outside the root hair cylinder by two, since area is directly related to radius. Hence, on the basis of diffusion, P uptake during the first minutes after development of the root and its root hairs may be related to the increase in surface area caused by the root hairs, which in this case, would be an increase by a factor of 3.75. After the P in the root hair zone begins to be depleted, the uptake due to the greater effective radius would be greater by a factor of two.

The effect of root hairs on P depletion in the soil in the rhizosphere has been investigated by Bhat et al. (1976). They computed the theoretical distribution of P about rape roots where root hairs were present and found experimental depletions were greater than predicted using the relationship between P in soil and in solution obtained from a desorption isotherm.

Root hairs were reported to increase P uptake by Barley and Rovira (1970). However, Bole (1973) found little difference in P uptake among wheat varieties that varied widely in root hair density. Possibly rate of diffusion was fast enough that the low root hair density still was able to absorb much of the diffusible P in the soil within the root hair cylinder.

Bhat and Nye (1974a) observed that in soils low in P, root hairs were longer. This may be a method used by the plant to partially compensate for low P soil levels. Powell (1974) found that root length per unit weight of roots decreased as the P level in the soil increased. So the plant may both increase root length and root hair length in order to obtain more P when P levels in the soil are low.

In addition to root hairs, infection of the roots with mycorrhizae increases the P uptake rate of the root. Rhodes and Gerdemann (1975) have demonstrated that increased P uptake occurred in roots infected with endomycorrhizae. This subject is discussed in detail in Chapter 22.

E. Uptake of P as Influenced by Proportion of the Root System Supplied with P

Phosphorus uptake characteristics of roots have been described by Michaelis-Menten kinetics. Following this system, Claasen and Barber (1974) used I_{max}, K_m, and efflux to describe the relation between I_n and C_l, concentration of the ion in solution. Uptake of P by the root may be influenced by demand of the shoot for P. Plants usually develop a particular shoot/root ratio so that, by supplying only part of the root system with P, demand of the shoot per unit of root supplied with P will be increased.

Jungk and Barber (1975) studied the effect of supplying only part of the root system with P on P influx into corn roots and found no effect of increasing shoot demand when plants had the roots split between no-P and P-containing solutions for 2 days. They also found that trimming the roots to reduce the amount of roots supplied with P gave the same results as splitting the root system between a P-containing and a P-deficient solution. Trimming was done 1 day before P influx measurement. An example of the results obtained is shown in Table 2. They conducted experiments to compare uptake of trimmed and untrimmed plants of ages ranging from 12 to 82 days and found that trimming did not influence P influx at any plant age.

Edwards and Barber (1976) studied the effect of root trimming on P uptake by soybeans ranging in age from 18 to 74 days. Trimming was done 2 days before P influx measurement. Phosphate influx was evaluated by the procedure of Claassen and Barber (1974). For 18- and 35-day-old plants trimming increased I_{max} such that uptake per plant of trimmed and untrimmed plants was similar. Hence, it appears that reducing the amount of roots supplied with P influences P influx differently with different species.

Table 2—Effect of supplying part of the root system with P on P influx by that portion of the root system.

Shoot dry wt	Root length†	P influx µg atoms
g	m	m^{-1} day^{-1}
1.09	22.0	2.35
1.01	31.3	2.12
0.76	28.6	2.61
1.06	42.6	2.87
0.99	59.6	2.33
1.01	57.4	2.46

† Initial root length was approximately 58 m. Lengths shown are those present after trimming 1 day before uptake measurement.

F. Effect of Plant P Status on P Uptake Rate

Jungk (1975) found that P influx was increased when plants were starved for P by growing them several days in a P deficient nutrient solution. Tomato plants were transferred to a P-deficient nutrient solution for 1, 4, and 5 days before P influx measurements were made. After 1 day of no P, I_{max} was 0.088, and it was increased to 0.147 pmols cm^{-1} sec^{-1} for a 5-day no-P treatment. Percentage of P in the shoot was reduced from 1.08 to 0.78. This is similar to results obtained by Claassen and Barber (1976) for K uptake by corn roots where they observed that the increase in I_{max} for K was related to the decrease in percent K in the shoot. Neilsen (N. E. Neilsen, personal communication, Royal Veterinary and Agricultural University, Copenhagen, Denmark) grew corn for 7 days in eight solutions with P concentration varying from 1 to 100 μM. Influx of P was measured when the plants were 20 days old, and I_{max} varied from 0.97 pmol cm^{-1} sec^{-1} when pretreatment was 1 μM to 0.53 pmol cm^{-1} sec^{-1} when pretreatment was 100 μM. Shoot and root weights did not vary with treatment. It appears that reduction of P concentration in the shoot triggers some mechanism that causes an increase in the rate at which the roots can absorb P. This feedback mechanism appears to be a method for the plant to attempt to correct a deficiency of P.

G. Differences Between Species in P Absorption Rates

Plant roots vary between species in their morphology and it is likely they will vary in rate of P absorption. Loneragan and Asher (1967) compared P uptake by eight pasture species from solutions ranging in P concentration from 0.04 to 25 μM. They determined uptake rate in terms of microgram atoms per gram of fresh weight of root. The P uptake rate from 24 μM P solution varied from 9.8 to 20.6 μg atoms g(f.w.)$^{-1}$ day^{-1}. From 0.04 μM P solution the P uptake rates ranged from 0.10 to 0.49. Hence for eight species the range in uptake rate was from two- to fivefold.

Jungk (1975) compared the P uptake rates of corn, tomatoes, beans, and cucumbers. The I_{max} values were markedly different and they decreased in the order: corn > beans > cucumbers > tomatoes. There was a threefold range in values when calculated in terms of root length. However, K_m and C_{min} values were similar for the different species.

Phosphorus influx into soybean roots was compared with P influx into corn roots by Edwards and Barber (1976). The soybean plants ranged in age from 18 to 74 days. When expressed as uptake per centimeter of roots, the values of I_{max} for corn over a 4- to 6-hour period were five times those for soybeans; however, when net influx was calculated over a 60-day period, mean P influx for corn and soybean roots were similar. Possibly, the similarity in net influx is because many plant species have evolved on soils that are low in plant-available P, hence the supply of P to roots has been limited by rate of P diffusion through soil to roots rather than by the P absorption ability of the root. While root hairs may absorb additional P, they probably still do not provide the plant with enough P to allow it to absorb at near I_{max}. Since soil diffusion rate is often limiting, P supply to the root may have minimized the selection of species based on I_{max}.

H. Effect of Temperature on P Uptake

Phosphorus-deficient plants have sometimes been associated with low soil temperatures, suggesting that P supply to the root is less with low soil temperature. The reduction of P supply at low temperature could be due to one or more of the following factors (i) a reduction in I_{max} for P uptake by roots, (ii) a reduced rate of root or root hair growth relative to the increase in P requirement with shoot growth, or (iii) a reduction in the P level in soil solution and its replenishment by diffusion to root surfaces.

In a review of the effect of soil temperature on plant growth, Nielsen (1974) indicated that increasing soil temperature increased P uptake in soils low in P, but not in soils high in available P. Sutton (1969) reviewed research on the effect of soil temperature on P nutrition of plants and considered the effect of soil temperature on level of available P in soil. He concluded that most of the effect of temperature was on the inorganic rather than on the organic P, since the effect occurred too rapidly to have resulted from changes in the rates of microbiological activity. He reported an effect of temperature on total quantity of inorganic P that was potentially able to enter soil solution, since labile P was increased as soil temperature increased. Phosphorus concentration in the soil solution increased 1 to 2% for each 1°C increase in temperature. The amount of P extracted by shaking soil with an anion exchange resin increased with an increase in soil temperature. For one soil, the increase was 4% per 1°C over the rane 10 to 40°C. Since the concentration of solution P was increased, it would be likely that the rate of diffusion of P to the root would be increased with increasing temperature. Reduction in P uptake by plants growing at low soil temperature as compared to those growing at higher temperatures could be due to reduction in the rate of P supply to roots. When the effect of soil temperature on the proportion of P derived from the fertilizer was studied, a higher proportion of P came from fertilizer with low soil temperatures, indicating the soil P was less available to the plant relative to fertilizer P.

Temperature also affects the rate of P absorption from solution culture by plant roots. Using excised corn roots, Carter and Lathwell (1967) found the Q_{10} for I_{max} of P absorption was 1.8 to 2.0 between 20 and 40°C. Since P uptake requires respiration energy, when energy is limiting the rate of respiration will be influenced by temperature, and this in turn may influence uptake rate.

The rate of root growth relative to shoot growth is also affected by temperature. For many species, an increase in temperature from 5 to 35°C causes a reduction in root/shoot ratio (Cooper, 1973). This would mean that there would be relatively less roots per unit of shoot for plants growing at 35°C than for those growing at a lower temperature. Hence, the amount of roots present for P absorption at low temperature would be relatively larger. Corn had a somewhat different pattern for root/shoot growth, with a maximum ratio at 20°C. For plants in general, it appears that they attempt to compensate for the low rate of P supply from the soil at low temperatures by producing more roots.

From the results of experiments with soil, it appears that where low temperature is associated with P deficiency, the effect of temperature on the P level in solution, on total soil P supply, and on P uptake rate by the plant is greater than can be compensated for by increased root/shoot ratio at the lower soil temperatures.

V. MODELS FOR P UPTAKE BY PLANT ROOTS FROM SOIL

Mathematical and computer simulation models provide a useful tool for organizing ideas about the mechanisms of P uptake by plants roots growing in soil. The purpose of a model is to summarize current knowledge in an orderly fashion and also to permit prediction of the effects of changing a given set of conditions. The model can be used in conjunction with experiments to test whether the principles and relations used in developing the model are correct. Many systems are so complex that the interrelation of variables may be more conveniently studied by modeling.

The models which have been developed to describe P flux into roots growing in soil have been improved with time as more knowledge has been gained about the system. One of the early descriptions was by Bouldin (1961) where he described the diffusion of ions to plant roots. Olsen and Kemper (1968) reviewed the research on mass-flow and diffusion to plant roots. Nye and coworkers at Oxford University have contributed greatly to our knowledge about diffusion and mass-flow to plant roots (Nye & Marriott, 1969). Barley (1970) made a further review and provided information on the influence of root density on root growth. Helyar and Munns (1975) in developing their model for P uptake have provided the most recent review. Claassen and Barber (1976) have also developed a mathematical model to estimate ion uptake by the growing root systems. Many of the earlier researchers developed equations which predicted the concentration of P in the soil with distance perpendicular to the root surface. Most use similar models for the flux of P to the root by mass-flow and diffusion. The one used by Nye and Marriott (1969) is shown as Eq. [3]

$$\frac{dC_l}{dt} = \frac{1}{r}\frac{d}{dr}\left(rD_e\frac{dC_l}{dr} + \frac{v_o r_o C_l}{b}\right), \qquad [3]$$

where C_l is the concentration of P in the soil solution, r is the radial distance from the root axis, t is time, D_e is the mean effective diffusion coefficient, r_o is the root radius, b is the buffering capacity of the solid phase for C_l, and

v_o is the water flux into the root. The model assumes roots are smooth cylinderical sinks for P and hydrodynamic dispersion is unimportant.

When this equation was numerically integrated over the boundary conditions where

$$C_l = C_{li}; t = 0; r < r_o,$$

and

$$Db \frac{dC_l}{dr} + v_o C_{lo} = \frac{I_{max} C_{lo}}{K_m + C_{lo}}, t > 0; r = r_o,$$

the distribution of C_l with distance from the root as affected by time of absorption was obtained; C_{lo} is the concentration at the root surface.

This equation has been integrated numerically with the help of a computer. By changing the value of certain variables, it is possible to observe how each affects the theoretical distribution of P about the plant root.

Claassen and Barber (1976) have used the model approach to estimate K uptake and compare it with observed values. The model of Nye and Marriott (1969) gives C_l at $r = r_o$ with time. Michaelis-Menten kinetics were used to describe I_n as a function of C_l so that it was possible to get I_n per cm of root with time and by integrating over time obtain uptake per cm of root for various values of t. This relation was combined with rate of root growth to get uptake per plant. The variables required for the model are: I_{max}, K_m, E, C_{li}, D, B, rate of root growth, initial root length, and root radius.

All parameters except those describing rate of root growth and root radius can be measured independent of experiments conducted to test the model. This model assumes that root hairs do not contribute to nutrient uptake and that I_n is uniform with root and plant age, although, if known, the variation can be included. It also assumes that D_e and I_n are independent of r_o. The model was tested for K uptake from four different soils and several K levels and a remarkably close fit of $r^2 = 0.89$ was obtained between the calculated and observed K uptake. Calculated values were somewhat larger than observed uptake values which may have been due to root competition and reduction of I_n during the dark period. The same model was used to calculate P uptake by millet from five soils, each fertilized with P to give four levels of C_l for P in the soil (Adepetu & Barber, 1978). The observed uptake was much larger than the calculated uptake at low P levels in the soil. The difference between calculated and observed P uptake decreased as the P level in the soil was increased. With an initial soil solution P level of 0.02 µg/ml the observed P uptake was 14 to 25 times larger than calculated P uptake. While with an initial soil solution P concentration of 1.5 µg/ml, the observed uptake was 0.66 to 1.2 of the calculated uptake. With low soil solution P concentrations the buffering capacity is large and the diffusion coefficient small. Phosphorus diffuses for short distances and root hairs are important for P uptake since the average P diffusion distances may be less than the distance between root hairs.

Brewster et al. (1976) also obtained larger values for P uptake by rape than predicted from the model, but similar values when the test crop was onions which did not have root hairs. The greater uptake by roots having root hairs is much larger than that predicted from geometry and increased area for diffusion. Root hairs appear to influence the rhizosphere so as to increase P supply to the plant in a manner that is unknown at present.

A. Effect of Soil and Plant Factors on Predicted P Uptake

The effect of varying one factor frequently depends on levels of the other factors, so that in this discussion, only a few general statements will be made. Nye and Marriott (1969) and Helyar and Munns (1975) discuss the variation of factors in more detail.

The soil factors that are in the model are diffusion coefficient, D_e, P concentration in soil solution, C_l, and buffering capacity, b. These three are not independent of one another. As b increases, D_e decreases. As C_l increases, and b decreases, D_e increases. However, considering these independently, P uptake by the root will increase as C_l increases, as b increases, and as D_e increases. At one level of C_l, increasing b results in increasing C.

The plant root factors in the model include I_{max}, K_m, root radius, rate of water uptake, and presence of root hairs. The values of I_{max} and K_m describe the relation between P concentration in solution at the root surface, C_{lo}, and rate of uptake. When C_{lo} is high, uptake increases with I_{max}. However, when the rate of diffusion to the root is the limiting factor and C_{lo} is small, changes in I_{max} and K_m will have smaller effects on uptake. In general at low soil P levels, I_{max} and K_m are not the factors which limit uptake. However at high soil P levels or where P fertilizer is concentrated in bands, I_{max} and K_m may become limiting because C_{lo} is high.

When intact plant roots are allowed to deplete a solution of P, the P concentration will reach a minimum, C_{min}, which varies from 0.01 μM to 0.4 μM depending on species of plant. In soils low in P the value of C_{min} of a species may significantly affect the amount of P the plant root will absorb. Hence, C_{min} is a factor that may influence P influx, particularly in low P soils.

Reducing the root radius increases the relative P flux to the root per cm^2 of root surface because the diffusive flux is occurring radially. Rate of water flux usually has little effect on P absorption because the levels of P in the soil solution are low and the contribution of P by mass-flow is low.

If root weight remains constant, reducing root radius increases root length, and this will increase potential P uptake almost in proportion to the increase in root length per plant because P diffusion to the root is usually limiting. In solution culture where P concentration at the root is maintained at the root, uptake per gram of roots may be similar for fine and coarse roots.

VI. CHEMICAL EFFECT OF ROOTS ON SOIL ENVIRONMENT AND P UPTAKE

When plant roots grow in soil they change the concentrations of ions in the soil immediately adjacent to the root because of (i) mass-flow and diffusion of ions resulting from water and ion absorption; (ii) exudation of H^+ or OH^- (or HCO_3^-) as a result of imbalance between cation and anion absorption; (iii) exudation of organic substances from the root into the soil; and (iv) differential microbiological activity due to the presence of the root and its effect under (i), (ii), and (iii).

Accumulation of ions at the root surface when rate of supply by mass-flow exceeds the uptake rate has been theorized by Barber (1962) and Nye and Marriott (1969). The accumulated ions would cause a concentration gradient along which ions would diffuse away from the root and eventually a balance between mass-flow to the root and uptake plus back diffusion would be obtained. Riley and Barber (1970) measured salt accumulation at soybean root surfaces. The concentration of water soluble salts in the rhizocylinder (root plus adhering soil) was five to eight times that in soil not affected by the root. Barber and Ozanne (1970) demonstrated Ca accumulation about ryegrass and subterranean clover roots using autoradiographic procedures. The plants were grown in a sandy soil in which the Ca was labeled with ^{45}Ca. Fifteen-day-old plants showed a definite increase in Ca concentration near the root.

When both Ca^{2+} and SO_4^{2-} are present in the solution, accumulation of these ions can cause precipitation of $CaSO_4$ on the root surface. Malzer and Barber (1975) using both autoradiographic and petrographic procedures

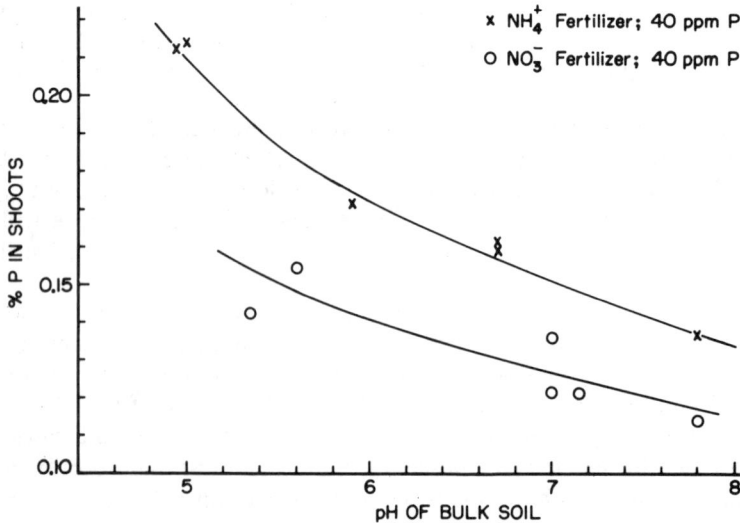

Fig. 4—Percent P in shoots of soybean seedlings fertilized with NH_4^+ or NO_3^- as related to bulk soil pH.

confirmed precipitation of $CaSO_4$ on root surfaces. In arid soils, $CaSO_4$ precipitation could be a common occurrence. Barber (1974a) found evidence in nature for $CaCO_3$ precipitation at root surfaces. In soils with a neutral or higher pH, Ca accumulation at the root combined with HCO_3^- released from the root could result in $CaCO_3$ formation on the root surface.

An imbalance between cation and anion absorption by plant roots results in a pH change at the root surface. Cation and anion absorption occurs independently so that when more cations are absorbed than anions, H^+ is released and the soil near the root surface becomes more acid. When anion absorption exceeds cation absorption the pH at the root increases. Since nonlegumes absorb more equivalents of N than any other nutrient, supplying the plant with NH_4^+ instead of NO_3^- will greatly change the ratio of cation-anion absorption and will result in release of H^+. The opposite will result in OH^- release. This will cause a change in the pH of the soil. Riley and Barber (1969) found that they could change the pH of the rhizocylinder of soybean roots from 6.2 to 7.2 by supplying all the N as NO_3^-, and where NH_4^+ was used as the N supply (Riley & Barber, 1971), from 6.3 to 5.6.

A. Effect of Ca Accumulation on P Availability

Accumulation of Ca at the root surface may reduce the corresponding solution P level if the pH remains constant. The accumulation of Ca may occur external to the epidermis, or it may occur in the free space of the roots (Eaton & Bernardin, 1964). If we consider the solubility product principle, increasing Ca activity will decrease P activity in solution, where a Ca phosphate solid phase is determining concentration of P in solution, provided pH remains constant.

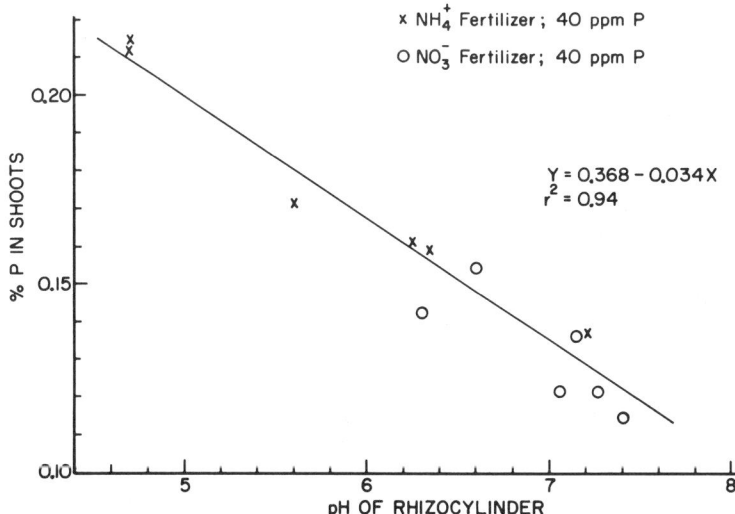

Fig. 5—Percent P in shoots of soybean seedlings fertilized with NH_4^+ or NO_3^- as related to rhizocylinder (root plus strongly adhering soil) pH.

Hoffmann and Barber (1971) investigated the effect of Ca accumulation on P uptake by wheat from four soils. One treatment was watered with deionized water and the other with 5 mM CaSO$_4$ solution. On soils which had a pH during development above 6.8, addition of CaSO$_4$ reduced P uptake; but, on more acid soils it did not. The higher pH soils had solid-phase Ca phosphate present, and the additional Ca probably reduced the P solubility by the common ion effect. One factor which tends to compensate for the effect of accumulated Ca on P level in solution is that as salt concentration increases, solution pH tends to decrease, and in the soils investigated by Riley and Barber (1971) this increased P solubility.

B. Effect of Changes in Soil pH

When cation uptake is more than anion uptake, pH of the rhizocylinder decreases and this increases P solubility. The reverse occurs when pH increases due to anion uptake becoming dominant. Riley and Barber (1971) investigated the effect of pH change of the rhizocylinder of soybean roots caused by applying N as NH$_4^+$ in one treatment and as NO$_3^-$ in another. The soil was limed to give four initial soil pH levels of 5.2, 6.3, 6.7, and 7.8. Equilibrating the soil with 0.01M CaCl$_2$ showed that reducing soil pH increased P concentration of the equilibrium solution. Soybeans were grown for 3 weeks, and at harvest pH of both rhizocylinder and bulk soil was determined as well as P concentration of the plant. Neither lime treatment nor form of N affected yields. The relation between pH of the bulk soil and percentage of P in the shoot is shown in Fig. 4. At the same soil pH, the NH$_4^+$ treatment gave considerably higher P levels in the shoot than did the NO$_3^-$ treatment. Increasing soil pH reduced the P level in the soybean plant.

The relation between pH of the rhizocylinder and percentage of P in shoots is shown in Fig. 5. The data indicated that P concentration in shoots decreased linearly with increased pH and N treatment had no differential effect. The pH of the rhizocylinder was reduced relative to that of the bulk soil when NH$_4^+$ was used and was increased when NO$_3^-$ was used. Hence, the close correlation of P concentration with rhizocylinder pH for both N sources indicated that the effect the root had on soil pH, caused by form of N used, influenced soil P availability. In this experiment the difference in P availability between the N forms used could be attributed to their indirect effect on pH of the soil in the rhizocylinder. There may be other influences of N form on P uptake. These have been reviewed by Miller (1974).

VII. EFFECT OF P DISTRIBUTION IN SOIL ON P UPTAKE BY ROOTS

The previous discussion dealt with P uptake where plant roots were assumed to be uniformly exposed to various concentrations of soil P. Frequently P fertilizer is applied as localized placement so that part of the root system is exposed to high concentrations of P and part to low concentrations of P. Localizing P application increases the rate of P per unit volume of P-fertilized soil, but reduces the portion of the root system that will be in P-fertilized soil. However, localizing P placement could be advantageous because frequently less applied P is "fixed," or made unavailable by the soil. The degree to which P localization in the soil is advantageous or disadvantageous depends on how localization of P affects the total P absorbed by the plant.

Gile and Carrero (1917) conducted research on P uptake by rice from solution culture. They found that when one-half of the roots were exposed to P, uptake per gram of roots exposed to P increased 28%, but P uptake per plant was only 76% of that where all of the roots were in P-containing solution.

Jungk and Barber (1974) investigated the relation between rate of P uptake by corn roots of 13-day-old plants and the proportion of roots exposed to P. They used two procedures; one was to divide the seminal roots between P-deficient and P-containing nutrient solutions, and the second was to trim off the roots 1 day before measuring P uptake. With both methods P uptake rate per unit of root length was not affected by supplying P to only part of the roots so that P uptake per plant was proportional to the amount of roots supplied with P.

Since P influx decreases with plant age, the effect of root trimming on P influx into corn roots was studied on corn ranging in age from 18 to 80 days (Jungk & Barber, 1975). Again P influx per unit of root growing in solution was not influenced. In these experiments roots were trimmed 2 days before P influx measurement, and P influx was measured over a wide range of P concentrations in solution. Phosphorus influx per unit of root was not influenced by reducing the proportion of root supplied with P. In the foregoing research, P influx was measured within 2 days of initiation of treatment. Jungk (1975) and Hoffmann (1968) found that when corn plants were starved for P, the P influx increased when they were subsequently placed in P-containing solutions. These results indicated that P level of the shoot influenced P uptake rate per unit of root. If this is the case, it would not be possible to supply as much P to the plant through part of the root system as through all of it.

The effect of distribution of fertilizer P in the soil on P uptake is dependent upon the rate of P diffusion through the soil to the root, the ability of the root to absorb P, and reactions of fertilizer P with soil. At lower rates of P the rate of supply to the root system determines P uptake rate of the plant. Hence, if four times as much P is applied to one-fourth of the soil volume, diffusive flux to the root would have to be increased fourfold. More than a fourfold increase could be obtained because usually less P is adsorbed tightly by the soil as the rate of P applied to soil increases and P is proportionately more available for diffusion to the root. Hence, as long as the rate of P supply to the root is less than about one-half the maximum rate that the root can absorb, localizing P could be advantageous. However, when the supply rate by diffusion and mass-flow from the soil for the local-

Table 3—Effect of P fertilizer placement on yield of corn and P concentration of the ear leaf at tasseling.

Treatment	Yield	Leaf P concentration	P soil test†
	kg/ha	%	kg/ha
Band by row	7,210	0.26	14
Broadcast and plowed under	7,590	0.27	21
Strip and plowed under	8,280	0.29	37
L.S.D. 0.05	363	0.02	9

† Soil sampled between rows at random during fifth year and tested by Bray P_1 procedure.

ized placement approaches and exceeds I_{max} for P uptake, localization of P in soil loses advantage over a more uniform distribution in soil permeated by roots.

Starting with a uniform mix of fertilizer P with the soil, the effect of localizing P will probably become more and more advantageous as the proportion of soil fertilized with P is reduced until a point is reached where P supply rate from P-fertilized soil to roots growing in that soil approaches I_{max} for P uptake by these roots. Localization can be very important at low rates of P application per unit of soil. Where sufficient P is added to obtain maximum yields, it is usually necessary to have most of the root system absorbing P at a rate near I_{max} so that restriction of the volume of soil fertilized cannot go very far.

Traditionally, P has been applied either by broadcasting and mixing with the surface plow layer by tillage operations, or by banding near the row when the crop is planted. For cereal crops the fertilizer and seed are often placed together; for row crops the seed and fertilizer are usually 5 cm apart. Barber (1974b) compared the effect on corn yield of three placements: (i) row application at planting; (ii) broadcast and plowing under to mix with the plow layer; and (iii) applying a 5- to 10-cm wide strip on the surface every 70 cm and then plowing. The last treatment should give a degree of mixing intermediate between (i) and (ii). In (i) the P fertilizer was mixed with only 2 to 3% of the soil, while with (ii) it was mixed with most of the soil. In treatment (iii) it probably mixed with 10 to 15% as much soil as in treatment (ii). The results from an experiment with corn conducted for 5 consecutive years on the same plots are shown in Table 3. The strip treatment was more effective than the other two treatments on this soil. Apparently in the band treatment, not enough roots were in the P fertilized soil. In the broadcast treatment more of the P was tied up in an unavailable form than for the other two treatments. The soil test on the band by the row was for soil not sampled near the row, and hence is the unfertilized soil level. The test for the strip is the mean of random samples from P-fertilized and unfertilized soil. The soil test on the strip treatment was higher, thus indicating that there was less "fixation" with this placement than with the broadcast placement. The strip treatment was a favorable compromise between the two.

VIII. FUTURE RESEARCH NEEDS

Additional research is needed to determine the root-soil interactions which influence P absorption kinetics of P uptake by plant roots with root hairs. The influence of soil temperature, soil pH, and form of P in the soil on P flux into the root and on P diffusion in soil is needed to increase our understanding of soil influence on P supply to the root.

A study of the plant root P absorption characteristcs, I_{max}, K_m, C_{min}, root radius, and root length per g of P required is needed to determine how these characteristics vary among species and within species and to determine their effect on P uptake rate by the plant root. Can new cultivars be developed that have more effective P foraging characteristics in the soil so that crops can produce high yields on soils low in P? Can we develop cultivars that use applied P fertilizer more efficiently? Other questions for which we do not have adequate answers include the following: What plant property controls the rate of P uptake per unit of plant root? What controls rate of root growth and root hair growth per unit of P needed by the shoot? How long do roots actively absorb P? What determines the effect of root age on P uptake rate?

The mathematical models for P uptake by plant roots need more elaboration so that we can include the realistic effect of the influence of root hairs; the change in P absorption by the root with age; the interaction of other ions such as Ca, K, Na, etc., on P influx; the effect of moisture gradients about the root; and the effect of increasing root density in the soil.

LITERATURE CITED

Adepetu, J. A. 1976. Evaluation of the kinetic processes involved in phosphorus availability to plant roots in soil. Ph.D. Thesis. Purdue Univ. Univ. Microfilms. Ann Arbor, Mich. (Diss. Abstr. 36:3152B).

Adepetu, J. A., and S. A. Barber. 1978. Soil phosphorus availability to millet roots. Comm. Soil Sci. and Plant Anal. 9:541–550.

Barber, S. A. 1962. A diffusion and mass-flow concept of soil nutrient availability. Soil Sci. 93:39–49.

Barber, S. A. 1974a. Influence of the plant root on ion movement in soil. p. 525–564. *In* E. W. Carson (ed.) The plant root and its environment. Univ. Press of Virginia, Charlottesville.

Barber, S. A. 1974b. A program for increasing the efficiency of fertilizers. Fert. Solutions 18(2):24–25.

Barber, S. A. 1974. Properties of the plant root that influence fertilizer practice. p. 25–33. Proc. 7th Int. Colloq. Plant Anal. Fert. Prob., Hanover, West Germany.

Barber, S. A., and P. G. Ozanne. 1970. Autoradiographic evidence for the differential effect of four plant species in altering the calcium content of the rhizosphere soil. Soil Sci. Soc. Am. Proc. 34:635–637.

Barber, S. A., J. M. Walker, and E. H. Vasey. 1963. Mechanisms for the movement of plant nutrients from the soil and fertilizer to the plant root. J. Agric. Food Chem. 11:204–207.

Barley, K. P. 1970. The configuration of the root system in relation to nutrient uptake. Adv. Agron. 22:159–201.

Barley, K. P., and A. D. Rovira. 1970. The influence of root hairs on the uptake of phosphate. Commun. Soil Sci. Plant Anal. 1:287–292.

Bhat, K. K. S., and P. H. Nye. 1973. Diffusion of phosphate to plant roots in soil. I. Quantitative autoradiography of the depletion zone. Plant Soil 38:161–175.

Bhat, K. K. S., and P. H. Nye. 1974a. Diffusion of phosphate to plant roots in soil. II. Uptake along the roots at different times and the effect of different levels of phosphorus. Plant Soil 41:365–382.

Bhat, K. K. S., P. H. Nye, and J. P. Baldwin. 1976. Diffusion of phosphate to plant roots in soil. IV. The concentration distance profile in the rhizosphere of roots with root hairs in a low-P soil. Plant Soil 44:63–72.

Bole, J. B. 1973. Influence of root hairs in supplying soil phosphorus to wheat. Can. J. Soil Sci. 53:169–175.

Bouldin, D. R. 1961. Mathematical description of diffusion processes in soil-plant system. Soil Sci. Soc. Am. Proc. 25:476–480.

Bowen, G. D. 1970. Effects of soil temperature on root growth and on phosphate uptake along *Pinus radiata* roots. Aust. J. Soil Res. 8:31–42.

Bowen, G. D., and A. D. Rovia. 1967. Phosphorus uptake along attached and excised roots measured by an automatic scanning method. Aust. J. Biol. Sci. 20:369–378.

Brewster, J. L., K. K. S. Bhat, and P. H. Nye. 1976. The possibility of predicting solute uptake and plant growth response from independently measured soil and plant characteristics. V. The growth and phosphorus uptake of rape in soil at a range of phosphorus concentrations and a comparison of results with the predictions of a simulation model. Plant Soil 44:295–328.

Carter, O. G., and D. J. Lathwell. 1967. Effect of temperature on orthophosphate absorption by excised corn roots. Plant Physiol. 42:1407–1412.

Claassen, N., and S. A. Barber. 1974. A method for evaluating the influence of concentration on the uptake rate of nutrients. Plant Physiol. 54:564–568.

Claassen, N., and S. A. Barber. 1976. Simulation model for nutrient uptake from soil by a growing plant root system. Agron. J. 68:961–964.

Cooper, A. J. 1973. Root temperature and plant growth. A review Commonw. Bur. Hort. Plant. Crops 4:1–73.

Eaton, F. M., and I. E. Bernardin. 1964. Mass-flow and salt accumulation by plants on water versus soil cultures. Soil Sci. 97:411–416.

Edwards, J. H., and S. A. Barber. 1976. Phosphorus uptake rate of soybean roots as influenced by plant age, root trimming, and solution P concentration. Agron. J. 68:973–975.

Ferguson, I. B., and D. T. Clarkson. 1975. Ion transport and endodermal suberization in the roots of *Zea mays*. New Phytol. 75:69–80.

Gile, P. L., and J. O. Carrero. 1917. Absorption of nutrients as affected by the number of roots supplied with the nutrient. J. Agric. Res. 9:73–95.

Hagen, C. E., and H. T. Hopkins. 1955. Ionic species in orthophosphate absorption by barley roots. Plant Physiol. 30:193–199.

Helyar, K. R., and D. N. Munns. 1975. Phosphate fluxes in the soil-plant system: a computer simulation. Hilgardia 43:103–130.

Hendrix, J. E. 1967. The effect of pH on the uptake and accumulation of phosphate and sulfate ions by bean plants. Am. J. Bot. 54:560–564.

Hodges, T. K. 1973. Ion absorption by plant roots. Adv. Agron. 25:163–207.

Hoffmann, W. E. 1968. Mineralstofftransport der Wurzel in Abhangigkeit von ikrem K- and P-status. Landwirtsch. Forsch. 21:203–212.

Hoffmann, W. E., and S. A. Barber. 1971. Phosphorus uptake by wheat (*Triticum aestivum*) as influenced by ion accumulation in the rhizocylinder. Soil Sci. 112:256–262.

Jungk, A. 1975. Phosphate uptake characteristics of intact root systems in nutrient solution as affected by plant species age and P supply. p. 185–196. Proc. 7th Int. Colloq. Plant Anal. Fert. Prob., Hanover, West Germany.

Jungk, A., and S. A. Barber. 1974. Phosphate uptake rate of corn roots as related to the proportion of the roots exposed to phosphate. Agron. J. 66:554–557.

Jungk, A., and S. A. Barber. 1975. Plant age and the phosphorus uptake characteristics of trimmed and untrimmed corn root systems. Plant Soil 42:227–239.

Lewis, D. G., and J. P. Quirk. 1967. Phosphate diffusion in soil and uptake by plants: III. ^{31}P movement and uptake by plants as indicated by ^{32}P autoradiography. Plant Soil 26:445–453.

Loneragan, J. F., and C. J. Asher. 1967. Response of plants to phosphate concentration in solution culture: II. Rate of phosphate absorption and its relation to growth. Soil Sci. 103:311–318.

Malzer, G. L., and S. A. Barber. 1975. Precipitation of calcium and strontium sulfates around plant roots and its evaluation. Soil Sci. Soc. Am. Proc. 39:492-495.

Mengel, D. B., and S. A. Barber. 1974a. Development and distribution of corn root systems under field conditions. Agron. J. 66:341-344.

Mengel, D. B., and S. A. Barber. 1974b. Nutrient uptake rates per unit of root of corn grown under field conditions. Agron. J. 66:399-402.

Miller, M. H. 1974. Effect of nitrogen on phosphorus absorption by plants. p. 643-668. *In* E. W. Carson (ed.) The plant root and its environment. Univ. Press of Virginia, Charlottesville.

Nielsen, K. F. 1974. Roots and root temperature. p. 293-334. *In* E. W. Carson (ed.) The plant root and its environment. Univ. Press of Virginia, Charlottesville.

Nissen, P. 1974. Uptake mechanisms: inorganic and organic. Ann. Rev. Plant Physiol. 25:53-79.

Nye, P. H. 1968. The use of exchange isotherms to determine diffusion coefficients in soil. Int. Congr. Soil Sci. Trans. 9th (Adelaide, Aust.) I:117-126.

Nye, P. H., and F. H. C. Marriott. 1969. A theoretical study of the distribution of substances around roots resulting from simultaneous diffusion and mass-flow. Plant Soil 30:459-472.

Olsen, S. R., and W. D. Kemper. 1968. Movement of nutrients to plant roots. Adv. Agron. 20:91-151.

Parsons, R. 1959. Handbook of electrochemical constants. Academic Press, Inc., New York.

Powell, C. L. 1974. Effect of P fertilizer on root morphology and P uptake by *Carex coriacea*. Plant Soil 41:661-667.

Rhodes, L. H., and J. W. Gerdemann. 1975. Phosphate uptake zones of mycorrhizal and non-mycorrhizal onions. New Phytol. 75:555-561.

Riley, D., and S. A. Barber. 1969. Bicarbonate accumulation and pH changes at the soybean [*Glycine max* (L.) Merr.] root-soil interface. Soil Sci. Soc. Am. Proc. 33:905-908.

Riley, D., and S. A. Barber. 1970. Salt accumulation at the soybean [*Glycine max* (L.) Merr.] root-soil interface. Soil Sci. Soc. Am. Proc. 34:154-155.

Riley, D., and S. A. Barber. 1971. Effect of ammonium and nitrate fertilization on phosphorus uptake as related to root-induced pH changes at the root-soil interface. Soil Sci. Soc. Am. Proc. 35:301-306.

Rovira, A. D., and G. Bowen. 1968. Anion uptake by plant roots: distribution of anions and effects of microorganisms. Int. Congr. Soil Sci., Trans. 9th (Adelaide, Aust.) II:202-217.

Sutton, C. D. 1969. Effect of low soil temperature on phosphate nutrition of plants—A review. J. Sci. Food Agric. 20:1-3.

Warncke, D. D., and S. A. Barber. 1972. Diffusion of zinc in soils. II. The influence of soil bulk density and its interaction with soil moisture. Soil Sci. Soc. Am. Proc. 36:42-46.

Warncke, D. D., and S. A. Barber. 1974. Root development and nutrient uptake by corn grown in solution culture. Agron. J. 66:514-516.

Weast, R. C. 1969. Handbook of chemistry and physics. 49th ed. The Chemical Rubber Co., Cleveland, Ohio. p. F-36.

Chapter 22

Role of Rhizosphere Microorganisms in Phosphorus Uptake by Plants

P. B. TINKER

Rothamsted Experimental Station
Harpenden, Herts, U.K.

I. INTRODUCTION

A. General

This paper discusses the role of microorganisms which are associated with the root and its immediate environment and which are dependent upon the presence of the root in some way, and the word *rhizosphere* is to be understood in this sense. The subject is extremely interesting because of the possibility that the rhizosphere microorganisms can be manipulated to allow crops to grow with less or cheaper P fertilizer than they would otherwise need, but there are still wide areas of uncertainty. There are two major topics: the free living microorganisms of the rhizosphere, including a vast array of bacteria, fungi, and other organisms, and the very limited number of species which form the symbiotic associations with roots called *mycorrhizae*. I exclude organisms which are normally regarded as direct root pathogens, though there are interesting interactions between P supply and root pathogen invasion.

B. Nutrient Uptake Processes

It is assumed in this paper that normal uptake of nutrients by roots is from the soil solution, and that prior to uptake nutrient ions move up to the root surface by mass flow and diffusion. The continuing supply of ions depends upon the sorption-desorption relationships of the ion on the soil colloid surfaces, and upon the mean distance between roots, which sets a limit to the exploitable soil volume. The general approach is embodied in the models described by Baldwin et al. (1973) and Nye and Tinker (1977) for uptake by whole root systems.

Copyright 1980 © ASA-CSSA-SSSA, 677 South Segoe Road, Madison, WI 53711, USA.
The Role of Phosphorus in Agriculture.

Phosphate is probably the most difficult nutrient ion to deal with in this way (apart from the trace metals) because of the complexity of its sorption relationships (Kafkafi et al., 1967; White & Taylor, 1977; White, 1980). Desorption isotherms are relevant to nutrient uptake since they can be used to find the buffer power, which determines the diffusion coefficient (Olsen & Kemper, 1968; Drew & Nye, 1970). Some of the suggestions of how bacteria improve plant P nutrition are equivalent to saying that the bacteria temporarily alter the sorption equilibrium, and thereby release sorbed P into free solution.

It is also assumed here that diffusion is nearly always a major rate-limiting step in P uptake from soil. This is shown by many autoradiographs of depletion zones close around roots (Lewis & Quirk, 1967; Bhat & Nye, 1974), and by the calculated P concentration gradients necessary to provide observed fluxes of P into roots (Bar Yosef et al., 1972; Brewster et al., 1975; Mengel & Barber, 1974). Root hairs help to circumvent this problem, since they absorb P from the soil between them, and in effect also give the root a larger radius, whereby the maximum theoretical uptake is increased (Nye, 1966).

The absorbing power of plant roots for P seems to be very large (Nye & Tinker, 1969), and many species can maintain maximum growth rate in $10^{-6}M$ solution culture if it is continually replenished (Lonergan & Asher, 1967). In this case uptake of P from soil by a root is likely to be limited by transport in the soil rather than by root absorbing power. If soil-grown roots behave in the same way as solution-grown ones (which is open to question), then we conclude that microorganisms are unlikely to enhance P uptake by significantly increasing the mean root demand coefficient, α, of the active root surface, though they could possibly diminish it so much that the uptake rate would be lowered (Nye & Tinker, 1977). Instead, present views tend to stress the great importance of root morphology for P uptake. An increase in total active root length per unit plant weight, which could be produced by changes in root-shoot weight ratio, root length per unit weight, or duration of active uptake, would reduce the mean uptake per unit root length (inflow) required for the normal growth of the plant (Brewster & Tinker, 1972).

To summarize, the mechanisms by which microorganisms could potentially alter P uptake rates seem to be:

1) Alteration of root morphology, in particular root hair length and density, or change in active root length.
2) Change of mean absorbing power of the root over all or part of its surface.
3) Displacement of sorption equilibria to produce higher local P concentrations in the soil solution, thereby allowing a higher flux toward the root surface and a higher uptake rate.
4) Facilitated transport of P to the root, again allowing a larger uptake rate, and possibly from a larger soil volume.

II. RHIZOSPHERE MICROORGANISM POPULATION

A. Composition and Distribution

The *rhizosphere* is the ill-defined region of soil around the root which is "influenced by" the root; if this referred to, for example, water uptake or presence of eelworm hatching factor, it could clearly extend for several centimeters, but it is normally restricted to the volume within which the microorganic population is appreciably increased. The rhizoplane is an elaborate word for the root surface. The terminology and definitions have been well discussed by Rovira and McDougall (1967).

There are several excellent reviews of observational and taxonomic work on rhizosphere microorganisms (Katznelson, 1965; Gams, 1967; Rovira & McDougall, 1967; Parkinson, 1967; Clark & Paul, 1970; Darbyshire & Greaves, 1973; Bowen & Rovira, 1976; Balandreau & Knowles, 1978). In general, it is always found that the mean population density at the surface rises to 10 or more times that in the bulk soil well away from the root surface, and that bacteria, actinomycetes, streptomycetes, protozoa, and fungi follow the same pattern, though to different degrees. There are always considerable uncertainties about whether the techniques of counting give correct absolute numbers, but the relative numbers at different distances should be reasonably dependable. Most authors who identify the species of microorganisms have noted a relatively large increase in *Pseudomonas* species, but the taxonomic identification of the organisms has as yet given little insight into the processes occurring. The root surface ecology is currently receiving much attention.

In addition to the free-living microorganisms, there are the symbiotic, N-fixing bacteria of the genus *Rhizobium* and some other N-fixing associations, symbiotic mycorrhizal fungi, and the smaller soil fauna such as nematodes, collembola, etc.

Recent studies on the distribution of microorganisms on the surface of the root (Fig. 1) have used electron microscopy (Darbyshire & Greaves, 1973; Foster & Rovira, 1976) and the techniques of quantitative ecology (Newman & Bowen, 1974; Rovira et al., 1974). These allowed the clumping and distribution of individual bacteria or colonies to be described accurately, showing that 4 to 10% of the grass root surface was covered by bacteria, with a further 3% by fungal hyphae. These results agree with earlier observational work (Bowen & Theodorou, 1973), and suggest that direct interference with the movement of ions to the root surface is unlikely. However, the entrance to the "free space" of the cortex is through the junctions between cells on the surface, and there is some indication that microorganisms tend to congregate there. For example, Warcup (1975) noted that *Endogone* hyphae tended to run along the junction between cells on the root surface. The whole topic of microbial colonization of plant roots has been reviewed by Bowen and Rovira (1976) and by Rovira (1979).

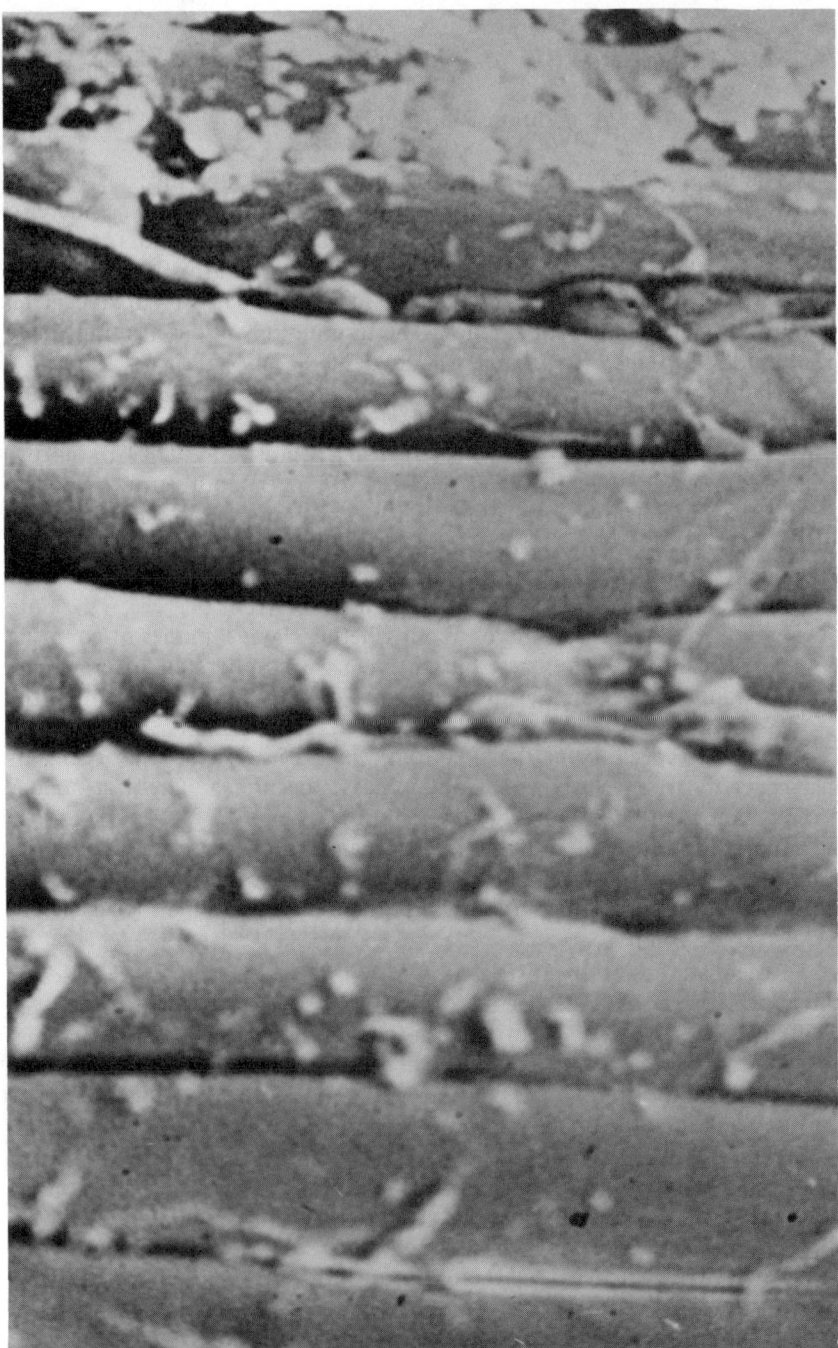

Fig. 1—(a) Scanning electron micrograph of onion root prepared by critical point drying, showing bacteria on surface, soil mineral particles, and hyphae of VA mycorrhizal fungi (Photo by G. Sparling).

Fig. 1—(b) VA hyphae in soil (Photo by K. Birkby).

The rhizosphere is probably best considered to include the "mucigel" (Rovira & McDougall, 1967; Mosse, 1975a). Many bacteria are embedded in this gel, which forms a layer some 1 to 5 μm thick around most roots (Jenny & Grossenbacher, 1963; Brams, 1965; Greaves & Darbyshire, 1972; Mosse, 1975a; Oades, 1978). It is composed mainly of material directly exuded from the root, especially the root tip, and some authors include breakdown products of epidermal cells and bacterially produced substances. Chemical analysis shows it to contain polymers of uronic acids and sugars. It is presumably a good habitat for bacteria, but there is no evidence that it has any direct stimulating or nutritional effect on them, and its persistence around individual bacteria argues against its being of great nutritional value. There are few good physical and chemical data on the properties of mucigel, and it is therefore rather difficult to decide its relevance to rhizosphere microorganisms or to plant nutrition (Rovira & McDougal, 1967). It may assist in maintaining root-soil contact, and the large production in the root cap suggests that it functions as a lubricant for the movement of the root tip through soil.

B. Substrate for the Rhizosphere Population

Nonsymbiotic microorganisms in the rhizosphere utilize material supplied by the root, in addition to the substrate present in the soil originally. The quantity of this supply is of considerable interest, because it sets the limits for the total amount of microorganic activity in the rhizosphere, and hypotheses about rhizosphere processes must be tested against this. This supply comprises soluble exudates, sloughed-off parts of the root cap, mucigel, root hair residues, and abraded epidermal cells. The terminology of exudation has become extremely complicated, and Rovira et al. (1979) have attempted to clarify it by a series of definitions. The components of this material are exceedingly diverse, and various authors have published lists of compounds found in the soluble exudates (Rovira, 1962; Rovira & Davey, 1974; Hale et al., 1971; 1978), the most common being organic acids, aminoacids, and sugars. It has often been suggested that hydroxycarboxylic or polycarboxylic acids could displace P from soil colloids; an analogous process is discussed below for bacterially produced chelating acids of this type. Apart from this, the precise composition of the exudates appears to have no particular significance. It can certainly vary with both species and with plant growth stage (Smith, 1970), but the consequences are not understood in detail.

The total amount of material available is probably of greatest importance. Very widely varying values have been published, probably due to differences in conditions, species, and techniques. Hale et al. (1978) and Rovira and Davey (1974) list the factors affecting exudation rate as plant species, plant age, nutrition, light, temperature, soil moisture, microorganisms, and the supporting medium. The last two factors imply that measurements should ideally be made in nonsterile soil, and that solution-derived data may be misleading (Barber & Gunn, 1973; Hale et al., 1973). In solid media it is particularly difficult to be certain whether it is the total C exported from the root, or simply the soluble exudate fraction, which is being measured. Griffin et al. (1976) found that the soluble exudate was less than 5% of the total C loss by peanut roots in solution culture. Bowen and Rovira (1973) concluded that the total loss was 1 to 2% of root dry weight and that 80% of total C export (i.e., all C lost from roots except as CO_2) was as solid material, so rhizosphere substrates should be considered more as solid material adjacent to the root rather than soluble compounds diffusing freely away. Other work has used ^{14}C-labeling methods which would ensure complete recovery of all solid material (plus possible dark fixation of $^{14}CO_2$ produced in the soil by root respiration). The mean export by seven forage species was found to be 7.9% (Shamoot et al., 1968), by barley to be 13 to 25% (Barber & Martin, 1976), and by wheat and mustard to be 13 to

17% (Sauerbeck & Johnen, 1976) of total plant dry weight. Such values are an order of magnitude larger than results of earlier work, and warrant careful attention. Barber and Martin's (1976) work gave evidence that the C export was increased greatly when microorganisms were present, and that the latter used the exported material rapidly. Simple observations of bacterial quantity around roots showed that there was more than could have been produced from the soluble exudates of sterile roots (Barber & Lynch, 1977). If such values are accepted, the question arises about the exact location at which this C becomes available. If much of it is as dead fine roots, or sloughed-off root cortex, it will not be in what is usually understood to be the rhizosphere. For example, Reynolds (1975) suggested that fine roots in a whole section of the root system of trees may die off at about the same time. However, most experiments referred to here are on young, annual plants, which should lose little material in this way.

Bowen and Rovira (1976) have discussed the problems of measuring and expressing microbial growth rates in the rhizosphere in relation to substrate supply. The distribution and quantity of the microbial biomass produced from this substrate will clearly vary with soil and plant conditions. Bowen and Rovira (1973) calculated a rough equivalence between observed numbers of bacteria, of 1 to 2×10^6/mg of dry root, and that predicted from a root exudation of 4 to 8 μg/g of dry root. Newman and Watson (1977) have used a mathematical model to predict distributions of bacteria, which agrees with observational work in finding the excess population to be concentrated closely around the root. There seems little doubt that most of the biomass is so close to the root surface that any solutes produced by the bacteria could easily reach the root (see Section III A).

However, at present the uncertainty over the actual amount and nature of exudation or export of material from the root is such that these calculations are highly speculative and are not further discussed here.

III. EFFECT OF FREE-LIVING MICROORGANISMS ON HIGHER PLANTS

A. Effects on Root Morphology

There seems little doubt that bacterial cultures may produce compounds with growth hormonal activity (Brown, 1974; Lynch, 1976). Effects similar to those produced by gibberellins and auxins have been reported, including differences in root morphology, but they are not yet well quantified. Bowen and Rovira (1961) determined the effects of bacteria on wheat roots in sand and agar culture, and found that root length, and root hair length and density were both reduced (Fig. 2). Even a light soil irradiation of 0.6 Mrad increased the fraction of grass root systems carrying root hairs from 41 to 85%.[1] Marked decreases in root growth, but increases in rate of

---

[1] G. Sparling. 1976. Effects of vesicular-arbuscular mycorrhizas on Pennine grassland vegetation. Ph.D. thesis, Univ. of Leeds, England.

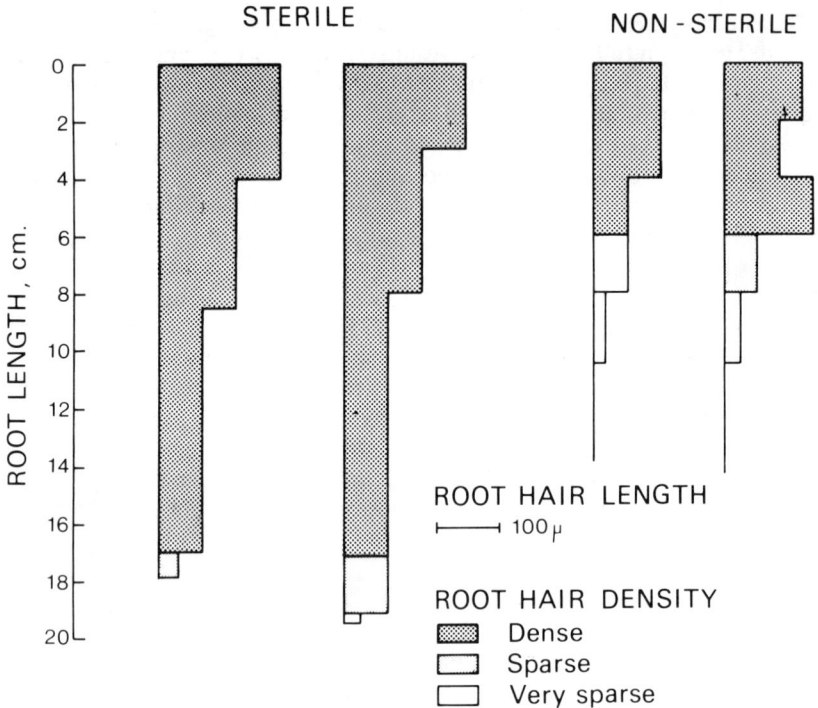

Fig. 2—Effects of bacteria on the growth of clover roots, and the production and length of root hairs, when grown in agar (after Bowen & Rovira, 1961).

K uptake, have also been caused by soil bacteria (Williamson & Wyn Jones, 1973). The production of intensely branched ("proteoid") roots by incubation with rhizosphere microorganisms (Malajczuk & Bowen, 1974) is particularly striking. A more detailed review has been given by Mosse (1975a).

B. Effects on Uptake of Phosphorus in Solution Culture

There is ample evidence that bacteria can alter the uptake rate of P by roots over such short periods that no morphological or developmental effects within the plant are likely (see Barber, 1978). Thus Barber (1969) found that nonsterile plants absorbed less than sterile ones from solution cultures with very low P concentrations, but at higher concentrations there was no difference. Bowen and Rovira (1966; 1969) found that nonsterile plants both absorbed more P and transported the largest fraction to the shoot. Recent work (Barber & Rovira, 1975) has shown that the effect of various microorganic inocula appeared to depend upon the plant age and the duration of the experiment. In 30-min experiments at 5 μM P concentration, nonsterile plants always absorbed and transported more P than sterile plants, but the effect decreased with age, and the effect on the P content of the shoot was small in older seedlings (Table 1). Mechanisms such as direct

Table 1—Effect of soil microorganisms on the uptake of phosphate by barley seedlings of different ages. The plants were grown in aerated, complete solution and uptake measured after treatment for 30 min in 0.005 mM-potassium phosphate labeled with ^{32}P (Barber & Rovira, 1975).

	Age (days)					
	6		8		12	
Condition of plants	Sterile	Nonsterile	Sterile	Nonsterile	Sterile	Nonsterile
Roots						
Dry weight (mg)	6.0	5.3	7.5	6.7	10.4	8.8
Phosphate absorbed ± S.E. (p mole·mg^{-1} dry weight)	599 ± 28	1,045 ± 53	631 ± 36	1,004 ± 61	584 ± 24	908 ± 48
Shoots						
Dry weight (mg)	13.2	15.1	17.5	19.6	27.0	26.9
Phosphate absorbed ± S.E. (pmole·mg^{-1} dry weight)	19 ± 1	40 ± 8	21 ± 1	31 ± 2	15 ± 1	12 ± 1
Total phosphate absorbed ± S.E. (nmole)	3.64 ± 0.30	6.34 ± 0.62	5.06 ± 0.29	7.23 ± 0.46	6.51 ± 0.46	8.33 ± 0.58
Percentage in shoot† ± S.E.	7.0 ± 0.3	9.1 ± 4.0	7.3 ± 0.5	8.3 ± 0.5	6.2 ± 0.3	3.8 ± 0.2

† Percent of total absorbed P found in shoot.

competition by bacteria for P can account for some results (Barber, 1978), but it seems probable that usually a hormonal mechanism must be involved. The present picture is thus rather daunting, since the uptake properties of the plant may vary in major ways depending upon the presence or absence of a microbial population. It would simplify the matter if it could be confirmed that all normal rhizosphere populations act in approximately the same way, but this appears rather unlikely since different cultures of common soil bacteria do not appear to contain the same range of growth-regulating compounds (Brown, 1974).

C. Effects of Free-Living Microorganisms on Plant Nutrition in Soil

For obvious technical reasons, it is impossible to measure P uptake rates by plants in soil over short defined periods in the way in which it can be done in solution culture. Experiments in soil are therefore necessarily longer-term, and the understanding of the results correspondingly less easy. It is often difficult to determine whether the plant size is increased because of a greater nutrient uptake rate, or vice versa.

1. SOIL STERILIZATION

There are many reports of plant growth improvements following soil sterilization (Powlson, 1975), and often these must result from the removal of pathogenic organisms, though the cause is rarely established with certainty. Benians and Barber (1974) showed that barley grown on a soil of basaltic parent material, and with a very low P level in the soil solution, was improved by 48% in weight and by 44% in P uptake if the soil were first sterilized with 5 Mrad of α-radiation. It appears to be generally agreed (Cawse, 1975) that irradiation is the sterilization method least likely to alter the chemistry of the soil, and in this instance no change in the E or L value of soil P was found. A large dose of P fertilizer removed the growth or uptake response to sterilization, and it was suggested that rhizosphere microorganisms were competing with the plants for soil P. The competition hypothesis is an attractive one, but it does assume that the microbial demand for P must be comparable to that of the plant, if it is to be of any importance (Tinker & Sanders, 1975).

From Bowen and Rovira (1973) it appears that the dry biomass of rhizosphere bacteria is about 1 to 3 $\mu g \cdot mg^{-1}$ of root. Even if the bacteria contain 3% P, the amount held is only about 10^{-7} $g \cdot mg^{-1}$ of dry root. According to Brewster and Tinker (1972) plant P uptake rates tend to be around 3×10^{-6} $g \cdot mg^{-1}$ dry root$\cdot$day^{-1}. If so the P held in the rhizosphere bacteria is only 3% of 1 day's uptake. Alternatively, if the plant dry matter contains 0.3% P, and the bacteria 3% P, then 1% of total plant mass exuded and converted to one half the mass of bacterial tissue represents 5% of total plant P uptake. Serious competition is thus hardly credible if exudation is only 1 to 2% of plant mass, but the much larger rates found recently suggest that an open mind should be kept on this point.

Phosphate-linked growth responses to incomplete soil sterilization have been obtained also (Kouchecki & Read, 1976). Sterilization was only slight and partial, and it seems unlikely that removal of competition by microorganisms could be the mechanism causing yield responses in such cases.

2. BACTERIAL INOCULATION

This type of work can be divided into laboratory experimentation, and the numerous field experiments associated with the bacterial fertilizer "phosphobacterin." This subject has been reviewed recently by Brown (1974), following earlier reviews by Cooper (1959), Swaby (1962), and Smith et al. (1962). In general, there has always been much doubt about the effects of bacterial fertilizer, and it is now generally considered that where these do have effects, they are not due to P "solubilization," which was the normal early claim for such inoculants. There is no doubt that appropriate bacteria can utilize mineral or organic P of low solubility in solution culture, but there is no evidence that this can happen in soil (Greaves & Webley, 1969). Rovira and Davey (1974), Tinker and Sanders (1975), and Hayman (1975a) considered the processes which "solubilization" would have to cover, and concluded that they were unlikely. Firstly, microorganisms would have to transfer P from an adsorbed or crystalline mineral form into solution. They could do so by producing sufficient acid to reduce the pH significantly, or chelating acids which either would dissolve low-solubility minerals by sequestering Ca, Mg, or Fe, or displace adsorbed P from soil mineral surfaces. These processes are feasible in unbuffered media, such as sand or agar, or even in microregions of the soil, such as between a mineral surface and a bacterial colony growing on it, but they are distinctly unlikely in bulk soil volumes. It has been postulated that 2-ketogluconic acid, which is produced by rhizosphere bacteria, plays a major part in P solubilization by complexing calcium, but doubt has now been cast upon its ability to do this (Moghimi & Tate, 1978). Secondly, it is not obvious how "solubilized" P could reach the root. If absorbed into bacteria, it is of no direct benefit to the root; if liberated into the bulk soil, it is simply readsorbed. A direct benefit would only seem possible if the soil immediately around the root were saturated with chelating acids, but Tinker and Sanders (1975) estimated that a large amount of chelating acids would be needed for the process to be significant. Similar considerations apply to the use of organic P produced by bacterial uptake and decay. Some of these are easily hydrolyzed by exoenzymes (Rogers et al., 1940), but the major soil organic P compounds are strongly adsorbed and thereby protected (Anderson et al., 1974).

Confusion is caused by the fact that some of the mechanisms postulated in "bacterial solubilization" processes could also operate directly from the plant root. Thus, plant roots can make their environment more acid or more alkaline, they can absorb or adsorb Ca, Fe, and Al, and they excrete chelating organic compounds. Johnston and Olsen (1972) concluded that Ca absorption allowed plants in solution culture to use P in fluorapatite. Chelating compounds may also form multiple complexes with Ca, Al, or Fe and phosphate ions, and may thus possibly increase the P mobility and availability in the soil (Ramamoorthy & Manning, 1974). Any "phosphate solubilizing" processes in the rhizosphere may thus be independent of the microbial population, or the effect of the latter may be indirect, by altering root exudation. There is some evidence that processes of this type occur, for example, Bhat et al. (1976) found that the P uptake of rape could not be accounted for by diffusion theory.

Much of the work in this field was stimulated by the experiments of Gerretsen (1948), but there have been few experiments since then which have explicitly confirmed this work; his experiments were carried out in sand culture, and hence proved nothing about the effects in soil. Lategan and Louw (1972) measured L-values for P in soil with monocotyledonous and dicotyledonous plants, finding them to be largest with the latter, but they were unable to relate this to any difference in "phosphate-dissolving" bacteria in the rhizospheres of these plants. Recently Barea et al. (1975) and Azcon et al. (1976) have reported that lavender grown on P-deficient soils responded to inoculation with a *Pseudomonas* and an *Agrobacterium* species selected for their ability to dissolve P in pure culture (Table 2). The authors considered it at least as likely that the growth results were due to plant growth substances produced by the bacteria as to any solubilizing action. The data indicated possible interactions between bacterial inoculants and vesicular-arbuscular mycorrhizae (see Sect. IV.C. below), and Bowen and Theoderou (1979) have shown that a range of bacteria alter the speed of formation of mycorrhizas on *Pinus radiata*.

3. CONCLUSION

There seems little doubt that an appropriate population of rhizosphere microorganisms can increase the growth of plants in some circumstances, and that bacterial inoculations can give yield responses. The mechanism is however highly uncertain, and on balance it appears that interactions between the microorganism and the plant are by growth hormones, rather

Table 2—Weight and P percentage of lavender grown in sterile and nonsterile conditions (Barea et al., 1975); C, control; B, bacterial inoculation; E, mycorrhizal inoculation.

Inoculation treatment		Shoots		Roots	
		Dry weight, mg	% P	Dry weight, mg	% P
Soil 3	C	340	0.34	162	0.30
	B	355	0.35	148	0.34
	E	360	0.42	166	0.41
	E+B	372	0.46	170	0.40
	L.S.D. at 5%	29	0.11	21	0.10
Soil 7	C	236	0.22	170	0.28
	B	290	0.27	197	0.31
	E	290	0.36	180	0.49
	E+B	290	0.35	188	0.48
	L.S.D. at 5%	28	0.06	12	0.11

than by any direct P solubilizing action. Over long periods there is, of course, no question that microorganisms break down soil organic P into inorganic orthophosphate and probably that they use low-solubility inorganic forms (Larsen, 1973), but there is no clear evidence that this is enhanced in the rhizosphere, or that such activities directly benefit higher plants. The prospects of utilizing free-living microorganisms to improve P nutrition are consequently not good.

IV. TAXONOMY AND BIOLOGY OF MYCORRHIZAL FUNGI

A. General

Mycorrhizae are a close, stable, and permanent association of a root of a higher plant and a fungus, where no apparent damage results to either partner, though there may be morphological changes. The arguments about whether benefits are obtained by one or both partners is somewhat academic; it is usual for the fungus to be strongly or wholly dependent upon the higher plant, whereas the plant may or may not benefit (Lewis, 1973). I only discuss here these major forms in which there is clear evidence that P nutrition of higher plants is improved by the mycorrhizal association. These are essentially the ectomycorrhizae or sheath-forming mycorrhizae, the vesicular-arbuscular (VA) mycorrhizae, and smaller groups such as the ericaceous mycorrhizae (Harley, 1969; Smith, 1974).

B. Ectomycorrhizae

Ectomycorrhizae are formed by a range of fungi with a number of tree species, the latter usually originating in the temperate zone and including oak, beech, pine, eucalyptus, birch, and larch (Harley, 1969; Marks & Kozlowski, 1973). The fungi are almost all Basidiomycetes. Some of the fungi certainly exist in the soil independently, but others may only grow in association with roots.

Mycorrhiza formation results from the penetration of the fungal hyphae between cortex cells. A loose weft of hyphae form round the root first, which thickens to produce a sheath or mantle, perhaps some 20 to 40 μm thick around the root (Fig. 3a). This is continuous with a network of hyphae lying between the cortical cells (Hartig net), but which never penetrates the endodermis. The exterior of the sheath carries mycelial strands, rhizomorphs, and a relatively sparse growth of external hyphae which penetrate the surrounding soil.

The formation of ectomycorrhizae is easily observed, both by the presence of the mantle and from changes in the root morphology (Harley, 1969). Frequently it results in characteristic dichotomously or pyramidally branched short lateral roots, which are all fully mantled. Main extension roots are more likely to remain nonmycorrhizal, possibly due to their speed of growth.

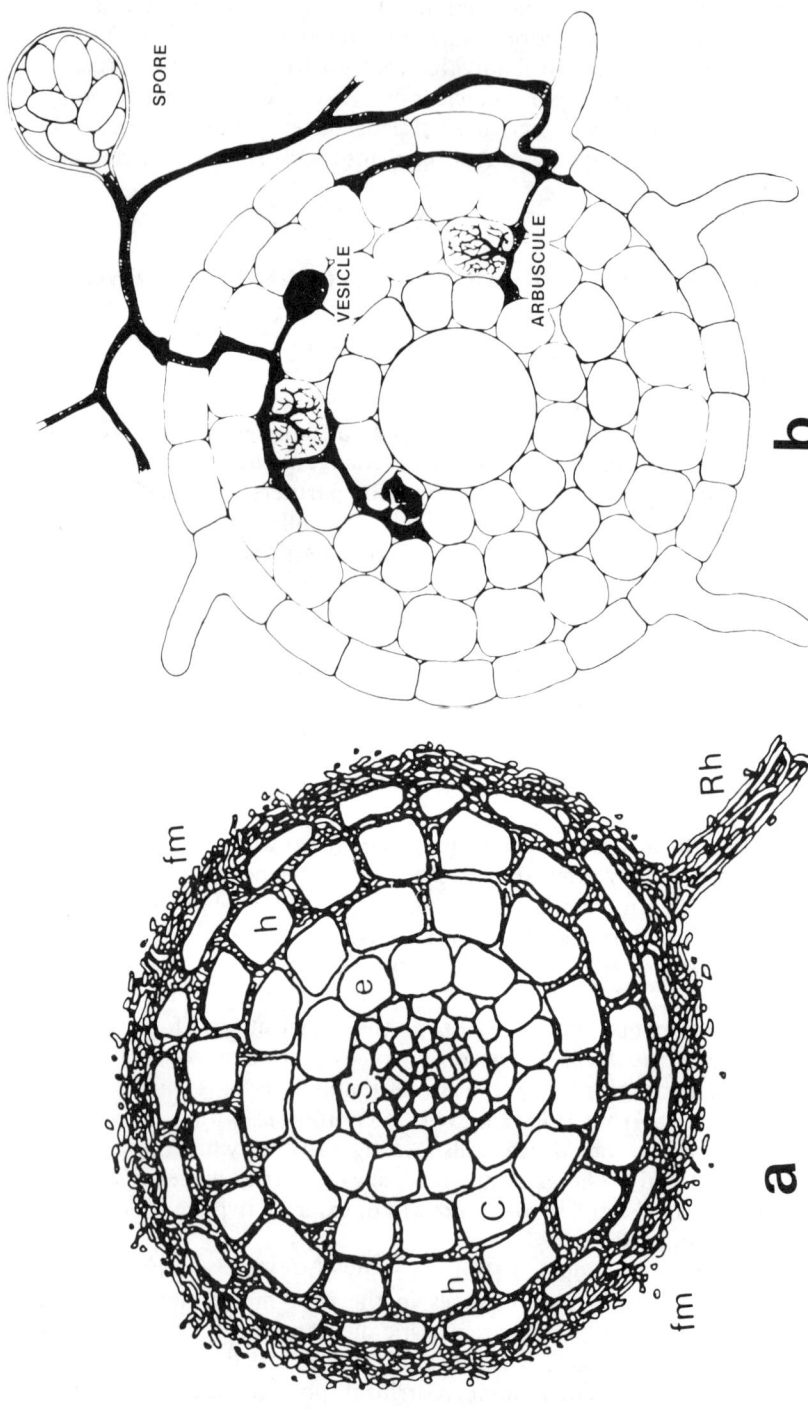

Fig. 3.—Diagrammatic representation of (a) ectomycorrhizae and (b) endomycorrhizae, both in section across the root. f.m. = fungal mantle; h = host cell; c = cortex; e = endodermis; s = stele; and RH = rhizomorph.

C. Endomycorrhizae

Endomycorrhizae have recently been discussed in detail (Sanders et al., 1975; Hayman, 1978). The most important group is undoubtedly the vesicular-arbuscular (VA) types (Fig. 3b) (Mosse, 1973; Tinker, 1975a; Gerdemann, 1975) followed by the ericaceous and lesser groups (Lewis, 1973). Morphologically the endomycorrhizae are quite distinct from the ectomycorrhizae, though intermediate groups are known (ectendomycorrhizae). In endomycorrhizae there is no external sheath, but only a loose mycelium which may not be obvious on inspection. It develops after initial infection of the root by a hypha or spore germ tube (Fig. 3b and 4). Whereas the ectomycorrhizae do not develop any structures within the root cells, the VA mycorrhizae in particular form extensive inter- and intracellular mycelium within the root cortex, and at least two characteristic structures, vesicles and arbuscules. The vesicles are rounded bodies of the order of 50 μm in diameter which form between cells, and probably act as stores of metabolites as lipid droplets. They become strongly radioactive if $^{14}CO_2$ is fed to the host (Cox et al., 1975). The arbuscules' form suggests that they are transfer organs; they originate from a "trunk" hypha, which branches from an intercellular hypha, and enters a cell (Cox & Sanders, 1974). Inside, this hypha branches dichotomously and repeatedly to form dense clumps of fine hyphae, which may fill a considerable part of the cell volume. These structures degenerate after a short time, lose their cytoplasm, and form agglomerations of fungal cell wall. This invasion is accompanied by up to a 20-fold increase in the host cell cytoplasm (Cox & Tinker, 1976), but the trend is reversed after the arbuscule degenerates. The cell then appears quite unchanged in function, and may be reinvaded. The morphology of different host-endophyte combinations may vary considerably (Gerdemann, 1975).

Vesicular-arbuscular mycorrhizae are found in all climatic zones, and so far as is known, in all soils carrying vegetation. There is a clear tendency for ectomycorrhizae to occur more frequently on tree species in temperate climates, whereas the VA mycorrhizae are found on the majority of the world's vegetation, including tropical tree species. Some host species may carry both endo- and ectomycorrhizae, even at the same time. Certain families are rarely if ever infected with VA fungi, particularly the Cruciferae and the Chenopodiaceae which contain many important crop plants, but recent observations have found infection in species previously thought to be immune, and it is possible that there is simply a wide range of susceptibility (Mosse, 1975b).

The taxonomy of the VA mycorrhizae is unsettled. None of these nonseptate phycomycetous fungi has been grown in pure culture yet (Warcup, 1975), and the taxonomic classification depends wholly upon the morphology of the large (40–400 μm) spores formed outside the root by the majority of such fungi. Recent work (Gerdemann & Trappe, 1975) has clarified the situation greatly. They proposed a classification system, which omits non-

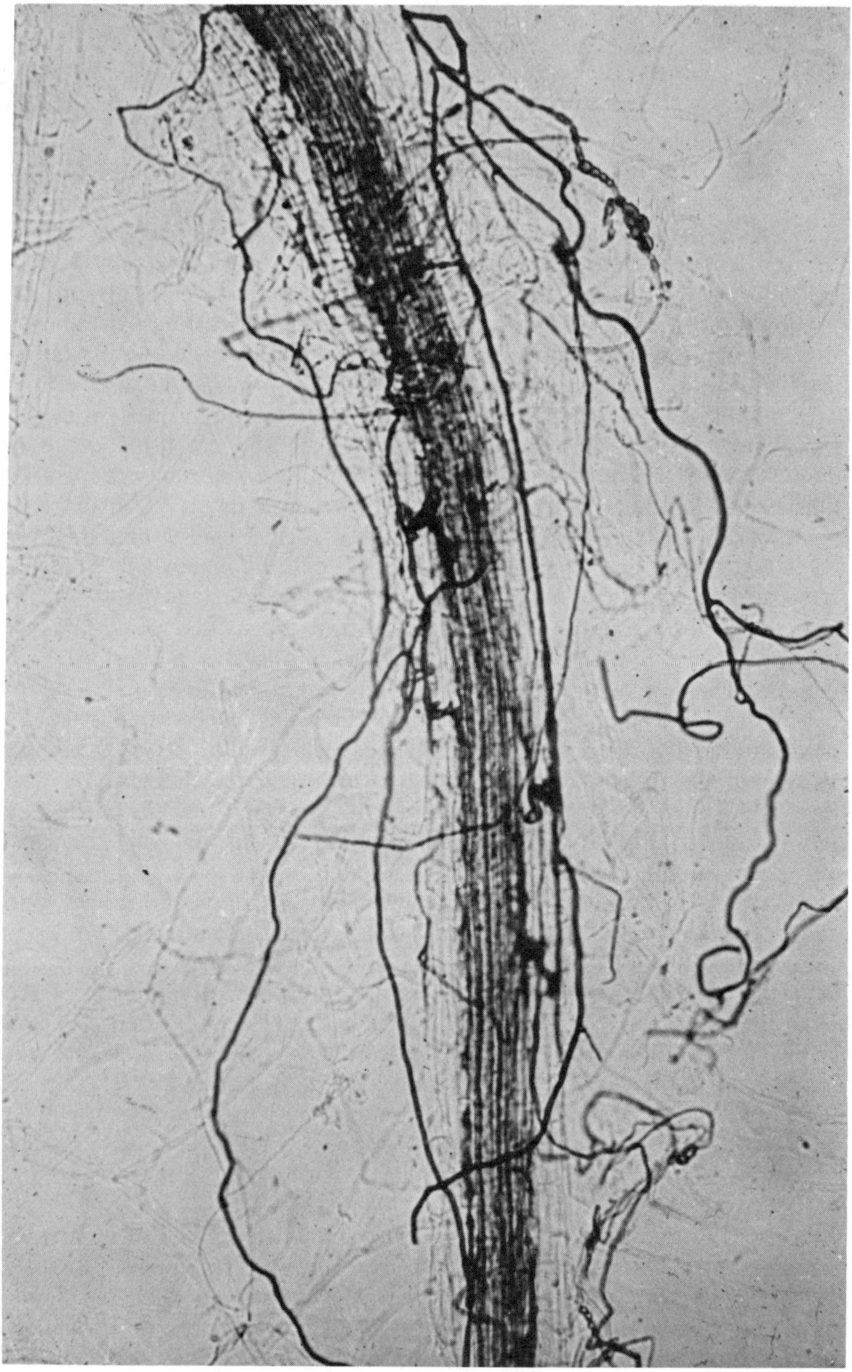

Fig. 4a—See opposite page for complete caption.

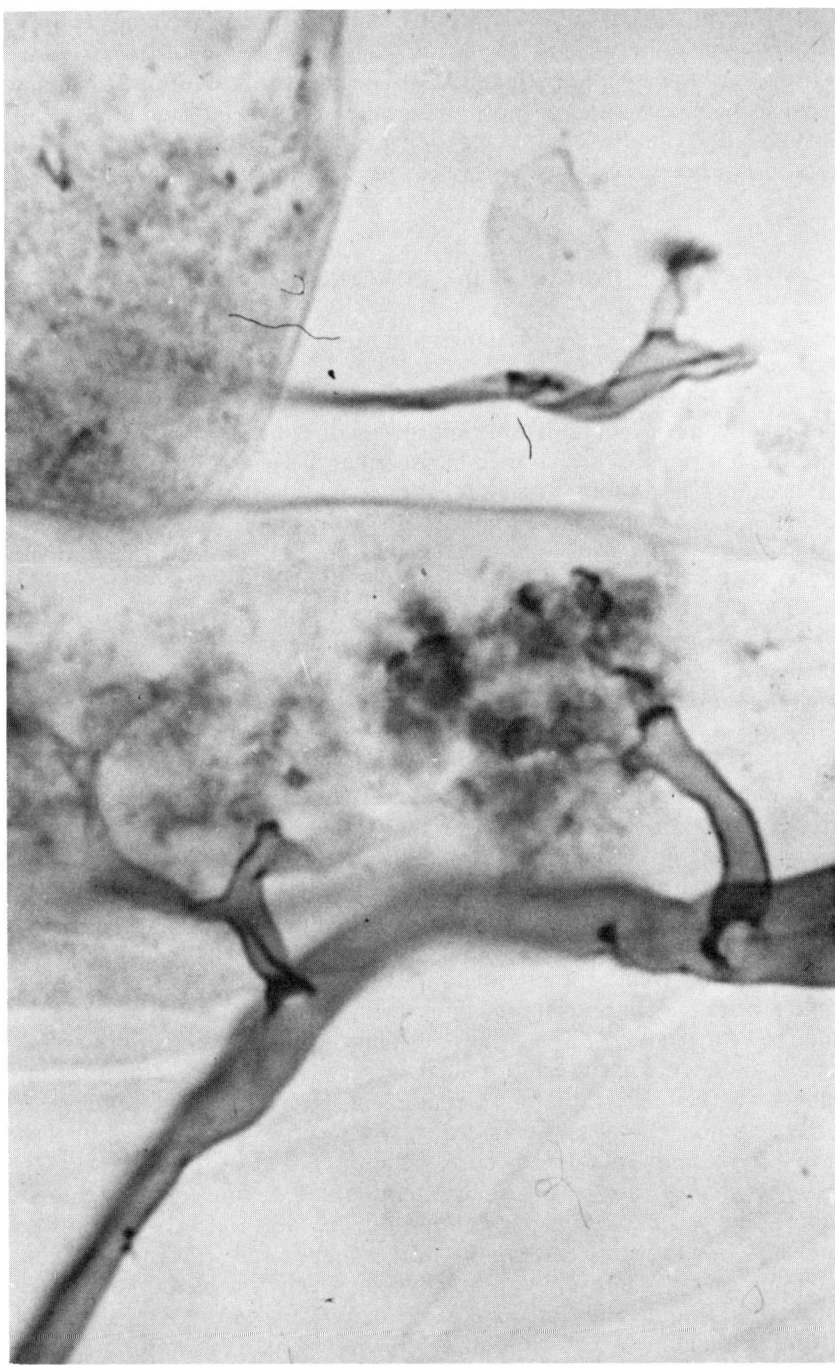

Fig. 4—Photomicrograph of VA mycorrhiza: (*a*) External mycelium and entry points and (*b*) Internal mycelium and arbuscules (Photos by F. Sanders).

sporing forms, because they will not be noticed unless their mycelium is distinctive in some way, as for example, with *Rhizophagus tenuis* which possesses narrow hyphae (Crush, 1973). The mycorrhizae formed by septate fungi in the Ericaceae are morphologically somewhat different (Read & Stribley, 1975) and exhibit a very intensive development of infection in appropriate hosts.

V. FACTORS INFLUENCING MYCORRHIZA FORMATION

A. Hosts and Fungi

In the ectomycorrhizae, most hosts can be infected by several fungi, and equally, most fungi can infect more than one host. Some fungi are quite specific, but precise comparative measurements seem to be lacking. There has been no report of any strict VA endophyte- or host-specificity, but there are reports of a considerable range in infectivity and susceptibility. For example, Moose (1975b) found three endophyte species infected respectively 26, 90, and 97% of the root length of *Centrosema pubescens*.

Laboratory or greenhouse experiments comparing fungi should distinguish between the rate of infection spread in relation to inoculum density, and the equilibrium level of infection reached under particular environmental conditions. Measurements with four species of VA endophyte on onion (Sanders et al., 1977) showed different rates and patterns of infection spread, so that different conclusions could have been reached depending upon date of harvest. Bevege and Bowen (1975) described the different times at which infection, arbuscules, and vesicles appeared with six different endophyte-host combinations and also the different rates of spread of infection and production of external mycelia.

Ectomycorrhizal hosts are normally perennial trees, and the local rate of spread of infection may not be of great importance. Many important VA hosts are annuals and the rate of mycorrhiza formation during periods of rapid growth may be crucial to plant growth. Different hosts growing in the same soil show widely differing susceptibility to VA infection, with cereals usually more heavily infected than potatoes (Hayman, 1975b). It is unknown whether this is due to a genuinely different root susceptibility, to differing amounts of external mycelium which carries the infection further, or to differences in root density or distribution.

The growth rate of mycelial strands of an ectomycorrhizal fungus along a root was 2 to 4 mm·day^{-1} (Bowen & Theodorou, 1973), or up to 1.5 mm·day^{-1} for hyphae. This compares reasonably well with the minimum extension rate of 2 mm·day^{-1} of VA hyphae which may be estimated from the work of Rhodes and Gerdemann (1975) and the measured spread rate of VA infection along a grass turf of 1 cm·week^{-1} (Sparling, 1976).[1] The dynamics of the development of infection in root systems need much further study, particularly on the mycorrhizal infection zone (Marks & Foster, 1973) and the relation between inoculum density and infection rate. Interpretation may use mathematical models of the type suggested by Tinker (1975a).

B. Soil Factors

There is only moderate agreement about the soil factors which determine the presence and numbers of mycorrhizal fungal propagules and the speed of development of infection and its effect upon the host. In general, waterlogged soils tend to have poorly infected roots. Ectomycorrhizae are few in anaerobic soils, possibly because of the high respiration rate of the fungal mantle. The rate of spread is slow in dry soils (Meyer, 1974). There is much evidence that ectomycorrhizal fungi grow best in a moderately acid environment of about pH 5, though mycorrhizal development may be greater in mull than in mor humus soils (Meyer, 1973). Mosse (1972) reported that *Glomus mosseae* infected hosts only in soils with pH greater than 5.5, though we have noted infection of grasses at lower pH. A survey during 1975 of root infection percentage in fields under continuous barley failed to show any strong effect of soil pH, texture, or other common soil analytical value (Black & Tinker, 1979).

Usually, increased soil fertility, in particular the supply of P, leads to lower infection with both ectotrophic (Marx et al., 1977) and VA mycorrhizae, though the reverse may happen if the P status is extremely low (Sanders & Tinker, 1973) and Meyer (1974) concluded that ectomycorrhiza were not especially favoured by low soil nutrient levels. This is clearly connected with the sensitivity of the fungus to the internal P concentration in the plant (Sanders, 1975). It has also been found that VA mycorrhizal spore populations are altered by added N (Hayman, 1975b), but the mechanism in this case is unknown. Reid and Bowen (1979) tested the effects of water stress on ectomycorrhizal development in *Pinus* and showed it to be greatest at moderate tensions. There is a large literature (see Marx & Krupa, 1978) on the establishments of mycorrhizal plants on spoil heaps and other polluted sites.

C. Light and Temperature Effects

The effect of other environmental factors undoubtedly requires further investigation under controlled conditions. Vesicular-arbuscular mycorrhizal infection is increased considerably by higher temperatures up to at least 25°C (Hayman, 1974; Furlan & Fortin, 1973). There is no information on the mechanism by which this occurs, but Dr. K. Cooper (private communication) found that P translocation decreased below 15°C in *Glomus mosseae* hyphae. If this indicates that translocation in general is lowered, it could explain fungal growth reduction. Bjorkman (Harley, 1969) developed a general theory suggesting that the rate of mycorrhizal infection depended upon the availability of free carbohydrate in the host roots. From this hypothesis we would expect more light to increase mycorrhizal infection by increasing the photosynthesis rate. As a general rule this appears to be true in both ectotrophic and endotrophic mycorrhizae, but it cannot be regarded as invariable, and the literature is somewhat confusing. Thus Hayman (1974) found the effects of light to differ depending upon the soil used.

Table 3—Effects of mycorrhizal infection on weight and nutrient percentage of *Pinus strobus*.

	Inoculated	Uninoculated
Dry weight/seedling (mg)	405	303
Root/shoot ratio	0.78	1.04
Nitrogen % dry weight	1.24	0.85
Phosphorus % dry weight	0.196	0.074
Potassium % dry weight	0.744	0.425
Nitrogen uptake, mg·g^{-1} dry weight root	29	16
Phosphorus uptake, mg·g^{-1} dry weight root	4.5	1.4
Potassium uptake, mg·g^{-1} dry weight root	17	8

VI. EFFECT OF MYCORRHIZAE ON PHOSPHORUS NUTRITION

Our interest in mycorrhizae is largely due to the frequent observation that plants grow better when infected by the appropriate fungi and that this increase may on occasion be very large. In almost every case which has been investigated, such growth stimulation has been traced to an improvement in the plants' mineral nutrition, and in the great majority of cases P has been the nutrient most affected. Other relatively immobile nutrients, such as Zn (Bowen et al., 1974), have also been implicated. Many experiments in the field or in controlled conditions showed clearly that plants infected with ectomycorrhizal fungi had higher P concentrations and uptake than uninfected plants (Harley, 1969; Bowen, 1973; Table 3). In general the growth response to infection decreased with P additions and, as noted above, the degree of mycorrhiza formation diminished also. Differences in plant morphology, such as the root-shoot ratio and leaf size and color, are often noted between infected and uninfected plants, but these morphological changes are probably also caused by different levels of P nutrition.

Similar work with VA mycorrhizae began only in the 1950's when it was fully appreciated that this type of infection also could improve growth (see Nicolson, 1967), and this again was soon associated with P nutrition (Table 4). More recently, there have been suggestions that the supply of Zn, S, K, and N could be improved (see Tinker, 1975a). Improvement in Zn nutrition leading to growth responses in field crops following mycorrhizal infection have been reported (Gerdemann, 1975).

Improved growth may itself lead to a more rapid uptake of all nutrients, and it is important to distinguish which one is the cause of the improvement. In general, adding P removes the mycorrhizal response. The best understanding of the effects of mycorrhizae is obtained from experiments in which the response curve to P is determined with and without mycorrhizae. The result of one such experiment was a large mycorrhizal growth effect in clover at the lowest level of soil P, which increased in absolute, but diminished in relative terms with increasing P additions, until the effect finally almost disappeared when the P supply was fully sufficient for the noninfected plants (Sparling & Tinker, 1978b) (Table 5). Yost and Fox (1979) have investigated a similar point in field plots on a tropical soil, using methyl bromide fumigation of soil maintained at different levels of P.

Table 4—Dry weight of shoots of onions grown in two soils, both nonsterilized and after γ-irradiation, and with and without inoculation with *Glomus mosseae* (Mosse, 1973).

Soil		\multicolumn{5}{c}{$Ca(H_2PO_4)_2 \cdot H_2O$ added (g/kg soil)}
		0	0.2	0.5	1.0	1.5
		\multicolumn{5}{c}{dry wt of shoots (mg) and *P content*, %}
		\multicolumn{5}{c}{Irradiated soil}
8C	Inoculated	221	517	604	574	511
		0.13	*0.23*	*0.33*	*0.78*	*0.71*
	Control	143	490	756	712	495
		0.09	*0.15*	*0.36*	*0.58*	*0.78*
7	Inoculated	1,480	2,147	2,218	2,087	1,860
		0.10	*0.12*	*0.21*	*0.34*	*0.38*
	Control	253	1,485	1,833	1,750	1,423
		0.09	*0.11*	*0.19*	*0.32*	*0.35*
		\multicolumn{5}{c}{Unsterile soil (different experiment)}
8C	Inoculated	338	656	470	--	--
		0.17	*0.22*	*0.43*		
	Control	99	506	657	--	--
		0.27	*0.35*	*0.32*		
11	Inoculated	623	795	739	--	--
		0.26	*0.25*	*0.36*		
	Control	158	413	371	--	--
		0.18	*0.18*	*0.24*		

They found that the soil solution P level at which mycorrhizal infection ceased to have an effect varied between 0.1 and 1.6 μg $P \cdot ml^{-1}$, depending upon species. This type of investigation, with the more fundamental approach described later (see Sect. VII, B), allow the mycorrizal function to be clearly related to soil properties. This was also found by Stribley et al. (1980) who determined responses of leeks to added P without mycorrhizas, with artificial inoculum, or with the natural soil inoculum. Excellent, but different, relationships with bicarbonate-soluble (Olsen) P in the soil were obtained over the whole range of soils tested when plants were uniformly nonmycorrhizal or heavily infected, but the relationship in the natural soils was much poorer, because degree of infection was variable. The predictive value of soil analysis for P may thus depend upon the state of mycorrhizal infection.

The most precise way of expressing improved root uptake is as mean uptake rate per unit weight, length, or surface area of root, since this defines the relative availability of P in a given soil to infected and uninfected roots. Bowen (1973) used early data of Hatch to estimate uptake per unit root weight for ectomycorrhizal *Pinus* (Table 3). Sanders and Tinker (1971, 1973) showed that VA infection increased the P uptake rate per unit root length (inflow) of onion by a factor of about 4. The uptake rates of other elements were also increased in these experiments, but to a smaller extent.

If the effect of mycorrhizal infection is to supply P to the plant, it is also possible to equate infection to a particular amount of fertilizer P for any one set of conditions, and values between 30 kg P•ha⁻¹ and 556 kg P•ha⁻¹ have been reported, with most around 100 kg P•ha⁻¹.

VII. MECHANISM OF PHOSPHATE UPTAKE BY MYCORRHIZAL ROOTS

A. Source of Soil Phosphate

The simplest way to determine the sources of an absorbed nutrient is to label the labile pool in the soil with an isotope, and to compare the specific activities (or the L value) of the plants with that in the soil solution. Sanders and Tinker (1971, 1973), Hayman and Mosse (1972b) and Powell (1975) measured the specific activity of onions growing in ^{32}P-labeled soil, and all concluded that the specific activity of P in mycorrhizal and nonmycorrhizal plants was the same, despite large differences in uptake and growth. Sanders and Tinker (1971) also found the specific activity of P in the plants to be close to that of soil solution P. In the absence of other results, it seems that mycorrhizae absorb P from the same immediate source as the uninfected roots, and that this is the soil solution (Tinker, 1975b; 1977) but further testing may be advisable (Swaminathan & Verma, 1977). In particular, this should be done with soils containing much organic P, in view of the frequently large quantities of the latter (see Chapt. 15 by G. Anderson, this book), its use by VA mycorrhizae in agar (Mosse & Phillips, 1971), and the phosphatases found on mycorrhizal surfaces (Bartlett & Lewis, 1973).

Much interest has been shown in the uptake of P from low-solubility sources by mycorrhizal plants, in the hope that fungal infection may render such materials more useful. Daft and Nicolson (1966) first tested the ability of tomatoes to utilize tricalcium phosphate (TCP), hydroxy apatite, phosphate rocks (PR), and other Ca phosphates mixed in sand culture, and found that mycorrhizal plants were much larger than nonmycorrhizal plants with TCP, less so with the very insoluble apatite, and least for the more soluble dicalcium phosphate. In their tests with bone meal, the mycorrhizal effect depended upon the amount added. Murdoch et al. (1967) conducted similar experiments with maize, and found the mycorrhizal effect greatest with TCP, less with PR, and least with soluble phosphate additions. Ross and Gilliam (1973) tested the growth of soybean with various P compounds, including Fe and Al phosphates, and showed that mycorrhizal plants grew better with all sources. The relative value of the phosphates did not alter greatly on infection, suggesting again that mycorrhizal and nonmycorrhizal plants utilize the same P source, the soil solution. Results of work by Hayman and Mosse (1972a) also suggested that mycorrhizae do not render wholly unavailable sources useful, but that they can accelerate the uptake rate from low-solubility sources which are of some value to uninfected plants. Azcon et al. (1975) reported that lavender plants responded to PR only after infection, but several comparisons in pots on the effect of PR on legume crops with and without mycorrhizas, in a range of soils, confirmed that VA mycorrhizas could improve, but not confer, the ability to use PR efficiently (Mosse et al., 1976; Mosse, 1977). Graw (1979) tested the relationships of soil pH and the availability of various phosphates to mycorrhizal and nonmycorrhizal plants. The results were extremely complex, with two plant species giving quite different results. With *Guizotia abyssinia,* growth was increased by mycorrhizal infection without added P or with ferric phosphate, but decreased when calcium monophosphate was given at all pH levels. With hydroxyapatite and aluminum phosphate, infected plants showed the best growth at pH 6.6, but noninfected plants were best at lower pH. The interaction of soil temperature and form of phosphate was tested by Nyabenda (1977)[2] who found that the mycorrhizal affect varied with both factors. The relative value of infection was always greatest with insoluble phosphates, but it varied greatly with temperature.

The detailed considerations affecting such uptake are discussed by Tinker (1975b). If the fungal hyphae simply absorb phosphate ions from the soil solution in the same way as the roots, they will promote dissolution of adjacent particles of mineral phosphate. The important factor should then be the mean distance between an absorbing surface and a mineral phosphate surface, and this could be decreased by subdividing the mineral, or by having hyphae or root hairs on the root. This view suggests greater efficiency of utilization by mycorrhizal plants, but no major difference in kind of ability to use such phosphates. Indeed, some host species with a particular reputation for using PR efficiently, such as mustard, are members of the Cruciferae, and are nonmycorrhizal. The possibility of utilizing low-solubility sources more efficiently is attractive, however, and deserves further investigation. It seems likely that mycorrhizae will not prove to be a universal means of improving uptake from PR and other low-solubility phosphates, but that they may have important uses where such P sources are only partially effective in promoting growth, or that they may allow the maximum growth response to be obtained with smaller applications (Table 5).

Utilization of poorly soluble phosphates by ectomycorrhizae has been

---

[2] P. Nyabenda. 1977. Einfluss der Bodentemperatur und organische Stoffe auf die Wirkung der vesikular-arbuskularen Mykorrhiza. Thesis, Univ. of Gottingen, Germany.

Table 5—Effect of different rates of phosphate rock in a brown earth soil (Malham series), on P uptake by clover with no, or one of three types of endophyte (after Sparling & Tinker, 1978b).

Phosphate rock	Endophyte			
	Nil	Local, "fine"	Local, "coarse"	*Glomus mosseae*
g·kg^{-1} soil		mg P per plant shoot		
0	0.005	0.3	0.4	0.005
0.1	0.01	0.2	0.25	0.04
1	0.4	0.9	0.9	0.35
2	0.6	1.8	1.3	1.2
5	2.6	5.5	4.3	5.1
20	4.9	5.6	6.1	6.5

discussed by Bowen (1973), though little work has been done on this subject. The fungi were grown in pure culture on agar media, and their use of poorly soluble mineral phosphate has been proven. As for equivalent work with bacteria (see Sect. III, C-2), this does not constitute evidence that the same process will occur at useful rates in soil. In tests with *Pinus radiata* (Bowen, 1973), mycorrhizal plants grown in soil responded well to PR but the relative increase in dry weight and P content was about equal in the mycorrhizal and nonmycorrhizal plants. Mejstrick & Krause (1973) tested various forms of P supply by ectomycorrhizae formed with two fungi. Solution and adsorbed phosphate were taken up similarly, but the mycorrhizal effect was much greater with humus phosphate.

B. Uptake and Translocation of Phosphate by Fungal Hyphae

There are now several lines of evidence showing clearly that mycorrhizal hyphae absorb and translocate nutrients. The simplest and most direct is that excised ectomycorrhizal mantles absorb phosphate (and other ions) more rapidly than the cortical roots, per unit weight. The excised mantle is a useful experimental material, and a wide range of studies have been made on its P relationships. The main results are that the uptake is stopped by low temperatures and metabolic inhibitors, hence it is active (Harley, 1969); that the fungal material can store comparatively large amounts of P; and that P passed on to the host plant is in some sort of interchange with various "pools" of P in the mantle. Ling-Lee et al. (1975) suggested that polyphosphate formed a storage pool for P. The comparatively large storage capacity for P in the sheath has caused suggestions (Harley, 1969) that this benefits the host by acting as reservoir for P during periods which are unfavorable for P uptake. However, the mobility of P in host tissues and the wide range of P contents in the latter suggest that the host could carry out this function itself if necessary. The important question is how the fungus allows the mycorrhiza to accumulate P at a faster rate than a normal root. From the slicing experiment of Harley and McCready (1952) it is clear that very little P will diffuse through the mantle under normal growth conditions in soil, and that substantially all the P taken up by an infected root will enter via the fungal partner. The relative uptake by the mantle itself and by the external hyphae and mycelial strands is still not clear, however. Melin and Nilsson (1958) showed, for ectomycorrhizal fungi on pine, that hyphae translocated P, N, and Ca readily. The ability of other fungal hyphae to act as extremely efficient translocating systems is well substantiated (Jennings et al., 1974). Uptake and translocation of ^{32}P by mycelial strands of *Rhizopogon luteus* mycorrhizae over distances of up to 12 cm was shown by Skinner and Bowen (1974). Sanders and Tinker (1971) and Tinker (1975b) calculated that the inflow of P to the uninfected roots of onion was near the maximum in that soil (Passioura, 1963), with the root functioning as a near-zero sink, and concluded that the additional inflow to infected roots must be carried by the VA hyphae. Other work (Hattingh et al., 1973; Rhodes & Gerdemann, 1975) has shown directly that VA mycorrhizal hyphae external to a root will absorb P from as far as 8 cm from the root and translocate it to the host. Direct measurements of translocation rates for P (flux of up to 1×10^{-9} moles·cm^{-2}·sec^{-1}) by Pearson and Tinker (1975) and for P (2×10^{-10}) S (1.6×10^{-11}) and Zn (2×10^{-12}) by Cooper and Tinker (1978) over distances of 1 to 2 cm have also been made.

Since fungal hyphae are able to increase the P uptake rate of roots, it is of interest to determine whether this is due to their greater absorbing power per unit surface area or to spatial and geometric factors. There is much evidence that plant roots can absorb P sufficiently for plants to grow well with a root surface concentration of around 10^{-6} M. In the experiments of Pearson and Tinker (1975), VA hyphae also absorbed P from a solution with this concentration. Responses to mycorrhizal infection of onion were found with the soil solution concentration 5×10^{-6} M (Sanders & Tinker, 1973), when the root uptake rate for uninfected plants was near the maximum by diffusion, and the concentrations of P at root surfaces were therefore much lower than in bulk soil. It was therefore probably the position of the ramifying external hyphae outside the root depletion zone which allowed them to absorb and translocate P, rather than any special absorbing power of the hyphae. It has been suggested by Mosse (1973) that plant roots may have a threshold concentration below which they do not absorb (Claassen & Barber, 1974), and that hyphae may have no or a lower threshold concentration, but this point has not been fully established yet (Nye & Tinker, 1977).

The above view has several corollaries (Tinker, 1975b). Responses to mycorrhizal infection can be expected whenever the maximum diffusive transfer of P to unit length of root (the inflow) is appreciably less than that required to maintain the growth of the plant. The required inflow (I) depends upon the relative growth rate (R_w), the length of active absorbing root (L), the total plant weight (W), and the nutrient percentage in the plant (N) by the following relation (Nye & Tinker, 1969).

$$I = R_w \frac{W}{L} \cdot \frac{N}{100}$$

The likelihood of a response to mycorrhizal infection thus depends upon plant, soil, and fungus variables. A sparse rooting system therefore calls for a large I, and Bowen (1973) has stressed the low root density of mycorrhizal trees compared to herbs. Mycorrhizal fungi can thus bypass the diffusive impedance around roots in a manner similar to root hairs, and it is believed that responses to infection are more likely in species without than in species with hairs (Baylis, 1970). In very dense root systems, mycorrhizal effects may be less likely because of (i) low I values and (ii) the distance between adjacent roots is so small that there is virtually no unexploited soil for hyphae to ramify into (Sparling & Tinker, 1978a). Mycorrhizae would also be beneficial if they absorbed for longer periods than uninfected roots, and thereby increased L. In ectomycorrhizae the mantle remains active in uptake for months, whereas the absorbing power of roots for P normally declines gradually over a period of days or weeks (Bowen, 1973). The increase in radius of roots by addition of the mantle may increase uptake slightly, but according to diffusion theory, the effect is relatively minor (Nye, 1966; Olsen & Kemper, 1968). The effect of root age on uptake has not yet been investigated for VA mycorrhizae, and much of the reported work is on seedling plants with young roots.

If fungal hyphae absorb P outside the depletion zone of root, root hairs, or mantle, the quantity and distribution of such hyphae become important. Bowen (1973) pointed out that mycelial strands penetrate up to 3.5 cm from a mycorrhiza, even though individual hyphae extended only up to 350 μm. By contrast, root hairs are commonly less than 1 mm long. For endomycorrhizas, it is known that the hyphae can extend several centimeters into soil (see Sect. VII, B), but more data are needed.

The total quantity of external hyphae has rarely been measured. Sanders and Tinker (1973) found quantities corresponding to about 80 cm (3 mg fresh weight) per cm of infected onion root in one soil type; under similar conditions Sanders et al. (1977) reported that three endophytes produced similar amounts of mycelium. The very large numbers of entry points found by Read and Stribley (1975) in the roots of *Calluna* suggest that hyphal lengths may be large. In ectomycorrhizae, no direct measurements have been made of total length yet, but Skinner and Bowen (1974) found very large differences in mycelial growth in different soils, and increases in growth due to added phosphate.

The uptake and translocation of P by external hyphae are therefore regarded as being the essential growth-promoting processes in mycorrhizae. The total transfer of labelled P from external solution to shoot of host plant was followed by Cooper and Tinker (1978). They noted a surprising 2-day lag in the uptake and translocation of ^{32}P, which could not be satisfactorily accounted for by isotopic dilution or by delay in transfer from fungus to host, and which suggests an enzymatic induction mechanism at some stage in the uptake or translocation process. Mention is made above of direct measurements of the mean flux of P through VA hyphae; Sanders and Tinker (1973) produced a much larger value (3.4×10^{-8} moles$\cdot$cm$^{-2}\cdot$sec^{-1}) by counting main entry hyphae into the root, and assuming that the extra inflow to infected, over that to the uninfected root, was carried by hyphae. All such values are too large to allow diffusion as the translocation mechanism within the hyphae, and Bowen (1973) and Tinker (1975a) considered that protoplasmic streaming was likely to be the important factor, with possibly some bulk flow of hyphal contents.

This topic gained further interest by the discovery of opaque granules in VA mycorrhizal hyphae (Cox et al., 1975) which probably contained polyphosphate. Polyphosphate granules have also been found in endo- and ectomycorrhizae by Ling-Lee et al. (1975). This opens the interesting possibility that the high P fluxes arise from the movement of polyphosphate granules by protoplasmic streaming (Tinker, 1975a), with hydrolysis of the polyphosphate in the arbuscules, and subsequent transfer to the host. Callow et al. (1978) confirmed the presence of polyphosphate in VA mycorrhizae by biochemical techniques.

C. Transfer of Phosphate from Fungus to Host

The mechanism of transfer from fungus to host in ectomycorrhizae has always been assumed to be by excretion of P compound(s) from the hyphae of the sheath and Hartig net, followed by reabsorption by the host cells. The storage capacity of the mantle causes a large buffer store, and most of the absorbed P is found in the mantle after uptake periods of a few hours. This is passed to the host over an extended period if no more P is supplied externally. The transfer process is oxygen-dependent, and may be affected by the transpiration rate of the shoot (Harley, 1969). In the VA mycorrhizae, the regular decay or digestion of the arbuscules (Cox & Sanders, 1974) has given rise to the hypothesis that P is transferred by this process into the host. Kinden and Brown (1975) and others have reported that hyphal contents are lost into the host cell on arbuscule degeneration, but Cox and Tinker (1976) measured fungal volume and other quantities, and suggested that this mechanism was not capable of sufficiently large rates of transfer. It therefore seems likely that transmembrane transfer occurs in the VA mycorrhizae also, and there is evidence of a specific alkaline phosphatase in the internal fungal material, which could be concerned with the transfer between fungus and host (Gianninazi et al., 1979). Woolhouse (1975) has suggested detailed mechanisms for transfer, but much further work is required on this topic.

VIII. OTHER EFFECTS OF MYCORRHIZAL FUNGI ON PLANTS

Effects other than those due to nutrient transfer have often been suggested, but rarely proved to occur. In the first instance, the fungus receives all its carbon from the host plant, and it might be expected that this would decrease host growth (Harley, 1975). There are occasional reports of decreased growth following infection with VA mycorrhizae, which may be related to their carbon demand (Tinker, 1977; Stribley et al., 1980). Potential losses due to carbon use of both endo- and ectomycorrhizas have been discussed by Bowen (1979), but the present situation is by no means clear. This process may reduce crop yields below that *theoretically* attainable, but mycorrhizal infection in practice normally increases yield.

It has been conjectured that the fungal partner must affect the host in some way by the production of hormonal compounds, and it seems certain that this must happen in the sheathing mycorrhizas (Shemakhanova, 1962; Slankis, 1973). Certainly, it is difficult to observe VA infections where the entire root cortex appears filled with fungal material without expecting some such effect on the host. The appearance may be deceptive, however, in that it is the projections of very fine hyphae which are seen and the total volume is much less than may appear. Direct nonnutritional interactions between host and fungus are, therefore, still a matter for speculation, but Gerdemann (1975) has reviewed several cases in which mycorrhizal infection affected susceptibility to disease or pests.

IX. PRACTICAL APPLICATIONS OF MICROORGANISMS IN IMPROVING PHOSPHATE NUTRITION OF HOST PLANTS

A. Theoretical Possibilities

It is helpful to consider first what possible benefits could reasonably be expected from association of microorganisms and roots of higher plants. Mycorrhizae might allow plants to make the same growth, but with a lower percentage of P in the tissues, thus leading to a net saving of P during cropping. In fact, however, all the evidence indicates the opposite and there are many instances where mycorrhizal plants require a higher P concentration to produce the same dry weight (Abbott & Robson, 1978; Stribley et al., 1980). The evidence discussed above (see Sect. VII, A) does not support the idea that highly insoluble mineral sources can be used, but suggests that sources of low, moderate, or high solubility can be used more efficiently. The most likely view at present is that mycorrhizae may allow plants to grow at their maximum potential relative growth rate at a lower P "availability," "intensity," or soil solution concentration in the soil than would otherwise be possible. This would have two practical advantages: first, less P need be added to deficient soils to bring them up to the optimum status; second, the continued maintenance of a lower "intensity" in regularly fertilized soils would cause a smaller rate of transfer from the labile pool P into nonlabile forms (Larsen, 1967). In this context, the interactions between root systems of different hosts cannot be ignored (Christie et al., 1978).

B. Practical Results in Field Tests

Field work with microorganisms has lagged behind laboratory and greenhouse experimentation, except in the one case of bacterial fertilizer, where the reverse is possibly true. However, the long and unsuccessful history of bacterial fertilizers for the improvement of P nutrition argues against any easy use of rhizosphere bacteria for this purpose. There is not only the fundamental question of whether appropriate bacteria do in fact release soil P for use by higher plants, but also the problem of establishing and maintaining such bacteria throughout a highly complex and competitive rhizosphere (Bowen & Rovira, 1976). With better understanding of the hormonal mechanisms which allow bacteria to affect plants, there may be scope for new developments, but there seems no particular reason to link all of these with P nutrition, or to expect rapid progress.

The possibilities seem brighter for VA mycorrhizae, since there is no doubt that they increase P uptake and that they form a stable symbiotic system. They are of particular importance because of the very wide range in host compatibility. However, there are field situations in which the mycorrhizae appear to confer no benefit on the host, and there is also the possibility that any potential benefit is already being obtained due to natural infection. So far, field results showing a clear yield response to inoculation with VA mycorrhizae are scanty. Two main possiblities are usually considered in this type of work (Gerdemann, 1975).

1) The naturally present inoculum in the soil is insufficient to give a useful level of infection to an annual crop early enough in the season (Hayman, 1970). This situation is most likely to develop after a sequence of noninfected crops or fallows on the same field (Winter, 1952; Black & Tinker, 1979), and the natural infectivity of soils can in any case vary widely (Mosse, 1977). Khan (1972, 1975) reported field yields of wheat and maize were approximately doubled after artificial inoculation with VA fungi; the fields were deliberately chosen to have small amounts of natural VA inoculum present. Yield increases have also been found in cotton (Rich & Bird, 1974) and in soybeans (Ross & Harper, 1970). Black and Tinker (1977) reported yield responses to inoculation in potatoes grown on a field kept free of mycorrhizal crops for 3 years. Where soil fumigants are used the native fungal population may be killed, and large responses to reinoculation of citrus nurseries have been found (Kleinschmidt & Gerdemann, 1972). Fumigation of experimental plots has also been used to show the field effects of natural mycorrhizal infection (Yost & Fox, 1979), the mean reduction in P uptake by fumigation being 25-fold where no P was given.

2) There is the possibility that the local endophytes may be relatively ineffective, and that introduction of other fungal species will cause a yield increase. This effect has been shown in greenhouse culture (Mosse, 1972; 1973, 1977), in which plants in unsterilized field soil grew much better after inoculation with *Glomus mosseae* and other spores (Table 4). Powell (1976) tested an introduced VA fungus against the indigenous fungi in 37 hill soils from New Zealand, and found it to give greater growth of clover in 24 of them. Mosse (1975b) has reported further experiments confirming this point, and has reviewed the whole question of differences in the efficiency of indigenous and introduced endophyte species in promoting host growth. Similar effects were reported by Abott & Robson (1978), who also found that 2 different isolates of *Glomus monosporus* gave different yield responses. This raises the whole question of differences in fungal strains, which is very difficult to clarify while the organism cannot be cultured. Sanders et al. (1977) analyzed growth promoting activity of four endophytes in terms of infection spread, external mycelium production, and ability of the mycelium to absorb P, and found one endophyte to be inferior in all three respects to the rest. The prospects of finding and using superior species of endophytes seem good. Direct tests with inoculated seedlings transplanted into the field have shown very promising responses for clover (Powell & Daniel, 1978; Hayman & Mosse, 1979) and for agricultural crops (Owusu Bennoah & Mosse, 1979).

There are quite complex interactions between soils, hosts, and endophytes, and the correct endophyte may have to be carefully selected. This may be particularly important when the environmental conditions have been drastically changed, for example, by liming of the soil or burning off vegetation. The indigenous endophytes then may not be the best adapted or the most useful under the new conditions.

Much more work is needed to determine how VA mycorrhizae can be used to improve crop growth, and it may be doubted whether full success is to be expected whilst our knowledge of the response of the fungi to changes in environmental conditions and state of host is so incomplete. However, with such striking and repeatable effects in partly controlled environments, it would be very surprising if further field effects of practical value could not be found. There are serious problems in producing effective and concentrated inoculum for field use (Crush & Pattison, 1975), but Hattingh and Gerdemann (1975) have shown that a mixed inoculum (spores, hyphae, and fine roots) can be attached to orange seeds with methyl cellulose, and that effective infection results from this. Very recently, Powell (1979) successfully tested the effects of pelleting seeds with heavily infected soil before sowing in the field.

There is a long and successful history of responses to ectomycorrhizal infection in the field (Harley, 1969; Mikola, 1973; Marx, 1977). Spores of many of these fungi are airborne, in contrast to those of VA mycorrhizae, and it is only the small range of hosts which prevents them from being universally present. Where efficient mycorrhizal associations have been established before, inoculation may be of little or no value, but for trees that will be planted in areas where no related species have been established before, inoculation may be essential. Henderson and Stone (1970) found good responses to inoculation in seedling conifers growing on soils which had been sterilized. Differences between the efficiency of ectomycorrhizal fungi in promoting host growth in pot experiments have been known for a long time. Theodorou and Bowen (1970) showed that artificially inoculated pine seedlings grew better than noninoculated ones when planted out into an area infected with indigenous ectomycorrhizal fungi, and that growth varied with the inoculum used. Infected soil is the usual inoculum, though the planting of infected seedlings may be most effective, and Theodorou and Bowen (1973) used freeze-dried basidiospores of *Rhizopogon luteolus* to infect *Pinus radiata*. Detailed comparisons of fungal species and methods of inoculation showed large differences (Marx et al., 1978) with *Pisolithus tinctorius* and *Thelephora terrestris* being the best.

X. CONCLUSION

This topic is at the interface of biology, chemistry, and biophysics, and collaboration between these disciplines will greatly assist the developments we may expect within the next decade. It seems to the author that the VA mycorrhizae offer the clearest and most immediate promise of improving P nutrition of crops because of their proven properties and wide host range. They may reduce the need for P fertilizers, or they may allow the use of cheaper and cruder forms. The greatest possibilities probably lie in the tropics, because of their large areas of P-deficient soils, large numbers of VA mycorrhizal host species, and rapid development of the endophytes. Current work already suggests two points which should be borne in mind. First, no field or pot work with P fertilizers or on P nutrition should be done without some consideration of the mycorrhizal condition of the crop. Second, any studies of crop rotation effects should consider the possibility that endophyte populations are altered by different sequences of crops. The exploitation of ectomycorrhizae on appropriate tree crops will undoubtedly develop further also. The chances of using bacteria to improve P nutrition seem slight, but work on the rhizosphere population and its effects on the host in general is extremely interesting and shows promise in several directions.

The ultimate aim must be to understand the rhizosphere microbial ecology so well that we can control its composition and quality as required, though this aim is obviously a distant one.

ACKNOWLEDGMENTS

I wish to thank Dr. F. E. Sanders and Dr. G. Bowen for comments.

LITERATURE CITED

Abbott, L. K., and A. D. Robson. 1978. Growth of subterranean clover in relation to the formation of endomycorrhizae by introduced and indigenous fungi in a field soil. New Phytol. 81:575-585.

Anderson, G., E. G. Williams, and J. O. Moir. 1974. A comparison of the sorption of inorganic orthophosphate and inositol hexaphosphate by six acid soils. J. Soil Sci. 25:51-62.

Azcon, R., J. M. Barea, and D. S. Hayman. 1976. Utilization of rock phosphate in alkaline soils by plants inoculated with mycorrhizal fungi and phosphate solubilizing bacteria. Soil Biol. Biochem. 8:135-138.

Balandreau, J., and R. Knowles. 1978. p. 243-268. The rhizosphere. In Y. R. Dommergues and S. V. Krupa (ed.) Interactions between non-pathogenic soil micro-organisms and plants. Elsevier, Amsterdam.

Baldwin, J. P., P. H. Nye, and P. B. Tinker. 1973. Uptake of solutes by multiple root systems from soil. III. A model for calculating the solute uptake by a randomly dispersed root system developing in a finite volume of soil. Plant Soil 38:621-635.

Barber, D. A. 1969. The influence of microflora on the accumulation of ions by plants. p. 191-199. In I. H. Rorison (ed.) Ecological aspects of the mineral nutrition of plants. Blackwell, Oxford.

Barber, D. A. 1978. Nutrient uptake. p. 131-162. In Y. R. Dommergues and S. V. Krupa (ed.) Interactions between non-pathogenic soil micro-organisms and plants. Elsevier, Amsterdam.

Barber, D. A., and K. B. Gunn. 1974. The effect of mechanical forces on the exudation of organic substances by the roots of cereal plants grown under sterile conditions. New Phytol. 73:39-45.

Barber, D. A., and J. M. Lynch. 1977. Microbial growth in the rhizosphere. Soil Biol. Biochem. 9:305-308.

Barber, D. A., and J. K. Martin. 1976. The release of organic substances by cereal roots into soil. New Phytol. 76:69-80.

Barber, D. A., and A. D. Rovira. 1975. Rhizosphere micro-organisms and the absorption of phosphate by plants. Ann. Rep. ARC Letcombe Laboratory, 1974. p. 27-28.

Barea, J. M., R. Azcon, and D. S. Hayman. 1975. Possible synergistic interactions between Endogone and phosphate-solubilizing bacteria in low-phosphate soils. p. 409–417. *In* F. E. Sanders, B. Mosse, and P. B. Tinker (ed.) Endomycorrhizas. Academic Press, London.

Bar Yosef, B., E. Bresler, and U. Kafkafi. 1972. Uptake of phosphate by plants growing under field conditions: II. Computed and experimental results for corn plants. Soil Sci. Soc. Am. Proc. 36:789–794.

Bartlett, E. M., and D. H. Lewis. 1972. Surface phosphatase activity of mycorrhizal roots of beech. Soil Biol. Biochem. 5:249–257.

Baylis, G. T. S. 1970. Root hairs and phycomycetous mycorrhizas in phosphorus-deficient soil. Plant Soil 33:713–716.

Benians, G. J., and D. A. Barber. 1974. The uptake of phosphorus by barley plants from soil under aseptic and non-sterile conditions. Soil Biol. Biochem. 6:195–200.

Bevege, D. I., and G. D. Bowen. 1975. *Endogone* strain and host plant differences in development of vesicular-arbuscular mycorrhizas. p. 77–86. *In* F. E. Sanders, B. Mosse, and P. B. Tinker (ed.) Endomycorrhizas. Academic Press, London.

Bhat, K. K. S., and P. H. Nye. 1974. Diffusion of phosphate to plant roots in soil. Depletion around onion roots without root hairs. Plant Soil 41:383–394.

Bhat, K. K. S., P. H. Nye, and J. P. Baldwin. 1976. Diffusion of phosphate to plant roots in soil. IV. Plant Soil 44:63–72.

Black, R. L., and P. B. Tinker. 1977. Interaction between effects of vesicular-arbuscular mycorrhiza and fertilizer phosphorus on yields of potatoes in the field. Nature 267:510–511.

Black, R. L., and P. B. Tinker. 1979. The development of endomycorrhizal root systems. II. Effect of agronomic factors and soil conditions on the development of vesicular arbuscular mycorrhizal infection in barley and on the endophyte spore density. New Phytol. 83:401–413.

Bowen, G. D. 1973. Mineral nutrition of ectomycorrhizae. p. 151–205. *In* G. C. Marks and T. T. Kozlowski (ed.) Ectomycorrhizae. Academic Press, New York.

Bowen, G. D. 1979. Dysfunction and shortfalls in symbiotic response. p. 231–256. *In* J. G. Horsfall and E. B. Cowling (ed.) Plant disease—an advanced treatice. Academic Press, New York.

Bowen, G. D., and A. D. Rovira. 1961. Effects of micro-organisms on plant growth. I. Development of roots and root hairs in sand and agar. Plant Soil 15:166–188.

Bowen, G. D., and A. D. Rovira. 1966. Microbial factor in short-term phosphate uptake studies with plant roots. Nature (London) 211:665–666.

Bowen, G. D., and A. D. Rovira. 1969. The influence of micro-organisms on the growth and metabolism of plant roots. p. 170–201. *In* W. J. Whittington (ed.) Root growth. Butterworths, London.

Bowen, G. D., and A. D. Rovira. 1973. Are modelling approaches useful in rhizosphere biology? p. 433–450. *In* E. Rosswall (ed.) Modern methods in the study of microbial soil ecology. Bull. 17, Ecological Research Committee, Stockholm.

Bowen, G. D., and C. Theodorou. 1973. Growth of ectomycorrhizal fungi around seeds and roots. p. 107–150. *In* G. C. Marks and T. T. Kozlowski (ed.) Ectomycorrhizae. Academic Press, New York.

Bowen, G. D., and C. Theodorou. 1979. Interactions between bacteria and ectomycorrhizal fungi. Soil Biol. Biochem. 11:119–126.

Bowen, G. D., M. F. Skinner, and D. I. Bevege. 1974. Zinc uptake by mycorrhizal and uninfected roots of *Pinus radiata* and *Araucaria cunninghamii*. Soil Biol. Biochem. 6:141–144.

Brams, E. 1965. The mucilaginous layer of citrus roots—its delineation in the rhizosphere and removal from roots. Plant Soil 30:105–108.

Brewster, J. L., and P. B. H. Tinker. 1972. Nutrient flow rates into roots. Soils Ferts. 35:355–359.

Brewster, J. L., K. K. S. Bhat, and P. H. Nye. 1975. The possibility of predicting solute uptake and plant growth response from independently measured soil and plant characteristics. Plant Soil 42:197–226.

Brown, M. 1974. Seed and root bacterization. Ann. Rev. Phytopath. 12:181–197.

Callow, J. A., L. C. M. Capaccio, G. Parish, and P. B. Tinker. 1978. Detection and estimation of polyphosphate in vesicular-arbuscular mycorrhizas. New Phytol. 80:125–134.

Cawse, P. A. 1975. Microbiology and biochemistry of irradiated soils. p. 213–267. *In* E. A. Paul and A. D. McLaren (ed.) Soil biochemistry, Vol. III. Marcel Dekker, New York.

Christie, P., E. I. Newman, and R. Campbell. 1978. The influence of neighbouring grassland plants on each others endomycorrhizas and root surface micro-organisms. Soil Biol. Biochem. 10:521–528.

Claassen, N., and S. A. Barber. 1974. A method for characterizing the relation between nutrient concentration and flux into roots of intact plants. Plant Physiol. 54:564–568.

Clark, F. E., and E. A. Paul. 1970. The microflora of grassland. Adv. Agron. 22:375–436.

Cooper, K. M., and P. B. Tinker. 1978. Translocation and transfer of nutrients in vesicular-arbuscular mycorrhizas. II. Uptake and translocation of phosphorus, zinc, and sulphur. New Phytol. 81:43–52.

Cooper, R. 1959. Bacterial fertilizers in the Soviet Union. Soils Ferts. 22:327–333.

Cox, G. C., and F. E. Sanders. 1974. Ultrastructure of the host-fungus interface in a vesicular-arbuscular mycorrhiza. New Phytol. 73:901–912.

Cox, G., and P. B. Tinker. 1976. Translocation and transfer of nutrients in VA mycorrhizas. I. The arbuscule and phosphorus transfer: a quantitative ultrastructural study. New Phytol. 77:371–378.

Cox, G. C., F. E. Sanders, P. B. Tinker, and J. Wild. 1975. Ultrastructural evidence relating to host-endophyte transfer in a vesicular-arbuscular mycorrhiza. p. 297–312. *In* F. E. Sanders, B. Mosse, and p. B. Tinker (ed.) Endomycorrhizas. Academic Press, London.

Crush, J. R. 1973. The effect of *Rhizophagus tenuis* mycorrhizas on ryegrass, cocksfoot and sweet vernal. New Phytol. 72:965–973.

Crush, J. R., and A. C. Pattison. 1975. Preliminary results on the production of vesicular-arbuscular mycorrhizal inoculum by freeze-drying. p. 485–493. *In* F. E. Sanders, B. Mosse, and P. B. Tinker (ed.) Endomycorrhizas. Academic Press, London.

Daft, M. J., and T. H. Nicolson. 1966. Effect of Endogone mycorrhiza on plant growth. New Phytol. 65:343–350.

Darbyshire, J. F., and M. P. Greaves. 1973. Bacteria and protozoa in the rhizosphere. Pestic. Sci. 4:349–360.

Drew, M. P., and P. H. Nye. 1970. The supply of nutrient ions by diffusion to plant roots in soil. III. Uptake of phosphate by roots of onion, leek and rye-grass. Plant Soil 33:545–563.

Foster, R. C., and A. D. Rovira. 1976. Ultra-structure of the wheat rhizosphere. New Phytol. 76:343–352.

Furlan, V., and J. A. Fortin. 1973. The formation of endomycorrhizae by *Endogone calospora* on *Allium cepa* under three temperature regimes. Naturaliste Canadien 100:467–477.

Gams, E. 1967. Mikro-organismen in der Wurzelregion in Weizen. Mitt. aus der biologisches Bundesanstalt fur Land und Forstwirtschaft. Berlin-Dahlem. Heft 123.

Gerdemann, J. W. 1975. Vesicular-arbuscular mycorrhizae. *In* G. D. Torrey and D. T. Clarkson (ed.) The development and function of roots. Academic Press, London.

Gerdemann, J. W., and J. M. Trappe. 1975. Taxonomy of the Endogonaceae. p. 35–52. *In* F. E. Sanders, B. Mosse, and P. B. Tinker (ed.) Endomycorrhizas. Academic Press, London.

Gerretsen, F. . 1948. The influence of micro-organisms on the phosphate intake by the plant. Plant Soil 1:51–81.

Gianninazi, S., V. Gianninazi-Pearson, and J. Dexheimer. 1979. Enzymatic studies on the metabolism of vesicular-arbuscular mycorrhiza. III. Ultrastructural localization of acid and alkaline phosphatases in onion roots infected by *Glomus mosseae*. (Nicol & Gerd), New Phytol. 82:127–132.

Graw, D. 1979. The influence of soil pH on the efficiency of vesicular-arbuscular mycorrhiza. New Phytol. 82:687–695.

Greaves, M. P., and J. F. Darbyshire. 1972. The ultrastructure of the mucilaginous layer in plant roots. Soil Biol. Biochem. 4:443–449.

Greaves, M. P., and D. M. Webley. 1969. The hydrolysis of myoinositol hexaphosphate by soil micro-organisms. Soil Biol. Biochem. 1:37–43.

Griffin, G. J., M. G. Hale, and F. J. Shay. 1976. Nature and quantity of sloughed organic matter produced by roots of axenic peanut plants. Soil Biol. Biochem. 8:29–32.

Hale, M. G., C. L. Foy, and F. J. Shay. 1971. Factors affecting root exudation. Adv. Agron. 23:89–109.

Hale, M. G., D. L. Lindsay, and K. M. Harmeed. 1973. Gnotobiotic culture of plants and related research. Bot. Rev. 39:261–273.

Hale, M. G., L. D. Morris, and G. J. Griffen. 1978. Root exudates and exudation. p. 163–203. *In* Y. R. Dommergues and S. V. Krupa (ed.) Interactions between non-pathogenic soil micro-organisms and plants. Elsevier, Amsterdam.

Harley, J. L. 1969. The biology of mycorrhiza. Leonard Hill, London.

Harley, J. L. 1975. Problems of mycotrophy. p. 1–24. *In* F. E. Sanders, B. Mosse, and P. B. Tinker (ed.) Endomycorrhizas. Academic Press, London.

Harley, J. L., and C. C. McCready. 1952. Uptake of phosphate by excised mycorrhizal roots of the beech II. New Phytol. 51:56–64.

Hattingh, M. J., and J. W. Gerdemann. 1975. Inoculation of Brazilian sour orange seed with an endomycorrhizal fungus. Phytopathology 65:1013–1016.

Hattingh, M. J., L. E. Gray, and J. W. Gerdemann. 1973. Uptake and translocation of ^{32}P labelled phosphate to onion roots by endomycorrhizal fungi. Soil Sci. 116:383–387.

Hayman, D. S. 1970. *Endogone* spore numbers in soil and vesicular-arbuscular mycorrhizae in wheat as influenced by season and treatment. Trans. Br. Mycol. Soc. 54:53–63.

Hayman, D. S. 1974. Plant growth responses to vesicular-arbuscular mycorrhiza VI. Effect of light and temperature. New Phytol. 73:71–80.

Hayman, D. S. 1975a. Phosphorus cycling by soil microorganisms and plant roots. p. 67–91. *In* N. Walker (ed.) Soil microbiology. Butterworths, London.

Hayman, D. S. 1975b. The occurrence of mycorrhiza in crops as affected by soil fertility. p. 495–509. *In* F. E. Sanders, B. Mosse, and P. B. Tinker (ed.) Endomycorrhizas. Academic Press, London.

Hayman, D. S. 1978. Endomycorrhizas. p. 401–442. *In* Y. R. Dommergues and S. V. Krupa (ed.) Interactions between non-pathogenic micro-organisms and plants. Elseview, Amsterdam.

Hayman, D. S., and B. Mosse. 1972a. The role of vesicular-arbuscular mycorrhizas in the removal of phosphorus from soil by plant roots. Rev. Ecol. Biol. Sol. 9:463–470.

Hayman, D. S., and B. Mosse. 1972b. Plant growth responses to vesicular-arbuscular mycorrhizas. III. Increased uptake of labile P from soil. New Phytol. 71:41–47.

Hayman, D. S., and B. Mosse. 1979. Improved growth of white clover in hill grassland by mycorrhizal inoculations. Ann. Appl. Biol. 93:141–148.

Henderson, G. S., and E. L. Stone. 1970. Interactions of phosphorus availability, mycorrhizae and soil fumigations on coniferous seedlings. Soil Sci. Soc. Am. Proc. 34:314–318.

Jennings, D. H., J. D. Thornton, M. F. Galpin, and C. R. Coggins. 1974. Translocation in fungi. Symp. Soc. Exp. Biol. 28:139–156.

Jenny, H., and K. Grossenbacher. 1963. Root-soil boundary zones as seen in the electron microscope. Soil Sci. Soc. Am. Proc. 27:273–277.

Johnston, W. B., and R. A. Olsen. 1972. Dissolution of fluorapatite by plant roots. Soil Sci. 114:29–36.

Kafkafi, U., A. M. Posner, and J. P. Quirk. 1967. Desorption of phosphate from kaolinite. Soil Sci. Soc. Am. Proc. 31:348–353.

Katznelson, H. 1965. Nature and importance of the rhizosphere. p. 187–209. *In* K. F. Baker and W. C. Snyder (ed.) The ecology of soil-borne plant pathogens. Murray, London.

Khan, A. G. 1972. The effect of vesicular-arbuscular mycorrhizal associations on growth of cereals. I. Effects on maize growth. New Phytol. 71:613–619.

Khan, A. G. 1975. Growth effects of VA mycorrhizae on crops in the field. p. 419–435. *In* F. E. Sanders, B. Mosse, and P. B. Tinker (ed.) Endomycorrhizas. Academic Press, London.

Kinden, D. A., and M. F. Brown. 1975. Electron microscopy of vesicular-arbuscular mycorrhizae of yellow poplar. II. Intracellular hyphae and vesicles. Can. J. Microbiol. 21:1768–1780.

Kleinschmidt, G. D., and J. W. Gerdemann. 1972. Stunting of citrus seedlings in fumigated nursery soils related to the absence of endomycorrhizae. Phytopathology 62:1447–1453.

Kouchecki, H. K., and D. J. Read. 1976. Vesicular-arbuscular mycorrhizae in natural vegetation systems. II. The relationship between infection and growth in *Festuca ovina* L. New Phytol. 77:655–666.

Larsen, S. 1967. Soil phosphorus. Adv. Agron. 19:151–210.

Larsen, S. 1973. Recycling of phosphorus in relation to long term soil reserves. Phosphorus Agric. 61:1–6.

Lategan, S., and H. A. Louw. 1972. The root region microflora of plants with reference to the utilization of rock phosphate by different plant species. Phytophylactica 4:119–126.

Lewis, D. H. 1973. Concepts in fungal nutrition and the origin of biotrophy. Biol. Rev. 48:261–278.

Lewis, D. G., and J. P. Quirk. 1967. Phosphate diffusion in soil and uptake by plants. Plant Soil 26:445–453.

Ling-Lee, M., G. A. Chilvers, and A. E. Ashford. 1975. Polyphosphate granules in three different kinds of tree mycorrhiza. New Phytol. 75:551–554.

Loneragan, J. F., and C. J. Asher. 1967. Response of plants to phosphate concentrations in solution culture. II. Rate of phosphate absorption and its relation to growth. Soil Sci. 103:311–318.

Lynch, J. M. 1976. Products of soil micro-organisms in relation to plant growth. CRC Critical Reviews in Microbiology. 5(1):67-107.
Malajczuk, N., and G. D. Bowen. 1974. Proteoid roots are microbially induced. Nature 251: 316-317.
Marks, G. C., and R. C. Foster. 1973. Structure, morphogenesis and ultra structure of ectomycorrhizae. p. 2-42. *In* Ectomycorrhizae. Academic Press, New York and London.
Marks, G. C., and T. T. Kozlowski (ed.). 1973. Ectomycorrhizae. Academic Press, New York and London.
Marx, D. H. 1977. The role of mycorrhizae in forest production. Tappi Conference Papers, Ann. Meeting Atlanta, Georgia. p. 151-161.
Marx, D. H., A. B. Hatch, and J. F. Mendicino. 1977. High soil fertility decreases sucrose content and susceptibility of loblolly pine roots to ectomycorrhizal infection by *Pisolithus tinctorius*. Can. J. Bot. 55:1569-1574.
Marx, D. H., and S. V. Krupa. 1978. Ectomycorrhizae. p. 373-400. *In* Y. R. Dommergues and S. V. Krupa (ed.) Interactions between non-pathogenic soil micro-organisms and plants. Elsevier, Amsterdam.
Marx, D. H., W. G. Morris, and J. G. Mexal. 1978. Growth and ectomycorrhizal development of loblolly pine seedlings in fumigated and nonfumigated nursery soil infested with different fungal symbionts. Forest Sci. 24:193-203.
Mengel, D. B., and S. A. Barber. 1974. Rate of nutrient uptake per unit of corn root under field conditions. Agron. J. 66:399-402.
Mejstrick, J. K., and H. H. Krause. 1973. Uptake of ^{32}P by *Pinus radiata* roots inoculated with *Suillus luteus* and *Cenococcum graniforme* from different sources of available phosphate. New Phytol. 72:137-140.
Meyer, F. H. 1973. Distribution of ectomycorrhizae in native and man-made forests. p. 79-105. *In* G. C. Marks and T. T. Kozlowski (ed.) Ectomycorrhizae. Academic Press, New York and London.
Meyer, F. 1974. Physiology of mycorrizae. Ann. Rev. Plant Physiol. 25:567-586.
Melin, E., and H. Nilsson. 1958. Translocation of nutritive elements through mycorrhizal mycelia to pine seedlings. Bot. Notis. 111:251-256.
Mikola, P. 1973. Application of mycorrhizal symbiosis in forestry practice. p. 383-412. *In* G. C. Marks and T. T. Kozlowski (ed.) Ectomycorrhizae. Academic Press, New York and London.
Moghimi, A., and M. E. Tate. 1978. Does 2-ketogluconate complex calcium in the pH range 2.4-6.4? Soil Biol. Biochem. 10:289-292.
Mosse, B. 1972. The influence of soil type and *Endogone* strain on the growth of mycorrhizal plants in phosphate-deficient soils. Rev. Ecol. Biol. Sol. 9:529-537.
Mosse, B. 1973. Advances in the study of vesicular-arbuscular mycorrhiza. Ann. Rev. Phytopath. 11:171-196.
Mosse, B. 1974a. A microbiologist's view of root anatomy. p. 39-66. *In* N. Walker (ed.) Soil microbiology. Butterworths, London.
Mosse, B. 1975b. Specificity in vesicular-arbuscular mycorrhizas. p. 469-484. *In* F. E. Sanders, B. Mosse, and P. B. Tinker (ed.) Endomycorrhizas. Academic Press, London.
Mosse, B. 1977. Plant growth response to vesicular-arbuscular mycorrhiza. X. Responses of *Stylosanthes* and maize to inoculation in unsterile soils. New Phytol. 78:277-288.
Mosse, B., and J. M. Phillips. 1971. The influence of phosphate and other nutrients on the development of vesicular-arbuscular mycorrhizae in culture. J. Gen. Microbiol. 69:157-166.
Mosse, B., C. L. Powell, and D. S. Hayman. 1977. Plant growth responses to vesicular-arbuscular mycorrhiza. IX. Interactions between VA mycorrhiza, rock phosphate and symbiotic nitrogen fixation. New Phytol. 76:331-342.
Murdoch, C. L., J. A. Jackobs, and J. W. Gerdemann. 1967. Utilization of phosphorus sources of different availability of mycorrhizal and nonmycorrhizal maize. Plant Soil 27: 329-334.
Newman, E. I., and H. J. Bowen. 1974. Patterns of distribution of bacteria on root surfaces. Soil Biol. Biochem. 6:205-209.
Newman, E. I., and A. Watson. 1977. Microbial abundance in the rhizosphere: a computer model. Plant Soil 48:17-56.
Nicolson, T. H. 1967. Vesicular-arbuscular mycorrhiza—a universal plant symbiosis. Sci. Progr. Oxf. 55:561-581.
Nye, P. H. 1966. The effect of the nutrient intensity and buffering power of a soil, and the absorbing power, size and root hairs of a root, on nutrient absorption by diffusion. Plant Soil 25:81-105.

Nye, P. H., and P. B. Tinker. 1969. The concept of a root demand coefficient. J. Appl. Ecol. 6:293-300.

Nye, P. H., and P. B. Tinker. 1977. Solute movement in the root-soil system. Blackwell, Oxford.

Oades, J. M. 1978. Mucilages at the root surface. J. Soil Sci. 29:1-16.

Olsen, S. R., and W. D. Kemper. 1968. Movement of nutrients to plant roots. Adv. Agron. 20:91-151.

Owusu Bennoah, E., and B. Mosse. 1979. Plant growth responses to vesicular-arbuscular mycorrhiza. XI. Field inoculation responses in barley, lucerne and onion. New Phytol. 83:671-679.

Parkinson, D. 1967. Soil microorganisms and plant roots. p. 449-473. *In* A. Burges and F. Raw (ed.) Soil biology. Academic Press, London.

Passioura, J. B. 1963. A mathematical model for the uptake of ions from the soil solution. Plant Soil 18:225-238.

Pearson, V., and P. B. Tinker. 1975. Measurement of phosphorus fluxes in the external hyphae of endomycorrhizas. p. 277-287. *In* F. E. Sanders, B. Mosse, and P. B. Tinker (ed.) Endomycorrhizas. Academic Press, London.

Powell, C. Ll. 1979. Inoculation of white clover and ryegrass seed with mycorrhizal fungi. New Phytol. 83:81-85.

Powell, C. Ll., and J. Daniel. 1978. Growth of white clover in undisturbed soils after inoculation with efficient mycorrhizal fungi. New Zealand J. Agric. Res. 21:675-681.

Powell, C. W. 1975. Plant growth responses to vesicular-arbuscular mycorrhizas. VIII. Uptake of phosphorus by onion and clover infected with different *Endogone* spore types. New Phytol. 75:563-566.

Powell, C. 1976. Mycorrhizal fungi stimulate clover growth in New Zealand hill country soils. Nature (London) 264:436-438.

Powlson, D. S. 1975. Effects of biocidal treatments on soil organisms. p. 193-224. *In* N. Walker (ed.) Soil microbiology. Butterworths, London.

Ramamoorthy, S., and P. G. Manning. 1974. Inorganic phosphate and the uptake of mineral nutrients by plants. Inorg. Nucl. Chem. Letters 10:623-628.

Read, D. J., and D. P. Stribley, 1975. Some mycological aspects of the biology of mycorrhiza in the Ericaceae. p. 105-117. *In* F. E. Sanders, B. Mosse, and P. B. Tinker (ed.) Endomycorrhizas. Academic Press, London.

Reid, C. P. P., and G. D. Bowen. 1979. Effect of water stress on phosphorus uptake by mycorrhizae of *Pinus radiata*. New Phytol. 83:103-107.

Reynolds, E. C. 1975. Tree rootlets and their distribution. p. 163-178. *In* J. G. Torrey and D. R. Clarkson (ed.) The development and function of roots. Academic Press, London.

Rhodes, L. H., and J. W. Gerdemann. 1975. Phosphate uptake zones of mycorrhizal and nonmycorrhizal onions. New Phytol. 75:555-561.

Rich, J. R., and G. W. Bird. 1974. Association of early season vesicular-arbuscular mycorrhizae with increased growth and development of cotton. Phytopathology 64:1421-1425.

Rogers, H. T., R. W. Pearson, and W. H. Pierre. 1940. The absorption of organic phosphorus by corn and tomato plants and the mineralizing action of exo-enzyme systems of growing roots. Soil Sci. Soc. Am. Proc. 5:285-291.

Ross, J. P., and J. W. Gilliam. 1973. Effect of *Endogone* mycorrhiza on phosphorus uptake by soybeans from inorganic phosphates. Soil Sci. Soc. Am. Proc. 37:237-239.

Ross, J. P., and J. A. Harper. 1970. Effect of *Endogone* mycorrhiza on soybean yields. Phytopathology 60:1552-1556.

Rovira, A. D. 1962. Plant root exudates in relation to the rhizosphere microflora. Soils Ferts. 25:167-172.

Rovira, A. D. 1979. Biology of the soil root interface. p. 145-160. *In* J. L. Harley and R. Scott Russell (ed.) The soil-root interface. Academic Press, London.

Rovira, A. D., and C. B. Davey. 1974. Biology of the rhizosphere. p. 153-204. *In* E. W. Carson (ed.) The plant root and its environment. Univ. Press of Virginia, Charlotteville.

Rovira, A. D., R. C. Foster, and J. K. Martin. 1979. Note on terminology: Origin, nature and nomenclature of the organic materials in the rhizosphere. p. 1-4. *In* J. L. Harley and R. Scott Russell (ed.) The soil-root interface. Academic Press, London.

Rovira, A. D., and B. M. McDougall. 1967. Microbiological and biochemical aspects of the rhizosphere. p. 418-463. *In* A. D. McLaren, and G. H. Peterson (ed.) Soil biochemistry. Edward Arnold, London.

Rovira, A. D., E. I. Newman, H. J. Bowen, and R. Campbell. 1974. Quantitative assessment of the rhizoplane microflora by direct microscopy. Soil Biol. Biochem. 6:211-216.

Sanders, F. E. 1975. The effect of foliar-applied phosphate on the mycorrhizal infections of onion roots. p. 261-276. *In* F. E. Sanders, B. Mosse, and P. B. Tinker (ed.) Endomycorrhizas. Academic Press, London.

Sanders, F. E., and P. B. Tinker. 1971. Mechanism of absorption of phosphate from soil by *Endogone* mycorrhizas. Nature (London) 233:278-279.

Sanders, F. E., and P. B. Tinker. 1973. Phosphate flow into mycorrhizal roots. Pestic. Sci. 4: 385-395.

Sanders, F. E., B. Mosse, and P. B. Tinker (ed.). 1975. Endomycorrhizas. Academic Press, London.

Sanders, F. E., P. B. Tinker, R. L. Black, and S. Palmerley. 1977. The development of endomycorrhizal root systems: I. Spread of infection and growth-promoting effects with four species of vesicular-arbuscular endophyte. New Phytol. 78:257-268.

Sauerbeck, D., and B. G. Johnen. 1976. Der Umsatz von Pflanzenwurzeln im Laufe der Vegetationsperiode und dessen Beitrag zur "Bodenatmung". Z. Pflanzenernahr. Bodenk. p. 315-328.

Shamoot, S., I. McDonald, and W. V. Bartholomew. 1968. Rhizo-deposition of organic debris in soil. Soil Sci. Soc. Am. Proc. 32:817-820.

Shemakhanova, N. M. 1962. Mycotrophy of woody plants. Acad. of Sciences of the USSR Inst. of Microbiology (Israel Program for Scientific Translation, 1967).

Skinner, M. F., and G. D. Bowen. 1974. The uptake and translocation of phosphate by mycelial strands of pine mycorrhizas. Soil Biol. Biochem. 6:53-56.

Slankis, V. 1973. Hormonal relationships in mycorrhizal development. p. 231-298. *In* G. C. Marks and T. T. Kozlowski (ed.) Ectomycorrhizas. Academic Press, London.

Smith, J. H., F. E. Allison, and D. A. Soulides. 1962. Phosphobacterin as a soil inoculant. USDA Tech. Bull. 1263.

Smith, S. E. 1974. Mycorrhorizal fungi. p. 273-313. *In* CRC critical reviews in microbiology. CRC Press, Cleveland, Ohio.

Smith, W. H. 1970. Root exudates of seedling and mature sugar maple. Phytopathology 60: 701-703.

Sparling, G. P., and P. B. Tinker. 1978a. Mycorrhizal infection in Pennine grassland. II. Effects of mycorrhizal infection on the growth of some upland grasses in γ-irradiated soils. J. Appl. Ecol. 15:951-958.

Sparling, G. P., and P. B. Tinker. 1978b. Mycorrhizal infection in Pennine grassland. III. Effects of mycorrhizal infections on the growth of white clover. J. Appl. Ecol. 15:959-964.

Stribley, D. P., P. B. Tinker, and R. C. Snellgrove. 1980. Effects of vesicular-arbuscular mycorrhizal fungi on the relations of plant growth, internal phosphorus concentration and soil phosphate analyses. J. Soil Sci. (in Press).

Swaby, R. J. 1962. Effect of micro-organisms on nutrient availability. Int. Soc. Soil Sci. Trans. (New Zealand) Comm. IV and V:158-172.

Swaminathan, K., and B. C. Verma. 1977. Symbiotic effect of the vesicular-arbuscular mycorrhizal fungi on the phosphate nutrition of potatoes. Proc. Indian Acad. Sci. 85B: 310-318.

Theodorou, C., and G. D. Bowen. 1970. Mycorrhizal responses of radiata pine in experiments with different fungi. Aust. For. 34:183-191.

Theodorou, C., and G. D. Bowen. 1973. Inoculation of seeds and soil with basidiospores of mycorrhizal fungi. Soil Biol. Biochem. 5:765-771.

Tinker, P. B. 1975a. Effects of vesicular-arbuscular mycorrhizas on higher plants. p. 325-349. *In* D. G. Jennings and D. L. Lee (ed.) Symbiosis. Symp. Soc. Exp. Biol. 29.

Tinker, P. B. 1975b. The soil chemistry of phosphorus and mycorrhizal effects on plant growth. p. 353-371. *In* F. E. Sanders, B. Mosse, and P. B. Tinker (ed.) Endomycorrhizas. Academic Press, London.

Tinker, P. B. 1978. Effects of vesicular-arbuscular mycorrhizas on plant nutrition and plant growth. Physiol. Veg. 16:743-751.

Tinker, P. B., and F. E. Sanders. 1975. Rhizosphere micro-organisms and plant nutrition. Soil Sci. 119:363-368.

Warcup, J. H. 1975. A culturable Endogone associated with eucalypts. p. 53-63. *In* F. E. Sanders, B. Mosse, and P. B. Tinker (ed.) Endomycorrhizas. Academic Press, London.

White, R. E. 1980. Retention and release of phosphate by soil and soil constituents. p. 000-000. *In* P. B. Tinker (ed.) Soils and agriculture: Critical reviews. Society of Chemistry and Industry, London.

White, R. E., and A. W. Taylor. 1977. Reactions of soluble phosphate with acid soils: the interpretation of absorption-desorption isotherms. J. Soil Sci. 28:314-328.

Williamson, F. A., and R. G. Wyn Jones. 1973. The influence of soil micro-organisms on growth of cereal seedlings and on potassium uptake. Soil Biol. Biochem. 5:569-575.

Winter, A. G. 1952. Zum Problem der Mykorrhiza bei landwirtschaftlichen Kulturpflanzen. Zt. f. Pflanzenernahr. Dung. Bodenk. 60:221-243.

Woolhouse, H. W. 1975. Membrane structure and transport problems considered in relation to phosphorus and carbohydrate movements and the regulation of endotrophic mycorrhizal associations. p. 209-239. *In* F. E. Sanders, B. Mosse, and P. B. Tinker (ed.) Endomycorrhizas. Academic Press, London.

Yost, R. S., and R. L. Fox. 1979. Contribution of mycorrhizae to P nutrition of crops growing on an Oxisol. Agron. J. 71:903-908.

Chapter 23

Interactions of Phosphorus with Other Elements in Soils and in Plants

FRED ADAMS

Auburn University
Auburn, Alabama

I. INTRODUCTION

Phosphate ions are involved in major chemical reactions in soils and in numerous metabolic reactions in plants. Consequently, P influences, or is influenced by, the availability or utilization of many other elements, both essential and nonessential. Such influences of P on other elements and other elements on P are called interactions. In defining "interaction," Russell (1973, p. 55) states:

> ... if two factors are limiting, or nearly limiting growth, adding only one of them will have little effect on growth, whilst adding both together will have a very considerable effect. Two such factors are said to have a large positive interaction in such circumstances, for the response of the crop to both together is larger than the sum of responses to each separately. If the crop response to the two factors together equalled the sum of its responses to each separately, we would say the two factors showed no interaction, or worked entirely independently of each other; and if the response to the two factors together was less than the sum of the responses to each factor separately, they are said to have a negative interaction with each other.

Under this definition, interactions may occur in the soil, in the plant, or in both. No specific mechanisms are implied in this definition.

There are numerous examples in the literature in which a response curve to applied P is obtained at low levels of another nutrient, while entirely different P response curves are obtained at higher levels of the other nutrient. Likewise, response curves to other nutrients change as soil P is changed from "deficient" to "sufficient" levels. While this may constitute "interaction" under Russell's definition if the effects deviate from additivity, it will not be treated in further detail in this review.

Interactions between P and other elements in the soil are the manifestations of specific chemical reactions, few of which have been quantitatively defined. Qualitatively, however, these reactions have found useful expressions in terms of nutrient availability and the efficiency of P fertilizers. Ef-

Copyright 1980 © ASA-CSSA-SSSA, 677 South Segoe Road, Madison, WI 53711, USA.
The Role of Phosphorus in Agriculture.

ficiency of P fertilizers has long been associated with chemical reactions involving phosphate and certain cations, especially Ca, Al, and Fe. Normal concentrations of these cations assures soil-solution P concentrations of less than 1 ppm in most soils (Adams, 1974).

Phosphorus plays a key role in many metabolic pathways within the plant. These various metabolic reactions are dependent on numerous other ions, either as activators of enzyme systems or as vehicles for electron transfers. The interaction of P with such ions within the plant is of importance under two separate conditions: (i) when adequate or excess levels of P affect the utilization of other ions by the plant and (ii) when adequate or excess levels of other inorganic nutrients affect the utilization of P.

Most mechanisms responsible for interactions between P and other elements in soils are reasonably well delineated, whereas those within plants are not. This is because the chemistry of soil P is simpler and better defined than the numerous metabolic pathways involving P in growing plants. In the discussion that follows, each element is examined separately for likely interactions with P, first within soils and then within plants.

II. NITROGEN

One of the more obvious interactions between N and P in soils is the coprecipitation of ammonium and phosphate when these two ions are applied in compound or mixed fertilizers. About 25 NH_4-P compounds containing either Al, Ca, Fe, or Mg have been suggested as possible soil-fertilizer reaction products (Lindsay et al., 1962; Lehr et al., 1967). It is difficult to generalize on the effects of these interactions on plant nutrition because the compounds vary from highly available to very slowly available as sources of N and P to plants (Lehr et al., 1967).

Some apparent interactions between N and P in soils are probably the result of other reactions, unrelated to N-P reactions per se. For example, there is considerable evidence that nitrification causes increased solubility of P compounds in alkaline soils and is responsible for the increased availability of P when NH_4 fertilizers are used. Chapman's early experiment (1936) with calcareous soils showed NH_4 salts to be superior to NO_3 salts, but inferior to HNO_3, in increasing soil P availability. Almost two decades later, Lorenz and Johnson (1953) reported similar findings for an alkaline, P-deficient soil, and concluded that "physiologically acid" fertilizers were responsible for the increased soil-P availability.

The dissolving action of concentrated fertilizers in fertilizer bands can also be expected to influence fertilizer-P reaction products and, consequently, their relative availability (Khasawneh et al., 1974; Blanchar & Caldwell, 1966a, 1966b). In fact, Bouldin and Sample (1958, 1959) concluded that the primary effect of intimate association of banded N and P fertilizers on P uptake was the result of increased P solubility.

The importance of N-P "balance" in fertilizers was demonstrated early when an application of $Ca(NO_3)_2$ was used to correct "P toxicity" in soybeans (*Glycine max* L.) (Shive, 1918). This adverse effect of P fertilizers on

N metabolism of legumes has received little attention since then (Rossiter, 1952, 1955), probably because it is not a widespread field problem. Instead, interest has focused on N-P interactions of fertilized nonlegumes.

Increased P absorption by plants is a common consequence of adding N fertilizers. In reviewing the effect of N on P uptake by plants, Grunes (1959) summarized the various explanations for this phenomenon: increased top growth, increased root growth, altered metabolism, and increased solubility of soil P. There is evidence to support each.

Yields of most nonlegumes are increased by N fertilizers, usually resulting in expected increases in demand for inorganic nutrients, e.g., P. The effect of N on P uptake in such cases may be best explained by physiological stimulation occurring within the plant as a consequence of the greater N supply (Bennett et al., 1962; Grunes et al., 1958; Simpson, 1961; Soltanpour, 1969).

Ohlrogge and coworkers at Purdue University in the 1950's stimulated much interest when they demonstrated the synergistic effect of conjoint placement of N and P fertilizers on root growth (Miller & Ohlrogge, 1958; Olson & Dreier, 1956; Robertson et al., 1954). A greater root mass generally resulted when N and P were banded together, and is believed to be responsible for the increased P uptake by the crop (Miller & Vij, 1962). Significantly, NH_4 fertilizers have a greater stimulating effect on P absorption than do nitrates.

Short-term absorption experiments suggest that the ammonium ion per se is not responsible for an enhanced P uptake; rather, it is the increase in N metabolites in the roots (Cole et al., 1963; Humble et al., 1969; Taber & McFee, 1972, 1974; Thien & McFee, 1970). In general, pretreating root systems with N increases subsequent P uptake and translocation. Ammonium and nitrate sources are equal in this regard, but both are without effect when present in the root medium along with the P.

Engelstad and Allen (1971) found that root yields and root P contents were increased by N fertilizer, with ammonium being superior to nitrate; they also found a lower N recovery from the nitrate fertilizer. Nitrate's inferiority in this case may have been caused by denitrification and nitrite toxicity, as evidenced by the somewhat stubby, brown-tipped roots.

In spite of the abundance of data supporting N-P interactions within plants, the metabolic pathways actually involved remain essentially unidentified. Present knowledge barely extends beyond the fact that increased tissue growth requires more of both N and P.

III. CALCIUM

A. Calcium-Phosphorus Interactions in Soil

Concepts about Ca-P interactions in soils have hardly changed since the excellent, comprehensive reviews by Kurtz (1953) and Olsen (1953). The general agreement that Ca and P react to form a solid-phase component has not been elucidated to the extent that we have an acceptable definition of

Ca-P relationships in soils that simultaneously encompasses all likely mechanisms. This is evidenced by the continued general use of such ambiguous expressions as "P fixation" and "P sorption."

Three basic experimental approaches have been used to study Ca-P interactions in soils: (i) fractionation of P compounds or components by extraction procedures, (ii) adsorption isotherms, and (iii) solubility of discrete compounds. With basically different objectives being pursued, different models being used, and chemical properties of systems being only partially defined, it is small wonder that little progress has been made in the last 25 years.

An extraction procedure by Dean (1938) to measure the amount of Ca phosphates in soil was modified by Chang and Jackson (1957) to include Al and Fe phosphates. Unfortunately, neither procedure measured what it was intended to measure. Adsorption isotherms for "P sorption" changed progressively from Freundlich's (Davis, 1935) to Langmuir's (Olsen & Watanabe, 1957) to multiple Langmuir's (Syers et al., 1973) without significantly altering the concept of Ca-P interaction. The view that Ca-P interactions in soils resulted from precipitation and dissolution of discrete compounds (Buehrer, 1932) was revived by the work of Aslyng (1954). The solubility approach has been hampered because data have been interpreted: (i) without a complete definition of the system, (ii) without adequate recognition of natural crystal defects and isomorphous substitution, and (iii) without adequate recognition of the dependence of solubility on exposed surface area and actual precipitate composition.

There is no argument that increasing the concentration of solution Ca decreases P solubility (Barrow, 1972; Lehr & van Wesemael, 1952; Mattson et al., 1951), irrespective of reaction mechanism. Conversely, increasing solution P by adding ammonium-P fertilizer lowers the concentration of soil-solution Ca (Adams, 1966; Bennett & Adams, 1970).

A concept encompassing the diverse experimental data on Ca-P interactions in soils remains elusive. However, a step toward possible unification appears in the work of Griffin and Jurinak (1973) in which Ca phosphate adsorbed on calcite surfaces fits the Langmuir isotherm, the BET equation, and the solubility products of a basic Ca phosphate. Significantly, they concluded that P was not present as an adsorbed monolayer on the calcite surface, but as adsorbed ions plus a nucleated crystalline species.

B. Calcium-Phosphorus Interactions in Plants

Considerable interest has developed during the last two decades concerning the interaction of Ca and P as plant nutrients. Methods for elucidating this interaction can be grouped into three experimental approaches: (i) short-term absorption periods with excised roots, (ii) short-term absorption periods by mitochondria, and (iii) absorption by intact root systems and translocation from the roots.

Using excised roots, Tanada (1955) showed that 1 mM Ca(NO$_3$)$_2$ greatly enhanced the absorption of P from a 0.1 mM KH$_2$PO$_4$ solution during a

10-min absorption period. Leggett et al. (1965) subsequently demonstrated that Ca-stimulated P uptake by excised roots actually continued for several hours in nutrient solutions. Using three different plant species, Franklin (1969) pretreated excised roots for 1 min with various dilute chloride salts before subjecting them to 0.02 mM KH$_2$PO$_4$ solution for different time intervals. He found that the valency of the pretreatment cation markedly influenced P uptake, the relative effectiveness being K$^+$ = NH$_4^+$ < Mg^{2+} = Ca^{2+} < Al^{3+} = Fe^{3+}.

A valid criticism of many short-term studies with excised roots, including those cited above, is the fact that the relative stimulatory effect of Ca was measured with control plants in solutions with Ca levels too low for plant growth or for membrane stability. Therefore, the reported Ca effects on P absorption based on such experiments should be validated by more rigorous conditions before they are accepted as conclusive.

It has been suggested that mitochondrial reactions control the uptake of P by roots (Jackson et al., 1962; Miller et al., 1972). The accumulation of Ca and P ions within mitochondria is believed to be mutually dependent (Elzam & Hodges, 1968; Hanson & Miller, 1967). It is further suggested that the coupled Ca and P accumulation occurs because Ca diverts the P from a phosphorylated high energy intermediate that would otherwise couple with ADP to produce ATP (Miller et al., 1972). Earnshaw and Hanson (1973) showed that a "postenergized" addition of P prevented the release of organic Ca, which follows the cessation of mitochondrial respiration, and it also led to an increased P uptake. It has been further demonstrated that Ca and P can accumulate inside mitochondria sufficiently to form deposits of inorganic Ca phosphates (Elzam & Hodges, 1968; Peverly et al., 1974), possibly as hydroxyapatite. Unfortunately, the mitochondrial experiments are generally subject to the same criticism as most short-term, excised-root experiments, namely, a Ca-deficient system.

Miller et al. (1972) claimed qualitative similarities of P uptake by corn roots (*Zea mays* L.) and isolated corn mitochondria. In an experiment with the apical 2 mm of intact corn roots (where appreciable ^{32}P translocation does not occur), they found mutual Ca and P synergism for absorption. In an effort to identify causal factors, roots treated with an inhibitor of ATP synthesis (oligomycin) continued to show Ca-stimulated P uptake, although at a lower P-uptake rate; roots treated with an inhibitor of "high energy intermediate formation" (dinitrophenol) failed to exhibit Ca-stimulated P uptake; and roots treated with an inhibitor of mitochondrial respiration (NaN$_3$) continued to show Ca-stimulated P uptake, although at a lower P-uptake rate. They concluded that Ca-induced P uptake is a response "to a Ca-stimulated P transport that may occur at mitochondrial membranes."

Experiments with intact root systems have given results substantially like those with excised roots. Greenwood and Hallsworth (1960) found no direct effect of P on Ca uptake (according to plant composition data), but found that earlier and more severe symptoms of Ca deficiency occurred where P levels were high. They concluded that the increased stress on Ca-deficient plants at high P levels was not the effect of a lowered Ca uptake. Bar-Yosef (1971), in fact, showed that higher P concentrations in solutions increased Ca flux into roots. Conversely, Edwards (1968) and Hyde (1966) showed that higher Ca concentrations in the nutrient solution caused a marked increase in P uptake.

The significance of the Ca-stimulated P-uptake phenomenon to plants growing in the field is still uncertain. In an effort to explain the observation that *Medicago* sp. are generally more sensitive to soil acidity and more tolerant of soil alkalinity than some *Trifolium* sp., Robson et al. (1970) measured Ca and P uptake by these species in a series of culture-solution experiments. Using Ca and P concentrations within ranges commonly found in soil solutions and absorption periods up to 10 days, they found that increasing Ca concentration in solution increased P absorption in all cases, with the increase being much greater with *Medicago* than with *Trifolium* and also relatively greater at 0.2 μM than at 5 μM P concentrations. Thus, the effect of Ca on P uptake would appear to be more important at deficient or near deficient P levels than at luxury P levels. The authors further showed that a Ca pretreatment of the root systems failed to stimulate P uptake, leading to the conclusion that the process of Ca absorption directly affects P uptake.

The mode of action of Ca in stimulating P absorption remains unresolved. One proposed mechanism claims that Ca increases the transport rate of P because of its effect on P "carriers." The other favored mechanism is a screening action of Ca (or other polyvalent cation) of electronegative sites, resulting in greater accessibility to absorption sites by $H_2PO_4^-$ ions.

IV. MAGNESIUM

The likelihood of Mg-P interactions in soils appears remote except under conditions that would favor $MgNH_4PO_4$ precipitation in or near fertilizer bands of NH_4 phosphate. In contrast, Mg has been ascribed the function of P "carrier" in plants (Truog et al., 1947), based on a positive correlation between the Mg and P contents of plants or between fertilizer P efficiency and the supply of available Mg.

The role of Mg as an activator of kinase enzyme systems has given it special importance with P metabolism because Mg activates practically all reactions involving phosphate transfers (Jackson et al., 1967). However, no quantitative assessment has been made as to how this relationship might affect absorption of Mg and P by plants. It has been shown by short-term absorption experiments with intact root systems (Edwards, 1968) and with excised roots (Franklin, 1969) that P uptake is increased in the presence of Mg.

V. POTASSIUM

Potassium has been shown to coprecipitate with P and a variety of associated cations when soluble phosphatic fertilizers are applied to soils (Cole & Jackson, 1950; Haseman et al., 1950; Lindsay et al., 1962).

Participation of K in P precipitation is more pronounced in soils with high exchangeable K or with easily decomposed K-bearing minerals. Potassium is probably a universal coprecipitant with P when mixed fertilizers containing both K and P are applied to soils. Lehr et al. (1967) list 12 K-P compounds, containing either Al, Ca, Mg, or Fe, that have been implicated as possible soil-fertilizer reaction products.

Although efforts to identify K-P interactions in plants have been reported (Adriano et al., 1971; Edwards, 1968; Franklin, 1969; Hyde, 1966; Rothstein, 1955; Shere & Jacobson, 1970; Ward et al., 1963), there is very little evidence to support the concept of interactions within the plant. If K-P interactions do play a role within the plant, it would appear to be as part of the cation-anion balance system in which organic acids play a significant role (Banwart & Pierre, 1975; Follett & Reichman, 1973; Watanabe et al., 1971). However, too little progress has been made in this area to propose viable interactions at this time.

VI. ALUMINUM

A. Aluminum-Phosphorus Interactions in Soils

The importance of Al-P interactions in acid soils has been recognized since the 1920's, when it was found that superphosphate fertilizer would alleviate symptoms of Al toxicity. The strong affinity between aqueous Al^{3+} and $H_2PO_4^-$ ions to form insoluble compounds confines their coexistence in soil solution to very low concentrations. The Al oxides and alumino-silicates of acid soils assure a suitable surface and/or supply of aqueous Al^{3+} ions to rapidly convert soluble P fertilizers into an insoluble form. There is no disagreement among researchers about the existence of this phenomenon, but there is about the mechanisms responsible.

Since Olsen and Watanabe (1957) empirically fitted the amount of soluble P that was converted to insoluble P to the Langmuir adsorption isotherm, the adsorption mechanism has received much favored attention. However, this disregards the earlier experiments of Cole and Jackson (1950) and Kittrick and Jackson (1955) in which they reported both the precipitation of variscite-type crystals from solution and the formation, within a few minutes, of variscite crystals as a separate phase on the surface of a synthetic Al oxide. Haseman et al. (1950) reacted electrodialyzed clays (Al-saturated) with phosphate solutions of K, NH_4, Na, and Mg and identified palmerite crystals (ideal formula: $HK_2Al_2(PO_4)_3 \cdot 7H_2O$), with Na and NH_4 freely substituting for K, and variscite (ideal formula: $AlPO_4 \cdot 2H_2O$), with Fe substituting for Al to yield barrandite, $(Al, Fe) PO_4 \cdot 2H_2O$. In addition, Lehr et al. (1967) compiled a list of other Al phosphates that vary in Al and P contents and in H, K, NH_4, Fe, Ca, and Mg contents. The ease with which some of these compounds form makes them likely candidates for reaction products between soluble P fertilizers and soil-supplied Al. It is not likely that all soil-precipitated Al phosphate compounds will have identical composition and crystallinity and will exhibit a single, invariant, thermodynamic solubility product. Yet, this criterion has been used to reject stoichiometric precipitation in favor of adsorption reactions (Hsu & Rennie, 1962; Kurtz, 1953).

A precipitation reaction between soluble P and Al occurs when their soil solution activities exceed a certain solubility product. The amount of soluble P immediately removed by Al, however, is a function of exchangeable Al because of the rapid rate of cation-exchange reactions in soils. In a study of 60 subsoils in North Carolina, Coleman et al. (1960) found a high correlation ($r = 0.84$) between "sorbed" P and exchangeable Al and concluded that the P and Al formed a variscite-type mineral.

B. Aluminum-Phosphorus Interactions in Plants

The propensity of Al and P to coprecipitate from soil solution and the drastic effect of Al toxicity on the absorbing root systems complicate the rationalization of Al-P interactions within plants. Soil solutions containing toxic concentrations of Al often contain sufficient P to allow adequate plant growth were the Al absent. Thus, the Al-P interaction in plant nutrition involves more than mere external precipitation of Al phosphates. It is not uncommon for plants to exhibit the characteristically stubby, gnarled root system of Al toxicity concomitantly with typical P-deficiency symptoms of aerial plant parts (Foy, 1974).

Aluminum toxicity is often manifested as a P deficiency (Foy & Brown, 1963, 1964; Lowe & Bortner, 1973; MacLean & Chiasson, 1966). Conversely, P is an effective agent for detoxifying excess Al (Wright, 1937, 1943). The proposed mechanisms to explain the Al-P interaction within plants fall into broad categories: (i) coreactions of internal adsorption or precipitation of Al and P and (ii) Al interference with normal P metabolism.

The precipitation-adsorption mechanism was suggested by Wright (1937, 1943) after observing excess P in Al-affected roots and a deficiency of P in meristematic tissue of barley (*Hordeum vulgare* L.). Similar results were obtained more recently by MacLean and Chiasson (1966), also with barley, when they noted that Al failed to suppress P uptake but did restrict P translocation from the root system, resulting in P-deficiency symptoms. Direct evidence of Al phosphate precipitation is offered by the electron microscopic technique of McCormick and Borden (1974) in which they identified "scattered globules" of Al phosphate within the mucilaginous layer along the surface of barley roots and in the intercellular regions of the root tip.

The adsorbed-precipitated Al phosphate appears to concentrate at the root surface, extending back a few millimeters from the root tip (McCormick & Borden, 1972). Within the root tissue itself, the Al phosphate seems to be confined to the "free space" (Rorison, 1965) and associated with cell walls outside the cytoplasmic membrane (Clarkson, 1966, 1967; McCormick & Borden, 1972).

Not all absorbed Al becomes adsorbed-precipitated at or near the root surface. Differential tolerances to Al toxicity, as well as differential Al accumulation in plants, suggest that organic acids act as chelating agents to prevent the precipitation of Al at physiological pH values (Jones, 1961). Such chelation would allow Al to move into cells, and possibly into the mitochondria, and allow Al to interfere directly with P metabolism.

Rorison (1965) studied ^{32}P uptake and metabolism by Al-treated excised roots and found that Al reduced P esterification to less than half of that in the absence of Al. He also found the ratio of nucleotide to hexosephosphate to be reduced significantly. The experiments of Randall and Vose (1963) with growth inhibitors led them to the conclusion that an Al-induced increase in P uptake was the result of some metabolic process, probably involving a cytochrome. However, Clarkson (1966) and Cheong and Chan (1972) found the higher P in Al-treated barley roots and sugarcane roots (*Saccharum officinarum* L.), respectively, to be present as inorganic P. Sugarcane roots showed no effect of Al on amount of P incorporated into phosphorylated compounds or on the size of the ATP pool. Barley roots, on the other hand, exhibited a sharp reduction in phosphorylated sugars and a larger pool of ATP when treated with Al. This led Clarkson (1966) to conclude that Al prevented the utilization of ATP in the glucose phosphorylation reaction, probably by inhibiting the enzyme hexokinase.

The reviews of Clarkson (1969) and Foy (1974) on Al toxicity emphasize the genetic variability of resistance to Al toxicity. Because the mechanism of Al toxicity and the nature of resistance to it are probably intimately associated with observed Al-P interactions in plants, the resolution of Al toxicity can be expected to precede a complete rationalization of Al-P interactions within plants.

Short-term absorption experiments with ^{32}P on Al-treated roots are frequently conducted in Ca-free solutions. Since such solutions tend to have a deleterious effect on membrane integrity, the observed effects of Al on P metabolism must be interpreted cautiously.

Solution experiments seeking to elucidate Al-P interactions within plants must be carefully examined. Frequently, solution levels claimed for Al and P are not thermodynamically possible because of the sparingly soluble nature of Al phosphates and Al hydroxides. There seems to have been little effort by several researchers to maintain constant Al, P, and pH levels in root media. In contrast, soil solutions by their very nature meet solubility criteria and tend to maintain solution composition. Probably related to the above criticism is the fact that claims for Al-stimulated growth and P uptake are strangely limited to solution experiments. Soils are most productive in the absence of measurable Al concentrations in the soil solution.

VII. IRON

Truog (1953) believed that hydrated Fe oxides were the principal substances in acid soils that combined with soluble phosphates and rendered them insoluble. Although current concepts may not assign Fe such a dominant role in determining P reaction products in soils, the affinity between Fe^{3+} and $H_2PO_4^-$ ions is too great for them to coexist in soil solution at more than trace levels. One view holds that $H_2PO_4^-$ is adsorbed by the Fe oxide surface; a second view believes the $H_2PO_4^-$ and Fe^{3+} to be stoichiometrically precipitated. Although natural minerals are rarely pure compounds, advocates of an adsorption reaction discount a precipitation reaction because an Fe-P compound of constant composition and solubility is seldom identified. The adsorption mechanism also implies that Fe oxides have negligible solubility (Kurtz, 1953). However, the similarity of Fe oxide and Fe phosphate surfaces provides a suitable environment for the precipitation of one compound on the surface of the other.

Since Fe oxides are major soil constituents and Fe phosphates are minor, it is not surprising that soil Fe oxides are often contaminated with P. The so-called "reductant-soluble" P method is intended to release this combined P. Bauwin and Tyner (1957) found powder diffraction patterns for geothite, but not for strengite and concluded that the reductant-soluble P in geothite nodules was "chemisorbed at random." Practically, of course, surface precipitation and chemisorption are not distinguishable.

The reaction of a dilute P solution with Fe oxides is dependent upon chemisorption of the P or the slight solubility of the oxide to provide Fe^{3+} ions for Fe phosphate precipitation. On the other hand, concentrated bands or pellets of P fertilizer may be powerful dissolving agents of Fe oxide (Lindsay et al., 1962). Similarly, the strong acidity developed by oxidation of exposed sulfide-containing soil material dissolves significant amounts of Fe oxides. Any reaction that increases the Fe^{3+} activity of Fe oxide will enhance the precipitation between Fe^{3+} and $H_2PO_4^-$, resulting in soil P of very low availability. A subsequent coating of Fe-oxide precipitate would effectively occlude P from the dissolving action of the soil solution. The Fe phosphates, whether precipitated, adsorbed, or occluded, are not generally adequate sources of P for agricultural crops.

Although reports persist of no measurable Fe-P interaction within plants (Adriano et al., 1971; Bingham, 1963; Bingham et al., 1958), there is overwhelming evidence to support the concept of Fe-P interaction and to conclude that it is genetically regulated. The interaction also appears to function at three separate loci: the absorption process, the translocation process, and the assimilation process.

When P is absorbed in excess of the plant tissues' "needs," the metabolic use of Fe and other ions is disturbed. In the words of Biddulph and Woodbridge (1952), "The passage of such ions through tissues rich in P is interfered with . . ." The interference of P with normal Fe metabolism is manifested as Fe chlorosis. However, chlorotic leaves often contain as much or more Fe than green leaves, and there may be no correlation between Fe and chlorophyll contents of leaves (Wallace & Lunt, 1960). Since the nature of Fe chlorosis is poorly understood, it is not surprising that much uncertainty accompanies the mechanistic explanations of P-induced Fe deficiency.

The Fe-P interaction has been explained as a precipitation reaction of Fe^{3+} phosphate externally at the root's surface (Rediski &Biddulph, 1953) or internally at vein-mesophyll junctions (Biddulph, 1953). The observation that P-induced Fe deficiency is aggravated by increasing pH (Biddulph,

1953) may be partially explained by the relative solubilities of $FePO_4 \cdot 2H_2O$ and $Fe(OH)_3$. Where these two solids coexist in equilibrium with solution, their solubility products may be written as

$$k_1 = (Fe^{3+})(OH^-)^2(H_2PO_4^-)$$

and

$$k_2 = (Fe^{3+})(OH^-)^3.$$

Dividing k_1 by k_2 yields

$$k_1/k_2 = (H_2PO_4^-)/(OH^-)$$

or

$$\log(H_2PO_4^-) = pH + \text{constant}.$$

Thus, solution $(H_2PO_4^-)$ increases with increasing pH in a $Fe(OH)_3$–$FePO_4 \cdot 2H_2O$ system.

The ineffectiveness of Fe salts (Fe^{2+} is rapidly oxidized to Fe^{3+}) as Fe sources for plants in nutrient solutions has long existed as a clue that Fe^{3+} reduction to Fe^{2+} was an integral part of the Fe absorption mechanism. The suggestion by Brown et al. (1961) that a Fe^{3+} reductive mechanism was involved has been strongly supported by the evidence of Chaney et al. (1972) that soybean roots reduced Fe^{3+} EDDHA to Fe^{2+} before the Fe was absorbed. More recently Brown and Jones (1975) studied the Fe-P interaction in four sorghum (*Sorghum vulgare* Pers.) lines and demonstrated that "reductants" released from the roots, which are genetically controlled, made the Fe available for uptake by the plant.

The release of Fe^{2+} from either $Fe(OH)_3$ or $FePO_4 \cdot 2H2O$ precipitates is a function of both redox potential and pH, according to the following thermodynamic equations. In the Fe^{2+}–$Fe(OH)_3$ system,

$$Fe^{2+} + 3H_2O \rightleftharpoons Fe(OH)_3 + 3H^+ + \text{electron} \qquad [1]$$

and

$$\log(Fe^{2+}) = \text{constant} - 3pH - 16.9Eh, \qquad [2]$$

where (Fe^{2+}) represents Fe^{2+} activity in solution and Eh is the redox potential of the system. Equation [2] shows that (Fe^{2+}) varies inversely with both pH and Eh. In the Fe^{2+}–$FePO_4 \cdot 2H_2O$ system,

$$Fe^{2+} + 2OH^- + H_2PO_4^- \rightleftharpoons FePO_4 \cdot 2H_2O + \text{electron} \qquad [3]$$

and

$$\log(Fe^{2+}) = \text{constant} - 2pH - 16.9Eh - \log(H_2PO_4^-). \quad [4]$$

Equation [4] shows that (Fe^{2+}) varies inversely with $(H_2PO_4^-)$ at each pH and Eh level. This thermodynamic relationship may explain the observation of Mikesell et al. (1973) that P appeared to interfere with the utilization of Fe by sorghum. It surely should apply to Brown's observation (1972) that Fe-efficient 'Hawkeye' soybeans solubilized and made available the precipitated Fe phosphate in response to Fe stress, whereas the Fe-inefficient PI line could not lower Eh enough to compensate for the higher $H_2PO_4^-$ activity.

Since phosphate ions compete with the plant for Fe, the effectiveness of P in causing Fe deficiency should vary with the $H_2PO_4^-$ activity, pH, and the redox potential created by the absorbing tissue. Thus, if a plant can compensate for higher $H_2PO_4^-$ activity by lowering the pH or Eh, Fe availability can be maintained at a suitable level.

Phosphorus may also interfere with the internal transport of Fe by forming Fe phosphates (Biddulph, 1953; Rediski & Biddulph, 1953). However, recent data suggest that Fe moves in plants as the Fe-citrate anion (Tiffin, 1970), which should shield the Fe from precipitation by P. In spite of this, plants showing P-induced Fe chlorosis may have a normal concentration of tissue Fe, but a higher than normal P/Fe ratio (DeKock et al., 1960), suggesting that more than just absorption and translocation is involved in Fe chlorosis. DeKock et al. (1960) believed that Fe and P metabolisms were closely connected, because Fe is bound to phosphoproteins as Fe^{3+} while other Fe is present as Fe^{2+}. Thus, the P/Fe ratio may be a measure of the Fe^{3+}–Fe^{2+} balance in cells, heme synthesis, and chlorophyll synthesis. The P/Fe ratio and Fe^{2+} content of leaves appear to be inversely related.

VIII. MANGANESE

Soil-solution concentrations of Mn^{2+}, as well as exchangeable Mn^{2+}, are governed by the solubilities of Mn oxides, such as Mn_2O_3 and MnO_2. The dissolution and precipitation of these oxides are governed by soil pH and redox potential, such as

$$Mn^{2+} + 2H_2O \rightleftharpoons MnO_2 + 4H^+ + 2\,\text{electrons} \quad [5]$$

and

$$\log(Mn^{2+}) = \text{constant} - 4pH - 35.7Eh \quad [6]$$

where (Mn^{2+}) represents Mn^{2+} activity in solution and Eh is the redox potential of the system. Equation [5] is dependent on soil microbial activity as well as on soil pH. Equation [6] shows that (Mn^{2+}) varies inversely with both pH and Eh, and significant concentrations of Mn^{2+} occur only at low pH or low Eh.

In a study of Mn-phosphate solubility in dilute solutions, Heintze (1969) found significant precipitation of such compounds only between pH 6 and 9. Because special precautions were required to prevent Mn-oxide precipitation above pH 6, it does not appear that $Mn_3(PO_4)_2$ precipitation plays a significant role in lowering the availability of either Mn or P. On the other hand, P fertilizers will increase the level of soil-solution Mn in some soils, probably because of the acidic reaction of dissolved superphosphate or because of the acidity produced by the nitrification of NH_4 phosphate. For example, increasing amounts of water-extractable Mn tends to accompany increasing rates of $Ca(H_2PO_4)_2$, especially on noncalcareous soils (Bingham & Garber, 1960; Racz & Haluschak, 1974).

The decreased soil pH associated with high P rates can account for numerous reports that P fertilizer increased Mn uptake by plants (Bingham & Garber, 1960; Bingham et al., 1958; Larsen, 1964; Page et al., 1963; Smilde, 1973). In a statistical evaluation of many field experiments in Great Britain, Page et al. (1963) found a significant correlation between the decreased soil pH and increased Mn content of oats associated with applications of P fertilizer.

Not all experiments, however, have reported an increased soil Mn availability associated with P fertilizer. Phosphorus fertilizer may have no effect (Bingham & Martin, 1956), or it may decrease Mn uptake by plants (Heintze, 1969; Racz & Haluschak, 1974). Since P fertilizers are not likely to affect the pH of calcareous or highly buffered soils, P should not increase solution Mn on such soils. At the same time, however, the cation in P fertilizer may increase the cationic antagonism associated with Mn uptake. This may be the reason that Heintze (1969) found that P fertilizer aggravated Mn deficiency in oats (*Avena sativa* L.) on a Mn-deficient, alkaline soil without measurably altering soil pH or exchangeable Mn.

Nutrient-solution experiments have failed to show a definite Mn-P interaction within plants. Bingham (1963) reported P rates were associated with a lower Mn uptake by citrus and tomatoes (*Lycopersicon esculentum* Mill.), but with a higher uptake by beans. In a study of Mn toxicity in tobacco (*Nicotiana tabacum* L.), Mehta and Patel (1969) found P intensified Mn toxicity symptoms without increasing leaf content of Mn more than 10%. Racz and Haluschak (1974) found that P in nutrient solutions had no effect on Mn uptake by wheat seedlings (*Triticum aestivum* L. sp.). Conversely, Ohki (1975) found P uptake by cotton seedlings (*Gossypium hirsutum* L.) to be unaffected by solution Mn above deficiency levels. The fact that Mn can activate enzymes in some P metabolic reactions (Jackson, 1967) has not been appreciably helpful in elucidating the overall Mn-P interactions within plants.

IX. ZINC

Olsen's (1972) review of Zn-P interactions surveyed results from many experiments conducted between 1936 and 1970. There are many apparent contradictions in those results, leaving the impression that the true interdependence of the two elements on plant growth is multifaceted or has eluded scientists.

Interest in this topic has been stimulated from time to time because P fertilizer sometimes induces or aggravates a Zn deficiency in a crop. Efforts to define or identify the "cause-and-effect" mechanism of this phenomenon have been frustrated, however, often because high P rates failed to curb Zn uptake or induce Zn deficiency (Armbruster et al., 1975; Bingham, 1963, Boawn et al., 1954; Ellis et al., 1964; Seatz et al., 1959). For example, Boawn et al. (1954) reported that attempts to induce Zn-deficiency symptoms by high P rates failed even where P contents in plant tissue were doubled.

The early suspicion that P-induced Zn deficiency resulted from precipitation of $Zn_3(PO_4)_2$ in the soil or on the roots was convincingly demolished by solubility experiments (Boawn et al., 1954; Jurinak & Inoye, 1962), by Zn-source experiments (Boawn et al., 1954; 1957), and by extraction experiments (Bingham & Garber, 1960; MacLean, 1974; Pauli et al., 1968; Racz & Haluschak, 1974). The solubility of $Zn_3(PO_4)_2$ is more than adequate to supply Zn to plants, $Zn_3(PO_4)_2$ is an effective Zn fertilizer, and higher P rates generally increase extractable soil Zn.

The contradictions of Zn-P interactions are shown by a series of experiments by Bingham and coworkers (Bingham, 1963; Bingham & Garber, 1960; Bingham & Martin, 1956; Bingham et al., 1958). Their earliest experiments suggested that P fertilizer reduced Zn uptake; their intermediate experiments showed a lowered Zn uptake only at excessive P rates; their last experiment led them to conclude that Zn uptake was unaffected by P rate.

The research of Boawn and coworkers illustrates another puzzling aspect of the Zn-P relationship, i.e., Zn content of leaves is not always an indicator of Zn deficiency. In early experiments with beans (*Phaseolus vulgaris* L.) (Boawn et al., 1954), they found no relationship between a P-induced lower yield and the Zn content of plant. Subsequently, they found that high P rates were associated with Zn deficiency symptoms on potatoes (*Solanum tuberosum* L.), but not with the Zn content of plant tissue (Boawn & Leggett, 1964). In a later experiment with potatoes (Boawn & Brown, 1968), they used a split-root technique, where Zn was supplied to part of the root system via soil and P was supplied to another part via solution, to show that high P levels could induce Zn deficiency symptoms in both beans and potatoes without decreasing Zn content of plants. They concluded that excess P interfered with "normal metabolism" of Zn.

Some of the confusion in the literature about the Zn-P interaction in plants is probably related to unrecognized differences in reference treatments. As pointed out by Olsen (1972), adding P to a P-deficient soil stimulates growth, thereby diluting the Zn content of the plant tissue. This dilution effect sometimes results in Zn deficiency (Brown et al., 1970; Ellis et al., 1964; Jackson et al., 1967; Warnock, 1970), but often does not (Armbruster et al., 1975; Ellis et al., 1964; Langin et al., 1962; Racz & Haluschak, 1974; Smilde, 1973), probably depending on the level of available soil Zn. Adding P to a P-sufficient soil does not stimulate plant growth, but usually stimulates P uptake and may result in Zn deficiency. The unpredictable effect of this phenomenon on Zn deficiency is illustrated by the 2-year field experiment of Ganiron et al. (1969), in which Zn deficiency was severe the first year and minor the second year. The Zn content of leaves was neither related to the appearance nor the severity of the deficiency.

The fact that Zn-deficiency symptoms were often unrelated to Zn content of plant tissue led Boawn and Leggett (1964) to suggest that Zn deficiency depended upon the P/Zn ratio in plant tissue. Later experiments, however, have tended to discount the concept of a critical P/Zn ratio (Giordano & Mortvedt, 1969; Sharma et al., 1968; Stukenholtz et al., 1966; Warnock, 1970).

Nutrient-solution experiments are often used to clarify concepts of mineral nutrition. In the case of Zn-P interactions, however, they have not been particularly enlightening. For example, experiments have shown that (i) P may have no effect on Zn uptake (Bingham, 1963), (ii) P may increase Zn uptake (Pauli et al., 1968; Wallace et al., 1973; Watanabe et al., 1965), or (iii) P may decrease Zn uptake (Adriano et al., 1971; Racz & Haluschak, 1974).

Such apparent anomalies suggest that the manifestation of Zn-P interactions is influenced by other growth factors, e.g., genetics (Ambler & Brown, 1969; Giordano & Mortvedt, 1969), soil compaction (Ward et al., 1963), available K (Stukenholtz et al., 1966; Ward et al., 1963), available N (Stukenholtz et al., 1966), Fe content of plant tissue (Ambler & Brown, 1969; Jackson et al., 1967; Warnock, 1970), and metabolic rate of plant growth (Ellis et al., 1964; Ganiron et al., 1969; Martin et al., 1965; Sharma et al., 1968).

The evidence seems clear that P-related Zn deficiency is sometimes associated with high Fe content of plant tissue. Jackson et al. (1967) and Warnock (1970) reported much higher Fe contents in plants suffering from P-induced Zn deficiency than in normal plants. Further evidence of this three-way relationship was presented by Ambler and Brown (1969), who found Zn deficiency to be aggravated by high levels of both Fe and P. They further noted that the bean variety which was more susceptible to Zn deficiency was also the one that absorbed more Fe and P. The determining factor appears to be the Fe content of the plant, and not the Fe concentration of the soil solution (Warnock, 1970).

The evidence is strong that P-related Zn deficiency is a metabolic malfunction of some sort, seemingly unrelated to the Zn content per se. Whereas Stukenholtz et al. (1966) suspected that P interfered with the metabolic processes responsible for Zn absorption by root cells and for Zn translocation to leaves, Smilde (1973) found no such evidence. Using different temperatures to vary the metabolic rate of plants, Martin et al. (1965) found P-induced Zn deficiency to be more acute at the lower temperature (10 to 15°C) even though Zn contents were higher. This was subsequently confirmed by Sharma et al. (1968), who also showed that the extent of Zn translocation to tops was directly proportional to temperature (between 15 and

30°C), being practically nil at 15°C, even though roots continued to absorb Zn at all temperatures. Ganiron et al. (1969) further noted that P accumulated in both roots and shoots of Zn-deficient plants at normal temperature, whereas roots of Zn-sufficient plants rapidly translocated P to the shoots.

Using 48-hour absorption periods for intact root systems developed under different regimes of P, Zn, temperature, and light intensity, Edwards and Kamprath (1974) found that shoot/root ratios were increased by either high P rate, high temperature, or low light intensity. However, P was without effect on Zn uptake or translocation, whereas temperature had a profound effect. Practically no Zn was absorbed or translocated at 14 to 18°C, but Zn was so effectively translocated at 26 to 30°C that low-temperature roots wound up with more Zn than high-temperature roots. Under a low-light regime, root growth was retarded and Zn absorption was so restricted that seed-contained Zn was about the only Zn available to the seedling.

The Zn-P relationship in plants is also manifested at excess Zn levels. In a tolerance-to-Zn evaluation of 18 economic plant species, Boawn and Rasmussen (1971) concluded that excess Zn was not directly toxic to plants, but that it upset normal P metabolism, thereby creating a growth response and visual symptoms typical of P deficiency. Smilde et al. (1974) in substantiating these results noted that P ameliorated Zn toxicity by increasing plant growth rather than by decreasing total Zn uptake.

X. COPPER

Copper concentrations in soil solutions are normally very low with a large proportion being present as organic complexes. Reports of P-induced Cu deficiency and P-alleviated Cu toxicity seem to be confined to citrus. Prolonged use of high P-fertilizer rates in Florida citrus groves had created Cu deficiencies by the 1940's (Forsee & Neller, 1944). Subsequent use of Cu in fertilizers and in foliar fungicidal sprays increased available soil Cu to toxic levels, which apparently caused Fe chlorosis in acid soils (Reuther & Smith, 1952). It was later demonstrated that high P rates for citrus would effectively detoxify Cu (Spencer, 1966).

Bingham and coworkers in California (Bingham & Garber, 1966; Bingham & Martin, 1956; Bingham et al., 1958), in a series of greenhouse experiments with citrus seedlings, confirmed that high rates of P fertilizer sometimes dramatically reduced Cu content of leaves, reduced growth, and caused Cu-deficiency symptoms. Since the amount of Cu absorbed by plants was actually less at the high P rates, the authors concluded that $H_2PO_4^-$ in soil solution interfered with Cu absorption. In a subsequent sand-culture experiment, Bingham (1963) was unable to show any Cu-P antagonisms, in sharp contrast to his earlier findings that "large soil applications of P tend to produce a marked reduction in growth and Cu content and prominent symptoms of Cu deficiency" on citrus. Adriano et al. (1971) also reported no evidence of Cu-P interaction in corn seedlings grown in nutrient solutions.

In contrast to most citrus experiments, the Cu-P interaction in subterranean clover (*Trifolium subterraneum* L.) was found to be synergistic instead of antagonistic (Greenwood & Hallsworth, 1960). The authors reported, "when Cu was limited. . ., both concentration and total uptake of Cu was also enhanced by P." Smilde (1973) reported that seedlings of some tree species had higher total Cu uptake at "normal" P rates, although concentrations tended to be lower, and concluded that P interfered with the root-to-shoot translocation of Cu rather than with absorption. In a soil and nutrient-solution experiment with wheat seedlings, Racz and Haluschak (1974) found that P reduced Cu uptake in most soils, even though the water-extracted Cu content of the soil was increased by $NH_4H_2PO_4$ additions.

XI. SULFATE

Experiments related to sulfate-phosphate interactions in soils have been more superficial than sophisticated. Sulfate ions are generally viewed as being retained in soils through adsorption mechanisms, especially by Al- and Fe-oxide surfaces, in spite of the fact that insoluble basic (Al, Fe) sulfates are known to precipitate rapidly from acid solutions (Bassett & Goodwin, 1949; Hsu & Bates, 1964) as well as being naturally occurring secondary minerals (Harvey & Vitalians, 1964; Hendricks, 1937; Hollingworth & Bannister, 1950).

The similarity in composition of the relatively insoluble Al or Fe sulfates and phosphates is shown by their formulae when written as $Al(OH)_2H_2PO_4$, $Al_4(OH)_{10}SO_4$, and $KAl_3(OH)_6(SO_4)_2$. The strong affinity of aqueous Al^{3+} and Fe^{3+} for $H_2PO_4^-$ and SO_4^{2-}, along with OH^- ions, suggests keen competition among the anions for adsorption-precipitation reactions with partially neutralized Al and Fe hydroxides. The stoichiometric composition and solubility of such compounds should be reflected in the composition of equilibrium soil solutions.

The effect of P fertilizer added to soils is to reduce the amount of "adsorbed" sulfate (Chao et al., 1962; Kamprath et al., 1956) and thereby increase its availability (Bailey, 1974). This concept has been applied to correlation experiments for soil testing, and P-containing solutions have been found to be highly effective sulfate extractants (Hoeft et al., 1973).

The interaction of absorbed P and S within the plant has received scant attention. The sulfate ion has been generally viewed as the most innocuous of the major nutrient ions and has consequently been the variable co-ion in most studies with cationic variables in nutrient solutions. The lack of attention, however, should not imply a lack of interaction.

XII. MOLYBDENUM

Available soil Mo probably exists in soil solution largely as MoO_4^{2-} and $Mo_2O_7^{2-}$ ions (Reisenauer et al., 1962). The pH-dependent solubility of Mo in soils, however, has not been sufficiently defined (Reyes & Jurinak, 1967) for a quantitative evaluation of chemical reactions in soils that involve Mo-P interactions. Since anions of both Mo and P react with Al and Fe oxides and other soil minerals in a somewhat similar manner (Barrow, 1970), competition between them for adsorption-precipitation reactions appears likely. Although Barshad (1951) and Gupta and Cutcliffe (1968) found that ordinary superphosphate and concentrated superphosphate, respectively, had no effect on water-extractable soil Mo, Barrow (1974) showed that $0.1M$ KH_2PO_4 was considerably more effective in extracting Mo than a comparable KCl-KOH solution. Still, the mechanism of replacement or ionic interaction in the soil remains essentially obscure.

The contradictory evidence for P fertilizers affecting soil Mo availability was effectively resolved by the rather comprehensive report by Stout et al (1951). They reported that SO_4-containing ordinary superphosphate reduced Mo uptake by plants, whereas concentrated superphosphate increased Mo uptake by as much as 10-fold. Phosphate was especially effective in stimulating Mo uptake in short-term absorption experiments with acid nutrient solutions. As long as sulfate was absent, P enhanced Mo uptake. This view has been generally supported by others (Bingham & Garber, 1960; Greenwood & Hallsworth, 1960; Gupta & Munro, 1969). However, Bingham (1963) reported that P lowered Mo uptake by some plants, but not others in a sand-culture experiment, and Bingham and Garber (1960) claimed P fertilizer enhanced Mo availability in acid soils, but suppressed it in alkaline soils.

Although convincing proof is lacking, it appears that the positive interaction of P on Mo uptake by plants is related to the plant's metabolic processes involving these two ions through a still-to-be-identified mechanism.

XIII. BORON

Possible B-P interactions in soils appear to have escaped the attention of soil scientists. However, since borate ions are removed from soil solutions by Al and Fe oxides and clay minerals through adsorption or precipitation reactions (Krauskopf, 1972), there is reasonable likelihood of at least minor interactions related to solution concentrations of borate and phosphate ions. Bingham and co-workers (Bingham & Garber, 1960; Bingham & Martin, 1956; Bingham et al., 1958) reported that applications of $Ca(H_2PO_4)_2$ to soils of southern California resulted in a lower availability of B, especially in acid soils, as measured by plant uptake and water extractability. No mechanism was proposed.

There is also a dearth of published data on B-P interaction within the plant. Robertson and Loughman (1974), noting conflicting reports describing the effect of varying B regimes on P absorption, attempted to determine whether B was inhibitory or stimulating on P uptake. They believed that the effect of B deficiency was too catastrophic on plant growth in some reported experiments to permit an unreserved interpretation of the effect of B on P uptake and metabolism. In short-term absorption experiments with intact root systems that suffered different levels of B deficiency, they were able to show the following: (i) B deficiency caused a thickening of roots and retarded root elongation, resulting in a smaller absorbing root surface and reduced P uptake; (ii) addition of B to B-deficient roots rapidly restored their ability to absorb and metabolize P, thus implicating B closely with P uptake and P metabolism; (iii) overall P metabolism in roots was not greatly affected by B deficiency because ^{32}P was equally distributed through similar pathways in severely deficient and nondeficient roots; (iv) partitioning of absorbed P between roots and shoots was similar in deficient and nondeficient roots; and (v) tissue formed with adequate B did not require a continuous supply of B to maintain normal P absorption and metabolism, even though root growth was being affected.

XIV. SILICATE

Use of silicates as soil amendments or liming materials has been beneficial in some instances, but has never been a widely accepted practice. Silicates not only raise pH, but they also increase the availability of soil and fertilizer P. Taylor's review (1961) cites the results of many early experiments in which silicate applications increased plant growth, the most famous being the one conducted at the Rothamsted Experiment Station without interruption since 1862 (Russell, 1973, p. 638-639).

An early explanation of the beneficial effect of silicates suggested that Si caused a more efficient utilization of P by plants. This was followed by the explanation that Si actually increased the availability of soil P (Taylor, 1961). It was further suggested that silicates were effective in increasing P availability only on soils of low or medium P levels. Recently, however, Roy et al. (1971) claimed significant growth increases for sugarcane in a pot experiment by adding silicate to Hawaiian soils that were very deficient in P.

There is considerable evidence that added Si increases water-soluble and easily extractable P (Laws, 1950; Reifenberg & Buckwold, 1954; Roy et al., 1971; Taylor, 1961; Toth, 1939). There is also evidence that added P increases solution Si (Bar-Yosef et al., 1969; Kafkafi, 1968; Reifenberg & Buckwold, 1954) and that prior-added Si decreases the amount of P that a soil "fixes" (Reifenberg & Buckwold, 1954; Roy et al., 1971). In spite of the evidence for Si-P interaction in soils, there is scant evidence on the actual mechanisms involved.

It has been suggested that silicate and phosphate ions compete for adsorption sites on clay and Al(Fe)-oxide surfaces (Bar-Yosef et al., 1969; Roy et al., 1971; Toth, 1939). This concept is supported by a considerable amount of data on anion adsorption by soil minerals (Hingston et al., 1967; Mekaru & Uehara, 1972).

A second proposal is that both silicate and phosphate ions form insoluble precipitates with such common ions as Al, Fe, and Ca, and that solution concentrations of these ions are governed by the solubility of the different precipitates (Jones & Handreck, 1967; Kafkafi, 1968). The problem of obtaining direct proof of precipitates and solubility products seems insurmountable at the moment, primarily because of the inability to acquire complete information about precipitate composition in a soil matrix. However, indirect proof through solubility data would appear to be accessible if adequate allowances can be made for the nonideal formulae of natural precipitates because of occluded ions, isomorphus substitutions, and solid solutions.

The precipitation concept has the advantage of considering solution ions other than silicate and phosphate and recognizing that all soil minerals are soluble to some extent. As a practical matter, of course, "chemisorption" and precipitation reactions are hardly distinguishable, and the argument becomes largely a matter of semantics.

Although most evidence has indicated that Si increases yields by increased P availability, there is recent evidence to support the earlier suggestion that Si results in a more efficient use of P by plants (Taylor, 1961). Roy et al. (1971) found that Si increased sugarcane growth while slightly decreasing the percentage of P in cane stalks. They concluded that the increased yield was associated with improved P nutrition and that Si decreased the plants' internal P requirements. Additional evidence was obtained for this view by short-term absorption experiments with sugarcane by Cheong and Chan (1972) who reported that Si did not increase P uptake, but significantly enhanced phosphorylation of glucose and fructose at the expense of ATP. They suggested that Si improved energy metabolism and sugar synthesis, possibly operating as a cofactor to the kinases, which is reflected in more growth.

LITERATURE CITED

Adams, Fred. 1966. Calcium deficiency as a causal agent of ammonium phosphate injury to cotton seedlings. Soil Sci. Soc. Am. Proc. 30:485–488.
Adams, Fred. 1974. Soil solution. p. 441–481. *In* E. W. Carson (ed.) The plant root and its environment. Univ. Virginia Press, Charlottesville.
Adriano, D. C., G. M. Paulsen, and L. S. Murphy. 1971. Phosphorus-iron and phosphorus-zinc relationships in corn (*Zea mays* L.) seedlings as affected by mineral nutrition. Agron. J. 63:36–39.
Ambler, J. E., and J. C. Brown. 1969. Cause of differential susceptibility to zinc deficiency in two varieties of navy beans (*Phaseolus vulgaris* L.). Agron. J. 61:41–43.
Armbruster, J. A., L. S. Murphy, L. J. Meyer, P. J. Gallagher, and D. A. Whitney. 1975. Field and growth-chamber evaluations of potassium polyphosphate. Soil Sci. Soc. Am. Proc. 39:144–150.
Aslyng, H. C. 1954. Phosphate potential and phosphate status of soils; the solubility and availability of phosphates. R. Vet. Agric. Univ. Yearb., Copenhagen. p. 1–50.
Bailey, J. M. 1974. Changes in the adsorbed sulphate status of a yellow-brown earth after phosphate fertilization and legume growth (soils). N.Z. J. Agric. Res. 17:257–265.
Banwart, W. L., and W. H. Pierre. 1975. Cation-anion balance of field-grown crops. II. Effect of P and K fertilization and soil pH. Agron. J. 67:20–25.
Barrow, N. J. 1970. Comparison of the adsorption of molybdate, sulfate, and phosphate by soils. Soil Sci. 109:282–288.
Barrow, N. J. 1972. Influence of solution concentration of calcium on the adsorption of phosphate, sulfate, and molybdate by soils. Soil Sci. 114:175–180.
Barrow, N. J. 1974. On the displacement of adsorbed anions from soil: 1. Displacement of molybdate by phosphate and by hydroxide. Soil Sci. 116:423–431.

Barshad, I. 1951. Factors affecting the molybdenum content of pasture plants. 2. Effect of soluble phosphates, available nitrogen, and soluble sulfates. Soil Sci. 71:387-398.

Bar-Yosef, B. 1971. Fluxes of P and Ca into intact corn roots and their dependence on solution concentration and root age. Plant Soil 35:589-600.

Bar-Yosef, B., U. Kafkafi, and N. Lahav. 1969. Relationships among adsorbed phosphate, silica, and hydroxyl during drying and rewetting of kaolinite suspension. I. Hydrogen-ion concentration. Soil Sci. Soc. Am. Proc. 33:672-677.

Bassett, H., and T. H. Goodwin. 1949. The basic aluminum sulfates. J. Chem. Soc. (London). p. 2239-2279.

Bauwin, G. R., and E. H. Tyner. 1957. The nature of reductant-soluble phosphorus in soils and soil concretions. Soil Sci. Soc. Am. Proc. 21:250-257.

Bennett, A. C., and Fred Adams. 1970. Calcium deficiency and ammonia toxicity as separate causal factors of diammonium phosphate injury to seedlings. Soil Sci. Soc. Am. Proc. 34:255-259.

Bennett, W. F., J. Pesek, and J. Hanway. 1962. Effect of nitrogen on phosphorus absorption by corn. Agron. J. 54:437-442.

Biddulph, O. 1953. The translocation of minerals in plants. p. 261-272. *In* E. Truog (ed.) Mineral nutrition of plants. Univ. of Wisconsin Press, Madison.

Biddulph, O., and C. G. Woodbridge. 1952. The uptake of phosphorus by bean plants with particular reference to the effects of iron. Plant Physiol. 27:431-444.

Bingham, F. T. 1963. Relation between phosphorus and micronutrients in plants. Soil Sci. Soc. Am. Proc. 27:389-391.

Bingham, F. T., and M. J. Garber. 1960. Solubility and availability of micronutrients in relation to phosphorus fertilization. Soil Sci. Soc. Am. Proc. 24:209-213.

Bingham, F. T., and J. P. Martin. 1956. Effects of soil phosphorus on growth and minor element nutrition of citrus. Soil Sci. Soc. Am. Proc. 20:382-385.

Bingham, F. T., J. P. Martin, and J. A. Chastain. 1958. Effects of phosphorus fertilization of California soils on minor element nutrition of citrus. Soil Sci. 86:24-31.

Blanchar, R. W., and A. C. Caldwell. 1966a. Phosphate-ammonium-moisture relationships in soils: I. Ion concentrations in static fertilizer zones and effects on plants. Soil Sci. Soc. Am. Proc. 30:39-43.

Blanchar, R. W., and A. C. Caldwell. 1966b. Phosphate-ammonium-moisture relationships in soils: II. Ion concentrations in leached fertilizer zones and effects on plants. Soil Sci. Soc. Am. Proc. 30:43-48.

Boawn, L. C., and J. C. Brown. 1968. Further evidence for a P-Zn imbalance in plants. Soil Sci. Soc. Am. Proc. 32:94-97.

Boawn, L. C., and G. E. Leggett. 1964. Phosphorus and zinc concentrations in Russett Burbank potato tissues in relation to zinc deficiency symptoms. Soil Sci. Soc. Am. Proc. 28: 229-232.

Boawn, L. C., and P. E. Rasmussen. 1971. Crop response to excessive zinc fertilization of alkaline soil. Agron. J. 63:874-876.

Boawn, L. C., F. G. Viets, Jr., and C. L. Crawford. 1954. Effects of phosphate fertilizers on zinc nutrition of field beans. Soil Sci. 78:1-7.

Boawn, L. C., F. G. Viets, Jr., and C. L. Crawford. 1957. Plant utilization of zinc from various types of zinc compounds and fertilizer materials. Soil Sci. 83:219-227.

Bouldin, D. R., and E. C. Sample. 1958. The effect of associated salts on the availability of concentrated superphosphates. Soil Sci. Soc. Am. Proc. 22:124-129.

Bouldin, D. R., and E. C. Sample. 1959. Laboratory and greenhouse studies with monocalcium, monoammonium, and diammonium phosphates. Soil Sci. Soc. Am. Proc. 23:338-342.

Brown, A. L., B. A. Krantz, and J. L. Eddings. 1970. Zinc-phosphorus interactions as measured by plant response and soil analysis. Soil Sci. 110:415-420.

Brown, J. C. 1972. Competition between phosphate and the plant for Fe from Fe^{++} ferrozine. Agron. J. 64:240-243.

Brown, J. C., and W. E. Jones. 1975. Phosphorus efficiency as related to iron inefficiency in sorghum. Agron. J. 67:468-472.

Brown, J. C., R. S. Holmes, and L. O. Tiffin. 1961. Iron chlorosis in soybeans as related to genotype of rootstock: 3. Chlorosis susceptibility and reductive capacity of root. Soil Sci. 91:127-132.

Buehrer, T. F. 1932. The physico-chemical relationships of soil phosphates. Arizona Agric. Exp. Stn. Tech. Bull. 34:155-212.

Chaney, R. L., J. C. Brown, and L. O. Tiffin. 1972. Obligatory reduction of ferric chelates in iron uptake by soybeans. Plant Physiol. 50:208-213.

Chang, S. C., and M. L. Jackson. 1957. Fractionation of soil phosphorus. Soil Sci. 84:133-144.

Chao, T. T., M. E. Harward, and S. C. Fang. 1962. Movement of ^{35}S tagged sulfate through soil columns. Soil Sci. Soc. Am. Proc. 26:27-32.

Chapman, H. D. 1936. Effect of nitrogenous fertilizers, organic mater, sulfur, and colloidal silica on the availability of phosphorus in calcareous soils. J. Am. Soc. Agron. 28:135-145.

Cheong, Y. W. Y., and P. Y. Chan. 1972. Incorporation of ^{32}P in phosphate esters of the sugarcane plant and the effect of Si and Al on the distribution of these esters. Plant Soil 38:113-123.

Clarkson, D. T. 1966. Effect of aluminum on the uptake and metabolism of phosphorus by barley seedlings. Plant Physiol. 41:165-172.

Clarkson, D. T. 1967. Interactions between aluminum and phosphorus on root surfaces and cell wall material. Plant Soil 27(3):347-356.

Clarkson, D. T. 1969. Metabolic aspects of aluminum toxicity and some possible mechanisms for resistance. p. 321-397. In I. H. Rorison et al. (ed.) Ecological aspects of the mineral nutrition of plants. Blackwell Scientific Publ., Oxford.

Cole, C. V., and M. L. Jackson. 1950. Solubility equilibrium constant of dihydroxy aluminum dihydrogen phosphate relating to a mechanism of phosphate fixation in soils. Soil Sci. Soc. Am. Proc. 15:84-89.

Cole, C. V., D. L. Grunes, L. K. Porter, and S. R. Olsen. 1963. The effects of nitrogen on short-term phosphorus absorption and translocation in corn (*Zea mays*). Soil Sci. Soc. Am. Proc. 27:671-674.

Colemn, N. T., J. T. Thorup, and W. A. Jackson. 1960. Phosphate-sorption reactions that involve exchangeable Al. Soil Sci. 90:1-7.

Davis, L. E. 1935. Sorption of phosphates by non-calcareous Hawaiian soils. Soil Sci. 40:129-158.

Dean, L. A. 1938. An attempted fractionation of the soil phosphorus. J. Agric. Sci. 28:234-246.

DeKock, P. C., A. Hall, and M. McDonald. 1960. A relation between the ratios of phosphorus to iron and potassium to calcium in mustard leaves. Plant Soil 12:128-142.

Earnshaw, M. J., and J. B. Hanson. 1973. Inhibition of postoxidative calcium release in corn mitochondria by inorganic phosphate. Plant Physiol. 52:403-406.

Edwards, D. G. 1968. Cation effects on phosphate absorption from solution by *Trifolium subterraneum*. Aust. J. Biol. Sci. 21:1-11.

Edwards, J. H., and E. J. Kamprath. 1974. Zinc accumulation by corn seedlings as influenced by phosphorus, temperature, and light intensity. Agron. J. 66:479-482.

Ellis, Roscoe, Jr., J. F. Davis, and D. L. Thurlow. 1964. Zinc availability in calcareous Michigan soils as influenced by phosphorus level and temperature. Soil Sci. Soc. Am. Proc. 28:83-86.

Elzam, O. E., and T. K. Hodges. 1968. Characterization of energy-dependent Ca^{+2} transport in maize mitochondria. Plant Physiol. 43:1108-1114.

Engelstad, O. P., and S. E. Allen. 1971. Effect of form and proximity of added nitrogen on crop uptake of phosphorus. Soil Sci. 112:330-337.

Follett, R. F., and G. A. Reichman. 1973. Ionic balance of barley as influenced by P fertility, water, and soil temperature. Agron. J. 65:477-482.

Forsee, W. T., Jr., and J. R. Neller. 1944. Phosphate responses in a valencia grove in the eastern Everglades. Proc. Fla. State Hortic. Soc. 57:110-115.

Foy, C. D. 1974. Effects of aluminum on plant growth. p. 601-642. In E. W. Carson (ed.) The plant root and its environment. Univ. Virginia Press, Charlottesville.

Foy, C. D., and J. C. Brown. 1963. Toxic factors in acid soils: I. Characterization of aluminum toxicity in cotton. Soil Sci. Soc. Am. Proc. 27:403-407.

Foy, C. D., and J. C. Brown. 1964. Toxic factors in acid soils: II. Differential aluminum tolerance of plant species. Soil Sci. Soc. Am. Proc. 28:27-30.

Franklin, R. E. 1969. Effect of adsorbed cations on phosphorus uptake by excised roots. Plant Physiol. 44:697-700.

Ganiron, R. B., D. C. Adriano, G. M. Paulsen, and L. S. Murphy. 1969. Effect of phosphorus carriers and zinc sources on phosphorus-zinc interactions in corn. Soil Sci. Soc. Am. Proc. 33:306-309.

Giordano, P. M., and J. J. Mortvedt. 1969. Response of several corn hybrids to level of water-soluble zinc in fertilizers. Soil Sci. Soc. Am. Proc. 33:145-148.

Greenwood, E. H. N., and E. G. Hallsworth. 1960. Studies on the nutrition of forage legumes. II. Some interactions of calcium, phosphorus, copper, and molybdenum on the growth and chemical composition of *Trifolium subterraneum* L. Plant Soil 12:97-127.

Griffin, R. A., and J. J. Jurinak. 1973. The interaction of phosphate with calcite. Soil Sci. Soc. Am. Proc. 37:847-850.

Grunes, D. L. 1959. Effect of nitrogen on the availability of soil and fertilizer phosphorus to plants. Adv. Agron. 11:369-396.

Grunes, D. L., F. G. Viets, Jr., and S. H. Shih. 1958. Proportionate uptake of soil and fertilizer phosphorus by plants as affected by nitrogen fertilization. I. Growth chamber experiment. Soil Sci. Soc. Am. Proc. 22:43-48.

Gupta, U. C., and J. A. Cutcliffe. 1968. Influence of phosphorus on molybdenum content of brussels sprouts under field and greenhouse conditions and on recovery of added molybdenum in soil. Can. J. Soil Sci. 48:117-123.

Gupta, U. C., and D. C. Munro. 1969. Influence of sulfur, molybdenum, and phosphorus on chemical composition and yields of brussels sprouts and of molybdenum on sulfur contents of several plant species grown in the greenhouse. Soil Sci. 107:114-118.

Hanson, J. B., and R. J. Miller. 1967. Evidence for active phosphate transport in maize mitochondria. Proc. Nat. Acad. Sci. USA 58:727-734.

Harvey, R. D., and C. J. Vitalians. 1964. Wall-rock alteration in the Golfield district, Nevada. J. Geol. 72:564-579.

Haseman, J. F., E. H. Brown, and C. D. Whitt. 1950. Some reactions of phosphate with clays and hydrous oxides of iron and aluminum. Soil Sci. 70:257-271.

Heintze, S. C. 1969. Manganese-phosphate reactions in aqueous systems and the effects of applications of monocalcium phosphate on the availability of manganese to oats in an alkaline fen soil. Plant Soil 29:407-423.

Hendricks, S. B. 1937. The crystal structure of alunite and the jarosites. Am. Mineral. 22:773-784.

Hingston, F. J., R. J. Atkinson, A. M. Posner, and J. P. Quirk. 1967. The specific adsorption of anions. Nature (London) 215:1459-1461.

Hoeft, R. G., L. M. Walsh, and D. R. Keeney. 1973. Evaluation of various extractants for available soil sulfur. Soil Sci. Soc. Am. Proc. 37:401-404.

Hollingworth, S. E., and F. A. Bannister. 1950. Basaluminite and hydrobasasluminite, two new minerals from Northampton. Mineral. Mag. 29:1-17.

Hsu, P. H., and T. F. Bates. 1964. Formation of x-ray amorphous and crystalline aluminum hydroxides. Mineral Mag. 33:749-768.

Hsu, P. H., and R. A. Rennie. 1962. Reactions of phosphate in aluminum systems. I. Adsorption of phosphate by x-ray amorphous "aluminum hydroxide". Can. J. Soil Sci. 42:197-209.

Humble, G. D., E. El Leboudi, and V. V. Randig. 1969. Effect of nitrogen on phosphorus absorption by excised barley roots. Plant Soil 31:353-364.

Hyde, A. H. 1966. Nature of the calcium effect in phosphate uptake by barley roots. Plant Soil 24:328-332.

Jackson, P. C., S. B. Hendricks, and B. M. Vasta. 1962. Phosphorylation by barley root mitochondria and phosphate absorption by barley roots. Plant Physiol. 37:8-17.

Jackson, T. L., J. Hay, and D. P. Moore. 1967. The effect of Zn on yield and chemical composition of sweet corn in the Willamette Valley. Proc. Am. Soc. Hortic. Sci. 91:462-471.

Jackson, W. A. 1967. Physiological effects of soil acidity. *In* R. W. Pearson and Fred Adams (ed.) Soil acidity and liming. Agronomy 12:43-124. Am. Soc. of Agron., Madison, Wis.

Jones, L. H. 1961. Aluminum uptake and toxicity in plants. Plant Soil 13:297-300.

Jones, L. H. P., and K. A. Handreck. 1967. Silica in soils, plants, and animals. Adv. Agron. 19:107-149.

Jurinak, J. J., and T. S. Inouye. 1962. Some aspects of zinc and copper phosphate formation in aqueous systems. Soil Sci. Soc. Am. Proc. 26:144-147.

Kafkafi, U. 1968. Hydrogen consumption and silica release during initial stages of phosphate adsorption on kaolinite at a constant pH. Isr. J. Chem. 6:367-375.

Kamprath, E. J., W. L. Nelson, and J. W. Fitts. 1956. The effect of pH, sulfate, and phosphate concentrations on the adsorption of sulfate by soils. Soil Sci. Soc. Am. Proc. 20:463-466.

Khasawneh, F. E., E. C. Sample, and I. Hashimoto. 1974. Reactions of ammonium ortho- and polyphosphate fertilizers in soil: I. Mobility of phosphorus. Soil Sci. Soc. Am. Proc. 38:446-451.

Kittrick, J. A., and M. L. Jackson. 1955. Rate of phosphate reaction with soil minerals and electron microscope observations on the reaction mechanism. Soil Sci. Soc. Am. Proc. 12:292-295.

Krauskopf, K. B. 1972. Geochemistry of micronutrients. p. 7-40. *In* J. J. Mortvedt, P. M. Giordano, and W. L. Lindsay (ed.) Micronutrients in agriculture. Soil Sci. Soc. Am., Madison, Wis.

Kurtz, L. T. 1953. Inorganic phosphorus in acid and neutral soils. *In* W. H. Pierre and A. G. Norman (ed.) Soil and fertilizer phosphorus in crop nutrition. Agronomy 4:59-88. Academic Press, Inc., New York.

Langin, E. J., R. C. Ward, R. A. Olson, and H. F. Rhoades. 1962. Factors responsible for poor response of corn and grain sorghum to phosphorus fertilization: II. Lime and P placement effects on P-Zn relations. Soil Sci. Soc. Am. Proc. 26:574-578.

Larsen, S. 1964. The effect of phosphate applications on manganese content of plants grown on neutral and alkaline soils. Plant Soil 21:37-42.

Laws, W. D. 1950. Water-soluble silicate application to a calcareous clay soil and effect on soil properties and nutrient uptake by plants. Soil Sci. Soc. Am. Proc. 15:89-92.

Leggett, J. E., R. A. Galloway, and H. G. Gauch. 1965. Calcium activation of orthophosphate absorption by barley roots. Plant Physiol. 40:897-902.

Lehr, J. J., and J. E. van Wesemael. 1952. The influence of neutral salts on the solubility of soil phosphate, with special reference to the effect of the nitrates of sodium and calcium. J. Soil Sci. 3:125-135.

Lehr, J. R., E. H. Brown, A. W. Frazier, J. P. Smith, and R. D. Thrasher. 1967. Crystallographic properties of fertilizer compounds. Tenn. Valley Authority Chem. Eng. Bull. no. 6, TVA, Muscle Shoals, Ala.

Lindsay, W. L., A. W. frazier, and H. F. Stephenson. 1962. Identification of reaction products from phosphate fertilizers in soil. Soil Sci. Soc. Am. Proc. 26:446-452.

Lorenz, O. A., and C. M. Johnson. 1953. Nitrogen fertilization as related to the availability of phosphorus in certain California soils. Soil Sci. 75:119-129.

Lowe, R. H., and C. E. Bortner. 1973. Effect of phosphorus nutrition and soil pH on "physiologic spotting" of L8 burley tobacco. Agron. J. 65:263-265.

MacLean, A. A., and T. C. Chiasson. 1966. Differential performance of barley varieties to varying aluminum concentrations. Can. J. Soil Sci. 46:147-153.

MacLean, A. J. 1974. Effects of soil properties and amendments on the availability of zinc in soils. Can. J. Soil Sci. 54:369-378.

Martin, W. E., J. G. McClean, and J. Quick. 1965. Effect of temperature on the occurrence of phosphorus-induced zinc deficiency. Soil Sci. Soc. Am. Proc. 29:411-413.

Mattson, S., E. Koutler-Andersson, R. B. Miller, and K. Vahtra. 1951. Phosphate relationships of soil and plants: VIII. Electrokinetics, amphoteric behavior and solubility relationships of calcium phosphates. Ann. R. Agric. Coll. (Sweden) 18:128-153.

McCormick, L. H., and F. Y. Borden. 1972. Phosphate fixation by aluminum in plant roots. Soil Sci. Soc. Am. Proc. 36:799-802.

McCormick, L. H., and F. Y. Borden. 1974. The occurrence of aluminum-phosphate precipitate in plant roots. Soil Sci. Soc. Am. Proc. 38:931-934.

Mehta, B. V., and N. K. Patel. 1969. Effect of different iron-, potassium-, phosphate-, and calcium-manganese relationships on the growth and chemical composition of aromatic strain of Bidi tobacco (*Nicotiana tabacum* L.). Plant Soil 30:305-315.

Mekaru, T., and G. Uehara. 1972. Anion adsorption in ferruginous tropical soils. Soil Sci. Soc. Am. Proc. 36:296-300.

Mikesell, M. E., G. M. Paulsen, R. Ellis, Jr., and A. J. Casady. 1973. Iron utilization by efficient and inefficient sorghum lines. Agron. J. 65:77-80.

Miller, M. H., and A. J. Ohlrogge. 1958. Principles of nutrient uptake from fertilizer bands: I. Effect of placement of nitrogen fertilizer on the uptake of band-placed phosphorus at different soil phosphorus levels. Agron. J. 50:95-97.

Miller, M. H., and V. N. Vij. 1962. Some chemical and morphological effects of ammonium sulfate in a fertilizer phosphorus band for sugar beets. Can. J. Soil Sci. 42:87-95.

Miller, R. J., J. H. Peverly, and D. E. Koeppe. 1972. Calcium-stimulated ^{32}P accumulation by corn roots. Agron. J. 64:262-266.

Ohki, K. 1975. Manganese and boron effects on micronutrients and phosphorus in cotton. Agron. J. 67:204-207.

Olsen, S. R. 1953. Inorganic phosphorus in alkaline and calcareous soils. *In* W. H. Pierre and A. G. Norman (ed.) Soil and fertilizer phosphorus in crop nutrition. Agronomy 4:89-122. Academic Press, Inc., New York.

Olsen, S. R. 1972. Micronutrient interactions. p. 243-264. *In* J. J. Mortvedt, P. M. Giordano, and W. L. Lindsay (ed.) Micronutrients in agriculture. Soil Sci. Soc. Am., Madison, Wis.

Olsen, S. R., and F. S. Watanabe. 1957. A method to determine a phosphorus adsorption maximum of soils as measured by the Langmuir isotherm. Soil Sci. Soc. Am. Proc. 21:144-149.

Olson, R. A., and A. F. Dreier. 1956. Nitrogen, a key factor in fertilizer phosphorus efficiency. Soil Sci. Soc. Am. Proc. 20:509-514.

Page, E. R., E. K. Schofield-Palmer, and A. J. MacGregor. 1963. Studies in soil and plant manganese. IV. Superphosphate fertilization and manganese content of young oat plants. Plant Soil 19:255-264.

Pauli, A. W., R. Ellis, Jr., and H. C. Moser. 1968. Zinc uptake and translocation as influenced by phosphorus and calcium carbonate. Agron. J. 60:394-396.

Peverly, J. H., R. J. Miller, C. Malone, and D. E. Koeppe. 1974. Ultrastructural evidence for calcium phosphate deposition by isolated corn shoot mitochondria. Plant Physiol. 54:408-411.

Racz, G. J., and P. W. Haluschak. 1974. Effects of phosphorus on Cu, Zn, Fe, and Mn utilization by wheat. Can. J. Soil Sci. 54:357-367.

Randall, P. J., and P. V. Vose. 1963. Effect of aluminum on uptake and translocation of phosphorus by perennial ryegrass. Plant Physiol. 38:403-409.

Rediski, J. H., and D. Biddulph. 1953. The absorption and translocation of iron. Plant Physiol. 28:576-593.

Reifenberg, A., and S. J. Buckwold. 1954. The release of silica from soils by orthophosphate anion. J. Soil Sci. 5:106-127.

Reisenauer, H. M., A. A. Tabikh, and P. R. Stout. 1962. Molybdenum reactions with soils and the hydrous oxides of iron, aluminum, and titanium. Soil Sci. Soc. Am. Proc. 26:23-27.

Reuther, W., and P. F. Smith. 1952. Iron chlorosis in Florida citrus groves in relation to certain soil constituents. Proc. Fla. State Hortic. Soc. 65:62-69.

Reyes, E. D., and J. J. Jurinak. 1967. A mechanism of molybdate adsorption on αFe_2O_3. Soil Sci. Soc. Am. Proc. 31:637-641.

Robertson, G. A., and B. C. Loughman. 1974. Reversible effects of boron on the absorption and incorporation of phosphate in *Vicia faba* L. New Phytol. 73:291-298.

Robertson, W. K., P. M. Smith, A. J. Ohlrogge, and D. M. Kinch. 1954. Phosphorus utilization by corn as affected by placement and nitrogen and potassium fertilization. Soil Sci. 77:219-220.

Robson, A. D., D. G. Edwards, and J. F. Loneragan. 1970. Calcium stimulation of phosphate absorption by annual legumes. Aust. J. Agric. Res. 21:601-612.

Rorison, I. H. 1965. The effect of aluminum on the uptake and incorporation of phosphate by excised sainfoin roots. New Phytol. 65:23-27.

Rossiter, R. C. 1952. Phosphorus toxicity in subterranean clover and oats grown on Muchea sand, and the modifying effects of lime and nitrate nitrogen. Aust. J. Agric. Res. 3:227-243.

Rossiter, R. C. 1955. The influence of soil type on phosphorus toxicity in subterranean clover. Aust. J. Agric. Res. 6:1-8.

Rothstein, A. 1955. Relationship of the cell surface to electrolyte metabolism in yeast. p. 65-100. *In* A. M. Shanes (ed.) Electrolytes in biological systems. Waverly Press, Inc., Baltimore.

Roy, A. C., M. Y. Ali, R. L. Fox, and J. A. Silva. 1971. Influence of calcium silicate on phosphate solubility and availability in Hawaiian latosols. *In* Proc. Int. Symp. Soil Fert. Eval. (Comm. II & IV, Int. Soc. Soil Sci.) New Delhi, India. 9-14 Feb. 1971. I:757-765.

Russell, E. W. 1973. Soil conditions and plant growth. 10th ed. Longman Group Ltd., London.

Seatz, L. F., A. J. Sterges, and J. C. Kramer. 1959. Crop response to zinc fertilization as influenced by lime and phosphorus applications. Agron. J. 51:457-459.

Sharma, K. C., B. A. Krantz, A. L. Brown, and J. Quick. 1968. Interactions of Zn and P with soil temperatures in rice. Agron. J. 60:652-655.

Shere, S. M., and L. Jacobson. 1970. The influence of phosphate uptake on cation uptake in *Fusarium oxysporum* f. sp. *vasinfectum*. Physiol. Plant. 23:294-303.

Shive, J. W. 1918. Toxicity of monobasic phosphates toward soybeans grown in soil and solution-cultures. Soil Sci. 5:87–122.

Simpson, K. 1961. Factors influencing uptake of phosphorus by crops in Southeast Scotland. Soil Sci. 92:1–14.

Smilde, K. W. 1973. Phosphorus and micronutrient metal uptake by some tree species as affected by phosphate and lime applied to an acid sandy soil. Plant Soil 39:131–148.

Smilde, K. W., P. Koukoulakis, and B. van Luit. 1974. Crop response to phosphate and lime on acid sandy soils high in zinc. Plant Soil 41:445–457.

Soltanpour, P. N. 1969. Effect of nitrogen, phosphorus, and zinc placement on yield and composition of potatoes. Agron. J. 61:288–289.

Spencer, W. F. 1966. Effect of copper on yield and uptake of phosphorus and iron by citrus seedlings grown at various phosphorus levels. Soil Sci. 102:296–299.

Stout, P. R., W. R. Meagher, G. A. Pearson, and C. M. Johnson. 1951. Molybdenum nutrition of crop plants: I. The influence of phosphate and sulfate on the absorption of molybdenum from soils and solution cultures. Plant Soil 3:51–87.

Stukenholtz, D. D., R. J. Olsen, G. Gogan, and R. A. Olson. 1966. On the mechanism of phosphorus-zinc interaction in corn nutrition. Soil Sci. Soc. Am. Proc. 30:759–763.

Syers, J. K., M. G. Bowman, G. W. Smillie, and R. B. Corey. 1973. Phosphate sorption by soils evaluated by the Langmuir adsorption equation. Soil sci. Soc. Am. Proc. 37:358–363.

Taber, H. G., and W. W. McFee. 1972. Nitrogen influence on phosphorus uptake by *Pinus radiata* seedlings. Forest Sci. 18:126–132.

Taber, H. G., and W. W. McFee. 1974. Kinetic analysis of nitrogen influence on phosphate absorption by excised roots of Monterey pine (*Pinus radiata*) seedlings. Forest Sci. 20:279–282.

Tanada, T. 1955. Effects of ultraviolet radiation and calcium and their interaction on salt absorption by excised mung bean roots. Plant Physiol. 30:221–225.

Taylor, A. W. 1961. Review of the effects of siliceous dressings on the nutrient status of soils. J. Agric. Food Chem. 9:163–165.

Thien, S. J., and W. W. McFee. 1970. Influence of nitrogen on phosphorus absorption and translocation in *Zea mays*. Soil Sci. Soc. Am. Proc. 34:87–90.

Tiffin, L. O. 1970. Translocation of iron citrate and phosphorus in xylem exudate of soybean. Plant Physiol. 45:280–283.

Toth, S. J. 1939. The stimulating effects of silicates on plant yields in relation to anion displacement. Soil Sci. 47:123–141.

Truog, E. 1953. Liming in relation to availability of native and applied phosphates. *In* W. H. Pierre and A. G. Norman (ed.) Soil and fertilizer phosphorus in crop nutrition. Agronomy 4:281–297. Academic Press, Inc., New York.

Truog, E., R. J. Goates, C. G. Gerloff, and K. C. Berger. 1947. Magnesium-phosphorus relationships in plant nutrition. Soil Sci. 63:19–25.

Wallace, A., and O. R. Lunt. 1960. Iron chlorosis in horticultural plants, a review. Proc. Am. Soc. Hort. Sci. 75:819–841.

Wallace, A., A. ElGazzar, and G. V. Alexander. 1973. High phosphorus levels on zinc and other heavy metal concentrations in Hawkeye and PI54619-5-1 soybeans. Commun. Soil Sci. Plant Anal. 4:343–345.

Ward, R. C., E. J. Langin, R. A. Olson, and D. D. Stukenholtz. 1963. Factors responsible for poor response of corn and grain sorghum to phosphorus fertilization. Soil Sci. Soc. Am. Proc. 27:326–330.

Warnock, R. E. 1970. Micronutrient uptake and mobility within corn plants (*Zea mays* L.) in relation to phosphorus-induced zinc deficiency. Soil Sci. Soc. Am. Proc. 34:765–769.

Watanabe, F. S., W. L. Lindsay, and S. R. Olsen. 1965. Nutrient balance involving phosphorus, iron, and zinc. Soil Sci. Soc. Am. Proc. 29:562–565.

Watanabe, F. S., S. R. Olsen, and C. V. Cole. 1971. Ionic balance and growth of five plant species in four soils. Agron. J. 63:23–28.

Wright, K. E. 1937. Effects of phosphorus and lime in reducing aluminum toxicity of acid soils. Plant Physiol. 12:173–181.

Wright, K. E. 1943. Internal precipitation of phosphorus in relation to aluminum toxicity. Plant Physiol. 18:708–712.

Chapter 24

Phosphate Nutrition of Corn, Sorghum, Soybeans, and Small Grains

J. J. HANWAY AND R. A. OLSON

Iowa State University, Ames, Iowa, and,
University of Nebraska, Lincoln, Nebraska, respectively

I. INTRODUCTION

The effectiveness and efficiency of P fertilizer use on corn, sorghum, soybeans, and small grains are of prime concern because: (i) these are the crops grown on most of the cropland and account for most of the crop production in the United States, and (ii) they are the crops for which about 60% of the P fertilizer is applied. These grain crops are grown on 85.5 million ha each year (Table 1) of which 29.7, 27.9, 24.6, 6.9, 6.5, and 4.4% is devoted to corn, wheat, soybeans, sorghum, oats, and barley, respectively. Of the total 261 million metric tons of grain produced, on the average, from 1972 to 1975, 52.4, 18.7, 14.3, 7.7, 3.7, and 3.2% was corn, wheat, soybeans, sorghum, oats, and barley, respectively.

Approximately 79% of the area devoted to grain crops and 83% of the total grain production in the United States are in 15 midwestern states. Corn and soybeans are produced primarily in Iowa, Illinois, and surrounding states. Sorghum is produced farther west and south, primarily from Nebraska to Texas. The area from Texas to North Dakota-Montana is the major wheat-producing region.

Much evidence has been accumulated to show that P deficiencies are common in these grain-producing areas. Although P availability in the soils varies widely, general use of P fertilizers is essential for the high yields now being produced. Approximately 61% of the total P fertilizer marketed in the U.S. is used in the 15 major grain-producing states.

Phosphate fertilization practices for these grain crops can be discussed by considering primarily practices for corn and wheat, the major grain crops. Fertilizer practices for these two crops are representative of the different types used for grain-crop production. They account for a major portion of the P consumption in the United States (Table 2). Soybeans are grown most extensively in a corn-soybean cropping system with the residual effect of the P fertilizer applied for the corn crop generally supplying adequate P for the soybean crop. Production and fertilization practices for

---

Copyright 1980 © ASA-CSSA-SSSA, 677 South Segoe Road, Madison, WI 53711, USA.
The Role of Phosphorus in Agriculture.

Table 1—Average grain production† of grain crops and average phosphate fertilizer consumption‡ in United States and individual states.

State	Grain production (1,000 metric tons)							Area of grain crops (10^6 ha)	Phosphate fertilizer consumption (1,000 metric tons P)
	Crop								
	Corn	Sorghum	Soybeans	Wheat	Oats	Barley	Total grain		
Iowa	28,046	42	6,083	39	1,149	--	35,359	8.1	158
Illinois	25,393	100	7,041	1,488	365	13	34,400	8.2	190
Nebraska	12,526	2,667	824	2,589	344	26	18,976	4.8	54
Minnesota	11,268	--	2,657	2,038	1,686	747	18,396	6.0	106
Kansas	3,510	4,551	682	9,279	88	60	18,170	6.9	60
Indiana	12,669	36	3,115	1,216	174	11	17,221	4.3	116
Texas	1,838	9,165	210	2,221	232	50	13,716	4.9	117
Ohio	7,023	--	2,360	1,430	396	13	11,222	3.5	107
Missouri	4,924	684	2,982	1,024	61	9	9,684	3.6	75
North Dakota	221	--	97	6,327	1,021	1,853	9,519	5.4	44
Michigan	3,358	--	3,979	795	266	22	8,420	1.5	59
South Dakota	2,978	230	214	1,626	1,342	388	6,778	3.4	24
Wisconsin	4,628	--	130	53	1,069	23	5,903	1.6	55
Oklahoma	158	662	123	3,692	81	123	4,839	2.8	42
Montana	19	--	--	3,214	154	1,161	4,548	2.5	24
Total (15 states)	118,559	18,137	30,497	37,031	8,428	4,499	217,151	67.5	1,231
% of U.S. Total	87	91	82	76	87	54	83	79	61
U.S. Total	136,903	20,021	37,378	48,850	9,642	8,353	261,147	85.5	2,009

† From USDA Statistical Reporting Service (1976) (1972–1975 average).
‡ From Hargett (1975).

sorghum are reasonably similar to those for corn, although sorghum is a somewhat better "farmer" for scavenging such nutrients as are available in the soil. Practices for oats and barley are similar to those for wheat except as they are influenced by the different soil and climatic conditions in the regions where they are grown.

Most of the P fertilizer used for grain crops is broadcast and plowed under or disked in before the crop is planted. This may be as a phosphate material or as NP, PK, or NPK fertilizer. Some fertilizer, usually a NP or NPK material, is applied in a band near the row at planting time as a "starter" fertilizer for corn, and concentrated superphosphate (CSP) without N or K is drilled with the seed as wheat is planted on P-deficient soils of

Table 2—Average P fertilizer use on United States grain crops.

Crop	Avg. kg of P/ha†	Avg. P fertilizer use (as % of total U.S. consumption)‡
Corn, sorghum	27	43
Wheat, oats, barley	8	14
Soybeans	6	6
Grain crop total	--	63

† Avg. of 1972–1974.
‡ Total U.S. consumption (avg. of 1972–1974) = 4,551,425 metric tons.

the western U.S. The starter for corn is most commonly employed where soil temperatures are low at planting time, especially in the northern part of the corn-growing region. However, farmers generally are planting earlier when soils are colder so the use of "starter" fertilizers is moving south.

II. PHOSPHORUS UPTAKE

A. Seasonal Demand Patterns

The seasonal pattern of growth and development generally is similar for all grain crops. After the seed germinates and the plant begins to grow, it then progresses through a period of vegetative growth followed by flowering, pollination, and subsequent grain formation. During the early period of vegetative growth, the rate of dry matter accumulation increases exponentially. This continues until enough leaves have been developed to intercept most of the sunlight, after which the daily rate of dry matter accumulation becomes essentially constant. This constant rate continues until later in the seed-filling stage when it decreases and then stops at maturity.

Phosphorus accumulation in the plants follows a seasonal pattern very similar to that for dry matter accumulation. The pattern of P accumulation in different plant parts differs from dry matter in that P accumulated in leaves, roots, and other vegetative plant parts is translocated to the seeds during the seed-filling period. This serves as an important source of P for the developing seeds, but results in severe depletion of P in the leaves and other vegetative plant parts during that period.

B. Rooting Pattern and Root Activity

The rooting pattern differs for the different crops. Early season root development of corn and wheat consists of a radical and a few seminal roots that develop from the seed soon after it is planted. These roots are most effective during the first 3 to 4 weeks of the season. During this time whorls of nodal roots develop from successive nodes of the stem near the soil surface. These roots develop in an "umbrella-like" configuration that spreads on all sides of the plants penetrating down into the soil to depths of 150–180 cm or more and branching extensively, especially in the surface plow-layer of soil. The root system of soybeans develops as a tap root with extensive branches and nodules, the latter developing primarily in the plow-layer and near the base of the plant.

Root growth and activity in nutrient uptake are dependent upon the supply of soluble sugars available to the roots. As the young plants develop for a short time after planting, these sugars are provided from carbohydrates stored in the seed. Later, as leaves on the plants develop, sugars are produced by photosynthesis in the leaves. During the period of vegetative growth of the plants, translocation of sugars to the roots is adequate and root growth is rapid. Later, as seeds develop, the sugars produced by photosynthesis in the leaves are channeled primarily to the developing seeds. Therefore, as the seed develops, the proportion of sugars translocated to the roots is limited so root growth slows and stops. In legumes nodules die and are sloughed off; N-fixation and much of the nutrient uptake by the roots slows and stops.

Environmental conditions—moisture, aeration, temperature, nutrient availability, and chemical and physical properties of the soil—influence root development and activity of all of these grain crops. Roots will not grow into a dry soil or absorb nutrients from a soil that has become dry. Excess water, or any other factor that limits aeration, restricts root development. Growth and nutrient uptake by roots are severely limited by cold temperatures or any nutrient deficiency. Although roots are not attracted to zones of high fertility, they do branch and develop profusely when they grow into zones of high concentration of available nutrients, but do not develop in zones of high salt concentration. Hard pans or other physical barriers limit root development in many soils. Some chemical conditions, such as strong acidity with associated Al toxicity, can also be inhibitive to root development. Thus, roots grow and develop most profusely in zones that are warm, moist, and well-aerated, and that have good physical conditions and high nutrient availability without excess salts. In most soils, these characteristics are most favorable in the surface 15 to 30 cm. Thus, a large proportion of the plant roots and their activity in nutrient uptake is concentrated in this area. Roots of grain crops may penetrate to depths of 2 m or more and this is often of great importance primarily for moisture uptake. The small proportion of roots found at these depths, however, can be responsible for a major portion of root activity for nutrient uptake in later growth stages with unfavorable conditions above of dried soil and depleted nutrient supply (Gass et al., 1971).

C. Foliar Absorption of Phosphorus

Although nutrient absorption into plants is primarily through the roots and fertilizers for field grain crops usually are applied to the soil, nutrients can be absorbed through the plant leaves and fertilizers can be applied as foliar sprays of fertilizer solutions. However, the amounts of salt (fertilizer) solution that can be applied at any one time as a foliar spray is limited because excessive amounts cause serious "leafburn" and many foliar sprayings are required to supply adequate P, and other nutrients, throughout the growing season. Therefore, fertilizer applications to the soil to supply adequate P for uptake through the roots is the most feasible method of assuring adequate P to the plants if applications are made when the roots can absorb the P.

There are times when plants do not effectively absorb P and certain other nutrients through the roos, even though the nutrients may be available in the soil. With grain crops, this occurs during the seed-filling period. During this period, photosynthate is channeled from the leaves to the developing seeds with little going to the roots to serve as a source of energy for nutrient uptake. During this period nutrients (especially N, P, K, and S) are translocated from the leaves and other plant parts to the developing seeds resulting in nutrient depletion of the leaves and a reduction in the rate of photosynthesis. Garcia and Hanway (1976) have shown that foliar fertilization during the seed-filling period can be an effective method of increasing soybean yields over and above yields obtained with adequate soil fertilization. The foliar spray solution must supply N, S, and K in addition to P. Barel (1975) has shown that greater amounts of P can be applied effectively as a foliar spray of polyphosphate than of orthophosphate. Preliminary results of Hanway at Iowa State University (unpublished) indicate that foliar fertilization with appropriate nutrient solutions during the seed-filling period can be effective on all of the grain crops considered in this chapter. However, other researchers have had little success with this technique. Obviously, more research is needed before foliar fertilization of field grain crops can become a practical practice for most farmers.

Other factors, such as dry surface soil, low soil temperature, or fixation of added nutrients by soils, may present other conditions where foliar fertilization of these grain crops is especially advantageous.

III. PHOSPHORUS REMOVAL AND PHOSPHORUS RECYCLING IN CROP RESIDUES

The amount of P removed in an average yield of harvested grain varies from 7 to 15 kg P/ha (16 to 34 kg P_2O_5/ha) with from 2 to 8 kg/ha taken up by the plants but returned to the soil in the crop residues left in the field (Table 3). Higher or lower crop yields will result in higher or lower amounts of P in the plants. With such P removal in the harvested grain, regular applications of P fertilizers are essential to maintain adequate levels of available P for optimum crop production on most soils except for those with large reserves of available P. Obviously, the amounts of P fertilizer required to maintain adequate levels of available soil P vary among soils and different crops, and will vary with yield of any one crop.

With recent expanded use of crop residues as forage for animals and as a potential source of energy in resolving national energy problems, it is apparent that annual soil P replenishment from crop residues will be less than in the past. Such use of the residues can be expected to result in greater fertilizer P requirements in the future.

Table 3—Phosphorus content of the above-ground portion of different grain crops.

Crop	Grain yield (kg/ha)	P content (kg P/ha)	
		Grain	Plant less grain
Corn	5,000	15	6
Sorghum	4,000	10	6
Soybeans	1,800	13	2 (+4 in leaf-fall)
Small grains—wheat	2,400	9	2
—oats	1,600	7	4
—barley	1,900	7	2

IV. YIELD—PHOSPHORUS CONCENTRATION RELATIONSHIPS

Since the P concentration in a plant reflects the adequacy of P for plant growth, many useful yield-P concentration relationships have been developed to serve as guides for recommending P fertilizer use. However, the P concentration varies among and within crop species, among different plant parts, and with age of the plants. Furthermore, since P exists in plants in many different forms, the amount of P found by any chemical analysis depends upon the forms measured in the determination.

Total P analysis of a given plant part at a given stage of plant development has been useful in establishing criteria of deficiency and sufficiency with many crop species. The part and time most commonly selected for the title grain crops have been a specified leaf sampled at or near anthesis—a leaf near the ear at silking for corn, a recently fully developed leaf near the top of the plant at anthesis (or full-bloom) for sorghum and soybeans, or the entire above-ground plant as the head is emerging from the boot for small grains. Generally, in such samples if the percentage of P (expressed on an oven-dry weight basis) exceeds 0.25%, the P concentration of the plant is considered sufficient. If the P concentration is less than 0.20% the plant is considered to be low in P, and if less than 0.15% the plant is considered to be very P deficient.

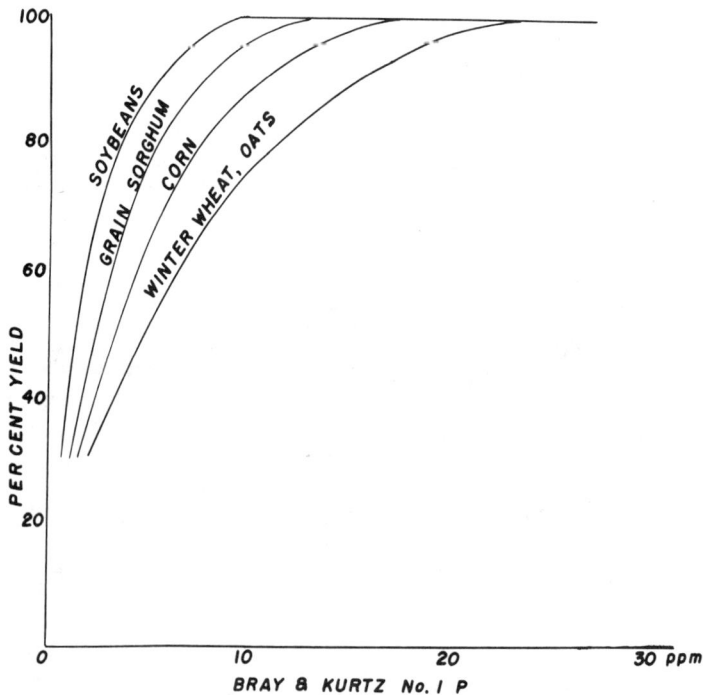

Fig. 1—Relation of soil P test levels to crop response to supplemental P under Nebraska environmental conditions (Olson & Sander, 1975)

(Percent yield = $\frac{\text{yield without applied P}}{\text{yield with applied P}} \times 100$).

V. FACTORS INFLUENCING PHOSPHORUS UPTAKE AND UTILIZATION

A. Species and Variety

The levels of P availability in soils required for optimum crop production vary among the different crops as shown in Fig. 1. Corn is significantly more responsive than soybeans to fertilizer P application (de Mooy et al., 1973), so in a corn-soybean cropping system it is advantageous to apply the P fertilizer for the corn crop. Similarly, wheat responds to P at higher soil P levels than required for maximum yields of corn, presumably related to the fact that wheat makes most of its growth under colder soil conditions than corn (Olson et al., 1962). Because the amounts of P removed vary among the different crops, the amounts of P fertilizer required to maintain optimum P levels in the soils also vary.

Differential varietal response to levels of P availability have also been recorded for the various species under consideration. Genotypic differences in rooting habit and activity along with varied yield potential presumably are responsible.

B. Other Nutrients

Applications of N fertilizer often have been shown to influence the P percentage in and total P uptake by nonleguminous plants (Bennett et al., 1962). For example, Bennett et al. (1963) reported increases in corn leaf P percentages and decreases in grain P percentages due to N fertilization on some soils. It appears that N deficiency can result in low P percentages in plants growing on soils with adequate available P.

The fact is reasonably well established that NH_4-N placed with fertilizer P in a band promotes crop uptake of the P, although reasons for the enhancement remain somewhat obscure (Blair et al., 1971; Fine, 1955; Miller & Ohlrogge, 1958; Olsen & Dreier, 1956). On the other hand, increased uptake of Zn and Ca from fertilizer and lime treatments commonly results in a decreased P content of the crop (Langin et al., 1962; Seatz et al., 1959).

C. Moisture

Phosphorus uptake is strongly influenced by soil moisture since plants cannot take up P from dry soil (Eck & Fanning, 1961). Maximum availability of the element is associated with a soil moisture tension of around 1/3 bar (Watanabe et al., 1960). In most soils P availability is highest in the surface plow layer and much lower in the subsoil. Therefore, if during a period of dry weather the surface soil becomes dry, plants suffer from a P deficiency even though moisture is still available in the subsoil and the

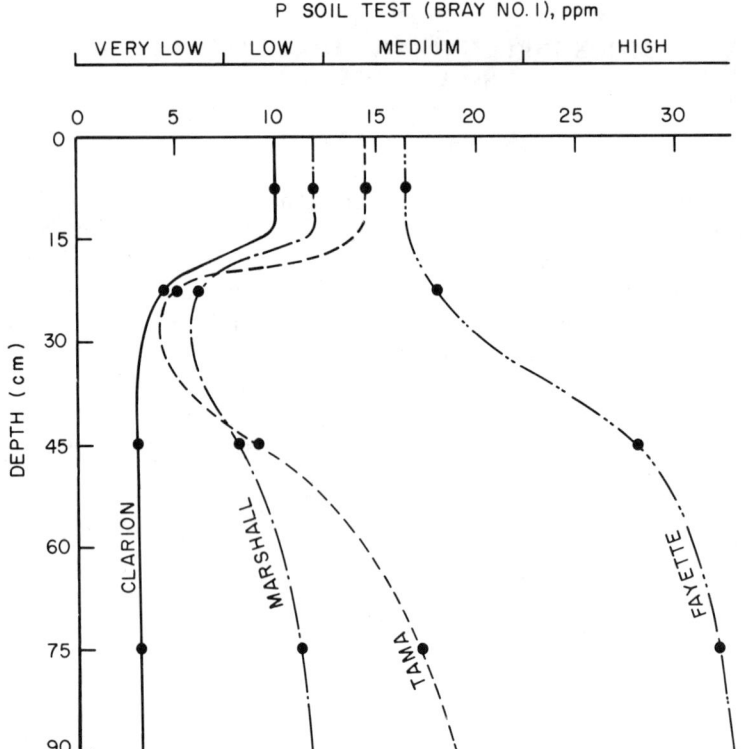

Fig. 2—Available soil P (Bray no. 1) of the 90-cm profile of major Iowa soils (Dumenil & George, 1968).

plants show no water stress. The situation cited is not axiomatic since, P availability is high in the subsoils of only a few soil series. For example, the Marshall, Tama, and Fayette series (Fig. 2) contain large reserves of available P in the deep subsoil below the characteristic grass depletion zone in the upper subsoil. This depletion zone is most apparent in the Marshall and Tama soils. The Clarion series, on the other hand, is much lower in available P in the subsoil. There are extensive areas of crop production in the 15 states (Table 1) where subsoil P is similar to that of the Clarion series. However, there are also large acreages in the western part of the Corn Belt where subsoil P reserves are so high that it limits, if not precludes, response of grain crops to fertilizer P after almost a century of cultivation since native sod breaking.

Excessive moisture which results in poor soil aeration also restricts P uptake by plants. Banding fertilizer near the plants to provide a zone of very high nutrient concentration, such as a row application of starter fertilizer for corn, results in very striking increases in plant growth on poorly drained soils.

Topdressed fertilizer P applications are not very effective for annual crops on medium- to fine-textured soils and are almost totally ineffective where the surface soil remains dry after application.

D. Temperature

In much of the area devoted to production of the grain crops under discussion, low soil temperatures early in the season restrict nutrient uptake by the plants. This becomes especially important in the northern parts of the Corn Belt, since research has shown that earlier planting, even though soil temperatures are lower, generally results in higher yields, and farmers are adopting this practice. As the planting date is moved earlier, greater responses from starter fertilizer containing P can be expected. The beneficial effects of banding P fertilizer as compared with mixed placement increases with decreasing temperature (Ketcheson, 1957).

Wheat is more responsive to applied P than are corn and soybeans, in part because of the temperature factor. It is planted in the fall with most of its tillering accomplished in the late fall and early spring when soil temperatures are much lower than those encountered during early growth of the other crops. Calibration ranges that have been effected in soil testing accordingly are notably higher for wheat than for the other crops even though total P removed by wheat is no greater than, if as great as, total P removed by the other crops (Fig. 1). The temperature differential at time of planting undoubtedly contributes to this difference as well as to the lower soil test requirement of soybeans and grain sorghum compared with corn. In some years the stimulation in root and top growth from P application in the more northerly regions can make the difference between a good stand of wheat in the following spring and one badly decimated by winter killing.

E. Efficiency of Fertilizer Phosphorus

Soil moisture status has a predominating influence on the availability to crops of any fertilizer P increment as has been elaborated above. Otherwise, from the standpoint of fertilizer P management, it is essential that any P applied to the soil for the benefit of annual grain crops be placed in the soil rooting zone at or before the time of planting. Later broadcast application to the soil surface after the crop is up will not be effective, except for sands, and will be responsible for serious root pruning if banding near the seed row is attempted.

Banding 4 to 5 cm to the side of the seed row has generally proved to be the most effective placement in deriving maximum "starter" benefit and greatest yield increase per unit of applied P in the case of corn (Nelson, 1956; Welch et al., 1966). The more water-soluble the P carrier the more advantageous is banding over mixed placement of the fertilizer (Lawton et al., 1956). Explanation for this response to banding is associated with the earlier planting dates employed by farmers. Early root development is primarily in a lateral direction in the surface few centimeters of soil until warming downward from this zone occurs as the season progresses (Mosher & Miller, 1972). Grain sorghum and soybeans planted at a later date do not respond the same as corn in this respect. They appear to derive quickest starter response from P placed below the seed, although the issue in this case is not as clearly defined, since the latter crops have proved less responsive to applied P than has corn. One recent investigation in Minnesota indicated best response of soybeans on a low P soil when moisture was favorable with a combined broadcast and row starter treatment (Ham et al., 1973). With low moisture and low soil P, broadcasting alone was superior, but with favorable moisture and very high soil P, neither placement was effective and both caused yield depression. Most studies on placement of fertilizer P for wheat have demonstrated decidedly superior results from application in or very near to the seed row (Duley, 1930; Lutz et al., 1961; Olson & Rhoades, 1953). Rather large application rates are possible without damage to germination with straight phosphatic carriers, but with any significant N and/or K component some separation of seed and fertilizer band becomes necessary (Miller & Ohlrogge, 1977).

Efficiency of fertilizer P in influencing yield is controlled to a very considerable degree by soil P availability in horizons below the surface soil. For example, many soils of the western Corn Belt that have developed on loess and are noncalcareous into the parent material have very high levels of available P in the lower rooting profile. Corn growing on these soils shows some visual response to applied P in early growth that disappears as the root system becomes active in the lower soil horizons (Olson et al., 1962).

In the grain-producing regions of the United States, however, there are soils very low in available P throughout the crop rooting profile that require something more than starter effects. In such cases, high production can be obtained only by incorporating P at sufficient rates to effect a modest equilibrium level of available P in the plow layer, supplemented by annual row applications of smaller starter rates.

The high efficiency of foliar-applied nutrients during the grain-forming stage of crop development as root activity decreases may modify somewhat earlier concepts of optimum timing and placement of P.

VI. RELATION BETWEEN PHOSPHORUS NUTRITION AND QUALITY OF HARVESTED PLANT PARTS

Phosphorus nutrition, except where P is extremely deficient, generally has no important effects on the quality of the grains produced.

Delayed maturity is one of the commonly observed characteristics of a crop seriously deficient in P. Appropriate application of fertilizer P will correct this condition, affording grain of low moisture content at harvest. This does not mean to say, however, that fertilizer P will hasten maturity of any one of these crops already functioning at a normal level of P nutrition.

LITERATURE CITED

Barel, Dirk. 1975. Foliar applications of phosphorus compounds. Ph.D. Thesis. Iowa State Univ., Ames, Iowa. Microfilm No. 76-1820.

Bennett, W. F., J. Pesek, and J. Hanway. 1962. Effect of nitrogen on phosphorus absorption by corn. Agron. J. 54:437-442.

Bennett, W. F., G. Stanford, and L. Dumenil. 1953. Nitrogen, phosphorus, and potassium content of corn leaf and grain as related to nitrogen fertilization and yield. Soil Sci. Soc. Am. Proc. 17:252-258.

Blair, Graeme J., C. P. Mamaril, and M. H. Miller. 1971. Influence of nitrogen source on phosphorus uptake by corn from soils differing in pH. Agron. J. 63:235-238.

deMooy, C. J., J. L. Young, and J. D. Kaap. 1973. Comparative response of soybeans and corn to phosphorus and potassium. Agron. J. 65:851-855.

Duley, F. L. 1930. Methods of applying fertilizers to wheat. J. Am. Soc. Agron. 22:515-521.

Dumenil, L., and R. George. 1968. Guard against drought induced nutrient hunger. Better Crops Plant Food Vol. L11, no. 3:2-6.

Eck, Harold V., and Carl Fanning. 1961. Placement of fertilizer in relation to soil moisture supply. Agron. J. 53:335-338.

Fine, L. O. 1955. The influence of nitrogen and potassium on the availability of fertilizer phosphorus. So. Dak. Agric. Exp. Stn. Bull. 453 (North Central Regional Publication 67).

Garcia, L. Ramon, and J. J. Hanway. 1976. Foliar fertilization of soybeans (*Glycine max* (1) Merrill) during the seed-filling period. Agron. J. 68:653-657.

Gass, W. B., G. A. Peterson, R. D. Hauck, and R. A. Olson. 1971. Recovery of residual nitrogen by corn (*Zea mays* L.) from various soil depths as measured by ^{15}N tracer techniques. Soil Sci. Soc. Am. Proc. 35:290-294.

Ham, G. E., W. W. Nelson, S. D. Evans, and R. D. Frazier. 1973. Influence of fertilizer placement on yield response of soybeans. Agron. J. 65:81-84.

Hargett, N. L. 1975. Fertilizer summary data, 1974. National Fertilizer Development Center, TVA, Muscle Shoals, Ala.

Ketcheson, J. W. 1957. Some effects of soil temperature on phosphorus requirements of young corn plants in the greenhouse. J. Soil Sci. 37:41-47.

Langin, E. J., R. C. Ward, R. A. Olson, and H. F. Rhoades. 1962. Factors responsible for poor response of corn and grain sorghum to phosphorus fertilization: II. Lime and P placement effects on P-Zn relations. Soil Sci. Soc. Am. Proc. 26:574-578.

Lawton, K., C. Apostolakis, R. L. Cook, and W. L. Hill. 1956. Influence of particle size, water solubility, and placement of fertilizer on the nutrient value of phosphorus in mixed fertilizers. Soil Sci. 82:465-476.

Lutz, J. A., Jr., G. L. Terman, and J. L. Anthony. 1961. Rate and placement of phosphorus for small grains. Agron. J. 53:303-305.

Miller, M. H., and A. J. Ohlrogge. 1958. Principles of nutrient uptake from fertilizer bands: 1. Effect of placement of nitrogen fertilizer on the uptake of band-placed phosphorus at different soil phosphorus levels. Agron. J. 50:95-97.

Miller, M. H., and A. J. Ohlrogge. 1977. Fertilizer factors. p. 31-53. *In* G. E. Richards (ed.) Band application of phosphatic fertilizers. Olin Corp., Little Rock, Ark.

Mosher, P. N., and M. H. Miller. 1972. Influence of soil temperature on the geotrophic response of corn roots (*Zea mays* L.). Agron. J. 64:459-462.

Nelson, L. B. 1956. The mineral nutrition of corn as related to its growth and culture. Advan. Agron. 8:321-375.

Olson, R. A., and A. F. Dreier. 1956. Nitrogen, a key factor in fertilizer phosphorus efficiency. Soil Sci. Soc. Am. Proc. 20:509-514.

Olson, R. A., A. F. Dreier, C. A. Hoover, and H. F. Rhoades. 1962. Factors responsible for poor response of corn and grain sorghum to phosphorus fertilization: I. Soil phosphorus level and climatic factors. Soil Sci. Soc. Am. Proc. 26:571-574.

Olson, R. A., and H. F. Rhoades. 1953. Commercial fertilizers for winter wheat in relation to the properties of Nebraska soils. Nebr. Agric. Exp. Stn. Res. Bull. 172.

Olson, R. A., and D. H. Sander. 1975. The effective use of fertilizers in wheat production. p. 256–261. *In* V. A. Johnson (ed.) Proc. 2nd Int. Winter Wheat Conf. 9–19 June 1975. USARS, USAID, Univ. of Nebr. IANR, Zagreb, Yug.

Seatz, Lloyd, F., Athan J. Sterges, and James C. Kramer. 1959. Crop response to zinc fertilization as influenced by lime and phosphorus applications. Agron. J. 51:457–459.

U.S. Department of Agriculture, Statistical Reporting Service. 1976. Crop production: 1975 annual summary. Crop Reporting Board, Stat. Rep. Serv., USDA, Washington, D.C. CrPr 2-1(76).

Watanabe, F. S., S. R. Olsen, and R. E. Danielson. 1960. Phosphorus availability as related to soil moisture. Int. Congr. Soil Sci., Trans. 7th (Madison, Wis.) III:450–456.

Welch, L. F., D. L. Mulvaney, L. V. Boone, G. E. McKibben, and J. W. Pendleton. 1966. Relative efficiency of broadcast versus banded phosphorus for corn. Agron. J. 58:283–287.

Chapter 25

Phosphorus Nutrition of Cotton, Peanuts, Rice, Sugarcane, and Tobacco

L. E. NELSON

Mississippi State University
Mississippi State, Mississippi

I. INTRODUCTION

Crops considered in this chapter occupy almost 14% of the world's cropland (FAO, 1974, 1975a, 1975b, 1977). Three of them, cotton, sugarcane, and tobacco, are normally well fertilized and, in some regions, these crops consume a large fraction of the fertilizers applied to crops. An attempt will be made to estimate the P requirements of these crops in terms of total uptake and of removal in the harvested portion of the crop. In addition, P composition and factors influencing it will be reviewed. Finally, a summary of foliar and tissue analyses will be presented along with critical concentrations where available.

Based on the 1974-1976 3-year average, cotton was produced on 31.9 million ha; the Soviet Union, China, and the United States accounted for 20.8, 18.6, and 17.0% of the world production of cotton lint, respectively (FAO, 1977). Other important producers are India, Pakistan, Brazil, Turkey, Egypt, Mexico, and Sudan. The area devoted to peanuts was 19.2 million ha; India and China accounted for .32.1 and 15.6% of the world production, respectively. Other important peanut-growing countries are the United States, Senegal, Nigeria, Sudan, Brazil, Argentina, and Burma. Rice was grown on 140 million ha; China and India accounted for 34.3 and 20.1% of the world production, respectively. Other important producers are Indonesia, Bangladesh, Japan, Thailand, Burma, Brazil, Korea, and the Philippines. Although the United States is 11th in world production, producing less than 2%, it is one of the largest exporters of rice (FAO, 1975b). Sugarcane was grown on 12.5 million ha; India and Brazil accounted for 21.2 and 14.4% of the world production, respectively. Other important producers are Cuba, China, Mexico, United States, the Philippines, Pakistan, South Africa, and Indonesia. Tobacco was grown on 43 million ha; China and the United States accounted for 18.6 and 17.3% of the world production, respectively. Other important producers are India, USSR, Brazil, Turkey, Bulgaria, Japan, Canada, and Rhodesia.

Copyright 1980 © ASA-CSSA-SSSA, 677 South Segoe Road, Madison, WI 53711, USA.
The Role of Phosphorus in Agriculture.

II. COTTON

A. Introduction

1. THE PLANT

Four species of cotton are cultivated. *Gossypium herbaceum* L. and *G. arboreum* L. are the Old World species and *G. hirsutum* L. (American Upland) and *G. barbadense* L. (Sea Island) are the New World species. The USA crop is 99% American Upland and it constitutes a high proportion of the commercial crop of most cotton-growing countries (Green, 1972).

Cotton is cultivated in both the tropics and subtropics and is grown under irrigated conditions as well as natural rainfall. It is herbaceous and is considered an annual, but in the tropics where the mean temperature of the coldest months does not fall below 18°C, it is a long-lived perennial. Cotton is indeterminate, carrying both flowers and fruit at the same time, which has important consequences in the nutrition of the plant.

2. PHOSPHORUS FERTILIZATION AND CROP RESPONSE

a. Total P in the Cotton Plant—According to Christidis and Harrison (1955), the dry matter distribution in the above-ground parts of mature cotton plants is: stems 25.4%, leaves 22.2%, burs 15.6%, seed 25.2%, and lint 11.6%. The production of 560 kg of lint/ha requires 17.4 kg of P/ha; 10.4 kg of P/ha for the vegetative parts, and 7.0 kg of P/ha for the seed. In addition, the roots require 1.0 kg of P/ha. The vegetative to seed ratio is not constant, however, and they illustrate with some data from Georgia that the higher the yield, the lower the amount of nutrients needed for the production of a certain amount of seed cotton. More recent work (Bassett et al., 1970) in California at yield levels of 1,178 to 1,628 kg of lint/ha bears this out. At these yield levels 560 kg of lint can be produced with 8.4 kg of P in the above-ground parts.

With a world-wide yield average of only 405 kg of lint/ha in 1976 (FAO, 1977), the figure of Christidis and Harrison gives an estimate of the total P required by the world's cotton crop in 1976 of 400,000 metric tons. Since the seeds contained 38% of total P in the plant, the P removed in the harvest was 152,000 metric tons.

If world yield levels were similar to those reported by Bassett et al. (1970), i.e., 1,400 kg/ha, then world production in 1976 could have been obtained from 9.1 million ha. The P required would have been 190,000 metric tons, and since the seeds contained 58% of the P, 110,000 metric tons would have been removed in the harvest. The difference in removal of P at harvest is probably because the value for lint turnout used by Christidis and Harrison was 31%, while that of Bassett et al. was 41%.

b. Crop Response—Cotton is grown on a wide variety of soils; thus, the response to P fertilization is quite variable. In the United States, good responses are obtained on the sandy, highly weathered upland soils in the southeast, whereas little or no response is obtained on the fertile alluvium of the Mississippi flood plain or the irrigated soils of New Mexico, Arizona, and California. Also, because cotton is a high-value crop, fertilization is a practice which is often adhered to, regardless of the need. At the usual rate of P fertilization in the past, soil P levels have been raised to the point that little or no response is obtained on farmers' fields and there is little justification for more than a maintenance application (Jones & Bardsley, 1968).

Variable responses are obtained also in other parts of the world, the responses depending on the soils and, particularly, the climatic and weather conditions (Geus, 1967). Ten reports in the literature representing 310 location-years show that 1 kg of P increased the yield of seed cotton 15.5 ± 4.0 kg. Application rates ranged from 10 to 31 kg of P/ha with a mean of 15.9. Increases in seed cotton ranged from 3 to 42 kg/kg of P for this range of application. As rates of application increased, yield increments became smaller.

The quantities of P applied in the United States have recently been reported by Jones and Bardsley (1968): in the southeast, 30 to 39 kg of P/ha; in the Mississippi Valley, 0 to 37; in the Plains States, 15 to 39; and in the west, 0 to 39. In India, on large tracts of black cotton soils, rainfall is so unpredictable and erratic that no fertilizer is used (Geus, 1967). In other areas, fertilizers are used to some extent and from 5 to 20 kg of P/ha is recommended. In Africa and the Near East, P is one of the main requirements for cotton and is recommended.

3. EFFECT OF PLACEMENT METHODS AND TIME OF APPLICATION

In a review of fertilization of cotton, Jones and Bardsley (1968) discussed the problem of placement and pointed out that band placement or the use of granular materials will reduce fixation of the water-soluble phosphates in acid soils while mixing the low-water-soluble materials will increase availability. Band placement or use of granular materials having a high degree of water solubility is important on calcareous or alkaline soils. Since P moves very little in the soil, it must be placed in the rooting zone.

They also point out that P applications should coincide as closely as possible with seedling utilization of P. This is particularly true when P is being applied to soils that are low in available P and that have a high P-fixing capacity. Thus, application just prior to planting is recommended. If, on the other hand, P is being applied to soils high in available P merely as a maintenance application, then time of application is immaterial.

B. Phosphorus Concentration

1. CONCENTRATIONS IN VARIOUS TISSUES

Cottonseed has a higher concentration of P than any of the other tissues in the plant and the percent P in the seed is remarkably constant, averaging around 0.6. Leaves, burs, stems, roots, and lint follow in order of decreasing concentration of P (Christidis & Harrison, 1955; Bassett et al., 1970).

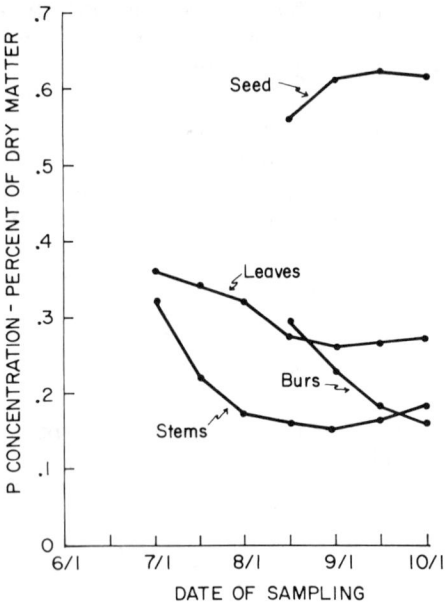

Fig. 1—The concentration of P in cotton plant parts. Mean of 6 sites (after Bassett et al., 1970).

As leaves and stems mature, concentration of P decreases (Fig. 1). In seeds, P concentration is very high at the beginning of development and increases slightly during maturation, while at the same time concentration in burs decreases.

2. EFFECTS OF OTHER NUTRIENTS AND ENVIRONMENTAL FACTORS

A. Other Nutrients—Effects of various nutrient regimes on the nutritional status of the cotton plant were investigated by Joham (1951). Plants were grown in sand cultures and low, medium, and high levels of macronutrients were applied. Changes in concentrations of soluble constituents were used as a measure of treatment effect. Increasing levels of P in the substrate generally increased soluble P in plants. At 90 and 145 days after emergence, increasing levels of N in the substrate decreased percent soluble P in petioles, and at 145 days, there was a negative correlation between substrate Ca levels and percent soluble P in the petioles. Samples were also obtained from a field experiment. The percent soluble P in petioles was negatively correlated with levels of N and K and positively correlated with levels of P. Decrease in percent soluble P due to N and K fertilization was associated with an increase in yield and may have been in part a dilution effect. Percent P in the cotyledons of seedling cotton has been found to decrease with increasing levels of Ca in the substrate (Nelson, 1971).

Skinner et al. (1944) found that variations in K fertilization had little effect on percent P in the whole cotton plant. Percent P tended to decrease as K application increased. However, the effect was much more apparent in 1939 than in 1940 and differed depending on the soil. The effect was more pronounced when yields were increased by K fertilization. In no case were the effects pronounced.

b. Environmental Factors—Fraps (1919) found the same variety grown in different locations to have different P concentrations in the tissues. This may be attributed to differences in soils and weather conditions. Kapp et al. (1953) presented evidence that a decrease in the availability of soil water decreased uptake of P. Concentration of P in cotyledons of seedlings has been found to increase with increasing root temperatures (Unpublished data, L. E. Nelson, Mississippi Agric. For. Exp. Stn.).

C. Phosphorus Uptake Demand Patterns

1. CUMULATIVE UPTAKE WITH TIME

Accumulation of P by cotton during growth is illustrated in Fig. 2 which is constructed from data derived from Bassett et al. (1970) and Olson and Bledsoe (1942). The former represents high-yielding, irrigated cotton grown in the western U.S.; the latter represents lower-yielding cotton grown under conditions of natural rainfall. In both cases, there is an initial period of very slow growth and P accumulation. Olson and Bledsoe (1942) compared the uptake of P by cotton grown on three different soils and found that only about 10% of total P taken up by the plant occurred during the first 90 days, i.e., from planting to early squaring. The pattern of P uptake was very similar to dry matter production. Bassett et al. (1970) found that a

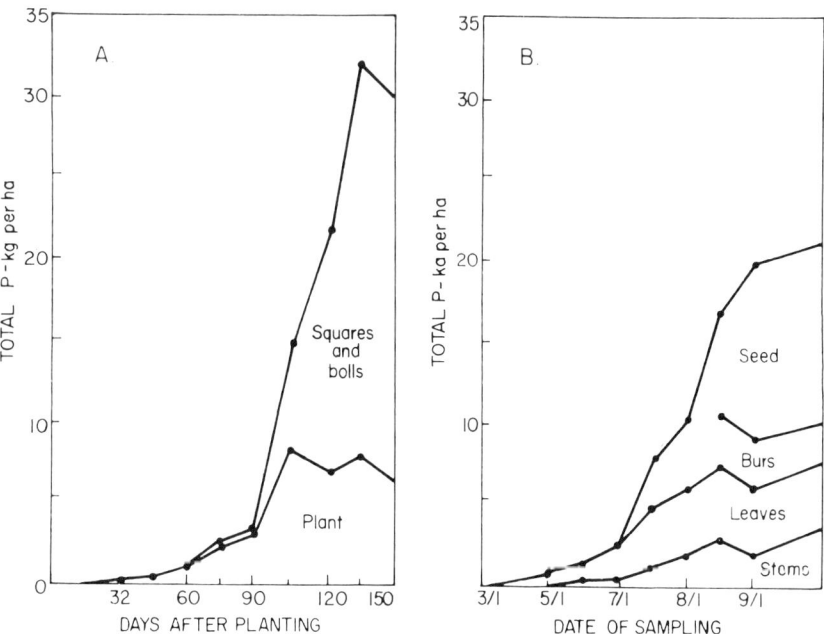

Fig. 2—Accumulation uptake and distribution of P by cotton: (*A*) Georgia (after Olson & Bledsoe, 1942); (*B*) California (after Bassett et al., 1970).

Table 1—Accumulation of dry matter and P by cotton in California expressed as a percent of the total accumulation. Average of six sites (after Bassett et al., 1970).

Item	Sampling date†						
	6/15	7/1	7/15	8/1	8/15	9/1	9/15
Dry matter	6	11	23	41	74	81	100
Phosphate	7	14	35	42	71	84	100

† Three planting dates. First flowers appeared on 25 June, 3 July, and 22 July for the 1 April, 1 May, and 1 June planting, respectively.

somewhat greater proportion of the P was taken up during the first part of the growing season (Table 1). About 35% of the P and 23% of the dry matter had been accumulated by July 15. Data obtained in India by Bhatt and Appukuttan (1971) suggest a similar pattern—29% of total P uptake at time of flowering and 20% of dry matter accumulation.

2. DISTRIBUTION AMONG VARIOUS TISSUES

Figure 1 shows that concentration of P in various tissues is different and that it also changes with time. If we look at the total P contained in a stand of cotton (Fig. 2), we see that at maturity most of it is in the reproductive tissue. This is further demonstrated in Table 2, which shows that from 48 to 82% of total P at maturity is in the reproductive tissue. The large variations reported in the literature are due to differences in yield and differences in morphology of the varieties represented.

There is little translocation of P from the vegetative tissues during fruiting so long as there is an adequate supply in the substrate (Eaton & Ergle, 1957; Ergle & Eaton, 1957; Fig. 2). The slight decrease in total P in the vegetative part of the plant in Fig. 2A may be due in part to leaf loss as a result of abscission. Bhatt et al. (1974) compared P accumulation of a normal cotton and two short-branch strains. Their data show that the vegetative portion of the plant contains 50% or less P at harvest than its maximum during the growing season. However, it is not possible to ascertain how much of the loss was due to translocation and how much to leaf abscission.

Table 2—The distribution of P among the various tissues of the cotton plant at maturity as reported in the literature.

Plant part	Percent of total seasonal uptake						
	Olson and Bledsoe (1942)	Bassett et al. (1970)	Bhatt and Appukuttan (1971)		Dastur and Ahad (1941)		McHargue (1926)
			MCU-1†	PRS-72‡	American	Desi§	
Stem	17.6¶	16.7	19.5	6.4	18.1	8.4	17.5
Leaves		19.0	26.2	22.7	34.2	24.6	23.6
Burs	82.4#	11.9	7.1	12.1	47.7#	67.0#	58.9#
Seed		52.4	47.2	58.7			

† Normal variety.   ‡ Short branch variety.   § Coarse, short staple variety.
¶ Includeds both stem and leaves.   # Includes both burs and seed

Perhaps because of its indeterminate nature, the cotton plant develops its reproductive tissues largely on the basis of continued uptake from the soil with little transfer from the leaves and stems. Under conditions of low P in the substrate, reproductive growth is limited more than vegetative growth, indicating that reproductive tissues do not have a higher priority for P in the plant than do the vegetative tissues (Ergle & Eaton, 1957).

3. EFFECT OF FERTILIZATION AND OTHER FACTORS

Ergle and Eaton (1957) compared the effect of low-P and high-P levels in sand and water cultures on the P nutrition of cotton. They found that P concentrations in the various tissues of plants grown in low-P solutions were one-fifth of that in plants grown in a high-P solution. In the field, however, where P levels in the soil are not extremely low, P fertilization may not affect greatly the P concentrations in the leaves (Samuels et al., 1959). Clark (1964) found no correlation between yield of lint and percent P in the plant, but found a good correlation ($r^2 = 0.67$) between yield of lint and total P uptake in kg/ha.

D. Yield-P Concentration-Age Relationships

1. TOTAL VS. SOLUBLE CONCENTRATIONS

Joham (1951), on the basis of the work of other researchers with various species, utilized a chemical test which determined the level of soluble P in the tissue. Samuels et al. (1959) in Puerto Rico, on the other hand, determined the total quantity of the element in the tissue. Since then, most analyses have determined the total P concentrations in the leaf blade (Anderson et al., 1971; Maples & Keogh, 1973; McClung et al., 1961; Sabbe et al., 1972). Determination of soluble constituents can be carried out using "quick" tests in the field, whereas analysis for total P has to be carried out in the laboratory.

Most work on tissue analysis of cotton has been directed towards assessing N and K status (Sabbe & McKenzie, 1973). In the development of tissue tests for cotton, Joham (1951), primarily on the basis of N and K studies, chose the leaf petiole as the most suitable tissue and analyzed for soluble constituents, while Samuels et al. (1959) in Puerto Rico found the concentration of total P in the leaf blade to be most suitable.

2. FACTORS AFFECTING CONCENTRATION-YIELD RELATIONSHIPS

Joham (1951) showed that the soluble P levels in the petioles of 90-day-old cotton were affected by varying the levels of N, P, and K applied to the soil (Table 3). Soluble P in petioles was increased by increasing P and decreased by increasing N and K in the substrate.

In the Sudan on an alkaline, clay soil with impeded drainage and high salts, percent P in the leaves at 5 weeks was negatively correlated in each of 3 years with yield of seed cotton (Jewitt, 1953), and there was a P concentration × season interaction. The yield differences were associated with higher

Table 3—Effect of varying fertilizer rates on the soluble P concentration in the petioles of cotton sampled at 90 days (from Joham, 1951).

Fertilizer nutrient varied	Correlation with soluble P in petiole, r*	Effect on yield	
		y_0	y_{max}
		%	
N	−0.79	52	100
P	0.56	23	100
K	−0.64	28	100

* All significant at $P = 0.01$ level.

presowing rainfall and with the presence or absence of a legume in the rotation. Undoubtedly the yield differences were due to differences in N supply, and high yields resulted in a dilution of the P. Thus, it appears that percent P × season interactions may make it difficult to establish relationships between percent P in the leaves and yield.

Stelly and Morris (1953) obtained an increase in yield and percent P in the whole plant from P fertilizer that had been applied 2 years earlier. There was no effect of lime when pH was increased from 5.5 to 6.5, but examination of their data suggests a slightly lower P concentration in plants growing on limed plots when yields were the same.

Differences in percent P in the leaves of several varieties were reported by Samuels et al. (1959). Their data suggest that varietal differences also will make it difficult to develop consistent relationships between percent P and yield.

3. CRITICAL P LEVELS

The quantities of fertilizer needed for optimum yields will in the forseeable future be determined by soil testing (Jones & Bardsley, 1968). Tissue analysis will still be useful for evaluation of a fertilizer program and as a diagnostic tool. In areas where the climate is consistently uniform or in climates where most of the water is supplied from irrigation, it may be possible to develop useful correlations.

Some of the common sampling methods and the foliar-P levels used to assess the P status of cotton are given in Table 4. It is evident from this table that there is a paucity of information on sampling methods and interpretation of tissue analyses.

E. Relationships Between P Nutrition and Plant Quality

1. PLANT CHARACTERISTICS

An adequate supply of P promotes earliness in cotton (Jones & Bardsley, 1968) but Lancaster and Savatli (1965) found that earliness was increased only if there was a yield increase.

Nelson and Ware (1932) reported that the size of boll and the lint in 100 bolls were increased slightly by P but that the weight of the seed, percentage of lint, and staple length were not affected. In another study Nelson (1949)

Table 4—Sampling methods for foliar diagnosis in cotton and the foliar-P levels used to assess the P status of the plant.

Geographical location	Plant part	No. of plants per sample and age	Determination	P in tissue	Reference
South Carolina	Petioles from third and fourth nodes from apex of main stem	Not given	Soluble in sodium acetate	Critical levels: 0.016% fresh wt (field grown) 0.019% fresh wt (greenhouse)	Joham, 1951
Puerto Rico, Sea Island variety	Leaf blades from third and fourth nodes from the apex of main stem.	12 plants per sample—45 days	Total P	No response when leaf P is above 0.40%.	Sanuels et al., 1959
Western U.S.	Petioles of most recent fully expanded leaf.	Not given	Soluble PO_4-P, ppm	Safe levels: First bloom 1,500–2,000 Peak bloom 1,200–1,500 First open boll 1,000–1,200 Maturity 800–1,000	Soil Improve. Comm., 1975
Arkansas, U.S.	Blades	1 August	Total P	Sufficiency range: 0.30–0.65%	Sabbe et al., 1972
Arkansas, U.S.	Blades of most recent fully expanded leaf	20 per plot	Total P	Critical levels: Squaring 0.31% Peak bloom 0.33% Late maturing 0.24%	Maples & Keogh, 1973
Brazil	Blades	Not given	Total P	Adequate—0.2%	Malavolta & Gomes, 1960

also observed an increase in yield and in weight of boll, but fiber length, fineness, strength, fiber weight per inch, percentage of mature fibers, and percentage of oil and N in seed were not affected by rates of P (0, 24.5, and 49.0 kg/ha). Eaton (1955) after reviewing the literature concluded that P fertilization usually has no effect on oil or N composition.

2. DISEASES

The effect of P fertilization on disease susceptibility of cotton has been summarized by Presley and Bird (1968). Phosphorus helps reduce the severity of *Phymatotrichum* root rot; and while it may have no influence on cotton susceptibility to wilts, P may increase the severity of infection in the presence of high rates of N.

III. PEANUT

A. Introduction

1. THE PLANT

The cultivated species of peanut, which is known outside of the United States as *groundnut*, consists of two subspecies, each containing two botanical varieties (Hammons, 1973). The subspecies *hypogea* includes the varieties *hypogaea* (in the United States, the Virginia and Runner market types) and *hirsuta* Kohler. The subspecies *fastigiata* Kohler includes the varieties *fastigiator* (the Valencia type) and *vulgaris* Harz (Spanish or bunch, erect or upright).

The peanut is an annual legume with a well-developed taproot system. Height varies from 15 to 30 cm. It requires a 100- to 140-day frost-free period, relatively high temperatures, and moderate moisture supplies. It is generally grown in very sandy soils since growth in medium- to fine-textured soils results in stained and dark-colored pods and difficulties in harvest.

2. PHOSPHORUS FERTILIZATION AND CROP RESPONSE

a. Total P in the Peanut Plant—The approximate P concentrations in the haulms, shells, and kernels of peanuts at harvest are 0.07, 0.03, and 0.36%, respectively (Bromfield, 1973; Bunting & Anderson, 1960; Collins & Morris, 1941). The values for the stems and shells are most variable while the value for the kernel is reasonably constant. The shelling percentage of the whole nut is around 70 or slightly less and kernel to haulm ratio is around 0.85. Based on world average yield of 958 kg of unshelled nuts/ha in 1976 (FAO, 1977), the quantities of haulms, shells, and kernels were 789, 287, and 671 kg/ha, respectively. The quantities of P contained in the haulms, shells, and kernels were 0.55, 0.09, and 2.42 kg/ha, respectively, for a total of 3.06 kg of P/ha. In the United States where the average yield was 2,763 kg/ha in 1976 (FAO, 1977), the total P removal in haulms, shells, and kernels was 1.59, 0.25, and 6.96 kg/ha, respectively, for a total of 8.8 kg P/ha. The quantity in the roots at harvest is quite small; Bunting and

Anderson (1960) found it to be less than 1% of the total contained in the plant. It can be seen that the P requirements of the peanut are rather small. Assuming that the haulms or vegetative portion is returned to the soil, the average P removal is roughly 2.5 kg/ha on a world-wide basis. If approximately 70% of the total P in the plant is removed when the nuts are harvested, the total P contained in the worldwide peanut crop in 1976 was about 68,500 metric tons, of which 48,000 metric tons is removed in the nuts. The remainder is returned to the soil and recycled.

b. Crop Response—In the United States peanuts are grown on light-textured, deep, well-drained soils. This is generally true throughout the world as well, but there are areas where they are grown on finer textured soils. The response of peanuts to P fertilization is quite variable, responses being quite good on soils newly put into cultivation or receiving little or no fertilization. This is to be expected since sandy soils are usually low in available P. However, on soils that have been well fertilized little or no response is observed. This, too, is to be expected, since the P requirement of peanuts is low and their roots are not sensitive to soluble Al (Adams & Pearson, 1970). Also, roots are quickly extended into the subsoil. Hall et al. (1953) observed that the taproot had extended to a depth of 60 cm within 3 weeks after emergence. Thus, the peanut should be able to exploit P present in the subsoil.

A review of 18 papers reporting on yield responses to P fertilization suggests that 1 kg of P will increase peanut yields 20.0 ± 4.3 kg. This figure is based on 722 location-years and an average rate of 21.5 kg of P/ha, the range in application being 10 to 41 kg of P/ha. As pointed out by Reid and Cox (1973), peanuts will respond to P, Ca, and S present in ordinary superphosphate (OSP) which has been the usual form of P applied in most fertilizer experiments. Thus, responses may have been to Ca or S present in the fertilizer rather than to P, since it is not always clear from the published reports whether or not the experiment was designed to separate P, Ca, and S responses. In the United States, where the use of gypsum is a common practice and where peanuts are grown in rotation with well-fertilized crops, responses are small or absent. For example, Scarsbrook and Cope (1956) reviewed the responses of peanuts to P in Alabama and showed that prior to 1940 good responses were obtained, while in 1952-1954, 18 tests showed no response. Still later in Alabama, 24 trials in 1967-1972 showed no response of peanuts to fertilization (Hartzog & Adams, 1973). Nye (1954) reported that residual effects of P fertilization in the Gold Coast were better the second year than either the direct effects of first-year residual effects. This may have been due in part to the presence of S in the fertilizer. Chesney (1975) observed no response to P on the second and third crops of peanuts and no beneficial effect of splitting the P; to the contrary, better response was obtained if all of the P was applied before planting, especially during periods of stress and at low rates of gypsum. Oram (1958) in a review of peanut production in Africa pointed out that large responses to P usually occurred on land recently brought into cultivation or on which an exhaustive cropping system was followed. Martin (1964) in a worldwide review of fertilization of peanuts pointed out that in most areas peanuts depended on residual P applied to other crops for their P supplies. Typical P recommendations are in the range of 15 to 50 kg of P/ha, depending on the soil test level of available P.

3. EFFECT OF PLACEMENT METHODS AND TIME OF APPLICATION

Early root growth in the peanut is primarily by the taproot. Lateral root growth contributes little to P absorption until the 11th week according to Hall et al. (1953). It is not surprising, therefore, that most studies have indicated that fertilizer placement in the row to the side and below the seed is most satisfactory (Futral, 1952; Goldsworthy, 1964; Oram, 1958; Sardone, 1960). It is essential that the fertilizer not come in contact with the seed (Futral, 1952; Sardone, 1960). In some cases little difference has been observed among broadcast, planting-line, or seed placement (Nye, 1954), and in other cases better response has been obtained by applying all to the preceding crop in the rotation rather than splitting it between the two crops (Hallock, 1962).

B. Phosphorus Concentrations in Peanuts

1. CONCENTRATIONS WITHIN THE VARIOUS TISSUES

Phosphorus concentrations in various tissues of the peanuts are shown in Fig. 3. The highest P concentration is found in the kernel, followed by

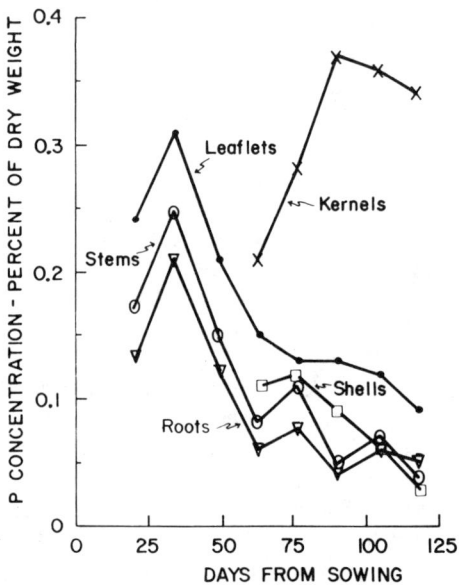

Fig. 3—The concentration of P as a percent of the dry weight in the various parts of peanuts grown in Tanganyika. Natal Common, an upright, bunch-type planted 19 Jan. 1950, first fruit appeared at 49 days (after Bunting & Anderson, 1960).

the leaflets, shell, stem, and roots, although at maturity concentrations in the roots, shell, and stems are quite similar.

As shown in Fig. 3 vegetative tissues reached their peak concentrations 34 days after sowing, then the concentrations decreased rapidly until 63 days after sowing, and then decreased more gradually to maturity. This decline in the foliage has also been observed by Hallock et al. (1969) and Cox et al. (1970). Concentration in the kernel, on the other hand, increases very rapidly until about 29 days before harvest with little change during the last 28 days.

2. EFFECT OF OTHER NUTRIENTS

a. Nitrogen—Nitrogen may have no effect or only a slight but statistically significant effect on P concentrations in peanut foliage (Prevot & Ollagnier, 1951, 1954). They suggest that there is an optimum ratio of percent N to percent P and that in peanuts it seems to be 20.

b. Potassium and Other Nutrients—Prevot and Ollagnier (1954) reported no effect of K fertilization on foliar P concentrations. There does not seem to be much information on the effect of other elements on P foliar levels. Prevot and Ollagnier (1954) suggested optimum rates between P and other elements: $P/K = 0.20$, $P/Ca = 0.10$ to 0.15, and $P/Mg = 0.3$.

C. Phosphorus Uptake Demand Patterns

1. CUMULATIVE UPTAKE WITH TIME

Accumulation of P by the peanut plant is illustrated in Fig. 4. Yields of peanuts and accumulation of P were as follows: Chamaye soil—1,280 and 4.5 kg/ha of peanuts and total P, respectively; respective values for the Norfolk soil without S were 1,320 and 6.8 kg/ha and for the Norfolk soil with S, 2,470 and 16.4 kg/ha. The Chamaye soil received 26 kg of P/ha before planting; on the other soil the prior crop was fertilized with P. Relative dry matter and P accumulation by peanuts depend in part on the level of available P in the soil. In the experiments illustrated in Fig. 4, when 50% of the dry matter had been accumulated, 50 and 33% of the P had been accumulated in the peanuts growing in the Norfolk and Chamaye soils, respectively. The Norfolk soil is high in available P; the Chamaye soil is deficient.

2. DISTRIBUTION AMONG VARIOUS TISSUES

The relative distribution of P among the various tissues changes during the growth of the plant. The amount of P in the vegetative portion of the plant can be rather small compared with that in the kernel, as in the case of the Chamaye soil, or relatively large, as in the S-deficient Norfolk soil. Roots generally contain small amounts relative to that in the tops, less than 1, 3, and 8% for the Chamaye, −S, and +S Norfolk soils, respectively (Fig. 4).

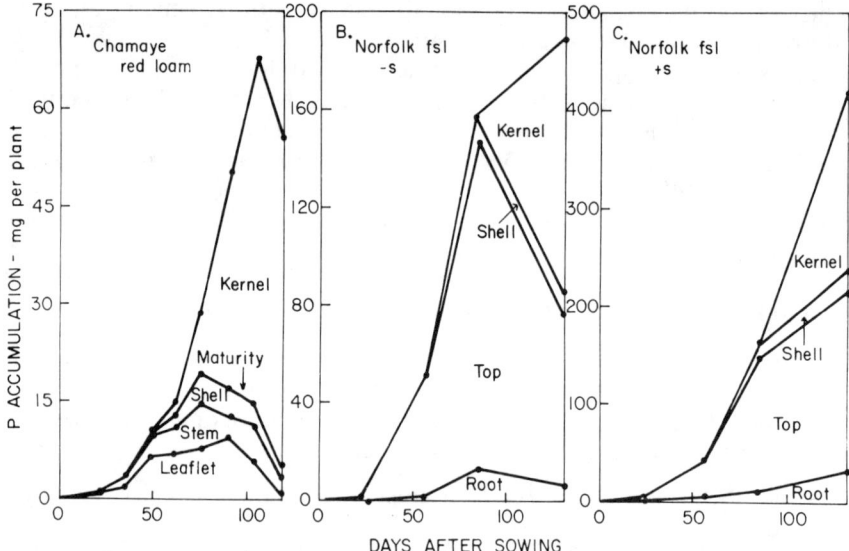

Fig. 4—Distribution of P among the various tissues of peanut during the growing season: (A) Tanganyika, responsive to P (after Bunting & Anderson, 1960); (B) Florida, Norfolk fine sandy loam, pH 6.3, no gypsum applied; (C) Same soil as (B), 67 kg S/ha as gypsum dusted on plants at flowering (after Killinger et al., 1947).

At maturity, a large fraction of the P is present in the fruit, ranging from 80% for the Chamaye soil, to 62% for the −S Norfolk soil, and to 49% for the +S Norfolk soil. The demand for P is also greatest during the fruiting period since much of the P required by the peanut is used for production of the kernel.

There is some evidence for translocation of P from vegetative tissues to the developing fruit (Fig. 4A). During the period of rapid accumulation in the kernel, the quantity of P present in the leaflets, stem, and shells decreased. Bunting and Anderson (1960) noted that after maturity most of the loss in P was due to abscission. There was considerable loss of P in the tops during fruiting (Fig. 4B). Part of this may have been due to translocation since the percent P in the tops decreased from 0.27 to 0.24, but most of it was probably loss due to abscission, since dry weight of tops decreased from 47 to 28 g/plant. However, when the soil was well supplied with P and S was not limiting (Fig. 4C), the plant was able to take up P from the soil to supply the needs of the developing fruit. Loss of P from the vegetative portion did not occur. Dry weight of tops actually increased and P concentration did not change. Translocation of P, then, may or may not occur in peanuts depending on the nutritional status of the plant and the supply of P in the soil.

3. EFFECT OF FERTILIZATION AND OTHER FACTORS

The yield level determines largely the amount of P taken up by the peanut since the kernel is relatively high in P and the concentration does not vary greatly, as illustrated by the data shown in Fig. 4. On a per-hectare-basis the −S treatment yielded 1,320 kg of peanuts/ha and the +S treatment 2,470 kg/ha. The respective P accumulations were 6.8 and 16.4 kg/ha. Bunting and Anderson (1960) noted a decrease in the rate of P uptake during a dry period.

D. Yield-P Concentration-Age Relationships

1. TOTAL VS. SOLUBLE CONCENTRATION

Burkhart and Page (1941) correlated soluble P extracted in boiling water with response to P additions in the "Virginia Bunch" variety. If soluble P in leaf petioles was less than 260 ppm, response could be expected.

The literature contains little information to indicate which tissue is most sensitive to changes in yield. Early workers (Burkhart & Page, 1941; Prevot & Ollagnier, 1954) concluded that the hot-water-soluble P in the petioles would give the best prediction of a P response. More recently the total P concentration in the main stem above the principal lateral branches (Hallock et al., 1969) or in the upper stem and leaves at early pegging stage (Small & Ohlrogge, 1973) has been recommended for evaluation of the nutrient status.

Although environmental and soil fertility status affect tissue-P levels and varieties differ in the P concentrations of their tissues, interactions with yield responses have not been reported in the literature, if indeed they have been investigated.

2. CRITICAL P LEVELS

Fertilizer recommendations for peanuts in the United States are usually based on soil tests which have been correlated with field trials. Foliar analysis is, however, useful for evaluating the adequacy of a fertilization program. The critical levels which have been established and the sampling and analytical techniques utilized are given in Table 5. As pointed out by

Table 5—Sampling methods and critical levels for assessing P status of peanuts.

Country	Plant part	Comments	Critical level	Reference
Madagascar	Leaves		0.2% total P	Roche et al., 1959
India	Leaves	Start of flowering	0.22% total P	Satyanarayana & Krishna Rao, 1962
United States	Upper stems and leaves	Early pegging stage —10 to 12 weeks after emergence	0.20–0.35% total P (sufficiency range)	Small & Ohlrogge, 1973
Africa	Fourth to sixth leaf on main stem	40 to 45th day after sowing	% total P: 0.20 if %$N=3.0$ 0.225 if %$N=3.5$ 0.250 if %$N=4.0$	Gillier & Prevot, 1960
United States	Lower blades	H_2O soluble P, fresh wt basis	260 ppm	Burkhart & Page, 1941

Small and Ohlrogge (1973) the information on foliar analysis of peanuts is quite inadequate.

E. Relationships Between Nutrition and Plant Quality

1. PLANT CHARACTERISTICS

In a sand culture experiment employing low, medium, and high levels of P in the nutrient solution, Nicholaides and Cox (1970) found no effect of P on days to flowering, total number of flowers per plant, or flowers per day per plant. A medium level (0.25 meq of P/liter) increased total pegs per plant, while a high level (1.0 meq of P/liter) decreased the percentage of flowers forming pegs. Satyanarayana and Krishna Rao (1962) found P fertilization increased number of flowers and pods and the shelling percentage. In a sand culture experiment, Reid and York (1958) imposed a P deficiency at the beginning of flowering and found that the number of flowers, pegs, pods, and reproductive branches decreased. The weight of seeds per pod was improved by P fertilization (Evelyn & Thornton, 1964). Phosphorus fertilization has been shown to increase root length and number of roots (Bhan & Misra, 1970) and also to decrease root length (Huber, 1956).

A common observation is that P fertilization increases the number of nodules (Geus, 1967; Huber, 1956), but Evelyn and Thornton (1964) reported no effect of fertilization on degree of nodulation in Gambia.

2. CHEMICAL COMPOSITION

The concentration of oil in peanuts has been increased by P fertilization. Arora et al. (1970) observed an increase from 46.8 to 49.8%, Kumar and Venkatachari (1971) from 46.6 to 49.3%, and Walker and Carter (1971) from 42.8 to 43.2%. Protein was found to increase about 2% with P fertilization (Arora et al., 1970; Kumar & Venkatachari, 1971). Arora et al. (1970) also found a decrease in total soluble carbohydrates but no effect on free fatty acids.

IV. RICE

A. Introduction

1. LITERATURE

The literature on rice is voluminous, but much of it is written in Japanese and thus is inaccessible to a large part of the world. Fortunately, we have the International Rice Research Institute 1964 Symposium on the Mineral Nutrition of the Rice Plant in which many authors have summarized much of this literature and the review of the physiology of the rice plant by Ishizuka (1971). I shall draw heavily on these two sources.

2. THE PLANT

Two species are cultivated, (i) *Oryza sativa* L., the common rice, and (ii) *O. glaberrima* Steud., which is cultivated to a limited extent in Africa (Purseglove, 1972). Three subspecies of *O. sativa* are usually recognized—*indica* Kato, *japonica* Kato, and *javanica* Kato.

Rice is a freely tillering annual grass. The roots are not typically aquatic as they are much branched and have a profusion of root hairs. Extensive aerenchyma develops in the cortex.

3. PHOSPHORUS FERTILIZATION AND CROP RESPONSE

a. Total P in the Rice Plant—The average P content of rough rice is 0.26%. Total quantity of P in the grain, based on the 1976 world-wide yield estimate (FAO, 1977), was 900,000 metric tons. The amount in the straw is variable since the grain-to-straw ratio varies with varieties. The *indica* varieties have a grain-to-straw ratio of 1.1, while the ratio for the *japonica* is around 0.55 (Ishizuka, 1971). At maturity the P concentration in rice straw is approximately 0.1%. Amounts of P in the above-ground parts of the rice plant have been estimated in Table 6. This suggests that from 60 to 74% of the P taken up is removed in the grain. The quantity in the straw may be returned to the soil as residue or in the ash if the straw is burned. In some areas, the straw is removed and utilized for various purposes, including feeding livestock. In the latter case, if the manure is returned to the soil, some of the P will be recycled.

b. Crop Response—Fine-textured soils are most suitable for rice since they permit puddling and the reduction in percolation reduces loss of water and nutrients. Nevertheless, rice is grown on most textural classes, provided there is adequate water from rain or irrigation. Rice is grown in many different ways, each of which may affect the supply of P to the plant differently. Ninety-three % is grown under flooded conditions, while 7% is grown under upland conditions as any other cereal is grown. In the tropics, where water supplies are adequate, usually two crops are grown per year. Under rainfed conditions rice is grown during the rainy season and another crop may follow rice and mature during the dry season.

Table 6—Estimated P contents of the above-ground parts of rice.

Country	Average yield of paddy†	Yield of straw‡	P content			Percent of total P in the grain
			Paddy	Straw	Total	
	— metric tons/ha —		——— kg/ha ———			
Japan	5.92	5.38	15.4	5.4	20.8	74
United States	4.98	5.53	12.9	5.5	18.4	70
China	3.27	5.95	8.5	6.0	14.5	60
India	1.67	3.04	4.3	3.0	7.3	60

† FAO (1975a).
‡ Grain/straw ratios: Japan 1.1, USA. 0.90, China, India 0.55.

Response to P fertilizers is uncertain under flooded culture because the anaerobic conditions which ensue after flooding increase the availability of P (Mitsui, 1956; Ponnamperuma, 1965). On soils low in active Fe or total P, sufficient P may not become available, and a response to P fertilization may be observed. Responses to P are more likely under upland conditions. Responses obtained on deficient soils have been summarized by Mukerjee (1965), FAO (1966), Geus (1967), Chu (1959), and Dewan and Mahdavi (1968). Analyses of these data indicate that on the average 1 kg of P would return 25 ± 3.1 kg of rough rice. It would be difficult to say what percentage of the world rice acreage would give this kind of response. Doyle (1966) reported that 1 kg of P would return 7.2 kg of rough rice, based on data from 433 responses in 25 countries and the correlation (r) between applied P and increase in yield of grain was only 0.45. The figure of 25 kg of rice/kg of P is for soils low in available P, while Doyle's figure is probably more representative of the overall response of rice to P.

Geus (1967) summarized the information on P recommendations. Rates vary from 10 to 66 kg of P/ha with the lower rates being recommended in the tropical areas and the high rates in northern Japan and Europe. In most cases, these rates were recommended on the basis of soil tests, soil type, or history of response.

4. EFFECT OF PLACEMENT METHODS AND TIME OF APPLICATION

According to Broeshart and Brunner (1964) P fertilizer applications to the main crop should be made in a single application at the time of transplanting. Under lowland conditions it should be applied on the surface or hoed into the surface of the soil. Ordinary superphosphate was equally effective placed on the surface or at a 5-cm depth on major Indian soils (Datta & Venkateswarlu, 1968). In a review of the literature, Davide (1965) concluded that there was little difference in sources except that certain phosphate rocks (PR) and bone meal (both finely ground) were better than OSP on extremely acid soils. Thompson et al. (1962) noted that P fertilization of rice in Arkansas encouraged excessive weed growth and that the P should be banded 7.5 cm deep just before planting. In a greenhouse study, Terman and Allen (1970) found P uptake decreased with a decrease in water solubility.

B. Phosphorus Concentrations in Rice

1. CONCENTRATIONS WITHIN THE VARIOUS TISSUES

There is no consistent relationship among the P concentrations in the various tissues of the rice plant. In Brazil, Gargantini and Blanco (1965) found that the aerial part of the plant was higher in percent P than were the roots. During the reproductive phase, the order was grain > leaves and stem > roots. Reyes et al. (1962), in the Philippines, found that the roots were higher in P during the vegetative phase, while in the reproductive

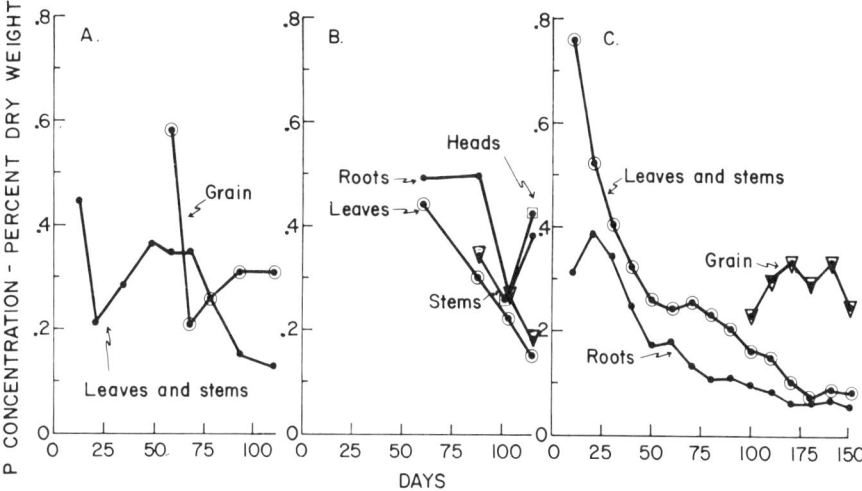

Fig. 5—Changes in the P concentration of various tissues in the rice plant: (A) Northern Japan first sample 12 June (after Ishizuka, 1965); (B) Philippines, first sample at tillering (after Reyes et al., 1962); (C) Brazil, upland rice in pots, first sample 10 days after emergence (after Gargantini & Blanco, 1965).

phase the order was heads = roots > stems > blades. Ishizuka (1965) found that grain was higher in P than stem and leaves. It appears, generally, that at maturity, the grain has the highest concentration of P.

The P concentration in the vegetative tissue decreases with age (Fig. 5) and is linked to maturation of the reproductive tissue. The magnitude of these changes in percent P is indicated by the following data from Murayama (1965):

	Leaf blade	Leaf sheath	Culm	Root	Ear
At heading	0.061	0.078	0.066	0.049	0.056
At harvesting	0.015	0.015	0.028	0.042	0.259

The pattern seems to vary, probably depending on variety, level of soil P, and environmental conditions that affect the rate of growth.

2. EFFECT OF OTHER NUTRIENTS AND ENVIRONMENTAL FACTORS

a. Nitrogen—Nitrogen applications tend to reduce the P concentrations in rice tissues (Coronel & Wallihan, 1971; Datta & Shinde, 1965; Enyi, 1969). However, Sims and Place (1968) found that percent P did not differ signfiicantly with respect to age of plant, N rate, and variety. It is more likely that if N results in large increases in dry matter under conditions of limited P supply, percent P will decrease as a result of dilution.

b. Other Nutrients—Coronel and Wallihan (1971) found that the percent P in the Y leaves at flower emergence was higher when the rice was deficient in K and Mn. Under conditions of limited P supply in nutrient solutions, percent P in rice was higher when the P was accompanied by sulfate rather than chloride (Mitsui, 1956).

c. **Environmental Factors**—Like other plants, the uptake of P is dependent on temperature (Ishizuka, 1964). Okajima (1965) found that P absorption was reduced 56% when the temperature was reduced from 30°C to 15°C.

Hydrogen sulfide in the nutrient solution caused exudation of P from roots (Mitsui, 1956).

An increase in the availability of P as a result of flooding would be expected to increase the percent P in the rice plant. Datta and Shinde (1965) reported that the percent P increased from 0.06 to 0.10 as a result of flooding. When fertilized with 40 kg of P/ha, the respective percent P levels were 0.07 and 0.16.

C. Phosphorus Uptake Demand Patterns

1. CUMULATIVE UPTAKE WITH TIME

Accumulation of P by the rice plant is shown in Fig. 6. Illustrated are the uptake patterns for four different conditions. Three are lowland rice grown in (i) northern Japan, representing the cool condition, (ii) southern Japan, the more temperate climate, and (iii) Los Baños, the Philippines, representing the tropical condition. The fourth one is the upland condition from the state of São Paulo in Brazil. At the two Japanese locations most of the P uptake occurred prior to the reproductive phase; at the Philippine location, P uptake was continuous throughout vegetative and reproductive phases. Upland rice from the Brazil location was intermediate, i.e., most of the uptake occurred during the vegetative phase, but there was also some uptake during the reproductive phase. At the time 50% of the dry matter had been accumulated, 80, 65, and 60% of the P had been taken up by the rice in the Japanese, Brazilian, and Philippine studies, respectively. A 150-day variety was found by Mikkelsen (1970) in California to have accumulated 50% of total P uptake when 50% of the dry matter had been accumulated.

2. DISTRIBUTION AMONG VARIOUS TISSUES

Data from the Philippines and Brazil indicate (Fig. 6) that P contained in the roots may be as high as 30% of the total in the plant at tillering, but decreases to about 10% of the total at maturity. The bulk of the P is contained in the aerial portion of the plant during the vegetative phase and flowering, but during the reproductive phase, it shifts to the reproductive tissues.

In all cases (Fig. 5) there is a decrease in total P contained in vegetative tissues during reproductive growth. Much of this can be attributed to translocation to the developing reproductive tissue (Ishizuka, 1965; Kasai & Asada, 1965). Undoubtedly part of the loss of P from leaf tissue is the result of senescence and death of the leaves. In most studies, such as those reported here, no attempt was made to account for dry matter and nutrient loss from leaves that dies. However, it appears that in the Japanese studies most of the P uptake occurred prior to ear filling and most of the P in the reproductive tissue came via translocation from the leaves and stems. In the Philippine study about 50% of the P in the ears came from the vegetative tissues. In the Brazilian study with upland rice, about 56% of the P in the ears came from the vegetative tissues.

It is difficult to say how typical these examples are and to what factors the observed differences can be attributed. Certainly the level of available P in the soil, the climate, the length of the growing season, and the nature of the variety are all important. It would be of considerable interest to know the pattern of uptake of some of the newer IRRI high-yielding varieties.

3. EFFECT OF FERTILIZATION AND OTHER FACTORS

Since a relatively large proportion of P in the rice plant is in the grain at maturity and P concentration of the grain is relatively constant, total uptake of P by the plant is affected by those factors that affect grain yield. Basak (1962) found that 70% of the variation in P uptake was associated with variations in grain yield. Sims and Place (1968) reported that total up-

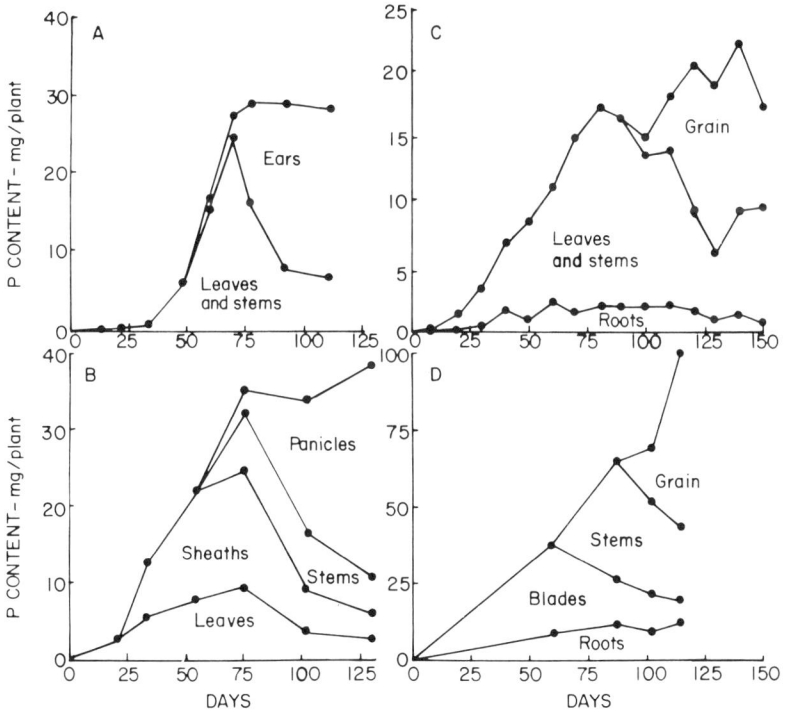

Fig. 6—The accumulation of P and its distribution in the various tissues of the rice plant. (A) Sapporo, Northern Japan, initial sample 12 June; (B) Kyushu, Southern Japan, initial sample 2 July (after Ishizuka, 1965); (C) Brazil, upland rice in pots, initial sample 10 days after emergence (after Gargantini & Blanco, 1965); (D) Philippines, initial sample at tillering (after Reyes et al., 1962).

take of P closely paralleled dry matter production. Both Datta and Shinde (1965) and Enyi (1969) reported increases in P uptake with applications of N and P.

D. Yield-P Concentration-Age Relationships

1. TISSUES FOUND MOST INDICATIVE OF YIELD LEVELS

Leaf tissue has been found most satisfactory for evaluating the nutritional status of the rice plant. Angladette (1965) found the flag leaf to be most suitable; Mikkelsen (1970) found the "Y" leaf, the most recently matured leaf during the active tillering stage, to be the most useful. Angladette (1965) also investigated the composition of the stalk in relation to nutritional status of the plant. Most workers determined total P, but Mikkelsen (1970) extracted the leaf tissue with 2% acetic acid.

2. FACTORS AFFECTING CONCENTRATION-YIELD RELATIONSHIPS

There seems to be little work on the P concentration-yield relationship, probably because at the low yield levels of the *indica* varieties sufficient P becomes available during flooding to take care of the needs of the plant and there has been little incentive to work on the problem. For example, Mikkelsen and Hunziker (1971) surveyed California rice fields, and based on plant analysis, only 5% of 400 samples tested low in P at maximum tillering and 11% at panicle initiation. Phosphorus levels in the soil and application of P fertilizer along with N and other nutrients may affect the P concentration in leaf tissue. Koyama and Chammek (1971) compared two local rice varieties in Thailand at rates of P ranging from 0 to 66 kg/ha. Phosphorus fertilization increased percent P markedly and the varieties differed considerably in percent P at tillering; however, varietal differences were not so great at flowering.

3. CRITICAL P LEVELS

Some of the critical levels that have been established (in some cases tentative only) are shown in Table 7. It seems clear that there is a need to refine critical levels in tissue-tests, particularly to evaluate such factors as nutrient levels, environment, and variety.

E. Relationships Between P Nutrition and Plant Quality

1. PLANT CHARACTERISTICS

According to Murayama (1965) P deficiency symptoms appear in the lower part of the plant and P deficiency decreases leaf number and leaf blade length, number of ears per plant, and number of filled grains per ear. Olsen (1958) found a marked decrease in number of tillers in a P-deficient nutrient solution. The improvement in tillering is greatest when P is added

Table 7—Sampling methods and the foliar-P levels used to assess the P status of the rice plant.

Plant part	Growth stage or age of plant	Critical level or range, %	Comments	Reference
Flag leaf	Onset of flowering	0.18 total P	If the second and third leaves are sampled the critical level is 0.12% P.	Angladette, 1965
"Y" leaf	Flower emergence	0.3 total P	Greenhouse study; critical level lower if N, K, or Mn deficient	Coronel & Wallihan, 1971
Leaf blade	Tillering	0.1 total P	P deficiency may be suspected if % P <0.2%. 1.0% P in straw at maturity toxic	IRRI, 1969
"Y" leaf	45 to 80 days after planting	0.06–0.08 (2% HOAc soluble PO_4-P)	Normal range 0.1–0.15%. Observed field range 0.03–0.27%.	Mikkelsen et al., 1970
"Y" leaf	Mid-tillering	0.14–0.27 total P	Range without P deficiency	Ward et al., 1973
	Panicle differentiation	0.18–0.29 total P	Range without P deficiency	

to a low-P soil (Terman & Allen, 1970). Takahashi et al. (1955) reported that increasing the P supply accelerated heading and increased the ratio of weight of ears to weight of straw.

2. DISEASE RESISTANCE

According to Tanaka and Akai (1963) excess P caused increased susceptibility to *Helminthosporium* leaf spot. Kozaka (1965) concluded that if P is limiting growth, rice blast disease can be reduced with P applications sufficient to produce normal growth. Excessive application, on the other hand, increased the disease, especially when plants received high rates of N. High rates of P also helped to control blast in Japan when the weather was cool or when cold water was used for irrigation.

V. SUGARCANE

A. Introduction

1. THE PLANT

Five species of sugarcane are known. *Saccharum officinarum* L. is the large-stalked, soft, juicy, sweet cane and varieties that come from it are called the "noble" canes. *Saccharum sinense* Roxb. and *S. barberi* Jeswiet are "thin" canes, and *S. spontaneum* L. and *S. robustum* Brandes and Jeswiet ex Grassl are wild types. Present-day varieties are complex hybrids derived from both cultivated and wild forms (Masefield, 1949). Sugarcane grows in both the tropics and subtropics between latitudes 35° north and south of the equator, where the annual rainfall is at least 1,524 mm and the temperatures are at least 24 to 27°C during most of the year. It is propagated vegetatively and its growing season varies from less than 10 months in Louisiana, to 2 years in Hawaii, Peru, and South Africa. Elsewhere, it is grown in 14- to 18-month plant crops and 12-month ratoons as stubble cane (Humbert, 1968). Generally, two ratoon crops are grown in the case of 14- to 18-month plant crops followed by 12-month ratoon, and the crop cycle consists of 4 years. Where 22- to 24-month plant and ratoon cane is grown the cycle is 8 years (Barnes, 1974).

2. PHOSPHORUS FERTILIZATION AND CROP RESPONSE

a. Total P in the Plant—According to Dillewijn (1952) the approximate average P contents in plant parts associated with a metric ton of millable cane are: millable cane 0.5 kg P, green tops 0.2, and abscised dry leaves 0.15. There is considerable variation among data from different parts of the world and this is attributed to differences in climate, weather, soil fertility, and length of growing season. The average values cited above suggest that, of the total P present in the above-ground portion of the cane, about 59% is removed during harvest. The trash remaining is either incorporated in the soil or burned so that the P it contains is recycled. Of the P in the millable cane, part of it goes into the molasses, some remains in the bagasse, and according to Barnes (1974), most of it ends up in the filter mud. Filter mud and sometimes bagasse are used as soil amendments, which would recycle considerable amounts of P. Most of the bagasse, however, is burned as a source of energy for processing the cane and the high temperatures attained in the furnaces convert the P in the bagasse to an unavailable form.

World cane production was 693 million metric tons in 1976 (FAO, 1977), suggesting (based on Dillewijn's figures) that the sugarcane crop contains in the above-ground portion 590,000 metric tons of P, of which 350,000 goes to the mill. Yield of molasses is approximately 3% of the crushed cane and contains approximately 0.9% P (Barnes, 1974). The quantity of P in the molasses which leaves the mills would then be 190,000 metric tons each year, leaving 160,000 metric tons that will either be rendered unavailable when the bagasse is burned or returned to the soil in the filter mud or in the bagasse if used as a soil amendment. On a per-hectare basis, since the average yield of sugarcane is 54.3 metric tons/ha, the total P requirement would be 46 kg/ha, of which 27 would go to the mill, leaving 19 to be returned to the soil directly in the form of green top and abscised leaves.

b. Crop Response—Sugarcane is grown on a wide variety of soils—Histosols in Florida, Entisols and Inceptisols in Louisiana and other parts of the world, Ultisols in the warm temperate and subtropical regions, and Oxisols in the tropics. It is not unexpected therefore that responses to P fertilization are variable.

Responses obtained to an average P application of 37 kg/ha on responsive soils in various parts of the world averaged 20 ± 2.3 kg of sugar/kg of P applied. As with other crops, responses vary with soils and decrease with subsequent increments. Response is usually greater with plant crops than with ratoon crops. Quantities of P applied to sugarcane in various parts of the world range from 0 to as much as 300 kg/ha, with the usual rate in the range of 20 to 60 kg/ha (Davies & Vlitos, 1969).

3. EFFECT OF PLACEMENT METHODS AND TIME OF APPLICATION

Placement of P fertilizer is affected by the nature of sugarcane root growth. On the basis of ^{32}P studies, Golden (1967) concluded that vertical roots develop in greater number and activity early in the growing season than do lateral roots. The ratoon crop has a shallower root system than the plant crop because the shoots originate at a higher level than those of a first-year plant. The old root system gradually dies and is replaced by a new one from the developing shoots (Dillewijn, 1952). Data from Hawaii have shown the importance of placement under the cuttings for maximum utilization (Humbert, 1968), and P fertilizer is almost always applied close to the roots (Davies & Vlitos, 1969). Thus, for the plant crop, P is placed in the furrow at time of planting; for the ratoon crop, it is placed in furrows on one side of and as near to the cane row as possible immediately after harvest of the plant crop (Barnes, 1974).

B. Phosphorus Concentrations in Sugarcane

1. CONCENTRATIONS WITHIN VARIOUS TISSUES

The most metabolically active tissues have the highest P concentrations. The meristem has the highest, followed by the elongating cane, the spindle cluster, young blades, etc., with the mature cane containing the least. Phosphorus concentrations are lower in cane grown on soils lower in available P and are higher in the ratoon crop than in the plant crop, although this is most evident in the cane grown on soils high in P (Clements, 1955).

As the tissues mature, P concentration decreases (Fig. 7). The rate of decrease in green leaves is rather constant over the growing period. In the stalks, the rate of decrease is very rapid during the first half of the growth period, reaching a minimum which is maintained during the last half.

2. EFFECT OF OTHER NUTRIENTS AND ENVIRONMENTAL FACTORS

a. Nitrogen—Nitrogen fertilization has been shown to reduce P concentration in cane tissues (Burr, 1955; Hartt, 1955; Lakshmikantham et al., 1963). Baver (1960) reported that this depression was confined to stalk tissues, and the decrease was greater on soils high in available P. Heavy applications of N have caused the concentrations of P in the 8–10 internode to drop below an initial level of 0.032% and have resulted in significant increases in sugar when additional P was applied (Humbert, 1968).

b. Potassium and Other Nutrients—Hartt (1955) found that increasing levels of K, Ca, and Mg in nutrient solution reduced total P concentrations

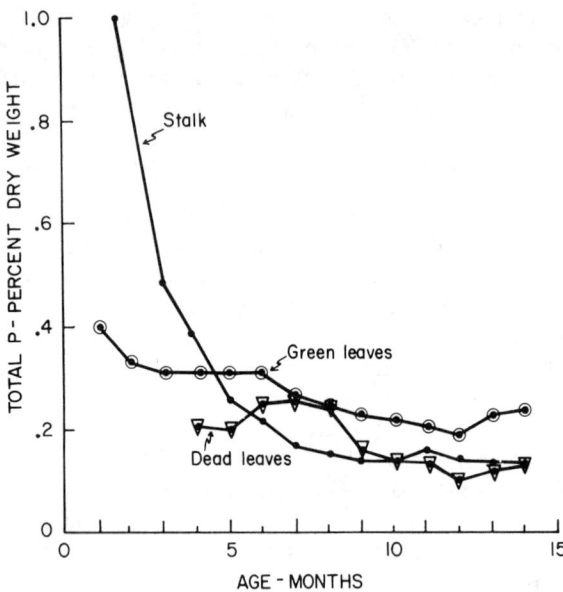

Fig. 7—Changes in P concentration of several tissues in the first ratoon crop of sugarcane during a 14-month period. Variety H1O9 grown on a high-P soil from August 1933 to October 1934 with 224 kg N/ha (after Ayres, 1936).

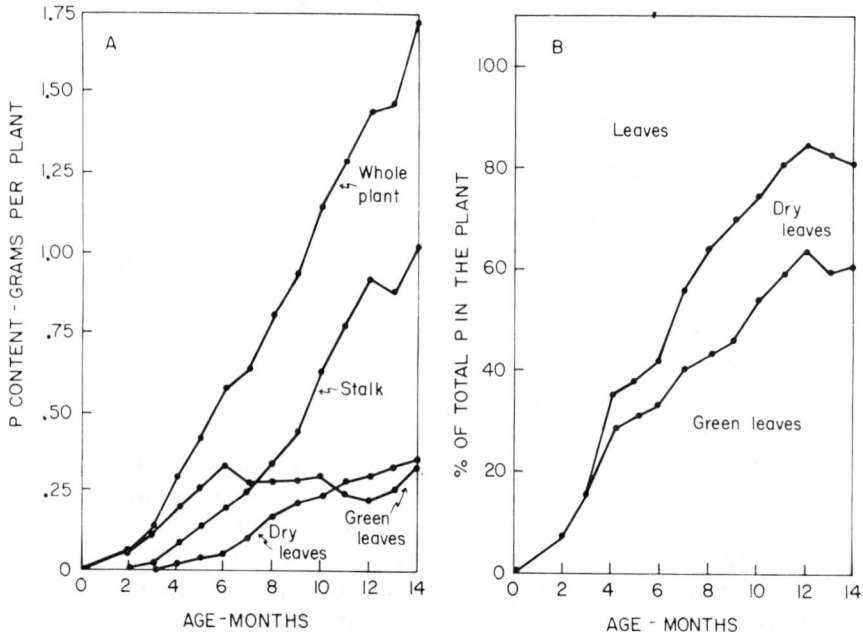

Fig. 8—Accumulation (A) and distribution (B) of P in sugarcane in Hawaii. In (B) the area between the curves represents the proportion of total P in the indicated tissue (after Ayres, 1936).

in the 8-10 internode. In a later nutrient solution study, it was found that plants with low supplies of K accumulated higher concentrations of P in the leaves, especially inorganic P. Deficiencies of N, Ca, and Mg, on the other hand, had no striking effect on P absorption (Hartt, 1958). Evans (1960) reported that excess Al found in extremely acid soils may prevent root extension and cause a P deficiency.

c. Effect of Environmental Factors—Low root temperatures reduce P uptake and plant growth to the same extent; consequently, percentage composition is not affected (Hartt, 1955). Optimum root temperature for both P absorption and growth is 26°C (Burr et al., 1957). The percent P in the 8-10 internode is increased by shade (Hartt, 1958). Samuels (1959) reported that the decrease in leaf P percentage of irrigated cane in Puerto Rico was not so rapid in the early months of the growing season as it was for nonirrigated cane. Evans (1963) pointed out that foliar P may drop drastically when drought and salinity conditions are present and recover ater 4 to 6 weeks of good rainfall.

C. Phosphorus Uptake Demand Patterns

1. CUMULATIVE UPTAKE WITH TIME

Accumulation of P by the sugarcane plant is illustrated in Fig. 8A, which is derived from data of Ayres (1936) on Hawaiian cane. Green leaves attained their maximum P content rather early, and this correlated well with dry weight of green leaves. Weight of dry leaves and of stalk increased steadily throughout the growth period, as did P accumulation. However, proportionally more of the total P accumulated during early growth than did dry matter. For example, 40% of the P had been accumulated when the plant was about 7.5 months of age, but only 30% of the dry matter had been accumulated.

2. DISTRIBUTION AMONG VARIOUS TISSUES

Concentrations of P in the various tissues are quite different and, as has been pointed out, concentrations change with age of the tissue. Looking at a stand of cane, for example, on 1 ha, we find that the distribution of total P contained in various tissues changes with time. Data of Ayres (1936) have been expressed on a relative basis and plotted against time in Fig. 8B. Early in the growth of a stand, most of the P is present in the green leaves. At 10 months, half is in the leaves and half in the stalk, but, of that in the leaves, about 40% is tied up in dead leaf tissue and is of no use to the plant. There is, of course, a considerable quantity of P tied up in the root system. According to Golden and Ricaud (1963), the amount in the stubble and roots is 14% of the total in the plant for cane grown in Louisiana, as compared with 18% for cane grown in other parts of the world.

Although Fig. 7 would suggest some loss of P from sugarcane leaves as they mature, Fig. 8A shows that P accumulated in dry leaves and that maximum accumulation in green leaves which occurred at 6 months coincided with maximum green-leaf weight.

3. EFFECT OF FERTILIZATION AND OTHER FACTORS

Since total accumulation is determined by percentage composition and total dry matter production, fertilizers and other growth factors can affect total P accumulation by either changing composition or dry matter production, or both. Data from Louisiana (Golden, 1961) show that increased yields due to N, P, or K fertilization increased P accumulation: N fertilizer increased accumulation of P by 21%, P fertilizer increased P accumulation 39%, and K fertilizer increased it 21%.

D. Yield-P Concentration-Age Relationships

1. METHODS OF FOLIAR DIAGNOSIS

Because sugarcane is grown under such widely different conditions—climatic as well as cultural—a number of methods of foliar analyses have been developed. These have been described by Samuels (1960) and consist of two major categories: the multiple plant analyses methods for crop-logging, and the selective plant analyses methods. Samuels (1960) states that crop- or stalk-logging is better suited for cane grown 18 to 24 months, while the Jamaican or Puerto Rican selective plant anlayses methods are better suited for cane cut at 12 months and receiving a high annual rainfall.

Hartt (1955) found that inorganic P concentrations in sugarcane leaves were better indicators of levels of available P in the soil than total P. Because of greater variability and difficulties in analysis, however, he concluded that total P was more suitable for routine use.

In Hawaii, stalk tissues have been found more responsive to applied P than the blades or sheaths (Table 8; Baver, 1960; Burr, 1955; Hartt, 1955; Humbert, 1968). Clements (1961) used P concentration in the sheath along with the concentration of P in the fifth mature internode to arrive at an 'Amplified P Index.' In other parts of the world, P concentration in leaf lamina has been found most suitable (Evans, 1963).

2. FACTORS AFFECTING CONCENTRATION-YIELD RELATIONSHIPS

Factors which affect the rate of growth may interfere with the concentration-yield relationship. Evans (1963) pointed out that foliar P may drop from 0.20% P to as low as 0.12% when drought and salinity conditions occur; it may return to 0.20% after several weeks of good rainfall. It is for this reason that sampling must be rigidly standardized and that certain indicators of the physiological status of the plant must be employed. In the crop-logging technique, a drop in sheath moisture indicates drought and plant vigor is reflected in the green weight of the sheath, the tissue moisture level, and the total sugar level (Clements, 1961). Where foliar analysis is practiced, certain growth indices have come into use, particularly cane

Table 8—Effect of P fertilization on P concentrations in various tissues of sugarcane grown in Hawaii. Fertilized August 1953, sampled 3 Feb. 1954 (Burr, 1955).

Treatment	Blade 4	Sheath 4	8–10 internode	Basal internode
kg/ha		% of dry wt		
0	0.248	0.085	0.029	0.031
98	0.266	0.097	0.053	0.048
Significance (P)	0.05	n.s.	0.01	0.01

elongation. It is assumed that if the plant is growing at the rate of 15 to 18 cm/week (near maximum growth rates on the steep part of the growth curve) and maintaining critical levels of P in the leaf, then this level of P should be more than adequate if the growth rate is slowed due to the effect of other factors (Evans, 1963).

Varieties differ in foliar P. Dillewijn (1952) reviewed data on 30 varieties of cane. The range was from 0.12 to 0.21% P with a mean of 0.16%. Presumably, the variations in foliar P are least when substrate P is low and greatest when substrate P is high. This was found to be the case for 8–10 internodes by Hartt (1958). Thus, foliar concentrations found to be adequate for one variety may not be adequate for another variety, and the general relationships published without identifying varieties must be checked whenever they are to be used for a new variety or for a variety for which they were not developed.

3. CRITICAL P LEVELS

Although the quantities of fertilizer required are generally determined by fertilizer trials (Davies & Vlitos, 1969) and the results of soil analyses (Du Toit, 1969), plant analyses are useful for checking the adequacy of the fertilizer program. Foliar-P levels used to assess the P status of the plant are given in Table 9.

E. Relationships Between P Nutrition and Plant Quality

1. PLANT CHARACTERISTICS

Application of P may stimulate initial tillering even on soils that are not deficient and on which no yield response would be obtained (Dillewijn, 1952). Plants adequately supplied with P have longer and thicker internodes and longer and broader but thinner leaves than those deficient in P. Plants deficient in P have poorly developed root systems (Humbert, 1968).

2. QUALITY OF CANE

Borden (1936) observed a direct relationship between purity and P concentration of crusher juices in H1O9 cane. Maximum P in the juice occurred at about 81% purity and decreased either below or above that purity. Humbert (1968) pointed out that in none of the 354 experiments conducted

Table 9—Phosphorus levels in sugarcane tissues used to assess the P status of the plant. Values are for leaves unless otherwise specified.

Country	Age at sampling	Crop†	Critical P levels	Reference
	Months		% dry wt	
Fiji	8–12	P	0.13–0.21	Holford, 1968
Guyana	4.5	P and R	0.21	Davies & Vlitos, 1969
Trinidad	4	R	0.16–0.24	Davies & Vlitos, 1969
Jamaica	4–5	P and R	0.22	Innes & Chinloy, 1955
Mauritius	5–7	R	0.21	Halais, 1963
Puerto Rico	3	P and R	0.18–0.25	Capó et al., 1955
Brazil	4	P	0.25–0.30	Malavolta & Gomes, 1960
Mexico	3–5	P	0.14	Portales, 1963
USA—Louisiana	3	P	0.18–0.22	Golden & Ricaud, 1965
USA—Hawaii	6–12	P and R	0.038–0.046‡	Hartt, 1960
USA—Hawaii	3–12	P and R	0.08§	Humbert, 1968
USA—Hawaii	3¶	P and R	1,800#	Clements, 1958

† P = plant, R = ratoon.
‡ Internodes.
§ Leaf sheaths.
¶ Begin crop logging.
# Amplified P Index based on P in sheath and internodes expressed as ppm (sugar-free dry wt).

during the 1940-to-1954 period was there a lowering of cane quality with P fertilization; 7 to 13% gave significant improvements.

The amount of P in cane juice has an effect on clarification (Honig, 1960) and should be in the range of 132 to 264 ppm of P when lime is used for clarification. A lower value may be needed in other methods of clarification.

VI. TOBACCO

A. Introduction

1. THE PLANT

Two species of tobacco are cultivated. *Nicotiana rustica* L. occupies only a small fraction of the tobacco acreage but, because it is high in nicotine, is grown mainly as a source of the drug for medicine and insecticides. *Nicotiana tabacum* L. provides the bulk of the leaf material for making cigars, cigarettes, and pipe tobaccos.

Tobacco is an herbaceous annual plant, at first rosette-like, but later attaining a height of 0.9 to 1.5 m. It produces a single main root which, if allowed to develop, becomes a strong taproot with extensive lateral development. Tobacco is established by first growing the seedling in a bed and then transplanting it into the field. During transplanting, the taproot is usually broken and a mass of fibrous roots develops. When the inflorescence appears, the plant is topped, which results in an increase in the size of upper leaves and of roots (Berthold, 1931; Petrie et al., 1939). Of the cultivated crops, tobacco has one of the widest top/root ratios which becomes wider with increasing temperature.

The plant is grown over a very wide range of climates—from lowlands to mountain regions, from the tropics to 50° latitude, from the highest light intensities to shade, and from very moist to relatively dry (Went, 1957).

2. PHOSPHORUS FERTILIZATION AND CROP RESPONSE

a. Total P in the Tobacco Plant—Average P concentration in cured tobacco leaf is 0.3%. The world tobacco crop in 1976 (FAO, 1977) removed in the harvested leaves about 17,000 metric tons of P. Since some tobaccos are harvested by stalk cutting, an additional quantity is removed in the stalk. Around one-third of the crop is harvested in this way. The stalk is about 40% of the above-ground part of the plant, and its average P concentration is 0.26% (Garner, 1946). This means an additional 3,300 metric tons of P is removed for a worldwide total of nearly 20,000. When tobacco is harvested by priming, i.e., only the leaves are removed, stalks are returned to the soil, and their P is recycled.

The average world yield level in 1976 was estimated at 1,270 kg/ha (FAO, 1977). As Garner (1946) pointed out, yields and leaf-P concentrations vary depending on tobacco types and soil properties. He estimated from 5.0 to 18.0 kg of P/ha in the above-ground part of the plant. If the crop is harvested by stalk cutting, then all of this will be removed; if harvested by priming, from 35 to 45% will be returned to the soil.

b. Crop Response—Tobacco is grown on well-aerated soils varying widely in texture, but predominantly on sandy soils. An examination of responses representing 172 location-years indicates that 1 kg of P will return 8.1 ± 0.9 kg of leaf at an average application rate of 30 kg of P/ha. These data represent responses on soils low in available P, mainly in the United States. Phosphorus responses have also been obtained in the production of transplants (McCants & Lamm, 1970).

According to Dierendonck (1959) the quantities of P applied to tobacco ranged from 0 to 132 kg/ha with the usual rate being 25 to 50. Variations in rates are due to differences in tobacco type, soil, climatic conditions, and cultural practices.

3. EFFECT OF PLACEMENT METHODS AND TIME OF APPLICATION

A large percentage of P uptake takes place in the early stages of growth (Woltz et al., 1949). So it is essential that the fertilizer be placed in the rooting zone. Most soils that have been cropped with tobacco for more than 5 years will be high in available P because rates of application are high in comparison to removal by crops. Swanbeck and Anderson (1952) found that broadcasting fertilizer and harrowing it in was as effective in high P soils as any other method. On soils low in fertility, it should be worked in well before transplanting (Dierendonck, 1959).

B. Phosphorus Concentrations in Tobacco

1. CONCENTRATIONS WITHIN THE VARIOUS TISSUES

In general, P concentrations are higher in leaves than in stalks (Fig. 9A), and higher in top leaves than in basal leaves. In a nutrient solution study, Kakie (1969) found that optimum growth occurred with 3.1 ppm of P and the percent P in the leaves, stems, and roots was 0.43, 0.38, and 0.44, respectively. At a very high level of substrate P the percent P in roots was very much higher than that in leaves and stems.

In a nutrient culture study, McEvoy (1951) found that the concentration of P in tobacco tissues decreased with age of the tissue (Fig. 9A). Grizzard et al. (1942) found that leaves at transplanting contained 0.45% P and at 63 days after transplanting, 0.29% P. Others who have found that P concentration decreased with age include Lockman (1970b), Chouteau (1962), Davidson (1895), and Watson and Petrie (1940).

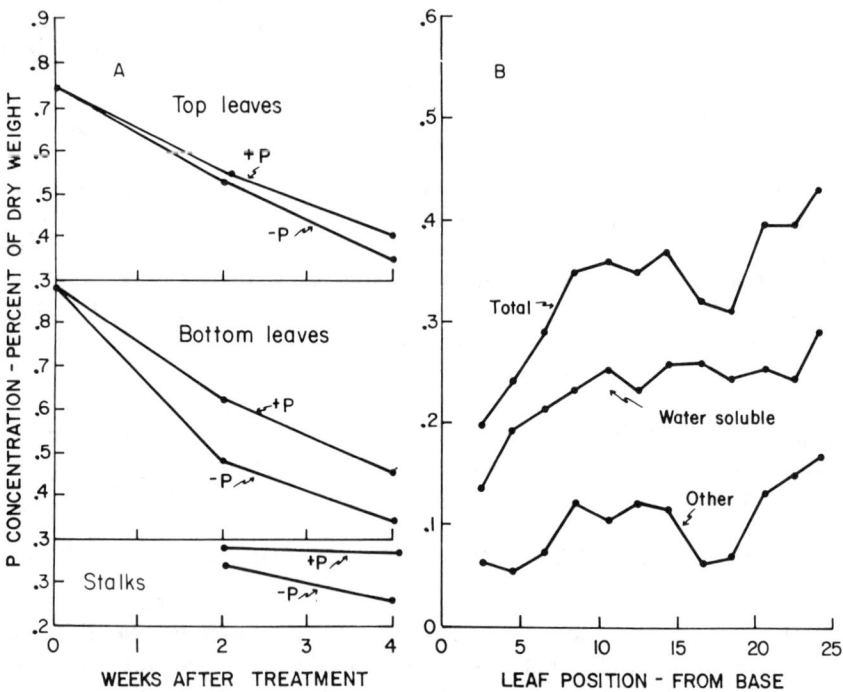

Fig. 9—The concentration of P in the tissues of tobacco. (A) Sand culture. The effect of removing P from the nutrient solution at the time of topping (8 weeks of age) on the composition of White Mammoth flue-cured tobacco (after McEvoy, 1951); (B) Strain 7D of Connecticut 49 sampled 49 days after transplanting. Each point represents a composite sample of 20 leaves, 2 each from 10 plants (after Vickery, 1961).

2. EFFECT OF OTHER NUTRIENTS AND ENVIRONMENTAL FACTORS

a. Other Nutrients—There are reports that N will increase (Bolton, 1956; McEvoy, 1951), decrease (Bolton, 1956; McEvoy, 1951; Nichols et al., 1958) or have no effect (Breland et al., 1967) on percent P. Probably the effect of N is related to its effect on dry matter. Elements other than N apparently have little or no effect (Bolton, 1956; Garner, 1939; Darkis et al., 1937). Skogley and McCants (1963) found that NO_3^- and Cl^- decreased percent P in the whole plant but NH_4^+ had no effect.

b. Environmental Factors—Low root temperatures decrease uptake of P in tobacco and lower P concentrations in leaves. McEvoy (1960) reported that P uptake increased when root temperature increased from 10 to 35°C. Parups and Nielsen (1960a, 1960b) found optimum soil temperature for tobacco to be 22°C. They obtained growth responses only at low soil temperatures and observed decreases in the P concentrations in the leaves as soil temperature decreased.

Moisture supplies may affect the concentration of P in the plant. Decreases due to drought or arid conditions have been observed by Garner (1939), Brown (1970), Bulkina (1969), and Parups and Nielsen (1960b). Lockman (1970b), however, observed an increase in the P concentration.

Maleic hydrazide, which reduces suckering, was found to increase percent P at first, then to cause a decrease with time (Peterson & Naylor, 1953).

Garner (1939) reported that flue-cured varieties grown for cigarette tobacco were lower in P than were Connecticut Valley varieties grown for cigar tobacco. The mean P concentration in the cured leaf of six flue-cured varieties grown at five locations for 3 years was found to be 0.23%, the means ranging from 0.20 to 0.25% (Collins et al., 1961). The differences were not significant.

C. Phosphorus Uptake Demand Patterns

1. CUMULATIVE UPTAKE WITH TIME

Accumulation of P by tobacco is illustrated in Fig. 10. The demand for P the first 3 weeks or so after transplanting is not very great, but once dry matter accumulation accelerates, then rate of P uptake accelerates also. For example, data by Grizzard et al. (1942) indicated that in 1939 about 50% of the dry matter had been accumulated at 6 weeks of age but 74% of the P had been taken up. In 1940, it took 7 weeks to accumulate 50% of the dry matter and 58% of the P. Rainfall was optimum in 1939 but in 1940 it was above average early and inadequate in the latter part of the growing season. Raper and McCants (1966) showed also that when 50% of the dry matter had accumulated, 58% of the P had been taken up.

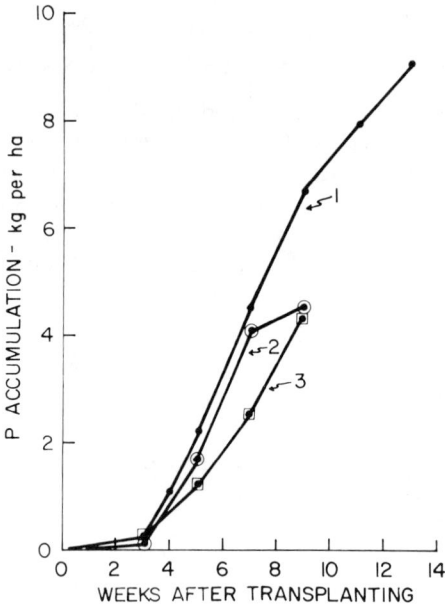

Fig. 10—Accumulation of P by flue-cured tobacco. *Curve 1*—NC-95 grown on Norfolk sl (after Raper & McCants, 1966); *Curves 2 and 3*—Yellow Mammoth grown on Granville sl in 1939 and 1940, respectively (after Grizzard et al., 1942).

2. DISTRIBUTION AMONG VARIOUS TISSUES

Distribution of P in the above-ground part of the tobacco plant is roughly 50-50 between the stalk and the leaves (Table 10; Bowman & Nichols, 1953). Within the leaf itself about one-third of the P is in the midrib. Data (not shown) indicate also that about 28% of the P is present in the bottom third of the leaves, 33% in the middle third, and 39% in the top third, primarily because both leaf size and percent P increase towards the top of the plant.

In the normal culture of tobacco, reproductive growth is prevented by

Table 10—Total P after curing and its distribution in the above-ground portion of two varieties of burley tobacco grown on Hermitage silt loam, Greeneville, Tenn., in 1949. Soil fertilized with 22.4 metric tons of barnyard manure and 896 kg of 3-4-5 (3-9-6) per ha. Plants set 10 June, sampled 9 September; 30 plants per sample (from Bowman & Nichols, 1953).

	Kentucky 16		Burley 1	
Plant part	Total P content	% of total	Total P content	% of total
	mg/plant		mg/plant	
Lamina	150	39	158	35
Midrib	44	12	52	12
Whole leaf	194	51	210	47
Stalk	187	49	238	53
Total	391	100	448	100

the practice of topping. Watson and Petrie (1940) compared the growth of topped and untopped tobacco plants in sand culture with four levels of P in the substrate. They found that topping, which takes place at the time the inflorescence begins to emerge, essentially stopped the net export of P from leaves 6 through 10 from the bottom, and at the higher P levels in the culture, led to increased accumulation. Thus, the normal pattern of translocation of P from the leaves to the inflorescence is prevented. The concentration of P in the leaves decreased, however. This was caused by a dilution due to increased production of dry matter. Since topping increased growth of leaves, the dilution effect was greater and percent P in the leaves was lower in the topped than in the untopped plants.

D. Yield-P Concentration-Age Relationships

Very little has been done to relate nutrient levels in the leaves of tobacco to yield. Much of the concentration data refers to cured leaf or, if not, it is not associated with fertilizer differentials and yield response. The limited information available has been summarized in Table 11. Lockman (1970a) compared P concentration in tobacco grown on a soil in which low and good fertility plots had been established. The P concentration was higher on the low than on the high fertility plots which may be a result of the Steenbjerg

Table 11—Literature values for P concentration in tobacco leaf and stalk tissues.

Region and soil	Fertilization	Response	Plant part	P as a % of dry matter	Reference
	kg P/ha				
North Carolina; Norfolk	0, 20, 39	None	Upper fully mature leaves at 10–12 leaf stage	0.29	Nelson et al., 1948
			Stalk at 10–12 leaf stage	0.28	
North Carolina; Hiwassee clay loam	44 to 1,408	None	Cured leaf	Increased from 0.21 to 0.26 0.21 = avg.	Whittey et al., 1966
Wisconsin; Waupun 1968 silt loam	0 to 135	Yes	Cured leaf	0.13–0.30 Plants with 0.13 showed deficiency symptoms	Peterson, 1968
Ohio; soils not specified	0 to 89	Yes	Upper leaf-bloom stage	0.27–0.30 Good fertility 0.30–0.37 Low fertility	Lockman, 1970a
Canada, and central and eastern U.S. states	Regular production practices	Yes	Upper fully developed leaves	0.27 (Canada) 0.50 (Central) 0.27 (Eastern)	Lockman, 1970a

Table 12—The P concentration in the leaves of tobacco sampled 12 Aug. 1965 when plants were in bloom. Means of three varieties, one burley and two flue-cured (from Lockman, 1970a).

Soil treatment	P concentration		T/B ratio†
	4th leaf from top	4th leaf from bottom	
	%		
Acid, low fertility	0.37	0.10	3.7
Neutral, low fertility	0.30	0.08	3.8
Acid, good fertility	0.27	0.11	2.5
Neutral, good fertility	0.30	0.11	2.7

† The ratio of P concentration in the top leaves to that in the bottom leaves.

effect. Lockman suggested that a comparison of P levels in top and bottom leaves might be useful since low P levels seemed to be reflected only in lower leaves. His results are shown in Table 12, which indicate that top/bottom ratios greater than 3.0 were associated with low soil test P. The burley tobacco had higher P levels than the flue-cured varieties, 0.37 and 0.12% for top and bottom leaves as compared with 0.24 and 0.08% and 0.30 and 0.10% for the other two varieties. Lockman (1970b) also showed that foliar-P concentrations were considerably higher in drought years than during years of adequate moisture. It is clear that at the present time there is insufficient information to define a sufficiency range of foliar P for tobacco.

E. Relationships Between P Nutrition and Plant Quality

1. PLANT CHARACTERISTICS

Maturity of tobacco is hastened by an adequate supply of P (Chouteau, 1968), although there may be exceptions to this if environmental conditions are unfavorable in the latter part of the growing season (McCants & Woltz, 1967).

Although heavy application of P fertilizers is not thought to have an effect on quality (Chouteau, 1968), there are reports indicating an adverse effect (Nichols et al., 1958). According to Tso (1972), phenolic compounds are known to play an important role in leaf color, quality, and physiological strength of smoke. Adequate P gives a lighter color, an important factor in cigarette tobacco, because it indicates low levels of phenols (Andersen et al., 1970).

The effect of P on nicotine content is not consistent. Woltz et al. (1948) found percent nicotine was negatively correlated, while percent sugar was positively correlated with percent P in the leaf. Matusiewicz (1962) found a decrease in percent nicotine as the P supply was increased. On the other hand, Hondt (1955), Komatsu (1951), and Parups and Nielsen (1960b) found no effect of P supply on percent nicotine. Avundzhyan (1963) found that the composition of the alkaloids changed with high P in the substrate; i.e., there was a greater proportion of nornicotine and anabasine present and a smaller proportion of nicotine.

The plant was found to be more susceptible to leaf spotting due to ozone when P supply was low (Hsieh & Kwan, 1973). Low soil pH, low available P, or both, caused 'physiological spotting' (Loche, 1969).

It had been thought that P fertilization did not have an effect on leaf burn or fireholding capacity (Bowling, 1959; Tso, 1972; Chouteau, 1968). However, Bowling (1967) reported that results of a 15-year test showed that duration of glow for the no-P treatment was 42 sec; for 15 and 44 kg of P/ha, it was 31 sec; and for 88 kg of P/ha, it was 22 sec.

A deficiency of Ca in cigar-wrapper tobacco results in a deformed leaf unsuitable for wrapper use. Rhoades (1972; 1974) found P fertilization decreased Ca levels in the tobacco plant. This was true for four varieties representing cigar-wrapper, cigar-filler, flue-cured, and Turkish type tobaccos.

Manganese toxicity reportedly affected the quality of the leaf. Bortner (1935) found that P reduced Mn toxicity, but Elliot (1969) found that superphosphate increased Mn, apparently because it lowered soil pH.

2. DISEASE SUSCEPTIBILITY

Losses of leaf due to downy mildew or blue mold (*Peronosphora tabacina*) and sporulation decreased with higher levels of P (Oczos, 1969). Infection by the tobacco leaf-curl virus was decreased from 64 to 47% in sand culture by increasing P to 93 ppm (Sastry & Nariana, 1962). Phosphorus was found to have no effect on *Cercospora* leaf spot disease (Stephen, 1958).

ACKNOWLEDGMENT

This is Journal Contribution no. 3315 of the Mississippi Agricultural and Forestry Experiment Station. Special thanks are due Dr. D. H. Bowman, Dr. J. D. Lancaster, and Mr. I. Spurgeon for critical reviews of the manuscript, and to Mr. D. M. Stewart and Ms. S. J. Williamson for their help in the preparation of the manuscript.

LITERATURE CITED

Adams, F., and R. W. Pearson. 1970. Differential response of cotton and peanuts to subsoil acidity. Agron. J. 62:9-12.

Andersen, R. A., J. F. Chaplin, R. E. Currin, and Z. T. Ford. 1970. Plant phenols in flue-cured tobaccos fertilized at different rates. Agron. J. 62:415-417.

Anderson, O. E., H. F. Perkins, R. L. Carter, and J. B. Jones. 1971. Plant nutrient survey of selected plants and soils of Georgia. Georgia Agric. Exp. Stn. Res. Rep. 102.

Angladette, A. 1965. Nutritional status as indicated by plant analysis. p. 355-372. *In* Int. Rice Res. Inst. The mineral nutrition of the rice plant. The Johns Hopkins Press, Baltimore, Md.

Arora, S. K., J. S. Saini, R. C. Gandhi, and R. S. Sandhu. 1970. Study of chemical composition and yield of groundnut as affected by *Rhizobium* inoculation. Oléagineux 25:279-280.

Avundzhyan, E. S. 1963. Effect of the nitrogen and phosphorus source on the alkaloid content of tobacco. Sov. Plant Physiol. (Fiziol. Rast.) 10:8-12.

Ayres, A. 1936. Effect of age upon the absorption of mineral nutrients by sugar cane under field conditions. Agron. J. 28:871–886.
Barnes, A. C. 1974. The sugarcane. 2nd ed. Halsted Press, John Wiley & Sons, Inc., N.Y. 572 p.
Basak, M. N. 1962. Nutrient uptake by rice plant and its effect on yield. Agron. J. 54:373–376.
Bassett, D. M., W. D. Anderson, and C. H. E. Werkhoven. 1970. Dry matter production and nutrient uptake in irrigated cotton (*Gossypium hirsutum*). Agron. J. 62:299–303.
Baver, L. D. 1960. Plant and soil composition relationships as applied to cane fertilization. Hawaii. Plant. Rec. 56:1–86.
Berthold, T. 1931. Effect of topping and suckering on development of the tobacco plant. Connecticut Agric. Exp. Stn. Bull. 326. p. 399–405.
Bhan, S., and D. K. Misra. 1970. Effects of variety, spacing and soil fertility on root development in groundnut under arid conditions. Indian J. Agric. Sci. 40:1050–1055.
Bhatt, J. G., and E. Appukuttan. 1971. Nutrient uptake in cotton in relation to plant architecture. Plant Soil 35:381–388.
Bhatt, J. G., T. Ramanujam, and E. Appukuttan. 1974. Growth and nutrient uptake in a short-branch strain of cotton. Cotton Grow. Rev. 51:130–137.
Bolton, A. 1956. Experiments on tobacco-seed production. II. The relationship between rate of applied nutrients, chemical composition of leaf and seed yield. Emp. J. Exp. Agric. 24: 161–166.
Borden, R. J. 1936. Some interesting cane-tonnage, purity, soil-analyses, juice-analyses relationships. Hawaii. Plant. Rec. 40:11–19.
Bortner, C. E. 1935. Toxicity of manganese to Turkish tobacco in acid Kentucky soils. Soil Sci. 39:15–33.
Bowling, J. D. 1959. Relation to fertilizer composition to growth and development of Maryland tobacco. Maryland Agric. Exp. Stn. Bull. A-101.
Bowling, J. D. 1967. Phosphorus, potassium, calcium and magnesium requirements of Maryland tobacco grown on Monmouth soil. Maryland Agric. Exp. Stn. Bull. A-151.
Bowman, D. R., and B. C. Nichols. 1953. Burley tobacco leaf composition according to position on the stalk. Tennessee Agric. Exp. Stn. Bull. 229.
Breland, H. L., W. L. Pritchett, and H. W. Lundy. 1967. Effect of fertilizer treatment on composition of flue-cured tobacco. Soil Crop Sci. Soc. Florida Proc. 27:235–242.
Broeshart, H., and H. Brunner. 1964. The efficiency of phosphate and nitrogen fertilization in rice cultivation. Int. Congr. Soil Sci., Trans. 8th (Bucharest, Romania) IV:209–218.
Bromfield, A. R. 1973. Uptake of sulphur and other nutrients by groundnuts (*Arachis hypogaea*) in northern Nigeria. Exp. Agric. 9:55–58.
Brown, P. W. 1970. Study of anatomical structure and lipid composition of Virginian type of tobacco grown in an arid environment. Tob. Sci. 14:91–94. Abstr. in Tob. Abstr. 14:1953 (1970).
Bulkina, E. V. 1969. Uptake of phosphorus by tobacco plants as dependent on weather conditions. Agrokhimiya no. 4:25–29. Abstr. in Field Crop Abstr. 22:3196(1969).
Bunting, A. H., and B. Anderson. 1960. Growth and nutrient uptake of Natal Common groundnuts in Tanzanyika. J. Agric. Sci. (Cambridge) 55:35–46.
Burkhart, L., and N. R. Page. 1941. Mineral nutrient extraction and distribution in the peanut plant. Agron. J. 33:743–755.
Burr, G. O. 1955. Plant analysis as indexes of nutrient availability and adequacy. Hawaii Plant. Rec. 55:103–109.
Burr, G. O., C. E. Hartt, H. W. Brodie, T. Tanimoto, H. P. Kortschak, D. Takahashi, F. M. Ashton, and R. E. Coleman. 1957. The sugarcane plant. Annu. Rev. Plant Physiol. 8: 275–308.
Capó, B. G., G. Samuels, P. Landrau, Jr., S.A. Alers, and A. Riera. 1955. The method of foliar diagnosis as applied to sugarcane. Puerto Rico Agric. Exp. Stn. Bull. 123.
Chesney, H. A. D. 1975. Fertilizer studies with groundnuts on the brown sands of Guyana. II. Effect of nitrogen, phosphorus, potassium, and gypsum and timing of phosphorus application. Agron. J. 67:10–13.
Chouteau, J. 1962. The manuring of French tobaccos: Results from the Experiment Institute for Tobacco, Bergerac. Fertilité 16:10–22. Abstr. in Hortic. Abstr. 33:5620 (1963).
Chouteau, J. 1968. Nutrition and mineral fertilization of tobacco. Phosphorus Agric. 51:27–32.
Christidis, B. G., and G. J. Harrison. 1955. Cotton growing problems. McGraw-Hill Book Co., Inc., New York. 633 p.

Chu, H. F. 1959. Fertilizer use in rice production in Taiwan. p. 44–49. *In* Rice improvement in Taiwan. Plant Ind. Ser. no. 15. Chinese-American Joint Comm. Rural Reconstruct., Taipai, Taiwan.

Clark, R. E. 1964. Direct and residual effects of two mixtures of nitrogen and phosphorus upon the growth of Acala cotton in the Pecos Valley. Agron. J. 56:18–20.

Clements, H. F. 1955. The absorption and distribution of phosphorus in the sugar cane plant. Hawaii Plant. Rec. 50:17–32.

Clements, H. F. 1958. Recent developments in the crop-logging of sugar cane-phosphorus and calcium. Hawaii Agric. Exp. Stn. Prog. Notes no. 114.

Clements, H. F. 1961. Crop logging of sugar cane in Hawaii. p. 131–147. *In* W. Reuther (ed.) Plant analysis and fertilizer problems. Pub. no. 8. Am. Inst. Biol. Sci., Washington, D.C.

Collins, E. R., and H. D. Morris. 1941. Soil fertility studies with peanuts. North Carolina Agric. Exp. Stn. Bull. 330.

Collins, W. K., G. L. Jones, J. A. Weybrew, and D. F. Matzinger. 1961. Comparative chemical and physical composition of flue-cured tobacco varieties. Crop Sci. 1:407–411.

Coronel, R. E., and E. F. Wallihan. 1971. The effects of nutrient deficiencies of nitrogen, potassium and manganese on the critical phosphorus concentrations in the rice plant (*Oryza sativa* L.). Philipp. Agric. 55:83–96.

Cox, F. R., J. J. Nicholaides, P. H. Reid, D. L. Hallock, and D. C. Martens. 1970. Nutrient concentrations in Virginia type peanuts during the growing season. North Carolina Agric. Exp. Stn. Tech. Bull. 204.

Darkis, F. R., L. F. Dixon, F. A. Wolf, and P. M. Gross. 1937. Chemical composition of flue-cured tobaccos produced on limed and nonlimed soils under varying weather conditions. Ind. Eng. Chem. 29:1030–1039.

Dastur, R. H., and A. Ahad. 1941. Studies on the periodic partial failures of Punjab-American cottons in the Punjab: III. The uptake and the distribution of minerals in the cotton plant. Indian J. Agric. Sci. 11:279–300.

Datta, N. P., and J. E. Shinde. 1965. Yield and nutrition of rice under upland and waterlogged conditions-effect of nitrogen, phosphorus and silica. J. Indian Soc. Soil Sci. 13:53–60.

Datta, N. P., and J. Venkateswarlu. 1968. Uptake of fertilizer phosphorus and nitrogen from different methods of application by lowland rice growing on major Indian soils. Int. Congr. Soil Sci., Trans. 9th (Adelaide, Australia) IV:9–18.

Davide, J. G. 1965. The time and methods of phosphate fertilizer applications. p. 255–268. *In* Int. Rice Res. Inst. The mineral nutrition of the rice plant. The Johns Hopkins Press, Baltimore, Md.

Davidson, R. J. 1895. Analyses of parts of tobacco plant at different stages of growth. Virginia Agric. Exp. Stn. Bull. 50.

Davies, W. N. L., and A. J. Vlitos. 1969. Fertilization of sugarcane. Proc. Int. Soc. Sugarcane Technol. 13:68–83.

Dewan, M. L., and A. F. Mahdavi. 1968. Soils in relation to rice production in Iran. Int. Congr. Soil Sci., Trans. 9th (Adelaide, Australia) IV:41–51.

Dierendonck, F. J. E. van. 1959. The manuring of coffee, cocoa, tea, and tobacco. Centre d'Etude de l'Azote, Geneva.

Dillewijn, C. van. 1952. The botany of sugarcane. Chronica Botanica Co., Waltham, Mass.

Doyle, J. J. 1966. The response of rice to fertilizer. FAO Agric. Studies no. 70, United Nations, Rome.

Du Toit, J. L. 1969. The basis of a fertilizer advisory service. Proc. Int. Soc. Sugarcane Technol. 13:800–808.

Eaton, F. M. 1955. Physiology of the cotton plant. Annu. Rev. Plant Physiol. 6:299–328.

Eaton, F. M., and D. R. Ergle. 1957. Mineral nutrition of the cotton plant. Plant Physiol. 32:169–175.

Elliot, J. M. 1969. Effect of applied manganese, sulfur, and phosphorus on the manganese content of oats and flue-cured tobacco. Can. J. Soil Sci. 49:277–285.

Enyi, B. A. C. 1969. Effect of varying supply of nitrogen and phosphorus on growth and nutrient uptake in upland and swamp rice varieties under 'dry' soil condition. Indian J. Agric. Sci. 39:180–195.

Ergle, D. R., and F. M. Eaton. 1957. Aspects of phosphorus metabolism in the cotton plant. Plant Physiol. 32:106–113.

Evans, H. 1960. Elements other than nitrogen, potassium and phosphorus in the mineral nutrition of sugarcane. Proc. Int. Soc. Sugarcane Technol. 10:473–508.

Evans, H. 1963. A review of recent developments and trends in sugar cane agriculture. Proc. Int. Soc. Sugarcane Technol. 11:47-68.
Evelyn, S. H., and I. Thornton. 1964. Soil fertility and the response of groundnuts to fertilizers in the Gambia. Emp. J. Exp. Agric. 32:153-160.
Food and Agriculture Organization. 1966. Rice—grain of life. World food problems no. 6, FAO, United Nations, Rome.
Food and Agriculture Organization. 1974. Production yearbook 1973. Vol. 27. United Nations, Rome.
Food and Agriculture Organization. 1975a. Production yearbook 1974. Vol. 28. United Nations, Rome.
Food and Agriculture Organization. 1975b. Cotton (lint) no. 9, p. 10, cottonseed no. 10, p. 9, and rice no. 2, p. 17. Monthly Bull. Agric. Econ. Stat. 24. United Nations, Rome.
Food and Agriculture Organization. 1977. Production yearbook 1976. Vol. 30. United Nations, Rome.
Fraps, G. S. 1919. The chemical composition of the cotton plant. Texas Agric. Exp. Stn. Bull. 247.
Futral, J. G. 1952. Peanut fertilizers and amendments for Georgia. Georgia Agric. Exp. Stn. Bull. 275.
Gargantini, H., and H. G. Blanco. 1965. Absorção de nutrientes pela cultura do arroz. Bragantia 24:515-528.
Garner, W. W. 1939. Some aspects of the physiology and nutrition of tobacco. Agron. J. 31:459-471.
Garner, W. W. 1946. The production of tobacco. The Blakiston Co., Philadelphia.
Geus, J. G. De. 1967. Fertilizer guide for tropical and subtropical farming. Centre d'Etude de l'Azote, Zurich.
Gillier, P., and P. Prevot. 1960. Fumures minérales de l'arachide au Sénégal. Oléagineux 15: 783-791.
Golden, L. E. 1961. Nutrient uptake by sugarcane in Louisiana. Sugar J. 23(11):22-24.
Golden, L. E. 1967. The uptake of fertilizer phosphorus by sugar cane in Louisiana as measured by radioisotope methods. Proc. Int. Soc. Sugarcane Technol. 12:640-646.
Golden, L. E., and R. Ricaud. 1963. The nitrogen, phosphorus and potassium contents of sugar cane in Louisiana. Louisiana Agric. Exp. Stn. Bull. 574.
Golden, L. E., and R. Ricaud. 1965. Foliar analysis of sugar cane in Louisiana. Louisiana Agric. Exp. Stn. Bull. 588.
Goldsworthy, P. R. 1964. Methods of applying superphosphate to groundnuts in northern Nigeria. Emp. J. Exp. Agric. 32:231-234.
Green, G. D. 1972. Principles of cotton production. Publ. by Mrs. Emily Berthea Green, Starkville, Miss.
Grizzard, A. L., H. R. Davies, and L. R. Kangas. 1942. The time and rate of nutrient absorption by flue-cured tobacco. Agron. J. 34:327-339.
Halais, P. 1963. The detection of NPK deficiency trends in sugar cane crops by means of foliar diagnosis run from year to year on a follow-up basis. Proc. Int. Soc. Sugarcane Technol. 11:214-221.
Hall, N. S., W. F. Chandler, C. H. M. van Bavel, P. H. Reid, and J. H. Anderson. 1953. A tracer technique to measure growth and activity of plant root systems. North Carolina Agric. Exp. Stn. Tech. Bull. 101.
Hallock, D. L. 1962. Effect of time and rate of fertilizer application on yield and seed-size of jumbo runner peanuts. Agron. J. 54:428-430.
Hallock, D. L., D. C. Martens, and M. W. Alexander. 1969. Nutrient distribution during development of three market types of peanuts: I. P, K, Ca and Mg contents. Agron. J. 61:81-85.
Hammons, R. O. 1973. Genetics of *Arachis hypogaea*. p. 135-173. *In* C. T. Wilson (Chm. ed. comm.) Peanuts—culture and uses, a Symp. Am. Peanut Res. Educ. Assoc., Inc., Stillwater, Okla.
Hartt, C. E. 1955. The phosphorus nutrition of sugar cane. Hawaii. Plant. Rec. 55:33-45.
Hartt, C. E. 1958. Total phosphorus in internodes 8-10 as a guide to the phosphorus fertilization of sugar cane. Hawaii. Plant. Rec. 243-270.
Hartt, C. E. 1960. The growth of sugar cane as influenced by phosphorus. The critical range of plant phosphorus. Proc. Int. Soc. Sugarcane Technol. 10:467-473.
Hartzog, D., and F. Adams. 1973. Soil fertility experiments with peanuts in 1972. Prog. Rep. Series no. 101, Alabama Agric. Exp. Stn.

Holford, I. C. R. 1968. Nutrient status of sugar cane in relation to leaf nutrient concentration. Aust. J. Exp. Agric. Anim. Husb. 8:606-614.

Hondt, H. A. de. 1955. The effect of phosphorus rates for Maryland tobacco on the content of nitrogen fractions in the plant. Ph.D. Dissertation. Univ. Maryland (Libr. Congr. Card no. Mic. 55-725). Univ. Microfilms, Ann Arbor, Mich. (Diss. Abstr. 15:14186).

Honig, P. 1960. The presence of phosphates in cane juices. Proc. Int. Soc. Sugarcane Technol. 10:356-361.

Hsieh, S. T., and K. H. Kwan. 1973. Phosphorus nutrition and susceptibility of tobacco leaves to ozone. Taiwan Tob. Wine Monop. Bur. Tob. Res. Inst. Annu. Rep. p. 131-136. Abstr. in Tob. Abstr. 18:293(1974).

Huber, A. 1956. Some observations on the correlated influence of fertilizers on peanut yields and vegetative development of the plants. Plant Soil 8:126-131.

Humbert, R. P. 1968. The growing of sugar cane, rev. ed. Elsevier Pub. Co., New York.

Innes, R. F., and T. Chinloy. 1955. Experiences with crop control in Jamaica. Hawaii. Plant. Rec. 55:149-159.

International Rice Research Institute. 1969. Nutritional disorders of the rice plant in Asia. Int. Rice Res. Inst. Annu. Rep. 1969. p. 152-155.

Ishizuka, Y. 1964. Absorption and utilization of phosphoric acid by wheat and rice plants at various growth stages. Int. Congr. Soil Sci., Trans. 8th (Bucharest, Romania) IV:459-470.

Ishizuka, Y. 1965. Nutrient uptake at different stages of growth. p. 199-217. *In* Int. Rice Res. Inst. The mineral nutrition of the rice plant. The Johns Hopkins Press, Baltimore, Md.

Ishizuka, Y. 1971. Physiology of the rice plant. Adv. Agron. 23:241-315.

Jewitt, T. N. 1953. Seasonal variations in the composition of young cotton leaves in the Sudan Gezira. J. Agric. Sci. (Cambridge) 43:89-91.

Joham, H. E. 1951. The nutritional status of the cotton plant as indicated by tissue tests. Plant Physiol. 26:76-89.

Jones, U. S., and C. E. Bardsley. 1968. Phosphorus nutrition. p. 213-254. *In* F. C. Elliot, M. Hoover, and W. K. Porter, Jr. (ed.) Advances in production and utilization of quality cotton: Principles and practices. Iowa State Univ. Press, Ames.

Kakie, T. 1969. Phosphorus fractions in tobacco plants as affected by phosphate application. Soil Sci. Plant Nutr. 15:81-85.

Kapp, L. C., R. J. Hervey, J. R. Johnston, and J. C. Smith. 1953. Response of evergreen sweet clover and cotton to phosphorus applications on Houston black clay. Soil Sci. 75: 109-118.

Kasai, Z., and K. Asada. 1965. Translocation of mineral nutrients and other substances within the rice plant. p. 75-92. *In* Int. Rice Res. Inst. The mineral nutrition of the rice plant. The Johns Hopkins Press, Baltimore, Md.

Killinger, G. B., W. E. Stokes, F. Clark, and J. D. Warner. 1947. Peanuts in Florida: I. Peanut growing. II. Chemical composition of the peanut plant. Florida Agric. Exp. Stn. Bull. 432.

Komatsu, N. 1951. The phosphorus compounds of tobacco leaves. J. Sci. Soil Man. Japan 22:19-22. Abstr. in Soils Fert. 15:1136(1952).

Koyama, T., and C. Chammek. 1971. Soil-plant nutrition studies in tropical rice (I) Studies on the varietal differences in absorbing phosphorus from soil low in available phosphorus (Part 1). Soil Sci. Plant Nutr. 17:115-126.

Kozaka, T. 1965. Control of rice blast by cultivation practices in Japan. p. 421-438. *In* Int. Rice Res. Inst. The rice blast disease. The Johns Hopkins Press, Baltimore, Md.

Kumar, M. A., and A. Venkatachari. 1971. Studies on the effect of intra row spacings and fertility levels on the yield and quality of two varieties of groundnut (*Arachis hypognea* L.). Indian J. Agric. Res. 5:67-73. Abstr. in Field Crop Abstr. 26:2706(1973).

Lakshmikantham, M., G. N. Rao, R. L. Narasimham, P. H. Rao, R. B. Subbarao, and T. S. Krishna. 1963. Studies on the influence of irrigation and nitrogenous fertilization on tissue composition as related to cane and sugar yields. Proc. Int. Soc. Sugarcane Technol. 11: 238-247.

Lancaster, J. D., and Z. A. Savatli. 1965. Foliar application of phosphorus for cotton. Mississippi Agric. For. Exp. Stn. Bull. 708.

Loche, J. 1969. Experiments on mineral fertilizers N, P and K for tobacco plants in various regions. Serv. Exploit. Ind. Tab. Allumettes. Annu. 6:11-49. Abstr. in Tob. Abstr. 14: 2000(1970).

Lockman, R. B. 1970a. Mineral composition of tobacco leaf samples: Part I. As affected by soil fertility, variety, and leaf position. Commun. Soil Sci. Plant Anal. 1:95-108.

Lockman, R. B. 1970b. Mineral composition of tobacco leaf samples: Part II. As affected by soil fertility, seasonal variation, and stage of growth. Commun. Soil Sci. Plant Anal. 1: 353-365.

Malavolta, E., and F. P. Gomes. 1960. Foliar diagnosis in Brazil. p. 180-189. *In* W. Reuther (ed.) Plant analysis and fertilizer problems. Pub. no. 8, Am. Inst. Biol. Sci., Washington, D.C.

Maples, R., and J. L. Keogh. 1973. Phosphorus fertilization experiments with cotton on delta soils of Arkansas. Arkansas Agric. Exp. Stn. Bull. 781.

Martin, G. 1964. La fumure de l'arachide dans le monde. Oléagineux 19:161-167.

Masefield, G. B. 1949. A handbook of tropical agriculture. Oxford Univ. Press, London. 196 p.

Matusiewicz, E. 1962. The effect of different phosphorus levels on the development and yield of leaves and seeds, and on the quality of seeds of tobacco. Roczn. Nauk rol., Ser. A, 86:209-219. Abstr. in Soils Fert. 26:511(1963).

McCants, C. B., and W. S. Lamm. 1970. Influence of fertilization practices on production of tobacco transplants. North Carolina Agric. Exp. Stn. Tech. Bull. 198.

McCants, C. B., and W. G. Woltz. 1967. Growth and mineral nutrition of tobacco. Adv. Agron. 19:211-265.

McClung, A. C., L. M. M. de Freitas, D. S. Mikkelsen, and W. L. Lott. 1961. Cotton fertilization on Campo Cerrado soils State of São Paulo, Brazil. IBEC Res. Inst. Bull. 27.

McEvoy, E. T. 1951. The physiological aspect of major element nutrition on the maturity of flue-cured tobacco. Sci. Agric. 31:85-92.

McEvoy, E. T. 1960. Influence of culture solution temperature on uptake of phosphorus by flue-cured tobacco plants. Can. J. Plant Sci. 40:211-217.

McHargue, J. S. 1926. Mineral constituents of the cotton plant. Agron. J. 18:1076-1083.

Mikkelsen, D. S. 1970. Recent advances in rice plant tissue analysis. Rice J. 73(6):2-5.

Mikkelsen, D. S., and R. R. Hunziker. 1971. A plant analysis survey of California rice. Agricchem. Age 14(6):18-19, 22.

Mikkelsen, D. S., M. D. Miller, M. Brandon, C. Wick, and J. Lindt. 1970. Plant analysis—A technique for diagnosing the nutritional status of rice. p. 56. *In* Proc. 13th Rice Tech. Working Group, Beaumont, Tex., 24-26 Feb. 1970.

Mitsui, S. 1956. Inorganic nutrition fertilisation and soil amelioration for lowland rice, 3rd ed. Yokendo, Ltd., Tokoyo.

Mukerjee, H. N. 1965. Fertilizer tests in cultivators' fields. p. 329-354. *In* Int. Rice Res. Inst. The mineral nutrition of the rice plant. The Johns Hopkins Press, Baltimore, Md.

Murayama, N. 1965. The influence of mineral nutrition on the characteristics of plant organs. p. 147-172. *In* Int. Rice Res. Inst. The mineral nutrition of the rice plant. The Johns Hopkins Press, Baltimore, Md.

Nelson, L. E. 1971. The effects of root temperature and Ca supply on the growth and transpiration of cotton seedlings (*Gossypium hirsutum* L.). Plant Soil 34:721-729.

Nelson, M., and J. O. Ware. 1932. The relation of nitrogen. phosphorus, and potassium to the fruiting of cotton. Arkansas Agric. Exp. Stn. Bull. 273.

Nelson, W. L. 1949. The effect of nitrogen, phosphorus, and potash on certain lint and seed properties of cotton. Agron. J. 41:289-293.

Nelson, W. L., B. A. Krantz, W. E. Colwell, W. G. Woltz, A. Hawkins, L. A. Dean, A. J. MacKenzie, and E. G. Rubins. 1948. Application of radioactive tracer technique to studies of phosphatic fertilizer utilization by crops: II. Field experiments. Soil Sci. Soc. Am. Proc. 12:113-118.

Nicholaides, J. J., and F. R. Cos. 1970. Effect of mineral nutrition on chemical composition and early reproductive development of Virginia type peanuts (*Arachis hypogaea* L.). Agron. J. 62:262-264.

Nichols, B. C., D. R. Bowman, and J. E. McMurtrey, Jr. 1958. Response of burley tobacco to fertilization in the Central Basin. Tennessee Agric. Exp. Stn. Bull. 280.

Nye, P. H. 1954. Fertilizer responses in the Gold Coast in relation to time and method of application. Emp. J. Exp. Agric. 22:101-111.

Oczos, A. 1969. Influence of mineral nutrition (NPK) on the chemical composition of leaves and the degree of susceptibility of two tobacco varieties (resistant and susceptible) to *Peronospora tabacina* Adam. Warsaw Inst. Uprawy, Nawozenia Gleboznawoze. Pamietnik Pulawski 36:251-280. Abstr. in Tob. Abstr. 14:603(1970).

Okajima, H. 1965. Environmental factors and nutrient uptake. p. 63-73. *In* Int. Rice Res. Inst. The mineral nutrition of the rice plant. The Johns Hopkins Press, Baltimore, Md.

Olsen, K. L. 1958. Mineral deficiency symptoms in rice. Arkansas Agric. Exp. Stn. Bull. 605.

Olson, L. C., and R. P. Bledsoe. 1942. The chemical composition of the cotton plant and the uptake of nutrients at different stages of growth. Georgia Agric. Exp. Stn. Bull. 222.

Oram, P. A. 1958. Recent developments in groundnut production, with special reference to Africa. Part 1 and 2. Field Crop Abstr. 11:1-6, 73-84.

Parups, E. V., and K. F. Nielsen. 1960a. The growth of tobacco at certain soil temperatures and nutrient levels in greenhouse. Can. J. Plant Sci. 40:281-287.

Parups, E. V., and K. F. Nielsen. 1960b. The growth, nicotine and phosphorus content of tobacco grown at different soil temperature, moisture and phosphorus levels. Can. J. Plant Sci. 40:516-523.

Petersen, E. L., and A. W. Naylor. 1953. Some metabolic changes in tobacco stem tips accompanying maleic hydrazide treatment and the appearance of frenching symptoms. Physiol. Plant. 6:816-828.

Peterson, L. A. 1968. Nitrate accumulation in tobacco leaves in relation to N, P, and K concentrations of the leaf. Agron. J. 60:26-29.

Petrie, A. H. K., R. Watson, and E. D. Ward. 1939. Physiological ontogeny in the tobacco plant. 1. The drifts in dry weight and leaf area in relation to phosphorus supply and topping. Aust. J. Exp. Biol. Med. Sci. 17:93-122.

Ponnamperuma, F. N. 1965. Dynamic aspects of flooded soils and the nutrition of the rice plant. p. 295-328. *In* Int. Rice Res. Inst. The mineral nutrition of the rice plant. The Johns Hopkins Press, Baltimore, Md.

Portales, R. S. 1963. Foliar analysis as a guide to sugar cane fertilization at San Cristobal, Mexico. Proc. Int. Soc. Sugarcane Technol. 11:222-232.

Presley, J. T., and L. S. Bird. 1968. Diseases and their control. p. 347-366. *In* F. C. Elliot, M. Hoover, and W. K. Porter, Jr. (ed.) Advances in production and utilization of quality cotton: Principles and practices. Iowa State Univ. Press, Ames.

Prevot, P., and M. Ollagnier. 1951. Application du diagnostic foliaire a l'arachide. Premiers résultats an Sénégal. Oléagineux 6:329-337.

Prevot, P., and M. Ollagnier. 1954. Peanut and oil palm foliar diagnosis interrelations of N, P, K, Ca, Mg. Plant Physiol. 29:26-34.

Purseglove, J. W. 1972. Tropical crops: Monocotyledons 1. John Wiley and Sons, Inc., New York.

Raper, C. D., Jr., and C. B. McCants. 1966. Nutrient accumulation in flue-cured tobacco. Tob. Sci. 10:109.

Reid, P. H., and F. R. Cox. 1973. Soil properties, mineral nutrition and fertilization practices. p. 271-297. *In* C. T. Wilson (Chm. ed. comm.) Peanuts—culture and uses, a Symp. Am. Peanut Res. Educ. Assoc., Inc., Stillwater, Okla.

Reid, P. H., and E. T. York, Jr. 1958. Effect of nutrient deficiencies on growth and fruiting characteristics of peanuts in sand cultures. Agron. J. 50:63-67.

Reyes, E. D., J. G. Davide, L. G. Orara, and R. A. Calixihan. 1962. Nitrogen, phosphorus, and potassium uptake by a lowland rice variety at different stages of growth. Philipp. Agric. 46:7-19.

Rhoads, F. M. 1972. Effect of phosphorus fertilization on calcium uptake by cigar wrapper tobacco. Commun. Soil Sci. Plant Anal. 3:87-95.

Rhoads, F. M. 1974. Response of five tobacco cultivars to sodium phosphate and ammonium polyphosphate. Commun. Soil Sci. Plant Anal. 5:557-563.

Roche, P., J. Velly, and B. Joliet. 1959. Utilization of foliar diagnosis with groundnuts in the soil conservation area of Vallée-Témoin. Agron. Trop. 14:165-197. Abstr. in Trop. Abstr. 14:2244(1959).

Sabbe, W. E., J. L. Keogh, R. Maples, and L. H. Hileman. 1972. Nutrient analysis of Arkansas cotton and soybean leaf tissue. Arkansas Farm Res. 21(1):2.

Sabbe, W. E., and A. J. MacKenzie. 1973. Plant analysis as an aid to cotton fertilization. p. 299-313. *In* L. M. Walsh and J. D. Beaton (ed.) Soil testing and plant analysis, rev. ed. Soil Sci. Soc. Am., Madison, Wis.

Samuels, G. 1959. The influence of the age of sugarcane on its leaf-nutrient (N-P-K) content. J. Agric. Univ. Puerto Rico 43:159-170.

Samuels, G. 1960. The relative merits of various methods of foliar diagnosis for sugarcane. Proc. Int. Soc. Sugarcane Technol. 10:529-537.

Samuels, G., J. P. Rodréguez, and P. Landrau, Jr. 1959. The response of cotton to fertilizers in Puerto Rico. J. Agric. Univ. Puerto Rico 43:89-102.

Sardone, L. T. 1960. The groundnut industry of Queensland. World Crops 12:351-354.

Sastry, K. S. M., and T. K. Nariani. 1962. Effect of host plant nutrition on growth and susceptibility of tobacco plants to infection with tobacco leaf curl virus. Ind. J. Agric. Sci. 32:288-293.

Satyanarayana, P., and D. V. Krishna Rao. 1962. Mineral nutrition of peanuts by the method of foliar diagnosis. Andhra Agric. J. 9:329-343. Abstr. in Soils Fert. 27:1848(1964).

Scarsbrook, C. E., and J. T. Cope, Jr. 1956. Fertility requirements of runner peanuts in southeastern Alabama. Alabama Agric. Exp. Stn. Bull. 302.

Sims, J. L., and G. A. Place. 1968. Growth and nutrient uptake of rice at different growth stages and nitrogen levels. Agron. J. 60:692-696.

Skinner, J. J., J. G. Futral, and N. McKaig, Jr. 1944. The uptake of nutrients by the cotton plant when fertilized with acid forming and non-acid forming fertilizers combined with different rates of potash. Georgia Agric. Exp. Stn. Bull. 235.

Skogley, E. O., and C. B. McCants. 1963. Ammonium and chloride influences on growth characteristics of flue-cured tobacco. Soil Sci. Soc. Am. Proc. 27:391-394.

Small, Jr., H. G., and A. J. Ohlrogge. 1973. Plant analysis as an aid in fertilizing soybeans and peanuts. p. 315-327. In L. M. Walsh and J. D. Beaton (ed.) Soil testing and plant analysis, rev. ed. Soil Sci. Soc. Am., Madison, Wis.

Soil Improvement Committee. 1975. p. 148. In Western fertilizer handbook, 5th ed. California Fert. Assoc., Sacramento.

Stelly, M., and H. D. Morris. 1953. Residual effect of phosphorus on cotton grown on Cecil soil as determined with radioactive phosphorus. Soil Sci. Soc. Am. Proc. 17:267-269.

Stephen, R. C. 1958. The influence of nitrogen, phosphorus, and potassium nutrition of flue-cured tobacco on the incidence of Cercospora leafspot disease. Emp. J. Exp. Agric. 26:64-69.

Swanback, T. R., and P. J. Anderson. 1952. Fertilizer placement for Connecticut tobacco. Connecticut Agric. Exp. Stn. Bull. 561.

Takahashi, J., M. Yanagisawa, M. Kôno et al. 1955. Studies on nutrient absorption by crops: VIII. Effect of temperature on the growth, yield and nutrient absorption of rice plants. 2. Study of the relationship between temperature and phosphorus supply. Bull. Natl. Inst. Agric. Sci. Tokyo B4:63-71. Abstr. in Soils Fert. 19:1690(1956).

Tanaka, H., and S. Akai. 1963. Effect of some nutrient elements on the susceptibility of rice plants to *Helminthosporium* leaf spot. Annu. Phytopath. Soc. Japan 28:144-152. Abstr. in Soils Fert. 27:481(1964).

Terman, G. L., and S. E. Allen. 1970. Fertilizer and soil P uptake by paddy rice, as affected by soil P level, source and data of application. J. Agric. Sci. (Cambridge) 75:547-552.

Thompson, L., R. Maples, and J. Wells. 1962. Recommendations for rice fertilization in southern states. Arkansas Rice J. 65(1):5-6.

Tso, T. C. 1972. Physiology and biochemistry of tobacco plants. Dowden, Hutchinson, and Ross, Inc., Stroudsburg, Pa.

Vickery, H. B. 1961. Chemical investigations of the tobacco plant. XI. Composition of the green leaf in relation to position on the stalk. Connecticut Agric. Exp. Stn. Bull. 640.

Walker, M. E., and R. L. Carter. 1971. The effect of fertilization and storage temperatures on percent nitrogen, oils and germination of Spanish and runner peanut seed. Georgia Agric. Exp. Stn. Res. Bull. 88.

Ward, R. C., D. A. Whitney, and D. G. Westfall. 1973. Plant analysis as an aid in fertilizing small grains. p. 329-348. In L. M. Walsh and J. D. Beaton (ed.) Soil testing and plant analysis. rev. ed. Soil Sci. Soc. Am., Madison, Wis.

Watson, R., and A. H. K. Petrie. 1940. Physiological ontogeny in the tobacco plant: 4. The drift in nitrogen content of the parts in relation to phosphorus supply and topping, with an analysis of the determination of ontogenetic changes. Aust. J. Exp. Biol. Med. Sci. 18:313-340.

Went, F. W. 1957. The experimental control of plant growth. Vol. 17, Chron. Bot., Waltham, Mass.

Whitty, E. B., C. B. McCants, and L. Shaw. 1966. Influence of width of fertilized band of soil on response of burley tobacco to nitrogen and phosphorus. Tob. Sci. 10:17-22.

Woltz, W. G., N. S. Hall, and W. E. Colwell. 1949. Utilization of phosphorus by tobacco. Soil Sci. 68:121-128.

Woltz, W. G., W. A. Reid, and W. E. Colwell. 1948. Sugar and nicotine in cured bright tobacco as related to mineral element composition. Soil Sci. Soc. Am. Proc. 13:385-387.

Chapter 26

Phosphorus Nutrition of Vegetable Crops and Sugar Beets

O. A. LORENZ AND M. T. VITTUM

University of California-Davis, and New York State Agricultural Experiment Station, Cornell University, Geneva, N.Y., respectively

I. INTRODUCTION

A. Production Areas

Vegetables and potatoes are grown commercially in every state, but major production is centered in a relatively few. California is the leading state, and in 1974 accounted for 43% by value of vegetables for fresh market and over 50% for processing (USDA, 1976). California, Florida, Texas, Michigan, and Arizona account for more than two-thirds of the total area devoted to fresh market production, while Wisconsin, California, and Minnesota predominate in production for processing. Potatoes are grown extensively in Idaho, Maine, North Dakota, Washington, and California; these states account for about half of U.S. production. Sugar beets are grown predominately in five states, namely, California, Minnesota, North Dakota, Colorado, and Idaho. The United States currently farms nearly 2,100,000 ha of vegetables and potatoes and 600,000 ha of sugar beets.

B. General Use of P Fertilizer

Phosphorus fertilizers are almost universally applied to vegetables and sugar beets. Vegetables receive high rates of P as compared with other crops. Even so the total amount of P applied to vegetables in the United States is less than 100 million kg, or less than 5% of that used on all crops. The rate of application of P varies greatly between crops, but probably averages about 45 kg/ha. Rates as high as 200 kg/ha are often applied to high-value vegetables. The average application of P to sugar beets approaches 45 kg/ha.

Copyright 1980 © ASA-CSSA-SSSA, 677 South Segoe Road, Madison, WI 53711, USA.
The Role of Phosphorus in Agriculture.

C. Status of P in Soils

Due to high fertilization rates of P and to low crop removal, contents of both available and total P have increased greatly in many soils, especially those devoted to the production of high-value market crops. This is particularly evident in such intensive vegetable-producing areas as the Salinas and Imperial Valleys in California, and in the market garden areas in New York, Florida, Michigan, and Wisconsin. In these areas it is not uncommon to find available soil P values several times those of the virgin soils. It is evident that these soils should not continue to receive P fertilizer at the same rate as in the past. Peck et al. (1965a) stated that many soils used for vegetable production in New York have been heavily fertilized for many years and, therefore, had accumulated large amounts of available P. Peech (1946) reported a marked accumulation of readily soluble P in soils in the important potato-producing areas along the Atlantic Coast. In light-textured soils with large amounts of readily soluble P in the surface layer, there was appreciable movement of P into the subsoil, but in heavier soils Peck et al. (1965a, 1976) found very little downward movement of P. Unlike N which is quite mobile, moving up and down with soil moisture, P is relatively immobile, remaining essentially where it is placed in the soil.

II. PHOSPHORUS COMPOSITION OF VEGETABLES AND SUGAR BEETS

A. Phosphorus Removal

Compilations of yields, percentage P, and P absorption by the harvested portion of many vegetables are presented in Table 1. Yields were estimated from the average yield for the United States for the years 1969–1973 multiplied by a factor of 1.3 to better approximate good yields. In most cases tissue concentrations of P are the average values for many samples as reported by Chatfield and Adams (1940) and Beeson (1941). On a dry-wt basis, P commonly varies from 0.35 to 0.50%, although it is less than 0.3% in table beet roots, eggplant, onions, potatoes, watermelons, and sugar beet roots and is over 0.7% in broccoli and cauliflower.

The amount of P removed by the harvested portion of the plant is usually less than 10 kg/ha. Celery and garlic have the highest uptake of about 30 kg/ha. Less than 3 kg/ha of P is removed by asparagus, cucumbers, and watermelons. Absorption of P by the entire plant is commonly in the range of 15 to 20 kg/ha, but approaches 30 for tomatoes and celery and is less than 10 for beans, cucumbers, and watermelons. These values for P absorption agree well with those of vegetables grown in the United Kingdom as reported by Greenwood et al. (1974), except for onions and carrots where they reported higher values. They found the uptake of P to vary from a low of 4 and 8 kg/ha for radish and lettuce, respectively, to over 40 kg/ha for leeks.

Table 1—Good commercial yield, percent P, and P absorption by vegetable crops and sugar beets.

Crop	Good yield, fresh wt†	Plant part harvested	P, dry wt	P uptake Harvested portion	P uptake Total plant
	kg/ha		%	kg/ha	
Artichoke	9,100	Bud	0.56	8.3	--
Asparagus	3,500	Spear	0.73	2.0	--
Beans (snap)	7,200	Pod & seed	0.48	4.8	7.8
Beans (lima)	3,500	Immature seed	0.41	4.8	7.8
Beets	37,900	Root	0.27	11.2	25.2
Broccoli	10,500	Flower stalk	0.79	8.3	24.9
Brussels sprouts	15,500	Bud	0.45	10.5	20.3
Cabbage (fresh market)	23,500	Head	0.38	8.2	12.0
Cabbage (processing)	54,500	Head	0.38	15.7	23.5
Cantaloupe	17,800	Fruit	0.53	6.6	11.1
Carrots	36,200	Root	0.33	14.1	19.6
Cassava	41,000	Root	0.14	21.6	--
Cauliflower	14,000	Flower bud	0.76	8.9	26.7
Celery	70,000	Stalk	0.64	28.2	31.4
Corn (sweet)	13,800	Kernels	0.26	7.9	15.8
Cucumber	13,700	Fruit	0.45	2.4	7.4
Eggplant	23,500	Fruit	0.27	4.7	14.7
Endive	18,300	Leafy head	0.50	6.6	9.9
Honeydew melon	23,400	Fruit	0.29	6.4	12.7
Garlic	18,700	Bulb	0.38	29.7	52.1
Lettuce	30,900	Head	0.50	8.9	12.0
Onion	42,700	Bulb	0.26	13.9	17.3
Green peas	3,800	Immature seeds	0.57	5.5	12.4
Green peppers	13,000	Immature fruit	0.44	7.9	17.9
Potatoes	33,300	Tubers	0.25	18.5	22.3
Spinach	17,800	Leaves	0.50	9.0	9.1
Sweet potato	15,500	Roots	0.12	5.8	7.8
Taro	17,500	Corn	0.28	11.1	--
Tomato	60,100	Fruit	0.55	19.5	29.5
Turnip	34,900	Root	0.36	11.4	25.4
Watermelon	14,700	Fruit	0.14	2.1	7.6
Sugar beets	57,200	Root	0.13	14.8	31.8

† Avg. U.S. yields 1969-1973 × 1.3 (source: USDA, 1974).

Some of the most comprehensive research on P nutrition of vegetables has been carried out by Peck and coworkers at Geneva, N.Y. (Peck et al., 1965b; Peck & MacDonald, 1969, 1972; Peck & Stamer, 1970; Peck, 1975a, 1975b). Four rates of fertilizer P and four rates of fertilizer K were applied to the same field plots each year, thus establishing wide ranges in the levels of P and K in the soil. Over a period of years, five different vegetable crops were grown on these plots. Crops were sampled at different stages of growth, ranging from 3 sampling dates for peas to 11 for sweet corn. At each sampling date, plants were separated into different morphological parts, and each part was analyzed for P, K, Ca, Mg, Mn, Zn, Fe, and Na. In general their values for P uptake agree with those cited above.

Draycott (1972) summarized P absorption by sugar beets as determined by various researchers in the United Kingdom and reported that total P absorption varied from 10 to 33 kg/ha with an overall average of 23 kg/ha. Absorption of P by vegetables and sugar beets is very small when measured as percentage of that normally applied. Most crops absorb from 5 to 15% of the applied P in the first year, but absorption may vary from as low as 1 to as high as 20%.

B. Critical P Levels in Selected Plant Parts

Plant analyses have been used by many investigators as an aid in determining fertilizer needs (Danielson, 1953; Tyler & Lorenz, 1964). Plant analyses for P are simple, accurate, and can be performed at very low cost. The methods often used are those in which either fresh or dried material is extracted with 2% acetic acid (Johnson & Ulrich, 1959). Usually some conductive tissue such as petiole or stem is used, but leaf blades can be used equally well. It appears that plant analysis has been used most successfully in warm, naturally dry areas where water is supplied by irrigation and other environmental factors are more nearly optimum.

Deficiency and sufficiency levels for many crops as determined largely from field experiments by Lorenz and Tyler (1971) and from various sources as reported by Geraldson et al. (1973) are presented in Table 2. These levels are related to definite stages of growth and to specific parts of the plant at time of sampling. Note that 2,000 ppm acid-soluble PO_4-P can be used as the critical or deficiency level for many plants. Only broccoli, cauliflower, and cabbage exhibit higher critical values. Crops for which lower values can be used include sweet corn, beans, and asparagus. In sugar beets, 750 ppm PO_4-P represents the deficiency level in petiole tissue; while 1,000 ppm is suggested as the deficiency value for the blade of the first mature leaf (Ulrich et al., 1959). Where the P concentrations remained above these values for the entire season, the crop did not respond to P fertilizer even though more P was absorbed.

If the concentrations of P in plant tissues approach the sufficiency level, there is very little chance that crops would respond to any additional P fertilization. Between the deficient and sufficient levels, a yield response to P fertilization might or might not be obtained, depending upon other environmental factors. Concentrations much higher than those classified as sufficient are commonly observed in field-grown vegetables as noted in numerous field surveys.

It is often necessary to relate the age of the plant to the deficiency level, because in most plants the concentration of P decreases from early growth to maturity. This concept is illustrated with peas (Peck & MacDonald, 1969) cabbage (Peck & Stamer, 1970), table beets (Peck & MacDonald, 1972), snap beans (Peck, 1975b), and sweet corn (Peck & MacDonald, 1975). In California, concentrations of 2,000 ppm in the petiole tissue in early growth, 1,500 ppm at early fruit set, and 1,000 ppm at the time of first mature fruits, have been suggested as the deficient levels for samples of canta-

Table 2—Plant analysis guide for sampling time, plant part, and nutrient levels of vegetable crops and sugar beets.

Crop	Time of sampling	Plant part	2% acetic acid soluble PO$_4$-P, dry weight	
			Deficient	Sufficient
			ppm	ppm
Asparagus	Midgrowth of fern	10-cm tip section of new fern branch	800	1,600
Bean, bush snap	Midgrowth	Petiole of fourth leaf from tip	1,000	3,000
Bean, bush snap	Early bloom	Petiole of fourth leaf from tip	800	2,000
Broccoli	Midgrowth	Midrib of young, mature leaf	2,500	5,000
Broccoli	1st buds	Midrib of young, mature leaf	2,000	4,000
Brussels sprouts	Midgrowth	Midrib of young, mature leaf	2,000	3,500
Brussels sprouts	Late growth	Midrib of young, mature leaf	1,000	3,000
Cabbage	At heading	Midrib of wrapper leaf	2,500	3,500
Cantaloupe	Early growth (short runners)	Petiole of sixth leaf from growing tip	2,000	4,000
Cantaloupe	Early fruit set	Petiole of sixth leaf from growing tip	1,500	2,500
Cantaloupe	1st mature fruit	Petiole of sixth leaf from growing tip	1,000	2,000
Carrot	Midgrowth	Petiole of young, mature leaf	2,000	4,000
Cauliflower	Buttoning	Midrib of young, mature leaf	2,500	3,500
Celery	Midgrowth to maturity	Petiole of newest fully elongated leaf	2,000	4,000
Cucumber, pickling	Early fruit set	Petiole of sixth leaf from tip	2,500	2,500
Lettuce	At heading	Midrib of wrapper leaf	2,000	4,000
Lettuce	At harvest	Midrib of wrapper leaf	1,500	2,500
Pepper, chili	Early growth	Petiole of young, mature leaf	2,000	3,000
Pepper, chili	Early fruit set	Petiole of young, mature leaf	1,500	2,500
Pepper, sweet	Early growth	Petiole of young, mature leaf	2,000	4,000
Pepper, sweet	Early fruit set	Petiole of young, mature leaf	1,500	2,500
Potatoes	Early season	Petiole of fourth leaf from growing tip	1,200	2,000
Potatoes	Midseason	Petiole of fourth leaf from growing tip	800	1,600
Potatoes	Late season	Petiole of fourth leaf from growing tip	500	1,000
Spinach	Midgrowth	Petiole of young, mature leaf	2,000	4,000
Sweet corn	Tasseling	Midrib of first leaf above primary ear	500	1,000
Sweet potato	Midgrowth	Petiole of sixth leaf from growing tip	1,000	2,000
Tomato (canning)	Early bloom to first mature fruits	Petiole of fourth leaf from growing tip	2,000	3,000
Watermelon	Early fruit set	Petiole of sixth leaf from growing tip	1,500	2,500
Sugar beets	Any time after seedling stage	Petiole of young mature leaf	750	2,000
		Blade of young mature leaf	2,000	4,000
Sugar beets	Seedling	Petiole	1,500	3,000
		Blade	3,000	6,000

loupe (Tyler & Lorenz, 1964). In potatoes, 1,200 ppm PO$_4$-P is the suggested deficiency level at early growth, 800 at midseason, and 500 in late season, for petiole tissue of the fourth oldest leaf (Tyler et al., 1961). Most studies have shown that samples taken early in the growth of plants are more reliable than those taken at a later stage for estimating the plant's nutritional requirement for P. Greater differences in tissue concentration of P between deficient and sufficient plants are noted in samples taken during early growth. Tremblay and Baur (1952) emphasized the great effect of P on early growth and absorption of P by peas. During the latter part of the growing season there was often no correlation between yield of peas and P concentration of the tops and leaves. The best time to sample pea plants to determine their P status was between the four to eight-node stage. In sugar beets, if the youngest mature leaf is selected the same critical level can be used for plants at all ages, except for seedlings where a higher concentration is required.

Plant analyses have definite limitations; The growing period for many vegetables is short and by the time the plant is large enough to sample effectively, it is too late to apply additional fertilizer. Nevertheless, this information is of value in determining fertilizer programs for future crops.

It is often desirable to analyze plant tissue for total P rather than for some soluble fraction. Many analyses of total P concentration in the aboveground portion of vegetables are available (Table 3). The lower value in many cases approaches the critical concentration which is associated with reduced yields, while the higher values represent concentrations above those required for maximum yields. As with the soluble fraction of P, values for total P do not vary as much as those for many of the other nutrients. In leaves or petioles of most crops, concentrations as low as 0.30% total P approach the critical level, while 0.50% is higher than required for maximum yields. Crops which seem to be exceptions to this generalization are cauliflower, cabbage, lettuce, and tomatoes. High concentrations of P are observed in these crops.

C. Soluble and Total P for Estimating Critical Levels

Most researchers have found P soluble in 2% acetic acid to be a better measure of the P status of the plant than total P. As long as concentrations remain above 2,000 ppm soluble P, most plants have sufficient P to produce maximum growth. The relationship between total P and soluble P concentrations in potato petioles is shown in Fig. 1. These data were obtained from many fields ranging from soils deficient in P to soils containing ample quantities of P for maximum plant growth (Lorenz et al., 1964a). There was a high correlation between soluble and total P and either form would serve as a good measure for predicting the P status of the plant. Approximately two-thirds of the total P was soluble in 2% acetic acid. Similar information for cantaloupes from a large number of fields is shown in Fig. 2 (Tyler & Lorenz, 1964). Leaf blade tissue did not give as high a correlation between soluble P and total P as petiole tissue. Correlation coefficients were 0.65 and 0.86 for the blades and petioles, respectively. In petioles the soluble P was slightly more than two-thirds of the total P. The range in concentration

Table 3—Common P levels in leaf tissue of vegetable crops and sugar beets.

Crop	Growth stage	Plant part	P, dry wt %
Asparagus	Mature fern	Fern from 45 to 90 cm	0.30–0.35
Beans, snap	Bud	Mature trifoliate leaf	0.30–0.50
Beet	Mature	Young mature leaf	0.20–0.30
Broccoli	At heading	Young mature leaf	0.30–0.50
Brussels sprouts	Midgrowth	Young mature leaf	0.25–0.50
Cabbage	Heads half grown	Young wrapper leaf	0.30–0.50
Cantaloupe	Fruit set	Leaf blade	0.25–0.40
Carrots	Midgrowth	Young mature leaf	0.20–0.35
Cauliflower	At heading	Young mature leaf	0.50–0.70
Celery	Half grown	Young mature leaf	0.30–0.50
Cucumbers	Fruit set	Young mature leaf	0.25–0.40
Kale	Almost mature	Young mature leaf	0.30–0.60
Lettuce	Heads at half size	Wrapper leaf	0.40–0.60
Onion	Midgrowth	Young mature leaf	0.25–0.40
Peas	Midgrowth	Young mature leaf	0.25–0.35
Peppers, bell	Midgrowth	Young mature leaf	0.20–0.30
Potatoes	Tubers half grown	Young mature leaf	0.20–0.40
Potatoes	Tubers half grown	Petioles of young mature leaf	0.20–0.40
Spinach	30–50 days old	Young mature leaf	0.40–0.55
Spinach	Mature	Young mature leaf	0.30–0.50
Sweet corn	Silking	Ear leaf	0.20–0.30
Sweet potatoes	Midseason	Young mature leaf	0.20–0.30
Turnip	Midgrowth	Young mature leaf	0.35–0.60
Tomato (MH)	First mature fruit	Young mature leaf	0.55–0.80
Tomato staked	First mature fruit	Young mature leaf	0.80–1.00
Watermelon	Midgrowth	Young mature leaf	0.20–0.30
Sugar beets	Mature	Young mature leaf	0.20–0.30

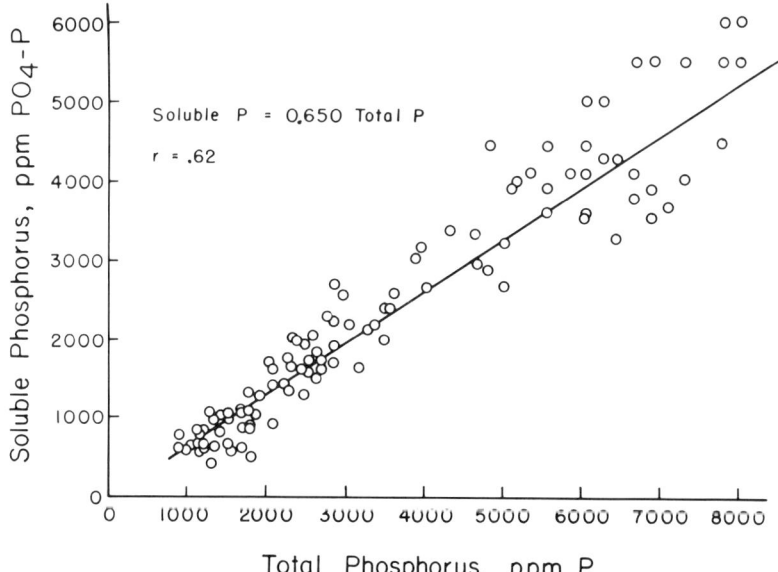

Fig. 1—Comparison of soluble P, extracted with 2% acetic acid, with total P in potato petioles.

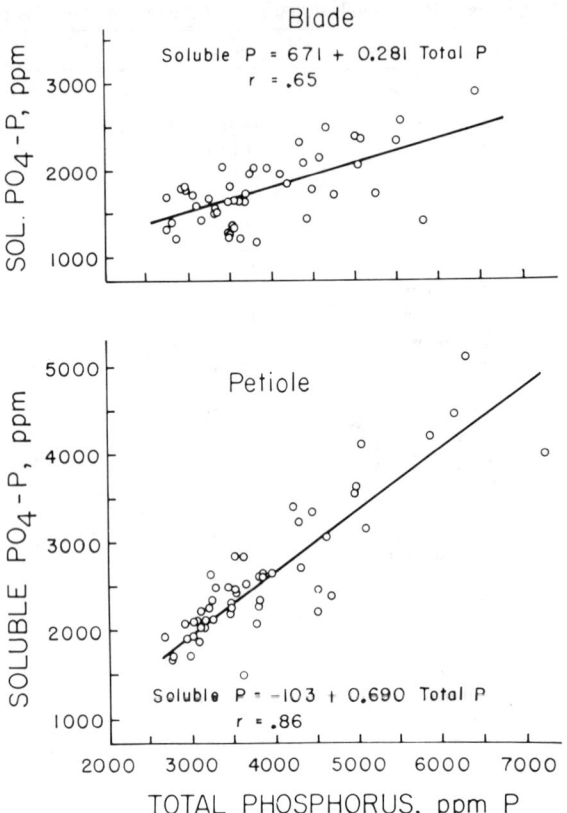

Fig. 2—Comparison of soluble P, extracted with 2% acetic acid, with total P in cantaloupe leaf blades and petioles.

for soluble P was much less in the blade than that in petiole tissue, thus petiole tissue would provide a better estimation of the P nutritional status of the plant.

Peck and MacDonald (1972) found that for table beets the petiole is a better indicator of P nutrition than the leaf blades. In contrast, they reported (1975) that for snap beans the concentration of total P in leaf blades is often higher than in petioles.

III. PHOSPHORUS UPTAKE DEMAND PATTERNS

Numerous researchers have reported on the course of P absorption by various crops. Notable is the information for celery (Zink, 1963), lettuce (Zink & Yamaguchi, 1962), spinach (Zink, 1965), onions (Zink, 1966), potatoes (Carpenter, 1957), melons (Tyler & Lorenz, 1964), peas (Wolf, 1945; Peck & MacDonald, 1969), peppers (Cochran & Olson, 1941), snap beans (Peck, 1975b), cabbage (Peck & Stamer, 1970), table beets (Peck & Mac-

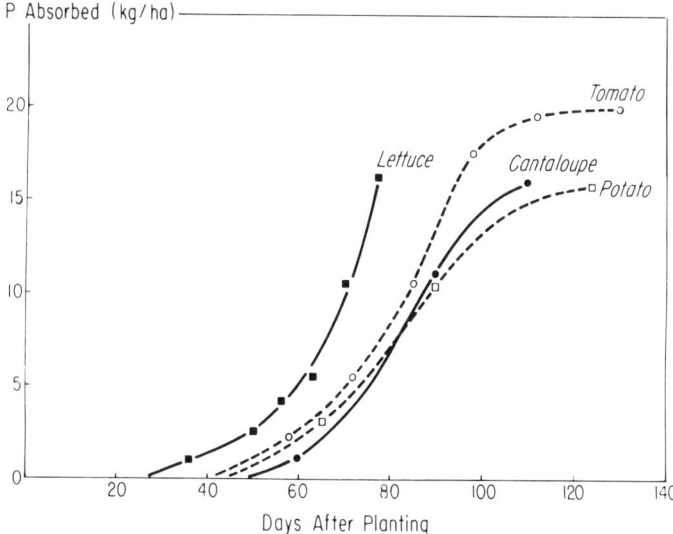

Fig. 3—Phosphorus absorption by vegetable crops in California. Data for lettuce redrawn from Zink and Yamaguchi (1962).

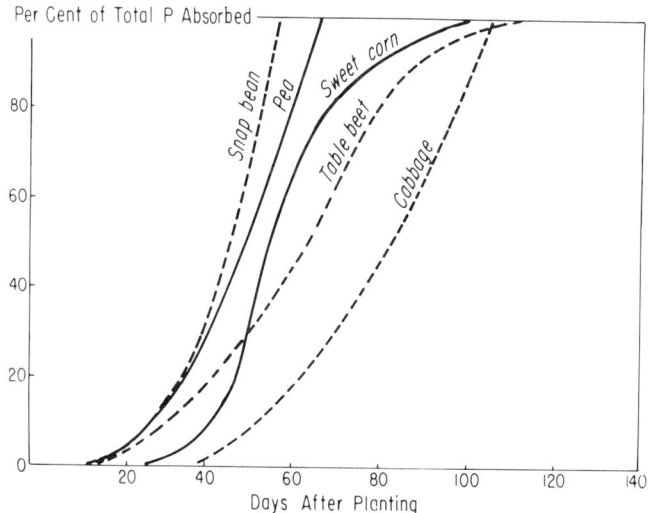

Fig. 4—Phosphorus absorption by vegetable crops in New York. Data redrawn from Peck (1975a).

Donald, 1972), sweet corn (Peck & MacDonald, 1975), and various vegetables (Hester et al., 1951). Demand uptake patterns for various vegetables were summarized by Lorenz and Bartz (1968) and by Peck (1975a). Many vegetables are grown for a relatively short period and most are harvested before full maturity. Carrots, celery, lettuce, snap beans, peas, sweet corn, and many other vegetables are harvested at an immature stage of growth since highest product quality is attained while the crop is still growing vigorously. With these crops most of the P is absorbed during the last one-fourth of the growing period, but even so the greatest need for P fertilization as determined from many observations is during the seedling stage.

The relative accumulation of P at different stages of growth of several vegetables is shown in Fig. 3 and 4. Uptake patterns are illustrated for crops maturing in less than 60 days to those maturing after 120 days. Snap beans which were harvested 56 days after seeding were still growing actively and were rapidly absorbing P at the time of harvest (Fig. 4). The same was true of lettuce which was harvested less than 80 days after seeding (Fig. 3). The greatest rate of P accumulation for both crops occurred in the last 2 weeks before harvest. Similar demand patterns would be exhibited by peas and by foliage vegetables, such as spinach, celery, broccoli, and endive. This is contrasted with table beets, potatoes, and processing tomatoes which were not harvested until the crops had begun senescence. In these crops, maximum rate of P accumulation occurred during the grand period of growth and then decreased as the crops matured. Cantaloupes present a third uptake pattern in which fruits mature and are harvested over a period of several weeks. In this case, P absorption slows down during the last days but the plant is still absorbing P actively when harvest is discontinued. Similar uptake patterns occur with crops having indeterminate fruit-set, such as cucumbers, tomatoes, summer squash, okra, etc.

Uptake of P during the growth of sugar beets was described by Draycott (1972). During the first few weeks after seeding, P absorption is slow and then increases during the grand period of growth. The rate of accumulation prior to harvesting depends on the growth stage at harvest. If the crop is harvested early, P is still being accumulated in significant amounts, but if it is not harvested and is allowed to remain in the soil and even overwintered, as in some parts of California, absorption of P is very low in the immediate preharvest stage.

IV. CROP RESPONSES AND PHOSPHORUS FERTILIZATION

A. Phosphorus Deficiency Symptoms

Phosphorus deficiency on most vegetables is not readily detected by visual symptoms. It is usually typified by stunting of the plant and often by dark green and/or purple leaves and stems. The stems are thin and shortened in growth. Poor growth and retarded development are often the only symptoms of P deficiency. In potatoes, leaves are often paler rather than darker. Under severe deficiency leaf veins may become reddish or purplish as observed with tomatoes, cabbage, cauliflower, and some cultivars of sweet corn.

Maturity is usually delayed in plants which are P-deficient. In foliage crops such as lettuce, P-deficient plants often reach harvestable size as much as several weeks later than plants receiving ample quantities of P. Consequently, early yields may be greatly reduced even though there may be no effect on subsequent total yields. Lingle and Wight (1964) showed that the response of cantaloupes to P was most evident in early maturity of the crop. Phosphorus fertilizer hastens maturity of sweet corn. Furthermore, high P holds sweet corn at good processing quality for a longer period of time, thus giving growers and processors more time in which to harvest the crops (Peck & MacDonald, 1975). Applications of P beyond those required for optimum growth do not further hasten maturity.

Symptoms of P-deficiency of sugar beets were described by Draycott (1972). In severely deficient plants, cotyledons and primary leaves in the seedling stage are dark green and brown-pitted. Petioles curl upwards and a reddish-brown necrosis develops from the tips and the edges of the leaves. In mature, severely deficient plants, a brown netted veining forms in the tissues of fully expanded leaves when they dry up and die. On slightly deficient plants, leaves are dark green and plants appear as though they are stunted and delayed in growth.

B. Rates of Fertilizer P

Rates of P required for optimum yields of vegetables vary greatly, due in part to the methods of measuring yields. Crops such as lettuce and celery must attain a certain size and quality before they are acceptable for marketing. Fruit size is also important for vegetables such as cantaloupes and fresh market tomatoes. With other crops, yield differences are more quantitative than qualitative. A review of experiments and recommendations for many states shows that P fertilizer is usually recommended for commercial vegetable production. Usually the recommended rates range from 50 to 80 kg/ha of P but vary from a low of about 25 to a high of 200 kg/ha for some high-value crops. While there are probably some varietal differences in response to P fertilization, these would be minor as compared to the large differences between species.

Compared to field crops, rates of P applied to vegetables are usually high. Recommendations are usually grouped according to crops and soil test values for P. In Massachusetts, for example, Maynard and Thomson (1970) recommended 60 to 90 kg/ha of P for most vegetables grown on low-fertility soils. Rates from 30 to 50 kg/ha were recommended for beans, carrots, peas, peppers, and radish, while 100 to 130 kg/ha were recommended for celery, potatoes, and sweet corn. When grown on high-fertility soils the recommended rates of P applications were approximately half those for low-fertility soils. Similar adjustment of recommended amounts of P fertilizer, depending on soil fertility level, was made for New York State by Peck (1975b).

Some species are more responsive to P than others as shown in field experiments conducted by Lorenz (unpublished) in California on a Meloland fine sandy loam soil analyzing 6 ppm bicarbonate-extractable PO_4-P (Table 4). Lettuce and cabbage gave much larger yield increases from P fertilization than did carrots, onions, and cantaloupes. Maximum yields of cabbage

Table 4—Relative response of vegetables to rates of concentrated superphosphate.†

P application	Yields as percent of maximum				
	Cabbage	Lettuce	Carrots	Onion	Cantaloupe
kg/ha					
0	0	13	73	79	85
30	99	83	78	90	87
60	100	94	81	98	100
90	100	100	100	100	100

† Lorenz, unpublished data (3-year avg.); soil—Meloland fine sandy loam, pH 7.8, $NaHCO_3$ extractable PO_4-P 6 ppm; fertilizer—applied broadcast and folded into bed previous to planting crops.

were obtained with 30 kg/ha of P, while three times this amount was required for lettuce. Without P fertilizers cabbage and lettuce grew very poorly while onions and cantaloupe yielded about 85% of the maximum.

Cleaver and Greenwood (1975) in the United Kingdom found that spinach and lettuce gave the greatest relative yield increase to P; cabbage, cauliflower, Brussels sprouts, and radish gave the least; and potatoes, carrots, beans, and onions were intermediate (Table 5). Spinach yields were increased by P applications to soils analyzing as high as 150 ppm bicarbonate-extractable P, while yields of cabbage and cauliflower did not increase by raising soil P values above 20 ppm P. Nishimoto and Fox (1972) in Hawaii reported that lettuce and Chinese cabbage gave larger increases to P fertilization than sweet corn, sweet potatoes, and cucumbers. They also related P response to the concentration of P in the soil solution. Three-g soil samples were equilibrated with 30 ml of 0.01 M $CaCl_2$ containing graded amounts of CSP and P determined in the supernatant solution. At 0.003 ppm P in the soil solution, lettuce produced about 1% of maximum yield and at 0.025 ppm about 14% of the maximum. Optimum concentrations of P in soil solution ranged from 0.04 ppm for cabbage to 0.3 for lettuce (Table 6).

Ware and Johnson (1949) compared the yield response of 26 vegetables to applied P in bins containing P-deficient soils. Lettuce, endive, and onion gave the greatest relative response, while sweet potatoes, beans, turnips, and radish exhibited the least. The other vegetables were intermediate.

It is evident that recommendations for P fertilization should be related to the amount of available P in soil as determined by a soil test. Cleaver and Greenwood (1975) and Greenwood et al. (1974) established soil response values for many vegetables (Table 5). Based on expected yields, recommendations are made as to the amount of P to apply. As an example, to produce the maximum yield lettuce grown on a soil testing 10 ppm $NaHCO_3$ soluble P should receive 300 kg/ha of P, while on a soil testing 75 ppm of P, the recommended rate of P would be only 100 kg/ha. Winter cabbage grown on a soil testing 10 ppm P would require only 60 kg/ha of P and, on soils testing more than 20 ppm P, no P would be required. For most vegetables grown on soils low in available P, their recommendation is to apply about 150 kg/ha of P, and on soils testing over 50 ppm available P, only 50

Table 5—Predicted phosphate levels for vegetables grown on soils with various levels of exchangeable P.†

Crop	Response values of crops to P ppm PO_4-P in soil (sodium bicarbonate extraction)					
	10	20	30	40	50	60
	kg/ha of P					
Broad beans	40	20	10	10	10	5
Brussels sprouts	10	5	5	0	0	0
Calabrese	25	10	10	5	5	0
Carrots	50	25	15	15	10	5
French beans	45	25	15	10	10	5
Leeks	25	25	15	15	10	5
Lettuce	140	70	50	35	25	10
Onions	60	30	20	15	10	5
Parsnips	25	15	10	5	5	5
Peas	10	5	5	5	0	0
Potatoes	35	20	10	10	10	5
Radish	10	5	5	5	0	0
Red beet	25	10	10	5	5	0
Spinach	350	180	120	90	70	35
Summer cabbage	10	5	5	0	0	0
Summer cauliflower	10	5	5	0	0	0
Swedes	15	5	5	5	5	0
Turnips	30	15	10	10	5	5
Winter cabbage	5	5	0	0	0	0

† Source: T. S. Cleaver and D. J. Greenwood. 1975. Natl. Vegetable Res. Stn., Wellesbourne, Warwick, U.K.

Table 6—Relative yield of several vegetable crops at specified P levels in soil solution.†

Crop	P in soil solution, ppm							
	0.003	0.006	0.012	0.025	0.05	0.1	0.2	0.4
	% of maximum yield							
Head lettuce	1	2	6	14	26	52	81	100
Cucumber	20	32	45	58	72	83	97	--
Tomato			43	70	80	89	94	99
Chinese cabbage	27	44	58	70	81	90	97	100
Soybean	46	55	70	75	83	91	96	100
Sweet potato	72	74	77	82	87	94	99	100
Headcabbage			87	91	96	99	100	100

† Source: R. K. Nishimoto and R. L. Fox. Sept. 1975. Univ. of Hawaii, Honolulu.

kg/ha of P should be used. Lettuce and spinach should be fertilized at rates of about 150 kg/ha even on soils high in available P.

Tolman et al. (1956) reported 60 kg/ha of P as the optimum rate for sugar beets on soils testing between 15 and 45 ppm bicarbonate-soluble P and none for soils above 75 ppm P. Based on the bicarbonate-soluble P in soils, Draycott (1972) in the United Kingdom recommended 180 kg/ha of P for soils with less than 10 ppm P, 120 kg for soils with 11 to 15 ppm, 30 for soils within 26 to 45 ppm, and none for soils above 45 ppm.

Peck et al. (1976) applied various rates of P to soils in New York and determined that annual applications of about 70 kg/ha were required to maintain an adequate level of available soil P as determined by Morgan's sodium acetate-acetic acid extracting solution. Below this the soil P levels became depleted with time, while applications of 280 kg/ha of P showed large accumulation of available soil P.

C. Placement and Method of Application

It is evident that no single method of P application can be applied to all vegetable crops grown under all conditions. Many factors must be considered, such as rooting characteristics of the plant, length of growing season, crop rotations, soil moisture and irrigation, soil characteristics, rate of P application, and source of P fertilizer. These interact to determine the best practice to use at a particular time.

Over 50 years ago, Sayre (1934) showed that applying fertilizer in bands for tomato transplants was more efficient than broadcasting it on the surface. Many placement experiments were conducted throughout the United States between 1925-1958, and results were summarized and recommendations for fertilizer placement were prepared by the National Joint Committee on Fertilizer Application (1958). For practically all vegetable crops and sugar beets, it was recommended that fertilizer be placed in bands 5 to 8 cm to the side and 3 to 5 cm below the seed level. In most situations small amounts of N and K are recommended with the P in the band. Large amounts of N and K at this time are likely to harm the young seedlings. If only P is applied, the band may be placed 5 cm deep directly under the seed. A recent review by Lucas and Vittum (1976) summarizes the present status of fertilizer placement for vegetables.

Although band placement of P is theoretically better than a broadcast application because some fertilizer is placed near the seedling to stimulate early growth and reduce P fixation, experiments comparing band placement vs. broadcasting have occasionally given conflicting results. Exceptions are explained by the fact that vegetables usually receive high rates of fertilizer P, and even if applied broadcast, a sufficient quantity of P is available for growth of young seedlings, thus overcoming the benefit of closely banded fertilizers. This is especially true in soils that have received repeatedly high rates of P fertilizers. When low rates of P have been applied, banding has been shown to be superior to broadcasting as illustrated by Lingle and Wight (1964) with cantaloupes.

Numerous workers have shown that fertilizer for potatoes applied at planting should be in bands 5 cm to each side of the seed piece on a level slightly below the seed piece.

Schmehl et al. (1955) with sugar beets in Colorado found little difference in the final yield between band-placed fertilizers and those placed under or applied broadcast and disked in. If banded, the fertilizer should be applied as near the seed as possible without burning the young seedlings. Grunes et al. (1958) obtained significantly higher yields of sugar beets from

P applied in bands as compared with broadcasting. Ulrich et al. (1959) in California recommended placement of P in close proximity to the seed. Placement several cm below the seed was very effective in overcoming early-season P deficiencies.

Tomatoes which are direct-seeded respond to P applied at seeding, even at relatively low rates of application. In California (Sims et al., 1968), it is a common practice to apply about 30 kg/ha each of N and P about 5 cm directly below the seed. This is usually a neutral liquid mix. The tomato plant has a high P requirement early in the season when temperatures are low and the root system is small. Wilcox (1967) in pot experiments found that P placed 5 cm under the seed row was superior for early seedling growth to placement 4 cm to the side and 5 cm below the seed. For the first 5 weeks of growth, the starter band effectively met the P requirement of the plant.

The importance of P as a starter fertilizer is widely recognized. Diammonium phosphate and other liquid mixes high in P are used. Transplants often respond favorably if solutions are used at the time of field setting. Starter fertilizers are useful with seeded as well as transplanted crops since high amounts of P are required in early growth. Response to starter fertilizer is especially marked early in the season when soils are apt to be cold.

Lingle and Wight (1964) compared band placement vs. broadcasting of P fertilizer at rates of 12 and 60 kg/ha, and reported that banding of P fertilizer for cantaloupes increased P efficiency. Banding was slightly more beneficial than broadcasting at the lower rate of application, but yields were increased by the higher rate of P with both methods of application.

MacKenzie and Stockinger (1956) showed that close placement of P for lettuce resulted in high yields and early maturity. Placement 10 cm below the seed resulted in delayed growth as compared with broadcast application. A large part of lettuce grown in California and Arizona receives up to 100 kg/ha of P applied broadcast and folded into the bed prior to planting. This results in a localization of the P supply and provides ample P close to the developing seedlings. Onions, carrots, and many other vegetables also show the need for some P placed close to the seedlings.

Foliar applications of P have not proven economical and are seldom recommended. It is possible to produce vegetable crops with foliage-applied P provided it is applied several times per week. Foliar application seems to be of little benefit when the level of available soil P is adequate. The greatest need for P is during early stages of growth but at this time leaves are small, few in number, and there is inadequate surface area for absorption of P by foliar application. Silberstein and Wittwer (1951) dipped leaves of tomato, bean, and sweet corn in solutions of several P materials and noted slight responses in plant growth. Early but not total yields of tomatoes were increased by 4 weekly sprays of a 25-mM solution of orthophosphoric acid. Using foliar-applied, radioactive orthophosphoric acid, they found that P was readily absorbed by leaves of tomatoes, corn, bean, and squash plants and translocated to the root tips and other centers of high metabolic activity. Teubner et al. (1962) found that about 12% of the total P in the plant parts harvested for food of field-grown tomatoes, potatoes, and beans could be supplied through multiple foliar sprays. Foliar sprays did not increase the amount of total P absorbed nor did they increase yields.

D. Time of Phosphorus Application

The time of application of P is closely related to the method of application. Application by broadcasting or banding under the row is obviously made at or immediately prior to seeding. It has been repeatedly demonstrated that the greatest response to P fertilizer is observed in young seedlings. Application of P late in the growing season has not increased yields above those resulting from banding all of the P at planting time. Lingle and Wight (1964) showed that sidedressing cantaloupes with P after emergence of the seedlings was of no value to the crop, but good responses were obtained from P applied at seeding. Haddock and Linton (1957) demonstrated that P was more effective when applied immediately preceding a pea crop than when applied 1 to 2 years before. Potatoes normally receive all of the P at planting time, placed in a band on both sides of the row. Some growers split the application, applying about half by broadcasting and then incorporating it into the soil before planting. The remainder is then applied in the planting operation. Vittum and Hulburt (1963) reported that transplanted tomatoes produced maximum yields from placing part of the fertilizer in bands and the remainder at the bottom of the plow furrow at plowing.

Grunes et al. (1958) showed that percentage of fertilizer P absorbed by sugar beets decreased as the time of application was delayed. Olsen et al. (1950a) reported P placed near sugar beet seed at planting increased uptake of P markedly as compared to band placement at thinning time.

Data by all researchers suggest that the total application of P should be made at the time of seeding or transplanting and rarely, if ever, are additional P applications recommended. The need for P is critical during the early stages of growth.

E. Seasonal Influences on P Fertilization

It is commonly observed that vegetables grown during cool seasons give greater response to P fertilization than crops grown under warmer conditions. Locascio and Warren (1960) noted that P deficiency symptoms of field-seeded tomatoes are often more pronounced in early plantings than in later plantings. Lingle and Davis (1959) reported that field-seeded tomatoes planted early in cool weather generally respond to P fertilization while later plantings may not respond at all. Lettuce grown during warm weather in the Imperial Valley of California requires very little P fertilizer, but crops grown during the coldest winter months respond to very high rates of P fertilization and often to rates as high as 300 kg/ha (Lorenz et al., 1964b).

In the petiole tissue of tomatoes, Tiessen and Carolus (1963) noted an increase in P concentration as the air temperature increased from 10 to 22°C, but increasing the soil temperature from 8 to 22°C had no pronounced influence on petiole P concentration. Locascio and Warren (1960) found increases in growth of tomato plants as P fertilization was increased from 70 to 550 kg/ha at 13°C, while, at 21 and 30°C, growth was increased up to only 140 kg/ha of P and with little increase about 70 kg/ha. The uptake of P per plant increased as the soil temperature increased from 12 to 21°C. Apple and Butts (1953) reported more pronounced P deficiency symptoms in early planted pole beans than in later plantings. The percentage of P in pole beans was increased more by P fertilization at 17°C soil temperature than at 37°C. Mack et al. (1964) found increased concentrations of P in snap beans and peas as soil temperature was increased from 13 to 26°C. However, application of P at very high rates did not compensate for the growth-retarding effect of soil temperatures.

F. Sources of Phosphorus for Vegetables

The agronomic effectiveness of phosphate fertilizers as related to management practices is discussed in detail by Engelstad and Terman in Chapter 12. It is generally agreed that all of the water-soluble forms of P are of equal value in fertilizing vegetable crops and that a major amount of the P should be water soluble when used in a band or seed placement. When applied as a broadcast application, the degree of water solubility is of less importance and there is little influence on P uptake by the crop (Lorenz & Bartz, 1968). There is a strong interaction between granule size and water solubility which must be considered when making comparisons of agronomic effectiveness of P sources.

Most data show that, on the basis of crop yields, powdered dicalcium phosphate (DCP) is as good a source of P as ordinary superphosphate (OSP) on neutral and acid soils. On a calcareous soil in Imperial Valley, California, Lorenz et al. (1964b) found minor differences in yield, P uptake, and available soil P between concentrated superphosphate (CSP), DCP, and liquid phosphoric acid. Liquid phosphoric acid or neutral mixes of aqua ammonia and phosphoric acid, applied to the soil surface as a spray and later incorporated, were as effective as dry granular materials applied broadcast and incorporated into the soil. Lingle (1960) reported no differences in yields of field-grown tomatoes from ammonium phosphate, diammonium phosphate (DAP), or nitric phosphate (NP) (20-20-0 with 50% water-soluble P) banded at planting.

Olsen et al. (1950b) reported that potatoes on calcareous soils in Colorado absorb equal amounts of P from OSP and from calcium metaphosphate, ground to pass a 100-mesh screen. Mattingly (1963) in England showed that yields of potatoes on an acid soil were the same with granular nitrophosphates, potassium metaphosphate, high soluble basic slag, and granular OSP. Phosphate rock gave lower yields. Murphy and Goven (1966) in Maine reported no differences in yield or specific gravity of potato tubers from OSP, CSP, monoammonium phosphate (MAP), and ammonium polyphosphate (APP).

Schemhl et al. (1955) with sugar beets grown on calcareous soils in Colorado found ammonium metaphosphate, calcium metaphosphate, and NP to be unsatisfactory when applied in bands, but were comparable to OSP when mixed with the soil. Olsen et al. (1950b) found the absorption of P was highest from OSP and calcium metaphosphate followed by DCP and least from tricalcium phosphate. Yields of sugar beets were usually higher from OSP than the other sources. Crops absorbed from 10 to 12% of the fertilizer P.

V. PHOSPHORUS NUTRITION AND CROP QUALITY

A. Market Quality

Most information on the effect of P on quality of vegetables is related to the effect on market quality, particularly size and grade. Small sizes usually suffer price discrimination as compared with the larger sizes. Likewise, the very large sizes are often not desired. Using potatoes as an example, without adequate P the plants produce small tubers which do not meet the accepted grade or size. With adequate fertilization, larger tubers are produced and a high yield of U.S. no. 1's results. With still higher fertilization, the tubers may be too large and often show growth cracks, knobs, or other defects which make them unacceptable for premium grades.

The effect of P fertilization on hastening the maturity of vegetables and the subsequent effect on quality and price should not be overlooked. As indicated previously, in crops such as lettuce the time of maturity is often hastened by as much as 2 weeks by P fertilization, as compared with crops grown on P-deficient soils. Various internal disorders such as tipburn, russet spotting, and pink rib often develop in older plants. Fertilizer practices that hasten maturity and allow early harvesting may eliminate these disorders. Greenwood et al. (1974) conducted 109 experiments in England on 19 vegetable crops and reported the effects of P on quality were negligible. Phosphorus fertilizers above the amount required for optimum yield have not affected quality. Phosphorus fertilization has not been shown to affect the storage or shelf life of vegetables except through its effect on maturity and quality at the time the produce is stored. The effect of P on the quality of vegetables was reviewed by Lorenz (1963).

B. Nutritive Value

Although vegetables are usually rather colorful and attractive on the dinner table, they are not a major source of most of the nutrients essential for humans in current American diets. They supply a negligible amount of fat, only 7% of the body's energy requirements, and 8% of the protein needs. Other foods, such as cereals, meat, milk, and eggs are more efficient sources of these vital constituents. On the other hand, vegetables are an important source of the fiber and bulk which is necessary in the proper functioning of the human digestive system, and, as a group, contribute over 40% of our vitamin A and vitamin C requirements (V. R. Boswell. 1961. What we need in quality of vegetables. p. 7-16. *In* Proc. 23rd Eastern States Agron. Conf., Springfield, Mass. Mimeo).

Phosphorus fertilization has been associated with higher sugars in some fruit vegetables. Bradley and Fleming (1959) reported a shortage of P resulted in a slight reduction in sugar content of watermelons. When yields were not changed by fertilizer rate differences, sugar content was not affected. Lingle and Wight (1964) found P fertilization to increase the soluble solids of cantaloupe fruits.

Stevens and Paulson (1973) reported that P concentration is an important factor in the quality of tomato fruits because of its effect on the relationship between pH and titratable acidity. There has been little success, however, in increasing the P concentration in tomato fruits by fertilization. Kattan et al. (1957) found no effect of P on firmness, color, pH, viscosity, or solids of fresh or canned tomatoes.

Most information on the effect of P on quality of vegetables is related to environmental factors such as soil, location, and season—and management factors—such as variety, stage of maturity at harvest, and posthandling procedures. These have a far greater effect on the nutritive value of vegetables than do fertilizers, provided the crop is grown without a serious deficiency of any of the essential elements. A well-fertilized crop is usually a high-quality crop (Vittum, 1963).

Sugar percentage and purity of the juice are commonly used to measure quality of sugar beets. Sugar percentage is used as the basis for payment to growers. Draycott (1972), after reviewing the results of numerous experiments on the effect of P on quality of sugar beets, concluded that P fertilization increased sugar percentage slightly on severely deficient soils. On moderately fertile soils P fertilization had no effect on sugar percentage. Juice purity was not affected by P fertilizer on any soil. Herron et al. (1964) in Kansas found no effect of P on sugar percentage, while Ogden et al. (1958) in Minnesota reported a decrease in sugar percentage from high rates of P fertilization, but little effect on juice purity.

VI. RELATION OF SOIL P LEVELS TO CROP RESPONSE

A. Soil Tests

Vegetables have high nutrient requirements and many have shallow root systems, thus making them very sensitive to soil nutrient supply. Most of the information relating vegetable crop responses to P in calcareous soils has been evaluated by extraction of soil with sodium bicarbonate as described by Olsen et al. (1954). Knudsen (1975) states the Bray procedure for P is used by all state soil testing laboratories in the north central region except North Dakota. This method gives good correlation with crop yield on most acid and neutral soils. Most soil testing laboratories in the northeastern states use the sodium acetate-acetic acid extraction solution developed by Morgan and perfected by Peech and English (1944).

B. Crop Response

There is great variation in the deficiency levels of available soil P as recommended by various researchers. Evidently the crop, season, and cultural conditions greatly affect the response to P fertilization. Lorenz and Bartz (1968), after reviewing much published data, concluded that levels above about 20 ppm of bicarbonate-extractable P were adequate for vegetables grown on calcareous soils. On acid soils, when the Bray extractant of $0.025N$ HCl is used, levels of less than 17 kg of P/ha are considered low, 17 to 34 medium, and above 34 high. Reisenauer et al. (1976) set the critical range for cool-season vegetables at 12 to 20 ppm bicarbonate-extractable P and for warm-season vegetables at 5 to 9 ppm. Tyler et al. (1961) reported on the results of many field experiments with potatoes grown on alkaline soils. Soil levels of bicarbonate-extractable P as low as 6 ppm and as high as 70 ppm were observed in potato soils in California. They showed concentrations below 15 ppm P to be deficient, 15 to 30 ppm as intermediate with a response to P fertilization uncertain, and above 30 ppm as sufficient, where there would be no response to P fertilizers. Haddock and Linton (1957) in Utah recommended that the soil contain 40 ppm bicarbonate-extractable P for peas.

Cleaver and Greenwood (1975) in the United Kingdom stated that spinach and lettuce required the highest levels of available soil P. They obtained responses to P additions on soils testing as high as 75 ppm bicarbonate-extractable P for lettuce and 100 ppm for spinach. Other vegetables gave little or no response on soils testing above 50 ppm and the *Brassicas* did not respond beyond 35 ppm. These levels are considerably higher than levels recommended for vegetables in the United States.

Nishimoto and Fox (1975), using acid soils in Hawaii, related the levels of P in the soil solution to the response of various vegetables (Table 6). Concentration of P in the soil solution varied from 0.003 to 1.6 ppm. For yields at 95% of the maximum, the P requirement in the soil solution was slightly less than 0.1 ppm for sweet corn and sweet potatoes, 0.2 ppm for Chinese cabbage and cucumber, and 0.4 for lettuce.

Peck (1975a) concluded that 45 kg/ha of available P by the sodium acetate-acetic acid extraction method was sufficient in New York soils for snap beans, sweet corn, cabbage, and table beets, and that a response to fertilizer P would be unlikely except in cold, wet soils.

Sailsberry et al. (1968) recommended P fertilization for sugar beets on soils testing below 10 ppm bicarbonate-extractable P. Marked responses in early plant growth were observed on soils testing below 8 ppm. James et al. (1967) found that Washington soils responded to P fertilization only on soils below 20 ppm P. Tolman et al. (1956) compared the results from many sugar beet field tests and reported that response to P fertilization was obtained when bicarbonate-extractable P was below 15 ppm, while, in soils testing between 15 and 45 ppm, yield response was obtained in about 75% of the cases.

C. Disease Resistance

The effect of P on disease resistance seems to be most closely related to its effect on plant growth, particularly during the seedling stages. Phosphorus applications have a positive effect on seedling vigor, which increases the number of plants that survive the attack of various seedling diseases. Draycott (1972) reported that sugar beet seedlings on severely P-deficient soils die from attacks of black leg (*Pythium* spp.) and other fungal infections.

VII. EFFECTS OF EXCESS P

The chief effect of excessive levels of P fertilizer seems to be reduced uptake of certain micronutrients, particularly Zn as well as internal micronutrient imbalance. This subject is discussed in detail by Fred Adams in Chapter 23 in this book. Peck and MacDonald (1972), for example, found that high soil P reduced the Zn concentration in leaf blades of table beets throughout the season; it had little effect on Zn in the petioles until late in the season, at which time it lowered Zn concentration; and it increased Zn in the roots throughout the season. Some of Peck's results are summarized in Table 7. High levels of soil and fertilizer P invariably increased the P concentration of plant tissues, and it reduced the concentrations of Zn in practically all tissues except roots. In snap beans, high P reduced K and Fe concentrations but tended to increase Mg. From a practical point of view, applications of P at rates normally recommended for vegetables should have very little effect on micronutrient availability and utilization except possibly on high pH soils where a Zn deficiency could develop.

VIII. FUTURE CONSIDERATIONS

Yields of most vegetables have continually increased due to improved varieties and better cultural and harvesting practices. These higher yields increase the P needs of the crops. Efforts should continue towards a more efficient use of P fertilizers and the conservation of phosphate rock as a natural resource. Since P is being applied to many soils at high rates, and soil P levels are being increased, the relation of P to other soil nutrients will become more important. It is especially necessary to evaluate the effect of high soil P concentrations on the availability of Zn, but other nutrient imbalances should not be ignored.

The development of new forms of fertilizer N, such as urea-formaldehyde

Table 7—Significant effects of high soil and fertilizer P on mineral concentration of plant tissue at harvest maturity.†

Crop and part of plant sampled	High P increased									High P decreased								
	P	K	Ca	Mg	Mn	Zn	Fe	Na	P	K	Ca	Mg	Mn	Zn	Fe	Na
Beets																
Leaf blades	**	NS	--	--	--	--	--	NS	--	*	*	**	**	**	--	--
Leaf petioles	**	--	--	**	--	--	--	**	--	NS	**	--	NS	**	--	--
Roots	**	--	--	NS	--	**	--	**	--	NS	NS	--	NS	--	--	--
Sweet corn																
Stalks	**	NS	**	**	--	--	NS	*	--	--	--	--	NS	**	--	--
Leaves	**	**	***	NS	NS	--	**	--	--	NS	--	--	--	**	--	--
Husks	**	**	NS	*	--	--	NS	NS	--	--	--	--	NS	NS	--	NS
Cobs	**	--	--	--	--	--	--	*	--	NS	NS	NS	NS	**	NS	NS
Kernels	NS	--	--	--	--	--	--	--	--	NS	NS	NS	NS	**	NS	NS
Snap beans																
Roots	**	--	--	--	--	--	--	--	--	**	**	NS	NS	NS	NS	**
Stems	**	--	--	**	--	--	--	--	--	**	**	--	*	**	**	**
Old petioles	**	--	--	**	--	--	--	--	--	**	**	--	*	**	**	NS
Young petioles	**	--	--	NS	--	--	--	--	--	**	*	--	NS	**	**	NS
Pedundes	**	NS	--	NS	--	--	--	--	--	**	**	--	NS	**	**	**
Old blades	**	--	**	**	--	--	--	--	--	**	--	--	NS	**	**	--
Young blades	**	--	NS	**	--	--	--	--	--	*	--	--	NS	**	**	**
Large pods	**	--	--	**	--	--	--	--	--	**	**	--	NS	**	**	*
Small pods	**	--	NS	**	--	--	--	--	--	**	--	--	**	*	**	NS

**, *, NS—Highly significant, odds exceed 99:1; significant, odds exceed 19:1; and nonsignificant, odds <19:1, respectively.
† Data taken from Peck (1975b) and Peck and MacDonald (1972, 1975).

sulfur-coated urea, and N-P combinations, such as MAP and DAP, require additional research. What effect do these materials have on germinating seedlings? Is a localized zone of high pH immediately surrounding the fertilizer granule detrimental to the root that penetrates this zone? Can precision equipment be developed so that the fertilizer band is always in the same location with respect to the seed? Why is a P starter fertilizer so important in the growth of young seedlings?

Soil testing procedures need revision. We need tests that not only identify P and K deficiencies in soil, but will provide quantitative indices for soils having deficient, adequate, and excessive levels of these elements.

LITERATURE CITED

Apple, S. B., Jr., and J. S. Butts. 1953. The effect of soil temperature and phosphorus on growth and phosphorus uptake by pole beans. Proc. Am. Soc. Hortic. Sci. 61:325-332.

Beeson, K. C. 1941. The mineral composition of crops with particular reference to the soils in which they were grown. A review and compilation. USDA Misc. Pub. 369.

Bradley, G. A., and J. W. Fleming. 1959. Fertilization and foliar analysis studies on watermelons. Arkansas Agric. Exp. Stn. Bull. 610.

Carpenter, P. N. 1957. Mineral accumulation in potato plants. Maine Agric. Exp. Stn. Bull. 562:3-5.

Chatfield, Charlotte, and Georgian Adams. 1940. Proximate composition of American food materials. USDA Circ. 549.

Cleaver, T. S., and D. J. Greenwood. 1975. Ready reckoner to predict best fertilizer levels for vegetables. Grower 83:1269-1271.

Cochran, H. L., and L. C. Olson. 1941. Uptake of nutrients in the perfection pimiento plant under field conditions and its relation to fertilizer practices in Georgia. Georgia Agric. Exp. Stn. Bull. 208.

Danielson, L. L. 1953. Rapid chemical plant tissue tests for diagnosing fertilizer deficiencies in growing vegetable crops. Virginia Truck Exp. Stn. Bull. 112.

Draycott, A. P. 1972. Sugar-beet nutrition. John Wiley and Sons, New York. 250 p.

Geraldson, C. M., G. R. Klacan, and O. A. Lorenz. 1973. Plant analysis as an aid in fertilizing vegetable crops. p. 365-379. In L. M. Walsh and S. D. Beaton (ed.) Soil testing and plant analysis. Soil Sci. Soc. Am., Madison, Wis.

Greenwood, D. J., T. J. Cleaver, and M. K. Turner. 1974. Fertilizer requirements of vegetable crops. Natl. Veg. Res. Stn. Proc. 145, Wellesbourne, Warwick, UK.

Grunes, D. L., H. R. Haise, and L. O. Fine. 1958. Proportional uptake of soil and fertilizer phosphorus by plants as affected by nitrogen fertilization: Field experiments with sugar beets and potatoes. Soil Sci. Soc. Am. Proc. 22:49-52.

Haddock, J. L., and D. C. Linton. 1957. Yield and phosphorus content of canning peas as affected by fertilization and irrigation regime and sodium bicarbonate soluble soil phosphorus. Soil Sci. Soc. Am. Proc. 21:167-171.

Herron, G. M., D. W. Grimes, and R. E. Finkner. 1964. Effect of plant spacing and fertilizer on yield, purity, chemical constituents, and evapotranspiration of sugar beets in Kansas. J. Am. Soc. Sugar Beet Technol. 12:669-714.

Hester, J. B., F. A. Shelton, and R. L. Isaacs, Jr. 1951. The rate and amount of plant nutrients absorbed by various vegetables. Proc. Am. Soc. Hortic. Sci. 57:249-251.

James, D. W., G. E. Liggett, and A. I. Jaw. 1967. Phosphorus fertility relationships of central Washington irrigated soil, with special emphasis on exposed calcareous subsoil. Washington Agric. Exp. Stn. Bull. 668.

Johnson, C. M., and Albert Ulrich. 1959. Analytical methods for use in plant analysis. California Agric. Exp. Stn. Bull. 766, Berkeley.

Kattan, A. A., F. C. Stark, and A. Kramer. 1957. Effects of certain preharvest factors on yield and quality of raw and processed tomatoes. Proc. Am. Soc. Hortic. Sci. 69:327-342.

Knudsen, D. 1975. Recommended phosphorus soil test procedures. North Dakota Agric. Exp. Stn. Bull. 499. p. 16-19.

Lingle, J. C. 1960. The effect of sources of phosphorus on the growth and phosphorus uptake of tomato seedlings. Proc. Am. Soc. Hortic. Sci. 76:495-502.

Lingle, J. C., and R. M. Davis. 1959. The influence of soil temperature and phosphorus fertilization on the growth and mineral absorption of tomato seedlings. Proc. Am. Soc. Hortic. Sci. 73:312-322.

Lingle, J. C., and J. R. Wight. 1964. Fertilizer experiments with cantaloupes. California Agric. Exp. Stn. Bull. 807, Berkeley.

Locascio, S. J., and G. F. Warren. 1960. Interaction of soil temperature and phosphorus on growth of tomatoes. Proc. Am. Soc. Hortic. Sci. 75:601-710.

Lorenz, O. A. 1963. Effect of mineral nutrition on quality of vegetables. Proc. 39th Annu. Meet. Counc. Fert. Applic., Amherst, Mass. Natl. Plant Food Inst., Washington, D.C.

Lorenz, O. A., and J. F. Bartz. 1968. Fertilization for high yields and quality of vegetable crops. p. 327-352. *In* L. B. Nelson (ed.) Changing patterns in fertilizer use. Soil Sci. Soc. Am., Madison, Wis.

Lorenz, O. A., and K. B. Tyler. 1971. Plant tissue analysis of vegetable crops. p. 44-49. *In* Statewide Conf. Soil and Tissue Testing. Div. of Agric. Sci., Univ. of California, Davis.

Lorenz, O. A., K. B. Tyler, and F. S. Fullmer. 1964a. Plant analyses for determining the nutritional status of potatoes. p. 226-240. *In* C. Bould (ed.) Plant analysis and fertilizer problems, IV. Am. Soc. Hortic. Sci., Geneva, N.Y.

Lorenz, O. A., K. B. Tyler, and O. D. McCoy. 1964b. Phosphate sources and rates for winter lettuce on a calcareous soil. Proc. Am. Soc. Hortic. Sci. 84:348-355.

Lucas, R. E., and M. T. Vittum. 1976. Fertilizer placement for vegetables. p. 75-88. *In* G. E. Richards (ed.) Phosphorus fertilization—principles and practices of band application. Olin Corp., St. Louis, Mo.

Mack, H. J., S. C. Fang, and S. B. Apple. 1964. Effects of soil temperature and phosphorus fertilization on snap beans and peas. Proc. Am. Soc. Hortic. Sci. 84:332-338.

MacKenzie, A. J., and K. R. Stockinger. 1956. How phosphorus placement increases lettuce yields. Western Grower and Shipper 27:14-15.

Mattingly, G. E. 1963. The agricultural value of some water and citrate soluble phosphate fertilizers. An account of recent work at Rothamsted and elsewhere. Proc. Fert. Soc. (London) 75:57-98.

Maynard, D. N., and C. L. Thomson. 1970. Nutrition of vegetable crops in Massachusetts. Massachusetts Coop. Ext. Serv. Pub. no. 63.

Murphy, H. J., and M. J. Goven. 1966. A comparison of rate and source of phosphorus for potato fertilizers. Maine Farm Res. 14:29-30.

National Joint Committee on Fertilizer Application. 1958. Methods of applying fertilizer. Natl. Plant Food Inst., Washington, D.C.

Nishimoto, R. K., and R. L. Fox. 1972. Comparative phosphorus needs of lettuce and Chinese cabbage. Hawaii Farm Sci. no. 3:8.

Nishimoto, R. K., and R. L. Fox. 1975. Comparative external and internal phosphorus requirements of vegetable crops growing in tropical soils. HortScience 10:325.

Ogden, D. B., R. F. Finkner, R. F. Olson, and P. C. Hanzas. 1958. The effect of fertilizer treatment upon three different varieties in the Red River Valley of Minnesota for: 1. Stand, yield, sugars, purity, and non-sugars. J. Am. Soc. Sugar Beet Technol. 10:265-271.

Olsen, S. R., C. V. Cole, F. S. Watanabe, and L. A. Dean. 1954. Estimation of available phosphorus in soils by extraction with sodium bicarbonate. USDA Circ. 939.

Olsen, S. R., R. Gardner, W. R. Schmehl, F. B. Watanabe, and C. O. Scott. 1950a. Utilization of phosphorus from various fertilizer materials by sugar beets in Colorado. Proc. Am. Soc. Sugar Beet Technol. 6:317-331.

Olsen, S. R., W. R. Schmehl, F. S. Watanabe, C. O. Scott, W. H. Fuller, J. V. Jordan, and R. Kunkel. 1950b. Utilization of phosphorus by various crops as affected by source of material and placement. Colorado Agric. Exp. Stn. Tech. Bull. 42.

Peck, N. H. 1975a. Vegetable crop fertilization. New York's Food and Life Sci. Bull. no. 52. New York Agric. Exp. Stn., Geneva, N.Y.

Peck, N. H. 1975b. Plant response to concentrated superphosphate and potassium chloride fertilizers. V. Snap beans (*Phaseolus vulgaris* var. *humilis*). SEARCH Agric. 5(2). New York Agric. Exp. Stn., Geneva.

Peck, N. H., and G. E. MacDonald. 1969. Plant response to concentrated superphosphate and potassium chloride. I. Pea (*Pisum sativum* L.). New York Agric. Exp. Stn. Bull. 825, Geneva, N.Y.

Peck, N. H., and G. E. MacDonald. 1972. Plant response to concentrated superphosphate and potassium chloride fertilizers. IV. Table beet (*Beta vulgaris* L.) SEARCH Agric. 2(14). New York Agric. Exp. Stn., Geneva.

Peck, N. H., and G. E. MacDonald. 1975. Plant response to concentrated superphosphate and potassium chloride fertilizers. VI. Sweet corn (*Zea mays* L. var *rugosa*). SEARCH Agric. 5(3). New York Agric. Exp. Stn., Geneva.

Peck, N. H., G. E. MacDonald, M. T. Vittum, and D. J. Lathwell. 1965a. Accumulation and decline of available P and K in a heavily fertilized Honeoye silt loam soil. Soil Sci. Soc. Am. Proc. 29:73-75.

Peck, N. H., G. E. MacDonald, M. T. Vittum, and D. J. Lathwell. 1976. Effects of concentrated superphosphate and potassium chloride on residual available P, K, and Cl in three depths of soil derived from calcareous glacial till. Agron. J. 68:504-506.

Peck, N. H., and J. R. Stamer. 1970. Plant response to concentrated superphosphate and potassium chloride fertilizers. III. Cabbage (*Brassica oleracea* var. *capitata*). New York Agric. Exp. Stn. Bull. 830, Geneva, N.Y.

Peck, N. H., M. T. Vittum, and G. E. MacDonald. 1965b. Response of table beets (*Beta vulgaris* L.) to banded fertilizer phosphorus and potassium at different levels of soil phosphorus and potassium in a heavily fertilized Honeoye silt loam soil. Soil Sci. Soc. Am. Proc. 29:417-420.

Peech, Michael. 1946. Nutrient status of soils in commercial potato-producing areas of the Atlantic and Gulf Coast. II. Chemical data on the soils. Soil Sci. Soc. Am. Proc. 10:245-251.

Peech, M., and L. English. 1944. Rapid microchemical soil tests. Soil Sci. 57:167-195.

Reisenauer, H. M., J. Quick, and R. E. Voss. 1976. Soil test interpretive guides in soil and plant-tissue testing in California. p. 39-40. *In* H. M. Reisenauer (ed.) Univ. of California Bull. 1879, Berkeley.

Sailsbery, R. L., F. J. Hills, and B. A. Krantz. 1968. Sugar beet yields increased by phosphorus fertilization. Calif. Agric. 22:19.

Sayre, C. B. 1934. Fertilizer placement for vegetable crops. Proc. Natl. Joint Comm. on Fert. Applic. 10:52-54. Natl. Plant Food Inst., Washington, D.C.

Schmehl, W. R., S. R. Olsen, R. Gardner, S. D. Romsdal, and R. Kunkel. 1955. Availability of phosphate fertilizer materials in calcareous soils in Colorado. Colorado Agric. Exp. Stn. Tech. Bull. 58.

Silberstein, O., and S. H. Wittwer. 1951. Foliar application of phosphatic nutrients to vegetable crops. Proc. Am. Soc. Hortic. Sci. 58:179-190.

Sims, W. L., M. P. Zobel, and R. C. King. 1968. Mechanized growing and harvesting of processing tomatoes. Univ. California Agric. Ext. Serv. AXT-232, Berkeley.

Stevens, M. A., and K. N. Paulson. 1973. Phosphorus concentration in tomato fruits: Inheritance and maturity effects. J. Am. Soc. Hortic. Sci. 98:607-610.

Teubner, F. G., M. J. Bukovac, S. H. Wittwer, and B. K. Guar. 1962. The utilization of foliar-applied radiophosphorus by several vegetable crops and tree fruits under field conditions. Michigan Agric. Exp. Stn. Quar. Bull. 44:455-465.

Tiessen, H., and R. L. Carolus. 1963. Effects of soluble starter fertilizer and air and soil temperatures on growth and petiole composition of tomato plants. Proc. Am. Soc. Hortic. Sci. 82:403-413.

Tolman, B., R. Johnson, and R. S. Gaddie. 1956. Comparison of CO_2 and $NaHCO_3$ as extractants for measuring available phosphorus in the soil. J. Am. Soc. Sugar Beet. Technol. 9:51-55.

Tremblay, F. T., and K. E. Baur. 1952. Plant analysis. A method of determining the phosphorus requirement of peas. Agron. J. 44:614-618.

Tyler, K. B., and O. A. Lorenz. 1964. Diagnosing nutrient needs of melons through plant tissue analysis. Proc. Am. Soc. Hortic. Sci. 85:393-398.

Tyler, K. B., O. A. Lorenz, and F. S. Fullmer. 1961. Plant and soil analyses as guides in potato nutrition. California Agric. Exp. Stn. Bull. 781, Berkeley, Calif.

Ulrich, Albert, David Ririe, F. S. Hills, A. G. George, and M. D. Morse. 1959. Plant analysis —a guide for sugar beet fertilization. California Agric. Exp. Stn. Bull. 766, Berkeley, Calif.

U.S. Department of Agriculture. 1976. Agricultural statistics. U.S. Government Printing Office, Washington, D.C.

Vittum, M. T. 1963. Effect of fertilizers on the quality of vegetables. Agron. J. 55:425–429.

Vittum, M. T., and W. C. Hulburt. 1963. Fertilizer placement and rates for tomatoes. New York Agric. Exp. Stn. Bull. 797, Geneva, N.Y.

Ware, L. M., and W. A. Johnson. 1949. Phosphorus studies with vegetable crops on different soils. Alabama Agric. Exp. Stn. Bull. 268.

Wilcox, G. E. 1967. Effect of phosphorus fertilization on tomato seedling growth rate. Proc. Am. Soc. Hortic. Sci. 90:330–334.

Wolf, B. 1945. Amounts of fertilizer elements removed by peas at three stages of growth. J. Am. Soc. Agric. 37:292–296.

Zink, F. W. 1963. Rate of growth and nutrient absorption of celery. Proc. Am. Soc. Hortic. Sci. 82:351–357.

Zink, F. W. 1965. Growth and nutrient absorption in spring spinach. Proc. Am. Soc. Hortic. Sci. 87:380–386.

Zink, F. W. 1966. Studies on the growth rate and nutrient absorption of onion. Hilgardia 37:203–218.

Zink, F. W., and M. Yamaguchi. 1962. Studies on the growth rate and nutrient absorption of head lettuce. Hilgardia 32:471–500.

Chapter 27

Phosphorus Nutrition and Fertilization of Forest Trees[1]

RUSSELL BALLARD

North Carolina State University
Raleigh, North Carolina

I. INTRODUCTION

Forest Tree nutrition received little attention until the 20th century and only in the last 2 or 3 decades has the large scale application of fertilizers to forests become an accepted silvicultural technique (Hagner, 1971; Baule, 1973a). Interest in tree nutrition has developed along with the transition in forestry practices from the traditional exploitation of native forests to the intensive culture and management of forest resources (Stone, 1975). Increased demand for wood products in conjunction with increased agricultural, industrial, and social pressures on land resources has forced this transition on most countries over the last 50 years.

Less exploited native forests in most parts of the world have maintained, and still do maintain, a high degree of productivity without artificial nutrient inputs; however, man's disturbance of forest ecosystems in quest for wood and by intensive silvicultural practices has enhanced nutrient losses from forested areas (Stone, 1973). Thus the maintenance of productivity in intensively managed forests of many parts of the world is dependent on fertilizer inputs. As in agriculture, the most commonly required nutrient elements are N, P, and K (Bengtson, 1976). Nitrogen is used predominantly in native coniferous forests on mineral soils in the cooler northern latitudes where N becomes progressively immobilized in the organic debris of the forest floor. Phosphorus, alone or in conjunction with N and K, is applied principally to exotic and native coniferous plantations on organic soils or mineral soils depleted of P by intense weathering and leaching or erosion. The total area of forest land fertilized with P up to 1977 is probably only in the vicinity of 500,000 ha world-wide, although the extent of P deficiencies in forests is undoubtedly several times this amount (Pritchett, 1976), and is likely to increase as intensive forest management places greater demands on soil nutrient reserves.

---

[1] Contribution prepared while the author was at the Forest Research Institute, Rotorua, New Zealand.

Copyright 1980 © ASA-CSSA-SSSA, 677 South Segoe Road, Madison, WI 53711, USA.
The Role of Phosphorus in Agriculture.

Although vigorous tree stands have a gross annual uptake of nutrients close to that of many agricultural crops (see Table 7), the need for fertilizer inputs to maintain tree productivity is seldom as great as that of agricultural crops. This is largely due to the ability of forest trees to renew their own nutrient supply through cycling of nutrients within the ecosystem. Ebermayer (1867) in his treatise on forest litter was perhaps the first to document the importance of nutrient cycling in meeting the nutritional needs of forest trees. Nutrient cycling, deep rooting habit, long life span, and sheer size of forest trees are responsible for most of the unique features and problems of forest tree nutrition.

Since the bulk of P fertilizers are applied to man-made exotic forests and intensively managed indigenous coniferous forests, this review concentrates on aspects of tree nutrition in these forest ecosystems.

II. PHOSPHORUS NUTRITION IN FOREST NURSERIES

A. Fertilizer Practices

The primary objective of forest nurseries is to produce seedlings with a high survival capability and good growth potential once they are planted in the forest. Sound nutritional practices play an important part in attaining this objective (Aldhous, 1972; Armson & Sadreika, 1974; Stoeckler & Arneman, 1960).

Nutrition of tree nursery crops differs little from many agricultural crops in that the plants are grown from seed, nutrient cycling between crop and soil plays an insignificant part in the nutrition of the crop, the rooting zone is shallow, and the crop is harvested after a relatively short period in the ground. Dry matter production and nutrient drain on the site of tree nursery crops (Table 1) are also similar to those for many agricultural crops

Table 1—Estimates of P removed (tops + roots) from nursery soils by a range of tree crops.

Species	Age	P removed	Dry weight	Reference
	years	—————kg/ha—————		
Radiata pine	1	8	5,980	Knight, 1978
Radiata pine	1.5	13	12,750	Knight, 1978
Loblolly pine	1	27	16,900	Switzer & Nelson, 1956
Ponderosa pine	2	12	7,840	Youngberg, 1958
Red pine	2	27	14,600	Lunt, 1938
White pine	2	12	6,450	Lunt, 1938
Jack pine	2	10	4,190	Stoeckeler & Jones, 1957
Sitka spruce	2	12	4,010	Benzian, 1966
White spruce	2	8	4,000	Lunt, 1938
Norway spruce	2	13	5,750	Lunt, 1938
Spruce	3	13	pp†	Baule & Fricker, 1970
Pines	2	4	--	Baule & Fricker, 1970
Beech	3	8	--	Baule & Fricker, 1970
Oaks	3	11	--	Baule & Fricker, 1970

† Not reported.

(Fried & Broeshart, 1967). For these reasons tree nursery fertilizer practices, particularly for P, are very similar in terms of quantities and sources used to those for agricultural crops. Thus only cursory attention will be paid to them in this review.

Tree nursery crops do however differ from most agricultural crops in one respect. Unlike agricultural crops, which usually pass through a series of development phases (seedling, reproductive, maturity) during their crop life, nursery crops only have a vegetative seedling phase. Thus, nutrient uptake by nursery crops tends to closely parallel dry matter accumulation, and nutrient ratios remain fairly constant throughout the crop's history (Armson, 1965; Baule & Fricker, 1970), although manipulation of these ratios to alter the physiological quality of the nursery crop at different stages during its development is fairly common practice (Aldhous, 1972; Brix & van den Driessche, 1974).

Phosphorus fertilizer prescriptions for nursery beds are usually based on soil analysis. However, the art of nursery soil testing is generally not well developed; the extraction techniques and associated fertility ratings used are normally those developed for agricultural crops (Aldhous, 1972; Knight, 1978). Where soil tests have been specifically calibrated for nursery crops, the methods employed have often been fairly coarse, using soil test results from soils either supporting productive natural stands of the species (Wilde, 1958) or nursery beds of good growth (May, 1957) as a guide to adequate fertility levels. Soil P test standards for a wide range of tree species, soils, and soil test methods have been published (Aldhous, 1972; Stoeckeler & Jones, 1957; Stoecheler & Slabaugh, 1965; Switzer & Nelson, 1956; van den Driessche, 1969; Wilde, 1958).

While not generally used for determining routine fertilizer prescriptions for nursery crops, visual deficiency symptoms and tissue analysis are used as back-up techniques in nursery fertility management. These techniques assume greater importance with nursery crops grown for more than one growing season. Descriptions of P deficiency symptoms in seedlings of most conifers planted in the Northern Hemisphere can be found in the nursery manuals by Aldhous (1972), Armson and Sadreika (1974), Stoeckeler and Jones (1957), and Stoeckeler and Slabaugh (1965), and in publications by Baule and Fricker (1970) and van Goor (1970). Visual P-deficiency symptoms in radiata pine, slash pine, and loblolly pine have been described by Truman (1972), and those in a number of *Eucalyptus* species have been reported by Will (1961) and Kaul et al. (1966, 1968, 1970a, 1970b). Phosphorus concentrations in the foliage and/or tops of nursery seedlings which can be used for diagnostic purposes have been published for a wide range of tree species (Aldhous, 1972; Armson & Sadreika, 1974; Benzian & Smith, 1973; Ingestad, 1960; Knight, 1978; van den Driessche, 1969; Will, 1964a[2]). In general, the critical P concentrations found in foliage of 1-year-old nursery crops closely approximate those reported for forest-grown trees using the same age foliage (Knight, 1978; Terman & Bengtson, 1973).

---

[2] G. M. Will. 1964. Growth rates and foliage analyses, eucalypt fertiliser trial, FRI nursery, 1961–63. N.Z. For. Serv., For. Res. Inst., Silvicultural Rep. no. 20.

Soil analysis and other diagnostic techniques are used to determine the need for P fertilizers in nurseries, but the rates of P applied are usually based on experience in individual nurseries. They vary considerably among nurseries, crops, and years according to the P-retention capacity of the soil, rainfall, and depletion by the crop. In general, application rates of between 50 and 250 kg P/ha are used when applying inorganic fertilizers (Aldhous, 1972; Armson & Sadreika, 1974; Knight, 1978; Mikola, 1959; Stoeckeler & Arneman, 1960). As P is less mobile than N and K in most soils, P fertilizers are usually applied only in a single annual dressing to nursery crops, by incorporation into the seedbed before sowing for first-year crops, followed by annual spring topdressings if 2- or 3-year-old crops are being produced.

With recent evidence of the importance of timing the application of different fertilizer elements in relation to crop development for controlling the physiological quality of seedlings (Brix & van den Driessche, 1974), nursery workers now tend to rely on more than a single annual P dressing in order to obtain greater control over the nutrient status and development of the crop (Knight, 1978). This trend towards more frequent side or topdressings of P, applied in varying ratios with N and K, has been facilitated by the production of high-analysis mixed NPK fertilizers with a wide range of nutrient ratios. The close control of nutrient inputs in the production of seedling stock has reached its greatest development in containerized seedling systems (Brix & van den Driessche, 1974).

A wide variety of P fertilizer sources is used in forest nurseries. Many nurseries are deliberately located on sandy soils, which are low in organic matter, nutrient retention, and water-holding capacity. Nursery workers have thus been traditional users of a wide variety of organic amendments (Stoeckeler & Arneman, 1960) which contribute to varying degrees to the P status of nursery soils. Inorganic P fertilizers, however, are used most extensively for increasing the P status of forest nursery soils. Ordinary and triple superphosphate are the most commonly used P fertilizers, particularly in basal dressing before seeding (Aldhous, 1972; Armson & Sadreika, 1974; Knight, 1978). On acid soils, particularly those with low P-retention capacities, water-insoluble fertilizers, such as phosphate rocks, basic slag, and reverted forms of superphosphate are used also (Baule & Fricker, 1970; Knight, 1978).

B. Effect on Seedling Quality

The quality of seedlings used for field planting largely determines early growth and survival in the field. Nursery fertilization, through its effect on seedling size and sturdiness as well as nutrient concentrations, has a well-documented effect on growth and survival of field-planted seedlings (Anderson & Gessel, 1966; Smith et al., 1966; Switzer & Nelson, 1967). Most of these effects, however, have been associated with N treatments, this being the element manipulated to the greatest extent in nursery fertilizer regimes. Nevertheless, P fertilization, particularly where it contributes to a balanced nutrition, is important to the survival and growth of out-plants (Aldhous, 1972; Baule & Fricker, 1970).

Shirley and Meuli (1939) reported that N and P fertilization increased the drought resistance of red pine, and Allen and Maki (1955) noted that a balanced NPK fertilization, but not N alone, improved the drought resistance of longleaf pine. The mechanism of drought resistance is not fully understood; however, P nutrition probably ensures high carbohydrate reserves, a root system with a high regenerative capacity, and a sturdy, hardened seedling. Brix and van den Driessche (1974) pointed out that reducing the N and increasing the P and K supply in late summer hastens the onset of dormancy, a process which reduced water loss.

Phosphorus nutrition has seldom been implicated directly in decreasing resistance to frost damage. Malcolm and Freezaillah (1975), however, reported that high P fertilization increased frost damage of Sitka spruce seedlings by extending the growth season into the frost danger period. Current opinion appears to indicate that a balanced nutrition, particularly of N and K, is a prerequisite for frost-resistant stock (Timmis, 1974).

III. PHOSPHORUS NUTRITION OF FOREST TREES

A. P Status of Forest Soils

Soils supporting coniferous forests and plantations have a much lower P status than most soils used for agricultural production. There are several reasons for this. Trees differ considerably in their site requirements. The more exacting hardwoods (as compared to conifers) normally are found on inherently more fertile sites (Remezov & Pogrebnyak, 1969). "Hardwood sites" have traditionally been exploited for agricultural purposes and the less fertile "pine sites" have been left under forest cover (Wilde, 1958). Where similar soils are used for both agriculture and forest production, those under use in agriculture have generally been subject to much better soil husbandry—including P fertilization and liming—than those used for forest production. Until very recently production forestry has involved only soil exploitation with no fertilizer input to compensate for the nutrient drain in forest products. Traditional land use policy, particularly in Southern Hemisphere countries involved in establishment of exotic forests, has been to relegate for forestry use only that land considered unsuitable for agricultural production—often because of its inherent low fertility.

Despite forest soils being generally lower in P than most agricultural soils, P deficiencies are not particularly widespread in forests. Deficiencies of P are found principally in managed pine forests on leached acidic sands in Europe and in the southeastern U.S., on peats in Europe, and on strongly weathered clays and sands in exotic pine forests in Southern Hemisphere countries such as Australia, New Zealand, and South Africa (Baule, 1973a). The ability of coniferous species to thrive at relatively low levels of soil P is reflected in the critical levels of soil P for pines, being several magnitudes lower than those for agricultural crops (see Section IVb). This ability is usually attributed to the efficient use of P through recycling within the ecosystem.

B. Accumulation and Cycling of P in Forest Ecosystems

1. ACCUMULATION OF P IN FOREST STANDS

The pattern of accumulation of P and biomass in forest monocultures is practically the same for all tree species and is illustrated in Fig. 1 for a loblolly pine plantation. In young stands, the rate of P accumulation increases relatively more rapidly than the rate of dry matter accumulation. Once full site occupancy has been achieved, at which stage foliage biomass stabilizes (Switzer et al., 1968), the rate of accumulation of P declines rapidly while dry matter accumulation, principally in stems and branches, continues.

For different sites, management practices, and species, the time scale and weights shown in Fig. 1 will differ. The influence of age, species, and, to a limited extent, site on the amounts of P and dry matter accumulated in the above-ground components of coniferous stands is shown in Table 2. Despite the relatively high accumulation of dry matter in many forest stands,

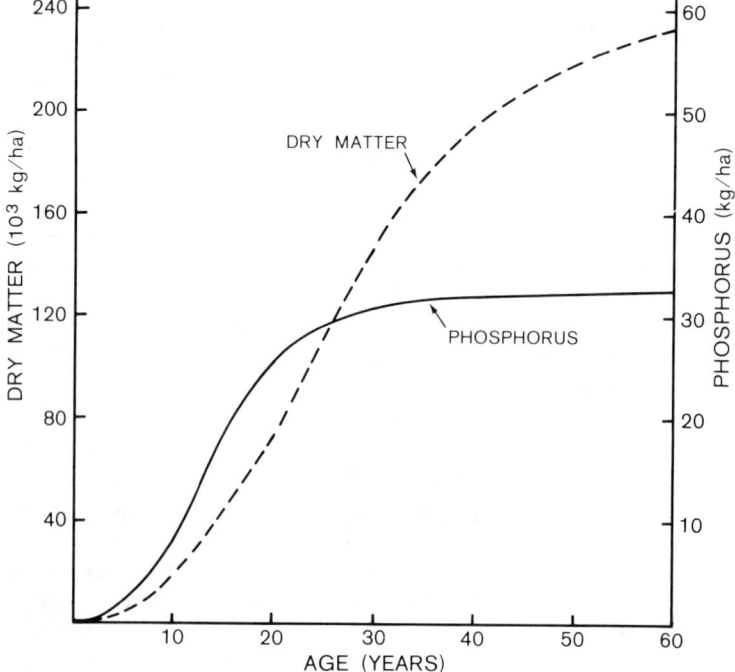

Fig. 1—Accumulation of dry matter and P in the above-ground components of a loblolly pine plantation (adapted from Switzer et al., 1968).

the weights of P accumulated are small and, by agricultural standards (Fried & Broeshart, 1967), the annual rates of accumulation are generally very low. Within any age group there is a strong correlation between dry matter and P content. Data in Table 2 suggest, and specific studies (Table 3) confirm, that mean annual accumulation rates of P are highest during the early development stages of the tree stand. The maximum rate of dry matter accumulation occurs at a later stage. The maximum rate of P accumulation tends to coincide with the period of maximum canopy development, while that of dry matter tends to coincide more with the period when maximum stem-volume increment occurs.

As a forest stand develops, it influences the accumulation of nutrients and dry matter in not only the tree but also other components of the forest ecosystem (Table 4). During the early years of stand development, other herbaceous plants often contain a high proportion of the system's biomass. As the stand develops and a closed canopy situation is reached, most other plants are shaded out of the ecosystem. Corresponding with this and the increased litterfall of the tree component at this stage (discussed later), organ-

Table 2—Dry matter and P content of the above-ground part of coniferous forest stands of comparable age in different parts of the world.

Species	Age	Location	Dry matter	Phosphorus Total content	Accumulation rate	Reference
	years		— kg/ha —		kg/ha per year	
Radiata pine	4	New Zealand	22,200	16.5	4.1	Madgwick et al., 1978
Radiata pine	6	New Zealand	52,400	40.2	6.7	Madgwick et al., 1978
Slash pine	5	USA	42,100	19.8	4.0	White & Pritchett, 1970[†]
Slash pine	5	USA	57,800	26.8	5.4	White & Pritchett, 1970
Loblolly pine	5	USA	23,600	12.8	2.6	White & Pritchett, 1970[†]
Loblolly pine	5	USA	30,600	16.2	3.2	White & Pritchett, 1970
Loblolly pine	5	USA	15,700	9.4	1.9	Nelson et al., 1970
Radiata pine	17	New Zealand	284,000	70.7	4.2	Madgwick et al., 1978
Loblolly pine	16	USA	156,000	30.9	1.9	Wells et al., 1975[†]
Loblolly pine	20	USA	90,000	19.3	1.0	Switzer & Nelson, 1972
Maritime pine	14	Australia	123,000	20.8	1.5	Keay & Turton, 1970
Scots pine	18	Scotland	54,900	16.6	0.9	Wright & Will, 1958
Corsican pine	18	Scotland	25,800	12.0	0.7	Wright & Will, 1958
Jack pine	20	Canada	16,000	3.5	0.2	Morrison, 1973
Douglas fir	15–20	Canada	64,800	31.0	1.8	Webber, 1977
Radiata pine	26	New Zealand	223,000	28.0	1.1	Orman & Will, 1960
Loblolly pine	30	USA	176,000	31.0	1.0	Switzer et al., 1968
Scots pine	28	Scotland	94,000	25.0	0.9	Wright & Will, 1958
Scots pine	28	Finland	17,900	6.2	0.2	Malkonen, 1974[†]
Scots pine	32	Scotland	150,000	30.0	1.1	Ovington & Madgwick, 1959[†]
Corsican pine	28	Scotland	68,500	14.0	0.5	Wright & Will, 1958
Jack pine	30	Canada	81,300	13.0	0.4	Foster & Morrison, 1976[†]
Red pine	37	USA	83,000	25	0.7	Wittwer et al., 1975
Virginia pine	27–31	USA	--	8	0.3	Madgwick, 1970
Douglas fir	36	USA	205,000	66	1.8	Cole et al., 1967[†]

[†] Also provide information on root P content and biomass.

Table 3—Estimates of mean annual accumulation of P and dry matter in the above-ground part of the tree crop over different age periods for coniferous forests.

Species	Age period	Mean annual accumulation		Reference
		P	Dry matter	
	years	kg/ha per year		
Radiata pine	0–10	3.5	17,000	Will, 1968
Radiata pine	10–35	0.5	15,600	Will, 1968
Radiata pine	2–4	7.9	10,700	Madgwick et al., 1978
Radiata pine	4–8	4.3	15,100	Madgwick et al., 1978
Loblolly pine	0–5	0.5	1,200	Switzer & Nelson, 1972
Loblolly pine	5–10	1.4	4,400	Switzer & Nelson, 1972
Loblolly pine	10–15	1.3	7,000	Switzer & Nelson, 1972
Loblolly pine	15–20	0.7	5,400	Switzer & Nelson, 1972
Loblolly pine	1–16	1.9	9,700	Wells & Jorgensen, 1975
Loblolly pine	15–16	0.9	7,500	Wells & Jorgensen, 1975
Jack pine†	0–20	0.6	2,600	Foster & Morrison, 1976
Jack pine	20–30	0.2	2,900	Foster & Morrison, 1976
Jack pine	30–65	0.1	900	Foster & Morrison, 1976
Corsican pine	0–18	0.6	--	Wright & Will, 1958
Corsican pine	18–28	0.2	--	Wright & Will, 1958
Corsican pine	28–48	0.4	--	Wright & Will, 1958
Scots pine	28–29	1.3	2,400	Malkonen, 1974

† Includes roots.

ic matter starts to accumulate on the forest floor. In warmer temperature climates the weight of the forest floor tends to stabilize at about the same time as foliar biomass (Forrest & Ovington, 1970; Switzer & Nelson, 1972); in cooler climates it tends to increase steadily with age (Turner & Long, 1975). The forest floor in most coniferous stands contains a significant proportion of the above-ground P content, about 30% being a common value for a wide range of climatic conditions and species (Cole et al., 1967; Lamb & Florence, 1975; Malkonen, 1974; Switzer & Nelson, 1972).

Table 4—Changes in the P content and dry matter of the herbaceous understory, forest floor, and tree components of a loblolly pine plantation ecosystem during the first 20 years (Switzer & Nelson, 1972).

Component	System age (years)				
	0	5	10	15	20
	kg/ha				
	P content				
Herbaceous understory	7.8	3.3	0.0	0.0	0.0
Forest floor	0.0	1.1	6.9	8.2	9.1
Trees	0.0	2.3	9.5	15.8	19.3
Total	7.8	6.7	16.4	24.0	28.4
	Dry matter				
Herbaceous understory	11,000	8,300	0	0	0
Forest floor	0	1,700	12,700	15,700	17,000
Trees	0	6,500	28,600	64,000	90,000
Total	11,000	16,500	41,300	79,700	107,000

The pattern of P accumulation by forest stands within a single growing season has received little attention because of the difficulty of sampling and a general feeling that in such long rotation crops, a single season's pattern is unimportant relative to the overall accumulation pattern. However, Nelson et al. (1970) studied a seasonal pattern in a stand of 5-year-old loblolly pine. They reported that P accumulated faster than dry matter with 100% of the uptake having occurred by September. There was some loss after this which they attributed to needle fall or translocation into the root system. Within tree components, P accumulated preferentially in foliage, and once the foliage biomass had been accumulated by August/September there was an increased accumulation in bark and wood.

2. DISTRIBUTION OF P AMONG STAND COMPONENTS

The distribution of P and biomass between stand components varies with age and stand development (Table 5). In young stands prior to canopy closure, when the foliage biomass is expanding rapidly (Switzer et al., 1968), foliage usually contains the greatest proportion of both the biomass and P in the standing crop. As stand development proceeds and canopy closure occurs, the relative contribution of foliage to the overall biomass and P content declines rapidly, while that of woody tissue (stemwood and branches) shows a concomittant increase. Once canopy closure occurs and foliage biomass stabilizes, the relative contribution of foliage steadily declines with the age of the stand while that of stemwood increases.

At all stages in stand development, foliage contains a much greater proportion of the stand's P than its biomass; the reverse is true of stemwood, while branches and stembark make approximately equal contributions to both (Table 5; Foster & Morrison, 1976; Wells & Jorgensen, 1975). The different concentrations of P in the various components account for this. A search of the literature on P concentrations in the components of conifer stands reveals that, with few exceptions, the range in concentrations found in any one component is largely independent of stand age and species: concentration ranges are 0.05-0.30% for foliage (all ages),

Table 5—Effect of age on the relative distribution of P and dry matter (DM) among the aboveground components of radiata pine stands (adapted from Madgwick et al., 1978, and Orman & Will, 1960).

Component	2 years		4 years†		8 years		17 years		26-29 years	
	P	DM	P	DM	P	DM	P	DM	P	DM
					%					
Foliage	68.4	50.7	60.7	32.2	29.4	7.2	24.8	5.6	24	2.5
Live branches	14.5	16.9	24.1	29.6	34.9	28.8	14.8	7.7	8	5.5
Dead branches	--	--	--	--	0.8	1.9	2.3	3.2		
Cones	--	--	--	--	--	0.1	0.4	1.8	ND	ND
Stembark	7.9	8.4	5.1	5.7	9.4	6.4	15.0	7.7	19	12
Stemwood	9.2	23.9	10.2	32.4	25.4	55.7	42.7	75.7	40	80
Total wt (kg/ha)	0.76	710	16.5	22,170	33.8	82,630	70.7	283,650	28	222,500

† Canopy closure occurred between 4 and 8 years.

0.01–0.08% for live branches, 0.004–0.08% for stembark, and 0.002–0.04% for stemwood. Within foliage, P concentrations generally decrease with the age of foliage (Madgwick et al., 1978; Malkonen, 1974) and increase towards the base of the crown (Will, 1957). Within bark, concentrations increase inwards from the outer dead bark to the inner living bark, while within stemwood, they decrease markedly from the outer sapwood to the inner heartwood (Orman & Will, 1960). It is the very low P concentrations found in stemwood and the major contribution of this tissue to the accumulative biomass after canopy closure that accounts for the very low rate of net accumulation of P (especially in relation to dry matter accumulation) in older conifer stands (Fig. 1).

Insufficient studies have been carried out on the biomass and nutrient content of root systems of coniferous stands to gain a clear picture of species and site differences and the relative contribution of roots to total stand biomass and P content at different stages of stand development. Data indicate that roots can contribute from about 10% (Cole et al., 1967; Foster & Morrison, 1976) to 40% (Malkonen, 1974) of the total P content of stands; most published data suggest a value of 30% as a reasonable "ballpark" figure for most stands (Malkonen, 1974; Ovington & Madgwick, 1959; Wells & Jorgensen, 1975; White & Pritchett, 1970). There is limited evidence to suggest that after site occupancy (and full exploitation of the rooting zone) the relative contribution of the root system to total stand P content declines with age (Malkonen, 1974). Phosphorus concentrations within the root system decrease with increasing diameter of the individual root (Malkonen, 1974; Ovington & Madgwick, 1959). White and Pritchett (1970) found in 5-year-old slash pine that P concentrations in fine roots (<0.5-cm diam) closely approximated those in the foliage, while those in the laterals and stumps were close to those in the branches and stemwood, respectively.

3. CYCLING OF P IN FOREST ECOSYSTEMS

The cycling of nutrients in forest systems can be examined over a number of time periods—daily, seasonal, annual, or crop development stages. The annual cycle has been studied most intensively and will be concentrated on in this review. Switzer and Nelson (1972) recognized three major nutrient cycles which are of importance to the nutrition of forest trees:

a. Geochemical Cycle—This cycle covers the import and export relationships of the forest ecosystem. In an undisturbed forest stand the geochemical cycle is usually only of minor importance to the P nutrition of forest stands. Annual inputs of P in precipitation and dryfall are normally negligible and seldom >0.4 kg/ha per year (Cole et al., 1967; Foster & Morrison, 1976; Malkonen, 1974; Switzer & Nelson, 1972; Turner & Singer, 1976; Weetman & Webber, 1972; Will, 1959). Such inputs usually represent <5 or 6% of the annual P requirements of forest stands (Foster & Morrison, 1976; Malkonen, 1974; Switzer & Nelson, 1972). However, as a proportion of the annual accumulation of P (net uptake) in forests, particularly in older stands, even input rates < 0.4 kg/ha per year can be very significant (see Table 3).

Annual losses of P from established forest stands are usually restricted to those leaching below the rooting zone. As P is strongly retained in most soils, annual losses below the rooting zone in forest ecosystems are normally very small and usually represent less than the annual input in rainfall (Cole et al., 1967; Weetman & Webber, 1972; Wells & Jorgensen, 1975; Will, 1968), thus leading to a net accretion of P in the ecosystem. Of greatest concern on the output side of the nutrient budget in managed forest ecosystems is the major effect of harvesting and other silvicultural operations on nutrient losses from the site. This aspect is covered in a later section of this review.

b. Biogeochemical Cycle—This cycle encompasses the transfer of nutrients between plant and soil. There are three major transfer paths of nutrients from plant to soil: (i) in rainwater (stemflow or throughfall) following washing and leaching of the canopy; (ii) in litterfall to the forest floor and thence to the soil following decomposition; and (iii) in mortality and decomposition of fine roots.

Stemflow and throughfall are fairly minor pathways for transfer of P from trees to soil. The amount of P leached from forest canopies is usually small in relation to amounts of other elements, seldom exceeding 1 kg/ha per year (Cole et al., 1967; Foster & Morrison, 1976; Malkonen, 1974; Switzer & Nelson, 1972; Turner & Singer, 1976; Will, 1959). The amounts involved in any one stand usually represent only a relatively small fraction of the stands annual P requirement (Table 6), although in some stands stemflow and throughfall have been observed to contribute as much as litterfall (Malkonen, 1974; Will, 1959).

Litterfall is usually the major pathway by which P is transferred from the tree to the forest floor. Litterfall in even-aged coniferous plantations tends to be slightly out of phase with, but closely parallel to, canopy development. Very little litterfall occurs during the first 2 to 5 years of stand development but increases rapidly as crown competition begins, eventually stabilizing 3 to 4 years after foliar biomass stabilizes (Bray & Gorham, 1964; Forrest & Ovington, 1970; Will, 1959). There are reports of litterfall slowly declining with age, but most data indicate no inherent tendency towards higher or lower litterfall with increasing age beyond canopy closure, although disease conditions may alter this (Bray & Gorham, 1964). Litter production, and hence P return, is closely related to stand productivity. Less productive stands in cool temperate regions return around 0.2 to 2.0 kg/P per year in litterfall (Table 6; Foster & Morrison, 1976), while productive stands in warm temperate regions return from 2 to 8 kg/P per ha per year (Wells & Jorgensen, 1975; Will, 1959). In closed canopy stands these values are usually close to the annual requirement of new foliage production (Table 6), and can represent a significant proportion of the total P content of the forest biomass (Switzer & Nelson, 1972).

Prior to becoming available for tree uptake, P in litterfall must first be mineralized in the forest floor. The rate of mineralization varies among

Table 6—Estimated annual P requirements and transfer rates in selected conifer forest ecosystems.

	Loblolly pine		Scots pine§		Pacific silver fir¶
Component	16-year-old†	20-year-old‡	28-year-old	45-year-old	Old stand
	kg/ha				
Requirements of trees					
Foliage	6.3	5.0	1.15	2.29	
Branches	1.0	0.2	0.36	0.43	
Stems	0.6	0.3	0.10	0.16	
Roots	12.7	--	0.56	0.31	
Total	20.6	5.5	2.17	3.19	1.7
Transfer within trees	0	3.3	0.11	0.50	1.0
Transfer to forest floor					
Litterfall	7.8	(4.6)	0.22	0.85	2.0
Throughfall + stemflow	0.5	0.8	0.22	0.22	0.5
Total	8.3	(5.4)	0.44	1.07	2.5
Transfer to soil					
From forest floor	4.0	2.1	--	--	1.1
From roots	12.3	--	--	--	--
Total	16.3	2.1			1.1
Net demand on soil	4.3	0.1	1.62	1.62	0
Leaching loss from soil	0.03	--			0.5

† Wells and Jorgensen, 1975.
‡ Switzer and Nelson, 1972.
§ Malkonen, 1974.
¶ Turner and Singer, 1976.

species and with climatic and soil conditions (Miller, 1974). Irrespective of the obvious effect of these conditions on the rate of mineralization, litter decomposition studies in conifer stands tend to show certain common trends. Mineralization of P is rapid in the first year after litterfall. Will (1967) and Wells and Jorgensen (1975) have reported release of 60 to 80% of the P from litter during the first year after deposition.

After the first year, when the litter becomes part of the O2 horizon, release of P drops dramatically (Will, 1959), and some workers report an actual net gain in litter P (Wells & Jorgensen, 1975). Mineralization of P usually occurs at a somewhat slower rate than decomposition of organic matter, producing an increase in P concentration with age of litter and a more rapid accumulation of P than organic matter in the forest floor. Most studies on accumulation of P in the forest floor show a rapid build-up with age associated with canopy closure followed by a slow accretion with age after this point (Foster & Morrison, 1976; Switzer & Nelson, 1972). Where decomposition and mineralization rates are relatively low, the amount of P immobilized in the litter, even during the period after canopy closure, can be greater than that immobilized in the growing crop (Foster & Morrison, 1976). However, despite immobilization of P in the litter layer, that released by mineralization usually represents a major proportion of the annual requirements of the above-ground portion of well-established conifer stands (Table 6).

Minimal work has been done on the amounts of P released by the death and decomposition of components of conifer root systems. Wells and Jorgensen (1975) estimated that 12.3 kg P/ha per year were released by decomposition of roots in a stand of loblolly pine. Although it was estimated that this release was balanced by root uptake of P, the amount involved represents such a high proportion of the total annual turnover of P in the system that it must be treated with suspicion, since the root system of established stands normally contains only about 30% of the total P in coniferous forests. The role of root systems in nutrient cycling and immobilization is obviously in need of more detailed examination.

c. Biochemical Cycle—This cycle encompasses the internal nutrient cycle in the tree—the translocation of nutrients from older to young tissues for reuse in biosynthetic processes. Estimates of the importance of internal transfer in meeting the P requirement of new tissue vary considerably (Table 6). Switzer and Nelson (1972) and Turner and Singer (1976), working in very different forest ecosystems, estimated that internal transfer accounted for about 60% of the P requirements of new tissue, while Wells and Jorgensen (1975) estimated that internal transfer of P was nonexistent in a stand of the same species and similar age to that studied by Switzer and Nelson (1972). Malkonen (1974) estimated internal transfer accounted for 5 to 16% of the P requirements of current tissue in stands of Scots pine. These marked differences are undoubtedly partly due to the methods and assumptions used by the various authors in calculating internal transfer. However, so little is known about factors controlling remobilization of nutrients in tree tissues that the possibility of these estimated differences being real cannot be discounted. Florence and Chuong (1974), for instance, reported that the gradient in foliar P concentrations from 1- to 4-year-old needles in radiata pine was much greater on infertile than on fertile soils. This suggests that the contribution of internal transfer to the P requirements of current tissue is partially controlled by the ability of current uptake to meet these requirements. Nevertheless, the recorded reductions with age in P concentrations and quantities in foliage on a wide variety of sites (Florence & Chuong, 1974; Madgwick et al., 1978; Malkonen, 1974; Wells & Metz, 1963) indicate that internal transfer of P is a significant cycle in coniferous trees. The transfer of P from older to younger tissue is not restricted to foliage; reductions in P concentrations have also been reported with increasing age of wood and bark (Orman & Will, 1960; Wright & Will, 1958).

Although mean annual accumulations of P in the biomass of trees is small (Tables 2 and 3), gross annual uptakes are many magnitudes larger and approach those of agricultural crops (Table 7). The large differences between uptake and retention in the biomass are due to the efficiency of the P cycle. The magnitude and importance of P cycling in coniferous forests are well illustrated by the data of Switzer and Nelson (1972), who estimated that during the 20th year of a loblolly pine stand, 19% of the total P in the

Table 7—Estimates of gross annual P uptake and dry matter production in the above-ground part of the tree crop for coniferous forests.

Species	Age period	Gross annual P uptake	Gross annual dry matter production	Reference
	years	kg/ha per year		
Radiata pine	0–12	4.5	164,000	Will, 1964
Radiata pine	4–8	13.8	--	Madgwick et al., 1978
Loblolly pine†	19–20	5.5	11,000	Switzer & Nelson, 1972
Loblolly pine†	15–16	8.8	153,000	Wells & Jorgensen, 1975
Jack pine	29–30	1.7	--	Foster & Morrison, 1976
Scots pine	28–29	1.9	2,400	Malkonen, 1974
Black spruce	0–65	3.7	--	Weetman & Webber, 1972
Douglas fir	35–36	7.2	10,000	Cole et al., 1967
Conifers (European)	--	2.7–5.1	--	Duvigneaud & Denaeyer-De Smet, 1970

† Overestimated by amount recycled annually within the tree.

above-ground part of the system was cycling, and of the annual requirement of new tissue only 7% was retained and the remaining 93% recycled.

4. IMPACT OF SILVICULTURAL OPERATIONS

Various silvicultural operations practiced in intensively managed forests have a pronounced effect on both the rate of cycling of nutrients within the forest ecosystem and on the input/output nutrient balance. Operations which have the most pronounced effect, and on which this review will concentrate, are harvesting, site preparation (including burning), and fertilization.

a. Harvesting—Clearfelling, the traditional harvesting practice in intensively managed conifer forests, has been the subject of much debate with regard to its depletive effect on the nutrient status of forested land (Stone, 1973). Clearfelling influences nutrient depletion through removal of harvest products from the site and by accelerating leaching and erosional losses.

The common harvesting technique of extracting only the merchantable bole from coniferous forests removes only relatively small amounts of P from forest sites, ranging from 8 to 30 kg/ha every 20 to 100 years (Table 8). On an annual basis this usually amounts to a depletion of <1 kg/ha, a quantity which is not much above natural inputs to forest ecosystems in dust and precipitation (Table 6; Weetman & Webber, 1972). Rennie (1955), in an early review of European data, reported mean annual losses in stemwood and bark (thinnings plus clearfelling) of 0.18 and 0.35 kg P/ha per year for pines and other conifers, respectively. These low losses, despite the considerable quantities of dry matter removed (Table 8), can be attributed to the harvesting of only the nutrient-poor components of the forest stand (see Section IIB2).

The need to obtain larger yields from forest land has recently led to an interest in greater utilization of the standing crop, shorter rotations, and more intensive cultural inputs. Greater utilization (whole-tree harvesting) involves extraction of the nutrient-rich components of the stand, such as

Table 8—Influence of harvesting method and rotation length on P and dry matter removal from coniferous forests.

Species	Age	Harvesting method	Dry matter removed	P removed		Reference
				Total	Annual	
	year		—— kg/ha ——		kg/ha per year	
Radiata pine	26	Logs	177,000	15	0.6	Bunn & Will, 1973
		Whole-tree	221,000	28	1.1	
Black spruce	65	Logs	54,000	12	0.2	Weetman & Webber, 1972
		Whole-tree	107,000	42	0.6	
Norway spruce	70	Logs	121,500	8	0.1	Malkonen, 1973
		Whole-tree†	197,500	41	0.6	
Loblolly pine	16	Logs	116,000	14	0.9	Wells et al., 1975
		Whole-tree	156,000	30	1.9	
		Whole-tree†	185,000	37	2.3	
Loblolly pine	20	Logs	--	17	0.8	Switzer & Nelson, 1973
		Whole-tree	--	27	1.4	
	40	Logs	--	28	0.7	
		Whole-tree	--	40	1.0	

† Includes stumps and large roots.

foliage and branches, as well as stems. While this practice increases biomass yield by 25 to 100%, it has a much more substantial effect on P removal, increasing it by 100 to 400% (Table 8). Reducing rotation lengths also enhances nutrient removal with the effect being more pronounced if whole-tree harvesting is practiced (Table 8; Switzer & Nelson, 1973). Although increased utilization and shorter rotations markedly increase the removal of P from forest sites, the quantities removed on an annual basis (Table 8) are still well below those removed off-site by harvesting of agricultural crops (Fried & Broeshart, 1967).

Because clearfelling increases mineralization rates and effectively removes the intercepting root mass, the impact of this silvicultural operation on leaching losses has received considerable attention. However, as P is a relatively immobile element in most soils, it is not surprising that many catchment and lysimeter studies have reported little impact of clearfelling on leaching losses of P (Stone, 1973; Tamm et al., 1974). Cole and Gessel (1965) working with Douglas-fir on a gravelly soil found that although clearfelling increased the movement of P through the forest floor from 0.84 to 2.31 kg/ha over a 10-month period after treatment, the movement below 90 cm (rooting depth) only increased from 0.03 to 0.11 kg/ha. Several catchment studies in the Pacific Northwest examining the impact of clearfelling Douglas-fir on nutrient losses have all shown minimal losses of P following treatment (Brown et al., 1973; Fredriksen, 1971; Fredriksen et al., 1975). The greatest loss was reported by Fredriksen (1971) in a felled and burned catchment from which an additional 0.4 kg P/ha were lost in the year following treatment. On sandy sites with soils of very low P-retention capacity, losses greater than these may be expected.

Harvesting, including partial harvesting (thinning), not only affects the input/output balance, but has a dramatic effect on the distribution and cycling of nutrients within forest ecosystems. Clearfelling temporarily interrupts nutrient cycling, dramatically increases the biomass of the forest floor (where tree crowns are left on site), and speeds up the transfer of nutrients from the forest floor to the soil. Thinning, in addition to initially increasing forest-floor biomass, also alters the cycling of P by temporarily reducing litter return to the forest floor (as a result of crown expansion and longer retention of foliage on remaining trees). It also enhances decomposition of the forest floor which tends to improve the P status of the remaining trees (Wells & Jorgensen, 1975; Will, 1968; Wollum & Schubert, 1975).

b. Site Preparation—Cultivation techniques (including bedding) used as a prelude to planting increase the mineralization of P from organic debris thereby improving the P status of young planted stock (Haines et al., 1975). In soils of very low P retention capacity this increased mineralization rate in conjunction with the limited development of root systems of recently planted trees could enhance leaching losses (W. L. Pritchett, personal communication).

The forest floor developed under the previous crop and the logging debris from harvesting often contain a considerable proportion of the site's P reserves and so play an important role in the nutrition of the following crop (Will, 1968). Site preparation techniques [e.g. KG blading (shearing stems at ground level), windrowing, and rootraking] used to remove debris from the site which interferes with further cultivation or planting, can have a detrimental effect on the site's nutrient status. Unfortunately these operations frequently remove topsoil from the planted area, which combined with the removal of organic debris can have a very adverse effect on the P status and productivity of the following crop (Ballard, 1978a; Glass, 1976).

Burning, either as a site preparation technique or to reduce the accumulation of fuel in established stands, has a well-documented effect of increasing the availability of P to the following or existing crop (Boyle, 1973; Viro, 1974; Wells, 1971). This is usually attributed to the mineralization of P from the organic layers, although it has been attributed to a calcination of soil P minerals under very hot fires (Humphreys & Lambert, 1965). Most studies indicate that while fire influences the distribution of P between the forest floor and the mineral soil, very little loss of P occurs from the system (Lewis, 1974; Wells, 1971). However, Harwood and Jackson (1975) recorded a loss of 10 kg P/ha in particulate matter from a slash burn in Tasmania.

c. Fertilization—The addition of P fertilizers to forest stands has an obvious effect on cycling of P in the forest ecosystem; this is considered in detail in Section IV. Nitrogen fertilizers, both urea based and ammoniacal forms, have been shown to increase the rate of movement of P from the forest floor to the underlying soil (Beaton, 1973; Cole & Gessel, 1965). Bengtson (1970a) discussed the possible mechanisms responsible, including solubilization of organic matter by high pH and ammonium ions, and stimulation of the microbial population. Where N fertilization produces a biomass response in a forest stand, the quantities of annual cycled P are usually increased. Miller et al. (1976) reported increases of up to 37% in the amount of P in total litterfall over a 6-year period following fertilization of Corsican pine with ammonium sulphate, and this was despite a decrease in P concentration in the needle litter.

Intensification of culture in forest plantations (site preparation, improved genetic stock, tending, fertilization) can dramatically increase both productivity and mean annual nutrient demand. Switzer and Nelson (1973) calculated that by increasing the productivity of a 10-year-old stand of loblolly pine from 35 to 166 metric tons/ha by intensive cultural practices, the mean annual demand for P would be increased from 1.3 to 5.2 kg/ha. They speculated that natural P inputs and the reserves of most forest sites are unlikely to be able to sustain such increased demands if short rotations with complete tree utilization are practiced.

C. Role of Mycorrhizae

All conifers are ectomycorrhizal and the advantages to trees of this association have been long recognized. One of the principal benefits attributed to mycorrhizae is the improved nutrition of the host plant, particularly the increased uptake of P (Bowen, 1973; Harley, 1970; Marx & Bryan, 1975; Tinker, 1978; Wilde, 1968). The increased uptake of P by mycorrhizal as compared to nonmycorrhizal trees has been attributed to a number of factors including production by mycorrhizal roots of acids and chelating compounds capable of solubilizing poorly soluble soil P compounds, a greater phosphatase activity on the root surface, and a vastly increased root surface area involved in nutrient absorption.

The importance of acid or chelate production in increasing P uptake under field conditions is questionable (Bowen, 1973). Recent work by Mosse et al. (1976) with vesicular-arbuscular mycorrhizae has convincingly shown that mycorrhizae increase P uptake from insoluble phosphate rock only under soil conditions (acid) capable of slowly solubilizing the P mineral. Nevertheless, the possibility of phosphate-solubilizing bacteria in the mycorrhizosphere enhancing the dissolution of insoluble phosphates cannot be discounted (Azcon et al., 1976; Ralston & McBride, 1976).

Bowen (1973) questioned the phosphatase hypothesis on the grounds that few studies had been able to distinguish between the role of the fungus and host tissue in phosphatase production and the complete lack of quantitative studies on the relative phosphatase activity of mycorrhizae and nonmycorrhizal short roots. However, Williamson and Alexander (1975) recorded a two to eight times greater phosphatase activity in mycorrhizal as compared to nonmycorrhizal beech roots, with the phosphatase activity being detected in the fungal sheath. Surface phosphatase activity on mycelial strands and hyphae of mycorrhizal fungi in conjunction with an ability to produce antibiotics against soil microorganisms (Marx, 1969) could be of considerable importance to the P nutrition of established stands which rely heavily on P recycled through the litter layer to meet their annual requirements. These two mechanisms, plus the ability of some mycorrhizae to grow at low water potential, could provide mycorrhizal fungi with a competitive advantage over other microorganisms in the litter and organic soil layers and in effect enhance the rate of transfer from the litter back to the tree. Gadgil and Gadgil (1975) have suggested that mycorrhizal control of litter decomposition and nutrient release could protect the nutrient pool against leaching losses.

An abundance of research indicates that the greatly increased absorption surface of infected roots is perhaps the major mechanism whereby P uptake is enhanced (Bowen, 1973). Tree root systems are invariably coarse with large inter-root distances, a feature which tends to limit the availability to trees of immobile elements, such as P, within the large inter-root soil volumes. The effective extension of the root system into these "unexploited" volumes by growth of mycelial strands and the increased radii of diffusion of ions about mycorrhizal roots due to their greater longevity contribute greatly to the increased uptake of P by trees (Bowen, 1973; Harley, 1970; Marx & Bryan, 1975).

An aspect of mycorrhizal research which is assuming greater importance now that research results are starting to be put to practical use, is defining differences in effectiveness of various species of mycorrhizal fungi in enhancing P uptake. Differences have been documented (Bowen, 1973) and introduction of selected fungi have already been successfully achieved in some situations (Marx & Bryan, 1975). A cautionary note on selection for specific properties, such as P uptake, has been sounded by Bowen (1973). He pointed out that mycorrhizal fungi fulfill many important roles other than just P uptake and that efficiency of a fungus in P uptake does not mean high performance in other respects, such as disease resistance, water uptake, persistence, and growth factor production. It is also probable that strong interactions between fungal effectiveness and site conditions will exist which will further complicate selection of suitable fungi for practical inoculation.

IV. PHOSPHORUS FERTILIZATION OF FOREST STANDS

A. Application of Fertilizer

Phosphatic fertilizers are applied to forest stands principally for the purpose of increasing wood production. The timing, methods and rates of application, and fertilizer source are selected to optimize the returns from wood production per unit area of forest stand.

1. TIME OF APPLICATION

Fertilizers are usually applied to forests at two principal stages of stand development: (i) at or near time of stand establishment, and (ii) at or after canopy closure. On P-deficient forest soils, fertilizer applications near time of planting are usually required to obtain satisfactory establishment and early growth of conifers. Establishment fertilization of conifers is practiced on sandy soils in the southeastern U.S. (Pritchett & Smith, 1975), France (Mauge, 1972), and Central Europe (Baule, 1971); on peats in the United Kingdom (Everard, 1974), Ireland (Farrell, 1976), and Finland (Salonen, 1967); and on a range of P deficient sites in Japan (Kawana, 1969), New Zealand (Ballard, 1978b), Australia (Waring, 1973), South Africa (Schutz, 1976), and a number of tropical countries (Ojo & Jackson, 1973). Establishment applications are usually made at times ranging from just prior to planting, with the fertilizer incorporated into the soil by cultivation (Pritchett & Smith, 1975), to several months after planting, with fertilizer applied either by hand or mechanized means. Most research indicates that, provided the P fertilizer is applied within 12 months of planting, the long-term response is very much the same (Ballard, 1978b). Delaying P fertilizer applications on P-deficient sites beyond the first year after planting usually results in a loss inproductivity which cannot be recaptured by later fertilization (Waring, 1973).

Phosphorus fertilization of older stands is practiced where establishment applications were not made on P-deficient sites, due to the relatively recent operational use of fertilizers in forests, and where establishment applications are insufficient to sustain maximum productivity throughout a crop rotation. Establishment applications, particularly those applied in a localized spot, tend to have a fairly short-term effectiveness. This is attributed to the relatively small quantities applied to each seedling at establishment, and the restricted availability of localized fertilizer applications to rapidly expanding root systems (see below). The timing of P fertilizer applications to established stands is usually determined by monitoring foliar P levels: once levels decline below those established as suboptimal for growth, fertilizer is applied (Ballard, 1977; Bevage & Richards, 1972; Everard, 1974; Kawana & Leaf, 1973). The season of application is not critical with P fertilizers, although they are normally applied during the period of peak growth in late spring (Pritchett, 1976). Unlike response to N (Woollons & Will, 1975), responses to P are not as sensitive to stocking and crown competition levels (Mead & Gadgil, 1978), and thus applications are not usually timed to coincide with silvicultural operations such as thinning.

2. METHODS OF APPLICATION

Methods of application vary with the stage in stand development at which fertilizers are applied. The various techniques currently used in forestry have been reviewed and thoroughly discussed by Bengtson (1973, 1976), and Hagner (1971). Briefly, pre- and immediate post-planting applications are usually made either by hand or tractor-mounted spreaders, while applications to established stands are made principally from the air, and to a lesser extent by ground spreading using tractor-mounted spreaders.

Establishment applications are usually localized to some extent to improve accessibility to the restricted root system of seedlings, restrict weed competition, and limit immobilization of P on highly sorptive soils (Ballard, 1978b; Gentle & Humphreys, 1968; Schutz, 1976). With fertilizers applied by hand this is achieved by a spot application either on the soil surface or in a slit in close proximity (5-20 cm) to the seedling. Mechanized spreading utilizes special delivery mechanisms to restrict the fertilizer distribution to bands down the planted row (Bengtson, 1973). Mechanized spreading by tractor is used principally where the soil has been intensively cultivated prior to planting, as this restricts weed competition and enables incorporation of the fertilizer into the soil (Terry & Hughes, 1975)—a practice which is crucial to the effectiveness of certain fertilizer materials such as phosphate rock (Bengtson, 1970b; Brendemuehl, 1970).

Mechanized establishment applications are usually expected to be adequate to meet the P needs of a tree crop throughout a rotation if the correct P source is used (Pritchett, 1976); however, recent evidence suggests this is highly unlikely for hand-applied spot applications (Ballard, 1978b; Humphreys, 1977). The rapid decline in effectiveness of spot-applied fertilizer, irrespective of rate of application, on a soil of medium P-retention capacity is shown in Fig. 2. It will be noted that the spot applications still gave a more rapid initial response over the first 3 to 4 years. The rapid decline in effectiveness of spot applications has been attributed to a decreasing proportion of the expanding root system in contact with the fertilizer-enriched zone and the disproportionate "drying out" of the fertilizer-enriched zone due to intense rooting activity (Ballard, 1978b; Humphreys, 1977). Since increasing rates of application appear to have little influence on the long-term effectiveness of spot applications, further applications at a later stage of stand development are usually anticipated (Ballard, 1978b; Schutz, 1976).

Aerial applications to established stands are made by both fixed-wing aircraft and helicopters. Fixed-wing aircraft still dominate the scene in Scandinavian and the Southern Hemisphere countries (Bengston, 1973; Hagner, 1971), but helicopters are gaining in popularity with increases in their lifting capacity. Ground spreading by tractor or hand is practiced to a limited extent in Europe and Japan (Hagner, 1971). One advantage of ground spreading is that finely ground material such as phosphate rock, which cannot be effectively spread from aircraft, can be broadcast. Distribution of fertilizer from both fixed-wing aircraft and helicopters has often been variable (Armson, 1972; Ballard & Will, 1971); thus careful control of aerial operations is required as poor distribution patterns can have a marked effect on the variability of stands because of the infrequency of topdressing of forest stands.

3. RATES OF APPLICATION

Application rates of P fertilizers to coniferous forests are remarkably similar throughout the world, despite a wide range in soil and climatic conditions and the use of a variety of P sources. Spot application rates at establishment fall in the range of 10-20 g P/seedling (Ballard, 1978b; Baule, 1973a; Everard, 1974; Humphreys, 1977; Kawana, 1969; Ojo & Jackson, 1973; Schutz, 1976). Broadcast rates at establishment and rates to established stands mostly fall in the range of 50-100 kg P/ha (Baule, 1973a; Baule & Fricker, 1970; Binns, 1974; Boardman, 1974; Mead & Gadgil, 1978; Pritchett, 1976).

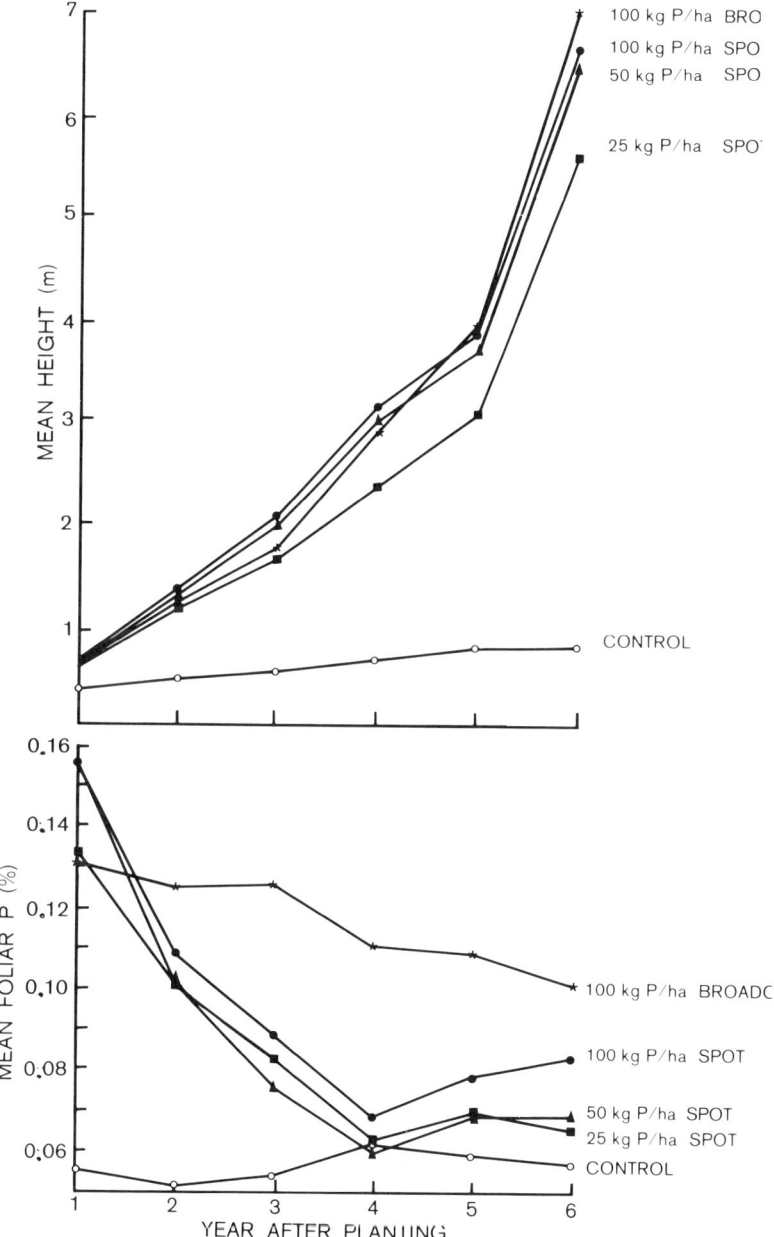

Fig. 2—Effect of superphosphate placement on growth response and foliar P concentrations of radiata pine. Cpt. 30 Mullions Range S.F., N.S.W. (F.R. Humphreys, personal communication).

The remarkable similarity in rates used is probably related to (i) the practice of adjusting the frequency of fertilizer applications rather than rates of individual applications to sustain the response throughout a rotation, (ii) the similar P requirements of most conifers, and (iii) the ability of established tree crops, irrespective of soil and climatic conditions, to conserve nutrients within their ecosystem by means of nutrient cycling.

4. FREQUENCY OF APPLICATION

Single broadcast applications of P fertilizers at rates little different from those used for agricultural crops will often give sustained responses lasting for 15 to 20 years (Pritchett & Smith, 1974; Pritchett & Swinford, 1961; Waring, 1973; Gentle et al., 1965). Ballard (1978c) reported that an application of 224 kg P/ha to a first rotation stand of radiata pine had a considerable residual effect on the second rotation crop planted 20 years later. The efficient use of applied P through recycling and the apparent ability of trees to utilize less available forms are normally cited as reasons for long-term responses of tree crops to P fertilizer (Pritchett & Smith, 1974); the careful selection of appropriate P sources to suit particular site conditions has undoubtedly also contributed (see below).

Despite the many examples of long-term responses, repeat applications of P fertilizers are often required during a rotation to sustain growth on P-deficient soils. These are not only required following establishment spot applications, but also following broadcast applications to established stands. Mead and Gadgil (1978) reported an example in New Zealand on extremely P-deficient soils of moderate P-retention capacity where foliar analysis indicated a repeat application was required just 3 years after an application of 120 kg P/ha as OSP.

5. UTILIZATION

Will (1965) estimated that 15% of the fertilizer P applied (224 kg P/ha) existed in the crop biomass (trees and litter) 8 years after application to a severely deficient stand of a 25-year-old radiata pine. At the time of this estimation the treated stand was still healthy and increasing in volume relative to the untreated stand. Pritchett and Smith (1974) calculated that the above-ground biomass (trees and litter) of a 15-year-old stand of slash pine contained 30% of the 118 kg P/ha applied at establishment. As was the case in the Will (1965) study, the stand was still responding at time of sampling. Both of these studies were in stands on soils of moderate P-retention capacity. In a biomass study examining the effectiveness of spot applications at establishment of radiata pine, Ballard (1978b) recorded utilization figures ranging from <1 to only 4% of the P applied at 11 and 22 g P/seedling 3 years previously on P-responsive sites.

Although the above studies were done on a limited range of sites, the relatively low recovery values, despite large biomass responses, suggest that the long response duration to many fertilizer P treatments can be attributed to applying amounts of P in excess of those actually required to initially rejuvenate the system. There is obviously considerable scope for manipulation of methods, rates, and frequencies of application as well as fertilizer sources to obtain improved efficiency of utilization. In this respect the intensive fertilization practices recently adopted in South Australia are of interest (Woods, 1976).

B. Fertilizer Sources

Characteristics of P fertilizers used in forest fertilization operations, their relative effectiveness, and soil and site factors requiring consideration in selection of a suitable P source have been comprehensively examined in a number of recent reviews (Baule & Fricker, 1970; Beaton, 1973; Bengtson, 1973, 1976; Binns, 1975; Pritchett & Smith, 1969; Terman, 1968). Only a few of the major points of particular significance to forestry will be touched on in this review.

In contrast to agriculture, various water-insoluble P sources, in particular certain phosphate rocks (PR), have been used most effectively in forestry. On forest soils of moderate acidity and P-retention capacity, water-soluble phosphates such as ordinary superphosphate (OSP) and triple superphosphate (TSP), are usually of equal or superior effectiveness to water-insoluble sources (Bengtson, 1976; Boardman, 1974; Mead, 1974); the superiority of the soluble sources is usually more pronounced when applied to seedlings (Gentle & Humphreys, 1968). On these types of soils OSP and TSP are the principal P sources used in most countries (Ballard & Will, 1977; Bengtson, 1976; Pritchett & Gooding, 1975; Schutz, 1976). Acid soils of both very low (peats, Spodosols, and siliceous sands) and very high (clays and soils derived from basic parent material) P-retention capacities are commonly used for forestry purposes. On both of these types of soils water-insoluble P sources such as reactive PR frequently show a superiority

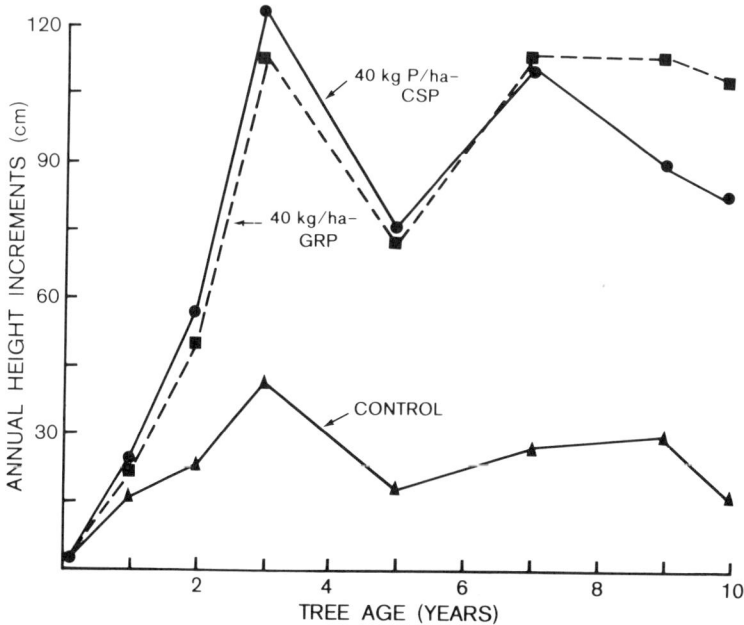

Fig. 3—Annual height increments of slash pine as affected by application of 40 kg P/ha as either TSP or Florida PR at time of planting on a soil with a high capacity for P fixation (Pritchett, 1976).

over soluble sources such as OSP and TSP, particularly in terms of duration of response (Bengtson, 1976; Binns, 1975; Hopkins, 1960; Huikari, 1973; Hunphreys & Pritchett, 1971; Pritchett & Gooding, 1975; Pritchett, 1976; Fig. 3). However, even on these soils the effectiveness of PR's is strongly determined by their citric acid solubility (Binns, 1975; Bengtson et al., 1974; Huikari, 1973) and the degree of contact with the soil afforded by the method of application (Brendemuehl, 1970; Bengtson, 1970b; Dickson, 1971) Recent advances in techniques of granulating ground PR's, which make them amenable to aerial broadcasting, have considerably increased the potential for use of these sources in forestry (Bengtson, 1976).

As intensification of forestry cultural practices continues, multiple deficiencies are becoming more common. The practice of applying compound NP, KP, or NPK fertilizers, or blended mixes of fertilizers containing these elements is increasing (Ballard & Will, 1977; Bengtson, 1976; Kawana, 1969; Pritchett & Gooding, 1975). Application of P in conjunction with N and K fertilizers is likely to increase the efficiency of utilization of N and K not only by stimulating root activity, but also by increasing the cation retention capacity of some soils (Ryden et al., 1977).

C. Interaction with Other Management Operations

1. SITE PREPARATION

Reports of interactions between response to P fertilization and site preparation techniques are common in the literature (Ballard, 1978b; Mauge, 1972; Pritchett & Smith, 1974; Waring, 1973; Woods, 1976). The interaction is normally attributed to the elimination of other growth-limiting factors such as excess moisture, restricted rooting depth, high soil bulk density, and weed competition. The magnitude of the additional response obtainable from P fertilization following amelioration of adverse site conditions is illustrated in Fig. 4 for 2-year-old radiata pine on a Spodosol in New Zealand. These data also indicate the futility of applying fertilizer on sites where adverse soil physical conditions restrict growth.

2. NITROGEN FERTILIZATION

Strong interactions between N and P fertilizers when applied at planting on P-deficient soils are common (Ballard, 1978b; Maftoun & Pritchett, 1970; Pritchett, 1972; Richards & Bevege, 1967; Waring, 1973). The major cause of the interaction, a suppression of growth following fertilization with N alone, has been attributed to the N fertilizer restricting mycorrhizal formation and thus P uptake, acidification of the soil by the N fertilizer reducing availability of soil P, and N additions stimulating microbial activity causing biological immobilization of P and possibly increased incidence of root disease. Where N is combined with P, N fertilization tends to enhance the response to P fertilization (Ballard, 1978b; Pritchett & Smith, 1972; Waring, 1973).

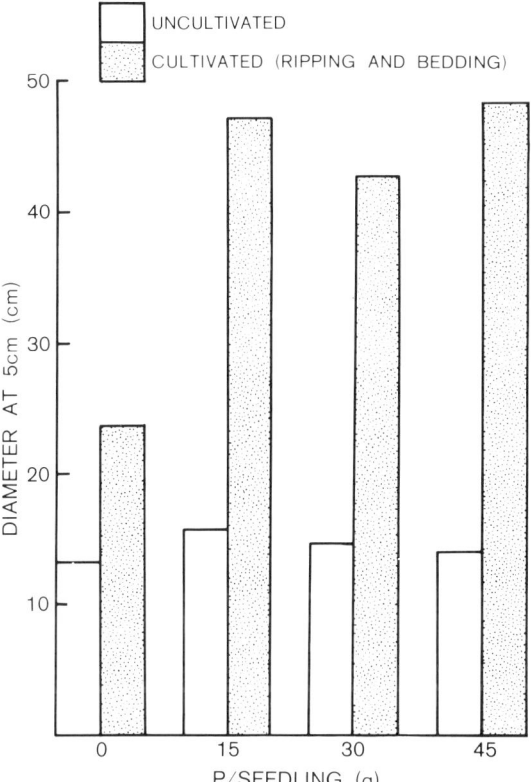

Fig. 4—Effect of cultivation and P fertilization on diameter growth of radiata pine 2 years after treatment and planting on a Spodosol.

3. TREE BREEDING

Foresters have long recognized and exploited differences in nutrient requirements among species. However, they have been slow to exploit differences within species (Smith & Goddard, 1973). Mounting evidence suggests that genotypes within conifers used for production forestry purposes respond differentially to fertilizers and often show a marked fertilizer × genotype interaction (Bengtson, 1968; Jahromi et al., 1976; Smith & Goddard, 1973). There is some evidence, however, that the fertilizer × genotype interaction is not as strong as frequently suspected (Matziris & Zobel, 1976).

For practical application, the ideal genotype is one that is capable of sustaining good growth under marginal nutritional conditions and that also responds well to fertilizer additions. Fertilizer × genotype interactions indicate that this ideal may not be all that common; the most responsive genotypes are those that fare relatively poorly in the absence of fertilizer addition, and the lines most efficient in the absence of fertilizer show only little response to fertilizer addition (Jahromi et al., 1976). Breeding strategies will obviously need to be carefully defined with the knowledge that

Table 9—Effect of five P sources on the basal area growth of three half-sib radiata pine families, Blue Range, S.F., Australian Capital Territory (H. D. Waring, personal communication).

Fertilizer source	Half-sib family		
	948	408A	954
	cm^2/plot		
Ordinary superphosphate	17.6	31.9	43.1
Phosphoric acid	13.5	16.6	18.2
Phosphoric acid + lime	14.9	15.8	19.0
Christmas Island calcined PR	10.7	18.5	6.6
Phosphate rock	11.6	7.8	5.5

selection and breeding of trees for response to fertilizer may eventually lead to fertilizer-dependent trees. Data provided by H. D. Waring (personal communication) which show a genotype × P fertilizer source interaction (Table 9) also illustrate the danger of basing selections on response to one fertilizer source and assuming the relative gain will be true for other sources. Such interactions are also likely to vary from site to site depending on soil conditions which influence the effectiveness of different P sources.

Concentration of P in the foliage is commonly used to diagnose the P status of trees (see below). Burdon (1975) reported, however, that although foliar analysis reflected the nutrient status of a population, P concentrations in foliage among clones were unrelated to growth rates. Jahromi et al. (1976) reported a poor relationship between foliar nutrient concentrations and growth parameters. Thus foliar analysis, although it may be a powerful tool for defining fertilizer needs of stands of trees, must obviously be used with caution as a guide to the nutrient efficiency of individual trees.

D. Nonvolumetric Effects

1. WOOD QUALITY

The early fears of reductions in wood density and timber quality following volume response to P fertilization have been largely allayed by findings of recent research (White, 1973; Baule & Fricker, 1970). Large volume responses to P fertilizer have been associated mostly with small, inconsequential decreases in density, mainly as a result of an increase in the proportion of early wood produced (Bamber, 1971; Cown, 1977; Gentle et al., 1965; Gooding & Smith, 1972). Further, there is some evidence to indicate that pulp yield per unit weight can be somewhat greater in lower density wood (Gagnon & Hunt, 1975), which could compensate for any reductions in wood density.

There appears to be little or no adverse effect of P fertilization on the quality of pulp and wood (Baule & Fricker, 1970; Gooding & Smith, 1972; Schmidtling, 1973; White, 1973). Baule and Fricker (1970) cautioned against fertilization of young trees with mostly juvenile wood, but pointed out that, in general, fertilization increases not only growth but also timber quality, particularly of pruned and older trees.

2. RESISTANCE TO PESTS AND DISEASE

The effect of fertilization on susceptibility of trees to attack by pests and diseases has been examined in several recent reviews (Barker, 1977; Baule, 1973b; Foster, 1968; Hesterberg & Jurgesen, 1972; Shigo, 1973); most studies have been concerned with effects of N and K fertilization. The limited amount of work with P suggests that P fertilization may reduce the degree of damage by chewing insects (Pritchett & Smith, 1972), but may enhance damage by sucking and boring insects (Baule, 1973b) and fungal diseases (Hollis et al., 1975; Smith et al., 1977). There are reports which conflict with these generalizations, and it is apparent that effects will be modified by rates and timing of application, the pre- and post-treatment condition of the tree (vigor, nutrient balance of organic constituents), and local site and climatic conditions (Baule, 1973b).

3. WATER QUALITY

The P retention capacities of most forest soils and the ability of tree crops to intercept mobile elements (Bengtson & Kilmer, 1975) are such that little of the P added to forest catchments is detectable in streamflow. In North Carolina, Sanderford (1975) observed no detectable increase in streamflow P outputs following application of 240 kg/ha of OSP to 13% of a southern pine catchment. Neary and Leonard (1977) reported the results of several catchment studies in New Zealand where losses of P in streamflow, following aerial P fertilization of radiata pine catchments with OSP, amounted to <0.1% of that applied. These authors observed that nearly all the output occurred soon after the topdressing operations and could be attributed to direct input into water courses within the catchments; the use of untreated riparian strips could alleviate even this small loss.

There are, however, two types of sites where considerable leaching can be expected following P fertilization—leached acid sands and acid organic soils which possess extremely low P-retention capacities. Humphreys and Pritchett (1971) and Tamm (1973) have reported considerable downward movement of water-soluble P fertilizers on acid sand and organic soils, respectively, although no loss was detected from either site. In both of these studies, however, movement of P from water-insoluble P sources was found to be negligible. In the interest of efficient use of fertilizer, it is usually recommended that such soils be delineated and fertilized only with water-insoluble or partially-soluble P fertilizers (Ballard & Pritchett, 1974).

4. FAUNAL AND FLORAL HABITAT

Several reviews on the impact of forest fertilization on the understory vegetation and wildlife indicate that the effects of P fertilization may be both beneficial and harmful (Hilmon & Douglass, 1968; Behrend, 1973; Leaf et al., 1975). Understory vegetation often increased in both quantity and quality following P fertilization, which increased the wildlife (and domestic stock) carrying capacity of the fertilized forests. The effect may be transient with stimulated canopy development shading out the understory. Unfortunately fertilization also tends to increase the palatability of the tree crop to animals resulting in increased browse damage.

5. SEED PRODUCTION

Most specific work on the effect of fertilization on seed production in conifer trees has been concerned with N (Jackson & Sweet, 1973). However, in many studies where fertilizers have increased seed production, NPK fertilizers have been used (Barnes & Bengtson, 1968; Baule & Fricker, 1970; McLemore, 1975). The effect of fertilization on stimulating seed production is often largely confounded by site and genotype differences, and the apparent importance of fertilizer timing and source (Barker, 1977).

V. DETERMINATION OF P FERTILIZER REQUIREMENTS

A number of techniques have been used to diagnose nutritional problems in forests. These include visual symptoms, pot trials, field trials, indicator plants, and plant tissue and soil analyses (Ballard, 1977; Tamm, 1964; Zottl, 1973). Of these techniques only plant tissue and soil analyses offer practical and effective means of routinely determining the fertilizer requirements of forest stands; consequently, discussion will be limited to these two techniques.

A. Tissue Analysis

Forest tree tissue analysis for diagnostic purposes is almost synonymous with foliar analysis, although other tissues, including bark, buds, and roots, have been suggested as potentially useful for diagnostic purposes (Leaf, 1973). The principles involved and specific problems associated with the use of foliar analysis in forestry have been the subject of several in-depth reviews (Armson, 1973; Leaf, 1973; Qureshi & Srivastava, 1966; van den Driessche, 1974; Zottl, 1973) and will only be outlined in this review, with special emphasis on P.

1. SAMPLING

Foliar P levels within any tree vary with age of the foliage, position in the crown, and time of the year, which necessitate rigid standardization of sampling procedures for comparative purposes. Foliar P levels also vary with crown class and genotype, which require sampling from a number of trees on any one "site" to obtain an index of its P status. Soil heterogeneity also contributes to between-tree variation on any one site. Procedures adopted for routine sampling still vary somewhat between countries, agencies, and species. For most coniferous crops, however, sampling is done during the period from early autumn to early winter by collecting foliage off the previous spring's flush (6- to 9-month-old needles) in the upper whorls of the crown (no light competition) from a number of trees (average of about 10-20 trees according to a 1970 Int. Union of Forest Res. Organ. survey). This sampling procedure tends to utilize that period of the year when P levels in the foliage are relatively stable, good resolution between sites (sensitivity) is obtained, and between-tree variability is at a minimum (Mead & Will, 1976).

2. INTERPRETATION

Routine use of foliar analysis for diagnostic purposes relies upon prior calibration of the concentration of P in foliage against either tree growth or responsiveness to fertilizer. Relationships between foliar P concentrations and tree growth or response parameters are used to identify critical levels which are then used for interpretative purposes. The *critical level* has been variously defined, but the most common usage is that concentration above which no measurable response to fertilizer addition will be achieved or which is associated with 90% of maximum yield (Bevege & Richards, 1972; Pritchett, 1968). As most calibrations are based on statistical inferences, P concentrations for any one species are often divided into three ranges: (i) low range, associated with poor growth and a high probability of response to fertilizer, (ii) marginal range, associated with marginal growth and only a moderate probability of obtaining a response to fertilizer, and (iii) high range, associated with good growth and a low probability of response to fertilizer. These ranges for a number of conifer tree crops are given in Table 10. For conifer species these values are reasonably independent of both the age of the tree (Swan, 1972a) and local soil and climatic conditions.

Precise interpretation of foliar analysis data, even where soundly calibrated relationships are available, is difficult because of the effects of nutrient interactions, multiple deficiencies, and environmental effects which can produce year-to-year and location-to-location variations in relationships (Leaf, 1973; van der Driessche, 1974). In many cases soundly calibrated relationships useful for predicting P requirements are not available, as foliar P levels of only a limited number of species in limited localities have been statistically calibrated against response to fertilizers (Ballard, 1977; Pritchett, 1968; Wells et al., 1973). Considerable caution should be employed in extrapolating beyond the soil, site, and crop conditions under which a calibration was derived. However, critical values derived from some species, e.g., radiata pine (Ballard, 1977; Raupach, 1967) and loblolly pine (Bevege & Richards, 1972; Wells et al., 1973), do agree well between countries, but those for others do not (see Norway spruce, Table 10).

A major shortcoming of foliar analysis when used for determining P fertilizer requirements is that, while it may indicate the need for P fertilizer, it gives little indication of suitable rates of application, duration of response, or suitable P source, all of which management requires information on. In most regions where operational P fertilization is practiced, foliar analysis is used only to diagnose the need for P fertilizer; rates and sources are based on results of field trails on similar soils (Ballard, 1977; Pritchett & Gooding, 1975). Periodical monitoring of foliar P levels (crop-logging) is

Table 10—Foliar P concentrations associated with poor, marginal, and good growth of conifer tree crops (guidelines only).†

Species	Foliar P associated with			Reference
	Poor growth	Marginal	Good growth	
		% oven-dry wt		
Radiata pine	<0.12	0.12–0.14	>0.14	Ballard, 1977
Loblolly pine	<0.10	0.10–0.13	>0.13	Wells et al., 1973
Slash pine	<0.09		>0.10	Pritchett, 1968
Ponderosa pine	<0.09		>0.09	Vlamis & Biswell, 1974
Caribaean pine			>0.10	Raupach et al., 1975
Lodgepole pine	<0.12	0.12–0.15	>0.15	Everard, 1973
	<0.10	0.10–0.17	>0.17	Swan, 1972c
	<0.07	0.07–0.10	>0.10	Swan, 1972d
Red pine	<0.12	0.12–0.15	>0.15	Swan, 1972b
Jack pine	<0.14	0.14–0.18	>0.18	Swan, 1970
Corsican pine	<0.13	0.13–0.18	>0.18	Everard, 1973
Scots pine	<0.14	0.14–0.17	>0.17	Everard, 1973
Norway spruce	<0.09		>0.09	Armson et al., 1975
	<0.15	0.15–0.20	>0.20	Swan, 1972a
	<0.14	0.14–0.24	>0.24	Everard, 1973
Black spruce	<0.14	0.14–0.18	>0.18	Everard, 1973
Sitka spruce	<0.14	0.14–0.18	>0.18	Everard, 1973
Douglas fir	<0.14	0.14–0.20	>0.20	Everard, 1973
Western hemlock	<0.20	0.20–0.35	>0.35	Everard, 1973
Larch	<0.20	0.20–0.35	>0.35	Everard, 1973

† The reader is advised to consult the individual references to determine the sampling methods employed and the site conditions under which these values were derived before attempting any extrapolation.

practiced in parts of Australia and New Zealand (Ballard, 1977; Bevege & Richards, 1972) to enable detection of the onset of a deficiency or decline in effectiveness of a previous application. This practice enables the fertilizer response to be sustained throughout the rotation by adjusting the frequency of application rather than the rate of a single application.

B. Soil Analysis

The use of soil analysis in forestry has been reviewed by Tamm (1964), Leaf (1968), Armson (1973), Zottl (1973), and Ballard (1977). Although soil testing is the principal diagnostic method used by agriculturists, its use in forestry has been somewhat limited. The success achieved with foliar analysis, especially in comparison with soil analysis (Ballard & Pritchett, 1975a; Wells et al., 1973), and a mistrust of agricultural techniques have relegated soil analysis to a secondary role in diagnosis of forest site fertility. The superior success of foliar analysis is attributed to the ability of trees to integrate the large number of soil and site factors controlling the uptake of P (Leaf, 1973). Difficulties with the use of soil analysis in forestry generally have been attributed to the problems of obtaining representative soil samples from throughout the tree's rooting zone and devising soil tests which reflect the fraction of soil nutrients that is available to trees.

1. SAMPLING

The deep rooting habit of trees has always been of concern to forest soil scientists attempting to obtain soil samples representative of the soil volume contributing to the nutrient requirements of trees. Most studies indicate, however, that levels of P in the surface horizon are much more closely correlated with the P status or responsiveness of trees than those in lower horizons (Alban, 1972; Ballard & Pritchett, 1975b; Pawluk & Arneman, 1961; Wells, 1965). Nevertheless, improved predictions have been reported when P-enriched lower horizons have been taken into consideration (Kessell & Stoate, 1938) and available P has been determined on samples from within the effective rooting zone rather than from just the surface 7.5 cm (Jackson & Gifford, 1974). Ballard and Pritchett (1975b) also reported that the critical soil P level for slash pine in the southeastern U.S. Coastal Plain was lower for soils with no impenetrable horizon within the surface 75 cm than for those with a smaller effective rooting volume.

The presence of substantial surface organic layers under older stands contribute significantly to the P supply of the trees. The chances of successfully calibrating P levels in the surface mineral horizon with stand P status are limited, particularly under circumstances where the P supply from organic layers is determined more by the rate of decomposition than the concentration of P. It is perhaps not by chance that most successful calibrations of soil P tests in forestry have involved young stands (Ballard, 1974; Ballard & Pritchett, 1975b; Wells et al., 1973).

2. TEST METHODS

Early work in Australia indicated that total soil P provided a good delineation of P-deficient and P-sufficient sites (Kessell & Stoate, 1938; Young, 1948). This early work with soil analysis helped foster the concept that soil tests developed for agricultural purposes were likely to be of little value in forestry. However, much subsequent work has shown that total P tests can only be successfully calibrated within groups of genetically similar soils (Ballard, 1970; Baur, 1959; Humphreys, 1964). A wide range of agricultural soil tests have been evaluated for use in forestry; P extracted by most of those successfully used in agriculture (Truog, Bray 1 and 2, North Carolina, Olsen, NH_4OAc, and Lactate methods) has been found to be significantly correlated with tree growth or response under some conditions (Ballard, 1974; Ballard & Pritchett, 1975a, b; Baule & Fricker, 1970; Pritchett, 1968; Wells et al., 1973).

The success of particular extractants, as in agriculture, undoubtedly varies from region to region because of differences in soil properties which influence the suitability of extractants (Thomas & Peaslee, 1973). In forestry the time dimension apparently is important also in determining the suitability of particular soil tests. Wells et al. (1973) and Ballard (1974) observed improved prediction of response to P fertilizer as the response period increased from 1 to 3 years after fertilization at planting. In a closer examination of this phenomenon, Ballard and Pritchett (1975a) found that early response (first year) in recently planted (or potted) seedlings was best predicted by mild extractants which essentially removed water-soluble P, while response over longer periods (3 to 5 years) was best predicted by stronger extractants which removed both water-soluble P and part of the solid phase components in equilibrium with it.

3. INTERPRETATION

As with other advisory practices, soil tests in forestry should not be used outside the range of conditions (both soil and crop) over which it was calibrated. Critical values for soil tests should be determined for each species and over a specified range of conditions. It is imperative that critical values derived for other crops not be used for tree crops. For instance slash pine will grow successfully in soils testing 5 ppm P extractable in $0.05N$ HCl + $0.025N$ H_2SO_4 (Ballard & Pritchett, 1975b), yet most agricultural crops on similar soils require a minimum of 16-37 ppm P (Thomas & Peaslee, 1973).

Very few soil tests have been calibrated against degree of response expected from fertilizer addition or quantity of fertilizer required to achieve maximum productivity. This has been done only in certain regions where extensive P fertilization at establishment is carried out (Ballard, 1974; Ballard & Pritchett, 1975b; Wells et al., 1973). For most other areas this has not been possible because of a lack of trials suitable for calibration purposes.

Despite the obvious limitations of soil analysis in forestry, it has advantages which make it a useful if not indispensable aid in formulating complete fertilizer programs:

1) It is the only reasonably sensitive diagnostic method available for predicting fertilizer requirements at time of planting. This is fortuitous, since in the first few years after crop establishment the mode of nutrition of trees is similar to that of agricultural crops—topsoil feeding with minimal contribution from nutrient cycling.
2) Soil analysis is an important back-up technique to foliar analysis and, when used in conjunction with it, can help prevent misinterpretation of effects caused by nutrient interactions, dilution effects, and multiple deficiencies.
3) Background soil analysis data are useful in selecting appropriate P sources, i.e., the value of pH and P retention in selecting P sources of varying solubility (Ballard & Pritchett, 1974) and choosing rates of application and predicting the likely duration of response (Ballard, 1978d).

The art of diagnosing fertilizer requirements for forest crops is still very much at an early stage of development compared to both agriculture and horticulture. Fortunately, the problems facing foresters are not as different as originally thought from those encountered and in many cases resolved, by agriculturists. Thus, there is obvious potential for rapid advancement in both the precision of calibrations and in the administering of diagnostic and corrective services.

VI. FUTURE RESEARCH NEEDS

Forest fertilization is fairly recent and pertinent research is consequently insufficient; thus there is much scope for future investigation. The weaknesses in the research effort to date are probably self-evident from this review. Some particular areas in which the author feels further work would yield results of considerable practical significance include:

a. Diagnosis—As operational forest fertilization gains momentum, the era of identifying severely deficient stands is rapidly passing. In the future, the need will be to detect incipient deficiencies which will call for considerably more refined calibrations than those shown in Table 10, with a wide range of P levels in the marginal region. Soil analysis data will need to give specific indications of the most effective P sources and rates of application in order to conserve a nonrenewable resource.

b. Nutrient Cycling—Except for the rather neglected root system, there is an abundance of quantitative information on tree and forest biomass. Data are generally lacking or imprecise, however, especially on the dynamics of nutrient cycling between the various pools in the ecosystem and the impact of intensive cultural practices on both the size of the pools and the rates of transfer between them.

c. Root Systems—The key to efficient utilization of soil P reserves probably lies in the root system. Because of the immobility of P, genotypes with fine, intensive root systems should be more efficient exploiters of soil P than those with coarse extensive root systems. This aspect warrants further attention, particularly in species evaluation and breeding programs. The potential for inoculation with specific mycorrhizal fungi suited for particular sites warrants further attention, also.

d. Disease—The influence of P fertilization on susceptibility of tree crops to disease and insect attack needs constant attention. The quest for fertility-responsive forest stands will inevitably reduce their variability and thus increase the risk of epidemics.

e. Fertilizer Efficiency—Biomass studies indicate that relatively inefficient use is being made of P fertilizers applied to trees. Means of improving this utilization, such as matching P source to site, and manipulation of methods, rates, and frequency of application, will require constant research.

f. Response Assessment—As energy becomes an overriding factor in society, traditional methods of assessing response to fertilizer, such as biomass or wood-volume increment, may become less meaningful in certain production forestry. In anticipation of the future energy role of some production forests, research is required on the energy balance of forest fertilization.

g. Breeding—There is obviously tremendous scope for manipulating the genetic makeup of forests to improve nutrient utilization and/or maximize productivity. Interactions between genotype and fertilization and other cultural practices need to be further examined as exploitation of these interactions offers some of the most exciting prospects of greatly increasing the productivity of forests (Stone & Bengtson, 1975).

ACKNOWLEDGMENTS

Thanks are expressed to Messrs. H. D. Waring and F. R. Humphreys for making available unpublished data.

LITERATURE CITED

Alban, D. H. 1972. The relationship of red pine site index to soil phosphorus extracted by several methods. Soil Sci. Soc. Am. Proc. 36:664-666.

Aldhous, J. R. 1972. Nursery practice. Bull. For. Comm. Lond. 43.

Allen, R. M., and T. E. Maki. 1955. Response of longleaf pine seedlings to soils and fertilizers. Soil Sci. 79:359-362.

Anderson, H. W., and S. P. Gessel. 1966. Effects of nursery fertilisation on outplanted Douglas fir. J. For. 64:109-112.

Armson, K. A. 1965. Seasonal patterns of nutrient absorption by forest trees. p. 65-76. In C. T. Youngberg (ed.) Forest-soil relationships in North America. Proc. 2nd North Am. Forest Soils Conf., 26-31 Aug. 1963, Oregon State Univ. Oregon State Univ. Press, Corvallis, Oreg.

Armson, K. A. 1972. Fertilizer distribution in the aerial fertilization of forests. Faculty of Forestry, University of Toronto, Tech. Rep. no. 11.

Armson, K. A. 1973. Soil and plant analysis techniques as diagnostic criteria for evaluating fertiliser needs and treatment response. p. 155-166. In A. L. Leaf and R. E. Leonard (ed.) Proc. Symp. on Forest Fertilization. State Univ. of New York. 22-25 Aug. 1972. USDA For. Serv. Gen. Tech. Rep. NE-3.

Armson, K. A., and V. Sadreika. 1974. Forest tree nursery soil management and related practices. Ministry of Natural Resources, Ontario.

Armson, K. A., H. H. Krause, and G. F. Weetman. 1975. Fertilization response in the northern coniferous forest. p. 449-466. In B. Bernier and C. H. Winget (ed.) Forest soils and forest land management. Proc. 4th North Am. Forest Soils Conf., 20-25 Aug. 1973, Land Univ., Quebec, Can. Les Presses de l'Universite Laval, Quebec.

Azcon, R., J. M. Barea, and D. S. Hayman. 1976. Utilization of rock phosphate in alkaline soils by plants inoculated with mycorrhizal fungi and phosphate solubilizing bacteria. Soil Biol. Biochem. 8:135-138.

Ballard, R. 1970. The phosphate status of the soils of Riverhead Forest in relation to growth of radiata pine. N. Z. J. For. 15:88-99.

Ballard, R. 1974. Use of soil testing for predicting phosphate fertilizer requirements of radiata pine at time of planting. N.Z. J. For. Sci. 4:27-34.

Ballard, R. 1977. Predicting fertiliser requirements of production forests. p. 33-44. In R. Ballard (ed.) Use of fertilisers in New Zealand forestry. N.Z. For. Serv., For. Res. Inst., Symposium no. 19. 7-10 Mar. 1977. Rotorua, New Zealand.

Ballard, R. 1978a. Effect of slash and soil removal on the productivity of second rotation radiata pine on a pumice soil. N.Z. J. For. Sci. 8:248-258.

Ballard, R. 1978b. Use of fertilisers at establishment of exotic forest plantations in New Zealand. N.Z. J. For. Sci. 8:70-104.

Ballard, R. 1978c. Effect of first rotation phosphorus applications on fertiliser requirements of second rotation radiata pine. N.Z. J. For. Sci. 8:135-145.

Ballard, R. 1978d. Use of the Bray soil test in forestry. I. Predicting phosphate retention capacity. N.Z. J. For. Sci. 8:239-247.

Ballard, R., and W. L. Pritchett. 1974. Phosphorus retention in Coastal Plain forest soils: II. Significance to forest fertilization. Soil Sci. Soc. Am. Proc. 38:363-366.

Ballard, R., and W. L. Pritchett. 1975a. Evaluation of soil testing methods for predicting growth and response of *Pinus elliottii* to phosphorus fertilization. Soil Sci. Soc. Am. Proc. 39:132-136.
Ballard, R., and W. L. Pritchett. 1975b. Soil testing as a guide to phosphorus fertilization of young pine plantations in the Coastal Plain. Univ. Fla. Agric. Exp. Stn. Tech. Bull. 778
Ballard, R., and G. M. Will. 1971. Distribution of aerially applied fertiliser in New Zealand forests. N.Z. J. For. Sci. 1:50-59.
Ballard, R., and G. M. Will. 1977. Past fertiliser use in New Zealand forests. p. 1-20. *In* R. Ballard (ed.) Use of fertilisers in New Zealand forestry. N.Z. For. Res. Inst., Symposium no. 19. Rotorua, New Zealand. 7-10 Mar. 1977.
Bamber, R. K. 1971. Some studies of the effects of fertilizers on the wood properties of Pinus species. p. 366-379. *In* R. Boardman (ed.) The Australian forest-tree nutrition conference. For. and Timber Bureau, Canberra, Australia.
Barker, J. E. 1977. Non-volumetric aspects of forest fertilisation. p. 267-290. *In* R. Ballard (ed.) Use of fertilisers in New Zealand forestry. N.Z. For. Serv., For. Res. Inst., Symp. no. 19. Rotorua, New Zealand. 7-10 Mar. 1977.
Barnes, R. L., and G. W. Bengtson. 1968. Effects of fertilization, irrigation, and cover cropping on flowering and on nitrogen and soluble sugar composition of slash pine. For Sci. 14:172-180.
Baule, H. 1971. Fertilisation in Midwest Europe. Background paper at I.U.F.R.O. Section 32 Meeting, Gainesville, Fla.
Baule, H. 1973a. World-wide forest fertilization: its present state, and prospects for the near future. Potash Rev. no. 6/1973.
Baule, H. 1973b. Effect of fertiliser on resistance to adverse agencies. p. 181-214. *In* Proc. Int. Symp. on Forest Fertilization. Paris. 3-7 Dec. 1973. Publ. by FAO/IUFRO.
Baule, H., and C. Fricker. 1970. The fertilizer treatment of forest trees. (Translation by C. L. Whittles) BLV, Verlagsgesellschaft mbH, Munich.
Baur, G. N. 1959. A soil survey of a slash pine plantation, Barcoongere, New South Wales. Aust. For. 23:78-87.
Beaton, J. D. 1973. Fertiliser methods and applications to forestry practice. p. 55-71. *In* A. L. Leaf and R. E. Leonard (ed.) Proc. Symp. on Forest Fertilization, State Univ. of New York. 22-25 Aug. 1972. USDA For. Serv., Gen. Tech. Rep. NE-3.
Behrend, D. F. 1973. Wildlife management-forest fertilization relations. p. 108-110. *In* A. L. Leaf and R. E. Leonard (ed.) Proc. Symp. Forest Fertilization, State Univ. of New York. 22-25 Aug. 1972. USDA For. Serv., Gen. Tech. Rep. NE-3.
Bengtson, G. W. 1968. Progress and needs in forest fertilization research in the South. p. 234-241. *In* Forest fertilization—theory and practice. Gainesville, Fla. 18-21 Apr. 1967. TVA, Muscle Shoals, Ala.
Bengtson, G. W. 1970a. Forest soil improvement through chemical amendments. J. For. 68: 343-347.
Bengtson, G. W. 1970b. Placement influences the effectiveness of phosphates for pine seedlings. p. 51-63. *In* C. T. Youngberg and C. B. Davey (ed.) Tree growth and forest soils. Proc. 3rd North Am. Forest Soils Conf., 5-10 Aug. 1968, North Carolina State Univ. Raleigh. Oregon State Univ. Press, Corvallis, Oreg.
Bengtson, G. W. 1973. Fertilizer use in forestry: materials and methods of application. p. 97-153. *In* Proc. Int. Symp. on Forest Fertilization, Paris. 3-7 Dec. 1973. Publ. by FAO/IUFRO.
Bengtson, G. W. 1976. Fertilizers in use and under evaluation in silviculture: a status report. Paper presented at 16th IUFRO World Congress in Oslo, Norway. 20 June-2 July 1976.
Bengtson, G. W., E. C. Sample, and S. E. Allen. 1974. Response of slash pine to P sources of varying citrate solubility. Plant Soil 40:83-96.
Bengtson, G. W., and V. J. Kilmer. 1975. Fertilizer use and water quality: Considerations for agriculture and forestry. p. 245-265. *In* B. Bernier and C. H. Winget (ed.) Forest soils and forest land management. Proc. 4th North Am. Forest Soils Conf., 20-25 Aug. 1973, Laval Univ., Quebec, Can. Les Presses de l'Universite Laval, Quebec.
Benzian, B. 1966. Manuring young conifers: experiments in some English nurseries. Proc. Fert. Soc. no. 94:5-37.
Benzian, B., and H. A. Smith. 1973. Nutrient concentrations of healthy seedlings and transplants of *Picea sitchensis* and other conifers grown in English forest nurseries. Forestry 46:55-69.
Bevege, D. I., and B. N. Richards. 1972. Principles and practice of foliar analysis as a basis for crop-logging in pine plantations. II. Determination of critical phosphorus levels. Plant Soil 37:159-169.

Binns, W. O. 1974. Fertilisers in forestry, 2. Nutrient elements, soils and climate. For. Home Grown Timber, April/May, 1974:35-36.

Binns, W. O. 1975. Fertilisers in forests: a guide to materials. (British) For. Comm. Leaflet no. 63.

Boardman, R. 1974. Pine stand improvement in the South-eastern region of South Australia. A review of phosphorus nutrition studies. South Aust. Woods For. Dep. Bull. no. 21.

Bowen, G. D. 1973. Mineral nutrition of ectomycorrhizae. p. 151-206. In G. C. Marks and T. T. Kozlowski (ed.) Ectomycorrhizae—their ecology and physiology. Academic Press, New York.

Boyle, J. R. 1973. Forest soil chemical changes following fire. Commun. Soil Sci. Plant Anal. 4:369-374.

Bray, J. R., and E. Gorham. 1964. Litter production in forests of the world. Adv. Ecol. Res. 2:101-157.

Brendemuehl, R. H. 1970. The phosphorus placement problem in forest fertilization. p. 43-50. In C. T. Youngberg and C. B. Davey (ed.) Tree growth and forest soils. Proc. 3rd North Am. Forest Soils Conf., 5-10 Aug. 1968, North Carolina State Univ., Raleigh. Oregon State Univ. Press, Corvallis, Oreg.

Brix, H., and R. van den Driessche. 1974. Mineral nutrition of container-grown tree seedlings. p. 77-84. In R. W. Tinus, W. I. Stein, and W. E. Balmer (ed.) Proc. North Am. Containerized Forest Tree Seedling Symp. Denver, Color. 26-29 Aug. 1974. Great Plains Agric. Council Publ. no. 68.

Brown, G. W., A. R. Gahler, and R. B. Marston. 1973. Nutrient losses after clearcut logging and slash burning in the Oregon Coast range. Water Resour. Res. 9:1450-1453.

Bunn, E. H., and G. M. Will. 1973. Management operations affecting nutrient cycling and fertilizer response in forest stands. p. 33-54. In Proc. Int. Symp. on Forest Fertilization, Paris. 3-7 Dec. 1973. Publ. by FAO/IUFRO.

Burdon, R. D. 1975. Foliar macronutrient concentrations and foliage retention in radiata pine clones on four sites. N.Z. J. For. Sci. 5:250-259.

Cole, D. W., and S. P. Gessel. 1965. Movement of elements through a forest soil as influenced by tree removal and fertilizer addition. p. 95-104. In C. T. Youngberg (ed.) Forest-soil relationships in North America. Proc. 2nd North Am. Forest Soils Conf., 26-31 Aug. 1963, Oregon State Univ. Oregon State Univ. Press, Corvallis, Oreg.

Cole, D. W., S. P. Gessel, and S. F. Rice. 1967. Distribution and cycling of nitrogen phosphorus, potassium and calcium in a second-growth Douglas fir ecosystem. p. 197-232. In H. E. Young (ed.) Symp. on Primary Productivity and Mineral Cycling in Natural Ecosystems. New York. 27 Dec. 1967. College of Life Sciences and Agric., Univ. of Maine, Orono, Maine.

Cown, D. J. 1977. Summary of wood quality studies in fertiliser trials. p. 307-310. In R. Ballard (ed.) Use of fertilisers in New Zealand forestry. N.Z. For. Serv., For. Res. Inst. Symp. no. 19. 7-10 Mar. 1977. Rotorua, New Zealand.

Dickson, D. A. 1971. The effect of form, rate and position of phosphatic fertilisers on growth and nutrient uptake of Sitka spruce on deep peat. Forestry 44:17-26.

Duvigneaud, P., and S. Denaeyer-de Smet. 1970. Biological cycling of minerals in temperate deciduous forests. p. 199-225. In D. E. Reichle (ed.) Analysis of temperate forest ecosystems. Springer-Verlag, New York.

Ebermeyer, E. 1867. Die gesammte Lehve der Waldstreu. Berlin.

Everard, J. E. 1973. Foliar analysis. Sampling methods, interpretation and application of the results. Quart. J. For. 67:51-66.

Everard, J. E. 1974. Fertilisers in the establishment of conifers in Wales and Southern England. (British) For. Comm. Booklet 41.

Farrell, E. P. 1976. Forest fertilization in Ireland. For. Chron. 52:194-196.

Florence, R. G., and P. H. Chuong. 1974. The influence of soil type on foliar nutrient concentrations in Pinus radiata plantations. Aust. For. Res. 6:1-8.

Forrest, W. G., and J. D. Ovington. 1970. Organic matter changes in an age series of Pinus radiata plantations. J. Appl. Ecol. 7:177-186.

Foster, A. A. 1968. Damage to forests by fungi and insects as affected by fertilizers. p. 42-46. In Forest fertilization—theory and practice. Gainesville, Fla. 18-21 Apr. 1967. TVA, Muscle Shoals, Ala.

Foster, N. W., and I. K. Morrison. 1976. Distribution and cycling of nutrients in a natural Pinus banksiana ecosystem. Ecology 57:110-120.

Fredriksen, R. L. 1971. Comparative chemical water quality—natural and disturbed streams following logging and slash burning. p. 125-137. In Forest land uses and stream environment. Oregon State Univ. Press, Corvallis, Oreg.

Fredriksen, R. L., D. G. Moore, and L. A. Norris. 1975. The impact of timber harvest, fertilization, and herbicide treatment on streamwater quality in western Oregon and Washington. p. 283-313. *In* B. Bernier and C. H. Winget (ed.) Forest soils and forest land management. Proc. 4th North Am. Forest Soils Conf., 20-25 Aug. 1973, Laval Univ. Quebec, Can. Les Presses de l'Universite Laval, Quebec.

Fried, M., and H. Broeshart. 1967. The soil-plant system in relation to inorganic nutrition. Academic Press, New York.

Gadgil, R. L., and P. D. Gadgil. 1975. Suppression of litter decomposition by mycorrhizal roots of *Pinus radiata*. N.Z. J. For. Sci. 5:33-41.

Gagnon, J. D., and K. Hunt. 1975. Kraft pulping and specific gravity in the uppermost stem of fertilised Balsam fir. Can. J. For. Res. 5:399-402.

Gentle, S. W., and F. R. Humphreys. 1968. Experience with phosphate fertilizers in man-made forests of *Pinus radiata* in New South Wales. Proc. 9th Commonwealth Forestry Conf., India. Forestry Commission of New South Wales, Sydney, Australia.

Gentle, S. W., F. R. Humphreys, and Marcia J. Lambert. 1965. An examination of a *Pinus radiata* phosphate trial fifteen years after treatment. For. Sci. 11:315-324.

Glass, G. G. 1976. The effects of root raking on an upland Piedmont loblolly pine (*P taeda* L.) site. N.C. State Univ. School. For. Resc. Tech. Rep. no. 56.

Gooding, J. W., and W. H. Smith. 1972. Effects of fertilization on stem and wood properties and pulping characteristics of slash pine. p. 1-18. *In* The effects of growth acceleration on the properties of wood. USDA For. Serv., Forest Products Lab., Madison, Wis.

Hagner, S. 1971. Techniques in sylvicultural operations with main emphasis on mechanization of fertilization. Proc. 15th IUFRO Cong., Sect. 32. Gainesville, Fal. 15 Feb. 1971.

Haines, L. H., T. E. Maki, and S. G. Sanderford. 1975. The effect of mechanical site preparation on soil productivity and tree (*Pinus taeda* L. and *P. elliottii* Engelm. var. *elliottii*) growth. p. 379-396. *In* B. Bernier and C. H. Winget (ed.) Forest soils and forest land management. Proc. 4th North Am. Forest Soils Conf., 20-25 Aug. 1973, Laval Univ., Quebec, Can. Les Presses de l'Universite Laval, Quebec.

Harley, J. L. 1970. Mycorrhiza and nutrient uptake in forest trees. p. 163-179. *In* Physiology of tree crops. Proc. Symp. at Long Ashton Research Station, University of Bristol, England. Academic Press, London.

Harwood, C. E., and W. D. Jackson. 1975. Atmospheric losses of four plant nutrients during a forest fire. Aust. For. 38:92-99.

Hesterberg, G. A., and M. F. Jurgensen. 1972. The relation of forest fertilisation to disease incidence. For. Chron. 48:92-96.

Hilmon, J. B., and J. E. Douglass. 1968. Potential impact of forest fertilisation on range, wildlife, and watershed management. p. 197-202. *In* Forest fertilization—theory and practice. Gainesville, Fla. 18-21 Apr. 1967. TVA, Muscle Shoals, Ala.

Hollis, C. A., W. H. Smith, R. A. Schmidt, and W. L. Pritchett. 1975. Soil and tissue nutrients, soil drainage, fertilization and tree growth as related to fusiform rust incidence in slash pine. For. Sci. 21:141-148.

Hopkins, E. R. 1960. The fertilizer factor in *Pinus pinaster* plantations on sandy soils of the Swan coastal plain, Western Australia. W. Aust. For. Dep. Bull. 68.

Huikari, O. 1973. Use of different N, P and K fertilizers in forests growing on drained peat. p. 391-404. *In* Int. Symp. on Forest Fertilization, Paris. 3-7 Dec. 1973. Publ. by FAO/IUFRO.

Humphreys, F. R. 1964. The nutrient status of pine plantations in central New South Wales. Appita 18:111-120.

Humphreys, F. R. 1977. Current fertilizer practice, New South Wales state forests. p. 123-124. *In* R. Ballard (ed.) Use of fertilisers in New Zealand forestry. N.Z. For. Serv., For. Res. Inst., Symp. no. 19. 7-10 Mar. 1977. Rotorua, New Zealand.

Humphreys, F. R., and Marcia J. Lambert. 1965. An examination of a forest site which has exhibited the ash bed effect. Aust. J. Soil Res. 3:81-94.

Humphreys, F. R., and W. L. Pritchett. 1971. Phosphorus adsorption and movement in some sandy forest soils. Soil Sci. Soc. Am. Proc. 35:495-500.

Ingestad, T. 1960. Studies on the nutrition of forest tree seedlings, III. Mineral nutrition of pine. Physiol. Plant. 13:513-533.

Jackson, D. S., and H. H. Gifford. 1974. Environmental variables influencing the increment of radiata pine (1) Periodic volume increment. N.Z. J. For. Sci. 4:3-26.

Jackson, D. I., and G. B. Sweet. 1973. Flower initiation in temperate woody plants. A review based largely on the literature of conifers and deciduous fruit trees. Hort. Abstr. 42:9-24.

Jahromi, S. T., R. E. Goddard, and W. H. Smith. 1976. Genotype × fertilizer interactions in slash pine: growth and nutrient relations. For. Sci. 22:211-219.

Kaul, O. N., P. B. L. Strivastava, and N. K. S. Bora. 1966. Nutrition studies on *Eucalyptus*. I. Diagnosis of mineral deficiencies in *Eucalyptus* hybrid seedlings. Indian For. 92:264-268.

Kaul, O. N., P. B. L. Srivastava, and V. N. Tandon. 1968. Nutrition studies on *Eucalyptus*. III. Diagnosis of mineral deficiencies in *Eucalyptus grandis* seedlings. Indian For. 94:831-834.

Kaul, O. N., P. B. L. Srivastava, and V. N. Tandon. 1970a. Nutrition studies on *Eucalyptus*. IV. Diagnosis of mineral deficiencies in Eucalyptus *globulus* seedlings. Indian For. 94:453 453-456.

Kaul, O. N., P. B. L. Srivastava, and J. D. S. Negi. 1970b. Nutrition studies on *Eucalyptus*. V. Diagnosis of mineral deficiencies in *Eucalyptus citriodora* seedlings. Indian For. 96:787-790.

Kawana, A. 1969. Forest fertilization in Japan. J. For. 67:485.

Kawana, A., and A. L. Leaf. 1973. Maintenance of productivity under short rotations in Japan. Voluntary paper at the Int. Symp. on Forest Fertilization, Paris. 3-7 Dec. 1973. Publ. by FAO/IUFRO.

Keay, J., and A. G. Turton. 1970. Distribution of biomass and major nutrients in a maritime pine plantation. Aust. For. 34:39-48.

Kessell, S. L., and T. N. Stoate. 1938. Pine nutrition. W. Aust. For. Dep. Bull. 50.

Knight, P. J. 1978. The nutrient content of *Pinus radiata* seedlings: a survey of planting stock from 17 NZ forest nurseries. N.Z. J. For. Sci. 8:54-69.

Lamb, D., and R. G. Florence. 1975. The influence of soil type on the nitrogen and phosphorus content of radiata pine litter. N.Z. J. For. Sci. 5:143-151.

Leaf, A. L. 1968. K, Mg and S deficiencies in forest trees. p. 88-122. *In* Forest fertilization—theory and practice. Gainesville, Fla. 18-21 Apr. 1967. TVA, Muscle Shoals, Ala.

Leaf, A. L. 1973. Plant analysis as an aid in fertilizing forests. p. 427-454. *In* L. M. Walsh and J. D. Beaton (ed.) Soil testing and plant analysis (revised ed.). Soil Sci. Soc. Am., Madison, Wis.

Leaf, A. L., R. E. Leonard, and N. A. Richards. 1975. Forest fertilization for non-wood production benefits in northeastern U.S. p. 435-448. *In* B. Bernier and C. H. Winget (ed.) Forest soils and forest land management. Proc. 4th North Am. Forest Soils Conf., 20-25 Aug. 1973, Laval Univ., Quebec, Can. Les Presses de l'Universite Laval, Quebec.

Lewis, W. M., Jr. 1974. Effects of fire on nutrient movement in a South Carolina pine forest. Ecology 55:1120-1127.

Lunt, H. A. 1938. The use of fertilizer in the coniferous nursery with special reference to *Pinus resinosa*. p. 723-766. *In* Conn. Agric. Exp. Stn. Bull. 416.

Madgwick, H. A. I. 1970. The nutrient contents of old-field Virginia pine. p. 275-282. *In* C. T. Youngberg, and C. B. Davey (ed.) Tree growth and forest soils. Proc. 3rd North Am. Forest Soils. Conf. 5-10 Aug. 1968, North Carolina State Univ., Raleigh. Oregon State Univ. Press, Corvallis, Oreg.

Madgwick, H. A. I., D. S. Jackson, and P. J. Knight. 1977. Above-ground dry matter, energy and nutrient contents of trees in an age series of *Pinus radiata* plantations. N.Z. J. For. Sci. 8:445-468.

Maftoun, M., and W. L. Prtichett. 1970. Effects of added nitrogen on the availability of phosphorus to slash pine in two lower Coastal Plain soils. Soil Sci. Soc. Am. Proc. 34:685-690.

Malcolm, D. C., and B. C. Y. Freezaillah. 1975. Early frost damage on Sitka spruce seedlings and the influence of phosphorus nutrition. Forestry 48:139-145.

Malkonen, E. 1973. Effect of complete tree utilization on the nutrient reserves of forest soils. p. 377-386. *In* H. E. Young (ed.) IUFRO Biomass Studies, University of Maine, Orono, Maine.

Malkonen, E. 1974. Annual primary production and nutrient cycle in some Scots pine stands. Commun. Inst. For. Fenn. 84.5.

Marx, D. H. 1969. The influence of ectotrophic mycorrhizal fungi on the resistance of pine roots to pathogenic infection. II. Production, identification and biological activity of antibiotics produced by *Leucopaxillus cerealis* var. *piceina*. Phytopathology 59:411-414.

Marx, D. H., and W. C. Bryan. 1975. The significance of mycorrhizae to forest trees. p. 107-118. *In* B. Bernier and C. H. Winget (ed.) Forest soils and forest land management. Proc. 4th North Am. Forest Soils Conf., 20-25 Aug. 1973, Laval Univ., Quebec, Can. Les Presses de l'Universite Laval, Quebec.

Matziris, D. I., and B. J. Zobel. 1976. The effect of fertilization on growth and quality characteristics of loblolly pine. For. Ecol. Manage. 1:21-30.

May, J. T. 1957. Technical assistance program for southern forest nurseries. South. Lumberman 195:95–96.

Mauge, J. P. 1972. Etudes et experimentations sur Pin maritime. 1965–70. Association Foret-Cellulose, Nangis, France.

McLemore, B. F. 1975. Cone and seed characteristics of fertilized and unfertilized longleaf pines. USDA For. Serv. Res. Pap. SO-109.

Mead, D. J. 1974. Response of radiata pine to superphosphate and Christmas Island "C" phosphate fertilizers. N.Z. J. For. Sci. 4:35–38.

Mead, D. J., and R. L. Gadgil. 1978. Fertiliser use in established radiata pine stands in New Zealand. N.Z. J. For. Sci. 8:105–134.

Mead, D. J., and G. M. Will. 1976. Seasonal and between-tree variation in nutrient levels in *Pinus radiata* foliage. N.Z. J. For. Sci. 6:3–13.

Mikola, P. 1959. Studies on soil properties and seedling growth in Finnish forest nurseries. Comm. Inst. For. Fenn. 49:1–78.

Millar, C. S. 1974. Decomposition of coniferous leaf litter. p. 105–126. *In* C. H. Dickinson and G. J. F. Pugh (ed.) Biology of plant litter decomposition. Vol. 1. Academic Press, London.

Miller, H. G., Jean M. Cooper, and J. D. Miller. 1976. Effect of nitrogen supply on nutrients in litterfall and crown leaching in a stand of Corsican pine. J. Appl. Ecol. 13:233–248.

Morrison, I. K. 1973. Distribution of elements in aerial components of several natural jack pine stands in northern Ontario. Can. J. For. Res. 3:170–179.

Mosse, B., C. L. Powell, and D. S. Hayman. 1976. Plant growth response to vesicular-arbuscular mycorrhiza. IX. Interactions between VA mycorrhiza, rock phosphate and symbiotic nitrogen fixation. New Phytol. 76:331–342.

Neary, D. G., and J. H. Leonard. 1977. Environmental aspects of forest fertilisation. p. 247–266. *In* R. Ballard (ed.) Use of fertilisers in New Zealand forestry. N.Z. For. Serv., For. Res. Inst. Symp. no. 19. Rotorua, New Zealand. 7–10 Mar. 1977.

Nelson, L. E., G. L. Switzer, and W. H. Smith. 1970. Dry matter and nutrient accumulation in young loblolly pine (*Pinus taeda* L.). p. 261–274. *In* C. T. Youngberg and C. B. Davey (ed.) Tree growth and forest soils. Proc. 3rd North Am. Forest Soils Conf., 5–10 Aug. 1968, North Carolina State Univ., Raleigh. Oregon State Univ. Press, Corvallis, Oreg.

Ojo, G. O. A., and J. K. Jackson. 1973. The use of fertiliser in forestry in the drier tropics. p. 351–364. *In* Int. Symp. on Forest Fertilization, Paris. 3–7 Dec. 1973. Publ. by FAO/IUFRO.

Orman, H. R., and G. M. Will. 1960. The nutrient content of *Pinus radiata* trees. N.Z. J. Sci. 3:510–522.

Ovington, J. D., and H. A. I. Madgwick. 1959. Distribution of organic matter and plant nutrients in a plantation of Scots pine. For. Sci. 5:344–355.

Pawluk, S., and H. F. Arneman. 1961. Some forest soil characteristics and their relationship to jack pine growth. For. Sci. 7:160–172.

Pritchett, W. L. 1968. Progress in the development of techniques and standards for soil and foliar diagnosis of phosphorus deficiency in slash pine. p. 81–87. *In* Forest fertilization—theory and practice. Gainesville, Fla. 18–21 Apr. 1967. TVA, Muscle Shoals, Ala.

Pritchett, W. L. 1972. The effect of nitrogen and phosphorus fertilizers on the growth and composition of loblolly and slash pine seedlings in pots. Soil Crop Sci. Soc. Fla. Proc. 32:161–165.

Pritchett, W. L. 1976. Phosphorus in forest soils. Phosphorus Agric. 67:27–35.

Pritchett, W. L., and J. W. Gooding. 1975. Fertilizer recommendations for pines in the Southeastern Coastal Plain of the United States. Univ. Fla. Agric. Exp. Stn. Bull. 774.

Pritchett, W. L., and W. H. Smith. 1969. Sources of nutrients and their reactions in forest soils. Soil Crop Sci. Fla. Proc. 29:149–158.

Pritchett, W. L., and W. H. Smith. 1972. Fertilizer response in young pine plantations. Soil Sci. Soc. Am. Proc. 36:660–664.

Pritchett, W. L., and W. H. Smith. 1974. Management of wet savanna forest soils for pine production. Univ. Fla. Agric. Exp. Stn. Tech. Bull. 762.

Pritchett, W. L., and W. H. Smith. 1975. Forest fertilization in the U.S. Southeast. p. 467–476. *In* B. Bernier and C. H. Winget (ed.) Forest soils and forest land management. Proc. 4th North Am. Forest Soils Conf., 20–25 Aug. 1973, Laval Univ., Quebec, Can., Les Presses de l'Universite Laval, Quebec.

Pritchett, W. L., and K. R. Swinford. 1961. Response of slash pine to colloidal phosphate fertilization. Soil Sci. Soc. Am. Proc. 25:397–400.

Qureshi, I. M., and P. B. L. Srivastava. 1966. Foliar diagnosis and mineral nutrition of trees. Indian For. 92:447-460.

Ralston, D. B., and R. P. McBride. 1976. Interaction of mineral phosphate—dissolving microbes with red pine seedlings. Plant Soil 45:493-507.

Raupach, M. 1967. Soil and fertilizer requirements for forests of *Pinus radiata*. Adv. Agron. 19:307-353.

Raupach, M., A. R. P. Clarke, B. G. Gibson, and K. M. Cellier. 1975. Cultivation and fertiliser effects on the growth and foliage nutrient concentrations of *P. radiata, P. pinaster* and *P. caribaea* on three soil types at Anglesea (Victoria). CSIRO, Div. of Soils. Tech. Pap. no. 25.

Remezov, N. P., and P. S. Pogrebnyak. 1969. Forest soil science. Israel program for scientific translations, Jerusalem.

Rennie, P. J. 1955. The uptake of nutrients by mature forest growth. Plant Soil 7:49-95.

Richards, B. N., and D. I. Bevege. 1967. Effect of cultivation and fertilizing on potential yield of pulpwood from loblolly pine. Aust. For. 31:202-210.

Ryden, J. C., J. R. McLaughlin, and J. K. Syers. 1977. Mechanisms of phosphate sorption by soils and hydrous ferric oxide gel. J. Soil Sci. 28:72-92.

Salonen, L. K. 1967. Evolution of forest fertilisation in Finland. p. 39-40. *In* Forest fertilisation. Proc. 5th Colloquim of the Int. Pot. Inst., Jyvaskyla, Finland. 1967.

Sanderford, S. G. 1975. Forest fertilisation and water quality in the North Carolina Piedmont. N.C. State Univ. School of For. Resources, Tech. Rep. no. 53.

Schmidtling, R. C. 1973. Intensive culture increases growth without affecting wood quality of young southern pines. Can. J. For. Res. 3:565-573.

Schutz, C. J. 1976. A review of fertilizer research on some of the more important conifers and eucalypts planted in subtropical and tropical countries with special reference to South Africa. S. Afric. Dep. For. Bull. 53.

Shigo, A. L. 1973. Insects and disease control: forest fertilization relations. p. 117-121. *In* A. L. Leaf and R. E. Leonard (ed.) Proc. Symp. on Forest Fertilization. USDA For. Serv. Gen. Tech. Rep. NE-3.

Shirley, H. L., and L. J. Meuli. 1939. The influence of soil nutrients on drought resistance of two-year-old red pine. Am. J. Bot. 26:355-360.

Smith, J. H. G., A. Kozak, O. Sziklai, and J. Walters. 1966. Relative importance of seedbed fertilization, morphological grade, site, provenance, and parentage to juvenile growth and survival of Douglas fir. For. Chron. 42:83-86.

Smith, W. H., and R. E. Goddard. 1973. Effects of genotype on the response to fertilizer. p. 155-168. *In* Int. Symp. on Forest Fertilization, Paris. 3-7 Dec. 1973. Publ. by FAO-IUFRO.

Smith, W. H., C. A. Hollis, and J. W. Gooding, III. 1977. Influence of soil factors on fusiform rust incidence. p. 81-88. *In* R. J. Dinus and R. A. Schmidt (ed.) Management of fusiform rust in Southern pines. Proc. Symp. Univ. Fla., Gainesville, Fla. 7-8 Dec. 1976.

Stoeckeler, J. H., and H. F. Arneman. 1960. Fertilizers in forestry. Adv. Agron. 12:137-195.

Stoeckeler, J. H., and G. W. Jones. 1957. Forest nursery practice in the lake states. USDA For. Serv. Agric. Handb. no. 110.

Stoeckeler, J. H., and P. E. Slabaugh. 1965. Conifer nursery practice in the Prairie-Plains. USDA For. Serv., Agric. Handb. no. 279.

Stone, E. L. 1973. The impact of timber harvest on soils and water. p. 427-467. *In* Report of the President's advisory panel on timber and the environment. U.S. Govt. Printing Office, Washington, D.C.

Stone, E. L. 1975. Soil and man's use of forest land. p. 1-19. *In* B. Bernier and C. H. Winget (ed.) Forest soils and land management. Proc. 4th North Am. Forest Soils Conf., 20-25 Aug. 1973, Laval Univ., Quebec, Can. Les Presses de l'Universite Laval, Quebec.

Stone, E. L., and G. W. Bengtson. 1975. Towards all out fiber production. p. 55-67. *In* W. P. Martin (ed.) All-out food production: strategy and resource implications. Am. Soc. of Agron. Special Publ. no. 23, ASA, Madison, Wis.

Swan, H. S. D. 1970. Relationships between nutrient supply, growth and nutrient concentrations in the foliage of black spruce and jack pine. Woodl. Pap. Pulp Res. Inst. Can. Pap. 19.

Swan, H. S. D. 1972a. Foliar nutrient concentrations in Norway spruce as indicators of tree nutrient status and fertiliser requirement. Woodl. Pap. Pulp Res. Inst. Can. Rep. 40.

Swan, H. S. D. 1972b. Foliar nutrient concentrations in red pine as indicators of tree nutrient status and fertiliser requirement. Woodl. Pap. Pulp Res. Inst. Can. Rep. 41.

Swan, H. S. D. 1972c. Foliar nutrient concentrations in lodgepole pine as indicators of tree nutrient status and fertiliser requirement. Woodl. Pap. Pulp Res. Inst. Can. Rep. 42.

Swan, H. S. D. 1972d. Foliar nutrient concentrations in shore pine as indicators of tree nutrient status and fertiliser requirements. Woodl. Pap. Pulp Res. Inst. Can. Rep. 43.

Switzer, G. L., and L. E. Nelson. 1956. The effect of fertilization on seedling weight and utilization of N, P, and K, by loblolly pine (*Pinus taeda* L.) grown in the nursery. Soil Sci. Soc. Am. Proc. 20:404-408.

Switzer, G. L., and L. E. Nelson. 1967. Seedling quality strongly influenced by nursery soil management, Mississippi study shows. Tree Planters' Notes 18:5-14.

Switzer, G. L., and L. E. Nelson. 1972. Nutrient accumulation and cycling in loblolly pine (*Pinus taeda* L.) plantation ecosystems: the first twenty years. Soil Sci. Soc. Am. Proc. 36:143-147.

Switzer, G. L., and L. E. Nelson. 1973. Maintenance of productivity under short rotations. p. 365-390. *In* Int. Symp. on Forest Fertilization, Paris. 3-7 Dec. 1973. Publ. by FAO/IUFRO.

Switzer, G. L., L. E. Nelson, and W. H. Smith. 1968. The mineral cycle in forest stands. p. 1-9. *In* Forest fertilization—theory and practice. Gainesville, Fla. 18-21 Dec. 1967. TVA, Muscle Shoals, Ala.

Tamm, C. O. 1964. Determination of the nutrient requirements of forest stands. Int. Rev. For. Res. 1:115-170.

Tamm, C. O. 1973. Effects of fertilizers on the environment. p. 299-317. *In* Int. Symp. on Forest Fertilization, Paris. 3-7 Dec. 1973. Publ. by FAO/IUFRO.

Tamm, C. O., H. Holmen, B. Popovic, and G. Wiklander. 1974. Leaching of plant nutrients from soils as a consequence of forestry operations. Ambio 3:211-222.

Terman, G. L. 1968. Fertilizer, soil, and plant properties affecting crop response to P fertilizers. p. 77-80. *In* Forest fertilization—theory and practice. TVA, Muscle Shoals, Ala.

Terman, G. L., and G. W. Bengtson. 1973. Yield-nutrient concentration relationships in slash and loblolly pine seedlings. Soil Sci. Soc. Am. Proc. 37:445-450.

Terry, T. A., and J. H. Hughes. 1975. The effects of intensive management on planted loblolly pine (*Pinus taeda* L.) growth on poorly drained soils of the Atlantic Coastal Plain. p. 351-378. *In* B. Bernier and C. H. Winget (ed.) Forest soils and forest land management. Proc. 4th North Am. Forest Soils Conf. 20-25 Aug. 1973, Laval Univ., Quebec, Can. Les Presses de l'Universite, Laval, Quebec.

Thomas, G. W., and D. E. Peaslee. 1973. Testing oils for phosphorus. p. 115-132. *In* L. M. Walsh and J. D. Beaton (ed.) Soil testing and plant analysis. Soil Sci. Soc. Am., Madison, Wis.

Timmis, R. 1974. Effect of nutrient stress on growth, bud set and hardiness in Douglas fir seedlings. Great Plains Agric. Counc. Publ. 68:197-193.

Tinker, P. B. 1980. Role of rhizosphere microorganisms in phosphorus uptake by plants. p. 000-000. *In* F. E. Khasawneh et al. (ed.) The role of phosphorus in agriculture. Am. Soc. Agron., Madison, Wis.

Truman, R. 1972. The detection of mineral deficiencies in *Pinus elliottii, Pinus radiata* and *Pinus taeda* by visual means. For. Comm. of New South Wales Tech. Pap. 19.

Turner, J., and J. N. Long. 1975. Accumulation of organic matter in a series of Douglas fir stands. Can. J. For. Res. 5:681-690.

Turner, J., and M. J. Singer. 1976. Nutrient distribution and cycling in a sub-alpine coniferous forest ecosystem. J. Appl. Ecol. 13:295-301.

van den Driessche, R. 1969. Forest nursery handbook. B.C. For. Serv., Victoria, Canada. Res. Note 48.

van den Driessche, R. 1974. Prediction of mineral nutrient status of trees by foliar analysis. Bot. Rev. 40:347-394.

van Goor, C. P. 1970. Fertilization of conifer plantations. Irish For. 27:68-80.

Viro, P. J. 1974. Effects of forest fire on soil. p. 7-45. *In* T. T. Kozlowski and C. E. Ahlgren (ed.) Fire and the ecosystem. Academic Press, New York.

Vlamis, J., and H. H. Biswell. 1974. Growth and phosphorus uptake by pine seedlings grown on phosphate deficient and phosphate fixing soils. Soil Sci. 118:374-379.

Waring, H. D. 1973. Early fertilisation for maximum production. p. 215-242. *In* Int. Symp. on Forest Fertilization, Paris. 3-7 Dec. 1973. Publ. by FAO/IUFRO.

Webber, B. D. 1977. Biomass and nutrient distribution patterns in a young *Pseudotsuga menziesii* ecosystem. Can. J. For. Res. 7:326-334.

Weetman, G. F., and B. D. Webber. 1972. The influence of wood harvesting on the nutrient status of two spruce stands. Can. J. For. Res. 2:351-369.

Wells, C. G. 1965. Nutrient relationships between soils and needles of loblolly pine (*Pinus taeda*). Soil Sci. Soc. Am. Proc. 29:621-624.

Wells, C. G. 1971. Effects of prescribed burning on soil chemical properties and nutrient availability. p. 86-96. *In* Proc. Symp. on Prescribed Burning. 14-16 Apr. 1971. Southeastern Forest Exp. Stn., Asheville, N.C.

Wells, C. G., and L. J. Metz. 1963. Variation in nutrient content of loblolly pine needles with season, age, and soil and position on the crown. Soil Sci. Soc. Am. Proc. 27:90-93.

Wells, C. G., D. M. Crutchfield, N. M. Berenyi, and C. B. Davey. 1973. Soil and foliar guidelines for phosphorus fertilization of loblolly pine. USDA For. Serv. Res. Pap. SE-110.

Wells, C. G., and J. R. Jorgensen. 1975. Nutrient cycling in loblolly pine plantations. p. 137-158. *In* B. Bernier and C. H. Winget (ed.) Forest soils and forest land management. Proc. 4th North Am. Forest Soils Conf., 20-25 Aug. 1973, Laval Univ., Quebec, Can. Les Presses de l'Universite, Laval, Quebec.

Wells, C. G., J. R. Jorgensen, and C. E. Burnette. 1975. Biomass and mineral elements in a thinned loblolly pine plantation at age 16. USDA For. Serv. Res. Pap. SE-126.

White, D. P. 1973. Effect of fertilization on quality of wood and other forest products. p. 271-298. *In* Int. Symp. on Forest Fertilization, Paris. 3-7 Dec. 1973. Publ. by FAO/IUFRO.

White, E. H., and W. L. Pritchett. 1970. Water table control and fertilization of a flatwood soil for pine production. Univ. Florida Agric. Exp. Stn. Tech. Bull. no. 743.

Wilde, S. A. 1958. Forest soils. The Ronald Press Co., New York.

Wilde, S. A. 1968. Mycorrhizae: their role in tree nutrition and timber production. Wis. Agric. Exp. Stn. Res. Bull. no. 272.

Will, G. M. 1957. Variations in the mineral content of radiata pine needles with age and position in tree crown. N.Z. J. Sci. Tech. Bull. 38:699-706.

Will, G. M. 1959. Nutrient return in litter and rainfall under some exotic conifer stands in New Zealand. N.Z. J. Agric. Res. 2:184-193.

Will, G. M. 1961. Some changes in the growth habit of *Eucalyptus* seedlings caused by nutrient deficiencies. Emp. For. Res. 40:301-307.

Will, G. M. 1963. Dry matter production and nutrient uptake by *Pinus radiata*. Comm. For. Rev. 40:57-70.

Will, G. M. 1965. Increased phosphorus uptake by radiata pine in Riverhead forest following superphosphate applications. N.Z. J. For. 10:33-42.

Will, G. M. 1967. Decomposition of *Pinus radiata* litter on the forest floor. Part I. Changes in dry matter and nutrient content. N.Z. J. Sci. 10:1030-1044.

Will, G. M. 1968. The uptake, cycling and removal of mineral nutrients by crops of *Pinus radiata*. Proc. N.Z. Ecol. Soc. 15:20-24.

Williamson, B., and I. J. Alexander. 1975. Acid phosphatases localized in sheath of beech mycorrhiza. Soil Biol. Biochem. 7:195-198.

Wittwer, R. F., A. L. Leaf, and D. H. Bickelhaupt. 1975. Biomass and chemical composition of fertilized and/or irrigated *Pinus resinosa* Ait. plantations. Plant Soil 42:629-651.

Wollum, A. G., II, and G. H. Schubert. 1975. Effect of thinning on the foliage and forest floor properties of ponderosa pine stands. Soil Sci. Soc. Am. Proc. 39:968-972.

Woods, R. V. 1976. Early silviculture for upgrading productivity on marginal *Pinus radiata* sites in the south-eastern region of South Australia. S. Aust. Woods and Forests Dep. Bull. no. 24.

Woollons, R. C., and G. M. Will. 1975. Increasing growth in high production radiata pine stands by nitrogen fertilisers. N.Z. J. For. 29:243-253.

Wright, T. W., and G. M. Will. 1958. The nutrient content of Scots and Corsican pines growing on sand dunes. Forestry 31:13-25.

Young, H. E. 1948. The response of loblolly and slash pine to phosphate manures. Queensl. J. Agric. Sci. 5:77-105.

Youngberg, C. T. 1958. Uptake of nutrients by western conifers in forest nurseries. J. For. 56:337-340.

Zottl, H. W. 1973. Diagnosis of nutritional disturbances in forest stands. p. 75-96. *In* Int. Symp. on Forest Fertilization, Paris. 3-7 Dec. 1973. Publ. by FAO/IUFRO.

Chapter 28

Phosphorus Nutrition of Forages

D. A. MAYS

Tennessee Valley Authority
Muscle Shoals, Alabama

S. R. WILKINSON

Science & Education Administration, AR, USDA
Southern Piedmont Conservation Research Center
Watkinsville, Georgia

C. V. COLE

Science & Education Administration, AR, USDA
Colorado State University
Fort Collins, Colorado

I. INTRODUCTION

The term *forage* encompasses a broad spectrum of annual and perennial plants. The importance of forages to U.S. and world agriculture has been discussed by Turner (1972), Hodgson (1968), Hodgson (1974), Hanson (1974), and others. In the United States alone, there are at least 30 legumes, over 100 species of grasses, and numerous forbs which are used for forage.

Several annual crops, including corn, sorghum, soybeans, and the small grains, are commonly ensiled or harvested for hay and can also be classified as forage crops. However, these species will not be dealt with in this chapter since their fertility requirements are discussed by R. A. Olson and J. J. Hanway in Chapter 24.

Grasslands occupy more than 400 million ha of land in the United States. Of this amount about 60 million ha is in hay and cropland pasture or improved pasture. The remainder is in permanent pasture, range, and in public grazing land. A summary of native grasslands by type is shown in Table 1.

Forages provide about 65% of the nutrients consumed by dairy cattle, 75% of those consumed by beef cattle, and more than 90% of the diet of

---

Copyright 1980 © ASA-CSSA-SSSA, 677 South Segoe Road, Madison, WI 53711, USA.
The Role of Phosphorus in Agriculture.

Table 1—Types of native grasslands in the United States (USDA For. Serv., 1970).

Type	Area, million ha
Forest grasslands†	146.4
Shortgrass	80.2
Semidesert and desert grasslands	63.6
Sagebrush grasslands	39.1
Pacific bunchgrass	17.2
Mountain grasslands and meadows	13.8
Tallgrass	7.5
Total	367.8

† Consisting of: open conifer 24%, pinyon juniper 21%, oak hickory 16%, loblolly pine, shortleaf pine, and hardwood 15%, Woodland chapparral 9%, longleaf slash pine 7%, open aspen 1%, and oak pine 7%.

sheep, goats, and horses. If the trend toward growing calves to heavier weights on grass before feedlot finishing continues, grasslands will be an even more important source of nutrients for cattle.

Hodgson (1974) estimated that forages in the United States produce nutrients equivalent to 210 million metric tons of corn annually. The wild swings in agricultural prices experienced in recent years make crop value or profitability comparisons of limited use in general discussions such as this. However, crop value comparisons indicate that the value of all forages based on the value of equivalent nutrients from corn would have exceeded $9 billion at the lowest corn price experienced since 1970 and would have been about three times that amount at the highest corn price. A valuation based on some percentage of the sale price of cattle also would have resulted in wide fluctuations during the same period. Forages and grains are not necessarily directly exchangeable in value since efficient forage utilization is largely limited to ruminants and horses, while grains can be utilized by all classes of livestock as well as by humans.

Forages are of varying importance on a worldwide basis. Several countries, including Australia, New Zealand, Uruguay, and Argentina, have agricultural economies based largely on forage and livestock production. At the opposite end of the spectrum, many of the heavily populated countries in Asia produce little forage for animal consumption.

On the whole, forages always have been less adequately fertilized than have cash crops. Turner (1972) calculated that only 15% of the hay and cropland pasture, 18% of the improved permanent pasture, and 1.5% of the permanent pasture and range in the United States is fertilized on a routine basis. These percentages are likely to vary from year to year in response to different fertilizer cost/product value ratios. Beaton and Berger (1974) reported that the total P used on forages in the United States in 1970 was 452,000 metric tons. They calculated the potential for P utilization to be 1,869,000 metric tons.

In some regions of western Europe, Australia, and New Zealand, where the agricultural economy depends almost entirely on high-producing grasslands, fertilization is a routine production practice on a large percentage of the land devoted to forage crops. Consequently, high outputs of animals and animal products per hectare result.

Reasons for the limited use of fertilizer on forages include the following:

1) Forages are considered to be low value crops and thus not worth fertilizing.
2) Reduced yields due to low fertility are not always obvious, particularly under grazing.
3) It is difficult for the farmer to measure a fertilizer response in term of dollars of profit per unit area of land.
4) The level of management practiced by many farmers does not result in full utilization of the extra forage produced by fertilization.

Areas of P deficiency are extensive on a worldwide basis, except in North America and western Europe where economics and/or technology have favored the application of large amounts of fertilizer.

Beeson (1945) delineated the areas of P deficiency in the United States. However, as a result of extensive federal cost-sharing programs for fertilizer application on grasslands and educational programs sponsored by public agencies and the fertilizer industry, large contiguous areas of P deficiency no longer exist in the United States. The P status can change almost on a field-by-field basis as a result of past differences in fertilization. Because of this wide variability in soil P supply, many of the more than 500 published results of P application experiments with forages are of value only when used in conjunction with valid soil tests. One cannot assume that a field will be responsive or nonresponsive to P fertilization from knowledge of the P status of similar fields. In contrast, N is universally needed by grasses and the response to a certain application rate can be fairly well predicted for a given grass species and environmental situation.

While yield-limiting P deficiency can be found on many soil types, extreme deficiencies which preclude plant growth are usually now found only on drastically disturbed areas such as strip mine spoils, highway cuts, and other scalped areas where no topsoil is present. Some land only recently brought into cultivation is also extremely P deficient. Mays and Bengtson (1974) reported results from a strip-mined site in northeast Alabama which was so P deficient that a fall planting application of 28 kg of P/ha was insufficient to insure winter survival of new clover and grass seedlings. A 56-kg/ha application resulted in good survival and subsequent yield; 112 kg/ha produced even higher yields. Such extreme deficiencies should be identified and characterized by soil tests. Fertilizer recommendations suitable for low-fertility farmland are not sufficiently high to assure successful revegetation of these disturbed sites.

While this chapter will deal only with forages grown to support livestock, grasses are also extremely valuable for use in land reclamation; roadside stabilization; as ground cover on playgrouns, parks, and home lawns; and in other situations where aesthetics rather than profit provides the motive for establishment. At least moderate applications of fertilizer are necessary to keep these special purpose plantings healthy and attractive.

II. RELATIONSHIPS OF PHOSPHORUS APPLICATION TECHNIQUES TO ROOT MORPHOLOGY

A. Morphological Differences Between Grasses and Legumes

Phosphorus immobility in soil is a phenomenon which can have both good and bad effects. Loss of P from the root zone through upward or downward movement is not a problem in most soils. However, because of this immobility root growth and extension are important factors in the P supply pattern of all plants. Black (1967) estimated that almost all of the P is supplied to plants from a soil layer about 1 mm in thickness around each root.

The morphological differences between grass and legume roots affect their capacities for P uptake from soil. Most grasses produce a dense network of fine fibrous roots which completely ramify into the surface soil layers to a depth of 25 to 50 cm. This root network extensively explores the soil mass and is able to extract sufficient P from soils that would not supply adequate P for legumes or other plants with less pervasive root systems.

Forage legumes are taprooted or have heavy, branched roots which are much less extensive and fibrous than grass roots. A number of legumes, including alfalfa and sericea lespedeza, have extremely deep-growing taproots which may extend 5 m or more into deep soils. However, a deep-rooting habit is of limited help in P nutrition as most P is usually found in the top 20 to 30 cm of soil.

Lamba et al. (1949) found significant differences in the depth and distribution of the root systems of alfalfa, red clover, bromegrass, and timothy (Table 2). However, even the deepest rooted of these species had most of the roots in the top 40 cm of soil, indicating a strong capability for nutrient removal from the upper soil layers.

Schwendiman et al. (1966) compared the root production of alfalfa with that of several dryland grasses in Washington and Oregon. The grasses produced two to three times as much root by weight as did alfalfa in the surface 20 cm of soil. Considering the proportion of alfalfa taproot to total root weight in the surface 15 to 20 cm, the grasses must have had a tremendous advantage in total nutrient absorbing surface.

Table 2—Root distribution of two legumes and two grasses after two growing seasons on Miami silt loam (Lamba et al., 1949).

Sampling depth	Alfalfa	Red clover	Bromegrass	Timothy
cm		% of total root system		
0– 20	57.3	85.6	64.1	91.3
20– 40	11.8	9.5	13.1	6.1
40– 60	10.5	4.4	10.6	2.6
60– 80	9.2	0.5	10.1	--
80–100	7.5	--	10.1	--
80–100	7.5	--	2.1	--
100–120	3.8	--	--	--

Keay et al. (1970) compared the rates of P absorption of several annual pasture species in Australia. They found that ryegrass and silvergrass had greater ability grow at low soil-P concentrations than did several annual clovers and forbs. However, the grasses exhibited lower maximum P uptake rates at high soil-P concentrations.

B. Band Seeding

Duell (1974) reviewed the literature on P fertilization for forage establishment, and found general agreement that legume seedlings are usually less able than grasses to obtain P from the low soil-P concentrations associated with broadcast fertilizer applications. Seedling growth of both grasses and legumes is often enhanced by placing P in concentrated bands directly underneath the seed row (band seeding). Moving the fertilizer band as little as 2 or 3 cm to the side of the seed row is often sufficient to significantly reduce early growth of legume seedlings. Favorable responses to banding of P were reported more often with alfalfa and the clovers than with birdsfoot trefoil.

The data of Sheard et al. (1971, Fig. 1) show that, while the P concentrations in alfalfa and bromegrass were both increased by banding of P, the magnitude of alfalfa response was greater. The effects of banding on tissue

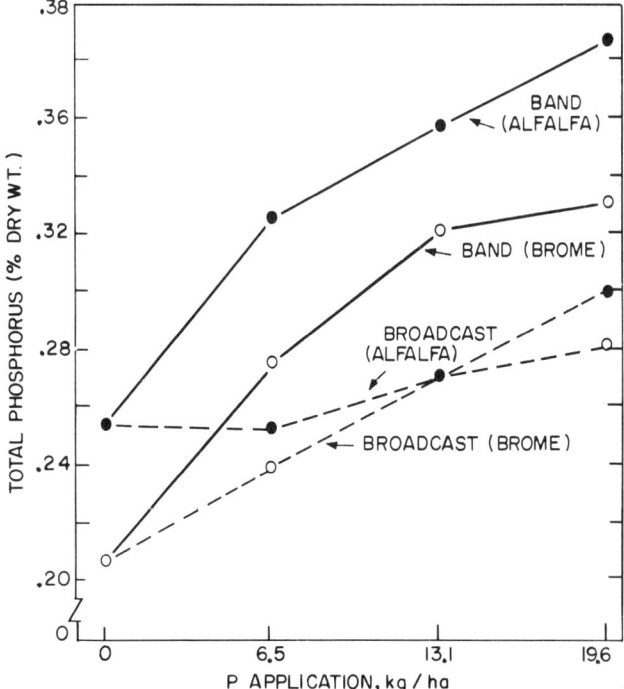

Fig. 1—Total P concentrations in alfalfa and bromegrass seedlings resulting from rate and placement of fertilizer P (Sheard et al., 1971).

P concentrations were large in seedlings 6 weeks old but small in seedlings 12 weeks old. Tissue P concentration was higher in alfalfa at 6 weeks but higher in brome at 12 weeks. Placing the fertilizer band 9 cm to one side of the row prevented fertilizer P uptake by the seedlings.

Brown (1959) seeded alfalfa with triple superphosphate banded and broadcast at rates ranging from 100 to 800 kg/ha. He found that all banded rates except the lowest resulted in a doubling of alfalfa seedling size, while only the highest broadcast rate increased the rate of seedling growth.

Robinson et al. (1959) studied the effect of temperature on response of red clover to banded P on a P-deficient soil. They noted a yield response to band seeding of 272% at 10°C but only 34% at 27°C. They found no concentration of roots around the P bands. These findings agree with observations on the relationship between planting date and response to starter fertilizer on corn where "pop-up fertilizers" have usually given responses only on cold soils. The data suggest that band seeding might be more important with early spring than with autumn seedings, which are usually made on warmer soils.

It has been shown in pot experiments with established plants that all roots do not have to be in contact with fertilized soil for near maximum response to occur. Burton (1966) reported that common and coastal bermudagrass and Pensacola bahiagrass produced 89% of maximum top growth and 100% of maximum root growth when only 18% of the root zone was fertilized, provided the whole root zone was watered. However, failure to water part of the root zone resulted in large reductions in both root and top growth.

In a split root experiment with corn, Engelstad and Allen (1971) showed that P applied to one side of the root system was translocated throughout the entire root system and was effective in promoting root and top growth. They found that the presence of ammonium N enhanced the uptake of P from a band, but had no effect on uptake of P mixed throughout the soil.

Duell's review (1974) cited instances of both positive and negative effects of N on seedling legume response to banded P, with general agreement that any positive responses were confined to the early growth period. Many forage workers consider band seeding to be good insurance when adequate soil fertility levels or favorable weather are at all questionable.

C. Effectiveness of Surface Application of P

The relative immobility of P has caused many people to question the effectiveness of surface application on grasslands, but there has been considerable evidence for several decades that no serious problem exists.

Sprague (1933) showed that Kentucky bluegrass could obtain adequate P from surface-applied superphosphate. O'Donnell and Love (1970) showed by use of labeled P that 'Merion' Kentucky bluegrass has > 30% of its total root weight in the top 5 cm of soil, and that it can obtain a significant amount of P from this soil layer (Fig. 2).

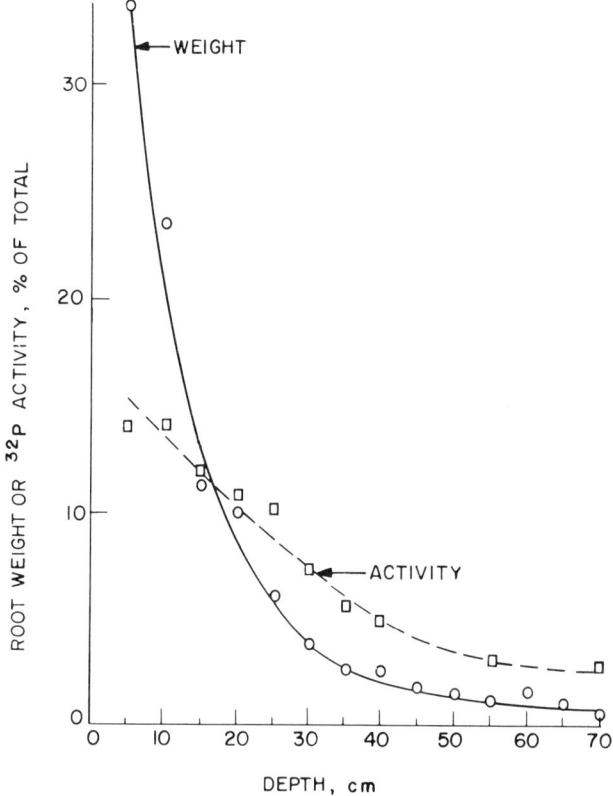

Fig. 2—Comparison of root weights and root activities with depth, as measured by ^{32}P injections in Merion bluegrass cut at 4.4-cm height (O'Donnell & Love, 1970).

Massey and Sheard (1970) reported that alfalfa was equal to bromegrass in utilizing surface-applied P. They indicated that many of the actively absorbing roots are in the top 3 cm of soil. In contrast, Lawton et al. (1954) showed that alfalfa obtained somewhat less P from shallow placement and more from deep placement of P than did bromegrass (Fig. 3). However, both crops obtained almost all of their P from the top 30 cm of soil.

Stanford et al. (1950) compared uptake of surface-applied radioactive P by alfalfa, ladino clover, and orchardgrass (Table 3) and found little difference in their capacity to obtain P from the soil surface. However, tissue concentrations indicated that perhaps this P supply was marginal for maximum production, even at the highest application rate.

Riley et al. (1975) reviewed the literature on availability of surface-applied P for no-till systems and concluded that it was readily absorbed. Reasons for this included proliferation of roots very close to the soil surface under a mulch, and movement of P into the 2.5- to 5.0-cm layer by freezing and thawing, wetting and drying, and action of soil fauna. Similar factors should result in good uptake of surface-applied P in pastures. (See also Moschler et al., 1972.)

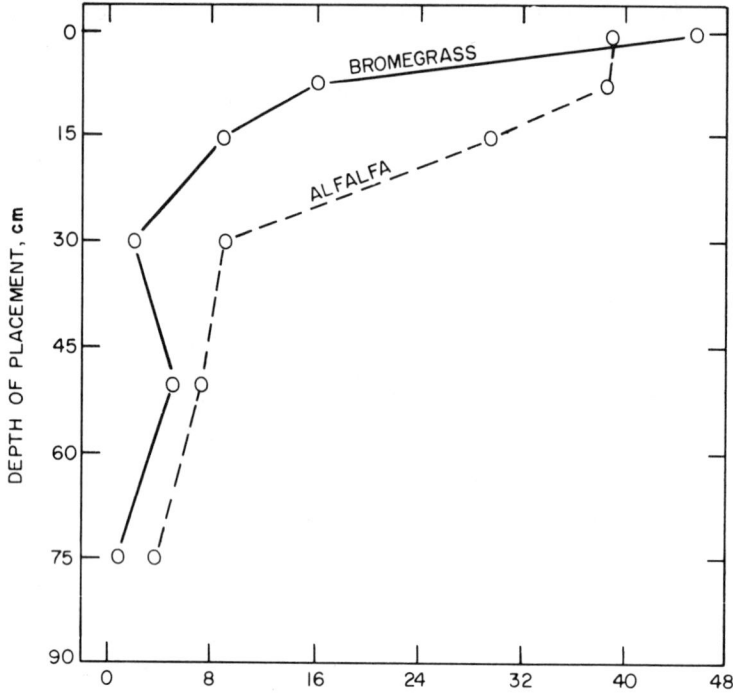

Fig. 3—Plant P in first cutting alfalfa-bome hay derived from superphosphate placed at various depths (Lawton et al., 1954).

Table 3—Total P in plants and proportion of total P supplied by surface-applied ordinary superphosphate (Stanford et al., 1950).

Rate	P from superphosphate			P in forage		
	Alfalfa	Ladino clover	Orchard-grass	Alfalfa	Ladino clover	Orchard-grass
kg OSP/ha	%					
0	0	0	0	0.234	0.226	0.216
224	20.0	17.8	15.8	0.260	0.230	0.218
1,120	45.9	49.5	43.9	0.300	0.277	0.232

III. PHOSPHORUS NUTRITION OF COOL-SEASON GRASSES AND LEGUMES

A. Grasses

The effect of initial soil P status on the magnitude of response to applied P by a given crop is well illustrated by Lunt et al. (1965). They grew alta fescue in the greenhouse on six different soils having HCO_3^--extractable P levels ranging from 2.8 to 17.5 ppm. Their data (Table 4) show an 80% yield increase to application of 66 ppm of P to soil having 2.8 ppm of ex-

Table 4—Yield and P concentration of alta fescue forage as affected by level of soil and applied P (Lunt et al., 1965).

HCO$_3$-extractable P in soil	P added, ppm		
	0	6.6	66
ppm	Yield, g/pot		
2.8	2.7	3.3	4.8
4.2	4.4	5.3	6.3
4.6	2.7	4.0	5.5
8.2	5.2	5.2	6.2
13.8	5.1	5.3	6.1
17.5	5.5	5.9	6.3
	P concentration, %		
Avg., low P soils	0.26	0.30	0.39
Avg., high P soils	0.35	0.36	0.46

tractable P, but only a 19% increase for soil testing 8.2 ppm or higher. They found that the 66-ppm application increased tissue P by 50% on low P soils but only 31% on high P soils.

1. SPECIES DIFFERENCES

Grasses exhibit marked species difference in response to P supply. Jones et al. (1970a) grew Italian ryegrass, hardinggrass, tall fescue, orchardgrass, and tall oatgrass with 0, 100, 200, 300, and 400 kg of P/ha. Tall fescue, ryegrass, and tall oatgrass produced the highest yields without P, but tall fescue and ryegrass also showed the greatest response to P fertilization.

Reid and Jung (1974) reported that the P requirement for lactating dairy cows producing 20 to 30 kg of milk daily is 0.35% of the ration, while beef cows need about 0.18%, and rapidly growing steers and heifers need as much as 0.43% P in the ration. The P needs of dairy cows and growing animals are at or above the critical levels in the herbage for maximum forage production and well above the P concentrations of forages grown at P-deficient levels. Direct feeding of P compounds in salt blocks and mixes or in concentrate supplements is so low cost and easy that fertilizing grasslands solely to increase the forage P concentrations for the benefit of the animals is rarely, if ever, justified. However, adequate soil and/or applied P usually results in much higher forage yields.

Reid et al. (1970) fertilized several grasses with 50 kg of P and 112 kg of N/ha and analyzed clippings taken at different stages of maturity for several plant nutrients. The P concentrations are shown in Table 5. All species in the vegetative state were adequate in P for grazing animals. Phosphorus concentration declined most rapidly in timothy, but by late bloom, P in all species was too low to meet the full needs of lactating dairy cows and of steers or heifers.

In the strip mine experiment of Mays and Bengtson (1974) orchardgrass seedlings were more sensitive than tall fescue to P deficiency. A complete loss of the orchardgrass stand occurred overwinter with a P fertiliza-

Table 5—Concentrations of P in several grasses as affected by stage of maturity (Reid et al., 1970).

Stage of growth	Potomac orchardgrass	Lincoln bromegrass	Kentucky 31 fescue	Common timothy
	%			
Vegetative	0.55	0.44	0.62	0.46
Boot	0.55	0.46	0.47	0.34
Headed	0.45	0.40	0.47	0.34
Early bloom	0.34	0.30	0.26	--
Late bloom	0.30	0.28	0.30	--
Mature	0.25	0.21	0.25	0.21
Vegetative regrowth	0.55	0.38	0.43	0.41

tion rate of 28 kg/ha, while a satisfactory stand of fescue persisted at this P rate.

Crossley and Bradshaw (1968) found that cultivated strains of orchardgrass were more responsive to P fertilization than were wild strains. This response probably reflects the results of natural ecotype selection under conditions of low P fertility as compared with the grass breeder's selection under conditions of fairly adequate P fertility in nurseries.

2. EFFECT OF TEMPERATURE AND MOISTURE

Finn and Mack (1964) showed that both soil temperature and soil moisture affected P response of orchardgrass (Fig. 4). At a 10°C soil temperature, maximum yield of the S-143 cultivar occurred with 35 ppm of applied P; at 20°C, yields were still increasing with 70 ppm. Yields were higher at 75 than at 25% of the moisture-holding capacity, with the differences being similar for both soil temperatures. The Chinook cultivar (data not shown) exhibited a much smaller effect of both temperature and moisture variables, with the optimum P rate being near 35 ppm for both temperatures. Orchardgrass forage yields were greater at 20°C but root yields were greatest at 10°C.

3. EFFECT OF N

Maximum P response occurs only when other nutrients are not limiting. Templeton et al. (1969) conducted a pot test on very low fertility soil and showed the effects of N on P response by Boone orchardgrass (Fig. 5). Without applied N maximum yield was reached with a P application of 22 kg/ha, while yields with 112 kg of N/ha were still increasing at the 44 kg/ha P rate. Phosphorus fertilization hastened leaf emergence, produced longer individual leaves, and greater total length of expanded blades. Added P produced higher shoot/root ratios in early growth but this difference did not persist.

4. RANGE IN P CONCENTRATION

Wedin (1974) reported that P concentrations in cool-season grasses range from about 0.14 to 0.50%. In most situations concentrations below

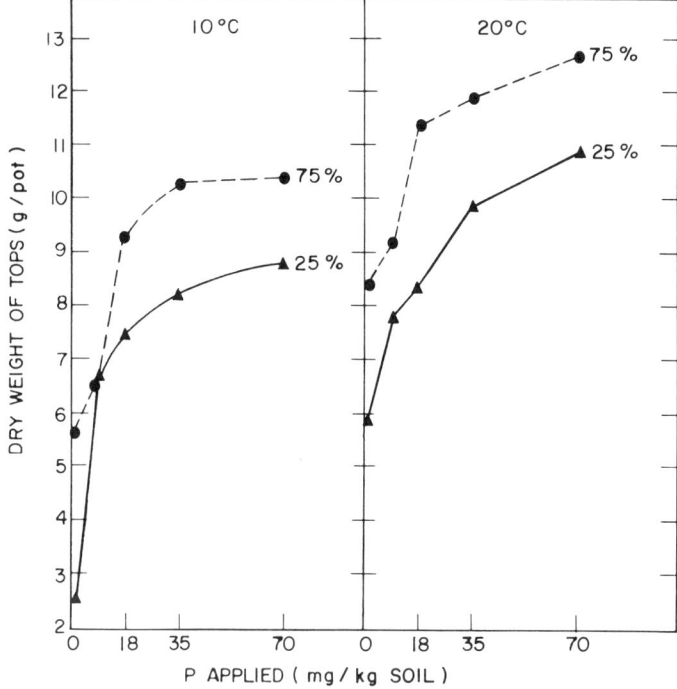

Fig. 4—Total yields of S-143 orchardgrass grown at 10 and 20°C soil temperature and 25 and 75% moisture, as affected by applied P (Finn & Mack, 1964).

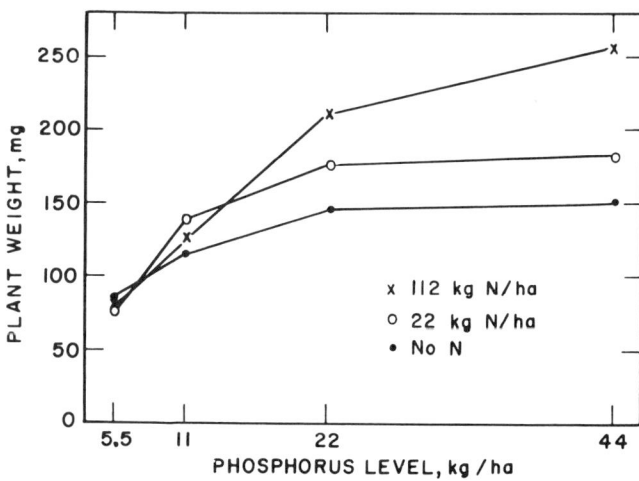

Fig. 5—Dry weights of 80-day-old Boone orchardgrass plants grown under three N and four P levels (Templeton et al., 1969).

about 0.20% indicate a deficiency for plant growth while 0.30 to 0.35% is usually necessary for optimum yields. Unlike N and K, a liberal supply of soil P does not usually result in luxury use.

Using National Research Council values for P concentration, Wedin (1974) calculated that the maximum annual P uptake for cool-season grasses at several yield levels ranged from 4 to 56 kg/ha (Table 6). Even if the P concentrations with minimum yields were somewhat higher, total P uptake would still be low because of low yields. These uptake values also represent removal from the system if grass is harvested as hay. However, under grazing at a maximum forage consumption of only 60 to 65%, a significant amount of ingested P is returned to the soil in the dung of grazing animals. Removal of P would be less than 25 kg/ha even at the highest yield levels and would be negligible at low yield levels.

Reid and Jung (1965) reported the effect of P fertilization on the

Table 6—Calculated annual P uptake by cool season grasses at several assumed yields (Wedin, 1974).

Yield, metric tons/ha	Assumed P concentration, %	P uptake, kg/ha
Minimum		
2	0.20	4
4	0.20	8
Optimum		
8	0.35	28
12	0.35	42
16	0.35	56

Table 7—Mineral and organic constituents in tall fescue as affected by P fertilization (Reid & Jung, 1965).

	First cutting		Aftermath	
Constituent	No P	220 kg P/ha	No P	220 kg P/ha
	%			
K	2.29	2.50	2.18	2.08
P	0.34	0.38	0.37	0.37
Ca	0.50	0.46	0.62	0.62
Mg	0.17	0.17	0.23	0.19
	ppm			
Mn	75	62	93	113
Cu	17	14	4	3
Zn	28	26	22	22
Mo	0.9	0.7	0.6	0.8
Co	1.1	1.7	0.5	0.4
	%			
Cellulose	31.1	31.9	29.9	30.4
Acid detergent fiber	33.6	34.7	36.5	36.8
Acid insoluble lignin	3.2	3.0	4.0	4.6
Cell wall constituents	65.2	66.1	65.1	64.7
Soluble carbohydrates	8.8	13.0	11.7	10.5

mineral and organic constituents of tall fescue (Table 7). Apparently the experiments were conducted on soil with good P fertility since forage P concentrations were affected only slightly by applied P in the first harvest and not at all in the aftermath. Concentrations of the other mineral constituents did not appear to be affected to an important degree by P fertilization. Among the organic constituents, soluble carbohydrates in first-cutting forage were increased by almost 50% by P fertilization. This change could improve palatability and digestibility.

B. Legumes

Much of the world's research on P fertilization of forage crops has been done with legumes. In those areas where economic analysis dictates minimum cash inputs into grassland agriculture, legumes are relied upon to supply most of the N requirements of the system. Low native P fertility and very low or nonexistent past applications have resulted in P being the most yield-limiting factor in such production systems. Thus, P has received tremendous research attention. In such situations response to the first or second increment of P is usually noted.

The opposite end of the spectrum of P supply is represented by work reported by Markus and Battle (1965). In a 9-year alfalfa experiment in New Jersey applied P did not affect total yield. They calculated that the soil supplied 19 kg of P/ha annually. Soil P was gradually depleted in those plots without applied P, while P accumulated with 50 kg/ha applied annually.

As is true with grasses, there are wide species and varietal differences in P response of legumes. Jones et al. (1970b) applied P for 21 cultivars of 5 species of annual clovers. Yield differences among different cultivars of the same species were almost as great as yield differences among species. Relative differences in yield were much greater than in P uptake. Snaydon and Bradshaw (1962) found that cultivated strains of white clover were more P responsive than native strains. Helyar and Anderson (1970) fertilized five forages with superphosphate at rates ranging from 112 to 3,300 kg/ha. They found that alfalfa and white clover responded to higher application rates while ryegrass, hardinggrass, and subterranean clover did not respond at all to the highest application rates. This emphasizes the need for plant breeders to conduct progeny tests at more than one fertility level.

While it was noted earlier that soil temperature influences the effectiveness of band seeding, Biddiscombe et al. (1969) also related soil temperature to competition among pasture species. They found that, when P is limiting, several grasses and forbs have higher early season growth rates than clovers, and thus appear to be more competitive than clover in the spring.

When relating fertilizer application rates to critical nutrient concentrations in plant tissue, factors such as growth stage and sampling position on the plant are of utmost importance, since valid comparisons cannot be made with tissue of different ages or from different positions on the plant.

Table 8—Concentrations of P in alfalfa and red clover at various growth stages (Reid et al., 1970).

Stage of growth	Narragansett alfalfa	Dollard red clover
	%	
Vegetative	0.33	0.43
Bud	0.30	0.35
Early bloom	0.31	0.32
Late bloom	0.25	0.32
Seed formation	0.24	0.25
Regrowth early bloom	0.31	0.29

Jones et al. (1972) studied the effects of age and time of recovery from clipping on P concentrations in subterranean clover. The critical P level of leaves declined from 0.52% in 48-day-old plants to 0.11% in 120-day-old plants when both were clipped only once. In multiple clipping situations critical concentrations were higher in regrowth clippings than in the first harvest. Phosphorus concentrations were lower in stems than in leaves.

Reid et al. (1970) analyzed well-fertilized alfalfa and red clover at several stages throughout the growth period and also found a gradual decrease in P concentration with advancing maturity (Table 8). In the vegetative stage red clover was somewhat higher in P, but no difference existed by seed formation time. This result shows that legumes as well as grasses must be harvested at a fairly immature stage if the P concentration is to be high enough to meet the needs of growing or lactating ruminants.

Chambliss et al. (1970) assayed alfalfa plants to find a sampling technique sensitive to treatment differences yet relatively insensitive to differences due to growth cycles. They found that analysis of midstem sections gave a high degree of uniformity among sampling dates within a given year and a fairly good reproducibility between years.

As mentioned above, legumes are often relied upon to supply most of the N in pasture mixtures which include grass. For this to be successful, legumes must be provided with a suitable environment, including the proper pH and an adequate supply of mineral nutrients. Wolfe and Lazenby (1973) grew mixtures of tall fescue, ryegrass, and hardinggrass with white clover on land that had never received any fertilizer (Fig. 6). During the first growing season only the clover responded to P, but in subsequent years both grass and clover growth was increased by each increment of P. The most logical explanation seems to be that improved P supply increased the growth and N fixation rate of white clover; the additional N was then responsible for more luxuriant grass growth in subsequent years.

The observation of Reid and Jung (1965) that P fertilization increases soluble carbohydrates in forage implies that P fertilization may improve palatability. Ozanne and Howes (1971) studied the effect of P fertilization on utilization of subterranean clover by sheep in a system where forage is accumulated during the moist growing season then grazed continuously during a subsequent summer dry period. Yields (Fig. 7) were increased by P applications as high as 40 kg/ha. Utilization of the accumulated forage by sheep as shown by the difference in October and March yields increased for

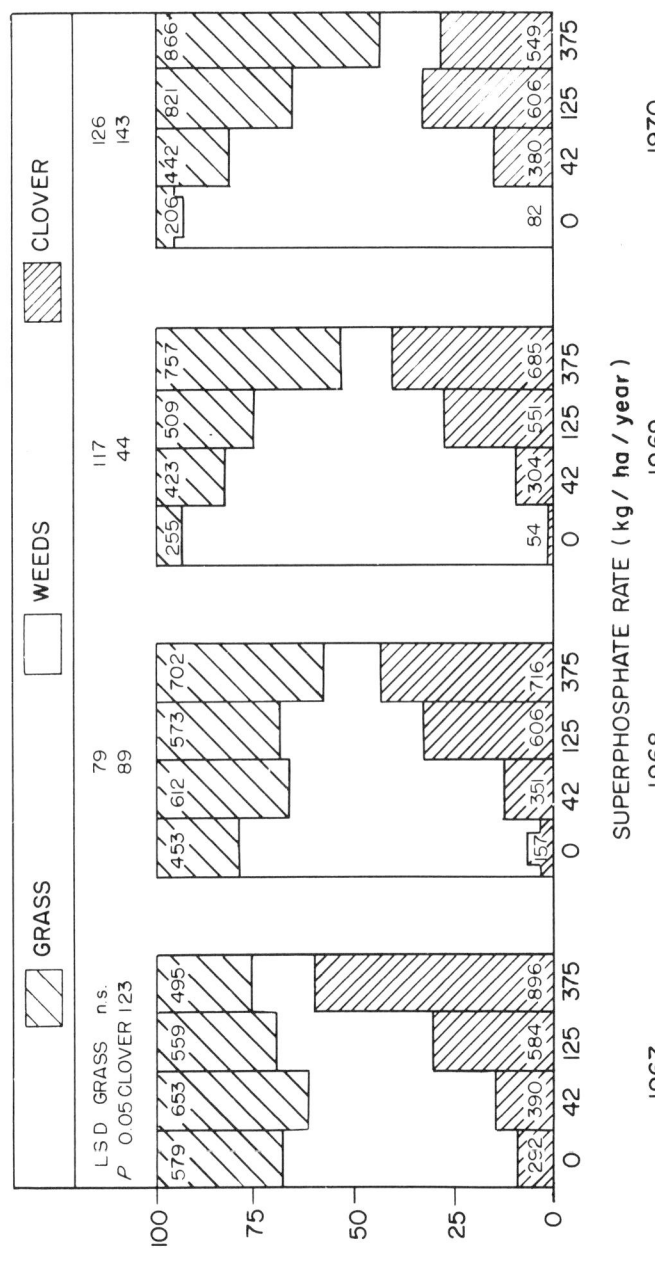

Fig. 6—The botanical composition of grass-white clover pasture as affected by annual applications of ordinary superphosphate (Wolfe & Lazenby, 1973).

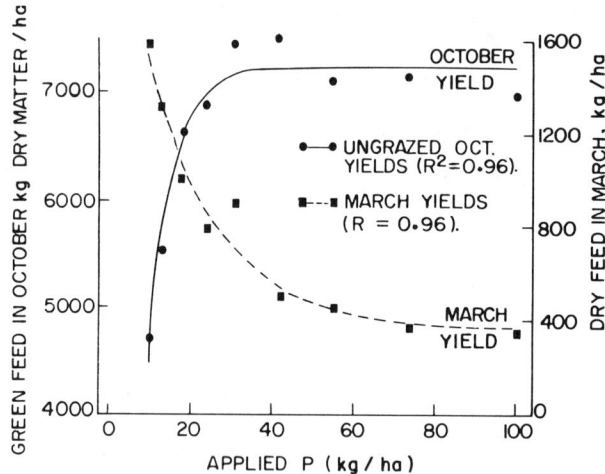

Fig. 7—Green pasture yields of subclover in October from ungrazed plots and dry feed residues from the same plots after grazing from October until March, as affected by applied P (Ozanne & Howes, 1971).

application rates up to about 80 kg/ha; this indicates that the applied P must have improved palatability.

IV. PHOSPHORUS NUTRITION OF WARM-SEASON GRASSES AND LEGUMES

The response to P of improved grasses and legumes adapted to the humid tropical and subtropical areas, including the southern United States, will be discussed in this section. These species may yield 45 metric tons/ha of dry forage under the long growing seasons and high rainfall of these regions (Vicente-Chandler et al., 1974). Such yield levels require intensive management practices, including careful selection of high-yielding species, heavy fertilization, favorable soil pH, weed control, and intensive utilization. These grasses may be established from stem cuttings, sprigs, or seeds, and are grown on a wide variety of soil types.

Soil fertility is a major factor affecting forage yields in the subtropics and tropics (Rotar & Plucknett, 1973). Fertilizer requirements are high because of heavy removal of nutrients in high yields of forage and the potential for loss of nutrients by leaching and soil fixation. Except where legumes are present, high rates of N are required to obtain maximum yields and high protein levels with rapidly growing genera such as *Pennisetum, Panicum, Digitaria, Cynodon,* etc.

Mosse (1973) found bahiagrass, molassesgrass, and the tropical legume centro to be responsive to inoculation with vesicular-arbuscular mycorrhiza. She indicated that roots infested with mycorrhizal hyphae are more proficient in absorbing P from very low P soil-solution concentrations than are noninfected roots.

A. Grasses

1. PHOSPHORUS REQUIREMENTS FOR ESTABLISHMENT

Establishment requirements for P have not been separated from production and maintenance requirements. However, evidence suggests that P levels for establishment on new, disturbed, or previously unfertilized land are critical, and that mature plants use P more effectively and are more efficient in obtaining P from deficient soils. Rotar and Plucknett (1973) recommended 200 kg/ha of a 12-5-10 grade N-P-K fertilizer at planting, followed by 30 kg/ha 2 to 3 months later. Younge and Moomaw (1960) found that an application of 672 kg of P/ha was needed for rapid grass establishment on bauxite strip-mine spoil 2 to 5 m below original soil surface compared to 336 kg of P/ha for previously unfertilized topsoil. Heavy initial applications of P to acid, highly weathered latosols may be necessary for rapid establishment of improved grass species.

2. YIELD RESPONSES OF HARVESTED FORAGE TO APPLIED P

Coastal bermudagrass responds to P fertilization at high levels of N and K inputs (Table 9). Jackson et al. (1959) reported that coastal bermudagrass responded to P fertilization only after the 4th year of cropping at 986 kg of N/ha on a Tifton loamy sand. The induced P deficiencies were corrected by applying sufficient P to raise the soil level to approximately 37 kg of available P/ha by the modified Truog method, and then applying a 4-1-2 grade ratio of complete fertilizer for all N rates from 224 to 896 kg/ha. Fisher and Caldwell (1959) reported that plant and soil analyses suggest that a 5-1-2 grade ratio should be applied to maintain high production levels. Response to P at relatively high levels of N fertilization has been shown by Welch et al. (1963) for coastal bermudagrass grown in the Southern Piedmont of Georgia and by Taliaferro et al. (1975) for midland bermudagrass grown in Oklahoma.

Coastal bermudagrass yields of 27.2 metric tons of dry matter/ha have been reported in south Georgia and in Texas. With a plant P concentration of 0.25% such a yield would remove 68 kg of P. This level of P removal re-

Table 9—Yield, P concentration and uptake, and recovery of applied P in coastal bermudagrass as affected by P fertilization (3-year avg.).†

P applied, kg/ha	Yield, metric tons/ha‡	P, %§	Uptake, kg/ha§	Recovery, %
0	13.2	0.18	23.1	--
24.7	14.7	0.22	30.5	30
49.3	15.1	0.23	33.4	21
98.6	16.4	0.24	38.7	16

† Together with 448-186 kg N-K/ha annually.
‡ Welch et al. (1963).
§ Adams et al. (1966).

quires high maintenance levels of P fertilization, or soil with high P-supplying capability.

Single annual applications of P are adequate as long as adequate levels of available P are maintained in the soil. Woodhouse (1969) and Jordan et al. (1966) reported that about 25 ppm of soluble P (dilute double-acid extractant) is adequate for coastal bermudagrass in North Carolina and Alabama (Fig. 8). Lunt et al. (1965) indicated that about 14 ppm of $NaHCO_3$-extractable P was adequate for common bermudagrass harvested frequently as turfgrass; Bruce and Bruce (1972) indicated that 17.5 ppm of P extractable with $0.01N\ H_2SO_4$ was necessary for high production of various tropical grasses and legumes. Field responses to P fertilization have not been documented in the United States for bahiagrass or dallisgrass. However, Rodulfo and Blue (1970) noted a bahiagrass response to 8 ppm of applied P in a greenhouse study using a low-P virgin soil.

The standard fertilizer recommendation for high grass production in Puerto Rico is 4.5 metric tons/ha of 15-2.2-8.3 annually (675-98-374 kg N-P-K/ha). Vicente-Chandler et al. (1974) concluded that tropical species such as guineagrass, napiergrass, and stargrass require fertilization with at least 73 kg of P/ha annually when cut every 40 to 60 days. They found that pangolagrass responded strongly to P fertilization on two soils, only slightly on four soils, and not at all on four additional soils of Puerto Rico. They reported that under Puerto Rican conditions, P need be applied only once

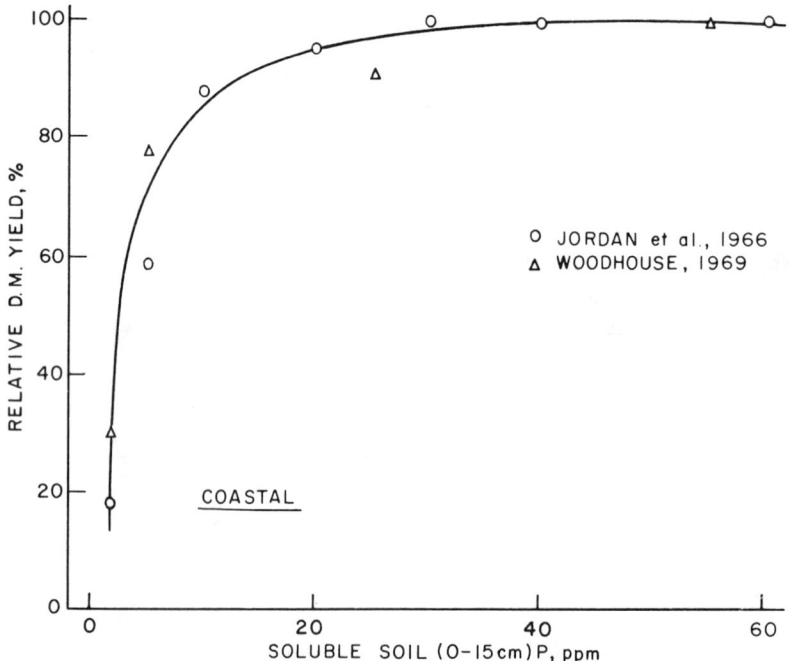

Fig. 8—Relative yield of coastal bermudagrass on loamy sand and sandy loam soils as affected by dilute double acid-extractable soil P (data of Jordan et al., 1966, and Woodhouse, 1969; reproduced from Wilkinson & Langdale, 1974).

yearly for the tropical grasses, which need heavier applications for soils having a high P-fixing capacity. Tropical grasses appear to have little tendency toward luxury consumption of P; concentrations of 0.17% P in these grasses cut every 60 days are considered adequate for optimum growth.

Phosphorus responses of kikuyugrass and pangolagrass in Hawaii were phenomenal. Check plots yielded 7,510 kg of dry matter/ha, while plots receiving 1,120 kg of P/ha initially yielded a 3-year average for both grasses of 20,200 kg/ha (Tamimi, 1972). The greatest response was noted with the first 560 kg of P applied; high rates of P were required on low P soils high in fixing capacity before a significant response was obtained.

Lucas and Blue (1973) found extremely low P concentrations in pangolagrass in a greenhouse study. However, they reported responses to 390 ppm of applied P for limed and 348 ppm for unlimed soil. Most of the pangolagrass growth was in stolons, of which the internodes contained about 0.04% P. Node-leaf tissue averaged about 0.17% and whole tops 0.10% P. Forage yields decreased with increase in lime rates regardless of applied P level. Lime appeared to have an unfavorable unexplained effect on root growth.

West and Prine (1974) concluded that P and K fertilization offers little hope for preventing winterkill of pangolagrass in the northern part of the pangola belt in Florida.

Concentrations of P in tops of tropical grasses decrease with age (Table 10), as with temperate grasses. Pangolagrass had the lowest, while molassesgrass and kikuyugrass had the highest P concentrations. Strickland (1973) found that P concentrations of bermudagrass, pangolagrass, and *Digitaria didac* did not differ although yields harvested were different.

3. CRITICAL P CONCENTRATIONS

Critical P concentrations for maximum growth (Table 11) decrease with maturity of the grass; within the same species, researchers disagree as to the critical value. The critical value for buffelgrass determined by Andrew and Robins (1971) was appreciably higher than that reported by Smith (1975), who attributed this to an effect of stage of growth. Smith (1975) also found that critical P concentrations of the youngest expanded leaves of *Setaria* declined with age. Critical P concentrations for pangolagrass may be lower than for many of the other grasses at similar stages of growth. Such data emphasize that diagnosis of P deficiency based only on analyses of tissue should be used with caution. Availability of soil P, seasonal conditions, stage of growth, past fertilization, local experience, and crop characteristics and behavior are also important in assessing the need for P fertilization.

The decline in P with stage of growth is commonly a dilution effect of a greater rate of dry matter accumulation than of P absorption and/or redistribution to younger growth. Such redistribution of P may result in established plants performing relatively better under P-deficient conditions than

Table 10—Concentrations of P in six tropical grasses after three growth periods (Gomide et al., 1969).

Tropical grass	Growth period, weeks		
	4	12	36
	%		
Molasses	0.31	0.20	0.12
Pangola	0.16	0.11	0.11
Napier	0.33	0.15	0.08
Kikuyu	0.24	0.21	0.14
Suwanee bermuda	0.22	0.15	0.11
Semipreverede	0.29	0.19	0.12
Mean	0.26	0.16	0.12

do seedlings. Christie (1975b) reported that buffelgrass was dependent on the external supply of P after the two-leaf stage of growth.

Ramkrishnan and Gupta (1973) reported that differential responses of three edaphic ecotypes of bermudagrass to P were correlated with the fer-

Table 11—Critical P concentrations in some subtropical and tropical grasses.

Grass identification	Age or stage of growth	Critical P concentration, %	Reference
Bermudagrass	4-5 weeks	0.20-0.26	Martin & Matocha, 1973
Midland and common bermudagrass	4-5 weeks	0.24-0.28	
Panicum Maximum, VAR Trichoglume, CV Petrie	3-4 leaf	0.55	Smith, 1975
	4-5 leaf	0.32	
	6-7 leaf	0.15	
	I.P.F.† (57 days)	0.20	Andrew & Robins, 1971
Kikuyugrass	I.P.F.† (45 days)	0.22	
Pearl millet	4-5 weeks	0.16-0.20	Martin & Matocha, 1973
Setaria spacelata, CV Nandi	4-leaf	0.46	Smith, 1975
	5-leaf	0.36	
	6-leaf	0.24	
	7-leaf	0.14	
Setaria anceps, CV Nandi	I.P.F. (57 days)	0.21	Andrew & Robins, 1971
Dallisgrass	I.P.F. (57 days)	0.25	
Pangolagrass	I.P.F. (45 days)	0.16	
	4-weeks	0.12-0.16	Martin & Matocha, 1973
Buffelgrass	I.P.F. (57 days)	0.25	Andrew & Robins, 1971
	(whole plant tops inflorescences just appearng)	0.16	Smith, 1975
	youngest expanded leaves sampled	0.17	
Buffelgrass	I.P.F. (39 days)	0.26	Christie, 1975a
	46 days	0.30	Christie & Moorby, 1975
Rhodesgrass, CV pioneer	I.P.F. (57 days)	0.22	Andrew & Robins, 1971
Molassesgrass	I.P.F. (57 days)	0.18	
Johnsongrass	boot stage	0.16-0.20	Martin & Matocha, 1973
Sorghum-sudangrass and sudangrass	4-5 weeks	0.14-0.20	

† I.P.F. means immediate preflowering stage of growth. Most analyses apply to whole plants unless otherwise specified.

tility status of the natural habitats of these ecotypes. It is likely that evolution of plants has favored those ecotypes with P requirements which could be satisfied by the natural fertility of the habitat soils. Asher and Loneragan (1967) reported great differences among eight annual pasture species in their ability to absorb P from low concentrations in nutrient culture. Christie and Moorby (1975) found that mulgagrass required a lower external P concentration for optimal growth than did mitchellgrass and buffalograss. Buffalograss yielded best at 30 ppm, mitchellgrass at 3 ppm, and mulgagrass at 0.3 ppm of P in the external solution. However, critical P concentrations were 0.30, 0.34, and 0.48% for these grasses, respectively. Buffalograss was the highest yielding and mulgagrass the lowest yielding. Plants with a low growth rate adapt to conditions of P deficiency more readily than those with a high growth rate.

4. PHOSPHORUS UPTAKE—DEMAND RELATIONSHIP

The capacity of the warm season grasses to sustain high growth rates under favorable conditions implies a high daily nutrient demand. Growth rates of coastal bermudagrass in the Southern Piedmont of Georgia have averaged 165 kg/ha per day during the period of most favorable growth (Table 12). Yields reported by Fisher and Caldwell (1959) of 27.2 metric tons/ha over a growing season of 205 to 260 days in Texas averaged a daily growth rate of about 105 kg/ha. With a critical P concentration of 0.22% P, the annual P uptake in harvested tops would be about 60 kg/ha. Daily P uptake of coastal bermudagrass having a growth rate of 155 kg/ha and a critical P concentration of 0.22% would be 0.34 kg/ha, or 9.5 kg/ha for a 4-week interval. This does not include P fixed in the soil nor that used to satisfy the P requirements of the other parts of the plant.

Measured P uptake during the season is shown in Table 13 for coastal

Table 12—Effect of growth interval, season, and irrigation on the growth rate of coastal bermudagrass in the Southern Piedmont, Georgia (avg. 1964-1967). (Unpublished data from Southern Piedmont Conservation Research Center.)

Interval	Growth Period†	Not irrigated	Irrigated
weeks		kg/ha per day	
4	1	36	50
	2	84	120
	3	138	155
	4	110	118
	5	34	64
	6	36	46
6	1	59	110
	2	137	165
	3	118	135
	4	33	58

† Growth periods start from 19 April-3 May and end 4 October-17 October over a 4-year period with annual season being 168 days each year.

Table 13—Seasonal P uptake of coastal bermudagrass as affected by rate of P-K fertilization (1955-1956 avg.).†

P-K fertilization	Approximate harvest date				Total
kg/ha	1 June	5 July	10 August	15 October	
			kg P/ha		
0	5.0	9.8	7.7	3.1	25.5
49– 93	7.5	13.3	10.2	4.4	35.4
99–186	10.1	14.5	11.0	5.7	41.2

† Data of W. E. Adams (now deceased), Southern Piedmont Conservation Research Center, Watkinsville, Ga.

bermudagrass growing in the Southern Piedmont. Such levels of P uptake require large maintenance applications. Phosphorus fertilization increased P uptake relatively more at the beginning and end of the growing season, but over 66% of the total uptake occurred during midseason when growth rate was highest. With an estimated growing season for coastal bermudagrass of 168 days, the daily demand for P was about 0.24 kg/ha. In a similar situation, Knauer (1973) found the rate of daily uptake was closely related to rate of increase in dry matter production.

During the early stages of growth, uptake of P proceeds more rapidly than does dry matter accumulation (Fig. 9). At a high N/K fertilization level, up to 77% of the P uptake in 6 weeks had taken place by the 4th week but only 61% of the growth. This relationship suggests that redistribution of P within the plant may, at times, substitute for uptake in meeting the P requirements for growth. For this reason, P uptake by forage harvested every 4 weeks may be relatively more than that harvested every 6 weeks. Demand on the soil supply of P probably is most severe during early stages of regrowth of perennial forages, and probably declines with increasing age because of this intraplant cycling of P.

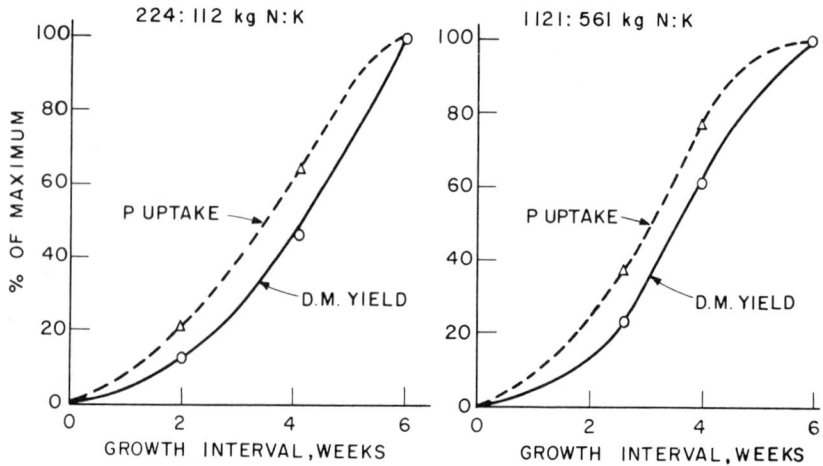

Fig. 9—Relative growth rate and total P uptake of coastal bermudagrass as affected by length of growth interval and N/K level (unpublished data, S. R. Wilkinson).

B. Tropical and Subtropical Legumes

These legumes are considered nutritious for ruminants, and under proper management are capable of producing excellent animal gains per hectare (Rotar & Plucknett, 1973). In general, levels of dry matter production are similar to those of tropical grasses receiving low to moderate levels of n fertilization.

1. PHOSPHORUS REQUIREMENTS FOR ESTABLISHMENT

Phosphorus deficiency is a major factor limiting the establishment of legumes in many acid, highly weathered, previously unfertilized soils (Plucknett, 1970). Keya and Kalangi (1973) found that topdressing with superphosphate was essential to establishment of silverleaf desmodium by oversowing in uncultivated grasslands of western Kenya. Werner and DeMattos (1972) found that P deficiency was the most limiting factor for development and N fixation of centro on a Red Latosol in the state of Sao Paulo, Brazil. For many soils within the region of adaptation of legumes from the *Stylosanthes, Centrosema,* and *Desmodium* genera, P fertilization at the time of preparation of the seedbed is necessary. If rainfall and management practices allow high levels of dry matter production, then initial P fertilization at relatively high rates may be justified. The cost of fertilization for establishment should be treated as a capital investment ratherthan as an annual production cost. Plucknett (1970) suggested this concept has merit for some Hawaiian soils.

2. YIELD RESPONSE TO APPLIED P

Fisher and Campbell (1972) found that Townsville stylo growing on a P-deficient red earth soil in western Australia responded to initial applications of up to 750 kg/ha of superphosphate (60 kg P/ha). First-year yields were 1,750 kg/ha without superphosphate and 7,420 kg/ha with 500 kg of superphosphate. Although there was a fairly high residual value, additional annual applications of 250 kg of superphosphate/ha provided a growth response. Concentrations of P of 0.12% or greater were associated with the higher yields.

Stylo pastures are prone to invasion by native grasses. Woods and Dance (1970) suggested that the grasses prevent the stylo from fully responding to superphosphate. Pure stylo pastures are desirable for maximum animal performance and maximum response of the legume to superphosphate.

Shaw et al. (1966) concluded that Townsville stylo was adapted to low fertility conditions, but that it also was responsive to higher levels of fertility. Increasing P applications increased both yields and N concentrations of stylo (Table 14).

Bryan and Andrew (1971) found that stylo and lotononis were more efficient in using phosphate rock than were silverleaf desmodium and phasey

Table 14—Response of *Stylosanthes humilis* to P fertilization, Australia (Shaw et al., 1966).

Rate of OSP		Yield of dry matter		Chemical composition, March 1962		
First year	Second year	19 Apr. 1961	23 Jan. 1962	N	P	K
		kg/ha			%	
None		1,450	2,750	2.36	0.06	0.75
112	56	1,730	4,870	2.73	0.09	0.80
224	112	2,040	5,840	3.20	0.16	0.89
448	224	1,920	6,500	3.28	0.21	0.85

bean. Stylo and lotononis appear to require relatively small amounts of P for maximum dry matter production while desmodium and phaseolus may require much larger amounts.

Dradu (1974) obtained large P responses with desmodium on Buganda loam soil and Kyebe red loam soil in Uganda. These soils were deficient in P, S, Mo, and Cu, with P being the major deficiency. Adding these nutrients increased yields from 5.9 to 8.9 g of dry matter/pot.

Andrew (1966) found that P absorption rate of excised roots of Townsville stylo much greater than that of phasey bean, silverleaf desmodium, alfalfa, and barley. These legumes may have evolved in soils lower in fertility, particularly P, and may be more efficient in uptake of P and Ca. Tropical legumes, generally, may have a lower optimum soil pH than the temperate legumes.

3. CRITICAL P CONCENTRATIONS

Published critical P concentrations for these legumes are presented in Table 15. Most of these values were derived from greenhouse or laboratory studies, with some field corroboration. There is considerable variation in the critical P concentrations, some of which is related to differences in stages of growth, or physiological maturity. Concentrations range from a low of 0.12% for *Stylosanthes Guyanensis* to values as high as 0.30% for *Desmodium* and siratro.

Robinson and Jones (1972) reported that Townsville stylo transports appreciable P from leaves to inflorescences as flowering commences. They found a rapid decline in P concentrations both before and after flowering. This may have a significant impact on the actual critical P concentration. Adequately fertilized Townsville stylo contained only 0.02% P in the mature leaf, while seeds contained 0.36% P. Bruce (1974) indicated that rainfall in the period prior to sampling, as well as the plant part sampled and sampling intensity, may affect critical P percentage.

Jones (1974) found large differences in growth and P uptake among groups of stylo accessions with progressive increase in P supply. Differences in growth and P uptake among accessions were small under conditions of severe P deficiency. However, accessions of stylo appeared to vary in ability to produce dry matter at a given P concentration in their tops. At moderately high levels of applied P (96 to 192 kg/ha equivalent), stylo accessions

Table 15—Critical P concentrations of some tropical legumes.

Species	Growth stage or age	Critical P concentration, %	Reference
Phasey bean, cv. Murray	I.P.F.†	0.20	Andrew & Robins, 1969a
Siratro	I.P.F.	0.24	
Siratro	40 days	0.30	White, 1972
P. Phaseoloides	65 days	0.23	Falade, 1973
Townsville stylo	I.P.F.	0.17	Andrew & Robins, 1969a
Townsville stylo	41 days	0.24	White, 1972
Stylosanthes gracilis	65 days	0.27	Falade, 1973
S. Guyanensis, cv. Schofield	25 cm tips	0.16	Bruce, 1974
	whole plants	0.12	
Stylosanthes (30 accessions)		0.20–0.25	Jones, 1974
Centro	I.P.F.	0.16	Andrew & Robins, 1969a
Centro	65 days	0.29	Falade, 1973
C. plumeris	65 days	0.28	
Greenleaf desmodium	I.P.F.	0.22	Andrew & Robins, 1969a
Greenleaf desmodium	42 days	0.30	White, 1972
Silverleaf desmodium	I.P.F.	0.23	Andrew & Robins, 1969a
Glycne wightii, cv. cooper	I.P.F.	0.23	
Lotononis	I.P.F.	0.17	
Vigno luteola	I.P.F.	0.25	
Alfalfa cv. Hunter broadleaf‡	I.P.F.	0.24	

† I.P.F. means immediate preflowering stage of growth (47–70 days old).
‡ Not considered a tropical legume, but included for comparative purposes.

in several groups developed foliar symptoms resembling P toxicity, had depressed yields of dry matter, and accumulated high P concentrations in their tops (in excess of 1.0%). In the groups of accessions showing only slight declines in relative yields with highest P applications, P concentrations rarely exceed 0.6%. Relative yields associated with the high P concentrations were as low as 30% of their maximum yields. The visual symptoms reported varied from marginal necrosis on cotyledons during the first 10 days to plant death. In plants mildly affected, the first three or four leaves developed irregular chlorosis, and subsequent necrosis, but younger leaves were affected less and plants survived.

These symptoms were similar to those found for greenleaf desmodium by White (1972), who reported similar relative responses to applied P for Townsville stylo and siratro. Plant P concentrations in *Desmodium* where yield depressions occurred were 0.86, 0.66, 0.63, and 0.52% for harvests made 21, 28, 35, and 42 days from sowing, respectively. The P concentrations associated with yield reductions decreased with increasing plant age. Mean P absorption and relative plant growth rates averaged over all P levels fell in the order greenleaf desmodium > stylo > siratro. The efficiency of P utilization fell in the order siratro > stylo > desmodium.

4. EFFECTS OF P ON N FIXATION AND STRESS TOLERANCE

Several workers have found that P fertilization increased N concentration in the tropical legumes (Andrew & Robins, 1969b; Dradu, 1974; Shaw et al., 1966), while others found no change (Falade, 1973). The increase in N concentrations in stylo with P fertilization reported by Shaw et al. (1966) was greatly enhanced by the addition of S. Total N fixed also increased as P fertilization increased yield. Dradu (1974) found that the application of 625 kg/ha of single superphosphate increased desmodium dry matter yield and N uptake by 75 and 99% on Buganda loam and by 198 and 372% on Kyebe red loam (Uganda soils), respectively. He also reported that omission of P on these soils depressed nodulation, N yield in nodules, P concentrations in tops, seedling vigor, and seedling growth.

Gates (1974) found that P fertilization of N- and P-deficient soil decreased the time interval for initiation of first nodules, and enhanced nodule development by increasing the number, volume, dry weight, and nodule growth rate on seedlings of stylo. He presented evidence that greater nodule mass and more effective symbiosis were obtained by improved P supply.

Effects of P on concentration of other nutrients in plants, such as K, Ca, and Mg, are complicated by the addition of one cation or another with the P source. Andrew and Robins (1969a) and Falade (1973) reported that cation concentration is affected by P source, soil and its relative proportion of cations, growth dilution, and suitability of the growth environment.

Gates et al. (1973) in a study of the possible relationships between cold stress, age, and P nutrition in lotononis found that its productivity and N-fixation capacity were favored by high levels of P fertilization. This effect of P nutrition was most important in mature plants. Increases in N concentrations with P fertilization seemed to be enhanced under cold stress.

V. PHOSPHORUS NUTRITION OF SEMIARID GRASSLANDS

Water availability is the most limiting factor for forage production in approximately one-half of the U.S. grasslands and in much of the rest of the world. In many semiarid areas fertilization increases production and water-use efficiency. Greater use of N and P fertilizers may be expected in reclamation and revegetation of depleted rangeland in the Great Plains of the United States (Taliaferro et al., 1975). The role of P in management of large areas of grassland needs to be carefully evaluated in response to pressures for more efficient use of this resource (Lorenz & Rogler, 1973).

While native grasses can be high in quality, much of the forage actually produced by semiarid grasslands is of low nutritional value, particularly with respect to P. Phosphorus concentrations in several legume and grass species grown under irrigated and dryland conditions at approximately 300 locations on the Columbia Plateau (U.S.) were determined by Boawn and Allmaras (1974). They found that P concentration of grasses ranged from 0.15 to 0.30% during immature to full-bloom stages but dropped to 0.03 to 0.09% during dormancy.

A major part of the grassland in semiarid regions of the United States is still populated by native species with suites of perennial grasses and forbs that are adapted to localized moisture and temperature conditions and to specific levels of biologically active P and other mineral nutrients. Nitrogen levels reflect the proportions of native leguminous species which are more sensitive than grasses to P supply. This probably accounts for the significant correlation between total N and P in virgin sods of the Great Plains (U.S.) found by Haas et al. (1961). Thus, the long-term effects of P on productivity of semiarid grasslands are probably larger through indirect effects on N fixation than through direct yield responses. On Australian grasslands with generally low soil P levels, the practice of P fertilization coupled with introduction of legumes has been widely accepted, with superphosphate accounting for 96% of the fertilizer applied (Williams & Andrew, 1970).

Rangeland fertilization studies in the United States have been well covered in reviews by Vallentine (1971), Heady (1975), and Stoddard et al. (1975). Phosphorus studies in Canadian grasslands are included in Sadler and Stewart's review (1974) of residual fertilizer P. These reviewers found that only marginal initial or residual responses to P applied along have been observed even in soils of low plant-available P. Next to moisture availability, N is usually the most critical factor limiting grassland responses to P.

Lorenz and Rogler (1972) investigated the response of mixed prairie vegetation to annual applications of 0, 45, 90, and 180 kg of N/ha and 0, 20, and 40 kg of P/ha over an 8-year period in North Dakota. Response to P was often not significant during the first 3 years, however, over the 8-year period each increment of P produced a significant yield increase. In the absence of applied N, response to P was small, but as N level increased, P response increased.

Phosphorus fertilization by itself provided very little yield increase at selected sites of native grassland in western Canada (Kilcher et al., 1965). The combination of N and P at 112 and 29 kg/ha resulted in yield increases slightly greater than those from N alone at 8 of the 10 sites over a 3-year period. Similar results were obtained by Black (1968) in southeastern Montana. Nitrogen and N plus P fertilization consistently increased forage yields of native and crested wheatgrass, but there were no yield responses to P applied alone. Plant P contents were increased substantially for at least 4 years following P application, regardless of N treatment, and P fertilization stimulated uptake of soil N by both grasses by about 20%. The stimulation of N uptake was attributed to increased root weights in the upper 60 cm of soil and to possible increased N mineralization in this severely P-deficient soil. Cosper et al. (1967) and Smika et al. (1961, 1965) reported that the application of P resulted in more efficient N utilization by grasses.

Taliaferro et al. (1975) compared the performance of midland bermudagrass, Morps weeping lovegrass, and Plains bluestem, which are all warm-season perennial grasses widely used in revegetation of depleted rangelands in the Southern Plains (U.S.). The introduced grasses and native range all responded to 90 kg of P/ha, with the response being greatest at the highest N levels. Average forage yields of midland bermudagrass receiving 90 kg of P/ha were 31% higher than where no P was applied.

Bowns (1972) investigated low-level fertilization of high-elevation ranges in southwestern Utah, and found that the most effective rates were 67 kg/ha each of N and P. Residual yield increases from a single application carried over for two growing seasons, while increases in P concentration were noted for three seasons. All P treatments resulted in a considerable increase in plant P. Yield responses to N and P applications in factorial combinations on six Utah range and meadow sites were subjected to economic analysis by Workman and Quigley (1974). Nitrogen applications were of borderline profitability, but there was no yield response to P.

Fertilizer N and P are immobilized by the soil-plant system to such an extent in semiarid grasslands that low application rates often fail to supply these nutrients in adequate amounts for plant growth. Thus nutrient availability has been underestimated as a growth-limiting factor in rangelands. Recent studies have utilized high application rates of N and P to establish maximum production levels for rangeland ecosystems. Johnston et al. (1967) increased yields of mixed prairie vegetation in Alberta, Canada, 300% with graduated rates up to 185 and 155 kg/ha of N and P, respectively, and 900% with rates up to 1,090 and 820 kg/ha of N and P. Increased yields were accompanied by major changes in botanical composition at the higher application rates. In a subsequent study, Johnston et al. (1968) examined the effects of five fertilizer application rates up to 1,120 kg of N/ha and 860 kg of P/ha on production of native fescue range and range seeded to bromegrass and creeping red fescue. Yield responses to NP applications were large, with greater responses in the second and third year than in the first year. The seeded range was less productive than the native range on unfertilized plots, but responded better to fertilization, with yields two to four times the native range at higher levels of N and NP applications.

Baldwin et al. (1974) reported 400% responses in herbage production to heavy applications of 27-12-0 fertilizer on native rangeland in Oregon. Substantial responses to a single application continued through the first 4 years of the study. Fertilization extended the green-forage season by 6 weeks or more each summer and increased the utilization of mature herbage by livestock.

Black and Wight (1972) applied high levels of N and P to examine production limits in the absence of nutrient limitations in southeastern Montana. They braodcasted factorial combinations of ammonium nitrate at 0, 112, 336, and 1,008 kg of N/ha and concentrated superphosphate at rates of 0, 112, and 224 kg of P/ha on a native range site. In 2 years the addition of a high rate of N and P increased total forage production by 330%, total crude protein by 67%, and plant N and P concentrations by about 200%. When compared within the same level of added N, P increased total crude protein production about 30%. Percentage recovery of N fertilizers was substantially increased by P fertilization. Wight and Black (1972) compared the productivity of various cropping systems with two levels of management. Native range and Russian wildrye compared favorably with fallow or continuous culture dryland wheat with above-average management. Above-average management for native range consisted of a single application of 1,008 kg of N/ha and 224 kg of P/ha. Water-use efficiencies were increased from 49 kg of dry matter/ha·cm of precipitation on the unfertilized range to 148 kg/ha·cm on the high NP treatment plots.

Halm found that the application of 20 kg of P/ha to a very P-deficient clay soil in Saskatchewan produced no initial or residual yield response for native grassland, but resulted in a 25% increase in grain yield for wheat grown on a recently cultivated adjacent site (B. J. Halm. 1972. The phosphorus cycle in a grassland ecosystem. Ph.D. Thesis. Univ. of Saskatchewan, Saskatoon, Canada). When 168 kg of N/ha was applied with the P, the yield of above-ground material in the grassland plots doubled in the year of application and in the subsequent year. The increases were greater than the sum obtained by separate N and P applications.

Phosphorus cycling in this grassland was examined in further studies by Halm et al. (1972). They obtained detailed information on P concentrations in above- and below-ground plant material, microbial P, and soil P fractions as influenced by season to construct a conceptual diagram of P flow in a grassland ecosystem. Stewart et al. (1973) and Cole et al. (1977) used this information as a basis for development of a simulation model for P cycling in semiarid grasslands. The model allowed integration of the effects of moisture, soil properties, plant phenology, and microbial turnover in a way not previously possible. Application of the model to two grassland sites and the comparison of P flows with actual field measurements pinpointed gaps in information on key processes, such as mineralization of organic P, which have not been fully researched.

High mountain meadows accunt for significant cattle production and have special management problems involving short seasons, cool temperatures, and organic matter accumulation, all contributing to slow nutrient cycling rates. While most of the fertility research in mountain meadows has involved N, recent studies have indicated that responses to P are also significant. Ludwick and Rumburg (1976) investigated the effects on grass production of topdressing N and P on a mountain meadow in Colorado with a mixed stand of bromegrass, timothy, orchardgrass, bluegrass, and red clover. Hay yields the first year were significantly incrased from 2.5 metric tons/ha with no fertilizer to 10.8 metric tons/ha with 358 and 59 kg/ha of N and P, respectively. Yields increased with increasing rates of both nutrients, and yields from fall-applied NP combinations were significantly greater than yields from similar spring-applied treatments. Residual P effects were significant through the fourth year after application, although yields declined each year. Phosphorus reapplied the fourth year significantly increased both yield and P uptake over all residual levels.

Responses to P in flood meadows of eastern Oregon were attributed to increased establishment and maintenance of white-tip clover in a mixed clover-rush-sedge hay (Cooper & Hunter, 1959). Phosphorus fertilization increased crude protein indirectly as a result of an increased proportion of clover in the hay. Similar indirect effects of P fertilization through increase in proportion of legume species have been reported for annual grasslands (Jones & Evans, 1960; Ofer & Seligman, 1969; Luebs et al., 1971).

VI. PHOSPHORUS CYCLING IN GRASSLANDS

An understanding of the quantities and rates of movement of P among various pools in the ecosystem is essential to the establishment and maintenance of biological and economic productivity of grasslands, effective use of P fertilizers, and of other management inputs. The P cycle describes these transfers, accumulations, and losses of P among major components within ecosystems.

Intensive pasture ecosystems are young, productive growth systems, and people manage them to obtain economic productivity and stability. Rangeland ecosystems are governed by similar principles particularly in maintaining biological and economic stability. The ecosystem concept permits all systems to be evaluated for productivity and stability by examining interrelationships among components of the ecosystem and the environment. Forages grown and mechanically harvested have P cycling characteristics similar to row crop agricultural systems, with allowances made for the perennial nature of many forage crops. Removal of P by harvesting creates most of the P requirement of such systems, whereas in grazed situations the return of P through plant and animal residues results in a nearly closed P cycle. Grazing introduces additional variables such as selectivity, treading, and partial defoliation.

The total ecosystem pool of P includes "unavailable" soil P, "available" soil P, residue P, plant P, and livestock P. Atmospheric P is relatively minor, with very little P arriving or leaving the ecosystem by deposition or volatilization processes. In terms of relative sizes of subpools, the soil fraction is by far the largest. However, the unavailable soil P, whether immobilized organic P or chemically fixed inorganic P, is not part of the active cycle. The desired objective of pasture management is to keep the transfer rate of P from residues and unavailable soil P to the available soil P pool high enough to supply adequate P to maintain a sufficient plant growth rate for good livestock grazing. The return of P in plant residues and animal excreta completes the cycle. Losses of P which occur with moving water either as runoff or percolate will be small but proportional to the loss of water from the ecosystem. There is no evidence that P is lost in gaseous form, even under anaerobic conditions with an abundance of decomposable organic matter (Alexander, 1977). Phosphorus cycling, therefore, is accomplished through the plant and animal components of the ecosystems. The need for P fertilization is determined by the turnover rate of P within the ecosystem, as well as by the total P removed.

Factors affecting P availability in the soil for plant uptake, requirement for and utilization of P in the plant, and chemistry of P important to understanding the P cycle are discussed in detail in other chapters. Therefore, we shall discuss only features of the P cycle which are unique to grasslands. These include the return of P in plant residues and animal excreta and the availability of the P in these residues for subsequent plant uptake. We shall also review P responses under grazed conditions.

A. Return of P Through Plant and Animal Residues

The proportion of P recycled by animals depends primarily on the amount of herbage utilized by grazing ruminants. Animals rarely graze more than 60% of the herbage (Minderhoud et al., 1975). Carter and Day (1970) found that herbage utilization from a ryegrass-subterranean clover pasture was about 53% at a stocking rate of 15 sheep/ha and 77% at a stocking rate of 25 sheep/ha. Herbage P not consumed is recycled through plant residues, and that consumed is recycled through animal excreta. Table 16 shows a simplistic relationship between percentage of herbage utilized and the probable proportions of P recycled via plant or animal residues. Greater herbage utilization results from high stocking rates, and more P is recycled via feces but less through plant residues. Phosphorus is excreted mainly in the feces with only traces occurring in the urine (Barrow & Lambourne, 1962). The amount of P removed in animal products is generally low. Wilkinson and Lowrey (1973) reported contents of P in animals and animal products of 6.71, 6.76, 4.93, 4.53, 0.31, and 1.03 kg/1,000 kg for calves, steers, lambs, sheep, unwashed wool, and cows' milk, respectively.

The important question now becomes: Is P recycled more effectively from animal excreta or in plant residues? Major considerations include the availability of P in the plant residues and excreta, as well as the physical return of P in space and time. Return from plant residues (roots, litter, etc.) is quite uniform with no net redistribution of P within the grazed area. On the other hand, P return in feces is subject to uneven return in time and space because of the mobility of the animals. Factors affecting the time and space distribution of excreta include stocking rate, camping tendencies, grazing patterns, type of animal (species, breed, and sex), and area affected by each fecal deposit. Wilkinson and Lowrey (1973) found that cows defecate about 12 times/day with an average area covered by each excretion of about 0.09 m², or a daily area coverage of 1.1 m². They estimated that the area covered with excreta from a cow-calf pair was 4 to 12% of the grazing area for 1 year, depending on various assumptions about overlap of excreta (stocking rate of 2.47 cow-calf pairs/ha). During and Weeda (1973) indicated that the

Table 16—Hypothetical effects of herbage utilization on recycling pathway of P from animal excreta or plant residues.†

Degree of utilization	Plant uptake	Animal product P	Return as plant material	Return as feces
%	kg/ha			
25	30	0.88	22.5	6.61
50	30	1.77	15.0	13.23
75	30	2.66	7.5	19.84

† Assumptions: Net primary productivity is 12,000 kg of dry herbage/ha containing 0.25% P. It requires 9,000 kg/ha to be consumed to produce a 400-kg calf containing 2.66 kg of P. Therefore, at a herbage utilization efficiency of 25%, 3 ha will be required; at a utilization efficiency of 75% only 1 ha is required.

area on which dung influences P uptake may be as much as five times the area physically covered by it. They commented that, at a stocking rate of 4 dry cattle/ha and an overlap similar to that characterized by Peterson et al. (1956), 40% of the pasture area would be affected by dung the first year and 75% affected in 3.5 years. They suggested that P uptake is likely to be affected by dung spots over half the grazing area at any one time with a heavy stocking rate of 3.4 cows/ha. MacDiarmid and Watkins (1972) reported that dairy cows defecated about 14 times/day with each defecation covering an area of 0.07 m^2, or an approximate coverage of 1.0 m^2/day. They grazed 40 Jersey cows on 0.73 ha over a 2-day period, and found 0.7% of the paddock area covered with dung. Assuming that the paddock was completely grazed during the 2 days, P in the grass consumed from the entire paddock was excreted on less than 1% of the original surface. These authors assumed that three times the area covered by dung is affected, and that the effect lasts through several grazings. They calculated that in 2.5 years an area equal to that of the paddock will be affected, and in 10 years all of the paddock would have been covered by dung. In low-intensity grazing systems, dung effects are much diluted and may be noticeable primarily around shade and water areas.

Because of the lateral spread of roots, stolons, and rhizomes, washing action of rainfall, and mechanical disturbance by grazing animals, it is reasonable to suppose that P nutrition may be affected in an area two to five times the original area covered by dung.

Sheep may have greater camping tendencies than cattle (Hilder, 1966). The increased number of sheep required to give grazing pressure equivalent to cattle should result in an increased potential for spreading and mixing of feces with soil by hoof action. Sheep may also graze the pasture more completely and more fully utilize herbage than cattle. Hilder (1966) found that 83% of the feces was deposited in 50% of the area of a 37.2-ha rotationally grazed paddock. Gillingham and During (1973) mapped a 12.5-ha, long-established, moderately steep hill pasture grazed by sheep and occasionally beef cattle into five strata according to occurrence, color, and vigor of pasture species and distribution of excreta. They concluded that large quantities of P were transferred into camp areas, with much of the transfer occurring from the least productive areas of the paddock.

The available evidence suggests considerable redistribution of P may be associated with P return in feces, but it does not permit a generalized quantitative assessment of this redistribution.

1. AVAILABILITY OF P IN PLANT RESIDUES

Phosphorus losses from live plant tissues are probably small. However, 69 to 80% of the total P may be leached from dormant or dead vegetation (Harley et al., 1951; Jones : Bromfield, 1969; Kline, 1969; Timmons et al., 1970). Microbial activity in such tissues substantially reduces the amount of water-soluble P. The intensity and duration of rain, as well as the interval between tissue dormancy or senescence and the first precipitation, affect the amounts of P returned to soil, or lost to runoff.

When herbage is ungrazed and the plant matures, P is redistributed to new shoots or to roots for growth and metabolism (Biddulph et al., 1958; Greenway & Gunn, 1966; Bouma, 1967). Ozanne and Howes (1971) considered intraplant cycling to be the cause for a lower P requirement for ungrazed than for grazed subterranean clover in Australia. Floate (1970d) found that monthly cut red fescue-colonial bentgrass and *Nardus stricta* pastures removed 1.78 and 1.54 times as much P as did these same pasture species cut annually. Since the yield of the annually cut herbage was greater, part of this difference in P yield represented intraplant redistribution. In perennial pastures with low intensity of herbage utilization, intraplant recycling may be an important quantitative factor in the P cycle.

2. AVAILABILITY OF P FROM FECES

Almost all P is excreted in the feces (about 0.06 g of organic P excreted per 100 g of feed eaten); the remainder is excreted as inorganic P. Barrow (1975) found evidence that P in sheep feces is present as dicalcium phosphate.

Floate (1970a) found that inorganic P increased by 62 to 78% as annually and monthly cut *Nardus* and fescue-bentgrass herbage was converted to feces. Bromfield and Jones (1970) found that up to 80% of the organic P in green feed was mineralized in the animal; the amount decreased with forage maturity. Inorganic P in sheep dung was reported to be soluble in dilute acid, but not very water soluble. Inorganic P from plant materials was highly water soluble (Bromfield, 1961; Bromfield & Jones, 1970). Sheep feces retained 40% of the initial total P after 2 years of exposure to weathering and the leaching action of more than 100 cm of rain; 90% of this residual P was organic (Bromfield & Jones, 1970).

Whether or not net mineralization of P occurs from residues of plant or animal origin depends on spatial relationships to the soil, initial organic P, total P, and water contents of the residues, and on temperature and time.

Floate (1970a) reported that from 3 to 30% of the original organic P in sheep feces and 2 to 15% of organic P in plant materials was mineralized. Net amounts mineralized were positively correlated with the initial levels of organic P. He suggested that positive net mineralization is unlikely to occur when organic P contents are less than 0.09% during the first few weeks of decomposition. Organic P in both plant and animal residues appears to be more of a "sink" than a source of P for cycling (Floate, 1970a, b, c, d; Bromfield & Jones, 1970). Decomposition of residues which contain less than 0.2% total P probably results in little net mineralization of P (Fuller et al., 1956; Swaby, 1962).

Floate (1970c) suggested that variations in temperature and moisture are more important than continuous moisture in their effects on mineralization. Heating or drying the soil and then rewetting often results in a flush of organic matter decomposition several times greater than the rate associated with continuously moist conditions (Alexander, 1977). This may be important in the release of P from surface-deposited residues, particularly in areas of erratic rainfall, or in areas with regular dry or rainy periods.

Organic P mineralized from feces was 10% at 30°C, 2.5% at 10°C, and −12% at 5°C (net immobilization of P), while mineralization of organic P from *Nardus* and fescue-bentgrass herbage was 0.2% of the original P at 30°C and −41% at 5°C (Floate, 1970b). Increased mineralization at temperatures greater than 30°C indicates that the thermophillic range is more favorable than the mesophillic range for mineralization of P (Alexander, 1977).

Rumen bacteria operate at temperatures of 39 to 41°C, and have P contents of the order of 5% on a dry-weight basis. Rumen fluids normally contain about 10 mmoles of P per liter. The higher temperatures in the rumen, plus higher P contents in rumen bacteria than in soil bacteria (1.5 to 2.5% P), suggest that, when bacteria move from the rumen to the lower intestinal tract, considerable mineralization of P may occur. Ruminants thus may enhance dissolution and mineralization of P. The process of digestion will also result in narrowing C/P ratios because of the greater loss of C than of P during herbage digestion by ruminants.

The availability of P in feces mixed with the soil is high (Bromfield, 1961; Bornemissza & Williams, 1970). Dung beetles, earthworms, and other soil fauna increase the availability of P in feces by burying and mixing of soil and feces. Phosphorus recoveries by millet from cattle feces applied to the soil surface were 3% without dung beetles, 17% with confined dung beetles, and 26% for feces mixed with the soil (Bornemissza & Williams, 1970). Similar effects could be expected from earthworm action. Soil fauna enhance nutrient cycling by providing foci of high nutrient concentrations which increase root growth and nutrient uptake. They also may serve to inoculate soil with microflora which hasten decomposition of residues (Macfadyen, 1961).

Cole et al. (1977) have indicated that rates of mineralization of organic P, decomposer P requirements, and better knowledge of the activity and morphology of roots were critical areas of information for development and operation of models simulating P cycling in semiarid grasslands.

B. Response to P in Grazing Studies

Numerous trials have been conducted on small plots where excreta has been returned mechanically or naturally, or has not been returned. Many of these have been reviewed by Wilkinson and Lowrey (1973). In general, the results suggest that only in the long term will the return of P in excreta significantly affect fertilizer P requirements. We shall review grazing trials in which P fertilization was a variable, and in which outputs were measured in animal products such as meat, wool, etc. Information published since the review by Wilkinson and Lowrey (1973) is emphasized.

Phosphorus deficiency is widespread in Australia and New Zealand, and consequently much of the published information is derived from studies in these countries. Most of the grazing trials have involved sheep as the grazers, and have had stocking rates as a variable.

Kirk et al. (1970) reported on comparisons of cattle performance on pangolagrass pastures in Florida with annual P rates of 0, 25, or 49 kg of P/ha over the period 1951 to 1965. Several P sources were compared. Annual N inputs to all pastures stocked at rates from 0.6 to 1.2 ha/cow unit were 56 to 112 kg of N/ha. Pastures were limed to pH 5.5. They concluded that all P sources increased establishment rate and persistence of pangolagrass stands, that the average stocking rate of phosphated pastures was 58% above the nonphosphated, and that the average gains for the period 1951 to 1965 were 129 kg/ha for the check not receiving P and from 215 to 251 kg for P-fertilized pastures. Average weaning weights and calf crop percentage were not significantly affected. Phosphorus fertilization decreased consumption of nonphosphatic mineral supplements and improved bone density and strength. Gross P deficiency symptoms appeared in a high-producing, 12-year-old cow grazing unfertilized pangolagrass pasture. They recommended annual application rates of 12 to 25 kg of P/ha as single or triple superphosphate or basic slag.

Cumberland et al. (1971) conducted a 2-year grazing trial in New Zealand on a mixed grass-clover pasture using three rates of superphosphate and three stocking rates with 8-month-old Romney wether hoggets. They reported responses to superphosphate at rates up to 1,000 kg/ha annually, and felt that wool production and weight gain were statistically more sensitive indexes of fertilizer response than rate of pasture growth. Their results suggested that high rates of superphosphate (1,000 kg/ha) may help buffer the effect of drought on forage availability. Drought reduced herbage dry matter yields 50% in low-P plots, (125 kg superphosphate/ha in the first year only), but only 30% on high-P plots (1,000 kg for both years), while hogget gains were reduced 70% in low-P paddocks and 30% on high-fertilized paddocks.

Scott (1968) investigated the effect of ordinary superphosphate (OSP) at rates of 112 and 336 kg/ha on herbage dry matter production and on lamb and wool production over a 7-year period. Sward dry matter of mixed grass and clover responded to P at the lower rate, but the response was not expressed in animal products at stocking rates considered near optimum or slightly excessive.

Cannon (1972) reported that wool production from Merino wethers was not increased when annual fertilization was increased from 56 to 280 kg of OSP/ha. Gross marginal returns were enhanced more by increased stocking rates. Kohn (1975) found similar results on an annual subterranean clover-barley grass pasture which had a history of P fertilization. Superphosphate fertilization increased the amount of herbage available, but also increased barley grass dominance of the pasture.

Carter and Day (1970) indicated that P fertilization was as necessary for increased herbage availability at high stocking rates of 25 Merino wethers/ha as at lower stocking rates. At the high stocking rate there was a marked dependence on OSP applications of 188 or 282 kg/ha to ensure satisfactory levels of pasture production, pasture availability, wool production, sheep survival, and profitability.

Simpson et al. (1974a), Bromfield and Simpson (1974), and Simpson et al. (1974b) conducted an experiment over a 5-year period to determine the effects of heavy and light grazing pressure at suboptimal, adequate, or luxury rates of OSP (43, 129, and 387 kg/ha, respectively) on growth, composition, and nutrient status (N, P, S), and availability of *Phalaris*-subterranean clover-annual grass leys. Heavy grazing pressures and suboptimal P weakened *Phalaris* and increased annual grass growth. This resulted in a greater accumulation of subsoil nitrates. In glasshouse trials with topsoil removed at the end of the fourth and fifth years of grazing, the suboptimally fertilized treatments were still responsive to additional P fertilization. At high P levels, the great variability of the *Phalaris*-clover pasture availability observed at suboptimal P levels was reduced. In microplot studies within the grazing plots, withholding P fertilization was shown to increase grass dominance. These workers found that heavy grazing pressures increased N availability for cropping in the surface soil, but that grazing pressures had no effect on P or S availability. Phosphorus availability increased only with application of superphosphate. Organic P accumulation was not affected by a ninefold change in superphosphate application rate. Consequently, the percentage of soil P in the organic form decreased with the application of inorganic P. The amount of organic P which accumualted over the experimental period of about 6 years was 21.4 ppm of P (3.6 ppm of P/year). Phosphorus was apparently redistributed from below the 7.5-cm depth, since recovery of P in the 0- to 7.5-cm soil layer exceeded the amount applied.

These authors concluded that there was no convincing evidence that P availability was directly affected by changing grazing pressure. The results also indicated that P fertilization increased carrying capacity when an attempt was made to fully utilize the herbage produced under an environment of medium but erratic rainfall (45 to 89 cm/year over the experimental period).

LITERATURE CITED

Adams, W. E., M. Stelly, R. A. McCreery, H. D. Morris, and C. B. Elkins, Jr. 1966. Protein, P, and K composition of coastal bermudagrass and crimson clover. J. Range Manage. 19: 301–305.

Alexander, M. 1977. Introduction to soil and microbiology. 2nd Ed. p. 333–349. John Wiley & Sons, Inc., New York.

Andrew, C. S. 1966. A kinetic study of phosphate absorption by excised roots of *Stylosanthes humilis, Phaseolus lathyroides, Desmodium unicanatum, medicago sativa,* and *Hordeum vulgare.* Aust. J. Agric. Res. 17:611–624.

Andrew, C. S., and M. F. Robins. 1969a. The effect of phosphorus on the growth and chemical composition of some tropical pasture legumes. I. Growth and critical percentages of phosphorus. Aust. J. Agric. Res. 665–674.

Andrew, C. S., and M. F. Robins. 1969b. The effects of phosphorus on the growth and chemical composition of some tropical pasture legumes. II. Nitrogen, calcium, magnesium, potassium, and sodium contents. Aust. J. Agric. Res. 20:675–685.

Andrew, C. S., and M. F. Robins. 1971. The effect of phosphorus on the growth chemical composition and critical phosphorus percentages of some tropical pasture grasses. Aust. J. Agric. Res. 22:693–706.

Asher, C. J., and J. F. Loneragan. 1967. Response of plants to phosphate concentration in solution culture. I. Growth and phosphorus content. Soil Sci. 103:225-233.

Baldwin, D. M., N. H. Hawkinson, and E. W. Anderson. 1974. High-rate fertilzation of native rangeland in Oregon. J. Range Manage. 27:214-216.

Barrow, N. J. 1975. Chemical form of inorganic phosphate in sheep feces. Aust. J. Soil Res. 13:63-67.

Barrow, N. J., and L. J. Lambourne. 1962. Partition of excreted nitrogen, sulphur, and phosphorus between the feces and urine of sheep being fed pasture. Aust. J. Agric. Res. 13: 461-471.

Beaton, J. D., and J. Berger. 1974. Present and potential use of fertilizer for forage production in temperate zones. p. 17-37. *In* D. A. Mays (ed.) Forage fertilization. Am. Soc. of Agron., Madison, Wis.

Beeson, K. C. 1945. The occurrence of mineral nutritional diseases of plants and animals in the United States. Soil Sci. 60:9-13.

Biddiscombe, E. F., P. G. Ozanne, N. J. Barrow, and J. Keay. 1969. A comparison of growth rates and phosphorus distribution in a range of pasture species. Aust. J. Agric. Res. 20: 1023-1033.

Biddulph, O., S. Biddulph, R. Cory, and H. Keontz. 1958. Circulation pattern for phosphorus, sulphur, and calcium in the bean plant. Plant Physil. 33:293-300.

Black, A. L. 1968. Nitrogen and phosphorus fertilization for production of crested wheatgrass and native grass in northeastern Montana. Agron. J. 60:213-216.

Black, A. L., and J. R. Wight. 1972. Nitrogen and phosphorus availability in a fertilized rangeland ecosystem of the northern Great Plains. J. Range Manage. 25:456-460.

Black, C. A. 1967. The role of phosphorus in plant growth. Hortic. Sci. 4(4):314-320.

Boawn, L. C., and R. R. Allmaras. 1974. Mineral concentrations in animal feedstuffs grown in the Columbia Plateau and adjacent valleys. Washington State Univ. Res. Center Bull. 799. 11 p.

Bornemissza, G. F., and C. H. Williams. 1970. An effect of dung beetle activity on plant yield. Pedobiologia 10:1-7.

Bowns, J. E. 1972. Low level nitrogen and phosphorus fertilization on high elevation ranges. J. Range Manage 25(4):273-276.

Bouma, D. 1967. Nutrients uptake and distribution in subterranean clover during recovery from nutritional stress. I. Experiments with phosphorus. Aust. J. Biol. Sci. 20:601-612.

Bromfield, S. M. 1961. Sheep feces in relation to the phosphorus cycle under pastures. Aust. J. Agric. Res. 12:111-123.

Bromfield, S. M., and O. L. Jones. 1970. The effect of sheep on the recycling of phosphorus in hayed-off pastures. Aust. J. Agric. Res. 21:699-711.

Bromfield, S. M., and J. R. Simpson. 1974. Effects of management on soil fertility under pasture. II. Changes in nutrient availability. Aust. J. Exp. Agric. Anim. Husb. 14:532-535.

Brown, B. A. 1959. Band versus broadcast fertilization of alfalfa. Agron. J. 51:708-710.

Bruce, R. C. 1974. Growth response, critical percentage of phosphorus and seasonal variation of phosphorus percentages in *Stylosanthes Guyanensis,* CV Schofield topdress with superphosphate. Trop. Grassl. 8:137-144.

Bruce, R. C., and I. J. Bruce. 1972. The correlation of soil phosphorus analyses with response of tropical pastures to superphosphate on some North Queensland soils. Aust. J. Exp. Agric. Anim. Husb. 12:188-194.

Bryan, W. W., and C. S. Andrew. 1971. The value of Nauru rock phosphate as a source of phosphorus for some tropical pasture legumes. Aust. J. Exp. Agric. Anim. Husb. 11: 532-535.

Burton, G. W. 1966. Significance of the underground parts of several southern grasses. p. 30-39. *In* Proc. Annu. Meet. Am. Forage Grassl. Counc., 1-4 Feb. 1966, New Orleans, La.

Cannon, D. J. 1972. The influence of rate of stocking and application of superphosphate on the prouction and quality of wool, and on the gross margin from merino wethers. Aust. J. Exp. Agric. Anim. Husb. 12:348-354.

Carter, E. D., and H. R. Day. 1970. Interrelationships of stocking rate and superphosphate on pasture as determinants of animal production. I. Continuously grazed old pasture land. Aust. J. Agric. Res. 21:473-491.

Chambliss, C. G., D. A. Miller, and J. A. Jackobs. 1970. Selection of an alfalfa plant part for phosphorus analysis. Agron. J. 62:294-296.

Christie, E. K. 1975a. Physiological responses of semi-arid grasses. II. The pattern of root growth in relation to external phosphorus concentration. Aust. J. Agric. Res. 26:437-446.

Christie, E. K. 1975b. A study of phosphorus nutrition and water supply on the early growth and survival of buffelgrass grown on a sandy red earth from Southwest Queensland. Aust. J. Exp. Agric. Anim. Husb. 15:239-249.

Christie, E. K., and J. Moorby. 1975. Physiological responses of semi-arid grasses. I. The influence of phosphorus supply on growth and phosphorus absorption. Aust. J. Agric. Res. 26:423-426.

Cole, C. V., G. S. Innis, and J. W. B. Stewart. 1977. Simulation of phosphorus cycling in semi-arid grasslands. Ecology 58:1-15.

Cooper, C. S., and A. S. Hunter. 1959. A legume for native flood meadows: II. Phosphorus fertilizer requirements for maintaining stands of white-tip clover (*Trifolium variegatum*). Agron. J. 51:350-352.

Cosper, H. R., J. R. Thomas, and A. Y. Alsayegh. 1967. Fertilization and its effect on range improvement in the northern Great Plains. J. Range Manage. 20(4):216-222.

Crossley, G. K., and A. D. Bradshaw. 1968. Differences in response to mineral nutrients of populations of ryegrass (*Lolium perenne* L.), and orchardgrass (*Dactylis glomerata* L.). Crop Sci. 8:383-387.

Cumberland, G. L., C. B. Dyson, and E. N. Honore. 1971. A comparison of phosphorus responses in pasture and sheep, and effects of increasing grazing pressure in a dry season. Proc. N.Z. Soc. Anim. Prod. 31:66-73.

Dradu, E. A. A. 1974. Soil fertility studies on loam soils for pasture development in Uganda pot experiments. East Afr. Agric. For. J. 40:126-121.

Duell, R. W. 1974. Fertilizing forage for establishment. p. 67-93. *In* D. A. Mays (ed.) Forage fertilization. Am. Soc. of Agron., Madison, Wis.

During, C., and W. C. Weeda. 1973. Some effects of cattle dung on soil properties, pasture production, and nutrient uptake. I. Dung as a source of phosphorus. N.Z. J. Agric. Res. 16:423-930.

Engelstad, O. P., and S. E. Allen. 1971. Effect of form and proximity of added N on crop uptake of P. Soil Sci. 112:330-337.

Falade, J. A. 1973. Effect of phosphorus on the growth and mineral composition of four tropical legumes. J. Sci. Fed. Agric. 24:795-802.

Finn, B. J., and A. R. Mack. 1964. Differential response of orchardgrass varieties (*Dactylis glomerata* L.) to nitrogen and phosphorus under controlled soil temperature and moisture conditions. Soil Sci. Soc. Am. Proc. 28:782-785.

Fisher, F. L., and A. G. Caldwell. 1959. The effects of continued use of heavy rates of fertilizers on forage production and quality of Coastal bermudagrass. Agron. J. 51:99-102.

Gates, C. T., K. P. Haydock, and W. T. Williams. 1973. A study of the interaction of cold stress, age, and phosphorus nutrition on the development of *Lotononis bainessi* Baker. Aust. J. Biol. Sci. 26:87-103.

Gillingham, A. G., and C. During. 1973. Pasture production and transfer of fertility within a long-established hill pasture. N.Z. J. Exp. Agric. 1:227-232.

Gomide, J. A., C. H. Noller, G. O. Mott, J. H. Conrad, and D. L. Hill. 1969. Mineral composition of six tropical grasses as influenced by plant age and nitrogen fertilization. Agron. J. 51:120-123.

Greenway, H., and A. Gunn. 1966. Phosphorus retranslocation in *Hordeum vulgare* during early tillering. Planta 71:43-67.

Haas, H. J., D. L. Grunes, and G. A. Reichman. 1961. Phosphorus changes in Great Plains soils as influenced by cropping and manure applications. Soil Sci. Soc. Am. Proc. 25:214-218.

Halm, B. J., J. W. B. Stewart, R. L. Halstead. 1972. The phosphorus cycle in a native grassland ecosystem. p. 571-589. *In* Isotopes and radiation in soil-plant relationships including forestry. IAEA Proc. Ser. STI/PUB/292, Vienna.

Hanson, A. 1974. The importance of forages to agriculture. p. 1-16. *In* D. A. Mays (ed.) Forage fertilization. Am. Soc. of Agron., Madison, Wis.

Harley, C. P., H. H. Moon, and L. O. Reguimbal. 1951. The release of certain nutrient elements from simualted orchardgrass mulch. Proc. Am. Soc. Hortic. Sci. 57:17-23.

Heady, H. F. 1975. Rangeland management. McGraw-Hill, New York. 460 p.

Helyar, K. R., and A. J. Anderson. 1970. Response of five pasture species to phosphorus, lime, and nitrogen on an infertile soil with a high phosphate sorption capacity. Aust. J. Agric. Res. 21:677-692.

Hilder, E. J. 1966. Distribution of excreta by sheep at pasture. p. 977-981. *In* Proc. X Int. Grassl. Congr., Helsinki, Finland.

Hodgson, H. J. 1974. Forages, cattle, and the consumer. p. 56–65. *In* Proc. 7th Am. Forage Grassl. Counc. Res.-Ind. Conf., Bossier City, La.

Hodgson, R. E. 1968. The place of forages in animal production—now and in the years hence. p. 8–22. *In* Proc. 1st Am. Forage Grassl. Counc. Res.-Ind. Conf., Chicago, Ill.

Jackson, J. E., M. E. Walker, and R. L. Carter. 1959. Nitrogen, phosphorus, and potassium requirements of coastal bermudagrass on a Tifton loamy sand. Agron. J. 51:129–131.

Johnston, A., A. D. Smith, L. E. Lutwick, and S. Smoliak. 1968. Fertilizer response of native and seeded ranges. Can. J. Plant Sci. 48:467–472.

Johnston, A., S. Smoliak, A. D. Smith, and L. E. Lutwick. 1967. Improvement of southeastern Alberta range with fertilizers. Can. J. Plant Sci. 47(6):671–678.

Jones, M. B., and R. A. Evans. 1960. Botanical composition changes in annual grassland as affected by fertilization and grazing. Agron. J. 52:459–461.

Jones, M. B., L. M. Freitas, and K. H. de Mohrdieck. 1970a. Differential response of some cool-season grasses to N, P, and lime. IRI Res. Inst. Bull. no. 37, New York, N.Y. 24 p.

Jones, M. B., P. W. Lawler, and J. E. Ruckman. 1970b. Differences in annual clover responses to phosphorus and sulphur. Agron. J. 62:439–442.

Jones, M. B., J. E. Ruckman, and P. W. Lawler. 1972. Critical levels of P in subclover (*Trifolium subterraneum* L.). Agron. J. 64:695–698.

Jones, O. L., and S. M. Bromfield. 1969. Phosphorus changes during the leaching and decomposition of hayed-off pasture plants. Aust. J. Agric. Res. 20:653–663.

Jones, R. K. 1974. A study of the phosphorus responses of a wide range of accessions from the genus *Stylanthes*. Aust. J. Agric. Res. 25:847–862.

Jordan, C. W., C. W. Evans, and R. D. Rouse. 1966. Coastal bermudagrass response to application of P and K as related to P and K levels in the soil. Soil Sci. Soc. Am. Proc. 30: 477–479.

Keay, J. F., E. F. Biddiscombe, and P. G. Ozanne. 1970. The comparative rates of phosphate absorption by eight annual pasture species. Aust. J. Agric. Res. 21:33–44.

Keya, N. C. O., and D. W. Kalangi. 1973. The seeding and superphospahte rates for the establishment of *Desmodium uncinatum* (JACQ) D.C. by oversowing in uncultivated grasslands of Western Kenya. Trop. Grassl. 7:319–325.

Fisher, M. J., and N. A. Campbell. 1972. The initial and residual responses to phosphorus fertilizers of Townsville stylo in pure ungrazed swards at Katherine, N.T. Aust. J. Exp. Agric. Anim. Husb. 12:488–494.

Floate, M. J. S. 1970a. Decomposition of organic materials from hill soils and pastures. II. Comparative studies on the mineralization of carbon, nitrogen, and phosphorus from soil. Soil Biol. Biochem. 2:173–185.

Floate, M. J. S. 1970b. Decomposition of organic materials from hill soils and pastures. III. The effect of temperatures on the mineralization of carbon, nitrogen, and phosphorus from plant materials and sheep feces. Soil Biol. Biochem. 2:187–196.

Floate, M. J. S. 1970c. Decomposition of organic materials from hill soils and pastures. IV. The effects of moisture content on the mineralization of carbon, nitrogen, and phosphorus from plant materials and sheep feces. Soil Biol. Biochem. 2:275–283.

Floate, M. J. S. 1970d. Mineralization of nitrogen and phosphorus from organic materials of plant and animal origin and its significance in the nutrient cycle in grazed upland and hill soils. J. Br. Grassl. Soc. 25:205–302.

Fuller, W. H., D. R. Neilson, and R. W. Miller. 1956. Some factors influencing the utilization of phosphorus from crop residues. Soil Sci. Soc. Am. Proc. 20:218–224.

Gates, C. T. 1974. Nodule and plant development in *stylosanthes humilis* H.B.K.: Symbiotic response to phosphorus and sulphur. Aust. J. Bot. 22:45–55.

Kilcher, M. R., S. Smoliak, W. A. Hubbard, A. Johnston, A. T. H. Gross, and E. V. McCurdy. 1965. Effects of inorganic nitrogen and phosphorus fertilizers on selected sites of native grassland in Western Canada. Can. J. Plant Sci. 45:229–237.

Kirk, W. G., R. L. Shirley, E. M. Hodges, G. K. Davis, F. M. Peacock, J. F. Easley, and F. G. Martin. 1970. Production, performance, and blood and bone composition of cows grazing pangolagrass pastures receiving different phosphate fertilizers. Florida Agric. Exp. Stn. Tech. Bull. 735. 55 p.

Kline, J. R. 1969. Soil chemistry as a factor in the function of grassland ecosystems. p. 71–88. *In* R. L. Dix and R. E. Beidleman (ed.) The grassland ecosystem. A preliminary "Synthesis." Range Sci. Dep. Sci. Series no. 2, Colorado State Univ., Fort Collins.

Knauer, N. 1973. The importance of the dynamics of nutrients in the soil in the determination of the nutrient requirements of grassland. Phosphorsaeure 30:27–42. (Cited from Herb. Abst. 4118, 45, 1975).

Kohn, C. D. 1975. Superphosphate utilization in clover by farming. I. Effects on pasture and sheep production. Aust. J. Agric. Res. 25:525-535.

Lamba, P. S., H. L. Ahlgren, and P. J. Muckenhirn. 1949. Root growth of alfalfa, medium red clover, bromegrass, and timothy under various soil conditions. Agron. J. 41:451-458.

Lawton, K., M. B. Tesar, and B. Kawin. 1954. Effect of rate and placement of superphosphate on the yield and phosphorus absorption of legume hay. Soil Sci. Soc. Am. Proc. 18:428-432.

Lorenz, R. J., and G. A. Rogler. 1972. Forage production and botanical composition of mixed prairie as influenced by nitrogen and phosphorus fertilization. Agron. J. 64:244-249.

Lorenz, R. J., and G. A. Rogler. 1973. Interaction of fertility level with harvest data and frequency on productiveness of mixed prairie. J. Range Manage 26:50-54.

Lucas, L. N., and W. G. Blue. 1973. Effect of lime and phosphorus on selected alluvial entisols from eastern Costa Rica. 2. Forage plant responses. Trop. Agric. 50:63-74.

Ludwick, A. E., and B. C. Rumburg. 1976. Grass hay production as influenced by nitrogen and phosphorus. Agron. J. 68:933-937.

Luebs, R. E., A. E. Laag, and M. J. Brown. 1971. Effect of site and rainfall on annual range response to nitrogen and phosphorus. J. Range Manage. 24:366-370.

Lunt, O. R., R. L. Branson, and S. B. Clark. 1965. Response of five grass species to phosphorus on six soils. p. 419-423. In Proc. 10th Int. Grassl. Congr., Sao Paulo, Brazil.

MacDiarmid, B. N., and B. R. Watkin. 1972. The cattle dung patch. 3. Distribution and rate of decay of dung patches and their influence on grazing behavior. J. Br. Grassl. Soc. 27:48-54.

Macfadyen, A. 1961. Metabolism of soil invertibrates in relation to soil fertility. Ann. Appl. Biol. 49:215-218.

Markus, D. K., and W. R. Battle. 1965. Soil and plant responses to long-term fertilization of alfalfa (*Medicago sativa* L.). Agron. J. 57:613-616.

Martin, W. E., and J. E. Matocha. 1973. Plant analysis as an aid in the fertilization of forage crops. p. 393-425. In L. M. Walsh and J. D. Beaton (ed.) Soil testing and plant analysis. Rev. Ed. Soil Sci. Soc. Am., Madison, Wis.

Massey, D. L., and R. W. Sheard. 1970. Utilization of surface-applied phosphorus by established stands of alfalfa and bromegrass. Can. J. Soil Sci. 50:9-16.

Mays, D. A., and G. W. Bengtson. 1974. Fertilizer effects on forage crops on strip-mined land in Northeast Alabama. TVA, Natl. Fert. Dev. Center Bull. Y-74. 23 p.

Minderhoud, J. W., P. F. J. Van Burg, B. Deinum, J. G. P. Dirven, and M. L. 't Hart. 1975. Effects of high levels of nitrogen and adequate utilization on grassland productivity and cattle performance with special reference to permanent pastures in the temperate regions. Stikstof 18:2-11.

Moschler, W. W., G. W. Shear, D. C. Martens, G. D. Jones, and W. R. Wilmouth. 1972. Comparative yield and fertilizer efficiency of no-tillage and conventionally tilled corn. Agron. J. 64:229-231.

Mosse, B. 1973. Advances in the study of vesicular arbuscular mycorrhiza. Annu. Rev. Phytopathol. 11:170-196.

O'Donnell, J. L., and J. R. Love. 1970. Effects of time and height of cut on rooting activity of Merion Kentucky bluegrass as measured by radioactive phosphorus uptake. Agron. J. 62:313-316.

Ofer, Y., and N. G. Seligman. 1969. Fertilization of annual range in northern Israel. J. Range Manage. 22:337-341.

Ozanne, P. G., and K. M. W. Howes. 1971. The effects of grazing on the phosphorus requirement of annual pasture. Aust. J. Agric. Res. 22:81-92.

Peterson, R. G., W. W. Woodhouse, Jr., and H. L. Lucas. 1956. The distribution of excreta by freely grazing cattle and its effect on pasture fertility. I. Excretal distribution. Agron. J. 48:440-444.

Plucknett, D. L. 1970. Productivity of tropical pastures in Hawaii. p. A38-A49. In Proc. 11th Int. Grassl. Congr. Queensland, Australia.

Ramakrishnan, P. S., and N. Gupta. 1973. Nitrogen, phosphorus, and potassium nutrition of the edaphic ecotypes in *Cynodon dactylon* L. Pers. Ann. Bot. 37:885-894.

Reid, R. L., and C. A. Jung. 1965. Influence of fertilizer treatment on the intake, digestibility, and palatability of tall fescue hay. J. Anim. Sci. 24:615-625.

Reid, R. L., and G. A. Jung. 1974. Effects of elements other than nitrogen on the nutritive value of forage. p. 395-435. In D. A. Mays (ed.) Forage fertilization. Am. Soc. of Agron., Madison, Wis.

Reid, R. L., A. J. Post, and G. A. Jung. 1970. Mineral composition of forages. West Virginia Agric. Exp. Stn. Bull. 589T.

Riley, D., J. Coutts, and M. A. Gowman. 1975. Placement, mobility, and plant uptake of nutrients in no-tillage systems. p. 15-28. *In* No tillage forage. Ohio State Univ. and Ohio Agric. Res. and Dev. Center, Columbus.

Robinson, P. J., and R. K. Jones. 1972. The effect of phosphorus and sulphur fertilization on the growth and distribution of dry matter, nitrogen, phosphorus, and sulphur in Townsville Stylo (*Stylonanthes humilis*). Aust. J. Agric. Res. 23:633-640.

Robinson, R. R., V. G. Sprague, and C. F. Gross. 1959. The relation of temperature and phosphate placement to growth of clover. Soil Sci. Soc. Am. Proc. 28:225-228.

Rodulfo, S., and W. G. Blue. 1970. The availability to forage plants of accumulated phosphorus in Leon fine sand. Soil Crop Sci. Soc. Fla. Proc. 30:167-174.

Rotar, P. P., and D. L. Plucknett. 1973. Tropical and subtropical forages. p. 358-371. *In* M. E. Heath, D. S. Metcalfe, and R. F. Barnes (ed.) Forages. Iowa State Univ. Press, Ames.

Sadler, J. M., and J. W. B. Stewart. 1974. Residual fertilizer phosphorus in western Canadian soils: A review. Saskatchewan Inst. of Pedology Pub. no. R136. 37 p.

Schwendiman, J. L., R. B. Foster, and O. K. Haglund. 1966. The influence of climate, soils, and management on the root development of grass species in western states. p. 40-57. *In* Proc. Annu. Meet. Am. Forage and Grassl. Counc., New Orleans, La.

Scott, P. S. 1968. Animal and pasture production as indices of fertilizer maintenance requirements. Proc. N.Z. Soc. Anim. Prod. 28:53-64.

Shaw, N. H., C. T. Gates, and J. R. Wilson. 1966. Growth and chemical composition of Townsville lucerne. 1. Dry matter yield and nitrogen content in response to superphosphate. Aust. J. Exp. Agric. Anim. Husb. 6:150-156.

Sheard, R. W., G. J. Bradshaw, and D. L. Massay. 1971. Phosphorus placement for establishment of alfalfa and bromegrass. Agron. J. 63:922-927.

Simpson, J. R., S. M. Bromfield, and G. T. McKinnery. 1974a. Effects of management on soil fertility under pasture. 1. The influence of experimental grazing and fertilizer systems on the growth, composition, and nutrient status of the pasture. Aust. J. Exp. Agric. Anim. Husb. 14:470-478.

Simpson, J. R., S. M. Bromfield, and O. L. Jones. 1974b. Effects of management on soil fertility under pasture. 3. Changes in total soil nitrogen, carbon, phosphorus, and exchangeable cations. Aust. J. Exp. Agric. Anim. Husb. 14:487-494.

Smika, D. E., H. J. Haas, J. F. Power. 1965. Effects of moisture and nitrogen fertilizer on growth and water use by native grass. Agron. J. 57:483-486.

Smika, D. E., H. J. Haas, G. A. Rogler, and R. J. Lorenz. 1961. Chemical properties and moisture extraction in rangeland soils as influenced by nitrogen fertilization. J. Range Manage. 14:213-216.

Smith, F. W. 1975. Tissue testing for assessing the phosphorus status of green panic, buffelgrass and setaria. Aust. J. Exp. Agric. Anim. Husb. 15:383-390.

Snaydon, R. W., and A. D. Bradshaw. 1962. Differences between natural populations of trifolium repens L. in response to mineral nutrients. I. Phosphate. J. Exp. Bot. 13:422-434.

Sprague, H. B. 1933. Root development of perennial grasses and its relation to soil conditions. Soil sci. 36:189-209.

Stanford, G., C. McAuliffe, and R. Bradfield. 1950. The effectiveness of superphosphate topdressed on established meadows. Agron. J. 42:423-426.

Stewart, J. W. B., B. J. Halm, and C. V. Cole. 1973. Nutrient cycling: I. Phosphorus. Can. Comm. Int. Biol. Prog. (Matador Project) Tech. Rep. no. 40. Univ. Saskatchewan, Saskatoon, Canada.

Stoddard, L. A., A. D. Smith, and T. W. Box. 1975. Range improvements for increasing forage production. Chapt. 14. *In* L. A. Stoddard, A. D. Smith, and T. W. Box (ed.) Range management. McGraw-Hill, New York.

Strickland, R. W. 1973. A comparison of dry matter yield and mineral content of three *Cynodon Dactylon* with *Digitaria decumbens* and *Digitaria* didac. Trop. Grassl. 7:313-317.

Swaby, R. J. 1962. Effects of microorganisms on nutrient availability. p. 159-172. *In* Trans. Joint Comm. IV and V Int. Soil Sci., Palmerston, New Zealand.

Taliaferro, C. M., F. P. Horn, B. B. Tucker, R. Totusek, and R. D. Morrison. 1975. Performance of three warm-season perennial grasses and a native range mixture as influenced by N and P fertilization. Agron. J. 67:289-292.

Tamimi, Y. N. 1972. Response of kikuya and pangolagrass to rates of nitrogen, phosphorus, and potassium. II. Effect of high rates. Hawaii Univ. Ext. Misc. Publ. 81:85-92.

Templeton, W. C., J. L. Menees, and T. H. Taylor. 1969. Growth of young orchardgrass (*Dactylis glomerata* L.) plants in different environments. Agron. J. 61:780-782.

Timmons, D. r., R. F. Holt, and J. S. Latterell. 1970. Leaching of crop residues as a source of nutrients in surface runoff water. Water Resour. Res. 6:1367-1375.

Turner, J. R. 1972. Maximizing use of fertilizer in forage crop production. p. 5-11. *In* Proc. 5th Am. Forage Grassl. Counc. Res.-Ind. Conf., Louisville, Ky.

USDA Forest Service. 1970. Range ecosystem research. Agric. Inf. Bull. no. 346. Washington, D.C.

Vallentine, J. F. 1971. Range developments and improvements. Brigham Young Univ. Press, Provo, Utah.

Vicente-Chandler, J., F. Abruna, R. Caro-Costas, J. Figarella, S. Silva, R. W. Pearson. 1974. Intensive grassland management in the humid tropics of Puerto Rico. Univ. Puerto Rico AES Bull. 233. 164 p.

Wedin, W. F. 1974. Fertilization of cool season grasses. p. 95-144. *In* D. A. Mays (ed.) Forage fertilization. Am. Soc. of Agron., Madison, Wis.

Welch, L. F., W. E. Adams, and J. L. Carmon. 1963. Yield response surface, insoquants, and economic fertilizer optima for coastal bermudagrass. Agron. J. 55:63-67.

Werner, J. C., and H. B. DeMattos. 1972. studies on the nutrition of centrosema (*Centrosema pubescens* Benth). Bol. Ind. Anim. 29:375-391. (Herb. Abstr. 44:2843).

West, S. H., and G. M. Prine. 1974. Winterkill and hay yield of pangola digitgrass as affected by fall application of N, P, K. Proc. Soil Crop Sci. Soc. Fla. 33:16-20.

White, R. E. 1972. Studies on mineral ion absorption by plants. I. The absorption and utilization of phosphate by *Stylosanthes-humilis, phaseolus atropureus,* and *Desmodium intortum.* Plant Soil 36:427-447.

Wight, J. R., and A. L. Black. 1972. Energy fixation and precipitation use efficiency in a fertilized rangeland ecosystem of the northern Great Plains. J. Range Manage. 25:376-380.

Wilkinson, S. R., and G. W. Langdale. 1974. Fertility needs of the warm season grasses. p. 119-145. *In* D. A. Mays (ed.) Forage fertilization. Am. Soc. of Agron., Madison, Wis.

Wilkinson, S. R., and R. S. Lowrey. 1973. Cycling of mineral nutrients in pasture ecosystems. p. 248-390. *In* G. W. Butler, R. W. Bailey (ed.) Chemistry and biochemistry of herbage, Vol. 2. Academic Press, New York.

Williams, C. H., and C. S. Andrew. 1970. Mineral nutrition of pastures. Chapter 21. *In* C. H. Williams and C. S. Andrew (ed.) Australian grasslands. Australian Natl. Univ. Press, Canberra.

Wolfe, E. C., and A. Lazenby. 1973. Grass—white clover relationships during pasture development. I. Effect of superphosphate. Aust. J. Exp. Agric. Anim. Husb. 13:567-574.

Woodhouse, W. W., Jr. 1969. Long-term fertility requirements of coastal bermudagrass. II. Nitrogen, phosphorus, and lime. Agron. J. 61:251-256.

Woods, L. E., and R. A. Dance. 1970. Seed and nutritional aspects of grass-Townsville Stylo competition. J. Aust. Inst. Agric. Sci. 36:45-57.

Workman, J. P., and T. M. Quigley. 1974. Economics of fertilizer application on range and meadow sites in Utah. J. Range Manage. 27(5):390-393.

Younge, O. R., and J. C. Moomaw. 1960. Revegetation of strip-mined bauxite lands in Hawaii. Econ. Bot. 14:316-330.

# Chapter 29

Relationship Between Phosphorus Nutrition of Plants and the Phosphorus Nutrition of Animals and Man

R. L. REID

West Virginia University
Morgantown, West Virginia

I. INTRODUCTION

The significance of phosphorus (P) and calcium (Ca) in the nutrition of humans and of farm animals has been recognized for centuries and has been the subject of an extensive literature which it would be virtually impossible to review. The disease of rickets in children may have been first described in the 17th century (Leitch, 1964) and the complex interactions of diet, climate, and socio-economic status on the incidence of Ca and P disorders in humans were elucidated gradually over the following 300 years. Similarly, observations relating to a condition of aphosphorosis in cattle date back to the late 18th century, and the widespread occurrence of this problem throughout the world was reviewed by, for example, Theiler and Green (1932) and Beeson (1941). These authors pointed out the close association between P status of the soil and bone disorders of grazing animals and suggested control of the problem either by the provision of P supplements in solid or liquid form, or by changing the P content of the pasture and forage by soil amendment.

The clinical condition of rickets in children has now almost been eliminated as a major nutritional problem; the parallel conditions of osteoporosis and osteomalacia are less well understood and considerable research is still being directed to the significance of dietary levels and ratios of Ca and P on the incidence of bone disorders in the adult human. Without doubt, however, the main impact of P deficiency conditions at the present time is on ruminant livestock; the situation has been encapsulated by Underwood (1966) in the observation that: "Phosphorus deficiency is predominantly but not exclusively a condition of grazing ruminants, especially cattle, whereas calcium deficiency is more a problem of hand-fed animals, especially pigs and poultry." Major attention in this review will therefore be directed to the P requirements of ruminant animals fed forage-based diets,

---

Copyright 1980 © ASA-CSSA-SSSA, 677 South Segoe Road, Madison, WI 53711, USA.
The Role of Phosphorus in Agriculture.

and to the problems of maintaining adequate P nutrition under conditions of limited P supply. Further objectives will be to consider the possible effects of lower P inputs on the nutritional quality of grain and other food crops consumed directly by monogastric species, and to evaluate the potential of unconventional P sources as components of animal diets.

As indicated by both Underwood (1966) and Allaway (1962), the question of whether to attempt to modify the mineral nutrition of livestock by soil amendment practices or by provision of appropriate supplements in the diet is an economic one and will depend upon (i) the mineral involved, and (ii) the nature of crop response to soil amendment. The latter in turn will relate to nature of the soil, climate, and plant species. Allaway (1971) pointed out that on some soils the major response to P fertilization may be simply an increase in crop yield, with little or no change in the P concentration of the plant. The situation under grazing conditions may be further complicated by the fact that animal response to P fertilization may be primarily conditioned by the change in botanical composition brought about by fertilizer treatment, rather than by an increase in the P content of the herbage. There is, then, no apparent simple unifying principle relating the P nutrition of forage and food crops to that of animals consuming the diet. Under certain conditions, which will be examined, alleviation of P deficiency in animals, or improvement of animal output, may be brought about simply by fertilization and soil amendment practices. In other conditions, direct supplementation may be the more effective and feasible approach.

What does appear to be certain, however, is that both routes to supplying P as a nutrient to animals are subject to much the same kinds of economic pressure. Essentially the same factors apply to the supply and cost of feed phosphates as apply to the fertilizer industry. Approximately 84% of the demand for P in the United States is consumed in the manufacture of fertilizer, 10% is used in animal feeds, and 5% goes to the production of detergents and cleaning agents (USDA-ERS, 1974). This led, in 1974, to the development of a shortage of P supplements for use in animal feedstuffs and to a dramatic increase in the cost of feed P. A report prepared by a National Research Council (1974b) committee for the USDA indicated that, in relation to an annual requirement of 1.45 million metric tons of P, there was an available supply of 1.18 million tons. This reflected a fivefold increase in the domestic demand for P supplements since 1951, as indicated in Fig. 1, taken from data of Highton (1974). The committee recognized that the demand for feed P supplements might exceed production capacity for a period of 6 to 24 months and prepared a series of guidelines for the adequate nutrition of livestock and poultry during conditions of P shortage. While phosphate production capacity increased markedly in 1976, the continuing high cost of P supplements for feeding purposes warrants a current evaluation of the P requirements of different classes of animals, including man, and of the various options available to farmers and to the food industry for supplying these requirements in the most economic and effective fashion.

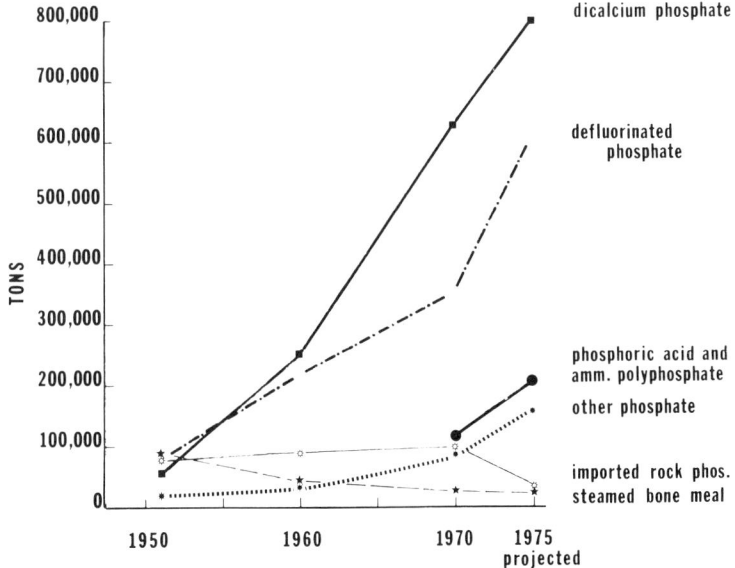

Fig. 1—Changes in U.S. production of P feed supplements during the period 1950–1975. Data for 1975 are estimated (from Highton, 1974).

II. PHOSPHORUS REQUIREMENTS OF ANIMALS

The mineral requirements of all classes of livestock and of man are affected by a variety of factors. As defined by Underwood (1966), these include (i) species or breed; (ii) age, sex, and rate of growth; (iii) type and rate of production desired; (iv) level and chemical form of mineral in diet; (v) overall nutrient balance and adequacy of diet; (vi) hormonal and other physiological processes; (vii) climate or nondietary environment; (viii) criteria of adequacy employed. The effects of certain of these factors and their interactions on mineral requirements can be defined with reasonable accuracy; the effects of others are relatively unknown.

A. Establishment of Requirements

The criteria used to determine adequacy, and the experimental methods used to estimate requirements, are particularly important. In general, estimates of mineral needs are made by one of three general techniques (Leitch, 1964). One is the *experimental depletion method*, in which animals are deprived of a nutrient and the amount required for repletion or to offset deficiency signs is determined. The second is *physiological*, and may be based on the classical balance trial or on an estimation of mineral retention by body analysis under various conditions of growth and production. The validity of balance techniques in the determination of mineral requirements has been examined by a number of authors; Duncan (1966) concludes that for large animals and for man "their validity remains to be established." In an earlier paper (Duncan, 1958), this author identified the occurrence of systematic error in the estimation of P requirement by balance trial in ruminants. Similar problems in the interpretation of Ca and P balance trials conducted with humans have been discussed by Hegsted (1973, 1976).

A final approach to determining requirements is the *epidemiological* or *diet survey method* in human populations, or alternatively, in the case of experimental animals or livestock, the better controlled feeding trial method relating mineral intake on a variety of diets to such criteria as rate of growth, feed conversion, milk production, bone density, etc.

Each of these techniques, or different combinations of them, has been used in making recommendations for P allowances, and recommended values from different authors, agencies, or countries may differ markedly depending on the criteria adopted and the methods used for establishing requirement. An example of differences in interpretation of requirement is seen in Table 1, taken from a review by Little (1970) of the P nutrition of beef cattle in Australia. Recommendations by the National Research Council (NRC) (1963) and Mitchell (1947) were higher at low body weights, and lower at high body weights, than values suggested by the Agricultural Research Council (1965). It may be pointed out at this stage that part of this divergence of opinion stems from a dearth of knowledge concerning the availability of P in natural feeds, and this problem will be discussed again. Similar differences are seen in published values for recommended dietary intakes of P and Ca by human populations. Hegsted (1973), in commenting upon the significantly lower levels of calcium intake recommended by the World Health Organization (1962) as compared to recommendations of the Food and Nutrition Board (1968), states that there are yet no definitive studies associating health with higher levels of mineral intake.

B. Recommended Levels of Phosphorus in the Diet

Within these general constraints it is possible to make recommendations on amounts or concentrations of dietary P which will allow for reasonable health and various levels of animal production. A summary of P requirements for the more important productive functions in different classes of domestic livestock is provided in Table 2, together with representative values for the P content of feeds and supplements commonly used in the maintenance of the different species. A more detailed statement of requirements may be found in the NRC (1974b) report *Feed Phosphorus Shortage: Levels and Sources of Phosphorus Recommended for Livestock and Poultry,* or in the series of NRC publications (1971b, 1973, 1975, 1976, 1978) on the *Nutrient Requirements of Domestic Animals.* The P requirements of individual species will be considered at greater length in later sections, but certain generalizations may be made from these data.

Table 1—Phosphorus requirements for growth of cattle at various bodyweights (Little, 1970).

Body weight, kg	P requirements and source					
	ARC (1965)		NRC (1963)		Mitchell (1947)	
	g/day†	% of dry ration‡	g/day§	% of dry ration‡	g/day‡	% of dry ration§
100	7.3	0.29	10.1	0.40	10.8	0.43
200	9.8	0.20	11.1	0.22	13.5	0.27
300	14.6	0.20	12.3	0.16	15.0	0.20
400	23.7	0.24	13.4	0.13	16.0	0.16

† Published values for maintenance and growth at 0.5 kg/day.
‡ These values were calculated from the corresponding figures, assuming a daily dry matter intake of 2.5% of body weight.
§ Published values interpolated graphically for these particular body weights; rates of growth marginally higher than for those considered by the Agricultural Research Council (1965).

1) Species differ in their P needs, and greatest requirement is associated with periods of high metabolic activity, e.g., rapid growth and heavy lactation.
2) Due to the fairly generous "insurance" factor built into most feeding stuff recommendations to allow for individual animal variation, a reduction of P intake is probably feasible without marked impairment of performance; the degree of this reduction will relate both to species and to function, with greatest reductions possible in fattening animals and in mature animals not producing milk, or in early stages of pregnancy.
3) Protein concentrate feeds are good sources of P, grain crops are intermediate, and forage crops tend to be low and variable in their P content. The necessity to provide P supplements routinely in the diet of the productive monogastric species is apparent. The ability of forages to meet the P requirements of ruminants will relate directly to physiological state of the animal, to class of forage, and to agronomic management of the pasture or forage crop.

For humans, a distinction is usually drawn between requirements and *recommended dietary allowances* (RDA) of a nutrient. By definition, *RDA*, as used in the United States, are the levels of intake of essential nutrients considered by the Food and Nutrition Board to be adequate to meet the known nutritional needs of almost all healthy persons; RDA, therefore, allow for individual variation and are higher than estimates of requirement. Recommended allowances for P and Ca for humans are given in Table 3, taken from data by Hegsted (1973), and from the National Academy of Sciences publication, *Recommended Dietary Allowances* (NRC, 1974a).

The general lack of emphasis on P allowances for humans stems from the fact that P is present in most foods and, at least in western diets, is normally adequate to meet requirements. The Food and Nutrition Board, in fact, did not make specific recommendations for P intake until 1968. In the American diet, more than 60% of the P is derived from dairy products, meat, and fish; some 12% from flour and cereals; and the rest from foods

Table 2—Phosphorus requirements and concentrations of P in some major dietary sources for classes of livestock.

P requirements and sources	Beef cow Growing-finishing	Beef cow Dry preg.	Beef cow Lactating (avg. milk)	Dairy cow, lactating <20 kg	Dairy cow, lactating >30 kg	Sheep Fattening lambs	Sheep Preg. ewe	Sheep Lactating with twins	Poultry Broiler starter	Poultry Broiler grower	Poultry Layer	Swine Growing finishing	Swine Bred gilts and sows	Swine Lactating gilts and sows
Body weight, kg	400	454	400			40	70	70				35-100	100-250	110-250
							%P							
NRC (1974b) recommendations†														
Present	0.21-0.26	0.18	0.28	0.33	0.39	0.19	0.20	0.31	0.7	0.7	0.6	0.4	0.5	0.5
Minimal	0.19-0.23	0.16	0.25	0.30	0.36	0.15	0.16	0.25	0.65	0.5	0.45	0.4	0.5	0.5
Emergency	0.18-0.22	0.15	0.24	0.28	0.34	0.12	0.12	0.17	0.6	0.4	0.4	0.38	0.46	0.46
Forages														
Pasture‡	0.2-0.5	0.2-0.5		0.2-0.5			0.2-0.5							
Grass hay§	0.21	0.21		0.21			0.21							
Legume hay§	0.30	0.30		0.30			0.30							
Corn silage§	0.22	0.22		0.22			0.22							
Concentrates¶														
Corn, dent yellow, grain	0.27	0.27		0.27			0.27			0.27			0.27	
Barley, grain	0.36	0.36		0.36			0.36			0.36			0.36	
Wheat, grain	0.36	0.36		0.36			0.36			0.36			0.36	
Soybean meal	0.65	0.65		0.65			0.65			0.65			0.65	
Peanut meal	0.65	0.65		0.65			0.65			0.65			0.65	
Inorganic¶														
Defluorinated phosphate	18.0	18.0		18.0			18.0			18.0			18.0	
Dicalcium phosphate	18.2	18.2		18.2			18.2			18.2			18.2	
Steamed bonemeal	14.3	14.3		14.3			14.3			14.3			14.3	

† Prepared by Committee on Animal Nutrition Task Force on Phosphorus Requirements of Livestock and Poultry (NRC, 1974b). Present levels are those defined by NRC for acceptable animal performance and health; minimal levels can be fed for extended periods with little or no harmful effect on performance and health; emergency levels should be fed only under emergency supply conditions and for short periods of time.
‡ Whitehead (1966).   § Stout et al. (1977).   ¶ NRC (1971a).

Table 3—Recommended intakes of Ca and P (Hegsted, 1973; NRC, 1974a).

Age	Ca[†]	P[†]	Ca[‡]	Ca[§]
		mg/day		
0-12 months	360-540	240-400	500-600	500
1-9 years	800	800	400-500	700-1,000
10-15 years	1,200	1,200	600-700	1,200
16-19 years	1,200	1,200	500-600	900
Adult	800	800	400-500	500
Pregnancy and lactation	1,200	1,200	1,000-1,200	1,200

[†] From NRC (1974a).
[‡] From World Health Organization (1962).
[§] From Canadian Council on Nutrition (1964).

such as eggs, beans, vegetables, and potatoes (Bogert et al., 1973). It has been pointed out (Davidson & Passmore, 1969) that the average daily intake of P by adults in the United Kingdom is approximately 1.5 g., a level considerably higher than the recommended allowances. In practice, therefore, and probably also from lack of experimental evidence, recommendations for dietary intakes of P have generally been based upon those for Ca, assuming a desirable ratio for Ca/P of 1:1. The situation in developing countries, where meat and dairy products are replaced largely by vegetable foods, is obviously more marginal in relation to both Ca and P supply. Under these conditions, factors affecting the content and availability of these elements in the staple diet may have considerable bearing on bone structure and formation and on the general health of the population.

III. FACTORS AFFECTING THE PHOSPHORUS CONTENT OF PLANTS

The effects of soil amendment practices and agronomic management on plant growth and composition are reviewed by other authors in this symposium and discussion of these areas will therefore be brief and will emphasize the relationship of plant composition to animal requirement. Fleming (1973) summarized the main factors affecting the mineral composition of forage plants as follows: (i) soil (parent material, development, moisture, and reaction); (ii) genus, species, and variety of plant; (iii) stage of maturity; (iv) season and temperature effects; (v) fertilizers and management; and (vi) mineral distribution in the plant.

A. Soil

The influence of soil parent material and location on plant mineral composition may be illustrated by a number of recent studies in Pennsylvania (Adams, 1975; Stout et al., 1977)[1] relating the mineral content of

[1] D. P. Belesky, W. L. Stout, and G. A. Jung. 1974. Mineral status of Pennsylvania forages. Agron. Abstr. p. 104.

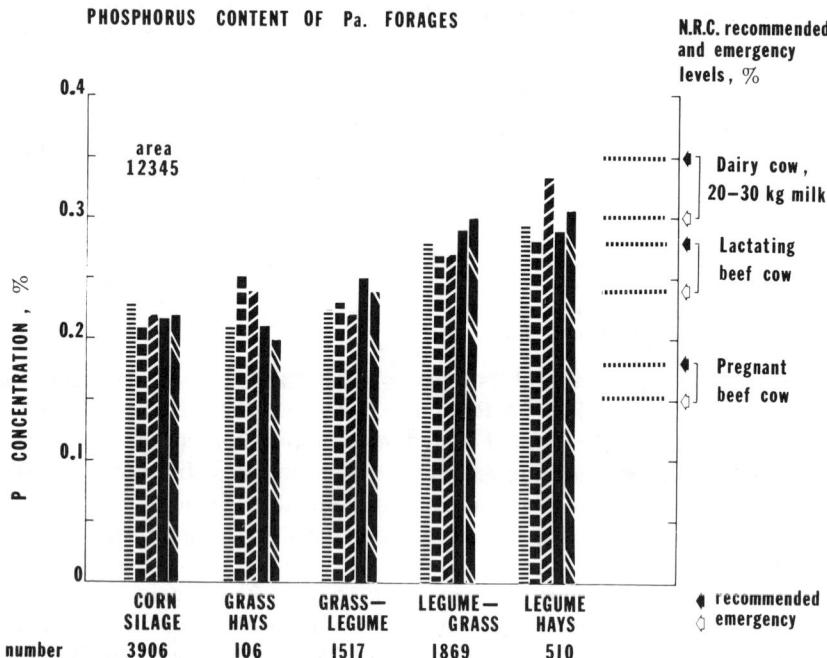

Fig. 2—Phosphorus content of forage crops from different areas of Pennsylvania, in relationship to NRC recommended and emergency levels for dairy and beef cows (NAS, 1974b) (data from Belesky[1] and Stout et al., 1977).

feeds to problems of dairy herd health and productivity. Figure 2 compares the P content of five major forage classes produced in five physiographic areas of the state, in relationship to the requirements of dairy and beef cattle at recommended and emergency levels. Variations in P concentration by region and parent material were less evident than for elements such as Mg, but significant regional differences were noted for corn silage, grass-legume, and legume-grass hays. The data also indicate clearly that by NRC standards, the main harvested or preserved forage crops provided adequate levels of P for the pregnant beef cow, were marginal for the lactating beef animal, and, with the exception of the class of legume hays, did not provide sufficient P to meet the needs of the dairy animal producing moderate to high levels of milk. In a study of the mineral composition of plant species on mountain pastures in central West Virginia (Baker & Reid, 1977), soil series did not have a consistent effect on mineral content of the herbage, but did appear to have considerable influence on the species present. The P content of the forage was more consistent among soil series than was either Ca or Mg. Obviously, the nature of the relationship between soil properties and composition and mineral levels of the plant population will depend a great deal on such factors as climate, history and nature of cultivation, and fertilization practices. The comprehensive review by Beeson (1941) does, however, indicate a frequent association between available soil P and the concentration of P in vegetable, cereal, and forage crops.

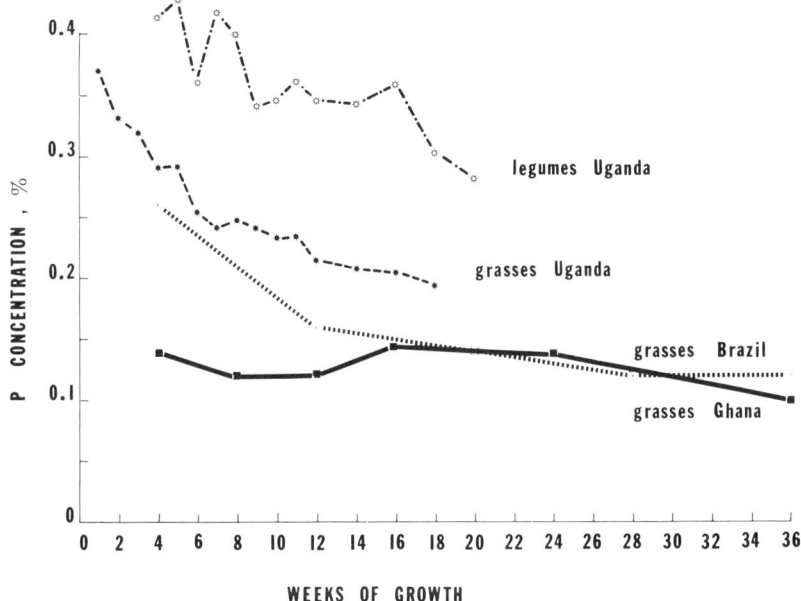

Fig. 3—Changes in P concentration of tropical grasses and legumes with age. Data on Uganda forages (Reid et al., 1975a) are mean values for 42 species and varieties of grass and 11 legumes, with fertilization. Data from Brazil (Gomide et al., 1969) are mean values for six grasses, with fertilization. Results from Ghana (Sen & Mabey, 1965) are mean values for 25 indigenous grasses, without fertilization.

B. Genus, Species, and Variety

Generally, in comparing the mineral composition of economically important forage classes, the most striking differences are to be found between grasses and legumes. For P, there is no particularly consistent evidence for such a difference in the temperate forages. The earlier work of Thomas et al. (1952) indicated that legumes contained higher levels of Ca, P, and Mg than the grasses, and the data from Pennsylvania summarized in Fig. 2 for classes of legume and grass hays would support this difference. Whitehead (1966), however, in a review of the literature, concluded that grasses and legumes contained approximately equal amounts of P, and studies in West Virginia (Reid et al., 1970; Baker & Reid, 1977), both with pure stands and with a range of species collected from pastures, showed no significant differences between grasses and legumes in P content.

Cooper (1973) and Fleming (1973) have referred to the generally lower levels of minerals in tropical forages than in temperate species. Relatively few data are available to permit comparisons of the mineral composition of tropical grasses and legumes. Figure 3 summarizes changes in the mean P concentration of grasses and legumes grown under the same fertilization and management conditions in Uganda (Reid et al., 1975a) and indicates also the effect of growth stage on P content. It is apparent that the legumes contain consistently higher levels of P than the grasses and that, under

Table 4—Heritability estimates for nutritional constituents, derived from a 6 by 6 diallel analysis of contrasting ryegrass varieties† (Cooper, 1973).

Constituent	Heritability (narrow-sense)	Range between extreme varieties, % of dry matter
In vitro digestibility	0.42	72.9 –76.7
Water-soluble carbohydrates	0.84	22.4 –47.0
N	0.63	1.96– 3.78
P	0.68	0.32– 0.47
Mg	0.86	0.14– 0.21
Ca	0.78	0.34– 0.56
K	0.80	2.65– 3.85
Na	0.55	0.07– 0.15

† Young vegetative material in seeding year.

reasonable levels of soil fertility, the P concentrations are comparable to those of legumes in temperate areas.

This study indicates also that there were marked species and varietal differences in mineral accumulation among tropical grasses and legumes. Similar differences in temperate forages have been reviewed by Fleming (1973), and the physiological basis for differential nutrient uptake by plants has been discussed by Vose (1963). It was found, for example, that the efficiency of P uptake may relate to root type, root development, and the proportion of secondary to primary roots in the plant. The possibility of exploiting genetic variation in selection programs for nutritionally superior forage plants is reviewed by Cooper (1973), who provides heritability estimates for components of Italian and perennial ryegrass varieties (Table 4). Hill and Jung (1975) showed significant variability in mineral accumulation between genotypes of Saranac alfalfa and concluded that it would be possible to develop populations with levels of P adequate to meet the requirements of a dairy cow, and with Ca/P ratios of the order of 3.0 or 3.5. It is, as Cooper points out, technically feasible to select and develop varieties with different levels of specific minerals; the problem is to assess the priority of this approach in relation to the options of modification of plant composition by fertilization or of supplying the mineral directly to the animal in the feed.

C. Stage of Maturity

A decline in the P content of a range of plant species with age has been demonstrated by many workers (e.g. Beeson, 1941; Thomas et al., 1952; Kivimae, 1959; Van Riper & Smith, 1959; Kirchgessner et al., 1967; Fleming & Murphy, 1968; Reid et al., 1970; Fleming, 1973). Changes in the concentration of P with increasing maturity are illustrated for tropical forages in Fig. 3. The very low levels of P in the Ghana study (Sen & Mabey, 1965) with indigenous grasses may be noted, and the fact that in these conditions there is little change in concentration with maturation of the plant. A summary of the effects of aging on the P content of selected temperate species is

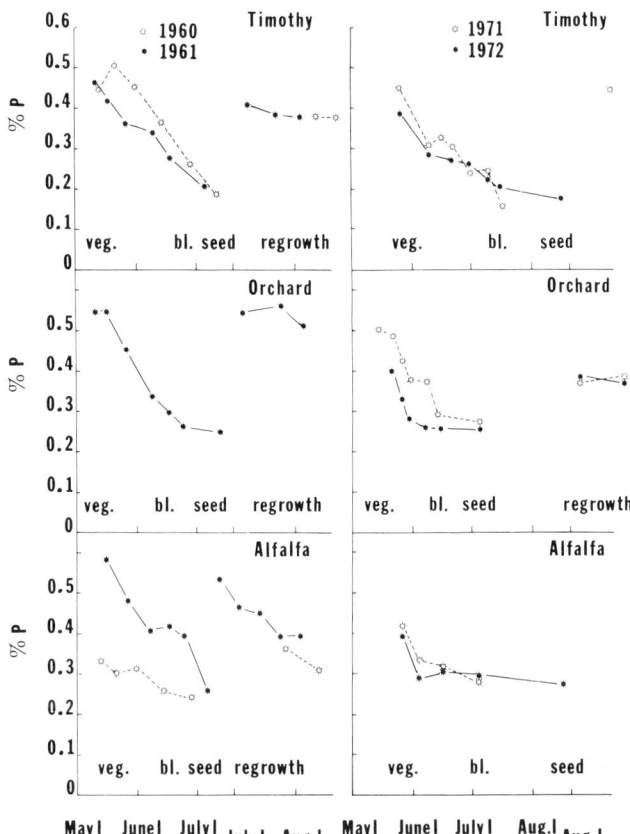

Fig. 4—Effect of stage of maturity in the first growth and regrowth on the P content of timothy, orchardgrass, and alfalfa in West Virginia trials (Reid et al., 1970; Baker & Reid, 1977). Samples in 1961 were taken from pure stands; samples in 1972 from grazed pastures.

given in Fig. 4, taken from West Virginia trials (Reid et al., 1970; Baker & Reid, 1977). The 1960-1961 results were obtained in pure stand trials and the 1971-1972 data under a pasturing program on different soils. The general similarity of effect of the maturation process on P levels in the plant is, however, evident. The decline in P content generally parallels that of protein concentration. It may be seen also that the P content of regrowth material is relatively high and would, in general, be adequate to meet the requirements of ruminant livestock.

The problem, then, in terms of feeding requirements lies mainly with the class of first-cutting hays which are harvested at a late stage of maturity. The prevalence of delayed harvesting, at least in the northeastern U.S., may be illustrated by referring again to the Pennsylvania data summarized in Fig. 1. The mean P concentrations for classes of first-cut grass hays, grass-legume, and legume-grass hays were 0.20, 0.22, and 0.27%, respectively. The first two classes represent the major part of the hay crop harvested in the temperate areas of this country; on an average, this type of forage might supply sufficient P for the maintenance of adult cattle and sheep and for the finishing of slaughter animals. It would not provide the levels of P required for lactation or for rapid growth. The same would be true for corn silage.

D. Season and Temperature Effects

Sullivan (1969) concludes that there is "little and inconsistent" seasonal variation in the P concentration of foliage, since the amount of available P in the soil is the primary factor regulating uptake by the plant. In examining such effects, where it is necessary to eliminate confounding due to maturation by clipping or grazing management, a further problem is introduced by the necessity to fertilize to stimulate growth after defoliation (Fleming, 1973). In cutting trials with grasses and white clover, Fleming and Murphy (1968) found a general pattern of seasonal decline in P and K content of the herbage. A definite seasonal pattern of mineral composition in herbage has been noted also in grazing studies with groups of ewes maintained on orchardgrass pasture and hay over a 5-year period in West Virginia (Reid et al., 1974a, 1974b). In these trials, herbage from the pastures was sampled at approximately monthly intervals, and the results for P concentration as they relate to changes in serum Ca and P values in the ewes are illustrated in Fig. 5. The data represent mean concentrations for four levels of N fertilizer applied to the pastures; maintenance P and K fertilizer applications were made in 1967 and 1969. Fairly consistent changes in P content of the herbage were observed from year to year, with primary peaks in April-May in the spring vegetative growth, a subsequent decline in midsummer, and a secondary peak in the fall regrowth. Changes in blood pat-

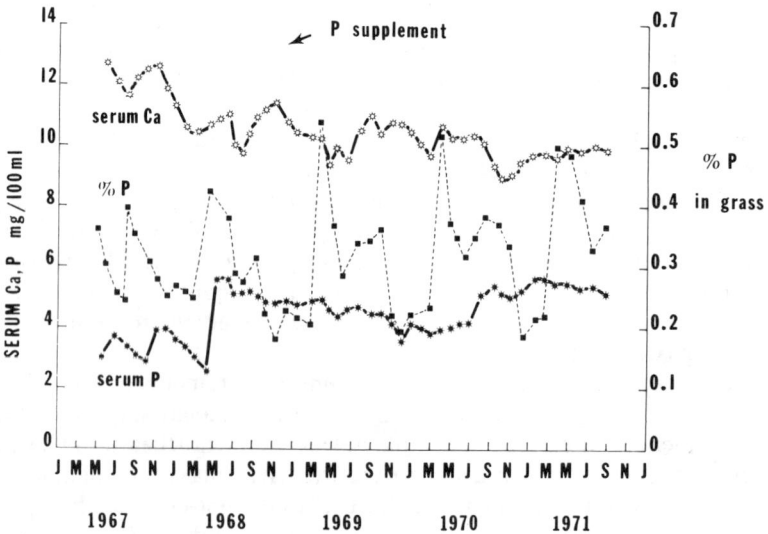

Fig. 5—Seasonal changes in herbage concentration of P, and in blood Ca and inorganic P concentrations of ewes maintained on orchardgrass over a 5-year period. Data are mean values for four N fertilizer treatments (from Reid et al., 1974a).

terns in the ewes are confounded by the fact that no dietary P supplement was provided until the animals had been on treatment for 1 year and were supporting their first lamb crop. At that time, serum inorganic P had declined to low levels (<2.5 mg/100 ml). With provision of dicalcium phosphate in a mineral mixture, the blood P values returned to a normal range and remained within this range for the remainder of the trial period. The results indicate the frequently borderline status of P nutrition in grazing ruminants, even when maintained on pastures of apparently adequate feeding value.

Fleming (1973) points out also that the decline in the P content of perennial grasses in summer is often accompanied by an increase in Ca, thereby creating the conditions for a Ca/P imbalance for grazing animals, and that the Ca/P ratio may also vary with the species of grass. In a New Zealand study, Metson and Saunders (1978) found that while the concentration of P in grass-legume pastures did not change markedly with season, there was considerable fluctuation in Ca/P ratios around an "ideal" value of 1.5; they did not consider that the variations were large enough to constitute a danger to animal health. The effects of temperature per se on the mineral composition of plants appear to vary with species (Smith, 1970, 1971). Balasko and Smith (1971) found in growth chamber studies that mineral concentrations were generally lowest at the temperature regime for optimal growth; concentrations of P in switchgrass decreased with increasing temperature, while P levels in timothy increased.

E. Fertilizers and Management

It is probably not surprising that reports in the literature on the effects of fertilization on the mineral composition of plants are frequently conflicting. Figure 6, adapted from Underwood (1966), indicates some possible relationships between soil nutrient status and plant growth and composition, and defines factors which may affect plant response to nutrient amendment. Allaway (1971) concludes that the effect of adding a nutrient in fertilizer may range from no increase in the concentration of the element in the plant—although marked yield increases may be obtained—to significant increases in the level of the element in plant tissue without any change in yield.

In a review of P fertilization studies in the western U.S., Peterson et al. (1953) extracted a number of general trends from what they define as "masses of heterogeneous and frequently contradictory data." They found that P fertilization consistently increased the yield and P composition of alfalfa, generally improved the response of small grain crops on deficient soils, and frequently increased the P content of the grain and straw. Fertilization often improved the quality of truck crops, increased the P content of leaves of tree fruits, and had a variable effect on the yield and P content of pastures and hay crops. Subsequent studies have confirmed the variable response of forage crops, in terms of composition, to the use of phosphatic

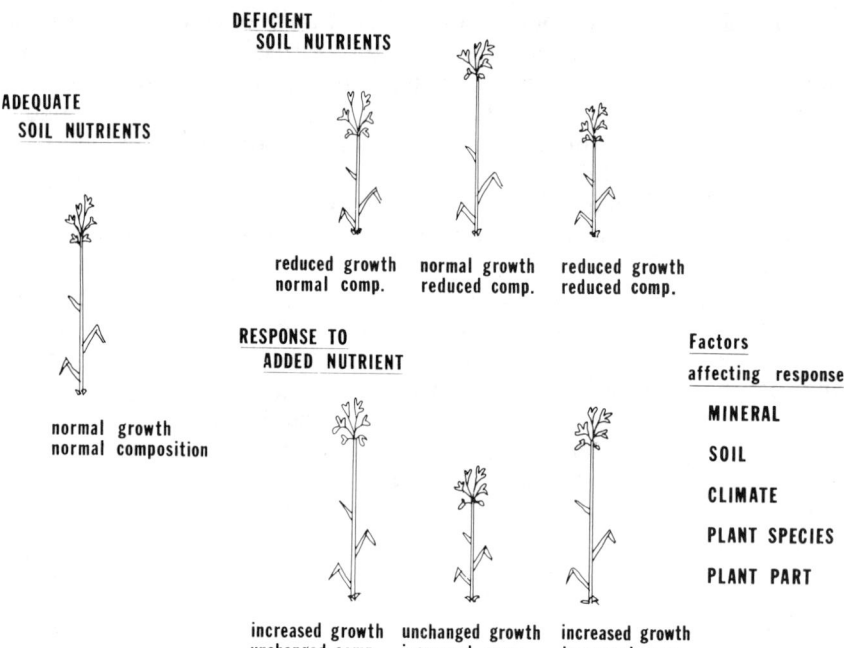

Fig. 6—Some possible relationships between nutrient status of soil and plant growth and composition, and factors affecting plant response to nutrient amendment (adapted from Underwood, 1966).

fertilizers (Reid & Jung, 1974). Whitehead (1966) concludes that alfalfa and red clover show a greater response (by an increase in P concentration) to P fertilization than do most forage species. Under hill-farming conditions, where deficiencies of Ca and P are common, Reith (1972) referred to major increases in both yield and P concentration of herbage obtainable with moderate applications of superphosphate. Similar responses have been noted with tropical grasses and legumes grown on low P soils (Andrew & Robins, 1969; Kirk et al., 1970). In many circumstances, however, only very moderate increases in P concentration have been obtained in response to P fertilization (Reith, 1965; Brown & Apgar, 1969).

The effects of N fertilization on P content have also been variable, and apparently relate to P availability status of the soil. Stewart and Holmes (1953) and Reith et al. (1964) found that N had relatively little effect on the P content of herbage, while other studies (Whitehead, 1966, 1970) have shown a considerable depression. Molloy et al. (1978) found that fertilizer N applied to ryegrass-white clover pastures in New Zealand did not appreciably change the concentrations of P, Ca, K, or Mg in herbage. The form of N used has been found to influence mineral uptake. MacLeod (1965) showed that heavy applications (672 kg of N/ha) of ammonium nitrate decreased P concentrations in grasses, and Nielsen and Cunningham (1964) noted that the nitrate form of fertilizer tended to reduce the levels of P, S, and chloride in ryegrass, while the ammonium form decreased the percentage of Ca. Similar effects have been observed for corn by Blair et al. (1970). In reviewing mineral balance studies carried out with dairy cows in the Netherlands, de Groot[2] concluded that when N was applied as the ammonium salt to pastures, the uptake of P, S, chloride, and Si was increased.

Since acidity of the soil may have a direct effect on the availability of soil P through the formation of insoluble Fe and Al phosphates, liming practices might be thought to bring about major changes in P accumulation by plants. Liming an acid soil generally results in increases in plant Ca, P, and Mg contents under temperate conditions, but the degree of change in P concentration is often modest. The response varies with soil pH; Doll et al. (1963), for example, found no increase in the P content of red clover when the pH was raised above 6.2. Reith (1972) has commented on the large increases in Ca content and minor effects on P in reseeded grass pastures brought about by higher levels of liming on acid soils (pH 4.8); at higher soil pH's (5.6), liming had little influence on composition. In mixed stands, marked changes in Ca content of the herbage with liming may result from a shift in botanical composition towards a higher content of legumes. The implications of such changes in Ca/P ratios of pastures are of considerable interest in relation to animal health, although cause-and-effect relationships are difficult to establish. Hignett (1950) and Crichton (1968), however, have suggested that the altered Ca/P ratios in improved pastures may be factors involved in an increased incidence of milk fever and reproductive problems in dairy cattle.

Finally, the influence of management on the P content of plants may be quite significant. Fleming (1973) has discussed the effects of grazing and cutting management on the mineral composition of forages and has attributed observed changes brought about by grazing to alterations in leaf/stem ratios and to returns to the pasture in the form of excreta. Whitehead (1970) found that the P content of N-fertilized grass decreased from 0.42 to 0.30% when the grass was harvested and removed, while no change was observed under a grazing system. However, while there is evidence for an increase in P yield of pastures due to return of animal excreta, it has been less easy to demonstrate consistent increases in the P content of the forage. The subject of nutrient recycling in soil-plant-animal systems has been reviewed by Wolton (1963), Barrow (1967), Mott (1974), and Frissel (1978).

F. Mineral Distribution

The partitioning of P between different parts of a plant will obviously be of nutritional significance for the grazing animal, for animals maintained on crops harvested at different stages of maturity, and for animals fed greater or lesser proportions of plants which have been physically separated into, for example, grain and straw fractions. Allaway (1971) used data from Larson et al. (1952) to illustrate the marked differences in P concen-

---

[2] T. de Groot. 1967. Mineral balance of milking cows on high nitrogen swards. Mimeo. Brit. Soc. Anim. Prod., Sept. 1967.

Table 5—Distribution of P in flowering shoots of six temperate grasses at early anthesis (Smith, 1973).

Plant part	P content in dry matter (%)					
	Timothy	Brome-grass	Orchard-grass	Canary-grass	Tall fescue	Quack-grass
Inflorescence	0.39	0.35	0.39	0.46	0.34	0.39
Leaf blade	0.18	0.25	0.27	0.23	0.21	0.23
Leaf sheath	0.13	0.15	0.19	0.15	0.16	0.17
Internode	0.20	0.17	0.21	0.12	0.21	0.15
Herbage (shoot-stubble)	0.21	0.22	0.26	0.16	0.22	0.20
Stem base (stubble)	0.20	0.14	0.20	0.10	0.15	0.15
Shoot	0.21	0.21	0.25	0.15	0.22	0.20

tration between oat grain and oat straw. Even with P fertilization, oat straw would be an inadequate source of P for ruminants, whereas oat grain would provide sufficient P (0.3 to 0.4%), even without fertilization. The relatively high levels of P in grain crops are shown in Table 2. The distribution of mineral elements in forage species has been examined by a number of workers (Fleming, 1963; Pritchard et al., 1964; Davey & Mitchell, 1968). For legumes, P is present at higher concentrations in the flowering heads than in other parts of the plant. Smith (1973) sampled flowering shoots of a number of temperate grasses at the stage of anthesis and a summary of the P content of the shoot parts is given in Table 5. While species differences are apparent, there is a common pattern for highest concentrations of P to be found in the inflorescence, with a marked decline in the leaf blade fraction, and generally lowest levels in the leaf sheath, internode, and stem base parts. Fleming (1973) pointed out that the availability of minerals to the animal may vary with the fraction of the plant in which the mineral is located, and this may be particularly important in the case of P. Phosphorus occurs in plants in inorganic and phytate forms. Phytate P is poorly utilized by monogastric species and may reduce the availability of other minerals. Phytic acid is normally stored in seeds and there is little evidence for significant concentrations in vegetative growth. Cohen (1975), however, suggests that the phytate/inorganic P ratio may be of practical importance under conditions where cattle are maintained during the winter on senescent pasture of low P content and a high Ca/P ratio.

IV. AVAILABILITY OF PHOSPHORUS IN DIETS

So far, the adequacy of plant sources to provide the P requirements of animals has been discussed only in terms of total P concentration of the diet. It will be apparent that only a fraction of the mineral content of any feed will be absorbed from the alimentary tract and utilized for productive purposes. It is therefore necessary to consider briefly the question of "availability" of P, as this may have a profound effect on dietary recommendations. That the question is a complex one is illustrated by the number of factors cited by Peeler (1972) as affecting utilization: type of ration fed; chemical form of the element; the Ca/P ratio; age of animal; sex; fat and energy levels; plane of nutrition; environment; hormones; disease and parasites; protein and microelement levels; interactions with other minerals and nutrients; chelating agents; the physical nature of P sources and other feedstuffs in the diet (particularly particle size); feed processing; and others. The problem is further compounded by the fact that there is no unanimity among animal scientists in the definition of availability, nor in the use of techniques for its determination. Among the more common methods are those discussed by Thompson (1965):

% apparent digestibility (or apparent absorption) $= I - F$

% true digestibility $= I - (F - Fe)$

% net retention $= I - (F + U)$

% availability $= I - (F - Fe) - (U - Ue)$

where I is intake of dietary element, F is total excretion of element in feces, U is excretion of element in urine, Fe is net excretion of element of body origin into intestinal tract, and Ue is excretion of element in urine at zero net retention. All of these terms have been used as measures of availability. The interpretation of availability data therefore requires considerable care; they may, however, serve to illustrate the relative extent of P utilization in the major classes of feedstuff or supplement.

A. Forages

Probably less is known about the availability of P in forage crops than in any other feed. Butler and Jones (1973), reviewing the literature on availability of herbage minerals for herbivores, cite the studies of Lofgreen and Kleiber (1953) indicating an apparent digestibility of 22% and a true digestibility of 91% for P in lambs fed alfalfa. Hill (1961) and Underwood (1966) state that there is evidence for a more efficient absorption of P than of Ca from forages. Kemp (1966), in experiments to determine mineral balance in dairy cows fed fresh herbage, obtained mean apparent availability values of 30, 17, and 27% for Ca, Mg, and P, respectively. In balance trials with sheep fed four species of grasses and legumes harvested at different growth stages in several growth cycles, Gueguen and Demarquilly (1969) found little variation in the amounts of P and Ca absorbed by the animals, but considerable differences in the amounts of mineral ingested. Calculated true availabilities for P were generally in the range of 40 to 90% and the availability of P increased markedly with maturation in the first growth cycle. Interestingly, in terms of balance, only alfalfa and meadow fescue were found to provide positive or zero balances of P and Ca during the year; perennial ryegrass gave variable results, and balances on timothy were always negative.

In similar trials with wether lambs fed a variety of grass and legume hays grown on different soils in Pennsylvania and West Virginia, it was found that P balances on grass hays cut at conventional harvesting dates were frequently negative, while balances on legume hays (red clover and alfalfa) were consistently positive (Reid et al., 1978). The apparent absorption values for P in these trials were low—a mean of 4.3% for 28 grass hays and of 17.0% for 10 legume hays. In trials with lambs fed different grasses in the fresh form, significant species and growth stage effects on the apparent absorption and retention of P were obtained (Reid et al., 1975b). Both of these studies showed a significant positive correlation between the concentration of P in the forage and its apparent absorption (%) and retention (g/day) by lambs. Joyce and Rattray (1970) in New Zealand ran a series of mineral balance trials with growing lambs fed white clover and perennial ryegrass and obtained a mean apparent absorption or digestibility figure for P of 10.2%; using an estimated value for endogenous fecal P, true availability was calculated to be approximately 74%.

Whatever the limitations of the balance trial approach, these recent studies show that: (i) forage crops and pastures provide only marginal levels of available P for ruminant species, even where concentrations of P in the forage appear adequate by recommended standards; and (ii) there are significant plant species and growth stage differences in P availability, with legumes apparently supplying more available P than the grasses. The further investigation of possible species and varietal differences in P availability among the grasses would merit attention.

B. Grain and Animal and Plant Protein Feeds

More than 50% of the P in grains and seeds may be present in the form of phytin, a compound made up of Ca, Mg, and K salts of phytic acid. McGillivray (1974) states that phytin P can constitute 75% of the total P of soybeans, followed by a phosphatide or phospholipid fraction (12%), then by nucleic acid P (6%) and, finally, inorganic P at only 4.5% of the total. The availability of phytic acid P is low in poultry and swine; Peeler (1972) summarized the findings of a number of authors using different techniques to estimate phytate availability in pigs (Table 6). Considerable disagreement has appeared in the literature dealing with the availability of phytate P to poultry. General evidence, however, indicates that availability to the chick is low (values of 10 and 20% have been reported), and that the ability to utilize phytate P may increase with age (Nelson, 1967). The Ca content of the diet may influence the ability of the laying bird to utilize phytate, and the P requirement may be related directly to level of egg production and to management of the flock (Singsen et al., 1962). Few comparable data are available for the human. The existence of phytase enzymes in the alimentary tract of man has not been shown, although Leitch (1964) concludes that "it

Table 6—Biological value of phytate P for swine (Peeler, 1972).

Reference	Weight of pig	Biological availability
	kg	%
Bayley and Thomson (1969)	27.2	20–30
Woodman and Evans (1948)	22.7–40.9	30–40
Besecker et al. (1967)	22.7	18–24
Noland et al. (1968)	Growing	30–60
Avg.		25–40

seems likely that those who habitually eat large quantities of undermilled cereals have the means to break down the phytate complex." The major concern in human nutrition is whether the phytate present in cereal flours may impair absorption of Ca. Irving (1973) refers to experiments carried out in England during World War II indicating that bread made from high-extraction flours contained enough phytic acid to slowly demineralize the skeleton, and points out that in England chalk is added to the flour to offset the action of phytate, while in the United States dried milk and improvers are used as a supplement. It is concluded, however, that for most diets normally consumed by man there is no danger from anticalcifying factors in cereals. A recent study (Reinhold et al., 1976) in this area demonstrated that Iranian subjects developed negative balances of P, Ca, Mg, and Zn when white bread in the diet was replaced by a wheaten wholemeal bread, even though consumption of the wheaten bread increased P intake significantly. The authors considered that the effects on mineral balance might have been due either to the phytic acid or the fiber content of the diet.

In contrast to the situation in the monogastric animal, early studies reported by Tillman and Brethour (1958) showed that Ca phytate was as available to sheep as monocalcium phosphate, and Dutton and Fontenot (1967) found no differences in the absorption of P from phytic acid or monosodium dihydrogen orthophosphate, although retention of P from the organic source was lower. Sullivan (1969), in fact, concludes that all forms of P are available to the ruminant. In an extensive study of the availability of the major minerals, Lomba et al. (1969) conducted balance trials on dairy cows fed a wide range of diets containing both forages and concentrates, but the data do not show clear effects of ration composition on P utilization. This work, in fact, illustrates the frequent difficulty of interpretation of balance trial results—P balances were very variable, ranging from -42.7 to 40.9 g/cow per day for dry cows, and from -23.5 to 47.7 g/cow per day for lactating cows. Endogenous P excretion was variable and high and there was no significant relationship between P intake and output in the feces. The authors also concluded that P balance may depend on the previous P status of the animal.

The P in animal protein sources seems to be as available as that in inorganic form. McGillivray (1974) gives values of 95 to 102 for biological availability of P (based on bone ash) in fish meal, poultry byproduct meal, and meat and bone meal when compared with inorganic supplements such as dicalcium phosphate or monosodium phosphate.

C. Inorganic Supplements

A great deal of work has been directed to examining the availability of P in mineral supplements for all classes of livestock. Underwood (1966) comments that it is still widely held that minerals derived from natural feedstuffs will be better utilized than those from inorganic sources, although most experimental data show the reverse to be true. In an earlier review, Hill (1961) concluded that the availability of P from supplements for ruminant animals was appreciably greater than that of Ca, but depended on the type of phosphate present; when the P was in an insoluble form such as Fe-Al-phosphate, or as meta- or pyrophosphate, utilization was essentially nil. Peeler (1972) reviewed the more recent literature and ranked the biological availability (based on a range of criteria) of several inorganic P sources for swine as follows: sodium phosphate = phosphoric acid = monocalcium phosphate > dicalcium phosphate > defluorinated phosphate = bone meal > low fluorine phosphate rock > soft phosphate. Much the same order of ranking has been obtained with chickens and laying hens and, indeed, with ruminants such as beef cattle, dairy cattle, and sheep. Apparently the only consistently unsatisfactory source of P is soft phosphate with colloidal clay. Cohen (1975) mentions problems encountered in Australia with monosodium phosphate as a supplement for sheep, and with phosphoric acid for sheep and cattle, but the evidence is not conclusive. With the development of a P shortage, other inorganic sources have been investigated. The P in fertilizer phosphates (e.g., triple superphosphate, ammoniated phosphates) appears to be highly available when compared, for example, with sodium phosphate, and the use of such material for livestock feeding would appear to be limited primarily by its fluoride content rather than by considerations of availability.

V. MAINTAINING ADEQUATE PHOSPHORUS NUTRITION IN ANIMALS

Animal species differ in their requirements for minerals; in their tolerance of or susceptibility to deficiencies, excesses, or imbalances of specific elements; in their ability to digest and absorb minerals from various dietary forms and complexes; and, frequently through management systems imposed by man, in the choice of dietary components from which minerals are derived. These differences need to be taken into account in considering the possibility of amendment of existing dietary practices or the use of new forms of mineral supplement to maintain optimal performance and health. The literature on the P and Ca nutrition of livestock and man is extensive, and no attempt will be made to review the subject in detail. Rather, selected references will be used to define for a limited number of economically important animal species the nature and consequences of P deficiency, dietary factors which may affect P status, the establishment of "minimal" and "optimal" requirements, and the feasibility of modifying dietary P levels without impairment of productivity.

A. Ruminants

Accounts of cattle suffering from clinical aphosphorosis in the earlier South African studies (Theiler & Green, 1932) included descriptions of depraved appetite, low milk yields, depressed growth, reproductive failure, and bone fractures. Unfortunately, certain of these conditions are not

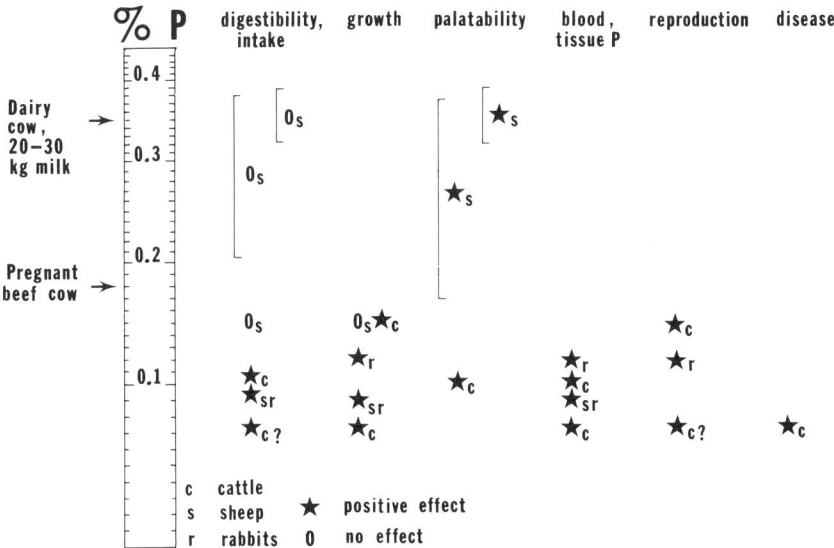

Fig. 7—Diagrammatic representation of relationship between P concentration in diet and demonstration (or otherwise) of P deficiency signs in cattle, sheep, and rabbits. References are drawn from the American, Australian, European, and South African literature and have been discussed by Reid and Jung (1974).

unique to aphosphorosis and may result in part from simultaneous deficiencies of other nutrients, such as energy and protein. Little (1970) considers that a basic effect of P deficiency is a reduction in feed intake, upon which other disturbances inevitably follow, and Underwood (1966) states that P deficiency has a specific effect on interruption of the estrus cycle, resulting in delay or prevention of conception. Reid and Jung (1974) reviewed the literature relating the concentration of P in forage crops or pasture to various criteria of P deficiency in cattle, sheep, and rabbits. A summary of these data is presented in Fig. 7, which shows a general trend for detectable deficiency signs in these species at a concentration of 0.07 to 0.15% P in the diet, although effects on palatability of forages have been observed at considerably higher levels. Little (1970) also stressed the need to consider *subclinical aphosphorosis,* which he defines as "a syndrome of insidious onset causing nonspectacular depressions in rate of production, but nonetheless of serious economic consequence." While such effects are probably very real, it has not yet been possible to define them in any quantitative fashion.

1. BEEF CATTLE

Much of the natural range and grazing land on P-deficient soils throughout the world may produce forage containing less than the average 0.1% associated with deficiency conditions (Fig. 7). This is particularly the case for mature forage under arid conditions.

The NRC (1976) recommendations for P in the diet of beef cattle are based upon estimates of body requirements and of requirements of rumen microorganisms (Table 2). For growing-finishing animals (weight 400 kg) the requirements are given as 0.18 to 0.26% P in the ration; for the pregnant beef cow (454 kg), 0.18%; for the lactating cow producing moderate amounts of milk (approx. 7.5 kg), 0.28%; for the cow producing high levels of milk, 0.38%. It is considered that these levels are close to the minimum required for optimal performance. An NRC (1974b) report states that it may be possible to reduce the intake of P as much as 10% below the NRC recommendations without depressing animal performance. Under more severe conditions of P shortage it is suggested that pregnant and milking cows on forage-based diets be given priority over growing and fattening cattle under feedlot conditions where grains make up a higher proportion of the diet. It may be noted that the NRC (1976) recommendation for the beef cow in the last third of pregnancy is 15 g/day, compared to the ealier (NRC, 1970) recommended level of 12 g/day. Hemingway (1978) found that beef cows fed an oat straw-urea-beet pulp diet providing 12 g/day in late pregnancy and early lactation showed a depression in straw intake and a decreased digestibility of the whole diet.

The question of whether to correct what amounts to a normal deficit of P in beef cattle under grazing conditions by direct supplementation or by soil amendment is mainly an economic one. In low-rainfall areas of sparse vegetation, direct supplementation would be the preferred course. With higher rainfall, P fertilization frequently results in both higher yields and increased concentrations of P in the herbage. Hodges et al. (1964), for example, refer to Florida trials in which the P content of pangolagrass was increased from 0.08 to 0.27% by annual fertilization with 56 kg/ha of superphosphate. They found that beef cows stayed in good health and at moderate levels of production while deriving all their energy, protein, Ca, and P from pasture alone. Under similar conditions in Australia, Underwood (1966) found that application of 56 kg/ha of superphosphate doubled the pasture yield but increased the P concentration only from 0.08 to 0.12%. Cattle maintained on such pastures would require further direct supplementation.

If plant response to soil treatment with P is variable, so also is the response of animals to direct supplementation with P compounds. Cohen (1975) pointed out that cattle in Australia on P-deficient pastures show much less pronounced responses to supplements than have been reported in South African and American studies. In trials in New South Wales, Cohen (1972), for example, found no effect of P supplementation on the liveweight gains of steers grazed or fed carpet grass having a P content ranging from 0.04 to 0.11%, nor was there any effect on the apparent digestibility of dry matter, N, or energy. The author suggests that such discrepancies in the literature from different geographic areas may result from interactions of P nutrition with other dietary factors such as vitamin D, protein, available energy, S, or Ca. Call et al. (1978) examined the effects of raising beef heifers for a 2-year period on diets containing 0.14% P as compared to 0.36% P, with adequate levels of energy, protein, vitamins, and other minerals. They found no significant effect of P concentration on intake, feed efficiency, age of puberty, or pregnancy rate.

The significance of the Ca/P ratio of diets for both beef and dairy cattle has received a great deal of attention during the last several years, and it has been suggested that the better responses to P supplementation in American and South African studies may be due to the maintenance of more favorable Ca/P ratios rather than to the provision of P alone. In contrast to species such as the rat, pig, and chick, which require rather narrow Ca/P ratios for optimal growth, the ruminant is less sensitive to the ratio of Ca/P in the diet. For beef calves, Wise et al. (1963) found that nutrient conversion and performance were depressed at Ca/P ratios lower than 1, that ratios between 1:1 and 7:1 gave similar responses, and that performance decreased again, but to a lesser degree, when the ratio exceeded 7:1. Ricketts et al. (1970) fed growing dairy steers diets containing Ca/P ratios of 1:1, 4:1, and 8:1. They found that animals receiving the 8:1 diet gained less daily than the other two groups, and that steers fed the 1:1 ratio had higher concentrations of serum inorganic P, lower blood Ca values, and a narrower blood Ca/P ratio than steers fed the other two Ca/P ratios.

It should also be stressed that adequate supplies of vitamin D are required for the absorption and utilization of both Ca and P and that, as stated by Underwood (1966), "Dietary ratios (Ca/P) found unsatisfactory under conditions of low vitamin D supply could be quite satisfactory when supplies of the vitamin are generous." On most types of diet fed to beef cattle, with the possible exception of a ration consisting mainly of alfalfa hay, it would be highly unlikely that a dietary Ca/P ratio of 7:1 or 8:1 would be exceeded.

Under typical feedlot fattening conditions there would usually be little problem of P supply, although the NRC (1970) bulletin illustrates a case in which deficiency occurred on a diet consisting of wet beet pulp, alfalfa hay, and beet molasses (0.12% P). Where the ration is based on corn grain, corn silage, and soybean meal, the P concentration normally ranges from 0.28 to 0.35%, and Preston and Stone (1974) found no advantage to adding supplemental P to such a diet. Stone (1974), in fact, comments that feedlot rations for beef animals tend to be oversupplemented with P, and that P should only be added to the ration for finishing beef cattle when urea is the source of supplemental N.

2. DAIRY CATTLE

The quantitative requirements for P of the lactating dairy cow are, obviously, much higher than those of the beef animal. The mature beef cow producing moderate amounts of milk requires approximately 25 to 28 g of P/day; the dairy cow giving 30 kg of milk requires 75 to 80 g of P/day by NRC (1978) standards. In terms of percentage requirement, however, the difference is not as great as might be expected, since the high-producing dairy cow is able to adjust its intake of food substantially to meet its energy and nutrient requirements (Underwood, 1966). Thus, the recommended NRC percentage of P for the dairy cow producing less than 20 kg of milk per day is 0.33; for the cow producing more than 30 kg of milk it is 0.39 (Table 2). Normally, the high-producing dairy cow at peak of lactation will not be able to meet the metabolic demand for Ca and P from dietary sources alone, and will mobilize bone reserves, resulting in a negative Ca and P balance. It is then necessary to ensure that the supply of these minerals in the ration is maintained at a high level during late lactation and the dry period to allow for repletion of the skeleton. Current research, as reviewed by Todd (1976), suggests the desirability of feeding a prepartum diet low in Ca (<0.50% Ca) but adequate in P and vitamin D for the prevention of milk fever.

In the dairy animal, deficiency of P has been found to result in (i) loss of appetite; (ii) a decline in blood inorganic P concentrations to values of the order of 2 mg/100 ml in mature animals and 3 mg/100 ml in calves; (iii) fragility and fracture of bones; (iv) decline in milk production; (v) depraved appetite, e.g., chewing of bones; and (vi) reproductive disturbances, e.g., low conception rates and failure to come into estrus. There is still considerable debate as to whether metabolic and health problems associated with P supply in dairy animals are due to outright deficiency of the element or to aberrations of the Ca/P ratio in the diet. The Dutch workers (Comm. on Miner. Nutr., The Hague, 1973) state that the ratio has no particular significance and that, as long as the cow receives adequate Ca, P, and vitamin D, a wide range of ratios will not affect productivity or health. Smith et al. (1966) found that ratios of Ca/P of 1:1, 4:1, and 8:1 in the diet of lactating dairy cows had no effect on feed utilization or production, while other studies (Colovos et al., 1958; Kendall & Byers, 1959; Littlejohn & Lewis, 1960; Manston, 1967) have shown that an excess of Ca over P, as found, for example, in diets based on legume forages or supplemented with limestone, may lower growth rate or reduce the utilization of P in the ration. It is also possible that an imbalance of Ca/P may decrease the availability of microelements in the diet. Buchanan-Smith (1978) concludes that, with the exception of milk fever prevention, it is reasonably safe to allow up to a 4:1 Ca/P ratio provided that minimum P requirements are met from readily available sources.

Dietary P and the ratio of Ca/P have been implicated in the disease of milk fever (parturient paresis or hypocalcemia) in dairy cows. Serum inorganic P, as well as blood Ca, is generally depressed in the paretic cow. Blood values, as summarized by Todd (1976), show the following differences:

Blood value	Normal cow	At calving	
		Normal	Milk fever
		mg/100 ml	
Blood serum Ca	10	7	3–5
Blood inorganic P	5–6	3–4	1–3
Blood serum Mg	2.5	3.2	3.4†

† 20% of cases may show low serum Mg concentrations of 0.5 to 1.5 mg/100 ml.

Stott (1963) found that adjusting the dietary Ca/P ratio to approximately 1:1 to 1.6:1 reduced the incidence of parturient paresis in cows with a previous history of the condition, and present evidence indicates that a Ca/P ratio in the region of 1:1 to 2:1 in the prepartum diet may be optimal for the prevention of milk fever in dairy cattle. A number of workers (e.g., Gardner, 1970, 1971; Gardner & Park, 1973; Kendall et al., 1968, 1970) found that cows fed a 2.3:1 Ca/P ratio had a lower incidence of milk fever than cows receiving diets with either wider or narrower ratios, although this finding is not supported by the studies of Beitz et al. (1974). The latter group concluded that factors other than Ca/P ratio must be involved in the development of milk fever in dairy cattle.

Phosphorus deficiency, or imbalances of dietary Ca and P, has also been associated with problems of dairy herd fertility, but the evidence for such relationships is, at best, equivocal. Morrow (1970) defined infertility losses as due to retained placenta, metritis, anestrus, silent estrus, cystic follicles, repeat breeding, and abortions, and concluded that a primary factor in a herd with a history of repeat breeding was a deficiency of dietary P in high-producing cows during early lactation, although there was a simultaneous deficit of dietary energy. The interaction between energy and P is further emphasized in a study by Carstairs et al. (1975), in which it was shown that Holstein heifers fed high-energy and low-P, or low-energy and high-P, diets had a longer interval to first estrus after calving than animals maintained on high-energy, high-P or low-energy, low-P rations.

The NRC (1974b) report indicates that in the event of a shortage of inorganic phosphorus, the earlier NRC recommended levels for P could safely be reduced by about 10%, with the highest priority given to the lactating cow to meet the requirement for milk and normal reproduction, and the lowest priority to dry cows and to growing animals over 100-kg liveweight. Such reductions seem reasonable, as under most modern dairy feeding programs where an inorganic source of P is normally added to a ration based on grain, a protein supplement and corn silage, or a high-quality forage, the P supply may frequently exceed the animal's requirement. Noller et al. (1977) found that the lower level of P recommended in the NRC (1974b) report was adequate to meet the requirements of growing dairy heifers, as evidenced by weight gains, efficiency of feed conversion, conception rate, and concentrations of minerals in blood serum. Pritchard (1974) also concluded from a ration evaluation program on a large number of Ohio herds that the average cow producing approximately 20 kg of milk was receiving 44 g less Ca and 17 g more P daily than required. Concern has frequently been expressed that dairy rations which include a high-quality alfalfa hay or haylage as the main forage source may be imbalanced in terms of the Ca/P ratio, and that this may result in a variety of metabolic problems in the dairy animal. While there is no convincing experimental evidence for such effects, it may be desirable to adjust the mineral supplementation of the ration to provide Ca/P ratios between 1:1 and 2:1. Guidelines for such adjustment are provided in Table 7 by Jacobson et al. (1972).

Table 7—Guide for including the indicated amounts of Ca and P in a mixture containing primarily corn and soybean meal† (Jacobson et al., 1972).

Ration	Kind of roughage fed	Amount	Concentrate fed	Feed grade limestone	Dicalcium phosphate	Monosodium phosphate
		kg		%		
1	Excellent alfalfa hays§	13.6	8.6	0	0	0.7
2	Good alfalfa hay¶	11.4	10.9	0	0	0.6
3	Avg. alfalfa hay#	9.1	12.3	0	0	0.45
4	Poor alfalfa hay	6.8	14.1	0	0	0.3
5	Alfalfa-grass hay	11.4	10.9	0	0.1	0.65
6	Good alfalfa hay†† and corn silage	9.1 15.0	8.6	0	0	0.85
7	Avg. alfalfa hay and corn silage	4.5 25.9	8.6	0	0.25	0.6
8	Good orchardgrass hay	9.1	12.3	0.85	0.8	0
9	Good orchardgrass hay and corn silage	6.8 20.9	8.6	1.2	1.2	0
10	Avg. grass hay and corn silage	4.5 26.4	8.6	1.25	1.15	0
11	Corn silage only	35.0	8.6	1.9	1.1	0
12	Corn silage only	29.1	10.9	1.0	0.7	0

Heading above mineral columns: Amount of mineral‡ to be included in the concentrate to provide adequate P and a 1.5 to 1.0 Ca/P ratio in total diet (90% DM)

† Assuming 20% soybean meal and 75% corn in concentrate mixture, 30 kg milk.
‡ Computations based upon feed grade ground limestone (38% Ca, 0% P), dicalcium phosphate (26% Ca, 18% P), and monosodium phosphate (0% Ca, 25% P).
§ Ca/P ratio of total diet would be 4.1 to 1.0 (higher than desired).
¶ Ca/P ratio of total diet would be 2.8 to 1.0 (higher than desired).
# Ca/P ratio of total diet would be 2.1 to 1.0 (still somewhat higher than desired).
†† Ca/P ratio of total diet would be 2.4 to 1.0 (higher than desired).

3. SHEEP

Underwood (1966) and McDonald (1968) stated that P deficiency occurs more frequently and more severely in grazing cattle than in grazing sheep, due both to the fact that sheep have higher energy requirements per unit of bodyweight and to the superior ability of sheep to select leafy material of high P content. Cohen (1974) also considers that sheep retain P (on a unit of metabolic body weight basis) more efficiently than do cattle when the intake of P is low. Australian researchers have claimed that grazing sheep do not respond to P supplementation. This was not the case in the grazing trials summarized in Fig. 5; here, even though herbage P concentrations appeared adequate, young lactating ewes responded immediately to the provision of a dicalcium phosphate–salt mixture by changes in blood P and Ca concentrations (Reid et al., 1974a). McMeniman and Little (1974) demonstrated a marked interaction between supplemental P and energy in the response of sheep fed mulga in Queensland; P supplements (monosodium orthophosphate in water) increased liveweight and wool growth only when fed with molasses. This type of interaction has been noted previously with dairy cattle.

The current NRC recommendations for P have been increased to allow for high levels of production. Recent trends in this direction, as summarized by Hogue (1974), include increased confinement, early weaning, increased multiple births, decreased lambing intervals, and earlier first breedings. Thus, the NRC (1968) standard for a 70-kg ewe in the first 8 to 10 weeks of lactation was 0.20%; in the 1975 publication the corresponding percentage is 0.34%. Recommendations for the fattening lamb have not been changed substantially and range from 0.23% P for a 30-kg lamb to 0.16% for a 55 kg-animal. Bushman et al. (1965) obtained no improvement in liveweight gain in lambs with rations containing more than 0.25% P when fed at normal Ca concentrations. In practice, feedlot lambs are generally fed rations containing 0.35 to 0.40% Ca, and 0.25 to 0.30% P (Church, 1972). This level of P is easily provided by most high-concentrate diets.

Phosphorus supplementation in lamb rations requires care, as several studies have shown a relationship between P intake and the incidence of urinary calculi. Bushman et al. (1965) and Emerick and Embry (1964) found that concentrations of 0.80 and 0.60% P in the diet resulted in a higher frequency of calculi in lambs than concentrations of 0.35% or less. More recent studies in South Dakota (Hoar et al., 1970a, 1970b) showed a reduction of calculi formation by increasing the level of Ca in the diet. Phosphorus at a concentration of 0.47% reduced weight gains, but this effect was overcome by the addition of $CaCl_2$ or ground limestone; Ca as $CaCl_2$ was more effective in reducing the incidence of calculi than Ca as ground limestone. The highest incidence of calculi (approximately 50%) was obtained at dietary levels of 0.47% P and 0.31% Ca or a low level of ground limestone (0.56% Ca). Church (1972) noted also that the feeding of base-forming supplements such as disodium phosphates, sodium tripolyphosphate, or dipotassium phosphate may cause an increased incidence of calculi.

With the exception of this specific metabolic problem, sheep appear to be as tolerant as cattle to fairly wide variations in Ca/P ratios in the diet. Young et al. (1966) found that a diet with adequate P and a Ca/P ratio of 9.9:1 did not result in decreased absorption of dietary P; however, when the diet was deficient in P (0.076%), the availability of P was reduced. The authors considered that a greater acidity at the site of absorption (the upper small intestine) might account for the greater tolerance to wide dietary Ca/P ratios in ruminant species than in the monogastric animal.

The NRC (1974b) report recommends the following order of priorities for P use in sheep in the event of limitations of supply: (i) ewes in the first 8 weeks of lactation; (ii) ewes in the last 8 weeks of lactation, particularly when suckling twins; (iii) lambs weaned at 1 month; (iv) lambs 3 to 6 months old; (v) ewes in late lactation, suckling singles; (vi) replacement lambs and yearlings; (vii) fattening lambs 6 to 10 months old; (viii) ewes at maintenance and in early pregnancy; and (ix) rams. It is considered that under conditions of P shortage the grazing ewe should be given a P supplement only when the area is known to be P deficient. This situation may, however, be more prevalent than is generally recognized.

B. Monogastric Species

As indicated by Underwood (1966), P deficiency in the economically important monogastric species is less of a problem than is Ca deficiency, due to the nature of the diets ingested. Corn grain contains approximately 0.03% Ca, compared to 0.27% P. Also, the Ca requirements for growth, and for the production of eggs by the laying bird, are much higher in monogastric animals than in ruminants. However, in view of the high cost of P supplements and of the effect of dietary Ca concentration on phytate P utilization, it is important that "the fullest use should be made of the phosphorus in cereals and protein concentrates and that the margin of safety allowed in dietary formulations should be minimal" (Taylor, 1975). The dietary requirements for P will be considered for poultry, for swine, and for humans.

1. POULTRY

Unfortunately, much of the P in dietary plant sources for poultry is in unavailable form, particularly for the young bird. There is considerable variation within feed ingredients of plant origin in the proportion of P present in the phytic acid form. Nelson et al. (1968a) found, for example, that phytate P as a percentage of total P ranged from 56 to 89 within the grains and grain byproducts. Later work (Nelson et al., 1968b, 1971) indicates that it may be possible to increase phytate P utilization by the incorporation of a mold-produced phytase enzyme in the diet of chicks. Thayer and Jackson (1975) reported that the inclusion of a live yeast culture at a level of 2.5% of the ration markedly increased phytate P utilization by stimulation of phytase production in the alimentary tract of the growing chick; they calculated that use of the culture might reduce the need for inorganic P supplementation of a corn-soybean oil meal ration by 40 to 60%.

Until such techniques become technically feasible, poultry will continue to require considerable supplementary P. Bird (1960), in fact, calculated that P use by layers might account for 22% of the total supplies of P available. The NRC (1971b) recommends that the diet of the chick should contain 0.7% P, and that 0.5% should be derived from inorganic sources. The older bird has a greater ability to utilize phytate P, but apparently there is little general agreement as to the degree of inorganic supplementation required. Studies reviewed by Harms (1974) indicate that the NRC recommendations for P requirements might be safely reduced for certain classes and ages of poultry. The requirement for maximum growth of chicks was found to be not more than 0.5% when fed the optimal amount of Ca, although a higher level was required for maximum bone ash. Similarly, in comparison with the NRC recommendation of 0.6% P for the laying hen, it was established that the actual requirement for hens maintained on the floor was between 0.4 and 0.5%; the higher level of 0.6% was required for caged birds to prevent cage layer fatigue. Couch (1975), in a review of trials conducted at different stations, concluded that the laying hen does not require more than 0.51% P but does require more than 2.5% Ca; further, caged layers could be fed a diet containing as little as 0.35% P "without serious detriment to various production parameters."

The level of P required for maximum hatchability apparently cannot be obtained from plant sources alone. Waldroup et al. (1967) found that highest hatchability was obtained when a level of 0.18% inorganic phosphate was added to a basal diet containing 0.34% P. There is no indication that lowering the P content of the diet will depress eggshell quality; rather, as noted by Harms (1974), the reverse may be true. Trials by Taylor (1965), Singh et al. (1971), and Scott et al. (1975) indicate an inverse relationship between dietary P and eggshell quality. In the last study it was found that egg production was not different between laying hens receiving either 0.55 or 0.26% available P in the diet; further, the breaking strength of the eggs during the sixth month of egg production was significantly higher in the birds on the lower levels of P.

It would also seem possible to reduce the level of P required by finishing and replacement birds. Waldroup et al. (1974) found that the addition of more than 0.12% supplemental P to a corn-soybean meal diet containing 0.42% total P resulted in little improvement in performance of broiler chickens of 4 to 8 weeks in age. Studies by Nelson et al. (1963) indicated that the P requirement of the developing-finishing turkey was no more than 0.49% (compared to the NRC recommended level of 0.7%), and that this could be supplied completely from plant sources. Similarly, Berg et al. (1964) found that the P requirement of 8- to 21-week-old pullets was not more than 0.3% with dietary Ca/P ratios of 1:1 to 2:1, and that this requirement could be met without inorganic supplementation.

Research in the area of Ca and P requirements of poultry is still extremely active. The divergence of opinion that exists on the specific needs of the bird for P appears to relate to the marked interactions with Ca, vitamin D, and other dietary components, variation in methods of bioassay, and considerable question as to the availability of P in natural and supplementary components of the ration. It would appear, however, that the P levels recommended by NRC incorporate a significant insurance factor. The NRC (1974b) report suggests that, in periods of P shortage, the P in layer diets should be reduced first, followed by broiler-finishers, pullet-growers, and turkey-grower diets. The extent of such reductions for certain classes of poultry is summarized in Table 2.

2. SWINE

Swine face much the same problem as domestic fowls in the derivation of P from plant sources. Harmon (1972) estimated that utilization of P in unsupplemented corn-soybean meal diets ranges from 23% for the finishing pig to approximately 38% for the weanling pig and nursing sow (compare with data in Table 6). Phytate utilization is improved at low levels of Ca in the diet, and other factors of importance are the vitamin D and Zn contents of the ration, the Ca/P ratio, and pH of the alimentary tract.

Consequences of P deficiency in swine include slow growth, decreased feed intake, lameness and stiffness, weakened bone structure, impaired breeding or gestation, and lowered levels of blood inorganic P (NRC, 1973). Evidence for a relationship between low levels of Ca or P and incidence of atrophic rhinitis (Brown et al., 1966) has not been confirmed (Cromwell et al., 1970; Harmon et al., 1970).

In most modern swine feeding systems it is common practice to include Ca, P, and microelements in the basal diet at levels to meet NRC requirements. In addition, as indicated by Cunha (1972), the same minerals are also self-fed in a mineral box. The result is frequently an oversupply, and the NRC (1974b) report on feed P shortages estimates that P supplementation in excess of NRC requirements may vary from as high as 50% for the bred or lactating sow to 100% for the growing-finishing pig. One immediate solution to the problem of limited supplies and/or high costs of feed P would therefore be to reduce supplementation of diets to levels recommended by the NRC. Other alternatives suggested by Illinois workers (Harmon et al., 1974) would be to (i) further reduce P feeding levels for swine of different ages, (ii) include animal-byproduct sources of P in the ration, (iii) use inorganic P sources other than dicalcium phosphate for swine feeding (including fertilizer P), and (iv) increase the utilization of P by grain processing.

Considering the first of these alternatives, the present NRC (1973) standards represent percentages required in a complete diet for maximum gain in body weight, but it is recognized that they may not be adequate for optimal bone development. Cunha (1972), in comparing NRC (1968) recommendations for Ca and P nutrition of the growing pig with those of the Agricultural Research Council (1967), concluded that the NRC levels were too low, particularly for P. However, several studies with the growing-finishing pig (Combs et al., 1962; Cromwell et al., 1970, 1972; Bayley & Thomson, 1969; Harmon et al., 1974; Newman & Elliott, 1976) have confirmed that the P requirement of this animal is not appreciably greater than the 0.4% recommended in the present NRC allowance. Similarly, recent studies (Kornegay et al., 1973; Harmon et al., 1975) have shown that low levels of dietary P do not appreciably depress reproductive performance. The latter workers found that an unsupplemented corn-soybean diet containing 0.33% P was adequate during gestation, while a level of 0.45% was required for lactation. These findings support the claim (NRC, 1974b) that the present NRC recommendations are close to minimal values. It is interesting to note, however, that in an evaluation of the Ca and P requirements of swine as determined by empirical and factorial methods, Braude (1978) suggests the possibility of a lower recommended level of P based on the higher availability values demonstrated in recent trials.

The use of alternative sources of P supplement for swine feeding has received a great deal of attention. Dicalcium phosphate has been most frequently used in feed formulation, but, as seen in Fig. 1, the use of defluorinated phosphate and phosphoric acid derivatives has increased rapidly in recent years. The utilization of P from defluorinated phosphate appears to be as efficient as from dicalcium phosphate. Harmon et al. (1974) found that soft phosphate was an effective supplement for finishing pigs at a total dietary P concentration of 0.46%. These workers also tested fertilizer P sources (superphosphate, ammoniated phosphate) containing levels of fluoride greater than 2.0% and concluded that growth rates and performance of finishing pigs were as good as on dicalcium phosphate at dietary concentrations of 0.50% P. It is, however, recommended that P supplements of this nature be used only with species (swine and poultry) having a high fluoride tolerance, and then only in the finishing phases of production.

The possibility that the effective utilization of plant P by swine may be increased by processing is of particular interest. Summers et al. (1967) found that steam pelleting increased the availability of P in corn-soybean diets for growing chicks, and Bayley and Thomson (1969) confirmed this observation for growing pigs. Steam pelleting a nonsupplemented corn-soybean ration increased net absorption of plant P from 19 to 29% and improved animal gains and bone development. In balance trial studies with weanling pigs, Cornelius and Harmon (1974) demonstrated that the digestibility and retention of P in animals fed high-moisture ensiled corn was significantly higher than in pigs fed regular corn. Further work in this area is certainly indicated.

3. HUMANS

As already mentioned, problems of P nutrition in human populations have not received much attention, mainly because of the widespread distribution of this element in components of man's diet. The principle, as stated by Davidson and Passmore (1969), appears to be "take care of the calcium and the phosphorus will look after itself." At the present time, recommendations on P are made only by the United States, Canada, and the German Democratic Republic (Int. Union of Nutr. Sci., 1975). In addition to the dietary intakes given in Table 3, the Food and Nutrition Board (1974) recommends that in early infancy the Ca/P ratio in the diet be 1.5:1. The German standards recommend a Ca/P ratio of 1:1 for children up to 3 years, of 1:1.2 for adolescents, and of 1:1.5 for adults. Canadian recommendations call for P allowances at least equal to Ca. These recommendations may be compared with the forthright statement by Davidson and Passmore (1969) that "Ca/P ratios in food can therefore be forgotten, along with other once popular but now outmoded scientific fashions." This opinion was based on experiments with adult males and infants, indicating that the level of P intake had no effect on Ca absorption.

There may, however, be cause for concern in the more extreme Ca/P ratios which tend to characterize contemporary "western" diets. Krook et al. (1972a, 1972b) commented that the production of Ca-containing foods in the United States 20 years ago supplied an average of 383 g of Ca and 1,070 g of P per person per year. Between 1955 and 1965 the U.S. production of milk and milk products declined by 10%, while during the same time meat and meat products (with a Ca/P ratio of about 1:20) increased by 10%. The authors calculated that the present Ca/P ratio in American diets may be approximately 1:3 and considered that an imbalance caused by dietary Ca deficiency and/or excess P may be involved in the conditions of osteoporosis and periodontal disease in the adult human, perhaps as a result of secondary hyperparathyroidism. They found that patients with low dietary Ca intakes showed improved alveolar bone structure in response to Ca supplementation at the rate of 1 g/day.

In contrast to the situation in the more highly developed countries, where diets are derived from a variety of sources and generally contain rather high proportions of meat and dairy products, people in developing countries may have to rely to a much greater extent on cereal and vegetable crops grown locally on soils of variable nutrient status. In this respect, their nutritional condition may be closer to that of the grazing ruminant. It has, however, been extremely difficult to relate the consumption of such diets (generally high in phytic acid) to outright manifestations of Ca and P deficiency. It is well known that whole wheat flour is rachitogenic, and that the content of phytic acid as a proportion of total P increases with the extraction grade. The early studies of McCance and Widdowson (1942), based on balance trials with humans, indicated that the phytic acid content of brown bread impaired Ca absorption, when compared with white bread. Later work with children (Widdowson & McCance, 1954), however, showed no effects on health associated with the long-term feeding of diets based primarily on vegetables and bread made from wheat flour of different degrees of extraction. The explanation, as discussed by Hegsted (1973), may lie in the phenomenon of adaptation; Walker et al. (1948) found that individuals consuming diets high in whole wheat were temporarily in negative mineral balance but gradually returned to an equilibrium state. Hegsted et al. (1952) demonstrated Ca equilibrium in individuals in Peru consuming diets which provided only 100 to 200 mg of Ca/day. Similarly, in trials with children fed either a poor-quality vegetarian diet or one supplemented with Ca, P, and vitamins, Kantha et al. (1957) observed a positive P retention of 76 mg on the basal diet supplying only 421 mg of P/day, compared to a retention value of +164 mg on the fortified diet providing 796 mg of P/day.

Within the limits of the balance trial technique, however, it has been found that P absorption and retention may vary appreciably with composition of the diet. Tewell et al. (1973) determined the P balance of adults consuming rice alone or various combinations of rice, milk, and wheat flour. They found that intakes of P between 1,120 and 1,150 mg/day did not ensure positive retention in all subjects; balances ranged from a mean of -81 ± 110 mg on rice alone to -6 ± 119 mg on the different combinations. Similar differences were noted by Moon et al. (1974). With diets including beans and maize, all P balances were negative despite intake levels of 1.19 to 1.49 g of P/day. Reinhold et al. (1973) and Reinhold (1975) commented on the state of depletion of P, Ca, and Zn in Iranian villagers due to the habitual consumption of whole wheat bread. This was found to contain twice the phytate content of bread eaten in the cities. They recommended that a wheat extraction rate lower than 95 to 100% should be used in rural Iran prior to a fermentation treatment.

Beyond the particular problems of Ca/P ratios and of the availability of P and Ca from plant sources, there is the more general question of whether the "quality" of man's food may be affected by changes in fertilization practice, and whether nutritional adjustments are desirable in response to greater or lesser inputs of specific nutrients to food crops. The complexity of this problem is considerable. Obviously, with respect to P, there are built-in physiological buffer mechanisms in plants which prevent drastic changes in composition; the immediate response to a soil deficit of P will be a reduction in yield rather than a change in mineral content. For example, Cöic (1974) found in studies with wheat on low-P soils that a fourfold decrease in dry matter yield was accompanied by only a one-third decrease in the P concentration of the grain, when compared with control treatments. The author considered that supplying a P fertilizer to correct a P deficiency might result in an increase in the P content of the grain, but would simultaneously decrease its protein content.

Allaway (1971), in reviewing the relationship of fertilizer use to human health, pointed to the positive correlations which can be derived from statistics of fertilizer consumption data and human mortality. There is, however, no evidence for any causative relationship between the two factors. Allaway concludes that the most important effects of fertilizer use upon national diets will result from responses in crop yield and total food production. Or, as expressed by a participant in a conference on fertilizers and crop quality: "Instead of trying to produce a perfect quality grain by correct use of fertilizers, I suggest it is more sensible simply to let the farmers concentrate on producing wheat and let other people concentrate on making up the deficiencies in the product by breeding and appropriate additives." These issues are still being debated. At present there is little evidence to suggest that lowering P intakes would have any adverse effects on man's health or well being.

LITERATURE CITED

Adams, R. S. 1975. Variability in mineral and trace element content of dairy cattle feeds. J. Dairy Sci. 58(10):1538-1548.

Agricultural Research Council. 1965. The nutrient requirements of farm livestock, no. 2. Agric. Res. Counc., London.

Agricultural Research Council. 1967. The nutrient requirements of farm livestock, no. 3. Pigs. Technical reviews and summaries. Agric. Res. Counc., London.

Allaway, W. H. 1962. Relation of soil to plant and animal nutrition. p. 13-23. In Cornell Nutr. Conf. Proc., Cornell Univ., Ithaca, N.Y.

Allaway, W. H. 1971. Feed and food quality in relation to fertilizer use. p. 553-556. In R. A. Olson, T. J. Army, J. J. Hanway, and V. J. Kilmer (ed.) Fertilizer technology and use. 2nd ed. Soil Sci. Soc. Am., Madison, Wis.

Andrew, C. S., and M. F. Robins. 1969. The effect of phosphorus on the growth and chemical composition of some tropical pasture legumes. I. Growth and critical percentage of phosphorus. Aust. J. Agric. Res. 20:665-674.

Baker, B. S., and R. L. Reid. 1977. Mineral content of forage species grown in central West Virginia on various soil series. West Virginia Univ. Agric. Exp. Stn. Bull. 657.

Balasko, J. A., and Dale Smith. 1971. Influence of temperature and nitrogen fertilization on the growth and composition of switchgrass (*Panicum virgatum* L.) and timothy (*Phleum pratense* L.) at anthesis. Agron. J. 63:853-857.

Barrow, N. J. 1967. Some aspects of the effect of grazing on the nutrition of pastures. J. Aust. Inst. Agric. Sci. 33:254-262.

Bayley, H. S., and R. G. Thomson. 1969. Phosphorus requirement of growing pigs and effect of steam pelleting on phosphorus availability. J. Anim. Sci. 28:484-491.

Beeson, K. C. 1941. The mineral composition of crops with particular reference to the soils in which they were grown. A review and compilation. USDA Misc. Pub. 369:1-164.

Beitz, D. C., D. J. Burkhart, and N. L. Jacobson. 1974. Effects of calcium to phosphorus ratio in the diet of dairy cows on incidence of parturient paresis. J. Dairy Sci. 57:49-55.

Berg, L. R., G. E. Bearse, and L. H. Merrill. 1964. The calcium and phosphorus requirements of white leghorn pullets from 8-21 weeks. Poult. Sci. 43:885-896.

Besecker, R. J., Jr., M. P. Plumlee, R. A. Pickett, and J. H. Conrad. 1967. Phosphorus from barley grain for growing swine. J. Anim. Sci. 26:1477.

Bird, H. R. 1960. Trends in phosphorus for laying hens. p. 10-13. In Trends in phosphorus. Smith-Douglass Co., Inc., Norfolk, Va.

Blair, G. J., M. H. Miller, and W. A. Mitchell. 1970. Nitrate and ammonium as sources of nitrogen for corn and their influence on the uptake of other ions. Agron. J. 62:530-532.

Bogert, L. J., G. M. Briggs, and D. H. Calloway. 1973. Nutrition and physical fitness. 9th ed. W. B. Saunders Co., Philadelphia-London-Toronto.

Braude, R. 1978. Calcium and phosphorus requirements of pigs. p. 39-49. In First Annu. Minerals Conf., St. Petersburg Beach, Fla. Int. Minerals and Chem. Corp., Skokie, Ill.

Brown, C. S., and W. P. Apgar. 1969. Effects of phosphorus fertilization on the quality of timothy hay. Res. Life Sci., Maine Agric. Exp. Stn. 17:18-23.

Brown, W. R., L. Krook, and W. G. Pond. 1966. Atrophic rhinitis in swine. Etiology, pathogenesis and prophylaxis. Cornell Vet. 56. Supp. 1.

Buchanan-Smith, J. G. 1978. Calcium and phosphorus requirements of dairy cattle. p. 51-60. In First Annu. Minerals Conf., St. Petersburg Beach, Fla. Int. Minerals and Chem. Corp., Skokie, Ill.

Bushman, D. H., R. J. Emerick, and L. B. Embry. 1965. Incidence of urinary calculi in sheep as affected by various dietary phosphates. J. Anim. Sci. 24:671-675.

Butler, G. W., and D. I. H. Jones. 1973. Mineral biochemistry of herbage. p. 127-162. In G. W. Butler and R. W. Bailey (ed.) Chemistry and biochemistry of herbage. Vol. 2. Academic Press, London and New York.

Call, J. W., J. E. Butcher, J. T. Blake, R. A. Smart, and J. L. Shupe. 1978. Phosphorus influence on growth and reproduction of beef cattle. J. Anim. Sci. 47:216-225.

Canadian Council on Nutrition. 1964. Dietary standards for Canada. Canadian Bull. of Nutr. 6, No. 1.

Carstairs, J. A., D. A. Morrow, and R. S. Emery. 1975. Energy and phosphorus influence on postpartum estrus and metabolites. J. Dairy Sci. 58:750-751.

Church, D. C. 1972. Finishing lambs in the feedlot. p. 234-249. In D. C. Church (ed.) Digestive physiology and nutrition of ruminants. Vol. 3. Oregon State Univ. Bookstores, Inc., Corvallis, Oreg.

Cohen, R. D. H. 1972. Phosphorus nutrition of beef cattle. I. Effect of supplementation on liveweight of steers and digestibility of diet. Aust. J. Exp. Agric. Anim. Husb. 12:455-459.

Cohen, R. D. H. 1974. Phosphorus nutrition of beef cattle. 4. The use of faecal and blood phosphorus for the estimation of phosphorus intake. Aust. J. Exp. Agric. Anim. Husb. 14:709-715.

Cohen, R. D. H. 1975. Phosphorus and the grazing ruminant. World Rev. Anim. Prod. XI: 27-43.

Cöic, Y. M. 1974. Mineral fertilization and quality of the crops. p. 591-617. In V. H. Fernandez (ed.) Fertilizers, crop quality, and economy. Elsevier Sci. Publ. Co., Amsterdam-Oxford-New York.

Colovos, N. F., H. A. Keener, and H. A. Davis. 1958. Effect of pulverized limestone and dicalcium phosphate on the nutritive value of dairy cattle feed. J. Dairy Sci. 41:676-682.

Combs, G. E., J. M. Vandepopuliere, H. D. Wallace, and M. Koger. 1962. Phosphorus requirement of young pigs. J. Anim. Sci. 21:3-8.

Committee on Mineral Nutrition, The Hague. 1973. Tracing and treating mineral disorders in dairy cattle. Centre for Agric. Pub. and Document., Wageningen, The Netherlands.

Cooper, J. P. 1973. Genetic variation in herbage constituents. p. 379-417. In G. W. Butler and R. W. Bailey (ed.) Chemistry and biochemistry of herbage. Vol. 2. Acad. Press, London and New York.

Cornelius, S. G., and B. G. Harmon. 1974. Phosphorus digestibility in high moisture and regular corns. J. Anim. Sci. 39:180 (Abstr.).

Couch, J. R. 1975. Review of research reported at 1975 Poultry Science Assoc. Meet. Feedstuffs 47(43):30–38.

Crichton, C. 1968. Observations on mineral deficiencies in dairy cattle. J. Br. Grassl. Soc. 23:186–193.

Cromwell, G. L., V. W. Hays, C. H. Chaney, and J. R. Overfield. 1970. Effects of dietary phosphorus and calcium level on performance, bone mineralization, and carcass characteristics of swine. J. Anim. Sci. 30:519–525.

Cromwell, G. L., V. W. Hays, and J. R. Overfield. 1972. Effects of phosphorus levels in corn, milo, and wheat base diets on performance and bone strength of pigs. J. Anim. Sci. 35:1103 (Abstr.).

Cunha, T. J. 1972. Mineral and vitamin requirements of the growing pig. p. 225–242. *In* D. J. A. Cole (ed.) Pig production. Pennsylvania State Univ. Press, University Park and London.

Davey, B. G., and R. L. Mitchell. 1968. The distribution of trace elements in cocksfoot (*Dactylis glomerata*) at flowering. J. Sci. Food Agric. 19:423–431.

Davidson, L. S. P., and R. Passmore. 1969. Human nutrition and dietetics. 4th ed. Williams and Wilkins Co., Baltimore, Md.

Doll, E. C., H. F. Miller, and J. R. Todd. 1963. Effect of phosphorus fertilization and liming on yield and chemical composition of corn, wheat, and red clover. Kentucky Agric. Exp. Stn. Bull. 682.

Duncan, D. L. 1958. The interpretation of studies of calcium and phosphorus balance in ruminants. Nutr. Abstr. Rev. 28(3):695–715.

Duncan, D. L. 1966. The balance trial and its limitations. p. 51–80. *In* J. T. Abrams (ed.) Recent advances in animal nutrition. Little, Brown and Co., Boston.

Dutton, J. E., and J. P. Fontenot. 1967. Effect of dietary organic phosphorus on magnesium metabolism in sheep. J. Anim. Sci. 26:1409–1414.

Emerick, R. J., and L. B. Embry. 1964. Effects of calcium and phosphorus levels and diethylstilbestrol on urinary calculi incidence and feedlot performance of lambs. J. Anim. Sci. 23:1079–1083.

Fleming, G. A. 1963. Distribution of major and trace elements in some common pasture species. J. Sci. Food Agric. 14:203–208.

Fleming, G. A. 1973. Mineral composition of herbage. p. 529–566. *In* G. W. Butler and R. W. Bailey (ed.) Chemistry and biochemistry of herbage. Vol. 1. Academic Press, London and New York.

Fleming, G. A., and W. E. Murphy. 1968. The uptake of some major and trace elements by grasses as affected by season and stage of maturity. J. Br. Grassl. Soc. 23(2):174–184.

Food and Nutrition Board. 1968. Recommended dietary allowances, 7th Rev. Ed., Natl. Res. Counc. Pub. 1694, Washington, D.C.

Frissel, M. J. 1978. Cycling of mineral nutrients in agricultural ecosystems. Elsevier Sci. Publ. Co., Amsterdam-Oxford-New York.

Gardner, R. W. 1970. Effects of calcium to phosphorus ratios prepartum and protein levels postpartum on parturient paresis and lactation responses of Holstein cows. J. Dairy Sci. 53:682 (Abstr.).

Gardner, R. W. 1971. Responses of Holstein cows to varying calcium to phosphorus ratios prepartum, and protein sources and percent postpartum. J. Dairy Sci. 54:794 (Abstr.).

Gardner, R. W., and R. L. Park. 1973. Effects of prepartum energy intake and calcium to phosphorus ratios on lactation response and parturient paresis. J. Dairy Sci. 56:385–389.

Gomide, J. A., C. H. Noller, G. O. Mott, J. H. Conrad, and D. L. Hill. 1969. Mineral composition of six tropical grasses as influenced by plant age and nitrogen fertilization. Agron. J. 61:120–123.

Gueguen, L., and C. Demarquilly. 1969. Influence of the vegetative cycle and the growth stage on the mineral vlaue of some herbage plants for adult sheep. p. 745–754. Int. Grassland Congr., Proc. 9th (Sao Paulo, Brazil).

Harmon, B. G. 1972. Calcium and phosphorus for confinement fed swine. 23rd Minnesota Nutr. Conf., Proc. Univ. of Minnesota, Minneapolis. p. 25.

Harmon, B. G., A. H. Jensen, D. H. Baker, and G. R. Carlisle. 1974. Phosphorus sources and phosphorus utilization by swine. Univ. of Illinois Pork Industry Day, 10 Dec. 1974. Univ. of Illinois, Urbana.

Harmon, B. G., C. T. Liu, A. H. Jensen, and D. H. Baker. 1975. Phosphorus requirements of sows during gestation and lactation. J. Anim. Sci. 40:660–664.

Harmon, B. G., J. Simon, D. E. Becker, A. H. Jensen, and D. H. Baker. 1970. Effect of source and level of dietary phosphorus on structure and composition of turbinate and long bones. J. Anim. Sci. 30:742-747.

Harms, R. H. 1974. The phosphorus crisis—what can we do about it? Distill. Feed Res. Counc. Proc. 29:41-54.

Hegsted, D. M. 1973. Calcium and phosphorus. p. 268-286. In R. S. Goodhart and M. E. Shils (ed.) Modern nutrition in health and disease. Lea and Febiger, Philadelphia.

Hegsted, D. M. 1976. Balance studies. J. Nutr. 106:307-311.

Hegsted, D. M., I. Moscoso, and C. C. Collazos. 1952. A study of the minimum calcium requirements of adult men. J. Nutr. 46:181.

Hemingway, R. G. 1978. Phosphorus nutrition of beef cattle. p. 61-71. In First Annu. Minerals Conf., St. Petersburg Beach, Fla. Int. Minerals and Chem. Corp., Skokie, Ill.

Highton, H. 1974. Outlook-trends in phosphorus feed supplements. p. 3. In Trends in phosphorus. Borden Chemical, Norfolk, Va.

Hignett, S. L. 1950. Factors influencing herd fertility in cattle. Vet. Rec. 62:654-663.

Hill, R. 1961. The provision and metabolism of calcium and phosphorus in ruminants. World Rev. Nutr. Diet. 3:129-148.

Hill, R. R., and G. A. Jung. 1975. Genetic variability for chemical composition of alfalfa. I. Mineral elements. Crop Sci. 15:652-657.

Hoar, D. W., R. J. Emerick, and L. B. Embry. 1970a. Influence of calcium source, phosphorus level and acid-base-forming effects of the diet in feedlot performance and urinary calculi formation in lambs. J. Anim. Sci. 31:118-125.

Hoar, D. W., R. J. Emerick, and L. B. Embry. 1970b. Potassium, phosphorus and calcium interrelationships influencing feedlot performance and phosphatic uriolithiasis in lambs. J. Anim. Sci. 30:597-600.

Hodges, E. M., W. G. Kirk, F. M. Peacock, D. W. Jones, G. K. Davis, and J. R. Neller. 1964. Forage and animal response to different phosphatic fertilizers on pangolagrass pastures. Florida Agric. Exp. Stn. Bull. 686.

Hogue, D. E. 1974. Formula feeds for intensive sheep management systems. p. 58. In Proc. Cornell Nutr. Conf., Cornell Univ., Ithaca, N.Y.

International Union of Nutritional Sciences, Report of the Commission on International Dietary Allowances. 1975. Nutr. Abstr. Rev. 45(2):89-111.

Irving, J. T. 1973. Calcium and phosphorus metabolism. Academic Press, New York and London.

Jacobson, D. R., R. W. Hemken, F. S. Button, and R. H. Hatton. 1972. Mineral nutrition, calcium, phosphorus, magnesium, and potassium interrelationships. J. Dairy Sci. 55:935-944.

Joyce, J. P., and P. V. Rattray. 1970. Nutritive value of white clover and perennial ryegrass. III. Intake and utilisation of calcium, phosphorus, and magnesium. N.Z. J. Agric. Res. 13(4):800-807.

Kantha, J., M. Narayanarao, M. Swaminathan, V. Subrahmanyan. 1957. The metabolism of nitrogen, calcium and phosphorus in undernourished children. 3. The effect of a supplementary multipurpose food on the metabolism of nitrogen, calcium and phosphorus. Br. J. Nutr. 11:388-391.

Kemp, A. 1966. Mineral balance in dairy cows fed on grass, with special reference to magnesium and sodium. p. 411-415. Int. Grassland Congr., Proc. 10th (Helsinki, Finland).

Kendall, K. A., and J. H. Byers. 1959. Blood and excretion levels of calcium and phosphorus associated with varied mineral supplements fed. J. Dairy Sci. 42:933 (Abstr.).

Kendall, K. A., K. E. Harshbarger, R. L. Hays, E. E. Ormiston, and S. S. Spahr. 1970. Responses of dairy cows to diets containing varied levels of calcium and phosphorus. J. Dairy Sci. 53:681-682.

Kendall, K. A., R. L. Hays, and E. E. Ormiston. 1968. Postpartum serum calcium and phosphorus levels associated with calcium carbonate and monosodium phosphate. J. Dairy Sci. 51:978 (Abstr.).

Kirchgessner, M., E. Pahl, and G. Voigtländer. 1967. The influence of the stage of vegetative growth on the mineral contents of red clover (*Trifolium pratense* L.) and lucerne (*Medicago varia* Martyn). Wirtschaftseigene Futter 13:173-188.

Kirk, W. G., R. L. Shirley, E. M. Hodges, G. K. Davis, F. M. Peacock, J. F. Easley, and F. G. Martin. 1970. Production performance and blood and bone composition of cows grazing pangolagrass pastures receiving different phosphate fertilizers. Florida Agric. Exp. Stn. Bull. 735.

Kivimae, A. 1959. Chemical composition and digestibility of some grassland crops. Acta Agric. Scand. Suppl. 5.

Kornegay, E. T., H. R. Thomas, and T. N. Meacham. 1973. Evaluation of dietary calcium and phosphorus for reproducing sows housed in total confinement on concrete or in dirt lots. J. Anim. Sci. 37:493-500.

Krook, L., L. Lutwak, J. P. Whalen, Per-Ake Henrikson, G. V. Lesser, and R. Uris. 1972a. Human periodontal disease. Morphology and response to calcium therapy. Cornell Vet. 62:32-53.

Krook, L., J. P. Whalen, G. V. Lesser, and L. Lutwak. 1972b. Human periodontal disease and osteoporosis. Cornell Vet. 62:371-391.

Larson, W. E., L. B. Nelson, and A. S. Hunter. 1952. The effects of phosphate fertilization upon the yield and composition of oats and alfalfa grown on phosphate-deficient Iowa soils. Agron. J. 44:357-361.

Leitch, Isabella. 1964. Calcium and phosphorus. p. 261-307. In G. H. Beaton and E. W. McHenry (ed.) Nutrition: a comprehensive treatise. Vol. I. Academic Press, New York and London.

Little, D. A. 1970. Factors of importance in the phosphorus nutrition of beef cattle in Northern Australia. Aust. Vet. J. 46:241-248.

Littlejohn, A. I., and G. Lewis. 1960. Experimental studies of the relation between the Ca/P ratio of the diet and fertility in heifers. Vet. Rec. 11:33-44.

Lofgreen, G. P., and M. Kleiber. 1953. The availability of the phosphorus in alfalfa hay. J. Anim. Sci. 12:366-371.

Lomba, F., R. Paquay, V. Bienfet, and A. Lousse. 1969. Statistical research on the fate of dietary mineral elements in dry and lactating cows. III. Phosphorus. J. Agric. Sci. (Cambridge). 73:215-222.

McCance, R. A., and E. M. Widdowson. 1942. Mineral metabolism of dephytinized bread. J. Physiol. 101:304-313.

McDonald, I. W. 1968. The nutrition of grazing ruminants. Nutr. Abstr. Rev. 38:381.

McGillivray, J. J. 1974. Biological availability of phosphorus in feed ingredients. p. 15-23. In Proc., 35th Minnesota Nutr. Conf., 16-17 Sept. 1974, Univ. of Minnesota, Bloomington, Minn.

MacLeod, L. B. 1965. Effect of nitrogen and potassium on the yield and chemical composition of alfalfa, bromegrass, orchard grass and timothy grown as pure species. Agron. J. 57: 261-266.

McMeniman, N. P., and D. A. Little. 1974. Studies on the supplementary feeding of sheep consuming mulga (*Acacia aneura*). 1. The provision of phosphorus and molasses supplements under grazing conditions. Aust. J. Exp. Agric. Anim. Husb. 14:316-321.

Manston, R. 1967. The influence of dietary calcium and phosphorus concentration on their absorption in the cow. J. Agric. Sci. (Cambridge) 68:263-268.

Metson, A. J., and W. M. H. Saunders. 1978. Seasonal variations in chemical composition of pasture. I. Calcium, magnesium, potassium, sodium, and phosphorus. N.Z. J. Agric. Res. 21:341-355.

Mitchell, H. H. 1947. The mineral requirements of farm animals. J. Anim. Sci. 6:365-377.

Molloy, L. F., R. Ball, T. W. Collie, and D. J. Ross. 1978. Influence of fertilizer nitrogen on higher fatty acids and on Mg, Ca, K, and P in grazed grass-clover herbage. N.Z. J. Agric. Res. 21:57-64.

Moon, W. H., J. L. Malzer, and H. E. Clark. 1974. Phosphorus balances of adults consuming several food combinations. J. Am. Diet. Assoc. 64:386-390.

Morrow, D. A. 1970. Diagnosis and prevention of infertility in cattle. J. Dairy Sci. 53:961-969.

Mott, G. O. 1974. Nutrient recycling in pastures. p. 323-339. In D. A. Mays (ed.) Forage fertilization. Am. Soc. Agron., Madison, Wis.

National Research Council. 1963. Nutrient requirements of beef cattle. Pub. 1137. Natl. Res. Counc., Washington, D.C.

National Research Council. 1975. Nutrient requirements of domestic animals. 5. Nutrient requirements of sheep. 5th ed. Natl. Acad. of Sci., Washington, D.C.

National Research Council. 1970. Nutrient requirements of domestic animals. 4. Nutrient requirements of beef cattle. 4th ed. Natl. Acad. of Sci., Washington, D.C.

National Research Council. 1971a. Atlas of nutritional data on United States and Canadian feeds. Natl. Acad. of Sci., Washington, D.C.

National Research Council. 1971b. Nutrient requirements of domestic animals. 1. Nutrient requirements of poultry. Natl. Acad. of Sci., Washington, D.C.

National Research Council. 1973. Nutrient requirements of domestic animals. 2. Nutrient requirements of swine. 7th ed. Natl. Acad. of Sci., Washington, D.C.

National Research Council. 1974a. Recommended dietary allowances. 8th ed. Natl. Acad. of Sci., Washington, D.C.

National Research Council. 1974b. Feed phosphorus shortage. Levels and sources of phosphorus recommended for livestock and poultry. Natl. Acad. of Sci., Washington, D.C.

National Research Council. 1975. Nutrient requirements of domestic animals. 5. Nutrient requirements of sheep. 5th Natl. Acad. Sci., Washington, D.C.

National Research Council. 1976. Nutrient requirements of domestic animals. 4. Nutrient requirements of beef cattle. 5th ed. Natl. Acad. of Sci., Washington, D.C.

National Research Council. 1978. Nutrient requirements of domestic animals. 3. Nutrient requirements of dairy cattle. 5th ed. Natl. Acad. of Sci., Washington, D.C.

Nelson, F. E., L. S. Jensen, and J. McGinnis. 1963. Influences of previous calcium and phosphorus intake and plant phosphorus on the requirement of developing turkeys for calcium and phosphorus. Poult. Sci. 42:579–585.

Nelson, T. S. 1967. The utilization of phytate phosphorus by poultry—A review. Poult. Sci. 46:862–871.

Nelson, T. S., L. W. Ferrara, and N. L. Stover. 1968a. Phytate phosphorus content of feed ingredients obtained from plants. Poult. Sci. 47:1372–1374.

Nelson, T. S., T. R. Shieh, R. J. Wodzinski, and J. H. Ware. 1968b. The availability of phytate phosphorus in soybean meal before and after treatment with a mold phytase. Poult. Sci. 47:1842–1848.

Nelson, T. S., T. R. Shieh, R. J. Wodzinski, and J. H. Ware. 1971. Effect of supplemental phytase on the utilization of phytate phosphorus by the chick. J. Nutr. 101:1289–1294.

Newman, C. W., and D. O. Elliott. 1976. Source and level of phosphorus for growing-finishing swine. J. Anim. Sci. 42:92–98.

Nielsen, K. F., and R. K. Cunningham. 1964. The effects of soil temperature and form and level of nitrogen on growth and chemical composition of Italian ryegrass. Soil Sci. Soc. Am. Proc. 28:213–218.

Noland, P. R., M. Funderburg, and Z. Johnson. 1968. Phosphorus availability in a practical diet for swine. J. Anim. Sci. 27:1155.

Noller, C. H., A. G. Castro, W. E. Wheeler, D. L. Hill, and N. J. Moeller. 1977. Effect of phosphorus supplementation on growth rate, blood minerals, and conception rate of dairy heifers. J. Dairy Sci. 60:1932–1940.

Peeler, H. T. 1972. Biological availability of nutrients in feeds: Availability of major mineral ions. J. Anim. Sci. 35:695–712.

Peterson, H. B., L. B. Nelson, and J. L. Paschal. 1953. A review of phosphate fertilizer investigations in 15 western states through 1949. USDA Circ. no. 927, Washington, D.C.

Preston, R. L., and R. L. Stone. 1974. Role of supplemental phosphorus and frequency of DES implantation in finishing steers. Ohio Agric. Exp. Stn. Res. summary: Beef cattle research. Ohio Agric. Exp. Stn.

Pritchard, D. E. 1974. Research in brief. Ohio Rep. Res. Dev. 59:118.

Pritchard, G. I., W. J. Pigden, and L. P. Folins. 1964. Distribution of potassium, calcium, magnesium, and sodium in grasses at progressive stages of maturity. Can. J. Plant Sci. 44:318–324.

Reid, R. L., K. Daniel, and J. D. Bubar. 1974a. Mineral relationships in sheep and goats maintained on orchardgrass fertilized with different levels of nitrogen, or nitrogen with microelements, over a five-year period. p. 565–575. Int. Grassland Congr., Proc. 12th, (Moscow, USSR).

Reid, R. L., and G. A. Jung. 1974. Effects of elements other than nitrogen on the nutritive value of forage. p. 395–435. In D. A. Mays (ed.) Forage fertilization. Am. Soc. Agron., Madison, Wis.

Reid, R. L., G. A. Jung, A. J. Post, F. P. Horn, E. B. Kahle, J. D. Bubar, and K. Daniel. 1974b. Effects of nitrogen and micro-element fertilization on quality of pasture and on the health, nutritional status and reproductive performance of sheep. J. Anim. Sci. 38:163–171.

Reid, R. L., A. J. Post, and G. A. Jung. 1970. Mineral composition of forages. West Virginia Agric. Exp. Stn. Bull. 589T.

Reid, R. L., A. J. Post, and F. J. Olsen. 1975a. Chemical composition and digestibility of tropical forages. J. Anim. Sci. 40:182 (Abstr.).

Reid, R. L., Karen Powell, and J. A. Balasko. 1975b. Performance and mineral utilization in lambs grazing four perennial grass species. J. Anim. Sci. 41:338 (Abstr.).

Reid, R. L., G. A. Jung, I. J. Roemig, and R. E. Kocher. 1978. Mineral utilization by lambs and guinea pigs fed Mg-fertilized grass and legume hays. Agron. J. 70:9-14.

Reinhold, J. G. 1975. Phytate destruction by yeast fermentation in whole wheat meals. J. Am. Diet. Assoc. 66:38-41.

Reinhold, J. G., B. Faradji, P. Abadi, and F. Ismail-Beigi. 1976. Decreased absorption of calcium, magnesium, zinc and phosphorus by humans due to increased fiber and phosphorus consumption as wheat bread. J. Nutr. 106:493-503.

Reinhold, J. G., H. Hedayati, A. Lahimgarzadeh, and K. Masr. 1973. Zinc, calcium, phosphorus and nitrogen balances of Iranian villagers following a change from phytate-rich to phytate-poor diets. Ecol. Food Nutr. 2:157-162.

Reith, J. W. S. 1965. Mineral composition of crops. NAAS Q. Rev. 68:150-156.

Reith, J. W. S. 1972. Soil conditions and nutrient supplies in hill land. p. 5-12. In P. A. Gething, P. Newbould, and J. B. E. Patterson (ed.) Hill pasture improvement and its economic utilisation. Potassium Inst. Ltd., Henley-on-Thames, Oxfordshire, England.

Reith, J. W. S., R. H. E. Inkson, W. Holmes, D. S. MacLusky, D. Reid, R. G. Heddle, and J. J. F. Copeman. 1964. The effects of fertilizer on herbage production. II. The effects of nitrogen, phosphorus and potassium on botanical and chemical composition. J. Agric. Sci. 63:209-219.

Ricketts, R. E., J. R. Campbell, D. E. Weinman, and M. E. Tumbleson. 1970. Effect of three calcium/phosphorus ratios on performance of growing Holstein steers. J. Dairy Sci. 53:898-903.

Scott, M. L., A. Antillon, and P. A. Mullenhoff. 1975. The effects of levels of calcium, phosphorus and vitamin D on bone development and eggshell quality in modern laying hens. p. 77-80. In Cornell Nutr. Conf. Proc. Cornell Univ., Ithaca, N.Y.

Sen, K. M., and G. L. Mabey. 1965. The chemical composition of some indigenous grasses of coastal savanna of Ghana at different stages of growth. Int. Grassland Congr., Proc. 9th (Sao Paulo, Brazil) 1:763-771.

Singh, R. D., J. K. Bletner, and O. E. Goff. 1971. Bone density index, bone breaking strength, egg production, and egg specific gravity as affected by dietary Ca and P. Poult. Sci. 50:1631 (Abstr.).

Singsen, E. P., A. H. Spandorf, L. D. Matterson, J. A. Serafin, and J. J. Tlustohowicz. 1962. Phosphorus in the nutrition of the hen. 1. Minimum phosphorus requirements. Poult. Sci. 41:1401-1414.

Smith, A. M., G. L. Holck, and H. B. Spafford. 1966. Symposium: Re-evaluation of nutrient allowances for high-producing cows. Calcium, phosphorus, and vitamin D. J. Dairy Sci. 49:239-243.

Smith, D. 1970. Influence of temperature on the yield and chemical composition of five forage legume species. Agron. J. 62:520-523.

Smith, D. 1971. Levels and source of potassium for alfalfa as influenced by temperature. Agron. J. 63:497-500.

Smith, D. 1973. Distribution of dry matter and chemical constituents among the plant parts of six temperate-origin forage grasses at early anthesis. Univ. of Wisconsin Res. Rep. R2552, Madison, Wis.

Stewart, A. B., and W. Holmes. 1953. Nitrogenous manuring of grassland. I. Some effects of heavy dressings of nitrogen on the mineral composition. J. Sci. Food Agric. 4:401-408.

Stone, R. L. 1974. Research in brief. Ohio Rep. Res. Dev. 59:118.

Stott, G. H. 1963. Parturient paresis related to dietary phosphorus. J. Dairy Sci. 46:635 (Abstr.).

Stout, W. L., D. P. Belesky, G. A. Jung, R. S. Adams, and B. L. Moser. 1977. A survey of Pennsylvania forage mineral levels with respect to dairy and beef cow nutrition. Pennsylvania State Agric. Exp. Stn. Progr. Rep. 364.

Sullivan, J. T. 1969. Chemical composition of forages with reference to the needs of the grazing animal. USDA-ARS 34-107.

Summers, J. D., S. J. Slinger, and G. Cisneros. 1967. Some factors affecting the biological availability of phosphorus in wheat by-products. Cereal Chem. 44:318-323.

Taylor, T. G. 1965. Dietary phosphorus and eggshell thickness in domestic fowl. Br. Poult. Sci. 6:79-87.

Taylor, T. G. 1975. Perspectives in mineral nutrition. Proc. Nutr. Soc. 34:35-41.

Tewell, J. E., H. E. Clark, and J. M. Howe. 1973. Phosphorus balances of adults fed rice, milk and wheat flour mixtures. J. Am. Diet. Assoc. 63:530-535.

Thayer, R. H., and C. D. Jackson. 1975. Improving phytate phosphorus utilization by poultry with live yeast culture. Oklahoma Agric. Exp. Stn. Anim. Sci. and Ind. Res. Rep. p. 131-139.

Theiler, A., and H. H. Green. 1932. Aphosphorosis in ruminants. Nutr. Abstr. Rev. 1(3):359-385.

Thomas, B., A. Thompson, V. A. Oyenuga, and R. H. Armstrong. 1952. The ash constituents of some herbage plants at different stages of maturity. Emp. J. Exp. Agric. 20:10-22.

Thompson, A. 1965. Mineral availability and techniques for its measurement. Proc. Nutr. Soc. 24(1):81-88.

Tillman, A. D., and J. R. Brethour. 1958. Utilization of phytin phosphorus by sheep. J. Anim. Sci. 17:104-112.

Todd, J. R. 1976. Calcium, phosphorus and magnesium metabolism, with particular reference to milk fever (parturient hypocalcemia) and grass tetany (hypomagnesaemic tetany) in ruminant animals. Nuclear techniques in animal production and health. Int. Atomic Energy Agency, Vienna.

Underwood, E. J. 1966. The mineral nutrition of livestock. Commonw. Agric. Bur., The Central Press (Aberdeen) Ltd., United Kingdom.

U.S. Department of Agriculture, Economic Research Service. 1974. United States and world fertilizer outlook, 1974 and 1980. Agric. Econ. Rep. no. 257. USDA, Washington, D.C.

Van Riper, G. E., and D. Smith. 1959. Changes in the chemical composition of the herbage of alfalfa, medium red clover, ladino clover, and bromegrass with advance of maturity. Wisconsin Agric. Exp. Stn. Res. Rep. 4.

Vose, P. B. 1963. Varietal differences in plant nutrition. Herb. Abstr. 33(1):1-13.

Waldroup, P. W., R. J. Mitchell, and K. R. Hazen. 1974. The phosphorus needs of finishing broilers in relationship to dietary nutrient density levels. Poult. Sci. 53:1655-1663.

Waldroup, P. W., C. F. Simpson, B. L. Damron, and R. H. Harms. 1967. The effectiveness of plant and inorganic phosphorus in supporting egg production in hens and hatchability and bone development in chick embryos. Poult. Sci. 46:660.

Walker, A. R. P., F. W. Fox, and J. T. Irving. 1948. Studies in human mineral metabolism. 1. The effect of bread rich in phytate phosphorus on the metabolism of certain mineral salts with special reference to calcium. Biochem. J. 42:452-461.

Whitehead, D. C. 1966. Nutrient minerals in grassland herbage. Commonw. Agric. Bur. Mim. Pub. no. 1, Farnham Royal, Bucks, England.

Whitehead, D. C. 1970. The role of nitrogen in grassland productivity. Commonw. Agric. Bur. Bull. 48, Commonw. Bur. of Pastures and Field Crops, Hurley, Berkshire, England.

Widdowson, E. M., and R. A. McCance. 1954. Great Britain Med. Res. Counc. Spec. Rep. Ser. no. 287.

Wise, M. B., A. L. Ordoveza, and E. R. Barrick. 1963. Influence of variations in dietary calcium/phosphorus ratios on performance and blood constituents of calves. J. Nutr. 79:79.

Wolton, K. M. 1963. An investigation into the simulation of nutrient returns by the grazing animal in grassland experimentation. J. Br. Grassl. Soc. 18:213-219.

Woodman, H. E., and R. E. Evans. 1948. Nutrition of the bacon pig. XIII. The minimum level of protein intake consistent with the maximum rate of growth. J. Agric. Sci. 38:354.

World Health Organization. 1962. Calcium requirements. WHO Tech. Rep. Ser. no. 230, WHO, Geneva.

Young, V. R., W. P. Richards, G. P. Lofgreen, and J. R. Luick. 1966. Phosphorus depletion in sheep and the ratio of calcium to phosphorus in the diet with reference to calcium and phosphorus nutrition. Br. J. Nutr. 20:783-794.

Glossary of Common and Scientific Names of Plants and Other Organisms

Alfalfa	*Medicago sativa* L.
Aspen	*Populus tremuloides* Michx.
Bahiagrass	*Paspalum notatum* Flugge
Banksia	*Banksia* sp.
Barley	*Hordeum vulgare* L.
Barleygrass	*Hordeum leporinum* Link
Barrelmedic	*Medicago tribuloides*
Beans	
broad	*Vicia faba*
bush	*Phaseolus vulgaris* L.
field	*Phaseolus vulgaris* L.
phasey	*P. lathroides* L.
Beech	*Fagus sylvatica* L.
Beets	
red	*Beta vulgaris*
sugar	*Beta saccharifera*
Bentgrass, colonial	*Agrostic tenuis* L.
Bermudagrass	*Cynodon dactylon* L. Pers.
Bigsting nettle	*Urtica dioica*
Birch	*Betula* sp.
Birdsfoot trefoil	*Lotus corniculatus* L.
Black currant	*Ribes nigrum*
Bluegrass	*Poa pratensis* L.
Bluestem	*Bathriochloa ischaemum* L.
Bromegrass, smooth	*Bromus inermis* Leyss.
Browntop grass	*Agrostis tenuis* Sibth
Buckwheat	*Fagopyrum estilentum*
Buffalograss	*Buchloe dactyloides* Nutt. Engelem.
Buffelgrass	*Cenchrus siliaris*
Cabbage	*Brassica oleracea* L. var. *capitata* L.
Chinese	*Brassica pekinensis*
Capeweed	*Arctotheca calendula*
Carpetgrass	*Axonopus affinis*
Carrot	*Daucus carota* L.
Cassava	*Manihot esculenta* Crantz
Cauliflower	*Brassica oleracea*
Centro	*Centrosema pubescens* Benth.
Clover	*Trifolium* sp.
crimson	*T. incarnatum* L.
cupped	*T. cherleri*
ladino	*T. repens* L. var. *ladino*
red	*T. pratense* L.
subterranean	*T. subterraneum* L.
white	*T. repens* L.
whitetip	*T. variegatum* L.
Cocksfoot (orchardgrass)	*Dactylis glomerata* L.
Cocoa	*Theoboroma cacao*
Corn	*Zea mays* L.

GLOSSARY OF SCIENTIFIC NAMES

Cotton	
old world species	*Gossypium herbaceum* L. and *G. arboreum* L.
American upland	*G. hirsutum* L.
Sea Island	*G. barbadense* L.
Cucumber	*Cucumis sativus*
Curlycress	*Lapidium sativum* L.
Dallisgrass	*Paspalum dilatatum* Poir.
Desmodium	
greenleaf	*Desmodium intortum* Mill
silverleaf	*D. unicinatum*
Douglas-fir	*Pseudotsuga menziesii* (Mirb.) Franco
Elephantgrass	*Pennisetum purpureum* Schum.
Eucalyptus	*Eucalyptus* spp.
Fescue	*Festuca* spp.
meadow	*F. elatior* L.
native	*F. scabrella* torr.
red	*F. rubra* L.
tall	*F. arundinacea* Schreb.
Fir	
Douglas	*Pseudotsuga menziesii* (Mirb.) Franco
Pacific silver	*Abies amabalis* (Doug.) Forb
Flatweed	*Hypochoeris glabra*
Flax	*Linum* spp.
Guineagrass	*Panicum maximum* Jacq.
Hardinggrass	*Phalaris tuberosa* L.
Heather	*Calluna vulgaris* (L.) Hull
Hemlock, western	*Tsuga heterophylla* Sarg.
Johnsongrass	*Sorghum halepense* L. Pers.
Juniper, pinyon	*Juniperus* spp.
Kentucky bluegrass	*Poa pratensis* L.
Kikuyugrass	*Pennisetum dandestinum* Nodist.
Larch	*Larix* sp.
Lavender	*Lavendula officinalis,* Chaix.
Lettuce	*Lactuca sativa* L.
Lotononis	*Lotononis bainsii* Baker
Lovegrass	*Eragrostis*
weeping	*E. curvula* Schrad.
Lucerne	*Medicago sativa* L.
Lupin	
narrow leafed	*Lupinus angustifolius*
sand plain	*Lupinus consentinii*
Maize	*Zea mays* L.
Mesquite, velvet	*Prosopis juliflora*
Millet	*Setaria* spp.
foxtail	*Setaria italica*
Broomcorn (proso)	*Panicum miliaceum*
Mitchellgrass	*Astrebla elymoides*
Molassesgrass	*Melinis minutiflora* Beau V.
Mulga	*Acacia aneura*

GLOSSARY OF SCIENTIFIC NAMES

Mulgagrass	*Thyridolepsis mitchelliana*
Mustard	*Sinapis alba* L.
Napiergrass	*Pennisetum purpureum* schum.
Oak	*Quercus* spp.
Oat	*Avena sativa* L.
Oatgrass, tall	*Arrhenatherum elatius* L. Presl.
Onion	*Allium cepa* L.
Orange	*Citrus sinensis* Osbeck.
Orchardgrass, cocksfoot	*Dactylis glomerata* L.
Pangolagrass	*Digitaria decumbens* Stent.
Pangola digitgrass	*Digitaria decumbens* Stent.
Peanuts	*Arachis hypogea* L.
Millet, pearl	*Pennisetum americanum* L. Leeke
Pine	*Pinus* spp.
caribaean	*P. caribaea* var. *hondurensis*
corsican	*P. nigra* var. *maritima* (Ait.) Melv.
eastern white	*P. strobus* L.
jack	*P. Banksiana* Lamb
loblolly	*P. taeda* L.
lodgepole	*P. contorta* Dougl.
longleaf	*P. palustris* Mill.
maritime	*P. pinaster* Alt.
Monterey	*P. radiata* D. Don.
ponderosa	*P. ponderosa* Laws.
radiata	*P. radiata* D. Don.
red	*P. resinosa* Alt.
scots	*P. sylvestris* L.
shortleaf	*P. echinata*
slash	*P. elliottii* var. *elliottii* Engelm.
Virginia	*P. virginiana* Mill.
western white	*P. monticolla* Dougl. ex D. Don
Pond snails	*Heterogen longispira*
Potatoes	*Solanum tuberosum* L.
Potatoes, sweet	*Ipomoea batatas* Lam.
Protozoa of the rumen	*Dasytricha* sp.
	Diplodinium sp.
	Entodinium sp.
	Isotricha sp.
	Ophyrosocolex sp.
Radishes	*Raphanus sativus* L.
Rape	*Brassica napus* L.
Reed canarygrass	*Phalaris arundinacea*
Rhodesgrass	*Chloris gayana* Kunth.
Rice	*Oriza* sp.
common	*O. sativa* L.
African	*O. glaberrima* Steud.
Russian wildrye	*Elymus junceus* Fish.
Rye	*Secale cereale* L.
Ryegrass	*Lolium* sp.
Italian	*L. multiflorum* Lam.
perennial	*L. perenne*
wimmera	*L. rigidum* Gand.

Sea anemone	*Anthopleura elegantissima*
	Metridium dianthus
Sedge	*Carex globularis*
Semipreverede	*Panicum maximum* Jacq.
Sericea lespedeza	*Lespedeza cuneata* (dumont) G. Don
Silvergrass	*Vulpia myuros*
Siratro	*Phaseolus atropurpureus* Mot. and Sesse
Sorghum	
grain	*Sorghum bicolor* L. Moench
forage	*S. bicolor* L. Moench × *S. sudanense* P. Stapf.
Sorghum sudangrass hybrid	*S. bicolor* L. Moench × *S. sudanense* P. Stapf.
Soybeans	*Glycine max* L. Merrill
perennial	*Glycine wightii*
Speargrass	*Heteropogon contortus*
Spinach	*Spinacea oleracea* L.
Spirodela	*Spirodela* spp.
Spruce	*Picea* spp.
black	*P. mariana* (Mill.) B.S.P.
Norway	*P. abies* (L.) Karst.
sitka	*P. sitchensis* (Bong.) Carr.
white	*P. glauca* (Moench) Voss.
Squash, zucchini	*Cucurbita pepo* L. var. *medullosa* A lef.
Stargrass	*Cynodon plectostachyus* (K. Schum) Pilger
Stylo, Townsville	*Stylosanthes humilis*
Sudangrass	*Sorghum vulgare sadanense*
Sugarcane	*Saccharum* spp.
noble cane	*S. officinarum* L.
thin canes	*S. sinense* Roxb. and *S. barberi* Jeswiet
wild types	*S. spontaneum* L. and *S. robustum* Brandes and Jeswiet ex Grassl
Switchgrass	*Panicum virgatum* L.
Tannier	*Xanthosoma saggitifolia*
Tickclover	*Desmodium*
Timothy	*Phleum pratense* L.
Tobacco	*Nicotiana* spp.
forleaf	*N. tabacum* L.
forleaf nicotine	*N. rustica* L.
Tomato	*Lycopersicon esculentum* Mill.
Turnips	*Brassica rapa* L.
Wavy hairgrass	*Deschempsia flexuosa*
Wheat	*Triticum vulgare* L.
fall	*Triticum aestivum* L.
Wheatgrass, crested	*Agropyron cristatum* L.
Yam	*Dioscorea* spp.

Glossary of Mineral Compositions

Mineral	Composition
Apatite: Fluorapatite	$Ca_{10}(PO_4)_6F_2$
Hydroxyapatite	$Ca_{10}(PO_4)_6(OH)_2$
Francolite	$Ca_{10-a-b}Na_aMg_b(PO_4)_{6-x}(CO_3)_xF_{0.4x}(F,OH)_2$
Anatase	TiO_2
Ankerite	$(Fe,Mg)_2(CO_3)_2$
Attapulgite	$Mg_5(OH)_2(OH_2)_4Si_8O_{20} \cdot 4H_2O$
Barite	$BaSO_4$
Calcite	$CaCO_3$
Chalcedony	SiO_2
Chlorospodiosite	Ca_2ClPO_4
Crandallite	$CaAl_3(PO_4)_2(OH)_5 \cdot H_2O$
Dolomite	$CaMg(CO_3)_2$
Fluorite	CaF_2
Goethite	$Fe_2O_3 \cdot H_2O$
Hematite	Fe_2O_3
Hilgenstockite	$Ca_4O(PO_4)_2$
Millisite	$(Na,K)CaAl_6(PO_4)_4(OH)_9 \cdot 3H_2O$
Nagelschmitite	$Ca_7(PO_4)_2(SiO_4)_2$
Nepheline	$(Na,K)AlSiO_4$
Opal (silica hydrogel)	$SiO_2 \cdot nH_2O$
Perovskite	$CaTiO_3$
Pyrite	FeS_2
Pyroxene group	$(Na,Ca,Mg,Fe)(Mg,Al,Fe)Si_2O_6$
Quartz	SiO_2
"Rhenania" phosphate	$Ca(Na,K)PO_4$ (α, β-forms)
Silicocarnotite	$Ca_5P_2SiO_{12}$
Strengite	$FePO_4 \cdot 2H_2O$
Variscite	$AlPO_4 \cdot 2H_2O$
Wavellite	$Al_6(F,OH)_6(PO_4) \cdot 9H_2O$
Wardite	$NaAl_3(PO_4)_2(OH)_4 \cdot 2H_2O$
Whitlockite	$(Ca,Mg)_3(PO_4)_2$

SUBJECT INDEX

Absolute citrate solubility (ACS), 65
Absolute solubility index (ASI), 65
Absorption of P by plants
  movement into root, 569-570
  zone of root, 568
Acid soils, P fixation, 471-514
Acrisols, 471
Activity of phosphate ions
  $H_2PO_4^-$ and HPO_4^{2-} role in determining P absorption, 366-369
Adenosine triphosphate (ATP), 559, 574
Adsorption isotherms, 299-302, 447
  uses and limitations, 388-390
Adsorption of phosphate, 264-274, 285, 288, 291, 299-303, 339, 384
  by soil constituents, 264-272
    alumino-silicates, 269
    calcium carbonate, 270-271
    clay minerals, 268
    gibbsite, 269, 272
    goethite, 272
    illite, 268
    kaolinite, 268, 272
    montmorillonite, 268
    pseudoboehmite, 269
  characterization of, 299-302
    Freundlich equation, 300
    Langmuir equation, 268-270, 285, 300
    other models, 301
    Temkin equation, 300
  effect on negative charge, 267
  hydroxyl release and, 266-269
  irreversibility, 385
  maximum, Langmuir, 266, 300
  mechanisms of
    anion exchange, 264, 268
    chemisorption, 265, 273
    physical adsorption, 264, 273
    specific adsorption, 265-269
  organic matter effect on, 271
  silicate release and, 269
  sulfate release and, 269
Adsorption parameters, soil P and plant growth, 376
Adsorption-desorption isotherms
  soil P characterization
    Freundlich equation, 382-383
    Langmuir equation, 383-384
    Temkin equation, 384
Adsorption-desorption rate, for soil P characterization, kinetic methods, 391-393
Alabama, phosphate deposits, 35
Alaska, phosphate deposits, 38
Alfalfa. *See* Forages, Legumes, Phosphorus
Alfisols, 472, 475
Alum process, 125
Alumina crystallization, 125-126

Aluminum
  crystallization, 125
  effect on P metabolism, 663
  exchangeable, 482
  exchangeable and soluble phosphate, 662
  induced P deficiency, 662
  interaction with P, 661
  in plants, 662-663
  in soils, 661-662
  phosphates, 661
  saturation, 492, 501
  toxicity in plants, 662
Aluminum oxides, 480
Aluminum phosphates, 438-439
  soil phosphate reaction products, 264-269, 281, 284-288
Aluminum-iron phosphates, 43, 44, 51, 71
Ammonium nitrate, crop response, 317
Ammonium phosphate supply, 232
Ammonium phosphate-sulphate fertilizer, 125
Ammonium phosphates, 315
  production, environmental problems, 223
  products and processes, 204-209
    diammonium phosphate, granular, 205
    monoammonium phosphate, nongranular, 207-209
    monoammonium phosphate, granular, 207
Ammonium polyphosphates
  crop response, 312, 317
  energy needed, 245
  granular, process for, 213
  history of, 196
Ammonium sulfate, sulphur use in manufacture of, 135
Amorphous product, 124
Andepts, high P fixation, 472, 473
Andosols, 473
Animal production
  grasses and legumes, 805-807
  P cycling, 834-840
  P in manure, 837-838
  P requirement, 813
Animal wastes (manure, etc.)
  composition, 519
  constraints on use, 532-535
  crop response, 528-530
  decomposition, 522
  effect of management, 521
  effect of species, 519-520
  forms of P, 521-522
  metals, 520
  N losses, 521
  N, P, and K, 520
  P cycling in grasslands, 835-840
  source of agricultural P in the environment, 554

893

SUBJECT INDEX

Animal wastes *(continued)*
  trace elements, 520
Anion exchange resins, 341, 346, 348
Apatite
  composition, 45, 53-55, 60-61
  deposits of igneous origin, 8-10
  mineralogy of, 1
  occurrence, 45
  refractive index, 59
  types, 45, 56
  unit-cell parameters, 55-56, 60-61
Application of P, 562
  central role, 559-560
Arbuscules, 629-630, 634
Arkansas, phosphate deposits, 34
Availability of P, 561
  definition of, 863
  in feedstuffs, 862-865
  in forages, 863-864
  in supplements, 865-866
  to animals, 862-866, 874-878

B-butanol solvent, 125
Bacteria, 619, 621-622, 627-628
Bacterial competition, 626, 627
Bacterial inoculation, 627-628
Band application, 290
Band seeding, forages, 809
Banding P fertilizer, 337, 495-496
Barley, P nutrition of, 681-692
Barrandite, 122
Basic slag, 312, 496, 499
  properties and use, 218
Beef cattle, P requirements of, 850, 852, 867-869
Beneficiation/upgrading B-zone, 123
Biological P cycle, 411-412
Blood phosphorus, factors affecting, 858, 867, 869, 870-872, 876
Bone meal, as feed supplement, 866
Bone phosphate of lime (BPL), 4, 44
Boron
  effect on P uptake, 672
  interaction with P, 672
Broadcast P application, 322, 496
Buffer capacity of a soil P system, 389
Buffering capacity, 479, 489
  effect on P diffusion in soil, 594, 595, 606
Bulk blending, 210
Bulk density, soil, 594

C-zone utilization, 125
Cadmium from waste materials, 533-535
Calcareous soils, 339
Calcination
  high temperature (over 1,000°C), 123
  lower temperature (under 550°C), 123
Calciphos (bulk granulated), 124

Calcium, 657
  adsorption of P, 658
  and urinary calculi, 873
  availability, 863, 865, 866, 878
  interactions with P, 657
  in plants, 658-660
  in soils, 657-658
  P solubility, 658
  P uptake, 658
  requirements, 853, 870, 871, 873, 875-878
Calcium accumulation at root surface, 609, 610
Calcium phosphates, 315, 438-439
  soil phosphate reaction products, 264, 271, 279, 277-288
Calcium phosphorus ratios, significance of in nutrition, 853, 856, 859, 861, 862, 869-878
Calcium polyphosphate, 312, 317
Calcium sulphate, 126
Calcium-iron-aluminum phosphates, 44
Calcium-magnesium phosphates, fused, process for, 218
California, phosphate deposits, 37, 38
Capacity factor, plant growth and soil P, 374
Carbonate fluorapatite, mineralogy of, 1
Carbonatites, 49
Cation exchange capacity, 491-493
Caustic soda leach C-zone, 126
Chemical conversion, 125
Christmas Island, 121-127
Citraphos (fines), 124
Citrate solubility, 497-499
  AOAC, 312, 318
Clay content, 480, 481
Clay mineralogy, 480, 481
Clover. *See* Forages, Legumes, Phosphorus
Coal, source of S for P fertilizer, 145
Coated fertilizers, granular water-soluble, 318
Compost
  amount, 536
  composition, 517
  crop response, 524-525
Compound fertilizers (NPK)
  products and processes, 210-217
    bulk blending, 210
    liquid fertilizers, 211-214
    potassium phosphates, 215-217
    urea-ammonium phosphates, 215
Condensed phosphates, 352
Consumption
  Canada, 231-232
  forecast
    Canada, 233
    United States, 233
    World, 235
  United States, 231-232
Copper, 670
  availability as affected by P, 670
  interaction with P, 670

SUBJECT INDEX

Corn, P nutrition of, 681-692
Corundum formation, 124
Cotton
 fertilization of, 694-695
 P accumulation by
  concentration in tissues, 695
  critical levels in, 700
  nutrition of, 695
  requirement for P, 694
 P accumulation by, 697
Crandallite, 122
Critical concentrations of P in the plant, 575, 577-578
Crop logging, 791
Crop quality, relation to P nutrition, 690
Crop removal of P, 335
Crop residues, P removal and recycling in, 685
Crop response to applied P
 chemical compounds present, 312
 chemical evaluation, 313
 crop species, 327, 687
 granule size-solubility relation, 341
 P source effects, 315-320
Crop yield response to added P
 corn, 254-256, 686-690
 cotton, 256-258
 P concentration relations, 686
 soybeans, 254-257, 686-690
 wheat, 255-257, 686-690
Crop yields. *See* Yields
Cultivar effect, critical concentration of P in plant tissue, 458
Cultivation and residual value, 337

Dairy cattle P requirements of, 852, 869-872
Davies equation, 365
Debye-Huckel equation, 364
Deficiencies
 nutrient, 559-560
Defluorinated phosphate, as a feed supplement, 852, 866, 876
Demand. *See* consumption
Desorption of P, 346-348, 350, 385
 by anion exchange resins, 346
 effect of cations, 347
Diammonium phosphate
 crop response, 316, 318, 327
 energy needed, 243-245
 history of, 195
 products and processes, 205-209
Dicalcium phosphate, 362, 476
 as a feed supplement, 852, 866, 876-877
 crop response, 312, 315-316
 soil phosphate reaction product, 270-271, 277-288
Diffusion coefficient of P, 600-601
Diffusion of P to plant roots, 592-594, 600-601, 604, 607-608, 611

Direct appliation of calcined C-zone ore, 124
Disassociation constant, selected ion pairs, 364
Disease resistance, effect of P on
 cotton, 702
 rice, 715
 tobacco, 729
Dissolution of phosphate fertilizers, 274-277, 302, 339
 effect of nonphosphatic salts, changes with time, 276-277
 effect of soil moisture, 274
 effect on nonphosphatic salts, 276-277
 solution formed in soil, 274-277
  chemical compositions, 275-277
  concentration of, 275-277, 285, 302
  dissolution of soil constituents by, 268-272, 279-282, 302
  metastable triple-point solution, 274, 280, 289
  movement through soil, 274, 279-286
  osmotic gradient and, 274, 276
  pH of, 275-277, 302
  phosphatolysis of minerals by, 280
  triple-point solution, 275, 280
  vapor transport and, 274
Dissolution of soil constituents, 268-272, 279-282, 302
 aluminum oxide, 268, 280
 attapulgite, 269, 280
 carbonates, 270-271, 280
 illite, 269, 280
 iron oxide, 268, 280
 kaolinite, 269, 280
 montmorillonite, 269, 280
 organic matter, 272
Distribution of P within the plant, 571-572, 576
Drying and air classification, 123
Dystric Nitosols, 474

Economic rate of P fertilizer, optimum, 584
Economics, energy output and input, 258-261
Ectomycorrhizae (sheathing mycorrhizae), 629-630, 634-635, 645-646
Edgington's formulation, isotopic exchange methods, 395
Efficiency of agriculture
 conversion, 259
 definition, 251
Efflux, 596
Egg production, effects of P on, 874-875
Elovich equation, isotopic exchange methods, 395-396
Emergency P levels of animals, 852, 854
Endodermis, 569
Endomycorrhizae, 631-634

SUBJECT INDEX

Energy
  agricultural
    phosphate fertilizer, input spectrum, 252–254
    phosphate fertilizer, output spectrum, 254–261
  biological
    output of crops as influenced by fertilizers, 255–257
  conservation, P fertilizer production, 249
  consumption
    calcination, 124
    U.S. agriculture, 251–252
  economics of output and input, 258–261
  phosphate fertilizer applications, 251–262
  process consumption, 243
  requirements
    pollution abatement, 246–247
    total fertilizer, 249
  returns, food, 251–262
  solar, conversion efficiency, 251–259
  supply curtailment, 241
Environment, agricultural P in, 545–557
Environmental factors, effect of on P concentrations
  cotton, 697
  rice, 711
  sugarcane, 717
  tobacco, 725
Environmental problems, P fertilizer industry, 223–225
Environmental protection, P fertilizer production, 246–249
Equilibrium methods, soil P characterization, 379–390
Ericaceous mycorrhizae, 629
Erosion, source of P in the environment, 546, 549, 552, 553
Erosion losses of P, 336
Escorias Thomas, 499
Eutric Nitosols, 475
Eutrophic red yellow podzolic, 475
Exchangeable Al, 480, 482
Excretion of ions by roots, 347, 350
External hyphae, 629
Exudates, 622

Feedback mechanism, 603
Ferralitic soils, 474
Ferralsols, 474
Ferrophosphorus, 74
Ferruginous soils, 475
Fertilizer applications, agricultural P in the environment, 551–553
Fertilizer efficiency, P nutrition of corn, sorghum, soybeans, small grains, 689–690
Fertilizer phosphorus
  as feed supplement, 877
  effects on P concentration, 859–861
  recommendations, 442–444, 463

Fertilizer solution. See Dissolution of phosphate fertilizers
Fertilizer × genotype interactions, 787
Fertilizers, ultra high analysis
  N-P compounds, crop response, 317
  process technology, 222
Fines separation, 123
Fixation of phosphates, 318, 326, 328
  See also Retention of phosphates
Flooded soils, 482
Florida, phosphate deposits, 27–30
Fluid fertilizer vs. solid forms, 325–326
Fluorapatite, 312
Fluoride, significance in animal feeding, 877
Foliar analysis. See Tissue analysis
Foliar fertilization, 326, 562
  P nutrition of corn, sorghum, soybeans, small grains, 684
Forages
  contribution to livestock diets, 805
  definition, 805
  economic value, 806
  effects of soil on P composition, 853–854, 859–861
  extent of fertilization, 806
  factors affecting P concentration, 853–862
  importance to agriculture, 805, 806
  mineral composition of, 852–862
  phosphorus concentration in
    animal products, 835
    bermudagrass, 821
    cool season grasses, 813–816
    cool season legumes, 818
    native grasses, 830
    rumen bacteria, 838
    tropical grasses, 824
    tropical legumes, 828–829
  phosphorus deficiency
    areas of, 807
    on disturbed land, 807, 813
    symptoms in plants, 829
  phosphorus effects on
    animal performance, 839
    botanical composition, 819
    mineral constituents, 816
    N fixation, 818
    organic constituents, 818, 829
    palatability, 818, 820
  phosphorus response
    bermudagrass, 821–822
    cool season grasses, 813–817
    cool season legumes, 817–819
    effect of N on, 814
    effect of temperature and moisture on, 814–815
    flood meadows, 833
    mountain meadows, 833
    pangolagrass, 822–823
    prairie grasses and legumes, 830–833
    species differences, 813
    tropical legumes, 827–830

SUBJECT INDEX 897

under grazing, 838-840
warm season grasses, 821-824
potential for P utilization, 806
Foreign trade, exports and imports, 233
Forest fertilization with P
application, 780-784, 786
disease and pest incidence, 789, 795
faunal and floral habitat, 789-790
frequency, 784
genotype interactions, 787, 795
methods, 781-782
N-P interactions, 786
nutrient cycling, 772-776
rates, 782-784, 794
research needs, 795
response magnitude, 783, 785, 787, 795
seed production, 790
timing, 780-781
utilization of applied P, 784, 795
water quality, 789
wood quality, 788
Forest floor, 769, 770, 773, 774, 778
Forest harvesting and P removal, 776-777
Forest nurseries
deficiency symptoms, 765
dry matter production, 764
fertilizer application, 766
phosphorus uptake, 764
seedling nutrition, 764
seedling quality, 766
Forest site preparation
response to P fertilization, 786
soil P status, 778
Forest tree nutrition, 763-804
Francolites, 57
Frasch process, 142
Free space, outer, 569
Freundlich equation
adsorption-desorption isotherms, 382-383
quantity factor, 376
Fungal sheath (mantle), 630, 640
Fungal species, 629
Fungal spores, 631
Fungi, 619, 629

Gangue materials, 23-24
Genetic inheritance, role of P, 327, 573, 687
Geology, Christmas Island, 121
Georgia
phosphate deposits, 29-30, 35
tertiary rocks, 35
Gibbsihumox, 481, 488, 492
Grade of ore, definition, 4
Granular fertilizers
application, 290
granule size of P fertilizers, 314, 316
products and processes
ammonium phosphates, 205-209

compound fertilizers (NPK), 213
ordinary superphosphate, 199-200
triple superphosphate, 200-201
Granulation, 124
Grasses
cool season
composition of, 853-862
mineral availability, 863-864
tropical, composition of, 855-856
Grasses. *See also* Forages and Phosphorus
band seeding, 809
cool season, 812-817
phosphorus cycling, 834-840
phosphorus requirement, 821, 827
root morphology, 808-809
semiarid, 830-833
surface-applied P, 810-812
warm season, 820-826
Grasslands
extent in U.S., 805, 806
semiarid, 830-833
Guano, deposits of, 11, 13

$H_2PO_4^-$, 596
H_3PO_4, 312, 319
HPO_4^{2-}, 595-596
Hapludult, 480-481
Haplustox, 481, 492
Hartig net, 629
Heavy metal accumulation from waste materials, 533-535
Hematite, 124
High analysis fertilizers
crop response, 317
new processes, 222
Hot air classification, 123
Humans, P nutrition
effect on health, 877-879
P requirements of, 851, 853, 877-878
Hydrochloric acid, use in manufacture of P fertilizers, 138
Hydrochloric acid leach, C-zone, 125
Hydrometallurgical, 125
Hydroxyapatite, soil phosphate reaction product, 271, 281, 284, 288
Hyphal uptake and translocation, 640-643

Idaho, phosphate deposits, 36
Illinois, phosphate deposits, 34
Immobilization of P, 774
Infrared analysis, 293
Inheritance, 573
Inositol hexaphosphates
amounts in soil, 420
complexities, 419
sorption, 425
Inositol phosphates, 418-421
Intensity capacity factor, 479

Intensity factor of soil solution P
  activity calculations, 363
    activities and potentials, 366–369
    electrical conductivity method, 364–366
    iterative method, 363
    relation of plant growth to P intensity, 370–374
  definition, 363
  high P fixation, 479
  measurement, 363
  molar concentration of P, 363
Ion absorption, kinetics of, plant growth and P intensity, 370
Ion exchange processes, 125
Ion-pair concentrations, 364
Iowa, phosphate deposits, 34
Iron, 663
  absorption by roots, 665
    as affected by P, 666
  chlorosis as affected by P, 666
  deficiency induced by P, 664
  interactions with P in soils and plants, 663–664
  oxidation-reduction reactions, 665
  oxide and phosphate solubilities, 665
  oxides and phosphate adsorption, 664
  phosphate compounds in soil, 664
  precipitation in roots as phosphate, 664
  transport in plants and P interference, 666
Iron oxides, 480
Iron phosphates, soil P reaction products, 264–269, 281, 284–288
Iron-aluminum phosphates, 44
Isotopic exchange
  for soil P characterization
    kinetic methods, 394–395
    residual P, 351
    sites, 375

Kansas, phosphate deposits, 35
Kaolinite, 481
Kentucky, phosphate deposits, 32
Kinetic methods
  soil P characterization, 390–398
    adsorption-desorption rate, 391–393
    desorption by resins, 393–394
    isotopic exchange methods, 394–395
    time-temperature interactions, 396–398

L-value, 334, 341, 345
Labile phosphorus
  definition, 362–374
  net- and surface-exchangeable fractions, 377
Laboratory evaluation, 124
Langmuir equation
  adsorption-desorption isotherms, 383–384
  quantity factor, 376
Latosols, 474

Leaching losses of P, 336
Legumes. See also Forages and Phosphorus
  band seeding, 809
  cool season, 817–819
  mineral availability, 863–864
  mineral composition of, 855–857
  root morphology, 808–809
  surface applied P, 812
  tropical and subtropical, 827–830
Light effect, 635
Lime addition, 123
Lime soda sinter, 126
Liming, 498, 501
  and P losses, 336
  and residual value, 352
  effects on P concentration in forages, 861
Liquid fertilizers
  products and processes, 211–214
    granular ammonium polyphosphate, 213
    pipe reactor, 212–213
    suspension fertilizers, 214
Litterfall, forest fertilization, 773
Localizing P, placement in soil, 610–612

Magnesium interactions with P, 660
Magnesium phosphates, soil P reaction products, 271, 281–287
Magnesium-ammonium phosphate, 126
Manganese, 666
  availability as affected by P, 667
  interactions with P, 667
  oxides, 666
  uptake as affected by P, 667
Manure. See Animal wastes
Mass flow, 592, 605, 608, 611–612
Matrix, definition, 4
Maximum influx (I_{max}), 596, 597, 602–604, 606–607, 611–612
McKay equation, isotopic exchange methods, 395
Melt-type granulation, TVA process, 219
Metallurgical grade Al_2O_3, 125
Metaphosphates, 352
Metastable triple-point solution
  composition, 275
  formation in soil, 274, 289
  reactions in soils, 280, 289
Michaelis constant (K_m), 596, 598, 606–607
Michaelis-Menten kinetics, 596–597, 602, 606
Microorganism-produced compounds, 622, 626–627
Milk fever (parturient paresis), 861, 870–871
Milk production, effects of P on, 869–871
Millisite, 122
Mineral supplements, inorganic, nutrition of animals and man, 865
Mineralogy, Christmas Island phosphates, 122
Mining rights, 121
Mississippi, phosphate deposits, 35

Missouri, phosphate deposits, 35
Mitscherlich equation, crop response to applied P, 581-582
Moisture stress, 562
Mollisols, 494
Molybdenum, 671
 availability as affected by P, 672
 interaction with P, 672
Monoammonium phosphate
 crop response, 316
 energy needed, 246
 history of, 196
 products and processes, 207
Monocalcium phosphate, 476, 479
 as feed supplement, 866
 crop response, 327
Montana, phosphate deposits, 36
Montmorillonite, 481
Mucigel, 621
Municipal wastes
 compost, 517, 524-525, 536
 refuse, 517, 536
 sewage sludge, 517-519, 525-527, 530-535, 537
 waste water, 516-517, 523-524, 530, 533, 536
Mycorrhizae
 endogone, 563
 forest trees, 779
 formation (factors affecting), 634-635
 in acid soils with high P fixation, 506
 incubation, 644-646
 infection by, 566
 infection spread rate, 566, 634
 mechanism of P uptake, 638-643
 P uptake, 563-564, 636-642
 plant use of residual P, 350
 taxonomy and biology, 629-634
 vesicular arbuscular (va), 563-564
 vesicular arbuscular
  role in P nutrition of plants, 563, 566, 630-635, 639, 640-643, 645-647

Natural gas, source of S for P fertilizers, 144
New York, phosphate deposits, 34
Nitrate leaching from waste materials, 533
Nitric acid, use in manufacture of P fertilizers, 138
Nitric phosphates
 crop response, 312
 history of, 196
 production, 223
 products and processes, 209
Nitrogen
 comsumption
  Canada, 232
  United States, 232
  world, 235
 effect on P concentration, 445-456
 effects on P concentration of forages, 814, 860-861
 fertilization of forest trees, 786
 fertilizer and P uptake, 657
 fertilizer sources and P availability, 656
 fixation, 564, 575, 818, 829
 interactions with P, 656
North Carolina
 Pungo River area deposits, 28
 tertiary rocks, 35
Nucleic acids, RNA and DNA, 421-422
Nutrient
 cycling in forest, 772-775
 imbalances, 580
 ratios, 577, 579
Nutrition of plants. *See* Phosphorus nutrition
Nutritional value of feed, P content, 579-580

Oats, P nutrition of, 681-692
Octocalcium phosphate, 339
 soil phosphate reaction product, 270-271, 278, 281, 284-288
Oil, "sour" crude, source of S, 144
Oklahoma, phosphate deposits, 35
Ordinary superphosphate. *See* Superphosphates
Ore, definition, 4
Organic matter, soil,
 acid soils and P fixation, 482
 complex ion formation and, 297
 dissolution by fertilizer solution, 272
 effect on P retention, 271
Organic P in soils
 amounts of, 416
 availability of, 424-427
  experiments with pure esters, 424
  experiments with indigenous soil materials, 425
 biological immobilization in virgin soils, 411
 changes in, 413-415
 deoxyribonucleic acids, 421
 equilibrium level, factors affecting, 415-416
 extraction, 417
 immobilization of P, 336, 350
 inositol hexaphosphates, 420
 inositol phosphates, 418
 measurement, 418
 mineralization of, 774, 777, 778
 nucleic acids, 421
 other esters, 423
 phospholipids, 422
 phytic acid, 418
 rates of accumulation, 336
 ribonucleic acids, 421
 teichoic acid, 423
 temperature and P uptake, 604
Orthophosphate, crop response, 317
Osteomalacia, 847
Osteoporosis, 847, 878
Oxisols, 472-474, 488, 492

Peanuts
  fertilization of, 702–704
  P accumulation by, 702
  concentration in, 704
  critical levels in, 707
  nutrition of, 704
  requirement for, 702
Pennsylvania, phosphate deposits, 34
Periodontal disease, in humans, 878
pH, 483, 493
  and crop response to P, 315–316, 321
  soil, effect on P uptake by roots, 610
  soil acidity, 635
Phloem, transport in, 570, 572
Phosphate compounds
  identification of, 291–299
    chemical analysis, 296
    infrared methods, 293
    microscopic methods, 292
    phosphate potentials, 297
    solubility products, 281, 296
    thermal analysis, 295
    X-ray diffraction analysis, 294
  in fertilizers, 275
  in soil phosphate reaction products, 271, 277–285
  solubility products of, 281, 296–297
    activity corrections for, 296–297
Phosphate deposits of the USA
  Atlantic and Gulf Coastal Plains
    geology, 26
    hardrock phosphate, Fla. and Ga., 30
    land-pebble district, Fla. and Ga., 27–29
    mining and beneficiation, 30–31
    Pungo River area, N.C., 28, 30
    river-pebble deposits, Fla., Ga., S.C., 30
    Savannah River area, Ga. and S.C., 29
  central area
    blue rock deposits, Tenn., 33
    brown rock deposits, Tenn., 32
    brown rock deposits, Ky. and Ala., 32
    Cason Shale, Ark., 34
    Chattanooga Shale, 35
    Cretaceous rocks, Ga., Ala., Miss., and Tex., 35
    Maquoketa Shale, Iowa, Ill., Wis., 34
    mining and beneficiation, 36
    Oriskany Sandstone, N.Y., Pa., and Va., 34
    Pennsylvanian rocks, Kan., Okla., and Mo., 35
    Pitkin Limestone and Hale Formation, Ark., 34
    tertiary rocks, N.C., S.C., Ga., and Tex., 35
    white rock deposits, Tenn., 33
  offshore
    California, 38
    eastern USA, 39
  western area
    mining and beneficiation, 38
    Miocene rocks, Calif., 37
    Permian and Triassic, Alaska, 37
    Phosphoria Formation, Idaho, Wyo., Mont., and Utah, 36
Phosphate fertilizers. *See also* Dissolution; Fertilizer solution
  compounds in, 275
  dissolution of in soils, 274–277
  energy input spectrum
    farm use of fertilizer, 253
    manufacturer to farm, 252
  energy output spectrum
    biological energy output, 255–258
    crop response, 254
    economics, 258–261
  history of, 195–196
  processes, objective of, 196
  product trends
    agricultural needs, 109–111
    ammonium phosphates, 114, 115
    ammonium phosphates, WPA intermediates, 114
    ammonium polyphosphates, 119
      pipe-reactor process, 119
    concentrated superphosphates, 113, 114
    direct application uses, 116, 117
    direct application, granulation, 117
    direct application, suspensions, 117
    direct application, Fe-Al phosphate rock types, 117
    economic factors, 110, 111
    nitric phosphates, 115, 116
    NP suspension fertilizers, 118
    ordinary superphosphates, 112, 113
    P-solubility of superphosphates, 112–114
    phosphate rock quality, 110–112
    potassium polyphosphates, 119
    technology limitations, 111
    thermal phosphates, 117
    urea-phosphoric acid compositions, 118, 119
  reactions in soils. *See* Adsorption; Precipitation
Phosphate fertilizers and process technology
  current products and processes, 196–218
    acidulation processes, 197–210
    compound (NPK) fertilizers, 210–217
    thermal processes, 217–218
  environmental problems, 223–225
  new, 218–222
    longer term, 222
    present and near term, 218–222
Phosphate fractionation methods, 398–401
Phosphate industry of the U.S., 19–42
  Atlantic and Gulf Coastal Plains, 26–31
  central USA, 32–36
  conservation, 40

SUBJECT INDEX

economic factors, 21-26
  gangue materials
    carbonate minerals, 23
    clay minerals, 23
    iron minerals, 24
    organic material, 24
    quartz, 23
  geographic location, 25
  geologic setting, 22
  phosphate content, 22
  product use, 25
  raw materials other than phosphate, 25
  thickness of the phosphorite bed, 22
  tonnage, 22
  water, 25
  weathering, 24
environmental considerations, 39
future outlook, 40-41
offshore deposits, 38-39
western USA, 36-38
Phosphate ions
  H_2PO^- and HPO_4^{2-}
    activity calculations, 363-366
  $H_2PO_4^-$ and HPO_4^{2-}
    activity and potentials, 366-369
    intensity measurement, 363
Phosphate reserves and resources, 1-18, 5-13
  apatite deposits of igneous origin, 8-10
  by type of phosphate deposit, 6-13
  common rock deposits, 14
  deposits of guano, 13
  economic reserves, 14
  environmental considerations, 16
  estimating, 5-6
  historical overview, 2
  low-grade deposits, 14
  marine phosphorite deposits, 6-8
  other resources, 14-15
  past estimates, 13-14
  phosphate mining industry, 2
  phosphatized rock
    from guano, 11
    from phosphatic sedimentary rocks, 12
  residual deposits, 10
  subeconomic deposits, 14
  variables that influence reserve tonnage, 15
Phosphate rock
  accessary minerals, 45-49, 51-52, 59, 62-63, 71-72, 75
  acid soils with high P fixation, 497
  apatite mineral
    compositions, 81, 84
    minor elements, 84
  as feed supplement, 866, 876
  beneficiation, 47, 62-64
  characterization methods, 52-61
    chemical analysis, 53-55
    infrared, 57-59
    interpretation, 60-61
    petrographic, 59-60
    X-ray, 55-57
  chlorides, 48, 76
  commercial ores
    accessory minerals in, 83, 99
    Francolite apatites, 84, 91, 93-96
    igneous-metamorphic types, 83
    mining sources, 88
    resources-reserves, 88-89
    sedimentary, carbonate types, 83
    sedimentary, Fe-Al types, 83
    sedimentary, siliceous types, 83
    variability, 82-89
  crop response, 312, 318, 326
  definition, 3, 43
  energy needed, 242-244, 247-249
  first prouction, 2
  granulated, 320
  history of mining, 2
  impurities and phosphoric acid manufacture, 184-186
    consumption of sulfuric acid, 184
    corrosion and scaling, 184
    crystal growth and habit, 185, 186
    foaming, 185
  marginal ores
    characterization of, 74-78
    use of, 222
  mineralogy, 44, 62
  mineralogy of for phosphoric acid manufacture, 183-184
  mining and beneficiation, environmental problems, 223
  minor and trace elements, 74, 77
  mycorrhizal effect on P uptake, 638-640
  organic matter, 48, 59, 68, 77
  partially acidulated, crop response, 319, 320
  phosphoric acid manufacture, 186-191
    "ideal" phosphate rock, 186
    rocks from Africa, 188-190
    rocks from Australia, Nauru and India, 187
    rocks from Christmas Island, 187
    rocks from middle east and eastern Europe, 190-191
    rocks from South America, 187
    rocks from the United States, 186
  phosphorite, 43
  reactivity solubility, 64-67
  reserves, 230
  residual value, 342, 351, 352
  suitability for phosphoric acid, research trends, 191-192
  texture, 48-52, 60, 62
    igneous rocks, 49
    metamorphosed rocks, 49
    sedimentary rocks, 50-52
    weathering influence, 52
  utilization and treatment, 64, 67, 68-73
  weathering, 52
  world production, 2

Phosphate rock products
  commercial concentrates
    P grades, 86, 87
    BPL trade notation, 86
    calcination treatments of, 98, 99
    influence of mining practices, 88, 89
    quality factors, 85–87
    quality, future trends, 109–112, 120
    sources of variability, 82
  fertilizer uses
    calcined product defluorinated types, 98–100
    compositions of, 100
    defluorinating reagents, 99
    chemical products effect of phosphate rock impurities, 101, 102
    ammonium orthophosphates, 103–107
    ammonium polyphosphates, 107
    compositions of nitric phosphate, 108, 109
    Fe-Al impurities in superphosphate, 107
    impurity phases in nitric phosphate, 109
    solid impurities in NP grades, 106
    solid impurities in WPA, 102, 103
    superphosphate types, 106–108
    superphosphate compositions, 106, 107
    nitric phosphate types, 108
    WPA intermediates, 102, 103
  direct application
    effect of binding agents, 92, 93, 97
    Fe-Al phosphate rock types, 97, 98, 117
    granulation methods, 92
    granulation trends, 116, 117
    phosphate rock suspensions, 117
    reactivity scales, 93–96
    selection principles, 91, 92
    solubility indexes, 92
    X-ray analysis of phosphate rock, 91–92
  low-temperature calcination, 98
Phosphate-soil reactions
  adsorption reactions. *See* Adsorption of phosphates
  identification of reaction products, 291–299
  methods of study, 288–291
  precipitation reactions. *See* Precipitation of phosphates
Phosphate solutions
  dilute, 291
  saturated, 289
Phosphatic fertilizers, sulphur in manufacture of, 129–130, 133–135
Phosphatic particles, definition, 4
Phospho-cristobalite, 124
Phospholipids, 423
Phosphonitrilic hexaamide
  crop response, 317
  production of, 222

Phosphoric acid
  capacity
    Canada, 234
    United States, 234
    world, 237
  feed supplement, 866, 876
  manufacture, suitability of phosphate rocks for, 183–192
  marginal materials, 125–126
  products and processes, 203–204
Phosphoric acid processes
  future development, 178–183
    calcium sulphate, use of, 180–181
    capacity, 178
    Gulf Research & Development, 180
    IMC Bonnie Plant, 179
    Oak Ridge National Laboratory, 180
    pollution abatement, 181–182
    two-stage hemihydrate-dihydrate, 179
    uranium recovery, 179
    W.R. Grace & Co., 180
  landmarks in development, 162
    Dorr Strong Acid Process, 162
    Dorr Weak Acid Process, 162
    Kunstdunger Patent Verwertungs a.g. (k.p.v.), 162
    Landskrona, 163
    Larsson, 163
    Lehrecke, 163
    Leijenroth, 163
    Nordengren, 163
    Vercelli, 163
  modern commercial processes, 164–165
    Central Prayon, 174
    Dorr HYS, 175
    Fisons dihydrate, 168
    Fisons single-stage hemihydrate, 171, 172
    Fisons, HDH, 175, 177
    Kellogg-Lopker, 169, 170
    Mitsubishi, 173, 174
    Nissan, 173–175
    NKK, 173, 174
    Prayon dihydrate, 167
    Saint Gobain/Rhone-Progil, 164
    Singmaster and Breyer, 175, 176
    Swenson Isothermal, 169, 170
    Trepca, 175
    UCEGO filter, 167
    Windmill Holland, 171
  wet process
    chemistry of, 152–159
      anhydrite, 157
      calcite, 153
      calcium sulphate, hydrated forms, 156
      cationic impurities, 155
      crystallization, crystal habit, 159
      filter cloth, 157
      fluorapatite, 152
      fluorine recovery, 155, 161
      fluosilicates scaling, 155

SUBJECT INDEX

Phosphoris acid processes, wet (*continued*)
  foam, 154
  francolite, 152
  gypsum, 157
  hemihydrate, 157
  hydrofluoric acid reactions, 155
  inhibition, 152
  magnesium fluosilicate, 155
  octahedral crystals, 156
  silicon, 155
  sludge, 155
  specific surface area, 152
  consumption, 131
  energy needed, 243–245, 247
  equipment and processes, 203–204
  production, environmental problems, 223
  sulfur use in manufacture of, 131
  unit operations, 159–161
    calcium sulfate, 160, 161
    evaporation, type of, 161
    filtration, types of, 160
    particle size distribution, grinding, 159
    reaction, agitated vessels, 160
    recrystallization, 160
    rock handling, 159
    sulfuric acid, 161
Phosphorite, definition, 3
Phosphorus
  adsorption capacity, 445
  application level and residual value, 341, 353
  availability to animals, 850, 862–866
  balance in animals, 850, 863–865, 877, 878
  concentration and yield, 578
  concentration in feed
    animal preferences, 580
    deficiencies, 479
  concentration in solution
    minimum (C_{min}), 596, 607, 612
  consumed as fertilizer, 136
  content of feeds, 852, 864–865
  content of forages, 852–862
  content of plants, 853–861
  cycling
    from animal excreta, 835–837
    from plant residues, 836–837
    in grasslands, 833, 834
    in the ecosystem, 412, 834
    role of soil fauna, 838
  deficiency in animals, 866–879
  desorption, 595, 601
  diffusion, 618, 628, 638–642
  diffusion rates in soil, 593–596
  distribution within the plant, 571–572, 576
  effects on yield, 859–860
  feed supply, 848–849
  fixation, factors affecting, 480
  forms of in soil, 439
  in human diets, 851, 853, 877–879
  influx, 597, 602, 603, 605, 611, 638–640
  interactions with other elements in soils and plants, 655–674
  organic in soils
    availability of, 424–427
    biological immobilization of P, 411–417
    characterization of, 417–423
    research needs, 427–428
  placement, 612
  requirement in acid high P fixing soils
    external, 483–485, 504
    internal, 504–505
  requirements
    beef cattle, 850–852, 867–869
    dairy cattle, 852, 869–872
    general, 849–853
    humans, 851, 853, 877–879
    of applied P, 567, 582, 584
    poultry, 852, 874–875
    sheep, 852, 872–873
    swine, 852, 875–877
  soil analyses for available P
    evaluation of soil test extractants, 434–442
    forest trees, 792–794
    soil solution as a measure of available P, 444–447
    sugar beets, 749
    vegetables, 749, 755
  solubilization, 627–629
  sorption index, 389
  sorption isotherms, 478, 480, 483
  status in soils, 401–403, 635, 636, 638
  supplements, 848–849, 852, 865–866, 868, 872, 873, 875, 876–877
  uptake and utilization by crops
    compost, 525
    cool season grasses, 816
    crop quality, 690
    efficiency of fertilizer P, 689
    foliar asorption, 684
    moisture, 687
    municipal wastes, 524–527
    other nutrients, 687
    predicted, 607
    removal and recycling in crop residues, 685
    rooting pattern and root activity, 683
    seasonal demand patterns, 683
    sewage sludge, 526
    soil tests and forms of P, 436, 440
    species and variety, 687
    temperature, 689
    tissue P concentrations, 458
    warm season grasses, 821
    waste water, 523–524
    yield-P concentration relationships, 686
  uptake patterns in
    cabbage, 745
    cantaloupes, 745
    lettuce, 745

Phosphorus, uptake patterns (*continued*)
  peas, 745
  potato, 745
  snap beans, 745
  sugar beets, 739, 744
  sweet corn, 745
  table beets, 745
  tomatoe, 745
  vegetables, 739, 744
  uptake rate, 624, 638, 640–642
Phosphorus (agricultural) in the environment, 545–557
  biosphere
    distribution in, 545
  effects of P additions
    fertilizer applications and soil management, 551–553
    in animal wastes, 554
    urban runoff, 555
  erosion losses, 546
Phosphorus fertilizers
  application methods
    broadcasting, 322
    fluid vs. solid forms of P, 325
    foliar, 326
    phytotoxic effects, 327
    row and band placement, 323
    strip application, 325
    surface, 325
  forest trees
    basic slag, 766
    organics, 766
    phosphate rock, 766, 782, 785–786, 788
    reverted superphosphate, 766
    triple superphosphate, 766, 783, 785, 788
  placement
    band seeding, 809
    effect of immobility on, 808–809
    effect on uptake, 812
    effectiveness of surface application, 810
    surface, 810–812
    temperature relationships, 810
Phosphorus fixation in acid soils
  amounts, 480–482
  capacity, 484, 485, 490, 501
  concentration in solution, 483–484
  distribution of soils with, 473–475
  management alternatives
    high input, 487–495
    low input, 495–508
  mechanisms involved, 475–478
  methods of estimating, 485–486
Phosphorus fractionation, soil, 658
Phosphorus in forest ecosystems
  accumulation, 768–771
  cycling, 772–775, 795
    biochemical cycle, 775
    biogeochemical cycle, 773–775
    geochemical cycle, 772
  distribution, 771–772
  uptake, 776, 780

Phosphorus in the environment
  background levels
    analytical problems, 548
    concentration in subsurface and surface flow, 546–548
    concentration in snow melt, 548
    geological, 546
    organic P, 547
    sediment, transport by, 549–551
    subsurface flow, 546
    surface flow, 547–548
    transport on sediments, 549–551
Phosphorus nutrition of plants
  buffering capacity, 560
  chemical composition
    cotton, 699
    peanut, 706
    rice, 713
    sugarcane, 720
  compounds in tissues, 572–573
  concentration yield relationships in
    cotton, 699
    peanut, 707
    rice, 717
    sugarcane, 720
    tobacco, 727
  cool season grasses, 812–817
  cool season legumes, 817–819
  critical levels in
    cotton, 700
    peanut, 707
    rice, 714
    sugarcane, 721
    tobacco, 727
  crop responses
    sugar beets, 757
    vegetables, 748, 749, 756
  distribution in
    cotton, 698
    peanut, 705
    rice, 712
    sugarcane, 719
    tobacco, 726
  distribution within soil profile, 562
  fertilization placement and response
    cotton, 694–695
    peanut, 703–704
    rice, 709–710
    sugarcane, 716–717
    tobacco, 723
  fertilizer placement
    sugar beets, 750
    vegetables, 750
  fertilizer rates
    sugar beets, 737, 749
    vegetables, 737, 747
  forms in soil, 561
  movement in soil, 561
  plant diseases
    Cercospora leaf spot of tobacco, 729
    cotton wilt, 702

SUBJECT INDEX 905

Phosphorus nutrition of plants
  plant diseases (*continued*)
    downy mildew of tobacco, 729
    Helminthosporium leaf spot of rice, 715
    phymatotrichum root rot, 702
    rice blast, 715
    tobacco leaf curl virus, 729
  seasonal influences, 752
  semiarid grasslands, 830-833
  sources
    sugar beets, 754
    vegetables, 753
  species differences, 813
  timing
    cantaloupes, 752
    peas, 752
    potatoes, 752
    sugar beets, 752
    tomatoes, 752
  tropical and subtropical legumes, 827-830
  warm season grasses, 820-826
Photosynthesis, role of P in the plant, 574
Physical properties, 493
Phytase activity, 424
Phytate P significance in nutrition, 864-865, 874-875, 878
Phytic acid, 418
Pipe reactor, liquid phosphate fertilizers, 212
Placement of P fertilizer
  in acid soils with high P fixation, 495
  row, 612
  split root, 602, 611
  strip, 612
Plant analyses. *See* Tissue analyses
Plant growth
  P utilization model, 462
  physical-chemical characteristics of soil P, 362-379
  rate, effect on tissue P concentration, 460
Plant nutrition. *See* Phosphorus nutrition
Plant quality
  effect of P on
    cotton, 700
    peanut, 708
    rice, 714
    sugar beets, 755
    sugarcane, 721
    tobacco, 728
    vegetables, 754-755
Plant residues, P cycling in grasslands, 835, 836
Plant species and varieties, tolerance to low P availability, 504-508
Plant tissue age, importance in critical concentration of P, 457
Plant tissue analyses. *See* Tissue analyses
Plants, phosphorus content of, 853-861
Pollution abatement, 246
Polyphosphates in fertilizers
  crop response, 316-317, 326
  hydrolysis in soils, 282-283
    factors affecting, 283
    pyrophosphate, 282
    tripolyphosphate, 282
  reactions with soil constituents, 269, 271-272, 282
    pyrophosphate, 269, 271-272, 282
    reaction products, 279, 284-285
    tripolyphosphate, 269, 271, 282
  residual P, 352
  solutions formed from, 275
Pop-up application, 323, 324
Potassium
  consumption
    Canada, 232
    United States, 232
    world, 235
  interactions with P, 660
Potassium phosphates, products and processes, 215-217
Potassium polyphosphate, crop response, 312, 317
Poultry, P requirements of, 852, 874-875
Precipitation of phosphates in soils
  by soil constituents, 268-274, 280-286
    alumino silicate minerals, 269
    aluminum oxides, 268
    calcium carbonates, 271
    iron oxides, 268
    magnesium carbonate, 271
  initial reaction products, 271, 280-286
    changes with time, 286-288
    dissolution and reprecipitation, 288
    dissolution and adsorption, 288
    from ammonium polyphosphate, 282
    from ammonium pyrophosphate, 282
    from ammonium tripolyphosphate, 271, 282
    from diammonium phosphate, 271, 281
    from dipotassium phosphate, 271, 286
    from monoammonium phosphate, 271, 281
    from monocalcium phosphate, 271, 281, 286
    from monopotassium phosphate, 271, 281, 286
  identification of, 291-299
  methods of studying, 288-291
  residual, 339
  stable reaction products, 287-288, 303
  within fertilizer application site, 277-279, 302
    dicalcium phosphate, 277-279
    effect of associated salts, 277-279
    octocalcium phosphate, 278
    with micronutrient elements, 278, 279
Prices
  concentrated superphosphate, 229
  relationship of crop and fertilizer, 228
  selected agricultural commodities, 229

Production capacity. *See also* Supply
  P fertilizers, 30, 36, 38, 101, 109-115, 120, 164, 178, 195-196, 242
Production. *See* Supply
Profitability, determination of, 584
Pyrometallurgical, 125
Pyrophosphates, residual P, 352
Pyrophosphates in fertilizers. *See* Polyphosphates

Quantity factor, soil P and plant growth
  adsorption parameters, 376
  chemical extractants, 378
  definition, 374
  high P fixation, 479
  isotopic exchange, 375
  resin exchange, 377

Recommended dietary allowances (RDA), 851
Red mud residue, insoluble, 126
Red Yellow Podzolic, 474
Redox potential, Fe-P interaction in soil, 665
Refuse (municipal)
  amount, 536
  composition, 517
Reproduction in animals, effects of P, 866-876
Reserves (world phosphate), definition, 5
Residual P
  in acid soils with high P fixation, 487-490
  relative effectiveness, 334, 353
Residual value of P
  and buffering capacity for P, 334
  and difference between soils, 345-346
  and difference between plants, 349-352
  index, 333
  measured by
    L values, 334
    response curves, 334, 351
    soil tests, 335
    sorption isotherms, 334
    uptake, 334
  measured in the field, 334
  plant use, 349-352
  rate of change, 341-345
  simulation model, 353-355
Resin desorption, for soil P characterization, kinetic methods, 394-395
Resin exchange
  quantity factor, soil P and plant growth, 377
  soil P characterization, quantity factor in, 377
Resin regeneration, 125
Resources (world phosphate), definition, 5
Response curves
  applied P, 581
  curvature of, 583

P concentration, 578
Retention of phosphates by soils. *See also* Adsorption of phosphates; Precipitation of phosphates
  by soil constituents, 264-274
    alumino silicate minerals, 268
    aluminum oxides, 264-268
    iron oxides, 264-268
    soil carbonates, 270
  definition, 263
  mechanisms, 264-274
  P concentration and, 273
  soil organic matter and, 271
    cation complexing, 271
    humic acid, 271
    humus, 271
    manure, 271
Reversibility of P reactions with soil, 346
Rhenania phosphates
  crop response, 312
  in acid soils with high P fixation, 499-500
  process for, 217-218
  raw materials, 71
Rhizobia, 563-564, 566
Rhizocylinder, 608, 612
Rhizosphere, 601, 607, 617, 619, 622, 623-627, 640
Rhodic Luvisols, 475
Ribonucleic acid, 559, 572
Rice
  fertilization of, 709-710
  P accumulation by, 709, 712
    concentration in, 710
    critical levels in, 714
    nutrition of, 709
    requirement for, 709
Rickets, 847
Root absorbing power, 618, 626
Root hairs, 565, 618, 623-624, 634, 640
  P uptake by, 600, 601, 603, 606, 607, 613
Root radius, 600, 601, 605-607
Root soil contact, 622
Root trimming, 602, 611
Rooting pattern and activity, P nutrition of corn, sorghum, soybeans, small grains, 683
Roots
  age and P uptake, 598-599, 611
  competition, 565-566
  density, 605, 640-642
  development, 564-565
  difference between plant species, 603
  distribution down soil profile, 562
  growth response, 562
  length, 563, 598, 601, 611
  mechanisms of P supply, 592
  models for P uptake, 605-607
  morphology, 603, 618, 623, 808
    effect on P uptake, 809
  movement of P into, 569

SUBJECT INDEX

P uptake characteristics, 596–605
  influence on tissue P concentrations, 458–460
  suberized, 598, 601, 611
  temperature effect on P uptake, 604
  zone of P entry, 568
Row and band placement, P application, 314–315, 323–324, 327
Ruminants, P nutrition of, 866–873
Runoff, transport of agricultural P in the environment, 552–553, 555

Salt accumulation from waste materials, 535
Sampling for tissue analysis,
  reasons for, 575
  time of, 576
Schofield's phosphate potential, 373
Sediments, transport of P in the environment, 549–551
Serpentine phosphate, 312
Sewage sludge
  cadmium, 535
  composition, 517–519
    forms of P, 518–519
    inositol phosphates, 519
  constraints on use, 532–535, 537
  crop response, 525–527
  effect on soil, 530–531
  metals, 518
  N, P, and K, 518
  trace elements, 518
Sheep, P requirements of, 852, 872–873
Shoot demand, 602
Shoot-root ratio, 602, 604
Silicate
  effect on P availability, 673
  in acid soils with high P fixation, 501, 503
  interactions with P, 673
Silverman solution, 53
Simulation models for P uptake, 605–613
Slow reactions between soil and P, 339–349
  differences between soils, 345
  effects of temperature, 340, 351
  mechanism, 348
  rate, 341
  reversibility, 346–348
Sodium aluminate, 126
Sodium carbonate, 126
Sodium phosphate as feed supplement, 866
Soil depth, P nutrition of plants, 599
Soil factors in P nutrition of plants, 560–561
Soil P characterization
  equilibrium methods
    adsorption-desorption isotherms, 382–386
    solubility isotherms, 379–382
    two-surface equations, 386–388
    uses and limitations, 388–390
  fractionation methods, 398–401

  kinetic methods
    adsorption-desorption rate, 391–393
    desorption by resins, 393–394
    isotopic exchange methods, 394–395
Soil P levels, 311, 321, 323, 327, 328
Soil P retention capacity, 766, 781, 785, 789, 794
Soil pH. *See* pH
Soil phosphate reactions. *See* Phosphate soil reactions
Soil phosphorus solution
  characteristics of, 362
  labile and nonlabile P, 362
  labile P, 374
Soil solution
  definition, 368
  electrochemical potential of P, 371
Soil solution culture studies
  excised roots and whole plants, 371
  P buffering capacity, 560
  plant growth and P intensity, 372–374
Soil solution P
  as a measure of available P, 444–447, 451, 452
Soil solutions and P buffering, 560
Soil sterilization, 622, 626, 645
Soil testing for P, 335, 347
  acid soils with high P fixation, 485
  conventional, 433–469
  correlations, 440–441, 452
  critical P levels, 442–443
  extractants, evaluation of, 436, 437, 439, 440, 452
  forest trees, 765, 788, 792–795
  methods of analysis, 793–794
  status of P in soils, 361–410
Soil texture, 503
Solid fertilizers vs. fluid, 325–326
Solubility isotherms, uses and limitations, 379–382
Sorghum P nutrition of, 681–692
Sorption, 477
Sources of P and residual value, 352
South Carolina
  phosphate deposits, 29–30
  tertiary rocks, 35
Soybeans, P nutrition of, 681–692
Species, differences in P requirement, 567–568
Steam, from sulfuric acid manufacturer, 244–245
Strengite, soil phosphate reaction product, 281, 284, 287–288
Strip application, 325
Struvite, soil phosphate reaction product, 271, 281–286
Sugar beets
  market quality, 755
  P application, 749, 750, 752
  P uptake patterns, 739, 744
  plant analyses, 739, 741, 743

Sugar beets (*continued*)
  production areas, 737
  soil testing, 749
  sources of P, 754
Sugarcane
  fertilization of, 716-717
  P accumulation by, 716, 719
    concentration in, 717
    critical levels in, 721
    nutrition of, 716
    requirement for, 716
Sulfate, 671
  adsorption as affected by P, 671
  availability as affected by P, 671
  interactions with P, 671
  precipitation by Al and Fe, 671
Sulfur
  acid and nonacid uses, 135
  consumption by type, 132
  consumption for P forecast to, 1980, 139
  energy needed, 244, 249
  phosphate rock on acid soils, 498
  production by Frasch process, 142
  production from
    coal, 145
    involuntary sources, 148
    petroleum, 144
    pyrites, 142
    sour natural gas, 144
    stack gases, 146
  reserves, cost-price relationship, 141-142
  resources
    outside U.S., 141
    U.S. only, 141
  supply
    factors affecting, 149
    involuntary sources, 143-144
    voluntary sources, 142
  supply forecast
    for phosphatic fertilizers, 149-150
Sulfuric acid
  advantage in manufacture of P fertilizers, 137
  energy requirement, 244
  from metal smelting and refining, 147-149
  leaching of calcined C-zone ore, 125
Superphosphate
  ammoniated concentrated, crop response, 312, 316
  ammoniated ordinary, crop response, 312, 316
  coated, crop response, 318
  concentrated
    consumption, 132
    crop response, 312, 316, 317
    prices, 229
    sulphur use in manufacture of, 132
    supply, 233
  history of, 195
  in acid soils with high P fixation, 476, 499
  normal
    energy needed, 243, 245, 246
    supply, 233
  ordinary, 312, 316
    consumption, 131
    sulphur use in manufacture of, 131
  ordinary (single) products and processes, 197-200
    batch type mixer, 197
    chemical composition and properties, 199
    continuous dens, 198
    operating conditions, 198
    TVA cone mixer, 197-198
    use in producing granular fertilizers, 199-200
  production
    environmental problems, 223
  triple, 200-201
    chemical analysis, 200
    continuous processing, 200-201
    energy needed, 242, 243, 245, 246
    granular products, 200-201
    run-of-pile, energy needed, 245, 246
Superphosphoric acid
  crop response, 316
  energy needed, 245, 247
Supplements
  inorganic P, 848-849, 852, 865-866
Supply
  ammonium phosphate, 233
  concentrated superphosphate, 233
  forecast
    Canada, 234
    United States, 234
    world, 237
  normal superphosphate, 233
  phosphoric acid, 237
Surface application (no-till), 325
Suspension fertilizers, 326
  process for, 214
Swine, P requirements of, 852, 875-877
Symplasm, 569

Taranakites, soil phosphate reaction products, 281, 284, 287
Teichoic acids, 423
Temkin equation
  adsorption-desorption isotherms, 384
  quantity factor, 376
Temperature
  effect on adsorption of P, 340
  effect on mycorrhiza, 635
  effect on P diffusion in soil, 595
  effect on P uptake, 595, 599, 604
  effect on slow reaction, 340
Tennessee, phosphate deposits, 32-33
Terra roxa estruturada, 475
Texas, phosphate deposits, 35
Thermal analysis, 295

SUBJECT INDEX

Thermal and chemical conversion, 125
Thermal processes for phosphate fertilizers, 217-218
  basic slag, 218
  fuse-Ca-Mg-phosphate, 218
  rhenania phosphate, 217-218
  sulfur requirements, 139
Thermophosphate, 499
Throughfall, forest fertilization, 773
Tissue analyses, 575-577, 765, 781, 783, 790-792
  acetic acid soluble P
    cantaloupes, 744
    potatoes, 743
    sugar beets, 741
    vegetables, 741
  assessing soil P status, 447-463
    research needs, 447, 464
  choice of, 576
  conventional, 433-469
  critical P concentrations, 448-449, 451, 453, 455
  for P in cotton, 699
  forest trees, 790-792
  reasons for, 575
  sampling time, 576
  total P
    cantaloupes, 744
    potatoes, 743
    sugar beets, 739, 743
    vegetables, 739, 743
  use of, 577
  what to sample, 576
Tissue analysis for P in
  peanuts, 707
  rice, 714
  sugarcane, 720
  tobacco, 727
Tobbaco
  fertilization of, 723
  P accumulation by, 723, 725
    concentration in, 724
    critical levels in, 727
    nutrition of, 724
Toxicity of P in the plant, 573, 578, 579
Translocation of P in plants, 569-570, 775
  cotton, 698
  peanut, 706
  rice, 712
  tobacco, 727
Tricalcium phosphate, 312
Triple-point solution
  composition, 275
  reaction in soils, 280
Triple superphosphate. See Superphosphates
Tripolyphosphates in fertilizers. See Polyphosphates
Trisodium phosphate, 126
Tropical and subtropical legumes, P nutrition, 827-830

Tropical areas, P sources, 328-329
Tropical forages
  composition of, 855-856
Tropical soils, acid, P fixation, 471-514

U.S. agriculture, energy use, 251-252
Ultisols, 472, 474, 479, 481
Ultra-high analysis fertilizers. See Fertilizers
Urban runoff, source of P in the environment, 555
Urea, crop response, 316
Urea phosphate, process and properties, 219-222
Urea-ammonium phosphate
  crop response, 316, 327
  new process, 218
  products and processes, 215
Urea-ammonium polyphosphate, crop response, 312, 316
Urinary calculi, effects of dietary P, 873
Utah, phosphate deposits, 36

Variscite
  Al-P interactions in soils, 661
  soil phosphate reaction product, 281, 284, 287-288
Vegetable crops
  market quality, 754
  P application, 747-752
  P uptake patterns, 744-745
  plant analyses, 740-744
  production areas, 737
  soil testing, 749, 755
  sources of P, 753
Vertisols, 494
Virginia, phosphate deposits, 34

Waste water as a source of P, 516-517, 523-524, 530, 533, 535-536
  amount, 536
  application rate, 523-524
  composition, 516
  crop response, 523-524
  disposal, 516-517, 530, 533
  municipal, 516
  processing, 516
Wastes
  application rates
    metals, 533-535
    nitrate leaching, 533
    salinity, 535
  composition, 513-523
  constraints on use, 532-535, 537
  crop response, 523-530
Water flux into roots, 605, 607
Water quality, agricultural P in the environment, 545, 556

Water solubility, and crop response to P, 314–316, 323
Water-holding capacity of soil, 334
Weight units, definition, 5
Wet process phosphoric acid. *See* Phosphoric acid processes
Wet washing and screening, 123
Wheat, P nutrition of, 681–692
Whitlockite (Ca-phosphate), 126
Whitlockite Formation, 124
Wisconsin, phosphate deposits, 34
Wyoming, phosphate deposits, 36

X-ray diffraction analysis, 294
X-amorphous colloids, 473, 475, 480, 481

Xylem, movement in, 569

Yield determinants
  mathematical models, 581
  maximum yield, 582
  responsiveness, 583

Zero point of charge, measurement of P fixation, 477, 478
Zinc, 667
  deficiency induced by P, 668
    as affected by iron, 669
    as affected by temperature, 669
  interactions with P in plants and soils, 667